Geology and the Environment: Living with a Dynamic Planet

Eighth Edition

Paul Bierman is a Professor of Environmental Science and Natural Resources at the University of Vermont. Now in his thirtieth year at UVM, Paul's areas of expertise include understanding how humans and landscapes interact using his expertise in hydrology, geochemistry, and geomorphology. He is particularly interested in the impact of humans on the built and natural landscape as well as science education at all levels. Paul teaches a variety of courses, including Earth Hazards, Climate Change, and Science Communication. He has a BA degree from Williams College and his Ph.D. from the University of Washington. Research interests include measuring the rate at which Earth's surface changes and climates of the past, which involves field work in such locations as central Australia and Greenland. Bierman directs UVM's Cosmogenic Nuclide Extraction Lab—one of only a handful of laboratories in the country dedicated to the preparation of samples for analysis of 10-Be and 26-Al from pure quartz. He manages the Landscape Change Program, a National Science Foundation-supported digital archive of historic Vermont Landscape images used for teaching and research, Paul's research is funded by the National Science Foundation, the U.S. Geological Survey, the National Geographic Society, and the U.S. Army. In 1996, Paul was awarded the Donath medal as the outstanding young scientist of the year by the Geological Society of America; he has since received a CAREER award from the National Science Foundation specifically for integrating scientific education and research. In 2005, Paul was awarded the NSF Distinguished Teaching Scholar award in recognition of his ongoing attempts to integrate these two strands of his academic life. Together, Paul, his graduate and undergraduate students, and collaborators have nearly 200 publications in refereed journals and books including a modern textbook, *Key Concepts in Geomorphology*. In his spare time, Paul enjoys walking, Nordic skiing, cooking local food, and growing flowers.

Richard W. Hazlett is a Professor Emeritus from Pomona College, Claremont, California, where he was a four-time Wig Distinguished Teaching award winner, former Chair of the Geology Department, and Coordinator of the interdisciplinary Environmental Analysis Program. *The Princeton Review* ranked him one of The Best 300 Professors in 2012. Presently, he is a research affiliate of the U.S. Geological Survey Hawaiian Volcano Observatory and an adjunct faculty member in earth sciences at the University of Hawai'i, Hilo. His areas of environmental interest include soils and the movement of nutrients and pollutants in natural systems. He has many years' experience working in volcanology, researching and mapping active volcanoes in Alaska, Italy, Central America, and Hawai'i. His published books include *The American West at Risk: Science, Myths, and Politics of Land Abuse and Recovery* (2008; co-authors Wilshire and Nielson); *Volcanoes: A Global Perspective* (2022; co-authors Lockwood and De la Cruz-Reyna); and *Roadside Geology of Hawai'i* (2022; co-authors Gansecki and Lundblad). He was lead editor for the Oxford University Press Research Encyclopedia of Agriculture and the Environment (2020). In addition to his academic career, Professor Hazlett is a former employee of the U.S. National Park Service, where he assisted with exhibit planning and illustrations and wrote interpretive brochures for visitors. He credits much of his teaching ability to the "on-the-job-training" of this early career experience.

Dr. D. D. "Dee" Trent has worked at or taught geology since 1955. After graduating from college, he worked in the petroleum industry, where his geologic skills were sharpened doing projects in Utah, Arizona, California, and Alaska. When one company offered to send him to Libya, he decided it was time to become a college geology teacher. He earned a Ph.D. from the University of Arizona and for twenty-eight years taught geology, physical oceanography, and physics at Citrus Community College. He has also served as an adjunct professor at the University of Southern California, taught minicourses for the University of California at Riverside Extension Division, worked for the National Park Service, did field work on glaciers in California and Alaska, appeared on several episodes of the PBS telecourse "The Earth Revealed," authored or co-authored papers on various geologic and mining topics, and co-authored (with Richard Hazlett) *Joshua Tree National Park Geology*. Now retired, Dee enjoys playing banjo and guitar in jazz combos, and does *en plain air* oil painting.

Geology and the Environment: Living with a Dynamic Planet

Eighth Edition

Paul Bierman
University of Vermont

Richard Hazlett
Pomona College

D. D. Trent
Citrus College

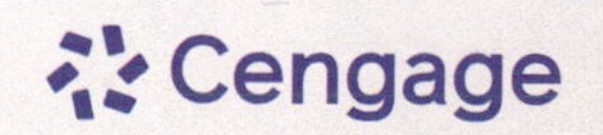

Australia • Brazil • Canada • Mexico • Singapore • United Kingdom • United States

Cengage

Geology and the Environment: Living with a Dynamic Planet, **Eighth Edition**
Paul Bierman, Richard Hazlett, and D. D. Trent

SVP, Product: Cheryl Constantini

VP, Product Management & Marketing: Thais Alencar

Portfolio Product Director: Maureen McLaughlin

Sr. Portfolio Product Manager: Vicky True-Baker

Product Assistant: Jeanette Ames

Content Manager: Anupam Bose, MPS Limited

Digital Project Manager: Kristin Hinz

Director, Product Marketing: April Danae

Product Marketing Manager: Andrew Stock

Content Acquisition Analyst: Ann Hoffman

Production Service: MPS Limited

Designer: Chris Doughman

Cover Image Source: Abstract Aerial Art/ DigitalVision/Getty Images

Library of Congress Control Number: 2023903086

ISBN: 978-0-357-85165-4

Cengage
200 Pier 4 Boulevard
Boston, MA 02210
USA

Cengage is a leading provider of customized learning solutions. Our employees reside in nearly 40 different countries and serve digital learners in 165 countries around the world. Find your local representative at **www.cengage.com.**

To learn more about Cengage platforms and services, register or access your online learning solution, or purchase materials for your course, visit **www.cengage.com.**

Printed at CLDPC, USA, 06-23

Brief Contents

Section I: Our Planet

Section II: The Solid Earth

Section III: The Surface Earth

Section IV: Earth and Society

Ahmad Sahel Arman/AFP/Getty Images

Contents

Section I: Our Planet

Mainichi Shimbun/Reuters

ZUMA Press, Inc./Alamy Stock Photo

Hemis/Alamy Stock Photo

iStock.com/Bierchen

Albertog33/Dreamstime.com

Bjoern Wylezich/Shutterstock.com

iStock.com/StockByM

Greg Ward NZ/Shutterstock.com

iStock.com/ClarkandCompany

Section III: The Surface Earth

Robert & Barbara Decker/USGS

iStock.com/F11photo

iStock.com/Saro...photo

iStock.com/BeyondImages

Our Changing Planet

Environmental geology is the study of the relationship between humans and their geological environment. This relationship goes both ways. Not only do naturally occurring geological phenomena affect the lives of people each day, but also human activities affect geological processes, sometimes with tragic consequences. Given the growing global population, the potential risk for experiencing abrupt, if not catastrophic, geological and climatic changes has never been greater for more people. During your lifetime, there is a strong possibility that you will either indirectly or directly experience at least one of many possible phenomena of interest to environmental geologists; perhaps an earthquake, flood, landslide, volcanic activity, extreme climate event, release of hazardous materials, or other significant environmental incident that impacts your local community. A basic, practical understanding of environmental geology is essential for understanding such occurrences and for accepting and moving beyond them proactively.

Overview of Organization

This book leads you through an understanding of Earth from an environmental perspective. Unlike many introductory geology texts, the book does not deal extensively with the distant past, but rather concentrates on the here, recent, and now. Its contents are divided into four general sections. The first three chapters "set the stage," providing a context in terms of time, space, and process, for all of the chapter material to follow. This includes an introduction to environmental geology and geologists, the large-scale systems that make Earth livable, and our planet's rapidly changing climate.

The book's second section considers the solid Earth, including the materials that make up the bedrock and soil beneath your feet. We follow with a pair of chapters teaching you about two formidable products of plate tectonics: earthquakes and volcanoes. As you'll learn throughout the book, threats to human welfare come not so much from the unleashed forces of nature, but rather from the placement of people, quite often unintentionally or through no personal choice, in the path of generally avoidable threats. Sometimes disaster is simply a matter of poor building and home construction and planning. Earthquakes and volcanism have always been a part of the Earth scene and always will be. We can ease our experience of them by learning as much as we can about their triggers and potential impacts.

The third section considers processes active on Earth's dynamic surface: the formation and development of soils, and the dynamics of the hydrological cycle, including the behavior of streams and rivers. We consider mass movements, landslides and debris flows and the havoc they can cause to people and society. Finally, we consider waves acting on coasts–both those driven by wind and storms and tsunami caused by earthquakes, volcanic eruptions, and landslides.

The last two chapters of this book are more focused on human affairs: energy production and waste management. We consider how resources are extracted from the Earth and the environmental ramifications of fossil fuel production and transport. We then consider how we deal with the waste that we all generate. As you'll see, effective waste management requires a keen awareness of how pollutants can move through the environment when something goes wrong. Indeed, we must prepare for things occasionally not to work out as well as we would like. This brings us back to the basic premise of this book–that a keen understanding of geology and how the natural world "works" is our best hope for living well today and building a better future as citizens of our planet.

In a culture where science and technology are interwoven with economics and political action, an understanding of the sciences is increasingly important. The National Science Foundation, the National Center for Earth Science Education, and several other prestigious Earth science organizations are promoting Earth science literacy in the United States for students from elementary school through college. The hope is that, through education, today's students will become better stewards of our planet than we have been. Earth science literacy helps us to appreciate Earth's beauty while also recognizing its limitations. One thing is certain: the planet can sustain just so much life, and it is now being pushed to the limit. The fragility of Earth was eloquently described by Apollo 14 astronaut Edgar Mitchell:

> It is so incredibly impressive when you look back at our planet from out there in space and you realize so forcibly that it's a closed system—that we don't have unlimited resources, that there's only so much air and so much water. You get out there in space and you say to yourself, "That's home. That's the only home we have, and the only one we're going to have for a long time." We had better take care of it. We don't get a second chance.

For you, the reader—don't be spooked! This book is meant to inform (we'd like to say "enlighten!") readers who are taking a one-semester undergraduate introductory-level geology course. As authors, each of us has taught undergraduate students extensively during our careers. Each of us is passionate about our efforts to reach out and provide insights that we have found exciting, even emotionally engaging. We attempt, we hope with reasonable success, to reduce the complex, beautiful mechanism of the geological world to terms that stimulate your interest and desire to learn more. Wherever feasible, we also try to make the topical material directly relevant. Not all is "gloom and doom," nor should it be. Some great, unsung success stories in adaptation, engineering, and shear human ingenuity also shine through. But be prepared to face some possibly strange, if not disturbing, new facts and perspectives about Planet Earth—all of which, we hope you'll agree, can be potentially useful to know.

This book is really a "work in progress." Science is an iterative, continually evolving pursuit. New areas of importance continually appear, while people lose interest in others and let them recede into the past—never fully lost, but dormant or meaningless in the present day. Simply compare this textbook with similar ones written 10, 20, or even 40 years ago and perhaps you'll be amazed, both by what hasn't changed and what has in terms of emphasis and basic understanding.

Science, unlike many other human endeavors that make the world so busy today, is also an intensely self-correcting process, an effort to arrive at the "truth" about the raw, natural state of the world (and cosmos) in as impartial and nonjudgmental a way as possible. Scientific progress is driven both by new technology and by ordinary, very human curiosity. Although it is possible to tell a lie as a scientist, one can't get away with doing that for long, because science requires intensive peer review and public correction. Science is all about the sharing of carefully nurtured insights with others.

Recently, one of us was confronted by a student who said, matter-of-factly, "I'm a humanities major; I don't do science." We must acknowledge that (however inadvertently) teachers sometimes fail to make science as clear and easy as it can and should be. The ongoing student refrain in a well-run science class should be, "Wow, that makes sense!"

To close one's mind to learning about how science works in the most basic way, and to blind oneself to the awesome insights it provides about simply being alive in this world, is a sad thing. We expect that you will have to work hard to grasp some ideas and concepts, especially in the first few "orientation" chapters. We believe that there is simply no other way to get to a higher level of understanding without some hard thinking. Learning is a dance with two partners: the instructor to provide the information as skillfully as possible and the student to seek out its meaning, irrespective of personal background. It cannot be merely a passive data dump–a simple matter of memorization and regurgitation. If we sometimes fail to make things as clear as we might in our writing, we offer our apologies. Know at least that we have tried to place ourselves in your shoes, well aware too of just how deep and broad this field is.

We attempt to present a portrait of environmental geology as honestly as we can. We certainly take full responsibility for any scientific errors in our text (and indeed expect there will be a few that crop up in the normal course of evolving research). May your class instructor be quick to jump in with fresh, critical insights as you learn! In the end, this is all about you.

Distinctive Features of the Book

At the start of each chapter, we provide a series of **learning objectives** that let you know what you'll be learning as you read on. We organize the chapter around these learning objectives, one for each major chapter heading.

We provide a streamlined set of features to aid your learning. **Galleries** of photos illustrate many geological wonders and the ways in which humans and Earth interact. The galleries are intended to stimulate your curiosity about and appreciation for natural geological wonders, and the ever-dynamic engagement of people with the environment. **Case Studies** highlight the relevance of the text discussion. These cover a broad spectrum of subjects and geographical areas, but many of them focus on the causes and aftereffects of bad environmental and geological planning.

At the end of each chapter, we provide is a list of **Key Terms** introduced in the chapter and a **Summary** of the chapter in outline form linked, for ease of studying, to the learning objectives.

The Eighth Edition

- The eighth edition of *Geology and the Environment* is unlike any earlier edition of this book. We have retained a focus on case studies, added copious new illustrations, and attempted to streamline the text in an easy-to-read style while updating every facet of the text and its many figures and tables.
- Every chapter of the book has been substantially, if not thoroughly rewritten. Several chapters have been combined and shortened in order to emphasize key concepts and only the most pertinent examples–the result, a user-friendly book aiming to keep and build student interest. Two new chapters, "Earth as a System" and "Climate Change" now lead the book, providing students an up-to-date context for each of the chapters to follow.
- Almost every photograph in the book has been replaced to facilitate student understanding and appreciation of geological phenomena and related human impacts. A newly formatted Photo Gallery section at the end of each chapter helps illustrate chapter content with current examples.

- All data in graphs, tables, figures and text has been updated to reflect the most recent understanding of our environment, as of the fall of 2022.
- Learning Objectives now lead each chapter and form the basis for chapter organization via major headings allowing students to readily find and master specific concepts. The structural framework of this edition is far more simplified than that of previous editions, adding to ease of use.

Supplements

Instructor Resources

Instructor Companion Site

Everything you need for your course in one place! This collection of book-specific lecture and class tools is available online via www.cengage.com/login. Access and download PowerPoint presentations, images, instructor's manual, and more.

Cengage Learning Testing Powered by Cognero

Cengage Learning Testing Powered by Cognero is a flexible online system that allows you to:

- Author, edit, and manage test-bank content from multiple Cengage Learning solutions
- Create multiple test versions in an instant
- Deliver tests from your Learning Management System, your classroom, or wherever you want

Acknowledgments

We are deeply indebted to Sonya Vogel for her vast contributions to this project. Her attention to detail and research efforts to ensure all the data presented are current and accurate added tremendous value. We also wish to thank the team of people from Cengage Academic without whom this edition would not have been possible, particularly Vicky True-Baker, Senior Portfolio Product Manager, and Abby DeVeuve, Project Manager. We also wish to thank Anupam Bose, Senior Project Manager at MPS Limited, for keeping us on task and managing all the moving pieces. Finally, we want to thank our families for their patience as we worked on this project.

We gratefully acknowledge the thoughtful and helpful reviews by a great number of individuals, some of whom also reviewed earlier editions. We also acknowledge the contributions of other reviewers and those who have generously contributed published and unpublished materials and photographs as this book evolved into its eighth edition. These have truly helped us build this book. The list is long and we offer our thanks to all:

Herbert G. Adams, *California State University, Northridge*
Jürg Alean, *Kantonsschule Zücher Unterland, Bülach, Switzerland*
Richard M. Allen, *University of Wisconsin–Madison*
Thomas B. Anderson, *Sonoma State University*
Kenneth Ashton, *West Virginia Geological Survey*
James L. Baer, *Brigham Young University*
Ed Belcher, *Wellington, New Zealand*
William B. N. Berry, *University of California, Berkeley*
Robert Boutilier, *Bridgewater State College*
Tom Boving, *University of Rhode Island*
Kathleen M. Bower, *Eastern Illinois University*
David Bowers, *Montana Department of Environmental Quality*
Lynn A. Brant, *University of Northern Iowa*
T. K. Buntzen, *Alaska Department of Natural Resources*
Don W. Byerly, *University of Tennessee, Knoxville*
Susan M. Cashman, *Humboldt State University*
Elizabeth Catlos, *Oklahoma State University*
Dan Cayan, *U.S.G.S.*
Ward Chesworth, *University of Guelph*
Christopher Cirmo, *State University of New York, Cortland*
Robert D. Cody, *Iowa State University*
Kevin Cornwell, *California State University–Sacramento*
Jim Cotter, *University of Minnesota, Morris*
Rachael Craig, *Kent State University*
John Dassinger, *Maricopa Community College*
Charles DeMets, *University of Wisconsin, Madison*
Terry DeVoe, *Hecla Mining Company*
Lisa DuBois, *San Diego State University*
Greg Erickson, *Sullivan County Community College*
Mark W. Evans, *Emory University*
Edward B. Evenson, *Lehigh University*
Larry Fegel, *Grand Valley State University*
John Field, *Western Washington University*
Lydia K. Fox, *University of the Pacific*
Earl Francis
Robert B. Furlong, *Wayne State University*
Marion M. Gallant, *Colorado Department of Public Health and Environment*
John Gamble, *Wellington, New Zealand*
Josef Garvin, *Eden Foundation, Falkenberg, Sweden*
Rick Giardino, *Texas A&M University*
Gayle Gleason, *SUNY, Cortland*

Raymond W. Grant, *Mesa Community College*
John E. Gray, *U.S.G.S.*
Bryan Gregor, *Wright State University*
Tark S. Hamilton, *Consultant*
Gilbert Hanson, *State University of New York, Stony Brook*
Edwin Harp, *U.S.G.S.*
Raymond C. Harris, *Arizona Geological Survey*
Douglas W. Haywick, *University of South Alabama*
Eric Henry, *University of North Carolina, Wilmington*
Lynn Highland, *U.S.G.S.*
Barbara Hill, *Onondaga Community College*
Roger D. Hoggan, *Ricks College*
Bryce Hoppie, *Mankato State University*
Alan C. Hurt, *San Bernardino Valley College*
Pam Irvine, *California Division of Mines and Geology*
David D. Jackson, *UCLA*
Gaoming Jiang, *China Academy of Sciences*
Randall Jibson, *U.S.G.S.*
Bill Kane, *University of the Pacific*
Steve Kenaga, *Grand Valley State University*
Marcie Kerner, *Arco, Anaconda, Montana*
Hobart King, *Mansfeld University*
Joe Kirschvink, *California Institute of Technology*
Peter Kresan, *University of Arizona*
Robert Kuhlman, *Montgomery County Community College*
Kevin Lamb
Kenneth A. LaSota, *Robert Morris College*
Douglas J. Lathwell, *Cornell University*
Rita Leafgren, *University of Northern Colorado*
Joan Licari, *Cerritos College*
Rick Lozinski, *Fullerton College*
Lawrence Lundgren, *University of Rochester*
Michael Lyle, *Tidewater Community College*
Berry Lyons, *University of Alabama*
Harmon Maher, *University of Nebraska, Omaha*
Alex K. Manda, *East Carolina University*
Peter Martini, *University of Guelph*
John Maurer, *CIRES, University of Colorado*
Larry Mayer, *Miami University*
David McConnell, *University of Akron*
Garry McKenzie, *Ohio State University*
Lisa McKeon, *U.S.G.S.*
Matthew L. McKinney, *University of Tennessee*
Robert Meade, *California State University, Los Angeles*
Chuck Meyers, *U.S. Department of the Interior, Surface Mining Reclamation and Enforcement*
Siddhartha Mitra, *East Carolina University*
William Mode, *University of Wisconsin, Oshkosh*
Marie Morisawa, *State University of New York, Binghamton*
Jack A. Muncy, *Tennessee Valley Authority*
George H. Myer, *Temple University*
William J. Neal, *Grand Valley State University*
James Neiheisel, *George Mason University*
Jennifer Nelson, *Indiana University–Purdue University Indianapolis*
Michael J. Nelson, *University of Alabama, Birmingham*
Ed Nuhfer, *University of Colorado, Denver*
June A. Oberdorfer, *San Jose State University*
Lloyd Olson
Alberto E. Patiño Douce, *University of Georgia*
Darryll Pederson, *University of Nebraska–Lincoln*
Libby Pruher, *University of Northern Colorado*
Kristin Riker-Coleman, *University of Wisconsin Superior*
Albert J. Robb III, *Mobil Exploration and Producing U.S., Inc., Liberal, Kansas*
Charles Rovey, *Missouri State University*
Robert Sanford, *University of Southern Maine*
Steven Schafersman, *Miami University*
James L. Schrack, *Arco, Anaconda, Montana*
Feride Schroeder, *Cuesta College*
Robert Schuster, *U.S.G.S.*
Geoffrey Seltzer, *Syracuse University*
Conrad Shiba, *Centre College*
Jennifer Shosa, *Colby College*
Edward Shuster, *Rensselaer Polytechnic Institute*
Kerry Sieh, *Earth Observatory of Singapore*
Susan C. Slaymaker, *California State University, Sacramento*
Joe Snowden, *University of Southeastern Missouri*
Frederick M. Soster, *DePauw University*
Neptune Srimal, *Florida International University*
Konrad Steffen, *CIRES, University of Colorado*
Dean Stiffarm, *Environmental Control Officer, Fort Belknap Reservation*
Hongbing Sun, *Rider University*
Terry Swanson, *University of Washington*
Siang Tan, *California Division of Mines and Geology*
Glenn D. Thackray, *Idaho State University*
Peter J. Thompson, *University of New Hampshire*
Joan Van Velsor, *California Department of Transportation*
Adil M. Wadia, *University of Akron, Wayne College*
Peter W. Weigand, *California State University, Northridge*
Todd Wilkinson, *Bozeman, Montana*
Nancy S. Williams, *Missouri State University*
Nathaniel W. Yale, *Pomona College*
Simon Young, *Montserrat Volcano Observatory*
Ning Zing, *University of Maryland*

About the Cover Stuðlagil Canyon, in eastern Iceland, is a dramatic landscape of rock and water. The canyon exposes near perfect columns of basalt, formed as molten lava slowly cooled and contracted. Iceland is an island of volcanos and marks the fiery boundary between two tectonic plates. The pale green water is opaque with fine sediment, ground to a powder by the glaciers upstream. Over millennia, water carved the canyon, taking advantage of every weakness in the rock to pluck one more piece and send it tumbling downstream. Water and rock: together they define much of environmental geology.

Abstract Aerial Art/DigitalVision/Getty Images

"We have altered the physical, chemical and biological properties of the planet on a geological scale. We have left no part of the globe untouched."

—David Suzuki

1

Why Environmental Geology Matters

Learning Objectives

LO1 Explain what environmental geology is and why it matters

LO2 Give examples of major problems approached by environmental geologists

LO3 Describe the tools used by environmental geologists doing science

Cars cross a bridge in Iceland. The black sand is from volcanoes that built the island but threaten its people. The water is thick with silt from melting glaciers.

Bim/E+/Getty Images

Deadly Mud

iStock.com/anamejia18

Figure 1.1 In 1985, this was a street in the thriving town of Armero. Today, it's abandoned, the victim of mud pouring from the nearby Nevado del Ruiz volcano.

The students were on a university geology field trip, planning to spend the night in the city of Armero about 85 kilometers (50 miles) west of Bogotá, Columbia, on November 13, 1985. With a population of 29,000 in 1985, Armero consisted mostly of one- or two-story buildings stretching along the Lagunillas River (**Figure 1.1**). It was an agricultural hub surrounded by small farms, ranches, and strips of undeveloped forest in gently rolling foothills; much quieter than the bright-lights, big-city of Bogotá.

Forty-some kilometers (approximately 25 miles) west of Armero is Los Nevados National Natural Park, a destination for adventuresome tourists, featuring stately wax palms, hot springs, and snowy volcanic peaks–the "spine" of the Andes Mountains at this latitude. The park is a major attraction for geology students. Most prominent of the volcanoes is Nevado del Ruiz, which rises to 5,320 meters (17,460 feet). Visitors can climb to the rugged top, though the hike there is strenuous, to say the least. The volcano is active, with a steaming summit crater cloaked in snow, glacial ice, and volcanic ash (**Figure 1.2**).

Nevado del Ruiz had reawakened from a 70-year-long volcanic slumber on September 11, 1985, just 2 months before the geology students came to town. Soon after, a large volcanic blast covered the peak with hot volcanic ash and steam that melted large amounts of snow and ice capping the summit. A vast flow of meltwater poured into the headwaters of streams, forming thick, dangerous slurries of mud and rock that travelled as far as 30 kilometers (20 miles) from the summit crater. The government of Colombia was properly worried by this event and requested the assistance of a team of United Nations volcano scientists to assess the situation. They did–and warned repeatedly of worse things to come!

The scientists made no effort to contact local residents directly about their concerns, however. That was the job of the civil government. The authorities did not communicate warnings in a coordinated way that made sense to the population. Some local newspapers suggested that such warnings were a plot and accused officials of deliberately driving down property values for personal gain. In addition to these socio-political problems, some locals did not take the scientists seriously, many of whom were foreigners who'd been in the country for only a short time.

A few hours before the geology field class settled in to sleep at their Armero hotel on 13 November, Nevado del Ruiz once again erupted, though this went largely unnoticed in the lowlands. A light fall of powdery ash settled over the city in the afternoon, but it was a stormy day, and most people were unable to hear, let alone see, the volcano as it rumbled many kilometers away. The storm impeded efforts to warn and evacuate

iStock.com/Nicolas Socadagui

Figure 1.2 Steaming summit crater, Nevado del Ruiz, in 2022.

people from threatened zones. Certainly, no one expected that this time a dense flood of muddy debris (known as a lahar by geologists) would come roaring at high speed down the Lagunillas River channel. But it did, a river of mud emerging out of the darkness moving at 50 kilometers (30 miles) per hour. This nasty surprise buried the city (**Figure 1.3**). More than 20,000 people were killed by the mud, which was so thick it entombed people, who were unable to move. Many of those rescued were covered in mud and badly injured from the debris carried by the lahar (**Figure 1.4**).

Afterward, student Jose Luis Restrepo recounted what happened to U.S. Geological Survey scientist John P. Lockwood:

> [At] About 11:30[pm] all the lights went out as Armero's generating plant was inundated and the flow arrived a few minutes later. It was only about 2 meters deep as it entered town, rushing down streets, and the only lights were from the headlights of cars being tumbled about by the flow.
>
> We ran to our hotel and three of us climbed to the third-floor terrace, thinking this would a safe place since the walls were [made] of concrete. Lower-standing buildings around us were swept away as the flood kept rising. I felt great shocks as rocks slammed into the hotel and suddenly it all crashed down. The screams of people were terrible, and I was sure I was going to die a horrible death, but with four companions we climbed into a concrete water cistern that had come from the hotel roof.
>
> We were swept along with debris and screaming people, as burning gas tanks lit up the terrible scene. We saw a high wave of mud coming at us, but somehow our tank stayed afloat. Some mud was warm—but most was cold. We crashed against a standing tree. There were many people around us stuck in the mud, but we were unable to help them. We dragged one man into the tank but his legs had been amputated and he was in great pain. When morning came, we saw that we were only 50 meters (175 feet) from the edge of the mud, but when I saw the terrible devastation behind us, I lost all control and cried—Armero was gone—except for our little island of trees. About 300 people were able to struggle through the mud to a small hill—the Armero cemetery—where we waited among the tombstones for rescue.

Of the 32 students in the university group, 13 died.

questions to ponder

1. What factors limited the ability of scientists to communicate with the public prior to the Armero disaster?
2. Imagine that you are asked by a government to put together a team of people to prevent future volcanic disasters such as Armero from taking place. What type of skills would you seek to have represented on your team, and why include this skill set in particular?

Langevin Jacques/Sygma/Getty Images

Figure 1.3 A view from above shows that much of Armero was buried deeply in mud.

Pool Bouvet/Duclos/Hires/Gamma-Rapho/Getty Images

Figure 1.4 A man rescued from the lahar in Armero is rushed to medical attention after being evacuated by helicopter.

1.1 What is Environmental Geology and Who Are Environmental Geologists? (L01)

Environmental geology is an applied science that matters to billions of people around the globe because it concerns real-life problems like flooding, natural resource availability and extraction, landslides, volcanic eruptions, coastal erosion, waste management, groundwater, and earthquakes. No place on Earth is safe from the processes that shape our planet and our lives, and thus understanding how the Earth works is critical for mitigating hazards and improving people's futures. Environmental geology is science that matters to you and to the 8 billion people now populating Planet Earth.

The Armero account clearly illustrates the close connection people have with the land and natural geological processes. By having the knowledge to change our ordinary behaviors, being better prepared, and planning wisely, we can control outcomes and reduce both personal and societal risk as the Earth beneath us changes on time scales that matter to us as people. The disaster at Armero could have been averted through better communication between scientists, diverse government agencies, and local citizens. A better appreciation of potential volcanic activity, of glaciers and their melting, and of streams, floods, and the debris they carry, not to mention foresight in planning for crises, could have saved lives. If following the afternoon ash fall on November 13, citizens had moved to higher ground at the edge of town, a short walk from most homes and businesses, many more would have survived (**Figure 1.5**).

Langevin Jacques/Sygma/Getty Images

Figure 1.5 Armero in the days after the lahar. The lowlands are buried in mud, but the hills escaped unscathed. If people had reached those hills, their chances of survival would have been far higher.

What Do Environmental Geologists Do?

Environmental geologists study the solid Earth, the surface Earth, water, and the oceans with a variety of tools; some are high tech, but some are as low tech as a shovel and rock hammer (**Figure 1.6**). Some environmental geologists use microscopes; others use chemical analyses, mathematical models, and historical photographs to understand how the Earth works (**Figure 1.7**).

The solid Earth is a remarkably active place for what we consider most days to be stable bedrock. Volcano monitoring is a critical task that helps build both a fundamental understanding of how volcanoes work and a means to better predict eruptions (**Figure 1.8**). Analysis of earthquake damage is key for mapping faults and understanding where, when, and how earthquakes are likely to strike (**Figure 1.9**). Environmental geologists increasingly use high technology means like global positioning systems (GPS) to measure and

Evgeny Haritonov / Alamy Stock Photo

Figure 1.6 Geologist in a quarry with a highly useful favorite tool, a rock hammer.

Paul Bierman

Figure 1.7 Environmental geology students measure water quality in the laboratory.

Environmental geologists are professionals who apply understanding of the natural environment–at least its geological dimension–to identify potential problems for people, and to help solve those problems. This doesn't simply mean learning how to recognize impending disasters to prevent or avoid their occurrence, although much of that is considered in the following text. Environmental geologists are also involved with the development and extraction of natural resources such as oil, coal, water, and minerals, but in ways that minimize or avoid developing threats such as land degradation, pollution, and unnecessary resource depletion.

In today's world of hyperspecialization, the job description "environmental geologist" is deliberately vague. Scientists doing environmental geology come from many different backgrounds: hydrologists (water), pedologists (soil), seismologists (earthquakes), mine and waste managers, physical oceanographers, field geologists, geochemists, volcanologists, and climate researchers. An environmental geologist is anyone whose study and work involves one or more components of our dynamic geological landscape (Chapter 2). The bottom line is that what determines "a better future for all" is what defines "a better way of living on Planet Earth." We've written this book with that spirit in mind.

1.2 Environmental Geology: Why It Matters to You and The World (LO2)

In the following examples we examine a selection of incidents or situations that illustrate what environmental geologists do professionally and the problems they help solve. Most of these examples concern water and climate change, either directly or indirectly. That is no accident: It turns out that water in one way, shape, or form, is a primary concern for many, if not most, environmental geologists; and climate change is a major global process impacting everyone at this point in history.

This is not to say that other environmental phenomena aren't just as important; they are, depending upon when and where they occur, including hurricanes, landslides, and threats posed by coal mining. We'll get to these and other topics later in this book. The examples below pertain throughout the field of environmental geology. Each illustrates specific points that we summarize at the end of this section.

1.2.1 The Newark Contaminant Plume

The eastern end of the Ontario Plain, a suburb of Los Angeles, is home to over a half million people who depend on water drawn not only from the Colorado River and

USGS

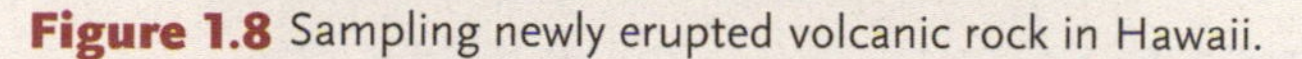

Figure 1.8 Sampling newly erupted volcanic rock in Hawaii.

monitor Earth movements as faults slip, volcanoes bulge, and landslides creep downslope (**Figure 1.10**).

Resource extraction, be it energy, minerals, or construction materials, is a field in which the skills of environmental geologists are in high demand. Expertise in how both rocks and the surface Earth behave allows resource extraction with the least environmental footprint and meaningful remediation so that the legacy of extraction is as benign as possible (**Figure 1.11**). The siting and maintenance of geothermal energy production facilities requires geologic expertise to find hot rocks and manage the crystallization that occurs where hot salty fluid from deep in the Earth cools and depressurizes at the surface (**Figure 1.12**).

Ken Hudnut/USGS

Figure 1.9 Measuring offset on road line after it was displaced by an earthquake.

Measuring environmental change at and near Earth's surface keeps environmental geologists busy and outdoors in all kinds of weather at all kinds of places. Stream gauging is a "must do" to determine how much water moves through a

(*Continued*)

Arctic-Images/Stone/Getty Images

Figure 1.10 Using high-precision GPS to measure the location of points on Earth's surface.

iStock.com/LordHenriVoton

Figure 1.11 Geologist photographing a rock sample in a quarry.

Roger Ressmeyer/Corbis/VCG/Getty Images

Figure 1.12 Sampling volcanic gases in a geothermal field.

Isabella Bennett

Figure 1.13 Measuring the speed at which water flows in a stream in order to calculate discharge.

stream over time—a key aspect of water management and flood prediction (**Figure 1.13**). Many environmental geologists are employed in sampling, collecting soil and water near facilities that contain hazardous materials, including those used for energy and chemical manufacturing as well as waste disposal (**Figure 1.14**). Soil analysis is key to maximizing agricultural productivity and siting septic systems so that they function properly (**Figure 1.15**). Waste management is another area of major concern in environmental geology (**Figure 1.16**): how do we safely store and manage the waste that we

iStock.com/Sorapop

Figure 1.14 Sampling surface water near a plant that uses hazardous materials.

BartCo/E+/Getty Images

Figure 1.15 Using a soil auger to collect soil samples.

USGS

Figure 1.16 USGS scientist samples a well to determine groundwater purity. Such wells encircle active and closed landfills to detect whether contaminants are moving off site.

generate through all the resource extraction we've undertaken to support our global civilization?

Earth materials have different strengths. Some, like granite, are hard and strong, stoically resisting erosion. Others, like ocean clay, are weak and soft, failing in landslides when provoked by an earthquake or even heavy equipment. Environmental geologists measure the strength of Earth materials and the movements of landslides (**Figure 1.17**). Along coastlines, they monitor waves and the movement of sand on the beach (**Figure 1.18**), while also tracking coastal erosion so people can be warned well before roads and homes collapse in the ocean (**Figure 1.19**).

Environmental geologists are deeply involved in understanding Earth's climate system, how it has changed over time, and how to mitigate and adapt to the impacts of climate change in the future. Some of the most extreme fieldwork an environmental geologist might do is on the world's giant ice sheets in Greenland and Antarctica, monitoring to determine where and how quickly they are melting (**Figure 1.20**).

USGS

Figure 1.17 Instrumenting a landslide to study how rapidly and when it moves.

USGS

Figure 1.18 Using GPS to map a beach.

Chris J Ratcliffe/Getty Images News/Getty Images

Figure 1.19 Geologists inspect home undermined by coastal erosion.

Arctic-Images/Stone/Getty Images

Figure 1.20 Collecting a shallow ice core.

Figure 1.21 The city of Riverside, California.

California aqueducts (surface supplies in this semi-arid region) but from groundwater also. It's a stunning place with mountains towering over a wide, heavily developed valley bottom (**Figure 1.21**). **Groundwater** is the technical term for water that collects in underground pore spaces and fractures within rock (Chapters 2 and 8). We extract it by pumping from wells. Large bodies of groundwater in completely saturated bedrock are termed **aquifers**.

The Bunker Hill Basin aquifer underlies 285 kilometers^2 (110 miles^2) of the eastern Ontario Plain. The underground water body is up to 200 meters (650 feet) thick and provides approximately 1.4 million acre feet (1,750 billion liters, or 467 billion gallons) of freshwater each year to residents, from over 200 extraction wells. This includes 25% of the residential water supply in the City of San Bernardino and 75% in nearby Riverside (**Figure 1.22**).

In 1980, as water district officials implemented a new water supply monitoring program, managers discovered to their horror that a potent plume of toxic pollutants was seeping into the aquifer from near its northwest corner. They dubbed the apparent source area, where the pollutants first percolated into the ground, the Newmark Groundwater

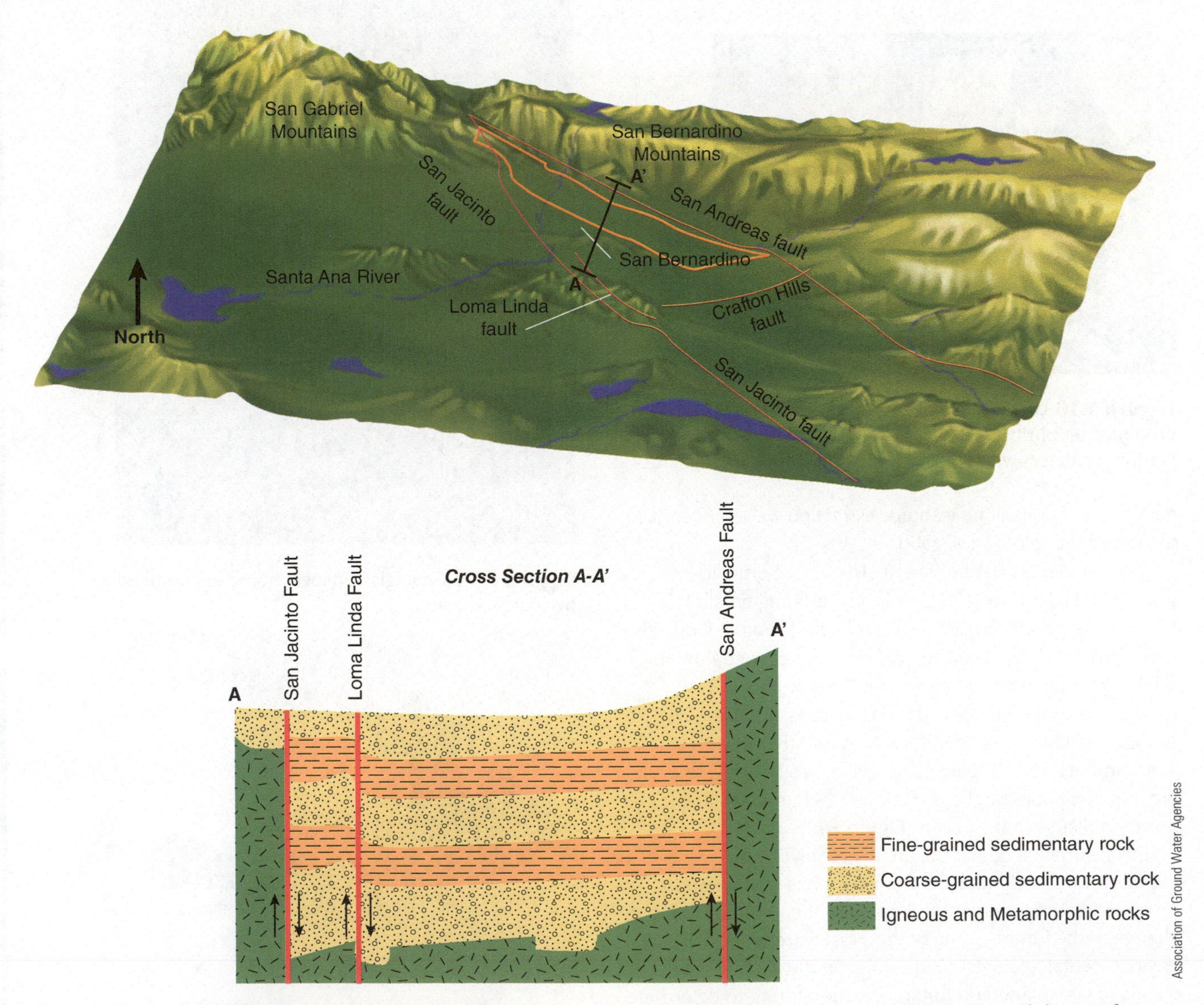

Figure 1.22 Map and cross-section of the geology of the Bunker Hill Basin. The basin is surrounded by faults, and the aquifer hosted in sand and gravel.

Table 1.1 Contaminants of Chief Concern, Newmark Contaminant Plume

Chemical	Use	Human Health Hazards
1,2-Dichloroethane (DCE)	Used in production of vinyl and molded plastic toys such as soldiers and holiday decorations. Also used as a paint remover and degreaser. Produces chlorine gas in a laboratory as needed by chemists for their work.	Irritates eyes, burns skin, is hard on lungs, causing chronic bronchial infections, and nervous system and liver damage. One of the most caustic of synthetic chemicals.
Tetrachloroethylene (PCE)	Sweet odor. Used in dry cleaning and for degreasing metal parts in automotive industry. Found in some paint strippers and spot removers.	Probable carcinogen. Attacks the human central nervous system, causes skin irritation, and may lead to Parkinson's disease.
Trichloroethylene (TCE)	Produced as an industrial solvent. Once used in vegetable oil extraction and coffee decaffeination and to flavor hops and spices.	Readily depresses the human central nervous system, and is anesthetic. Acute exposure can produce symptoms resembling alcohol intoxication. Produces liver cancer in mice and kidney cancer in rats. Possible link to Parkinson's disease.

Contamination Site. The pollutants included 1,2-DCE (dichloroethane), PCE (tetrachloroethylene), and TCE (trichloroethylene). **Table 1.1** summarizes the source and human impacts of these pollutants. People were at risk of drinking this stuff, never mind using it to clean cuts and bruises, shower, wash their cars, water lawns, and other domestic applications.

The Newmark contaminant plume extended 8 kilometers (5 miles) downslope into the aquifer. It was not alone. There are several other contaminant plumes in the aquifer (**Figure 1.23**). Twenty water supply wells within a 10-kilometer (6-mile) radius of the impacted area were quickly shut down, and soon the State and Federal governments joined the County in a years-long effort, working with environmental geologists to map the contamination identify its source, and discuss its cleanup. The U.S. Environmental Protection Agency (EPA) quickly designated the area a **Superfund site**–that is, an area in need of priority environmental cleanup according to the **Comprehensive Environmental Response, Compensation, and Liability Act (CERCLA)**, passed by Congress in 1980 (see Case Study 1.2).

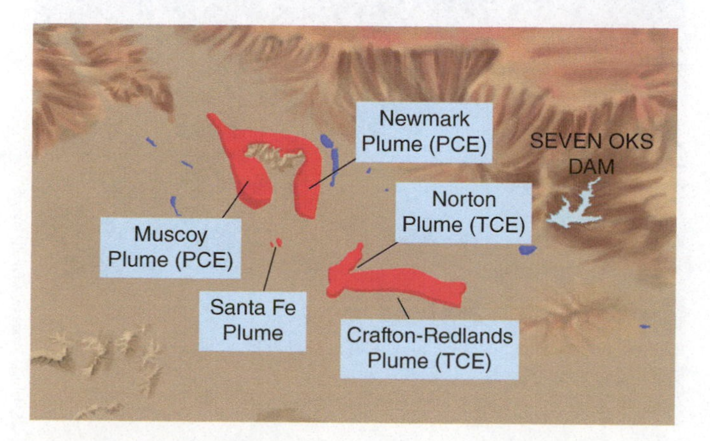

Figure 1.23 Five different plumes of contaminants spreading through the aquifer.

Who was responsible? Ultimately, the finger pointed in part toward the U.S. Army, which operated the 650-hectare (1,600-acre) San Bernardino Engineering Depot near the apex of the Newmark plume at the bottom of Cajon Pass. The Depot was just below the mountains at the northwestern side of the Ontario Plain and operated from 1940–1947.

A lot went on in the Depot. Almost 500 Italian prisoners of war were housed there during WWII and provided labor to pick fruit in the nearby orange groves as well as do maintenance work for the Army. The San Bernardino Engineering Depot was a supply source for materials headed to camps around the western United States, where American citizens of Japanese descent were forced to live during World War II (WWII) after being removed from their homes by the U.S. Government. The Depot was also the center of logistical support for training operations in the Mojave Desert during WWII, and that's where the problems may have started.

The Depot stored and used quantities of oil and many petrochemicals. These substances, most of them toxic, were used for much of the work that reports show went on there. Vehicle maintenance, and dry cleaning before tents and uniforms were repaired, were the most important jobs. Waste management practices in the 1940s (operating in a climate of haste and national emergency) were likely to blame, including chemical spills, incomplete cleanup of maintenance yards, and perhaps intentional ground disposal of liquid chemical waste. The EPA had not yet established that the Depot was *definitely* the source of the Newmark plume, but in a court settlement the U.S. government agreed in August 2004 to pay the City of San Bernardino $69 million, implicitly to help start cleaning up the mess.

Ongoing cleanup includes the pumping out of contaminated groundwater for storage in drums at secured landfills elsewhere, the extraction of gases by using sophisticated wellhead air suction devices, and the use of carbon filters–all tools of the trade in environmental geology (**Figure 1.24**). The solid

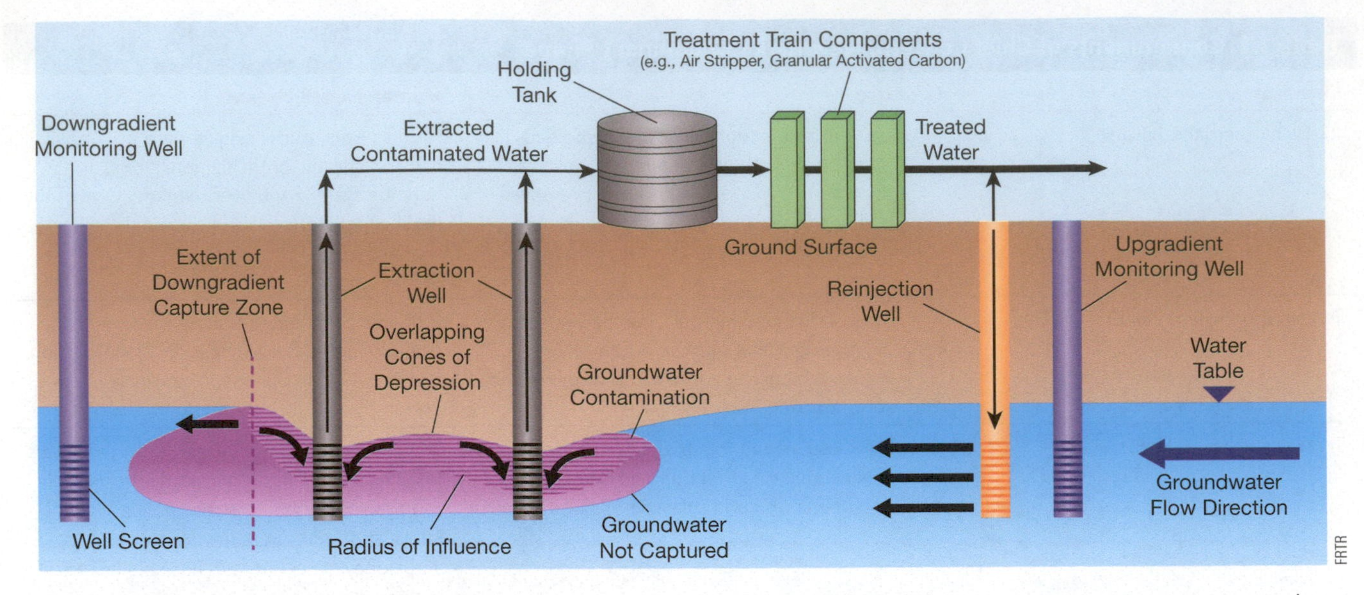

Figure 1.24 This cross-section shows how pumping and treating contaminated groundwater can remove contaminants and slow their spread.

and liquid contaminants, once removed from the aquifer, are isolated at public expense. Several EPA Five-Year Review Reports suggest that the remedial strategy taken seems to be working insofar as containing further expansion of the plume and isolating it from residents and water users.

There are several important takeaways from this example. Environmental problems of a geological nature can develop slowly. They may not be identified for many years, even decades, after they begin–in this case, almost 40 years! By then, it may also be unclear who ultimately is to blame. Quite often the public (taxpayers) end up footing the bill for cleanup. (A responsible party, even if identified, may have passed away or no longer be in business). In the Newmark case, environmental geologists were critical at two stages: (1) discerning the problem and (2) advising the solution; that is, the process of Superfund clean-up. They did their work by applying theory and observations to understand how groundwater accumulates, flows beneath the surface, and carries contaminants.

1.2.2 Fukushima: A Geological Nuclear Disaster

Much of Japan is a highly crowded, industrialized country. It is also highly mountainous, meaning that most of the population tends to crowd onto coastal plains with wide, fertile valley floors. Mountain areas are subject to landslides and flash flooding, making them unsuitable for nuclear power plant development. Hence, Japan's nuclear stations are positioned in coastal areas, where most of the country's food also tends to be produced. This makes good geographical and technological sense, but from a geological standpoint this choice can be problematic.

The coasts of Japan are especially vulnerable to storm surges and **tsunami**–giant seismic sea waves. Engineers plan for these natural hazards by construction of sea walls. But are these engineering works adequate? The Tohoku-Oki Earthquake of March 11, 2011 put that question to the test (Chapter 6). The tremor caused a major tsunami (**Figure 1.25**), and right within its path of travel on the northeast coast of Honshu stood the Tokyo Power Company (TEPCO) Fukushima (Daiichi) nuclear power station (**Figure 1.26**).

Nuclear power plants operate by releasing heat through controlled nuclear reactions (see Chapter 11). That heat in turn is used to make steam, which spins turbines to generate electricity. Important to this operation is a means to control the heat released by the reactors, where the nuclear reactions are taking place. A coolant such as cold water circulates through the system to keep the reactor core from getting too hot. Although removal of the nuclear fuel rods from reactors can slow down or stop the reactions taking place, the heat keeps coming as radionuclides decay. If cooling water does not circulate, reactors and stored fuel rods risk melting down

Sadatsugu Tomizawa/AFP/Getty Images

Figure 1.25 The 2011 tsunami coming ashore in Fukushima, overwhelming tall trees and tossing cars in the surf.

TEPCO / Alamy Stock Photo

Figure 1.26 The Fukushima Daiichi Nuclear Power Station February 21, 2007, 4 years before it was severely damaged by the tsunami of 2011.

from overheating, spreading deadly radioactive particles into the surrounding environment.

The Fukushima power station consisted of six separate boiling water reactors. Two had been shut down for planned maintenance at the time of the earthquake, while one had had its fuel rods removed and so was harmless. The three that were in service (Reactors 1, 2, and 3) shut down automatically with the shaking, and emergency generators immediately switched on to provide electricity to keep water moving through the reactor systems to cool them down safely.

Engineers had built a seawall to protect the plant from the largest tsunami thought possible for this part of the Japanese coast; but the 4.5-meter (15-foot)-wall wasn't high enough (**Figure 1.27**). The incoming wave crested at 13.5 meters (45 feet)! To make matters worse, the diesel-fuel generators were placed in low-lying rooms, and when the sea

Kurita KAKU/Gamma-Rapho/Getty Images

Figure 1.27 Photo from the Tokyo Electric Power Company shows the March 11, 2011, tsunami just breaking over a crumbling sea wall and rushing toward tanks of heavy oil needed for plant operation at the Fukushima nuclear complex.

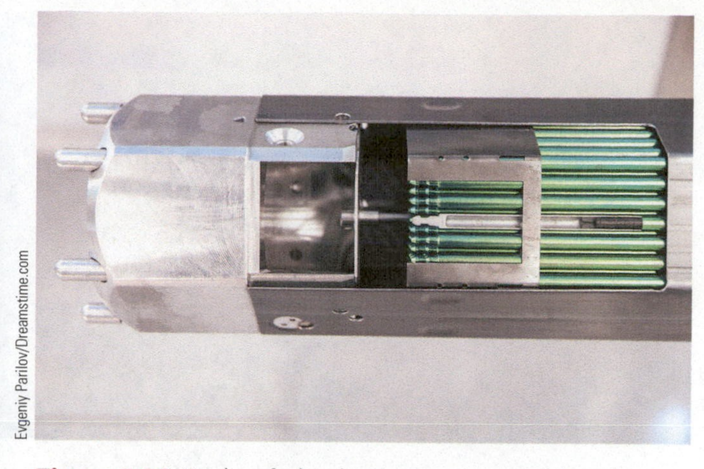

Evgeniy Parilov/Dreamstime.com

Figure 1.28 Nuclear fuel rod with zirconium cladding shown in green, with cutaway showing grey uranium fuel.

came surging in, these rooms quickly flooded, shorting out these emergency electricity sources. Six of twelve generators servicing the plant immediately stopped. Fuel rods could not be removed, and cooling water in the pipes entering the reactors stagnated. The reactors immediately began heating up. As the workers struggled, the fuel rods grew dangerously hot.

The standard zirconium cladding that sheathed each rod imparted strength and inhibited the release of radioactivity from the uranium fuel (**Figure 1.28**). But zirconium is a fairly volatile substance. It catches fire quite easily if exposed to oxygen at high temperature. And at high pressures, water or steam that is heated to a temperature above 482°C (900°F) will react to form pure hydrogen gas, an extremely flammable substance:

$$Zr + 2H_2O \rightarrow ZrO_2 + 2H_2 \text{ (gas)}$$

The hydrogen reacts explosively when it comes in contact with free oxygen in the air. (This type of explosion destroyed the airship *Hindenburg* infamously as it arrived at Lakehurst, New Jersey, in 1937; see Chapter 11).

Unknown to emergency responders, the reactors at Fukushima vented a large amount of hydrogen to the surrounding building spaces. The gas also filled the piping between Reactors 3 and 4. In short order, gas explosions blew open the pipes and tore off the roofs of four of the six power buildings (**Figure 1.29**), allowing radiation to escape over the surrounding countryside–the second-worst nuclear plant disaster in history, following the ruinous 1986 meltdown of the Chernobyl nuclear reactor in the U.S.S.R. Problems worsened as the three reactor cores active at the time of the quake melted down during the next few days, releasing radioactivity (**Figure 1.30**).

Under these circumstances, it seems almost miraculous that no one was killed or severely injured at the Fukushima power station, though decades may pass before full health impacts are known among the population exposed downwind of the failed reactors. This includes the 60,000 or so persons still evacuated from the "exclusion zone" 20 kilometers (12 miles) surrounding

YouTube, LLC

Figure 1.29 Television cameras captured the hydrogen explosion that damaged the reactor containments.

hasty decision-making involving expensive commitments. You can find many examples in modern history of people "leaping (disastrously) before looking" when it comes to land-use practices.

After the fact, environmental geologists became de facto members of the scientific group evaluating soil and groundwater contamination and the spread of radionuclides (radioactive particles) across the Pacific Ocean basin. They will remain important players in the decades to come as the short-lived radionuclides decay, long-lived nuclides are diluted by soil mixing, and people begin to reoccupy the fertile coastal region surrounding the TEPCO facility at Fukushima.

the plant, along with the residents of four small cities who will not be permitted to return to their homes except for brief periods each for the next two decades. This population remains, in effect, wards of the state, which will not provide the resources needed for property owners (unable to sell for obvious reasons) to relocate by repurchasing new homes elsewhere.

Not only was the land contaminated with radioactivity but radioactive waste was released into the surrounding ocean as cooling water leaked from the plant. Radioactivity in the water was quickly and massively diluted but functioned as a tracer of water movement across the Pacific Ocean–as did tuna that took up radioactivity near Japan and then swam east. Radioactivity was also blown across the Pacific toward the Americas. In air, fish, and in water, radioactivity levels were extremely low, far lower than levels of radioactivity in the 1980s that persisted from atmospheric testing of nuclear weapons in the 1950s and 1960s. Levels of radioactivity in Japanese air was about 1,000 times higher than in the air of North America (**Figure 1.31**).

It took the ever-so-slightly radioactive air about 5 days to arrive in California. It took the tuna about 4 months to swim across the Pacific Ocean before they were caught (and sampled by scientists) near San Diego, California, and it took about 2 years for water bearing traces of radioactivity from Fukushima to reach North America. This was an odd, unplanned, but very important demonstration about the speed at which different environmental transport processes happen on Earth.

There were operational and design errors that caused Fukushima disaster. You may be able to identify a few of them just from the brief description given above. In terms of environmental geology, better understanding of the potential sizes of tsunami along this coast was needed, and such knowledge should have been developed prior to construction of the power station (see Chapters 6, 10 and 11). Insufficient information about geology and geological processes vexes much

1.2.3 Warming Climate, Shifting Shores; Kiribati

Kami Rita Sherpa has climbed Mount Everest 24 times over many years. Several years ago, he noticed that his experience on the mountain was changing. He was more frequently seeing bodies of long-dead climbers, their bones and old climbing gear scattered over the rocks (**Figure 1.32**). More than 100 bodies of climbers who died have never been recovered from Everest; it's just too dangerous to find and remove the dead from such high elevations. In the past, the corpses were well preserved by the cold dry air and often buried in snow and ice. Now, the climate is warming and snow and ice cover on Everest, and in fact over most of the Himalayan Mountains, sometimes dubbed the third pole, is shrinking. Models suggest that by 2040, the Himalayas could be 2.1°C (3.8°F) warmer than they are today. By 2100, a third of Himalayan glaciers are likely to have vanished–melted away.

Water from the ice that's melting in the Himalayas (and everywhere else on Earth's lands) ends up in the ocean, increasing the volume of water and flooding shorelines. Over the past 100 years, glaciers, which are melting and receding in just about all the world's mountain ranges, have raised sea-level 0.75 millimeters per year. Add to that melting of polar (high-latitude) ice and we are in trouble. The Greenland Ice Sheet is now adding 0.8 millimeters per year to sea level, and models suggest Antarctica has begun to melt at similar rates. Today, global sea level is rising on average about 3.4 millimeters per year, a rate that is expected to increase rapidly over the next few decades as the climate heats up, the oceans warm, and seawater thus expands.

What happens in cold places, whether high in latitude or altitude, affects coastlines around the globe as sea level rises. Low-lying lands, particularly ocean islands (like coral atolls), river deltas, and barrier islands, are particularly at

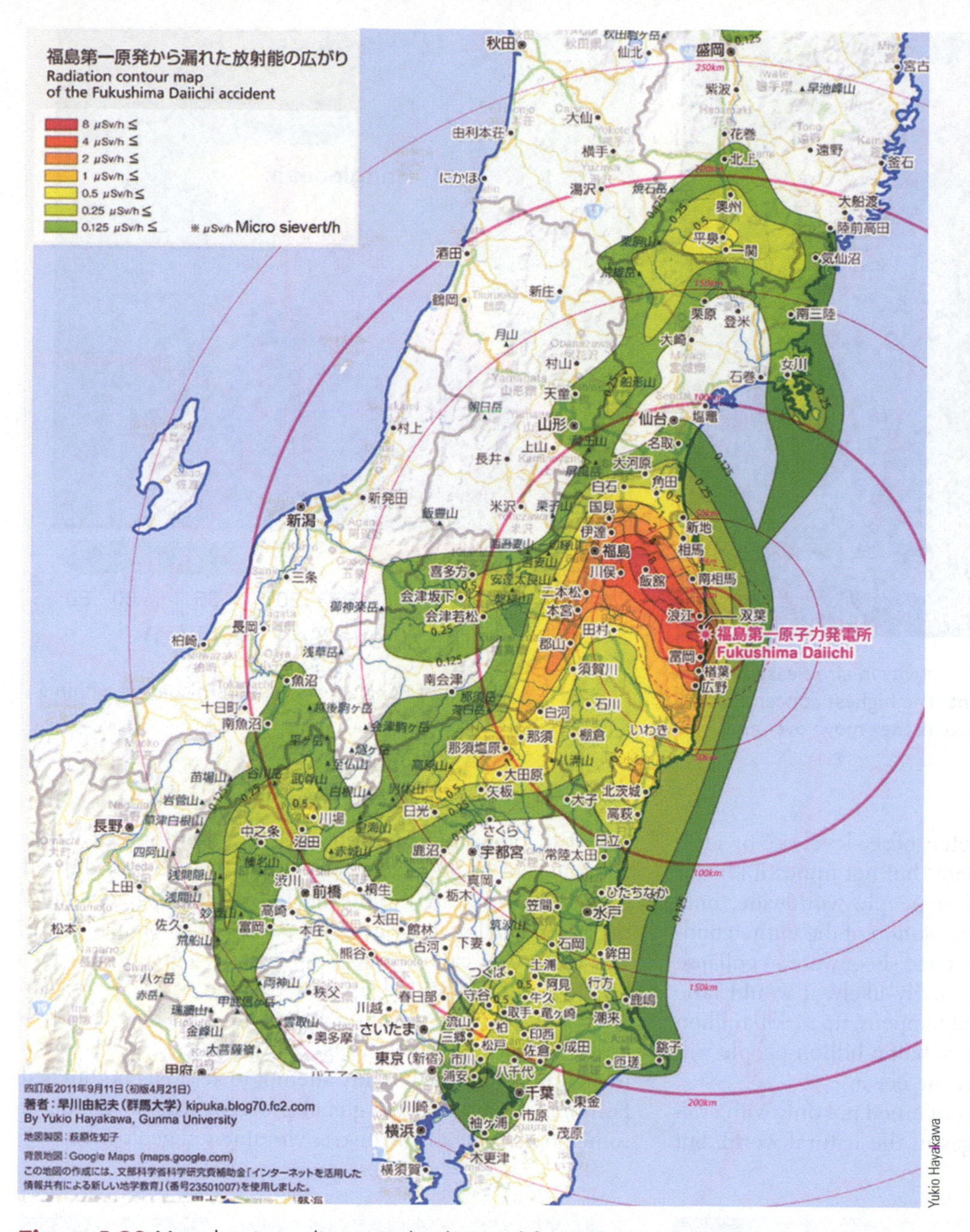

Figure 1.30 Map showing radioactivity levels in soil for 2020. Samples were collected by citizens. The map was published to educate people attending the 2020 Olympics.

risk of flooding as melting glaciers pour water once held on land into the ocean, raising sea level. On gently sloping coasts, 1 meter of sea-level rise can spread water a kilometer or more inland.

One region of concern is the country of Bangladesh, located on a delta at the head of the Bay of Bengal. The delta is massive, built of sediment transported by the Ganges and Brahmaputra Rivers all the way from the Himalayan Mountains, many hundreds of kilometers away and towering as high as 8,800 meters (29,000 feet) above sea level, to the Indian Ocean. The delta extends inland 200 miles (320 kilometers) from the coast (**Figure 1.33**).

The average elevation of the entire country is about 10 meters (32 feet) above sea level, but that is misleading: most of the land area of Bangladesh, supporting many of its 163 million citizens (about half the population of the United States), lies at lower elevation (**Figure 1.34**). Estimates suggest that by 2100, sea-level rise will have permanently flooded at least 1,000 square kilometers of low-lying land on which 800,000 people live today (other estimates suggest that sea level rise will displace upwards of 2 million people by 2100). More than twice that area and several million people will be further exposed to flooding from storm surges. It's reasonable to ask where these people will go.

Bangladesh is not alone. Kiribati is a Pacific Island nation (**Figure 1.35**). Composed of 33 islands (of which about two-thirds are inhabited), it's small, with a total land area of 313 square miles (811 square kilometers). The country's high point is a mere 3 meters (9.8 feet) above sea level. Most inhabitable land is lower. The atoll of Tarawa, which holds the nation's capital, is also home to most of the nation's 112,000 citizens.

Several of Kiribati's smaller islands have already washed over and are no longer above the waves. To make matters worse, owing to rising sea level, groundwater beneath the islands is becoming increasingly salty as seawater infiltrates precious freshwater wells (see Chapter 8). Coastal erosion increases as the ocean laps ever higher. (**Figure 1.36**). Many people wonder how the world will deal with a nation once it loses the land where its people live–the land that defines it. It seems certain that there will be many **environmental refugees**, people escaping drastic, long-lasting environmental change, in this case related to the climate. Kiribati is on the way to becoming the first nation in the world to vanish because of climate change.

NOAA

Figure 1.31 NOAA model of radioactivity in air released from Fukushima 9 days after the accident. The highest concentrations are in red over Japan and are diluted as they move over and around the Pacific Ocean.

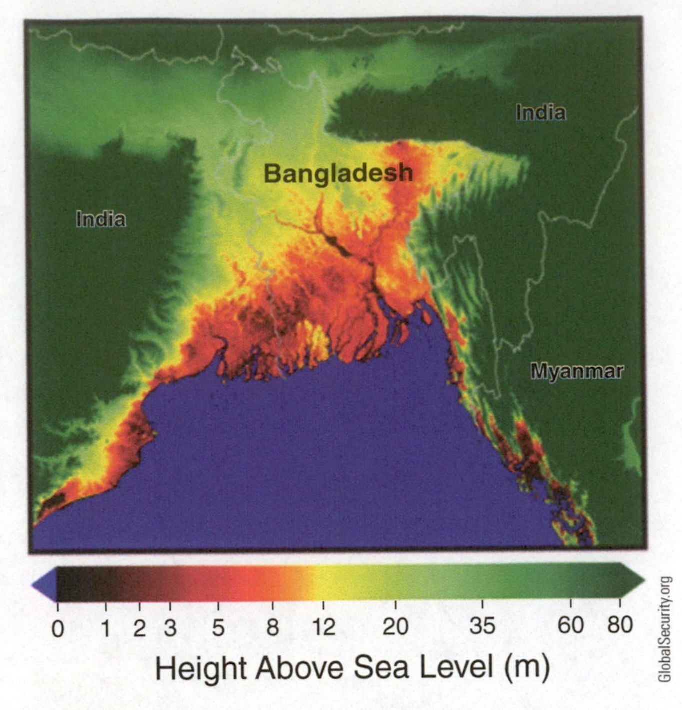

GlobalSecurity.org

Figure 1.33 Much of Bangladesh's southern coast is within a few meters of sea level.

Mapping suggests that 2 meters of sea-level rise by 2100 (that's on the high end of estimates but not impossible) will displace as many as 187 million people worldwide, only about 20% less than the 2022 population of the entire country of Brazil. If the West Antarctic ice sheet were to collapse and melt, which appears increasingly likely, it would raise sea level 5 to 6 meters. When that happens, up to 430 million people will need to move. Almost half a billion people will be flooded out of their homes by the ocean!

Environmental geology is concerned not only with evaluating the magnitude of changes in the natural world, but also with proposing solutions. **Climate mitigation** is the effort to rein in the degree of climate change already well underway, primarily through technological upgrades ranging from small-scale (e.g., solar panels on rooftops) to the very large (e.g., capture of carbon dioxide, a greenhouse gas, then pumping it back deep underground where it can dissolve and be stored in saline groundwater). There are obviously serious social, economic, and political aspects to climate mitigation too, but any attempt to act in a coordinated fashion to deal with this global problem will no doubt, at some point, crucially involve environmental geologists.

Dawa Finjhok Sherpa/Seven Summit/Redux

Figure 1.32 On Mount Everest, bodies of dead climbers are becoming exposed more frequently as snow and ice melt in our warming climate. Here, one of those bodies is removed by a team of climbers.

SOPA Images Limited / Alamy Stock Photo

Figure 1.34 With their fields and homes flooded by Cyclone Amphan, Bangladeshis take shelter on the only high ground—levees.

Raimon Kataotao / EyeEm/Getty Images

Figure 1.35 Everywhere in Tarawa, the capital of Kiribati, is close to sea level.

Jeremy Sutton-Hibbert/Alamy Stock Photo

iStock.com/JohnHodjkinson

Figure 1.36 High tides, even without storms, are flooding people's homes in Kiribati. (a) Tiaon Bwere watches waves of a "king tide" demolish his home in the village of Betio. As sea level rises, such tides bring the sea ever higher and into people's homes. (b) Waves at high tide crash into and over a seawall meant to protect this home.

1.2.4 Radon: The Silent Killer

Radon (^{222}Rn) is a radioactive gas formed by the natural disintegration of uranium as it transmutes step by radioactive step, a process that eventually leads to stable, nonradioactive lead. The immediate parent element of radon is radium (^{226}Ra), which emits an alpha particle (^{4}He) to form ^{222}Rn. ^{222}Rn has a half-life (the time it takes half the material to radioactively decay) of only 3.8 days, but it, in turn, breaks down into other elements, many of which emit dangerous radioactivity and are solid not gaseous (**Figure 1.37**). Inhaling radon is hazardous to your health as the elements into which it breaks down can lodge in your lungs, slowly damaging respiratory tissues.

Radon emanates from soils that are derived from uranium-bearing rocks, including fine-grain dark shale, a sedimentary rock, and coarser-grained granite, an igneous rock. The connection to rocks means that environmental geologists have an important role to play in assessing and reducing the radon risk because they know where such rocks are common and thus where radon is likely to be a problem (**Figure 1.38**).

Radon-generating rocks have uranium concentrations as low as a few parts per million; far less than what is found in the vicinity of uranium mines, but all the more hazardous because the danger may be unsuspected. As radium disintegrates in the top few meters of ground, radon seeps into the atmosphere or into houses through cracks in concrete slabs, basement walls, or around openings in pipes (**Figure 1.39**). The type of residential construction matters, as does the movement of air within each house. That's why careful testing for radon in your home's air is key for reducing risk.

The measure of radon activity is *picocuries per liter* (pCi/L)–the number of nuclear decays per minute per liter of air. One picocurie per liter represents 2.2 potentially cell-damaging disintegrations per minute. There are 1,000 liters per cubic meter of air. Inasmuch as the

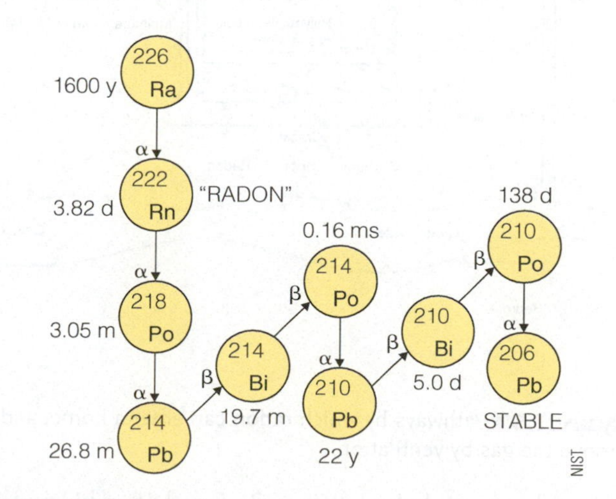

Figure 1.37 Decay chain that shows half-life (in years, days, minutes, and milliseconds) of each element on the decay chain of radium, which is derived from uranium.

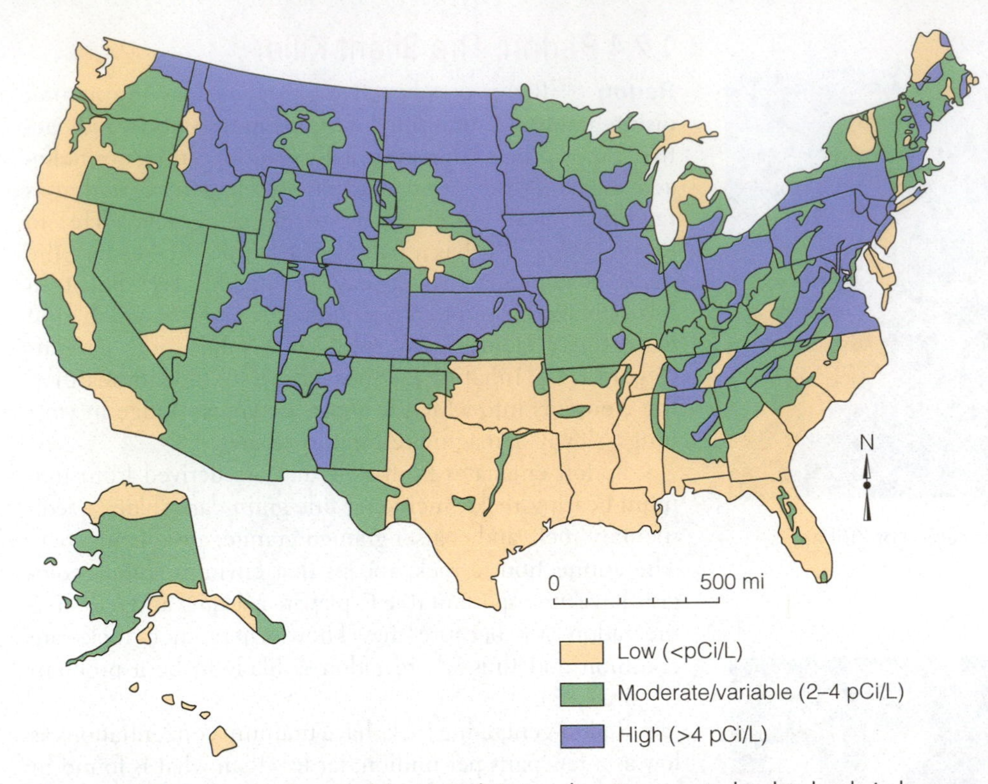

Figure 1.38 Generalized map of the United States showing expected radon levels in homes.

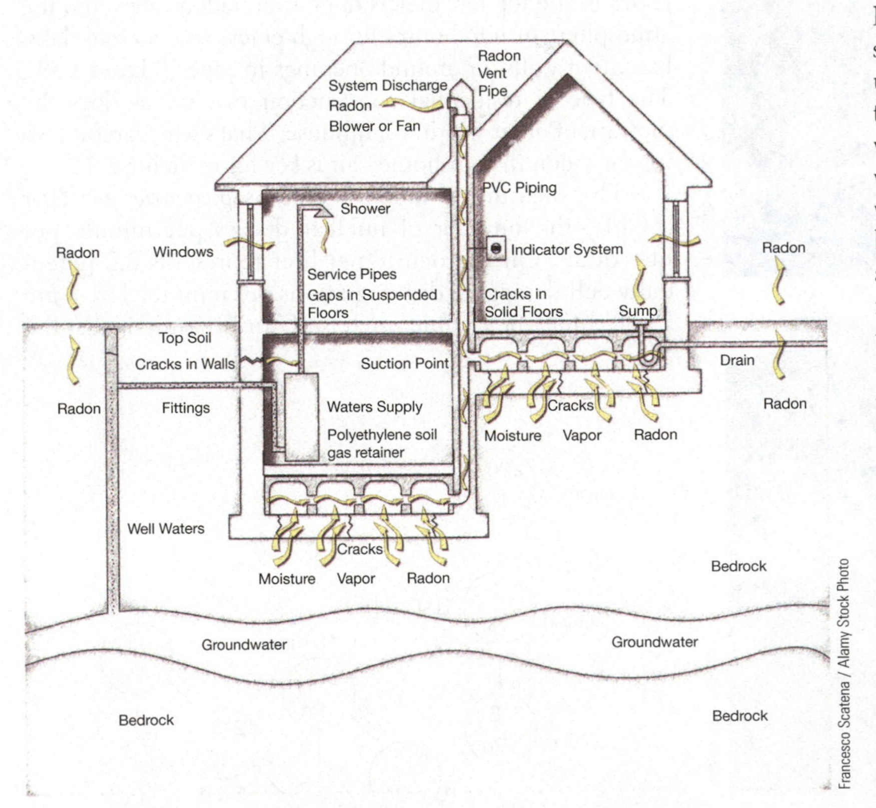

Figure 1.39 Pathways by which radon can enter a home, and some ways to remove the gas by ventilation.

average human inhales between 26.5 and 45 cubic meters (34 and 59 cubic yards) of air per day, high radon concentration in household air is a significant health hazard that can lead to lung cancer. Such cancer is rarely apparent before 5 to 7 years of exposure, and its incidence is rare in individuals younger than age 40; the issue is long term. Nevertheless, the EPA estimates (2022) that about 20,000 people in the U.S. die every year of lung cancer because they have inhaled radon and its decay products. The agency has established 4 pCi/L as the maximum acceptable indoor radon level. That is equivalent to smoking a half pack of cigarettes per day or having 300 chest X-rays in a year. **Figure 1.40** vividly illustrates the dangers of inhaling indoor radon (and of smoking).

Widespread interest in radon pollution first appeared in the 1980s with the discovery of high radon levels in houses in Pennsylvania, New Jersey, and New York. These areas are underlain by black shale, a sedimentary rock naturally high in uranium. The discovery was made accidentally when a nuclear-plant construction worker, Stanley Watras of Boyertown, Pennsylvania, set off the radiation-monitoring alarm when he arrived at work; this was very odd because the plant was still under construction and there was no nuclear fuel on site. Over the next few weeks, Watras repeatedly triggered the radiation alarm.

An investigation of his residence revealed very high levels of radon, about 1,000 times the safe limit and higher than most uranium mines (**Figure 1.41**). There was so much radium in Watras' home that his clothing was contaminated from radon decay products. This was the first time radon was recognized as a public menace in this way. The EPA moved in, and engineers dug up the Watras' basement and installed a radon removal system in the home. It worked: radon levels dropped below the 4 pCi/L limit. As of 2015, the Watras family still lived in their home and none of them had lung cancer. Remediation works.

It's simple to monitor your home for radon levels. Your local government is likely to supply a special charcoal canister that is placed in the household air for several months and then returned to the manufacturer for analysis (**Figure 1.42**). The charcoal absorbs radon in the air. This approach gives an

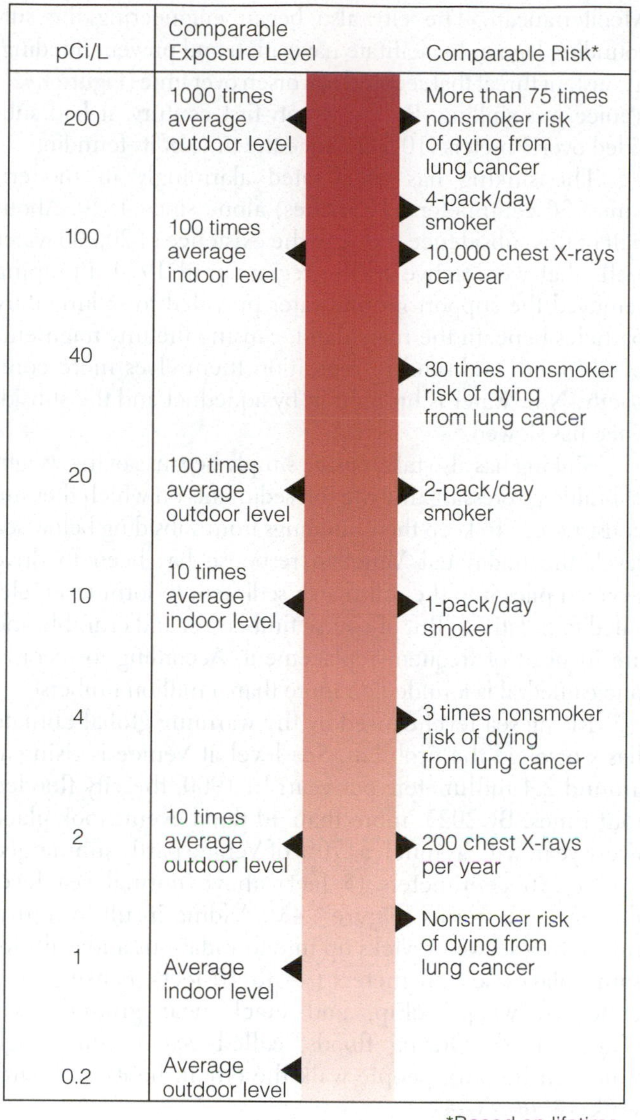

Figure 1.40 Lung cancer risk because of radon exposure, smoking, and X-ray exposure. A few houses have been found with radon radiation levels greater than 100 picocuries/liter, and one in Pennsylvania (Figure 1.41) had a level of 2,000 picocuries/liter.

average value across seasons (windows closed in the winter during heating season and open in the spring, fall, and perhaps summer). Any reading greater than 4 pCi/L requires follow-up measurements, but if the initial reading is between 20 and 200 pCi/L, you should consider immediate steps to reduce household radon levels. Common ways of dealing with the problem include increasing the natural ventilation in the basement or crawlspace beneath the structure, using forced air to circulate subfloor gases, and intercepting radon before it enters the house by installing gravel-packed plastic pipes below the floor slabs (**Figure 1.43**). For a home with marginal radon pollution, sealing all openings from beneath the house that go through the floor or the walls is usually effective.

Figure 1.41 Stanley Watras and his family in front of their Pennsylvania home where radon remediation was first applied—dropping airborne radon levels more than 1000 times.

1.2.5 Venice: A City Sinks

The mighty Po River, fed by numerous tributaries and draining the northern plain of Italy, enters the northern end of the Adriatic Sea, marked by marshlands, lagoons, and islands. On one of those low-lying islands is the legendary city of Venice (**Figure 1.44**). The heavy load of river sediment in this area causes Earth's underlying crust to respond by slowly sinking, though fresh inputs of sediment from periodic flooding add new land to replace older terrain that has subsided below sea level. Meanwhile the wetlands at the river's mouth support a thriving ecosystem for birds, fish, mussels, and other wildlife.

As the Western Roman Empire collapsed in the fourth and fifth centuries C.E., some citizens fled the invasions of Huns and Goths and settled in the wetland at the mouth of the Po River, eventually occupying 118 islets drained by canals within the main tidal lagoon. Water and dense vegetation protected them, and the City of Venice was born. By the twelfth century, Venice had become a major world power with a large navy and control of commerce throughout the eastern

Figure 1.42 The radon home-testing kit is a cannister you leave for several months in your living area, then return for analysis.

Mediterranean. The city also began engineering the surrounding lagoon to facilitate navigation and prevent flooding, a constant threat that seemed to worsen over time (**Figure 1.45**). Venice was sinking. By the twenty-first century, it had subsided over 3 meters (10 feet) since the time of its founding.

The sinking has accelerated alarmingly in modern times; 26 centimeters (10 inches) alone since 1870. About half of this subsidence is due to the existence of 20,000 water wells that were active in the region until 1970. Pumping removed the support groundwater provided to sedimentary particles beneath the marshland, causing the tiny fragments of silt, mud, and sand to reposition themselves more compactly. Now water is brought in by aqueduct and the subsidence has slowed.

Sinking has also taken place simply because of the weight of buildings pressing into the soft sediments on which they are constructed. To keep their buildings from subsiding below sea level, the traditional Venetian response has been to drive wooden piles into the soft marsh sediment to form level, elevated foundations. But of course timbers rot and crumble and are in need of frequent replacement. According to records, one cathedral is founded on more than a million timbers!

Rising sea level caused by the warming global climate has worsened the problem. Sea level at Venice is rising at around 2.4 millimeters per year; In 1900, the city flooded four times. By 2023, more than 40 flood events took place every year, with as much as 70% of Venice partly submerged up to 156 centimeters (5 feet) above normal sea level during some storms (**Figure 1.46**). Adding insult to injury, infiltrating seawater wicks up into foundations and walls, in some places several meters (~5 to 10 feet), causing concrete to "weep," chip, and crack near ground level (**Figure 1.47**). During floods, called *acqua alta* ("high water" in Italian), people walk the city in boots on boardwalks (**Figure 1.48**).

SM / Alamy Stock Photo

Figure 1.43 Active radon mitigation system pumps air from the basement to the outside of the home, taking radon with it. This system was installed in an older home in New Jersey.

iStock.com/Pavliha

Figure 1.44 Venice, laced by canals and sitting just above sea level.

Hemis / Alamy Stock Photo

Figure 1.45 Keeping your shop dry during high water in Venice is a challenge.

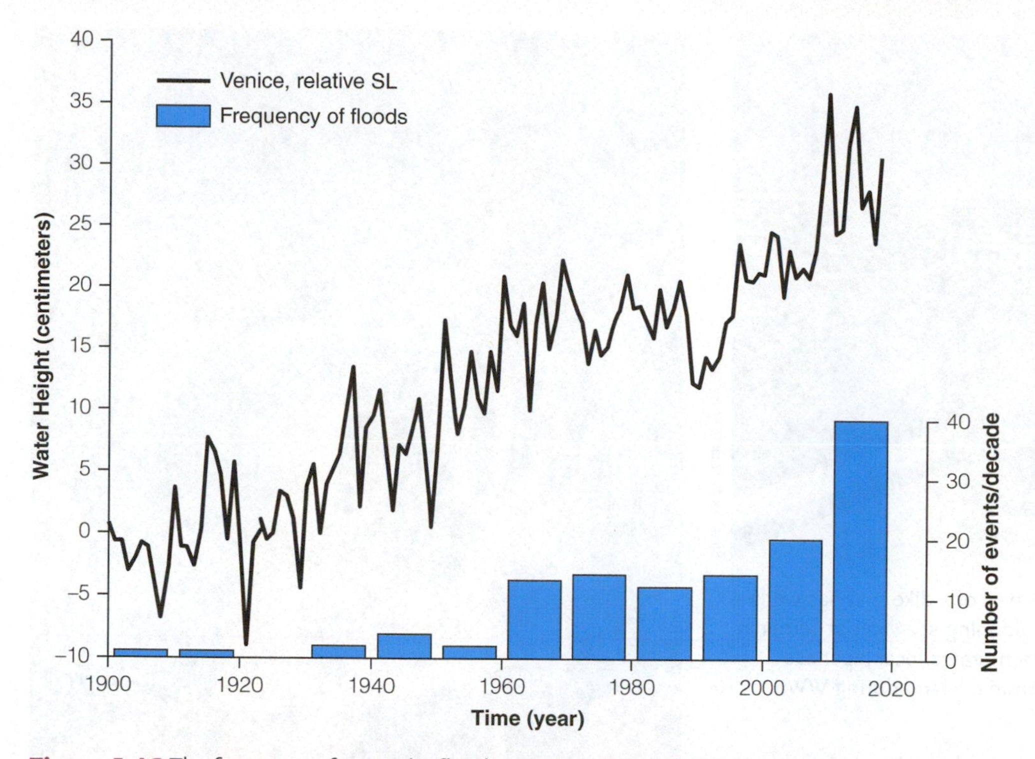

Figure 1.46 The frequency of *aqua alta* flooding in Venice since 1900. The increase in frequency is due to subsidence, settlement, and rising sea level, which is shown by the black line.

What is to be done? The city is a special site because of its historical importance, priceless antiquities, art, literature, and beauty, as well as its charming canals and gondolas.

Suggestions include a large dike or levee that once constructed would keep out the ocean (**Figure 1.49**). Terminal Island, a hub in the Port of Los Angeles, California now lying below sea level, is protected this way by a wall as much as several meters high. This approach would likely have serious consequences for the city, however. Urban wastewater would not be flushed out to sea, and the Venice lagoon would almost certainly become toxic with microbes and blue-green algae, collapsing the wetland ecology and related aquaculture industry.

Other solutions include pumping water back into the ground to restore volume to the delta sediment beneath Venice. This has been done with some success in other cases; for example, at the Port of Los Angeles, where severe subsidence caused by oil pumping throughout the first part of the twentieth century threatened to drown vital infrastructure. Seawater pumped underground stopped the sinking. Completely restoring original elevation to the land is impossible,

Figure 1.47 Damaged concrete on Venice home from seawater wicking up into the building.

Figure 1.48 Boots are more than a fashion statement when *acqua alta* floods St. Mark's Square in Venice.

Steve Cukrov / Alamy Stock Photo

Figure 1.49 This concrete seawall part of a dike that has worked to protect low, sinking ground from flooding seawater at Terminal Island, California. The statute commemorates local Japanese residents forcibly relocated from Terminal Island during WWII.

however, and won't prevent continued sea level rise due to climate change.

The solution for now is an expensive compromise. An ambitious scheme, the multibillion-dollar *Modelo Sperimentale Elletromeccanico* (MOSE), installed 78 hydraulic floodgates at the barrier-island entrances to the Venice lagoon in order to keep Adriatic storm surges and high tides from inundating the city (**Figure 1.50**). But at current rates of sea-level rise, within decades, the MOSE gates will have to be closed practically on a daily basis to prevent flooding. Doing so, the project will decrease tidal flushing of lagoon waters, allowing buildup of sewage released by the city. Many environmental geologists are concerned that this very expensive "fix" is not worth the cost and is temporary at best. The price tag for building MOSE was nearly $8 billion, and each time the barriers are raised the cost is over $300,000. MOSE has been busy. It was activated 13 times in 2020, and 20 times in 2021.

Piero Cruciatti / Alamy Stock Photo

Figure 1.50 People and their dogs watch as the yellow MOSE barriers are deployed for the first time (October 2020). The system kept Venice dry during what would have been an *acqua alta* flood.

Forskning & Framsteg

Figure 1.51 As the waters of Lake Nassar, impounded behind the Aswan High Dam, rose, masons were cutting apart and moving Abu Simbel to higher ground, one massive block of stone at a time.

Ultimately, the solution in the coming decades or centuries may be simply to cut the city up into movable blocks and transport it piece-by-piece to some safer inland location, as was done to avoid destruction of many famous Egyptian temples (e.g., Abu Simbel) due to building of the Aswan High Dam and filling of its reservoir in the 1960s (**Figure 1.51**).

Venice and the Port of Los Angeles are not the only urban areas threatened by subsidence and sea-level rise. Jakarta, Indonesia is sinking. So too is America's culturally iconic jazz capital, New Orleans. It too, someday, may have to pick up piecewise and move. In any event, environmental geologists will continue to be closely involved in efforts to save this, and other, well-loved heritage sites.

1.2.6 Pakistan: A Nation Drowns

Floods routinely damage communities such as Venice, but rarely do they affect entire countries. In August 2022, however, 30% of Pakistan went underwater (**Figure 1.52**) and 33 million Pakistanis were affected by flooding. Exceptionally heavy rains, linked to an extreme seasonal Asian monsoon and glacier melt tied to record summer warmth, generated rapid runoff. Falling for days on end, the rains poured off

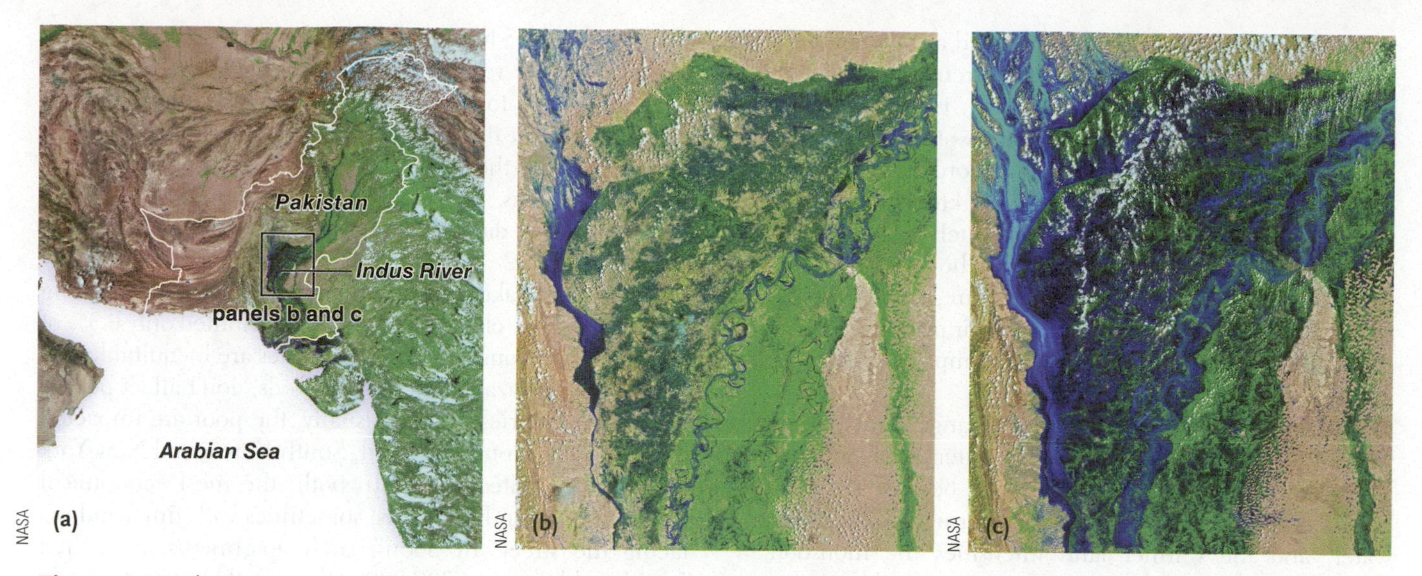

Figure 1.52 The 2022 monsoon floods in Pakistan were unprecedented in modern times. The Indus River filled with both glacial melt and the monsoon rainfall, both driven by the record summer heat. (a) Location map showing Pakistan and the Indus River. (b) NASA image showing Central Pakistan in April 2022; the Indus River is within its banks. (c) NASA image showing Central Pakistan in August 2022 with most of the area underwater as the Indus River has overflowed into the adjacent lowlands.

Figure 1.53 Flooding engulfs a village on a tributary to the Indus River in northern Pakistan, August 2022. Charsadda district.

Figure 1.54 With their homes underwater, residents of southern Pakistan sought shelter on the only high ground, levees topped with roads.

already saturated hillslopes, filling tributaries and rushing downstream. Rivers swelled until they rose and spilled out of their channels, turning much of the country into a shallow lake (**Figure 1.53**). Especially alarming was the behavior of the Indus River, which drains the western Himalaya Mountains and flows the length of Pakistan.

Flooding engulfed communities, driving people out of their homes to find refuge and a place to live on high ground (**Figure 1.54**). Sometimes, the only high ground was long, narrow levees along the banks of the raging rivers. One of every seven Pakistani's lost their homes. More than 5000 kilometers (3,000 miles) of roads were destroyed and more than 200 bridges throughout the country were damaged. (**Figure 1.55**). Without transportation networks, people were stranded, unable to obtain relief supplies including food, medicine, and clean water. Almost 8 million people were displaced, and over 1,700 died. Sherry Rehman, Pakistan's

Figure 1.55 A bridge was destroyed and a road damaged by flooding caused when heavy monsoon rains poured down on northern Pakistan's Swat Valley, August 2022.

minister for climate change, was quoted as saying, "An area bigger than the state of Colorado is currently submerged."

Pakistan is no stranger to flooding. Every summer, the monsoon brings heavy rains and rising water. And there have been big floods before. In 2010, record monsoon runoff swept through the country–this was known then as the "superflood." That event caused as much as $43 billion in damage and destroyed almost 2 million homes. The response was to rebuild with an eye to the future, raising roads and bridges in some cases meters above their prior level. It didn't help; many of those bridges were simply swept away by greater volumes of water in 2022.

Many scientists point to climate change as the culprit for these floods because it makes ocean water, the land, and the atmosphere warmer. This makes sense because the warmer ocean evaporates more water, the warmer air holds more water, and the warmer land intensifies the monsoon, a weather pattern caused by rising warm air over the continents (see Chapters 2, 3, and 8). Every year brings monsoon rains but the warmer the climate, the more intense the monsoon. Similarly, glaciers melt every summer, but as the climate warms, the ice melts more rapidly and extensively, releasing increasing volumes of water into streams and rivers. Northern Pakistan is glacier rich—at least for now (**Figure 1.56**).

Evidence for the climate connection is compelling: Nearly 400 millimeters of rain fell from June to August 2022, almost twice the 30-year average. In some parts of Pakistan, August rainfall was five to seven times higher than average! For every degree Celsius the atmosphere warms, the air can hold 6% to 7% more water vapor–and it's that vapor that condenses to form rainfall. From this simple analysis, it's logical to conclude that rains would have been less intense prior to the human-driven increase in atmospheric carbon dioxide over the past century.

Environmental geologists who study the extraction and use of fossil fuels report that Pakistan, in comparison to the rest of the world, uses little fossil fuel and releases scant amounts of carbon dioxide. Today, each Pakistani on average emits 0.87 tonnes [0.95 (short) tons] of carbon dioxide yearly, compared to 15.5 tonnes (17 tons) for Americans, 37 tonnes (41 tons) for the residents of Qatar and a global average of 4.8 tonnes (5.3 tons) per person annually. Looking back through time, the discrepancy is even more striking. Between 1750 and 2020, the United States was the global leader in carbon emissions, emitting 4.2×10^{11} tonnes (4.6×10^{11} tons) of carbon dioxide. Compare that to Pakistan which emitted only 5.2×10^{9} tonnes (5.7×10^{9} tons), nearly 100 times less. Pakistan is suffering dramatic and devastating effects of climate change; yet, it has contributed only 0.3% of all global emissions. Such discrepancies are inequitable.

Geologic hazards, including floods, don't affect people and their countries evenly. Usually, the poor are impacted most severely. Consider Seoul, South Korea, and New York City in the United States. In both, the most economical apartments are in basements, sometimes with tiny windows facing the street. In Seoul, such apartments are called *banjiha* and there are 200,000 of them in that one city alone.

When urban flooding hits, those living in these basements are particularly vulnerable as their apartments fill with water. To make matters worse, the pressure of the deepening water makes it impossible to open the entry door and many basement apartments have bars on their windows for security (**Figure 1.57**). Escape can be impossible. After devastating floods killed more than a dozen basement apartment dwellers in 2022, Seoul has vowed to phase out these residential units and turn them into warehouses and other uses less likely to endanger the city's citizens.

When the remnants of Hurricane Ida hit New York City in September 2021, they dropped 15 to 22 centimeters (6 to 9 inches) of rain, shattering records. That rain, and the

JSK / Alamy Stock Photo

Figure 1.56 Trekkers on Vigne glacier with K2 mountain and Godwin-Austin glacier in the background, Karakoram, Pakistan.

Anthony Wallace/AFP/Getty Images

Figure 1.57 People look through a street-level window of a *banjiha* in Seoul, Korea. The security grill has been removed so the apartment could be emptied. Three people drowned here, including a disabled woman and a teenager, after flooding, triggered by heavy rains in August 2022, filled the apartment with water. They could not escape.

Michael M. Santiago/Getty Images News/Getty Images

Figure 1.58 A flooded basement apartment in Passaic, New Jersey, a day after the remnants of Hurricane Ida dropped almost 23 cm (9 inches) of rain in September 2021.

Niday Picture Library / Alamy Stock Photo

Figure 1.59 Downtown Sacramento was a great place for boats in 1862, as water from more than a month of heavy rain filled the streets.

flooding it caused killed 11 people who lived in New York City basement apartments; more died in nearby areas of New Jersey. As in Seoul, occupants of these basement residences drowned, unable to escape as their underground homes filled with water (**Figure 1.58**). In New York, such apartments are illegal, so there is no official count but there are likely tens of thousands of such dwelling places in the city. For many new immigrants to the United States, they are the only affordable available housing in the city and for their landlords, a source of much-needed income.

It's sobering to think that we likely have not seen the biggest floods nature can deliver. A recent analysis, by geoscientists studying the California water cycle, suggests that megafloods are lurking in that state's future. Such floods have happened in California before. In late 1861, 5 meters (16 feet) of snow fell in the Sierra Nevada Mountains. That record snowfall was followed by 43 days of rain, the result of what is called an **atmospheric river**, a plume of moist air in the atmosphere that originates over the Pacific Ocean and comes ashore in California. After more than a decade of drought, the winter of 2023 looked a lot like 1861 with massive snow falls and more than a dozen atmospheric river events.

In late winter, 1862, an atmospheric river delivered enough rain to keep the capital of the state, Sacramento, under three meters (10 feet) of dirty water for months. With both the atmosphere and the Pacific Ocean warming, the chance of such a disastrous event has likely doubled. Estimates suggest it could cause a trillion dollars of damage and destroy the Central Valley fields that provide nearly 25% of the food Americans eat. Will downtown Sacramento once again become a thoroughfare for boats just as it was in 1862 (**Figure 1.59**)?

Flooding is a natural event, and it leaves a record that environmental geoscientists can readily decipher. Using a variety of tools including historic photographs, ancient flood debris high above modern rivers, and deposits of river sediment, environmental geologists can reconstruct giant floods of the past. This is a critical record that tells us how the Earth behaved before. Today, the frequency and size of floods are both increasing as climate change warms Earth's atmosphere and more rain falls from the sky. Environmental geoscientists use their expertise in how floods work and how they behaved in the past not only to understand the present but also to predict the future in a warming world.

1.2.7 Water Quality and Flooding in Jackson, Mississippi

Brown, turbid water with a funky smell dribbled into a Jackson, Mississippi kitchen sink. It was undrinkable. Jackson is the state capitol of Mississippi and for weeks in the fall of 2022, the tap water was unsafe to drink (**Figure 1.60**). The City of Jackson posted boil-water orders for those residents who had water, but many people had no water at all. Their taps were dry. How could this happen in the United States, one of the best-resourced nations in the world? History gives us some perspective. For millennia, civilizations have fallen both for lack of water and, at least in part, for what was in the water they drank.

Water is key to human survival and at the center of much of environmental geoscience. Ancient peoples, like the classic Maya civilization of Meso-America, appear to have been brought down by drought—a lack of water. Anthropologists tell us that major cities of the lowland Maya were abandoned around 760, 810, 860, and 910 C.E. Environmental

Figure 1.60 Brown water pours from a broken water main in Jackson, Mississippi, September 2022.

geologists working in both frigid Greenland (collecting cores from the ice sheet) and in warm tropics (collecting sediment cores from lakes and the Atlantic Ocean), find evidence for drought at that time. The Greenland data show that between 700 and 1000 C.E. it was cold and dry with slightly lowered output from the sun. The high-resolution sediment data are even more compelling—they suggest that a severe drought (perhaps the worst of the last 7,000 years) was most intense during the four time periods when the cities were abandoned (**Figure 1.61**). This is the geologic equivalent of a smoking gun. Without water, people die of thirst and crops fail.

One particularly telling example is the Mayan city of Tikal, which thrived in the Guatemalan jungle for more

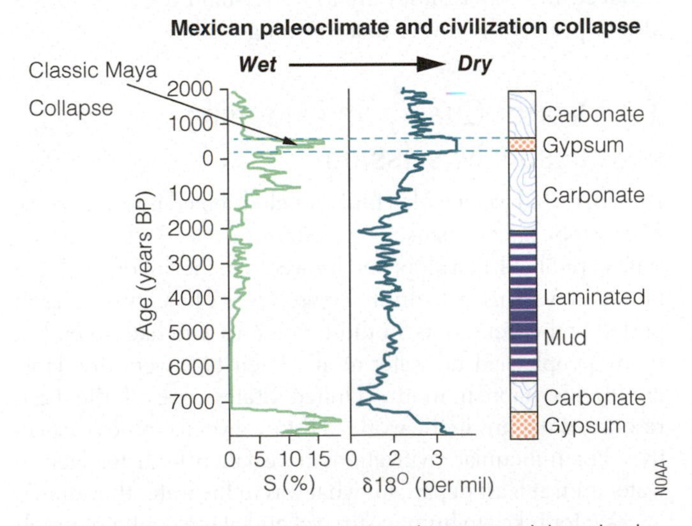

Figure 1.61 Sediment coring in central American lakes clearly shows changes in type of sediment over time as well as the percentage of sulphur, which increases during times of drought as the amount of gypsum ($CaSO_4$) in the sediment changes. Gypsum precipitates in lake waters when they become concentrated by evaporation (modified from Hodell et al., 1995).

Figure 1.62 Tikal National Park in Guatemala, a World Heritage Site. This ancient Mayan city was abandoned by 900 C.E.

than a 1,200 years (**Figure 1.62**). Slowly, over a century or two, the very large population dwindled. No new buildings were constructed. There is ample archeologic evidence for increased fighting between Mayan city-states and well as social unrest—all consistent with stress caused by increasing drought and resulting crop failures. About 900 C.E, Tikal was abandoned.

It wasn't just lack of water that brought Tikal down, however. Like all Mayan cities, Tikal managed water for its residents by means of reservoirs supplied by canals that concentrated runoff. Such reservoirs also collected sediment. In 2009 and 2010, scientists examined sediment from Mayan times preserved in the walls of excavations at the old reservoir sites (they were by now dry!). They did geochemical analysis on the material using modern, EPA-certified methods for mercury analysis and **PCR** (polymerase chain reaction) to determine DNA by species.

When the data were published in 2020, they created quite a stir. Reservoir sediment from the last years of Tikal were laced with mercury and the DNA of toxic cyanobacteria (**Figure 1.63**). Cyanobacteria DNA suggest major blooms of toxic blue green algae in the reservoirs–both mercury and cyanobacteria toxins are harmful to people when ingested with drinking water, which is exactly what the Maya were doing as Tikal withered away.

The mercury in the sediment most likely comes from cinnabar (HgS) a bright red mineral (**Figure 1.64**) used extensively for body and building decoration by the Maya. The reservoirs at Tikal got much of their water as run-off from impermeable surfaces including temples and courtyards where ceremonies involving mercury were common. The reservoirs used most by the ruling elite had the highest levels of mercury in their sediment. Perhaps the Mayan elite inadvertently poisoned themselves (and their citizens) by drinking mercury-tainted water and thus lost their ability to lead effectively. More likely, the collapse of the lowland Mayan civilization was brought on by several linked failures. Drought restricted Tikal's water supply, encouraged algal blooms, and concentrated mercury in shrinking reservoirs. Together these changes undercut the societal foundation of the great Mayan center.

Unlike Tikal, Jackson's water woes were not born of drought and water scarcity but from a lack of resources to maintain and modernize its aging water infrastructure. Many

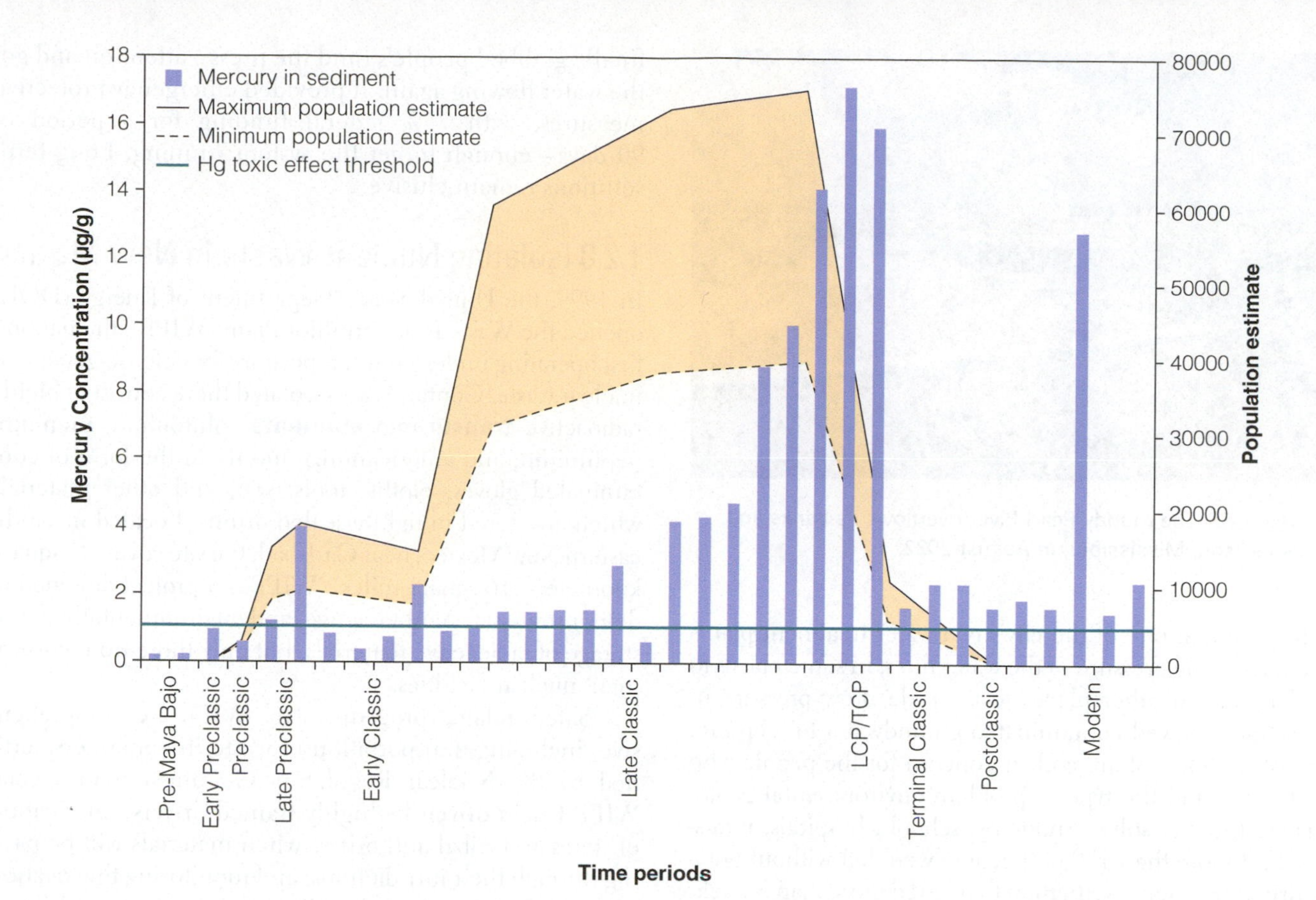

Figure 1.63 The highest levels of mercury (blue bars) were found in Tikal reservoir sediments just prior to and as the population of the city (in beige) crashed. The green line indicates the concentration of mercury at which sediment is considered likely to have toxic effects. (Adapted from Lentz, D.L., Hamilton, T.L., Dunning, N.P. et al. Molecular genetic and geochemical assays reveal severe contamination of drinking water reservoirs at the ancient Maya city of Tikal. Sci Rep 10, 10316 (2020).)

observers see this lack of investment as a clear manifestation of environmental racism and injustice (see Case Study 1.1). In 2022, the population of Jackson was 82.7% Black, 16.7% White, 0.7% multiracial, and 0.3% Asian; about a quarter of the population lived below the poverty line.

Figure 1.64 Cinnabar, a mercury-bearing mineral, and the pigment made by grinding it sit side by side in a German museum. The toxic material is safely displayed behind glass.

It has not always been that way: In 1969, a U.S. Supreme Court order desegregated public schools across the nation. That next fall, 10,000 students, mostly White, left the Jackson public schools and by year's end, the political, economic, and religious leaders of Jackson, many of whom were White, had fled for the suburbs. In 1 year, the Jackson school system lost 40% of its pupils. With the tax base shrinking, infrastructure investment dropped and there were no funds to properly maintain the water system. As a result, it had been degrading on and off for decades prior to the catastrophic failure in 2022. Because of Jackson's demographics, the impact of poor quality water and an intermittently functioning water system is predominantly on people of color. Those impacts can be significant, ranging from skin irritations caused by bacteria to a higher risk of developmental challenges for newborns and young children exposed to lead.

On August 29, 2022, the Pearl River, which runs through Jackson, poured over its banks (**Figure 1.65**). The flooding was the result of a very wet August capped off with exceptionally heavy downpours that caused the river, over just 2 days, to rise by nearly 8 meters (26 feet). Some of that water came from a reservoir upstream that was so full, it was about to overflow into its emergency spillway. At the same time, the

Chad Robertson Media/Shutterstock.com

Figure 1.65 The muddy Pearl River overflows its banks and floods Jackson, Mississippi, in August 2022.

main pumps at one of the city's two water treatment plants failed, reducing pressure in the water lines, in some places to nothing and in other places to a trickle. Low pressure in water lines allowed contaminated groundwater to seep into the distribution system; a clear concern for the people who lived there, and the type of problem environmental geologists are hired to solve. Suddenly, schools, hospitals, industries, and more than 150,000 people were left without potable drinking water—a situation that lasted more than 3 weeks before service was restored and clean water flowed from the city's taps (**Figure 1.66**).

Jackson had tried to improve the water system over the years but was stymied at the State level by politicians who failed to act, perhaps because they were not directly impacted by a lack of clean water. The city attempted to raise its sales tax to cover the cost of water system repairs but lawmakers in the state legislature said no. An effort to borrow funds using bonds and make repairs to the system died in a Mississippi State House committee without ever getting a vote. President Biden's emergency declaration

REUTERS / Alamy Stock Photo

Figure 1.66 The Jackson State University campus was without water. Delivery of bottled water allowed students to return to college.

finally grabbed people's (and the press') attention and got the water flowing again. It provided emergency protective measures, with 75% federal funding for a period of 90 days—enough to get the system running. Long term solutions remain elusive.

1.2.8 Isolating Nuclear Waste in New Mexico

In 1999, the United States Department of Energy (DOE) opened the Waste Isolation Pilot Plant (WIPP), the nation's first operating underground repository for defense-generated nuclear waste. Contaminants isolated there consist of highly radioactive **transuranic elements** (plutonium, uranium, neptunium, and americanum), mostly in the form of contaminated gloves, cloths, tools, soil, and other materials which are stored in tightly sealed drums. Located in southeastern New Mexico, near Carlsbad, the site covers 41 square kilometers (16 square miles). WIPP is a project designed to demonstrate safe, cost-effective, and environmentally sound storage of radioactive materials, but only those from government nuclear facilities.

Safety related procedures for WIPP are comprehensive, including transportation of waste in containers certified by the Nuclear Regulatory Commission in special WIPP trucks driven by highly trained drivers, notification of states and tribal authorities when materials will be passing through their jurisdictions, and monitoring the location of each shipment by a satellite tracking system while in transit (**Figure 1.67**). Oversight of the WIPP program is provided by numerous state and federal agencies and the Environmental Evaluation Group, an independent group that participates in, and comments on, various WIPP activities.

Storage of the radioactive materials is in excavated rooms 655 meters (2,150 feet–nearly a half-mile) underground in the Salado Formation, a 250-million-year-old, stable salt bed

Atomic / Alamy Stock Photo

Figure 1.67 A shipment of transuranic waste on its way to WIPP in a special, impact-resistant canister. This is the one-thousandth shipment of waste.

Nature and Science / Alamy Stock Photo

Figure 1.68 Mining salt at WIPP in 2007 so that geotechnical investigations could determine the site's suitability.

(**Figure 1.68**). The 610-meter (2,000-foot)-thick salt deposit was chosen as the site because there is very little earthquake activity and no flowing groundwater that could move waste to the surface or be contaminated by the waste, and because salt readily seals any fractures that form, given its ability to deform rapidly. Furthermore, squeezed by the high pressures that exist underground, the salt will slowly flow into and fill mined areas, further isolating the waste from the environment (**Figure 1.69**). From the standpoint of environmental geology, one could hardly ask for a better waste storage environment.

WIPP has operated fairly well since it opened, with waste containers placed one after another into cylindrical borings in the salt (**Figure 1.70**). But in 2014, a fire at a salt-haul truck followed by a slight radiation leak that exposed 22 workers forced temporary closure of the facility and a $73-million lawsuit from the New Mexico Environment Department. No traces of radioactivity escaped the site, fortunately, and by 2016 new safeguards were in place. A Safety Significant Confinement Ventilation System priced at $163 million is presently (2021) under construction. WIPP will be filled to capacity and closed for long-term storage sometime before 2035.

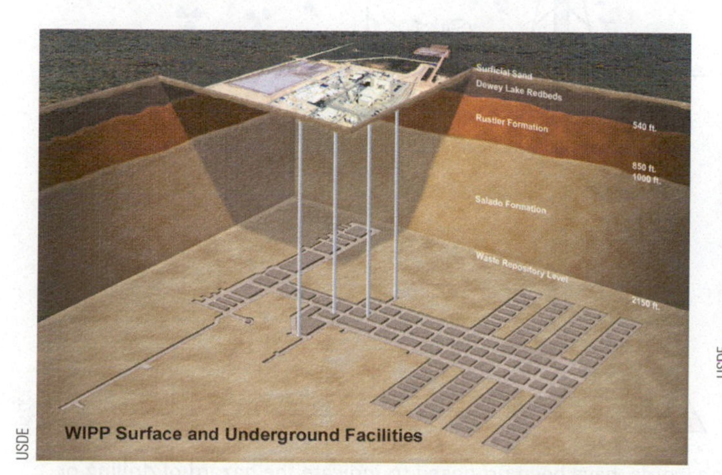

USDE

Figure 1.69 A schematic diagram of the WIPP facility.

Jim West / Alamy Stock Photo

Figure 1.70 Cylinders of nuclear waste fill these borings into the walls of salt nearly 700 meters (2300 feet) below the ground surface.

Then what? The contents of WIPP, sealed deep in a salt tomb, will remain toxic for hundreds of thousands of years, a time span that only geologists (and occasionally astronomers) are accustomed to discussing. A system of permanent markers was designed to inform our descendants that the area of the WIPP is not in its natural state, and that it has been marked for good reason (**Figure 1.71**). The site must remain inviolate for at least 10,000 years (an arbitrary, legal determination) and the warning components include a 9-meter (30-foot)-high berm, an information center, monuments along the perimeter, and a warning in seven languages printed on 9-inch-diameter discs made of granite, aluminum oxide, and fired clay.

Yet, a serious, largely academic question remains about just how one constructs a warning system that will both *endure* and *remain intelligible* to our descendants

USDE

Figure 1.71 Artist's rendition of a "spike field" sculpture designed to discourage people of the future from "messing" with WIPP. This is one of several designs considered to mark the site when it is closed.

through such a long span of time (**Figure 1.72**). To study this basic question, the DOE set up a group called the Expert Judgment Panel, in connection with its earlier planning for Yucca Mountain, a nuclear repository in southern Nevada that has never opened (see Chapter 12). After a decade of study, the group found that virtually nothing we can use as construction material will survive in satisfactory condition for so long, save, perhaps, a new synthetic product called **synroc**, made up of the rare minerals hollandite, zirconolite, and perovskite. If we leave a message rather than some sort of warning symbols, what should it say? Thomas Sebeok, lead author of the report *Communication Measures to Bridge Ten Millennia*, proposed that we initiate a transgenerational "Atomic Priesthood."

In this scheme, each successive generation of the Atomic Priesthood would be charged with the responsibility of seeing to it that our behest is heeded by the general public–if not for legal reasons, than perhaps for moral ones–with the veiled threat that to ignore our mandate [not to drill wells or dig in this area] would be tantamount to inviting some sort of supernatural retribution.

Adapting this authoritative overtone, the Expert Judgment Panel ultimately settled on the following message to be inscribed monumentally above the proposed Yucca Mountain nuclear waste site: *"This is not a place of honor. No esteemed deed is commemorated here. Nothing of value is buried here. This place is a message, and part of a system of messages. Sending this message was significant for us. Ours was considered an important culture."* It's certainly not a modest warning.

How would you interpret such a sign if you were to stumble across it in a desert wilderness? Furthermore, after 10,000 years, would the language even be recognizable? University of Iowa Professor John D'Agata provided the following translation using this very wording in Old English–direct ancestor of our modern language. How much of it do you understand?

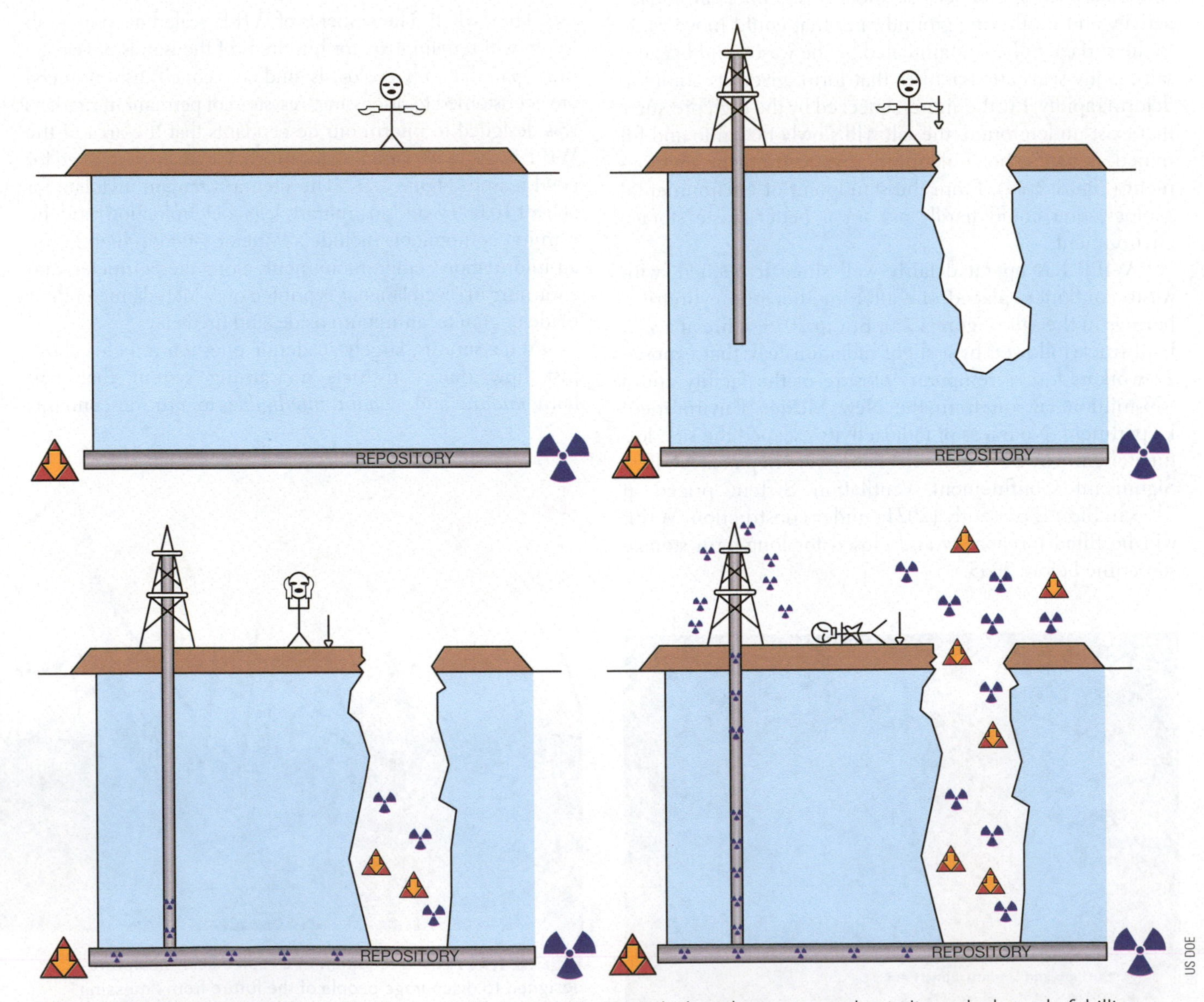

Figure 1.72 Expert-designed signs like this example, do not rely on words, have been proposed to indicate the hazard of drilling or digging into WIPP.

Nis weorðful stow. Nis last mære dæde na. Her nis naht geoweorðes bebyrged. Þeos stow bið ærend woruldes. GiemÞ wel! We sindon eornost! Þeos ærengiefu wæs niedmicel us. Hit Þuhte us Þuhte ys Þæt we wæron formicel cynn.

1.2.9 Lessons from These Situations

What do the above examples teach us about environmental geology and environmental geologists? One standout is that environmental geologists strive to anticipate physical problems that come from living on Planet Earth (Kiribati, radon); provide recommendations for how to avoid them (Venice, radon)–and when they occur, help society to heal and learn from them (Fukushima).

Here are some other reflections:

- We see that while some geological processes with disastrous consequences are triggered directly by human activity (Newmark, Kiribati, Venice, radon, and Jackson), others are not (Armero and Fukushima); they are simply part of natural geological change that spontaneously takes place everywhere on Earth.
- We often find ourselves in the way of what Nature is doing, though in most cases for easily avoidable reasons. To a certain extent, we can manage this conflict between people and nature. At other times, the conflict cannot be survived (Armero and Pakistan).
- We accept that sometimes the best way to avoid a bad geological situation is to plan for it (Fukushima), or simply back away (Armero, and perhaps Venice one day soon).

We learn that sufficient human understanding of how geological processes occur, perhaps especially for nonscientists, is urgently important, even lifesaving. Sometimes, what you don't know *can* hurt you. So, *know it*! (Radon, Armero, Fukushima, and WIPP.)

- We find that some harmful geological phenomena take place all at once (the tsunami at Fukushima and the flooding in Pakistan), while others build up gradually and persist indefinitely (Kiribati, radon, and the sinking of Venice). Environmental geologists can tell us how severe such problems are likely to be and what we can do about them.
- To everyone's dismay, we discover that the parties responsible for causing some environmental disasters may have long disappeared or not be held properly accountable, leaving expensive solutions for future generations (Newmark, Jackson, and to a large degree, Kiribati).
- Related to the previous point, we know that we often need to plan for transgenerational futures, whether these unfold over mere decades (Newmark, Love Canal, Kiribati) or many thousands of years–abstractly long periods of time (WIPP).
- We discover that tremendous inequities exist in the world between those who cause environmental problems and those who suffer from them. The "sins of the haves" tend to be visited on the "have-nots," even in economically well-resourced countries (Kiribati, Pakistan, Jackson). This is a major concern for **environmental justice**, the effort to assure that some people and societies do not suffer unfairly from environmental problems others escape or even trigger. (see Case Study 1.1).

Thinking over what you've just read, you may well have additional reflections to add to this list. It should also be clear that there are many avenues for environmental geologists to study and then apply their training and skills to providing service to humanity.

1.3 Tools of the Trade–Environmental Geology (LO3)

Environmental geologists rely on many different tools and technologies to study Earth at all levels: land, sea, sky, and the deep interior. It is beyond the scope of this book to describe in detail all the tools and how they work, but we will discuss some that environmental geologists frequently use to gather data and understand the Earth.

Mapping is fundamental to almost all aspects of environmental geology (**Figure 1.73**). It is the development of graphical representations that depict features of particular interest, from the roadmaps that we commonly use to get around, to detailed site studies of bedrock conditions beneath proposed building foundations, or the location of undersea volcanoes on a map of the deep ocean floor. At a small scale on land or on ice, maps may be produced using measuring tapes or laser range finders and devices known as total

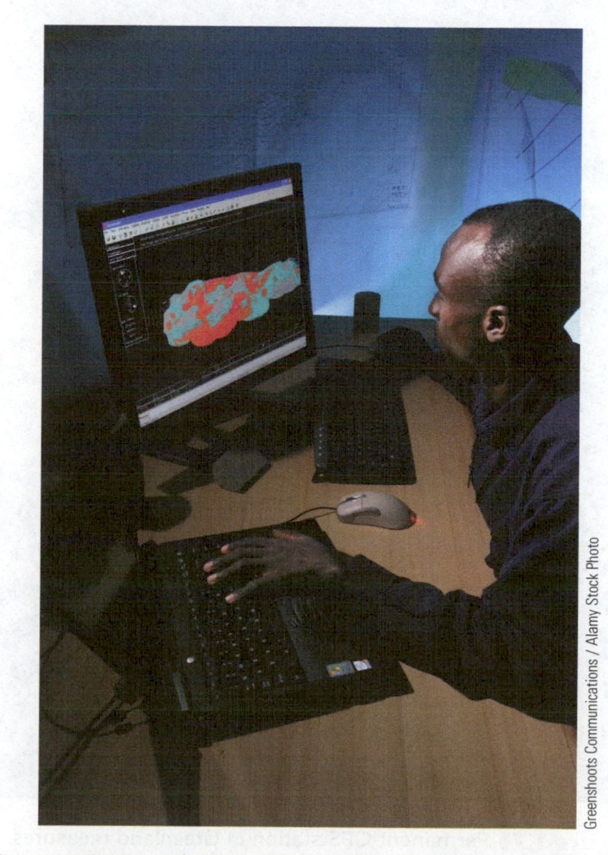

Greenshoots Communications / Alamy Stock Photo

Figure 1.73 A geologist uses computer graphics to map a goldmine in Eastern Ghana, West Africa.

Dmitriy Shironosov / Alamy Stock Photo

Figure 1.74 Geologists working in a quarry use a total station to survey the areas being mined.

stations that measure angles and distances between locations, both horizontally and vertically (**Figure 1.74**). Precise locations relative to Earth's geographical coordinates (longitude and latitude, or Universal Transverse Mercator–an alternative coordinate system) can be established using **global positioning system (GPS) receivers**, which receive signals from orbiting satellites. Some GPS receivers are so accurate and precise that they can locate specific sites to within a few millimeters (**Figure 1.75**).

UNAVCO

Figure 1.75 Permanent GPS station in Greenland measures the rise in bedrock elevation as the ice sheet melts and the land rebounds.

iStock.com/ChristopherBernard

Figure 1.76 GPS surveying of tar-sands mine in Fort McMurray, Alberta. Oil will be extracted from these sands.

An environmental geologist in the field can follow lines such as the boundaries between different rock formations, or earthquake ruptures, by walking along them while holding a GPS receiver to continuously locate positions (**Figure 1.76**). Data may be transferred easily to a digital map base afterward; a highly timesaving approach when there is much to be mapped. At sea, mapping is done remotely, via **sonar** (sound navigation and ranging) from ships (**Figure 1.77**). Acoustical signals are beamed from the ship to the seafloor, and the time it takes for the signals to return to the ship are a measure of the water depth. Because the sound energy used in sonar travels farther than do sounds in the air above, it is possible to map out great expanses of seabed (and such objects as shipwrecks) with single ship passes (**Figure 1.78**).

Space is the most exciting new area of technological development. **Satellites**, orbiting instrumental platforms, can examine the surface of our entire planet with great precision, collecting data ranging from air temperatures and ice

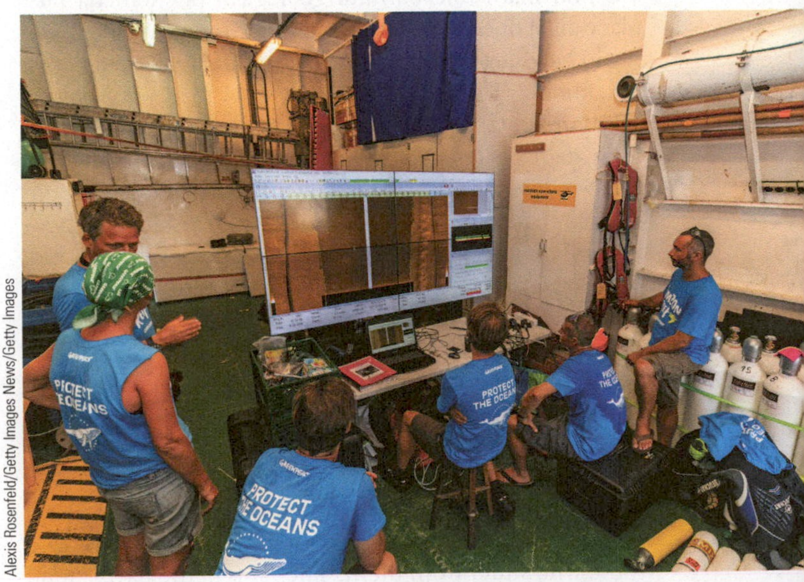

Alexis Rosenfeld/Getty Images News/Getty Images

Figure 1.77 Staff on an oceanographic vessel off of French Guyana monitor ongoing side scan sonar mapping as the data come in.

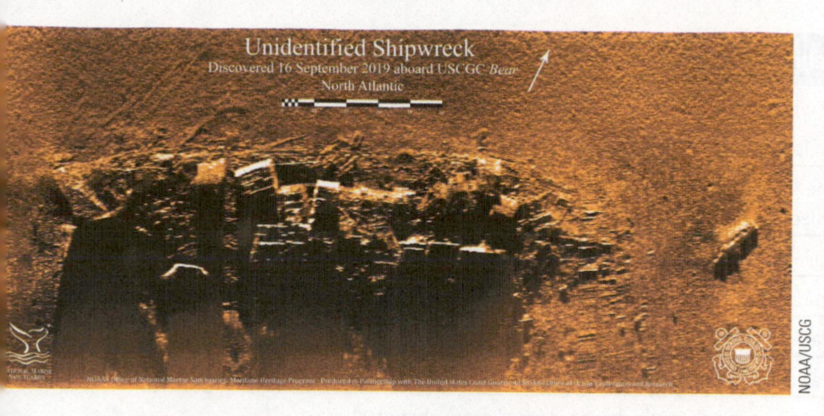

Figure 1.78 Side scan sonar shows a nearly intact shipwreck on the bottom of the North Atlantic. This is the Coast Guard cutter USRC Bear that went down in 1963, south of Cape Sable, Nova Scotia.

cover to the uplift of landscapes following major earthquakes (**Figure 1.79**). They carry into orbit cameras, infrared radiation sensors, synthetic aperture **radio detection and ranging (radar)** (using radio signals for determining elevations and mapping topography) (**Figure 1.80**), gravimeters, magnetometers, and other highly sensitive devices that observe Earth. Remote volcanic eruptions are easily detected, and the progress of their ash clouds tracked to warn people downwind. Weather forecasting, including the movement and sizes of hurricanes, depend heavily on satellite "eyes in the sky." We often see important events developing from outer space long before we know they're happening here on the ground!

Unlike data collected in most surface mapping, satellites have the potent ability to detect major, rapidly changing environmental conditions all over Earth's surface, including vegetation and ice cover, types of land use, progress of erosion, rising or falling temperatures, and atmospheric gas content. Such information could not be practically obtained

Figure 1.79 A series of Landsat satellites have mapped Earth's surface, detecting changes in land use and vegetation since 1972. Landsat data are an invaluable archive of planetary change over time.

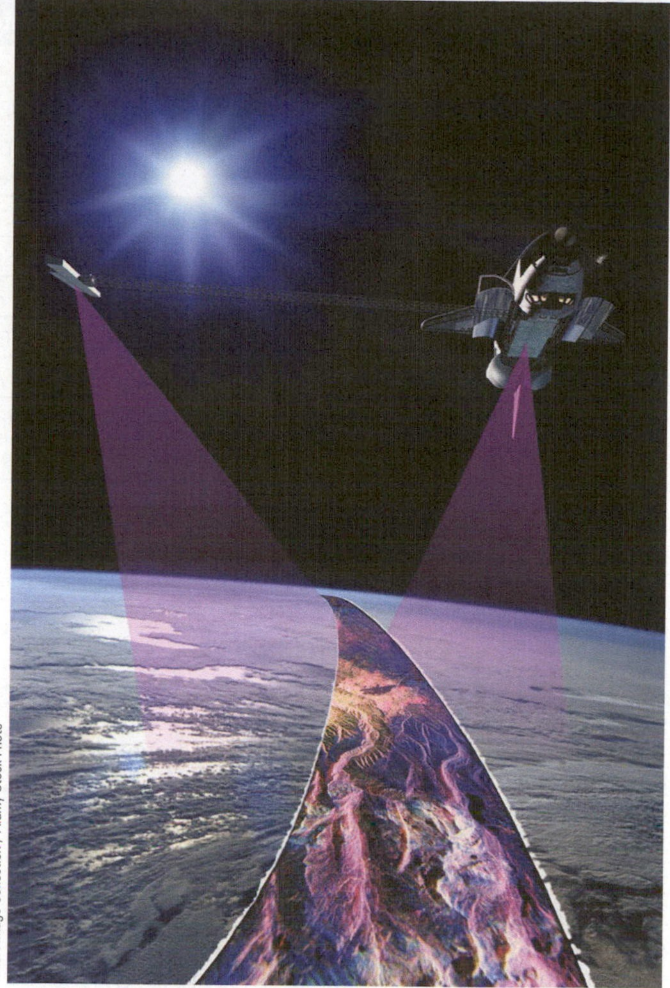

Figure 1.80 The Shuttle Radar Topography Mission or SRTM mapped most of the Earth's topography at high resolution using radar onboard the space shuttle. Its sensors see through the clouds.

otherwise. Examples of more environmentally important, recently launched satellites are described in **Table 1.2**.

Examining the inner Earth requires a range of tools. **Seismologists** study the pattern of energy waves generated by earthquakes to tease out the structure of Earth's deep interior (**Figure 1.81**). Their insights have allowed us to develop models for how the solid Earth "works" as a planet (see Chapter 2). Such investigation provides valuable information about the causes of earthquakes, where to expect them in the future, and even how to avoid their worst impacts.

At shallower levels, environmental geologists such as those who studied pollution at Newmark (Section 1.3.1) and Love Canal (see Case Study 1.2) use field sampling followed by laboratory analyses to detect toxic contaminants, ascertain the extent of groundwater pollution problems, and monitor the relative success of steps taken to clean underground water supplies (**Figure 1.82**). Surface water quality too is monitored using time-tested tools for establishing safe drinking-water standards. Often, environmental geologists work

Table 1.2 Some Currently Active (2022) Earth-Orbiting Satellites

Satellite Name	Launching Agency	Year Launched	Services Provided
CYGNSS	NASA	2016	Cyclone tracking for navigation
DSCOVR	NASA/NOAA	2015	The Deep Space Climate Observatory tracks sunlight reaching and reflecting from Earth's surface; a "barometer of global warming"
GOES 16, 17	NASA	2016	Collects weather observations
GPM	NASA & JAXA	2014	Studies rainfall and snowfall patterns
GRACE-FO	NASA	2018	Tracks changes in global sea levels, glaciers and ice sheets, soil moisture
GOSAT	JAXA	2009	Tracks changes in atmospheric greenhouse gases
ICESAT-2	NASA	2018	Measures ice sheet thicknesses for climate change studies
LANDSAT-9	NASA & USGS	2021	Images land surfaces and coastal areas at high resolution
Jason-3	NASA & CNES	2016	Radar altimeter used to measure sea surface height

Abbreviations:
NASA = (U.S.) National Aeronautical and Space Agency.
NOAA = (U.S.) National Oceanographic and Atmospheric Agency.
JAXA = Japan Aerospace Exploration Agency.
USGS = United States Geological Survey.
CNES = Centre National D'Etudes Spatiales.

with geophysicists and engineers to drill monitoring wells to identify the extent, structure, and purity of important aquifers (**Figure 1.83**). Intense interaction with various government agencies and even legal agencies and business interests may take place as scientists gather their data and share their conclusions.

In the last decade, drones have become all the rage for environmental geoscientists. First, they were used for mapping and photo-documenting the landscape (**Figure 1.84**). Now they fly all sorts of missions–some to places people fear to go, like the craters of active volcanos (**Figure 1.85**) where they monitor lava flows and sniff out volcanic gases. Other drones fly along the surface expression of faults using lasers to map the topography. The technology, termed **light detection and ranging (lidar)**, can also be used from airplanes and on the gorund to produce spectacular, detailed and very useful maps of Earth's surface in part because data processing allows the laser to "see through" the trees (**Figure 1.86**).

Environmental geology is a multidisciplinary scientific endeavor in service to humanity in which academic training is only the beginning, no more than an introduction. On-the-job experience, especially in a world with continuing technologies and scientific breakthroughs, assures that this field is continuously evolving, often in exciting ways providing greater capacity for data acquisition, precision, and insights.

dpa picture alliance archive / Alamy Stock Photo

Figure 1.81 Seismologists examine the seismological data from the 2011 earthquake that devastated Japan and destroyed the Fukushima nuclear complex.

USGS

Figure 1.82 Sampling a water well to test for chemicals dissolved in the water.

iStock.com/geogif

Figure 1.83 Drilling a monitoring well using augers and a track-mounted drill rig capable of working in uneven terrain without roads.

© Fiona D'Arcy

Figure 1.85 This drone is carrying instruments to measure gases coming from Turrialba volcano (in the background) The volcano is in Costa Rica.

UVM Spatial Analysis Laboratory

Figure 1.84 Drone image of a late fall landslide in Vermont with a dusting of snow on the ground and cars that provide scale. Blocks of debris (concrete) that used to be near the cars mark the toe of the slide by the river. Most of the trees are gone.

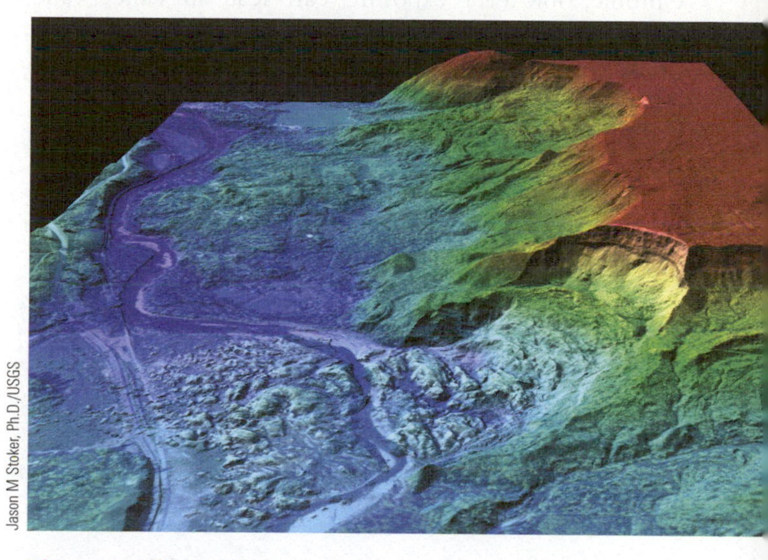

Jason M Stoker, Ph.D./USGS

Figure 1.86 The 2014 Oso, Washington landslide (in the foreground) captured in a lidar image (see Chapter 9). Can you find the road, the river, and an older landslide?

Case Study

1.1 Environmental Justice: The Role of Environmental Geoscience

In the summer of 1978, the U.S. government was considering a ban on **polychlorinated biphenyl compounds (PCBs)**. These are large, complicated, synthetic molecules, made of carbon, hydrogen, and chlorine. Nature does not make PCBs, chemists in labs do (**Figure 1.87**). PCBs were used in a wide variety of industrial applications, including circuit boards and transformers as well as window caulking, because they stayed flexible even at cold temperatures (**Figure 1.88**). Acute exposure to PCBs in the environment had been implicated in a number of illnesses, including Japan's infamous Yusho disease, where PCBs at high concentrations contaminated rice oil. The result was severe skin and eye lesions and numb limbs. Chronic, low-level exposure can lead to cancer and damage to body systems.

Seeking to rid itself of the expensive PCB waste disposal process, as mandated by the new U.S. Toxic Substance Control Act, passed in 1976, the Ward PCB Transformer Company secretly dumped 31,800 gallons of PCB-contaminated oil along 380 kilometers (235 miles) of highway shoulders in 14 counties and at the Fort Bragg Army Base in North Carolina. The crime took place over a period of 2 weeks. Drivers of a black truck drove along rural roads spraying the liquid toxin first along one shoulder one night, then the other shoulder the next. The press called the act "the Midnight Dumpings." In response, the governor of North Carolina had the Highway Department erect large yellow warning signs along the roads stating "Caution: PCB Chemical Spill on Shoulder" (**Figure 1.89**).

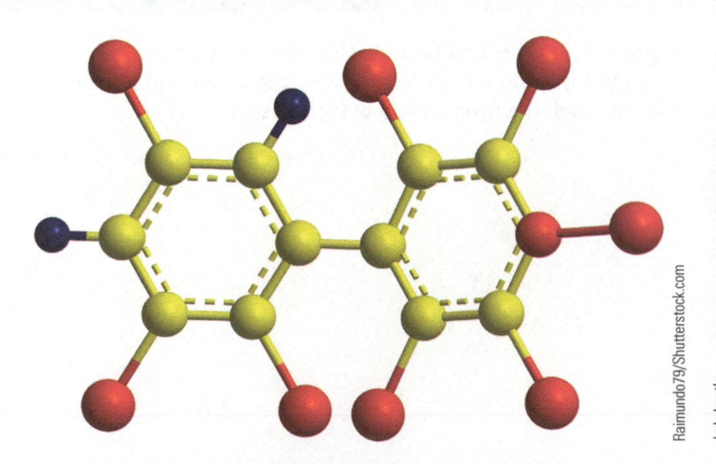

Raimundo79/Shutterstock.com

Figure 1.87 The structure of one specific PCB molecule. The yellow balls are carbon, the red chlorine, and the blue hydrogen.

US Army Corps of Engineers

Figure 1.88 Older electrical transformers were often filled with PCB-laced oil because it was resistant to high temperatures. Labelling allows for proper disposal.

The perpetrators were caught but served minimal time in jail. In the meantime, the state proposed to skim off the 54,500 tonnes (60,000 tons) of contaminated road shoulder soil (**Figure 1.90**) and put it in a new 7.7-hectare (19-acre) managed landfill in Warren County, then a poor, primarily agricultural backwater in North Carolina's Tidewater region. Warren County had a greater percentage of Black residents than any other county in North Carolina. The

Indy beetle

Figure 1.89 Signs along Warren county roads tell the story of the PCB hazard.

EPA

Figure 1.90 Removing PCB-contaminated soil from the shoulder of roads in North Carolina. Not only is the road crew wearing full hazmat protection, but so is the press corps in the back of the truck!

proposed landfill would be built on low ground, where groundwater is only few meters down. The local residents put up a fierce protest.

Groundwater tends to flow sluggishly in a downslope direction. Contaminants that are less dense than water float, including some PCBs, and tend to accumulate at the top of the groundwater. Groundwater moves far more slowly than water does in streams at the surface; but flow it does, and neighbors were concerned that any leakages could readily spread from the landfill underground to poison their shallow groundwater supplies.

Protests held off landfill construction for 4 years, and culminated in a 6-week-long nonviolent direct action campaign that was the largest civil protest in the American South since Martin Luther King Jr. marched through Alabama. The incident gave birth to the notion that *everyone* has a basic human right to live in a clean and safe environment, and moreover, that the burdens of environmental care and degradation otherwise should be shared with full disclosure as equally as possible–the very definition of environmental justice, a term introduced by protester Ben Chavis, who faced police harassment at the time (**Figure 1.91**). Pollutants should, in short, be treated as a common concern, not one concentrated on the poor, the uninformed, and the politically weak.

AP Images/Greg Gibson

Figure 1.91 Reverend Ben Chavis leads protestors in 1982. Their goal: to detoxify the landfill.

The residents were wise to be concerned. Despite their opposition, the need to sequester the contaminated soils somewhere proved overwhelming, and Warren County would just have to do, according to the state. But almost as soon as the landfill opened it began leaking PCBs, initially polluting the air as an odorless gas within a kilometer of the site (Figure 1.91). Shockingly, local residents were not informed of this situation for 15 years. The landfill operators eventually put a plastic liner atop the fill to seal in the gas. Finally, in 1993, North Carolina decided to detoxify the landfill after more protests and more arrests. Too much water was building up within it for the landfill to retain its poisonous content any longer. The wastewater could soon leak out above the base liner and allow contaminants to seep into community streams and wells. Detoxification continued until the end of 2003. It cost taxpayers $17 million.

The Warren County landfill case has a continuing legacy. It in large part inspired President Biden in 2021 to appoint two dozen advocate counselors to advise his Administration on environmental justice issues. The problem remains sadly significant: Black people in the United States remain four times more likely to die from pollution than White people. Thirty percent more Black Americans breathe heavily polluted air, and Black citizens are 75% more likely to live near a factory or plant that may be releasing pollutants.

Case Study

1.2 Love Canal: An Environmental Legislative Tipping Point

To most people, Buffalo, New York, is known for its chicken wings and nearby Niagara Falls. To geoscientists, it's the town where environmental geology got its start, at a place called Love Canal (**Figure 1.92**). A quiet middle-class neighborhood for decades, in 1978, Love Canal hit the national press as the place where people were being poisoned by industrial chemical waste buried in a poorly sited waste dump adjacent to their homes.

Love Canal itself was a failure–an attempt, in 1894, by William Love to divert some of the flow of the Niagara River to a site where he could generate hydropower (and make a fortune) some 11 kilometers (7 miles) away. The canal was never finished, and Love abandoned his plans after digging a 975-meter (3,200-foot)-long trench. It soon became a swimming hole as ground and rainwater filled the depression. The ditch proved tempting for other uses and Hooker Chemical Company, which had a large plant in Buffalo (**Figure 1.93**), purchased the canal. In 1942, Hooker started filling it with drums and tanks of chemical waste, including some of the most toxic organic chemicals of the time: dioxin, benzene, and PCBs. By 1953, the ditch was filled with 21,000 tons of hazardous waste and the company sold the land to the Buffalo Board of Education (which built an elementary school on the site). Hooker clearly stated in the deed that the site was contaminated. Nevertheless, in addition to the school, the City of Buffalo would soon build a park and baseball fields over the old canal and the buried toxic waste. Meanwhile, modest homes occupied the blocks around William Love's unfinished but by then filled canal (**Figure 1.94**).

Figure 1.93 Hooker Chemical Company plant in Buffalo, New York.

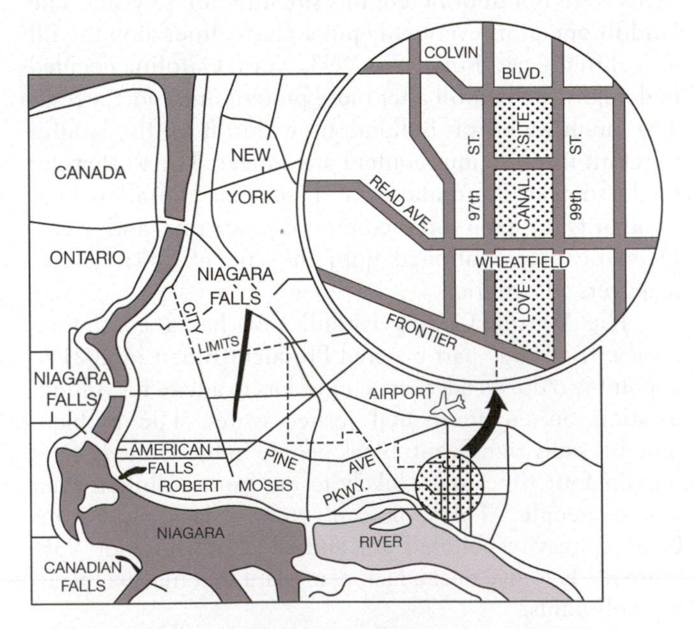

Figure 1.92 Love Canal is close to the Niagara River in southern Buffalo upstream from the falls.

Figure 1.94 Homes, some with swimming pools, surround the waste-filled Love Canal.

Underground, trouble was brewing. Beneath Love Canal was up to 40 feet of sediment left by the Laurentide ice sheet when it covered New York State during the last glacial period, perhaps 28,000 until about 14,000 years ago. The lowest sediment was glacial till; compacted by the immense weight of the overlying ice, till slowed the flow of groundwater. On top of the till was clay, likely deposited in a lake dammed by the retreating ice sheet. Clay, with low permeability, slows or altogether stops groundwater flow. The problem was the material at the top of the stack–sand with some silt and gravel. Although there were only a few meters of this material, it was permeable, allowing groundwater to move rapidly. The stack of glacial sediment was underlain by the Lockport Dolomite, a carbonate-rich rock prone to dissolution by groundwater (**Figure 1.95**).

University at Buffalo/Penelope Ploughman

Figure 1.96 Chemicals seeping through the cement wall of a basement in a home adjacent to Love Canal. This is one of many homes that in 1978, after a wet season, had dark, thick, strongly smelling liquids seep into their basement.

The stage was set for trouble because both Love Canal and the basements of the homes surrounding the canal were excavated into the same layer of sand with the clay lurking beneath. As contaminants leaked into the canal from rusting drums and tanks, groundwater, which could not flow down through the impermeable clay, flowed laterally, carrying toxic liquids toward the homes. It took several decades for the chemicals to migrate; then, the complaints began.

In the 1970s, foul-smelling, multicolored waste began seeping into people's basements (**Figure 1.96**). By 1976, several years of above-average rainfall had raised the water table and increased the flow of chemicals coming from the waste drums, some of which, after 20 plus years in the ground, had now rusted and ruptured. There were reports of drums and tanks poking out of the water-saturated soils of the canal (**Figure 1.97**). Puddles of colorful liquids dotted the surface. Eckhardt C. Beck (EPA Administrator for Region 2, 1977–1979) noted:

> "I visited the canal area at that time. Corroding waste-disposal drums could be seen breaking up through the grounds of backyards. Trees and gardens were turning black and dying. One entire swimming pool had been popped up from its foundation, afloat now on a small sea of chemicals. Puddles of noxious substances were pointed out to me by the residents. Some of these puddles were in their yards, some were in their basements, others yet were on the school grounds. Everywhere the air had a faint, choking smell. Children returned from play with burns on their hands and faces."

As state and federal agencies found ever more toxic materials, public protests began in earnest. In August 1978, President Jimmy Carter declared a National State of Emergency for Love Canal. The school that had been built over the filled canal was closed, and the 239 families who lived on the streets directly bordering the canal were evacuated (**Figure 1.98**). Another evacuation in 1979 removed 561 more families from a 10-block area around the canal. Their homes were bought out and demolished starting in 1982. In 1983, the school was demolished. Today, the waste has been removed from the canal and the area has been capped with both an impermeable fabric and clay. The site is well fenced.

Some of the first groundwater **numerical modeling** (a mathematical technique used to study

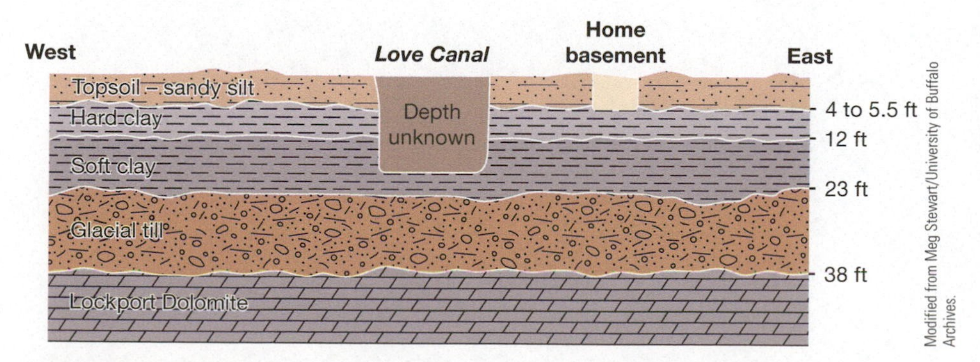

Modified from Meg Stewart/University of Buffalo Archives.

Figure 1.95 Diagram of the geologic materials underlying Love Canal. The contaminants mostly remained and travelled in the permeable, sandy topsoil.

Michel Philippot/Sygma/Getty Images

Figure 1.97 Rusting drums exposed at the surface of Love Canal.

University at Buffalo/Penelope Ploughman

Figure 1.98 Abandoned homes ringing Love Canal in 1982. They were demolished later that year. These homes were closest to the canal and were abandoned in 1978.

many physical processes) was done at Love Canal to understand how rapidly the plume of contaminated ground water would move and where it would go. The model was critical in designing a pumping and treatment system to keep the contaminated groundwater out of the Lockport Dolomite. This was a good thing because once the plume entered the porous dolomite, the model showed that contaminants would on average take less than 3 years to reach the Niagara River. To this day, Love Canal groundwater is aggressively pumped and treated to strip out chemicals which remain and to keep those chemicals from migrating offsite.

Activist citizens and scientists were key in bringing the environmental catastrophe that was Love Canal to light (**Figure 1.99**). Sampling by scientists, some working for the state and some working independently, identified the

AP Images/Anonymous

Figure 1.99 Lois Gibbs organized the citizen response to Love Canal when she was in her 20s and was raising two children. Gibbs led the Love Canal Homeowners Association. Here she trims a Christmas tree with decorations naming chemicals found in the Love Canal.

chemicals present in the canal and verified citizen's complaints. Protests by citizens, often involving children and teens, brought the issue of improperly disposed toxic waste into the public's consciousness (**Figure 1.100**). National concern was so great that, in 1980, the U.S. EPA's Superfund Law (otherwise known as the Comprehensive Environmental Response, Compensation, and Liability Act, CERCLA) was established to coordinate cleanup at waste sites around the United States. Love Canal and the evacuation of more than 800 families was a catalyst for this pivotal environmental legislation.

Bettmann/Getty Images

Figure 1.100 Protests forced the government's hand to deal with the hazardous materials buried in Love Canal.

Photo Gallery

Why Environmental

Geoscience is critical to understanding Earth and the environment in which we all live. As the human population has grown exponentially, so have human pressures on the environment and environmental pressures on people and society. Here are a few more examples of why geology matters!

iStock.com/SweetyMommy

1 Tough driving when a landslide drops the road out from underneath you, as it did here in Medicine Bow National Forest, Wyoming.

blickwinkel / Alamy Stock Photo

2 A rockfall in Uerguep, Turkey put an end to alfresco dining at this cafe.

James Davis Photography / Alamy Stock Photo

3 Recent eruptions mean that driving the speed limit on the Big Island of Hawai'i is going to be a challenge.

Hot springs, the result of geothermal activity are not only a favorite place for geologists, but in Nagano, Japan the snow monkeys have the warm water all to themselves.
Sean Pavone/Alamy Stock Photo

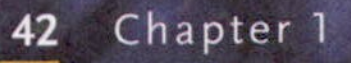

Everett Collection Historical / Alamy Stock Photo

4 Even being a famous geologist like Louis Agassiz gives you (well, OK, your statue) little protection against earthquakes as strong as the one that shook San Francisco in 1906. The statue fell from high on the zoology building at Stanford University.

iStock.com/Petr Kahanek

5 Sinkholes are usually something to avoid, but in Oman the collapse of the limestone provides people a place to swim and cool off in the desert at Hawiyyat Najm, the Bimmah Sinkhole pool.

Summary Why Environmental Geology Matters

1. What Is Environmental Geology and Who Are Environmental Geologists?

LO1 Explain what environmental geology is and why it matters

Environmental geology is the study of how geological forces impact, influence, and sometimes result from human activities. Environmental geologists work to improve human interactions with our physical planet. Environmental geology concerns issues such like flooding, earthquakes, volcanic eruptions, landsides, coastal erosion, water scarcity and contamination, and climate change. It also considers the impacts and potential impacts of resource extraction, such as mining and drilling for oil and water. Environmental geologists seek to understand the natural environment, identify problems, and find solutions by studying environmental change at or near the surface of the Earth. They use a variety of investigative approaches, ranging from detailed mapping of land and sea floors, to Earth-orbiting satellites, depending upon the needs of particular projects.

Study Questions

i. What distinguishes environmental geologists from ordinary environmentalists, and from ecologists (scientists who study living ecosystems?)

ii. Provide two examples of how environmental geologists can help improve the relationship of humanity to the physical world. What tools or approaches would they use in the examples that you cite?

2. Environmental Geology: Why It Matters to You and The World

LO2 Give examples of major problems approached by environmental geologists

a. The Newark Contaminant Plume

Beneath the eastern end of the Ontario Plain in California lies the Bunker Hill Basin aquifer, which supplies 25% and 75% of residential drinking water to San Bernardino and Riverside, respectively. In 1980, a contaminant plume was discovered in the aquifer, probably originating from U.S. Army operations of the San Bernardino Engineering Depot between 1940 and 1947. Ongoing cleanup includes pumping out contaminated groundwater, and the extraction of vapors at the expense of taxpayers.

b. Fukushima: A Geological Nuclear Disaster

The 2011 Fukushima tsunami exceeded estimations for tsunami hazard possible in this region of Japan, which is why Fukushima's nuclear plants were inadequately prepared for the sea wave. The wave completely overtook the seawall and flooded the nuclear power plants, resulting in the shutdown of backup generators used to continuously pump water and keep the nuclear reactors cool in the case of emergency. As a result, the reactors became dangerously hot, releasing hydrogen gas into the power plant. The buildup of this gas resulted in gas explosions that blew open pipes and tore off the roof of the plant. Radioactivity leaked from the plant into the surrounding atmosphere and ocean water, and eventually travelled to other parts of the world via wind, water, and fish.

c. Warming Climate, Shifting Shores: Kiribati

Water which has previously been trapped in glacial ice is melting at rapidly increasing rates as the climate warms. The meltwater flows into the oceans, resulting in sea-level rise. Bangladesh (among other nations around the globe) are predicted to be at risk as a result. Scientists estimate that by 2100, 1,000 square kilometers of low-lying land will be submerged displacing upwards of 800,000 people. Kiribati is a Pacific Island nation with several of its islands already flooded by seawater. The country's high point is 3 meters (10 feet) above sea level, the threat of sea-level rise is high.

d. Venice: A City Sinks

By the twelfth century, Venice had become a major world power with a large navy–and a major impending subsidence problem. The city began to engineer the surrounding lagoon to facilitate navigation and prevent flooding; nevertheless, the city continued sinking. In modern times, groundwater pumping has accelerated this issue immensely. Additional sinking has been caused by the weight of massive, heavy buildings on the soft sediments. With sea level rising and subsidence rate increasing, the risk of Venice one day becoming submerged grows greater and greater. The MOSE system, activated for the first time in 2020, isolates the lagoon and the city from the highest tides, greatly reducing flooding.

e. Pakistan: A Nation Drowns

In August 2022, the country of Pakistan was about one-third flooded as a result of exceptionally heavy rains linked to extreme seasonal monsoons, glacier melt tied to record summer warmth, and immense, rapid runoff. As a result, one in every seven Pakistani residents lost their homes and infrastructure throughout the country was destroyed. Many scientists point to climate change as the cause of these types of extreme floods.

f. Water Quality and Flooding in Jackson, Mississippi

On August 29, 2022, the Pearl River flooded due to heavy August rains in Jackson, Mississippi. At the same time, the main pumps at one of the city's two water treatment plants failed, reducing the pressure in the water lines, allowing contaminated groundwater to seep into the distribution system. Schools, hospitals, residences, and other water consumers in the region were left without potable drinking water. This situation persisted for 3 weeks before service was restored. However, this problem was avoidable. Jackson had tried to improve the water system for years, but these efforts had been stymied at the state level by politicians who failed to act. Many see this lack of investment as a clear manifestation of environment racism and injustice: as of 2022, Jackson's population was 82.7% Black, 16.7% White, 0.7% multiracial, and 0.3% Asian; about a quarter of the population lives below the poverty line.

g. Isolating Nuclear Waste in New Mexico

The Waste Isolation Pilot Plant (WIPP), opened in 1999, was the U.S. first underground repository for nuclear waste. WIPP has operated fairly well since opening. The site will be filled to capacity, then closed for long-term storage before 2035. After that, the site will be sealed permanently (because radioactive decay for some elements takes tens of thousands of years). A serious question still remains: how can a warning system be constructed at this site that will remain intelligible to descendants over a long span of time?

h. Lessons from These Situations

There are many lessons from these situations (and many more to learn), including: some geological processes are triggered directly by human activity, the best way to avoid a hazardous geological situation is to anticipate it and act accordingly, sufficient human understanding of how geological processes occur (for scientists and nonscientists) is urgently important, some parties responsible for environmental disasters are long gone and not held accountable, and tremendous inequities exist in the world that leave some people and communities at greater risk from geological and environmental hazards than others.

Study Questions

i. What is environmental justice, and how does it play a role in two of the situations described in the eight examples?

ii. Explain how two of the environmental issues in this chapter could have been avoided.

iii. Using examples from section 1.2 how has climate change increased the frequency and significance of environmental events?

3. Tools of the Trade–Environmental Geology

LO3 Describe the tools used by environmental geologists doing science

Many tools are utilized frequently by environmental geologists to understand the Earth. Mapping (using GPS, total stations, drones, sonar, and site studies) is a fundamental methodology employed by most environmental geologists to study land use, land (or sea) cover, topography, bedrock conditions, and locate undersea volcanoes. Seismologists use seismometers to monitor Earth's interior (plate) activity, whereas hydrologists use water samples, discharge rate measurements, and lab analyses to detect toxic contaminants and groundwater pollution, and to monitor drinking water purity in surface and near surface water. Drones have become popular among geoscientists to map landscape changes, monitor lava flows and active volcanoes, and map topography.

Study Questions

i. What are two examples of tools used by geoscientists, what do they measure, and how do they work?

ii. How are specific tools utilized in different specialties of environmental geology? Give two examples from the text.

Key Terms

- Atmospheric river
- Aquifers
- Climate mitigation
- Comprehensive Environmental Response, Compensation, and Liability Act (CERCLA)
- Environmental geology
- Environmental geologists
- Environmental justice
- Environmental refugees
- Global positioning system (GPS) receivers
- Groundwater
- Light detection and ranging (lidar)
- Mapping
- Numerical modeling
- Polychlorinated biphenyl compounds (PCBs)
- Radio detection and ranging (radar)
- Radon
- Satellites
- Seismologists
- Sound navigation and ranging (sonar)
- Superfund site
- Transuranic elements
- Tsunami

"Life did not take over the world by combat, but by networking."

—Lynn Margulis, American evolutionary biologist

2

Earth as a System

Learning Objectives

LO1 Define the Earth System and explain why it is a favorable environment for life

LO2 Explain the meaning of the "plates" in Earth's crust, and why their behavior is important to us

LO3 Explain why winds blow in the directions that they do on different parts of Earth

LO4 Describe the connection between oceanic circulation and global climates

LO5 Define the concept of biogeochemical cycling and provide one example

LO6 Describe the importance of the biosphere in the development of Earth's environment from earliest time

Earth from space – a view including Australia, Asia, the Middle East, and eastern Africa

iStock.com/Abrill_

Earth, One in a Universe of Planets

Earth is but one of countless planets in the known cosmos. Astronomers have located thousands of planetary bodies outside our Solar System so far, mostly orbiting stars in the nearby Milky Way, our home galaxy. They show tremendous diversity in their atmospheric compositions, sizes, average temperatures, and the shape of their orbits. Science-fiction is not so different from true science when it comes to imagining this multitude of distant worlds. We live in an exciting new age of planetary discoveries far from home! Many of these "exoplanets" very likely support life in one form or another (**Figure 2.1**).

One, planet Kepler-452b, is about the same distance from its host star as Earth is from the Sun, and even has a year length which nearly matches our own, 384 days. It lies in the constellation Cygnus, about 1,800 lightyears (1 lightyear is equal to approximately 9.5 trillion kilometers or 6 trillion miles) away from us. Kepler-452b is warm enough to retain oceans on its rocky crust, though its mass is about five times that of Earth's. As a result, gravity there is much stronger than it is here on Earth (**Figure 2.2**).

Promising as discoveries such as this are, most planets probably have environments hostile to life for multiple reasons. Their atmospheres are poisonous or overheated. They are too close or too far away from their host suns to have abundant liquid water at the surface, or they lack strong **magnetic fields**, the region of magnetic forces surrounding a planet, to protect them from deadly, charged cosmic-ray particles like protons that make up cosmic radiation.

Earth is blessed with a **goldilocks environment**–just the right distance from the Sun for abundant liquid water to exist at the surface, a robust magnetic field that intercepts damaging high-energy cosmic rays, and an atmosphere that life not only can tolerate but also exploit to renew itself and evolve. Earth's orbit around the Sun is steady enough and nearly circular, meaning that conditions do not change during the course of a year so much that life becomes unsustainable. As a species, we are byproducts of the environmental circumstances and events in our "nursery" world.

questions to ponder

1. Astronomers estimate that there are 1 to 10 trillion planets orbiting stars in our galaxy alone, and that there are between 100 and 200 billion galaxies in the observable universe. What do you think the likelihood of other goldilocks world's is, given these estimates?
2. The course and degree of life's development is closely tied to the world on which it is found. After you finish reading this chapter, return to this opener and consider the likelihood is of discovering creatures very much like human beings existing elsewhere in our galaxy.

NASA

Figure 2.1 Image from New Horizons spacecraft in 2015, flying past Charon, the largest moon of Pluto, about 5 billion kilometers (3 billion miles) from Earth. New Horizons found signs that Pluto has a liquid ocean beneath a thick crust of ice, as well as complex organic molecules. This suggests that even worlds as frigid as Pluto and Charon with almost non-existent atmospheres, could possibly sustain simple forms of life.

iStock.com/Darryl Fonseka

Figure 2.2 Artist's conceptualization of an Earth-like planet, Kepler 186f, about 490 lightyears from Earth. Imagined small, semi-circular seas fill meteor impact craters to right center of view.

2.1 Earth as a System (LO1)

A **system** is a set of components that work together as parts of a mechanism or an interconnected network or body. A watch is a good example. A simple wind-up watch typically consists of around 130 separate parts. We make watches to tell the time; that is their intended function. If one part of a system malfunctions, however (say that the spring breaks or battery fails), the whole mechanism stops. "Different parts working together" is the guiding definition of any system.

A system can also develop naturally *without any deliberate planning or intentional purpose*. **Natural systems** are products of physics and chemistry, operating according to principles such as the laws of thermodynamics, the structure of atoms, and the influence of gravity. They typically take very long spans of time to develop and can be quite complex. Natural systems are important products of the evolution of our universe as it has expanded and diversified since the Big Bang of 13.8 billion years ago. The Solar System, the set of planets including the Earth, the Moon, and other solid bodies in orbit around the Sun, is an example of a natural system. The Solar System has no "purpose": it is simply the consequence of the earlier explosion of a massive star that cast a vast amount of energy and gases into space. This explosion cloud later consolidated under the influence of gravity to form planets and suns with stable orbital relationships.

Another natural system is your own physical body. The "parts" that keep you functioning well (make you healthy) include your organs, skeleton, skin, and blood. If one part fails, the rest may feel the stress and suffer, often in ways that aren't immediately predictable given your natural complexity. Numb or tingling hands, for instance, may be a sign of diabetes, a blood sugar disease having no immediately obvious connection with hands. In this regard, physicians may be viewed as expert "systems specialists."

2.1.1 Earth as a System: The Spheres

Earth's life-giving environment–the **Earth System**–functions as a finely tuned machine; a natural, self-sustaining spaceship for its inhabitants. Just as the organs of the human body are connected and interdependent to enable a person to survive and stay healthy for many years, so too on Earth there are linkages between its rocky and molten interior (**geosphere**), the air (**atmosphere**), bodies of liquid water (**hydrosphere**), ice (**cryosphere**), and the living realm (**biosphere**) which have made our environment hospitable to life for great spans of time (**Figure 2.3**, **Table 2.1**).

We take for granted these linkages between the various parts of the Earth System because they are around us functioning all the time, usually (though not always) silently and continuously. Consider the geosphere; deep inside Earth's molten center, a powerful magnetic field has formed that you cannot feel, but that you can "see" whenever you use a compass. The needle points in the direction of strongest force in the magnetic field where you are standing.

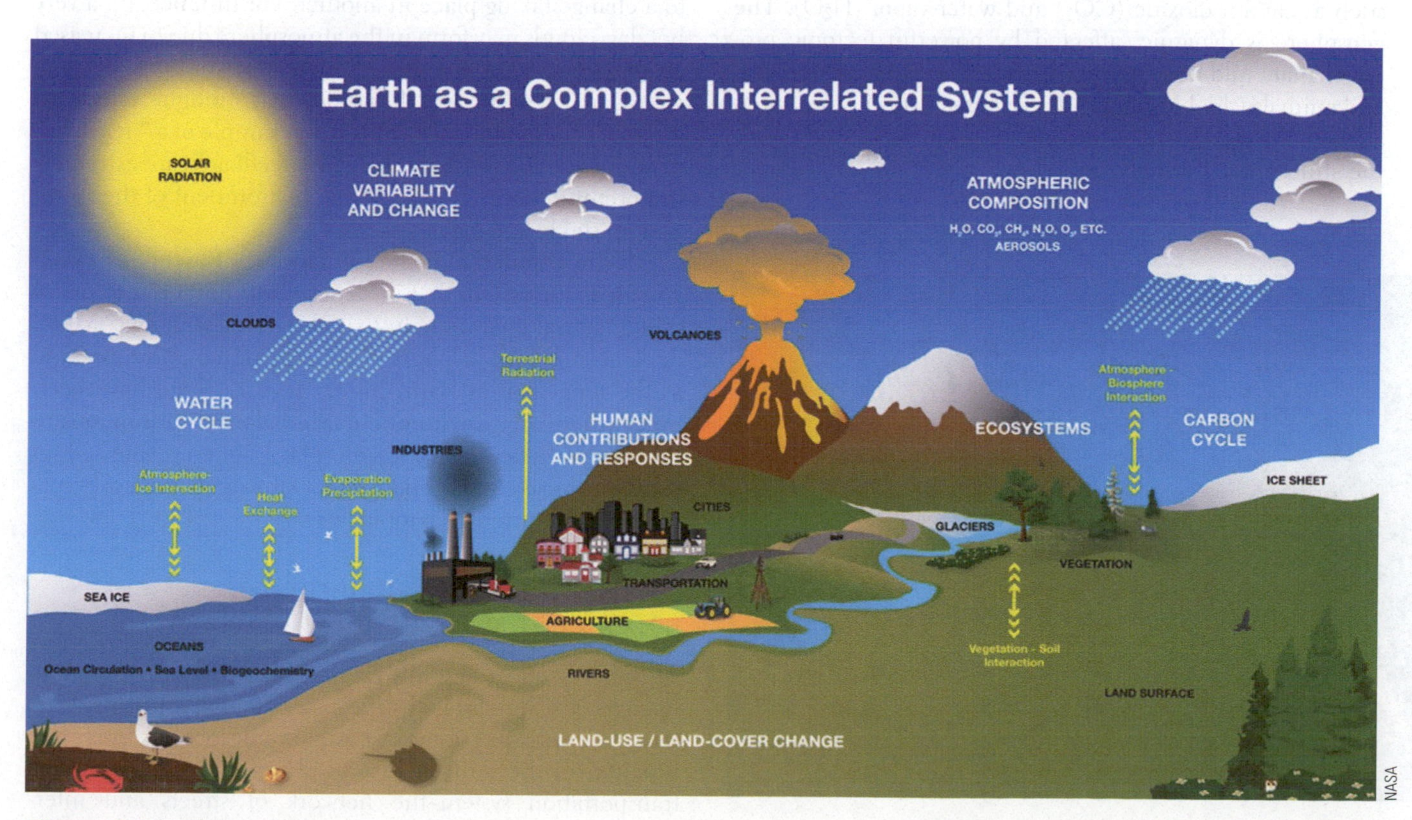

Figure 2.3 Earth as a system.

Table 2.1 Component Spheres of the Earth System

Component Name	Location	Some Key Contributions to the Overall System
Geosphere	Rocky body of the Earth, with a molten outer core	Generates magnetic field protecting terrestrial life forms; builds lands, mountain ranges, new ocean basins; influences ocean circulation, climates, and patterns of evolving life
Atmosphere	Gaseous envelop enclosing whole planet	Interacts with biosphere, hydrosphere, and cryosphere to determine concentrations of key greenhouse gases; protects terrestrial biosphere via ozone layer
Hydrosphere	Water present as oceans, rivers, lakes, and groundwater	Facilitates biogeochemical cycling; absorbs or releases atmospheric gases to moderate global temperatures; influences many regional climates; with biosphere, precipitates ancient, dissolved gases as limestone and gypsum (additions to geosphere)
Cryosphere	Ice cover near polar regions and at high altitudes	Impacts atmospheric circulation and temperature; can significantly impact biosphere
Biosphere	Living world of plants, animals, and other organisms	Interacts closely with atmosphere to determine greenhouse gas concentrations, with hydrosphere to capture circulating nutrients and water, and with geosphere to form soils (basis of the terrestrial food chain)

If you live at a high latitude, you can also view the magnetic field extending into the upper atmosphere to create colorful **aurora**, commonly known as the northern and southern lights (**Figure 2.4**). Earth's magnetism deflects most highly charged particles approaching from the Sun. These glowing displays of color in the sky on dark nights occur when the particles enter the upper atmosphere and collide with gases such as nitrogen and oxygen, which then glow brilliantly. If the particles reached us at the surface, however, they would destroy most surface life. The geosphere is linked to the atmosphere by supplying volcanoes with fresh gases such as carbon dioxide (CO_2) and water vapor (H_2O). The geosphere is dynamic, affected by powerful tectonic processes that build new mountains as well as habitats for plants and animals (the biosphere). Living organisms in turn interact with the geosphere by making soils at its surface, and with the atmosphere through respiration. And so on ...

iStock.com/RelaxFoto.de

Figure 2.4 The aurora flickers over the town of Alesund, Norway.

We explore these various interactions further in this chapter to develop a more detailed picture of how the Earth System fully works (Table 2.1). For the present, appreciate that what we call "the natural environment" is simply the blended outcome of various Earth System components operating at the surface. Their interactions maintain a rough balance with one another, leading to, or tending toward, environmental stability (homeostasis) by means of numerous feedbacks that are defined in the Box feature on the next page.

Think of feedback as the response of one part of a system to a change taking place in another. For instance, on a very hot day, clouds may form in the atmosphere due to increased evaporation (from the hydrosphere and biosphere). The clouds may tend to cool things down as sunlight reflects off their tops back into space. This is an example of a "negative," or counteracting feedback. Negative feedbacks are extremely important to keeping the natural environment of the Earth System stable and balanced.

If the system that is the human body functions to keep you alive, the system that is Planet Earth functions to maintain a natural environment that also keeps you alive. Much depends upon the balances of opposing forces, however; and as the long span of Earth history reveals, there have been times when large environmental imbalances–disruptions in the Earth System–have taken place, often expressed as extreme global climate change that kill large numbers of species. We'll return to that topic later.

2.1.2 Energy and Temperature in the Earth System

All systems, including the Earth System, require energy to operate. The human body–the system that is you–obtains its energy from eating food. An automobile obtains energy to function in the form of fuel or electricity. An automotive transportation system–the network of streets and interchanges necessary to permit movement of vehicles around a

Homeostasis and Feedback

The concept of **homeostasis** generally applies to living organisms and is defined as the ability of or tendency for a being to maintain internal equilibrium (e.g., a steady body temperature of 37°C or 98.6°F) by adjusting its physiological processes. When you are ill, your personal homeostatic system is strained, and your body temperature will typically be abnormally high. Recovery from illness is a good example of homeostasis at work. Your homeostatic system will stay in operation for the rest of your life.

The Earth System shows analogous homeostatic behavior by means of natural processes called **feedbacks** (**Figure 2.5**) Geoscientists identify two kinds of feedback: one positive and one negative. A **positive feedback** is a self-reinforcing set of reactions, increasing the intensity of the condition that set it into motion in the first place. (A nongeological example of a positive feedback is the simultaneous growth of human agriculture and population; both "feed" on one other, providing the facility for the other to increase). A **negative feedback** operates in the opposite sense, decreasing over time the intensity of the condition that originally set it into motion. (A nongeological example might be a swimmer; the faster a swimmer goes, the more resistance the water provides and the faster the swimmer becomes tired, more quickly slowing down to catch their breath and recharge energy).

In a perfectly balanced system, the effects of positive and negative feedbacks cancel out, and the net result is stability. But sometimes a feedback is so powerful that it overwhelms others opposing it and thus becomes a **runaway feedback.** A runaway feedback may continue being expressed indefinitely but quite often terminates in a new, stable set of conditions radically different from those with which it began. **Tipping points,** or thresholds, are conditions in which even small perturbations in a system—not necessarily runaway feedbacks—will cause it to change significantly. An example of a threshold is the boiling temperature of water: just a small increase in heat added to the water will cause it to transform suddenly from a liquid to a vapor state.

iStock.com/Paralaxis

Figure 2.5 Illegal deforestation in Jamanxim National Forest, Pará, Brazil. The deforestation sets in motion a positive (self-reinforcing) feedback: (1) removal of vegetation reduces the moisture evaporated back into the atmosphere above the region, which (2) further reduces local rainfall, hence (3) increasing drought conditions, which (4) increases the rate of forest die-back. Every link in this chain of consequences further increases the problem—drying out the land. Theoretically, the entire forest ecosystem in this part of the Amazon basin could collapse if a tipping point of rainfall reduction is exceeded. In other words, the feedback would "run away."

city—requires external energy inputs from routine street maintenance and the power for traffic signals to work.

The Earth System acquires its energy from two sources: one *internal*, the other *external*. Let's consider the internal energy source first. Inside our planet's deep interior, uranium, thorium, and to a lesser extent, potassium atoms, are continuously decaying into lighter-weight, more stable elements. One product of this **radioactive decay** is heat, which gradually escapes to Earth's surface. About half of the internally sourced energy at present is **radiogenic** in origin (made during radioactive decay); the rest results from gravitational squeezing and the heat left over from when Earth first formed at the birth of the Solar System, more than 4 billion years ago.

In general, the heat of rock inside Earth increases by about 25°C (45°F) each kilometer down into the planet you go, at least within the first few kilometers of the surface. The change in temperature with depth is called the **geothermal gradient**; a constant concern for deep-rock miners (**Figure 2.6**). The world's deepest mines, nearly

John Craven/Hulton Archive/Getty Images

Figure 2.6 Copper miner working deep in the Earth below France sweats in the heat.

4 kilometers (2.4 miles) down, have wall-rock temperatures as high as 60°C (140°F) at their base and must be cooled using huge, expensive air-conditioning systems (Chapter 4). Elsewhere, we see manifestations of our planet's escaping internal energy directly expressed as volcanoes, geysers, and hot springs (**Figure 2.7**). Some of the heat converts to kinetic energy, causing earthquakes, mountain building, and the slow shifting of the geosphere's massive tectonic plates.

More immediately noticeable to us on a daily basis, however, is the energy coming from Earth's external heat source, the Sun. Without the continuous input of solar energy, our planet would be a frozen, stony ice-ball, despite all the heat escaping from its interior. There would be no wind, rainfall, or circulating surface water. Because the biosphere depends on sunlight and liquid water for photosynthesis, life as we know it would be nonexistent.

The activities driven by the internal and external energy sources meet at Earth's surface. Any particular landscape is the result of forces in collision, some (largely internally generated) building up the land, others (mostly externally generated) tearing it back down. For example, wind, rain, ice, snow, and chemical weathering of rocks and minerals break down land that earthquakes, volcanic eruptions, sedimentation, and mountain building gradually build up (**Figure 2.8**).

Figure 2.7 Eruption of Giant Fountain geyser in Yellowstone National Park, Wyoming; a manifestation of intense heat energy from Earth's interior. The rock below the Yellowstone region is much hotter than most other parts of the world's shallow crust, owing to a massive, channelized upwelling of hot rock from depths of as great as 600 kilometers (400 miles).

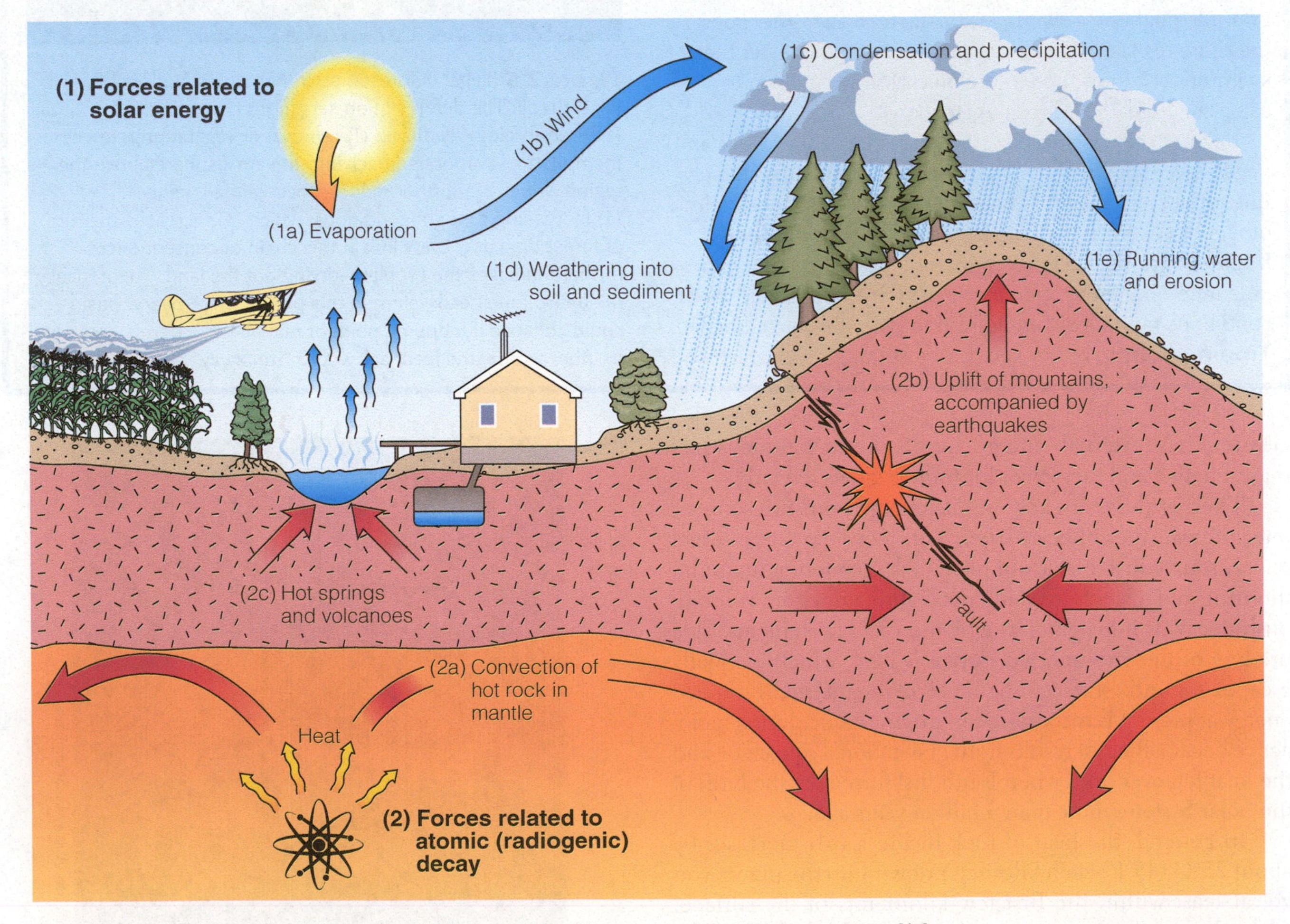

Figure 2.8 Earth's surface is a resolution of two sets of forces, modified by the activity of life.

2.2 The Geosphere: Land Building on a Large Scale (LO2)

Let's begin by taking a closer look now at the various parts of the Earth System, beginning with the geosphere, the ground under our feet. The geosphere is a fascinating place–and most of it is completely inaccessible to humans! In the recent past, though, Earth scientists with increasingly powerful tools have learned a lot about Earth's deep interior. Earthquake scientists (**seismologists**) have defined our planet's internal layering, density, and structure by studying the way that the energy released by earthquakes travels underground, much like the jiggling of Jell-O in a bowl that is suddenly slapped (Chapter 5). Geologists who study the composition and origin of rocks (**petrologists**) have been able to simulate conditions deep in Earth's interior through laboratory experiments (squeezing and heating rocks), learning much about what kinds of rocks must exist to account for the observations made by their fellow seismologists. Others have used chemical and physical techniques to date when rocks formed, what causes volcanoes to erupt where we find them (Chapter 6), and why the Earth shifts as it does along major faults. All make up a "Big Picture" of the geosphere that has only begun to come together since the mid-twentieth century.

2.2.1 How the Geosphere Came to Be

Astronomers have shed light on how it all began. The study of stellar nebula and young solar systems elsewhere in the galaxy indicate that our planet grew through the accretion of countless bodies of rocky, metallic debris condensing from a contracting, rotating cloud of hot gases left after the destructive explosion of an older star around 4.54 billion years ago (**Figure 2.9**). This process was fairly rapid, geologically speaking; mostly within a few tens of millions of years. The age of the human species, in comparison, goes back less than 400,000 years. We are truly late-comers on Planet Earth.

Early Earth was a largely molten body with "oceans" of churning, gaseous lava at the surface. Frequent heavy bombardment by meteorites and comets lashed the surface, adding new material to the growing planet, possibly including large quantities of water (H_2O) which added to that already being released by numerous volcanoes (**Figure 2.10**). One especially catastrophic strike completely detached a huge piece of Earth into orbit. This became our Moon. No life was possible in this unimaginably harsh birth-world environment.

Things changed slowly. Bombardments of new material diminished, and gravity gradually drew the densest material inside toward the center of the young, largely molten planet, forming a metallic iron-nickel-rich **core**. Less dense materials, including the elements silicon (Si), oxygen (O), and aluminum (Al), floated toward the top to form separate layers, the **mantle** and **crust**, wrapped like onion skins around the

Figure 2.9 Most meteorites striking Earth are fragments left from the condensation of the stellar birth nebula that gave birth to our planet roughly 4.54 billion years ago. Accretion of such matter provided the ingredients that differentiated under tremendous heat and pressure to build Earth's layers. (a) An iron meteorite cut open and etched with acid reveals blades of iron crystals. (b) A stony iron meteorite; a mixture of metallic and nonmetallic minerals, mostly silicates (c) A stony meteorite, almost entirely consisting of silicate minerals largely the same as what occurs in Earth's mantle today.

Science Photo Library/Alamy Stock Photo

Figure 2.10 Comets and meteorites frequently bombarded the Earth during its first few hundred million years; a lifeless world with little atmosphere for much of the time, and a planet that only gradually developed oceans.

core (**Figure 2.11**). This separation of matter based on density is called **differentiation**, and the three principal layers inside Earth–core, mantle, and crust–probably took several hundred million years to form this way. Yet, Earth's interior continues locally to differentiate up to the present in many places. Ultimately, our planet has also exhaled the lightest-weight elements and compounds as water and atmospheric gases, including nitrogen (N) and argon (Ar).

Figure 2.11 Cross-section of Earth's interior.

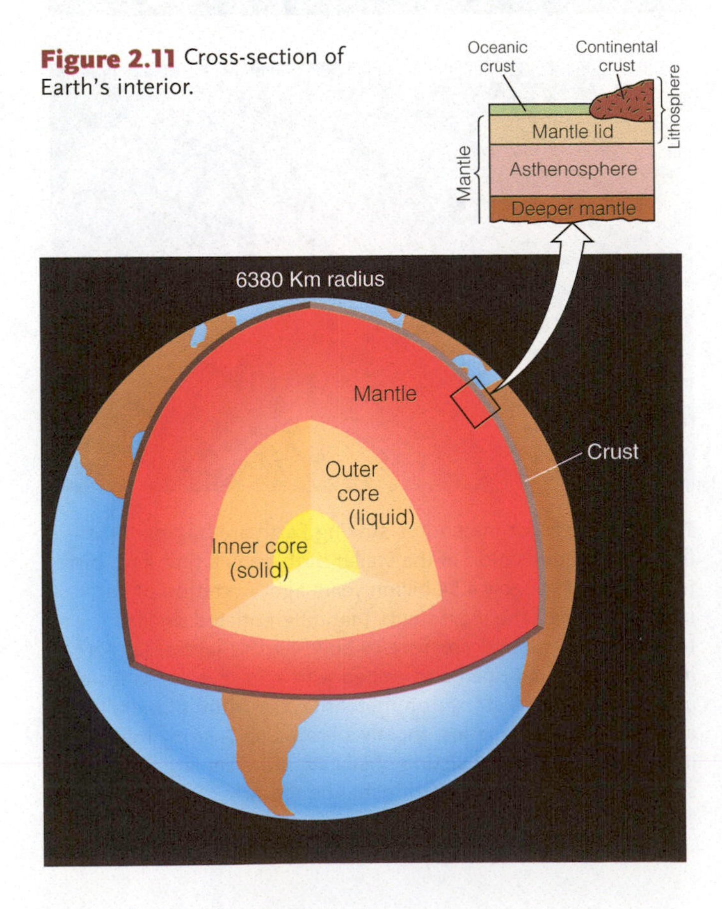

2.2.2 The Layered Earth

The outer part of the core, 2,900 to 5,100 kilometers (1,740 to 3,150 miles) deep, still remains molten, but the inner part (5,100 to 6,400 kilometers, or 3,150 to 3,950 miles deep) is solid, under too much pressure to melt despite having a temperature as high as 5,200°C (9,400°F), almost as hot as the surface of the Sun!

Earth's magnetic field originates by the movement of electrically charged currents in the molten outer core. The slow drift of these currents, probably related to Earth's rotation, accounts for the changing strength and position of the magnetic field over time. The magnetic field has **polarity**; that is, a North and South magnetic pole, much like the ends of a bar magnet (**Figures 2.12** and **2.13**). These do not exactly

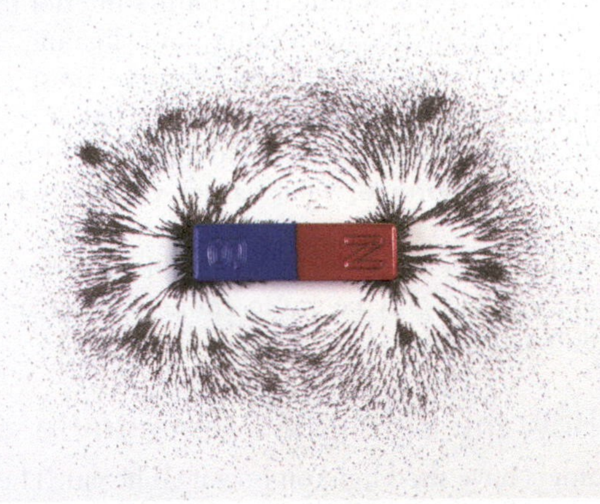

iStock.com/Wittyayut

Figure 2.12 Earth's interior acts like a giant bar magnet with north and south magnetic poles. Here, iron filings on a piece of paper are attracted to the two polar ends of a bar magnet in a classroom physics demonstration. Alignment of the filings defines the magnetic field created by the bar. Earth has a much larger magnetic field, of course: the magnetosphere, which extends on average between 60,000 to 300,000 kilometers (36,000 to 180,000 miles) into space from the planet's surface.

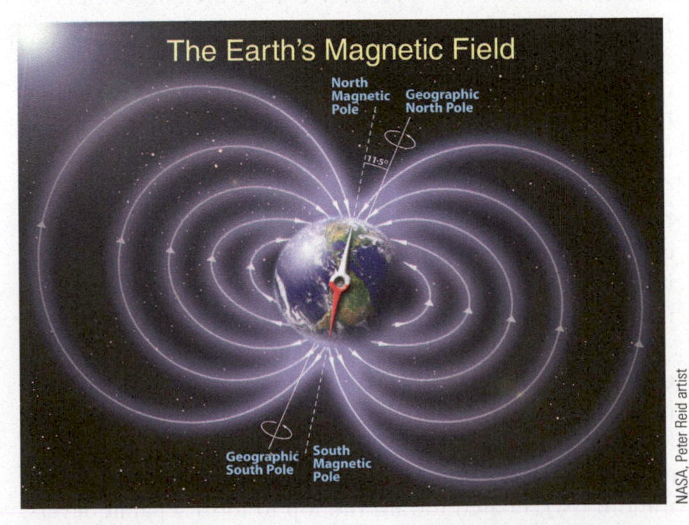

NASA, Peter Reid artist

Figure 2.13 Structure of Earth's magnetic field as defined by lines of magnetic field attraction. Our world is a giant, planetary "bar magnet."

match Earth's rotational North and South Poles, which are fixed geographical positions on a map. The mismatch can be as much as a thousand kilometers or more, though this is continuously changing. In 2019, the north magnetic pole was positioned only about 400 kilometers (250 miles) from the geographic North Pole, and for the first time in 360 years, compass needles in Britain pointed in the same direction as true geographical north.

Throughout the geologic past, the magnetic poles have rapidly (in a matter of thousands of years) swapped their locations when the overall magnetic field greatly weakens. No one knows why this happens, but these **magnetic reversals**, which can be detected in the rock record, have proven important for geologists who date and correlate groups of ancient rocks (**Figure 2.14**). This is because iron-rich minerals in rocks, especially lava flows, act like tiny bar magnets and record the orientation of the magnetic field at the time that they form. As far as we can see back into geologic history, Earth has had a magnetic field that every now and again swaps its poles.

Enclosing the molten hot core, a thick shell of magnesium and silicate-rich rock developed that we call the **mantle**. **Silicates** are rocks and minerals containing lots of the element silicon (Si) and oxygen (O) atoms, two of the most common elements making up the solid Earth. The silicon and oxygen are largely bonded together, forming silicate molecules, the building blocks of most minerals and almost half of some mantle rocks (Chapter 4).

The mantle is by far the thickest layer making up our planet, averaging about 2,900 kilometers (1,800 miles) from top to bottom. It makes up 84% of Earth's total volume. As far as we know, similar mantle layers also make up most of the interiors of other rocky planets. (In this feature, Earth is not unusual).

Overlying the mantle is a thin "skin" of relatively lightweight aluminum (Al) and alkali (sodium (Na) and potassium (K))-rich silicate rocks, termed the **crust**. Two basic types of crust exist: **oceanic crust** and **continental crust** (see Figure 2.11). Oceanic crust consists largely of dark basaltic rocks and is typically 5 to 10 kilometers (3 to 6 miles) thick. The continental crust, in contrast, consists largely of lighter-colored, silica-rich, igneous and metamorphic rocks (e.g., granite and gneiss) and related sedimentary cover, and ranges from 20 to 90 kilometers (12 to 56 miles) thick, with the greatest thicknesses found beneath mountain ranges like the Himalayas or Andes. We'll examine these various rock types in more detail in Chapter 4, but for now it is important to note that the oceanic rocks are characteristically denser (3.0 g/centimeters^3 or 0.108 pounds/inch^3) than continental rocks (2.7 g/centimeters^3 or 0.097 pounds/inch^3).

Earth's crust did not always consist of oceanic and continental fractions in the proportions we see today. Throughout Earth history, oceanic crust has dominated, supporting the notion that our planet is really an "ocean world." The 4-billion-year-old continental rocks found in central Canada, and even older rocks of the Jack Hills in Australia, however, show that continental material began forming very early in Earth history (**Figure 2.15**). The amount of continental mass around the world has only increased ever since, presently encompassing about 40% of Earth's surface and 70% of the crust's total volume.

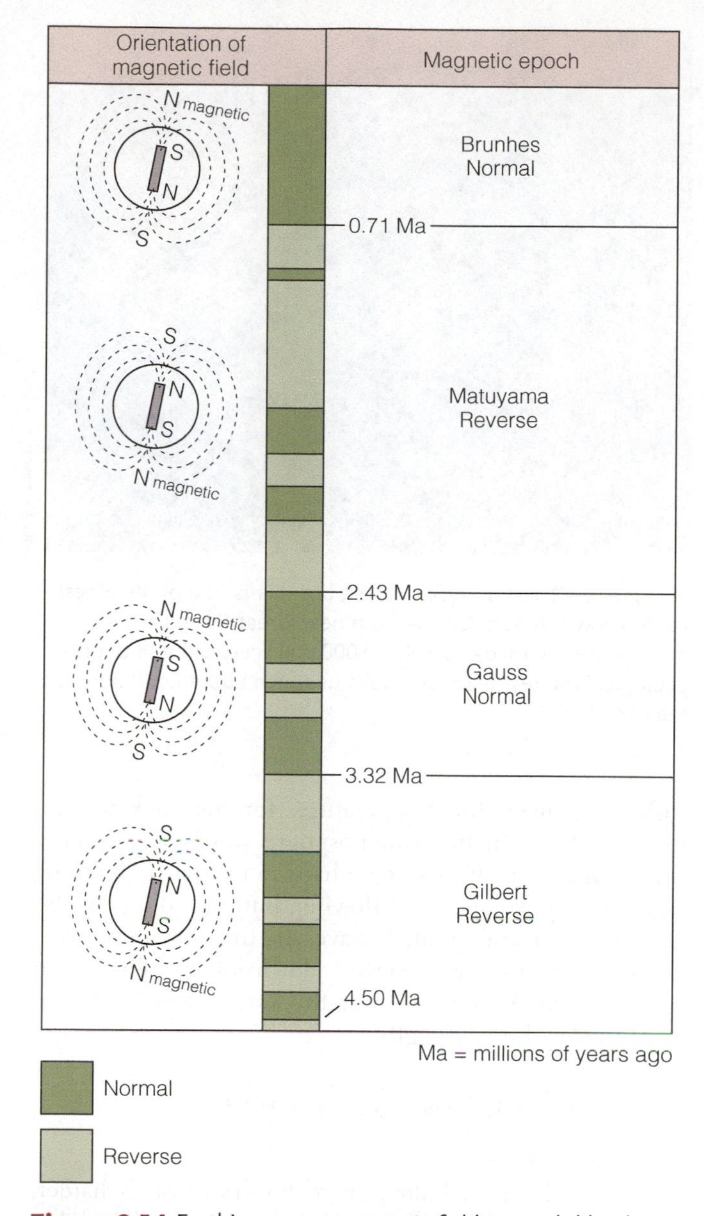

Figure 2.14 Earth's present magnetic field is much like the field that would be generated by a giant bar magnet inclined at 11° from Earth's rotational axis. Reversals of the field have occurred periodically, leaving "fossil" magnetism in rocks that can be dated. Each epoch is named after an important researcher of earth magnetism.

The crust and uppermost part of the mantle make up still another distinctive layer in Earth's interior, the **lithosphere** (the "rocky sphere"), on average about 100 kilometers (60 miles) thick. The lithosphere is defined as the outer layer of the Earth that behaves in a solid, rigid manner. It directly overlies a deeper layer made entirely of mantle rock, the **asthenosphere**, which extends as far as 250 kilometers (150 miles) beneath the surface. Pressures in the asthenosphere are very high, almost but not quite

Figure 2.15 It may look modest, but this is one of the oldest rocks known: Acasta Gneiss from near Great Slave Lake in northwestern Canada. It is 4,030,000,000 years old and of great geological interest because it tells us much about conditions on the early Earth.

Figure 2.16 Note that only a small part of the less dense ice cube stands above the water line. Most is below the denser water.

sufficient, given the temperature, for the rock to start melting. Rock in the asthenosphere slowly flows under these conditions. It may seem hard to think of something as solid as rock capable of flowing, but glacial ice, which is quite hard and solid, behaves in the same way as it slowly deforms and moves downslopes. The term "asthenosphere" derives from the Greek word *astheno*, meaning "without strength."

2.2.3 Ups and Downs of a Dynamic Geosphere

The lithosphere, including all of Earth's crust, is harder, stronger, and more brittle than the asthenosphere. The lithosphere literally floats atop the asthenosphere, just as ice floats on water. If you toss a cube of ice into a glass of water, the ice cube will bob up and down for a little bit before coming to rest with part of its mass afloat above the water level (**Figure 2.16**). Most of the mass of ice is submerged. You see only about one-ninth of it sticking into the air. It is floating in a state of equilibrium as long as it is undisturbed because the ice is *less dense* than the water. The percentage that continues above the water line is a function of the relative difference in the densities of the ice and water. Density is a function of mass divided by volume; thus, the greater the mass of the ice, the greater the volume of ice that is submerged underwater (**Figure 2.17**).

The geological analog of this phenomenon is called the concept of **isostasy**–or "floating equilibrium." Presuming that the lithosphere is everywhere roughly the same density, this means that the tallest mountains should also have the deepest foundational roots extending into the asthenosphere beneath them; and in general this is the observed to be the case (**Figure 2.18**). For instance, the **crustal root** (the thickness of low-density crustal rocks) beneath South America's Andes Mountains, which rise to 6 kilometers (4 miles) above sea level, extends to a depth of 50 to 80 kilometers (30 to 50 miles) beneath the surface.

Of course, the lithosphere, and certainly the crust, is not uniformly dense everywhere around the world. This, too, explains why some regions have significantly lower elevations than others. Oceanic crust is denser than continental crust; it lies at an average elevation of 3.7 kilometers

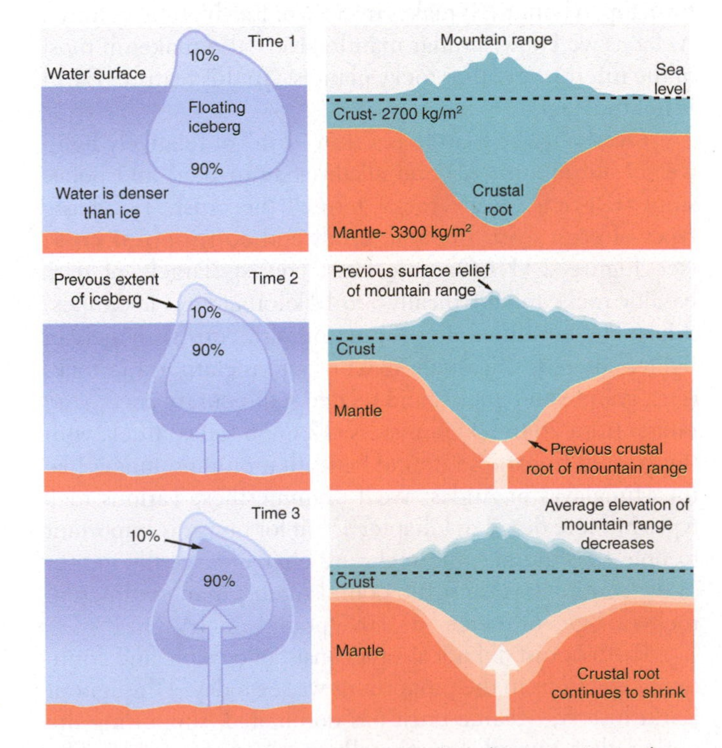

Figure 2.17 The isostatic response of Earth's crust is similar to an iceberg or an ice cube floating in water. (Adapted from Bierman P. R. Montgomery D. R. (2014). Key concepts in geomorphology, 2e. W.H. Freeman and Company Publishers.)

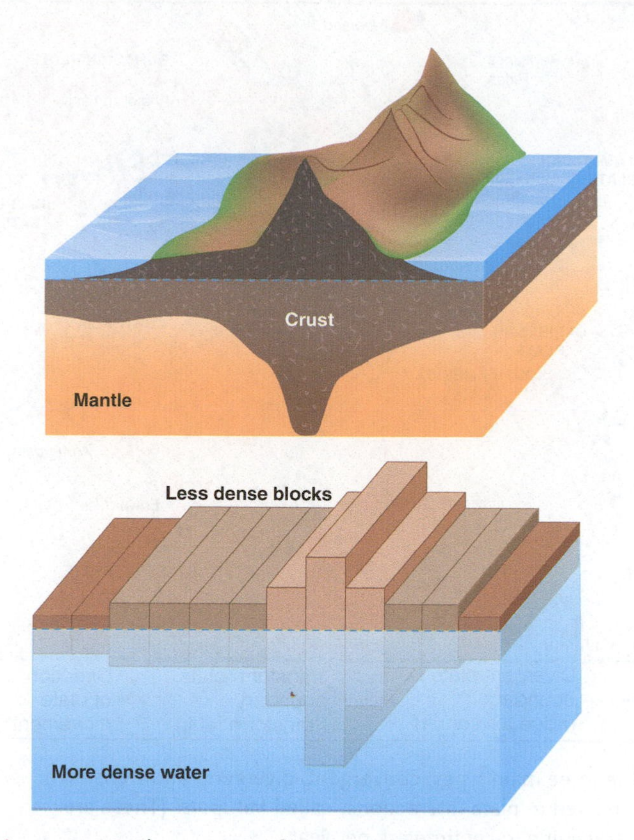

Figure 2.18 The concept of isostasy explains the presence of large, mountain root structures in the crust, showing that the lithosphere floats on the softer mantle (asthenosphere) below.

(2.3 miles) below sea level in contrast to continental platforms with an average worldwide elevation of 0.8 kilometers (0.5 miles) above sea level. So, it is relative differences in lithospheric density as well as thickness that explain why we have continents as well as ocean basins.

In many areas around the world, floating or **isostatic equilibrium** has not been attained because of various geological processes that throw the relationship of the lithosphere to the deeper mantle out of balance. For instance, widespread glaciation in Scandinavia, Alaska, and elsewhere until around 11,700 years ago buried and weighed down the underlying crust over broad areas, depressing the land surface hundreds of meters as the asthenosphere deep below deformed and flowed slowly outward, away from the ice. That ice has since largely melted away, and the unburdened land has been rebounding ever since (**Figure 2.19**). In the Gulf of Bothnia, the flooded saddle between Sweden and Finland (which was deeply buried under more than a kilometer of ice during the last glaciation), the seabed is rising at a rate of as much as 2 centimeters (almost an inch) each year. Unlike an ice cube suddenly tossed into a glass of water, it takes much longer for something as vast and ponderous as the asthenosphere to respond when it is in a state of isostatic disequilibrium: the long response time is the result of rock deforming much more slowly than bodies of water.

Figure 2.19 Isostatic uplift of Bathurst Island, northern part of Hudson Bay, Canada, following retreat of the great Laurentide ice sheet 20,000 to 6,000 years ago. The lines crossing the slope mark ancient beaches, oldest at the top, uplifted as the land continuously rises out of the sea.

The sluggish currents of nearly melted rock that flow like hot plastic in the asthenosphere are driven by the escape of heat from much deeper inside the planet. They may be set in motion at the core-mantle boundary, though the full pattern of mantle circulation is not well understood. One consequence of this internal stirring, nonetheless, is that the lithosphere has broken into large pieces, or **tectonic plates**, which may be compared to blocks of ice jostling on a frozen river, though far larger and denser. These same currents help shift those plates around the Earth at roughly the speed at which your fingernails grow. That is not a blinding rate in human terms, but millions of years of lithospheric restlessness have generated today's major geographical features, so it's important to give this phenomenon, plate tectonics, closer attention.

2.2.4 Plate Tectonics

The word *tekton* is from the Greek language, meaning "builder." The tectonic plates and their interactions define the shapes of our continents and major ocean basins, and the placement of many island groups, mountain ranges, and even major lakes and river valleys. Individual plates, which range from hundreds to thousands of kilometers across, are very active at their boundaries (**Figure 2.20**). In general, there are three types of plate interactions: plates sliding past one another (**transform boundaries**), plates pulling apart (**divergent boundaries**), and plates colliding (**convergent boundaries**):

- *Transform boundaries*. California's notorious San Andreas Fault system stretches 1,300 kilometers (800 miles) and has been the source of many destructive earthquakes that have killed (directly and indirectly) thousands of people (Chapter 5). The fault forms the boundary between two of Earth's largest tectonic plates, the North American Plate to the east and the Pacific Plate to the west. It is a

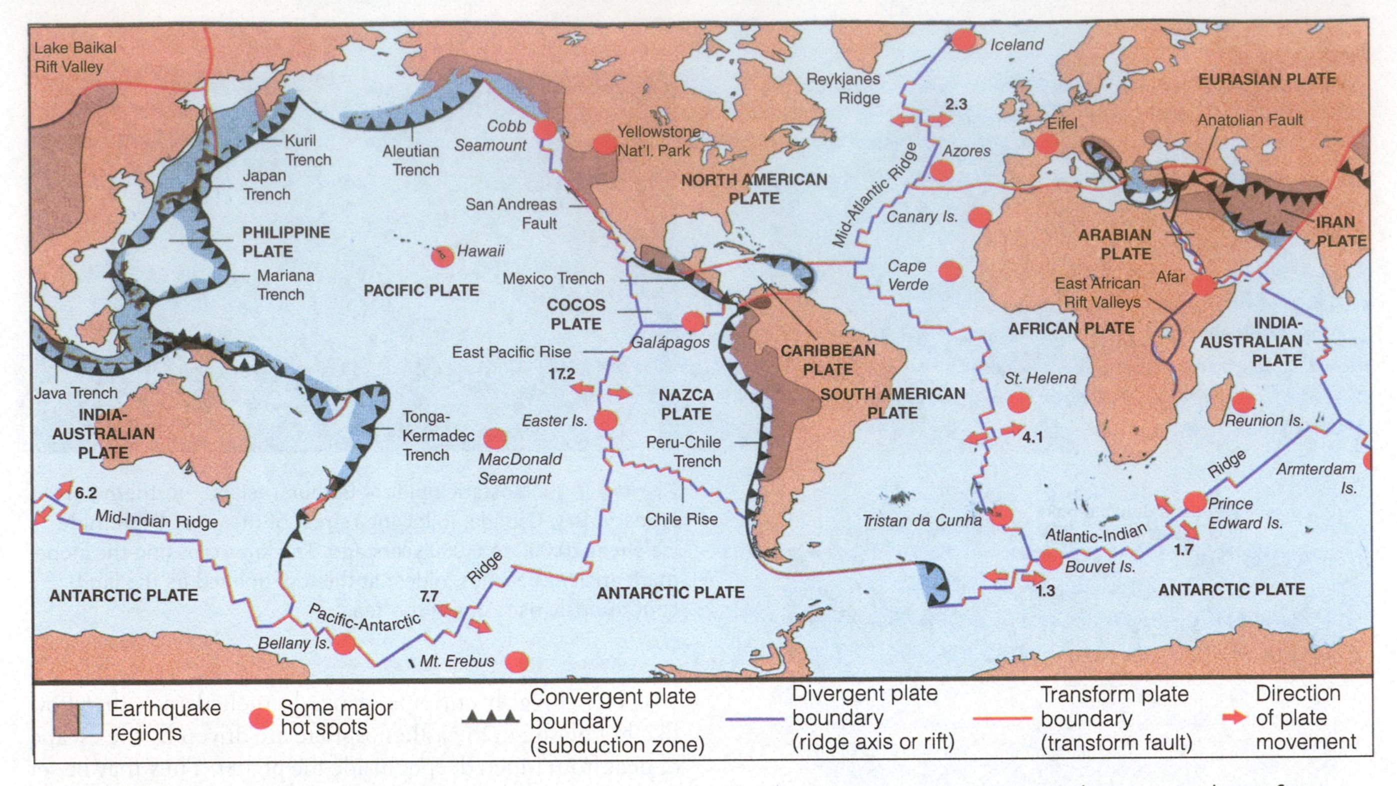

Figure 2.20 Map of the world's tectonic plate boundaries, showing the three main types: convergent, divergent, and transform. Also shown are areas of anomalous, intense volcanic activity, mostly unrelated to plate interactions, called *hot spots*. (These are discussed in Chapter 5.) Numbers next to arrows indicate rate of plate spreading in centimeters per year.

classic example of a **transform fault**, meaning that it links together two other boundary types at its ends–divergent margins in the Gulf of California and off the coasts of Oregon, Washington, and British Columbia (**Figure 2.21**). The plates slide past one another at an average rate of 40 to 60 millimeters (1.6 to 2.4 inches) per year, though they get hung up temporarily in many places, building up the stresses leading to massive earthquakes (Chapter 5). Transform fault boundaries in continental settings are a concern not only for residents of California, but also Turkey (North Anatolian Fault), New Zealand (Alpine Fault), Pakistan (Chaman Fault), Myanmar (Sagaing Fault), and a few other countries, though by far most transform faults lie on the ocean floor where they slice across divergent plate margins, rattling the deep ocean with earthquakes.

- At *divergent boundaries* plates move apart from one another under **tension** (stretching forces). The "tears" that develop in Earth's continental crust include rift valleys, of which the East African Rift Valley and Iceland are the most famous

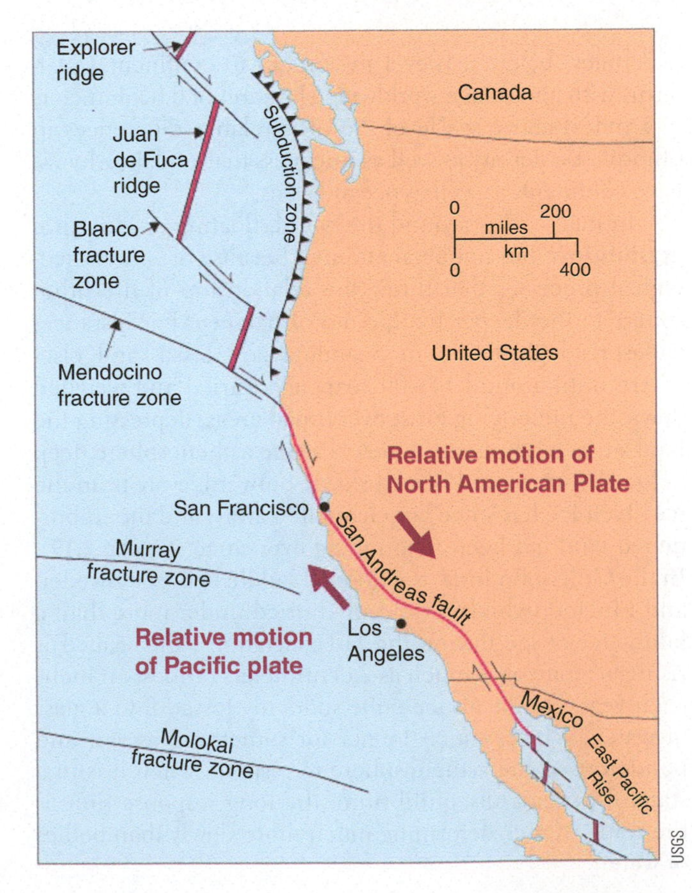

Figure 2.21 Tectonic map of San Andreas transform fault. It links two other plate boundaries: the East Pacific Rise, a divergent boundary in the Gulf of California, and the Juan de Fuca Ridge, another divergent boundary that itself is offset by shorter transform faults (e.g., the Blanco fracture zone).

iStock.com/Guentergoni

Figure 2.22 The western edge of the East African Rift Valley is a set of giant, eroded fault cliffs, one of which is shown here near Lake Turkana, next to the Kenya-Tanzania border. The rift valley floor (to right) is on average 600 to 900 meters (1,900 to 2,900 feet) deep, and the rift valley stretches 7,300 kilometers (4,310 miles): the longest continental rift valley on Earth.

Jose Antonio Peñas/Science Source

Figure 2.24 A "black smoker" seafloor volcanic gas and steam vent. Black smokers similar to this one are concentrated thousands of meters underwater all along the mid-ocean ridge system, and at other scattered submarine volcanoes not located on plate boundaries. They support rare forms of deep-sea life that take advantage of their heat and chemical emissions. They are also points of mineralization, including copper, gold, and other potentially valuable ores.

(Figures 2.22, 2.23). Along this axis, Africa is splitting apart at rates of 6 to 7 millimeters (0.25 to 0.3 inches) every year; sluggish, but impressive when you consider the size of the landmasses that are splitting apart from one another.

In the oceans, rifting occurs along the crest of **mid-ocean ridges**, with molten lava erupting through the fissures opened by the separating plates (see Figure 2.20). The lava cools to form new oceanic crust as the plates drift apart at rates from a few tens of millimeters per year (Mid-Atlantic Ridge) to as much as 170 millimeters (6.9 inches) per year (East Pacific Rise); much faster than observed in the case of East Africa (Figure 2.24). The process of rifting in the ocean crust, called **seafloor spreading**, not only creates new seafloor but rich deposits of minerals as well.

The various mid-ocean ridges are stitched together via transform faults to form a single, gigantic network: a global mid-ocean ridge system that winds around the surface of Earth like the seam on a giant baseball and links up with continental rift valleys in places. Its total length is more than 60,000 kilometers (37,000 miles)–by far the longest and one of the most circuitous mountain ranges on Earth, although it is mostly submerged and little noticed by land-bound humans (Figure 2.25).

iStock.com/Mlenny

Figure 2.23 Thingvellir, Iceland, location of an ancient Viking parliament, is also where plate divergence along the mid-Atlantic Ridge is tearing Iceland apart, as shown by enormous ground cracks such as this. The North American Plate is to the right of the tear, the Eurasian Plate to the left.

- At *convergent boundaries*, plates collide with tremendous **compressive** force. Where a plate capped with oceanic crust runs into a plate made up of continental material (for example, around much of the Pacific Rim), the continental plate thrusts over the oceanic plate, which is denser and sinks back into the mantle where it eventually is assimilated inside the deep Earth. Volcanic mountain ranges, such as the Andes, mark crumpling at the edge of the overriding plate along this type of convergent plate boundary.

  Weak, wet, seafloor sediment probably lubricates thrusting. This process is called **subduction** (meaning "under-moving"). The world's deadliest earthquakes and largest, most explosive volcanic eruptions are byproducts of subduction, often to the detriment of people who live in these regions (Chapters 5 and 6).

  Elsewhere, plates of oceanic crust collide. Subduction occurs in this situation, too, with the plate having the larger, generally older and slightly denser expanse of oceanic crust commonly subducting beneath the less dense, younger plate. Chains of islands created by explosive volcanoes, termed **island arcs**, are consequences of this type of plate

occur as these **intracontinental mountain chains** grow, but volcanic activity is rare (**Figure 2.27**).

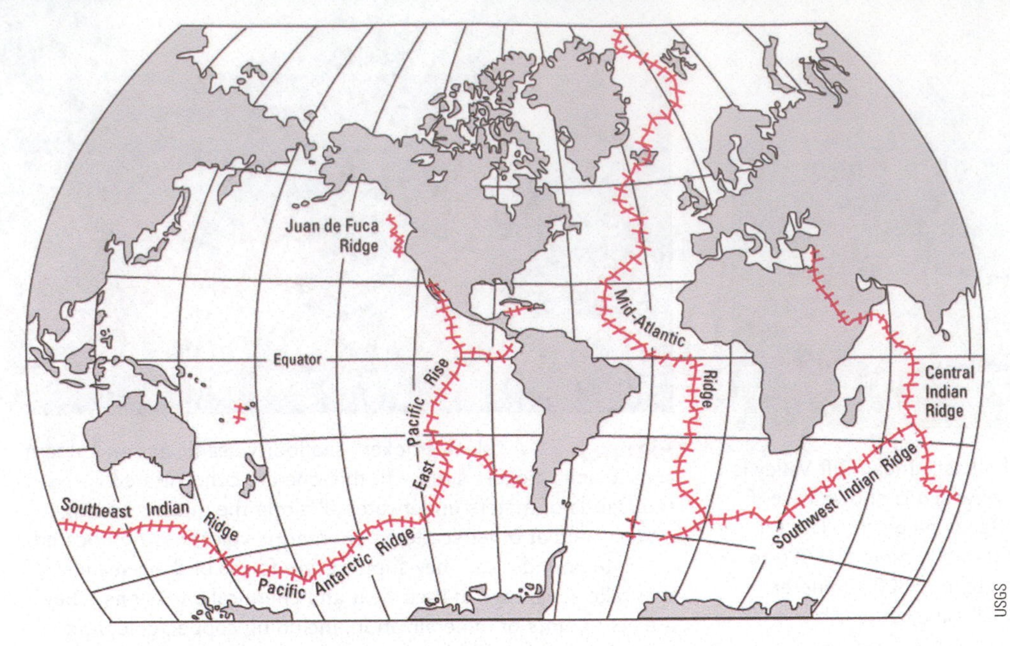

USGS

Figure 2.25 Map of mid-ocean ridge system, which corresponds to divergent plate boundaries trending across the ocean and younger seaway floors. The fastest rate of plate divergence is recorded along the East Pacific Rise. In comparison, divergence along the Mid-Atlantic Ridge is less than a quarter as fast.

convergence (**Figure 2.26**). The Antilles Islands fringing the Caribbean Sea are a good example. Here, the Atlantic oceanic lithosphere subducts beneath the Caribbean Plate at a rate of about 20 millimeters (0.8 inches) per year.

In another style of plate convergence, two plates made up of continental material collide, creating a great range of mountains because continental lithosphere is generally too thick and buoyant (relatively speaking) to subduct. The Himalaya Mountains and Alps formed in this way through the collisions of India and Africa, respectively, with Eurasia. Very powerful earthquakes often

Plate tectonics has played a key role in Earth's development, defining the broad outlines of Earth's surface–the arrangement of continents and ocean basins seen from space. The course of life's spread and evolution is linked closely to that ever changing world geography. Abundant new lifeforms have evolved and others have vanished due to plate tectonic activity. It is at plate boundaries that new landmasses develop, expanding the terrestrial environment available for life, including us. Even the very air we breathe is influenced by tectonism. Volcanic activity at plate boundaries contributes important gases such as carbon dioxide and water to the atmosphere. At the same time, rocks exposed by mountain ranges forced up by plate collisions weather rapidly, sucking carbon dioxide out of the air to help keep the buildup of this greenhouse gas in check (**Figure 2.28**). Astronomers have so far not found evidence for plate tectonics on any other known planet, underscoring one important if not perplexing aspect of Earth's "goldilocks" environment.

T.S. Miller/USGS

Figure 2.26 Pavlov and Pavlov Sister volcanoes, Alaska, two of over 130 large volcanoes lined up along the 2500 kilometer (1550 mile) long Aleutian island arc, continuing onto the Alaskan Peninsula, where the Pacific Plate under-thrusts the neighboring North American Plate, The two plates are colliding at a rate or around 6 centimeters (2.4 inches) per year.

iStock.com/Miljko

Figure 2.27 A woman stands atop 5600-meter (18,373-foot)-tall Kalu Patthar summit in the Central Himalaya of Nepal, gazing toward Mount Everest (center pyramid) and closer Lhotse, almost as tall (spiky peak to the right). These mountains began forming when India started colliding with Eurasia about 40 to 50 million years ago. They are still rising at a rate of around 1 centimeter (0.4 inches) every 10 years.

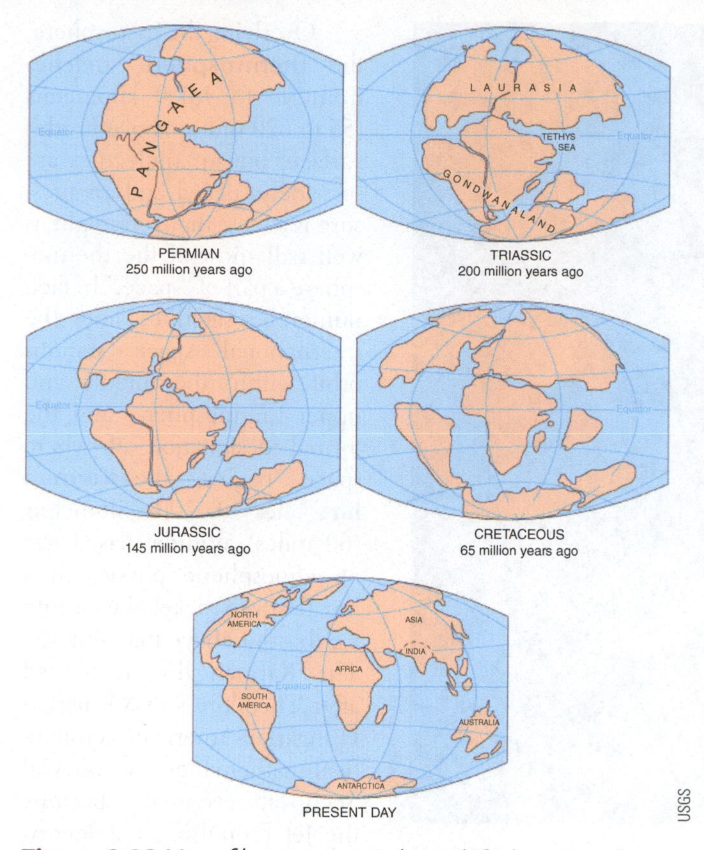

Figure 2.28 Map of how continents have shifted position due to plate boundary activity over the past quarter-billion years, during which time reptiles, mammals, and flowering plants have emerged as dominant life forms. The ancient supercontinent of Pangaea split apart to form the modern Atlantic and Indian Ocean basins.

2.3 The Atmosphere: Earth's Restless Outer Envelope (LO3)

Earth's atmosphere is a moving mass of gas that is vital to making planet Earth livable. Dominated by nitrogen and oxygen with only traces of other gases, the atmosphere is the birthplace of weather and integral to the movement of water around Earth. Thus, the atmosphere is tightly linked to the hydrosphere. Air currents move moisture and energy between the equator and the poles while trace gases such as carbon dioxide conspire with methane, nitrous oxide, and water vapor to keep our air sufficiently warm for liquid water, the elixir of life, to be the norm.

2.3.1 Atmospheric Structure

On a blue-skied, sunny day the atmosphere appears transparent and clear, but, like Earth's interior, it is also a complex, multilayered structure (**Figure 2.29**). The most restless and densest part of the atmosphere is the **troposphere**, which extends from the surface to an altitude of 8 to 18 kilometers (5 to 11 miles). The troposphere is thinner over the poles than the equator. It is also thinner during the winter months for the same reason–because cooler temperatures at that time of year make the air denser.

Seventy-five to 80% of the total mass of the atmosphere, and 99% of its water vapor, lie within the troposphere. Infrared radiation from the surface, felt as heat, warms the air near the base of the troposphere, which rises, spreads, and turbulently mixes with cooler air. The term "troposphere" literally means "a turn or change." Evaporated moisture condenses into clouds and storm systems. Our life-sustaining rainfalls are all tropospheric.

Mt. Everest, the tallest mountain on Earth, has an elevation of almost 9 kilometers (5.5 miles), high enough to penetrate the top of the troposphere. Air pressure at its peak is about a third of what it is at sea level. Most people cannot climb Everest without supplementary oxygen; humans generally have severe challenges breathing and risk life-threatening edemas (swelling) or hypoxia (lack of oxygen) when exerting themselves above 6 kilometers (4 miles) of elevation. By 19 kilometers (12 miles), air pressure is so low that blood inside a body begins to boil if one is exposed to the atmosphere without a pressurized suit or cabin, an instantly fatal situation. These numbers illustrate the thinness of the atmospheric envelope on which we depend. Drive a car 19 kilometers along a straight, level highway, and you will better appreciate how short this distance is in the grand scheme of things!

The top of the troposphere, the **tropopause**, marks the limit that the circulating, upwelling air can reach before stalling out due to loss of thermal energy. This is the base of the **stratosphere**, a layer of much more stable air (*strato* means layered). That is not to say that it is calm there, however. Stratospheric winds are furious, blowing as fast as 230 kilometers (130 miles) per hour. The notorious **jet streams** are an example (**Figure 2.30**). These "rivers of air" blow mostly west-to-east owing to Earth's rotation and are linked to the borders of warm and cold air masses in the troposphere below. Nevertheless, passenger planes and other aviation seek to fly in the lower part of the stratosphere to avoid the turbulence and potential storminess encountered at lower altitudes, and to save fuel by flying in the thinner air.

In the troposphere, air generally becomes cooler the higher you go. Not so with the stratosphere, because with increasing altitude the air is subject to increasingly intensive solar radiation, particularly in the ultraviolet (UV) spectrum–the short-wavelength, high-energy form of sunlight that leads to sunburns and, even worse, skin cancers. Fortunately for us, oxygen at the base of the stratosphere reacts with incoming sunlight to absorb and screen out the deadliest type of UV radiation, UV-C, through the formation of a heavy compound of oxygen called **ozone** (O_3). The less harmful forms of UV that pass through to the surface, UV-A and UV-B, can generally be managed using sunscreens and other skin protection. The layer of ozone production, the **ozone layer**, lies between 15 and 35 kilometers (9 to 22 miles) above the surface (**Figure 2.31**). Chemically, the reaction, termed the *oxygen-ozone cycle*, can be described as:

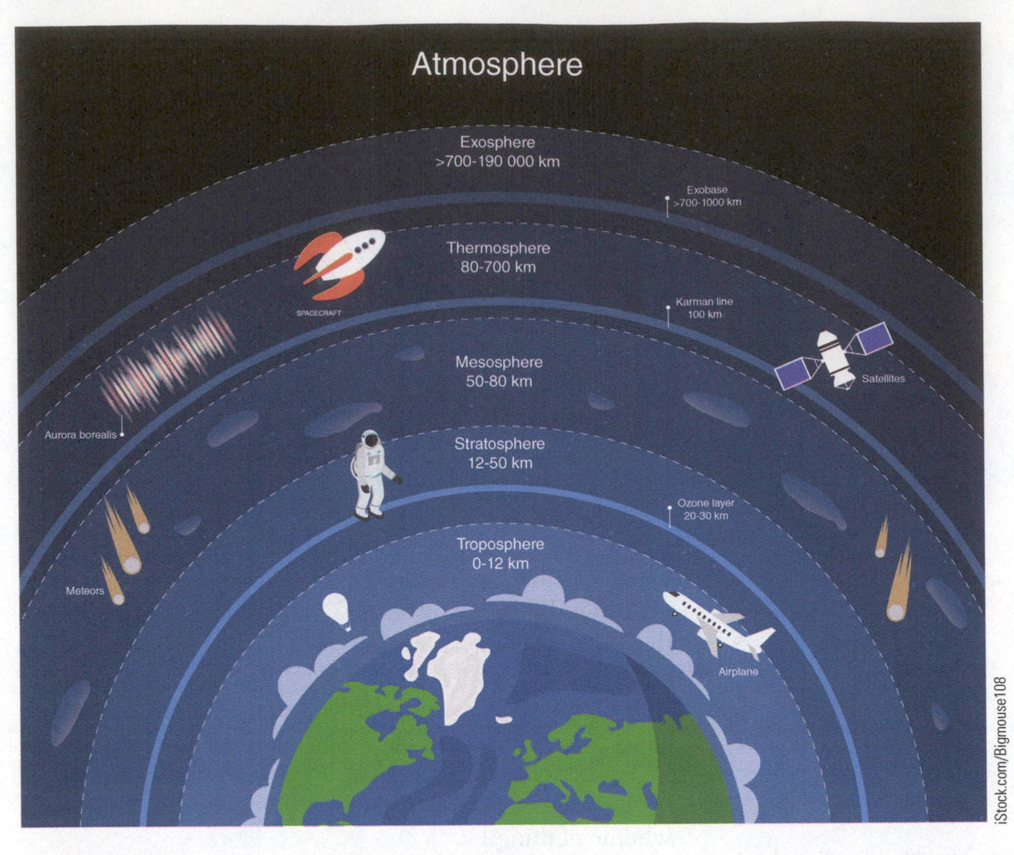

Figure 2.29 Layered structure of Earth's atmosphere.

O_2 (an ordinary atmospheric oxygen molecule) + *UV*
$\rightarrow 2O, O + O_2 \leftarrow\rightarrow O_3$ (ozone)

Ozone, an unstable molecule, quickly disintegrates back into the constituent oxygen molecules from which it is made. The oxygen-ozone reaction is continuous, however, because the sun continues to shine, assuring that we don't run out of stratospheric ozone, at least for natural reasons. As you may expect, ozone concentrations drop significantly at polar latitudes where winter nights can stretch through months of darkness. Come spring, however, the polar stratosphere is recharged with ozone in 24-hour daylight.

The stratosphere gives way upward to the **mesosphere**, a little understood part of the atmosphere from 50 to 85 kilometers (31 to 53 miles) altitude. This layer is too high for aircraft and research balloons, but too low for satellites and space vehicles; hence, it is not easy to study. We know though that most meteors vaporize as they heat up from friction at high speeds upon entering the mesosphere; it serves as a shield, protecting Earth's surface from as much as 90% to 95% of the natural debris headed our way from space. Vaporization also means that the mesospheric air is enriched in iron and other metal atoms related to meteor composition, an unusual form of "atmospheric pollution." While in the stratosphere, temperature increases upward thanks to the solar radiation. The reverse is the case with the mesosphere. The coldest part of the atmosphere lies at the top of the mesosphere, averaging around −90°C (−130°F).

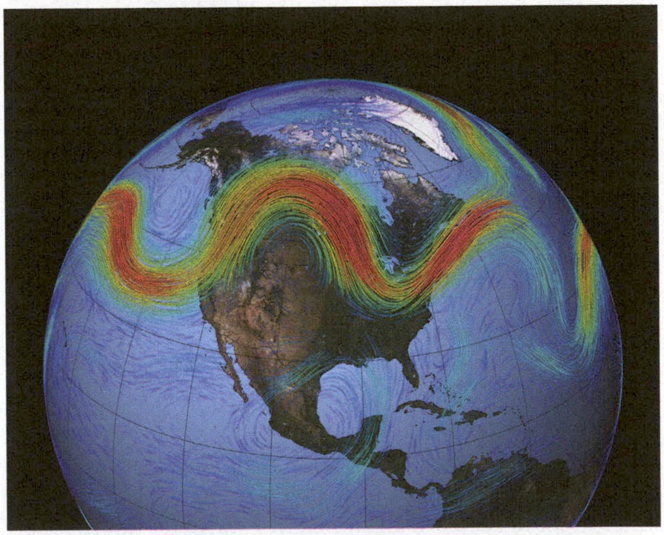

Overlying the mesosphere, the **thermosphere** stretches from 90 to 1,000 kilometers (55 to 620 miles) above Earth's surface, but air molecules are so widely spaced and air pressure is so low that you might as well call most of the thermosphere a part of "space." In fact, numerous satellites and the international space stations orbit within the middle and upper thermosphere, and the formal designation of where space "begins," the **Karman line**, lies at 100 kilometers (60 miles) altitude, based less on atmospheric physics than our ability to rocket objects into Earth orbit above that altitude. The Karman line is named after Theodore von Kármán, a Hungarian-American aeronautical mathematician whose Cal Tech lab eventually became the Jet Propulsion Laboratory,

NASA

Figure 2.30 The Northern Hemisphere polar jet stream is highlighted as the red and yellow band (warmer colors are a measure of higher wind speeds) where it zigzags above North America. Four such west-to-east blowing jet streams girdle our planet, one pair around the poles, the other closer to the Equator. They mark the upper boundaries of cells of overturning (convecting) air in the underlying troposphere and are largely a consequence of Earth's rotation. The zigs and zags result from air masses of different temperature coming together, influenced by the Coriolis effect, and from the presence of oceans and land masses below.

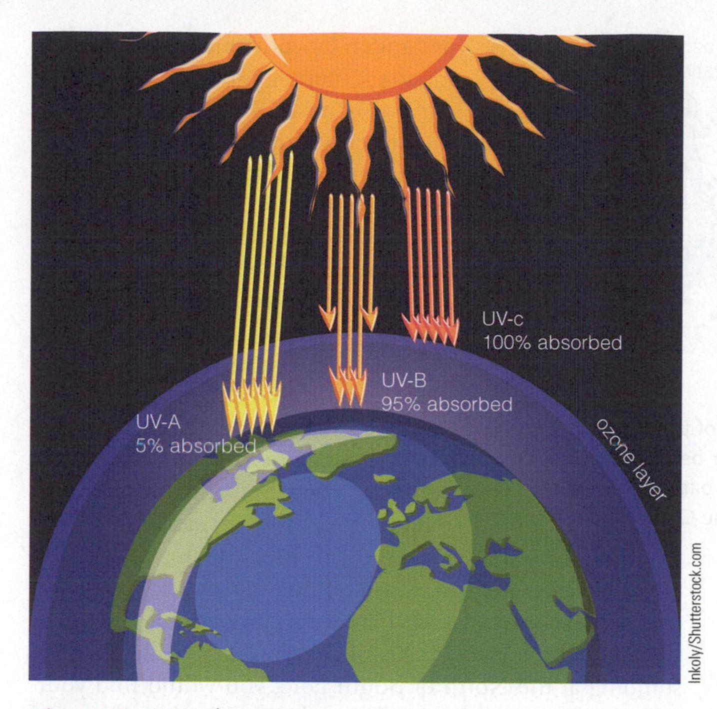

Figure 2.31 Earth's protective ozone sheath.

one of the most important space research centers in the world.

Traces of Earth's atmosphere may be detected out as far away as halfway to the Moon (192,000 kilometers, or 119,000 miles). Beyond that point, charged particles streaming from the Sun sweep away any ascending traces of gas and return them to deep space. As planetary atmospheres go, we do not have a very thick one. Then again, we are lucky to have one at all! Without it, we would not be here.

2.3.2 Patterns of Tropospheric Circulation

We are accustomed to think that we feel no pressure at all, standing casually in the open air at sea level. (You may feel even less pressure if you are standing on a beautiful beach with friends.) This is because we have developed as a species at the bottom of a very heavy atmosphere; we are simply used to and take for granted the pressure of all that air above us being pulled down by gravity toward Earth's surface. It is worth reflecting for a moment on one common standard for measuring that pressure: kilograms per square centimeter, or in the English system, pounds per square inch. At sea level, every square centimeter of your skin experiences a pressure from air of about one kilogram. That is equivalent to 14.7 pounds per square inch–almost all from the weight of 100 kilometers (60 miles) of air above that small patch of skin!

Just as air under high pressure escapes a balloon when you let it, so too does air flow naturally from areas of high pressure to low. We feel this movement and know it as **wind**. The atmosphere of course is not made up of multiple balloons refilled and emptied, although old Medieval maps often show in one corner the image of some big, full-cheeked fellow up in the heavens blowing through pursed lips–one of the Anemoi, or Greek wind gods, the source of the winds. Instead, pressure differences relate to temperature differences.

You may be accustomed to seeing smoky air escaping straight up above a chimney or smokestack on a cold winter morning (**Figure 2.32**). That air is rising because it is warm from the fire below; the individual air molecules, full of thermal energy from the fire, are agitated, become widely spaced in their molecular motion, and make the air less dense. Conversely, cold air sinks because the air molecules, lacking such energy, are less feisty, pack more closely together at the atomic level, and create a denser atmosphere. Dense, heavier air, like fluid, sinks and spreads out, displacing more buoyant, warmer air. This is a fundamental property for understanding how the atmosphere "works," at least in terms of winds and the weather.

Winds in the troposphere are predictable at the broadest scale and are driven by incoming solar radiation. Radiant solar energy warms Earth's surface and atmosphere unevenly, with the equatorial latitudes receiving more sunlight over the course of the year than the polar regions. Under most circumstances, warm tropical air rises and spreads laterally as it cools off at higher elevations, whereas cold polar air sinks. The troposphere is in a never-ending quest to equalize its temperature and pressure for any given elevation.

The latitude-dependent differences are significant. For example, the average annual temperature difference between the Philippines, lying in the sultry tropics, and the bitterly cold East Antarctic Plateau is a staggering 75.5°C (138°F). Because the Earth continues to rotate and the Sun continues to shine, such extreme differences persist–a perfect homogeneity of air temperature and density has never existed in our planet's history. Nevertheless, the Earth System, obeying the ordinary laws of physics, keeps trying to achieve it.

Given the natural tendency for the atmosphere to smooth out its physical differences, you might imagine that

Figure 2.32 Warm steam rising from smokestacks along the icy River Moskva on a cold, still winter morning in Moscow, Russia.

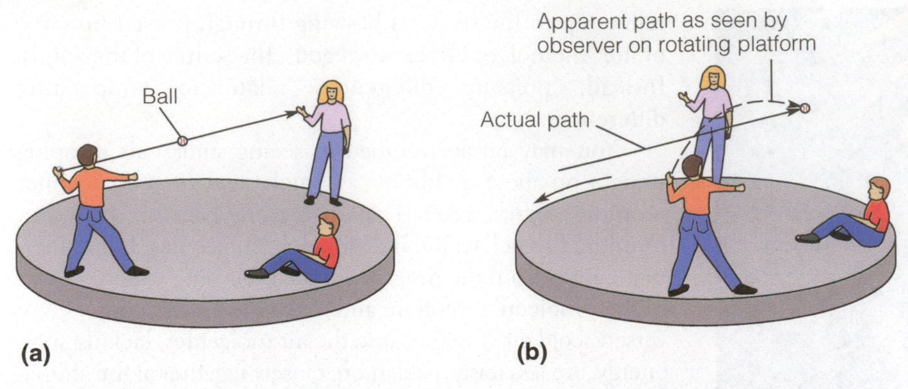

Figure 2.33 The Coriolis effect in a game of catch. (a) The path of the ball is straight when the game is played on a stationary platform, and an observer perceives it as straight. (b) On a rotating platform, the ball still follows a straight path, but to an observer on the platform, the path appears to be curved. This is the Coriolis effect.

winds blow in just two directions to be most efficient: hot air moving *directly* from the equator toward the poles and cold air moving in the *opposite* direction to mix uniformly somewhere in between. However, it doesn't quite happen that way, because of an important fact we haven't yet taken into consideration: the solid Earth and everything attached to it rotate beneath our atmosphere. Consequently, with respect to the surface, what are the actual paths taken by winds? They curve, in much the same way that a ball tossed to a bystander by someone standing on a rotating merry-go-round appears to curve (**Figure 2.33**). We term this illusionary curvature the **Coriolis effect,** and it applies not only to wind paths but also to ballistic missiles, ocean currents, passenger planes, and even fastballs traveling through the air (though at that scale the effect is hardly detectable) (**Figure 2.34**).

The Coriolis effect is one of the more complicated concepts to explain in earth science, but it is simple when grasped. Imagine several things happening at once. Specifically, the Earth rotates from west to east, with fixed locations at the equator having to cover a greater distance to get all the way around the planet during the course of a day than locations closer to the poles, through which Earth's axis of rotation passes. If you were standing at the North or South Pole, you would find yourself slowly turning around in place over the course of 24 hours. But at the equator, your locality would be moving through space at 1,670 kilometers (1,000 miles) per hour to return to your starting position 24 hours later, given Earth's 40,000-kilometer (24,000-mile) equatorial circumference.

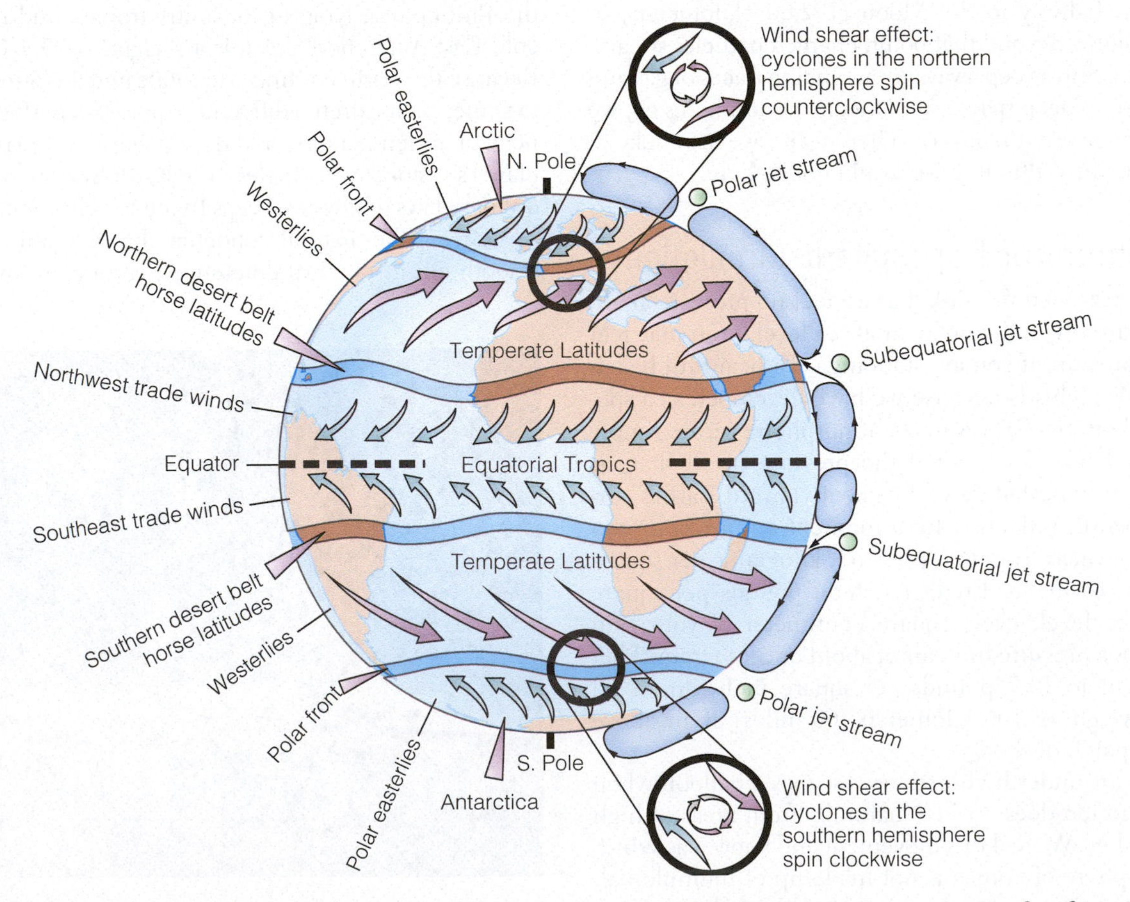

Figure 2.34 Pattern of global surface wind circulation. Note how Coriolis forces play a role in the circulation of surface winds in both hemispheres.

Because of the latitude-dependent change in rotational speeds, the winds bound for the pole tend to "race ahead" of the surface rotating beneath them, whereas equator-bound winds tend to "lag behind." To observers on the ground, the difference in relative motions of air and surface is apparent as wind paths tend to bend toward the right in the Northern Hemisphere and toward the left in the Southern Hemisphere. Coriolis deflection of the winds converging on, and escaping from, cyclones, hurricanes, and typhoons also helps explain the spin of these storms. Storms in the Northern Hemisphere turn counterclockwise, whereas those in the Southern Hemisphere rotate clockwise.

The takeaway point is this: using Earth's surface as a frame of reference few, if any, winds on Earth blow directly from the equator to the poles. In fact, quite often, winds tend to blow more west-east or east-west than in north-south directions. The same is true for the atmospheres of all other rotating planets; the bigger the planet, the stronger the winds. The highest wind speed ever documented on Earth is about 370 kilometers (230 miles) per hour, whereas on our giant neighbor Jupiter wind speeds as high as 1,450 kilometers (900 miles) per hour have been clocked.

Quite apart from the impact of Coriolis deflection, our atmosphere's circulation is not simple because winds moving toward the poles converge (just as Earth's longitudinal lines geometrically converge). Let's follow one hypothetical parcel of air at low elevation near the equator after sunlight heats it. Our air mass ascends, drawing in fresh air to replace it at low elevation from both north and south of the equator. The air is humid, thanks to evaporation over the ocean in the torrid tropical heat, but as the air rises, it cools, and the moisture it contains quickly condenses into clouds, from which heavy rain falls. By the time the air has moved away from the equator at elevations of 10 to 15 kilometers (6 to 9 miles), it has become much drier.

Moving toward the poles, the air converges with other high-altitude air masses, all headed in the same direction. The air grows denser because it cools and is squeezed by adjacent, northward-moving air masses. As a result, all this converging air sinks back to Earth roughly at latitudes 30° north and south of the equator. The air is compressed as it descends, and compression causes it to become warmer, even hot, a process called **adiabatic heating.** The parched descending air soaks up any moisture it can, desiccating the land and helping create the band of temperate-latitude deserts that circle the planet (**Figure 2.35**).

Figure 2.35 The Namib Desert, a horse latitude dry region in southwest Africa, Namib Naukluft National Park.

If the parcel of air descends to the ocean surface instead of land, the result is an "ocean desert," creating broad stretches of hot, still water–perilous to early mariners, who referred to them as **horse latitudes,** perhaps because of the death of horses (not to mention the sailors themselves), which could not survive the long, listless journey across these parts of the ocean (**Figure 2.36**). From the desert horse latitudes, the air parcel divides with some of it racing back toward the equator as **trade winds** (blowing from the east) and the rest continuing to angle poleward as westerly winds, or simply "**the westerlies**." (Wind directions are always given as the direction from

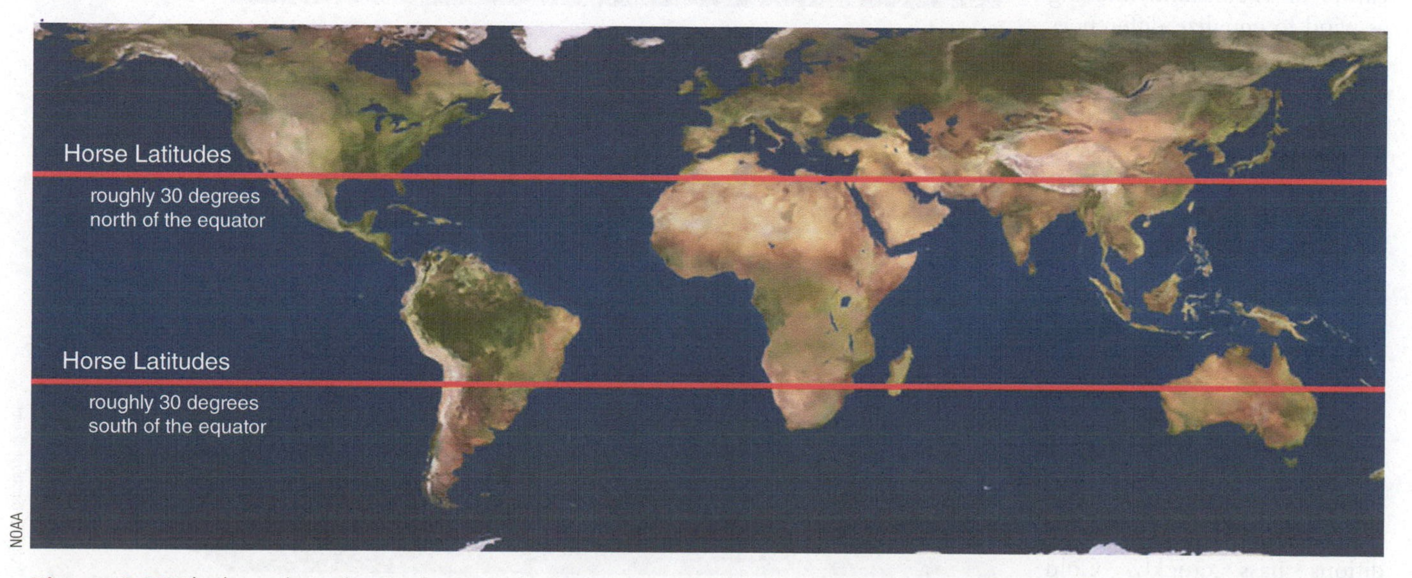

Figure 2.36 The horse latitudes; in the days of sailing ships, one of the most feared places on Earth. Note the correspondence of deserts (brown areas on continents) with horse latitudes both north and south of the equator.

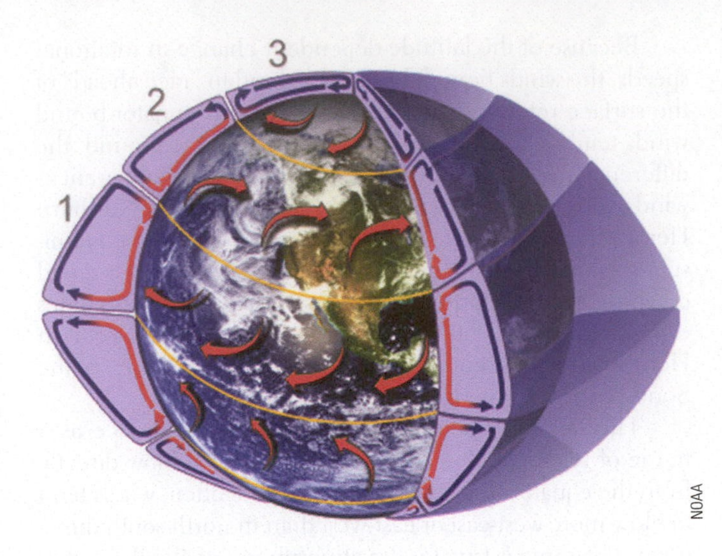

Figure 2.37 Three sets of circulation cells exist in Earth's near-surface atmosphere, each pair dependent on a particular latitude range. The cells closest to the equator (set 1) are termed Hadley cells. Set 2 is the mid-latitude (or Ferrel) cells, and set 3 the Polar cells. Note how the Coriolis effect influences the direction that warmer air in the lower parts of each cell blows across Earth's surface (shown by the curving red arrows). This plays an important role in determining weather patterns.

which the wind comes.) As the wind nears the poles, the warm air encounters a barrier of dense, frigid **polar air**. The colliding air masses partly mix, but some of the poleward-bound air, being warmer, simply avoids the barrier of polar air by rising and turning back toward the equator at high altitude. The zone of collision is called the **polar front**. The result of all this restlessness is an atmosphere divided into sets of overturning wind circulations called **circulation cells** (**Figure 2.37**).

2.3.3 Weather

The differences in heating of Earth's troposphere described above lead to the development of **air masses**. Cold, dry air is present over the poles, while the tropics are characterized by warm, moist air. Cold air masses are dense and tend to sink or stay close to the ground, while warm air masses are less dense, rise, spread buoyantly, and eventually cool down near the tropopause. The different air masses are separated by **fronts** (boundaries between air masses) that move through the atmosphere because, as mentioned earlier, air is constantly moving as wind to equalize differences in atmospheric pressure and temperature (**Figure 2.38**).

Where a cold air mass encroaches upon a mass of warm air, a **cold front** develops, with the cold air wedging in beneath and easily lifting the warm air in its way. This warm air rises quickly against the steep cold air front, and its high moisture content condenses into towering storm clouds (**Figure 2.39**). Heavy rainfall, snow, and hail often ensue, but these unstable and often destructive weather conditions pass quickly. Cold fronts are typically fast movers.

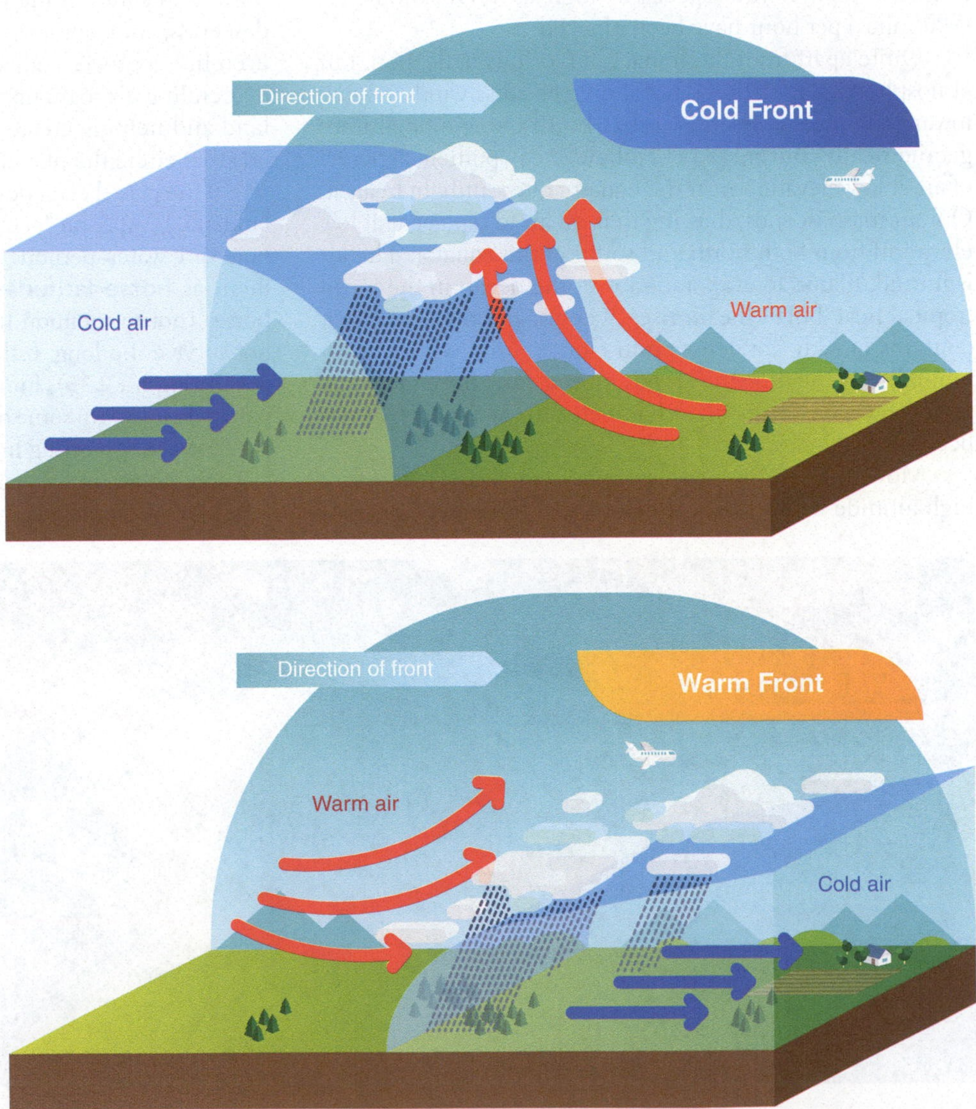

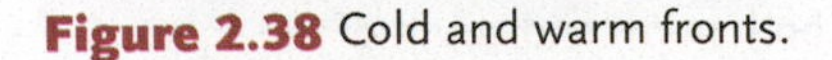
Figure 2.38 Cold and warm fronts.

Figure 2.39 An arriving cold front marked by the buildup of towering storm clouds and associated rainfall.

Figure 2.40 High-altitude cirrus clouds mark the approach of a warm front. These clouds will thicken, merge, and lower as the front approaches, leading to steady light rainfalls.

In contrast, encroaching **warm fronts** rise gradually above the cooler air masses they encounter. Warm fronts are slow movers. The zone of contact between the two air masses slopes gently. High-altitude clouds form, thicken, and lower to the surface over a period varying from many hours to a few days as the front approaches and produce long, steady rainfalls–a precious benefit for farmers (and the rest of us!) worldwide (**Figure 2.40**).

Fronts at temperate latitudes tend to travel west-to-east across Earth's spinning surface, and this has led to several popular folk sayings about the weather. *Red sky at morning, sailor's warning* refers to the approach of a typical weather front. Blown dust and other particles in the direction of the rising sun give the sky a ruddy color. To the west, the sky is dark and cloudy. It is time to get out your rain jacket or find shelter. *Red sky at night, sailor's delight* celebrates the passing of a front; skies are clearing to the west and fair weather is soon on its way. Centuries-old folk sayings help distinguish cold from warm fronts, including *"Rain foretold, long last–Short notice, soon will pass."* (You may be able to make up some of your own sayings with a sound understanding of how the weather works. Contrary to popular belief, there are many predictable aspects of the weather).

Where mountain systems block air flow, **orographic** conditions arise that may contrast greatly with those in the surrounding region (**Figure 2.41**). Winds speed up as air rises and sweeps across a mountain range, often leading to the development of smooth cloud caps (these are called **lenticular clouds**

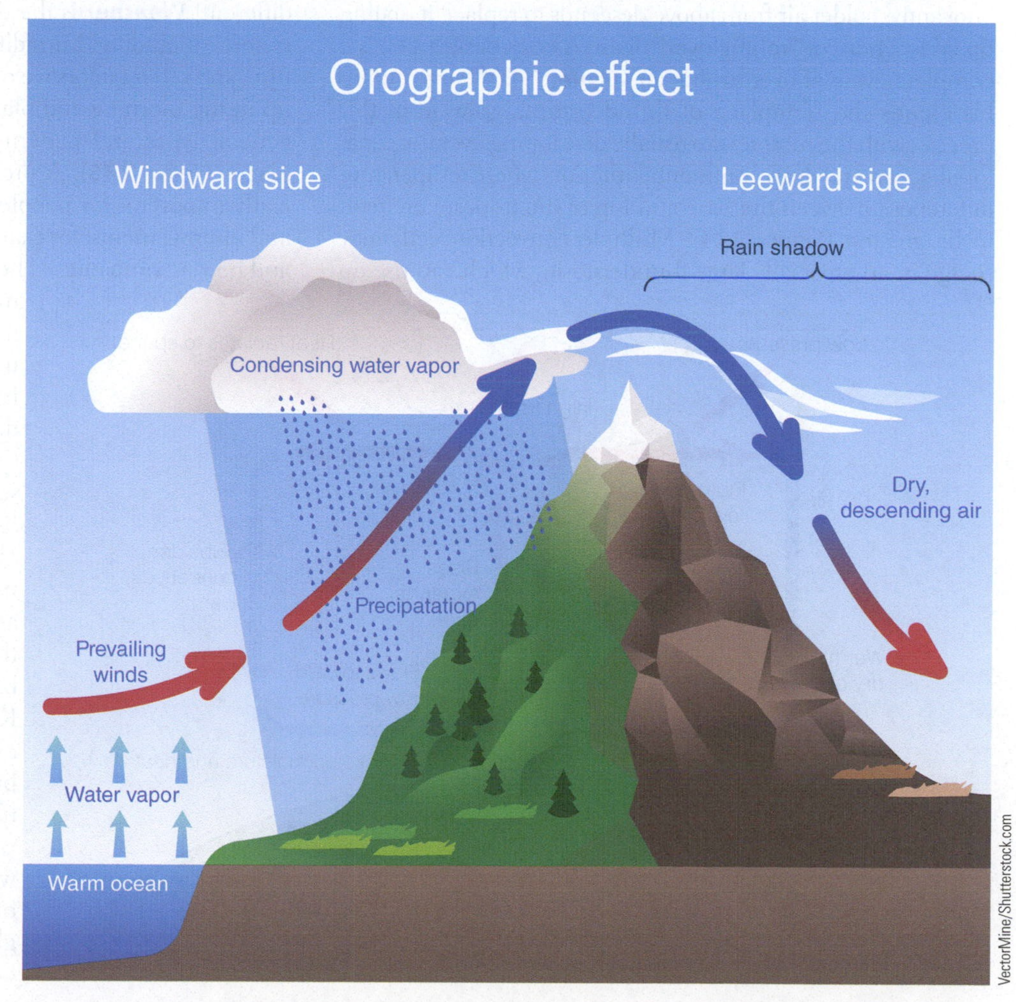

Figure 2.41 The orographic effect. Notice how rainfall changes as air flows across the high range divide. A dense forest may exist on the windward side of the divide, while high evaporation rates and adiabatic heating create a desert on the leeward side.

Figure 2.42 Lenticular and saucer shaped clouds form where winds speed up, cool and condense moisture crossing a high range divide, Pyrenees Mountains, Lleida, Spain.

Figure 2.44 A massive thunderhead, a cumulonimbus cloud that has risen to the top of the troposphere and is spreading out at the base of the stratosphere.

and indicate fast-moving air above the mountain range, **Figure 2.42**). When the fast-moving winds descend the far side, they heat up as the air compresses on reaching level ground at the downwind base of the range. Southern California's hot, high-pressure **Santa Ana winds** and the warm, blustery winter **chinooks** of Colorado are examples. As the air spreads downwind, it loses pressure and cools down.

In the troposphere, as warm air from the surface rises buoyantly, colder air from above descends to replace it, setting up convection, or "rolling-over" of air. A **convection cell** is a complete cycle of overturning air (**Figure 2.43**). Large cumulus clouds–the birthplace of thunderstorms–grow from this process, with the most severe usually developing over the equatorial ocean and deep continental interiors where temperature differences between the base and top of the troposphere tend to be greatest (**Figure 2.44**). Multiple convection cells may merge in an especially large thunderstorm, which can rise up into the base of the stratosphere as high as 22 kilometers (14 miles). Aircraft must carefully fly around such storms.

2.3.4 Climate

The term *climate* is often confused with "weather." They are different! **Weather** is the day-to-day conditions of the atmosphere, including humidity, rainfall, cloudiness, temperature, and other factors we often pay attention to when getting up in the morning and planning for the day. **Climate** is the general types and patterns of weather that characterize a place (**Figure 2.45**). A predictable (stable) climate is especially important for people who plan long term for projects and improvements; for example, farmers, water utility boards, and power companies. The effects of a changing climate are most severe in places where changing seasons are economically important. Take for example the ski industry; if winter snows go away, so do all the skiers and their wallets.

An international classification scheme for world climates, the **Köppen-Geiger Classification**, classifies world climates into five main groups: tropical, dry, temperate, continental, and polar. Within those groups, there are subgroups based largely on precipitation. The Köppen-Geiger Classification is closely related to the global distribution of ecosystems, as temperature and precipitation determine which plants are able to grow where. The worldwide distribution of climate groups is determined by global atmospheric circulation cells, latitude, and proximity to oceans and other large water bodies. As you might expect, with climate changing due to human

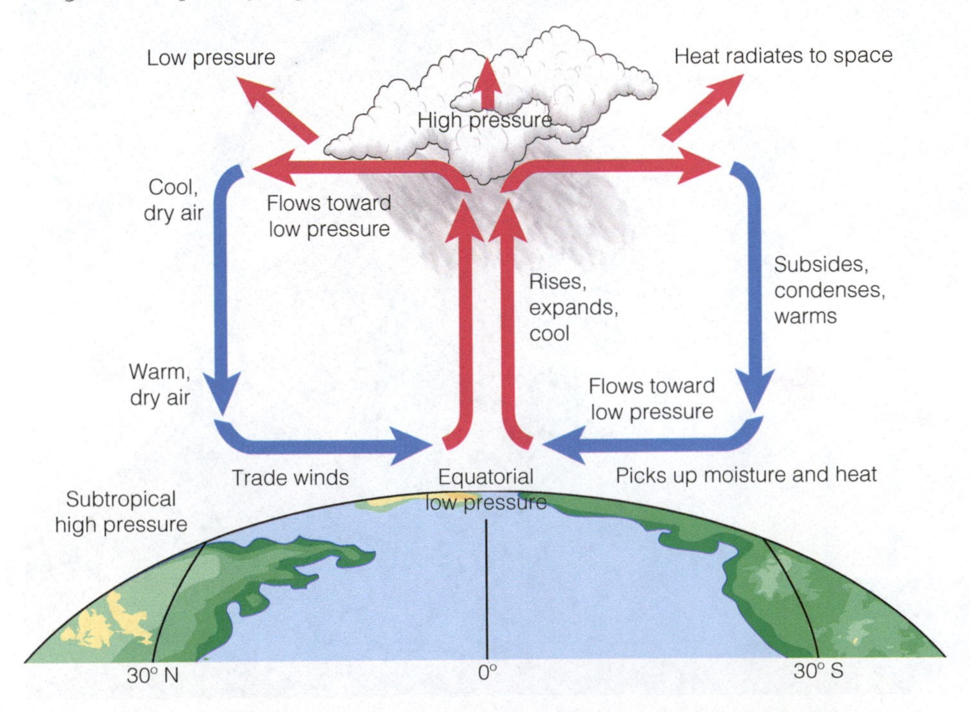

Figure 2.43 Convection cells form and warm air rises and expands, then cools and sinks. (Adapted from Kaufmann R. & Cleveland C. J. (2008). Environmental science. McGraw-Hill Higher Education.)

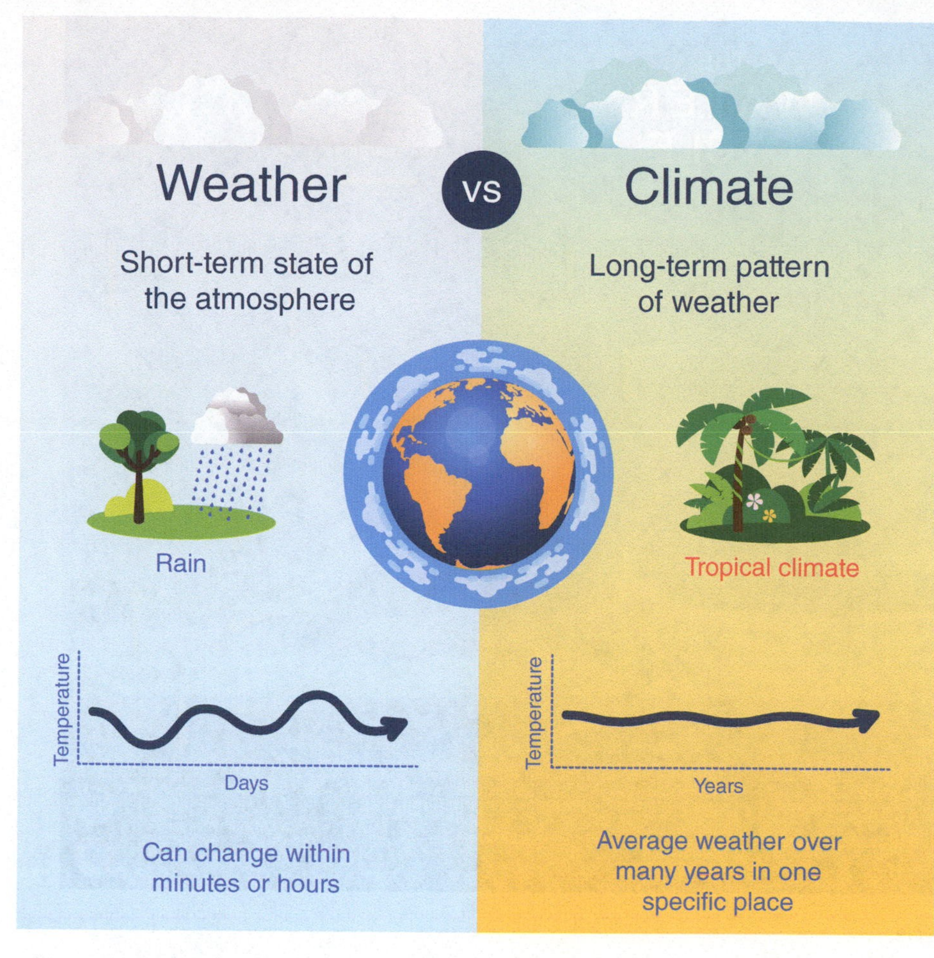

Figure 2.45 Weather and climate: two different but related concepts.

use of fossil fuels and land-use changes, many climates all over the world are being reclassified into warmer categories.

2.4 Hydrosphere and Cryosphere (LO4)

Molecules of water surround the Earth either in the form of liquid water, water vapor in the atmosphere, or ice. Water is one of the most common molecules in the universe, but it is especially active in liquid form on Earth, a "water world" teeming with life. It enables chemical reactions to take place for organisms to gain energy, make blood, thrive, grow, and eliminate waste. Because of its polar chemical nature (that means two ends of the molecule carry a positive charge while the center is negative), water dissolves and transports biologically essential nutrients in solution from place to place, including nitrogen (N), phosphorus (P), and potassium (K). As Earthlings, it is hard for us to imagine life anywhere in the universe existing without water, at least for very long. The body of a typical adult male human consists about 60% of water, and a female slightly less (55%). As much as 70% to 80% of the weight of newborn babies consists of water. In purely physical terms, newborns are largely containers of water wrapped in tissues.

2.4.1 The Hydrological Cycle

Water is always on the move, cycling throughout the Earth System. The **hydrological cycle** is driven by sunlight (**Figure 2.46**). Key to the hydrological cycle is evaporation from the surfaces of oceans, lakes, and other water bodies. Each day, heating from the sun evaporates roughly 1,400 cubic kilometers (336 cubic miles) of water worldwide. A cubic kilometer can be imagined as a volume equivalent to a giant cube measuring one kilometer (0.6 miles) on each side–a vast amount. Lake Erie holds 484 cubic kilometers, so imagine the lake evaporating entirely three times each day and you get the idea!

Water enters the atmosphere as vapor, adding to the air's **humidity** (water content). As moist air rises and cools, water vapor may condense to form liquid droplets–mist, fog, and clouds. Rain, snow, and hail may form and fall from the sky. Precipitation may return to the ocean directly, or indirectly by flowing as rivers and streams across the land. Some of the returning water may take a "side-trip," percolating into the earth to flow as slowly moving **groundwater** (underground water) for thousands of years. Groundwater flows slowly downslope, back toward the ocean or other nearby waterways and bodies, where it emerges as springs and seeps. Life taps into the hydrological cycle in many different ways, from plants developing roots to harvest water and nutrients from the soil to you simply drinking water from a glass. The cryosphere may be regarded as a "parking space" in the hydrological cycle: ice locks up water in solid form, though even ice eventually melts, of course, returning water to a more active, liquid state.

The hydrological cycle interacts not only directly with the biosphere and atmosphere, but also the lithosphere, by enabling the weathering of rocks at depth and the erosion of rocks at the surface, the deposition of sediments along coastlines, and the development of many minerals–including valuable ore deposits–as groundwater circulates through permeable crust sometimes heated by volcanic activity.

2.4.2 The Ocean, the Hydrologic Elephant in the Room

For all of its importance to life, only about 1% of Earth's water is fresh and available to the biosphere via streams, lakes,

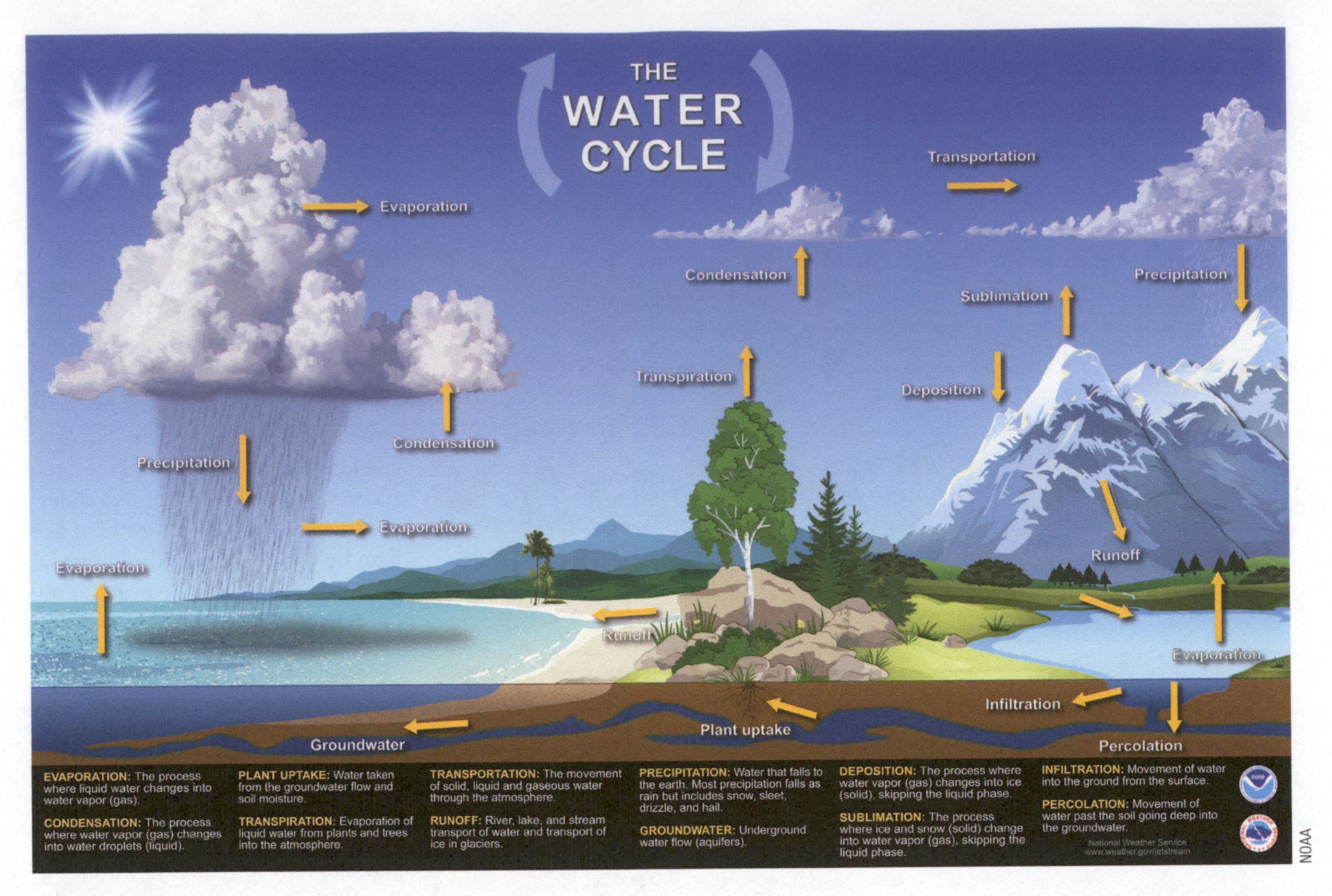

Figure 2.46 The hydrological (or water) cycle.

springs, and groundwater. About 2% is trapped as solid ice, mostly in the Antarctic Ice Sheet. The remainder, a whopping 97%, is found in the oceans. And it is quite salty–36 parts per thousand give or take. The sodium in salt is not bad for you, per se; your body can eliminate this important nutrient through the kidneys in urine. But there is simply too much of it in seawater for humans to process, about four times the concentration in your bloodstream. Drinking seawater means your kidneys are overworked and you end up passing more water to get rid of the excess salt than you can possibly drink. Ironically, people who drink seawater out of desperation die from thirst!

Most of that salt and other elements dissolved in the oceans have accumulated over billions of years from the weathering of rocks on land. Rainwater is slightly acidic and slowly dissolves rocks. Streams and rivers wash elements from these dissolved rocks into the ocean where they accumulate. When evaporation takes place, water enters the air, but everything else is left behind. It is no coincidence that the saltiest stretch of ocean in the world, the Red Sea, also has the highest evaporation rates. The oceans are the world's biggest, ultimate sump.

Though entirely liquid, the oceans are also *layered*, not unlike Earth's interior and atmosphere, but for very different reasons. Salt water is denser than fresh water (1020–1030 kg/m^3 versus 1000 kg/m^3) and cold water is denser than warm water. The saltiest, coldest water sinks to the bottom. Density is fundamentally responsible for ocean layering.

Solar-heated, generally less salty seawater forms the shallowest layer–the surface, or **epipelagic ocean**. Epipelagic waters are penetrated by filtered sunlight to at most about 200 meters (640 feet) deep and they contain about 90% of all ocean life, including most submarine plants. Life on Earth most likely originated in the epipelagic ocean billions of years ago.

The **deep ocean** includes many layers, defined by the different lifeforms that can tolerate these great depths (**Figure 2.47**). The border zone separating the epipelagic layer from deep ocean, called the **thermocline**, is defined by water temperature. It lies from around 200 to 1000 meters (640 to 3200 feet) below the surface. Above the thermocline, temperatures in the epipelagic zone do not change significantly with depth, thanks to nearly continuous mixing in the solar-heated and wave-churned, near-surface water. The epipelagic water is warm relative to the deep ocean, having a global average annual temperature of 12.8°C (55°F), though this varies across latitudes.

Beneath the thermocline, water temperatures continue to decline steadily with depth. Deep ocean floor temperatures typically hover in the range of 0°–3.9°C (32–39°F); just above the freezing point of pure water, although the freezing point of seawater is slightly lower than that of fresh water.

Despite the layering in the oceans, mixing and circulation takes place throughout, though these processes are usually sluggish at great depth. Nonetheless, just as winds blow to

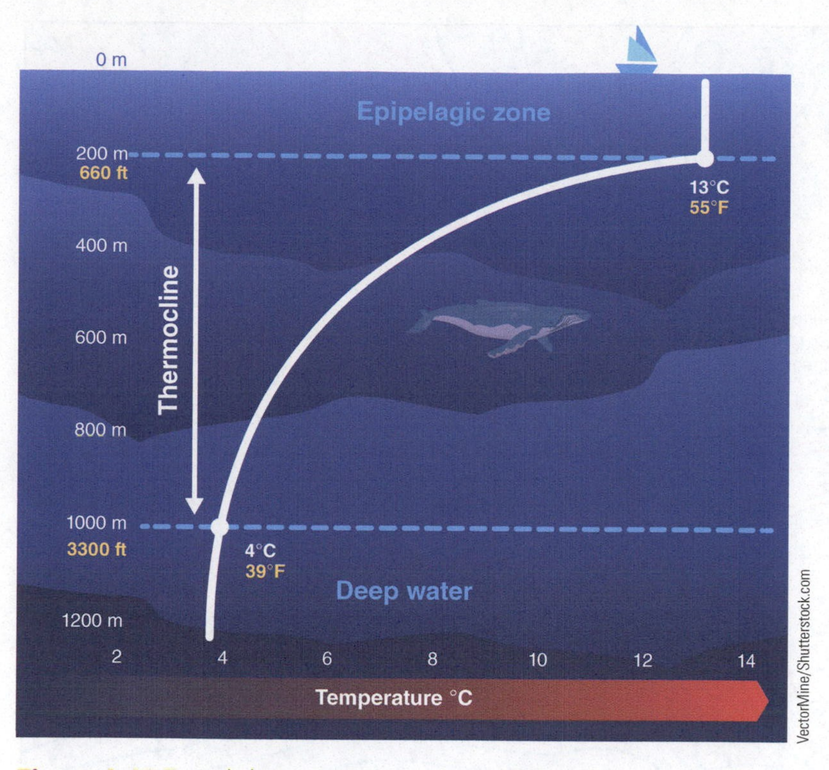

Figure 2.47 Typical change in temperature and layering of water with depth in the upper part of an ocean.

power walk. On an exceptional day, it can speed up to 9 kilometers per hour. The rate of flow may seem slow in comparison to, say, driving a car or flying an airplane, but given the mass of water on the move it is quite impressive. From its mainstream, numerous looping side-currents, termed **eddies**, branch off, some detaching as independently circling currents that generally don't last long (**Figure 2.50**). Gyre-related eddies are important for spreading nutrients from nearshore into parts of the sea that are nutrient-poor, encouraging the growth of many marine organisms, particularly plankton, across broad areas.

At the centers of the great gyre currents, making up some 40% of Earth's total surface, are large areas of calmer water, stirred only by occasional frontal systems or cyclones. Into these parts of the ocean much of the flotsam and jetsam drifts that people discard either accidentally or intentionally from ships and coastal communities, inspiring environmentalists to refer to them as **garbage patches.** In recent years, biologists have discovered that the garbage patches are severely harmful to marine life, especially birds who scavenge the flotsam (**Figure 2.51**).

equalize temperatures and pressures in the atmosphere, so too does the same physical driver apply to currents in the oceans. All also are tied ultimately to solar heating of Earth's surface.

2.4.3 Shallow Ocean Circulation

Wind drives the shallow circulation in Earth's oceans. If you have ever sailed a boat, you are familiar with the appearance of a fresh gust of wind across still water. The water roughens up and ripples develop and move downwind. Soon, the surface water is moving. Most waves at sea originate from wind transferring energy to water, though they can travel great distances beyond the area in which a current first begins moving (Chapter 10). On a much larger scale, the trade winds and westerlies drive global-scale ocean currents that travel thousands of kilometers and loop around to form **oceanic gyres**. Gyre patterns are shaped by three factors: (1) the placement of the continents, (2) the Coriolis effect on winds blowing across the waters, and (3) the Coriolis effect on water masses moving across latitudes (**Figure 2.48**). There are five gyres active in the world's oceans today, though in the deep geologic past, when the continents were positioned differently thanks to rifting and seafloor spreading, the numbers and sizes of oceanic gyres also no doubt were much different (**Figure 2.49**).

Gyres flow clockwise in the northern hemisphere and counterclockwise in the southern hemisphere. The fastest gyre-related current in the world, the Gulf Stream in the North Atlantic Ocean, flows at an average rate of 6.4 kilometers per hour (4 miles per hour), about the rate of a strong

2.4.4 Ocean Circulation and Climate

The hydrosphere and climate are tightly linked. Shallow ocean circulation is a great example by which to understand this linkage. The linkage requires understanding an important concept, **heat capacity**; that is, how much energy it takes to warm or cool a certain mass of material by a chosen number of degrees. Water has a high heat capacity, much higher than air – remember that, and the climate/ocean connection comes clear. Water can transport and hold a lot of heat!

Imagine yourself on the coast. During the day, the land warms more quickly than the water. Go inland and it's hot. Return to the coast and it's cooler. The land also cools off faster than coastal waters after dark; both release the absorbed solar energy as **infrared radiation** (warmth) but the land, holding less heat, cools more quickly than the water. You may recognize this from camping trips; nights can grow cold mighty fast once the sun goes down. Because of these differences in physical properties, coastal areas tend to remain cooler in summer and warmer in winter than places deep within continental interiors; extreme temperature swings during the course of a day are less common near, or right at, shorelines. In other words, oceans "moderate" climates for coastal inhabitants. Likewise, if an ocean current flowing along a coast is warm relative to average air temperature, it will tend to warm neighboring coastal areas through the slow release of heat to the atmosphere as it flows past. Cold marine currents, of course, have just the opposite impact.

Because of their great size and speed, oceanic gyres are especially important in transferring heat energy from lower to

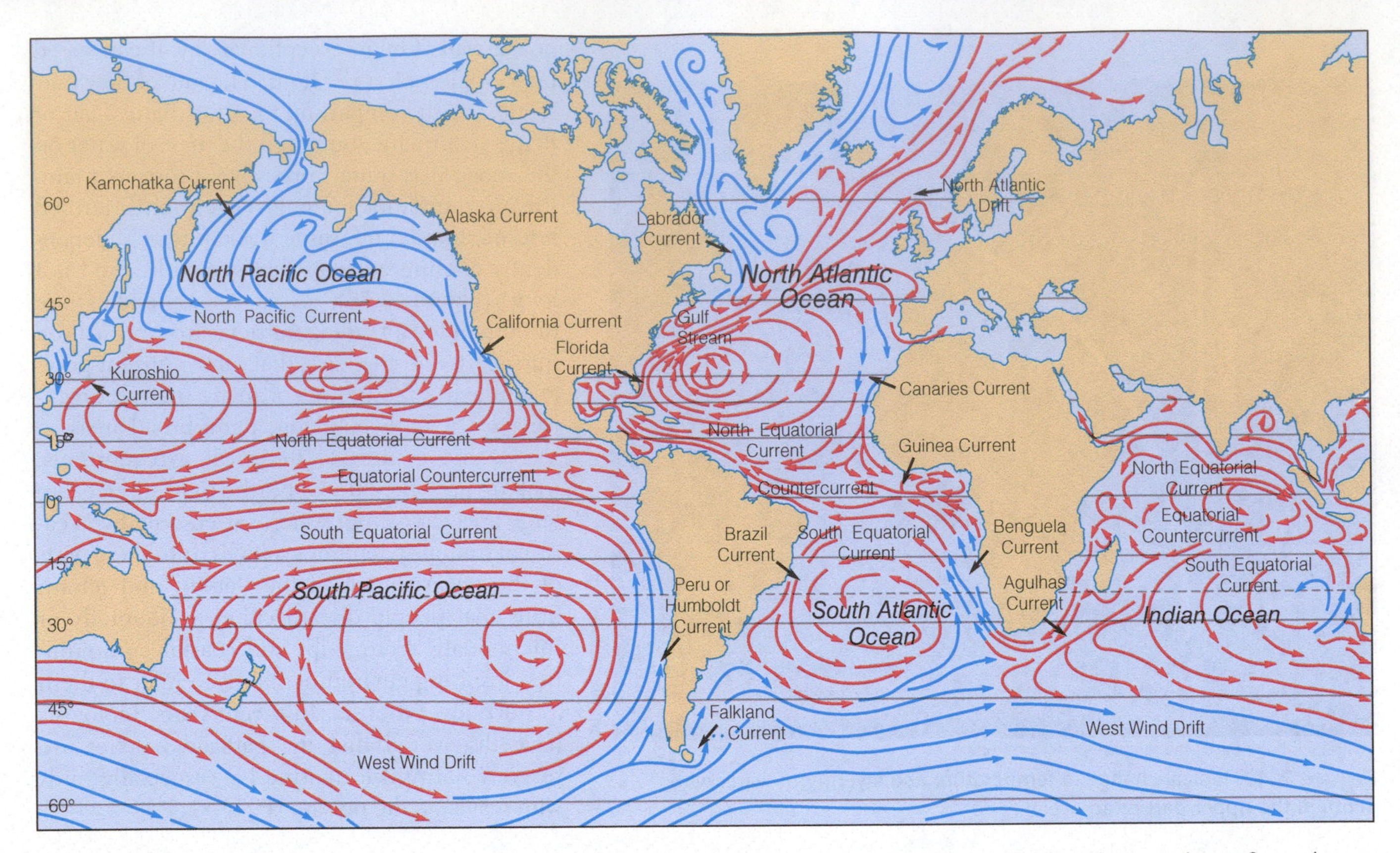

Figure 2.48 Major wind-driven surface currents. Warm currents are shown in red, cool currents in blue. The giant loops formed as the currents sweep around the rim of their respective ocean basins are called gyres. The gyres rotate clockwise in the Northern Hemisphere and counterclockwise in the Southern Hemisphere.

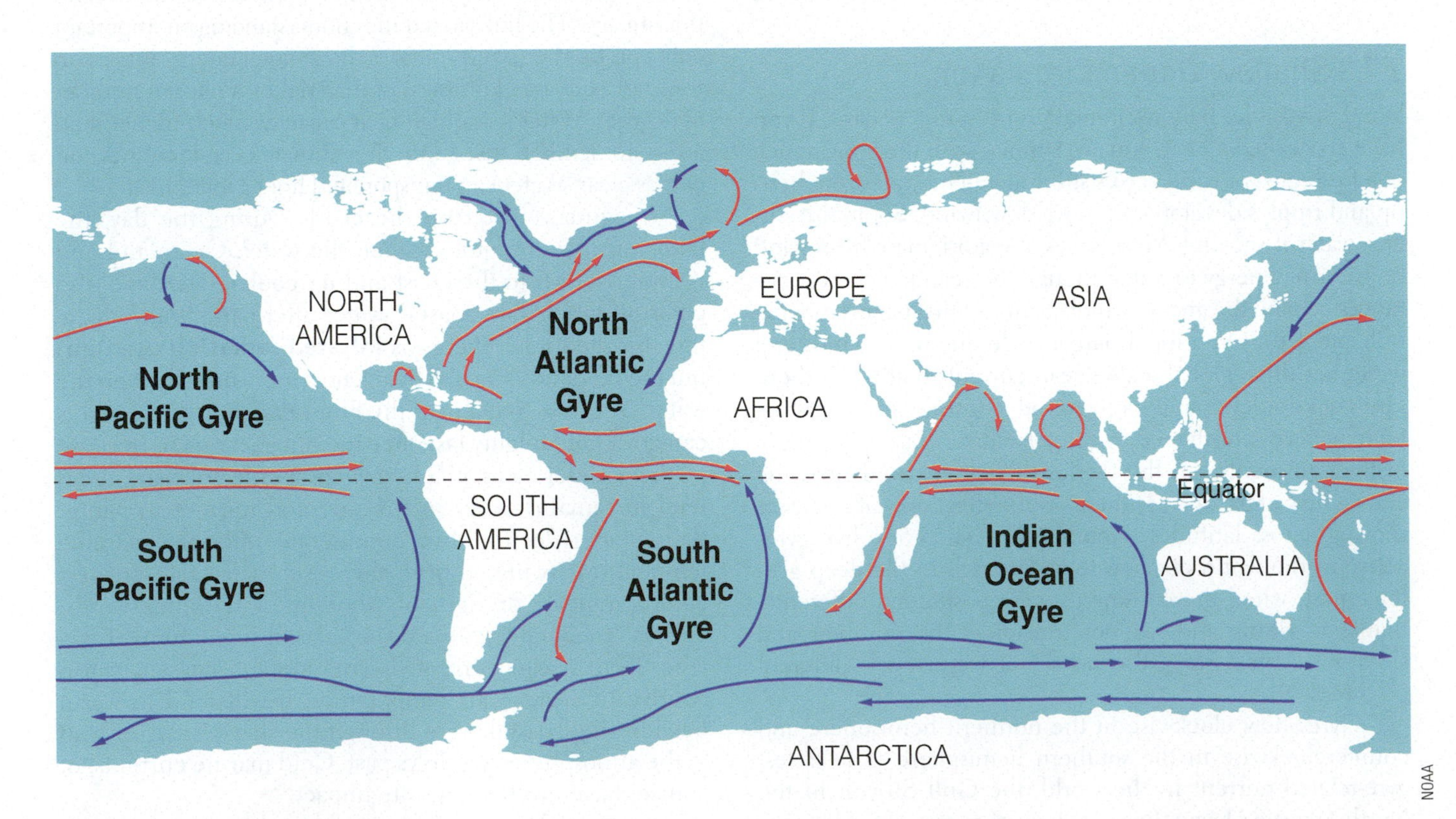

Figure 2.49 There are five ocean gyres.

Figure 2.50 High-resolution computer visualization of the many eddies spinning off the Gulf Stream and related currents, based on satellite observations and sea surface measurements.

higher latitudes. Warm-water currents such as the Gulf Stream and the Japanese Current in the North Pacific heat the air above to make climatic conditions more favorable for people living in such northerly locales as coastal Alaska, British Columbia, and Western Europe. At the same time, as these currents cool and head back toward the equator, they moderate coastal climates at lower latitudes that otherwise might be much hotter in the summer months.

Along shorelines on the western side of continents, persistent winds drive shallow waters away from the shore. To replace those waters, cold, deeper water rises (**upwells**), bringing with it important nutrients for marine life and chilling the moisture-laden offshore breezes to create coastal fogs. **Marine upwellings** occur over only about 0.2% of Earth's ocean surface, but the nutrients they provide sustain more than 20% of the world's total fishery catch. Seasonal fogs that result from such cold-water upwelling are important for ecosystems such as that of the Atacama Desert of northern Chile, South Africa's *fynbos* (from the Afrikaans, meaning "fine bush," the biologically important shrubland occurring in the Cape region), and California's redwood forests. The fogs often come as a great relief to seaside residents on otherwise hot summer days and for animals living in these areas provide a critical source of moisture. A beetle in the hyper-arid Namibian desert gets water by condensing fog on its hind legs–a good example of life specialized to this unique coastal environment (**Figure 2.52**).

Hence the oceans, coupled with the atmosphere, play an important role in determining Earth's climates. But they have an even greater, more subtle impact on keeping the planet hospitable for life, and this requires looking much deeper into their waters.

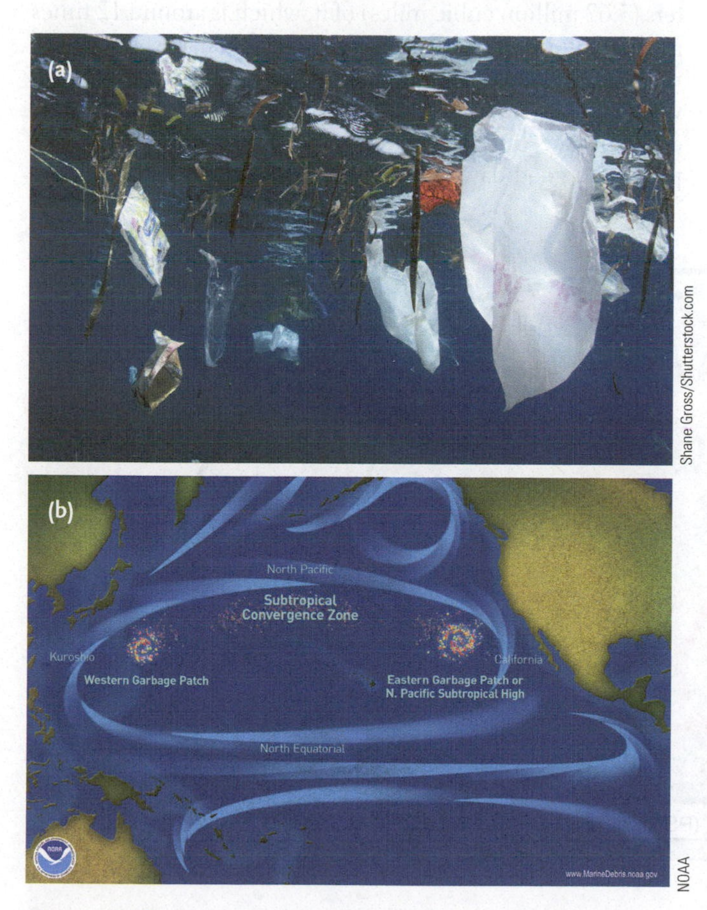

Figure 2.51 Humanity casts tremendous amounts of floating waste into the world's oceans–around 12.7 million tons every year–which currents sweep into huge fields of debris at the centers of gyres. About 80% of this waste is plastics (see Chapter 12). (a) Close up of plastic waste. (b) Pacific garbage patches.

2.4.5 Deep Ocean Circulation

Wind-driven ocean surface currents involve no more than about 10% of all ocean water at any given time, and in most places they stir the sea no deeper than the thermocline. At greater depths, the sluggish circulation that takes place is primarily **thermohaline**–that is, driven by density-related differences in saltiness and temperature. The major driver of

Figure 2.52 A Tenebrionid beetle, native to the dry Namibian desert is covered with condensation from the common coastal fog. In a place where years can go by without rain, only this water allows the beetle to live.

deep marine circulation in the modern world is the chilling of seawater along the coasts, in the North Atlantic, and under the ice shelves of Antarctica. An almost constant cascade of nearly freezing water slips into the deep along the margins of Antarctica, then spreads out northward across the seafloor, even crossing the equator into the Northern Hemisphere. This water is very salty because of the **brine exclusion effect**, which excludes dissolved salts in unfrozen water wherever ice forms; sea ice is salt-free, and the sea water that's left behind is saltier.

Sinking water must be replaced. As cold, dense, salty water sinks, it draws in generally warmer surface water from lower latitudes. In this way, deep marine and shallow marine circulations become connected. In the northern Pacific, the Antarctic deep water wells up as it warms, becoming incorporated into the shallow pattern of global wind-driven currents, in time mixing into the Gulf Stream, where, in the Norwegian Sea east of Greenland, cold temperatures prompt it to sink back to the seafloor as so-called North Atlantic deep water. The complete circulatory system has been named the **global conveyor** (**Figure 2.53**).

The global conveyor turns out to be of paramount importance to global climate because of a simple chemical property of carbon dioxide (CO_2), one of the most important greenhouse gases: CO_2 dissolves more readily in cold water than in warm water. Thanks to deep-water formation around Antarctica and near Greenland, billions of tons of CO_2 are transported by solution into storage in the cold, deep ocean, from which it takes, on average, more than 500 years for any given molecule of CO_2 to escape back to the atmosphere.

In addition to moderating atmospheric temperatures in coastal regions, the oceans also absorb as much as 30% of the CO_2 added to the atmosphere each year, greatly reducing the greenhouse warming worldwide that otherwise would take place. Notwithstanding the deep-water storage of CO_2 described in the previous paragraph, most of this gas is dissolved in shallow ocean waters where it may become available to organisms that make calcium carbonate-rich shells as well as carbonate reefs. There is exchange of mass and energy between the oceans, and the atmosphere clearly shows that the behavior of one part of the Earth System has important consequences for the other, with powerful feedbacks linking the two components of the system in lockstep. Note, too, how the biosphere is connected to this balance.

If Earth lacked oceans, it would probably lack life altogether, at least multicellular life. Imagine the extreme temperature differences, potential windiness, and heat that would characterize such a world.

2.4.6 The Hydrological Cycle, Underground

Groundwater is one of our most important natural resources. By far the largest amount of fresh water available to humanity is present underground; some 23.4 million cubic kilometers (5.62 million cubic miles) of it, which is around 12 times greater volume than all the world's surface lakes, reservoirs, rivers, and streams combined. Unlike surface water, groundwater does not evaporate unless within easy digging distance of the surface, but it does still creep seaward, albeit at a snail's pace on a good day, except where it can flow through cave

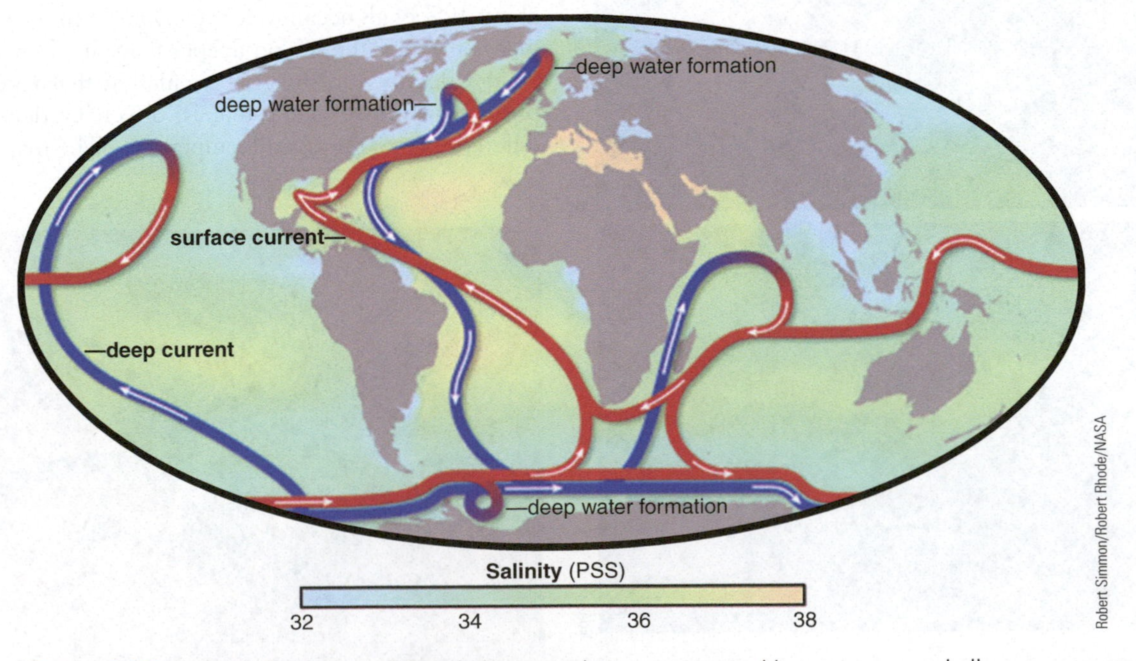

Figure 2.53 The global conveyor (thermohaline) circulation system. Red lines are warm, shallow currents which grow cold and sink as dense, salty water in polar seas. The deep, cold currents flow slowly across the ocean floor to resurface with gradual warming at lower latitudes. Conveyor waters mix with the Gulf Stream where it flows along the Atlantic Coast of North America. The entire overturning circulation takes about 1,500 to 2,000 years.

passages and open fractures in the rock. It is replenished (a process termed **recharge**) where rain, stream, or meltwater percolates (soaks into) into soil, sediment, or rock cracks.

The general layering observed in other components of the Earth System has its parallel in groundwater as well. Earth materials (rock, soil, sediment) with pore spaces only partly filled with water make up the near-surface **zone of aeration**. Deeper down, where water fills every available pore space or fracture, is the **zone of saturation**. If a body or layer of rock or sediment in the zone of saturation contains abundant extractable water, it is called an **aquifer**, a prized goal for water well drillers. Areas of rock or sediment that act as barriers for the downward percolation of water are termed **aquicludes** ("water precluders"). **Aquitards** are places where earth material inhibits but doesn't altogether preclude the percolation of water (**Figure 2.54**).

How deep does groundwater go? One can find evidence of flowing water present throughout the crust, many kilometers down. Much of this deep water, however, comes directly from Earth's interior; it is called **juvenile water**. While important for the creation of certain minerals, magmas, and ore deposits, juvenile water is of no practical interest for people seeking water at the surface. Nor is juvenile water part of the hydrological cycle except as a minor potential source for replenishment. The shallower water stored in aquifers that originates by percolation from the surface is said to be **meteoric water** instead–that is, "from the atmosphere" originally. Meteoric groundwater may infiltrate as deeply as 5 kilometers (3 miles) beneath continental surfaces, but most of it accumulates and flows at much shallower levels.

The presence of water and tiny openings to contain it far underground also implies the possibility of life existing in such extreme habitats, and indeed there is much evidence from deep drilling that microbial life is there, at depth. In fact, the total mass of simple lifeforms existing in this way far underground may exceed by two-fold the total biomass present in the world's oceans, and certainly the total biomass of the human species–roughly 15 to 23 billion tons of biological carbon! About 70% of the world's bacteria and Archaea (single-celled organisms lacking nuclei) appear to live underground, their ultimate habitat depth determined by how far meteoric water can seep down into the Earth.

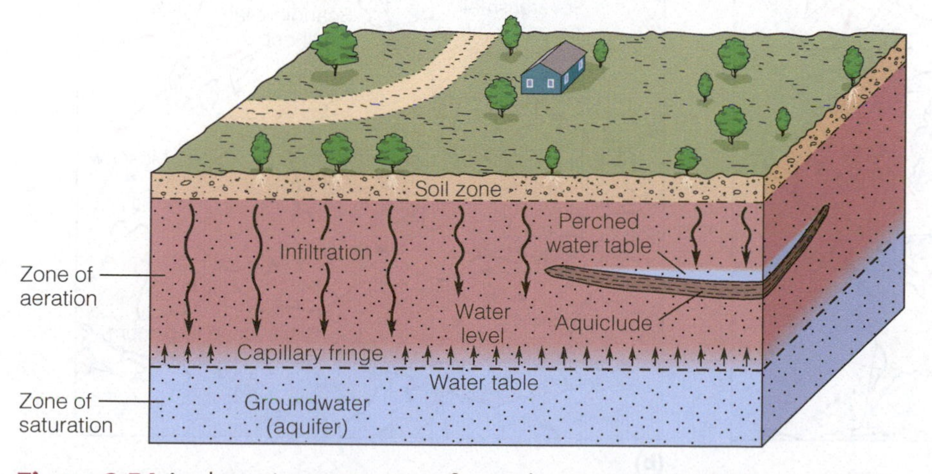

Figure 2.54 A schematic cross-section of groundwater zones. Perched water tables lie atop localized saturated zones supported by aquicludes. Capillary fringes are described in Chapter 8.

Figure 2.55 Arctic permafrost (frozen groundwater) cropping out in the bank of a gully.

2.4.7 Cryosphere–Frozen Water

More than double the amount of fresh water present in ground and surface water is locked up in solid form (aka, ice) in the form of glaciers, ice caps, and ice sheets. Considerable amounts of frozen groundwater, termed **permafrost**, also exist at high latitudes (**Figure 2.55**). Melting glaciers are an important source of fresh water for some regions, including the Ganges River Valley in India and parts of the Andes Mountains. Glaciers may be regarded as "frozen reservoirs" in such environments. They are critical for stabilizing the world's available freshwater supply.

Alpine glaciers grace high mountain valleys. They move downslope from elevations where more snow accumulates than can melt away. Snow that lasts more than a single year is termed **firn.** As the firn deepens, pressure at the bottom of the snowpack increases. The pressure eventually closes off pores, isolating bubbles of air. Fresh powder snow falls with a density typically in the range of 50 to 70 kilograms/meter^3 (84 to 117 pounds/yard^3). The delicate crystals quickly transform under pressure and some melting to more rounded and denser forms. When that density exceeds 830 kilograms/meter^3 (1,400 pounds/yard^3), the transition from snow to glacial ice is complete.

Ice is a weak solid and if stressed will flow. Ice is weak because at Earth's surface it is never far from the melting point. If a mass of ice grows thicker than a few tens of meters, the ice will flow under gravity, like a very slow-moving river, though solid. This is the definition of a **glacier**, flowing ice that

Figure 2.56 An alpine-style glacier exits the mountains at the shore of South Skjoldungen Fjord, Greenland. The dark ridge-like mound at the bottom of the glacier is an end (or terminal) moraine. Bands of rocky rubble stretching at the sides in up the interior of the glacier are also moraines.

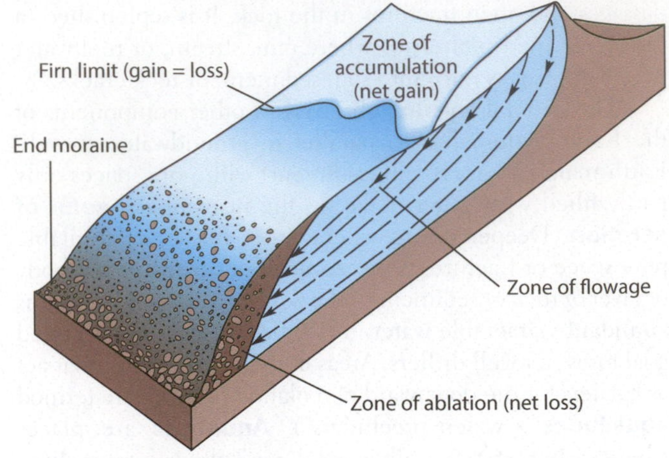

Figure 2.57 Typically, about two thirds of a glacier lies at elevations above the firn limit in the zone of accumulation, where more snow falls in winter than melts in summer. Below the firn limit is the zone of ablation, where glacier ice and the entrained rock debris are exposed by melting in the summer. The ridge of rock debris formed at the edge of the melting ice is an end moraine.

survives through summer after summer. Rocky debris falling atop the ice, and material eroded from the base, is dumped by the moving ice at the edge (**terminus**) of the glacier where melting takes place, leaving an irregular ridge of debris called a **moraine** (**Figures 2.56** and **2.57**).

Where alpine glaciers merge at their tops, the higher elevations of a mountain range or peak may be completely immersed in a broad, dome-shape **ice cap**. Each alpine glacier extending away from an ice cap shares a common source of fresh ice from upslope. Glaciers are continuously melting around their margins except, of course, when air temperatures plunge below freezing. But they may still advance, because their rate of downslope movement can exceed the rate at which their termini retreat by melting back upslope. If these rates are exactly matched, the front of the glacier appears stagnant.

On an even larger scale, ice can develop, climate permitting, across all sorts of terrain, including low, flat plains. The largest type of land-based glacier, a **continental ice sheet**, then grows. The scale of such features is mind-boggling (**Figure 2.58**). The great Laurentide Ice Sheet covered almost a third of North America during the last ice age, and most of Antarctica today is covered by two merging ice sheets, in places 3,000 meters (10,000 feet) thick. The edges of continental ice sheets are marked by sinuous moraines that may extend intermittently for thousands of

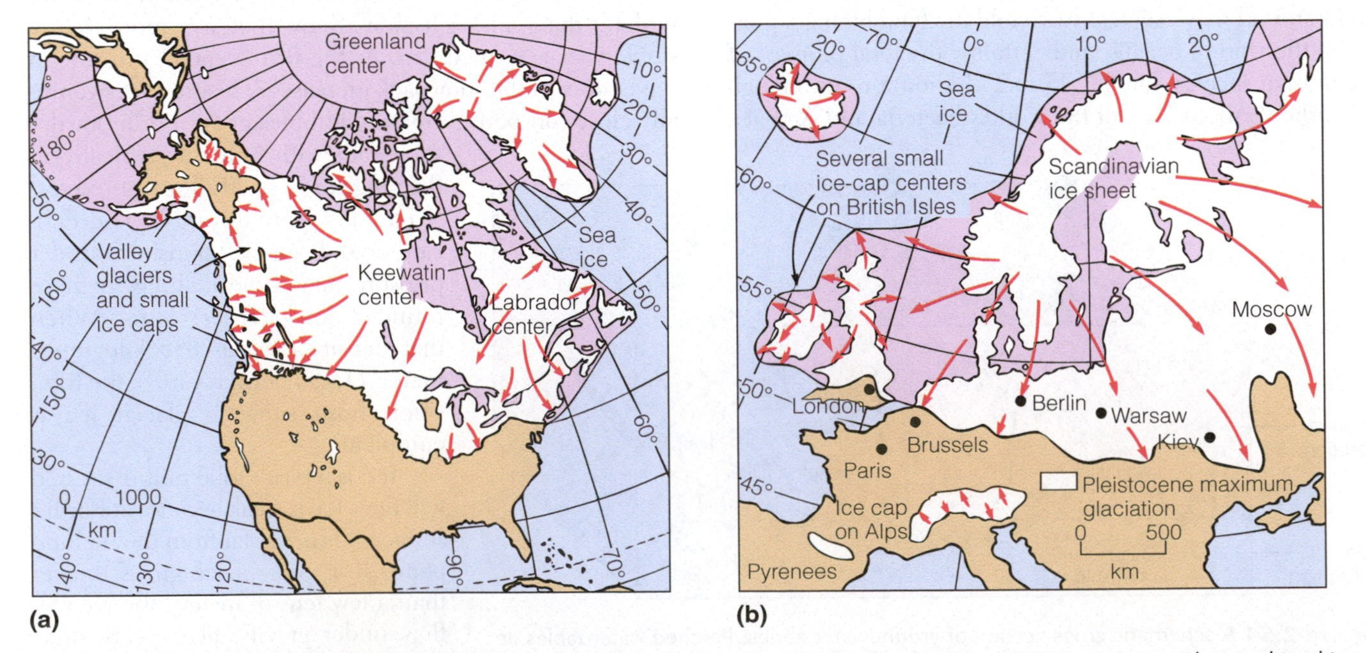

Figure 2.58 Giant ice sheets covered a large part of (a) North America and (b) Europe 20,000 to 25,000 years ago. The combined ice of two ice accumulation centers–the Keewatin, and Labrador–merged to make up the Laurentide ice sheet in North America.

Max Topchii/Shutterstock.com

Figure 2.59 Fjallsarlon, a proglacial lake, is fed by meltwater from neighboring Fjallsjökull glacier in Iceland.

kilometers, and by clay and silt deposited at the bottoms of vast glacial meltwater (or **proglacial**) lakes (**Figure 2.59**). The Great Lakes in North America were once part of an even larger proglacial lake network present along the southern edge of the Laurentide Ice Sheet at and after the climax of the last ice age. The largest, Lake Agassiz-Ojibway, was 1,500 kilometers (900 miles) long–a veritable inland sea. Rapid drainage of this monster freshwater lake via the St. Lawrence River Valley, owing to retreat of the ice sheet, had a major impact on the climatic transition from the past ice age to our warm modern epoch.

Where continental ice sheets meet the ocean, they may spread out as floating **ice shelves**, hundreds of meters thick, extending hundreds of kilometers from shore. They are the source of the world's largest icebergs. Elsewhere, the ocean surface itself can freeze directly to form a thin (meters thick) covering of floating ice. Such **sea ice**, for instance, seasonally covers much of the Arctic Ocean. Ice sheets cannot grow on open water but need the stability of an underlying landmass to thicken and spread. As ice goes, they are much more stable features than sea ice.

The cryosphere is a direct consequence of several factors: the amount of sunlight reaching Earth at high latitudes where glaciers (especially ice sheets) can grow, the freezing point of water, and the overall energy balance of the Earth. One of the most important controls on global energy balance is Earth's reflectivity, otherwise known as **albedo**–the percentage of incoming, shortwave solar radiation that is reflected back into space. In turn, the ice and snow influence the heat content of the oceans and atmosphere through the **ice-albedo effect**, a positive feedback in which snow and ice reflect sunlight directly back into space, keeping Earth cooler–and therefore more favorable to development of more snow and ice–than it would be if land and sea were exposed instead. Of course, this feedback can work the other way if ice and snow melt away, further warming the planet.

The cryosphere is sensitive to changing environmental conditions. You can appreciate that sensitivity by noting that at present ice covers about 11% of Earth's surface, mostly in Antarctica. Twenty-thousand years ago, when the Earth's atmosphere was on average 5°C (9°F) colder, ice coverage was close to 30%. Sea levels then stood about than 130 meters (440 feet lower) than at present. Conversely, during prior warm periods, such as 400,000 years ago, sea level was at least 6 meters (20 feet) higher than today, implying that at least alpine glaciers and the Greenland Ice Sheet shrank substantially. With the planet warming rapidly, our future world is likely to have less ice and more water than we are used to.

2.5 Biogeochemical Cycles (LO5)

The flow of matter and energy throughout the atmosphere, hydrosphere, lithosphere, and cryosphere is critical to life on Earth. It enables elements and compounds vital for life to circulate continuously through the world. These pathways, known as **biogeochemical cycles**, are a determining factor for where and how abundantly life exists on Earth. Biogeochemical cycling is a basic concern of scientists studying the environment. Here, we consider some of the most important cycles and underscore their essential role in the Earth System.

2.5.1 The Nitrogen Cycle

The **nitrogen cycle** is a complex exchange of the element nitrogen (N) in various forms between the atmosphere, hydrosphere, biosphere, and soils (**Figure 2.60**). Recall that nitrogen in the form of N_2 is the most abundant gas in Earth's atmosphere; the ultimate nitrogen "reservoir," inherited from when our planet first formed.

N_2 is an inert gas, unavailable in its elemental form to plants and animals despite its importance to us as a nutrient. Before nitrogen is of any use, it must be converted into molecular forms, primarily nitrite (NO_3^-) and nitrate (NO_2^-). Both are oxides of nitrogen and are capable of being taken up by complex creatures like us. This oxidation process, called **nitrification**, is readily enabled by energy-seeking microbes and bacteria in soils; these simple organisms take nitrogen directly out of the air, a process that is also called **nitrogen fixation**. Small amounts of biologically available nitrogen are also introduced to the environment by the infiltration of nitric acid (HNO_3), which in water becomes H^+ and NO_3^-, from rainfalls (acid rain) and by oxidation of atmospheric nitrogen when lightning super-heats small parts of the atmosphere breaking the N_2 bond, generating nitrate.

People have learned to make nitrogen-rich fertilizers artificially using the **Haber-Bosch process**, invented by the German chemists Fritz Haber and Carl Bosch. So important was their discovery that it earned them the Nobel Prizes in chemistry (despite the fact that Haber's research also enabled the Germans to manufacture poison gas to use against enemy troops during World War I). Haber won the award in 1918, whereas Bosch was honored (jointly with Friedrich Bergius) in 1931. And no wonder: such artificially fixed nitrogen is today applied to crops worldwide. More than half the world's rapidly growing population can eat because of it. Without artificially fixed nitrogen, our planet could not support 8 billion people. The widespread availability of nitrogen, which in the past was a limiting nutrient keeping plants from growing as quickly as they could, spawned the Green Revolution in agriculture. But fixing nitrogen is energy

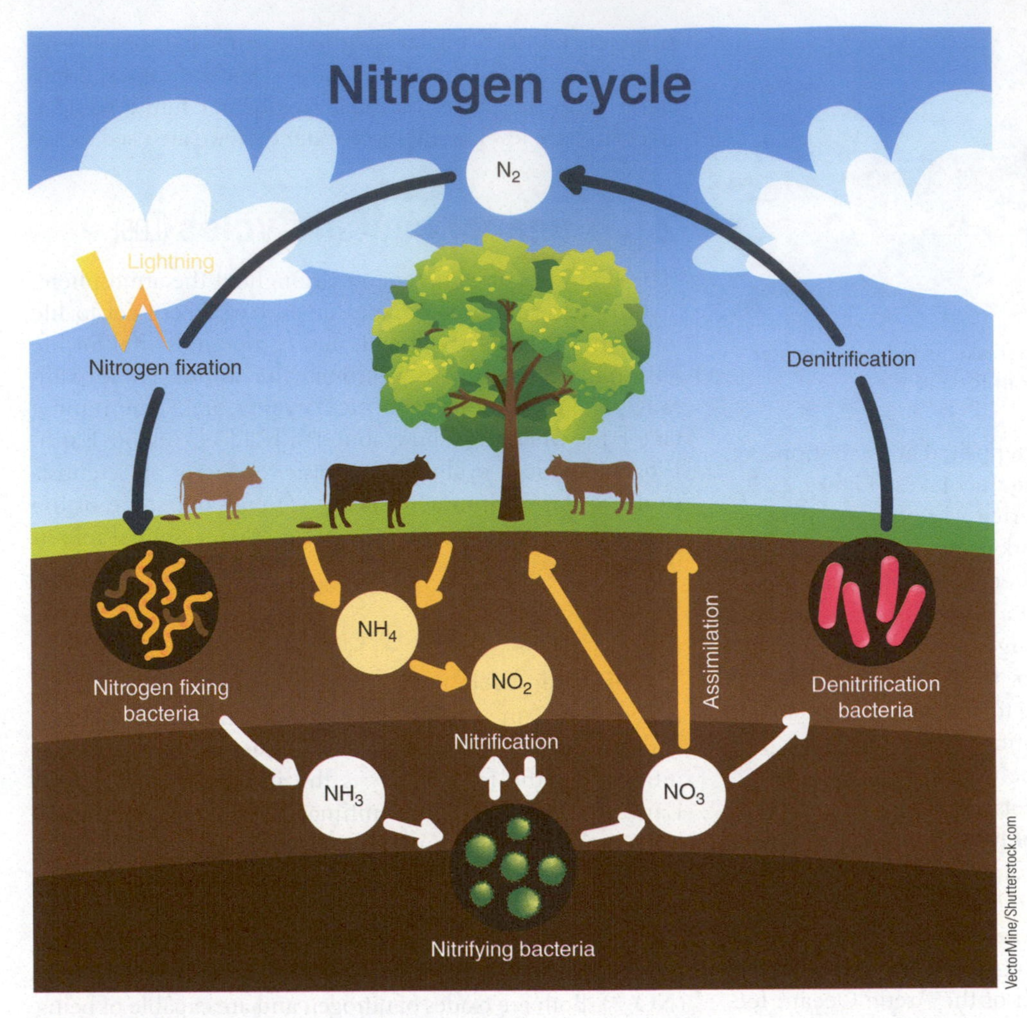

Figure 2.60 A simplified diagram of the nitrogen cycle on land.

intensive. Some estimates suggest that 1% of global human energy consumption is used to make nitrogen fertilizer.

Benefits sometimes come with consequences: although we've increased crop yields dramatically, so also have we harmfully, unintentionally fertilized waterways. Many streams and ponds near farmlands have become choked with **algal blooms** because nitrites and nitrates percolate from soils and fields into groundwater. The "lost fertilizer" can also wash out to sea both enriching and, where overly concentrated, poisoning waters for aquatic organisms as oxygen is taken up from the water by spreading algae. The sea can literally turn an unhealthy pale green (see Chapters 8 and 12).

When organisms die, bacteria and other microbes assist in decomposition and return some nitrogen to the soil in forms that new life can easily access. Other bacteria return the nitrogen to the air as N_2 in the process of **denitrification**. Denitrifying bacteria living in the ocean return the nitrogen back to the atmosphere in the form of N_2O (nitrous oxide) and NH_3 (ammonia) as evaporation and off-gasing takes place. The nitrogen cycle functions as a complete loop of interaction involving various spheres of the Earth System; it is far more complex than the water cycle to which it is linked.

2.5.2 The Phosphorous Cycle

Phosphorus (P), the eleventh most abundant element on Earth, is essential for making DNA, bones, teeth, and cell membranes. Unlike nitrogen and water, phosphorous does not require the atmosphere to complete its circulation cycle through the environment (**Figure 2.61**).

One source of phosphorus accessed by life is phosphorus-bearing minerals, notably the mineral **apatite** [$Ca_5[PO_4]_3$ (OH, F, Cl)], the material of which our bones are made. Chemical weathering of apatite releases phosphorus to soil, where organisms take it up. The growth, death, and decay of plants and animals recycle phosphorus for future generations, provided that erosion or other disturbances don't remove the soil and deposit it in lake or ocean sediments where it is inaccessible to most life.

Another source of phosphorus is **guano**, the excreta of birds and bats, in which phosphorus is highly concentrated. Mining of guano deposits on "bird islands" such as Nauru in the Pacific, and in bat caves such as Carlsbad Caverns in New Mexico, were major sources of commercial fertilizer for farmers before recent industrial production of synthetic fertilizers (**Figure 2.62**). Around 80% of the surface of Nauru has been strip mined for this purpose. Adding to the demand, the munitions industry has also sought phosphorus to make explosives. As recently as World War II (1939–1945), phosphorus derived from bird poop was used for military ordnance. The phosphorous mines of the West Indies, off the shore of South America, were guarded by U.S. soldiers to protect them from attack by Nazi U-boats.

Phosphorus is especially important for marine and lacustrine ecosystems. The amount of life possible in aquatic environments is often limited by dissolved phosphorus concentrations due to the scarcity of available phosphorus in water. As in the case of excess nitrogen, phosphorus added by runoff to aquatic ecosystems can cause tremendous damage by encouraging algal blooms.

2.5.3 The Carbon Cycle

The **carbon cycle** plays a pivotal role in determining global climate (Chapter 3) (**Figure 2.63**). Carbon dioxide (CO_2),

There is one carbon cycle, but it helps to consider two sub-cycles of importance to the Earth System: a shallow sub-cycle and a deeper one. The **deep carbon sub-cycle** provides the background needed to understand how CO_2–a biologically reactive gas–has persisted in Earth's atmosphere through billions of years. The **shallow carbon sub-cycle** is of greater importance, however, to grasp why human activity and atmospheric CO_2 concentration is such a major concern today. Humans have interconnected these sub-cycles by extracting fossil fuels, such as coal, gas, and oil from the deep carbon sub-cycle and introducing them, by combustion, into the shallow carbon sub-cycle.

The deep carbon sub-cycle begins deep inside the Earth, where carbon is quite abundant. Recent studies suggest that as much as 65% or more of Earth's carbon may be stored as high-pressure iron carbide minerals in the core. No doubt, based on measurements taken at volcanoes, abundant carbon dioxide is also present in the shallow mantle. CO_2 is the next most abundant gas after water vapor released by eruptions–as much as 280 to 360 million tons each year, especially where asthenospheric upwellings occur at divergent plate boundaries. If this number sounds impressive, however, consider that the annual human contribution of CO_2 to the atmosphere each year from burning oil, coal, and natural gas is almost 30 billion tons, 100 times greater than nature's emissions.

The release of CO_2 from the deep Earth replenishes carbon dioxide slowly extracted from the atmosphere as organisms use it for biological purposes. Where organisms die and sediment covers their organic remains, or their carbonate reefs and shells, the carbon will be buried and preserved as new rock added to the lithosphere. Examples include limestone, petroleum-bearing sandstone, and coal seams. Elsewhere, CO_2 in seawater can react with material on the deep ocean floor, where subduction carries that material back into the mantle for restorage or recycling. Other processes that remove carbon from the atmosphere include the weathering of rock and metamorphic reactions in which CO_2 is incorporated into minerals like calcite ($CaCO_3$) and dolomite [$Ca, Mg(CO_3)_2$]. Both solid carbonate in rock fragments and dissolved bicarbonate (HCO_3^-) can be washed out to sea,

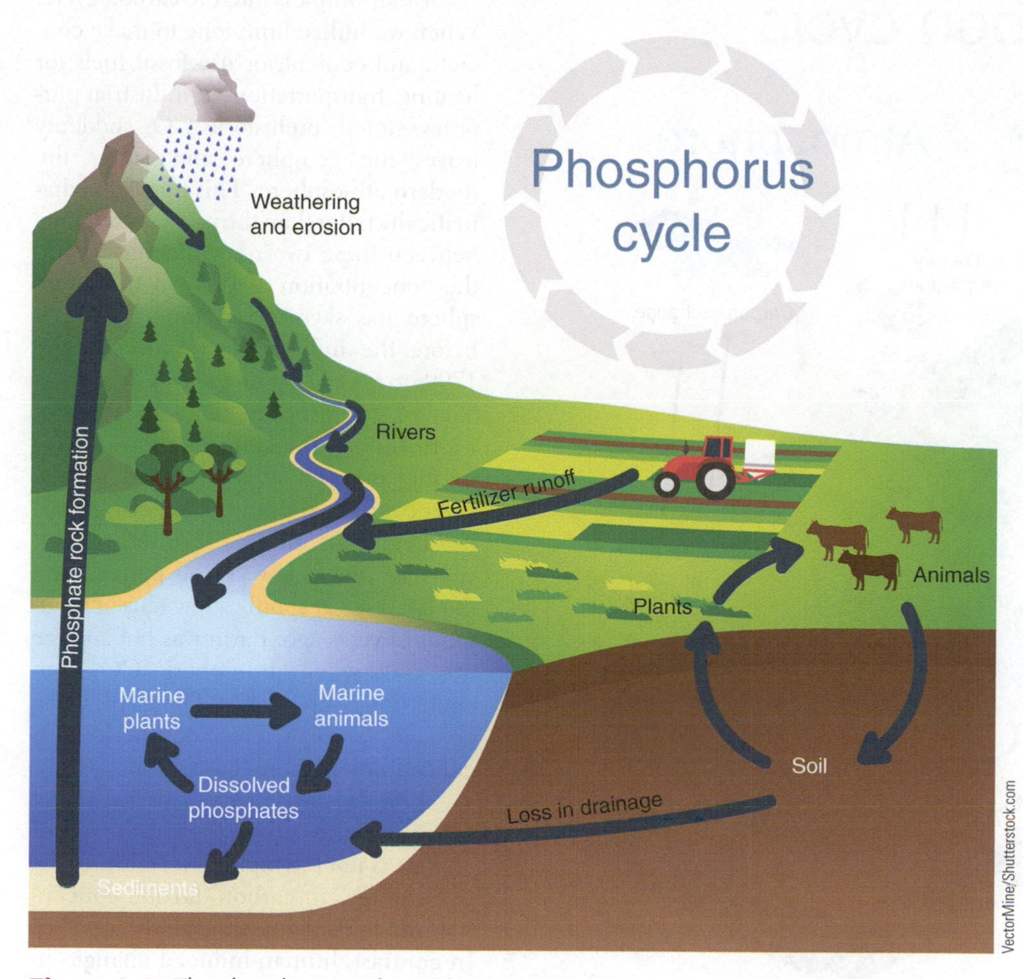

Figure 2.61 The phosphorus cycle, unlike other biogeochemical cycles, does not require the atmosphere as a nutrient source, although the transport of phosphorus through the Earth System does depend upon the hydrological cycle.

an important greenhouse gas (even at trace levels), has been present in the atmosphere since Earth's earliest days. Without CO_2 in the atmosphere, Earth would be a much colder planet. Indeed, Earth's average temperature over time is closely related to the atmospheric concentration of CO_2.

Figure 2.62 Cormorants leave a white coating of phosphorus-rich guano atop a rocky islet in Paracas National Park, Peru.

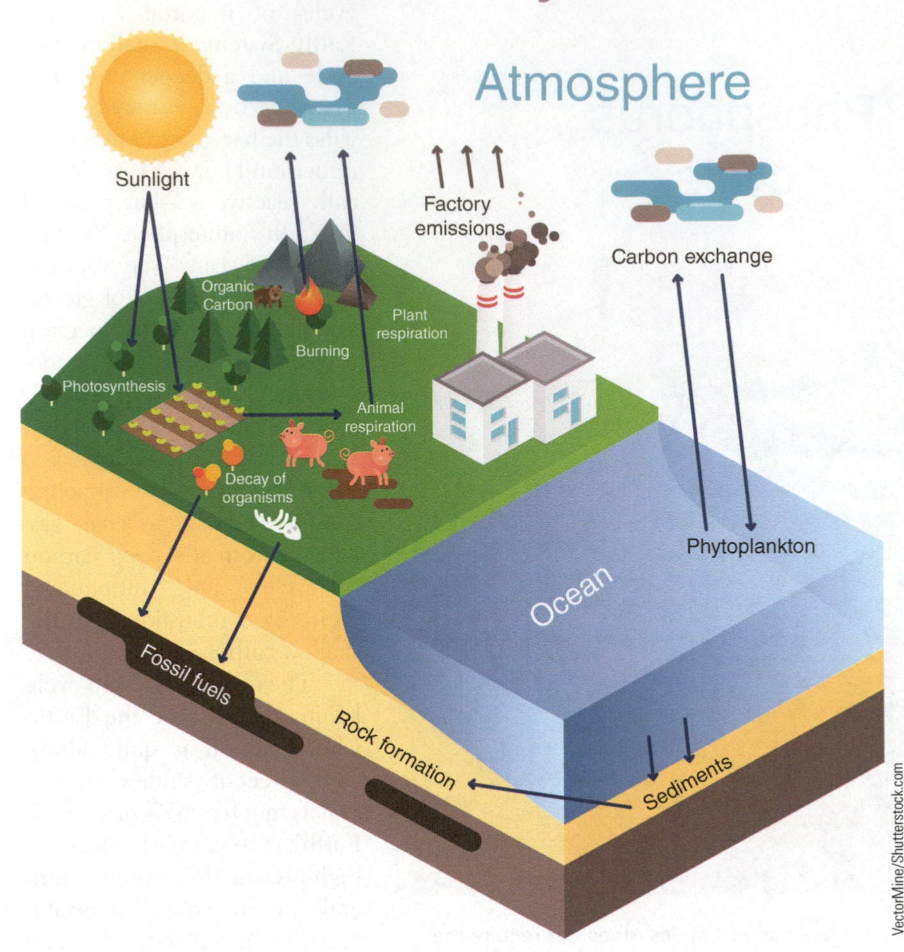

Figure 2.63 The carbon cycle on land and sea. Not included are processes related to the slow recycling of carbon from and back into the deep earth via volcanoes and plate tectonics.

where the rock ends up as sediment and the bicarbonate may precipitate and form fine crystals of calcite (which also settle to the seafloor). Much of this is mediated by marine organisms (the biosphere).

The shallow carbon sub-cycle involves all the different spheres (**Figure 2.64**). Plants take up CO_2 photosynthetically for biological purposes, returning it to the atmosphere when they die and decay or lose their leaves in the winter. Atmospheric carbon dioxide dissolves in seawater, returning to the air if water temperatures rise; or, if the seawater ends up in a deep ocean current, CO_2 is isolated from the atmosphere for hundreds to thousands of years because the deep ocean does not exchange gases with the atmosphere. Most of this air-sea gas exchange only takes place between the atmosphere and shallow ocean, where the water is well stirred by waves.

The geosphere is by far the largest natural reservoir (storage area) for CO_2 on Earth. The oceans are next, followed by the biosphere and then the atmosphere. Earth's atmosphere at present contains about 830 billion tons of CO_2. In comparison, permafrost (shallow, frozen soil) alone has trapped almost double this amount.

Enter humans into the carbon cycle. When we utilize limestone to make concrete and cement, or use fossil fuels for heating, transportation, or industrial purposes, stored, prehistoric CO_2 suddenly leaves the geosphere and enters the modern atmosphere. This transfer is dramatically faster than the natural exchange between these two reservoirs: the result, the concentration of CO_2 in the atmosphere has skyrocketed from 280 ppm before the industrial revolution in the 1700s to over 420 ppm today (Chapter 3). Viewed over the tremendous span of Earth history, the concentration of CO_2 in the atmosphere has ranged widely, from over 1,000 ppm (parts per million) in Cretaceous times, 145 to 66 million years ago, to as low as 180 ppm during the latest ice age, only a few tens of thousands of years ago. Earth was hot and ice was gone when atmospheric CO_2 concentrations were higher. The Earth was a cold, icy place when atmospheric CO_2 concentrations were lower.

Life clearly can survive great changes in the carbon cycle if the rate of change is not too rapid. Glacial/interglacial changes in carbon dioxide concentrations took many thousands of years. In contrast, human-induced changes in atmospheric carbon dioxide concentration are occurring at rates more than 10 times faster. There are certainly limits to what particular species and ecosystems can tolerate. On some planets (e.g., Venus), it appears that tipping points have been reached beyond which negative feedbacks no longer are effective in reversing the effect of high atmospheric CO_2

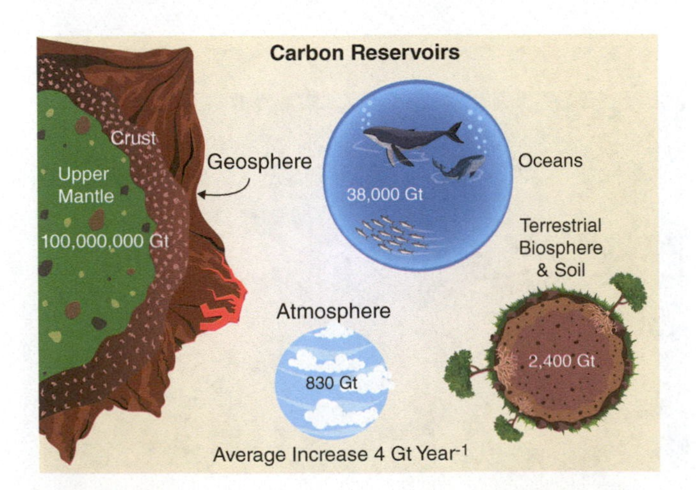

Figure 2.64 The geosphere is the largest carbon reservoir on Planet Earth. GT is a gigaton (billion tons).

concentrations. On Venus, carbon dioxide results in a hellishly hot world. Some planetary geologists suggest that as recently as 750 million years ago, our sister planet may have still been favorable for life.

2.6 Biosphere (LO6)

The biosphere–the mass of living matter on or near Earth's surface–functions as a regulator of the Earth System. Such regulation includes the negative feedbacks of ecosystem development and growth in response to changes in the amount of carbon dioxide and other greenhouse gases in the atmosphere; the **sequestration** (storing) of ancient CO_2 in limestone reefs, petroleum, coal, and natural gas; the albedo-related impact of biologically rich regions (dark as seen from space); and many other influences that are the principal concerns of ecology.

2.6.1 Earth's Evolving Atmosphere: The Living Connection

Today's atmospheric composition is largely a byproduct of life interacting with Earth's past atmospheres. The biosphere has had a profound impact on Earth's atmospheric composition, to which it is closely tied and well adapted. An astronomer looking for an exoplanet with abundant Earth-like conditions could identify such a world based on just a few key observations. That person would be looking for tell-tale indications of atmospheric gases using a **spectroscope**, a device that identifies the wavelengths of light characteristic of certain gases. For example, carbon dioxide, methane, water, and oxygen each have specific spectral lines and, together, are suggestive of life influencing atmospheric composition (**Figure 2.65**). Nitrogen, which is a nonreactive gas quite stable in sunlight, would be dominant if there are oceans, soils, and ecosystems there to dissolve or incorporate much of the other gases. The study of **biosignatures**, the distinctive spectroscopic signs of life and Earth-like conditions on exoplanets, is an exciting new branch of deep-space research.

In its most primitive state, our atmosphere consisted mostly of gases directly derived from the nebular cloud that formed the planet. These included abundant light elements such as hydrogen (H) and helium (He). This was Earth's Atmosphere 1.0. Radiation from the young Sun quickly drove this tenuous sheath away into space, leaving the nascent Earth with no substantial atmosphere, possibly for millions of years. Volcanic activity, however, gradually filled this void with fresh gases exhaled from inside the planet, including water vapor, CO_2, sulfur dioxide (SO_2) and nitrogen–all the ingredients of our modern air except for free oxygen. We could not have breathed such a concoction despite its mix of otherwise familiar gases.

Carbon dioxide and water vapor were dominant in Atmosphere 2.0 (volcanic atmosphere), but as ocean grew from continuous, intensive rainfalls, much of the carbon dioxide and noxious SO_2, dissolved in the accumulating seawater. Nitrogen became dominant because it was not dissolved in large quantities in the young seas; nor did it react with rocks and sediments as weathering took place on the bare continental landscapes. Consider oxygen, which is thousands of times more abundant in the Earth System than nitrogen. Most of Earth's oxygen is tied up in minerals (for instance, silicates), while only a small fraction of Earth's total nitrogen is incorporated these ways. In short, something resembling our modern air came into being.

Atmosphere 2.0 still lacked much oxygen. We could not have breathed it either. Atmosphere 3.0, our present air, required life. Its development was a slow process indeed. The first known lifeforms, primitive filamentous microbes discovered in the Apex Chert of western Australia, existed 3.5 billion years ago. For the next 2 billion years, Earth's dominant life form was cyanobacteria (blue-green algae); a gooey "pond scum" that thrived in shallow coastal waters. The cyanobacteria utilized CO_2 photosynthetically, much as plants do today, and this proved key to the development of Atmosphere 3.0 and all future life on Planet Earth (**Figure 2.66**).

2.6.2 Life and Oxygen

The Sun shone more weakly billions of years ago than it does today. The high concentration of CO_2 in the atmosphere

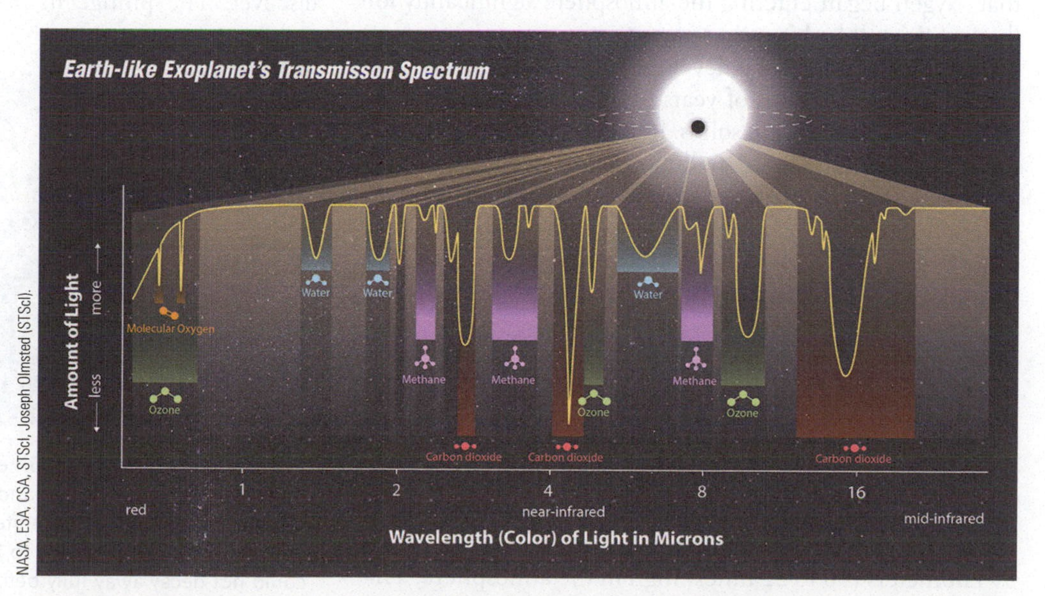

Figure 2.65 Spectrum of an Earth-like planet's atmosphere. Such spectra can be collected through telescopes to help us to evaluate whether other planets might harbor life.

Faviel_Raven/Shutterstock.com

Figure 2.66 Fossils of one of the earliest life-forms known, the species *Grypania spiralis*, which thrived 2.1 billion years ago. Grypania may have been a type of giant shallow-water bacterium, or a filamentous alga. Rock sample is from the Natural History Museum of Vienna.

PomInPerth/Shutterstock.com

Figure 2.67 The stony residues of modern stromatolites in Lake Thetis, Australia, whose algal ancestors thrived in shallow seas over a billion years ago.

kept the planet warm enough, however, for liquid water to persist at the surface and for simple life forms to evolve in the oceanic "soup," which was much less salty than it is today. The photosynthetic cyanobacteria (blue-green algae) flourished in great mats and mound-like structures called *stromatolites* (**Figure 2.67**) releasing oxygen as a highly reactive waste product:

$$6CO_2 + 6\ H_2O \leftarrow\rightarrow C_6H_{12}O_6 + 6O_2$$

[Carbon dioxide from the air + water yields algal tissue (glucose; a source of stored chemical energy) + waste oxygen]

The oxygen did not travel far before bonding with ferrous iron (Fe^{2+}) dissolved in the water to produce floating grains of the mineral hematite (Fe_2O_3), or with iron in rocks lining waterways and water bodies to give them a distinctive rust-red coloration. No oxygen was left to accumulate in the atmosphere … at first (Chapter 4). About 1.5 billion years ago, though, all available iron became sufficiently oxidized that oxygen began entering the atmosphere significantly for the first time. It had no other place to go.

Atmosphere 3.0 now developed fully, entirely because of hundreds of millions of years of quiet biological activity and chemical bonding in solids. In a sense, this was the environment's first great waste-pollution crisis, but the build-up of oxygen also enabled new life forms to evolve capable of utilizing free oxygen for metabolic purposes. The reaction they used, and which we use in our own bodies today, is just the opposite of photosynthesis:

$$C_6H_{12}O_6 + 6O_2 \leftarrow\rightarrow 6CO_2 + 6H_2O + \text{energy}$$

[Glucose (plant or animal matter consumed) + oxygen yields carbon dioxide + wastewater + energy needed for metabolism]

In addition to making possible a new aerobic form of life, rising concentrations of oxygen (O_2) reacted with ultraviolet radiation from the Sun to generate ozone (O_3) 15 to 35 kilometers (10 to 22 miles) high in the atmosphere. The ozone acted as a shield preventing deadly solar radiation from reaching land surfaces which, up until that time, had been inhospitable to life. Some 440 million years ago, when primitive funguses first appear on dry land, Earth's continents were harsh, windblown, and lifeless, perhaps resembling the present surface of Mars.

The amount of oxygen in the atmosphere did not increase steadily but fluctuated due to the appearance and disappearance of biomass that occasionally suffered severe extinction events, and recurrent opportunities to oxidize freshly exposed rocks and sediments. During Permian times (300 to 250 million years ago), oxygen levels may have ranged from a dangerously combustible high of 35% to a low of 15% atmospheric concentration (compare to today's 21%). Early and mid-Permian life was characterized by enormous insects adapted to living in oxygen-rich environments, including dragonflies the sizes of crows (**Figure 2.68**). Massive wildfires occasionally consumed whole forests that presently would be too wet to burn, leaving immense deposits of fossil charcoal (**fusain**) for future paleontologists to discover. The plunge in oxygen levels, corresponding to

Nicolas Primola/Shutterstock.com

Figure 2.68 Life on land in early Permian times was dominated by giant amphibians and winged insects, club mosses, ferns, and scale trees. Plant life was so lush in low wetland environments between 360 to 250 million years ago that it could not decay away fully before burial by sediment. Ninety percent of all coal beds today are residue of the biosphere that thrived during this time.

intense global cooling at the end of the Permian, could have contributed to major extinction of Permian faunas. In any event, it may be a shock to realize that Earth's surface has only become habitable for life as we know it *in just the past 10%* of the planet's history.

2.6.3 Geologic Time and the Importance of Biology

Changes in the biosphere serve as the basis for the **geologic timescale** geologists use to compare the relative ages of different suites of rocks worldwide. In 1815, British geologist William Smith established that fossil assemblages succeed one another in age-related sequences and began classifying various fossil-bearing rocks in age groups related to particular type-localities or regions. Ultimately, a hierarchy of age groupings emerged to generate the modern timescale shown in **Table 2.2**. Note the relationship of life forms, or ecological characteristics, to various geologic intervals. As you might expect, the detailed breakdown of these intervals shortens the closer we come to modern times as a reflection of the increased complexity and our improved understanding of past life. Of particular interest to many paleontologists (geologists who study fossils and past life forms) is the seemingly sudden emergence of complex multicellular organisms at the end of the Precambrian Era (505 million years ago) and the great mass extinction events that have periodically set back and restarted the clock of evolution in new directions since then. Life always seems to bounce back following ecological catastrophe due to a grand homeostasis synchronized with Earth's atmosphere and hydrosphere.

Table 2.2 Geologic Timescale

Era	Period	Epoch/Age	Timing (mya)	Characteristics
Cenozoic	Quaternary	Holocene	0–0.01	Humans dominant; stable, post-ice-age climate
	Neogene	Pleistocene	0.01–2.6	Ice-age; humans appear (0.38 mya)
		Pliocene	2.6–5.3	Warm; pre-ice-age world
		Miocene	5.3–23.7	Grasslands expand immensely; intense worldwide volcanism
	Paleogene	Oligocene	23.7–36.6	Large animals dominate, occupying niches previously held by large dinosaurs
		Eocene	36.6–57.8	Largely tropical world climate; ended by global cooling resulting in mass extinction, particularly in oceans
		Paleocene	57.8–65.5	First large mammals emerge; slow recovery from mass extinction in the End-Cretaceous period
Mesozoic	Cretaceous		65.5–144	Emergence of flowering plants; hotter world; shallower seas; roaming of large dinosaurs; period ends with catastrophic mass extinction
	Jurassic		144–208	First birds appear; dinosaurs emerge and begin to dominate; formation of the Atlantic Ocean; Pangaea supercontinent breaks apart
	Triassic		208–245	Pangaea supercontinent begins to break apart; small mammals begin to emerge; dinosaur ancestral reptiles predominate
Paleozoic	Permian		245–286	Global tropical climate; giant insects, ferns, and mosses; Pangea supercontinent forms; period ends in significant climate cooling and subsequent mass extinction
	Carboniferous		286–360	First reptiles and cartilaginous fish develop; great coal-forming swamps and forests cover a largely tropical world
	Devonian		360–408	First land animals emerge, dominated by amphibians; fish dominate in coastal oceans
	Silurian		408–438	Plants and fungi emerge and thrive on land; first insects and jawed fish appear
	Ordovician		438–505	First vertebrates appear and evolve; period ends with a major extinction event
	Cambrian		505–570	Trilobites and other invertebrates dominate in oceans; oxygen depletion suspected to be related to mass extinction events ending period
Precambrian	Proterozoic Eon		570–2500	First multicellular life appears; soft-bodied organisms begin to evolve near the end of the period; atmosphere 3.0 (oxygen-rich atmosphere) develops
	Archaean Eon		2500–3800	First unicellular life evolves; cyanobacteria, notably, appears
	Hadean Eon		3800–4600	Earth consolidated from nebular debris; atmosphere 1.0 and 2.0 emerge and evolve; oceans are formed

Case Study

2.1 Major Environmental Changes Over Earth's Long History

If Earth's present-day surface environment results from a fine-tuned homeostasis of well-balanced feedbacks related to air, sea, and land, why is the geological record full of evidence showing that wide variations in the steady state of our natural environment have taken place over time? Why has our world had fabulously warm, CO_2-rich intervals, such as the Cretaceous period (66 to 145 million years ago), when tropical forests grew as far north as Greenland and dinosaurs romped within a few hundred kilometers of the South Pole? Why, also, has Earth experienced episodic deep freezes until quite recently, in some instances practically smothering the entire world with thick sheets of ice?

One answer lies in the description of the carbon cycle. Periodically, episodes of intensive volcanic activity have occurred, greatly increasing the concentration of CO_2 in the atmosphere and oceans. This volcanism has accompanied increased rates of seafloor spreading in response to rapid overturning of hot mantle rock below, as was the case for instance during the Cretaceous period. Hotter ocean floors around divergent plate margins uplifted the seafloor (think isostasy) and that meant that global sea levels rose, flooding continental margins and even low-lying interiors. The American Midwest during the Cretaceous was largely submerged by a shallow ocean–the Western Interior Seaway–corresponding to rapid rifting and opening of the young Atlantic Ocean to the east.

Toward the close of Cretaceous times around 66 million years ago, intensive volcanism in what is now the Deccan Traps of southwestern India, one of the largest volcanic features on Earth, also contributed to rapid global warming. Much of the CO_2 from this event may not have come from erupting volcanoes directly, but rather leaked more slowly from widespread, underlying magma bodies. Paleontologists debate the extent to which Deccan volcanism led to the demise of the dinosaurs. Likely as not, the strike of the giant Chicxulub meteor on the opposite side of Earth (eastern Mexico) dealt the death blow in a single fiery day, throwing survivors into years of frigid darkness; but the Indian volcanic activity could well have destabilized ecosystems at this critical time and facilitated the development of new species to replace the dinosaurs once their dominion collapsed (**Figure 2.69**).

Plate tectonics can be a dramatic geospheric contribution to Earth's changing environments by repositioning continental landmasses. The circulation of currents in the world's oceans is constrained and guided in large part by the position of landmasses. Oceanic circulation was much different 250 million years ago, when the continents of the world came together incrementally via plate convergence to assemble a single, gigantic landmass, or supercontinent–**Pangaea** (pan-GUY-ah) (see Figure 2.28). No Atlantic Ocean and related global conveyor current existed then, and deep-water formation may not have taken place. The pattern of oceanic currents was much simpler and their movement notably slower than today, helping sustain a much warmer world. Meanwhile, the vast continental interior of Pangaea included an enormous wind-blown, hyper-arid desert larger and harsher than any presently existing.

The world has since cooled dramatically as rifting and seafloor spreading have shifted the continent of Antarctica, a piece of former Pangaea, into a critical new geographical location. Antarctica did not exist at the South Pole 180 million years ago. Only ocean existed there then, with polar water kept much warmer than it is now because of the very different marine circulation. Instead, Antarctica was located several thousand kilometers closer to the equator, near the center of a much larger supercontinent, **Gondwana** (the southern part of Pangaea), which has since split into the continental "fragments" we now know–in addition to Antarctica–as Africa, South America, India, and Australia.

As Gondwana broke up, Antarctica began drifting south. But 100 million years ago, it was still covered with luxuriant vegetation, fossil leaves of which can be found in sedimentary formations of the frigid Transantarctic Mountains. Around 35 million years ago, Antarctica began to get colder as its ponderous drift carried it closer to the pole. More

Esteban De Armas/Shutterstock.com

Figure 2.69 The Chicxulub asteroid impact at the end of Cretaceous times: a bad day for dinosaurs (and most other large species on Earth), but in retrospect, a good day for us!

importantly, it broke free of South America, allowing a **circumpolar marine current** to develop that isolated the continent from the climate-moderating influence of warmer ocean waters (**Figure 2.70**). The ice-albedo feedback activated with a vengeance, and Antarctica's dramatic refrigeration soon chilled Earth's climate worldwide, bringing us to modern, cooler climatic conditions. Today, about 29 million cubic kilometers (7 million cubic miles) of glacial ice almost completely cover this continent, in some places weighing down the underlying bedrock so much that it is pushed below sea level. Antarctica hosts 90% of Earth's current ice supply, enough to raise sea levels as much as 70 meters (230 feet) if it entirely melted. How different the world would be if this were the case!

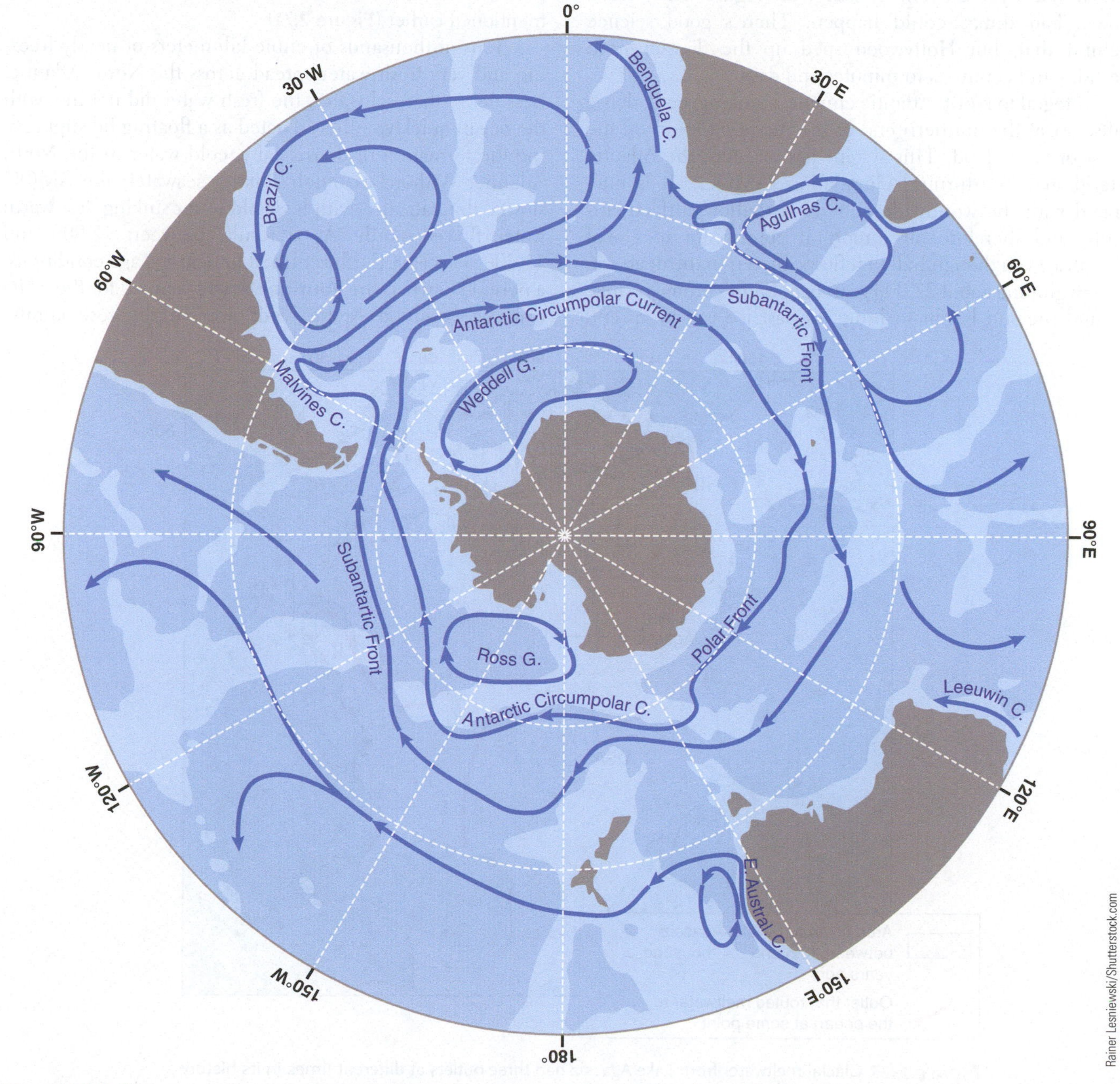

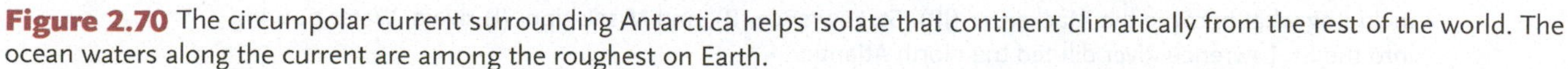

Figure 2.70 The circumpolar current surrounding Antarctica helps isolate that continent climatically from the rest of the world. The ocean waters along the current are among the roughest on Earth.

Case Study

2.2 Another Major Environmental Change About to Happen?

Have you seen the *Day After Tomorrow*, a popular sci-fi movie with the prerequisite tragedy and teen romance? Hollywood got the idea right and the science wrong. If the global conveyor current system (see Figure 2.53) slows down, bad things could happen. There's good science behind that, but Hollywood sped up the disaster from decades and centuries to minutes and days.

Integral to North Atlantic climate is sinking, cold, dense saltwater at the northern end of the Atlantic Ocean off the coast of Greenland. This system has an alias, the Atlantic Meridional Overturning Circulation (AMOC). It is integrated with the worldwide pattern of shallow surface currents, and therefore an important contributor to global climates, even though half of it flows unseen in the deep sea.

Beginning about 22,000 years ago, Earth's orbital change around the sun began to bring the last ice age to an end (Chapter 3). Earth grew warmer and marine currents such as the tropical Gulf Stream started flowing through the North Atlantic. The transition was not a steady one, however. As the ice sheets retreated in Eastern North America and Europe, the enormous proglacial lakes along their margins swelled in size and occasionally drained quite rapidly, perhaps over a period of months in the case of Lake Agassiz, the inland sea mentioned earlier (**Figure 2.71**).

Tens of thousands of cubic kilometers of nearly freezing and very fresh water spread across the North Atlantic. Less dense than saltwater, the fresh water did not mix with the ocean quickly; rather, it acted as a floating lid suppressing the formation of dense, salty, cold water in the North Atlantic. Without as much sinking seawater, the AMOC slowed dramatically–with less cold water sinking, less warm water flowed north. As a result, between 12,900 and 11,700 years ago, Earth returned to near-ice age conditions, a period known as the **Younger Dryas** (named for the *white dryas*, a common Arctic wildflower in the rose family)

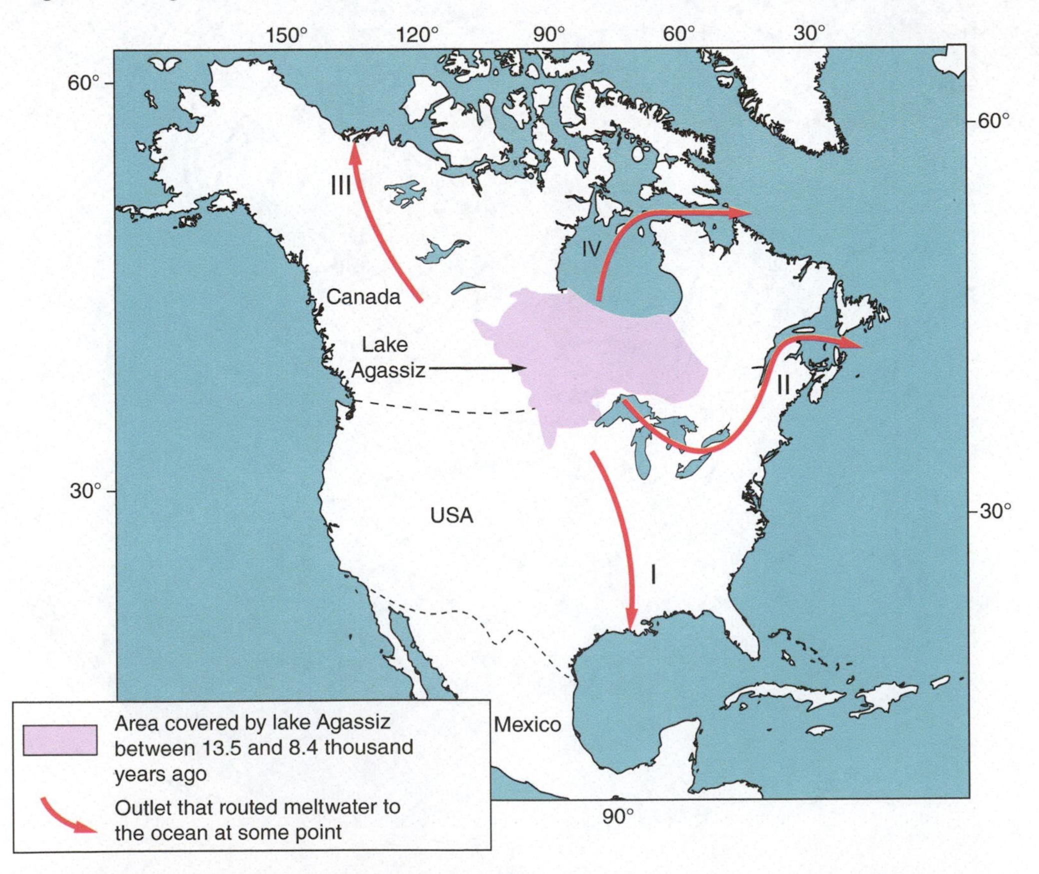

Figure 2.71 Glacial meltwater from Lake Agassiz had three outlets at different times in its history, including what became the Mackenzie (III), St. Lawrence (II), and Mississippi (I) rivers. Water pouring into the St. Lawrence River diluted the North Atlantic.

upwelling of deep, cold, replacement water during La Niñas. The wind-blown warm water flows toward Australia and East Asia, bringing with it increased chance of cyclone activity and heavy rainfalls in those regions. At the same time, periods of drought in western North America and South America intensify, a side-effect that brings great hardship to rural farmers and pastoralists, not to mention raising food prices for those who depend upon them. Recent research suggests that a greatly weakened global conveyor would throw the world into a more persistent La Niña condition. We seem to be entering unprecedented climate territory.

Figure 2.72 Dryas in flower on the Norwegian Arctic island of Svalbard.

(**Figure 2.72**). Only after 11,700 years ago did our modern warm climates establish themselves on both sides of the North Atlantic.

As global warming heats the atmosphere and rates of ice melting in Greenland accelerate, could the Younger Dryas scenario unfold once more? The Gulf Stream and global conveyor are both weakening at present, and many scientists are concerned, though there is much scientific debate about the degree to which we should worry and be prepared for major trans-Atlantic change in ocean currents. Remember, warm ocean currents that in our times contribute to moderate, hospitable climates in the Eastern United States, Canada, and Western Europe were displaced by the massive influx of frigid, fresh meltwater. Slow down the AMOC and, ironically, the North Atlantic region could get colder as the planet warms.

The concern is not simply restricted to North Atlantic nations, however; a weakened or stagnant global conveyor would spell trouble–certainly big changes–for people all over the world. A case in point is Australia, which experiences often catastrophic flooding during La Niña conditions. La Niñas are complex, wind-related weather patterns involving cooling of equatorial waters in the Eastern Pacific once every several years (**Figure 2.73**). They alternate with warm-water events called El Niños, which affect the same water. Winds blowing off South America drive shallow seawater from near shore, causing

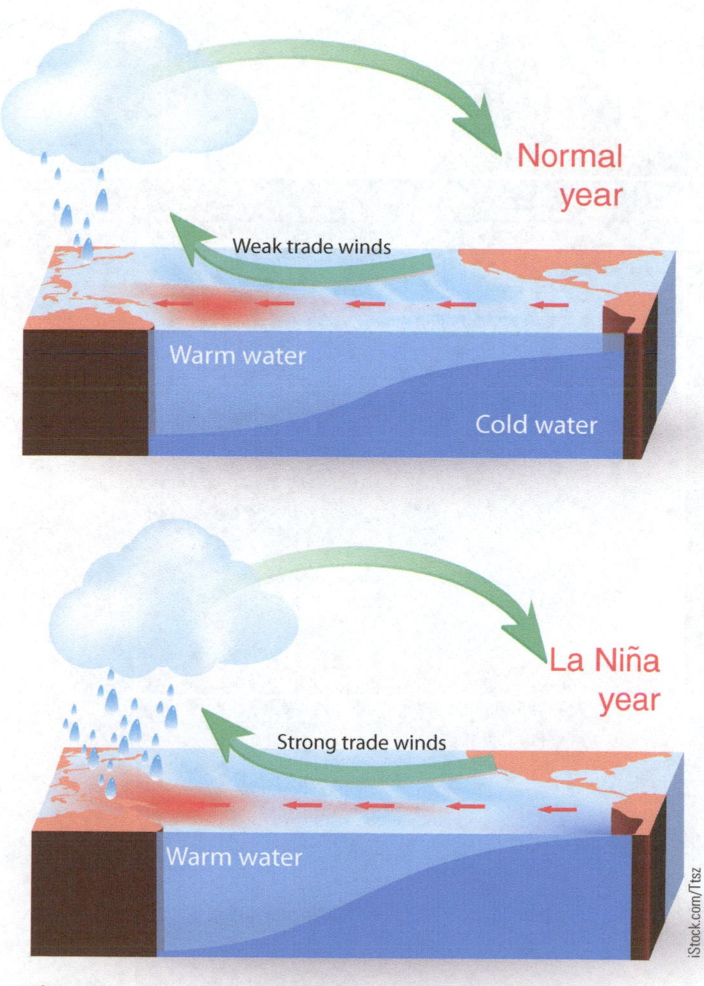

Figure 2.73 La Niña conditions promote upwelling of cold marine waters along South America's Pacific equatorial coast, with worldwide consequences.

Photo Gallery

Earth's Spheres

Earth is an amazing, dynamic system wherein the atmosphere, hydrosphere, geosphere, cryosphere, and biosphere never stand still. Rather, energy and mass are both stored in and flow through these systems, creating the planet we call home. Here are a few images of some of the most dramatic and important exchanges between these spheres that define Earth.

1 Cryosphere. Sea ice breaking up in spring near Kulusuk, Greenland. The white ice reflects most incoming sunlight. The dark water beneath does not.

iStock.com/Steve_is_On_Holiday

2 Atmosphere. A linear thunderstorm (known as a *derecho*, from Spanish for "straight") during a severe weather outbreak in Kansas.

iStock.com/Mdesigner125

3 Hydrosphere. Sometimes the hydrological cycle comes home to roost: a bathroom filled with sediment-laden floodwater.

iStock.com/Onurdongel

A thunderstorm approached Badlands National Park, South Dakota, United States.

Shawn Rundblade/Getty Images

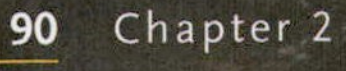

iStock.com/Gadaian

4 **Geosphere.** Calbuco Volcano in Chile erupts, spreading fertile volcanic ash over the countryside and into the atmosphere.

Fossil & Rock Stock Photos/Alamy Stock Photo

5 **Biosphere.** A fossil fern preserved in coal. Burning this coal allows ancient carbon to reenter our atmosphere.

Summary Earth as a System

1. Earth as a System

LO1 Define the Earth System and explain why it is a favorable environment for life

a. The Spheres

The Earth System is composed of five "spheres" that link the rocky and molten Earth interior (geosphere), the air (atmosphere), bodies of water (hydrosphere), structures of ice (cryosphere), and the living realm (biosphere). These linkages create a balance driven by numerous feedbacks which achieve environmental stability or homeostasis (though it is never completely achieved).

b. Energy and Temperature in the Earth System

There are two sources from which the Earth acquires energy. The first is internal, from within the Earth's deep interior, where heavy uranium and thorium atoms are continuously decaying to more stable elements (radioactive decay). Heat is released as a byproduct of these reactions and slowly escapes to the Earth's surface. The second, solar energy from the Sun, is external. These inputs of energy to the Earth System create livable, temperate conditions.

Study Questions

i. What are the five interconnected spheres which make up the Earth System? What does each sphere account for on Earth?

ii. What are the energy sources to the Earth System?

2. The Geosphere: Land Building on a Large Scale

LO2 Explain the meaning of the "plates" in Earth's crust, and why their behavior is important to us

a. The Origin of the Geosphere

Over 4.5 billion years ago, Earth was little more than a molten body with oceans of churning lava at the surface. Frequent bombardments by bolides added new materials like water and detached pieces of the Earth, from which our Moon would form. Gravity slowly pulled the densest materials toward the core. Less dense materials floated to the top forming the mantle and crust.

b. The Layered Earth

The Earth System is composed of many layers, with an outermost electromagnetic field originating from movement of electrically charged currents in the Earth's outer core. The innermost layer is the molten core, enclosed by the solid, magnesium- and silicate-rich rock mantle, the thickest layer making up Earth. The crust overlays the mantle, composed mostly of silica, alumina and alkali-rich rocks.

c. Ups and Downs of a Dynamic Geosphere

The uppermost mantle and crust, defined as the lithosphere or "rocky sphere," floats above the asthenosphere, or the deeper mantle rock. Isostatic equilibrium seeks to find a balance in the density of the floating lithosphere with the asthenosphere below; this explains the formation of deep ocean basins due to the high density of oceanic crust relative to continental crust.

d. Plate Tectonics

Interactions between the tectonic plates along their boundaries (divergence, convergence, and transform) result in the formation of mountains, volcanoes, and island groups, and cause earthquakes.

Study Questions

i. What is "isostasy" and how does it help explain the differing elevations on the surface of Earth?

ii. What are the Earth's layers and what are they each composed of?

3. The Atmosphere: Earth's Restless Outer Envelope

LO3 Explain why winds blow in the directions that they do on different parts of Earth

a. Atmospheric Structure

Like Earth's interior, the atmosphere is layered. The troposphere extends from the Earth's surface to an altitude between 8 to 18 kilometers. It holds 99% of the water vapor and 80% of the total mass of the atmosphere. Above is the stratosphere, separated by the tropopause. The stratosphere is much more stable than the troposphere, albeit much cooler and windier. Above the stratosphere is the mesosphere, then the thermosphere, where space "begins" at 100 kilometers above Earth's surface.

b. Patterns of Tropospheric Circulation

Air pressure differences, and the Earth System's quest to equalize atmospheric temperature and pressure globally, create familiar weather conditions in the troposphere. We experience the movement of air from high to low pressure as wind. Warm air rises as the movement of molecules increases, making the air less dense, while cold air sinks as the molecules slow down and the air becomes more dense.

c. Weather

Large air masses, called fronts, explain many changes in weather globally. Cold fronts wedge beneath warm air masses, causing them to rise forming clouds. Heavy rainfall, snowfall, and hail which come and go quickly are often the result of this phenomenon. Encroaching warm fronts, alternatively, rise gradually over cooler air masses, leading to prolonged steady rainfalls. Mountain systems block airflow, and this orographic effect often means that climate on the upwind and downwind side of mountain ranges differs.

d. Climate

Climate describes the general types and patterns of weather that characterize a region. Temperature, precipitation, latitude, and proximity to large bodies of water all influence a region's climate.

Study Questions

i. What is the difference between weather and climate?

ii. What are the differences between weather caused by encroaching cold fronts versus encroaching warm fronts?

4. Hydrosphere and Cryosphere

LO4 Describe the connection between oceanic circulation and global climates

a. The Hydrological Cycle

The hydrological cycle is driven by sunlight as solar energy produces evaporation. Water enters the atmosphere as water vapor, adding to the air's humidity. This water vapor can condense and fall as rain or snow, or form mist, fog, or clouds. Precipitation returns evaporated water to the earth, flowing into rivers, oceans, lakes, and streams. Some water percolates into the Earth's subsurface, replenishing groundwater supply.

b. The Ocean

Salt has accumulated in the oceans from weathering of rocks on land. This salt and its impact on density also contributes to the ocean's layering. The epipelagic ocean is the surface layer, which is penetrated by sunlight and contains about 90% of ocean life. Temperature doesn't change much with depth in this layer. The deep ocean is made up of many layers and is separated from the epipelagic ocean by the boundary called the thermocline. In the thermocline, ocean water temperature drops significantly with increased depth.

c. Shallow Ocean Circulation

Shallow ocean circulation is driven by the wind, forming waves locally and driving global-scale ocean currents called gyres. Eddies branch off from mainstream gyres, allowing nutrients to be distributed from nearshore regions into other parts of the ocean. The center of gyre currents are large areas of calmer water, stirred by frontal systems and cyclones.

d. Ocean Circulation and Climate

The ocean circulation impact on climate is directly related to the high heat capacity of water. Land has a relatively lower heat capacity, so it warms more quickly and loses heat faster as well. Thus, inland regions are typically hotter than coastal regions during the day, and relatively cooler in the evening. Wind off the cool water keeps coastal areas cooler than inland areas during the day, while at night, solar radiation is released from water slowly maintaining a higher temperature late into the evening. Ocean circulation patterns move cool and warm waters globally, impacting climate in other coastal regions around the world.

e. Deep Ocean Circulation

Deep ocean circulation is slow in comparison to shallow circulation and is driven by density-related differences in saltiness and temperature. Water cools along the icy Antarctic coast and sinks into the deep ocean where it begins to move northward. The cold, shallow water is replaced by warmer surface water from lower latitudes, connecting the shallow and deep marine circulations.

f. The Hydrological Cycle, Underground

Groundwater is water that moves slowly through rock fractures and sediment underground. It is recharged where rain, streams, and snowmelt percolate through the soil and sediment. Near the surface, pore spaces are partially filled with water in the zone of aeration. Beneath the water table is the zone of saturation, where water fills every pore space in the subsurface material. When the water in the zone of saturation is abundant (extractable), it is called an aquifer.

g. Cryosphere-Frozen Water

The majority of the Earth's freshwater supply is locked up in ice, in glaciers, ice sheets/shelves, and sea ice. The presence of the cryosphere is a consequence of less sunlight reaching high latitudes, the freezing point of water, and the concentration of greenhouse gases in the atmosphere. Ice and snow contribute to the ice-albedo effect, a negative feedback that reflects solar energy back into space, keeping Earth cooler.

Study Questions

i. Name two specific examples of interconnectivity between the five spheres mentioned in this chapter.

ii. Using one example from this chapter, explain how the Earth System's tendency toward homeostasis plays a role in making Earth a habitable planet.

5. Biogeochemical Cycles

LO5 Define the concept of biogeochemical cycling and provide one example

a. The Nitrogen Cycle

In order for nitrogen to become bioavailable to plants and animals, gaseous nitrogen (N_2) must be transformed to nitrite or nitrate, by nitrogen fixation. Nitric acid introduced to the environment by acid rain can be oxidized to form nitrate as well. Denitrification transforms nitrite and nitrate back to nitrogen gas. Denitrifying bacteria also return nitrogen back to the atmosphere in the form of nitrous oxide and ammonia.

b. The Phosphorus Cycle

The phosphorus cycle, unlike the nitrogen cycle, is not dependent on the atmosphere. Phosphorus is sourced from phosphorus-bearing minerals like apatite. When weathered, these minerals release phosphorus to the soil, where it can be taken up by organisms. Thus, the growth and decay of plants and animals primarily drive the phosphorus cycle.

c. The Carbon Cycle

The deep carbon sub-cycle begins far in the Earth's interior, where the majority of the carbon on Earth is stored. When volcanos erupt, carbon dioxide is released to replenish what is extracted from the atmosphere for biological purposes. The shallow carbon sub-cycle takes place in all of the Earth System spheres, starting with carbon dioxide uptake for photosynthesis and release from respiration. Atmospheric carbon dioxide dissolves in seawater and returns to the air if water temperatures rise. However, the whole of the carbon cycle is massively disrupted by humans, given the immense volume of carbon dioxide that is released anthropogenically.

Study Questions

i. What are some of the chemical reactions which drive the nitrogen cycle?

ii. In what ways have humans impacted the carbon cycle and what are the results of these impacts?

6. Biosphere

LO6 Describe the importance of the biosphere in the development of Earth's environment from earliest time

a. Earth's Evolving Atmosphere: the Living Connection

Biological processes play a role in the regulation and building of the Earth System we know today. An example of this connection is how warming weather and increased rainfall can result in greater volumes of vegetation growth. With more abundant vegetation performing photosynthesis, greater amounts of carbon dioxide are removed from the atmosphere, reducing the temperature increase that might take place otherwise.

b. Life and Oxygen

Originating from cyanobacteria, oxygen began to accumulate in Earth's atmosphere as a byproduct of photosynthesis, once all the available iron in the environment had been oxidized. This allowed new life forms to evolve, using free oxygen to metabolize glucose and make energy. Oxygen also reacted with sunlight to generate ozone and the ozone layer, which filters out the most dangerous solar radiation.

c. Geologic Time: the Importance of Biology

Changes in the biosphere, like population and species characteristics, ecological evolutions, and mass extinctions, serve as the basis for the geologic time scale.

Study Questions

i. How does the biosphere impact the regulation of the Earth System and its homeostasis?

ii. What are the characteristics of the three atmospheres, Atmosphere 1.0, 2.0, and 3.0?

Key Terms

Adiabatic heating
Air masses
Albedo
Algal blooms
Alpine glaciers
Apatite
Aquicludes
Aquifer
Aquitards
Asthenosphere
Atlantic Meridional Overturning Circulation (AMOC)
Atmosphere
Aurora
Biogeochemical cycles
Biosignatures
Biosphere
Brine exclusion effect
Carbon cycle
Chinooks
Circulation cells
Circumpolar marine current
Climate
Cold front
Compressive
Continental crust
Continental ice sheet
Convection cell
Convergent boundaries
Coriolis effect
Crust
Crustal root
Cryosphere
Deep carbon sub-cycle
Deep ocean
Denitrification
Differentiation
Divergent boundaries
Earth System
Eddies
Epipelagic ocean
Feedbacks
Firn
Fronts
Garbage patches
Geologic timescale
Geosphere
Geothermal gradient
Glacier
Global conveyor
Gondwana
Goldilocks environment
Guano
Haber-Bosch process
Heat capacity
Homeostasis
Horse Latitudes
Humidity
Hydrological cycle
Hydrosphere
Ice cap
Ice shelves
Ice-albedo effect
Infrared radiation
Intracontinental mountain chains
Island arcs
Isostasy
Isostatic equilibrium
Jet streams
Juvenile water
Karman Line
Köppen-Geiger classification
Lenticular clouds
Lithosphere
Magnetic fields
Magnetic reversals
Mantle
Marine upwellings
Mesosphere
Meteoric water
Mid-ocean ridges
Moraine
Natural systems
Negative feedback
Nitrification
Nitrogen cycle
Nitrogen fixation
Oceanic crust
Oceanic gyres
Orographic
Ozone
Ozone layer
Permafrost
Petrologists
Polar air
Polar front
Polarity
Positive feedback
Proglacial
Radioactive decay
Radiogenic
Recharge
Runaway feedback
Santa Ana winds
Sea ice
Seafloor spreading
Seismologists
Sequestration
Shallow carbon sub-cycle
Silicates
Spectroscope
Stratosphere
Subduction
Tectonic plates
Tension
Terminus
Thermocline

Thermohaline
Thermosphere
Tipping points
Trade winds
Transform boundaries
Transform fault
Tropopause
Troposphere
Upwells
Warm front
Weather
Westerlies
Wind
Younger Dryas
Zone of aeration
Zone of saturation

"The paleoclimate record shouts to us that, far from being self-stabilizing, the Earth's climate system is an ornery beast which overreacts even to small nudges. Climate is an angry beast and we are poking at it with sticks."

—Wallace Smith Broecker

3 Climate Change

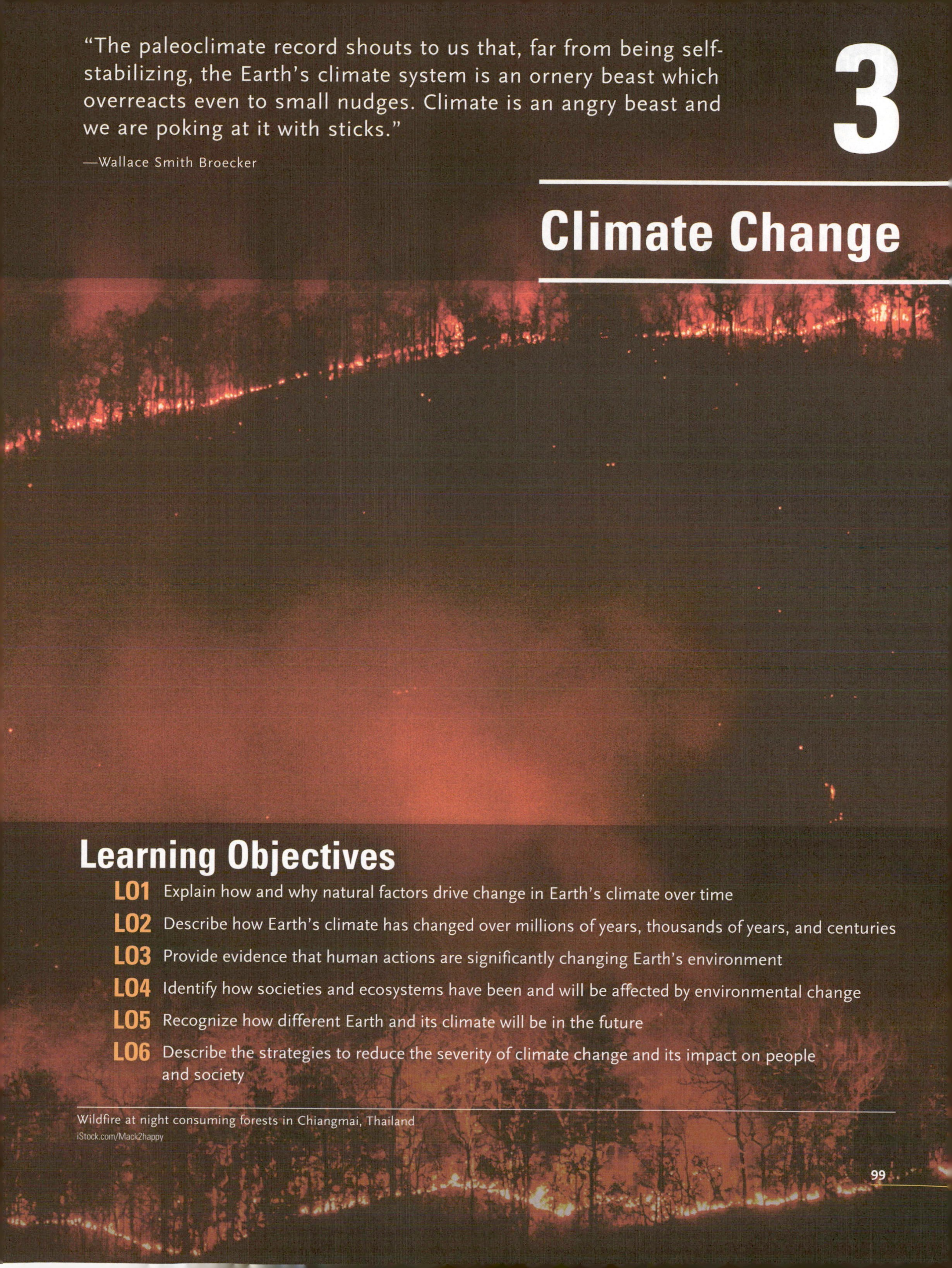

Learning Objectives

LO1 Explain how and why natural factors drive change in Earth's climate over time

LO2 Describe how Earth's climate has changed over millions of years, thousands of years, and centuries

LO3 Provide evidence that human actions are significantly changing Earth's environment

LO4 Identify how societies and ecosystems have been and will be affected by environmental change

LO5 Recognize how different Earth and its climate will be in the future

LO6 Describe the strategies to reduce the severity of climate change and its impact on people and society

Wildfire at night consuming forests in Chiangmai, Thailand
iStock.com/Mack2happy

Changing Climate and the Fate of Societies: Past and Present

The United States Bureau of Reclamation

Figure 3.1 A horseback rider in a prehistoric Hohokam canal.

In the mid-nineteenth century, when European American settlers first began cultivating the arid, irrigated plain of south-central Arizona, they quickly discovered the ruins of an ancient society–the Native American Hohokam culture. The Hohokam flourished between 300 and 1500 CE. They were master canal builders and engineers, tapping the sparse water supply of the Salt River to support widespread cultivation of squash, beans, agave, amaranth, and other crops in an arid landscape (**Figure 3.1**) The Hohokam dug more than 1,600 kilometers (1,000 miles) of irrigation channels, and many of these were simply cleaned out and repaired by the early American farmers to bring water to their fields. Hohokam communities were built of adobe, and many included ball courts and ceremonial platform mounds, like those found in the cities of the Aztecs. The Hohokam maintained an extensive trade in parrots, bells, and other commerce with Mexico. They prospered for more than four centuries.

But then their society mysteriously collapsed, leaving no direct testimony of what had happened. The metropolis of Phoenix has since risen in their homeland, building atop their ancient town sites and irrigation works (**Figure 3.2**). The name *Hohokam* is from the Pima language and means "all used up," perhaps indicative of an environmental catastrophe, although not necessarily one that occurred all at once. In fact, a gradual aggregation of environmental problems may have been more effective than any single event in their downfall. A period of steady, reliable,

iStock.com/Welcomia

Figure 3.2 The Phoenix metro area sprawls across the desert of southwest Arizona, the lands of the Hohokam.

iStock.com/DougVonGausig

Figure 3.3 Canals of the Central Arizona project bring water from the Colorado River to the Phoenix metro area.

annual rainfalls characterized the ascendancy of the Hohokam, who with great efficiency learned how to exploit this natural pattern to their benefit. The weather then became erratic, with intensive flooding and erosion destroying irrigation channels. There is some indication that the people began protecting themselves, migrating to a few large, multistory "fortress" communities. Sophisticated traditional ceramic art ended, and new canal works were built farther upslope in the Salt River watershed, depriving users downstream of their water supply. These changes signaled a society that was suffering from periodic resource scarcity. Research by anthropologist Donald Graybill of the University of Arizona suggests that climate change was a key factor.

Could modern-day Phoenix one day share a similar fate? With an exploding population, water supplies continue to be a major concern for the city. The onset of drought in recent years, exacerbated by forecasts of worse to come by climate change scientists, suggests history repeating itself in southern Arizona. Unlike the Hohokam, who depended on water from the Salt River, aqueducts have been constructed to support the much more populous modern city of Phoenix from watersheds as far as 600 kilometers (375 miles) away (**Figure 3.3**). Nevertheless, the differences with Native American predecessors are merely ones of technology and scale. As Phoenix archaeologist Tom Wright reminds us, "The Hohokam were a hydraulic society, and so are we ... We depend on the presence of water, the storage of water, the transport of water, and as long as the water is there and can serve our needs, we're fine. But if the water is not falling on the watershed, or if our needs start to outstrip what is available, that's a problem."

questions to ponder

1. How might urban growth, development, or governance be improved to provide for a flexible, successful social response in the face of increasing prospects for drought and a scarce water supply in the southwestern United States?
2. How should the "limit of growth" for a city such as Phoenix be defined and enforced?

3.1 Why Earth's Climate Changes (LO1)

Earth's climate has always changed. It has cooled and warmed, moistened and dried for billions of years before the first human walked this planet. Ice sheets came and went, and as a result sea level rose and fell as much as 130 meters (440 feet). Today, the 8 billion people who inhabit planet Earth are driving environmental change at a rate far faster than nature typically has in the past. On a global scale, the most widespread and pressing environmental change is occurring in our climate system as humans change the composition of Earth's atmosphere in a global experiment with no natural analog. In this chapter, we'll examine how and why climate changed before, how it is changing today, and what the future likely holds. Based on prior experiences of human societies, we examine the effects such changes are likely to have in the future.

3.1.1 Atmospheric Composition

Earth's atmosphere consists mostly of a colorless gas we can neither smell nor taste: nitrogen (N), which makes up 78.08% by volume of all the air we breathe. Oxygen (O) is the next most abundant gas in Earth's atmosphere, making up 20.95% of its volume, followed by argon (Ar; 0.93%), **carbon dioxide** (CO_2; 0.042%), and trace amounts of other gases (**Figure 3.4**).

Among many essential services the atmosphere provides for our natural environment, two other constituents are especially noteworthy: ozone (O_3), which screens out deadly ultraviolet radiation from the Sun high in the atmosphere; and **greenhouse gases**, including water vapor (H_2O), CO_2, and **methane** (CH_4), which keep Earth's atmosphere and surface warm enough for water to stay liquid and life to flourish. The explanation for this warmth is a phenomenon that has made much news in recent years: the **greenhouse effect** (**Figure 3.5**). Most of the media coverage of greenhouse gases is negative, but, in fact, we need these gases and the greenhouse effect for our survival. The problem the climate is experiencing now is "too much of a good thing."

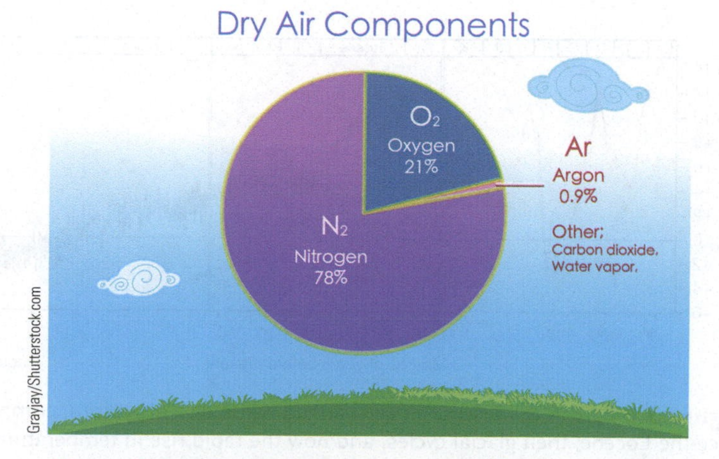

Figure 3.4 Earth's atmosphere is mostly nitrogen and oxygen, with lesser amounts of trace gases.

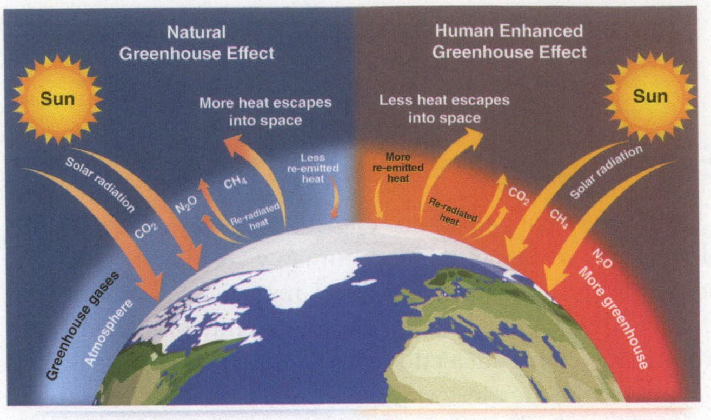

Figure 3.5 Human emissions of greenhouse gases have amplified the natural greenhouse effect trapping more energy in the Earth system by reducing the percentage of incoming radiation escaping back into space.

To understand how the greenhouse effect works, consider that the energy reaching Earth's surface from the Sun arrives mostly as short-wavelength radiation, some of which we characterize as visible light. On average, about a third of the short-wavelength solar radiation reaching Earth is reflected directly back into space, with the oceans reflecting much less (no more than about 15%) than fresh snow and ice (90% to 100%). Albedo is the term that quantifies this reflectivity (see Chapter 2). The darker an object, the less its reflectivity and lower its albedo (**Table 3.1**).

What energy is not reflected is absorbed, even by you standing under a bright Sun; then it is reradiated later, but mostly in a different and longer wavelength, the **infrared**, which we feel as heat. All that heat would escape back into space, except that it is intercepted in the atmosphere by greenhouse gases, which reabsorb it and re-radiate some of it back to Earth's surface–**greenhouse warming**. The extra boost of greenhouse warming makes a huge difference in overall air temperature. Without it, Earth's average air temperature would drop to between −22°C and −36°C (−9°F and −34°F), well below freezing. With it, Earth's average air temperature is a much more comfortable 14.5°C (58°F).

Table 3.1 Some Representative Albedos

Types of Surface	Albedo
Forests (coniferous and deciduous)	0.05–0.18
Grasslands	0.25
Sandy deserts	0.40
Fresh concrete	0.55
Earth, overall average	0.30
Moon	0.12

The most important greenhouse substances in the atmosphere are not the most abundant. Water vapor contributes most to greenhouse warming (60% to 70%), followed by CO_2 (15% to 20%), a family of industrial gases called **halocarbons** (7% to 10%), CH_4 (5% to 7%), O_3, and nitrous oxide (N_2O; 1% to 3%). Halocarbons are molecules with a backbone of carbon to which there are attached halogens such as chlorine, bromine, and fluorine. The human contribution to warming is indeed significant, but most greenhouse warming remains natural. Human activities contribute to the increasing concentrations of all the greenhouse gases.

3.1.2 Orbital Cycles

The geological record is full of evidence that climate has changed in the past. There were fabulously warm times, such as the Eocene period (about 50 million years ago), when tropical forests grew as far north as Greenland. About 2.7 million years ago, the most recent ice ages began with Earth's climate swinging between cold, icy, **glacial periods** and warm, stable, **interglacial** periods like the one we live in today (**Figure 3.6**). What controls the coming and going of glacial and interglacial climates?

Insofar as the ice ages and warm intervals of the past few million years go, it is not the Sun controlling these changes. The Sun's output has been steady, varying less than a percent, and bearing little relationship to Earth's temperature

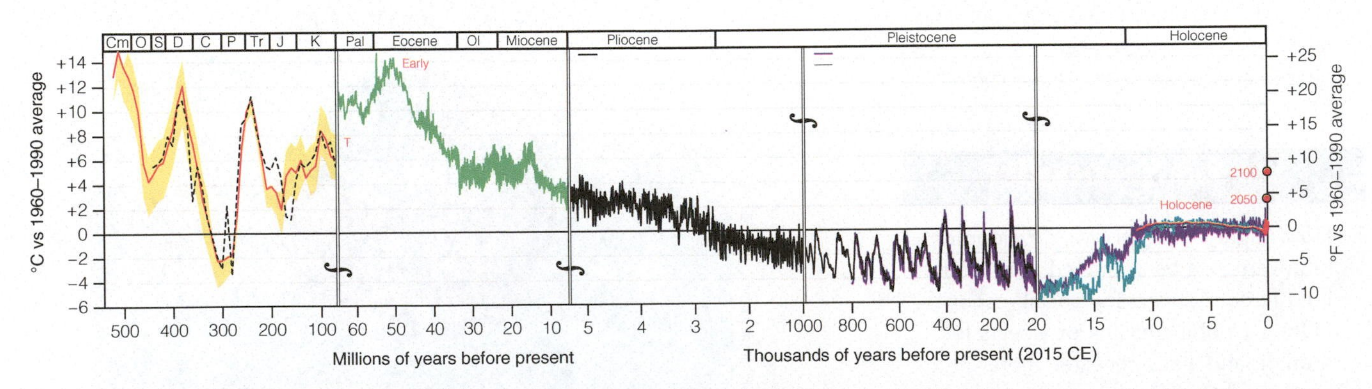

Figure 3.6 Earth's temperature has changed naturally. This record, compiled for the last half-billion years, clearly shows cooling since the Eocene, then glacial cycles, and now the rapid rise in temperature due to climate change.

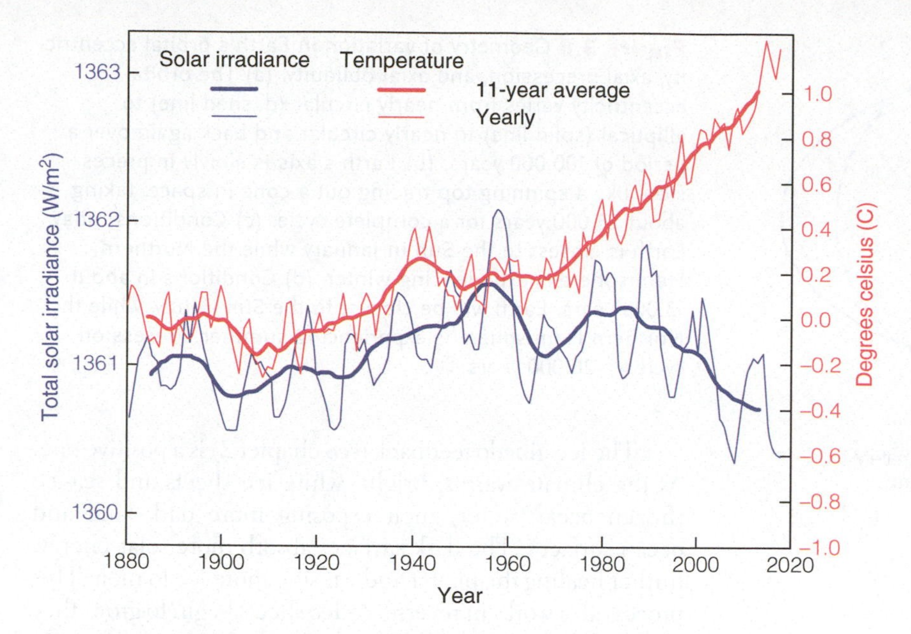

Figure 3.7 Solar activity, the amount of energy coming from the sun, has no relation to Earth's average temperature between 1880 and 2020. (Data from NASA-JPL/Caltech)

(**Figure 3.7**). Rather, it is the shape of Earth's orbit that matters. The shape of Earth's orbit changes cyclically over tens of thousands of years. These changes affect where and when solar radiation falls on the planet's surface (**Figure 3.8**).

Three variations occur as Earth follows its orbit:

- The **eccentricity**, or shape, of the elliptical path it takes in orbiting the Sun changes.
- The angle, or **obliquity**, of Earth's rotational axis relative to the planet's plane of revolution around the Sun changes.
- The rotational axis of the Earth **wobbles**, as a spinning top can wobble.

Let's explore the impact of these variations.

Earth's own orbit oscillates from a nearly circular path to a slightly drawn-out ellipse once every 100,000 years. Seasonal variation in temperature is greater when the orbit is more elliptical (or eccentric). Earth's present eccentricity is not great, resulting in only a 6% difference in the heat energy received between January and July. However, at Earth's most eccentric orbit, this difference is 20% to 30%. Think about how that change in energy could make the difference between snow melting every summer or sticking around to become a glacier, an accumulation of ice that flows under its own weight (Chapter 2). When such ice sticks around in the Canadian and Russian Arctic, at 65° north latitude, they become the birthplaces of ice sheets.

The obliquity or tilt of Earth's rotational axis varies between 21.8° and 24.5° over a 41,000-year cycle. The Earth's present tilt is in the middle of this range. This slight change causes variations in the temperature range between summer and winter. When Earth is at the minimum tilt, sunlight hits the polar regions at a higher angle, and the global seasonal temperature variation decreases worldwide.

Earth's axis presently points toward Polaris, the North Star, as generations of campers and wilderness scouts in the Northern Hemisphere have learned. But this is a coincidence that won't last long, geologically speaking. As Earth's axis wobbles, it will point toward other positions in space in years to come, gradually tracing out a circle, not to return to Polaris again for 26,000 years. The wobbling of the planetary axis means that the months in which summer and winter occur are constantly, slowly changing. About 13,000 years from now, the Northern Hemisphere summer will take place between December-February–the times northerners presently associate with freezing snow and short daylight hours, whereas conditions will be wintry in the Southern Hemisphere. Because the two hemispheres have different land-ocean configurations, they have very different average albedos, meaning that changes in seasonal timing greatly influence overall planetary climate.

Taken together, these three variations help explain the pacing of ice ages. But they are not the complete story. Feedbacks, which amplify or dampen changes in climate, come into play in this case, exaggerating the changes wrought by orbital forcings.

3.1.3 Climate Feedbacks

Feedbacks are critical to Earth's climate system. Recall from Chapter 2 that a positive feedback increases the intensity of the forcing that set it into motion in the first place. A negative feedback operates in the opposite sense, decreasing over time the intensity of the forcing that originally set it into motion. Sometimes a feedback is so powerful that it overwhelms others opposing it and thus becomes a runaway feedback. A runaway feedback quite often terminates in a new, stable set of conditions radically different from the ones with which we began. Tipping points are thresholds at which even small perturbations in a system will cause it to change significantly.

There are negative feedbacks that help explain why climate is generally stable. For example, heating of the atmosphere promotes more frequent and heavier rainfalls, and the growth of vegetation at higher latitudes and elevations in many parts of the world. As fresh plant growth incorporates CO_2 into plant tissues, CO_2 is taken out of the atmosphere, suppressing its role in global warming. If the atmosphere cools, this feedback can be thrown into reverse.

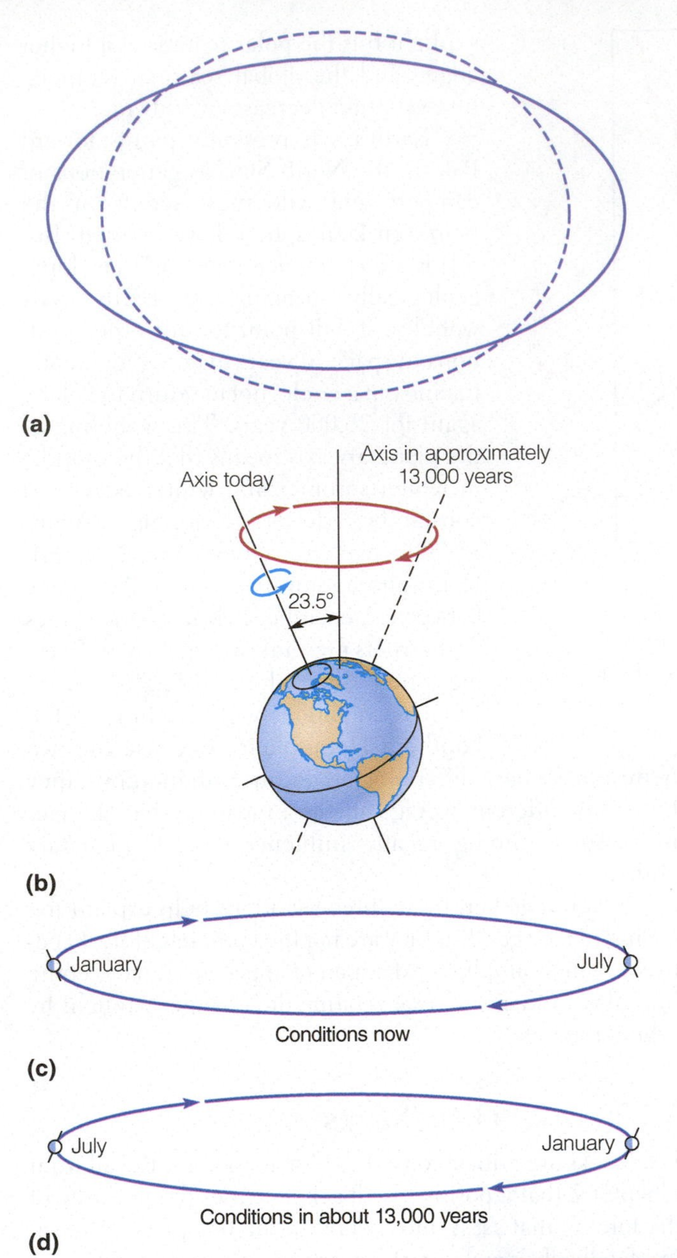

Figure 3.8 Geometry of variation in Earth's orbital eccentricity, axial precession, and axial obliquity. (a) The orbital eccentricity varies from nearly circular (dashed line) to elliptical (solid line) to nearly circular and back again over a period of 100,000 years. (b) Earth's axis is slowly in precession, like a spinning top tracing out a cone in space, taking about 26,000 years for a complete cycle. (c) Conditions today. Earth is closest to the Sun in January while the Northern Hemisphere is experiencing winter. (d) Conditions in about 13,000 years. Earth will be closest to the Sun in July, while the Northern Hemisphere is experiencing summer. Precession cycle = ~26,000 years.

The ice-albedo feedback (see chapter 2) is a positive one. As the climate warms, bright, white ice sheets and sea-ice (frozen ocean water) melt exposing more dark land and ocean surfaces. The dark surfaces absorb more solar energy, further heating the planet and causing more ice to melt. This process also works in reverse. As ice sheets begin to grow, they cool the climate, causing more snow to survive the summer, reflecting more sunlight and further cooling the planet. The ice-albedo feedback causes small changes in Earth's orbit to eventually generate massive ice sheets that cover much of the northern latitudes (**Figure 3.9**). In this way, an ice age begins.

3.2 Records of a Changing Climate (LO2)

The Earth records its climate history is a variety of places and ways. **Paleoclimatologists** (people who study past climates) spend a lifetime deciphering archives of Earth's history, including ice cores collected from glaciers around the world and muck dredged from the bottom of ponds. The data from thousands of climate archives, some stretching back millions of years, allow us to place human-induced climate change in the context of how Earth worked before humans.

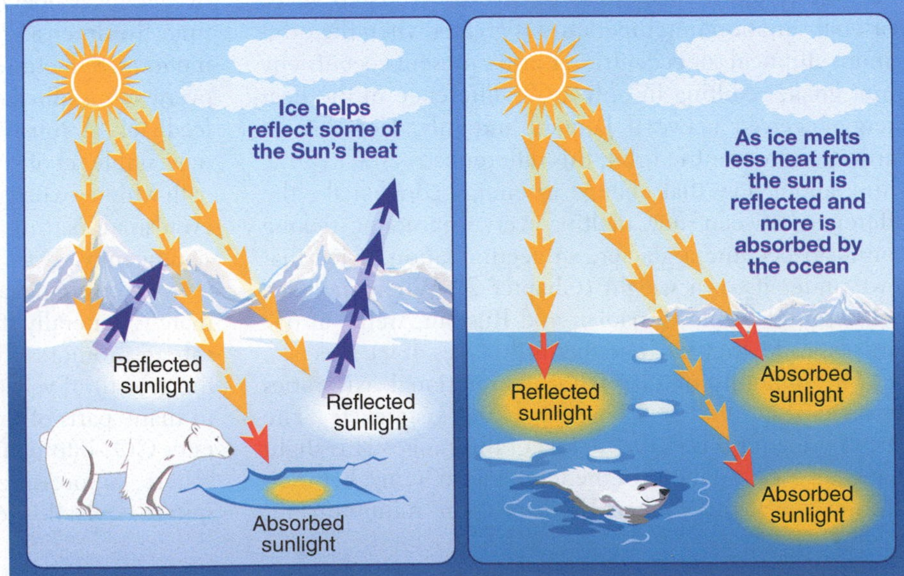

Figure 3.9 The ice-albedo feedback is all about reflectivity. Ice reflects most sunlight back to space. Seawater and land surfaces reflect less sunlight and thus, the planet warms, melting more ice. (Adapted from A Year in the Ice, University of Colorado Boulder)

3.2.1 The Retreat of Mountain Glaciers

Perhaps the most obvious example of current climate change is the worldwide retreat of **mountain glaciers**. Some 90% of Tanzania's Mount Kilimanjaro's ice has disappeared since the turn of the nineteenth century, and between 2014 and 2020, the summit's Furtwängler glacier lost 70% of its remaining ice mass. Of the 80 glaciers that graced Montana's Glacier National Park at the end of the Little Ice Age (~1850), only 26 remain based on aerial footage taken in 2015. Since these data were collected, some now may be too small to be considered glaciers anymore (**Figure 3.10**). Spring in the park starts 45 days earlier than in decades past, and warmer winters bring rain as well as snow. Mount Rainer National Park in Washington State, home of 29 glaciers and the most glaciated region within the contiguous United States, lost 20.1 square miles of glacial ice between 1896 and 2015.

Changes in mountain glaciers are significant because these bodies of ice are highly sensitive to temperature fluctuations. Their reaction time is much shorter than that of the vast ice sheets of Antarctica and Greenland; as a result, they provide significant information about twentieth-century climate change. At higher latitudes, Alaskan glaciers are in dramatic retreat. They are estimated to lose 66.7 gigatons (Gt) of ice per year, the most of any region on Earth. In fact, as of 2021, Alaska alone accounts for 25% of global ice mass loss. Using a laser altimeter attached to an airplane, a USGS research team flew over 67 glaciers in Alaskan mountains to compare their present surface elevations with those that were mapped in the 1950s. Since the 1950s, most glaciers show more than a hundred meters (330 feet) of thinning at low elevations and about 18 meters (60 feet) of thinning at higher elevations. Global glacial ice mass loss between 2000 and 2019 is estimated to be 267 ± 16 Gt per year in large part based on data collected by the **GRACE experiment** (Gravity Recovery and Climate Experiment), a pair of satellites traveling together. Together, these satellites measure minute changes in Earth's gravitational attraction as ice melts away and glaciers and ice sheets shrink. Ice loss between 2000 to 2019 equates to a rise in sea level of 0.74 ± 0.04 millimeters (0.03 inches) per year as glacier meltwater streamed into the global oceans.

Probably the best documented and most dramatic retreat of a glacier is found at Glacier Bay National Park, Alaska (**Figure 3.11**). When George Vancouver arrived there in 1794, he found Icy Strait, at its entrance, choked with ice and Glacier Bay only a slight dent in the ice-cliffed shoreline. By 1879, when John Muir visited the area, the glacier, by then named the "Grand Pacific," had retreated nearly 80 kilometers (50 miles) up the bay. The Grand Pacific Glacier had retreated another 24 kilometers (15 miles) by 1916, and today, a 104-kilometer (65-mile)-long **fjord**, an elongate, glacial-eroded valley that has been drowned by the ocean, occupies the area that just 200 years ago held a valley glacier that was as much as 1,220 meters (4,000 feet) thick.

Because of their small size and ice volume, mountain glaciers in the mid- and low latitudes are among the most sensitive indicators of climate change. They respond quickly even to small perturbations in climate. A worldwide survey of 160,000 mountain glaciers and ice caps reveals that the volume of the world's glacial ice is declining and the rate of loss is accelerating. Similar findings in midlatitudes are reported from Switzerland, where alpine glaciers have lost as much as half their mass since 1850; the Caucasus Mountains of Russia, where glacial ice decreased by about 50% during the twentieth century; and New Zealand, where, on average, the 127 glaciers in the Southern Alps have become 38% shorter and 25% smaller in area than when first studied in the twentieth century. At low latitudes, the largest glacier on Africa's Mount Kenya has lost 92% of its mass in the last few decades.

Figure 3.10 Glacier National Park's iconic glaciers are rapidly disappearing. Ice-covered areas in 1938 are now melted water in a lake, and the glacier has shrunk dramatically.

3.2.2 Ice Cores and Ocean Sediments

The best archives of past climates are those that preserve history in a way that changes in climate can not only be deciphered but also dated. Glacial ice and sediment preserved at the bottom of the ocean are examples that both provide invaluable archives of past climate history. **Ice cores** collected from both Greenland and Antarctica take us back hundreds of thousands of years, while **marine sediment cores** from the deep ocean preserve millions of years of climate history (**Figure 3.12** and **Figure 3.13**).

Ice cores provide evidence of changes in atmospheric chemistry, temperature, and dust, which correlate well with periods of cooling and warming during the last million years. As snow accumulates, it slowly compresses and transforms into glacial ice, preserving atmospheric

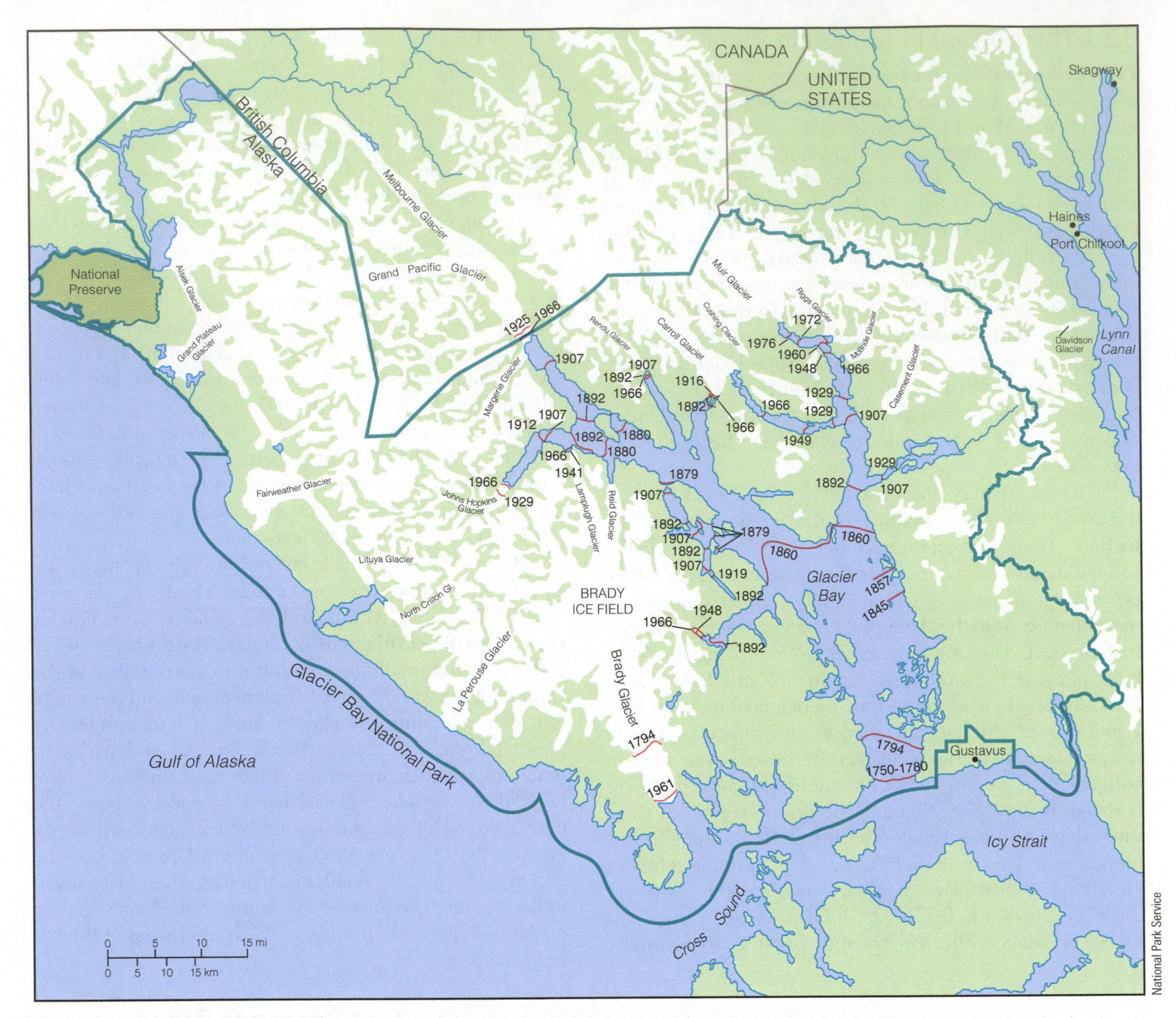

Figure 3.11 Map of Glacier Bay National Park, Alaska, showing the terminal positions of the retreating Grand Pacific and other glaciers from 1750 to the present.

information that becomes locked into the ice. The relative abundance of two slightly different oxygen atoms–the oxygen-18 and oxygen-16 isotope ratio (symbolized as $^{18}O/^{16}O$) in ice molecules indicates the atmospheric temperature when the snow fell from which the glacial ice formed (**Figure 3.14**). Other atmospheric gases such as CO_2, N_2O, and CH_4, dust, volcanic ash, and even traces of toxic pesticides and lead become trapped in glacial ice, awaiting analysis by a paleoclimatologist.

Figure 3.12 The Joides Resolution has traveled the world collecting marine sediment cores. These cores or mud and rock hold millions of years of climate history.

USGS

Figure 3.13 This Antarctic ice core preserves a dark band of volcanic ash from an eruption that occurred about 21,000 years ago.

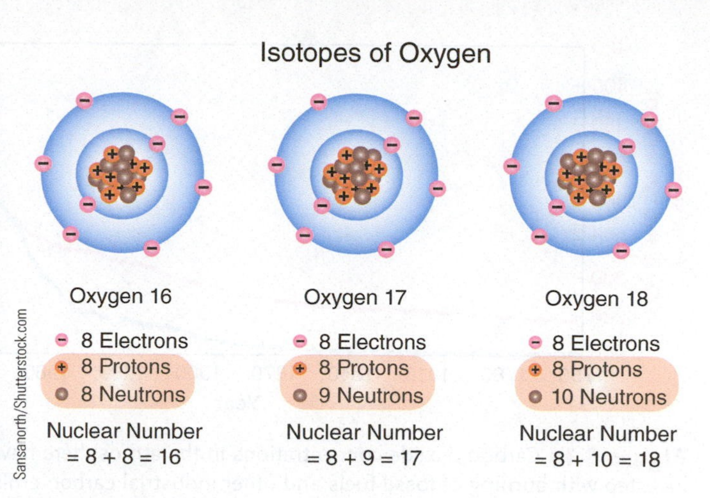

Figure 3.14 Oxygen has three different isotopes, each of which behaves differently during evaporation and condensation.

The bubbles of ancient atmosphere trapped and preserved in glacial ice obtained from 3,348-meter (10,981-feet)-deep cores drilled at Vostok Station and Dome C in Antarctica (Figure 3.15) together provide an 800,000-year record of variations on the atmosphere's temperature and composition through six full

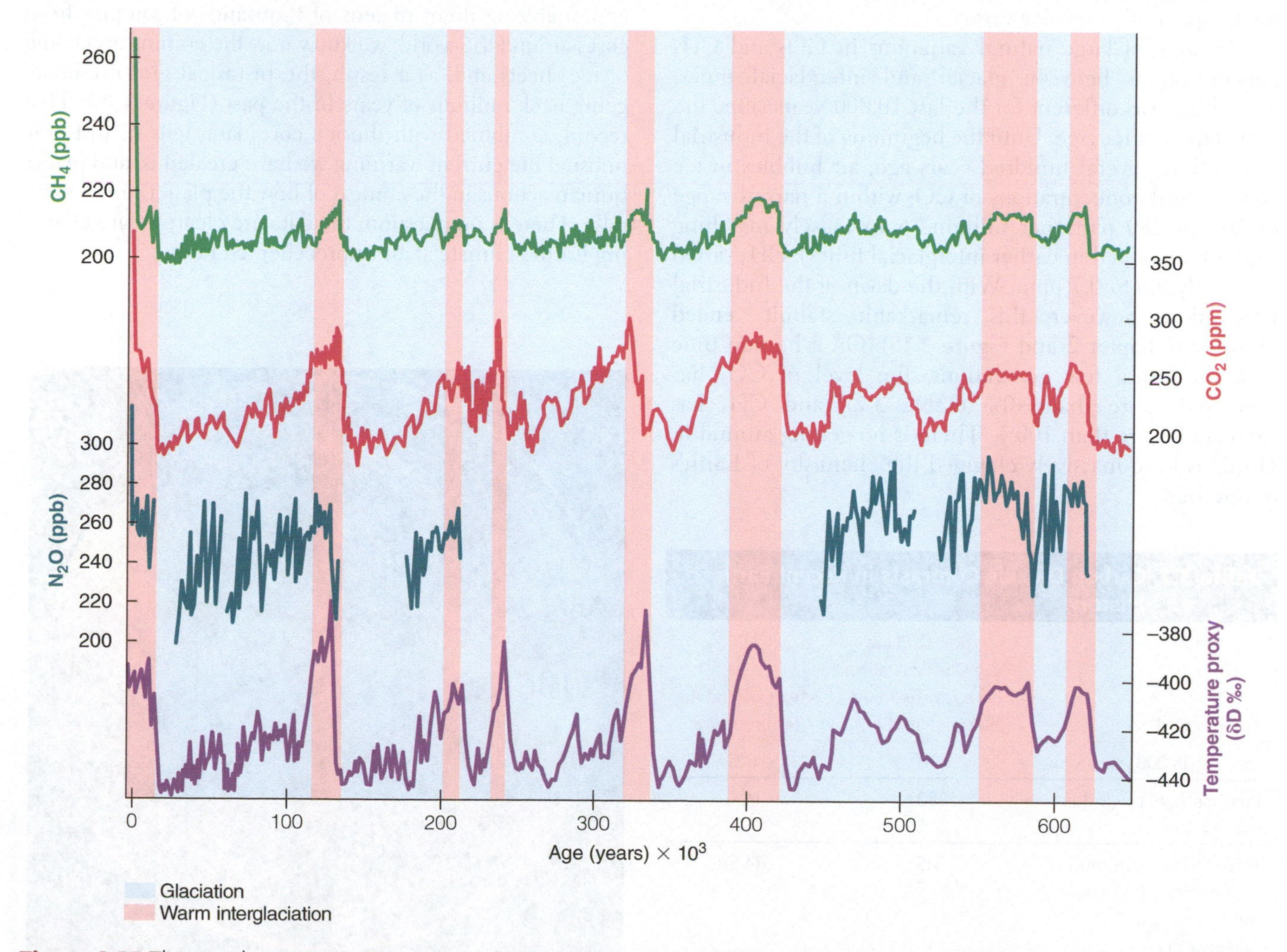

Figure 3.15 The greenhouse gases (CO_2, CH_4, and NO_2) and deuterium (dD) records for the past 650,000 years from the EPICA Dome C and the Vostok ice cores, Antarctica. A proxy for air temperature, dD is the deuterium/hydrogen ratio of the ice, expressed as a per mil deviation from the value of an isotope standard–the greater negative values indicate colder conditions. (Data from Brook, November 25, 2005, "The Long View," *Science* 310: 1285)

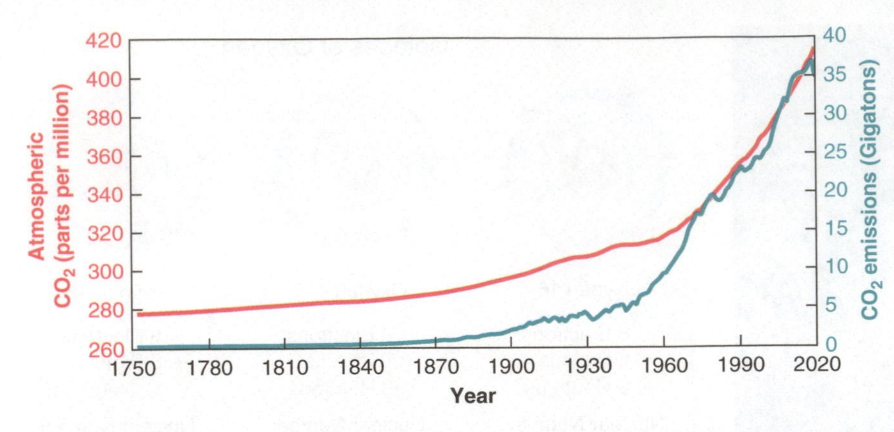

Figure 3.16 Carbon dioxide concentrations in the atmosphere have gone up in lockstep with burning of fossil fuels and other industrial carbon emissions since the late 1800s. (Data from NOAA)

glacial-interglacial cycles. The core clearly shows that the concentration of greenhouse gases (GHGs) in the atmosphere, especially CO_2, and temperature have gone up and down in lock step. When CO_2 concentration is high, so is the temperature, and vice versa.

In spite of large natural variations in CO_2 and CH_4 concentrations between glacial and interglacial times, something was different for the last 10,000 years since the end of the last Ice Age. Until the beginning of the Industrial Revolution, several hundred years ago, air bubbles in ice cores record concentrations of CO_2 within a narrow range of 260 to 280 parts per million (ppm closely matching values that existed in earlier interglacial times). CH_4 varied from only 0.6 to 0.7 ppm. With the dawn of the Industrial Revolution, however, this remarkable stability ended abruptly (Chapter 2 and **Figure 3.16**). On a human time scale of just a few generations, the level of CO_2 has increased more than 40% (**Table 3.2**), and CH_4 has increased more than 100%. There is no getting around it. Humans have massively changed the chemistry of Earth's atmosphere.

Table 3.2 Carbon Dioxide Contrasts in Recent Earth History

Time Period	Carbon Dioxide (CO_2) Concentration in the Atmosphere (parts per million)	Mean Global Air Temperature
Ice Age (18,000 years ago)	170	~6°C
Post-Ice Age, pre-Industrial Age	280	14°C
1958, First of continuous observations taken at the Mauna Loa Observatory	315	14.5°C
End of 21st century (2020)	370	15°C
2023	420	15.5°C

Deep ocean sediments accumulate more slowly than does snow on ice sheets and those sediments last a lot longer. Cores drilled in the deep ocean muds can be hundreds of meters deep and recover sediment that is many millions of years old. That sediment contains the shells of small, single-celled animals, plankton called **foraminifera** (forams for short) (**Figure 3.17**). These microscopic creatures grew and died in oceans of the past. Their remains, made of calcium carbonate ($CaCO_3$), sank to the bottom and were entombed in mud. $CaCO_3$ contains oxygen that can be analyzed for its isotopic composition, just like the frozen water in the ice cores.

The ratio of $^{18}O/^{16}O$ in the forams reflects both water temperature and how much water evaporated from the oceans and was then locked up in the ice sheets and glaciers of the world. After picking millions of forams from the mud and analyzing them in tens of thousands of samples from cores around the world, we know now the coming and going of ice sheets and, as a result, the historical global climate going back millions of years in the past (**Figure 3.18**). That record, combined with the ice core data, tells us just how unusual the current warming we have created is, and places human actions in the context of how the planet varied naturally. There is no question; the data are clear; we are changing Earth's climate at an unprecedented rate.

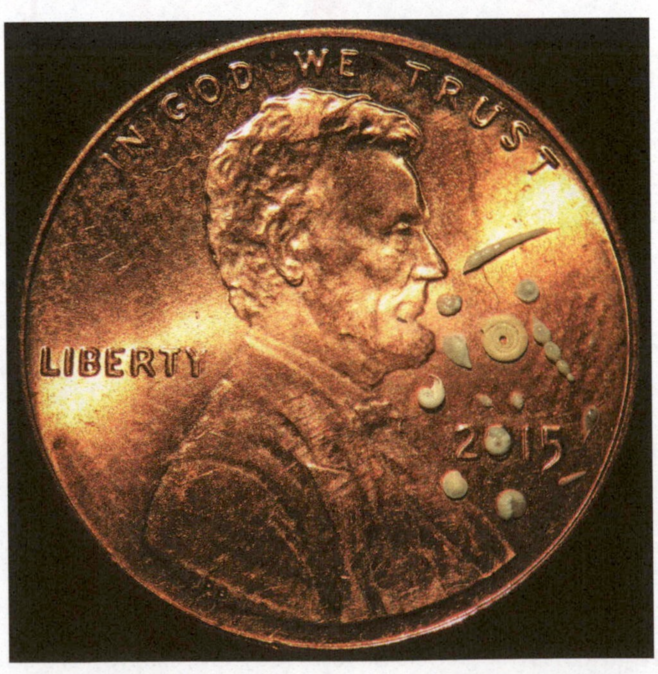

Figure 3.17 Forams on a penny for scale. These animals come in a variety of shapes and sizes.

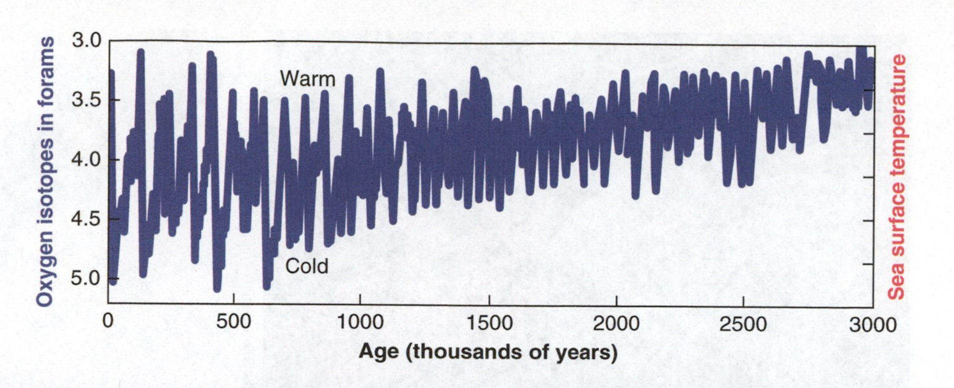

Figure 3.18 Oxygen isotopes measured in forams record the pulse of climate change over the past 3 million years. Peaks at the top of the graph indicate warm times and interglacials. Troughs at the bottom indicate times when the Earth was cold. (Adapted from International Ocean Discovery Program, Volume 303/306: North Atlantic Climate, 2006)

Isotopes: Climate Change Witnesses Tell Their Stories

Isotopes. Perhaps the word brings back memories of chemistry class. Each element on the periodic table has a set number of protons that makes it unique. But the number of neutrons in an element can vary, giving each **isotope** a different mass. For example, all carbon has 6 protons but there is carbon isotope with a mass of 12 (it has 6 neutrons), carbon 13 (with 7 neutrons), and carbon 14 (with 7 neutrons)– in shorthand, ^{12}C, ^{13}C, and ^{14}C. Although these isotopes are all carbon, and behave chemically the same way, there are subtle differences in the rate at which they react both in physical and biological systems. These slightly different rates of reaction lead to fractionation, the observation that some processes can enrich or deplete one isotope in relation to another (**Figure 3.19**).

Isotopes of carbon

Carbon-12 — 6 protons, 6 neutrons, 6 electrons

Carbon-13 — 6 protons, 7 neutrons, 6 electrons

Carbon-14 — 6 protons, 8 neutrons, 6 electrons

iStock.com/Ttsz

Figure 3.19 Carbon has three isotopes, each with the same number of protons but different number of neutrons.

The bias introduced by fractionation is tiny and depends on the difference in isotope masses. For example, ^{1}H and ^{2}H, two isotopes of hydrogen, differ in mass by 100%, but ^{12}C and ^{13}C differ in mass by only 8%. The mass of isotopes is often measured using a mass spectrometer, an instrument that weighs atoms by determining how well they can round a curve when moving rapidly. To understand mass spectrometry, imagine a truck and a bicycle entering a tight curve at high speed–the bike is likely to get around; the big truck, well, it might crash into the guard rail well before making the turn (**Figure 3.20**).

Isotopes are exceptionally useful tools for tracing processes and dating events in the geosphere, cryosphere, hydrosphere, and biosphere. Some isotopes are stable and do not radioactively decay. Stable isotopes of carbon, oxygen, and nitrogen are widely used as tracers of where and how mass moved on Earth, including water moving through the hydrosphere and carbon moving through the biosphere. Other isotopes are radioactive, decaying away at predictable rates and providing a clock by which we can learn about what happened in the past, and when. Carbon-14 is the most important of these radioactive isotopes. It has a **half-life** of 5730 years, which means that every 5730 years, half of the carbon 14 in a sample will decay radioactively and be gone. Measured in fossilized organic material, it has been critical in establishing the timing of glaciation, deglaciation, droughts, and floods over the past 40,000 years. That's an upper limit to its use for dating because after something has been dead for 40,000 years, there's not enough carbon-14 left to measure.

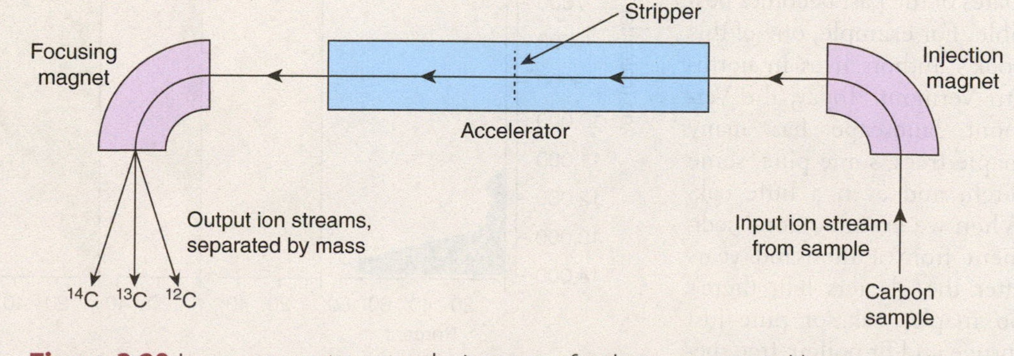

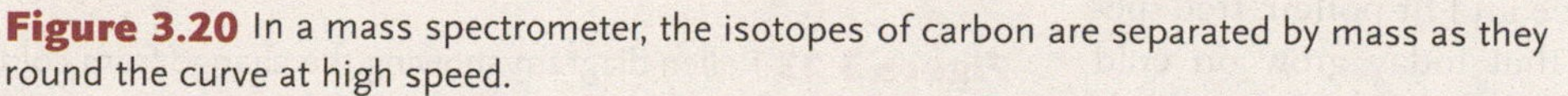

Figure 3.20 In a mass spectrometer, the isotopes of carbon are separated by mass as they round the curve at high speed.

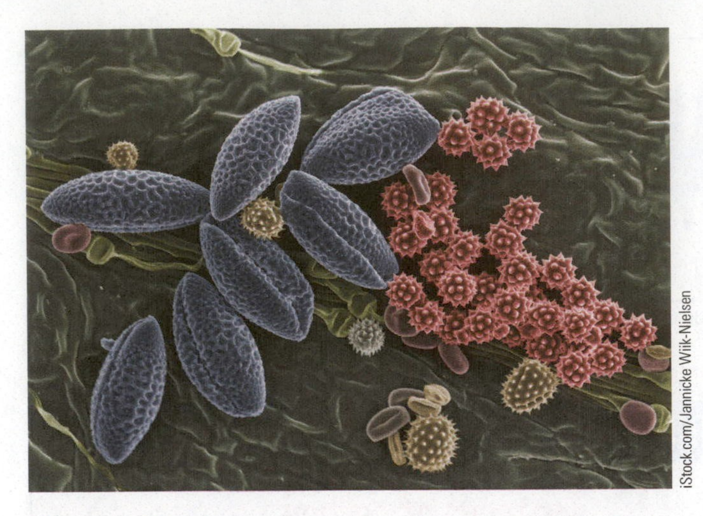

Figure 3.21 Pollen grains come in all shapes and sizes. Each species has a unique shape, which allows grains to be identified easily.

Figure 3.22 A twig recovered from sediment below the Greenland Ice Sheet. The sediment containing the twig was deposited at least 400,000 years ago. Having been frozen ever since, the twig is exceptionally well preserved.

3.2.3 Packrats and Pond Muck

Pond muck, the gooey black stuff that gets between your toes on a summer swim, is just as useful to paleoclimatologists as deep ocean mud. Buried in the mud is plant **pollen**. Flowering plants give off pollen as they reproduce. People with hay-fever know this all too well. Pollen settles onto the surface of ponds and lakes and soon sinks to the bottom, where it's encased in mud. Pollen is remarkably resistant to decay and widely useful in paleoclimate work because it is unique to every plant species. For example, sagebrush pollen looks very different than ragweed pollen and both look different than maple pollen (**Figure 3.21**).

By counting and identifying pollen grains, paleoclimatologists can identify plants that lived near the pond in the past. They even find fossil leaves and twigs and seeds (**Figure 3.22**) Once you know the climates in which different plants thrive, figuring out climates of the past becomes possible. For example, one of this book's authors lives in northern Vermont. Today, the Vermont landscape has many maple trees, some pine, some birch, and even a little oak. When we examine pond sediment from a thousand years after the glaciers left, there's no maple, oak, or pine–just spruce and fir pollen, tree species that today grow on cold mountaintops and hundreds of miles north. This tells us that it was much colder in Vermont 12,000 years ago than it is today. Similar pollen studies done in many parts of the world show the effect on plants at the end of the last cold glacial period and the warming that followed (**Figure 3.23**).

If you live in the desert, there are no ponds to collect pollen and bits of long-vanished plants, but there is something else to do the same work–**pack rats**. Pack rats are small, common desert creatures that scamper about the landscape collecting twigs and seeds that they use to make nests in rocky crevices. The packrats have an odd habit.

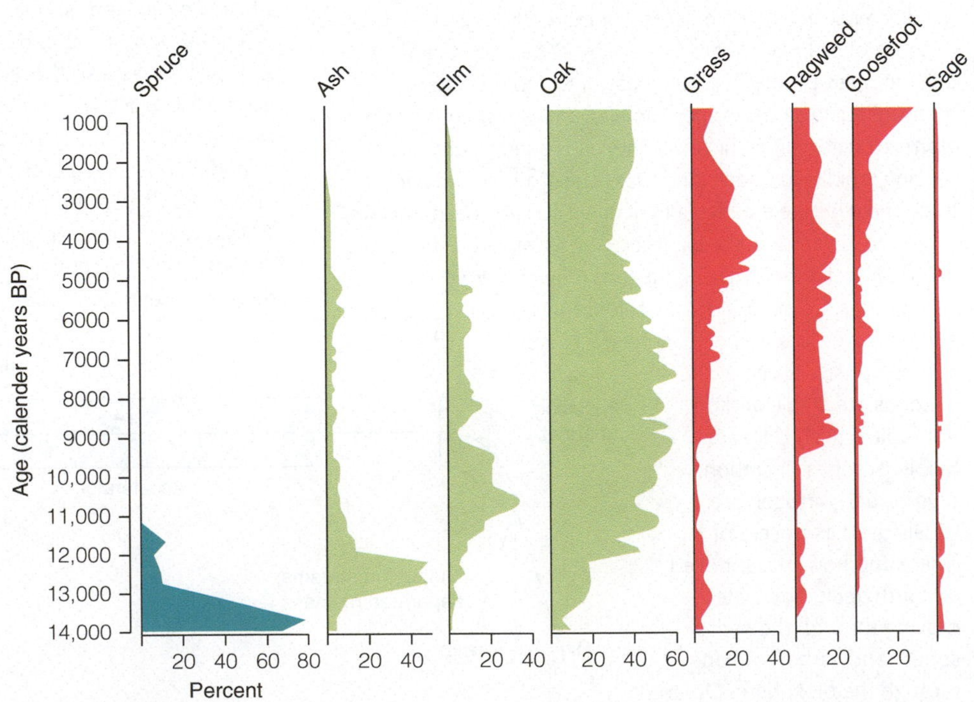

Figure 3.23 Pollen diagram showing the change from cold-loving species, such as spruce, at the end of the last glacial period to warmth-loving species, such as oak, today.

Lee Rentz/Alamy Stock Photo

Figure 3.24 Packrat nest (also called a midden) with rock hammer for scale. These are often founded in openings in rock as shown here.

They urinate on their nests (**Figure 3.24**). Because they live in the desert, they need to retain water and their kidneys have evolved to produce ultraconcentrated urine. The urine crystalizes and makes a remarkably tough, resin-like substance known as **amber-rat**.

The amber-rat encases the vegetation in the nest and preserves it. Thousands or tens of thousands of years later, geologists collect samples of the rat's nest. Back in the lab, they chip out fossil wood and twigs and date them using carbon-14. That organic material can be both identified and dated. Where today there is only sagebrush, at the height of the last glaciation, 25,000 years ago, the twigs and needles the rats collected tell us the landscape was covered by pinyon pines and juniper bushes. It was cooler and wetter then, and different plants thrived in that glacial climate.

3.2.4 Tree Rings

Tree rings are one of the most useful records of past climates. Each year, trees add a growth ring to their trunk–new wood (**Figure 3.25**). In good years, when there is lots of sun, warmth, and water, they add thick rings. In bad years, when there is drought, or the tree is stressed by cold or heat, they add only a thin ring. In some parts of the world, very slow growing trees, such as Bristlecone Pines, hold thousands of years of climate history in their rings (**Figure 3.26**).

To gather a climate record from trees, paleoclimatologists collect thin cores extending into the center of the tree (**Figure 3.27**). These cores are taken to a lab, polished, and their rings counted and widths measured. The count allows the rings to be dated (one each year) and the width tells the climate story. Tree rings have been used to characterize megadroughts in California that lasted for decades. In fact, those tree rings tell us that the current California drought is about as bad as any drought in the last 2000 years! In the arctic, rings from trees at the edge of their range show the effects of warming and cooling temperatures over the last few thousand years.

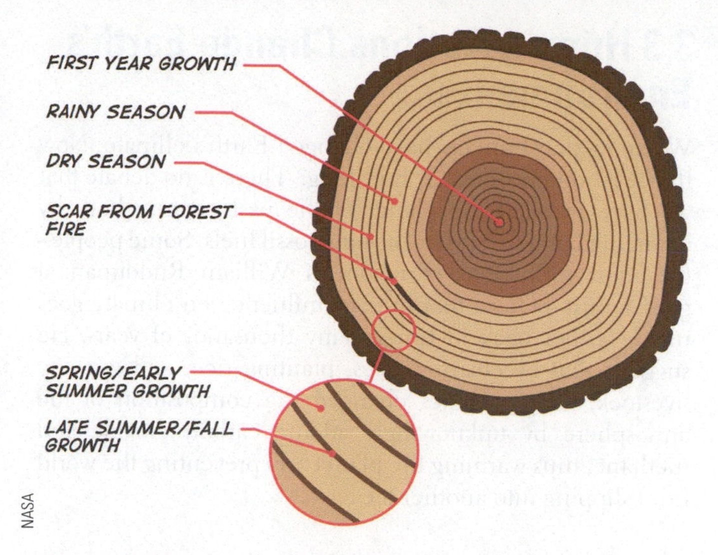

NASA

Figure 3.25 Each year, trees add a ring of new wood just below the bark; by counting those rings, the age of the tree can be determined.

NPS

Figure 3.26 Bristlecone Pines in the White Mountains of Southern California record thousands of years of climate history in their rings.

MediaNews Group/Los Angeles Daily News/Getty Images

Figure 3.27 Coring a tree to study its rings using a hand-held increment borer–a drill of sorts.

3.3 Human Actions Change Earth's Environment (LO3)

We know that humans have changed Earth's climate. Now the question becomes for how long? There is no debate that we have warmed the planet of the past century both by changing land use and by burning fossil fuels. Some people–the most influential of whom is William Ruddiman, a geoscientist–believe that human influence on climate goes much farther back in time, many thousands of years. He suggests that by clearing trees, planting rice, and growing livestock, early societies changed the composition of the atmosphere by unknowingly adding carbon dioxide and methane, thus warming the planet and preventing the world from slipping into another ice age.

3.3.1 Greenhouse Gas Emissions Changed Atmospheric Composition

In 1958, Charles Keeling, then a 30-year-old scientist, had an idea. He wanted to measure the CO_2 content of the atmosphere, not once, but over months and years to determine if it were changing. In Hawaii, near the top of Mauna Loa, at 3,400 meters (11,135 feet) above sea level, he set up his instruments and began taking data. In just a few years, he noticed that the concentration of CO_2 went up in December and January and down in June and July (that reflected plants growing in the northern hemisphere summer and not in the winter; the Northern Hemisphere dominates because it has more land area than the Southern Hemisphere). It took a few more years before his data showed a trend, upward.

Year after year, the measurements showed a steady increase in annually averaged atmospheric CO_2. In 1958, there were 316 (ppm) CO_2 in the atmosphere. In 2023, as we finished this book, the value is 420 ppm. The Keeling curve (**Figure 3.28**) was named for the man who started these measurements. Keeling died of a heart attack in 2005; today, his son continues the measurements, uninterrupted. The Keeling curve unambiguously shows the increase in CO_2 over more than half a century. Climate scientists quickly recognized its significance; the composition of Earth's atmosphere was rapidly changing. Even more alarming is the steepening upward curve of the graph, illustrating the accelerating rate of CO_2 emissions.

There is very strong isotopic evidence that burning coal and oil accounts for the increase in atmospheric CO_2. This is called the **Suess effect**, named for the Austrian chemist, Hans Suess, who first noted that the isotopic composition of carbon in the atmosphere was changing. Recall that carbon has three isotopes – ^{12}C, ^{13}C, and ^{14}C–the most of common of which is ^{12}C. Only one is radioactive, ^{14}C, made in the atmosphere by cosmic rays zipping in from the solar system. Fossil fuels have no ^{14}C because they are too old. ^{14}C decays quickly–in 40,000 years, it's mostly gone. Fossil fuels are millions of year old and thus devoid of all ^{14}C.

Suess noted two things, both directly related to the burning of fossil fuels. First, the ratio of $^{14}C/^{12}C$ was falling because the burning of fossil fuels pumped lots of ^{12}C into the atmosphere and no ^{14}C. Next, he noted that the $^{13}C/^{12}C$ ratio in the atmosphere changed because fossil fuels had less ^{13}C than atmospheric CO_2. Together, these isotope observations were the smoking gun of fossil fuel influence on atmospheric chemistry. There was no question left. The rise of CO_2 in the atmosphere measured by Keeling was caused by fossil fuel combustion. By the late 1960s, it was clear that humans were changing the composition of the atmosphere and, by inference, the temperature of the planet (**Figure 3.29**). The magnitude of that temperature change was unambiguous to scientists but has remained a point of strong disagreement in political and economic circles for decades.

3.3.2 Land-Use Change

Ever since they tamed fire, humans have been changing Earth's landscapes and thus influencing climate. At first the changes were subtle: burning grasslands and forest undergrowth to encourage plant growth and make hunting more efficient. As populations increased, so did human impact on landscapes. Thousands of years ago, people began clearing forests and cultivating land, and by doing so began

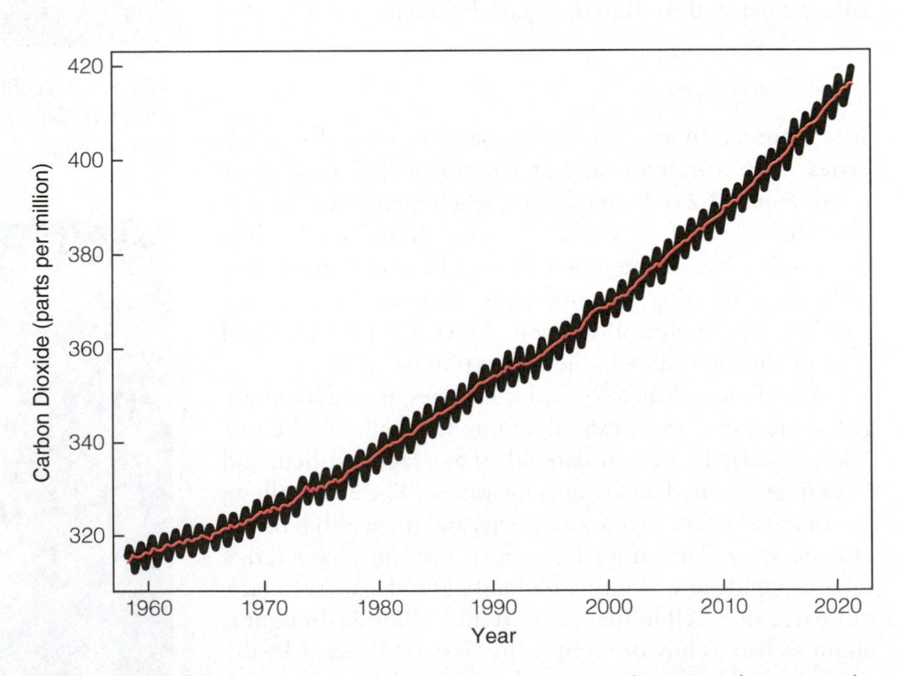

Figure 3.28 The continuous record of carbon dioxide in the atmosphere started by Keeling in 1958 at Mauna Loa in Hawaii. (Data from NOAA Global Monitoring Laboratory)

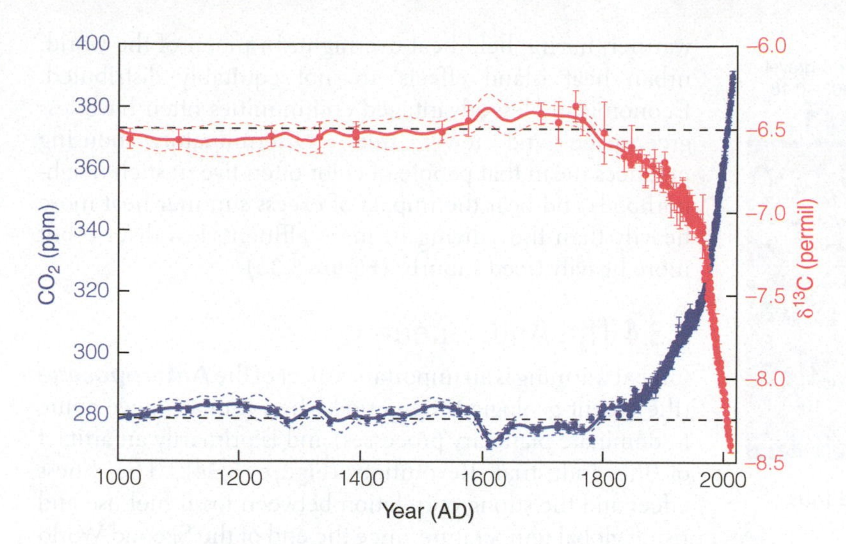

Figure 3.29 The Suess effect is clearly shown by the rapid change in the ^{13}C content of the atmosphere beginning after 1800, the start of the Industrial Revolution. As the CO_2 concentration in the atmosphere rises, the percentage of ^{13}C falls because fossil fuels are rich in ^{12}C. (Data from Rubino, M., et al. (2013). A revised 1000 year atmospheric $\delta^{13}C$-CO_2 record from Law Dome and South Pole, Antarctica. Journal of Geophysical Research: Atmospheres, 118(15), 8482–8499)

transferring carbon once held in soils and trees into the atmosphere (**Figure 3.30**).

With the coming of the Industrial Revolution, the rate and scale of landscape change grew exponentially. Whole forests were cleared, burned, and never regrew. Entire landscapes were plowed and carbon freed from the soil. That carbon moved into the atmosphere and wasn't recaptured by plants, as the land remained in agricultural use. Today, humans move more material over Earth's surface than do natural processes. Much of that movement is related to agriculture, and it has dramatically changed how carbon is stored near Earth's surface and the exchange of carbon between plants, soils, and the atmosphere.

Land-use change during the **Holocene** (the last 11,000 years of climate stability) has driven significant change in landscape albedo: a general increase where stands of trees are removed. Land-use change catalyzes transformation in patterns of cloud formation and precipitation, including drying out of landscapes in otherwise moist, tropical environments, where woodland **evapotranspiration** (the transfer of water from the ground to the air by plants) once played an important role in the cycling of water. Deforestation means a reduction in the biological carbon storage capacity of many regions because crops and grazing soils involve far less biomass than mature forests.

Figure 3.30 Burning slash after deforestation in the Amazon.

3.3.3 Urbanization

Have you ever wondered why cities tend to be warmer than the rural areas around them? The temperature difference between urban areas packed with buildings, pavement, and little greenspace, in contrast to leafy suburbs and rural areas can be stark–as much as 5 to 10 degrees C (9 to 18 degrees F) It's called the **urban heat island effect**. In a heat wave urban heat can be the difference between life and death (**Figure 3.31**).

The physics of hot cities is relatively simple. First, dark pavement has low albedo (**Figure 3.32**). Most of the solar energy striking asphalt is absorbed rather than being reflected back into the atmosphere, so the streets heat up. Then there's thermal mass. The brick, concrete, and asphalt warm during the day and hold that heat, slowly radiating it back to space at night. The lack of greenspace, parks, and trees reduces water evaporation from soil and evapotranspiration from trees. It takes energy to evaporate water and if there's no water to evaporate, then the energy simply remains as heat in the environment.

The heat island effect can be particularly striking at night. Overnight temperatures in cities can drop far less than in outlying areas. This sets people up for even more heat stress the following day as the city starts each new day

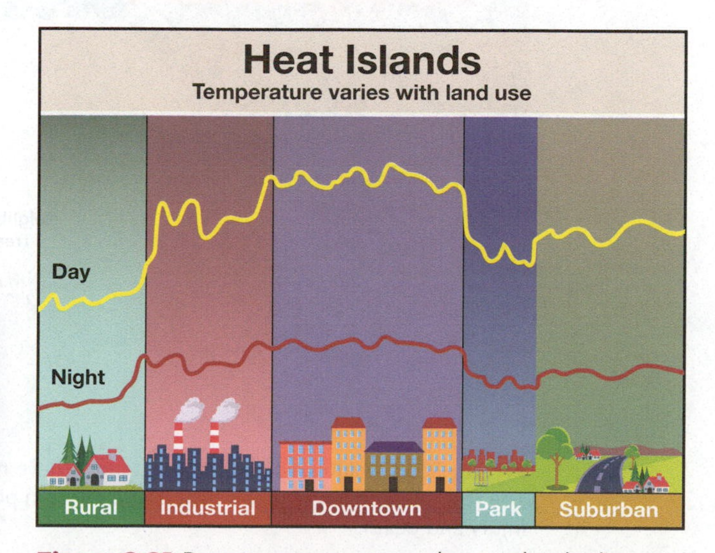

Figure 3.31 Downtowns are warmer than rural, suburban, and parkland areas, both day and night.

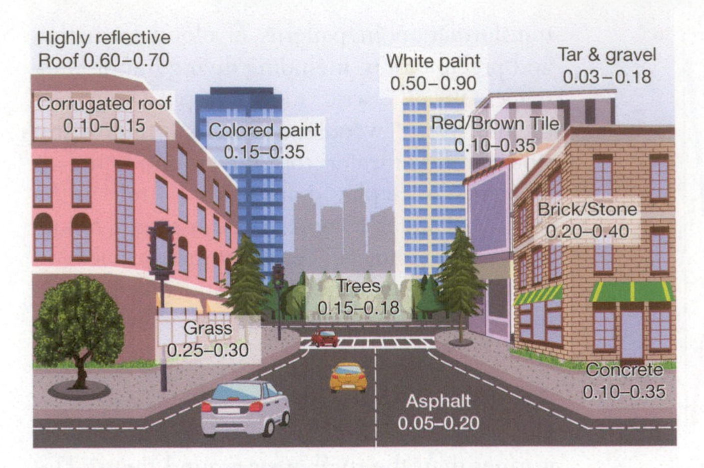

Figure 3.32 The albedo of cities tends to be low and thus they absorb rather than reflect most solar radiation.

warmer, having held heat overnight. In much of the world, urban heat island effects are not equitably distributed. Economically less-advantaged communities often have less greenspace and fewer trees. Discriminatory housing practices mean that people of color often live in such neighborhoods and bear the impact of excess summer heat more heavily than those living in more affluent, less diverse but more heavily treed suburbs (**Figure 3.33**).

3.3.4 The Anthropocene

Global warming is an important aspect of the **Anthropocene** (the recent geologic time period when humans have come to dominate planetary processes) and is primarily an artifact of the Industrial Revolution (**Figure 3.34**). The Suess effect and the strong correlation between fossil fuel use and rising global temperature since the end of the Second World War (**Figure 3.35**; **Table 3.3**) clearly show that CO_2 emissions

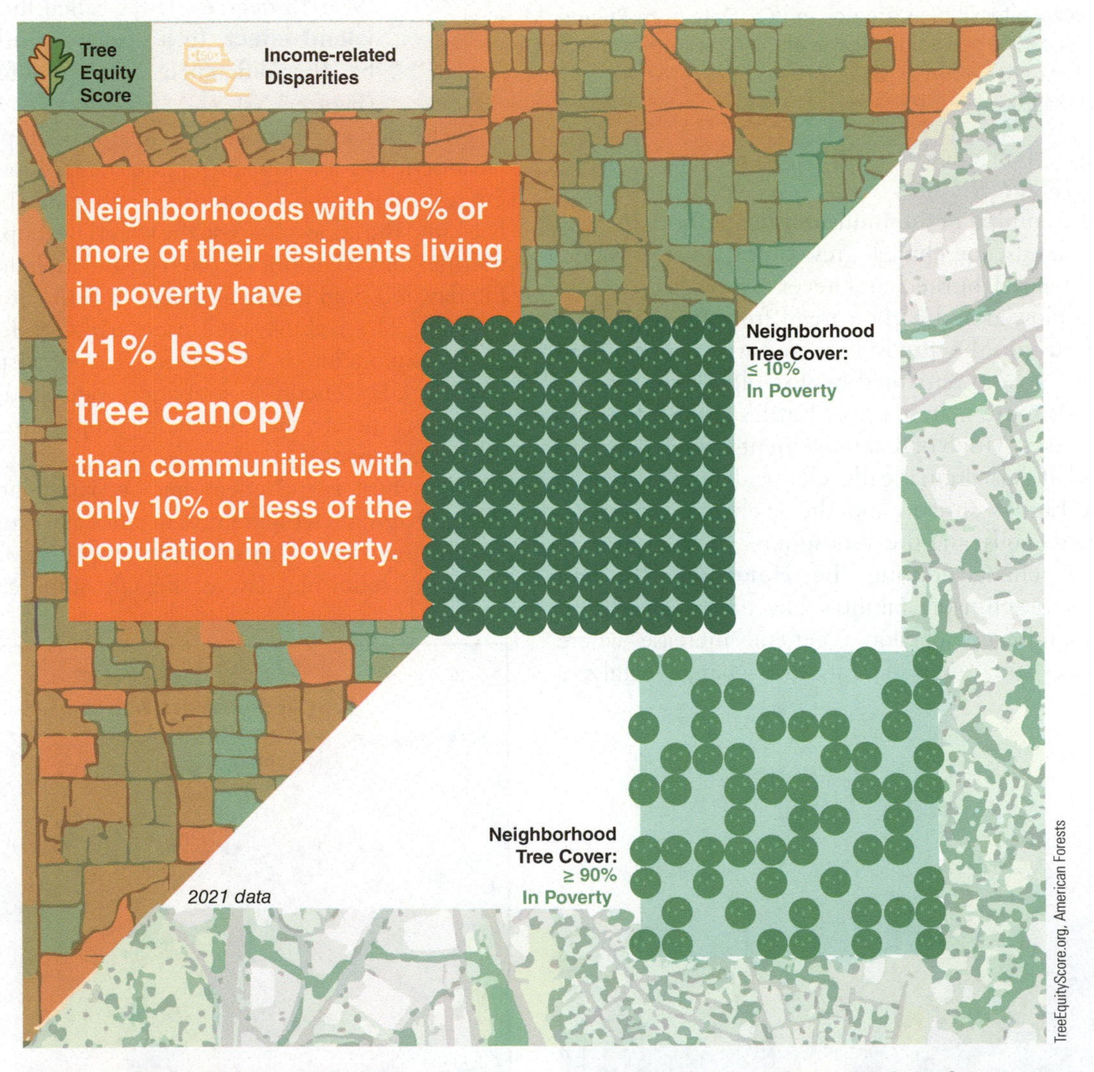

Figure 3.33 Wealthy neighborhoods tend to have more trees; poorer neighborhoods have fewer. The result is a stronger urban heat island effect in poor neighborhoods–and in heat waves, more injury and death.

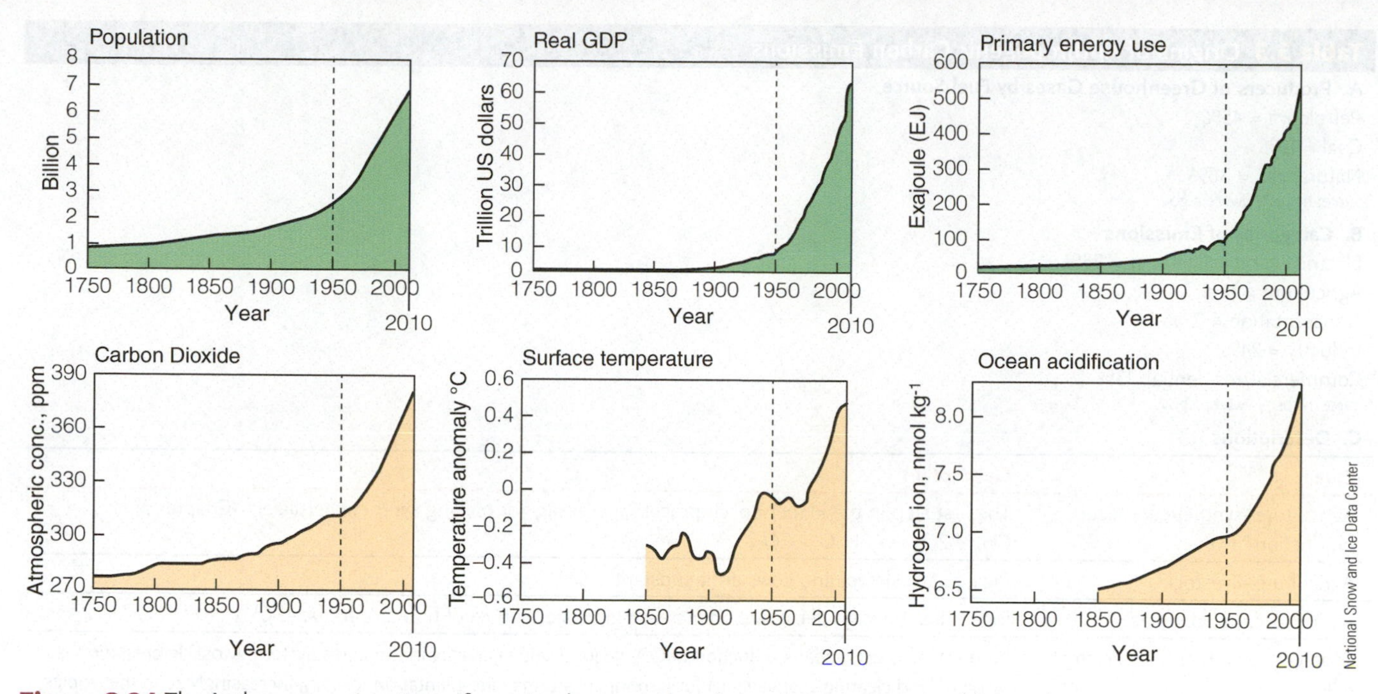

Figure 3.34 The Anthropocene is a time of massive, human-induced change in most parts of the Earth system.

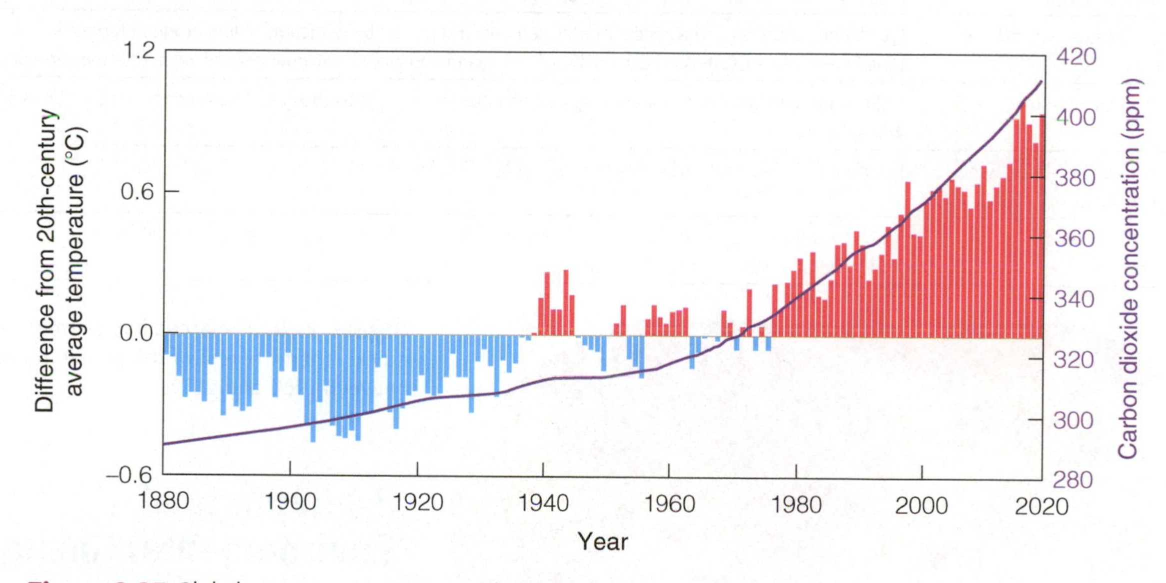

Figure 3.35 Global average temperature and carbon dioxide concentration in the atmosphere started increasing together after about 1960. (Data from NOAA)

related to gas, coal, and oil are to blame for climate change. Today's levels of atmospheric CO_2 exceed those which are characteristic of previous interglacial periods by a whopping 130 parts per million, or 40% (**Table 3.2**). Such high levels of CO_2 have not existed in Earth's atmosphere at any other time during the past few million years.

The related increase in temperature has not been uniform; regions of land at high northern latitude have shown the strongest change. This concentrated arctic warming is referred to as **arctic amplification**, which is driven by the ice-albedo feedback. Ocean areas have responded more slowly because of their large water mass (**Figure 3.36**). Industrial activity is not the only human activity contributing to this change. **Land conversion** is the biggest contributing factor after fossil fuel combustion–the clearing of forests and grasslands for farms, pastures, and urban areas, and many other manufacturing or commercial processes (see **Table 3.3**). In fact, changing the landscape of our planet is the hallmark of the Anthropocene.

Table 3.3 Origins of Anthropogenic Carbon Emissions

A. Producers of Greenhouse Gases by Fuel Source

Petroleum = 45%

Coal = 19%

Natural gas = 36%

Source: https://www.eia.gov

B. Categories of Emissions

Electricity and waste heat = 25%

Agriculture = 11%

Transportation = 27%

Industry = 24%

Commercial/residential: 13%

Source: https://www.epa.gov

C. Descriptions

Category	Description
Agriculture–synthetic fertilizer production	The first step in the Haber-Bosch process, necessary for making synthetic fertilizers (Chapter 2) is: $CH_4 + 2O_{2-} \rightarrow 2H_2O + CO_2$.
Agriculture–livestock	Pigs, goats, sheep, and cows emit substantial CH_4.
Agriculture–rice paddies	Rice fields, through decay, are an increasingly major source of human-produced CH_4.
Land conversion and associated deforestation	Removal of forests means reduction of CO_2 sequestration capacity in the Earth System. Most deforestation is a result of land clearing (conversion) for farming, pasturing, and plantation forestry, increasingly so in the tropics.
Soot-producing stoves in the lesser developed world	Dung and biomass-fueled stoves in developing countries create soot, which settles on snow and ice fields, most critically in the Himalayas. The result is significant albedo reduction.
Industrial processes–quicklime production	Quicklime (CaO) is used to make mortar, cement, and plaster by reduction of limestone in furnaces ($CaCO_3 \rightarrow CaO + CO_2$). As much as 4% of U.S.-generated global warming may be related to the cement industry.
Industrial processes–steel production	Coke (pure carbon from charcoal) is used to flux iron ore in foundries. The reaction is: $2Fe_2O_3 + 3C \rightarrow 4Fe + 3CO_2$.
Industrial processes–beer, wine, and cider production	Fermentation yields carbon dioxide: $C_6H_{12}O_6 \rightarrow 2C_2H_5OH + 2CO_2$.
Burning of fossil fuels	Transportation, manufacturing of medicines, clothing, plastics, and other products, heating and cooking, and general electrification all contribute.

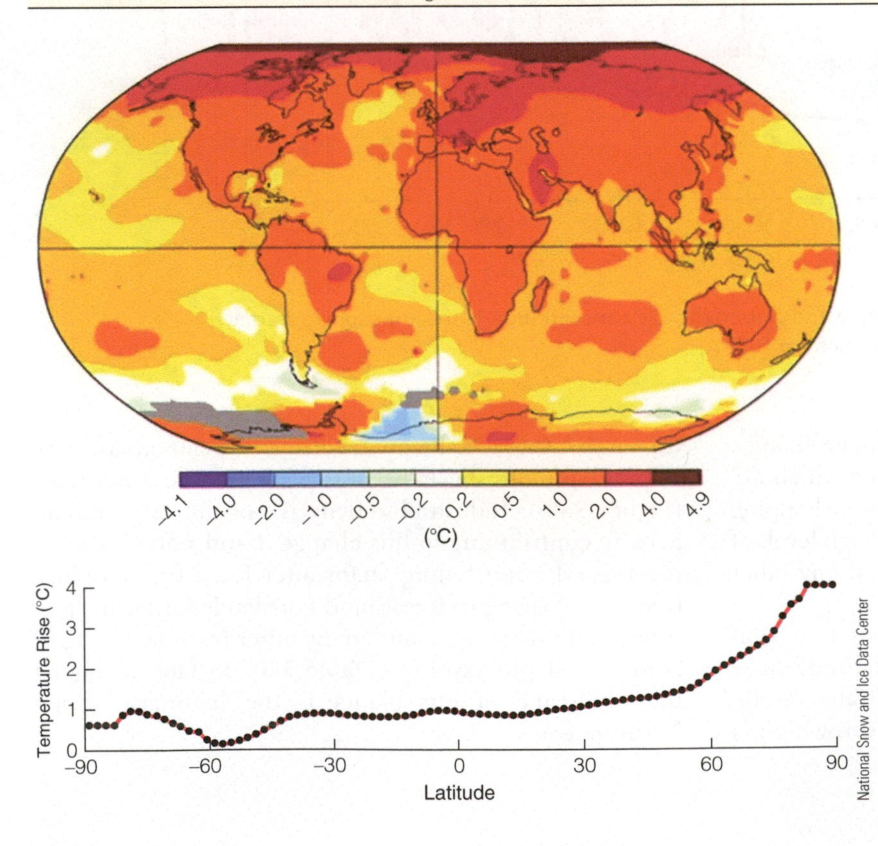

Figure 3.36 The arctic has warmed far more quickly than the rest of the planet in the years between 1960 and 2019.

3.4 Impacts of Environmental Change (LO4)

Human impacts on the environment of planet Earth are widespread and substantial. Temperatures are rising around the globe, precipitation patterns are changing, glaciers are shrinking, and sea level is rising. All of these changes threaten species around the world and, if left unchecked, will lead to political turmoil as water and goods become scarce and tens of millions of people are forced to migrate as temperatures spike and coastal zones flood.

3.4.1 A Baking Planet

Temperature records from land areas and oceanic shipping lanes reveal that as of

2021, Earth's average temperature had increased by 1.1°C (1.9°F) since 1880, with the majority of that increase occurring after 1975 (**Figure 3.37**). The rate of warming is roughly 0.15-0.2 C every decade. The year 2016 was the warmest year on record, followed closely by 2020 and 2019. 2021 tied with 2018 as the sixth hottest year on record. Nine of the ten hottest years globally have been within the last decade, with only 2014 failing to make the list.

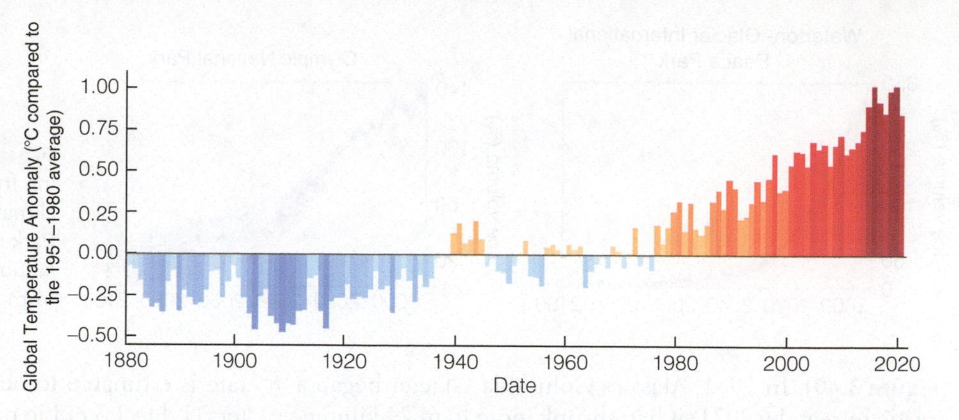

Figure 3.37 Earth is warming, and as a result, most of the warmest years on record have happened since 2010. (Data from NASA)

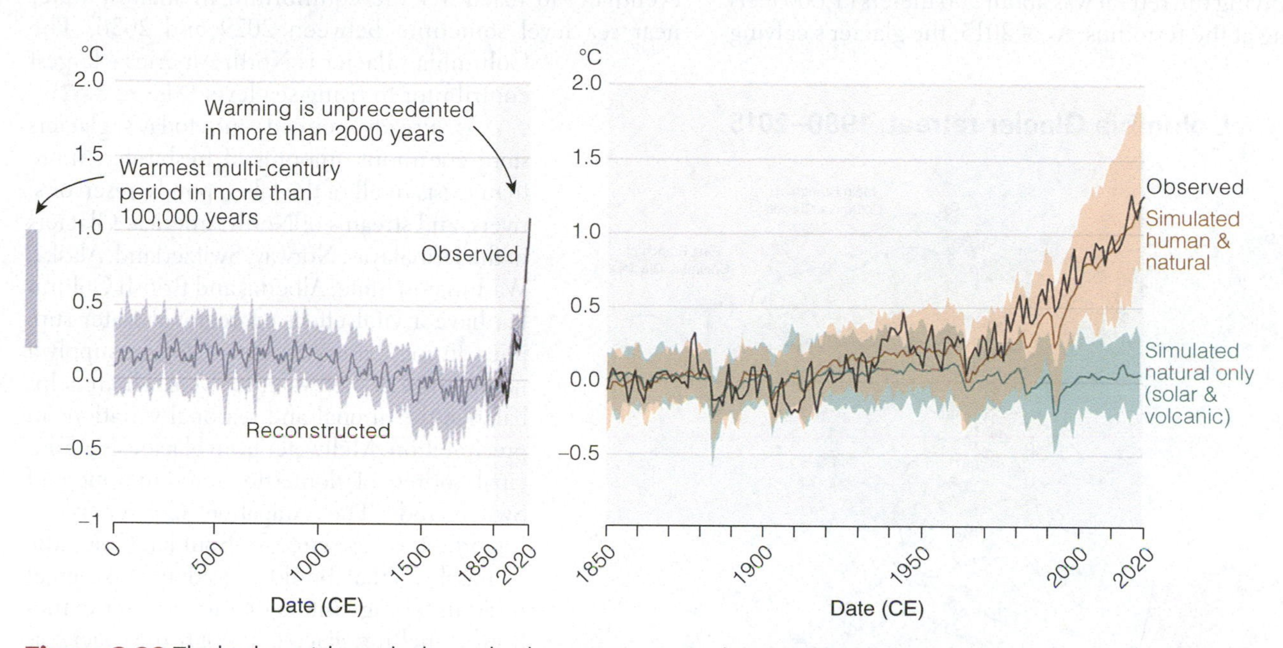

Figure 3.38 The hockey stick graph shows the dramatic rise in Earth's temperature over the last 150 years. The Earth is now warmer than it has been for more than 100,000 years. (Adapted from IPCC AR6 Working Group I report, 2021)

The greatest amount of warming occurred at night in the higher latitudes. Some areas of Canada, Alaska, and Eurasia warmed as much as 5.5°C (10°F) between 1965 and 2011. At high elevations in Glacier National Park, temperatures have risen three times as fast as the global average. By 2010, the average midwinter temperature at Palmer Station on the western shore of the Antarctic Peninsula had increased 6°C (10.8°F) since 1950; this is the highest rate of warming anywhere on Earth, about eight times the global average. Earth's present average temperature is the warmest it has been since the middle of the Holocene, thousands of years ago, a time when the Greenland Ice Sheet shrank back from today's margins by at least kilometers and perhaps tens of kilometers (**Figure 3.38**).

3.4.2 Shrinking Ice

About 97% of Earth's water is in the oceans, and three fourths of the remainder (2.25%) is in glaciers (**Figure 3.39**). These glaciers are shrinking, some very rapidly

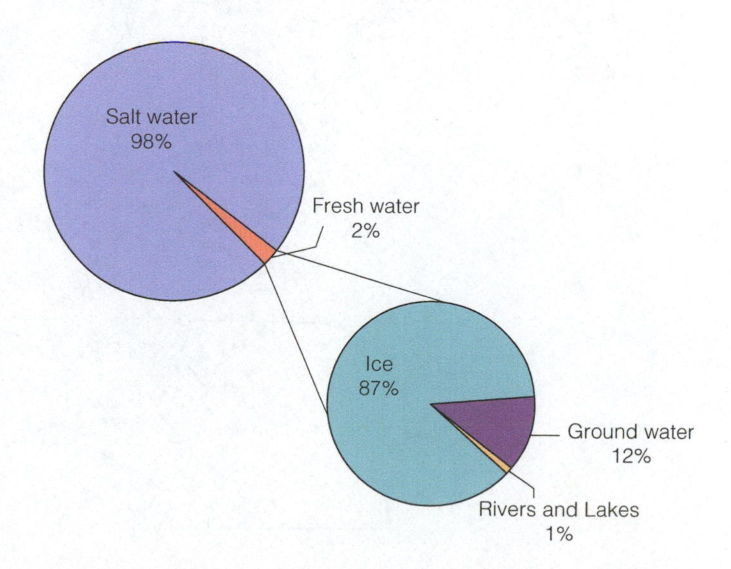

Figure 3.39 Most of Earth's water is in the ocean and of the water on land, most is help by glaciers.

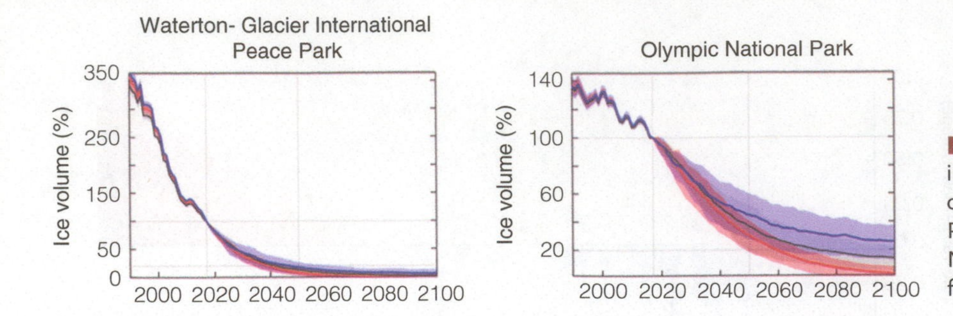

Figure 3.40 Glaciers are rapidly disappearing from America's National Parks due to climate change. By 2100, Glacier National Park is likely to be glacier-free and Olympic National Park will be close behind. (Data from USGS)

(**Figure 3.40**). In 1981, Alaska's Columbia Glacier began a drastic retreat; by 2021, it had shrunk more than 24 kilometers (15 miles) from its former terminus (**Figure 3.41**). Accompanying the retreat was about 500 meters (1,600 feet) of thinning at the terminus. As of 2015, the glacier's calving rate is estimated to be between 5.8 and 6.2 cubic kilometers (1.4 to 1.5 cubic miles) per year. It is expected to shrink back another 10 kilometers (6 miles) within a few decades, eventually to reach a static equilibrium in shallow water near sea level sometime between 2029 and 2036. The Columbia Glacier is North America's largest contributor to rising sea level (**Figure 3.42**).

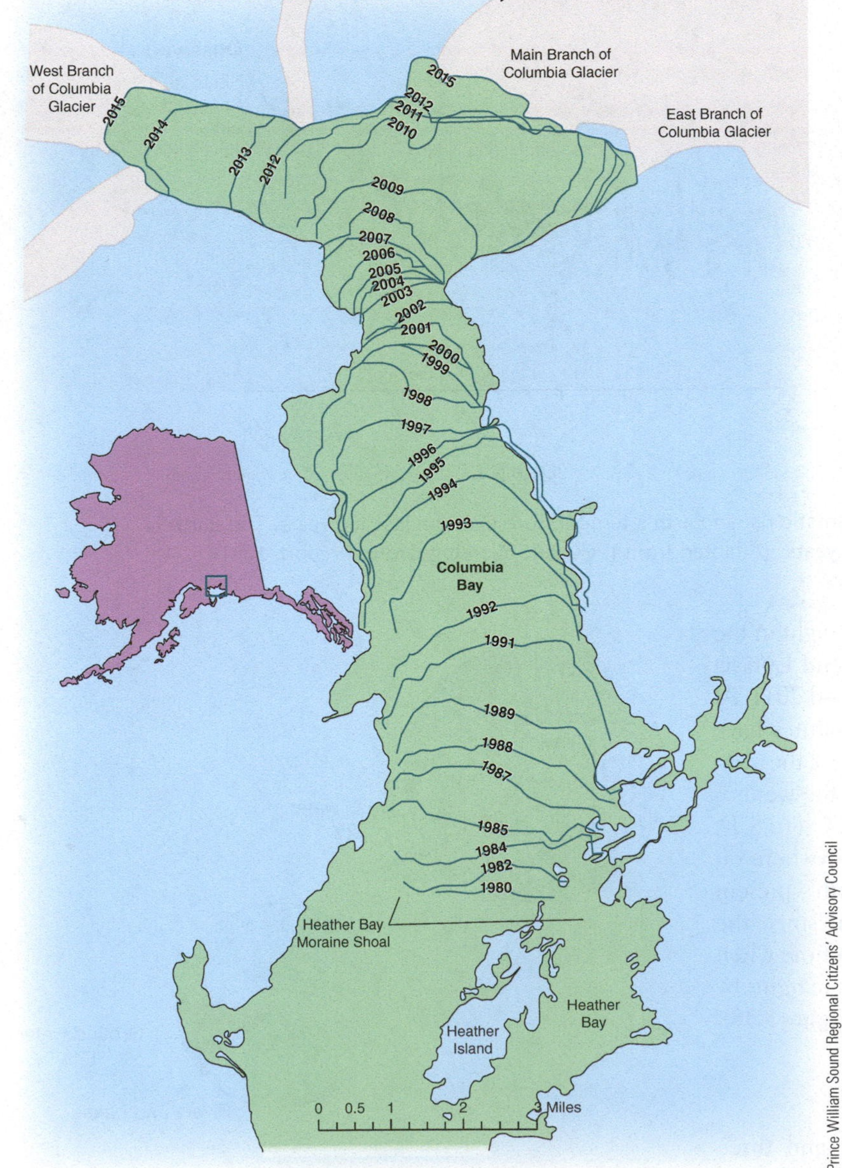

Figure 3.41 The Columbia Glacier is now rapidly retreating.

As already pointed out, today's glaciers store enormous amounts of freshwater, more than exists in all of the lakes, ponds, reservoirs, rivers, and streams of North America. Glaciers in the Himalayas, Norway, Switzerland, Alaska, Washington State, Alberta, and British Columbia have a vital effect on regional water supplies during the dry seasons. Glaciers supply a natural flow of meltwater to rivers, which helps balance the annual and seasonal variations in precipitation. Meltwater from glaciers is a principal source of domestic water in much of Switzerland. The Arapahoe Glacier is an important water source for Boulder, Colorado, so much so that Boulder residents take great pride in telling visitors that their water comes from a melting glacier (though the glacier is only part of a drainage basin of several tributaries that together supply the city's water.)

Many lowland areas along the base of the world's glacierized mountains are dependent on glacial meltwater. The desert regions of Pakistan, northwestern India, and parts of China count on water from the ice- and snow-covered Himalayas. Many agricultural regions in South America depend on water from the glacier-covered mountains of the Andes (**Figure 3.43**), as do the Swiss in agricultural areas in some of the large valleys that rely on meltwater from Alpine glaciers for irrigation.

Glaciers are equally important when it comes to hydroelectric power generation in mountainous regions such as Norway, Switzerland, and Austria (**Figure 3.44**). Tunnels and pipelines carry meltwater to lakes and reservoirs, where it is stored in summer to be released to hydroelectric power stations in winter when the demand for electricity is greatest. The potential impact on the food

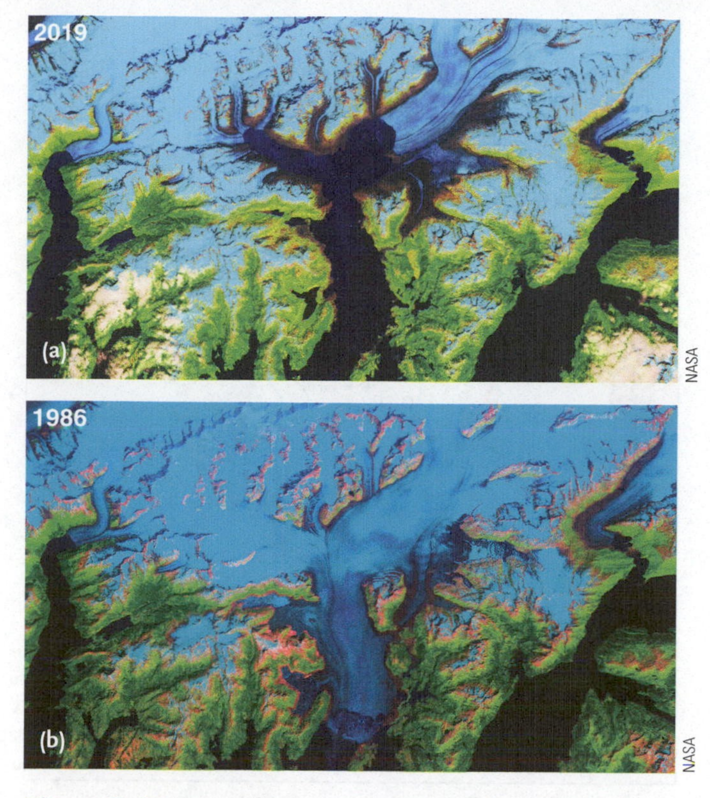

Figure 3.42 Images from space in 2019 (a) and 1986 (b) clearly show the dramatic retreat of the Columbia Glacier.

supply and power generation of the thinning and retreat of glaciers in mountainous regions due to global warming is a matter of concern and of global equity. Some of the world's least-resourced regions will be dramatically affected when the demise of glaciers ends the reliable summer stream flows, starving agricultural fields and people's taps of the water they need.

But mountain glaciers such as the Columbia contain only 6% of the world's ice. Of more concern are the enormous ice sheets of Greenland and Antarctica, which together contain about 90% of the world's freshwater (**Figure 3.45**). If only 10% of the water in these enormous ice sheets were to melt and be added to the ocean, sea level would rise by over 6 meters (20 feet), cause coastal devastation worldwide, and force the displacement of hundreds of millions of coastal dwellers. James Hansen, a highly respected senior climate scientist now retired from NASA, states that we are observing the disintegration of the Greenland and West Antarctica ice sheets, "both losing ice at the rate of about 150 million cubic kilometers per year," which is adding water to the ocean with "sea-level rise now going up about 3.5 centimeters (1.4 inches) per decade...that's more than double what it was 50 years ago." These ice sheets are responding directly to higher average global temperatures (**Figure 3.46**).

The Greenland Ice Sheet is a remnant of the vast continental ice sheets of the last Ice Age. Based on satellite data collected by the GRACE experiment managed by NASA's Jet Propulsion Laboratory, Greenland's glaciers shrank by a record 532 billion tons in 2019, the largest decrease ever recorded (**Figure 3.47**). This melt alone contributed 1.5 millimeters of global sea-level rise. Satellite radar measurements of ice speed reveal that the flow rates of several of Greenland's glaciers have quadrupled since the 1990s. In 2012, the Jakobshavn Glacier reached a record speed, flowing from land to sea at 17 kilometers (9.4 miles) per year. Combined, the effect has been a greater discharge of meltwater from the Greenland Ice Sheet as a whole, with meltwater draining from the Greenland Ice Sheet at double the average rate of flow in the Colorado River. Greenland holds about 2.5 million cubic kilometers, or 10%, of the world's ice–that's enough water to raise the global sea level more than 7 meters (23 feet) if it were all to melt.

Figure 3.43 In many parts of the world, people depend on glaciers for their water supply. Melting glaciers on Huascarán Sur, in Peru's Cordillera Blanca, supply most of the water for extensive farmlands along the base of the mountain. At an elevation of 6,748 meters (22,139 feet), Huascarán Sur is the highest peak in Peru and the fourth highest in the Western Hemisphere.

Figure 3.44 A major benefit of glaciers in developed, mountainous countries is their reliability for irrigation and hydroelectric power generation. A dam at Griesgletscher (Gries glacier) in Switzerland acquires and stores water in the summer melting season for use in generating power in the winter.

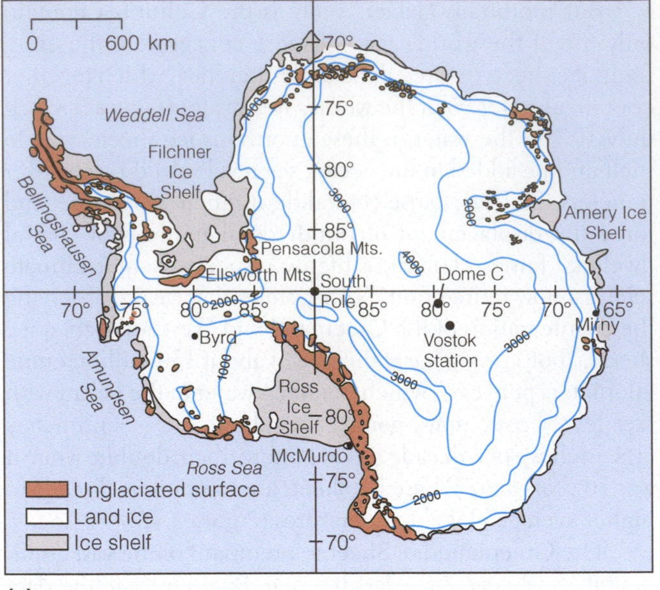

(a)

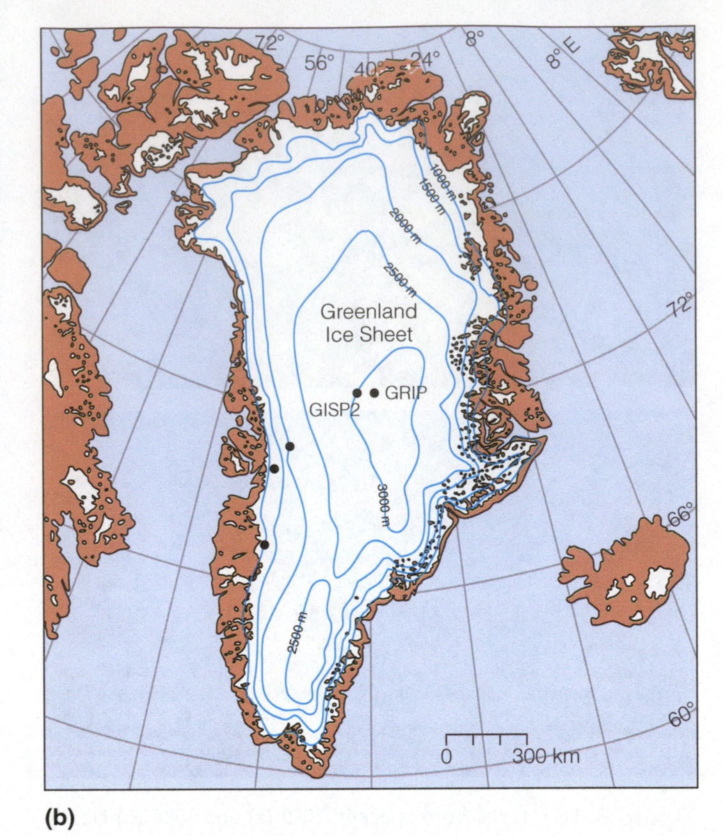

(b)

Figure 3.45 Earth's two continental glaciers. (a) The largest glacier complex covers practically all of Antarctica, a continent that is more than half again as large as the 48 contiguous states. The ice sheet's thickness averages about 2,160 meters (7,085 feet), with a maximum of 5,000 meters (16,500 feet). (b) Greenland's Ice Sheet has a maximum thickness of 3,350 meters (10,988 feet). Elevations are contoured in meters.

3.4.3 The Ups and Downs of Sea Level

Glaciers and ice sheets are a primary control on sea level as they hold frozen water on land that, when melted, increases the volume of water in the world's ocean. If all the polar ice were to melt, sea level would rise more than 60 meters (200 feet), and London, New York, Tokyo, and Los Angeles would be mostly under water (**Figure 3.48**). Tide gauges around the world reveal that sea level has risen by nearly 20 centimeters (8 inches) since 1880. Beginning in 1993, sea level has been accurately measured by satellites, which show sea level rising at the rate of 29.6 centimeters (1.17 inches) per decade on average around the world. (**Figure 3.49**).

The cause of sea-level rise is twofold, and both are consequences of global warming: freshwater entering the ocean from the melting of glacial ice and the expansion of warming seawater. It is estimated that 50% to 80% of the rise results from melting of land-based ice, such as the glaciers of mid-latitude mountains and the Greenland Ice Sheet, because their meltwater is added directly to the ocean. In contrast, a floating shelf ice that already displaces its own weight of water (Antarctica has many) has no impact on sea level. Thermal expansion of the global ocean is also a major factor. Sea level rises about 24 centimeters (9.5 inches) with each 1°C increase in seawater temperature. Although this effect seems quite minor for individual grams of water, there is quite a quantity of water in the ocean!

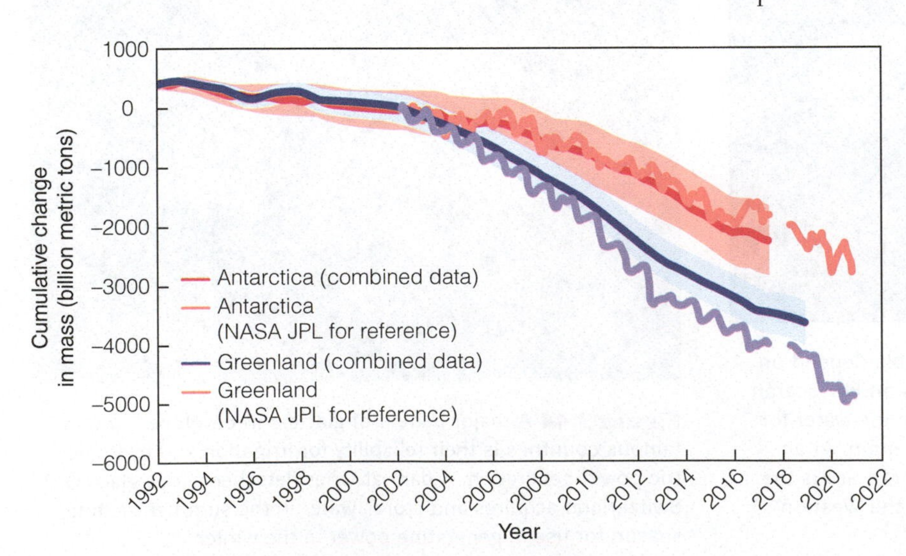

Figure 3.46 Both the Greenland and Antarctic ice sheets are in decline. After 2002, they both began rapidly losing ice mass. The combined data curves include information from over 20 different scientific studies. (Adapted from Climate Change Indicators: Ice Sheets, U.S. EPA.)

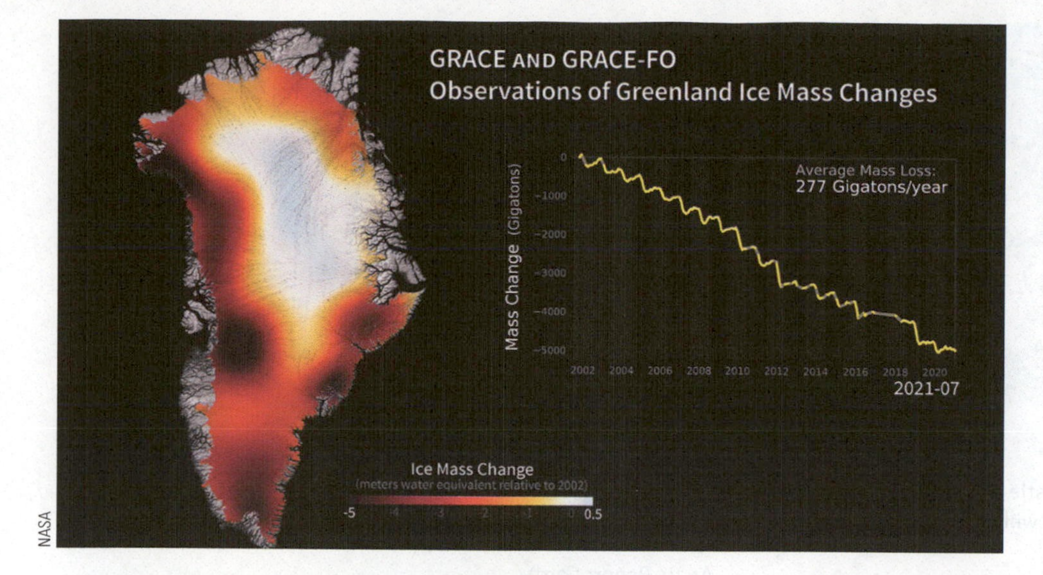

Figure 3.47 Greenland's ice is melting. Satellites orbiting far above the ice sheet measure the change in gravity as snow accumulates in winter and ice melts in the summer.

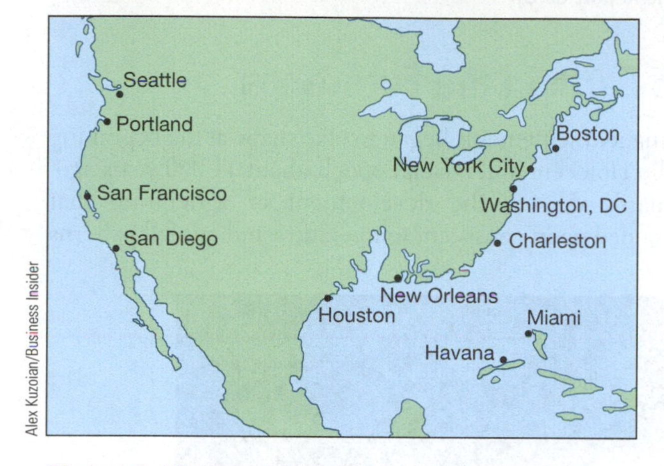

Figure 3.48 Many cities in North America would be underwater should all the polar ice melt. Map shows some potentially impacted major U.S. metro areas.

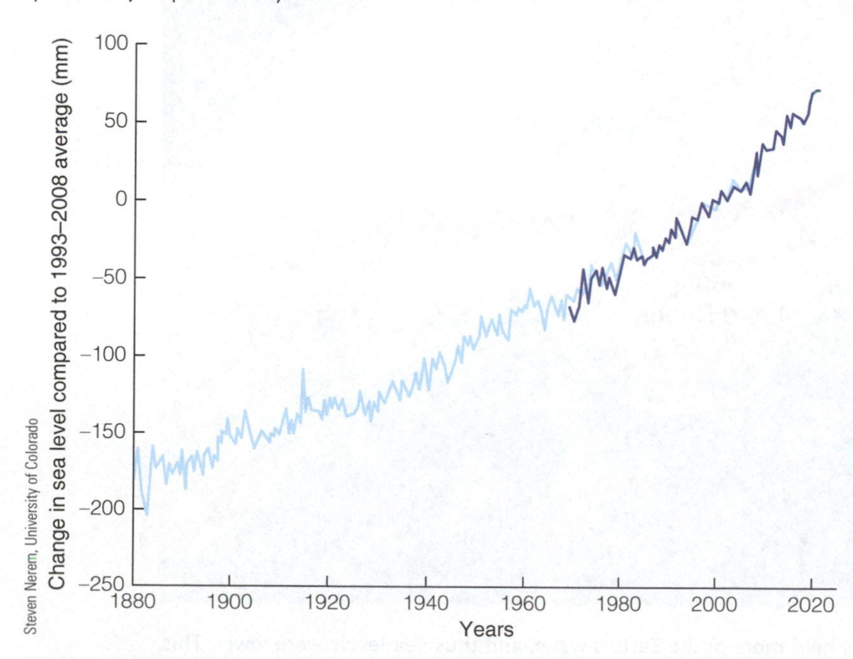

Figure 3.49 Two overlapping records of sea level. The values are the change in sea level compared to the 1993-2008 average.

Sea level has had a major influence on human societies in the past. When sea level dropped during the ice ages, rivers in coastal regions began downcutting across the exposed continental shelves as the rivers adjusted to the new conditions. When sea level rose to its present level, about 5,000 years ago, the mouths of the coastal rivers were drowned, forming what are now harbors and estuaries–culturally important parts of our landscape (**Figure 3.50**).

The formation of land bridges caused by the lowering of sea level during the ice ages was of major importance in establishing the current biogeography of Earth. Among the land bridges 25,000 years ago, the Bering Land Bridge (**Figure 3.51**) is especially significant because the two-way traffic across the bridge accounts for horses, mammoths, and other large mammals' crossing between North America and Asia. During the lowered sea level of the last major ice age, humans may have used the bridge to cross from Asia into North America. Rising sea level beginning about 20,000 years ago began to drown the bridge, and the 90-kilometer (54-mile) stretch of ocean between Alaska and Siberia now blocks most terrestrial organisms from migrating.

Predicting the rate of future sea-level rise is tricky. It will go up for sure; the question is, how fast and how much? Especially uncertain is how the large Antarctica and Greenland Ice Sheets will respond to warming temperatures, as their response involves complex processes of ice loss. For instance, a warming ocean destroys the floating tongues of ice that form where land-based glaciers float out to sea. The land-based tongues are anchored to rock, which holds back the glacier beyond. However, when the ice tongue breaks off, the moving glacier increases its speed. This is already happening to outlet glaciers in Antarctica and some glaciers in Greenland. Satellite measurements show

Figure 3.50 The harbor in Newcastle Australia is quite close to sea level–a slight rise in sea level will flood the harbor and the central business district.

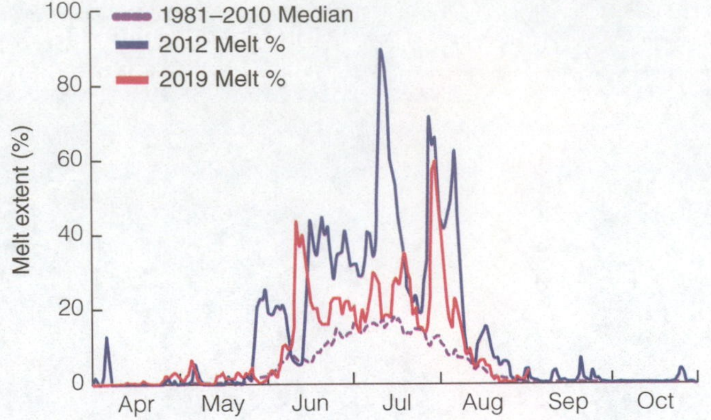

Figure 3.52 Melting in Greenland is increasing: 2012 and 2019 are so far the biggest melt years on record with more than half the ice sheet melting at one point or another during the summer. (Data from Tedesco, M. et al. (2019). Greenland Ice Sheet, NOAA Arctic Report Card)

that the rate of loss from Greenland's Ice Sheet has been increasing. During the summers of 2012 and 2019, almost the entire surface of the ice sheet melted during warm spells. In summer 2021, it rained instead of snowing for the first time in recorded history at the summit of the Greenland Ice Sheet (**Figure 3.52**).

3.5 Earth in the Future (LO5)

Permanent settlements began to take shape at the beginning of the Holocene interglacial epoch about 11,000 years ago, primarily due to the development of agriculture that depended upon consistent temperature and rainfall patterns

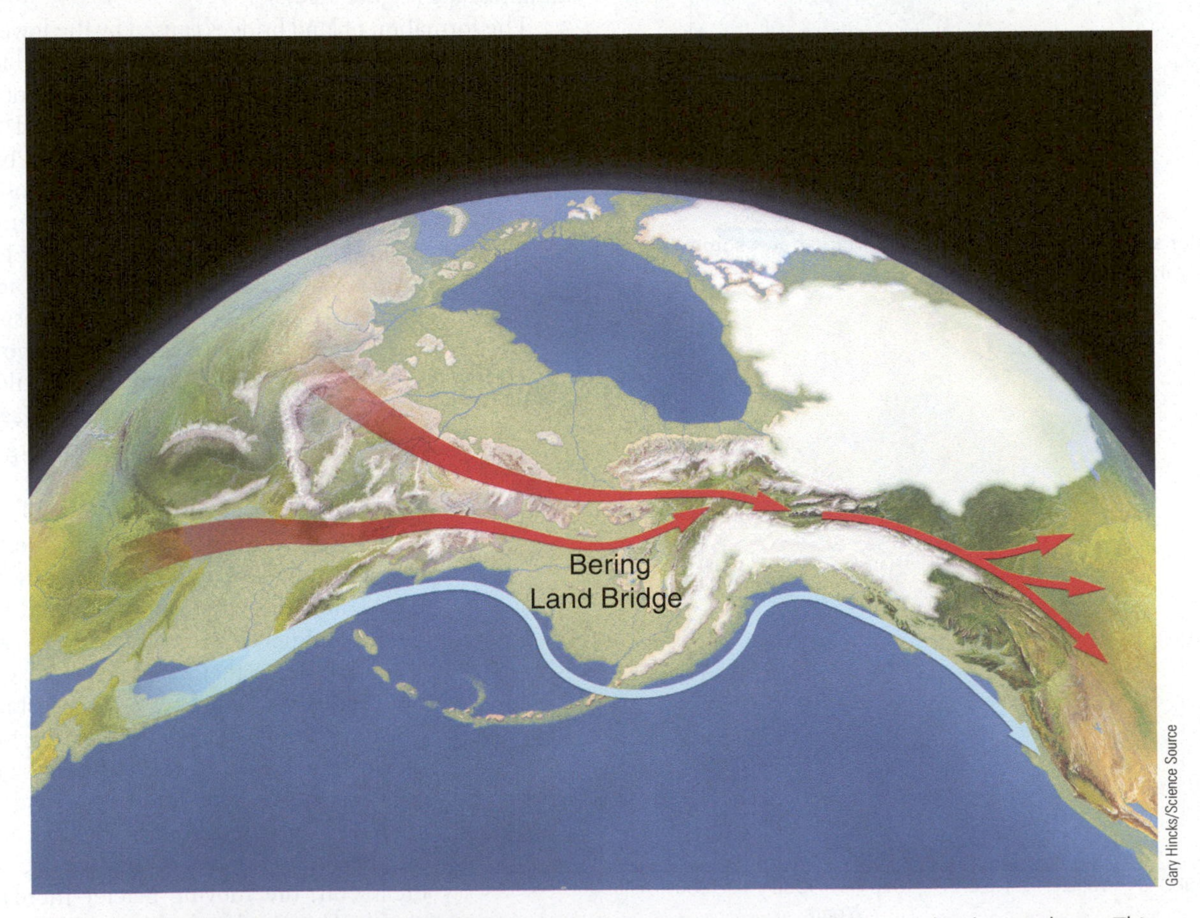

Figure 3.51 During glaciations, ice sheets held more of the Earth's water, and thus sea levels were lower. This allowed people and animals to migrate between continents over land bridge (possible migration routes represented by the red and light blue lines) now under the ocean.

© Ryan W Vachon

Figure 3.53 Ice coring in the high elevations of the tropics provides climate information unavailable in any other way.

that had not existed during the preceding ice age. This climate stability allowed farming to provide reliable amounts of surplus food for sedentary populations that began living in cities. Previously, human existence had consisted of nomadic hunting and gathering, with little established village life. Surplus food allowed new crafts and trades to develop and political and religious authorities to flourish. Likewise, stable sea levels were a must for developing worldwide maritime trade. Complex port cities with all their infrastructure could not thrive under conditions of rapidly fluctuating sea level. A stable high stand in sea level was important for the development of flood plains such as the Great Plain of China and the Nile Delta where fertile farmlands support millions of people. A dependable, reliable climate was essential in the past and is essential now to maintain stable societies.

3.5.1 A World With Little Ice

As the climate warms, we are headed to a world with little ice. Mountain glaciers will go first. Already, tropical glaciers high on Andean and African peaks have vanished. The only remaining bits of their ice are in freezers around the world–leftovers from cores taken by geologists rushing to gather samples before the ice melted in increasing tropical heat (**Figure 3.53**).

The Greenland Ice Sheet could be a mere shadow of itself in a few hundred to perhaps a few thousand years. If warming continues unabated, permanent ice there will likely cling only to mountainous uplands in the south and east, the result of heavy precipitation and cold temperatures at elevation. The Antarctic will fare a bit better. While the West Antarctic Ice Sheet is likely doomed to melt, ice in the East will remain, although shrunken.

Near the poles, sea ice (frozen sea water) forms on the surface of the ocean. In the Arctic Ocean, some sea ice persists from year to year. Both Antarctic and Arctic sea ice serve as critical habitats for marine birds and mammals, but Arctic sea ice appears to play another important role: regulating Northern Hemisphere climate. The sea ice in the arctic is thinning and less is surviving year to year.

Between 2018 and 2021, radar and laser altimeter (LIDAR) measurements show that Arctic sea ice lost 16% of its average winter volume, or half a meter (1.6 feet) of thickness in the duration of three years. For context, an 18-year record of Arctic ice thicknesses reveals that the volume of winter Arctic sea ice has decreased by 6,000 cubic kilometers (1,400 cubic miles), about a third of the total volume, as of 2021 (**Figure 3.54**). This is largely the result of reduced old (multiyear) ice and more thinner, seasonal ice. The summertime minimum extent of Arctic ice has been in steady decline since satellite measurements began in 1979. The record minimum was set in 2007, with the extent of summer ice in 2011 ranking second lowest of the 32-year record (**Figure 3.55**). Winter Arctic ice extent has also decreased since 1979, by about 4.2% per decade.

Combining satellite observations of sea-ice extent with conventional atmospheric observations reveals that decreasing summer Arctic ice impacts large-scale atmospheric phenomena during the following autumn and winter in regions well beyond the Arctic's boundary. Heat rising from the exposed Arctic ocean water in summer warms and destabilizes the lower atmosphere, causing increased cloudiness and weakening of the polar **jet stream**, the current of air that flows rapidly at the top of the troposphere between the Polar and Ferrel cells (see Chapter 2). The weak jet stream causes less winter precipitation in Scandinavia following the summers of reduced sea ice, resulting in less water in reservoirs

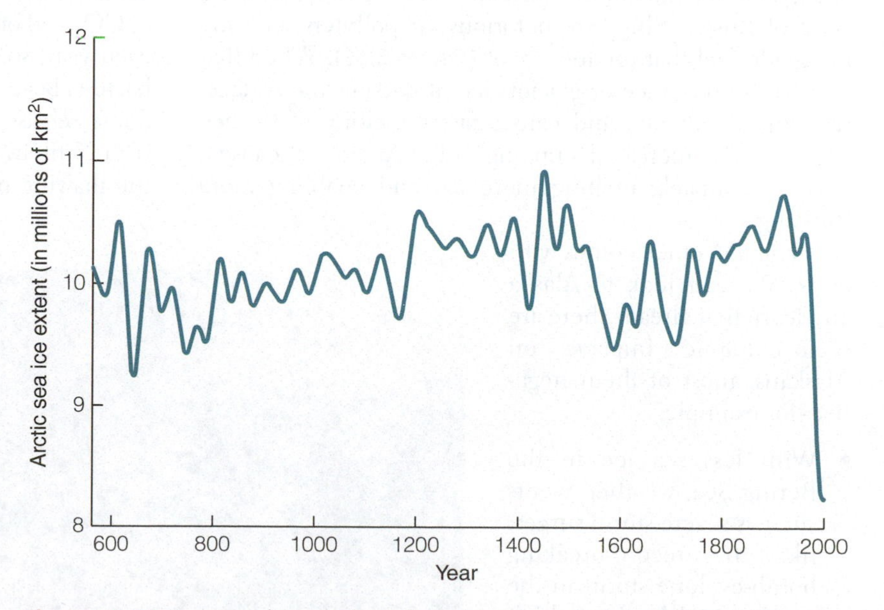

Figure 3.54 Sea ice volume today is less than at any time in the last 1,500 years. (Adapted from NOAA)

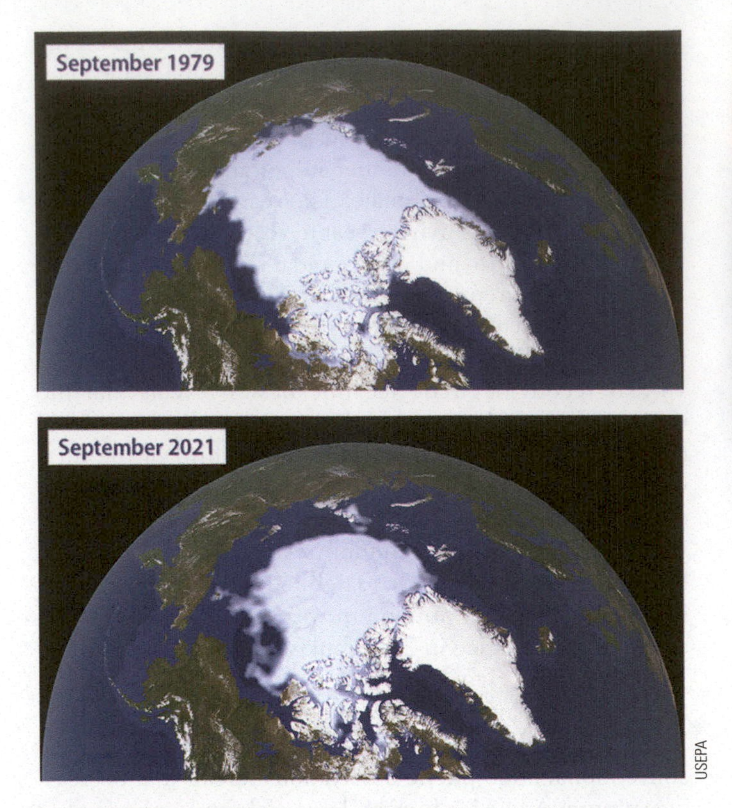

USEPA

Figure 3.55 Sea ice extent has shrunk dramatically since 1979, exposing much of the arctic ocean.

for hydroelectric power generation. Similarly, reduced sea ice aggravates droughts in the southwestern and southeastern United States.

A rapidly warming Arctic encourages more summer ship traffic through Canada's Northwest Passage and the Northern Sea Route. A Danish shipping company reports it saved a third of its usual fuel costs and half the time in shipping goods to China from Murmansk by following the Arctic coast of Russia. Ships are notorious air polluters, burning low-grade fuel that produces soot (**Figure 3.56**). When that soot lands on sea ice or glaciers ice, it darkens the surface, reducing its albedo, and causes faster melting of the ice (**Figure 3.57**). Increased shipping in the Arctic will cause a positive feedback, melting more ice and enabling more shipping.

The loss of ice comes with costs. We can look to Alaska and learn that already there are socioeconomic impacts on Alaskans, most of them negative–for example:

- With less sea ice in the Bering Sea, weather events such as severe storm surges, like the record-breaking bomb-cyclone storm in the Aleutian Islands which occurred in late 2020, have increased in frequency and severity, causing coastal erosion, inundation, and destruction of coastal villages and structures (**Figure 3.58**).
- Subsistence lifestyles of the native peoples in the region have been adversely affected by changes in sea-ice conditions. Hunting on the ice, for example, is more dangerous (**Figure 3.59**).
- Slope instability, landslides, and erosion are increasingly common in thawing permafrost terrain. This threatens infrastructure like roads and bridges, and worsens erosion (**Figure 3.60**).

iStock.com/EasyBuy4u

Figure 3.56 Oceangoing ship leaving a trail of soot in its wake.

3.5.2 Melting Permafrost

Permafrost thawing has caused major changes in the tundra, bog, and forest landscape features and wetland ecosystems in subarctic North America and Siberia. For example, the southern limit of permafrost in the region of James Bay, Canada, has moved 130 kilometers (81 miles) north since about 1960. As Earth warms up in response to millions of tons of CO_2 and other greenhouse gases entering the atmosphere each year, so does the permafrost. Once permafrost thaws, bacteria begin eating the organic matter it contains and in so doing release CO_2 and methane–another greenhouse gas, with 25 times the warming power of CO_2. Also released from the thawing permafrost are phosphates and nitrates, which

© Ben Pelto

Figure 3.57 Soot from wildfires covers glaciers in western Canada, reducing their albedo.

NOAA

Figure 3.58 Ocean waves undercut permafrost and erode the shoreline in Northern Alaska.

iStock.com/NiseriN

Figure 3.59 Polar bear stranded on an ice flow as the sea ice melts.

alter the ecosystem by helping novel plant types become established (see Chapter 2). The world's permafrost contains about twice as much CO_2 as the atmosphere. If even a small amount of that were released as CO_2 gas, along with the methane, it would contribute to a strong, positive feedback that would strengthen global warming.

In recent years, the incidence of tundra fires has increased; these too release carbon dioxide to the atmosphere that was once stored on the landscape. In 2020, forest fires in Siberia burned more than 155,400 square kilometers (60,000 square miles) of tundra, an area greater than that of the state of Florida (**Figure 3.61**). It is estimated that Siberian fires pumped about 800 million metric tons of CO_2 into the atmosphere in 2020. Scientists anticipate that Arctic fires will become more frequent as tundra and other permafrost plant communities become drier because of global warming.

iStock.com/Lyagovy

Figure 3.61 Wildfire burns boreal forest in Siberia releasing smoke and carbon dioxide.

Since about 1960, the number of days that oil company personnel can travel on Alaska's frozen North Slope to search for oil has shrunk in half. The tundra, which must be completely frozen before running vehicles on it, has been freezing later and thawing earlier. Building, road, and rail foundations are failing (**Figure 3.62**).

3.5.3 Extreme Events—Droughts, Floods, and Storms

A warming climate is likely to lead to more extreme weather events, which have already begun and are predicted to increase. Extreme heat is perhaps the most obvious result of a warming planet. Indeed, heat waves in Europe in 2019, 2021, and 2022 caused record fatalities, especially among the poor and the elderly who could not afford air conditioning and many of whom lived in urban area where the heat island effect exacerbated the warmth. In 2022, many places

Aleksandr Lutcenko/Shutterstock.com

Figure 3.60 Melting permafrost creates a sinkhole on the Yamal Peninsula in Siberia.

Nordroden/Shutterstock.com

Figure 3.62 Melting permafrost makes the ground unstable, here distorting a rail line in Siberia.

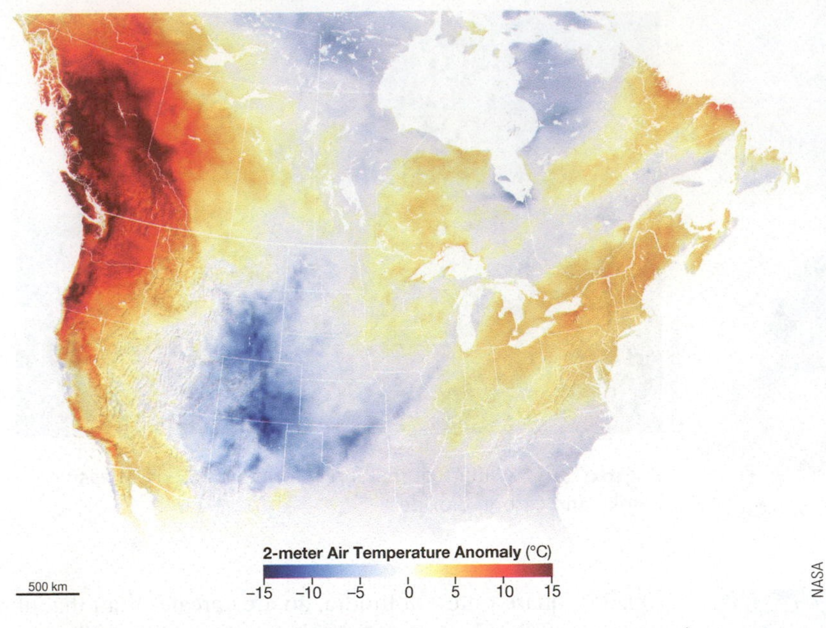

Figure 3.63 Red areas show how much warmer than average northwestern North America was during the June 2021 heat dome.

in the United Kingdom broke the 100°F (38°C) barrier for the first time ever, topping out at 104°F! A similar scenario played out in Oregon and Washington State in the summer of 2021, when a massive heat dome (a stagnant mass of hot atmosphere) enveloped the region, sending temperatures well over a 100°F (38°C) for days on end. The fatalities again were mostly older and poorer people who could not get to cooling stations.

Heat domes are the product of stagnant high pressure. They can linger for days as the sun warms the air. Such heat domes are becoming increasingly common as the climate warms – most likely because the path of the jet stream has gotten wavier. These waves, large meanders, move very slowly and sometimes stagnate, the stationary air just gets hotter and hotter. Adiabatic heating from the orographic effect can make heat domes even worse. That was the case in the U.S. Pacific Northwest in 2021 when temperatures Washington hit 49° C (120° F)! This event killed more than 120 people; almost 70% of those who died were over 65 years old. The hot air was restricted to the western United States and many places under the dome were more than 10° C (18° F) above normal summer temperatures (**Figure 3.63**).

Drought and fire are other expected outcomes of a warming climate. Warmer air temperatures increase evaporation, drying both soils and vegetation. And with more precipitation falling as snow and less as rain, snowpacks are thinner and melt faster in the spring, causing forests to dry out earlier in the spring. The result is more early-season fires and a longer fire season. Over the past several decades, drought has impacted water supplies throughout the western United States. The current drought in California rivals the greatest droughts on record, including those of 1200 years ago that devastated Native American populations (**Figure 3.64**).

Some effects of drought are not what you might think. Sure, reservoirs see water levels drop; some might even run dry. But in the 2021 and 2022, the news of drought from the southwestern United States included all sorts of things being

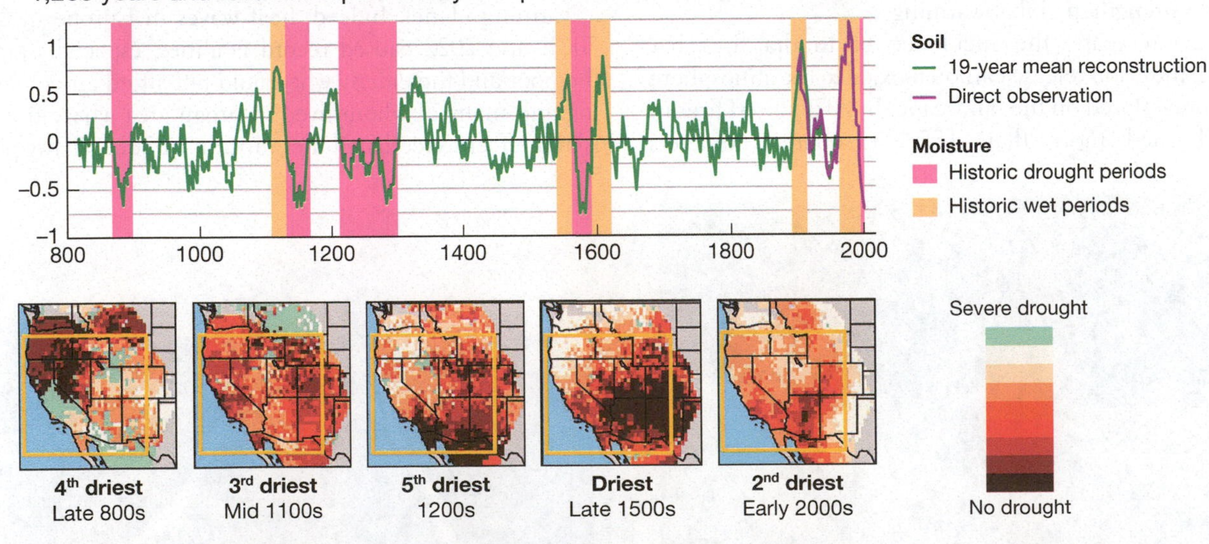

Figure 3.64 Today's drought in California is one of the most extreme–it is as dry today as it has been anytime in the last 1,200 years, according to paleoclimate records. (Adapted from Lamont-Doherty Earth Observatory of Columbia University)

exposed as water levels in some of western North America's biggest reservoirs drop to unprecedented lows (**Figure 3.65**). Sunken boats were no surprise, except when a 1945 WWII vintage landing craft emerged. It was once under more than 30 meters (100 feet) of water in Lake Mead. More awkward are the bodies. Four emerged from drying reservoirs in 2022. Most appeared to be drowning victims but one, found in a barrel with a gunshot wound, led police Lieutenant Jason Johansson to suggest that "Anytime you have a body in a barrel, clearly there was somebody else involved."

A warmer atmosphere holds more water, and more water can mean heavier rainfall and more intense storms. Take for example Hurricane Harvey, which stalled over south Texas in 2017. Over 4 days, it tapped warm, moist air from the Gulf of Mexico and flooded the flat landscape near Houston with over 1.5 meters (60 inches, or 5 feet of rain). The water had nowhere to go and the results were devastating (**Figure 3.66**): more than 100 people dead, nearly a quarter-million homes destroyed, and $125 billion in damage. Climate models suggest that climate change made a storm like Harvey about 40% stronger and more damaging than it would have been in the past.

Not only do such wet storms happen in humid regions, they can also happen in the driest parts of the world. Death Valley in southern California is not known for floods but rather for the highest temperature ever recorded in the lower 48 states, 58° C or 134° F. In an average year, the locale gets less than 5 centimeters of rain (2 inches). On Friday, August 5, 2022, that much rain fell in just a few hours and the landscape responded. Mud and rocks poured down dry creek beds and overwhelmed roads, cars, and tourist infrastructure in Death Valley National Park just six weeks after unusually intense storms did similar damage in Yellowstone National Park, 1,800 kilometers (1,000 miles) northeast (**Figure 3.67**). Noah Diffenbaugh, a Stanford University climate professor, clearly described the climate change science behind the massive rain: "What we're seeing with climate change consistently is that when the conditions that are well understood to produce intense precipitation do come together, the fact that there's more moisture in the atmosphere as a result of long-term warming means that those conditions are primed to produce more intense precipitation."

Damage from increasing precipitation does not require storms the size of Harvey. Indeed, relatively small changes in

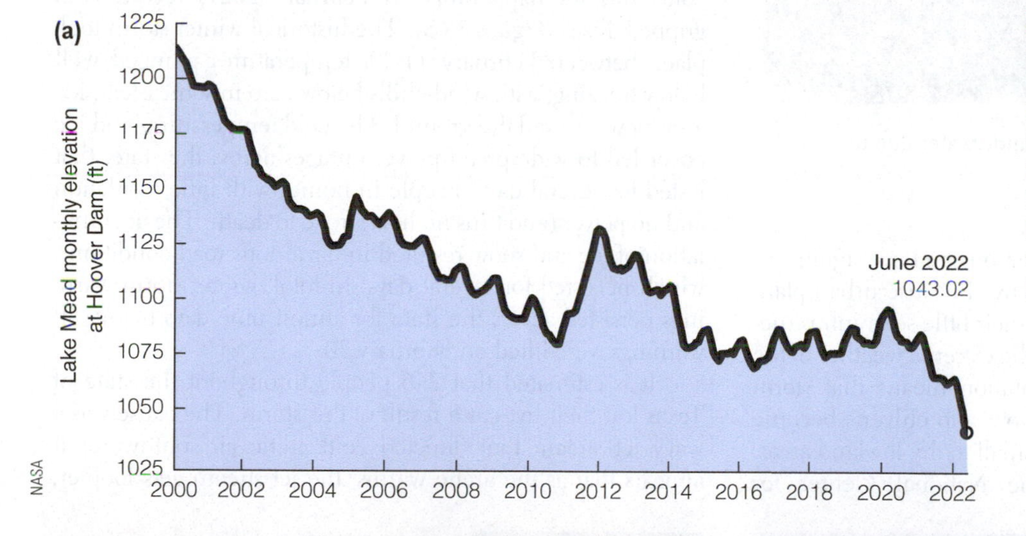

Figure 3.65 Lake Mead is running out of water. (a) The level of Lake Mead has fallen dramatically since 2000. (b) As the water level falls, the lakebed is increasingly exposed as the lake shrinks. (c) Behind the Hoover Dam, which holds in Lake Mead, is a bathtub rung of lake mud–quiet testimony to the lowering lake level.

2000 2021 2022
Virgin River
Overton Arm
(b)
NASA

Figure 3.66 Homes and roads are underwater due to flooding from Hurricane Harvey.

the amount and intensity of precipitation can have significant effects, especially in urban areas. Why? Because urban planners and engineers build cities and their little-seen infrastructure based on past climate records–the extreme weather of the past, not the future. More precipitation means that storm sewers fill, drainage channels overflow, and culverts become plugged. Pumping stations are too small to dry lowland areas. Andreas Prein, a scientist at the National Center for Atmospheric Research studies the most extreme storms and sums it up well, "The infrastructure we have is really built for a climate we are not living in anymore."

Determining whether and how much climate change is to blame for weather disasters, particularly extreme weather events, is termed **attribution science.** This field is growing more prevalent as costly weather events increase in frequency. Attribution science is based on statistics and computer modelling of what the climate has done in the past, what climate models predict it will do in the future, and the likelihood of any particular scenario (so much rainfall, so many hurricanes, so severe a drought) happening. A critical scientific advance of the last decade, it has finally allowed people to move past the idea that "you can't blame this storm or that flood on a changing climate." Now, it's possible to suggest the likelihood that a climate event is more extreme because of human changes to the atmosphere. Take the 2022 UK heat wave. Attribution science suggests such heat had a 1-in-1,000 chance of occurring in any given year and that as such the recorded high temperatures were more than 10 times more likely because of our greenhouse-gas emissions.

A less obvious outcome of climate warming is record cold. But it's happening. In February 2021, record cold gripped Texas (**Figure 3.68**). The historical winter storm took place between February 11-20; temperatures plunged well below freezing, with wind chills below zero in some areas; ice and snow covered the ground. The cold temperatures and ice cover led to widespread power outages across the state, that lasted for several days. People in homes with little insulation and no power (and thus no heat) froze to death. The accumulation of ice and snow resulted in hazardous road conditions, which persisted for several days. In total, winter storm warnings persisted across the state for almost nine days before all warnings were lifted on February 20.

It is estimated that 246 people throughout the state of Texas lost their lives as a result of this storm. The cause was a wavy jet stream that directed cold arctic air southward. It appears that as the arctic warms, the jet stream gets loopier,

Figure 3.67 Massive floods, far exceeding typical events, devastated two of America's favorite National Parks in 2022. (a) Mid-June flooding in Yellowstone took out many of the most important park access roads as floodwaters in swollen rivers undercut roadways. (b) Early August flooding in Death Valley sent torrents of mud and rocks down desert washes into tourist attractions, stranding thousands of visitors.

Figure 3.68 The Alamo the day after the 2021 arctic blast and snowstorm in Texas.

Figure 3.69 A wavy jet stream pulls arctic air towards the southern United States.

sending warm air north and cold air south with greater frequency than in the past (**Figure 3.69**). These storms and their effects are worsened when they impact areas underequipped to manage such weather, such as the disastrous impacts seen in the 2021 Texas storm.

3.5.4 A Long Warmth

The planetary climate for the next several hundred thousand years depends strongly on the carbon emissions we expel into the air over the coming century. Atmospheric CO_2 does not immediately return to the hydrosphere, biosphere, and soils, but lingers in the atmosphere. As much as a quarter of peak concentration in an emissions spike can persist for tens of millennia. Ultimately, silicate rocks like granite scrub out CO_2 from the air through natural weathering processes, but the timelines are too long (geological scale) to be humanly meaningful. In a more extreme super-greenhouse scenario, in which we load the atmosphere with 5,000 gigatons of CO_2, certainly within the realm of industrial possibility, the pollution tail stretches out more than 400,000 years–longer than the duration of the human species to date! Under these conditions, mean annual temperatures would rise as high as 6–7°C (11–13°F) in the tropics, and 10°C (18°F) at higher latitudes. Sea level could rise 50 meters (165 feet) or more. The chain of environmental consequences, in other words, is immensely long–and likely lethal for wildlife unable to shift ranges and adapt (**Figure 3.70**).

Of course, one could argue that deferring the next ice age or two is not such a bad thing. What would be worse for humanity, after all? A warmer climate that actually extends the range of human habitability across a larger area of Earth's surface–albeit including lands subject to dangerous heat waves or flooding at times–or the presence of grinding ice sheets that effectively obliterate whole regions, including Canada, northern Europe, and Eurasia? However one chooses to weigh this situation ethically and morally, we nonetheless find ourselves in a position of unprecedented power. Not since the Precambrian (half a billion years ago) has a lifeform (in that instance, primitive blue-green algae) had such immense influence over other forms of life as ours does today (see Chapter 2).

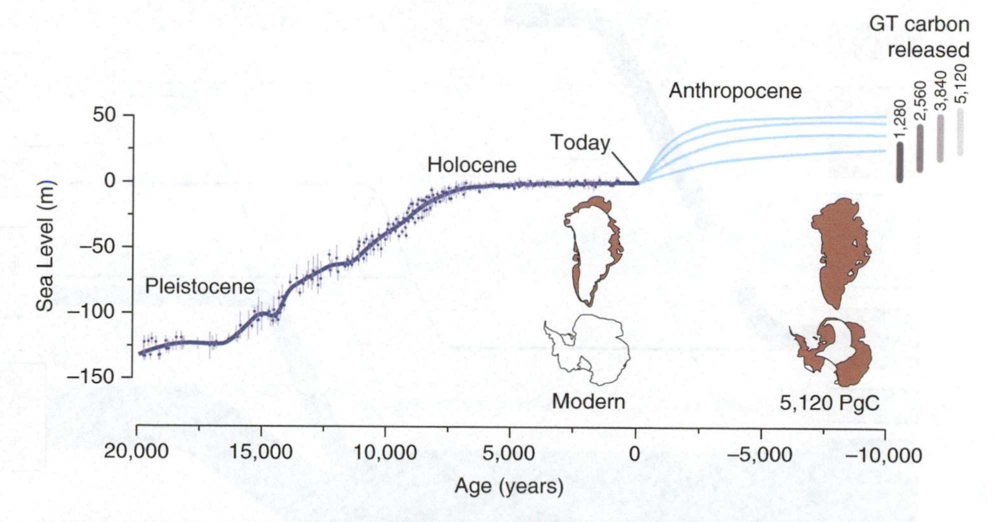

Figure 3.70 The more carbon dioxide we pump into the atmosphere, the warmer the planet will be, the less ice there will be at the poles, and the higher sea level will rise. (Adapted from Clark, P. et al. Consequences of twenty-first-century policy for multi-millennial climate and sea-level change. *Nature Clim Change* 6, 360–369 (2016))

3.6 Addressing Climate Change (LO6)

Greenhouse gas emissions and their effects on Earth's atmosphere, oceans, and living systems are global and could be devastating both to the natural world and to societies around the globe. There are steps we can take both as individuals and as citizens of the world's nations that can reduce the severity of climate change (**mitigation**), decrease the damage change will cause (**adaptation**), and perhaps alter the climate so the effects are less devastating (**geoengineering**). In this section, we consider each of these approaches.

3.6.1 Mitigation

The simplest and most energy efficient approach to climate change mitigation is lowering global emissions of greenhouse gases. The most efficient means to lower emissions is to use energy more efficiently–as a global society we are remarkably wasteful. Repeated analyses of United States energy use (and waste) by Lawrence Livermore National Laboratory shows just how much energy goes to waste (**Figure 3.71**). The 2021 data show that we reject (in other words, waste) twice as much energy as we actually put to good use. The only good news is that renewable energy (solar and wind) grew dramatically in 2021.

While each of us can play a role in mitigating climate change, the problem is so large and so systemic that solving it will take a global response at both the personal and governmental level. While you can buy an electric vehicle, that will only reduce your carbon footprint if the energy used to drive it is not derived from fossil fuels. Enter industry and the government, which build and regulate large-scale energy generation facilities (see Chapter 11). At such plants (your town likely has one) there are mitigation strategies that are only practical applied at a large scale. What makes sense at a big power plants very likely does not make sense attached to your individual tailpipe!

Consider **carbon capture and storage** (CCS), a process that traps CO_2 emissions and keeps them out of the atmosphere. It's been proposed for installation at power plants, refineries, cement plants, and other **point sources** (single smoke stacks) of carbon dioxide pollution. The idea is to trap CO_2 before it enters the atmosphere and bury it deep underground. Carbon dioxide in its supercritical form (greater than 31.1°C and pressure exceeding 72.9 atm), can be trapped in the geological subsurface by an impermeable rock layer (structural trapping), within the porous space between rocks (residual trapping), within salty brine in the pores of rocks (solubility trapping), or by forming carbonate minerals due to CO_2's chemical reaction with salt water (mineral trapping). Currently, the United States Department of Energy has 18 small-scale operations testing the economic viability of geologic carbon storage technologies in Michigan, Montana, Alabama, and several other states. Large-scale CCS is just beginning to occur around the world. Still, the technique remains in its infancy and capturing and storing carbon is energy intensive!

The best sites for potential geological storage (or "sinks") are depleted oil and gas fields, deep saline aquifers, and deep coal seams that cannot be mined (**Figure 3.72**). Ideally, the sink should be within 60 kilometers (37 miles) of one or more power plants, have a large reservoir volume, consist of porous and permeable reservoir rock, and be capped by a seal of impermeable shale to prevent leakage back into the atmosphere. The subsurface storage should be at a depth greater than 800 meters (2,600 feet), so that the overburden pressure will maintain the CO_2 as a high-density fluid instead of a gas. Once underground, the CO_2 will slowly react with mineral

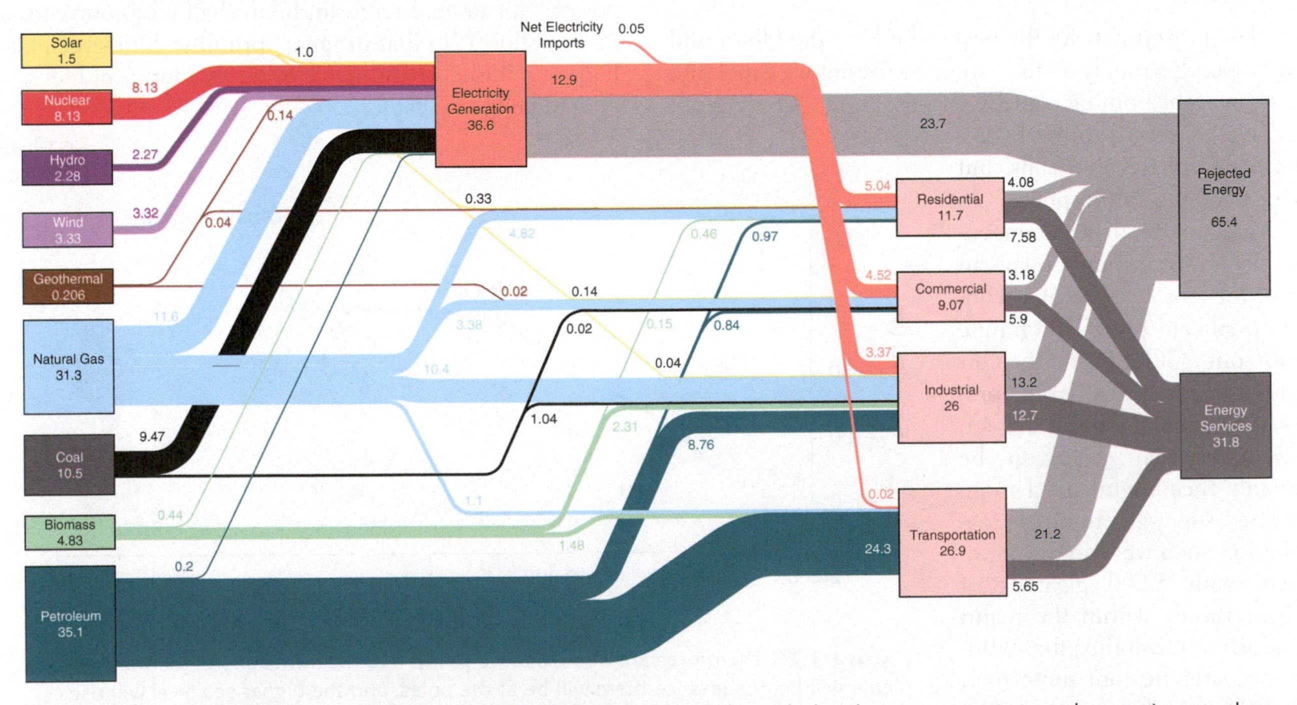

Figure 3.71 This diagram maps energy inputs and energy usage and concludes that we waste (rejected energy) more than twice as much energy as we use! (Data from Lawrence Livermore National Laboratory)

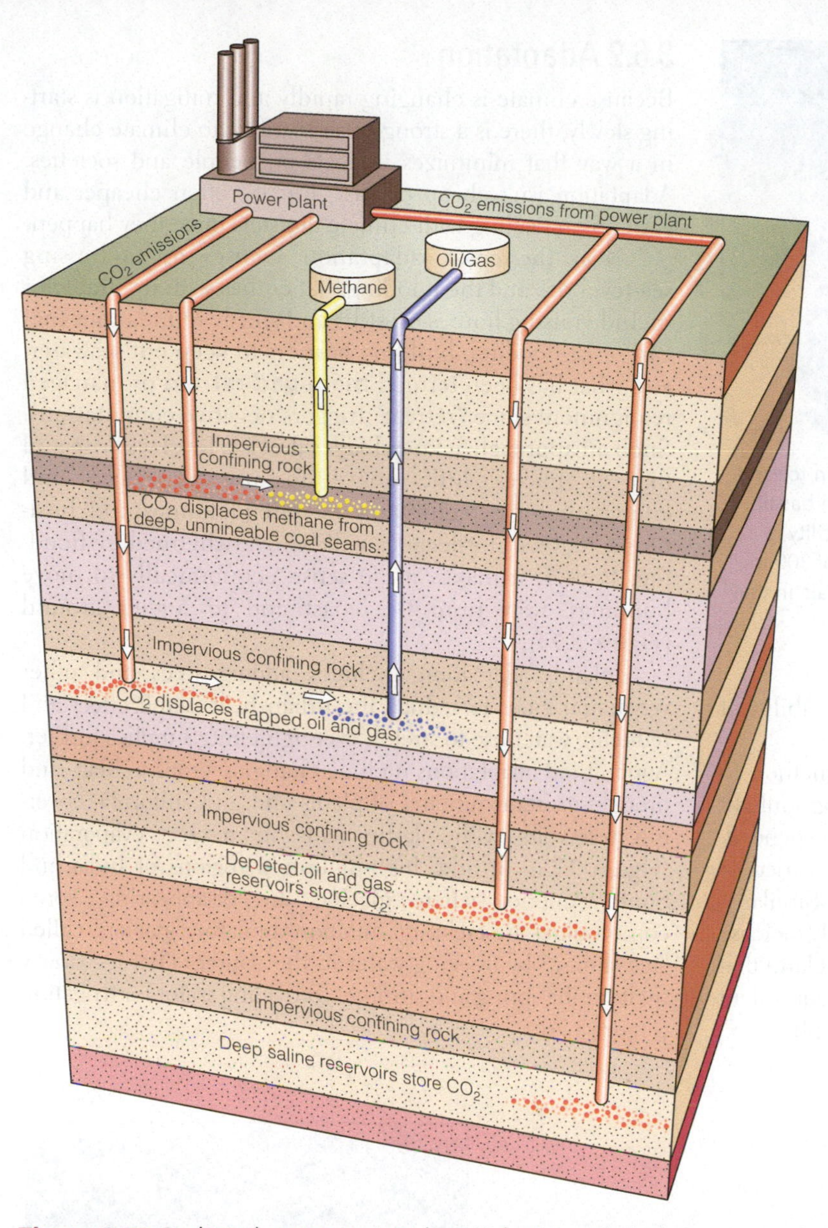

Figure 3.72 Geological storage options by CCS for CO_2 include deep, un-mined coal seams, depleted oil and gas reservoirs, and deep saline reservoirs. (Adapted from Ohio Department of Natural Resources, Division of Geological Survey)

material and organic matter to form stable minerals, primarily iron, calcium, and magnesium carbonates.

In the United States, more than 5,800 kilometers (3,600 miles) of pipeline already carry and store CO_2 underground, predominantly to oil and gas fields to stimulate additional production, a technology called **enhanced oil recovery** (EOR). The United States is the world leader in EOR and each year injects about 48 million tons of CO_2 to stimulate additional production from abandoned or depleted oil and gas fields. This is good practice for carbon capture and storage, although ironically it does enable the recovery and use of additional fossil fuels.

Norway's Statoil has been running an experiment since 1996 that reveals how CCS can play an important role in climate mitigation strategies. The CO_2 comes from the Sleipner natural gas field. The CO_2 is extracted from the natural gas before being piped to the mainland. This is the world's first commercial-scale operation for sequestering CO_2 in a deep saline reservoir. Since 1996, the project has injected some 2,800 tons of CO_2 per day, or about 1 million tons per year, but the demand for CCS is much greater–perhaps more than 10 billion metric tons per year. The question has been raised about potential leakage of CO_2, which is a highly reactive compound, from deep reservoirs. However, seismic monitoring of the Sleipner Project has shown that the CO_2 is safely stored beneath a thick layer of impermeable shale.

The capture of CO_2 from flue gases at fixed emission sites such as coal-fired power plants does nothing to reduce the CO_2 from mobile sources and that which we have already released to the air. If we could entirely halt anthropogenic CO_2 emissions today, significant climate impact from the remaining CO_2 would still be with us for hundreds to thousands of years. The other option–pull CO_2 out of the air–is energy intensive and expensive. If such direct air capture ever became practical, the CO_2 would need to be concentrated, compressed, and then moved deep in the crust or pumped deep in the ocean.

Another option is injecting CO_2 dissolved in water into thick flows of basalt lava such as those in the Columbia Plateau in the northwestern United States and the Deccan Traps in India. Basalts are rich in calcium silicate minerals, and CO_2, mixed with ground water will leach out the calcium and form the relatively stable mineral calcite. This chemical reaction, a form of geochemical weathering, goes on naturally all the time, and the technology would simply speed up the natural process.

A geothermal plant in Iceland is the site of a pilot study on the sequestration of CO_2 in basalt (**Figure 3.73**). Even better rocks for mineral sequestration of CO_2 would be the mantle-derived ultramafic rocks, peridotite and serpentinite. These rocks have been brought to the surface at several places on Earth by the collision of Earth's tectonic plates. Because they are more enriched in magnesium silicates than basalts, the CO_2 would react more readily and produce a higher concentration of stable minerals. To date, there is no economically feasible technology to accomplish mineral sequestration.

Sequestering carbon in the deep ocean has also been suggested. Below 3,000 meters (9,850 feet), liquid CO_2 is denser than seawater. If injected below that depth, it would sink to the seafloor. Some of the sequestered CO_2 would eventually escape to the atmosphere, but it would take centuries because of the slow speed of the oceanic conveyor belt (Chapter 2). By that time, we can only hope to have decarbonized our energy supply so that atmospheric CO_2 would be decreasing. Deliberate CO_2 storage in the deep ocean

Haraldur Stefansson / Alamy Stock Photo

Figure 3.73 The Hellisheidi geothermal power plant in Iceland. In a pilot study on the feasibility of CO_2 sequestration in basalt, the CO_2 released from the hot water that powers the facility is dissolved in water and injected into basalts at a depth of 300 to 800 meters (980–2,600 feet), where it reacts with minerals in the basalt to form stable new minerals.

would further acidify the ocean. This threatens the ability of many organisms to create shells and survive.

Large-scale forestation has been cited as a method to reduce carbon. Trees capture carbon naturally, using sunlight for photosynthesis. With a minimum of energy expended, trees could be grown, harvested, and then safely buried or burned with carbon capture to store the CO_2. Similarly, changing soil management practices can dramatically increase the volume of carbon stored in soils. Some people claim that repairing damaged soils worldwide could scrub the equivalent of 10 years of U.S. carbon emissions from the atmosphere.

3.6.2 Adaptation

Because climate is changing rapidly and mitigation is starting slowly, there is a strong push to adapt to climate change in a way that minimizes impacts on people and societies. Adaptation isn't cheap or easy, but it's often cheaper and easier than dealing with climate disasters once they happen.

Near the coast, adaptation focuses on addressing sea-level rise and the flooding that comes with it. Strategies include raising homes on stilts and sturdy foundations that can resist flooding, building barriers to keep out seawater, and what has been termed **managed retreat** or managed realignment since "retreat" triggers negative connotations. Such an approach acknowledges that sea level will rise and that it makes little sense to use expensive and energy-intensive techniques to defend homes and businesses that within decades will be overwhelmed by flooding. Rather, it makes more sense economically to move buildings away from the coast and to higher ground (**Figure 3.74**).

Inland, adaptation includes adding trees and other vegetation cities to reduce the urban heat island effect and reducing water use to stretch ever-tightening supplies further. Agricultural adaptations include changing crops to heat- and drought-resistant varieties of plants and, in California, covering reservoirs with plastic balls to reduce evaporation (**Figure 3.75**). In arid, high-mountain regions of India and South America where glaciers are disappearing, some communities are creating imitation miniature glaciers, called "ice stupas," by freezing water in the winter. That ice slowly melts in the spring and summer, providing water for irrigation (**Figure 3.76**).

Elena Hartley

Figure 3.74 Various adaptations to coastal sea-level flooding from sea-level rise, including managed retreat to reduce damage.

Figure 3.75 Workers release black plastic balls into a reservoir in California. The balls float on the surface, covering the water and reducing water loss by evaporation.

3.6.3 Geoengineering and Climate Change

Geoengineering is the intentional manipulation of Earth's climate in an effort to mitigate the effects of global warming from greenhouse gas (GHG) emissions. A number of wide-ranging and untested technologies have been suggested to counter the effects of increasing atmospheric CO_2 and global warming (**Figure 3.77**).

Oceanographers have been exploring the feasibility of dumping finely ground iron particles into the ocean to stimulate the growth of phytoplankton, which, like all plants, metabolize CO_2 by photosynthesis. Because in some parts of the ocean iron can be a limiting nutrient, this is called **ocean fertilization**. From initial experiments, it is estimated that spreading a half-ton of iron across 100 square kilometers (39 square miles) of a tropical ocean would stimulate enough plant growth to absorb some 350,000 kilograms (772,000 pounds) of CO_2 from seawater. Performed on a much larger scale, the iron fertilization of seawater could absorb billions of tons of CO_2, offsetting perhaps a third of global CO_2 emissions. The environmental side effects of such large-scale iron fertilization are difficult to predict, but the enhanced bloom of plants could enrich the entire oceanic ecosystem or send it into a tailspin.

Solar radiation management technologies use **sulfate aerosols** and **cloud reflectivity enhancement** to reduce incoming solar radiation. One way to turn down Earth's thermostat is to inject large amounts of sulfur dioxide aerosols (tiny particles) into the stratosphere, which would reflect

iStock.com/Tsewang Gurmet

Figure 3.76 A stupa grows in the winter in the Ladakh Region of the Himalayas, India. Come summer, the ice will melt, providing water for drinking and irrigation.

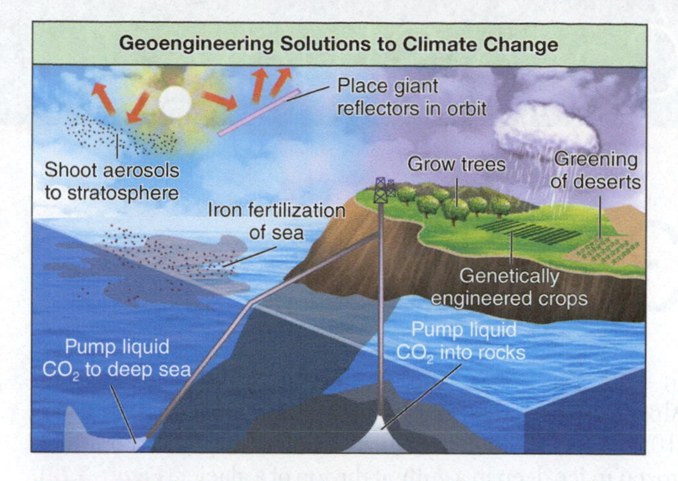

Figure 3.77 Some geoengineering ideas to reduce solar energy coming to Earth and remove carbon dioxide from the atmosphere. (Adapted from Mann & Kump (2015), Dire Predictions: Understanding Climate Change, 2nd Edition, Pearson Education)

sunlight back to space. You can think of this as volcano mimicry (see Chapter 5 for more on volcanic eruptions). Such a process would reduce global temperatures, much as happened following the eruption of Mount Pinatubo in 1991. This eruption put millions of tons of sulfate (SO_2^{-4}) into the air, causing a 0.5°C (1°F) drop in temperature for several months. Artificially injecting sulfate aerosols into the atmosphere would have the same effect, except that it would require continuous injection–dozens of planes a day, flying high in the atmosphere. Also, the technology would be expensive and come with the side effects of acid rain and probable damage to the ozone layer, not to mention a sky that would no longer be blue.

The goal of cloud reflectivity enhancement, which would create cloud condensation nuclei, is to increase the albedo of clouds so that they reflect more sunlight away from Earth. This might be done by spraying a mist of salty seawater from ships to form more low-lying clouds. According to two scientists at the University of Texas, the seawater mist could counteract a century's worth of warming for $9 billion, but not everyone agrees. Because the air over the ocean lacks dust and pollution, the technique would be more effective with cloud manipulation over the ocean than over land. None of these solar radiation techniques reduces the warming effects of CO_2 or other GHGs, nor would they reduce ocean acidification, which is already occurring as a result of increasing CO_2 levels–they simply change the reflectivity, albedo, of Earth.

Geoengineering is not a substitute for emissions control but is merely one of many schemes that could be considered along with emission-reduction efforts. Critics of geoengineering assert that the enormous sums of money required for most geoengineering technologies would be better spent on developing alternative power sources; that is, we need to focus our attention on the source of the problem, not its symptoms. Many critics voice concern that GHG mitigation by geoengineering will be self-defeating, leading to a false sense of security from the consequences of global warming and reducing popular and political pressure for emissions reduction and cutbacks on the use of fossil fuels.

Case Study

3.1 Ötzi – The Ice Man

In 1991, two hikers were climbing high up in the Austrian Alps near the Italian border at an elevation of 3,210 meters (10,530 feet). Suddenly, they stumbled upon a body half frozen in ice deep in a gully at the top of a glacier (**Figure 3.78**). The skin clung tightly to the ribs. The back was bare and the head was damaged. They reported their find to the authorities and the mummified human remains rapidly caught the world's attention. The mummy was named Ötzi, after the Ötztal (Ötz Valley) region where he was found.

The body was removed from the ice and it became clear this was no modern human. Years of work, including radiocarbon dating, showed that the man died more than 5,000 years ago not long after eating his last meal, the contents of which were identified by analyzing the food remaining in his stomach and guts. The "ice man" was remarkably intact, including his fur clothing, dagger, copper axe, and bow and arrows. The discovery proved to be a treasure trove of information for archaeologists about the life of a person who lived and died well before the time of the ancient Greek civilization (**Figure 3.79**).

Our warming climate likely exposed Ötzi, who had spent thousands of years frozen high up on a plateau at the top of an alpine glacier. Because he was so well preserved, ice there was likely not flowing but so cold that it was frozen to the rock below. Bits of plant material around the body were both older and younger than the body, suggesting that for perhaps a millenium after he died, snow and ice came and went in the gully where Ötzi lay frozen. Then, the climate cooled and for several thousand years he was immersed in ice. His melt-out and the discovery confirmed what glaciologists and climatologists already knew, ice in the Alps was disappearing, a consequence of climate warming. Such shrinkage of high-elevation glaciers in temperate regions is currently happening worldwide: in South America, the European Alps, central Asia, tropical Africa, the Himalayas, Irian Jaya, northwest America, California's Sierra Nevada, New Zealand, and Alaska. Are there more Ötzis out there waiting to be found?.

Paul Hanny/Gamma-Rapho/Getty Images

Figure 3.78 Climbers looking at the body of Ötzi soon after he was discovered melting out of the ice high in the Alps.

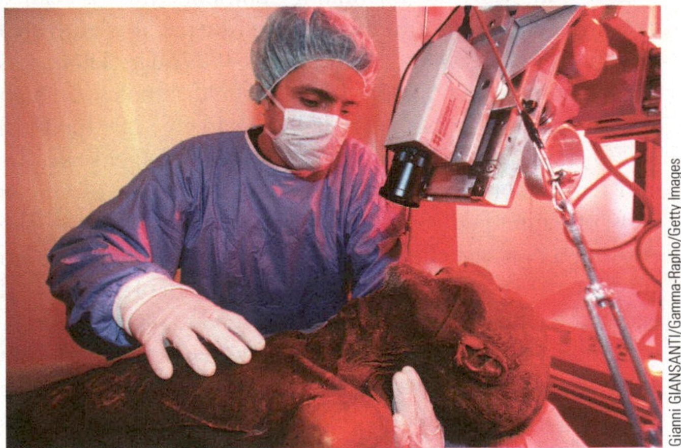

Gianni GIANSANTI/Gamma-Rapho/Getty Images

Figure 3.79 Scientists examine the mummified body of the ice man. He was dehydrated but otherwise exceptionally preserved because he spent most of his time in the ice, frozen.

3.2 Mitigation – What Individuals and Governments Can Do

Because more than 80% of the world's energy comes from burning fossil fuels–oil, coal, and natural gas–which release GHGs, mitigation involves cutting fossil fuel use. Currently, there are many cost-effective alternatives to fossil-fuel based energy (Chapter 11), but there is still much to be done. In the United States, individuals can do six major things:

1. Buy products that reduce energy consumption, such as front-loading, high-efficiency washing machines and Energy Star air conditioners, while replacing incandescent light bulbs with fluorescents or LEDs.
2. Use energy more efficiently. Add insulation to homes and businesses, seal gaps around windows, and replace old windows with new double- or triple-pane windows.

3. Forego fossil fuels. Use alternative energy when possible–biofuels, wind power, and photovoltaic (PV). Invest in companies with green practices and divest from oil company stocks.
4. Move closer to school or work. Use mass transit, carpool, or ride a bicycle. Cut down on long-distance plane travel, which is one of the fastest-growing sources of emissions.
5. Unplug electrical devices when not in use. It seems counterintuitive, but Americans consume more electrical energy to power devices when off than when on. This is so-called 'phantom load.'
6. Install smart meters that offer time-of-use incentives to encourage homeowner customers to shift electricity consumption for dishwashers electric vehicle charging, and laundry appliances to off-peak hours.

Governments can help meet the dual challenge of climate warming and ending the world's dependence on fossil fuels by providing:

1. Financial incentives for installing solar thermal (for heating water) and PV panels (for generating electricity) on roofs to increase the production of renewable energy.
2. Mandates that specify alternative energy or low-carbon requirements for companies supplying products under contract. This will promote more efficient and economic manufacturing processes and will lead to reduced costs for the rest of society.
3. An end to incentives for fossil-fuel extraction and distribution. The consumer would pay the true cost of energy upfront, and that would encourage the automobile and heating industries to manufacture high-efficiency vehicles.
4. Regulations that ensure newly built fossil-fuel-fired power plants include carbon-capture and storage (CCS) technology.

Table 3.4 summarizes the carbon impact of changes you, as an individual, can make to reduce your climate impact.

Table 3.4 Reducing Global Warming Begins at Home

Action to Take	Estimated Reduction of CO_2, Pound per Year
1. Insulate your home, install water-saver shower heads, clean and tune furnace.	2,480
2. Buy a fuel-efficient automobile with a rating of at least 32 mpg to replace your most-used car.	5,600
3. Do not drive your car two days a week.	1,590
4. Recycle all home waste metal, glass, newsprint, packaging, and cardboard.	850
5. Install solar heating to help provide hot water.	720
6. Replace washing machine with low-water-use, low-energy model.	420
7. Buy food and other products with recyclable or reusable packaging.	230
8. Replace refrigerator with a high-efficiency model.	220
9. Use a push lawn mower instead of a power mower.	80
10. Plant two trees.	20

Source: Oregon Energy Office.

Photo Gallery

Earth's Changing Climate

When we speak of "climate change," we are referring to changes in every climate zone of our planet. Arctic ice sheets melt, heat waves strike northern cities, lakes dry up, storms pound coastlines, forests burn, and tropical glaciers vanish.

iStock.com/Stan-Nascardi

1 Water pours off the Greenland Ice Sheet as arctic temperatures warm and ancient ice melts. This is a river of meltwater carving a channel through the ice.

iStock.com/MattGush

2 As California's drought worsens, reservoirs lower and marinas dry up, such as here at Folsom Lake.

Multiverse/Shutterstock.com

3 Hurricane Irma devastated the coastline of the Caribbean island of St. Maarten.

The 2022 drought in Bandai, India left this pond high and dry with nothing but cracked mud.
Prakash Singh/AFP/Getty Images

A'boursabroad/Shutterstock.com

4 The Alameda fire devastated communities in Oregon. Hotter, drier summers, the result of a changing climate, cause forests and communities to burn.

iStock.com/TheUntravelledWorld

5 As the climate warms, tropical glaciers are vanishing. Here on Mount Kilimanjaro in Africa, these stubs of ice are all that is left of one of the mountain's glaciers.

© Marika Massey-Bierman

6 The margin of the Mendenhall Glacier in Southeast Alaska has been rapidly retreating as the climate warms. The ice near the margin melts quickly as its darkened surface soaks up incoming solar radiation.

Summary Climate Change

1. Why Earth's Climate Changes

LO1 Explain how and why natural factors drive change in Earth's climate over time

a. Atmospheric Composition

Earth's atmosphere is composed primarily of nitrogen (78.08%), oxygen (20.95%), argon (0.93%), and carbon dioxide (0.042%), with trace amounts of other gases. The atmosphere contains greenhouse gases (e.g., carbon dioxide and methane) that trap and reradiate heat energy. This greenhouse effect heats the Earth's atmosphere and makes the planet livable.

b. Orbital Cycles

Earth's orbital cycles are a strong control on climate, influencing shifts between glacial and interglacial periods. The three variations of Earth's orbit are eccentricity, obliquity, and wobbles along the Earth's axis. Eccentricity fluctuates over 100,000 years, obliquity changes on a 41,000-year cycle, and the axis wobbles follow a 26,000-year cycle. While these changes in Earth's position relative to the Sun may have significant impacts on the Earth's climate (i.e., between glacial and interglacial periods), they do not explain the significant increases in temperature within the last 150 years driven by carbon dioxide emissions.

c. Climate Feedbacks

Positive feedbacks increase the intensity of forcing, such as the ice-albedo feedback in which warming temperatures lead to the melting of icesheets; the loss of reflective ice sheets decreases albedo, leading to increased warming. Negative feedbacks decrease the intensity of the initial forcing and often help explain climate stability.

Study Questions

i. Give an example of how the climate naturally changes and how we know that human-caused climate change is different, considering timescales.

ii. What is the composition of Earth's atmosphere?

2. Records of a Changing Climate

LO2 Describe how Earth's climate has changed over millions of years, thousands of years, and centuries

a. The Retreat of Mountain Glaciers

Changes in mountain glaciers are an excellent indicator of twentieth-century climate changes because these glaciers have short reaction times to temperature fluctuations. Mid- to high-altitude glaciers around the world have lost significant mass and volume since 1850 (the start of the Industrial Revolution).

b. Ice Cores and Ocean Sediments

Ice cores preserve evidence of changes in atmospheric chemistry, temperature, and dust in air bubbles trapped in the ice. Deep ocean sediment accumulates shells of single-celled animals called foraminifera composed of calcium carbonate, which can be dated and analyzed for many atmospheric and environmental characteristics of the Earth during the time they lived.

c. Packrats and Pond Muck

Pond muck preserves pollen (as well as fossilized twigs, leaves, and seeds), which allows paleoclimatologists to analyze when in recent geological history certain plants have thrived, allowing environmental conditions to be inferred during times in the recent past (tens of thousands of years). In deserts, the crystallized urine of packrats, called amber-rat, encases and preserves vegetation, which can be analyzed and dated.

d. Tree Rings

Climate records can be gathered from trees by taking thin cores and examining tree rings, which form each year. Thin tree rings indicate drought, temperature fluctuations, or other climate stresses on tree health. Thick tree rings indicate warm, sunny, growing conditions with sufficient moisture.

Study Questions

i. Give three examples of strategies used by paleoclimatologists to determine climates and environments throughout geological history.

3. Human Actions Change Earth's Environment

LO3 Provide evidence that human actions are significantly changing Earth's environment

a. Greenhouse Gas Emissions Changed Atmospheric Composition

Since the start of the Industrial Revolution, increased burning of fossil fuels has contributed to the increase of carbon dioxide in the atmosphere. The Keeling Curve provides evidence of a steady increase in atmospheric carbon dioxide since 1958. The Suess effect shows that the ratio of $^{14}C/^{12}C$ has decreased, as the burning of fossil fuels has released significant ^{12}C and no ^{14}C into the environment.

b. Land-Use Change

The increase in industrialization and the human population boom around the world led to an exponential rate and scale of landscape change. Forests are cleared for development and cultivation, releasing carbon once stored in the trees and soil into the atmosphere. Human impact on land-use has altered the way carbon is stored on Earth's surface, as well as how carbon exchanges within the atmosphere.

c. Urbanization

Urbanization significantly changes land cover, increasing dark brick, concrete and pavement coverage and reducing greenspace. Combined, these urban characteristics reduce albedo, as dark pavement absorbs heat (instead of reflecting solar energy back out of the atmosphere) and the lack of greenspace reduces evapotranspiration from trees and evaporation from soils, both of which take energy. These characteristics create excessive heat in urbanized areas.

d. The Anthropocene

The Anthropocene describes a new era of geological time in which the Earth is primarily changing because of human activity.

Study Questions

i. How have GHG emissions changed the composition of the atmosphere?
ii. What is the urban heat island effect, and what is the physical explanation for why it occurs?

4. Impacts of Environmental Change

LO4 Identify how societies and ecosystems have been and will be affected by environmental change

a. A Baking Planet

Since 1880, Earth's temperature has increased by 1.1°C, with most of the increase occurring after 1975. The greatest warming is occurring during the night at high latitudes.

b. Shrinking Ice

Glaciers store a significant volume of freshwater and supply a natural flow of meltwater to rivers, which many areas around the world depend on for drinking water and irrigation. This meltwater is also a significant source of flow for hydroelectric power in mountainous regions. Significant losses of these glaciers due to climate change has cascading impacts on societies that are dependent on these water resources.

c. The Ups and Downs of Sea Level

The primary controls on sea-level are glaciers and ice sheets, which, when melted, increase the volume of the world's oceans. Thermal expansion of the oceans also contributes to sea-level rise. Sea-level rise threatens coastal communities and infrastructure, and has impacts worldwide.

Study Questions

i. Explain the two factors controlling sea-level rise.
ii. Explain how the melting of glaciers impacts society in terms of energy production and drinking water availability.

5. Earth in the Future

LO5 Recognize how different Earth and its climate will be in the future

a. A World with Little Ice

Melting of sea ice has consequences that go far beyond the sea itself. Loss of sea ice has impacted places around the world by increasing the frequency and severity of storm surges, coastal erosion, forest fire frequency, and insect outbreaks.

b. Melting Permafrost

Melting permafrost releases carbon dioxide, methane. phosphates, and nitrates, which contribute to GHG emissions and alter nearby ecosystems. The emergence of drier tundra climates that result from permafrost melt has led to an increased frequency of tundra fires, causing environmental destruction and releasing tremendous volumes of carbon dioxide as vegetation and peat burn.

c. Extreme Events: Droughts, Floods, and Storms

Warming air is dangerous as heat waves become more common; warming air also increases evaporation, drying out soils and vegetation, causing droughts and forest fires. With warming air increasing evaporation and holding more water within the atmosphere, heavier rain is more likely resulting in more intense storms. Record cold is also possible, as the arctic warms and the jet stream becomes more wavy, carrying cold arctic air southward.

d. A Long Warmth

The future climate and its impacts on the biosphere, hydrosphere, and geosphere depend strongly on the actions we take now and the amount of greenhouse gases we continue to emit. Without change, humans have the capacity to cause environmental changes worldwide, lasting for millennia, even hundreds of thousands of years.

Study Questions

i. How does climate warming contribute to the increased frequency and severity of extreme weather events?

ii. What will Earth's climate be like in the future?

6. Addressing Climate Change

LO6 Describe the strategies to reduce the severity of climate change and its impact on people and society

a. Mitigation

Energy efficiency is the most effective climate change mitigation strategy. About two-thirds of energy used in the United States is wasted. Carbon capture and storage (CCS) describes a variety of methods that are currently being explored and used in a few moderate-scale operations to capture and trap carbon in the Earth's subsurface. Large-scale reforestation is another method of carbon sequestration that removes carbon already released into the atmosphere.

b. Adaptation

Around the world, people and communities have developed and implemented adaptation techniques to address the impacts of climate change. On coasts, homes are raised on stilts to resist flooding and sea walls are built to mitigate storm surges. Inland, cities increase greenspace to reduce the urban heat island effect.

c. Geoengineering and Climate Change

Geoengineering–which needs further consideration and research before large-scale implementation–could make options like aerosol injection, ocean fertilization, and cloud creation feasible. Critics argue that money invested in geoengineering may be better spent on emission reduction strategies.

Study Questions

i. List the three actions people could take now to mitigate climate change.

ii. List the two arguments for and two arguments against geoengineering.

iii. Explain the process of carbon capture and storage.

Key Terms

Adaptation
Amber-rat
Anthropocene
Arctic amplification
Attribution science
Carbon capture and storage
Carbon dioxide
Cloud reflectivity enhancement
Eccentricity
Enhanced oil recovery
Evapotranspiration
Fjord
Foraminifera
Geoengineering
Glacial periods
GRACE experiment
Greenhouse effect
Greenhouse gases
Greenhouse warming
Half-life
Halocarbons
Holocene
Ice cores
Infrared
Interglacial
Isotope
Land conversion
Managed retreat
Marine sediment cores
Methane
Mitigation
Mountain glaciers
Obliquity
Ocean fertilization
Pack rats
Paleoclimatologists
Point sources
Pollen
Suess effect
Sulfate aerosols
Tree rings
Urban heat island effect
Wobbles

"Geologists have a saying–rocks remember."

—Neil Armstrong

4 Minerals and Rocks

Learning Objectives

LO1 Describe why minerals are important to humanity

LO2 Define minerals, and explain what distinguishes a "mineral" from a "rock"

LO3 Recognize the importance of the rock cycle

LO4 Explain why valuable mineral deposits are in many cases "accidents of nature"

LO5 Identify the ways in which quarries and mines have negative environmental impacts and give examples of how we can remediate environmental problems caused by mining

Open pit iron mine in Turkey seen from above by a drone.
iStock.com/temizyurek

Little Things, Big Trouble!

Libby, Montana, population 2,775 (2020 census) rises on a floodplain of the Kootenay River and is surrounded by pine-covered slopes of the Rocky Mountains (**Figure 4.1**). Despite its beauty, Libby is the site of one of America's greatest environmental disasters, an incident that unfolded in near silence over many years. At the center of the problem was a typically benign mica-like mineral called **vermiculite**, which is familiar to organic farmers and compost managers because of its ability to retain water and nutrients in soil beds (**Figure 4.2**). Vermiculite is also highly resistant to heating and is prized as an insulator in the aluminum industry and for lining car brakes. It is even put to work as a bonding agent in some cements and was used in homes for insulation.

Vermiculite forms naturally under specific geological conditions. It crystallizes deep underground from slowly cooling magmas especially rich in silica and alkali elements (potassium and sodium). One such magma body formed in the crust about 11 kilometers (7 miles) east of what is now Libby during Cretaceous times, 94 million years ago (Chapter 2). Goldminers discovered this very large vermiculite deposit in 1881, and in 1939, the Zonolite Company began mining it, eventually selling out to another mining company, W. R. Grace, in 1963. Much of the vermiculite mined in Libby was used as insulation in homes, especially in attics, where it was packed between joists (**Figure 4.3**). At one point, the Libby mine supplied 80% of the world's vermiculite. Little surprise that the town downslope prospered.

There was a catch, though. The vermiculite deposit was intermixed with large amounts of **asbestos**–up to 25% by volume. Asbestos is not a specific mineral type, but rather a term given to any minerals that are naturally highly fibrous (**Figure 4.4**). Six types of minerals are commonly **asbestiform**, meaning they can crystalize in a way that results in long, skinny fibers, including chrysotile, amosite, crocidolite, anthophyllite, tremolite, and actinolite (**Figure 4.5**). Some types of asbestos are harmless, but other types are dangerous when people inhale the fine mineral fibers for long periods of time. Asbestos exposure is well linked to lung cancer, and **mesothelioma**, a serious cancer of the tissues lining the lungs, heart, abdomen and chest. **Asbestosis**, another debilitating non-cancerous, lung disease caused by scarring from the

Randy Beacham/Alamy Stock Photo

Figure 4.1 Spring in Libby, Montana, with the Cabinet Mountains in the background.

Fir Mamat/Alamy Stock Photo

Figure 4.2 Vermiculite, with a hand for scale.

iStock.com/Pixel-Productions

Figure 4.3 Vermiculite insulation used to insulate the floor of an attic.

Figure 4.4 Chrysotile asbestos fibers in a rock sample.

Figure 4.5 Asbestos fibers in vermiculite. The scale bar is 50 microns long; that is, 1/20th of a millimeter—so tiny!

fibers, is also a concern for people who breathe a great quantity of asbestos.

Decades of mining in Libby introduced to the surface environment large amounts of harmful tremolite asbestos. Asbestos fibers blew through the air as dust associated with mining, processing, and transport of the vermiculite. The fibers entered a tributary of the Kootenay River and infiltrated soils and sediment wherever they settled. Wildlife took them in. And so did people. Four hundred people died, and 2,400 others were diagnosed with asbestos-related diseases—more than the entire current population of Libby. Mining operations ceased in 1990, with damage already done to the town and its population.

In 2000, the U.S. Environmental Protection Agency (EPA) stepped in to clean up the mess in Libby, which otherwise could have continued to hurt people long into the future. By 2002, Libby was designated a major Federal clean-up (Superfund) site. The agency removed around 750,000 cubic meters (a million cubic yards) of soil at least partially contaminated with asbestos from the town and its environs, taking it to the old mine area where it could be safely isolated. EPA workers transported a further 750,000 cubic meters of asbestos-laced material from structures to a specially designated landfill for permanent storage (Chapter 12). Over 70 miles of sidings along railway tracks and highways were tested for asbestos and treated.

Collaborating with the U.S. Forest Service, the EPA assisted with fire suppression to prevent wildfires—common in this part of the Rockies—from spreading contaminants farther via fine particles in wood smoke. The agencies sampled animal and fish tissues to track ecological impacts. For human victims, the EPA declared its first-ever Public Health Emergency, freeing up Federal healthcare for persons in need. Clean-up objectives were achieved by 2018, except for the immediate vicinity of the mine and adjacent forest. As a result, the concentration of fibrous asbestos in the air of downtown Libby has dropped 100,000 fold from what it was when the mine and mill operated upslope. Libby is now regarded as a safe living environment for residents.

The story didn't simply play out in Libby, however, because asbestos-bearing vermiculite mined there was shipped all over North America to insulate homes—perhaps a million or more. Every time people renovated these homes, asbestos fibers had the potential to spread widely. Likewise, asbestos-bearing, dust-coated items stored in attics were a problem. When someone retrieved an item from storage, the dust came along with it, increasing the risk. Dealing with asbestos-contaminated vermiculite insulation is complicated and costly, requiring trained experts to remove the material safely. As part of its 2001-2012 bankruptcy settlement, W. R. Grace set up two trust funds totaling $4 billion; one trust pays homeowners half the removal cost for vermiculite shown by testing to come from Libby and for new insulation. The other fund compensates people injured by asbestos.

Not all asbestos-related issues can be related to Libby mine vermiculite. Until recent times, asbestos was applied in the same ways vermiculite had been. Unaware of the danger, contractors incorporated asbestos directly into insulation wraps around many heating pipes in houses and other buildings (**Figure 4.6**). It was used in roofing tiles, home siding, and wallboard for fireproofing; and floor tiles to increase strength and durability. It could also be found in crayons as a wax binder and in some cosmetics to inhibit caking. Structures of all sorts in the United States built before the 1970s still contain substantial amounts of asbestos, some benign, some not. Even the harmful stuff is no threat, though, as long as structural materials stay in good condition. But when these materials

iStock.com/Joe_Potato

Figure 4.6 Asbestos-containing insulation is shown here covering heating pipes in the basement of a home. The material is mostly wrapped, but the bar ends are releasing fibers into the home.

start to degrade or are damaged, then testing and safe removal of the asbestos-bearing material is important. Such remediation is expensive because the fibers are light, small and move through the air easily (**Figure 4.7**). Be ready for warning signs and tape, people in respirators and hazmat suits, as well as large fans with filters to trap particles (**Figure 4.8**).

The Libby vermiculite story illustrates the effects that even tiny pieces of minerals with certain shapes and composition, generally overlooked, can have on people. It also illustrates the unintended environmental consequences that can ensue when extracting resources from Earth's crust.

questions to ponder

1. Why did it take so long, or may have taken so long, to identify and treat the environmental problem caused by the mining of vermiculite at Libby?
2. Why were multiple external government agencies required to resolve the Libby vermiculite problem?

Albertog33/Dreamstime.com

Figure 4.7 Removing roofing material made with asbestos has to be done carefully to avoid spreading the fibers. Note that the workers are wearing protective suits and respirators.

iStock.com/BanksPhotos

Figure 4.8 Asbestos remediation releases fibers as the materials are being removed. You'll need to stay out of your home while the work is being done and until it's cleaned up.

4.1 Why Rocks and Minerals Matter to Us (LO1)

Civilization depends upon several types of natural resources; soils to produce food and fiber (see Chapter 7); fossil fuels and other natural sources of energy including heat from the deep Earth (geothermal power; Chapter 11), sunlight and wind (see Chapter 11); and the materials provided by rocks and minerals through mining and quarrying.

One product of minerals is **metals**, the typically hard, shiny, and easily formed substances we use to make everything from coins and hubcaps to electric wires and skyscrapers. Examples of metals include gold (Au), iron (Fe), copper (Cu), tin (Sn), and aluminum (Al). So important have metals been to humanity that we categorize stages of civilizational development by the important metals people learned first to use, including the Bronze Age ("bronze" is a blend of copper and tin) and Iron Age (**Figure 4.9**). Today, you could say that we live in the Nuclear Age, in which the toxic metal uranium (U) has become a major energy source and a component of weapons.

Dominic Gentilcore PhD/Shutterstock.com

Figure 4.10 White gypsum beds make up the desert landscape at Tule Fossil Beds National Monument, near Las Vegas, Nevada. While this deposit is not mined, elsewhere gypsum is extracted in vast quantities to make paper filler, textiles, fertilizers, plaster and other construction material, board products, Portland cement, tiles, and bricks.

If you stop and think about everything that you consume or depend upon to live comfortably on a day-to-day basis, you'll quickly appreciate how vital geological resources, including metals and fuels, have become for humanity. According to the Minerals Education Coalition and U.S. Geological Survey (2021 data), each year an individual in the United States directly or indirectly uses approximately 18,150 kilograms (40,000 pounds) of mineral and rock products to maintain a typical American standard of living. This includes 15 kilograms (34 pounds) of aluminum, 10 kilograms (23 pounds) of copper and lead, 2.7 kilograms (6 pounds) of zinc, and 270 kilograms (600 pounds) of other nonmetals such as gypsum (**Figure 4.10**). Common items, like automobiles, use large amounts of raw materials. (**Figure 4.11**).

The terms *rock* and *mineral* are equated by many persons, but they signify *two very different kinds* of earth materials. We'll explain this by beginning with basic "building blocks."

Dzuba Pavel/Shutterstock.com

Figure 4.9 Bronze Age copper ornaments, rings, chains, bracelets, discovered in Ukraine.

4.2 Elements, Minerals, and Rocks (LO2)

Elements are substances that cannot be changed into other substances by normal chemical methods. They are composed of extremely small particles called **atoms**. In 1870, Lothar Meyer published a table of the 57 then-known elements arranged in the order of their atomic weights. Meyer left blank spaces in the table wherever elements of particular weights were not known. About the same time, Dmitry Mendeleyev developed an alternative table–the basis of the periodic table that chemists use today–but his table was organized around grouping known elements having similar chemical properties. Mendeleyev sought to reorganize the table of elements because an element's placement based on atomic weight within a table did not clearly associate it with elements of similar properties. He also predicted the

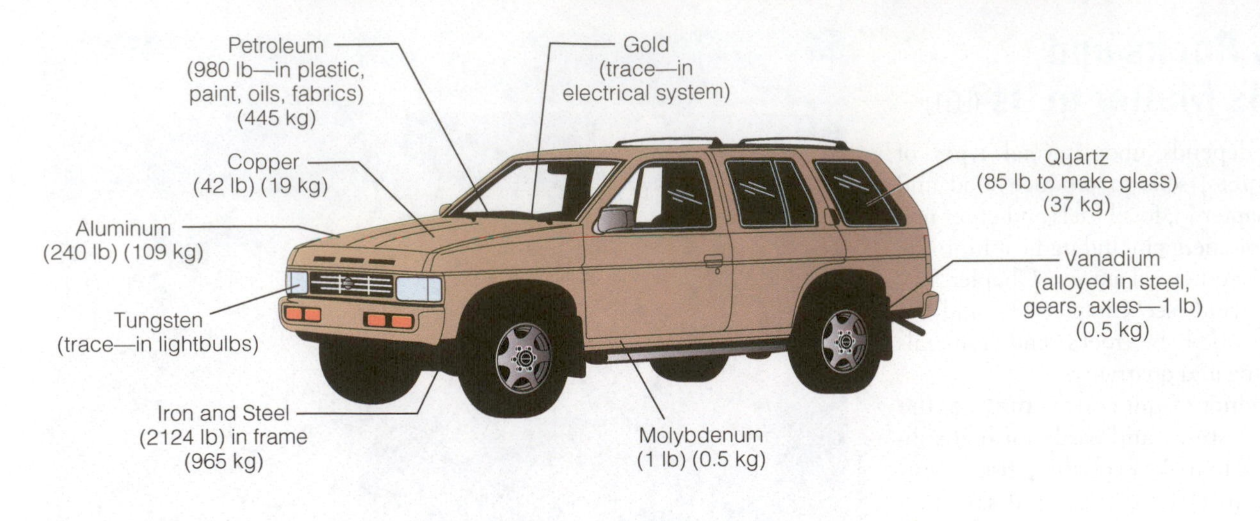

Figure 4.11 Estimated amounts of a few of the more than 30 mineral materials used in a midsize automobile.

properties of the "missing" yet-to-be-discovered elements, and shortly his predictions proved correct with the discoveries of the elements gallium (Ga), scandium (Sc), and germanium (Ge) (**Figure 4.12**).

The weight of an atom of any given element is contained almost entirely in its **nucleus**, which includes protons (atomic weight = "1" with an electrical charge of "+ 1") and neutrons (atomic weight = 1, but with an electrical charge = 0). An element's **atomic number** is the number of protons in its nucleus, and this number is unique to that element. The weight of an entire atom–its **atomic mass**–is the sum of its nuclear protons and neutrons. For example, $^{4}_{2}He$ denotes the element helium, which has 2 protons (denoted by the subscript) and 2 neutrons, for an atomic mass of 4 (denoted by superscript). Electrical neutrality of the atom is provided by balancing the positive proton charges by an equal number of negatively charged electrons orbiting at various distances, termed **shells**, around the nucleus (**Figure 4.13**).

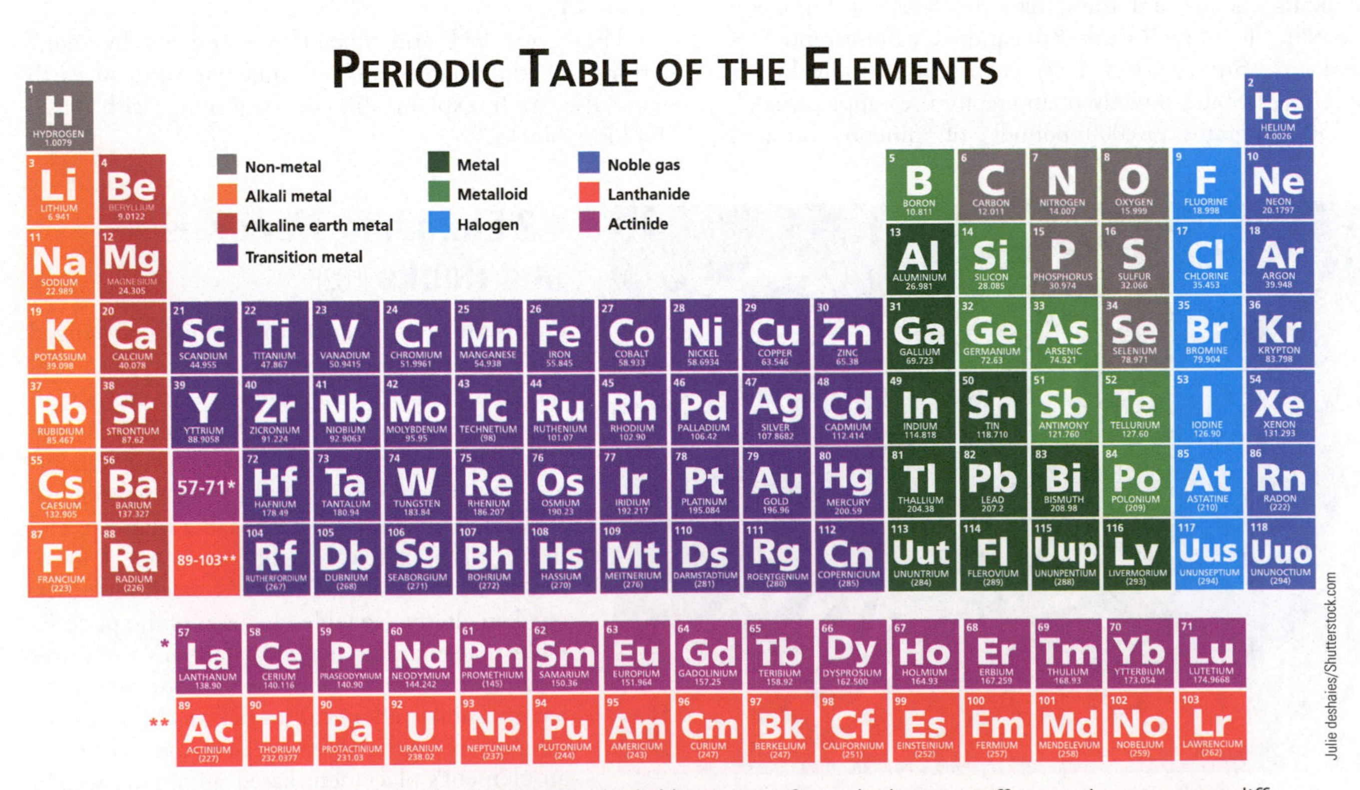

Figure 4.12 The periodic table of elements, using standard abbreviations for each element. Different colors represent different element groupings, as originally discerned by Dmitry Mendeleyev. The second to last row at the bottom (La to Lu) shows the lanthanide elements, including most of the rare earth elements discussed later in this chapter.

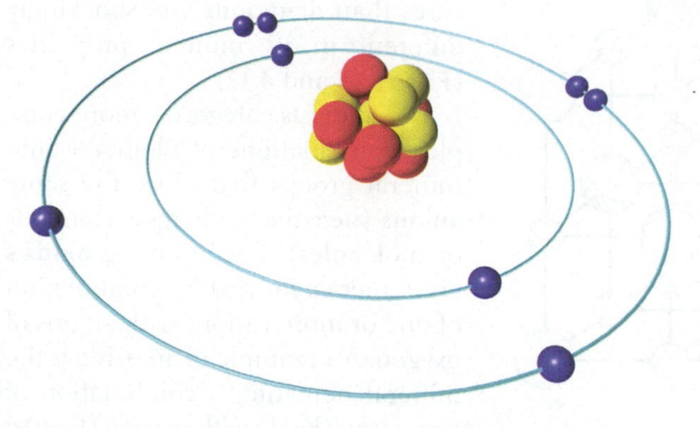

Emmily/Shutterstock.com

Figure 4.13 Simple model of an oxygen atom in which 8 protons (red balls) and 8 neutrons (yellow balls) are gathered in a nucleus around which 8 electrons (blue balls) orbit at various energy levels, or "shells." There may be an excess or deficit in the number of electrons present in the outermost shell, an unbalanced condition in which the atom is said to be an ion, or "charged."

iStock.com/MarcelC

Figure 4.14 The green, uranium-bearing copper phosphate mineral metatorbernite, from a deposit in Zaire, Africa. The uranium in this mineral is predominantly ^{238}U, though it is also somewhat radioactive, indicating that it also includes the more unstable isotope ^{235}U.

Ions are atoms that are positively or negatively charged because of a loss or gain of electrons in the outer electron shell. Thus, sodium (Na) may lose an electron in its outer shell to become a sodium ion (Na^+), and chlorine (Cl) may gain an electron to become a chloride ion (Cl^-). Positively charged ions are called **cations**, and negatively charged ions are called **anions**. **Valence** refers to ionic charge, and sodium (Na) and chlorine (Cl) are said to have valences of 1^+ and 1^-, respectively. The differences between the pure unionized forms of elements and their properties as ions in combination with other ions, may be profound; sodium is an unstable metallic solid that bursts into flame on exposure to oxygen, and chlorine is a deadly poisonous gas. When ionized, however, they may combine in an orderly fashion to form the mineral halite (NaCl), common table salt, a substance that is quite stable and safe in dry conditions, and vital for human existence.

Isotopes are forms (or species) of an element *that have different atomic masses*. The metallic element uranium, for instance, always has 92 protons in its nucleus, but varying numbers of neutrons. This means that uranium occurs in multiple isotopic forms. For example, uranium-238 (^{238}U) is the common naturally occurring isotope of uranium (**Figure 4.14**). It weighs 238 atomic mass units and contains 92 protons and 146 (238 *minus* 92) neutrons. Uranium-235 is the rare isotope that has 143 neutrons. Similarly, ^{12}C is the common isotope of carbon, but ^{13}C and ^{14}C also exist. Uranium-235 can be used to make atomic bombs for which Uranium-238 is useless.

4.2.1 Minerals

Atoms are the basic building blocks of **minerals**. Minerals, in turn are the basic building blocks of rocks. This hierarchy of increasing complexity is worth remembering: elements (types of atoms) → minerals → rocks → "stony" planets.

Most people understand minerals differently than what geologists consider them to be. We commonly speak of minerals as elemental supplements added to food to improve nutrition or crops to make them flourish. This is misapplied terminology, however. Strictly speaking, geologists define minerals as *naturally occurring, inorganic substances, each with a narrow range of chemical compositions and characteristic physical properties*. It's true that some minerals can be sources of elements that are nutritionally beneficial, but only a few types are consumed wholly and directly for that purpose. One example is dolomite [$CaMg(CO_3)_2$], a carbonate mineral that is a source of magnesium and calcium (good for heart and bone health). Individually, the magnesium and calcium are elements–the cations found in dolomite. Technically, they are not "minerals" in and of themselves, despite the labeling and advertisements that you otherwise may encounter in popular press.

Minerals also are **crystalline**. That means that the atoms that compose them are arranged in geometrically regular and orderly patterns. These patterns are visible using high-powered instruments such as x-ray diffractometers, but atomic orderliness can also be detected indirectly using polarizing microscopes. The regular spacing of atoms, and their links to one another, endows each mineral with a stable, long-lasting structure, and also imparts many of the physical characteristics of the crystals of these minerals that we see in nature, including color, hardness, and shape (**Figure 4.15**).

Given the limited numbers of elements identified by Meyer, Mendeleyev, and others, as well as the limited number of geometrically symmetrical arrangements possible for combining these elements in three dimensions, there

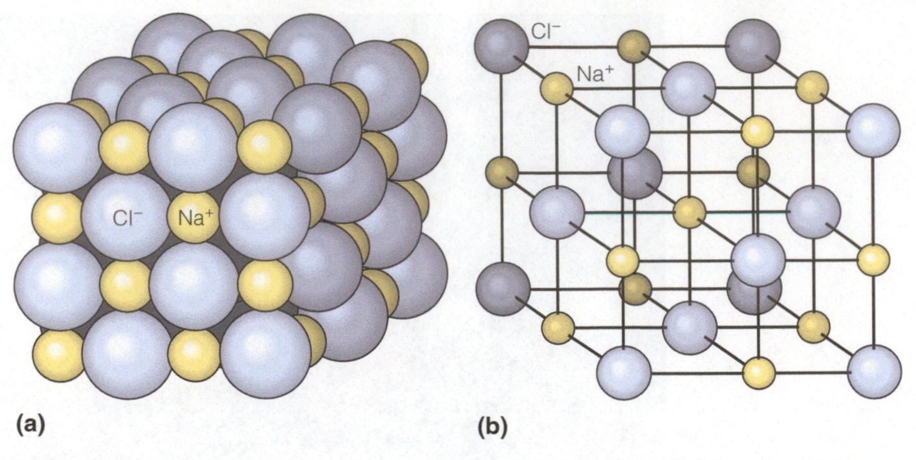

Figure 4.15 Two models of the atomic structure of the mineral halite (NaCl), common table salt: (a) a packing model that shows the location and relative sizes of the sodium and chloride ions; (b) a ball-and-stick model that shows the mineral's cubic crystal structure.

are only 230 types of crystals possible. That doesn't mean that there are only 230 kinds of minerals, however, because different combinations of elements can coincidentally share the same type of crystalline structure. All told, there are over 5,700 kinds of minerals known to date, with over a hundred new ones being discovered each year. However, out of this vast number, only about 20 minerals make up most of the solid Earth.

It is the chemical compositions of minerals, not their elegant crystal geometries, which are the basis of mineral classification. The simplest group is the **native elements**–that is, minerals consisting of only a single element compositionally. A good example is *gold* (chemical symbol Au) one of the most valuable minerals in the world. Carbon (C) is another example, although it can occur in two possible crystal forms: as diamond (a product of crystallization in high-pressure conditions), or as graphite–common pencil lead–formed at less extreme pressures than diamonds and shockingly different in its mineral properties (**Figure 16** and **4.17**).

Geologists categorize more complex combinations of elements into mineral groups that share the same anions (negatively charged elements or molecules). For instance, **oxides** are minerals formed by combination of one or more cations with anions of oxygen. An example of an oxide is the mineral hematite, a combination of ferric iron (Fe^{3+}) with oxygen (Fe_2O_3) (**Figure 4.18**). Hematite is easily spotted in some sandstones (and in household rust) due to its red color. Note that the net electrical charge of hematite is zero. You can do the math, combining 3 anions of oxygen (each oxygen carries a −2 charge) with two

Bjoern Wylezich/Shutterstock.com

Figure 4.16 Diamond, the same as graphite compositionally, formed under more extreme conditions of heat and pressure deep inside Earth. The difference between these two minerals is in the arrangement of their constituent carbon atoms.

Miriam Doerr Martin Frommherz/Shutterstock.com

Figure 4.17 Graphite, or native carbon, a metamorphic mineral that is used to make pencil leads.

vvoe/Shutterstock.com

Figure 4.18 A mass of hematite, an iron ore mineral that imparts a red color to many rocks.

ferric ion cations (each iron ion carries a 3+ charge), to see this:

$$[2(3+) + 3(-2) = 0].$$

This is important because hematite, and all other minerals, *won't form or remain stable for long* without such an internal charge balance present.

The **sulfide minerals**, much sought after by miners, are typically combinations of a metal with sulfur. Examples include galena (lead sulfide, PbS), chalcopyrite (copper-iron sulfide, $CuFeS_2$), and pyrite (iron sulfide, FeS_2), the "fool's gold" of prospectors (**Figures 4.19** through **4.21**).

Many minerals combine cations with negatively charged molecules (bundles) of ions, termed **radicals**. One of the most common such radicals in nature is the negatively charged $(CO_3)^{-2}$, the basis of the **carbonate** mineral group. Calcite ($CaCO_3$), the principal mineral of limestone and marble, is one example, as is the previously mentioned dolomite (**Figure 4.22** and **4.23**).

Minakryn Ruslan/Shutterstock.com

Figure 4.19 A clump of well-formed galena crystals.

Smallblackcat/Shutterstock.com

Figure 4.20 Chalcopyrite, a copper-iron sulfide mineral

Marco Fine/Shutterstock.com

Figure 4.21 A mass of well-formed crystals of pyrite, a pure iron sulfide mineral.

olpo/Shutterstock.com

Figure 4.22 An aggregate of calcite crystals. Superficially these may look like diamonds, but their crystal shapes are quite different, and calcite is a much softer mineral, its crystals easily scratched by diamond.

Eric Isselee/Shutterstock.com

Figure 4.23 Cliffs of limestone at Etretat, on the French coast.

The silica radical $(SiO_2)^{-4}$ has four excess electrons (hence the −4 charge), so it bonds readily to a wide range of anions and even other radicals. The result is the world's largest, most diverse group of minerals–the **silicates** (**Figure 4.24**). These range compositionally from such simple minerals as wollastonite ($CaSiO_3$, used in ceramics and car brakes) to tourmaline (a natural electrical conductor), the formula of which should earn you a prize to memorize:

$$(Ca, K, Na)(Al, Fe, Li, Mg, Mn)_3(Al,Cr,Fe,V)_6 (BO_3)_3(Si,Al,B)_6O_{18}(OH,F).$$

Note that it is hard to see how silicon and oxygen are important for building tourmaline until you near the end of its long formula (**Figure 4.25**). The silicons and oxygens combine in the same way within the crystals of both wollastonite and tourmaline, creating a structure termed a **silica tetrahedron**, an arrangement of ions that resembles a three-sided pyramid (four-sided–the "tetra"–if you include the bottom) (**Figure 4.26**). The highly charged silicon cation (+4) forms the center of each tetrahedron, while the lesser charged oxygen anions (−2 each) make up the corners, coordinated around the silicon. Because there are 4 oxygens for each silicon, the overall radical has a charge of −4, a glaring discrepancy which is addressed by nature as every crystal begins forming.

How does nature do it? One way is for the tetrahedra to link together three dimensionally so that each oxygen ion *shares* an electron with two silicons in adjoining tetrahedra, as happens in the common mineral, quartz (SiO_2). Or, the tetrahedra may only partly link, or can remain unlinked, simply relying on nearby cations of other elements to attach themselves to tetrahedral corners in order to neutralize charges.

Silicate minerals are subdivided into subcategories based on how their tetrahedra are arranged to deal with the "charge imbalance challenge" within them. Such arrangements impart important properties to specific minerals. Mica, for instance, is famous for its ability to peel apart (**Figure 4.27**). You can easily strip a mica crystal into thin, translucent sheets with your fingernails. Those sheets exist because at the atomic level, mica's silica tetrahedra are themselves linked into sheets separated by intervening cations whose bonds are easily broken. Micas therefore belong to the **sheet-silicate** group of silicate minerals. Clay minerals, so nutritionally important for growing crops, are another important sheet-silicate example. Some crucial elemental cations holding tetrahedral sheets together in clays, such as potassium (K), are also key plant nutrients and are easy for infiltrating water and plants to extract.

The silicate minerals also have chemically defined subcategories that become important in the study and classifications of rocks. Examples include the **feldspars** and **ferromagnesian** minerals (Box: Silicate "Solid Solutions" on page 155). Variation in the cation content of many silicate minerals in these groups' accounts for slight but important differences in their properties. Such variation arises because some cations have similar sizes (**atomic radii**) and can easily substitute for one another as a crystal grows, provided that they are both present together in the local environment. The parentheses in the formula for tourmaline, above, show that six different sets of elements can share available positions in this chemically complex mineral.

Students of mineralogy are often able to identify several hundred minerals without destructive testing. They are

Figure 4.24 Wollastonite, a calcium silicate mineral.

Figure 4.25 Multicolored "watermelon" crystals of tourmaline set in white (milky) quartz.

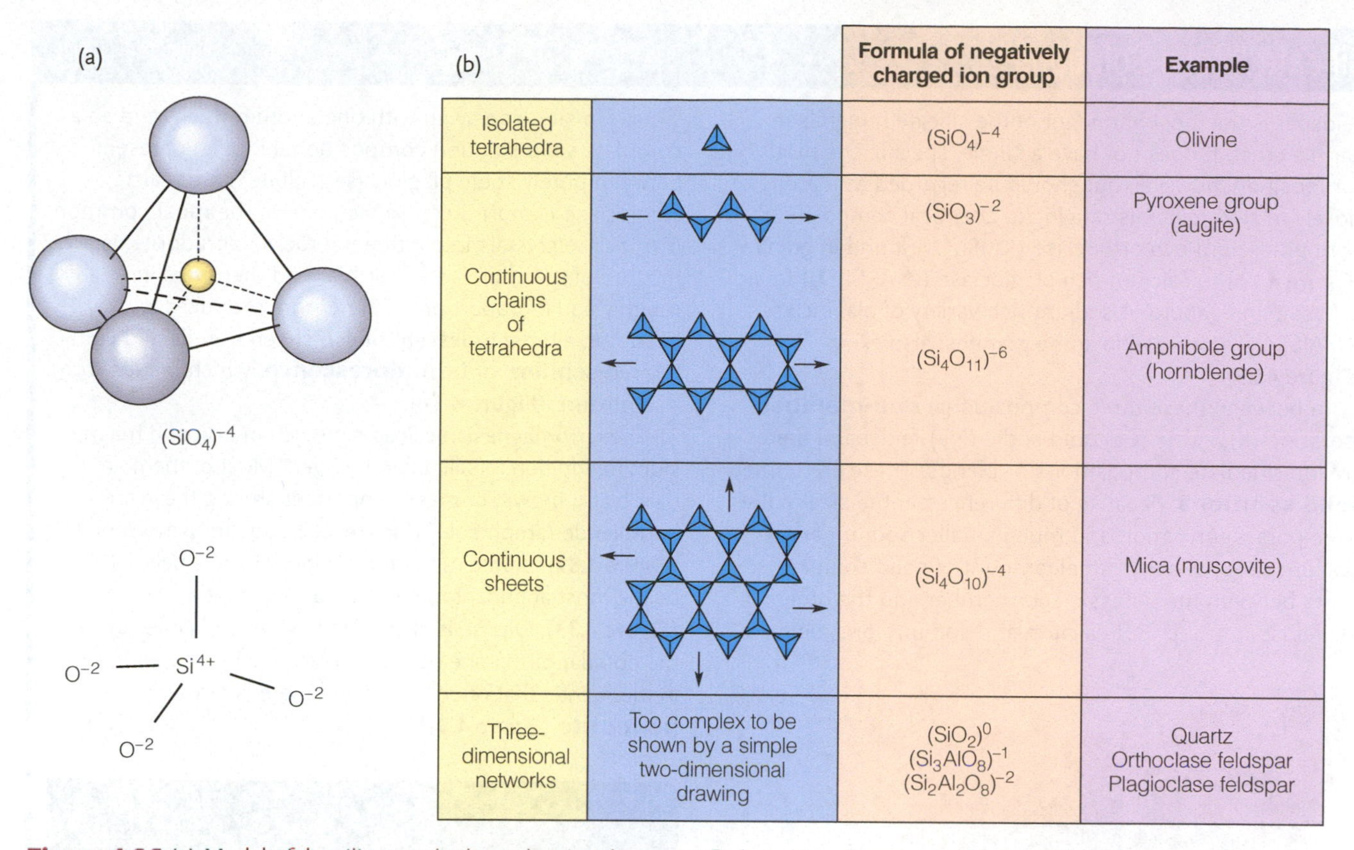

Figure 4.26 (a) Model of the silica tetrahedron, showing the unsatisfied negative charge at each oxygen that allows it to form (b) chains, sheets, and networks.

vvoe/Shutterstock.com

Figure 4.27 Muscovite mica showing sheet silicate layering.

able to do this mineral identification because minerals have distinctive physical properties, most of which are easily determined and associated with particular mineral species. With practice, one can build up a mental catalog of minerals, just like building a vocabulary in a foreign language. The most useful physical properties are hardness, cleavage, crystal form, and to lesser degrees, color (which is variable) and luster.

Mineralogists use **Mohs' hardness scale**, developed by Friedrich Mohs in 1812. Mohs assigned hardness values (H) of 10 to diamond (very hard) and 1 to talc, a very soft mineral (**Table 4.1**). Diamond scratches everything with which it comes in contact; hence its use on saw blades designed to cut rock. Quartz (H = 7) and feldspar (H = 6) can scratch glass (H = 5½–6), while calcite (H = 3) and your fingernail can scratch gypsum (H = 2).

Cleavage refers to the characteristic way particular minerals split along definite planes as determined by their crystal structure. Mica has perfect cleavage in one direction thanks

Table 4.1 Mohs' Hardness Scale

Hardness	Mineral Common	Example
1	Talc	Pencil lead, 1–2
2	Gypsum	Fingernail, 2½
3	Calcite	Copper penny, brass
4	Fluorite	Iron
5	Apatite	Tooth enamel Knife blade Glass, 5½–6
6	Orthoclase (potassium feldspar)	Steel file, 6½
7	Quartz	
8	Topaz	
9	Corundum	Sapphire, ruby
10	Diamond	Synthetic diamond

Silicate "Solid Solutions"

Feldspar is the most abundant of the silicate minerals in Earth's crust. It does not have a single, specific chemical composition, however, but should be regarded as a group of closely related minerals ranging in chemical composition from potassium-rich orthoclase ($KAlSi_3O_8$) found in granite (**Figure 4.28**) to calcium-rich plagioclase ($CaAl_2Si_2O_8$) found in basalt and gabbro. A sodium-rich variety of plagioclase ($NaAlSi_3O_8$), common in some granites, also exists (**Figure 4.29**).

In between these three compositional **end-members** [potassic (K), calcic (Ca), and sodic (Na) feldspars] there are intermediate compositions or blends. These are termed **solid solutions**. Because of differences in the size of the large potassium cation and much smaller sodium and calcium cations, only a small amount of solid solution exists between the potassic end member and the others in the feldspar group. But calcium and sodium ions, being similar in size, can swap with one another freely, and so a complete solid-solution compositional range is present between purely sodic plagioclase ("albite") and calcic plagioclase ("anorthite"). In fact, one of the most common plagioclase crystals in continental rocks, labradorite, is a mixture of the albite and anorthite end members in roughly equal proportions. Labradorite produces a beautiful, almost iridescent bluish sheen in reflected light, termed **schiller**, or **labradorescence**, which results from this mixture (**Figure 4.30**).

The ferromagnesian silicates are rich in iron and magnesium in addition to silicon and oxygen. Most of them are a dark black, brown, or green. Important among these are hornblende (amphibole) (**Figure 4.31**), augite (pyroxene) (**Figure 4.32**), biotite mica, and olivine, which is believed to be the most abundant mineral in Earth's mantle (**Figure 4.33**). Olivine is also prized by mineral collectors as the popular birthstone peridot. In fact, olivine is so common in the mantle that often rock found there is termed **peridotite** (**Figure 4.34**).

Fokin Oleg/Shutterstock.com

Figure 4.28 Orthoclase crystal.

PhotoChur/Shutterstock.com

Figure 4.29 Plagioclase. Note the closely spaced lines in the lower right side of this specimen. These are called twin planes, and they are a characteristic feature of this mineral group which develops during crystallization.

Alpha2473/Shutterstock.com

Figure 4.30 Labradorite crystal showing schiller.

Bjoern Wylezich/Shutterstock.com

Figure 4.31 The mineral hornblende; typically dark because of its iron and magnesium content, it forms elongate crystals that may have a nearly glassy luster but which are also in many cases quite dull.

Silicate "Solid Solutions" (*Continued*)

Figure 4.32 Augite, a common pyroxene group mineral, is dark like hornblende because of high iron content, but forms stubbier crystals that do not show the elongate cleavage so typical of hornblende and other amphiboles.

As in the case of the feldspars, solid solution compositional blends between cation end-members also occur within each of the main ferromagnesian mineral types mentioned above. For olivine, the range is from a glassy apple-green pure magnesium end-member (*forsterite;* Mg_2SiO_4) common in many basaltic lavas, to a shiny, black ferrous end-member (*fayalite;* Fe_2SiO_4), found only in some metamorphic rocks. Intermediate compositions of olivine show colors smoothly blended between these extremes, though the range of mixtures observed in nature is not complete, as it is in the plagioclase feldspars.

Figure 4.33 Olivine (peridot), characteristically with glassy luster and swirly fractures; one of the most common minerals on Earth.

Figure 4.34 Peridotite, a rock from Earth's mantle made up almost entirely of olivine and augite crystals. This specimen is embedded in dark-gray basalt lava, which brought it up from its place of origin in the upper mantle tens of kilometers down.

to its silica tetrahedral sheets; this is called *basal cleavage*, because it is parallel to the basal plane of the crystal structure. Feldspars split in two directions; halite, which has *cubic* structure, splits in three directions, as shown in **Figure 4.35**. Minerals that have perfect cleavage can be split readily by a tap with a rock hammer or even peeled apart, as in the case of mica. Some minerals do not cleave but have distinctive **fracture** patterns which help identify them. For instance, olivine and quartz fracture like glass, with smooth, curving breakage surfaces called **conchoidal fractures** (**Figure 4.36**). The common crystal shapes, presuming the crystals are well displayed, can also be useful for identifying minerals.

Color tends to vary within a mineral species, often due to the presence of solid-solution chemistry, as described in Box: Silicate "Solid Solutions" on pages 154–155. As a result, color may not be a reliable identifying property in many cases. **Luster**–how a mineral reflects light–often turns out to be a more dependable indicator. We recognize *metallic* and *nonmetallic lusters*, the latter being divided into descriptive types that you can easily imagine, such as glassy, greasy, and earthy (**Table 4.2**).

4.2.2 "Mineral-like," but Not Quite!

Not all inorganic solids found in nature are truly minerals. Some common "building-block" substances that we associate with Earth's surface lack crystalline structures necessary to be deemed a mineral. They are hodge-podge assemblages of atoms, including many ions, that never had time to become crystalline, or at least completely so. One common example is the black, shiny volcanic glass **obsidian**, valued by Native Americans for making projectile points and

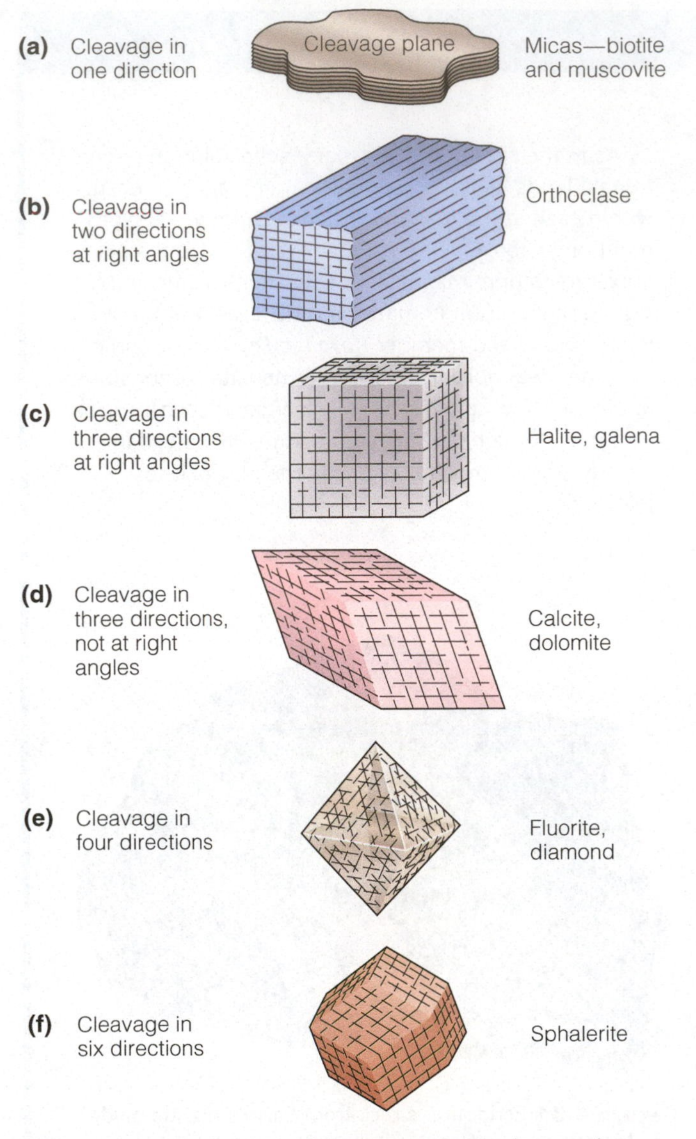

Figure 4.35 (a–f) Types of cleavage and typical minerals in which they occur.

Figure 4.36 Well-displayed conchoidal or "bottle glass" fracture in obsidian, a glassy volcanic rock. This distinctive type of fracture is also present in broken crystals of olivine, quartz, and some other minerals.

arrowheads (Figure 4.36). The silica-rich magma forming obsidian cools so quickly as it erupts that the atoms inside freeze randomly in place. Given many millions, or hundreds of millions, of years, however, the obsidian will lose its bright luster, as the glass finally begins crystallizing at a microscopic level, a process called **devitrification** (or "deglassing"). Obsidian is an example of a **metastable substance,** one that looks stable for long periods of time but is not.

Another example is the **mineraloid** ("close-to-being-a-mineral") opal ($SiO_2 + H_2O$), which may also be regarded as a common byproduct of volcanic activity, where heated groundwater circulates above shallow magma bodies

Table 4.2 Some Common Types of Mineral Lusters

Major Category	Subcategory	Notes	Ordinary Examples (including minerals and mineraloids)
Metallic		Having the appearance of a polished metal surface	Car fender or shiny metal pot (galena, gold, pyrite)
Submetallic		Like metallic, but less shiny	An iron skillet (sphalerite, cinnabar)
Nonmetallic			
	Glassy (vitreous) resinous	Highly reflective, shiny surface resembling glass	Fresh window or bottle glass (olivine)
	Greasy (oily)	Resembling a surface covered with a fresh coat of oil or grease	Fat or grease (halite)
	Earthy (dull)	Surface lacking any shininess or reflectivity	A piece of chalk (kaolinite clay)
	Pearly	A smooth surface that appears well polished and somewhat waxy	The surface of a pearl (opal)
	Resinous	Surface resembles smooth plastic	Moist chewing gum or resin seeping from a tree (amber)
	Adamantine	Like glassy, but even more shiny, even sparkly	Diamond

Cagla Acikgoz/Shutterstock.com

Figure 4.37 Opal showing a glassy luster and conchoidal fracture in reflected light. Opal can be milky white or translucent, as here. The play of light inside the mineraloid gives it the bright colors—so-called "fire opal."

(**Figure 4.37**). Nonvolcanic opal can also form when groundwater seeps into silica-bearing rocks buried deep below Earth's surface, then returns to shallow levels and evaporates. Opal is a prized gem, another birthstone, and one of the earliest substances people began to mine, perhaps as early as 6,000 years ago in Ethiopia. It is a chemically weak substance; a combination of water and silica that lacks strong ionic bonding. Thus, it does not stand up well with rough handling or even changes in climate. Rapid drying can cause it to split into fragments.

4.3 Rocks (LO3)

We progress up the hierarchy of Earth-building material to consider **rocks:** consolidated or poorly consolidated aggregates of one or more minerals, glass, or solidified organic matter (such as coal) that make up a significant part of Earth's crust. There are three categories of rocks, based on their origin: igneous, sedimentary, and metamorphic.

Igneous rocks (from the Latin *ignis*, "fire") crystallize from molten or partly molten material. **Sedimentary rocks** include both **lithified** (turned to stone) fragments of preexisting rock and rocks that were formed from chemical or biological activity. **Metamorphic rocks** are those that have been changed, in the solid state, by heat, fluids, or pressure within Earth.

To the various "cycles" of change that we described in Chapter 2, we now add in the present context an even larger, overriding planetary concept, the **rock cycle**. The rock cycle describes how the three distinct rock types transition from one to another, back and forth over time. The major drivers of this powerful, ponderous cycle are plate tectonics and interactions of the atmosphere and hydrosphere with the crust at or near Earth's surface (**Figure 4.38**).

For example, the rock cycle shows that an igneous rock may be eroded to form sediment, which subsequently becomes a **sedimentary rock**, which may then be metamorphosed by heat and pressure to become a metamorphic rock, and then melted to become an igneous rock once again. Even the reverse course of change is possible under certain conditions. Figure 4.38 is well worth spending the time to understand given your chapter reading up to this moment. We dissect the rock cycle more closely below.

4.3.1 Igneous rocks: A Closer Look

Igneous rocks are classified according to their textures and mineral compositions as well as whether they develop underground (intrusive igneous rocks), or via eruption at the surface (extrusive igneous rocks) (**Figure 4.39**). **Rock texture** is a function of the size and shape of its mineral grains. For igneous rocks, this is generally determined by how quickly or slowly a mass of molten rock cools.

If magma cools slowly, large crystals have the time to grow, and a rock with a coarse-grained **phaneritic texture** is formed (**Figure 4.40**). By definition, all the individual crystals making up a rock with phaneritic texture can be easily discerned with the naked eye. *Granite* is a good example of an intrusive rock, a light-colored phaneritic rock with which many people are familiar.

Large phaneritic rock bodies–crystallized magma chambers that cooled a few kilometers or more underground–are known as **batholiths** (from the Greek "deep rocks"). Greater than 100 square kilometers (39 square miles) in area, most batholiths now exposed by erosion at Earth's surface are granitic in composition. Batholiths show evidence of having intruded and pushed aside the surrounding older rocks, cooling and hardening where we see them today (**Figure 4.41**). Batholiths commonly occur in the cores of continental mountain ranges formed during subduction-related plate convergence (see Chapter 2).

If the exposed area of phaneritic igneous rock is less than 100 square kilometers, geologists refer to the rock body as a **stock**. Closely spaced batholiths and stocks may be linked to the same giant igneous body at depth, by chance only partly exposed in patches at the surface. Or a batholith may consist of multiple separate pulses of magma formation all merged to make a single, giant body of rock. Either way, geologists, collectively refer to these intrusive igneous masses as **plutons**, named for Pluto, the ancient Greek god of the underworld. Their phaneritic rocks are said to be **plutonic**.

Crystals within phaneritic rocks generally have roughly the same sizes, but some plutons can also include exceptionally large, well-formed potassic feldspar and quartz crystals (**megacrysts**) scattered throughout the mass of smaller mineral grains. The overall appearance of such rock is termed **porphyritic** (**Figure 4.42**). The origin of megacrysts is debated by geologists, but the commercial importance of **porphyries** (megacryst-rich rocks) is not; many host valuable metallic minerals, particularly copper.

If plutons form by slow crystallization of magma deep underground, why do we see them cropping out at Earth's surface today? Erosion is certainly one answer, but simultaneous uplift of the crust during mountain building–which

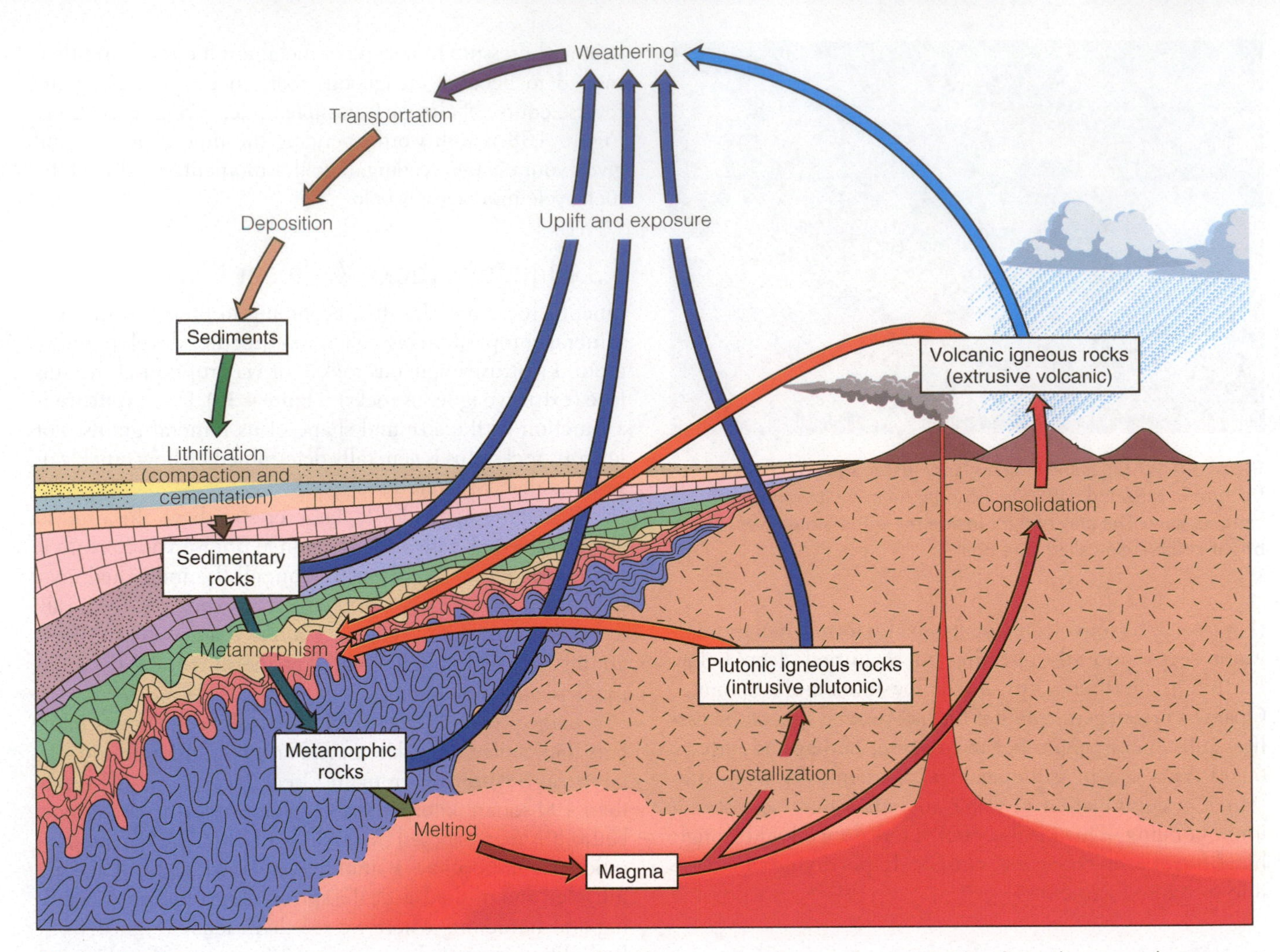

Figure 4.38 The rock cycle. The three rock types are interrelated by internal and external processes involving the atmosphere, ocean, biosphere, crust, and upper mantle.

greatly accelerates erosion, or "downcutting"–is also important. The overall process is called **tectonic exhumation,** the unroofing of rocks once deeply buried by tectonically driven processes. By studying the chemistry of minerals found in granitic rocks such as hornblende (amphibole), we have learned that some plutonic rocks exposed in outcrops today originated at depths as great as 30 kilometers (18 miles) below Earth's surface, a testament to the effectiveness of tectonic exhumation and the restlessness of Earth's crust (see Case Study 4.1).

Magma, because it is hot and therefore less dense, is typically more buoyant than solid crust, allowing some of it to rise all the way to the surface, where it erupts extrusively to form **volcanic rocks** (see Chapter 5). The erupting magma may pour out as liquid lava, or in a fierce spray of molten droplets and clots termed **pyroclasts**. In any case, the rocks produced this way typically lack crystals altogether. They are called **aphyric**, or crystal-free, because the magma was too hot to crystallize before erupting. In many places, however, you may discover some lavas and pyroclastic deposits including widely scattered, discernable crystals, especially of olivine and calcic plagioclases. These crystals form at higher temperatures than others and can appear suspended in a magma long before it erupts. The resulting volcanic rocks are called **phenocrystic**, and the individual crystals are **phenocrysts** (**Figure 4.43**).

Some magmas stall out on their ascent to the surface and crystallize rapidly at shallow depths in the crust. They are neither phaneritic (coarse-grained) nor aphyric, but contain numerous tiny crystals too small to distinguish with the unaided eye. The resulting rock texture is said to be **aphanitic**, from the Greek word for "invisible, "aphanēs." The rock may have a sparkly, sugary appearance. Some geologists apply this term to aphyric lavas and glassy volcanic rocks, too.

Chemically speaking, igneous rocks may also be classified according to their overall chemistry. Those that tend to be silica-rich, such as granite, are categorized as **felsic**, or silicic. Those that are silica-poor, like basalt, are called **mafic**, and those in between are **intermediate** (e.g., andesite). Together with whether they are plutonic or volcanic, this chemical variation serves as a basis for broadly categorizing all igneous rocks. The range of chemical compositions can be related to plate tectonic activity. We'll explore this further in Chapter 5.

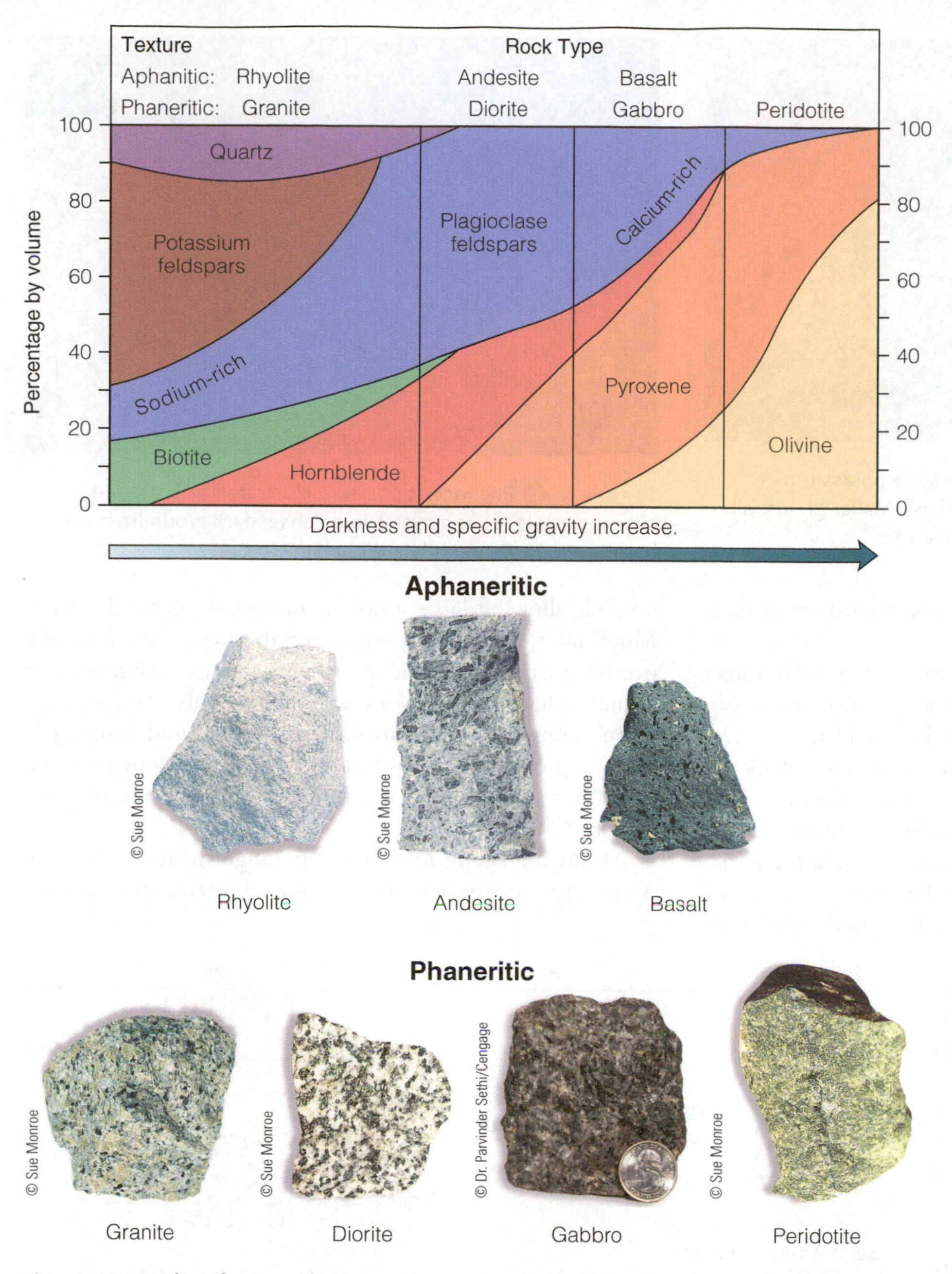

Figure 4.39 Classification of common igneous rocks by their mineralogy and texture. Relative proportions of mineral components are shown for each rock type. Fine-grained aphanitic texture results from rapid cooling of molten material. Coarse-grained phaneritic texture results from slow cooling.

Figure 4.40 Phaneritic texture in a granite, easily visible to the unaided eye. The reddish minerals are orthoclase (potassium feldspar), the white are plagioclase, and the gray are quartz crystals.

Israel Hervas Bengochea/Shutterstock.com

Figure 4.41 La Pedriza batholith, exposed in the core of the Guadarrama Mountains near Madrid, Spain.

4.3.2 Sedimentary Rocks

Sediment is particulate matter derived from the physical or chemical weathering of the materials of Earth's crust and by certain organic processes. It may be transported and redeposited by streams, glaciers, wind, or waves. We classify sediment according to particle sizes, ranging from clay granules too tiny to see without using a microscope, to boulders as large as cars (or larger).

Sediment accumulates wherever the transporting medium slows down and loses energy. For instance, when rushing water that is carrying or moving sediment enters a lake or pours into the ocean, it loses energy as it mixes into the larger, less active water body. It dumps its sediment load to form a bed of clay, sand, and occasionally larger particles on the bottom: a new sedimentary layer. The sediment can build deltas at river mouths, and spread along the shore to form beaches (see Chapter 10). As the accumulation of sediment increases, pressure hardens the lowest material in the stack into rock due to compaction. Filtration of groundwater that precipitates silica or carbonate cement between sediment grains also contributes to the process of **lithification** ("transformation into rock" **Figure 4.44**).

Only a particular category of sedimentary rocks forms this way, however. As a group, they are called **clastic sedimentary rocks**, where "clast" refers to

Figure 4.42 A porphyritic texture with large potassium feldspar megacrysts set in a groundmass of smaller grains, all coarse enough to be visible to the unaided eye.

Figure 4.43 Phenocrystic texture illustrated by scattered olivine crystals (phenocrysts) in a dense, dark groundmass of basalt lava, La Reunion Island, Indian Ocean.

individual sedimentary grains or pieces, regardless of their sizes (**Table 4.3** and **Figure 4.45**).

In contrast, **chemical sedimentary rocks** form by direct precipitation from water of solid material. For instance, ocean water contains high concentrations of dissolved calcium (Ca) and carbonate ions. Under the right circumstances–typically the presence of warm water and high evaporation rates–the calcium and carbonate may combine to become an oozy coating on the seafloor. The coating, layer after layer, hardens into *limestone*. To a greater degree, biological organisms also extract these chemicals from seawater to build shells and reefs (Chapter 2). After the organisms die, their hard, calcareous remains accumulate and cement together to form fossiliferous limestone. Diatoms, radiolarians, and other microbial organisms use silica instead of carbonate to make their hard parts, which sink to the seabed when life ends, forming widespread layers of smooth chert. These are examples of **biogenic sedimentary rocks**.

Other types of chemical sedimentary rocks include salt, beds of halite that are sometimes left by the evaporation of salty water. Deposits of gypsum ($Ca_2SO_4 \cdot 2H_2O$) also form this way where evaporation rates are quite high, such as in lagoons along desert coasts.

A distinguishing characteristic of most sedimentary rock is bedding, or **stratification**. The upper and lower boundaries of each bed, termed **contacts**, mark a time when the environmental conditions in which sedimentation took place changed. A river, for instance, may flow lazily for a long while, only bringing fine clay particles to a shoreline. Then, an abrupt change in climate may greatly increase regional rainfall, allowing large amounts of sand to enter the river. Much later, the contact separating the lower clay from the overlying sand can be read as "a record in rock" of that ancient climate change, acting on sediment supply. Sedimentary layers exposed in landscapes such as the Grand Canyon in Arizona provide wonderful insight into Earth's environmental history, stretching back hundreds of millions of years (**Figure 4.46**).

Individual sedimentary layers range in thickness from millimeters to tens of meters or more. The thickness of each

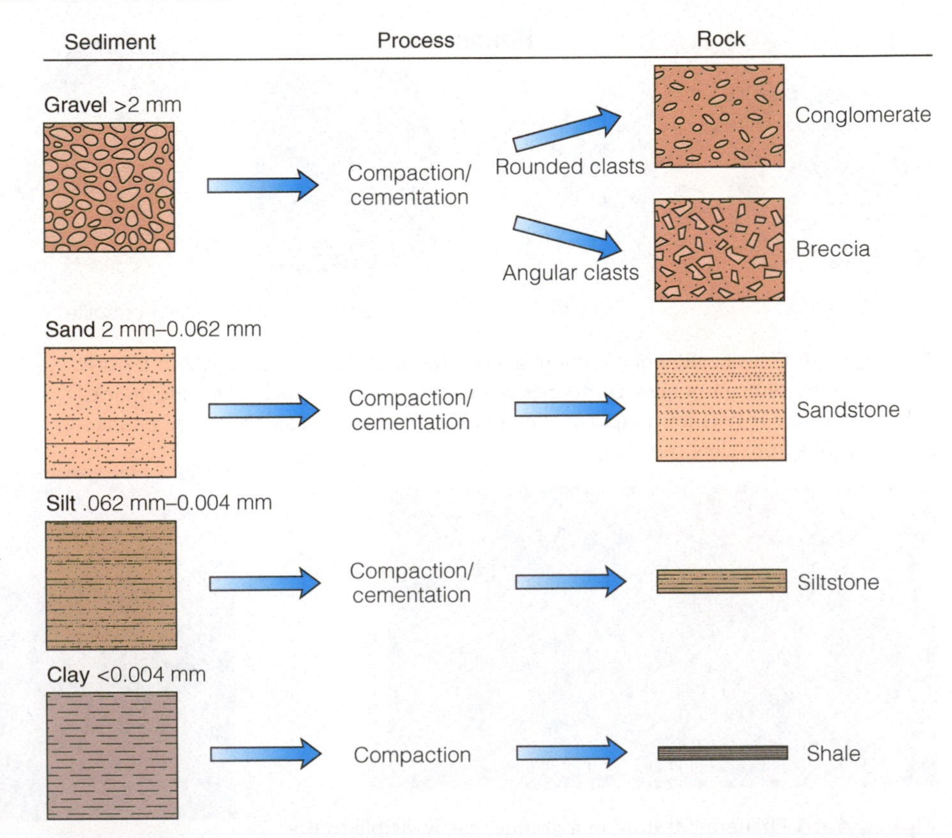

Figure 4.44 How sediments are transformed into clastic sedimentary rocks. The sequence of lithification is deposition to compaction to cementation to hard rock.

Table 4.3 Classification of Sedimentary Rocks

Detrital Sedimentary Rocks (Clastic Texture)		
Sediment	**Description**	**Rock Name**
Gravel (>2.0 mm)	Rounded rock fragments Angular rock fragments	Conglomerate Breccia
Sand (0.062–2.0 mm)	Quartz predominant, > 25% feldspars	Quartz sandstone Arkose
Silt (0.004–0.062 mm)	Quartz predominant, gritty feel	Siltstone
Clay, mud (<0.004 mm)	Laminated, splits into thin sheets Thick beds, blocky	Shale Mudstone
Chemical Sedimentary Rocks		
Texture	**Composition**	**Rock Name**
Clastic	Calcite ($CaCO_3$) Dolomite [$CaMg(CO_3)_2$]	Limestone Dolostone
Crystalline	Halite (NaCl) Gypsum ($CaSO_4 \cdot 2H_2O$)	salt gypsum
Biogenic Sedimentary Rocks		
Texture	**Composition**	**Rock Name**
Clastic	Shell calcite, skeletons, broken shells Microscopic shells ($CaCO_3$)	Limestone, coquina Chalk
Nonclastic (altered)	Microscopic shells (SiO_2), recrystallized silica Consolidated plant remains (largely carbon)	Chert Coal

layer is not a reflection of the time it took to accumulate. Some layers of gravel tens of meters thick form during single floods lasting only a few hours. In other places, layers of clay less than a centimeter thick may take centuries to accumulate. It is important to learn how sedimentary processes work and the time spans that they represent to appreciate the full meaning of these rocks when viewed in the landscape. This appreciation played a key role in the development of geology as a science beginning in the late eighteenth century.

Figure 4.45 Common clastic sedimentary rocks: (a) shale, (b) sandstone, and (c) conglomerate. Biogenic sedimentary rocks, (d) limestone composed entirely of shell materials (called coquina), and (e) coal.

Figure 4.46 Viewing the Grand Canyon from Mather Point, Grand Canyon National Park, Arizona.

Internal structures within sedimentary layers also provide hints about the environments in which they form. **Cross-bedding**–internal stratification that is inclined at an angle to layer contacts–show the influence of moving water that rippled the sediment as it accumulated. Especially thick cross-beds and frosted quartz grains in sandstone indicate an ancient desert sand dune environment. In this case, wind rather than water formed the cross-bedding (**Figure 4.47**).

Sedimentary rocks are the best places to look for fossils of ancient life (**Figure 4.48**). Some of the world's most instructive and spectacular fossil occurrences are in deposits left by massive floods or falling volcanic pyroclasts that quickly buried large tracts of land. Fine-grain sedimentary rocks like shale may preserve fossilized features as delicate as the veins in insect wings or ancient bird feathers–truly remarkable!

4.3.3 Metamorphic Rocks

The minerals found in many igneous and sedimentary rocks are generally not stable when subjected to tremendous pressures and/or temperatures such as those developed along the boundaries of tectonic plates. They break down or alter their crystal structures ("recrystallize") forming new, more stable crystals and minerals adjusted to the extreme change in crustal conditions. The entire **fabric** of a rock–its structures and overall appearance–can transform with heat and pressure. These alterations in crystal content and fabric are called **metamorphism**, literally "the process of changing form."

Figure 4.47 Windblown sandstone exhibiting large cross-beds; Zion National Park, Utah.

Figure 4.48 A fossilized dragonfly.

Metamorphism proceeds by stages, or **metamorphic grades**, from slight to intense. A rock that is only slightly metamorphosed (a "low-grade" of metamorphism) typically shows little difference from the original rock, or **protolith**, that existed before pressure and/or temperature in the surrounding crust changed. A high degree (or "high grade") of metamorphism, however, will make the metamorphosed rock look completely different from its protolith. New assemblages of minerals form in the rock, many of them stunningly beautiful and of considerable value to people. The pale-blue, aluminosilicate metamorphic mineral kyanite (Al_2SiO_5) is used to make molds for pouring molten metal in foundries because of its high melting point (**Figure 4.49**). Limestone that recrystallizes into larger grains of calcite ($CaCO_3$) forms the metamorphic rock *marble*, prized by sculptors the world over for many centuries. Likewise, sandstone recrystallizes during metamorphism to become *quartzite*, which is quarried to extract silicon, an element used with aluminum and iron to make electrical transformers, engine blocks, and machine tools.

Figure 4.49 Blue kyanite crystals set in clear gray quartz of an alumina-rich metamorphic rock.

Ryanclong photography/Shutterstock.com

Figure 4.50 Outcrop of slate with slaty cleavage included toward the left. This was once a deposit of clay and silt on the bottom of a water body.

If a rock is metamorphosed while temperatures and pressures build up in the surrounding crust, shouldn't it also "unmetamorphose" and return to its original state after these temperatures and pressures diminish? This can happen to a certain degree; during a process called **retrograde metamorphism**, as rocks cool, mineral assemblages can change. However, the mineral reactions required for retrograde metamorphism to proceed all the way take too much time and require too much energy–energy that is constantly being lost from the rock as it cools down and relaxes to surface conditions. In reality, the metamorphic fabric of the rock cannot be substantially undone or reversed. Because of this, a metamorphic rock typically preserves evidence of the *maximum grade* of metamorphism that it experienced during its long history.

A clastic sedimentary rock derived from erosion of silicate rocks undergoes a characteristic transformation with increasing metamorphic grade. A muddy seabed will harden into shale. The shale in turn transforms into fine-grained *slate* at low metamorphic grade. The slate may show a faint **foliation** (or "slaty cleavage") that is a flattening and layering of recrystallized or newly forming minerals that experience stress in the crust dominantly from one direction–for instance, the direction of tectonic compression (**Figure 4.50**). As nature ratchets up the temperature and pressure, the slate metamorphoses into shiny *phyllite* in which fine-grained white micas begin to grow (**Figure 4.51**). Individual crystals are too small to see with the naked eye, but at higher grade the crystals grow larger and new, readily visible minerals such as chlorite and garnet form. The rock becomes a *schist* (**Figure 4.52**). Foliation is especially conspicuous in schist because you can easily see the flattening of the mineral grains that define it. Ultimately, the schist transforms into *gneiss*, a hard, banded, coarse-grained rock; characteristic of deep, ancient continental crust (**Figure 4.53**). Gneisses that become so hot and under such high pressure at the extreme end of metamorphism begin to melt. They are preserved as *migmatites*–"mixed rocks;" partly metamorphic, partly igneous (Case Study 4.1; **Figure 4.54**). Many geologists regard schists, gneisses, and migmatites, despite their long, tortured histories, as among the most beautiful in the world.

KrimKate/Shutterstock.com

Figure 4.51 Phyllite; not as dull in luster as slate, with more pronounced foliation.

Bjoern Wylezich/Shutterstock.com

Figure 4.52 Mica-rich schist with a large garnet crystal included in the middle. The foliation is much coarser and less planar than it is in phyllite and slate.

Tyler Boyes/Shutterstock.com

Figure 4.53 Gneiss, having come a long way from once having been clay and silt on a seabed!

4.4 The Quest for Mineral Resources (LO4)

The mining industry is fond of claiming "If it isn't grown, it's mined." And it's true. Every material used in modern industrial society is either grown in or derived from Earth's natural mineral resources, which are usually classified as either "metallic" or "nonmetallic." Production and distribution of the thousands of manufactured products and the food we eat are dependent on the utilization of mineral resources. As

Figure 4.54 A migmatite, combining features of a gneiss with a felsic, phaneritic igneous rock. The gneiss had partly melted when temperatures and pressures started dropping back down to "freeze" it into the appearance seen here.

Zbynek Burival/Shutterstock.com

illustrated in the chapter introduction, the per capita U.S. consumption of mineral resources used directly or indirectly in providing shelter, transportation, energy, and clothing is enormous. With the value of nonfuel minerals produced in the United States in 2021 running about $90.4 billion, and the value of recycled scrap (glass, aluminum, steel, etc.) projected to reach almost $100 billion by 2031, minerals and mining are clearly a vital factor in the U.S. economy.

It is important to recognize that exploitable mineral resources occur only in particular places, having formed there because of unique geological conditions. Each of these deposits is exhaustible; there is only so much that recycling can do to replace it once it's been tapped out. To mine valuable mineral deposits is an enormously capital intensive, long-term enterprise, and there are always environmental trade-offs, sometimes painful to consider. Where do we draw the line between what we want as a society and what is best for the biosphere as a whole? It is never entirely clear, which is why questions like this quickly become political–beyond the scope of this book to consider. Science, however, does provide necessary context and focus for decision-makers and planners.

4.4.1 Economic Mineral Concentrations

Mineral reserves are deposits from which useful minerals are legally recoverable with existing technology. They may be nonmetallic (e.g., construction materials like gypsum) or metallic. If the minerals involved include important metals, we refer to them as **ores**. These include **precious metals** (e.g., gold, silver, platinum, and palladium) that are especially valuable because they are scarce, have represented wealth and power throughout much of history, and have played such practical roles as stabilizing currencies. They also have industrial significance. The more abundant **non-precious** (or **base**) **metals**, such as copper and tungsten (derived from minerals like chalcopyrite and scheelite), are of greater overall industrial importance–involved in the manufacture of wire, pipe, stainless steel, pots and pans, and many other products encountered daily.

Whether an ore deposit may be mined or not depends on two primary factors: the value that industry places on the metal that is sought, plus its geological concentration (or **concentration factor**) expressed as a ratio of a particular element's abundance in the deposit to its average continental-crustal abundance. **Table 4.4** illustrates that the minimum concentration factors for profitable mining vary widely for eight metallic elements. Note in the table that aluminum (Al), one of the most common elements in Earth's crust, has an average crustal abundance of 8%. An aluminum-ore deposit contains at least around 35% aluminum; thus, the present concentration factor for profitable mining of aluminum ore is about "4." In contrast, deposits of some rare elements, such as uranium (U), lead (Pb), gold (Au), and mercury (Hg), must have concentration factors in the thousands to be considered reasonably profitable for mining because elements like these are typically found in very low concentrations in rocks–parts per million, not percent.

A basic tenet of mining economics is that it is unlikely that we will ever run out of a useful mineral resource because there are always deposits of any mined substance that have lower concentrations than are currently economical to mine. If the supply of a particular resource drops, market forces will cause the price people are willing to pay for that resource to increase. What was too expensive to mine before then now becomes minable because we are now willing to dig further and spend more money extracting what we need. While some people dispute the correctness of this simplistic rationale, the point is that what we consider profitable for mining is always changing, due to the depletion of deposits, changes in market demand, altered public policies, and even the introduction of new means of mining.

For example, the exploitable reserves of gold increased dramatically in the United States in the late 1960s because of a combination of changes in government policies and advances in technology. Before 1971, the U.S. government

Table 4.4 Concentration Factors for Profitable Mining of Selected Metals

	Percentage Abundance (%)		
Metal	Average in Earth's Crust	In Ore Deposit	Concentration Factor
Aluminum (Al)	8.0	24–32	3–4
Iron (Fe)	5.83	20–69	4–14
Titanium (Ti)	0.86	22–86	25–100
Zinc (Zn)	0.00896	4	300–600
Copper (Cu)	0.0058	0.4–1.2	80–160
Silver (Ag)	0.00000896	0.00896	1000
Platinum (Pt)	0.000000596	0.00036	600
Uranium (U)	0.00016	0.08-0.16	500–1000
Gold (Au)	0.000000296	0.0012–0.0015	4000–5000

guaranteed that the dollar could be freely converted to gold at a rate of $35/troy ounce (1 troy ounce is equal to 31.103 grams, or 0.06 pounds). The gold was held in a national stockpile, and the size of that stockpile was a measure of the strength of the dollar in international trade. For various sound economic reasons, however, the government backed away from this arrangement in 1971, removing restrictions prohibiting private ownership of gold bullion (pure, raw gold). Now freely accessible for sale and purchase on the open market, gold prices rose to more than $800/ounce by 1980, and by 2022 had soared to almost $1,800 an ounce–an increase of over 5,000%! Many unworkable mining claims came to be considered economically profitable mines virtually overnight, thanks to market demand unleashed by a change in national policies regarding this element. A robust gold mining boom quickly began in Nevada; one of the biggest gold rushes in North American history. At the same time, gold exploitation became more cost-efficient with the development of cyanide heap-leaching technology in the 1980s, described later in this chapter.

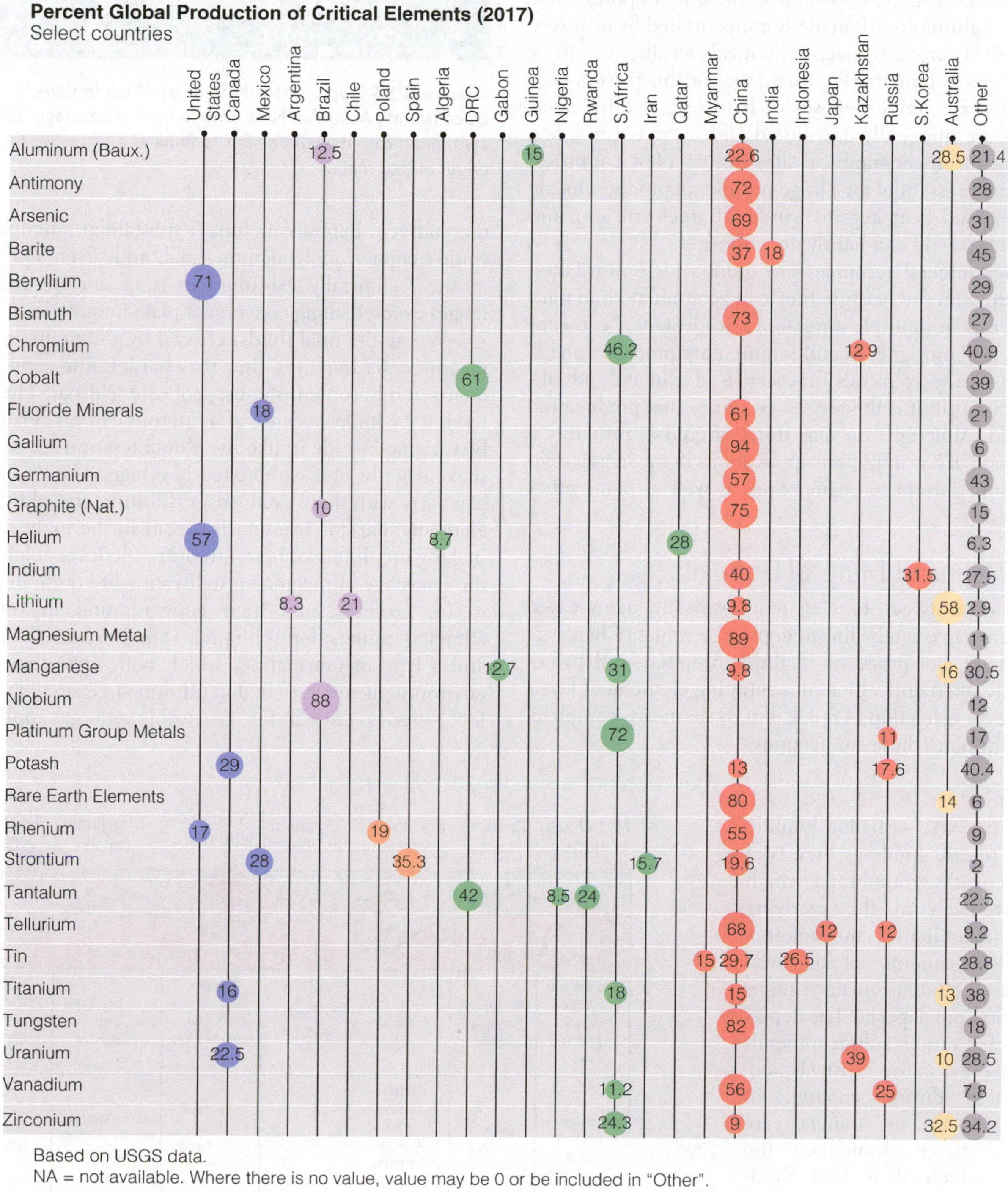

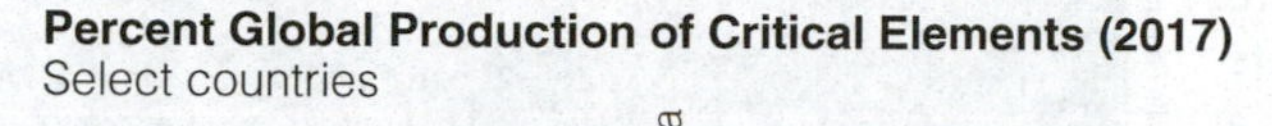

	United States	Canada	Mexico	Argentina	Brazil	Chile	Poland	Spain	Algeria	DRC	Gabon	Guinea	Nigeria	Rwanda	S.Africa	Iran	Qatar	Myanmar	China	India	Indonesia	Japan	Kazakhstan	Russia	S.Korea	Australia	Other
Aluminum (Baux.)					12.5							15							22.6							28.5	21.4
Antimony																			72								28
Arsenic																			69								31
Barite																			37	18							45
Beryllium	71																										29
Bismuth																			73								27
Chromium															46.2								12.9				40.9
Cobalt										61																	39
Fluoride Minerals			18																61								21
Gallium																			94								6
Germanium																			57								43
Graphite (Nat.)					10														75								15
Helium	57								8.7								28										6.3
Indium																			40						31.5		27.5
Lithium				8.3		21													9.8							58	2.9
Magnesium Metal																			89								11
Manganese											12.7				31				9.8							16	30.5
Niobium					88																						12
Platinum Group Metals															72									11			17
Potash		29																	13					17.6			40.4
Rare Earth Elements																			80							14	6
Rhenium	17						19												55								9
Strontium			28					35.3								15.7			19.6								2
Tantalum										42			8.5	24													22.5
Tellurium																			68			12		12			9.2
Tin																		15	29.7		26.5						28.8
Titanium		16													18				15							13	38
Tungsten																			82								18
Uranium		22.5																					39			10	28.5
Vanadium															11.2				56					25			7.8
Zirconium															24.3				9							32.5	34.2

Based on USGS data.
NA = not available. Where there is no value, value may be 0 or be included in "Other".

Figure 4.55 Global production of critical elements.

4.4.2 Distribution of Mineral Resources

Control of mineral deposits and reserves, plus other geological resources such as oil and water, has been sought by people and governments since the dawn of civilization. International trade agreements help keep some potential tension between nations at bay in modern times, but the catalog of global conflicts fought over mineral resources is a long one and remains a concern. The global distribution of valuable minerals continues to have no relation to the specific locations of political borders or technological capabilities (**Figure 4.55**).

Consider bauxite, for instance, the only important ore mineral of aluminum. Bauxite is concentrated by only one geochemical process: deep chemical weathering in a humid tropical climate (**Figure 4.56**). For this reason, economically exploitable deposits of bauxite and other such minerals are unequally distributed and restricted to a few highly localized geographical sites in just a few countries. Aluminum is essential for a huge number of products today, from cans and beer kegs to window frames and airplane bodies. The world can hardly do without it.

Other mineral resources are more widely distributed than aluminum ore because they can accumulate into minable reserves in multiple ways. Gold, for instance, can concentrate via both igneous and sedimentary processes and is mined in many countries irrespective of climate and latitude; as a result, it is the biggest countries that produce the most gold. Although you may think of gold as primarily a luxury good, it has important applications in medicine, electronics, and aerospace engineering, as well as many other industries.

Figure 4.56 Tropical bauxite mine at Weipa in Cape York , Queensland , Australia. Next time to drink a beverage from an aluminium can, it would be fair to think of a place like this as the origin of that metal.

4.4.3 Origins of Mineral Deposits

Mineral deposits can form in many ways. In many cases, their occurrence can be linked to plate tectonics (Chapter 2; **Figure 4.57**), but processes in the atmosphere and hydrosphere are often critical for concentrating elements at levels adequate for extraction. A quick, but by no means complete, review illustrates these connections.

Disseminated Ores and Ore Vein Deposits—Shallow granitic plutons release fluids as they cool and heat nearby groundwater, which then circulates in the overlying crust. The hot liquids can contain a tremendous amount of dissolved material, including metal cations which form ore deposits. The deposits consist of various sulfide minerals but sometimes native elements, too, precipitated during cooling. In some cases, the ore minerals are widely scattered throughout the rock in which their host fluids were active–"disseminated," in other words (**Figure 4.58**). Disseminated ore and vein deposits include a substantial amount of the world's copper, molybdenum, gold, and silver. They occur in the tectonically exhumed roots of ancient mountain ranges exposed along convergent plate boundaries.

Where the final fluids released by a crystallizing felsic magma enter fractures, they may harden into veins rich in quartz and in some instances, potassic feldspar. The quartz tends to be milky because of numerous carbon dioxide bubbles trapped inside it; like the cloudiness you see in a well-shaken bottle of a carbonated beverage. These veins also can carry with them gold, silver, lithium, and phosphorus; elements that don't ordinarily occur in the main minerals making up plutons. These unwanted "leftover" elements at the end-stage of magma crystallization are quite important to us, however, and their concentration in veins has attracted miners for millennia. Spodumene ($LiAlSi_2O_6$) and a type of tourmaline, which both crystallize in this environment, are enriched in lithium–an essential resource for batteries critical for renewable energy applications

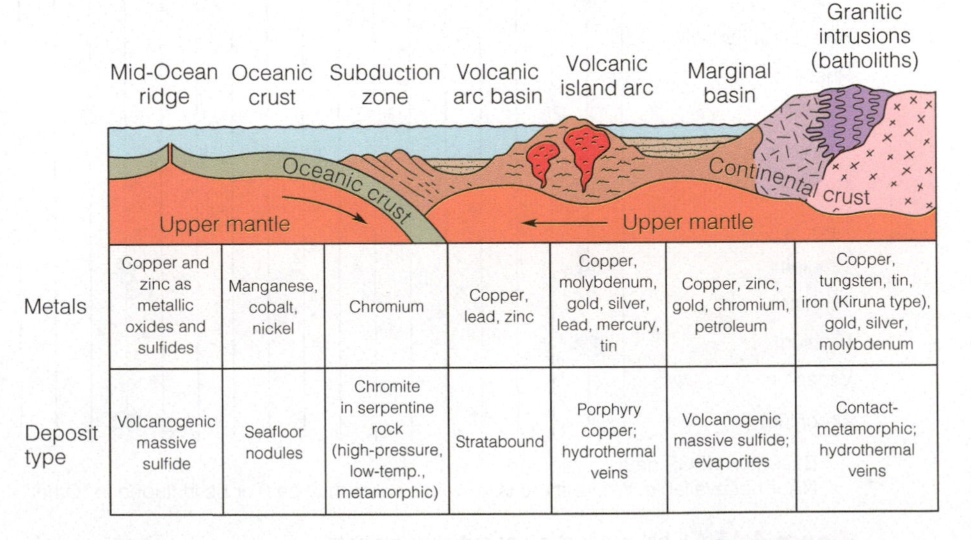

	Mid-Ocean ridge	Oceanic crust	Subduction zone	Volcanic arc basin	Volcanic island arc	Marginal basin	Granitic intrusions (batholiths)
Metals	Copper and zinc as metallic oxides and sulfides	Manganese, cobalt, nickel	Chromium	Copper, lead, zinc	Copper, molybdenum, gold, silver, lead, mercury, tin	Copper, zinc, gold, chromium, petroleum	Copper, tungsten, tin, iron (Kiruna type), gold, silver, molybdenum
Deposit type	Volcanogenic massive sulfide	Seafloor nodules	Chromite in serpentine rock (high-pressure, low-temp., metamorphic)	Stratabound	Porphyry copper; hydrothermal veins	Volcanogenic massive sulfide; evaporites	Contact-metamorphic; hydrothermal veins

Figure 4.57 Relationships between metallic ore deposits and tectonic processes.

Tim photo-video/Shutterstock.com

Figure 4.58 Disseminated copper-nickel ore in granitic rock. The abundant gold-colored mineral in this rock is chalcopyrite.

(**Figure 4.59**). Phosphorus-rich apatite, the importance of which we discussed in Chapter 2, also is abundant in mineral veins. Whether these minerals are sufficiently concentrated to be worth mining, however, depends upon nongeological, economic factors discussed above.

Magma Crystallization—Around 2.1 billion years ago, a massive asteroid slammed into Earth in what is now the eastern part of South Africa. It carved a huge, pear-shaped bowl in the crust, measuring 350 by 150 kilometers (200 by 90 miles) that rapidly filled with molten rock intruded through the shattered crust and erupted. The sea of primarily mafic lava and shallow magma cooled to crystallize phenocrysts of olivine, pyroxene, calcic plagioclase, and other minerals. Heavier metallic minerals settled into beds or layers, termed **reefs** (not to be confused by the reefs formed by marine organisms) toward the bottom of the molten mass. Eons later, Dutch African settlers named the region "Bushveld," or "landscape-plain of thorny bushes," and the name has stuck to designate the whole massive rock body, reefs of which include some of the richest ore deposits in the world. The Bushveld Complex supplies 75% of the world's platinum 50% of the world's palladium, and is the world's largest supplier of chromium (**Figure 4.60**). Vast quantities of vanadium, titanium, tin, and iron occur here, too. Everything from space satellites to automobiles, electronics, and jewelry utilizes ores largely from this single geological formation. How different would today's global economy and civilization be had that ancient Precambrian asteroid missed hitting Earth?

A few other similarly aged, possibly impact-related igneous masses also developed layered platinum and chromium ore deposits, including the Sudbury Complex in eastern Canada and Stillwater Complex in Montana. Sudbury is also a major source of nickel, at one time providing 80% of the world total. Of many uses in the present world, nickel is probably most important for making strong stainless steel. Both the Sudbury and Stillwater complexes are smaller than the Bushveld, but for people are also important "accidents of nature."

Massive Sulfide Deposits—Where volcanic activity breaks out on the deep ocean floor, particularly at mid-ocean ridges, ore deposits can form. At these submerged, divergent plate boundaries, seawater infiltrates fractures and faults overlying magma bodies, heats up, and reemerges as **hydrothermal fluid**. Sea water is full of dissolved chemicals including salts, and hot seawater can leach and transport additional elements from the oceanic crust through which it passes. Issuing from vents right along the boundary rifts, these chemicals react as the hot, chemically enriched water mixes with the colder ocean above to precipitate surface coats and flakes of metallic minerals and native elements. For good reason, marine geologists named these vents "black smokers;" and though they are patchy in distribution, there are countless thousands of them at the crest of the world's

Henri Koskinen/Shutterstock.com

Figure 4.59 Lithium-rich minerals from a Finnish pegmatite (coarsely crystalline igneous vein) mine, including red tourmaline, brownish spodumene, and a purplish lithium-rich mica, lepidolite. The modern battery industry depends on rocks like this one to provide some of its lithium.

iStock.com/Richard Meissner

Figure 4.60 Black layers of chromitite ore, the principal source of chromite, are interlayered with anorthosite, an igneous rock made up entirely of plagioclase, in the Bushveld complex near Steelpoort, Limpopo province in South Africa. Bands like this stretch for many kilometers and are termed "reefs" by the regional African miners.

Anna Kucherova/Shutterstock.com

Figure 4.61 Kokkinopezoula open pit copper mine in an ancient, massive sulfide deposit near Mitseros, Island of Crete.

mid-ocean ridges. The ore-bearing flakes they produce can settle like metallic snowfalls across surrounding landscapes, in time creating beds of rich ore; so-called **massive sulfide deposits** (**Figure 4.61**).

In a few places, later tectonic activity has uplifted slices of ancient oceanic crust, including black smoker vents and their related ore deposits. One historically important example is found on the Island of Cyprus. For more than 3,000 years, this eastern Mediterranean island supplied most of the world's copper, enabling the Bronze Age to take place. The hydrothermal activity that gave rise to this astoundingly rich concentration of copper occurred around 90 million years ago on the floor of the Tethyan Sea, a water body that has all but disappeared in the tectonic collisions of India, Africa, and Eurasia.

Placer Deposits—Most people are familiar with the image of a grizzled prospector with a mule and a pan for winnowing out flakes of gold from sediment in streams. This was once the only practical way for miners, operating as individuals, to "strike it rich" (though most did not, leading hardscrabble, if not short, lives). Today, much gold mining is organized as major corporate operations. Interest remains, however, in those gold-bearing stream sediments.

As weathering and erosion gradually expose and denude a pluton and related veins, the gold, tin, titanium, and other heavy metals inside are carried downslope by streams and rivers. These metals may initially be so disseminated in the host plutonic rocks that it cannot be extracted at a profit. But the process of sedimentary transport can concentrate them. Native gold, for example, is quite heavy compared to most sediment; at 19. 6 grams/cm^3, it's more than six times denser than the quartz and feldspar-rich sands with which it is associated. Gold settles out in still water along stream banks and at the downstream ends of sandbars. In time, a sedimentary bed develops with pockets of high-grade ore–a placer deposit (**Figure 4.62**).

Miners can recover the gold by panning in active streams, washing away the less dense minerals while

iStock.com/ilbusca

Figure 4.62 Placer deposit mining in California. Such hydraulic mining made a mess of the landscape, purposefully eroding large volumes of sediment which often choked rivers downstream.

leaving the bright, golden, dense flakes behind in the corrugated bottoms of their pans (**Figure 4.63**). Elsewhere, where the sediment has accumulated in thick deposits, mining and quarrying can extract much larger quantities of ore concentrated over many thousands of years (**Figure 4.64**). Historical **gold rushes** were urgent, sometimes lawless dashes to find treasure and fortune. They were launched by chance discoveries of gold along the banks of streams and rivers. Prospectors often follow these watercourses upslope to identify the source areas of the metal they seek, quite often cropping out as ore-bearing networks of plutonic veins termed **lodes** (**Figure 4.65**). A **mother lode** is the source area for gold and other metals found as placers throughout a broad region.

Rare Earth Element (REE) Deposits—The rare-earth elements (REE) include all 15 of what are termed by chemists the lanthanide elements; they are found in a row toward

iStock.com/Luftklick

Figure 4.63 Miner panning for gold in an active streambed placer deposit.

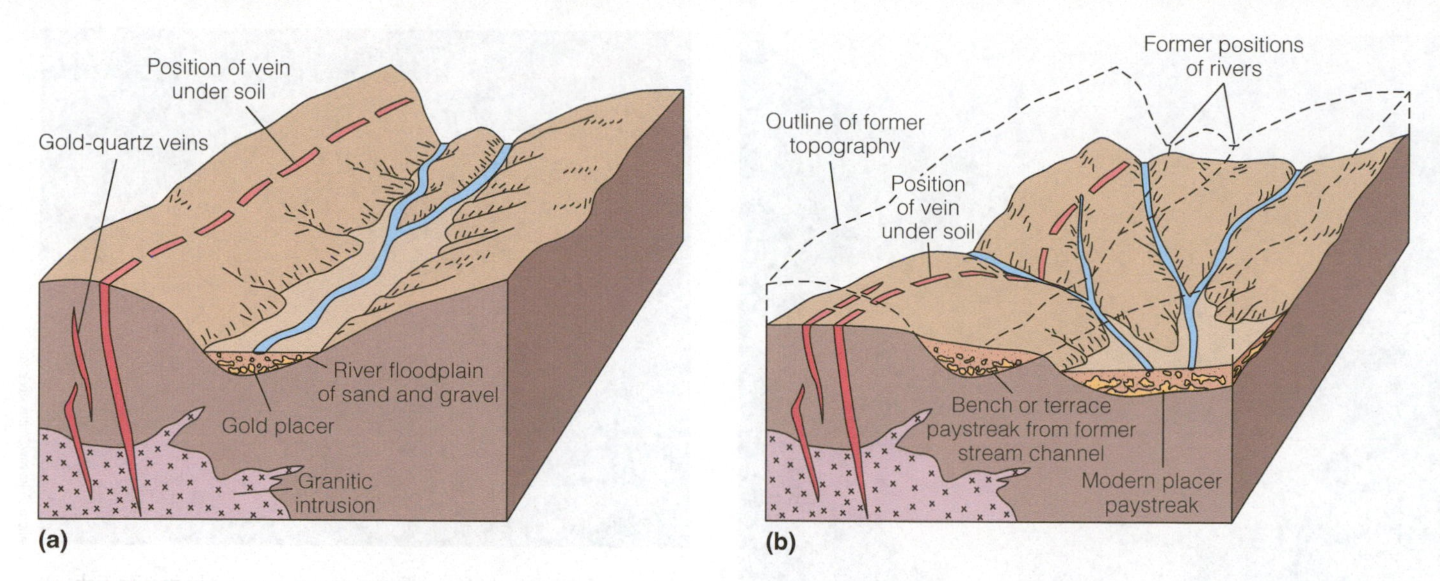

Figure 4.64 The origin of gold placer deposits. (a) An ancient landscape with a weathered and eroding gold-quartz vein shedding small amounts of gold and other mineral grains, which eventually become stream sediments. The gold particles, being heavier, settle to the bottom of the sediments in the channel. (b) The same region in modern time. Streams have eroded and changed the landscape and now follow new courses. The original placer deposits of the ancient stream channels are now elevated above the modern river valley as "bench" placers. (Adapted from Alaska's Oil/Gas & Mineral Industry, © 1982, with permission of Graphic Arts Books)

the bottom of the periodic table (containing all elements with atomic numbers between 57 and 71); cerium (Ce), terbium (Tb), samarium (Sm), and other uncommon elements that don't crystallize in the main rock-forming minerals (see Figure 4.12).

In fact, the REEs are not so rare; they are found in many places and in many rock types worldwide. However, concentrations of REEs in ore grade deposits are uncommon, and these elements, though needed only in very small amounts for any particular application, are vital for modern technologies including lasers and radars, magnets, hybrid and electric vehicles, and hard drives in desktop computers (**Figure 4.66**). Many devices we employ in the effort to achieve sustainability require REEs to manufacture.

Only a few rock types produce these ore deposits. **Carbonatites**, the products of rare, carbonate-based magmas erupted in continental rift zones, are the source of most REEs. Granitic rocks highly enriched with alkalis (potassium and sodium), and **pegmatites**, a type of quartz and feldspar-rich vein associated with some granites, also host REEs (**Figure 4.67**). REEs are particularly enriched in phosphate minerals such as apatite and monazite.

Elliot Lake, Ontario, is a good example of the special geologic conditions required to develop a rare earth ore

Figure 4.65 Copper and gold-rich lode ore exposed in the wall of a sub-surface Chilean mine.

Figure 4.66 Crystals of sphalerite (ZnS), rich in the rare earth element germanium. This element is used in many electronic devices and also enables the illumination of fluorescent lamps.

Mike Beauregard

Figure 4.67 A swarm of pegmatite veins (light-colored sub-horizontal streaks in mountainside) cross the landscape in northeastern Baffin Island, Canada, the products of Precambrian plate collision and mountain building in this region.

deposit. The REE minerals are present in placers within a quartz-pebble conglomerate, a coarse sedimentary rock derived from intense weathering and erosion of a nearby 1.4-billion-year-old granite. The placers also yield a large amount of uranium. The town of Elliot Lake, in fact, was once dubbed "The Uranium Capital of the World."

Almost all the world's REE mining and production is done now in China, an example of the geopolitical significance of mining. Much of the world's REE supply comes from the subtropical weathering of granites and the resulting accumulation of rare earth elements in soils and related crusts atop the aging rock. The greater the degree of weathering, the higher the REE enrichment as other minerals weather away.

Banded Iron Formation—In Chapter 2, you read that Earth's atmosphere became oxygen enriched in Precambrian times, thanks to the respiratory activity of blue-green algae and other early microbes. Cations of iron dissolved in sea water bonded to the oxygen atoms from the biological release of oxygen to form the minerals hematite (Fe_2O_3), magnetite (Fe_3O_4), and siderite ($FeCO_3$), which accumulated seasonally in widespread layers across the ocean floor. This went on from the time of life's inception–at least 3.7 billion years ago–until around 1.5 billion years ago, when the oceans finally became saturated with oxygen and could absorb no more.

Those ancient beds of iron oxide and related carbonates remain, however, in distinctive, **banded iron formations (BIFs)**. The beds are "banded" because the layers of iron-rich minerals, generally rust-red in color, alternate with darker layers of chert (silica-rich material) that probably formed every winter when microbial oxygen production was too low for the dissolved iron to oxidize (**Figure 4.68**). BIFs are the world's greatest source of iron ore, essential for the production of steel and other products. They are also among Earth's most unique rock types.

Andriy Solovyov/Shutterstock.com

Figure 4.68 Banded Iron Formation, showing hematite-rich bands alternating with darker cherts. Individual layers are typically no more than a few centimeters thick.

4.4.4 Types of Mining

Before the dawn of civilization, shortly after the end of the last Ice Age, people could walk around and simply pick up pieces of copper, gold nuggets, and other native elements off the ground. These were long undisturbed products of erosion that no doubt were interesting and attractive to wanderers. This "time of plenty" did not last long, however, once the potential importance of these resources became widely known. People began digging to find more minerals, and the need to organize digs was one factor driving early civilization, as in the case of the Phoenicians, who heavily mined silver in southern Spain, enabling them to compete with the Romans for control of the Western Mediterranean. The digging became deeper and more ambitious in the millennia since.

Quarries, also called **open-pit mines**, are pits or excavations in hillsides, open to the sky, from which rock is typically removed by explosions and the use of heavy hauling equipment. For many low-concentration, near-surface ore deposits, open-pit mining is the only economical way to extract ores. **Sub-surface mines** involve sinking shafts and tunneling underground to reach more concentrated mineral deposits otherwise too deep to quarry. Both types of excavation entail great physical challenges both for people and machinery; and the deeper the mining, the more dangerous the undertaking.

Among the oldest open-pit mines in the world is an 11,000-year-old site in central Israel, where blocks of limestone were extracted together with flint, a type of chert useful for making hard, sharp stone tools. The quarrying of rock continues today worldwide. About two thirds of the value of industrial resources mined in the United States are construction related, with **aggregates** (crushed stone, sand, and gravel) valued at over $17 billion, and cement (made from limestone and shale) valued at approximately $10 billion (**Figure 4.69**). Eighty percent of the aggregate is used in road building. Other major construction materials are clay for bricks and tile, and the mineral gypsum, mentioned earlier. It is the mining of aggregate that is probably the most familiar to urban dwellers, because quarries customarily are sited near cities, the major market for the product, to minimize transportation costs.

Sigur/Shutterstock.com

Figure 4.69 An aggregate quarry for construction materials.

The quantity of aggregate needed to build a 140 square meter (1,500 square foot) (139.35 square meters) house today is impressive: 51 m^3 (67 cubic yards) are required, each cubic meter consisting of 1.18 tonnes (1.3 tons) of rock and gravel and 0.8 tonnes (0.9 tons) of sand. This amounts to approximately 100 tonnes (110 tons) of rock, sand, and gravel per dwelling, including the garage, sidewalks, curbing, and gutters. Hence it is no wonder that aggregate has the greatest commercial value of all the rock and mineral products mined in the United States, ranking second only in those states that produce natural gas and petroleum. The annual per capita production of sand, gravel, and crushed stone in the United States amounts to around 5.4 tonnes (6 tons)!

The principal sources of aggregate are open-pit quarries in modern and ancient floodplains, river channels, and alluvial fans. In areas of former glaciation, aggregate is mined from glacial outwash and other deposits of sand and gravel that remained after the retreat of the giant Pleistocene ice sheets in the northern United States, Canada, and northern Europe.

The largest quarry in the world is the Kennecott Bingham Canyon open-pit mine near Salt Lake City, Utah (**Figure 4.70**). This great hole, cut into a shallow plutonic body and the surrounding limestone and quartzite rock, measures 4 kilometers (2.5 miles) across and is a nearly a kilometer (a half mile) deep. It is has yielded 15.3 million tonnes (17 million tons) of copper over its lifetime–one of the greatest single sources of copper on Earth. The rare element tellurium (Te), which is vital for the manufacture of photovoltaic solar panels, is presently being mined there, too.

Sub-surface mines in many cases have simple plans. Some consist of nearly level tunnels, termed **adits**, drilled sideways into hill and mountainsides (**Figure 4.71**). The adits are typically slanted gently toward mine entrances to drain water that may enter them through fractures in their walls and ceilings, and to expeditiously remove rocky rubble ("waste rock") cast out of the mine entrance across the neighboring landscape. Other mines begin as vertical **shafts** from which adits branch at depth, possibly at several depths or levels. As you may expect, elevators or cabled buckets are required to lower and raise miners, ore and waste rock in shafts. Shaft bottoms may be designed as **sumps**, or places for drainage water to collect. In some instances, tunnels are slanted at steep angles, with transit requiring underground tracks for ore carts and miners to get from one level to another.

Drifts are tunnels cut specifically to follow a valuable vein or mineral seam. They can slant at practically any angle depending upon the orientation of the material being mined. Drifts maximize access to some ore deposits underground, but their construction and maintenance can face even greater challenges than other kinds of mine passages. The world's largest underground mine, El Teniente, Chile, has over 3,000 kilometers (1,900 miles) of underground passageways, laid out practically like an underground city. (New York City for comparison has some 9,600 kilometers of paved streets). This operation intensively exploits a shallow plutonic, subvolcanic copper deposit but does so very differently than at the Kennecott Bingham Canyon open-pit mine.

YegorovV/Shutterstock.com

Figure 4.70 The Kennecott Bingham Canyon open pit copper mine in Utah, the world's largest excavation made by people.

iStock.com/sezer66

Figure 4.71 Sub-surface (underground) mining operation.

Don Johnston_ON/Alamy Stock Photo

Figure 4.73 Hydroseeding at work to establish grass on bare soil and thus slow erosion.

A special type of quarrying, **surface mining**, involves removing **overburden**, rock and soil that overlies the material of interest. Surface mining is generally less expensive, is safer for miners, and facilitates more complete recovery compared to sub-surface tunneling. Surface mining, however, causes more extensive disturbances to the land surface and has the potential for serious environmental consequences unless the land is carefully reclaimed, as you will later see. The three major methods of surface mining are contour mining, area mining, and mountaintop-removal mining. More often than not, coal is the target of surface mining although bauxite is often surface mined (see Figure 4.56).

Contour mining is typical in the hilly areas of the eastern United States, where coal beds occur in outcrops along hillsides (**Figure 4.72**). Mining is accomplished by cutting into the hillside to expose the coal and then following the coal seam around the perimeter of the hill. Successive, roughly horizontal strips are cut, enlarging each strip around the hillside until the thickness of the overburden is so great that further exposure of the coal bed would not be cost-effective. At each level of mining is a **highwall**, a cliff-like, excavated face of exposed overburden and coal that remains after mining is completed. Augers (giant drill bits) are used to extract coal from the areas beneath the highwall.

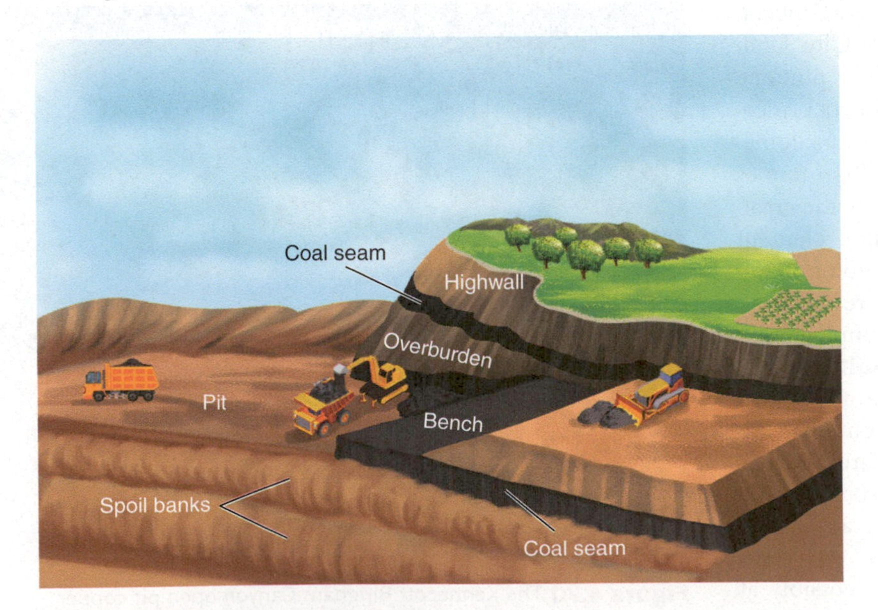

Figure 4.72 Contour mining of coal.

U.S. federal law requires that the surface mining sites be reclaimed–that is, returned to the original contour; and highwalls must be covered and stabilized after contour mining is completed. This is accomplished by backfilling **spoil**, the broken fragments of waste rock removed to mine the coal, against the highwall, and spreading and compacting as necessary to stabilize the reclaimed hillside. **Hydroseeding**, spraying seeds mixed with water, mulch, fertilizer, and lime onto regraded soil, is used on steep reclaimed slopes to aid in establishing vegetation to help prevent erosion (**Figure 4.73**).

Area mining is commonly used to mine coal in flat and gently rolling terrain, principally in the midwestern and western United States and parts of Europe (**Figure 4.74**). Because the pits of active area mines may be several kilometers long, enormous equipment is used to remove the overburden, mine the ore, and reclaim the land. Topsoil is stockpiled in special areas and put back in place when the mining is completed. After replacing, the soil may be tilled with traditional farming methods to reestablish it as cropland or pastureland.

Mountaintop-removal mining (MTR), perhaps the most destructive form of surface mining, is used primarily in the eastern United

iStock.com/hsvrs

Figure 4.74 Brown coal being area mined in Germany adjacent to a coal-fired power station.

States to recover coal that underlies the tops of mountains. It is now the dominant activity altering landscapes in the central Appalachian coalfields, where 300-million-year-old coal is currently being extracted (see Chapter 11). MTR mining involves clearing forests, stripping and disposal of overburden (soil and rock materials above the coal bed), and the use of explosives to shatter rocks to expose and mine the coal beds. It permits nearly complete removal of the coal. In some instances, whole communities and the people whose families lived for generations along the now-nonexistent stream banks have been displaced with minimal financial compensation (**Figure 4.75**).

After the coal is removed, the mined area is usually left as flat terrain that historically has been revegetated by planting a few herb and grass species. To date, less than 5% of the flattened land has been "restored" ecologically, and many reclaimed areas exhibit little or no regrowth of woody vegetation and minimal carbon storage even after 20 years. Proponents of MTR point to its ability to provide jobs, greater safety for miners than underground mining, its efficiency, and the resulting increase of flat land in a region where otherwise there is little. The coal industry claims that MTR mining brings prosperity; yet, regions where MTR mining is practiced have some of the highest poverty rates in the United States.

iStock.com/6381380

Figure 4.75 Mountaintop coal removal operation in West Virginia

4.5 Environmental Impacts of Quarrying and Mining (LO5)

For the people who do the mining, there are many immediate environmental concerns every day on the job.

Rock falls and bursts associated with reduction of pressure in quarry walls as quarrying continues can be extremely hazardous. Designing an open-pit mine to minimize this hazard is vital. Steel mesh nets, rock anchor bolts, and movable barriers can be emplaced to deal with the problem, but preventing disaster requires tailor-fitting the safety effort to each pit, because no two excavations are alike.

For sub-surface miners, air quality is an additional concern. The buildup of carbon dioxide and carbon monoxide can be serious, if not lethal, in underground mines which, unlike most natural caves, do not generally experience an efficient exchange of fresh air with the outside world. Methane concentration in coal mines often contribute to terrifying flash-fires (Chapter 11). One such explosive fire in a French mine in the early twentieth century killed more than 900 miners. Uranium mines experience the added threat of radon gas escaping from walls. Radon is an extremely toxic gas, a natural product of the radioactive decay of uranium (Chapters 1 and 12). Compounding the problem of contaminated mine air, the high temperatures far underground can be life-threatening.

To deal with all this, giant air-conditioning systems must be installed, including cooling towers at mine entrances, pumps, water sprayers, kilometers of electrical wiring, and for especially deep mines (>2 kilometers), secondary climate-control facilities placed deep within the mines themselves. The first such system was designed for a mine in 1937, in Canada. Before that time, miners simply labored and suffered (**Figure 4.76**).

A lesser concern for miners is infiltrating groundwater. Huge open-pit quarries may extend below the water table in the surrounding rocks, causing springs and seeps to issue from and weaken lower quarry walls. Likewise, mines in many places are flooded by groundwater. One of the drivers of the Industrial Revolution beginning 300 years ago was the need to develop steam-powered pumps to remove water pooling in deep British coal mines. Once these pumps became operational, the miners could dig even more deeply in their quest for coal with less fear of flooding and collapse of mine walls. It is no exaggeration to say that abundant fresh supplies of coal, together with the invention of new coal-fired steam engines, launched the world into modern times.

Figure 4.76 The many dangers of sub-surface coal mining including explosions, fires, and roof collapse.

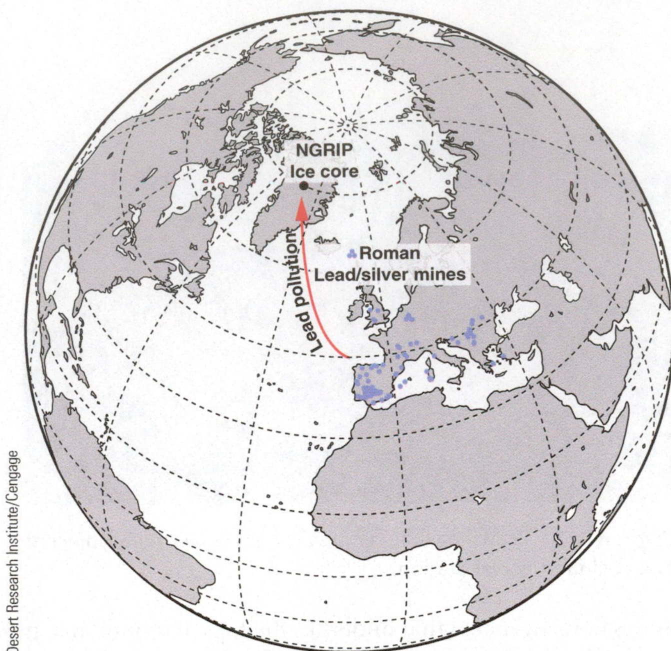

4.5.1 Air Pollution from Smelting and Processing

An active quarry on a windy day is a source of dust that can spread widely, coating surfaces and entering lungs for kilometers around. In general, however, the overall air pollution impacts of mining are much greater than this fairly localized annoyance. One of the early technological breakthroughs that spurred mining of ore deposits was the invention of **smelting**, using a furnace to extract pure metal from metallic minerals and rock. The technique was known to people in central Turkey as early as 8,000 years ago—before the invention of writing! Early Egyptians and Mesopotamians employed it to obtain copper and was independently developed by Native Americans 600 years before the arrival of Europeans. Air pollutants from distant ancient lead and silver smelting settled into ombitrophic (raised) peat bogs in Europe and accumulating Greenland ice, where they may still be found preserved today (**Figure 4.77**).

The technique of smelting is straightforward. The ore rock is placed in a furnace lined with material that can resist high temperatures without melting. A fire is lit beneath the rock, kept hot and active by blowing air into it using a long tube or nozzle called a *tuyère*. Also thrown into the mix is pure carbon in the form of charcoal or high-grade coal. As the ore rock melts, the carbon combines with oxygen in the rocks, which releases any metal to which the oxygen was bonded. The metal is

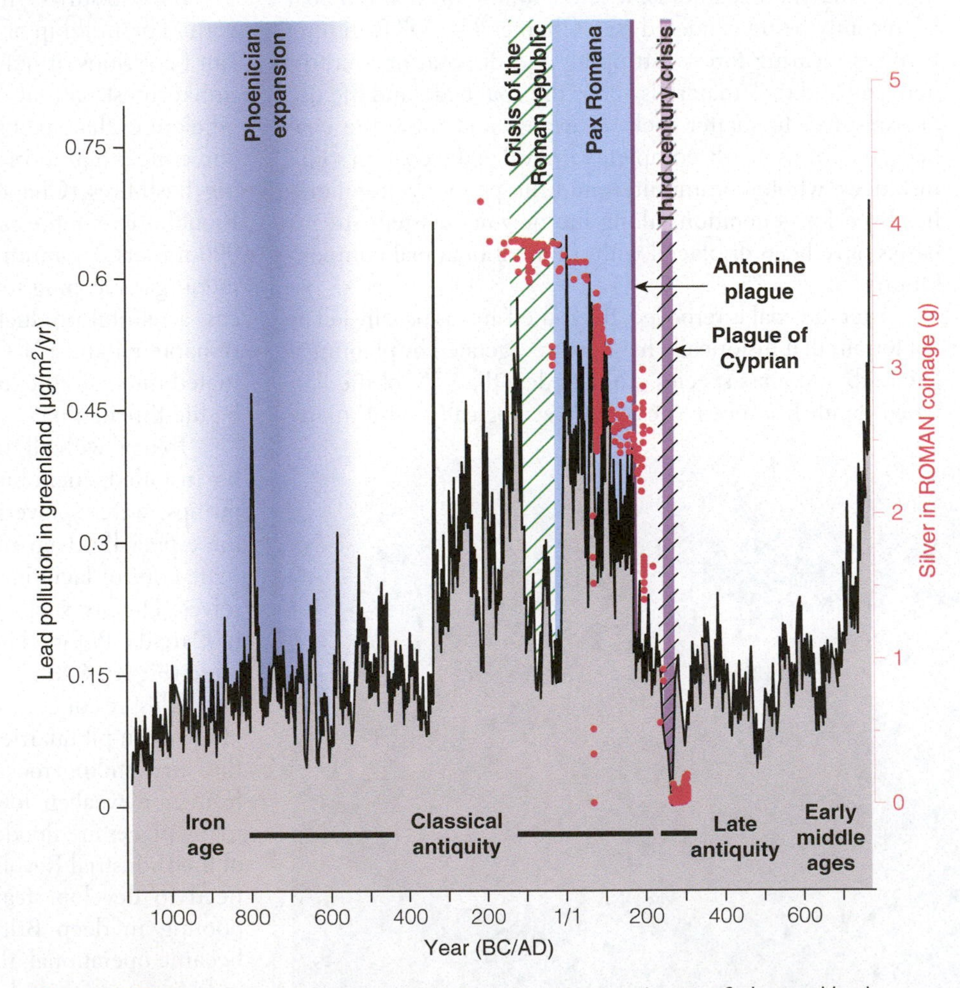

Figure 4.77 Atmospheric emissions (black line) from thousand years of silver and lead smelting are recorded in the peat bogs of Europe and the ice of the Greenland Ice Sheet. The red dots indicate the amount of silver in Roman coins. The globe shows the path of lead pollution (carried by the wind) from Europe to Greenland.

then collected as a molten fraction which drains into a bucket or well (**Figure 4.78**). Other elements in the decomposing rock escape as gases or run off separately as a human-made lava called **slag** (**Figure 4.79**). The coal is converted into mostly carbon dioxide gas that escapes to the air, contributing to the greenhouse gas effect and global warming (Chapter 3)–perhaps a percent or two our total greenhouse gas emissions.

Many smelters handle metallic sulfide ores. In this case, sulfur dioxide is also released, which downwind can combine with water vapor to form clouds–sources of sulfuric acid rain. Acid rains are defoliants that can strip vegetation of leaves across hundreds of square kilometers leach nutrients from soils, and impact regional water quality. While the Clean Air Act has reduced acid gas emissions in the United States, such emissions are not as well-regulated in other countries.

One of the modern world's greatest ore smelters is located at Sudbury, Ontario–site of the Precambrian meteorite impact mentioned earlier. Employing 20,000 miners, the Sudbury plant includes a towering smokestack, the 380-meter (1,250-foot)-tall "superstack." Engineers built this tower in 1972 subscribing to the idea that "the solution to pollution is dilution"–that is, spreading pollutants so widely that harmful concentrations wouldn't build up at any one location, or at least as often as otherwise (Chapter 12); **Figure 4.80**). The taller the smokestack, the more widespread the discharge, thanks to higher altitude windiness. The spread of sulfur dioxide emissions from Sudbury was so wide that large amounts of sulfur dioxide reached the United States and even Greenland, 2,700 kilometers (1,700 miles) away. Over 7,000 regional lakes were acidified, killing large numbers of trout and much other aquatic life. In cases like this, "dilution is not the solution."

On a happy note, both smelter management and local citizens have taken up a campaign of environmental restoration. **Sulfur scrubbers**, devices for removing sulfur gases from emission plumes, are now installed on smokestacks, and what was not long ago a landscape as bleak as the Moon's surface downwind from Sudbury is enjoying an intensive reforestation program–over 10 million new trees representing 80 different species have been planted in soils treated to neutralize their artificially boosted acidities. With money and minds put to it, people can help nature heal.

Less appreciated, perhaps, is the air pollution associated with making cement, the purpose of some of the world's largest open-pit mines which quarry limestone, the main raw ingredient of cement (**Figure 4.81**). In 2021, more than 7% of the CO_2 emitted by human activity every year came from cement production. China alone produces more cement-related carbon dioxide than all the rest of the world's countries combined.

Why is CO_2 such a significant waste product from this industry? The principal raw material going into cement is freshly quarried limestone, $CaCO_3$. Together with various clay minerals, bits of iron ore (which lowers the melting

Mentalmind/Shutterstock.com

Figure 4.78 Modern smelter operation: metallic ore is introduced to a furnace, the ore rock is heated and the metal in it separated as a molten fraction. The melted metal is collected in a bucket and poured out to be molded as it cools and hardens into final products.

iStock.com/Nenadpress

Figure 4.79 Fragments of slag from an iron smelting operation being disposed of on site.

WorldFoto/Alamy Stock Photo

Figure 4.80 Barren landscape downwind from a smelter is the result of acidic emissions. Sudbury, Ontario, Canada.

Tom Grundy/Alamy Stock Photo

Figure 4.81 Cement factory in Colorado. If you look closely, you'll see the mine for limestone as well as tailings piles and heaps of tires used for fuel.

point of limestone) and ash–much of it from coal-fired power plants (Chapter 12)—the crushed limestone is fed into giant rotating kilns with interior temperature of 1,400°C (2,500°F)–hotter than the hottest known lavas! The $CaCO_3$ breaks down at these temperatures into **lime** (CaO) and CO_2 gas. From the far end of the kiln emerges a material called **clinker,** in small gray balls that are then mixed with gypsum and additional limestone, crushed up, and sold to industry and big-box stores as "ready-mix concrete." Cement manufacturing produces more than three times the amount of CO_2 as all the world's airplanes flown each year (as of 2022).

4.5.2 Acid Mine Drainage

Acid mine drainage (AMD) is generally considered to be the most serious environmental problem facing the mining industry today. Uncontained, AMD can afflict waterways and soils around a mine for hundreds of years–long beyond the lifetimes of most companies responsible for creating it, never mind the miners and consumers of mining products themselves.

AMD typically ensues when high-sulfur coal or metallic-sulfide ore bodies are mined. In both cases, pyrite (FeS_2) and other metallic-sulfide minerals are prevalent in the walls of underground mines and open pits; in the **tailings**, finely ground, sand-sized waste material from the milling process that remains after the desired minerals have been extracted; and in other mine waste. The reaction of pyrite and other sulfide minerals with oxygen-rich water produces sulfur dioxide (SO_2) that combined with water forms sulfuric acid (H_2SO_4). The chemical reaction in turn, liberates free metallic ions, including iron, lead, zinc, copper, and manganese, into the environment (**Figure 4.82**). Such chemically charged substances kill aquatic life, much of which is far more sensitive to these pollutants than many people.

AMD runoff corrodes and destroys human-made structures, such as concrete drains, bridge piers, sewer pipes, and well casings. The problem is worsened by certain bacteria that have evolved to thrive in the artificially acidic environment, deriving energy from the conversion of one type of iron (ferrous, Fe^{+2}) to another (ferric, Fe^{+3}). In the case of the common ore mineral pyrite (FeS, or $Fe^{2+} + S^{2-}$), the biological oxidation of Fe^{2+} to Fe^{3+} causes the pyrite to dissolve, releasing both iron and acid-forming sulfur ions to the runoff. The bacteria accelerate the process of acid-production by almost a million times relative to what would happen in their absence. To make matters worse, AMD-related

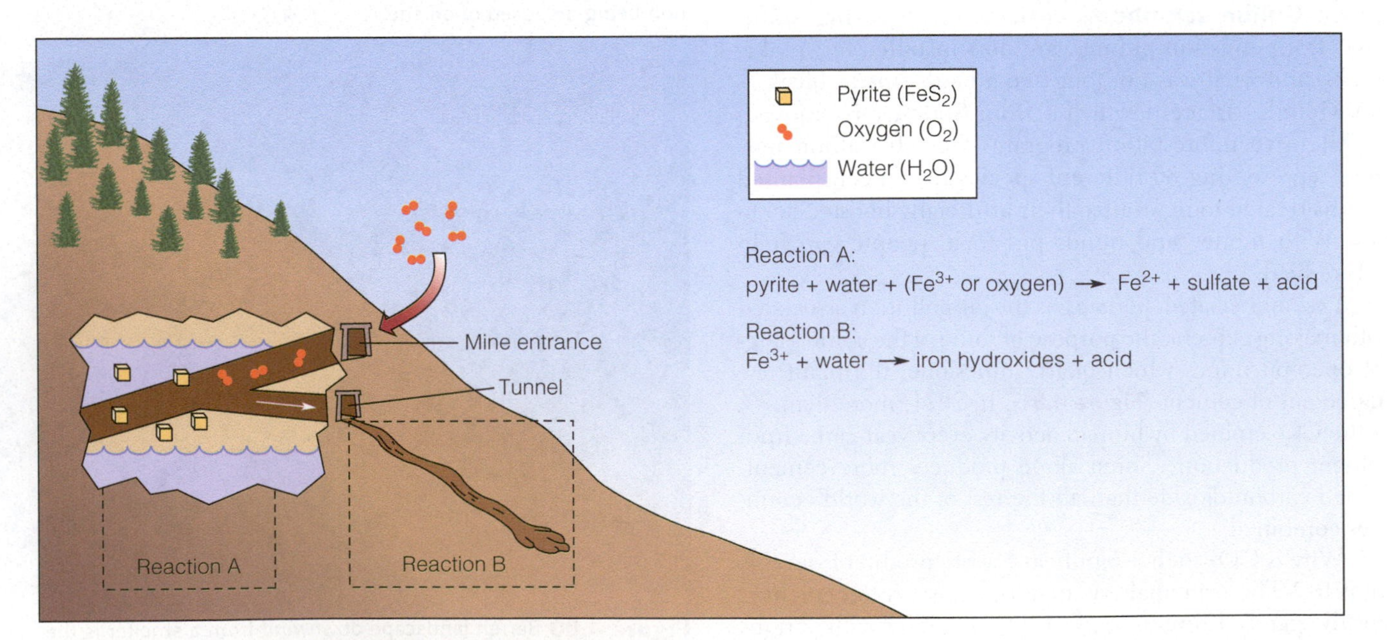

Figure 4.82 Pyrite reacts with air and water to produce sulfuric acid. (Modified from USGS)

chemical reactions are exothermic; heat produced provides positive feedback to the entire process by further speeding mineral disolution and the resulting acidification. Tell-tale reddening of brooks and streams from dissolved ferric iron ions can be observed for kilometers downstream of abandoned sulfide mines (**Figure 4.83**).

Through plant nutrient uptake, groundwater infiltration, and the food chain, toxins carried in AMD impact other components of the biosphere–including humans that unknowingly use contaminated groundwater–often with injurious effects. Cadmium found in AMD, for instance, can infiltrate water supplies and contribute to high blood pressure, liver damage, and cancer. Estimates suggest that between 8,000 and 16,000 kilometers (5,000 and 10,000 miles) of U.S. streams have been impacted by acid drainage. Where AMD-polluted water is used for livestock or irrigation, it may diminish the productive value of the affected land. Although expensive to implement, technologies do exist for preventing and controlling AMD. The problem is one of funding falling far short of environmental needs. There are 2,000 closed, hazardous AMD-producing mines on U.S. National Forest lands alone (as of 2020), many of them more than a century old. The cost to clean up and mitigate their ongoing pollution ranges between $2 billion and $4 billion.

Silvia B. Jakiello/Shutterstock.com

Figure 4.83 Acid mine drainge from a mine in Spain.

4.5.3 Hydraulic Mining

AMD is not the only significant water related problem associated with quarrying and mining. **Hydraulic mining**, or **hydraulicking**, involves directing a high-pressure jet of water through a nozzle, called a **monitor**, against hillsides containing ancient placers with certain valuable minerals, especially gold (**Figure 4.84**). The gold–and much more–washes out quite quickly, with little need for human labor. Hydraulicking requires the construction of ditches, reservoirs, a penstock (vertical pipe), and pipelines. A water-blasted open pit can yield thousands of cubic meters of gold-bearing gravel per day. The gold is recovered in **sluice boxes** (long, open-ended boxes with transverse slats on the bottom over which the slurry of muddy water flows).

Hydraulicking is efficient but destructive to the land. Because nineteenth-century hydraulicking was found to create river sediment that increased downstream flooding, clogged irrigation systems, and ruined farmlands, court injunctions stopped most such mining in the United States before 1900 (although its use continued in California's Klamath-Trinity Mountains until the 1950s). Hydraulicking is still being used in the northeast Baltic region for mining amber and to extract placer gold in remote areas of Brazil's Amazon Basin, as well as parts of Chile and Peru.

4.5.4 Mercury and Gold

Quite often, flakes of gold cannot be separated easily from other rock particles, making it difficult for miners to recover as much gold as they would like. This is especially true of very fine flakes of gold that can easily be overlooked. As early as 5,500 years ago, however, miners discovered that quicksilver (the element mercury, Hg) can be added to gold ore to create **amalgam**, an alloy, or mixture consisting of

Everett Collection/Shutterstock.com

Figure 4.84 Hydraulic mining of gold placer deposits in Alaska, early in the twentieth century.

both mercury and gold. The amalgam may be liquid, pasty, or solid, depending on the proportion of mercury present. It is easy to separate from other ore material by scraping and can then be heated in a metal container (retort) so that the mercury evaporates, leaving behind pure gold–easy to collect. The mercury may then be recovered in a condenser for re-use. Until the late nineteenth century, practically all gold mining operations used mercury amalgamation to extract the most gold possible from ore deposits, and in some parts of the world it is still used today; for instance, in many South American mines. The process is not 100% efficient, however. Some mercury always ends up escaping to the environment: as much as 100 tons enter the Amazon Basin each year alone.

The danger of freely circulating mercury in the natural world is that it takes on different geochemical forms, including a suite of organic compounds, the most important being **methylmercury**, which is readily incorporated into biological tissues and is highly toxic to humans; the Mad Hatter's disease of *Alice's Adventures in Wonderland*. In California's mother lode region, where hundreds of hydraulic placer gold mines operated from the 1860s through the early 1900s, the mercury loss to the environment from these operations is estimated to have been 12 million kilograms (26 pounds). Mercury persists in California ecosystems contaminating many streams and rivers, making the consumption of fish caught from those waterways hazardous to humans. Though methylmercury is also concentrated in the state's agricultural wetlands, there is no indication at present of significant contamination to regional food supplies.

4.5.5 Cyanide Heap Leaching

Ultimately, the use of mercury to separate gold gave way to a different chemical approach now routinely used in bedrock mining. The process is straightforward: once the gold ore rubble is gathered at the mouth of a mine, it is milled; that is, pulverized into fine powdery bits, spread out in piles over an impervious clay or high-density polyethylene liner, and sprayed with a dilute cyanide (CN) solution, commonly about 200 ppm (0.02%). The solution dissolves gold and silver (and several other metals) present in small amounts in the ore as it works its way through the heap to a neighboring pond. Finally, the gold and silver are recovered from the cyanide solution by adsorption on heat-treated (activated) charcoal, and the cyanide solution is re-used for additional extractions. The precious metals are removed from the charcoal by chemical and electrical techniques, melted in a furnace, and poured into molds to form ingots. The entire process is termed **cyanide heap-leaching** (**Figure 4.85**).

Although heap-leaching is efficient, it is controversial because open-pit mining, waste-rock dumps, and tailings piles are destructive to the landscape. Furthermore, cyanide is a hazard to wildlife and a contaminant if it escapes to ground and surface waters.

To deal with these concerns, monitoring wells are placed downslope from leach pads and ponds containing cyanide for detecting possible leakage. Maintaining the low concentrations of cyanide required by government regulations requires regular monitoring of the cyanide solution. The solution is kept highly alkaline by the addition of sodium hydroxide, a strong base, to inhibit the formation of lethal cyanide gas. No measurable cyanide gas has been detected above leach heaps where instrumental testing for escaping gas has been carried out. Unforeseen peculiarities of the ore chemistry and unusual weather have sometimes caused unexpected increases in cyanide concentrations at mines, and this has required an occasional shutdown of operations until the chemistry of the solution is corrected to established safe levels. Upon abandonment of a leach-extraction operation, U.S. government regulations require flushing and detoxification of any residual cyanide from the leach pile. Regular sampling and testing at groundwater-monitoring wells may also be required for several years following abandonment.

Securing the open cyanide ponds to keep wildlife from drinking the lethal poison is another environmental concern. After 900 birds died in one incident at a Nevada cyanide-tainted tailings pond, the company pleaded guilty to misdemeanor charges, paid a $250,000 fine, and contributed an additional $250,000 to the Nature Conservancy for preservation of a migratory bird habitat.

Various techniques for discouraging wildlife from visiting cyanide-extraction operations have been tried, including stringing lines of flags across leach ponds to frighten

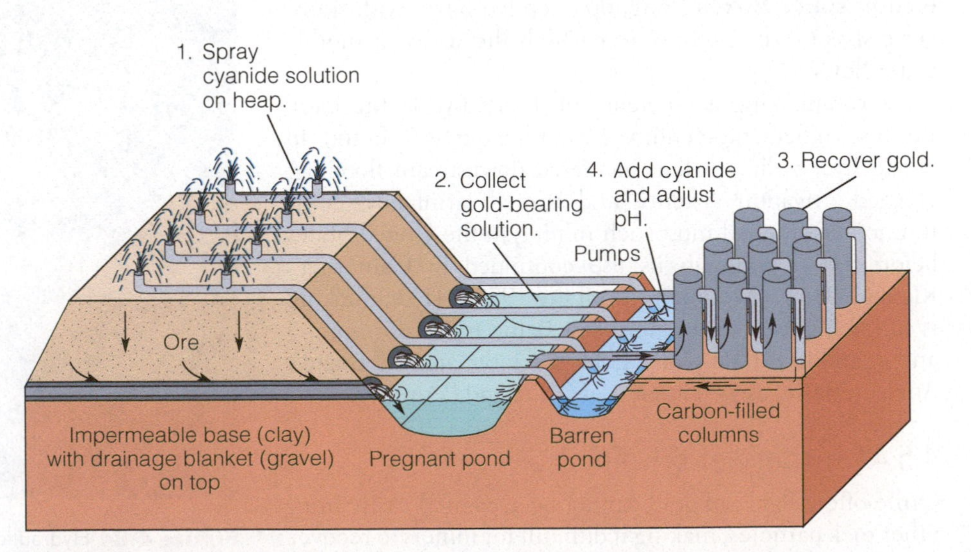

Figure 4.85 The major components of a cyanide heap-leach gold recovery.

birds away, covering cyanide ponds with plastic sheeting, and blasting recorded heavy-metal rock music from loudspeakers. At one mine, the heavy-metal recordings were effective in keeping migratory waterfowl away, but resident birdlife adjusted to the din and remained in the area. The current standard practice is to enclose the operations with chain-link fencing to keep out the larger animals and to stretch netting completely across the ponds to deter bird landings (**Figure 4.86**). The fences and netting have been successful in some 20 active heap-leach operations in the western United States: for example, bird losses have been reduced to only 50 to 60 a year in cyanide ponds. The number is trivial when compared with the estimated 1.9 to 4.8 billion birds killed in the United States every year. Most bird deaths are caused by cats (billions) followed by building collissions (millions).

D. D. Trent

Figure 4.86 A typical "pregnant pond" for collecting the cyanide solution containing dissolved gold from a leach heap. The face of the leach heap is at the upper right of the photograph. The pipes on the far side of the pond drain the "pregnant" solution from beneath the leach heap. The pond is lined with two layers of high-density polyethylene underlain by a layer of sand and a leakage-detection system of perforated PVC pipe. In the rare chance that leakage should occur, it is trapped under the pregnant pond and appropriate repairs are made. The pond is covered with fine-mesh netting to keep birds from the toxic solution.

4.5.6 The Widespread Scar of Mountaintop Removal

A major environmental problem with mountaintop removal mining is disposal of excess overburden and waste rock, which includes toxic heavy metals found in coal and former rock layers–now pulverized–such as mercury and arsenic. Millions of tons of waste are dumped into the source areas of streams and rivers, filling some smaller valleys completely and altering the landscapes of larger valleys forever, and in any case covering biologically rich terrain.

Some of the valley fills are kilometers long and many tens of meters high. For example, over 3,000 kilometers (2,000 miles) of valleys where coal mining in the eastern United States is intensive have been filled in. Over 5,600,000 hectares (14 million acres) of land have been impacted; an area roughly the size of the entire state of West Virginia where, in fact, much of this mining is concentrated. The levels of major chemical ions, such as iron, manganese, sulfate (SO_4), aluminum, and selenium, are often elevated downstream of fill sites, with selenium easily ingested by organisms reaching concentrations that are toxic to birds and fish. The degraded water can attain contamination levels that are lethal to laboratory test organisms, and macroinvertebrate and fish populations are impacted. Appalachia is home to the largest regional diversity of freshwater mussels in the world; more species live in the drainage of the Tennessee River alone than across the entire continent of Africa. MTR threatens this unique aquatic ecology.

The tale is not utterly grim. One citizen of eastern Kentucky, Patrick Angel, a scientist once involved in MTR mining and reclamation oversight, has been responsible with volunteer assistance for planting nearly 200 million trees on scarred mountaintops, helping restore biodiverse native woodland to around 2% of the total impact area. Professor of Appalachian History Kathy Newfont (University of Kentucky) commented, "When I first heard about these reforestation efforts, it was one of the most hopeful things I'd heard about the region in decades."

Case Study

4.1 Joshua Tree National Park–A Window Into The Deep Crust

Sprawling across a large swath of desert in Southern California, Joshua Tree National Park protects a range of geologic features beautifully exposed in the arid climate. Joshua Tree is famous for its bold, fractured, weirdly eroded outcrops of phaneritic and porphyritic rocks, including monzonites, granodiorites, and other granite-like ("granitic") masses; truly a rock climber's paradise (**Figure 4.87**). These rocks also have gold and other mineral deposits mostly associated with quartz veins and pegmatites developed as plutons cooled and injected their last fluids into the surrounding crust.

The highlands of Joshua Tree expose a large slice of Earth's crust ranging in age back to 1.7 billion years–some of the oldest rock in western North America. Extraordinarily, we can view it as a cross-section of Earth's ancient interior extending vertically from shallow depths (only a few kilometers deep when it formed) to as great as 30 kilometers (20 miles). Today, by fortunate chance, it is laid out almost horizontally thanks to extreme rotational movement during millions of years of fault slip and tectonic activity, combined with deep erosion. The national park lies next to a highly dynamic plate boundary, the San Andreas Fault (Chapter 2).

Extreme tectonic exhumation and erosion have revealed this cross section of ancient continental crust with remarkable completeness (**Figure 4.88**). We can literally hike "deep into the crust" that formed beneath this region during several episodes of ancient mountain building, almost to its base–all the while being careful, of course, to evade spiny cacti, mountain lions, and rattlesnakes that could be encountered along the way.

What is revealed is a remarkable picture of transitions. At the visible base of the section near the western boundary of the national park, the originally deepest rocks are gneisses formed from high-grade metamorphism of clastic sedimentary rocks (**Figure 4.89**). In places there are migmatites, meaning that the gneisses partly melted. The migmatites are the possible roots of the granitic rocks found higher in the crustal section, and an excellent example of an important link in the rock cycle. At originally shallower mid-crustal depths of 13 to 20 kilometers (8 to 13 miles), near the center of the park, the gneisses are largely replaced by numerous bands of granite-like rock up to tens of meters thick. Thin partitions of older gneiss separate some bands. In other places, they overlap. This is called the Bighorn

Figure 4.87 The iconic granitic rock formations of Joshua Tree National Park, California.

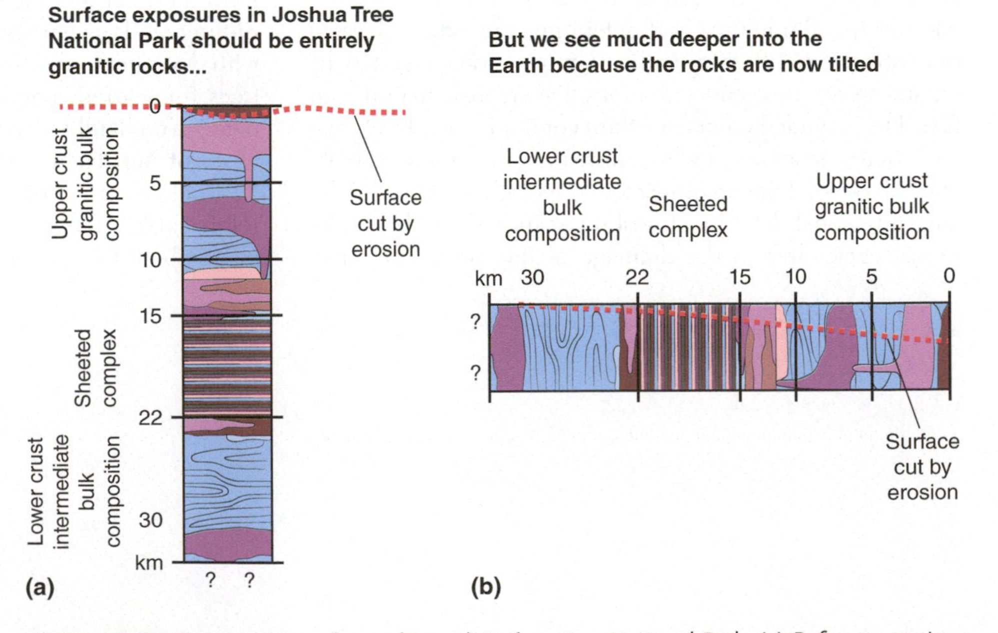

Figure 4.88 Cross-section of crust beneath Joshua Tree National Park. (a) Before tectonic deformation. (b) After tectonic deformation and erosion.

iStock.com/jenifoto

Figure 4.89 Extremely rugged, low mountains made of gneiss and, in places, migmatites, make up the oldest part of the landscape in Joshua Tree Park.

Sheeted Complex. Evidently the crust was shearing horizontally as the granitic magmas leaked up into it from the migmatites below. (Think of "horizontal shearing" as what happens when as you slide a deck of cards one over another with the palm of your hand across a tabletop). Such shearing commonly takes place as crust is compressed during the growth of large mountain ranges, although the mountains that once stood as the Bighorn Sheeted Complex developed have long since eroded away.

Above these depths, bodies of granitic magma from the sheeted complex ascended buoyantly into the shallow crust tens of millions of years ago, where they merged to form plutons that would slowly crystallize, and, thanks to slow but steady erosion, eventually become Joshua Tree's beloved climbing rocks.

4.2 Cleaning Up A Valley In Montana

The area of High Ore Creek in the Boulder River Basin in Montana was heavily mined beginning in the late nineteenth century. Metals extracted from here included gold, lead, silver, and zinc, taken from 26 now inactive or abandoned mining sites.

The biggest source of pollution was the Comet Mine and Mill (**Figure 4.90**). The mine was a major producer of gold, silver, lead, zinc, and copper, until 1941 when the ore finally ran out and operations shut down. The community that grew up around the mine became a ghost town and the Comet Mill was dismantled. However, mounds of waste rock, an abandoned open pit, and toxic heavy metals contained in a vast tailings dump remained. High Ore Creek wound its way through the 6-kilometer (3.7-mile) long, barren, and vastly altered floodplain (**Figure 4.91**). In the process, the creek flushed heavy-metals-loaded AMD and sediment downstream to the Boulder River at the nearby town of Basin, 8 kilometers (5 miles) to the southwest. Fish populations died off and sediment samples collected along the creek had metal concentrations greater than three times the natural background for arsenic, cadmium, copper, lead, manganese, mercury, silver, and zinc. Elevated levels of arsenic were measured in

© Helena/Montana Historical Society

Figure 4.90 Comet Mine and Mill, 1918. Notice the large pile of waste rock and tailings, to right of center.

Figure 4.91 High Ore Creek Valley, Montana, in (a) 1997, when the creek had eroded through a century's accumulation of metal-loaded tailings from the Comet Mine and Mill, washing them downstream to the river and enriching the water with heavy-metals some of which were toxic and accumulating in the aquatic food chain, and (b) 2006, the valley restored by the removal of toxic mine waste, regrading, the addition of topsoil, revegetation, and the rebuilding of the channel of High Ore Creek. The pollutants have been substantially reduced, and trout are returning to the creek after an absence of many decades.

streambank deposits at a school playground in Basin, containing levels that could lead to childhood cancers.

From 1997 to 2001, over a century after the mining began, the Federal Bureau of Land Management, the Montana Department of Environmental Quality, and the Montana Bureau of Mines and Geology stepped in to clean up the mess. (The responsible parties had passed away and/or were no longer in business). The effort consisted of removing 229,500 cubic meters (300,000 cubic yards) of tailings and waste rock that were returned to the Comet open pit from the surrounding landscape. The agencies collected a further 30,600 cubic meters (40,000 cubic yards) of the most polluting tailings and placed them in a separate repository, well-sealed off from the environment.

Next, water diversion channels were constructed to keep the mine site as dry as possible and to protect the area from erosion. Working downstream, the agencies also removed 47,430 cubic meters (62,400 cubic yards) of tailings-laced sediment and waste rock along about a 6-kilometer (4-mile) stretch of High Ore Creek. Waste rock slopes were graded to match natural contours in the adjoining terrain, and soil within the reclaimed area amended with a mix of compost and lime to support seeding with native plants. Netting to control erosion was spread over the landscape as new vegetation took hold. Toxic sediment settling ponds were in place to capture fresh pollutants that emerged during periods of high runoff from rainfall and snowmelt. Cost for the project was funded by a 35-cent tax on each ton of coal mined in the region.

Post-reclamation water analyses revealed that heavy-metal concentrations downstream in High Ore Creek dropped substantially. By 2006, the area was well on its way to recovery, with willows sprouting along the streambanks and grass and wildflowers growing on the reclaimed tailings sites. A trout fishery returned to High Ore Creek, too, after an absence of almost a century.

Montana state officials have identified hundreds of other abandoned mine sites in southwestern Montana that are threatening the environment and human health. The successful restoration techniques used in the Comet Mine cleanup–isolation of the worst pollutants, encapsulation of the remainder of the pollutants to the degree possible, keeping pollution sources area dry, and revegetation of the scarred landscape–point the way toward addressing similar problems elsewhere. To date (April 2022), the state has spent $17.5 million to clean up 31 miles of streams, seal 1,631 mine openings, and extinguish 16 underground coal fires. Sixty-four hectares (157 acres) of land impacted by mining operations have been reclaimed and are no longer ecologically problematic. Economically, the situation appears to be win-win. The Montana Department of Environmental Quality estimates that for every dollar spent on reclamation, three dollars currently enter the economy through contracts to work teams involved in restoring the land.

Photo Gallery

Rocks, Minerals, and Mining

Minerals are everywhere. Some are spectacular, some are colorful, and some are so valuable people will risk their lives to extract them from the Earth. Here are a few examples of how people and minerals interact.

Javier Trueba/Science Source

1 The Cave of Crystals (Cueva de los Cristales), a natural cavern intersected by the Naica Mine, Chihuahua, Mexico. These are the largest known mineral crystals in the world and are composed of gypsum (hydrated calcium sulphate). The cave is 290 meters (930 feet) deep. The crystals were discovered in 2000, when water was pumped from the mine.

Men extracting bright-yellow sulfur that has crystallized around vents from the Ijen volcano, Java, Indonesia.

Pascal Boegli/Alamy Stock Photo

Pulsar Imagens/Alamy Stock Photo

2 A miner in Ametista do Sul, Brazil, works to extract a large, amethyst-filled geode.

3 Large amounts of gold near the surface in the Democratic Republic of Congo catalyze informal mines and mining. Here, miners used a human chain to remove buckets of soil and hopefully gold at Iga Barriere.

4 The Berkeley Pit at Butte, Montana, is filled with water. The pit used to be a deep copper mine but today it holds a lake more than a kilometer long, almost a kilometer wide, and over 300 meters deep. The water is filled with heavy metals leached from the pit walls.

Summary Minerals and Rocks

1. Why Rocks and Minerals Matter to Us

LO1 Describe why minerals are important to humanity

Rocks and minerals matter to people and society. Geologic resources include metals used in everyday objects as different as electrical equipment and skyscrapers. On average, each person in the United States uses about 18,000 kilograms of mineral and rock products yearly.

i. What rocks, minerals, and elements are parts of your daily life, and where do you find them?

2. Elements, Minerals, and Rocks

LO2 Define minerals, and explain what distinguishes a "mineral" from a "rock"

a. Minerals

Geologists define minerals as naturally occurring, inorganic substances, each with a unique chemical composition and a characteristic physical structure, and which are crystalline. The chemical composition of minerals is the basis of mineral classification, including native elements (minerals consisting of a single element compositionally) and oxides (the combination of one or more cations with an oxygen anion). Mineral identification can be done without physical destruction because every mineral has distinctive physical properties (cleavage, color, luster, etc.).

b. "Mineral-like," but Not Quite!

Some common mineral-like substances, (like opal) are not truly minerals because they lack any orderly arrangement of atoms bonded together inside them, which is characteristic of crystals. We call them "mineraloids."

Study Questions

i. What are the five defining characteristics of a mineral?
ii. What are the three examples of mineral groups, and what characterizes the minerals in each group?
iii. How do geologists identify minerals without destroying their physical mineral structure?

3. Rocks

LO3 Recognize the importance of the rock cycle

a. Rocks

Rocks are consolidated aggregates of one or more minerals, glass, or solidified organic matter. The rock cycle describes how the three distinct rock types (igneous, metamorphic, and sedimentary) transition from one to another over time.

b. Igneous Rocks

Igneous rocks are crystallized from molten or partially molten material and are characterized by their textures and mineral compositions. Phaneritic, porphyritic, aphyric, aphanitic, and phenocrystic are major examples of rock textures. They describe the size and abundance of crystals that make up a rock based on how it was formed and cooled. Igneous rocks can also be classified by their overall rock chemistry: felsic rocks are silica-rich, mafic rocks are silica-poor, and those in between are classified as intermediate.

c. Sedimentary Rocks

Sedimentary rocks result from the accumulation of material at Earth's surface that hardens into rock. There are three main categories: Clastic sedimentary rocks are formed by the accumulation of sediment over time into layers, which compact and build pressure, resulting in lithification of the material into rock. Chemical sedimentary rocks form by precipitation from water of mineral grains or films, often calcium or carbonate rich. Or, when the process removing solids from water is initiated by a biological organism (such as in the formation of shells and reefs), the result is biogenic sedimentary rocks.

d. Metamorphic Rocks

Metamorphic rocks are formed when the minerals found in many igneous and sedimentary rocks are subjected to tremendous pressures and temperatures, leading to the breakdown of their crystal structures to form new crystals, minerals and rock fabrics. This transformation process is appropriately called metamorphism.

Study Questions

i. What are the three types of rocks and their defining characteristics?
ii. Briefly describe the rock cycle and how each rock type fits into it.

4. The Quest for Mineral Resources

LO4 Explain why valuable mineral deposits are in many cases "accidents of nature"

a. Economic Mineral Concentrations

Mineral deposits are locally high concentrations of potentially valuable minerals. The greater the quantity of an element or mineral, the richer the deposit and the more economically feasible it is to extract. An ore deposit is effectively the same thing as a mineral deposit, but the minerals involved include important metals in sufficient concentration to be worth mining. Concentration factor is the ratio of an element's abundance in a deposit to the crustal abundance. A specific ore mineral must be above a certain concentration factor in a deposit to be minable. That concentration factor changes according to what people are willing to pay to extract the ore. Hypothetically, if people are willing to pay "anything," the concentration factor can drop to zero.

b. Distribution of Mineral Resources

The global distribution of mineral deposits and reserves has no relation to political boundaries. While international trade agreements help alleviate some tension between nations, conflicts over resources have been and continue to be an issue between nations.

c. Origins of Mineral Deposits

Mineral deposits form in many ways, often connected to plate tectonics, hydrologic cycle flows, and atmospheric processes and conditions.

d. Types of Mining

Quarries are open-pit mines in which rock is removed by explosions and heavy hauling equipment to expose near-surface ore deposits. Sub-surface mines also use this type of mining; however, tunneling and sinking shafts are required to reach the deposits. Contour mining cuts into hillsides in rough, horizontal strips to expose coal. Area mining removes overburden using massive machinery generally to extract coal in places where the ground is level or nearly level. Mountaintop-removal mining is considered the most destructive sort of mining, in which forests are cleared, overburden is removed, and explosives are used to shatter summit rocks to expose coal beds. Drifts are tunnels cut into valuable veins or seams to mine valuable material directly.

Study Questions

i. The different types of mining are undertaken specifically for different reasons. Explain this using the examples of open pit, sub-surface, and mountaintop removal mining.

ii. Imagine that you are an engineer working for a mining company. You are planning a new mine. When would it make sense to propose excavating an adit as opposed to a shaft, or a drift to extract the ore?

5. Environmental Impacts of Quarrying and Mining

LO5 Identify the ways in which quarries and mines have negative environmental impacts and give examples of how we can remediate environmental problems caused by mining

a. Air Pollution from Smelting and Processing

Smelting is a process which uses a furnace to extract ore metal form metallic minerals and rocks. The problem is that, as ore rock melts away, carbon combines with oxygen in the rocks, releasing carbon monoxide and carbon dioxide into the atmosphere as a byproduct, contributing to the greenhouse gas effect. Smelters that contain sulfide ores release sulfur dioxide gas, which combines to water vapor in the atmosphere to form sulfuric acid and produce acid rain.

b. Acid Mine Drainage (AMD)

AMD occurs when high-sulfur coal or metallic-sulfide ores are mined and left uncontained. The tailings of these mining processes react with water to form sulfuric acid, liberating free metallic ions like iron, lead, zinc, copper, and manganese into the environment. These acidic and toxic sites kill aquatic life, corrode human-made structures, and infiltrate groundwater, amongst other serious effects.

c. Hydraulic Mining

Hydraulic mining directs a high-pressure jet of water at hillsides containing valuable minerals. These minerals wash out, requiring little human labor. However, hillsides are destroyed in the process, resulting in downstream flooding, clogging of irrigation systems, and ruined farmlands. In places where mercury amalgamation is used, mercury which escapes the mining process can become extremely toxic to humans and animals in organic forms like methylmercury.

d. Cyanide Heap Leaching

Pulverized gold ore rubble is sprayed with dilute cyanide solution to dissolve gold and silver present. Gold and silver are then recovered from the cyanide solution by adsorption on heat-treat charcoal. Problems occur when cyanide leaches out of mines and into the environment where it is a hazard to wildlife, and a contaminant in water sources.

e. The Widespread Scar of Mountaintop Removal

A major problem (amongst many) of mountaintop removal mines is the disposal of excess overburden and waste rock, which is pulverized when removed and dumped onto the source areas of many streams and rivers, not only ruining the tops of mountains but also covering other biologically rich terrain elsewhere.

Study Questions

i. What are the three examples of the environmental impacts of mining?

ii. Choosing one mining process example, explain in detail its environmental impact, and what can be done in the long run to reduce or eliminate it.

Key Terms

Acid mining drainage (AMD)
Aggregates
Anions
Aphanitic
Aphyric
Area mining
Atomic mass
Atomic number
Atomic radii
Banded iron formations (BIF)
Base metals
Batholiths
Biogenic sedimentary rocks
Carbonate
Carbonatites
Cations
Chemical sedimentary rocks
Clastic sedimentary rocks
Cleavage
Clinker
Color
Concentration factor
Contacts
Contour mining
Cross-bedding
Crystalline
Cyanide heap leaching
Devitrification
Drifts
Elements
End-members
Feldspars
Felsic
Ferromagnesian
Foliation
Fracture
Gold rushes
Highwall
Hydraulic mining
Hydraulicking
Hydroseeding
Hydrothermal fluid
Igneous rocks
Intermediate
Ions
Isotopes
Labradorescence
Lime
Lithification
Lithified
Lodes
Luster
Mafic
Massive sulfide deposits
Megacrysts
Metals
Metamorphic grades
Metamorphic rocks
Metamorphism
Metastable substance
Methylmercury
Mineral reserves
Mineraloid
Minerals
Mohs' hardness scale
Monitor
Mother lode
Mountaintop-removal mining
Native elements
Nucleus
Obsidian
Open pit mines
Overburden
Oxides
Pegmatites
Phaneritic texture
Phenocrystic
Phenocrysts
Plutonic
Plutons
Porphyries
Porphyritic
Precious metals
Protolith

Pyroclasts
Quarries
Radicals
Reefs
Rock cycle
Rock texture
Rocks
Schiller
Sediment
Sedimentary rock
Sedimentary rocks
Shafts
Shells
Sheet-silicate
Silica tetrahedron
Silicates
Slag
Sluice boxes
Smelting
Solid solutions
Spoil
Stock
Stratification
Sub-surface mines
Sulfide minerals
Sulfur scrubbers
Sumps
Surface mining
Tailings
Tectonic exhumation
Valence
Volcanic rocks

"Remind me that the most fertile lands were built by the fires of volcanoes."

—Andrea Gibson

5

Volcanoes

Learning Objectives

LO1 Describe how the location and composition of volcanoes relates to plate tectonics

LO2 Explain why volcanoes erupt

LO3 List the most common types of eruption products

LO4 State the basis for how volcanic eruptions are classified

LO5 Define the difference between volcanic hazards, risks, and benefits

LO6 Describe how geologists define active, dormant, and extinct volcanoes

LO7 Provide the examples of how volcanologists use modern technology to monitor active volcanoes

Early morning with steam rising from Mount Bromo, Indonesia.
iStock.com/StockByM

Ashes, Ashes, All Fall Down

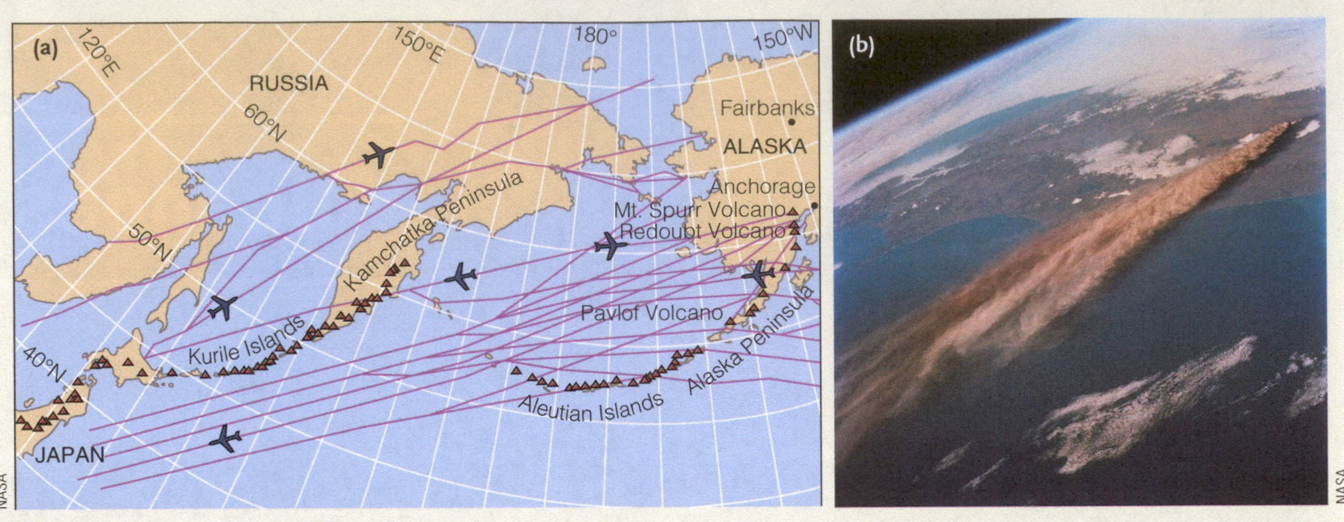

Figure 5.1 More than 10,000 passengers and millions of dollars in cargo are flown daily on the flight routes shown in red across the northwestern Pacific Ocean and Bering Sea. (a) The flight routes pass over or close by more than 100 active volcanoes (red triangles). (b) Air traffic to the Far East and Asia was disrupted in September 1994 when Russia's Mount Kliuchevskoi, the largest composite volcano in the world, erupted, spewing ash into high-altitude transportation routes.

One night in 1982, a British Airways Boeing 747 on a routine flight from Kuala Lumpur, Malaysia, to Australia cruised at an altitude of 12,000 meters (37,000 feet). Just before midnight, sleeping passengers were awakened by a pungent odor. Through the windows, they saw the huge plane's wings lit by an eerie blue glow. Suddenly, engine number 4 flamed out, followed almost immediately by the other three. The plane glided silently for an agonizing 13 minutes. At 4,500 meters (14,500 feet), engine number 4 was restarted, then numbers 2, 1, and 3. Nonetheless, an emergency was declared, and the plane landed in Jakarta, Indonesia, with only three engines operating.

This near-death experience gained the attention of the world's airline passengers and pilots. Before this, such a failure seemed virtually impossible in modern aircraft, which have backup systems for almost every contingency. Flying jet engines through clouds of **volcanic ash**–dust-like particles of exploded rock–however, turned out not to be one of them.

In December 1989, a KLM Boeing 747 encountered airborne ash from Redoubt Volcano at about 8,500 meters (28,000 feet) during a descent to land in Anchorage, Alaska. All four engines flamed out, and like the unfortunate British Air flight a few years earlier, the large aircraft suddenly became a glider (**Figures 5.1** and **5.2**). After a 4,100-meter glide (13,300 feet, greater than 2 miles), the pilots managed to restart the engines, and the plane went on to make what was described as an "uneventful" landing. All four engines required replacement, as did the windshield and the leading edges of the wings, flaps, and vertical stabilizer, which had all been "sandblasted." The plane's interior was filled with ash, and the seats and avionics equipment had to be removed and cleaned. It cost $80 million to return the aircraft to service. All these things considered, one may wonder what it takes to make a landing "eventful"!

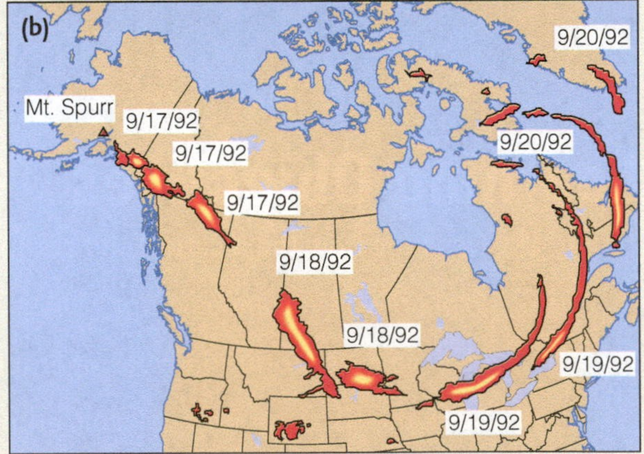

Figure 5.2 Mount Spurr volcano, a potential hazard to aviation. (a) Scientists from the Alaskan Volcano Observatory and the University of Alaska install seismometers at Mount Spurr to detect magma activity below the volcano. (b) The ash cloud from Mount Spurr's 1992 eruption traveled across Canada and the United States on prevailing westerly winds.

Volcanic ash clouds are difficult to distinguish from rain clouds, both visually and with radar; thus, they are challenging to map and to avoid. The ash cloud from Mount Pinatubo's 1991 eruption traveled westward more than 8,000 kilometers (5,000 miles) in 3 days from the Philippines to the east coast of Africa. Twenty aircraft were damaged by the cloud, most of them while flying a distance greater than 1,000 kilometers (600 miles) from the eruption.

North Pacific air routes are some of the busiest in the world. Since 1990, flying through ash clouds on those routes has caused damage to at least 15 commercial aircraft (including the KLM flight). Worldwide, almost a hundred large planes have been damaged this way over the past 30 years. There is nothing we can do to stop the hazard; the Alaskan Peninsula and the Aleutian Islands have 40 historically active volcanoes alone, and the Kamchatka Peninsula has even more. Satellite-based technologies and international cooperation, however, have allowed us to improve our ability to detect ash clouds as they develop, and to provide near real-time warnings to pilots through a new worldwide system of Volcanic Ash Advisory Centers. Flight disasters have been averted, but the economic disruption caused by volcanic ash eruptions remains severe.

In April 2010, the Eyjafjallajökull volcano in Iceland erupted huge ash clouds right in the middle of busy trans-Atlantic airline flightpaths (**Figure 5.3**). The ash-laden clouds only reached around 9 kilometers (6 miles) altitude, but they spread downwind over western Europe, shutting down roughly 300 airports in two dozen countries and forcing cancellations of over 100,000 flights during a 5-day period (**Figure 5.4**). Seven million passengers were impacted, but not a plane in flight was damaged, nor was a life lost. Volcano science and ash warnings work!

questions to ponder

1. Considering how volcanic ash can impact jet aircraft, how might ash also impact other devices or materials that people routinely use or rely upon? Think of some examples.
2. What are some ways that people can easily forecast where and how rapidly volcanic ash will spread from an eruption?

NG Images/Alamy Stock Photo

Figure 5.3 NASA satellite captures the ash cloud from Eyjafjallajökull streaming south from Iceland.

Owen Franken/Alamy Stock Photo

Figure 5.4 Passengers at Charles de Gaulle Airport in Paris are stranded by the eruption of Eyjafjallajökull. Every flight is canceled or postponed.

5.1 Location and Composition of Magma and Volcanoes (L01)

Volcanoes are the most dramatic expression of the great heat escaping Earth's interior. They also are conduits for pent-up gases that played such an important role in the evolution of Earth's early atmosphere and continue to influence the composition of the air and behavior of our global climate today (Chapter 2). Volcanoes are not merely terrifying agents of destruction; they also provide numerous benefits to people, as we'll explore later in this chapter. Some of the world's most beautiful landscapes have been created by volcanic activity. Think of the famous national parks and tourist attractions that feature volcanoes prominently. Examples include Yellowstone and Mount Rainier in the United States, Kilimanjaro in Tanzania, Tongariro in New Zealand, and Fujiyama in Japan.

5.1.1 Plate Tectonics and Volcanoes

Plate tectonics plays a role in the origin of the majority of–although not all–volcanoes. Most active volcanoes lie along plate edges, especially divergent and convergent plate boundaries (**Figure 5.5**). Some of the largest volcanoes, however, seemingly have nothing to do with plate movements, but form above anomalously heated areas in Earth's upper mantle and crust, appropriately termed **hot spots**. Hawai'i, Iceland, and Yellowstone are a few famous examples of these types of volcanoes.

Volcanoes, the **magma** (molten rock) that underlies them, and the **lava** that flows from them, vary dramatically around the planet. Some, found at subduction zones are tall, steep, and characterized by the emission of thick, explosive, lightly colored, silica-rich lava and ash. Others found at oceanic hot spots are larger, more gently sloping and typically erupt iron rich, dark, silica-poor lava. In the pages that follow, we will examine the how and why of these greatly different volcanoes.

The largest body of magma (molten rock) on Earth never erupts at all. That is the **outer core**, a mass of molten metal making up about 15% of Earth's total volume. The outer core is too deep (2,900 to 5,150 kilometers, or 1,785–3,700 miles) and the mantle (the rocky layer above the core) too strong a lid, however, to permit the escape of this metallic magma. The molten rock that feeds volcanoes all originates from a much shallower depth, mostly at the top of the mantle and in the lithosphere 70 to 100 kilometers (40 to 60 miles) below the surface of the Earth.

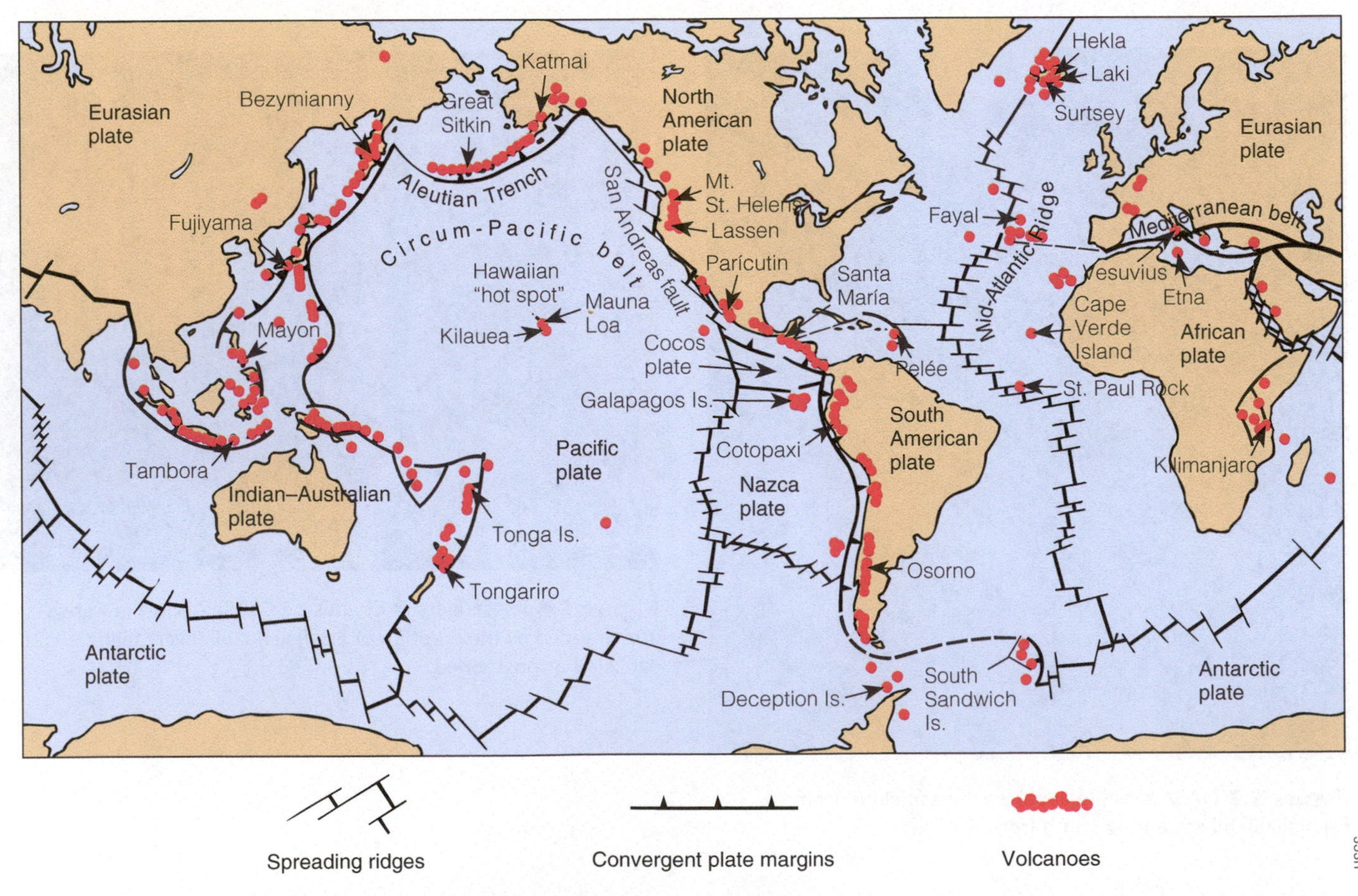

Figure 5.5 Distribution of Earth's active volcanoes at plate boundaries and hot spots. Note the prominent "Ring of Fire" around the Pacific Ocean, which contains 900 (66%) of the world's potentially active volcanoes. The remaining 450 are in the Mediterranean belt (subduction zones) and at spreading centers (divergent boundaries). A few important volcanic centers are related to hot spots, such as those in the Hawaiian and Galápagos islands.

5.1.2 Decompression Melting

Peridotite and other mafic (iron-rich) rocks (Chapter 4) making up Earth's mantle would certainly be hot enough to melt almost instantaneously if you somehow could scoop this material out and bring to the surface all at once. It can't melt where it occurs naturally, however, positioned tens of kilometers underground, simply because it is under too much pressure from the overlying rock. The tightly confined, individual atoms and molecules making up solid rock in the mantle and lower crust do not have the leeway to move around in a liquid state. If pressure drops significantly, though, this may induce at least partial melting–a process termed **decompression melting**.

Decompression of the underlying mantle, even just a few kilometers down, is greatest where tectonic plates separate at divergent plate boundaries. For this reason, the largest amount of magma erupted on Earth ordinarily comes to the surface as molten lava at the mid-ocean ridges (**Figure 5.6**). The zone of partial melting underlying ridges extends as deeply as 60 kilometers beneath the seafloor, and as much as 10% of the rock mass there may be molten. Erupted ridge crest lavas are among the hottest on Earth, up to 1,250°C (2,200°F), and as noted in Chapter 4, substantial ore deposits including copper and gold, may form around their undersea vents. Decompression melting also accounts for volcanic activity in continental rift valleys, (see Chapter 2) including eruption of some magmas that must have formed at depths exceeding 70 kilometers (40 miles), deeper than the partial melts underlying mid-ocean ridges.

Decompression melting also takes place in other geological settings. Hot spots originate as upwellings of extra-hot mantle rock that may start at the core-mantle boundary and ascend, like plumes of warm smoke rising through still, frigid winter air, right up to the base of the lithosphere. The hot rock experiences ever lesser pressure as it ascends, so in that sense it, too, undergoes decompression. At the base of the lithosphere, it begins to melt. For example, Hawaiian hot spot magmas form at a depth of around 90 to 130 kilometers (55 to 80 miles).

Decompression melting also takes place where fault motions (see Chapter 6) stretch ("extend") continental crust. In one extreme example, the collision of India with Eurasia, beginning 40 million years ago (Chapter 2), caused the interior landmass of Asia to stretch and tear, like the buckling of barrier hit by a truck. The associated decompression in the shallow mantle allowed basaltic volcanoes to erupt in central Mongolia, 2,300 kilometers (1,400 miles) to the north of the collisional plate boundary!

5.1.3 Melting Related to Fluid Release

Decompression can trigger partial melting, but so can the addition of **fluids** (such as H_2O, CO_2, and SO_2) to deeply buried hot rock. These are not "fluids" in the commonly understood sense of the word, but rather substances under high pressure that blend the physical properties of both pure gases and liquids. There is no analog at the surface to help explain this peculiar physical state, given the relatively low-pressure conditions we as humans experience at Earth's

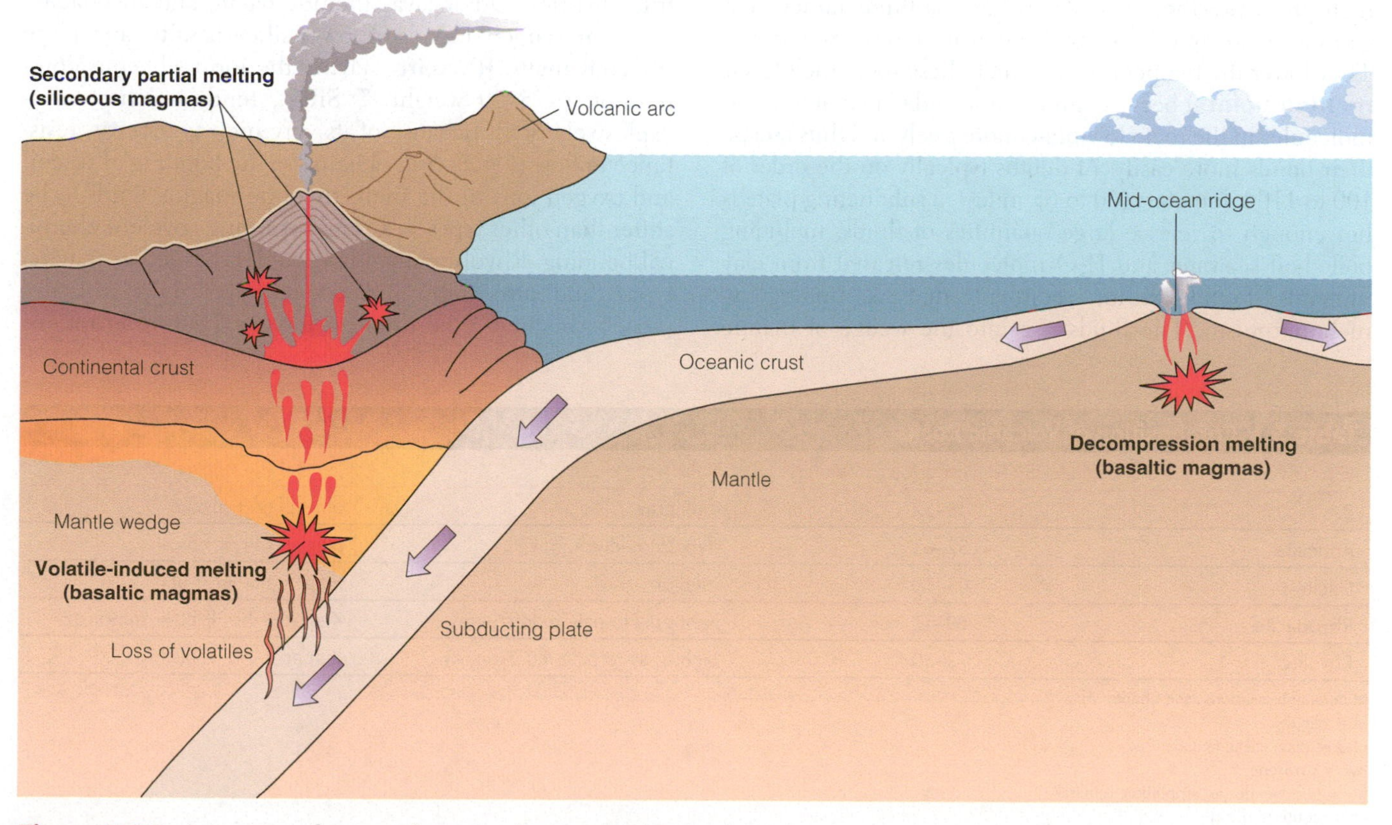

Figure 5.6 Environments of magma formation leading to volcanic activity.

Figure 5.7 A thick crust of sulfur crystals lines the rock surrounding the mouth of a fumarole in Hawaii. Such sulfur-producing fumaroles are called *solfataras*.

surface, but considering fluids in rock is important for understanding processes in the deep crust and upper mantle–including magma generation–as we explain below.

Volatiles are substances that do not fully participate in the formation of igneous rocks and minerals. They are the "leftovers" from rock and mineral formation, and include such molecules as H_2O, CO_2, and SO_2, largely the gases that we see escaping from active volcanic vents (**Figure 5.7**). At high pressures, these volatiles behave as fluids rather than gases and can travel widely through masses of solid rocks. They lower the temperature at which these rocks melt (their **melting points**) because fluids allow individual atoms and molecules in the rocks to vibrate more freely and thus escape their bonds more easily. At depths typically on the order of 100 to 110 kilometers (60 to 65 miles), a subducting plate is hot enough to release large quantities of fluids, including boiled-off seawater and H_2O molecules released from clay minerals found within sediment that is undergoing metamorphism. The fluids rise into the wedge of mantle separating the down-going plate from the overlying lithosphere, and the mantle rock responds by partly melting.

From the mantle wedge, the rising magmas enter the crust, bringing with them a tremendous amount of heat. This induces still another, shallower, round of magma generation, in this case within the crust itself, which is composed of rocks having much lower melting points than in the mantle. You can appreciate how complex this activity becomes; fluid infiltration from below, mantle-derived magmas rising buoyantly into the lower crust, secondary magma-genesis in response, taking place within intensely deformed crust caught up in plate collision. It would be an exciting place to visit but we'll never get there (except perhaps in the movies).

It is the sort of complexity that geologists love to investigate, but from the standpoint of most people, what matters is the ultimate product seen at the surface near subduction zones–an arcuate chain of volcanoes termed a **volcanic arc**. These volcanoes release great volumes of sulfurous and CO_2-rich gases (a reflection of the trapped seawater initially involved in magma formation) and have highly explosive, often destructive eruptions. The compositions of lavas and ash erupted here range from basaltic to rhyolitic, a complete spectrum of silica content (**Table 5.1**).

One of the most common products, almost unique to convergent plate volcanoes, is the lava **andesite** (named for the Andes Mountains volcanic arc; **Figure 5.8**). Andesite begins forming near the base of the crust and includes contributions from subducting plate fluids and partial melting in both the uppermost mantle and basal crust; a true "hybrid." Other lavas include dacite and rhyodacite, which are largely blends of lower-silica basaltic and more siliceous melts. **Rhyolite**, having the most siliceous composition (~75–80 weight % SiO_2), tends to erupt most explosively of all because of its very high **viscosity**, resistance to flow (**Figure 5.9**). Thanks to the bonding of silicon and oxygen ions in the melt, rhyolitic magma tends to be stiffer than other types as a result of strong covalent chemical bonding. Rhyolitic magma is cooler than other magma types, and also traps and concentrates more volatiles (10–15 weight %). Fortunately for us, rhyolitic eruptions

Table 5.1 Composition of Volcanic Rocks

Name	Silica (SiO_2) Content by % Weight	Color	Typical Phenocryst Minerals
Basalt	48–52	dark gray	ol + plag + pyx
Andesite	52–63	dark to medium gray	plag + minor hb
Dacite	63–69	medium gray	Plag + hb +/− bt
Rhyodacite	69–72	light gray to pale reddish gray	K-spar + hb + bt + minor qz
Rhyolite	72–80	Light gray to pale reddish-gray	K-spar + hb + bt + qz

Mineral abbreviations (see Chapter 4):
ol = olivine
plag = plagioclase feldspar
pyx = pyroxene
hb = hornblende (an amphibole mineral)
bt = biotite (a mica)
qz = quartz

vvoe/Shutterstock.com

Figure 5.8 Sample of typical andesite lava.

AlviseZiche/Shutterstock.com

Figure 5.9 Thick, rugged flows of rhyolite make up much of the landscape in Lanumannalaugar park, near Hekla Volcano in Iceland. Numerous hot springs and steam vents in the area indicate that this is still an active volcanic zone.

are much less common than other, less explosive types of volcanic activity: it simply does not erupt as easily as basalt or andesite owing to its viscous physical properties.

5.2 Magma Ascent, Why Volcanoes Erupt (LO2)

Magma chambers are important because they are commonly the sources of volcanic eruptions. Being molten, magma is less dense than the host rock from which it melts, and to equalize its density with the surrounding mantle or crust it rises toward the surface. As it does, however, the density contrast with the host rock steadily drops, and the magma finally stalls when it reaches a depth at which its density matches that of the enclosing lithosphere, a level termed the **neutral buoyancy zone (NBZ)**. Gradually, a substantial amount of molten material can accumulate at the NBZ, forming a single, large underground body, or magma chamber (**Figure 5.10**).

Most magma chambers lie at depths of several kilometers, but some very deep chambers (~10 to 30 kilometers; 6 to 20 miles) also exist. These may form not as a result of attempting to achieve neutral buoyancy, but because the rising magma encounters major changes in the structure or density of the lithosphere, such as the mantle-crust boundary. Other factors such as rapid loss of volatiles (gases and steam) can also cause magma to stop, concentrate, and form a large underground body. On a longer timescale, magma chambers are the places where, or from which, many metallic ore deposits form (Chapter 4), and these provide the heat needed for geothermal power plants (Chapter 11).

Eruptions ensue when conditions in and around a magma chamber drastically change. A magma chamber may be positioned for a long while beneath a volcano, unnoticed at the surface. A fresh pulse of magma entering the chamber from below can destabilize it, however, raising pressure sufficiently in the overlying rock to cause fracturing. This allows volatiles dissolved in the magma to rush out of solution, accelerating fracturing all the way to the surface to cause an eruption–somewhat akin to uncorking a champagne bottle.

The combination of intrusion of fresh melt into magma chambers and ensuing gas buildup is probably responsible for most eruptions. But seemingly minor external influences can also act to trigger volcanic activity, especially if a magma chamber is about ready to erupt anyway. For instance, the orbit of the Moon around the Earth not only raises substantial tides in the ocean, but also causes the solid crust to rise and fall, too, as much as 40 centimeters (16 inches). We do not notice this **solid earth tide** given its slow daily pace and planet-wide scale, but for rocks already stressed close to their breaking point, the tidal strain might be the "straw that breaks the camel's back."

Sets of eruptions have been recorded at Kīlauea in Hawai'i, Ilopango in El Salvador, and at Gamalama volcano in Indonesia correlated with times of maximum or minimum Earth-tidal strain. The most important components of the solid earth tide known to act as volcano triggers occur on a once every 12-hour (**semidiurnal**) or 14-day (fortnightly) basis–when the Moon is either full or in new Moon status. Once every 2 weeks (a fortnight) the Sun and Moon are in alignment with the Earth and the gravitational pull on Earth's crust from these other bodies in space is strongest.

Even sudden changes in atmospheric pressure can be significant enough to set off an eruption as the crust responds to the changing weight of the overlying air. In mid-June 1991, Typhoon Yunya passed within 75 kilometers (45 miles) of the Pinatubo volcano in the Philippines, beneath which fresh magma had intruded only a few months before. Within 3 hours of the greatest barometric pressure drop (6.3 mb) associated with the passing storm, Pinatubo produced one of the great eruptions of the twentieth century (**Figure 5.11**). Was this mere coincidence? Some volcanologists think not!

Most volcanoes are underlain by at least one magma chamber. Some have multiple magma chambers beneath them at different levels in the crust. They are linked to the surface by a pathway of chemically altered, fractured rock called a volcanic **conduit**. Some small volcanoes apparently have no

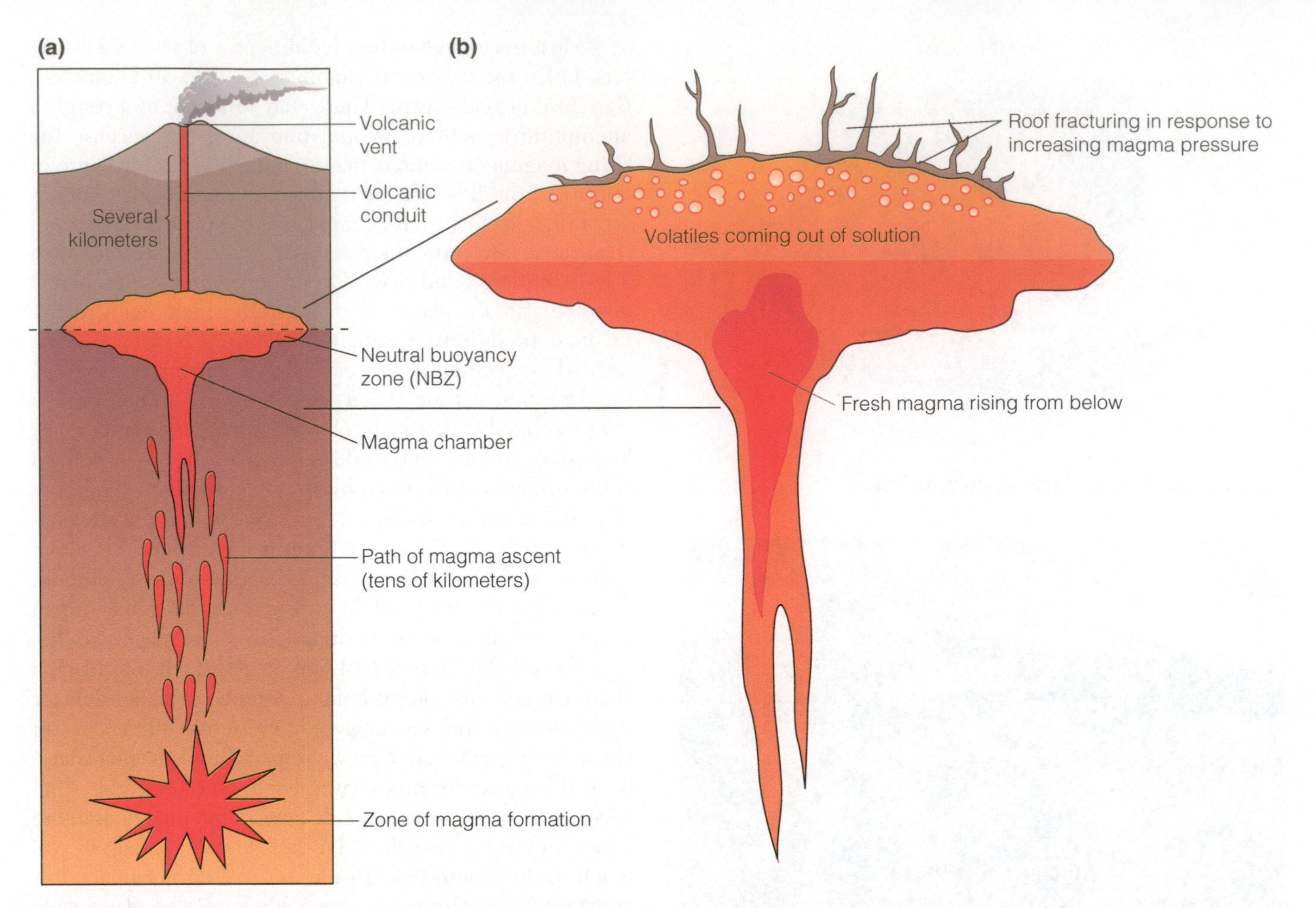

Figure 5.10 (a) Structure of magma supply beneath a typical volcano. (b) Major factors important in destabilizing a magma chamber, causing it to erupt.

Figure 5.11 Eruption of Pinatubo Volcano, Philippines, June 12, 1991.

underlying magma chambers at all. They tend to be quite short-lived however, forming during single eruptions. They are thus termed **monogenetic**; that is, made in one episode. Cinder cones, described later, are a great example of monogenetic volcanic landforms. The magma feeding them rises directly from the mantle without stopping on the way up. Larger volcanoes are all **polygenetic**, or formed through multiple eruptions.

5.3 The Products of Volcanic Eruptions (LO3)

Many volcanic eruptions, especially on basaltic volcanoes, are **effusive**, meaning they primarily produce flowing lava. In contrast, the magma erupted in **explosive eruptions** is thrown out in clots or great clouds of sprayed particles–the volcanic ash that so vexes airplanes, as described in the chapter opener. The collective term for debris thrown out by volcanic explosions is pyroclastic ("fire-broken") material (see Chapter 4). If pyroclastic particles become airborne for awhile after erupting, filtering out of the sky sometimes hours or even days later, we call them **tephra** (TEFF-ra), an Icelandic term.

In terms of total mass released, the largest product of any volcanic eruption is neither lava nor exploded fragments, but gases. The gases do not remain behind after activity ends, however, so we tend to ignore their very important role as eruption products. Where gases combine with water vapor, they can form strong acids, such as H_2SO_4 (sulfuric acid), which is a powerful natural defoliant and corroder of metals. Many visitors to active volcanoes must stay upwind of vents to avoid asphyxiation and, at worst, fatal seizures and acid burns.

Lava Type	Typical Lava Composition	Surface Appearances
Pāhoehoe	Basalt	Billowy, rippled, or smooth and platy
'A'ā	Basalt and andesite	Very rubbly, clinkery
Block	Dacite, rhyodacite, rhyolite	Blocks with smooth sides; may be very glassy (obsidian)

Table 5.2 Types of Lava Flows

etvulc/Shutterstock.com

Figure 5.13 Active 'a'ā flow from Mount Etna, Sicily.

5.3.1 Lava Flows

There are three main types of lava flows, based on their surface appearance upon hardening (**Table 5.2**). **Pāhoehoe** (pronounced PAH-hoy-hoy) has a smooth skin that may be wrinkled into rope-like structures. The name of the lava, translated from Hawaiian, means something like "the pattern of ripples made by passing a canoe paddle through water." A closely related type of lava, **'a'ā** (pronounced "ah-AH") has a very rough surface made up of largely loose fragments termed **clinkers** that would discourage anyone from walking across. 'A'ā is a Hawaiian word referring to a "combustible object." If the individual clinkers on the surface of the flow have smooth rather than jagged sides, then it is **block lava**.

In general, the most fluid lavas, typically basalts erupted at high temperature, flow out as pāhoehoe. Somewhat more viscous lava, which may be slightly cooler or disrupted by shearing during eruption, pour across a landscape as 'a'ā. Block lavas are typical of even more viscous eruptions such as those of andesitic or dacitic magmas, linked to cooler eruption temperatures and higher-silica contents (**Figures 5.12**, **5.13**, and **5.14**).

If an eruption of lava, particularly pāhoehoe, lasts for more than a few weeks, the lava typically develops a narrow channel in its interior that can crust over to form a pipe-like internal plumbing system, rather like the veins that convey blood in your hands. The often-used name for this feature is **lava tubes**, though some geologists object to this term on the grounds that, geometrically speaking, these features are rarely very tubular. In any case, the internal channels efficiently spread the lava far and wide downslope because they insulate the interior of the lava from rapid heat loss. Where the channels drain out at the end of an eruption, they may leave behind **lava caves** that people can explore, in some cases extending for kilometers (**Figure 5.15**). In cultural tradition, lava caves have been important places to collect water, seek shelter, practice religious rituals, and even bury the dead in sacred ways.

The rapid cooling of lava at Earth's surface causes it to develop a regular, almost geometrically symmetrical pattern of fractures in flow interiors, termed **columnar joining**. The fracture pattern is especially striking in the cores of thick 'a'ā flows, where the columns are oriented in the direction heat escaped as the flows cooled (**Figure 5.16**). Columnar joints and other fractures can be important avenues for groundwater and related fluids in the crust to move long after the flows have become solid rock (Chapter 12).

Channelized lava flows, erupted one after the other, build broad volcanoes with gently sloping sides: **shield volcanoes** (**Figure 5.17**). The largest mountains on Earth, the Hawaiian Islands, are all volcanic shields. From summit to base of the volcanic pile, Mauna Loa, the largest active volcano in the

www.sandatlas.org/Shutterstock.com

Figure 5.12 Active flow of pāhoehoe, Kīlauea Volcano, Hawai'i.

Frank Schulenburg/NPS

Figure 5.14 Block lava flow, Mount Lassen National Park, California.

Figure 5.15 Exploring a lava cave, Lava River, Arizona.

Figure 5.16 Vertical columnar joints in ancient lava flow cut by a modern river canyon, Studligal, Iceland. This same canyon is pictured from above on the cover of this book.

Figure 5.17 Common types of volcanic landforms.

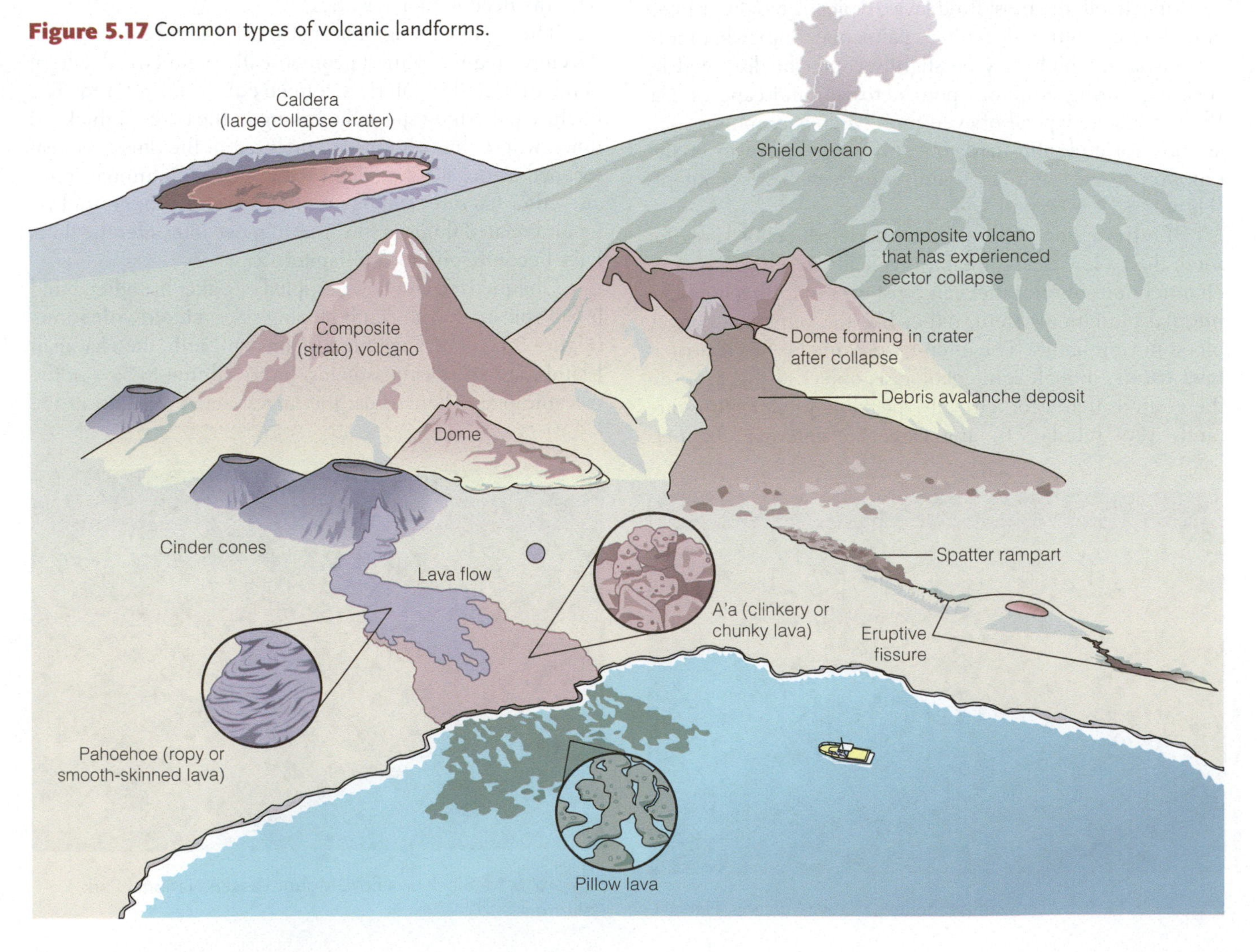

Figure 5.18 Mauna Loa, the world's largest active volcano, Island of Hawai'i.

world, has a height of 16,700 meters (55,000 feet), though given its great weight much of this giant mountain has subsided into the lithosphere, and its elevation above sea level is only about 4,150 meters (13,600 feet) (**Figure 5.18**).

5.3.2 Pyroclasts

Fragmental volcanic debris (pyroclasts) thrown out during explosive eruptions includes not only very fine particles of ash, but often coarser fragments that we classify according to their sizes and shapes (**Figure 5.19**). Volcanic ash and dust, formed by the most powerful explosions, consists of particles less than 2 millimeters in diameter. Fragments ranging from 2 to 64 millimeters (0.08 to 2.5 inches) in diameter are termed **lapilli**, while larger fragments are classified as volcanic **bombs** if they are rounded, or **blocks** if they are angular (**Figure 5.20**).

Pumice is a special kind of lapilli noteworthy for its foamy, high-porosity appearance. The name in fact derives for an ancient Greek word for "worm-eaten wood." It forms, like the foam on a carbonated beverage, by the rapid escape of gases and frothing of magma at the top of an eruptive conduit. So porous is pumice that it can float on water–the only known rock that can do so. Typically between 64% and 85% of the volume of a piece of pumice is air space (**Figure 5.21**).

Similarly sized, but less porous clots of lava are termed **cinder**, or **scoria** (**Figure 5.22**). Cinder can pile up around an explosive vent to build **cinder cones** up to a few hundred meters (a thousand feet) tall. Cinder cones are basaltic features, and often occur clustered together in continental interiors, in many places far from plate boundaries, as in the Basin and Range of the southwestern United States (**Figure 5.23**).

Volcanic bombs are not, as the name sounds, a natural form of explosive ordnance, but are large globs of lava pitched out by a blast that may be streamlined as they sail through the air to land with a thud on the ground up to several hundred meters from a vent. They can range in size up to a meter (several feet) across (**Figure 5.24**). Volcanic blocks are angular pieces of lava or other, older rock, torn from the walls of a vent during eruption and tossed across a landscape. Large, heavy, falling blocks can be extremely dangerous because they can travel at high speed in any direction away from a blast point.

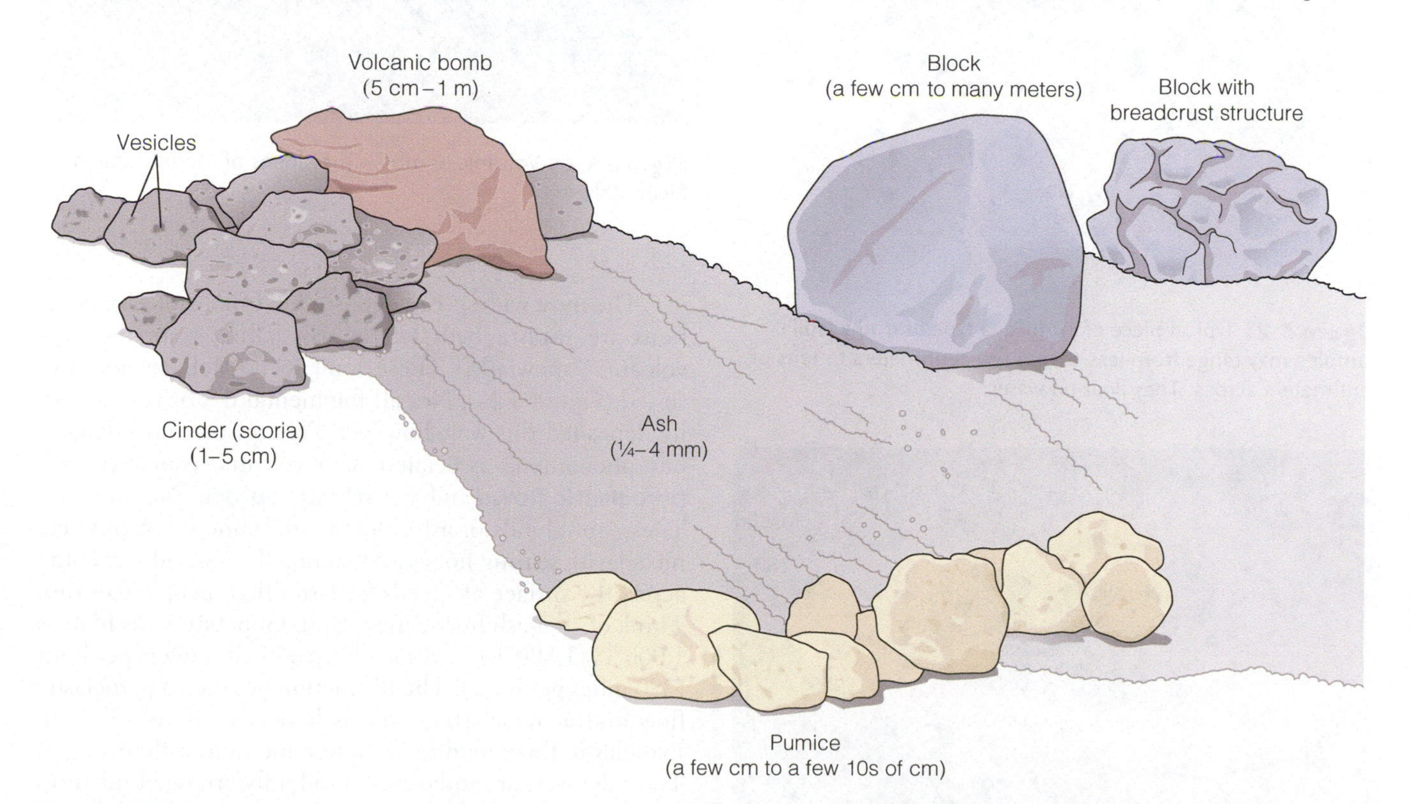

Figure 5.19 Common types of pyroclastic material with typical size ranges indicated. Note the high vesicularity (presence of gas bubbles) evident in cinder and pumice.

Siim Sepp/Alamy Stock Photo

Figure 5.20 Volcanic block from the summit of Mauna Kea. The width of the block is about 40 cm.

Stas Malyarevsky/Shutterstock.com

Figure 5.21 Typical piece of pumice, a type of frothy lapilli. Samples may range from less than a few centimeters to tens of centimeters across. They float on water.

lrosebrugh/Shutterstock.com

Figure 5.22 Volcanic cinder, also known as scoria.

iStock.com/Attila Adam

Figure 5.23 Amboy Crater in the Mojave Desert is a cinder cone. Isotopic dating suggests that the cinder cone erupted about 80,000 years ago.

iStock.com/jacquesvandinteren

Figure 5.24 Volcanic bomb on the slopes of Mount Etna in Sicily, Italy.

The most widespread deposits left from explosive eruptions are **tephra** beds left by ash falling from the sky—volcanic "snowfalls." These tend to be well layered and sorted (**Figure 5.25**). Not all fragmental debris is necessarily deposited this way, however. Perhaps the most dangerous phenomena associated with volcanic explosions are **pyroclastic flows** and **pyroclastic surges** (**Figure 5.26**). These are clouds of ash, blocks, and quite often pumice, mixed with searing-hot gases that hug the ground, sweeping across the surface at speeds far faster than people can run. Think of "hot-ash hurricanes" with temperatures as high as 1,000°C (1,800°F), and speeds up 450 kilometers per hour (280 miles per hour)! The distinction between a pyroclastic flow and a pyroclastic surge is basically one of dilution. Pyroclastic flows tend to be hotter and denser than surges. Their deposits are unlayered, randomly arranged mixtures of ash and other fragments, whereas pyroclastic surges show fine layering in many places gently cross-bedded.

Figure 5.25 Sixty meters thick, layered light-colored tephra deposit on the volcanic island of Santorini in the Aegean Sea of Greece. The island's last major eruption in 1600 B.C.E generated a massive tsunami that may have doomed the Minoan empire and the island of Crete.

Figure 5.26 A pyroclastic flow pours down the slope of Soufriere Volcano on the Island of Montserrat, in the Caribbean, January 2010.

Figure 5.28 The Ijen group of composite volcanoes. In the foreground is a lake-filled caldera, formed from the collapse of one of the cones into its underlying magma chamber. Banyuwangi Regency, East Java, Indonesia.

The cross-beds in this case did not form from wind or flowing water, but from hot, expanding volcanic gases.

Volcanoes built up of pyroclastic material with some subsidiary lava flows form steep-sided, gracefully symmetrical profiles. Geologists call them **composite** or **stratovolcanoes** (**Figure 5.27**). These are the typical volcanoes with which most people are familiar. Because they are commonly located along convergent plate margins with rich agricultural lands and commercial ports, such as Indonesia, the Pacific Northwest, and Japan, a large human population–about 10% of the world's total–also lives with these volcanoes virtually in their backyards (**Figure 5.28**).

5.4 Classification of Eruptions (L04)

Scientists first began studying volcanic eruptions systematically in the late nineteenth century. They observed that certain types of eruptions seemed characteristic of particular volcanoes, or volcanic areas, and named the eruption types after them accordingly. Although we now know that a single volcano can erupt in many ways, the general approach of naming eruption types for volcanoes or volcanic areas where they

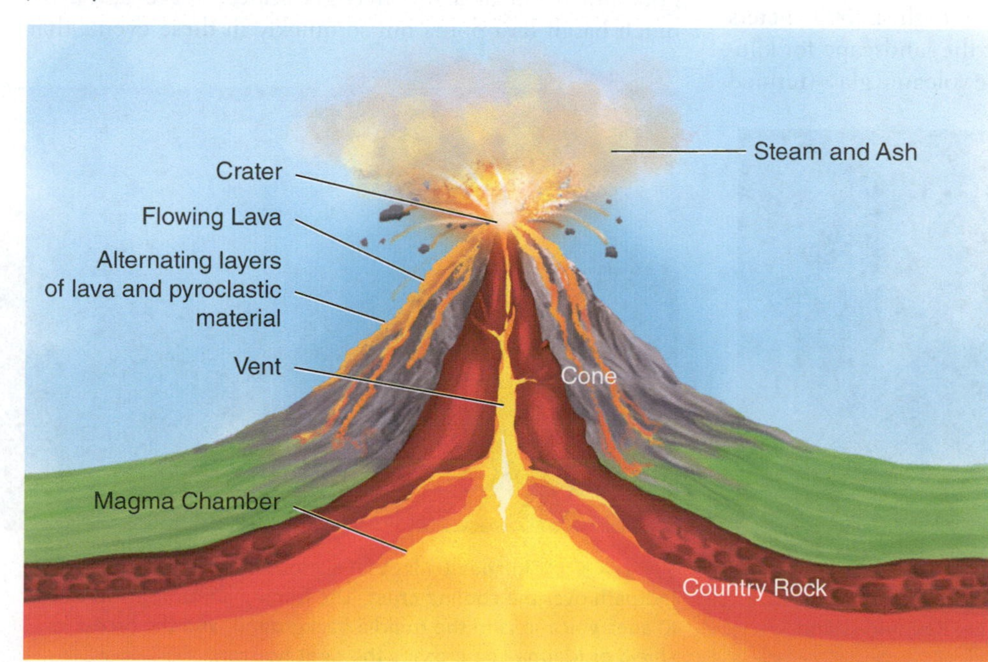

Figure 5.27 Cross-section of erupting composite or stratovolcano.

Pedro Carrilho/Shutterstock.com

Figure 5.29 Hawaiian-style fissure eruption, along the Holuhraun fissure system, Bárđabunga Volcano, Iceland, 2014.

Felix Nendzig/Shutterstock.com

Figure 5.31 Reticulite, a clot of spun glass which, like Pele's hair, is produced by high lava fountaining, then windborne sometimes for kilometers. Consisting of polygonal gas bubbles (vesicles) with very thin walls, it is more permeable than pumice, which is similar. It will sink in water.

have been observed continues up to the present. Let's consider these categories, going from most effusive to most explosive.

5.4.1 Hawaiian-Style Eruptions

Hawaiian-style eruptions are typical of oceanic hot spots and divergent plate boundaries. They are the typical eruption type responsible for building enormous shield volcanoes. They usually begin with the fountaining of basaltic lava from an eruptive fissure, or giant crack in the ground that opens directly above ascending magma (**Figure 5.29**). The lava travels away from the fissure as thin flows usually moving only a few tens of meters per hour. However, near a vent the lava can travel much faster. Very little pyroclastic material usually issues from these vents, though towering jets of lava, occasionally reaching as high as 600 meters (1,900 feet) can form which shower the landscape for kilometers downwind with threads of fine volcanic glass (termed **Pele's hair**, for the Hawaiian fire goddess) (**Figure 5.30**). Sometimes, pieces of foamy lava that resemble spun glass, called **reticulite**, also float away from these volcanic eruptions (**Figure 5.31**). Reticulite is even more porous than pumice, with up to 98% air space by volume. So foamy is it that, unlike pumice, it will sink in water because the pore spaces are all connected and open to air and water.

In some places, lava will pool above active vents or in craters to form **lava lakes**, which can take decades to cool and completely solidify after an eruption, depending upon how deep they are. A lava lake in 1.5-kilometer (1-mile)-long Kīlauea Iki Crater, Hawai'i, formed during a five-week long eruption in 1959 (**Figure 5.32**). The lava, 125 meters (415 feet) deep, developed a surface crust that by 1980 was 53 meters (173 feet) thick. Today, there is no liquid lava remaining, but the deep interior is still hot enough to boil water.

Flood basalt eruptions resemble Hawaiian-style eruptions but occur at a much larger scale (**Figure 5.33**). So much basalt lava pours out so quickly in these events that

Ouestusa/Shutterstock.com

Figure 5.30 A clump of Pele's hair, glassy threads spun by high lava fountaining, Pas de Bellecombe, Reunion Island, Indian Ocean.

Dmitri Kotchetov/Shutterstock.com

Figure 5.32 Kīlauea Iki lava lake, crossed by a straight footpath over the cooling crust. The fuming summit caldera of Kīlauea volcano is in the middle background, and the broad shield of Mauna Loa crosses the right horizon.

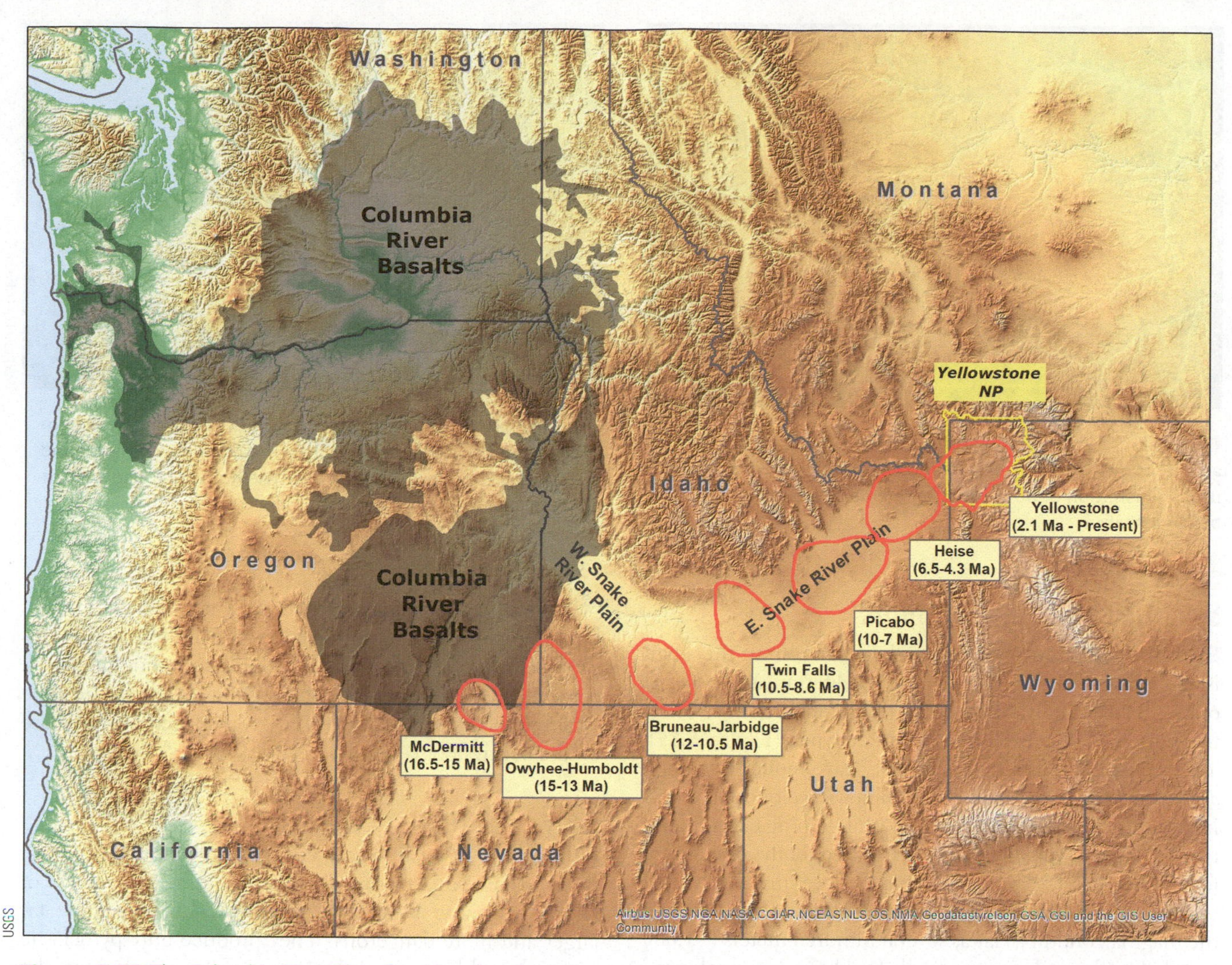

USGS

Figure 5.33 The Columbia River Plain flood basalts, shown in gray, cover most of eastern Washington State and southeastern Oregon. Also shown by red circles are smaller flood basalt and rhyolitic tuff (hardened ash and pumice) fields associated with the movement of the Yellowstone hot spot. The hot spot is shifting northeastward over time relative to the North American plate. Most of the Columbia River basalts erupted when the hot spot first appeared at the surface, 17 to 15 million years ago.

volcanic shields do not form. Instead, the landscape is completely smothered with fresh lava in a nearly level volcanic plain, dark and bleak, tens to hundreds of kilometers across. Although modern Hawaiian eruptions produce flows that range from less than 1 to a few cubic kilometers in volume, individual flood basalt lava flows have been measured up to 3,000 cubic kilometers (450 to 720 miles^3), with their source eruptive fissures 70 to 200 kilometers (40 to 125 miles) long! Flood basalt flows can be tens of meters thick–an order of magnitude more than ordinary basalt flows (**Figure 5.34**). Fortunately, flood basalt eruptions do not often take place on Earth. The latest such volcanic activity built the Columbia River Plain in the Pacific Northwest, between 5 and 17 million years ago. Flood basalt plains are common on a few of our planetary neighbors, notably Venus and Io, the volcanically active satellite of Jupiter. They make up the dark patches (**maria**) on the face of Moon and formed 3 to 4 billion years ago–the oldest lava you will ever see, albeit from a great distance.

iStock.com/Samson1976

Figure 5.34 Palouse Falls in eastern Washington tumbles over a stack of Columbia River flood basalt more than 100 meters thick.

iStock.com/Bierchen

Figure 5.35 A nighttime exposure of Stromboli shows the arcuate traces left by pyroclasts moving through the sky. It's a useful illustration of how these fragments move during Strombolian activity.

iStock.com/Ken Anderson

Figure 5.37 The symmetrical cone of the Stromboli volcano rises above the Mediterranean Sea. The cloud hovering around the summit is not entirely volcanic. The horizontal one is from convective updraft off the surrounding ocean.

5.4.2 Strombolian Eruptions

Strombolian eruptions derive their name from the small island volcano of Stromboli in the Mediterranean, just west of the "boot" of Italy. Strombolian eruptions are the smallest magnitude of the explosive eruption types (**Figure 5.35**). They consist of pulsating bursts that heave molten lava high above the crater rim together with billowing clouds of churning ash. Each explosion sounds like a booming cannon, but much louder. Typically, blasts occur frequently; once every few seconds. Most volcanologists believe that individual explosions are caused when a giant bubble of gas forms in the magma conduit as far as several hundred meters (~1,000 feet) below the surface, eventually building up enough pressure to hurl the overlying molten rock violently out of the vent. The sudden release of gas temporarily reduces the volatile pressure in the conduit–at least until a new bubble forms and starts the process all over again (**Figure 5.36**).

Marco Crupi/Shutterstock.com

Figure 5.36 Small Strombolian blasts produce puffs of ash atop Stromboli Volcano in the Mediterranean Sea, off the southwest coast of Italy.

While cinder cones are the most common landforms built by Strombolian eruptions, in some instances eruptions can take place repeatedly from a single vent, building a tall composite volcano over tens of thousands of years. The cone of Stromboli rises ~3,000 meters (~10,000 feet) from its base on the ocean floor (**Figure 5.37**).

5.4.3 Vulcanian Eruptions

Although people can safely view an ongoing Hawaiian or Strombolian eruption from just a few hundred meters or kilometers away, **Vulcanian eruptions** are generally far larger and more dangerous. They produce only pyroclastic material, and because of the power of their explosions, the pyroclastic debris consists of volcanic ash–fine dust to sand-sized particles much smaller than the cinders thrown out by Strombolian activity. Vulcanian blasts are typically continuous rather than pulsating, lasting up to 12 hours at a time. Rather than periodic cannon bursts, an eruption typically sounds like a jet engine turned on full blast. The clouds of roiling ash issuing from a crater are a mix of darker material fresh from the magma and lighter material that is pulverized older rock torn from the walls of the volcanic conduit.

Vulcanian ash clouds can rise kilometers above a vent, and because of the fine size of their particles, they develop strong **electrostatic charges**. As a result, forks of lightning frequently flash in and out of blast clouds, adding to the terror of their appearance, especially after dark. Volcanologists studying the eruption of Mount Redoubt in Alaska in 1992 correlated the energy of lightning bolts with the strength of eruption-related earthquake activity and volatile escape from the vent (**Figure 5.38**).

Not all pyroclastic material produced by Vulcanian eruptions consists of ash. Large blocks of rock also frequently blast in great numbers from vents, associated with the initial "throat-clearing" phase of activity. The blocks may be laced with a geometrical network of superficial

Mama Gipsy Olili/Shutterstock.com

Figure 5.38 Vulcanian eruption column laced with lightning, Eyjafjallajökull Volcano, Iceland, 2010.

DEA/G. ROLI/Getty Images

Figure 5.39 Bread-crust block erupted from a volcano in the Aeolian Islands of Italy.

cracks called, **bread-crust structure** (**Figure 5.39**). Bread-crust structure forms from the expansion of hot gases trapped in the blocks during their release, which tears open their brittle, chilled skins.

Some Vulcanian eruption clouds are so heavy with ash that they cannot rise far without collapsing back to the surface, running out across the landscape as hot, block-rich pyroclastic flows and surges. Some volcanologists recognize a distinct type of eruption, **Peléean eruptions**, that are essentially Vulcanian, but involve pyroclastic flows and surges released from the summit of a composite cone capped by a plug (dome) of hardened lava. One such event took place on May 8, 1902, on the Island of Martinique in the Antilles Arc. A pyroclastic flow swept through the port of St. Pierre from the neighboring volcano, Mount Pelée, and though it was quite diluted by the time it reached the city, the force of the blasting, searing gases ripped open buildings in its path and set everything combustible on fire (**Figure 5.40**). Twenty-eight thousand persons died in just a few minutes. The pyroclastic flow even continued spreading across the water, overturning and igniting ships anchored there. Today, the city has been rebuilt but the volcano remains a threat (**Figure 5.41**).

Volcanologist Frank Perret, writing in 1935, recalls the experience of one pyroclastic flow from Mount Pelée on the evening of April 15, 1931, during a small but potentially lethal Peleéan eruption. By this time, 29 years after the destruction of neighboring St. Pierre, a new dome with a craggy solid crust had extruded from the vent and filled the crater atop the volcano. A towering spine of hardened lava projected from one side of the dome. Precariously poised, it seemed ready to collapse without warning. Perret kept vigilant watch of the landscape at a cabin on an opposing ridge not far away:

> "Just at the close of the swiftly passing tropical twilight, in the dead calm between day and night winds, an unusual sound brought me to the cabin door. The whole mass [of the spine] had fallen, leaving a great scar on the dome from which poured forth an ash cloud of inky blackness, expanding upward as it rushed down the talus slope … At the same instant, an explosion on its eastern summit shot out a second avalanche with rising cloud as white as snow; two mighty parallel columns, ominous, terrifying, moving straight toward the station. I shall not attempt to describe all the sensations I felt, nor the thoughts that swept through my mind at that moment. My first thought was of instant flight. With a distance of 2.5 kilometers between me and the crater, it might be four minutes before the clouds reached me, but a moment's reflection indicated that this could not be enough to escape the wide path of the cloud. I decided to risk the protection of the frail shack. Doors, windows, cracks, and

Library of Congress

Figure 5.40 An old-fashioned stereo view shows the ruined city of St. Pierre, Martinique, shortly after the deadly eruption of Mount Pelée in 1902.

iStock.com/Damien Verrier

Figure 5.41 Mount Pelée towers over a rebuilt St. Pierre, Martinique.

holes were hastily closed and the onset awaited. Escapes from many former perils helped to allay my fears, but there was still the thought that this might be my last. I recall a sense of utter isolation; awe in the face of overwhelming forces of nature so indifferent to my feeble self. The track of the avalanche lay to the side of the station or these lines could never have been written. The chief dangers were heat and gas from the cloud. There was still a minute left. I peered out from the rear of the station. A sublime spectacle! Two pillars of cloud a thousand feet in height, apparently gaining in speed every instant and headed straight for my shelter. As I darted within, the blast was upon me–not a terrific shock as the reader might think, but swirling gusts of ash-laden wind, bringing a pall of darkness that might indeed be felt. The dusty air [that] entered every crevice of the shack was hot but not scorching. I felt the gases burning but not parching my throat and then came a feeling of weakness (carbon monoxide). It all lasted for half an hour, but it was nearly an hour before the feeling of suffocation was relieved by a kindly wind."

5.4.4 Plinian Eruptions

The most fearsome eruptions of all are named for the Roman naturalist and military commander Pliny the Elder. He died trying to help friends escape the eruption of Vesuvius in 79 C.E., which famously destroyed the coastal cities of Pompeii and Herculaneum. **Plinian eruptions** are explosive events that characteristically inject large amounts of volcanic gas and pyroclastic material into Earth's lower stratosphere. Like Vulcanian eruptions, they are too explosive to produce lava flows.

The eruptions consist of an initial phase of powerful gas release, which can create an **eruption column**, a pillar of ash and pumice fragments towering as high as 20 kilometers (12 miles) or more (**Figure 5.42**). At the level of stratospheric neutral buoyancy, the eruption cloud spreads out to form a capping umbrella, which may extend hundreds of kilometers in the downwind direction. Pliny's nephew, who documented the Vesuvius tragedy, called the combined column and umbrella a *pino* or "pine" cloud because he thought it resembled the shape of the long-trunked Italian pines with their spreading canopy that grew around his uncle's estate. Some powerful Vulcanian deposits can develop pinos, too, but Vulcanian eruptions are generally not as long-lasting as Plinian eruptions, which can go on for several days, nor do they impact the upper atmosphere quite so intensely (if at all). An extremely powerful Plinian eruption can produce as much as a hundred billion cubic meters of pyroclastic material–several orders of magnitude greater than the most powerful Vulcanian eruptions.

Large amounts of pumice are a distinctive product of Plinian eruptions. The pumice begins dropping from the Plinian umbrella the moment it begins spreading, raining down onto the landscape kilometers below. Downwind, well-sorted pumice accumulates in beds up to several meters thick. Citizens who fled Pompeii during the early hours of the Vesuvius eruption carried mattresses and other artifacts on their heads to protect themselves from the falling pumice. Some fragments were heavy enough to kill exposed individuals, as the broken skeletons of several unfortunate victims reveal. Approximately 20,000 persons, all but a few thousand of the city's population, nevertheless managed to escape, mostly on foot, under a shower of porous lightweight stones.

USGS

Figure 5.42 Plinian eruption column from Mount St. Helens, Washington, on May 18, 1980.

When the strength of a Plinian eruption begins to wane, the column cloud collapses and enormous pyroclastic flows and surges may race as far as tens of kilometers across the surrounding landscape. Their deposits consist mostly of ash and pumice, but also of blocks of vent-derived fragments and other stony materials scooped from the landscape across which the currents travel. The landscape around Vesuvius includes outcrops of 79 C.E. pyroclastic flow deposits rich in red roof tiles torn from Roman houses, in addition to ash, pumice, and other rock fragments. Pompeii itself was sealed beneath two pyroclastic surge beds 1.8 meters (6 feet) thick toward the end of the eruption, preserving the ruined city for future excavators to begin discovering 1,700 years later. Evidence of numerous victims are preserved as molds of their bodies (the flesh long-ago incinerated or fully decomposed) in these layers. Archaeologists have restored their shapes of many by filling their molds with plaster (**Figure 5.43**).

Collapse of a volcano's summit, and sometimes the entire volcano, generally occurs during a big Plinian eruption, as the weakening roof of the underlying, rapidly emptying magma chamber suddenly caves in. The term **caldera** (derived from the Spanish word for "cauldron") describes the resulting basin, which may be as wide as a few tens of kilometers. Backfilling by pyroclastic debris may fill the caldera as an eruption progresses, creating a steaming plain or volcanic tableland with no visible crater or basin. In some cases, enough of a physical depression exists to capture water over time, forming huge lakes within steep crater rims. The 10-kilometer (6-mile)-wide Crater Lake in Oregon is a famous example (**Figure 5.44**). It formed from the collapse of ancient Mount Mazama, a composite cone that experienced an intense Plinian eruption around 7,000 years ago, spewing ash that can be found today buried in soils and lakes across much of western North America (**Figure 5.45**).

Large calderas may continue to be areas of intense geyser and hot spring activity (**geothermal activity**) and eruption of silicic lavas long after they develop. An excellent case is Yellowstone caldera in Wyoming, preserved by the world's first national park. Yellowstone formed in three Plinian eruptions occurring between 1.8 million and 600,000 years ago. Future eruptions in the Yellowstone region almost certainly will take place; geologists monitor the crust of the park intensively for warning signs. The latest eruption within the caldera was mostly effusive; a thick, pasty rhyolite flow about 70,000 years ago.

Pung/Shutterstock.com

Figure 5.44 Crater Lake caldera, Oregon. The cinder cone at the center, Wizard Island, formed after caldera collapse.

Sulfuric acid aerosols from Plinian eruptions can reside in the lower stratosphere for several years. Because they reflect sunlight back into space, the aerosols can cool Earth's atmosphere by as much as 3°C (1.8°F to 5.4°F). In human terms, this can have a significant impact on growing seasons and agriculture. For example, the summer of 1816 in northern Europe, the northeastern United States, and eastern Canada was highly abnormal. Temperatures at this time of year in these regions are usually far too warm for frost and snow. However, that summer, temperature swings between 35°C (95°F) and near freezing took place in some cases within hours, and heavy snowfalls persisted well into June–prime agricultural season. Widespread crop failure ensued, and prices for oats, essential for feeding horses–a mainstay of

iStock.com/Flory

Figure 5.43 Cast of one of Pompeii's victims buried in volcanic ash. Early archeologists realized that although the bodies had decomposed, the molds created by the ash preserved people's last moments. The shapes were restored by filling each mold with plaster, a grisly but revealing task.

© HUGEfloods.com

Figure 5.45 Light-grey ash from the Plinian eruption of Mount Mazama exposed by a road cut along Interstate Highway 90 in Kittitas, Washington–500 kilometers from its source!

Figure 5.46 Standing on the brink of the caldera formed during the largest volcanic eruption in recorded history, at Tambora volcano, Sumbawa, Indonesia.

transportation in those days–rose from 12 cents to 92 cents a bushel. In ordinarily peaceful Switzerland, rising prices and food scarcities led to violence, whereas in the United States this climate catastrophe became known as the "Year Without a Summer" or the "Poverty Year," one of the worst famines of the nineteenth century.

In time, the cause would be linked to the massive eruption of Mount Tambora on the island of Sumbawa in Indonesia, the largest volcanic outburst to occur in more than 1,600 years and perhaps the deadliest, killing at least 71,000 people (**Figure 5.46**). The blasts from Tambora could be heard as far as 2,600 kilometers (1,550 miles) away, as the volcano explosively ejected some 160 cubic kilometers (38 cubic miles) of magma.

The Tambora eruption cast a dense pall of fine ash and sulfate gas aerosols into the stratosphere, dimming the sunlight reaching the surface below and cooling the global

The Volcano Explosivity Index (VEI)

In addition to the purely descriptive approach toward classifying volcanic eruptions, outlined above, the Volcano Explosivity Index (VEI), developed at Dartmouth College, New Hampshire, in 1982, categorizes volcanic activity on the basis of relative explosivity. Like the Richer magnitude scale for measuring earthquake strength (Chapter 6), the VEI is an open-ended logarithmic scale beginning with 0, running up to 9 or 10 for the most powerful eruptions (**Table 5.3**). The scale contrasts different parameters of volcanic eruptions, including height of eruption clouds (columns), duration of blasts, whether volcanic gases and ash penetrate the stratosphere, and overall volume of explosively ejected materials (**Figure 5.47**). One advantage of the VEI is that a geologist can infer much about the nature of an explosive eruption based on the size of its deposit long after the eruption took place, provided that erosion or burial by younger deposits has not progressed too far. The appearances of materials (e.g., whether pumice is abundant, presence of pāhoehoe and ʻaʻā) factor into making this inference too, of course.

The VEI also allows us to study and quantifiably estimate the frequency of different types of volcanic eruptions taking place worldwide. Resembling the example of earthquake magnitudes (Chapter 6), we find that eruption frequency tapers off with increasing VEI number in a nonlinear, geometric way. Although several Strombolian eruptions occur somewhere in the world every year, the incidence of eruptions having a VEI greater than or equal to 5 drops off to 1 every 10 years. Truly terrifying, caldera-forming eruptions (VEI > 8), in the meantime, are so infrequent that one shouldn't lose any sleep over their possibility.

Table 5.3 Selected Notable Worldwide Volcanic Eruptions*

Year	Volcano Name And Location	VEI	Comments
79	Vesuvius, Italy	5	Pompeii and Herculaneum buried; at least 3,000 killed
1783	Laki, Iceland	4	Largest historic lava flows; 9,350 killed
1792	Unzen, Japan	2	Debris avalanche and tsunami killed 14,500
1815	Tambora, Indonesia	7	Most explosive eruption in history; 92,000 killed; weather changed
1883	Krakatoa, Indonesia	6	Caldera collapse; 36,000 killed, mostly by tsunami
1902	Mount Pelée, Martinique, West Indies	4	Saint-Pierre destroyed; 30,000–40,000 killed; spine extruded from lava dome
1912	Katmai, Alaska	6	Perhaps largest 20th-century eruption; 33 km^3 tephra ejected
1914–1917	Lassen Peak, California	3	California's last historic eruption
1980	Mount St. Helens, Washington	4–5	Ash flow, 600 km^2 devastated
1991	Mount Pinatubo, Philippines	5–6†	Probably the second largest eruption of the 20th century; huge volume of SO_2 emitted
1983–	Kīlauea, Hawaii	2†	Longest continuing eruption, with more than 50 eruption events

*Historic lava or tephra volume ≥2 km^3, Holocene > 100 km^3, fatalities ≥ 1,500.
†Author's estimate.
VEI, Volcano Explosivity Index.
National Museum of Natural History, Smithsonian Institution, 2012.

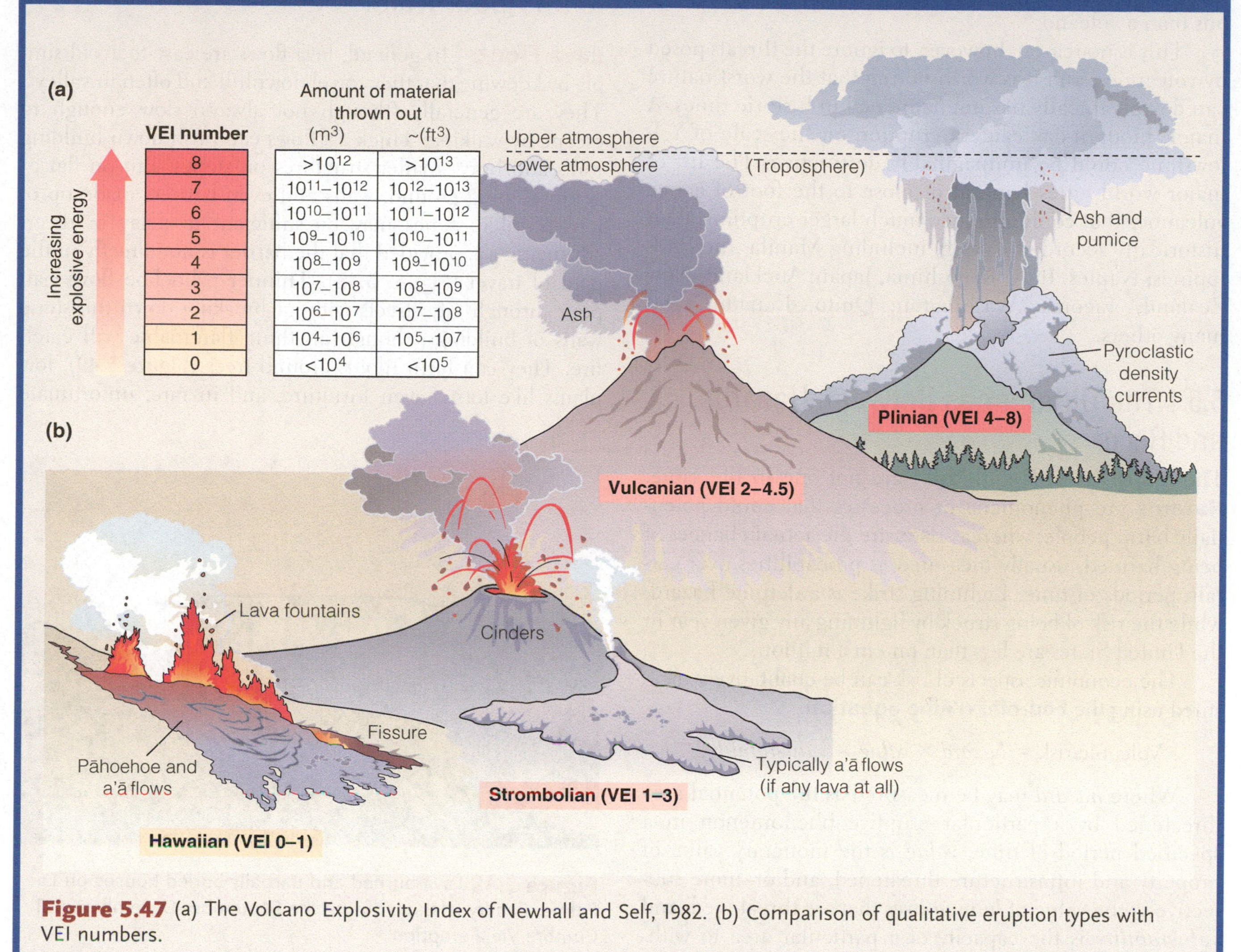

Figure 5.47 (a) The Volcano Explosivity Index of Newhall and Self, 1982. (b) Comparison of qualitative eruption types with VEI numbers.

climate by 0.4°C to 0.7°C (0.7°F to 1.3°F). Fine ash remained suspended 10 to 30 kilometers (6 to 18 miles) above the Earth worldwide for several years after the eruption. Volcanic blasts large enough to cool Earth's climate temporarily occur roughly once or twice every century.

In human terms, perhaps the severest case of volcano-induced global cooling took place 70,000 to 75,000 years ago, when a far more powerful eruption than that of Tambora created Lake Toba caldera–the world's largest volcanic collapse feature–on the island of Sumatra. This super-eruption generated on the order of 2,500 to 3,000 cubic kilometers (600 to 720 cubic miles) of ejecta, and as much as a hundred million metric tons of sulfuric acid escaped. Genetic evidence suggests that the world's human population, already struggling with ordinary ice-age instability, might have plummeted in response. The evidence for this catastrophic dieback is not conclusive, but surely volcanic activity 10 to 20 times greater than that causing the "Year Without a Summer" must have had significant global impact.

Around the year 1600, Huaynaputina volcano in Peru erupted, spreading a pall of ash that may have blanketed the entire planet. This Plinian explosion dumped 20 cubic kilometers (4.8 cubic miles) of tephra surrounding the volcano in just a matter of hours over an area of 300,000 square kilometers (16,000 square miles)–larger than the state of Nevada. Ice core records show that volcanic aerosols and finer ash settled over Antarctica and probably reached the Arctic, too–a worldwide impact. The eruption caused the coldest global summers in several centuries.

5.5 Volcanic Hazards, Risks, and Benefits (LO5)

Despite the fear that they can invoke, the chance of being injured or killed by active volcanoes is low compared to other natural hazards such as earthquakes and flooding. Since 1700 C.E., around 250,000 people have been killed by volcanic eruptions. Contrast this with the fact that each year approximately 1.3 million people die from traffic accidents alone. An automobile is on average far more dangerous than a volcano.

This is no reason, however, to ignore the threats posed by volcanoes, and it is worth noting that the worst nature can do volcanically has not happened in historic times. A major, modern-day caldera eruption on the scale of Yellowstone could be unimaginably destructive. The list of major world cities lying at or close to the foot of active volcanoes that could produce much larger eruptions than historically recorded is long, including Manila, the Philippines; Naples, Italy; Kagoshima, Japan; Auckland, New Zealand; Tacoma, Washington; Quito, Ecuador, and many others.

5.5.1 The Differences Between Hazards and Risks

The terms "hazard" and "risk" are not one in the same. **Hazards** are phenomena or processes that could potentially harm people, whereas **risks** are the actual chances of being harmed, usually measured as probabilities over certain periods of time. Lightning strike is a definite hazard, while the risk of being struck by lightning any given year in the United States are less than one in a million.

The economic aspects of risk can be qualitatively measured using the **Fournier d'Albe equation**:

$$\text{Volcanic risk} = \textit{hazard} \times \textit{value} \times \textit{vulnerability},$$

Where *hazard* may be measured as the potential area threatened by a particular eruptive phenomenon in a specified period of time; *value* is the monetary value of property and infrastructure threatened, and/or–more subjectively–the value of *human lives* that are threatened; and *vulnerability* is the capacity of a particular area to withstand the impact of the hazard. Vulnerability is tied to many factors, including local geography, wealth of an area and quality of construction, social stability, emergency services, and public preparation in advance for an eruption. It is easy to see how science relates to society by considering this formula.

Volcanic hazards should be regarded as chains of consequences. Primary, or first-order hazards, are the threats posed by immediate eruptive activity, like flowing lava, falling tephra, and pyroclastic flows. They are the obvious, direct hazards. They can trigger second-order hazards; hazards that wouldn't take place otherwise. For instance, heavy ash fall into streams on a mountainside will develop into lahars, deadly debris and mudflows originating on the slopes of active volcanoes and their environments. These in turn can generate third-order hazards, including disruption of transportation networks across an entire region, destruction of fisheries for many kilometers downstream, and economic hardship for people losing property in flood zones. Every eruption has its own chain of consequences, though these aren't necessarily difficult to anticipate using knowledge of past human experiences with eruptions.

5.5.2 Typical Eruption Hazards and What to Do About Them

Lava Flows—In general, lava flows are easy to avoid simply by knowing that they travel downhill and often in valleys. They are generally (though not always) slow enough to escape by walking. Thick a'ā flows can push down building walls and bury smaller structures on slopes. But on flat or gently sloping ground, walls and even barriers made up of loosely piled stones may be sufficient to delay or divert advancing a'ā, provided that the barrier is not directly in the path of travel (**Figure 5.48**). Thinner pāhoehoe flows can pour through city streets without breaking down the stone walls of buildings, though anything flammable will catch fire. They can form molds around trees (**Figure 5.49**), low plants like ferns, even furniture, and in rare, unfortunate

Anna LoFi/Shutterstock.com

Figure 5.48 Lava-burned and partially buried houses on La Palma, Canary Islands, during the September-December 2021 Cumbre Vieja eruption.

iStock.com/MNStudio

Figure 5.49 Tree trunks caught by an eruption in 1790 remain as pillars in Lava Tree State Monument on the Big Island of Hawai'i. The trees are moist and cold, and lava cools and hardens around them as it loses heat when the water in the tree turns to steam. The lava surrounding the hard case then drains away, leaving a pillar behind. The trunks burn or eventually rot away, leaving a hollow mold in each pillar.

instances, trapped animals, famously including groups of elephants in recent African eruptions.

Around some volcanoes, efforts have been made to construct barriers that divert lava flows or break up the channels that feed them, causing the lava to flow in other directions. In 1669 C.E., some daring citizens of Catania, Italy, threatened by lava from nearby Etna volcano, strapped cowhides onto their bodies for heat protection and pried open the flank of an approaching flow, allowing the liquid interior to pour away from their city. This did not work for long, however. Residents of another community downslope, now endangered instead, fought back by driving away the Catanians, and soon the lava returned to its original path, coming up against the city wall. That held for a while, turning the lava toward the sea, but after several days the ʻaʻā broke through and inundated part of Catania.

Since then, earth and rock lava diversion barriers have been constructed to turn the direction of a lava flow on several occasions, but with limited success, as in the Catania example above. However, in 1983, engineers hastily built a 14-meter (45-foot) diversion barrier to protect a ski resort and astronomical facility from a lava flow pouring down the flank of Etna, and it worked! Today, a 1.5-kilometer (1-mile)-long, 5 to 7 meters (16 to 23 feet) tall barrier made of bulldozed lava rubble wraps upslope of the National Oceanographic and Atmospheric Agency (NOAA) Observatory to protect it from lava flows high on the flank of Mauna Loa (**Figure 5.50**).

In 1936 and 1942, military aircraft bombed lava channels in other flows pouring down Mauna Loa, threatening the port of Hilo downslope. The working hypothesis was that if lava were to spread out upslope, thanks to an artificially clogged channel system, then the flow front would slow down or stop downslope. These efforts were not successful, however, and though the idea behind bombing remains sound, it is not long culturally accepted in Hawaiʻi, future risks notwithstanding.

RGB Ventures/Alamy Stock Photo

Figure 5.50 Lava diversion barrier constructed to protect the globally important Mauna Loa Observatory from lava flows coming from upslope (the bottom of the image).

FLPA/Alamy Stock Photo

Figure 5.51 White water hoses stand in stark contrast to black basalt as Heimay's residents attempt to cool and divert lava threatening their Iceland town. Eldfell volcano erupts in the background as the watered-down lava steams.

In 1973, as a thick ʻaʻā flow threatened to seal off the harbor at Heimay, Iceland, fireboats directed jets of water on the advancing flow front, cooling the lava to form a self-sealing dam of hardened crust (**Figure 5.51**). The project was successful; the flow thickened instead of lengthened. Massive water pumps helped douse around 6 million cubic meters (7.8 million cubic yards) across the flow during a 5-month time span. Observers discovered that each cubic meter (35 cubic feet) of sprayed water could immobilize about one cubic meter of lava at the steepening flow front. Today, much of the former town remains buried beneath ash and lava of the 1973 eruption, but archeologists have begun excavations–a modern-day Pompeii (**Figure 5.52**).

One of the most dangerous volcanoes in the world is Mount Nyiragongo, near the eastern border of the Democratic Republic of the Congo. The volcano looms above

Paul Bierman

Figure 5.52 Excavation of home buried in 1973 Heimay eruption from under basaltic cinders.

Ben Houdijk/Shutterstock.com

Figure 5.53 The city of Goma, Congo, at the northern shore of Lake Kivu, with Nyiragongo volcano steaming in the background not far from the suburbs.

the north end of Lake Kivu, filling a part of the East African Rift Valley 15 kilometers (9 miles) away. The city of Goma stands at the lake shore where it is closest to the volcano (**Figure 5.53**). Nyiragongo has a molten lava lake at its summit, fed continuously by a conduit linked to a central magma chamber below (**Figure 5.54**). The volcano is also positioned along a north-south rift (fracture) system through which lava can easily erupt. That rift system continues into the Goma urban area, which in 1977 had around 50,000 residents. On January 10, 1977, 20 to 22 million cubic meters (26 to 29 million cubic yards) of very fluid lava issued from the volcano's flank directly upslope from the city. For reference, an Olympic sized swimming pool has a volume of around 61,000 cubic meters (80,000 cubic yards), so imagine more than 300 swimming pools of lava. Given the high temperature and unusual chemical composition of the lava, together with steepness of slope, the flow poured especially fast downhill–as up to 60 kilometers (37 miles) per hour. Most of the residents of Goma fled on time, but 70 some people died. The summit lava lake drained in response to the flank eruption.

Eric Isselee/Shutterstock.com

Figure 5.54 Crater atop Nyiragongo, with a lava lake almost continuously active filling a central inner pit; the top of the volcano's magma conduit.

Nyiragongo wasn't done. While the hazard remained the same, the risk vastly increased. Between 1977 and 2002, the population of Goma ballooned. Civil war in the Congo and genocide in Rwanda (1994) swelled the city with refugees. By 2002, a half-million people lived there, many building new homes higher up the mountainside along the rift zone. On January 17, a new set of fissures opened, releasing 14 to 34 million cubic meters (18 to 44 million cubic yards) of very fluid lava once again partly supplied by drainage from the summit lava lake. It was as though a dam had burst. Eruptive fissures opened to within 1.5 kilometers (~1 mile) of the Goma Airport, and lava quickly covered a third of the runways, then destroyed the commercial and business heart of the city before entering the lake. Two-hundred-fifty people died and 350,000 fled. Almost a third of the city's population lost their homes during the 12-hour eruption.

Incredibly, things have since gone from bad to worse. By 2021, Goma's population reached around 2 million. Political corruption and unrest also greatly impeded operations of the local Goma Volcano Observatory. Guerilla fighters destroyed about 60% of the seismometers used to monitor the internal conditions of Nyiragongo, and the World Bank halted funding for volcanological operations owing to concerns about embezzlement. Geologists could not afford to maintain internet connections, essential for rapid data collection and response during emergencies. Nature didn't wait for people to solve their problems, and on May 22, 2021, the rift zone above Goma once again suddenly released a torrent of highly fluid lava, repeating the scenario that unfolded in 1977 and 2002. Fortunately, the flows stopped just short of the main city, destroying about a thousand houses. Thirty people died.

What is the moral of the Goma story? It is that volcanic risk is as much an outcome of human behavior as it is of nature, and that knowing about past mistakes is not necessarily the same thing as learning from them–or at least acting so as to prevent them from happening again.

Volcanic Gases—The gases from volcanic vents released by an eruption can be lethal, although they quickly disperse in the atmosphere. They pose no threat when people stay upwind of them. Beware otherwise; Pliny the Elder, who rushed to save friends downwind of the 79 C.E. Vesuvius eruption, likely died from gas inhalation not far from Pompeii. If caught in gases by a change in wind direction, a person can reduce the risk of poisoning by placing a damp cloth such as a handkerchief over the nostrils. Although not a long term solution, the moisture in the cloth absorbs the some of the gases and gives one time to escape.

The greatest threat posed by gases is not during eruptions themselves, but in the quiet times between them. Volcanoes continuously "breathe" gases, as long as there is a reservoir of magma beneath them. The gas can issue from steam vents or **fumaroles** (vents that issue gases other than just water vapor; **Figure 5.55**; or it can simply seep unseen through the soil all around a mountainside, a process called **soil efflux** (see Case Study 5.2).

For example, when the wind is calm, high concentrations of carbon dioxide can build up in areas of low ground because pure CO_2 is 1.5 times denser than normal atmosphere. These areas are termed mofettes. When concentrations exceed 3% to 4% of total air mass, animals–and people–wandering into them are lethally poisoned. Residents of the East African Rift Valley sometimes call mofettes "elephant graveyards." Volcanologist Thomas Jaggar, exploring near Yellowstone Park in the late nineteenth century, discovered an active mofette containing the bodies of several grizzly bears; and three skiers on Mammoth Mountain, a dormant California volcano, died in a CO_2-rich pocket in 2006. Visitors must stay mindful of low ground or cavities on calm days around degassing volcanoes.

One of the worst volcanic disasters in history occurred at Lake Nyos, a 1.8-kilometer (5,900-feet)-wide volcanic Crater Lake in Cameroon, in 1986. Although the eruption that formed Nyos volcano took place 400 years ago, magmatic carbon dioxide has continued to seep into the crater ever since, building up to high concentration in the cooler, deeper waters of the lake mostly below 180 meters (580 feet); a good example of a secondary volcanic hazard in the making. Carbon dioxide is more soluble in cold water than warm, and if a situation occurs in which cold water can mix readily with warm, a tremendous amount of CO_2 will suddenly come out of solution as a lethal cloud of gas–a so-called **limnic eruption**. The risk of such eruptions was not unknown to the local people. Bantu folklore is filled with stories of exploding lakes and "rains" of dead fish. According to oral tradition, at least three major degassing catastrophes occurred in the distant past. Many residents accept dead fish found floating on the water surface during minor events as gifts from their ancestors, whose spirits they believe live in the lake.

On the night of August 21, 1986, Lake Nyos suddenly discharged its asphyxiating load, possibly triggered by a small landslide from the crater wall. Scientists estimate that the gas rushed from the lake as fast as 100 kilometers (60 miles) per hour and rapidly filled the crater basin. It then silently poured over the low northern rim and flowed down a neighboring valley, overtaking livestock and a village of people who had, for the most part, gone to sleep. More than 1,700 villagers and hundreds of animals died of sudden respiratory failure (**Figure 5.56**). Since 2001, observers have used a 200-meter (650-feet)-long pipe to allow carbon dioxide to leave the lake without building up deadly concentrations. The gas generates a 40-meter (130-foot)-tall artificial geyser as it escapes (**Figure 5.57**).

Falling Ash—Ash falls darken skies, in some cases "blacker than night" with visibility reduced to only a few meters or less, making the movement of people and goods difficult at a time when emergency services may be needed. For most people who don't live so close to a volcano that other, more serious volcanic hazards are a risk, it is best to hunker down during an ash fall. Ash falls more than a few tens of kilometers from a source vent usually last no more than several hours.

People who decide to leave during an ash fall should bear in mind that it isn't just airplanes that are at risk from flying through ash clouds. Drifting ash particles will enter machinery and can cause engines to fail in any equipment,

Tkach Anastasiya/Shutterstock.com

Figure 5.55 Fumaroles and steaming ground, Tyatya Volcano, Kuril Islands, Russia.

Eric Bouvet/Gamma-Rapho/Getty Images

Figure 5.56 In 1986, carbon dioxide degassing from Lake Nyos killed more than 1,700 people and many farm animals around the lake. Here, a survivor stands in front of cattle that asphyxiated in the gas plume.

Louise Gubb/Corbis Historical/Getty Images

Figure 5.57 Collecting samples from the degassing system installed in Lake Nyos.

including power tools and even automobiles (**Figure 5.58**). Ash will cause expensive abrasive damage to both metal and glass, as many drivers who try to use windshield wipers during ash falls attest.

Those who stay behind quickly understand that ash is powdered rock, much heavier than snow. When soaking wet, ash is especially heavy; and unless roofs are steeply pitched, it can cause them to buckle and collapse under loads as thin as 15 centimeters (6 inches). During the 1982 eruption of Galunggung volcano on Java, many residents returned on a daily basis to their evacuated homes simply to sweep ash off their roofs (**Figure 5.59**). Ash can also enter sewer systems and drainage channels, clogging them and causing flooding. It provides abundant fine material for lahars (see Chapter 9) and can fill rivers with so much sediment that the passage of boats becomes impossible.

Ash particles are often coated with fine **sublimates**; water-soluble minerals deposited by aerosols of volcanic gas through which the ash falls. These can include ions of chlorine, fluorine, and other elements that re-dissolve as water infiltrates the tephra later, forming caustic acids. For this reason, it is a good idea for residents to keep emergency water supplies, at least until it is established whether sublimate contamination will be a problem or not. **Fluorosis** (ingesting fluorine), including bone and joint pain and deformation, has led to the deaths of livestock in Chile and Iceland that have consumed grasses contaminated with light ash coatings. That dieback has in turn caused human famine.

R. Hobblitt/USGS

Figure 5.58 Fresh ashfall covers cars and landscape at Clark Air Force, the Philippines, during the 1991 eruption of Pinatubo volcano.

USGS

Figure 5.59 Cleaning ash off a rooftop in Washington State following the May 18, 1980, Plinian eruption of Mount St. Helens.

Pyroclastic Flows and Surges—The only way to escape harm from these extremely dangerous phenomena is to evacuate a potentially threatened area well in advance. Frank Perret, whose experience with a pyroclastic flow at Mount Pelée we quoted earlier, was a lucky man. Most people who are in the path of a pyroclastic flow suffer severe burns not only externally but also deep into their respiratory systems: they cook from the inside out. The flows and surges move across the landscape with a deceptive, almost graceful softness; a gentle "shushing" sound, like a skier on snow. These flows in particular may be so hot at their bases that they glow dull red; hence the older French name for them, ***nuées ardentes***; "glowing clouds."

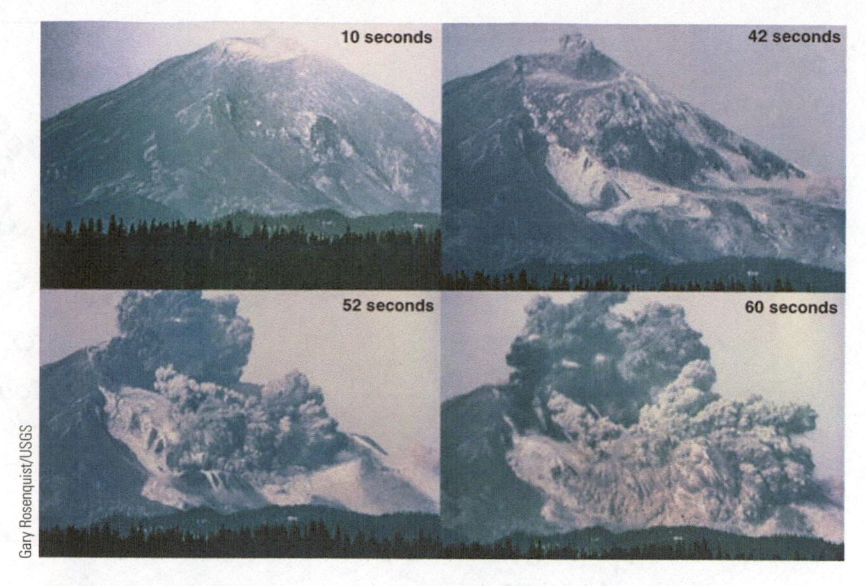

Figure 5.60 The first minute of the 1980 Mount St. Helens eruption, including the sector collapse and directed lateral blast.

Pyroclastic flows and surges typically travel downslope and tend to follow stream and river valleys. But staying on high ground or ridgelines is not necessarily sufficient protection. Some volcanic explosions eject pyroclastic clouds so powerfull that they race over high areas too, hill-and-dale, for many kilometers. This took place, for instance, during the 1980 Mount St. Helen's eruption.

On March, 20, 1980, a series of earthquakes in the same general area (termed an **earthquake swarm**) gave first indication that shallow magma was moving into the conduit beneath the volcano. Seven days later, small Vulcanian blasts began at the summit, but this did not relieve the pressure of the molten magma mass (it had the composition of dacite) building up viscously inside the mountain. A plug of older lava capping the summit caused the magma to begin expanding sideways at the top of the conduit, creating an enormous 2-square-kilometer (0.8-square-mile) bulge in the northern flank of Mount St. Helens. The mountainside stretched outward as much as 40 meters (130 feet). It didn't take much to destabilize this mass.

On the morning of May 18, a magnitude-5 earthquake shook the area, and the northern flank of the mountain gave way in a massive landslide (Chapter 9) now referred to as a **sector collapse**, because an entire sector of the volcano's cone collapsed. The magma chamber was "unroofed," but the collapsing summit prevented the full strength of the unleashed explosion from initially going straight up into the atmosphere to form a Plinian eruption column. Instead it was diverted almost horizontally to the north; a so-called **directed lateral blast** consisting of a mix of sliding avalanche debris and expanding pyroclastic material; ash, pumice, and blocks (**Figure 5.60**). Deadly.

The pyroclastic flow quickly overtook the head of the avalanche within about 15 seconds, at a speed of 90 to 110 meters per second (200 to 250 miles per hour), but the force of the blast combined with expanding gas pressure drove it even faster, to 235 meters per second (525 miles per hour), enabling it to overtop a 380-meter (1,250-foot)-tall ridge where a young volcanologist, David Johnston, was posted as an observer. He barely had time to radio in an alert ("Vancouver, Vancouver, this is it!") before being overtaken. His body was never found.

The pyroclastic cloud continued northward, devastating an area of over 600 square kilometers (230 square miles) in a matter of a few minutes. Thick stands of forest were plowed flat, like matchwood (**Figure 5.61**). Fifty-seven people died—almost all spectators outside the area that had been formally closed by authorities as the "red zone" of high volcanic danger (**Figure 5.62**). Many in the uplands were killed by extreme heat; others in the lowland were swept away or buried by the lahar. Had the blast not been a directed one, much of this area would have been spared this destruction, though heavy tephra fall would still have taken place.

Figure 5.61 A once-towering forest flattened by the 1980 Mount St. Helens eruption.

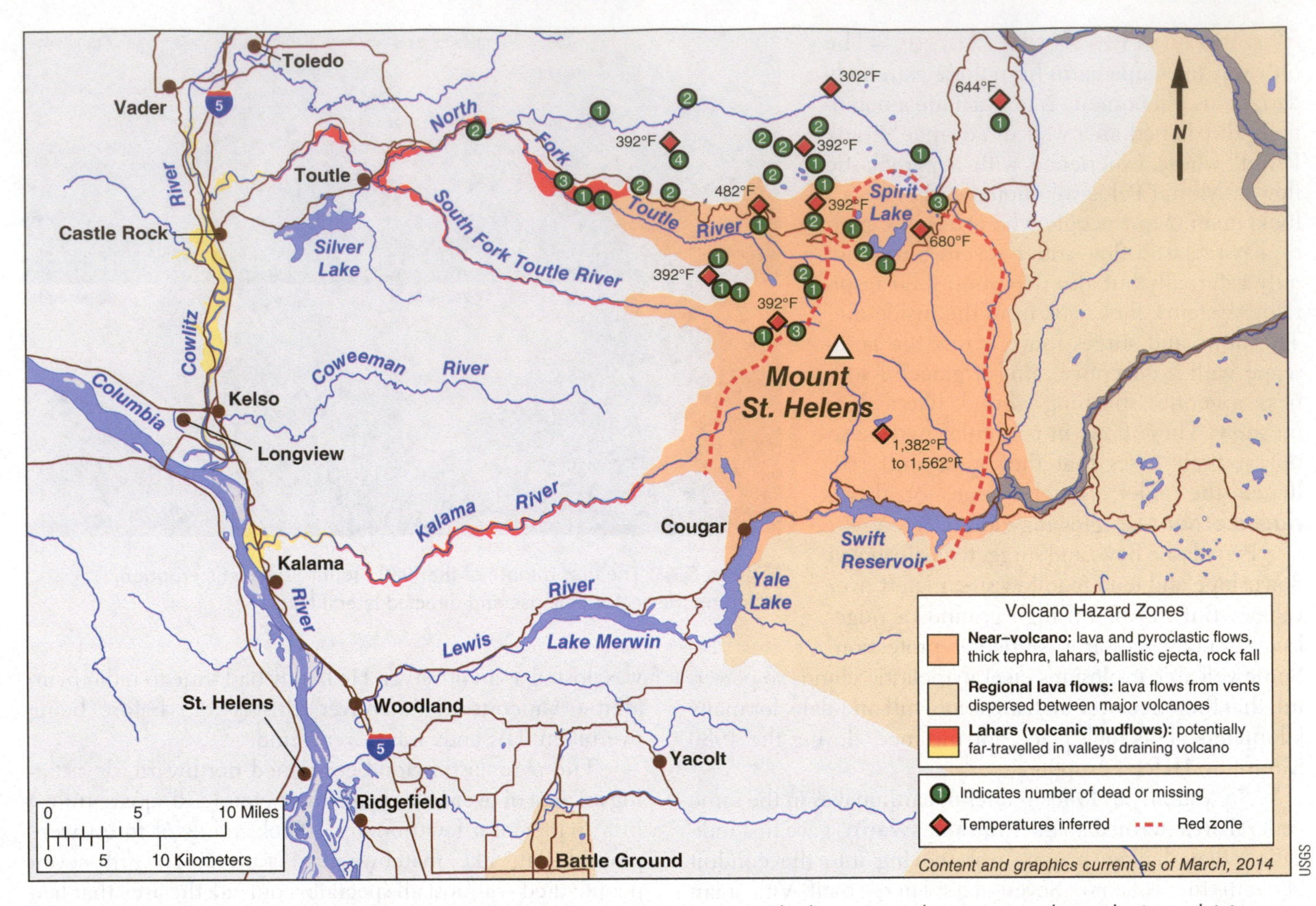

Figure 5.62 Most people killed by the Mount St. Helens eruption were outside the restricted zone set up by geologists advising emergency management specialists. Numbers indicate number of deaths and their location. Temperatures during the eruption are derived from the melting of plastic car parts.

One important point about the 1980 directed blast is that volcanologists and government authorities were taken largely by surprise. Directed blasts had been observed and described before, at Mount Lamington in New Guinea (1952) and Bezymianny in Kamchatka (1955–1956). Geologists had not seen warning signs at Mount Lamington, however, because it was not even recognized as a volcano before it erupted and is located in remote, inaccessible wilderness. In contrast, the Bezymianny eruption, which had been closely monitored and studied, occurred at the height of the Cold War when scientific exchange between East and West was minimal and much Russian scientific work remained untranslated in English. When it comes to volcanoes, what you don't know [or, rather what we haven't yet learned] can kill you.

Tsunami—**Tsunami**, giant sea waves (see Chapter 10), are secondary hazards resulting from a variety of geological events, most often undersea earthquakes (see Chapter 6). Shallow underwater eruptions or eruptions at coastal volcanoes have been responsible for some of the world's largest historical tsunami. In 1883, the collapse of the island of Krakatau during a Plinian eruption created an undersea caldera roughly 8 kilometers (5 miles) wide, simultaneously generating a monstrous sea wave that killed more than 36,000 people on both sides of Indonesia's Sunda Strait. At Unzen volcano in southern Japan in 1792, a giant landslide in the volcano flank generated a set of tsunami that destroyed 17 coastal communities and devastated 80 kilometers (50 miles) of shoreline. In some inlets, wave heights reached 60 meters (200 feet)! About 15,000 persons drowned with almost no warning because no precursory earthquake took place.

Pliny the Elder provided the first ever scientific description of a tsunami, formed in the Bay of Naples during intense earthquake activity while nearby Vesuvius erupted furiously. More than 90 volcanogenic tsunami have been reported since Pliny's time, killing more than 60,000 people worldwide.

Lahars—**Lahars** have been even more deadly than tsunami–and, in fact, are the most destructive of volcanic phenomena, historically speaking. The Geological Society of America defines a lahar as "a general term for a rapidly flowing mixture of rock debris and water (other than normal stream flow) from a volcano" (**Figure 5.63**).

Dr Morley Read/Shutterstock.com

Figure 5.63 Debris left by a lahar pouring down a valley from Tungurahua volcano, Ecuador.

Galih Sundoro/Shutterstock.com

Figure 5.64 Sabo dam and catchment basin meant to contain the impact of lahars in Japan.

They may be generated by hot ash falling or flowing over ice and snow on a mountainside, by the explosive ejection of volcanic lake water onto surrounding slopes laden with loose ash, or by heavy rain on weak volcanic soils or ash (see Chapter 1 and Chapter 9).

A lahar need not take place during an eruption, but can occur long afterward. Lahars continued to develop for several years following heavy rainfalls on thick ash deposits from Mount Pinatubo in the Philippines. The equivalent of 300 million dump-truck loads or 3 cubic kilometers (0.7 cubic miles) of rock, mud, silt, and other debris buried hundreds of square kilometers of populated agricultural land east of the volcano, including entire towns. Some 100,000 persons lost their homes, and an additional 100,000 remain threatened to this day, although since 1995 the risk of additional lahars has greatly subsided. The largest Pinatubo lahars, moving as fast as 32 kilometers (20 miles) per hour, were as much as 10 meters (~30 feet) thick and 100 meters (~300 feet) wide. They also arrived with little warning.

Having to contend with active volcanoes and tsunami, the Japanese are also no strangers to lahars. They have developed perhaps the world's most sophisticated lahar "self-defense" system, known as sabo engineering (**Figure 5.64**). This initially involves educating the public about the hazard along certain rivers and valleys near active volcanoes. Construction of new structures in potentially impacted lowlands is forbidden. Dams and catchment basins are built in the channels to strain out large rocks–the "battering rams" floated along by lahars that prove so destructive downstream. To keep the catchment basins open, accumulating sediment must be quarried, but it can have useful purposes elsewhere, including grading roads and making landfills.

5.5.3 Benefits of Volcanoes

Despite hazards associated with volcanoes, we owe them a great amount of gratitude, foremost being their roles in forming and replenishing the atmosphere we breathe (Chapter 2). Eruptive products build lands that people occupy by the tens of millions, and volcanoes are key to the development of ore deposits at their geological roots (Chapter 4), as well as heat for geothermal power (Chapter 11).

We should also thank volcanoes for food. Agricultural crops need 16 major and minor nutrients to thrive, including potassium, nitrogen, phosphorus, and iron. In many tropical environments, soil leaching rates are high; infiltrating water removes nutrients and washes them out to sea. Soils tend to be nutrient depleted, and the remaining nutrient supply is mostly taken up by plants and other organisms, only to be recycled via forest floors or savannah topsoils through death and decay. The Tropics is a tough region to farm, in other words. With the exception of nitrogen, which bacteria can readily scavenge from the atmosphere, volcanic ash supplies all the nutrients that crops need.

In tropical regions, farmers are quick to mix the soil with fresh ash after it falls to fertilize it naturally. Small eruptions of ash are widely welcomed! On Borneo, which lacks volcanoes, a farmer will be lucky to have a good harvest once every 3 years. But on neighboring Java, with its 20 recently active volcanoes, good harvests are possible 3 times a year thanks to frequent inputs of fresh volcanic ash, rich in nutrients and rapidly weathered in the tropical climate. What this means in terms of human numbers should be plain: Borneo has a population of 19 million in an area 60% larger than the State of California (population 40 million). Java, on the other hand, is less than a third as large as California and has a population of 145 million. **Andosol**, the technical name for volcanically derived soil,

em faies/Shutterstock.com

Figure 5.65 Java's population thrives largely because of the fertile volcanic soils supporting its agriculture. Agricultural lands in parts of California, Italy, Germany, New Zealand, Australia, Central America, Africa, Japan, and other regions are similarly enriched by volcanic ash.

is among the richest mediums for growing plants on Earth (**Figure 5.65**).

Volcanoes also have great ecological importance, contributing significantly to the development of biodiversity in many parts of the world. On the island of Hawaii, for instance, lava flows often surround patches of older terrain, called **kīpuka** (meaning "a variation or change in form" in the Hawaiian language, **Figure 5.66**). Biota within kīpuka are isolated by the fresh lava from similar biota elsewhere, and many insects and plants with highly localized reproductive ranges or mobility can evolve into distinctive varieties or entirely new species over short periods, biologically

C. Parcheta/USGS

Figure 5.66 Vegetated Kīpuka on Kīlauea, Big Island, Hawaii, isolated by lava flows.

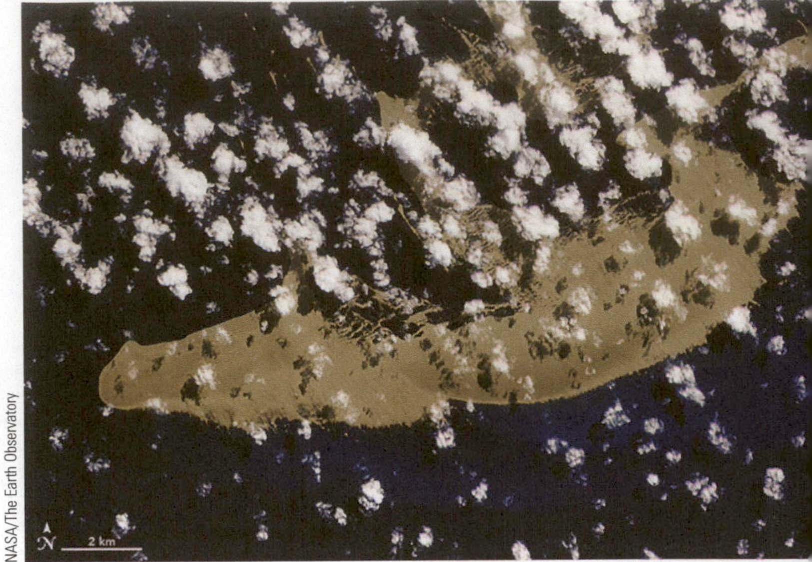

NASA/The Earth Observatory

Figure 5.67 A giant pumice raft, as large as Manhattan, floats in the ocean in Tonga in 2019.

speaking. A landscape rich in kīpuka quickly becomes a landscape rich in biodiversity–much richer, in fact, than comparable land areas that lack volcanic activity.

Giant rafts of pumice erupted from underwater volcanoes can bob at sea for up to 20 years, traveling hundreds or even thousands of kilometers in oceanic currents (**Figure 5.67**). The size of these floating islands can be impressive. In July 2012, an eruption of Havre Seamount in the Kermadec Islands, near New Zealand, was the probable source of an unbroken field of pumice that suddenly appeared on the overlying surface of the ocean, measuring 50 kilometers (30 miles) by 480 kilometers (300 miles)! Some larger pieces of pumice protruded as much as 60 centimeters (2 feet) above sea level. Boats can be damaged trying to navigate through such obstacles, which some mariners term "rubble slicks." Many nonplanktonic organisms attach themselves to floating pumice fragments or live sheltered around them. When the pumice washes up on a distant beach or degrades and finally sinks far from its point of origin, associated organisms become permanently isolated from their parent populations and, in turn, adapt to environments they have newly colonized.

5.6 Active? Dormant? Extinct? (LO6)

Living safely with volcanoes first requires general public understanding of them. Such understanding is not difficult to obtain, but for many reason volcanoes are not a priority concern for people or civil authorities; that is, until they erupt. One major challenge is for scientists to find ways of communicating more effectively with the populations they serve (see Chapter 1). This includes not only conveying the significance of current research and observations but also clarifying deep-seated historical misperceptions.

One common source of public confusion relates to the terms "active", "dormant", and "extinct". Many people think of active volcanoes as presently being in a state of eruption. Scientifically, however, an **active volcano** is one that has erupted frequently in historic time and will almost certainly continue to do so in the future, whether or not it is presently erupting. (Just what "frequently" means is subjective, however. Also, notice that the length of "history" differs from one region to the next around the world.) An active volcano must be regarded as threatening, whether it is actually erupting or not.

A **dormant volcano** shows signs of having erupted within the past few hundred or few thousand years but has not erupted in living memory (**Figure 5.68**). Many dormant volcanoes have only slightly weathered lava flows, appear not to be greatly eroded, and are sparsely covered with vegetation relative to the surrounding area. However, some dormant volcanoes are so heavily vegetated and eroded that people don't even know that they are volcanic without a careful geological study. Even if a volcano is dormant, it does not mean that it is safe to live nearby. We simply don't know if or when it will erupt again. Moreover, if an eruption does occur, it is likely to be dangerously explosive because of the cap of old, hardened rock that plugs the conduit of the volcano (as at Mount Pelée in 1902). Some dormant volcanoes that have returned to life catastrophically are El Chichón (Mexico), Mount Lamington (New Guinea), Mount Pinatubo, Vesuvius, and Mount St. Helens.

Volcanologists Chris Newhall and Steven Self published a study of 25 of the world's most violent historical eruptions. They warned that in 17 cases the eruptions were the first ever reported volcanic activity at the sites where they occurred. For others that had a record of previous activity, the mean time interval between eruptions was 865 years–long enough for people to forget the past and lose concern about the future. We do not know how many other dormant volcanoes exist that might one day spring back to life.

An **extinct volcano** is typically deeply eroded because it has not erupted in thousands, if not millions, of years. As you might guess, there are many more extinct volcanoes than dormant and active ones in the world today. Ultimately, erosion removes all traces of a volcanic cone, leaving perhaps the central plug of the former magma conduit standing behind as monumental landforms, such as Ship Rock, New Mexico (**Figure 5.69**).

Volcano scientists working through the United Nations and other organizations are eager to broadcast information about warning signs, potential hazards, and the steps that people can take to protect themselves during an eruption. They also produce detailed maps of the deposits left by past eruptions around active volcanoes. These deposits, as we've seen, can provide valuable information about the characteristic behavior and, therefore, the future anticipated hazards from a volcano. Careful, probability-based risk assessments based on this field work can give guidelines for developers, zoning planners, insurance companies, and civil authorities.

Perhaps the best way to save lives is not to put people in harm's way to begin with. Despite immense progress in eruption forecasting and monitoring in recent decades, the threat to human beings keeps growing because rising populations and pressure for land use around volcanoes has outstripped geoscientists' ability to protect people from nature's worst blows. The example of Goma, Congo, is a case in point. The trade-off in terms of general benefit may seem worth it in the long run, but for those people

DiskoDancer/Shutterstock.com

Figure 5.68 Just how dangerous is a volcano that isn't obviously erupting? That question isn't always easy to answer!

Sean Pavone/Shutterstock.com

Figure 5.69 Ship Rock, New Mexico, is a 500-meter (1,640-foot)-tall monogenetic volcanic conduit, or "neck," exposed by erosion. The volcano was active between 25 to 30 million years ago. It is now definitely extinct!

caught in the path of a volcanic eruption, such an argument has no meaning. The need for volcanologists to carry on the study and observation of volcanoes has never been stronger.

5.7 Tools of the Trade, Monitoring Technologies (L07)

In 1845, the King of Naples established the Vesuvius Volcano Observatory, the world's first such scientific facility. Since then, many other volcano observatories have come into existence, now forming an international community, the World Organization of Volcano Observatories (WOVO). Their mission is not only to help us better understand the workings of volcanoes, but to provide timely forecasts and warnings of impending eruptions, and advice to government authorities during volcanic eruptions that threaten large numbers of people.

Prediction is the ability to identify when and where something will happen. Although volcano scientists have gotten very close to predicting some eruptions within a few days of their occurrence, it is more accurate to say that they are forecasters; that is, they state in general terms the probabilities of eruptions taking place. The ability to forecast an eruption depends on how well a volcano is monitored with instruments and, to a certain extent, by how well its history of eruptions is known.

Almost every volcano has its own unique "personality," or eruptive behavior. Some show patterns of activity that can be useful in forecasting. Others are more chaotic, exhibiting great changes in their eruptive styles over time. The most effective tool for discerning and forecasting volcanic behavior is the **seismometer**, a device for measuring shaking in the ground caused by earthquakes (**Figure 5.70**).

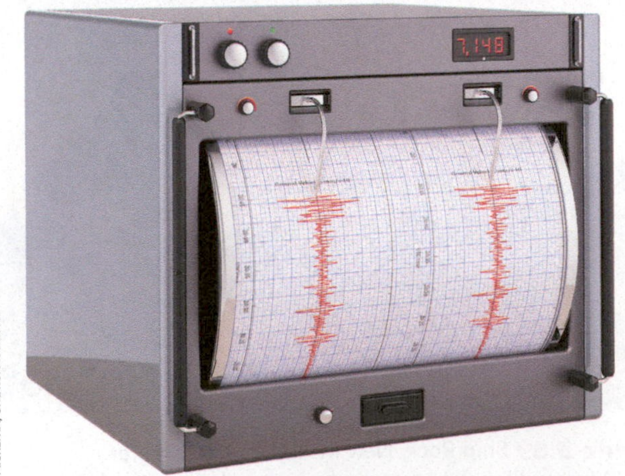

AlexLMX/Shutterstock.com

Figure 5.70 A seismograph is a device that records earthquake shaking. The swings in the recording needles are a measure of the strength of the shaking.

We shall explore the topic of earthquakes further in the next chapter, but it is worth focusing on volcano-related earthquakes here.

Seismologists distinguish two main categories of volcanic earthquakes: VT and LP. Volcanoes are inherently unstable structures, constantly adjusting and even collapsing under their own weights (Chapter 9). These adjustments can cause volcano-tectonic, or **VT-earthquakes**. During the May-September 2018 eruption of Kīlauea, swelling of the mountainside as it filled with intruding magma steepened the side of the volcano so much that a section of the flank covering roughly 700 kilometers^2 (270 miles^2) broke free and slid into deeper water. Fortunately, no tsunami ensued, but the strong shaking as so much rock sloughed off probably opened new fractures upslope inside the volcano, making its subsequent eruption much larger than it might have been otherwise.

Magma on the move within a volcano generates long-period, or **LP-earthquakes**. The "period" refers to seismic waves. If the energy waves lack high frequencies (that is, arrive less frequently than usual for most earthquakes), then they are said to be "long-period." When volcanologists detect a series of LP-earthquakes at shallow levels in the volcano, they prudently raise warnings about a possible eruption. Everything from fractures splitting open as magma intrudes, to collapsing gas bubbles in magma moving through a conduit, can generate this type of seismicity. In 2018, an intense **cluster** (series of seismic events) of earthquakes took place beneath Mount St. Helens, 6 kilometers (4 miles) down, in the roof of its current magma chamber. This disturbance did not lead to an eruption, though observers would not have been surprised if it had.

When magma comes within 1 to 3 kilometers of the surface, the character of the earthquakes changes. VT-earthquakes disappear, and a series of nearly continuous LP-earthquakes can appear, termed **volcanic tremor**. The onset of tremor is often the immediate signal that an eruption is about to begin; it may be only minutes away. Processing of tremor data to average the strength of seismic energy waves for particular frequencies and time intervals is done readily using a computer, a process called rectified (or real-time) seismic amplitude measurements (RSAM). The value of RSAM tremor analysis is that observers can monitor in real time the state of the magma feeding an eruption, and even detect new eruption activity before it is visually confirmed.

Another tool of the trade is the **tiltmeter**, a device for measuring very precisely the angle of slope of a volcano at a fixed station (**Figure 5.71**). As magma fills the interior of a volcano, it can swell, like a giant stone balloon. The swelling may not be much–perhaps only a few centimeters of uplift over a period of months. But that is a lot of pressure when you consider just how much solid crust is deforming! The swelling, termed inflation, can cause the slope of the volcano to steepen, though this is typically by mere microradians of angle. (A microradian is the change

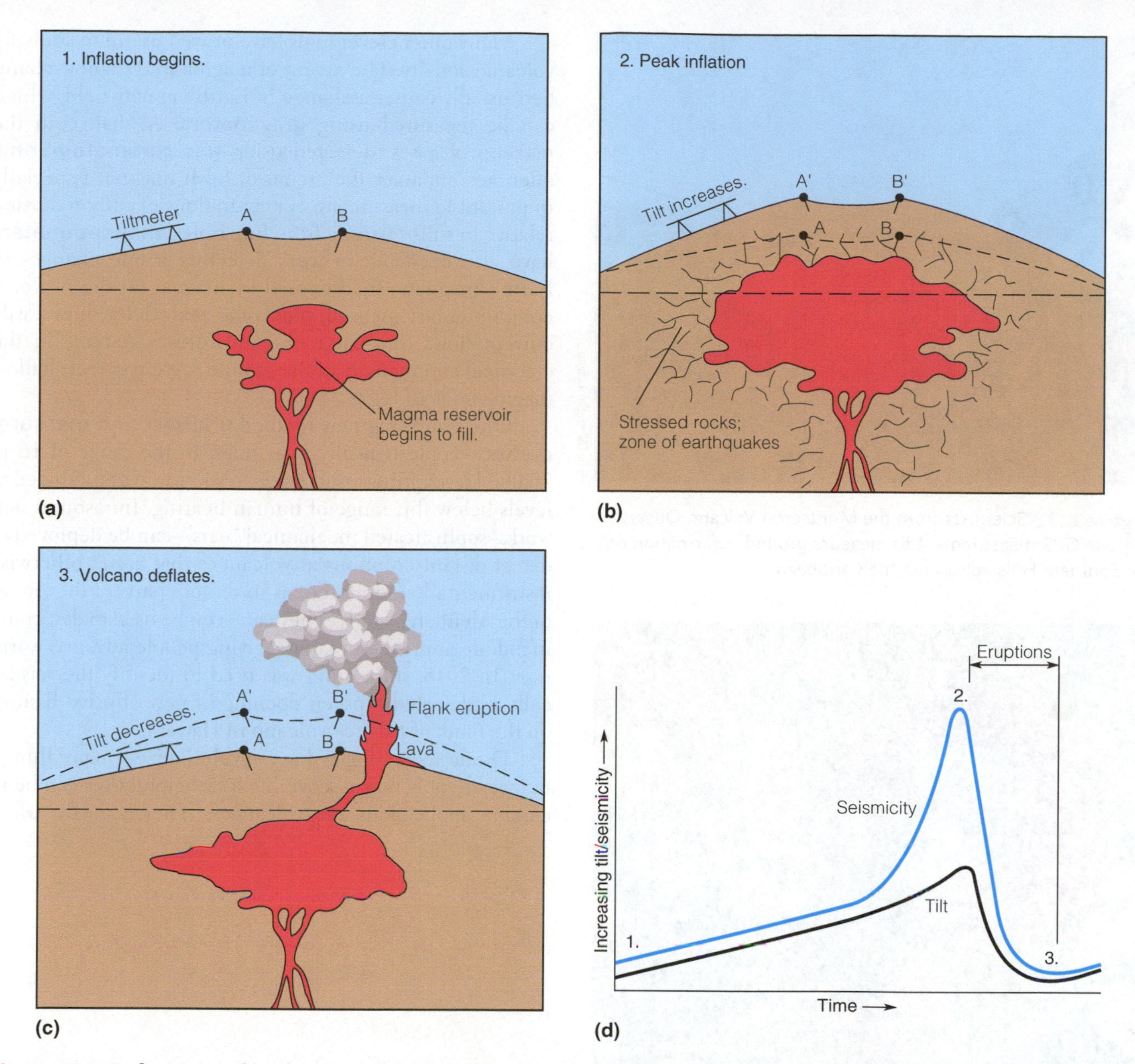

Figure 5.71 Deformation of a volcano is recorded by tiltmeters. They measure the angle of a volcano's slopes with great precision. Slopes steepen in the long run-up to an eruption as earthquake activity correspondingly increases, then sharply grow less steep (relax) immediately before and during the eruption. This pattern is caused by changes in the volume of magma in the underlying reservoir.

in angle you'd measure by lifting a kilometer-long board 1 millimeter at one end–not much.) Combined with earthquake activity, tilt is a powerful means of tracking the state of a volcano. Most tiltmeters, like most seismometers today, are automated, sending their signals by radio signal (telemetry) to distant observatory centers. This vastly improves the ability of volcanologists to observe a restless volcano over a large area safely and comprehensively.

Satellites have also been enlisted to help volcanologists. The Nimbus-7 and Meteor-3 satellites, using the moderate resolution imaging spectroradiometer (MODIS), have measured the total amount of gaseous aerosols injected into the stratosphere by large Plinian eruptions. This information is vital for understanding how such cataclysms impact global climate.

Global positioning system (GPS) stations linked to multiple satellites in orbit have revolutionized measurement of how a volcano changes its shape before an eruption, ever so slightly but tellingly. GPS monitors record with great precision not only positions on the surface, but also changes in elevation. Portable GPS stations are also easy to install on short notice at a volcano that is erupting or beginning to show restlessness (**Figure 5.72**).

Satellites using synthetic-aperture radars (SAR) can show how the shape of a volcano changes between successive orbits, even in areas that are too remote for ground-based instrumentation and observers. **Interferograms** (InSAR) constructed from these orbital data provide a map in broad color bands showing the deformation taking place in a volcano (**Figure 5.73**).

Hemis/Alamy Stock Photo

Figure 5.72 Scientists from the Montserrat Volcano Observatory use GPS measurement to measure ground deformation on the Soufriere Hills volcano in the Caribbean.

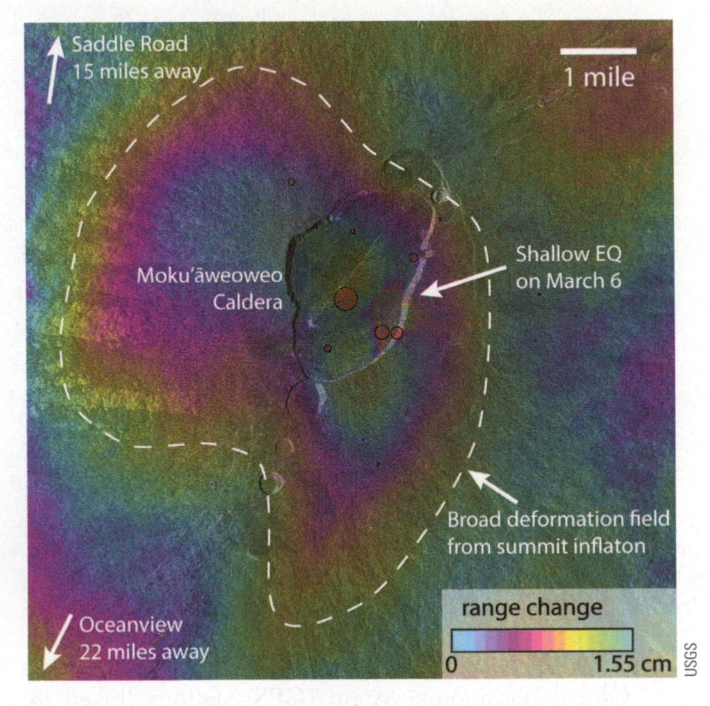

USGS

Figure 5.73 Satellite interferogram showing recent uplift at the summit of Mauna Loa Volcano.

Many other clever tools have proved useful in studying volcanic activity. The ascent of magma into shallow chambers usually causes a change in Earth's gravity field, which can be measured using **gravimeters.** A change in the makeup of gases, detected using **gas chromatography**, often accompanies the ascent of fresh magma. Especially important to measure are concentrations of carbon dioxide relative to sulfur issuing from fumaroles, **magnetometers** have also been put to work detecting subtle changes in Earth's magnetic field caused by magma on the move. A complimentary method, **electrical resistivity**, how easily current flows through rocks, measures variations in the electrical conductivity of the ground to help detect shallow magma bodies.

One promising new method is **infrasound measurements**. People typically hear noise in the range of 16 to 20,000 Hertz. Infrasound refers to soundwaves produced at levels below this range of human hearing. Infrasound networks–sophisticated mechanical "ears"–can be deployed to detect distant eruptions at volcanoes that aren't otherwise instrumentally monitored, as in remote parts of the Andes or the Aleutian Islands. They can even be used to detect the inaudible approach of lahars, giving people advance warnings. In 2018, infrasound was used to identify the seismically undetected, unseen opening of new effusive fissures on the flank of Kīlauea volcano in Hawaiʻi.

Drone technologies have revolutionized many things, but we are only now discovering how helpful they can be in certain applications on volcanoes. They can fly when helicopters are grounded, and hover long enough to collect valuable data about the speeds, directions, and development of active lava flows. They can carry cameras linked to satellites allowing observers to produce 3-D images of various volcanic features by merging hundreds of still photos, a process called **structure-from-motion** (**Figure 5.74**).

Urgent circumstances during an eruption also require that volcano scientists interact with concerned members of the public, the media, and government officials even if the scientists received little training in science communication. Volcanologists are the first people to whom governors, generals, the police, the public, media, and aid providers turn for advice on what to do next.

Volcanologists must be careful to convey information accurately and with consensus, so that they do not cause panic or costly, unwarranted flight. At the same time, they must determine the precise moment when it is important to urge civil authorities to organize evacuations.

Because few volcanologists are also educated in sociology, disaster management, economics, politics, and media studies, these are not easy calls to make. Experience matters greatly, with tremendous on-the-job-training. Statistics also play a vital role, enabling volcanologists to chart out pathways of projected volcanic behavior in terms of probabilities. These pathways are based on knowledge of past eruptions at an active volcano combined with current observations. They act as templates for guiding action as situations rapidly change. Throughout a period of crisis, the experts and authorities must always work with potentially life-threatening probabilities.

iStock.com/geyzer

Figure 5.74 A drone helps study a composite volcano on the Kamchatka Peninsula, Russia.

Case Study

5.1 New Zealand's Blast in the Past, A Ski Field, and Fluidization

The two islands of New Zealand are clearly distinguishable geologically. The North Island exhibits explosive volcanism and some of the most spectacular volcanic landforms in the world. The South Island lacks volcanic activity and is more alpine in nature. The North Island's volcanic zone includes Lake Taupo (**Figure 5.75**), the site of a cataclysmic Plinian eruption in 186 C.E. that was likely the largest eruption on Earth in 5,000 years (**Figure 5.76**).

Figure 5.75 Lake Taupo, New Zealand, fills a gigantic caldera. On the far shore in the distance are the volcanoes of Tongariro National Park.

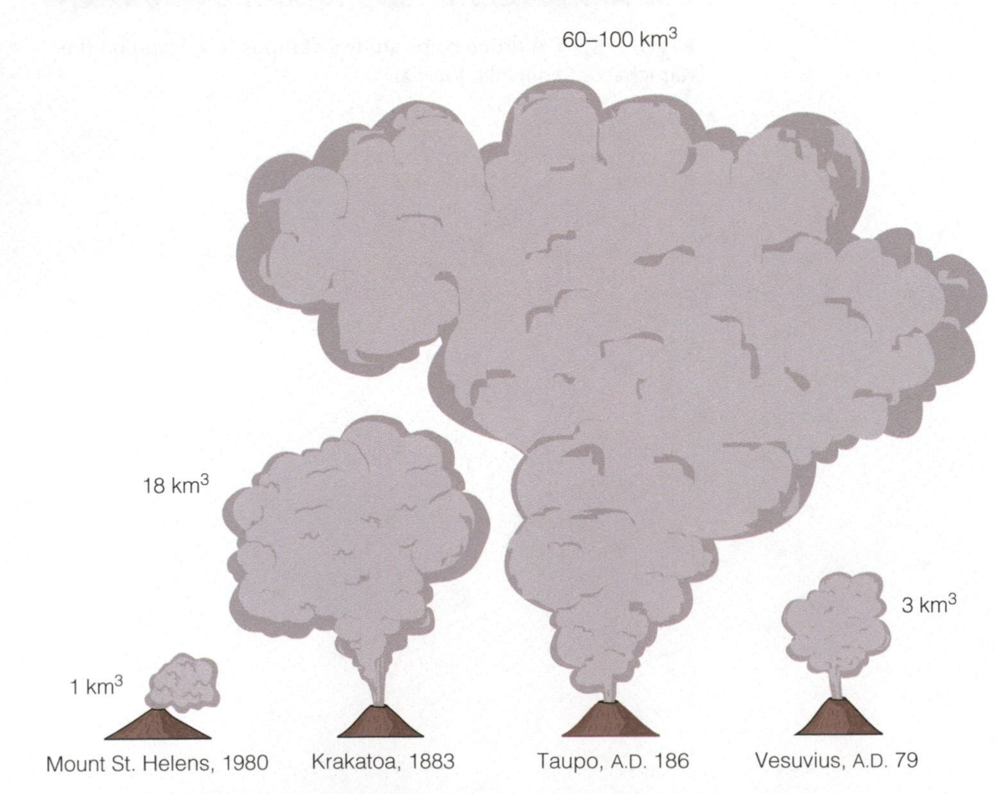

Figure 5.76 The relative sizes of four large eruptions of the past 2,000 years based on estimated volumes of ejected tephra.

A volcanic source, now beneath Lake Taupo, vented a column of hot gas, ash, and pumice more than 50 kilometers (30 miles) high, which spread downwind, forming light-yellow ashfall deposits more than 5 meters (16 feet) thick in some places. At 100 kilometers (60 miles) distance, the deposits are 25 centimeters (10 inches) thick; thick enough to cause ordinary roofs to cave in. Toward the end of the eruption sequence, the hot column became so heavy that it collapsed and flowed outward in all directions as a gigantic, pumice-rich pyroclastic flow. The speed of this moving cloud is estimated to 600 kilometers (375 miles) per hour, and the material traveled more than 80 kilometers (50 miles) from its source, overwhelming all the topography and vegetation in its path. Fortunately, the indigenous New Zealanders, the Maori, had not yet settled the island when the eruption occurred. When they did settle the island about 1000 C.E., they regarded the Tongariro-Taupo area as sacred. They believed that their gods and demons had brought volcanic fire with them from Hawaiki, the Maori's ancestral homeland.

How can a pyroclastic material move at such a phenomenal speed and travel such a great distance, seemingly defying friction and gravity? Some pyroclastic flows and surges are **fluidized**, meaning that the mix of high-pressure gas and fine particles within them moves like a giant liquid flood. Many chemical engineering professors demonstrate fluidization by forcing air through a powder on the bottom of a see-through box. As the air pressure is increased, the powder begins to quiver; then particles start dancing around, and eventually the whole mass is in motion. When the professor has it just right, they place a plastic duck (professor's choice) into the container, and it floats upon the air-solid mixture. If the duck is pushed down, it immediately pops up, proving that it is floating on a denser medium.

Scale this experiment up to that of a pyroclastic eruption, and you get the idea. When the pressure of the expanding gases ("forced air") in the fluidized flow is sufficiently great, the mass of hot gases, ash, and

Figure 5.77 Mount Ruapehu crater lake in spring time.

Figure 5.78 Ruapehu's June 1996 eruption buried the ski lifts in volcanic ash not snow - makes for some rough skiing!

pumice can overtop ridges in its way because the flow is spreading so fast.

At the south end of the Taupo volcanic zone is a cluster of composite volcanoes in Tongariro National Park. Two active volcanoes are popular playgrounds for New Zealanders, primarily for skiing and tramping (hiking). Mount Ruapehu, with a summit elevation of 2,797 meters (9,100 feet), is the highest point on North Island and has a lake in its summit crater (**Figure 5.77**). It is also the location of the Whakapapa ski field, a very popular, well-equipped ski area (**Figure 5.78**). The terrain at Whakapapa is bare, ragged lava and requires a lot of snow to be skiable. Throughout the ski field (or lava field, depending on the time of year), signs alert skiers and campers to stay on high ground and out of the canyons, should Ruapehu erupt, which it did in 1969, 1971, 1982, and sensationally in 1995 and 1996 and then again in 2006 and 2007.

The eruptions are **phreatic**; that is, superheated water at the bottom of the lake flashes to steam and overflows the crater. The lake's surface-water temperature, which ranges from 20°C to 60°C (68°F to 140°F), is monitored as a potential eruption predictor. When the water heats up, activity can be expected. A lahar warning system at Whakapapa is based on detection of earthquake swarms, which precede phreatic and magmatic eruptions. Tongariro (Maori tonga-, "south wind," and -tiro, "carried away") National Park is another case of humans both preserving and living wisely (but with some risk) within nature.

5.2 Carbon Dioxide, Earthquakes, and the Los Angeles Water Supply

The Eastern Sierra Nevada in California is a mecca for tourists who want beautiful scenery and outstanding recreation. Much of the area's appeal results from volcanic action. About 760,000 years ago, a huge eruption blew out 617 cubic kilometers (150 cubic miles) of magma that spread ash over nine states and as far to the east as present-day Nebraska (**Figure 5.79**). The surface above the magma chamber sank 2 kilometers (1.2 miles), forming the huge, oval caldera (16 × 33 kilometers; 10 × 20 miles) that is now called Long Valley. The eruption was 2,000 times the size of the 1980 Mount St. Helens eruption, and it left ash and pumice all over today's east-central California. The pyroclastic material was so hot as it accumulated that individual particles partly melded together, forming a hardened mass called **welded tuff** (**Figure 5.80**).

Volcanic activity in the region did not end with this event, however. Mammoth Mountain and Mono Craters, a linear chain of glassy domes stretching to the north, formed within the last 400,000 years (**Figure 5.81**). Steam-blast eruptions as recently as 620 years ago have taken place. The people of Long Valley are living with the caldera's continuing restlessness.

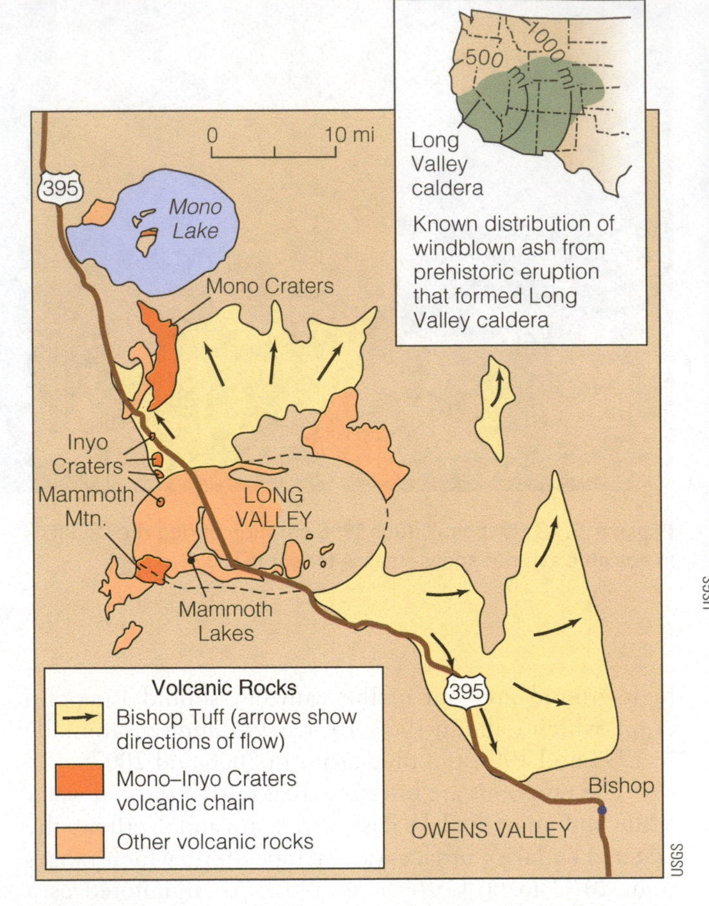

Figure 5.79 Mammoth Mountain and the Inyo-Mono Craters volcanic chain (red), California. The cataclysmic eruption 760,000 years ago that formed Long Valley ejected hot, glowing, pyroclastic flows, which cooled to form the Bishop Tuff (yellow). Inset map shows the distribution of Bishop Tuff and ash.

Figure 5.80 Outcrop of welded Bishop Tuff erupted from the Long Valley Caldera 760,000 years ago.

Figure 5.81 Mono Craters stretch in a line from near the shore of Mono Lake toward the Sierra Nevada.

The restlessness is manifest in hot springs, carbon dioxide (CO_2) emissions, and swarms of earthquakes, the most intense of which began in 1980 after two decades of quiescence. There were four strong, magnitude-6 earthquakes, three of them occurring on the same day. These quakes prompted studies by the USGS, which found uplift of 60 centimeters (about 2 feet) in the caldera center. Uplift, earthquake swarms, and gas emissions left little doubt that magma was rising and getting closer to the ground surface. Late in 1997, the volcano acted up again, with more than 8,000 earthquakes– most barely able to be felt, but a couple of them strong enough to catch people's attention. More than 1,000 earthquakes occurred one day, and many scientists believed the volcano was waving a red flag signaling danger. Again though, it did not erupt.

Much of the water for the city of Los Angeles comes from Sierra Nevada snowmelt and travels from Mammoth Lakes to the city via open aqueduct (see Chapter 8). Should the volcano erupt, it would jeopardize or totally cut off this resource on which the city depends. Furthermore, the area's main highway access is aligned in the direction of the prevailing wind from the mountains. For this reason, an "escape road" was built in the opposite direction, toward the north, to provide an alternate route for the area's residents and visitors.

Mammoth Mountain, a large rhyolitic volcano, is still hot. Tree-killing emissions of CO_2 were discovered at the base of the mountain in 1990. This indicated magma activity at shallow depth (**Figure 5.82**). By 1995, about 30 hectares (75 acres) of pine and fir trees had suffocated, and CO_2 concentrations of up to 90% of total gas were found in the trees' soil and root systems. Where CO_2 concentrations exceeded 30%, most of the trees died. One large campground had to be closed because park rangers found high levels of CO_2 in the restrooms and cabins after campers exhibited symptoms of asphyxia. It appears that CO_2 degases from the magma and migrates into a deep gas reservoir. When earthquake swarms cause fractures to develop in the reservoir rock, the gas, along with heat and sometimes hot water, finds its way to the surface (**Figure 5.83**). It is estimated that between 180 and 900 tonnes (200 and 1,000 tons) of CO_2 flow into the soil and atmosphere per day at Mammoth Mountain.

In 2017-2022, gas emissions declined, though this may simply have resulted from a 50% to 65% reduction in winter snowpack owing to drought conditions. The burden of winter snow and ice ordinarily helps "squeeze" stored carbon dioxide out of the volcano, illustrating a little studied link between climate and volcanic behavior. (Winter is the time of peak CO_2 emissions at Mammoth Mountain.) Beginning in 2017, when a regional drought began, there was a drop in overall CO_2 efflux by about 20%. Earthquake activity continued nevertheless: over 530 earthquakes shook the area between the spring of 2021 and the fall of 2022.

© B. Pipkin

Figure 5.82 Looking south through the Horseshoe Lake tree-kill area on the south side of Mammoth Mountain. The trees have died from high levels of carbon dioxide in the root zone.

Inyo National Forest/Jason D. Opliger

Figure 5.83 A small geyser of scalding water and gas erupts in a formerly popular bathing pool at Hot Creek, not far from Mammoth Mountain. Activity picked up here in the spring of 2006.

Photo Gallery Volcanic Wonders

Volcanoes present some of the strangest, most dramatic, most dynamic landscapes on Earth, as can be appreciated with these images.

1 Solfatara at the edge of an acidic volcanic lake, Kawah Ijen crater, Java.

Robert & Barbara Decker/USGS

2 Fossil fumaroles with incrustations of various iron oxides and other gas sublimate minerals, Valley of Ten Thousand Smokes, Katmai National Park, Alaska.

Adrienne Kettner/USGS

3 Hikers at the rim of Rinjani caldera on the island of Lombok, Indonesia, with the caldera lake below.

iStock.com/dislentev

Nyiragongo Lava Lake in the Democratic Republic of the Congo as night falls.

iStock.com/Olivier Isler

J. D. Griggs/Hawaiian Volcano Observatory/USGS

4 Call the automobile club!

5 Eroded basaltic lava flow probably less than 100,000 years old, with column-forming cooling fractures; Devil's Postpile National Monument, California.

USGS

Summary Volcanoes

1. Location and Composition of Magma and Volcanoes

LO1 Describe how the location and composition of volcanoes relates to plate tectonics

a. Plate Tectonics and Volcanoes

Most active volcanoes lie along plate boundaries, but not always. Some massive volcanoes form above hot spots, which have nothing to do with plate boundaries. Magma underlies volcanoes while lava flows from them, but magma and lava vary dramatically in different volcanoes.

b. Decompression Melting

Decompression melting is a process in which solid rock in the mantle and lower crust "melt" into a liquid state as the pressure of the overlying rock drops significantly. This process is common for volcanic activity in continental rift valleys, mid-ocean ridges, and sometimes hot spots.

c. Melting Related to Fluid Release

Partial melting can be triggered by the addition of fluids (volatiles) into deeply buried rocks, which decrease the rocks' melting point. Fluids released by subducting oceanic plates can trigger magma formation as they percolate up into the overlying mantle wedge. Fluids in the shallower crust also increase its chances of partial melting from heating caused by upwelling mantle-derived magma.

Study Questions

i. Explain the two mechanisms which facilitate partial melting.
ii. Where do volcanoes form and why?

2. Magma Ascent, Why Volcanoes Erupt

LO2 Explain why volcanoes erupt

a. Magma chambers form when less dense magma rises above denser mantle or crust. This occurs until the density of the rising magma and the host rock is the same; this is called the neutral buoyancy zone. The magma stalls out at this depth and forms a chamber. Eruptions are the result of drastic changes to the conditions in and around the magma chamber. Intrusion of fresh melt into magma chamber and ensuing gas buildup are likely responsible for most eruptions.

Study Questions

i. How do magma chambers form?
ii. Briefly explain two causes of volcanic eruptions.

3. The Products of Volcanic Eruptions

LO3 List the most common types of eruption products

a. Lava Flows

There are three main types of lava flows; Pāhoehoe, ʻaʻā, and block lava. Each differs in fluidity, temperature, viscosity, and movement.

b. Pyroclasts

Pyroclasts consist of fragmental volcanic debris including ash, dust, pumice, cinder, scoria, and old rock. Generally, any fragmental volcanic material that accumulates by falling out of the sky is called tephra.

Study Questions

i. Name and define the three major kinds of volcanoes resulting from eruptions.
ii. What are the three types of lava flows and how do they differ?

4. Classification of Eruptions

LO4 State the basis for how volcanic eruptions are classified

a. Hawaiian-Style Eruptions

This type of eruption is typical of oceanic hot spots as well as divergent plate boundaries, and they build enormous shield volcanoes. They typically begin with basaltic lava emerging from a fissure, which flows away sometimes quite rapidly. Erupting, lava sometimes spurts up into a towering lava jet or fountain, showering the surrounding area with Pele's hair and reticulite. Flood basalt eruptions resemble Hawaiian-style, but occur at a larger scale.

b. Strombolian Eruptions

Consisting of loud, frequent, pulsating bursts of molten lava high above the rim of the crater, Strombolian eruptions are the smallest magnitude primarily explosive eruption type. This type of eruption is likely the result of a giant bubble of gas that forms in the magma conduit until enough pressure is built up to cause an explosion.

c. Vulcanian Eruption

This highly violent type of explosive eruption can last up to 12 hours at a time, producing pyroclastic material such as volcanic ash clouds which develop strong electrostatic charges. A similar type of eruption is the Peleéan eruption, which involve pyroclastic flows released from the summit of a composite cone capped by a plug of hardened lava.

d. Plinian Eruptions

This highly explosive type of eruption releases large amounts of volcanic gas and pyroclastic material into the Earth's lower stratosphere. Eruptions consist of an initial phase of powerful gas release, which grows into an eruption cloud that spreads over several days. Large amounts of pumice drop from these umbrella clouds. The column cloud eventually collapses, and enormous pyroclastic flows surround the landscape. The infamous Mount Vesuvius eruption that buried Pompeii was Plinian eruption.

e. VEI Classification

Volcanic eruptions can also be categorized according to their Volcano Explosivity Index (VEI), with "0" being essentially nonexplosive (effusive), and 8 or 9 being most explosive. (The scale is open-ended.) The VEI of an eruption depends primarily upon the volume of material explosively erupted, the height of its eruption cloud (column), the duration of blasts, and whether the stratosphere is directly impacted.

Study Questions

i. What type of eruption at Mount Vesuvius destroyed Pompeii, why was it so devastating, and why is it so well studied today?

ii. Describe the two different types of volcanic eruptions and their unique or distinguishing characteristics.

5. Volcanic Hazards, Risks, and Benefits

LO5 Define the difference between volcanic hazards, risks, and benefits

a. The Differences Between Hazards and Risks

Hazards are phenomena that could potentially harm people. Risks are the actual chances of being harmed, typically measured as probabilities over certain periods of time. Volcanic risk is a function of hazard, value of what is threatened, and vulnerability of a population of people or possessions.

b. Typical Eruption Hazards and What to Do About Them

Lava flows, volcanic gases, falling ash, pyroclastic flows and surges, and tsunamis all present risks and hazards to communities impacted by volcanic activity. The best way to address these hazards and reduce risks is to monitor volcanic activity, stay aware, and use your best judgment to avoid areas impacted by volcanic eruptions.

c. Benefits of Volcanoes

Volcanic eruptions build lands, drive geothermal activity used for geothermal power, and are key to the development of ore deposits. Volcanic ash supplies most major nutrients that agricultural crops need to thrive, making volcanic soil great for farming, and volcanic ash can benefit nutrient-depleted soils.

Study Questions

i. How is volcanic risk measured and understood? How does this differ from volcanic hazards?

ii. What are the three benefits of volcanoes and volcanic activity and why are they beneficial?

6. Active? Dormant? Extinct?

LO6 Describe how geologists define active, dormant, and extinct volcanoes

a. Active Volcanoes

An active volcano is one that has erupted frequently in historic time and will almost certainly continue to do so in the future.

b. Dormant Volcanoes

A dormant volcano shows signs of having erupted within the past few hundred or few thousand years but has not erupted in living memory.

c. Extinct Volcanoes

An extinct volcano is deeply eroded because of non-eruption for thousands or millions of years.

Study Questions

i. Explain the differences between a dormant and extinct volcano.

7. Tools of the Trade, Monitoring Technologies

LO7 Provide the examples of how volcanologists use modern technology to monitor active volcanoes

a. Prediction

Volcano scientists study the historical and modern behaviors of individual volcanoes, enabling them to provide probabilities of eruptions taking place.

Seismometers can be used to help predict when an eruption is likely. Seismologists look for VT-earthquakes and LP-earthquakes to help determine when a volcano may erupt because of seismic activity. Other instruments used for predicting the probably of an earthquake include gravimeters, infrasound measurements, GPS, tiltmeters, and drones.

Study Questions

i. Choose two of the instruments mentioned in this chapter and explain how they are used in volcanic eruption prediction.

Key Terms

'a'ā
Active volcano
Andesite
Andosol
Block lava
Blocks
Bombs
Bread-crust structure
Caldera
Cinder
Cinder cones
Clinkers
Cluster
Columnar joining
Composite
Conduit
Decompression melting
Directed lateral blast
Dormant volcano
Earthquake swarm
Effusive
Electrical resistivity
Electrostatic charges
Eruption column
Explosive eruptions
Extinct volcano
Flood basalt
Fluidized
Fluids
Fluorosis
Fournier d'Albe equation
Fumaroles
Gas chromatography
Geothermal activity
Gravimeters
Hazards
Hot spots
Infrasound measurements
Interferograms
Kīpuka
Lahars
Lapilli
Lava
Lava caves
Lava lakes
Lava tubes
Limnic eruption
Lp-earthquakes
Magma
Magma chambers
Magnetometers
Maria
Melting points
Monogenetic
Neutral buoyancy zone (NBZ)
Nuées ardentes
Outer core
Pele's hair
Peléean eruptions
Pāhoehoe
Phreatic
Plinian eruptions
Polygenetic
Pumice
Pyroclastic flows
Pyroclastic surges
Reticulite
Rhyolite
Risks
Scoria
Sector collapse
Seismometer

Semidiurnal
Shield volcanoes
Soil efflux
Solid earth tide
Stratovolcanoes
Strombolian eruptions
Structure-from-motion
Sublimates
Tephra
Tiltmeter
Tsunami
Viscosity
Volatiles
Volcanic arc
Volcanic ash
Volcanic bombs
Volcanic tremor
VT-earthquakes
Vulcanian ash clouds
Vulcanian eruptions
Welded tuff

"At a primal level, we abhor randomness because it leaves us vulnerable."

—Dr. Lucy Jones, seismologist, California Institute of Technology

6 Earthquakes and People

Learning Objectives

LO1 Describe why and how earthquakes take place

LO2 Detail how the strength of an earthquake is measured

LO3 Identify some important ways that earthquakes can damage or destroy buildings and harm people

LO4 Explain the most important lessons that we've learned from past earthquake disasters

LO5 Discuss what we can do to forecast and prepare for future earthquakes

This freeway was well built but the ground beneath gave way in strong shaking from the 1995 Kobe, Japan, earthquake.

Pacific Press Service/Alamy Stock Photo

A Caribbean Catastrophe

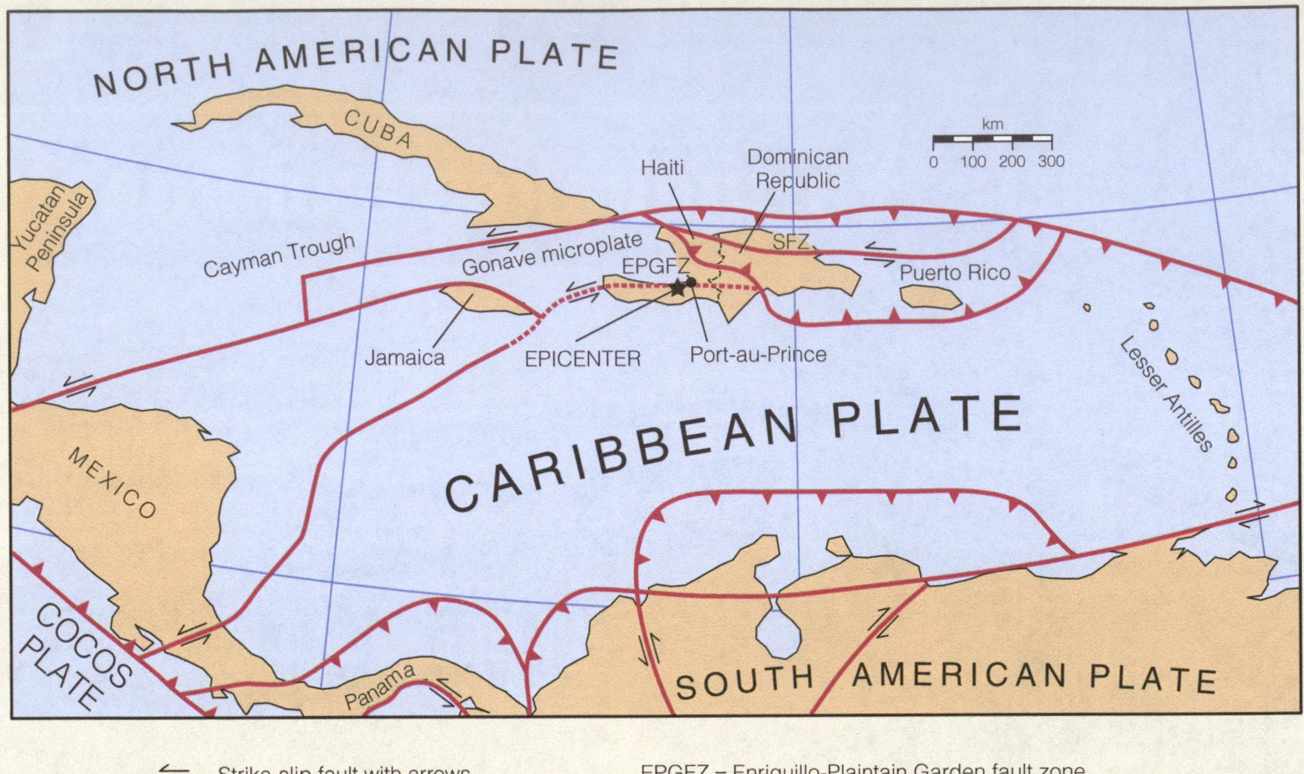

Figure 6.1 Generalized tectonic map of the northern Caribbean region.

The tropical Island of Hispañiola is the second-largest island in the West Indies (after Cuba), and most populous. Two countries share the island; the Dominican Republic to the east, Haiti to the west. Hispaniola is collage of many different rock types and pieces of crust uplifted at a transform plate boundary, along which the North American Plate slips past the Caribbean Plate at a rate of around 20 millimeters (0.8 inches) a year (**Figure 6.1**). As such, the island experiences many earthquakes. High levels of poverty and political instability, especially in Haiti, mean that larger earthquakes typically cause extensive damage and cost many lives.

On January 12, 2010, a magnitude (M) 7.0 earthquake suddenly struck along a completely buried ("blind") fault only about 15 kilometers (9 miles) south of the national capital, Port-au-Prince, at the base of the narrow, 200-kilometer (120-mile)-long Tiburon Peninsula. The earthquake was shallow; only about 10 kilometers (6 miles) deep. Given its proximity to the surface, location along many slopes composed of landslide-prone rock and soil, the weak standards of building construction, and insufficient infrastructure and emergency services, an extraordinary number of people died; greater than 300,000 according to one official estimate! Had this earthquake occurred in a better resourced country, the physical damage would still have been great, but fatalities would much likely have been much fewer (**Figure 6.2**).

Many of Port-au-Prince's 2 million residents lived in homes that cling precariously to steep hillsides. Deep valleys, some of which had been packed with homes, were swept bare or buried by quake-triggered landslides; many of those

Figure 6.2 Residents walk amidst the ruins of Port-au-Prince, the capital of Haiti, just a few days following its destruction by a large earthquake in 2010.

USGS

Figure 6.3 Three large landslides triggered by the 2010 Haitian earthquake dam the same river; if the dams were to break suddenly, they could endanger people living alongside the river downstream.

AP Images/Matias Delacroix

Figure 6.5 Just days after the 2021 earthquake, Haitian homes lay ruined in the foreground. In the background are landslides triggered by the heavy rains of Tropical Depression Grace that followed soon after the quake.

landslides dammed rivers, threatening people who lived downstream (**Figure 6.3**). The parliament building and part of the magnificent Presidential Palace collapsed. Tens of thousands of houses, schools, the national cathedral, hospitals, and the regional headquarters of the United Nations crumbled within a few deadly seconds. Damage to the city's main jail allowed prisoners to escape. Informal tent camps sprung up to house those whose homes had been destroyed (**Figure 6.4**).

The shocked world responded quickly. Within the first 3 years following the quake, $3.5 billion of assistance arrived, the United States being the biggest contributor. Slowly, Haiti rebuilt, albeit through a deadly cholera epidemic and political unrest.

Joe Raedle/Getty Images News/Getty Images

Figure 6.4 Tent city sprawls across a former golf course after the 2010 earthquake devastated Port-au-Prince, the Haitian capital.

Social media (for those with access to the Internet) and text messaging played important roles bringing people back together.

On August 14, 2021, another, even stronger earthquake struck (M 7.2), centered along a different fault system, 150 kilometers (90 miles) to the west. It was a complex event; one fault gave way, quickly followed by two others close by—a cascading rupture that merged into a single terrifying impulse of shaking. Thanks to a lower regional population density, fewer residents died this time. The earthquake was still shallow enough (10 kilometers deep), however, to be quite damaging; 2,500 people lost their lives and 12,200 were injured. Six hundred and fifty thousand surviving residents required assistance almost all at once. The beautiful, historic city of Les Cayes, Haiti's third largest, toppled into ruins. Four days later, Tropical Depression Grace, soon to become a powerful hurricane, moved over the quake-stricken area, dumping heavy rain (13 to 25 centimeters, or 5 to 10 inches), flooding tent camps, and triggering additional landslides (**Figure 6.5**).

Worldwide, M 7 earthquakes are not uncommon: about 20 quakes of this magnitude somewhere in the world each year. Haiti is not uniquely cursed by their occurrence, but recent events show that its national population remains poorly prepared to live with quakes. Many other parts of the world face similar challenges.

This chapter explores the relation of faults and plate tectonics to earthquakes and **seismicity** (ground shaking). We examine the causes of building collapse and the remedial measures that can be taken to avoid the costly devastation that accompanies large earthquakes. There is still much to learn, and any knowledge we gain and usefully apply (often beginning with abstract academic theory) can mean life or death for many people.

questions to ponder

1. What steps would you try to take as a local Haitian resident of limited means to reduce your risk from future earthquakes? (Think of the aftermath as well as the risk imposed by the event itself).
2. What can the government of a lesser-developed state such as Haiti do to prepare and better protect its citizens from severe earthquake impacts?

6.1 The Nature of Earthquakes (LO1)

Faults are fractures in Earth's crust along which great masses of rock slide past one another either so gradually that it is imperceptible, or more often, in violent, sudden jerks we call earthquakes. This movement leads to **displacement** (the offset of rock along the plane of rupture, which is termed the **fault plane**). The types of faults and the Earth forces that cause them are shown in **Figure 6.6**. The type of fault depends upon the conditions of stress present in the crust. When forces compress the crust to the point where rocks break, one mass of the crust slides up over another along a **reverse fault**. Under extreme compressive conditions, the angle of the reverse fault may be quite gentle; 15° inclination or less. Such "low-angle" reverse faults are termed **thrust faults** (**Figure 6.7**). In contrast, if crust stretches, or extends, beyond the breaking point, **normal faults** develop (**Figure 6.8**). The crust on one side of the normal fault slips down across the surface of the crust on the other side. Where shearing simply shifts one mass of crust past another in a horizontal direction, the resulting rupture is termed a **strike-slip fault**. Strike-slip faults may be classified as **right-lateral** if the crust viewed on the opposite side of a fault trace is displaced to the right relative to features on the side where the viewer is standing, or **left-lateral** otherwise (**Figure 6.9**).

Most earthquake activity is a result of plate tectonic interactions (see Chapter 2). Normal faults characterize earthquake ruptures at divergent plate boundaries, while reverse and thrust faults typify convergent plate boundaries. Strike-slip faults can also be found at these boundaries but are by far the most common fault types associated with transform plate interactions.

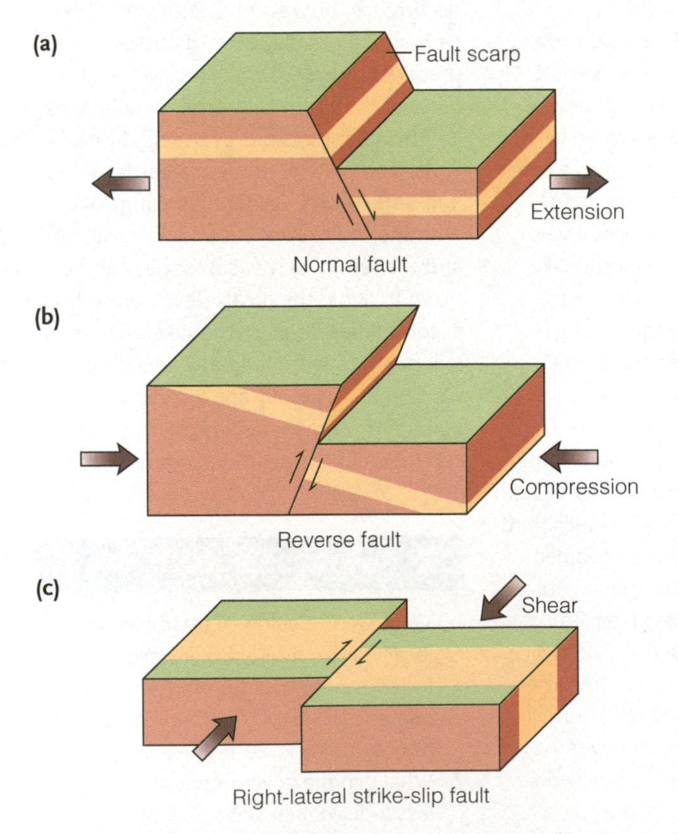

Figure 6.6 Schematic diagram of fault geometries including normal, reverse and strike slip.

Figure 6.7 A thrust fault, with fault trace highlighted in red. A rock hammer approximately 30 centimeters (1 foot) long is placed on the outcrop to provide scale. Note the gentle inclination of the fault trace, sloping off toward the left.

Figure 6.8 A normal fault offsets layers of sedimentary strata near Moab, Utah.

Figure 6.9 A plowed field displaced by a right-lateral strike-slip fault following an earthquake in the Imperial Valley, California, in 1979.

Figure 6.10 Most of downtown San Francisco burnt to the ground in the days immediately following the April 18, 1906, earthquake. Fatality estimates for the city and outlying areas range between 700 and 3,400 persons, many of whom may have died in this uncontrollable fire because water mains had been destroyed in the initial shaking.

Because the buildup of **stress** (force per unit area) that produces **strain** (deformation) in rocks can be transmitted long distances through the crust, an active fault does not necessarily occur exactly at a plate boundary, but it generally occurs in the vicinity (within a few hundred kilometers) of one. The mechanism by which stressed rocks store up strain energy along a fault to produce an earthquake was first explained by Harold E. Reid in a report to the State of California after the great San Francisco earthquake of 1906, which together with an ensuing fire, destroyed about 20% of the City by the Bay (**Figure 6.10**).

Reid proposed the **elastic rebound theory**, which states that when sufficient strain energy has accumulated in rocks, they may rupture rapidly–just as a rubber band breaks when it is stretched too far (**Figure 6.11**). The stored strain energy is released as vibrations that radiate outward in all directions (think of the shock that passes through your fingers when you release that rubber band). Most earthquakes are generated by movements on faults within the crust that do not produce ruptures at the ground surface. A fault rarely slips entirely during an earthquake, but the area of interface that does can cover an area of many square kilometers underground, called the **rupture zone**.

By studying the seismic energy released by an earthquake, we can pinpoint within the Earth where the rupture starts, a position called the **focus**, or **hypocenter** (**Figure 6.12**). Even easier to locate is the **epicenter**, or the location where the seismic energy first reaches Earth's surface and begins spreading outward across the landscape. **Seismology**, the branch of geology dedicated to studying seismic waves and energy release, allows us to determine the direction faults slip during earthquakes. The "how" of fault rupture is termed an earthquake's **focal mechanism**.

Because rock tends to flow plastically (think of Silly Putty) rather than accumulate elastic strain energy under the high-temperature and pressure conditions of Earth's deep interior (Chapter 2), most earthquake foci lie within the crust at depths of less than 15 kilometers (9 miles). The deepest foci in the world occur within subducted oceanic

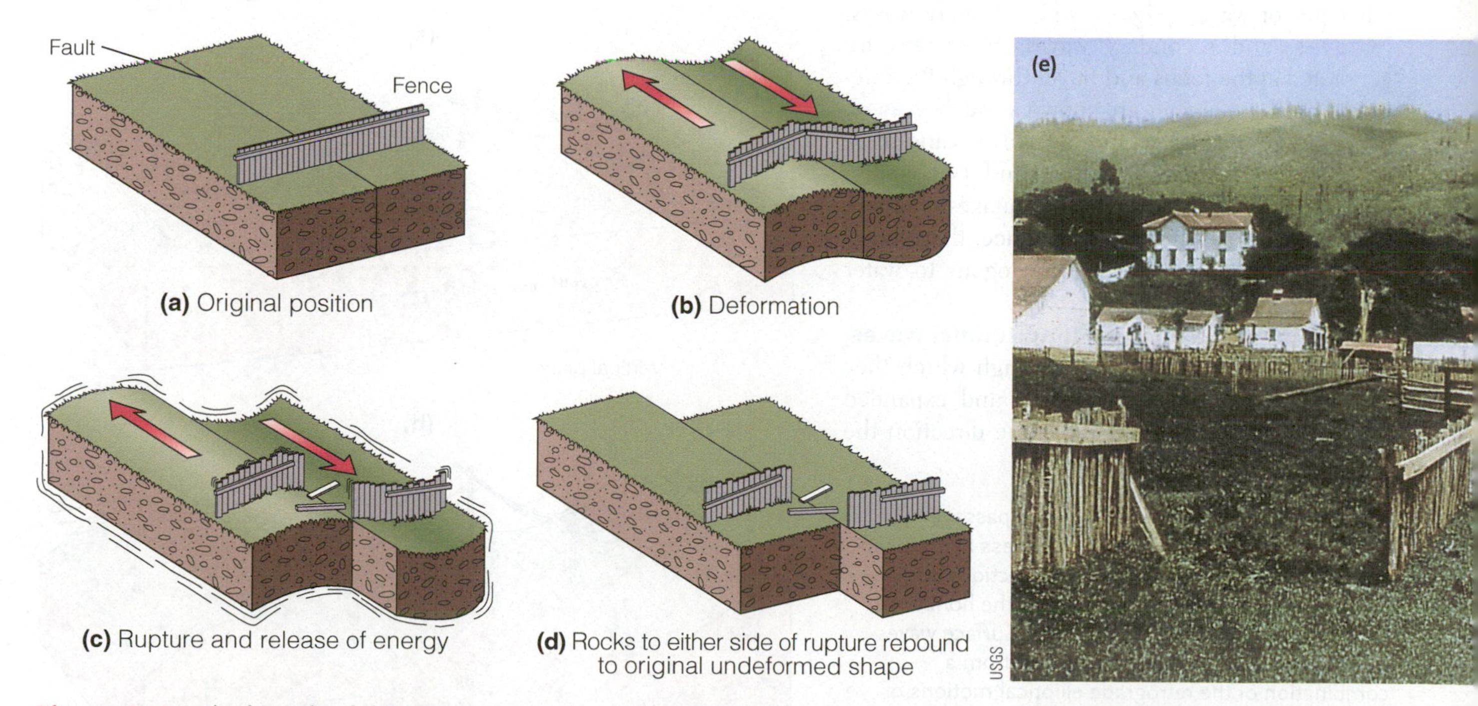

Figure 6.11 (a-d) The cycle of elastic-strain buildup and release for a right-lateral strike-slip fault according to Reid's elastic rebound theory of earthquakes. At the instant of rupture (c), energy is released in the form of earthquake waves that radiate out in all directions. (e) Right-lateral offset of a fence by 2.5 meters (8 feet) by displacement on the San Andreas Fault in 1906; Marin County, California.

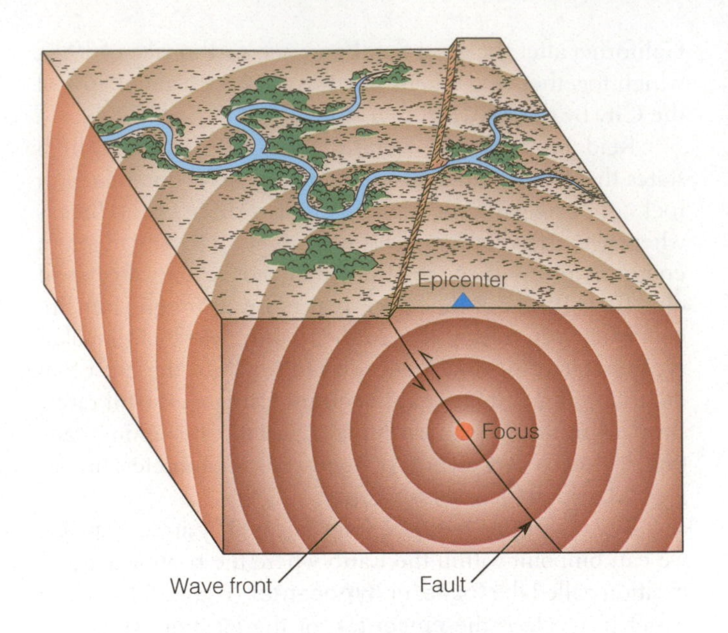

Figure 6.12 The focus of an earthquake is the point in the Earth where a fault rupture begins. The epicenter is generally, though not always, directly above the focus at the ground surface. Seismic-wave energy moves out in all directions from the focus.

slabs, which rupture as they heat up and sink into the mantle. **Subduction zone earthquakes** may originate at depths of as great as 700 kilometers (450 miles). These very deep earthquakes also tend to be more widely felt than shallower ones, though they rarely are very damaging because much of their seismic energy dissipates before reaching the surface.

6.1.1 Types of Seismic Energy Waves

The vibration produced by an earthquake is complex but can be described as three distinctly different types of waves (**Figure 6.13**). Primary waves, **P-waves**, and secondary waves, **S-waves**, are generated at the focus and travel through the interior of Earth; thus, they are known as **body waves**. They are designated P- and S-waves because they are the first (primary) and second (secondary) waves to arrive from distant earthquakes. As these body waves strike the planet's surface, they generate **surface waves** that are analogous to water ripples on a pond.

P-waves are examples of **longitudinal waves**; the solids, liquids, and gases through which they travel are alternately compressed and expanded ("pushed and pulled") in the same direction the waves move. Their speed depends on the resistance to change in volume (termed **incompressibility**) and compactness of the material through which they travel. Sound is a nonseismic equivalent of P-waves that people have evolved to sense with their ears. P-waves travel about 300 meters (1,000 feet) per second in air, 300 to 1,000 meters (1,000–3,000 feet) per second in soil, and typically around 6 kilometers (3.5 miles) per second in solid rock near the surface.

P-wave speed increases with depth because the materials composing the mantle and core become less compressible with distance inside the Earth. This stiffening of Earth's deep interior causes the waves' travel paths to bow downward as they move through the planet. Where they breach the ground surface, they have very small amplitudes (push-and-pull motion) and cause little property

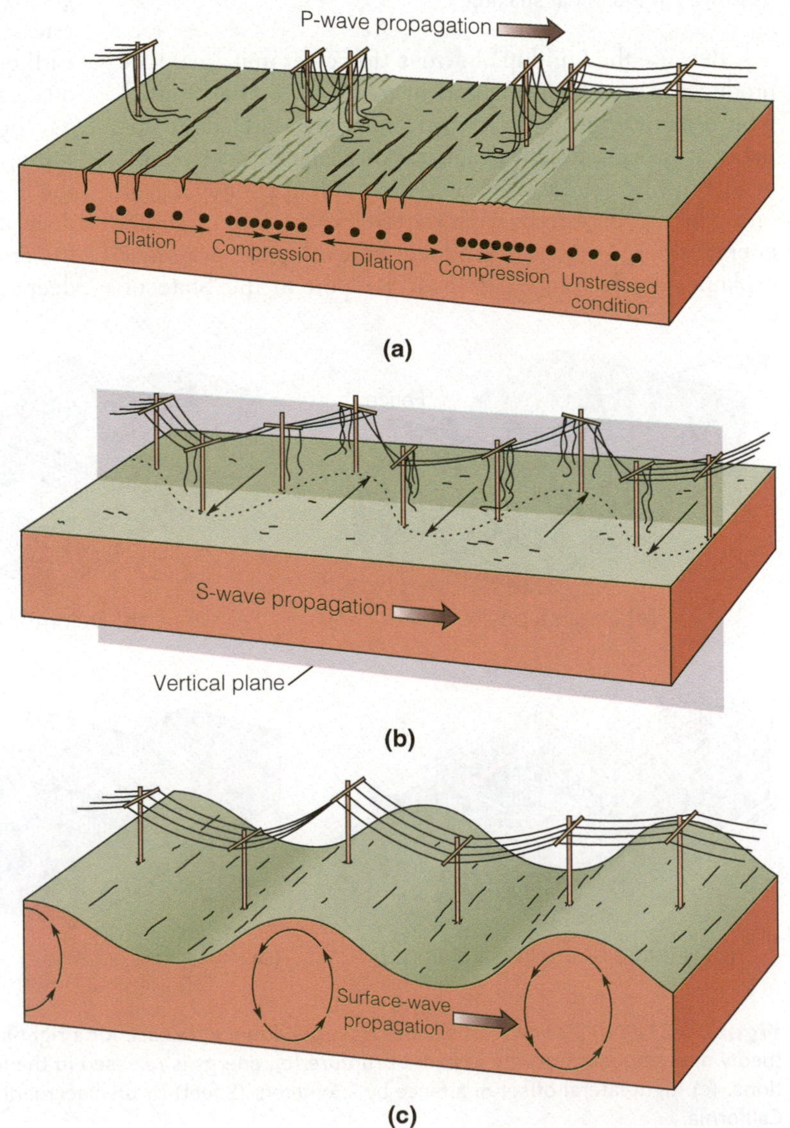

Figure 6.13 Ground motion during passage of earthquake waves. (a) P-waves compress and expand the Earth. (b) S-waves move in all directions perpendicular to the wave advance, but only the horizontal motion is shown in this diagram. (c) Surface waves create surface undulations that result from a combination of the retrograde elliptical motions of Rayleigh waves (shown) and Love waves, which move from side to side at right angles to the direction of wave propagation. (The S-waves shown in (b) are behaving as Love waves).

damage. Although they are physically identical to sound waves, they vibrate at frequencies below what the human ear can detect. The earthquake noise ("rumbling") that has been reported could be P-waves of a slightly higher frequency or other, related vibrations whose frequencies are in the audible range–at least for humans.

S-waves (see Figure 6.13) are **transverse (shear) waves**. They produce ground motion perpendicular to their direction of travel. This causes the rocks through which they pass to be twisted and sheared. These waves move much like energy does when a person takes one end of a garden hose or a loosely hanging rope and gives it a vigorous flip. They can travel only through a medium that resists shearing, the rupturing of a given mass into two parts that then slide past one another, each moving in the opposite direction.

S-waves travel only through solids because liquids and gases can't be "ruptured" and have no shear strength (just try piling water in a mound). This means that S-waves cannot move through the liquid outer core, creating a "shadow" and allowing seismologists to learn more about the structure of the deep Earth without ever taking a field trip there (see Chapter 2). Surficial S-wave ground motion may be largely in the horizontal plane and can result in considerable property damage. S-wave speed is typically around 4 kilometers (2.5 miles) per second from distant earthquakes; hence they arrive after the P-waves, with rare exception.

Unexpended P- and S-wave energy bouncing off Earth's surface generates a set of **surface waves** that cause a lot of damage. These waves produce a rolling motion in the ground. In fact, slight motion sickness is a common response to surface waves of long duration. Two major categories of surface waves are recorded by seismographs. **Love waves** are the most important. They exhibit horizontal motion normal to the direction of travel. **Rayleigh waves** exhibit retrograde (opposite to the direction of travel) elliptical motion in a plane perpendicular to the ground surface (see Figure 6.13 c). Surface waves have amplitudes that are greatest in near-surface unconsolidated layers and are the most destructive earthquake waves.

Love and Rayleigh waves also travel quite quickly through bedrock–in excess of 2 kilometers per second (1.2 miles per second), although not as fast as P and S waves. (The acronym to remember in terms of relative seismic wave speeds, from fastest to slowest, is "PSLR," pronounced "pleaser"). Their speeds, like those of body waves, vary according to the composition of the material through which they are moving, with some components traveling slowly enough in poorly consolidated materials to be easily visible during earthquakes–below the threshold of seismometer detection. Many observers (including your author Hazlett) have witnessed unforgettably creepy, wavelike motions rolling down streets and sidewalks once earthquakes are under way. During the 1964 Alaska earthquake, observers reported visible waves roughly a meter (yard) high that traveled through the town of Valdez, causing cracks to open and close with the passage of each wave. Jets of compressed groundwater squirted as high as 6 meters (20 feet) into the air as some fissures shut tight (**Figure 6.14**).

Figure 6.14 Large ground crack showing right-lateral strike-slip offset in highway lane dividing line, opened during the 2016 Kaikoura earthquake, New Zealand.

6.1.2 Seismic Wave Periods

Energy moving through any medium, be it water, magma, or solid earth, moves in pulses of force, or waves. You can easily visualize this thinking about or watching waves at sea approach a shoreline. **Wave period** is the time it takes for two successive wave crests (or troughs) to pass a particular point; for example, a buoy floating in the water (see Chapter 10). **Wave frequency** is a related concept referring to the number of wave crests, or troughs, that pass a particular point per unit interval of time (**Figure 6.15**).

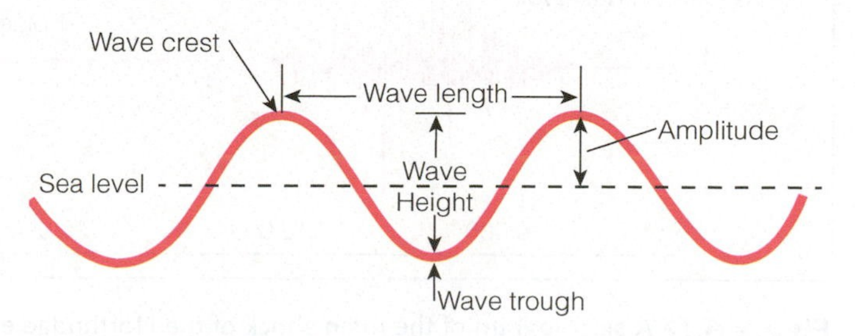

Figure 6.15 Labelled wave diagram with wavelength (period), amplitude, crest, trough, and wave height.

Figure 6.16 Seismological monitoring station on Mount Vesuvius, Italy, with wireless transmission so that scientists know immediately if an earthquake has occurred; perhaps related here to magma movement within or under the volcano.

The terms wave period and wave frequency apply to seismic energy, too, though the energy waves generated by earthquakes, of course, travel much faster than those at sea and generally have a much stronger punch at any given locality, as people who have experienced them can attest.

Seismic waves may be classified as **short** and **long period** (the LP earthquakes mentioned in Chapter 5). The shorter the period, the stronger the overall energy. As you might expect near the violent focus of an earthquake, short-periods predominate. Seismic energy dissipates with distance from source for many reasons, not the least being that it is expanding over a growing volume of Earth as seismic waves propagate. The energy may be absorbed by solid rock masses, bounced back toward the source from the boundaries between various rock layers, even dissipated to the atmosphere as sound waves and the oceans in the form of tsunami (described later). At greater distances, however, with diminishing overall energy, only longer seismic wavelengths and periods persist. The very-long-period seismic waves that pass all the way through the planet from big earthquakes to seismometers on the other side are called **teleseisms**. Arrival of teleseismic waves at a station signals to seismologists that a potentially disastrous earthquake has occurred somewhere far distant–at least hundreds and possibly thousands of kilometers away.

6.1.3 Locating the Epicenter of an Earthquake

A **seismometer** is an instrument designed specifically to detect vibrations in the Earth's crust (**Figure 6.16**). The actual record and measurement of those vibrations is made using a related device called a **seismograph** (see Chapter 5, Figure 5.70). Many seismometers are linked to seismographs using radio, wireless, or cell phones and thus can transmit critical data in real time to warn of earthquakes just as they begin. The actual record of shaking is a termed a **seismogram** that even in our digital age may be printed for initial interpretation and monitoring (**Figure 6.17**). Because seismometers are extremely sensitive to vibrations of any kind, they tend to be installed in quiet areas away from heavily trafficked roads and industrial centers, such as abandoned oil and water wells, cemeteries, and parks.

When an earthquake occurs, the distance to its epicenter can be estimated by computing the difference in P- and S-wave arrival times at various seismometers. Although the method used by seismologists is more accurate and determines the depth and location of the quake's focus and its epicenter, what is described here shows how seismic-wave arrival times can be used to determine the distance to an epicenter. Because it is known that the two kinds of waves are generated simultaneously at the earthquake focus and that P-waves travel faster than S-waves, it is possible to calculate where the waves started.

Imagine that two trains leave a station at the same time, one traveling at 60 kilometers (36 miles) per hour and the other at 30 kilometers (18 miles) per hour, and that the second train passes your house one hour after the first train goes by. If you know their speeds, you can readily calculate your distance from the train station of origin as 60 kilometers (36 miles). The relationship between distance to an epicenter and arrival times of P- and S-waves is illustrated in **Figure 6.18**. The epicenter will lie somewhere on a circle whose center is the seismometer and whose radius is the distance from the seismometer to the epicenter. The problem is then to determine where on the circumference of the circle the epicenter is. More data are needed to learn this–specifically, the

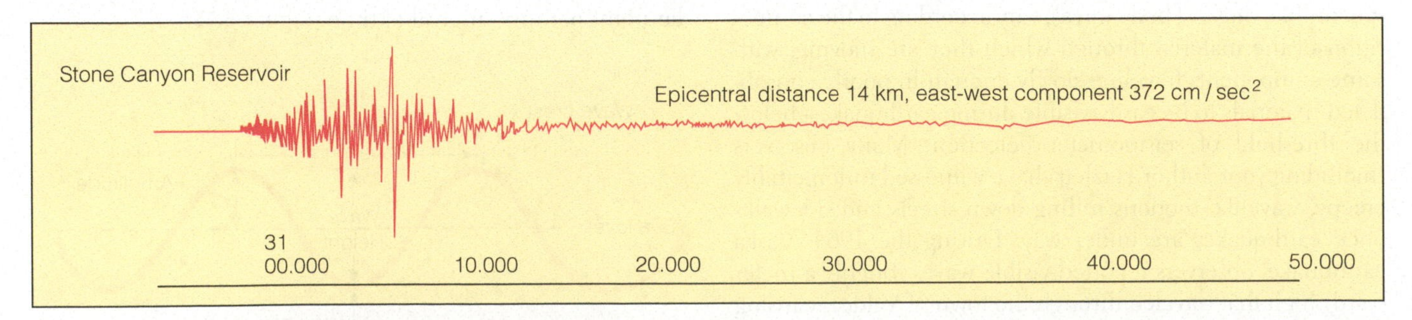

Figure 6.17 A seismogram of the main shock of the Northridge earthquake, January 17, 1994. The time in minutes and seconds after 4:00 a.m. appears at the bottom. Distance from the epicenter and direction and ground acceleration at the recording site are also indicated. (Acceleration is explained in the next section.)

distances to the epicenter from two other seismometers. Then the intersection of the circles drawn around each of the three stations–a method of map location called **trilateration**–specifies the epicenter of the earthquake (**Figure 6.19**).

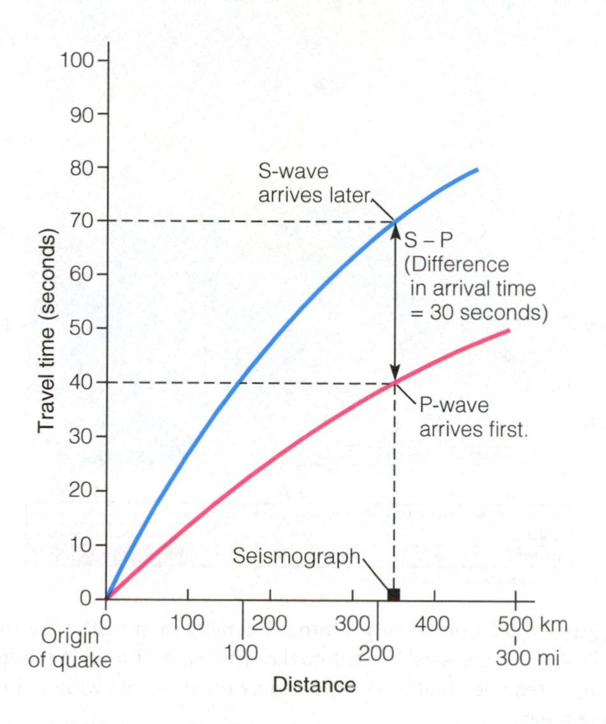

Figure 6.18 Generalized graph of distance versus travel time for P- and S-waves. Note that the P-wave has traveled farther from the origin of the earthquake than the S-wave at any given elapsed time. The curvature of both wave paths is due to increases in velocity with depth (distance from epicenter).

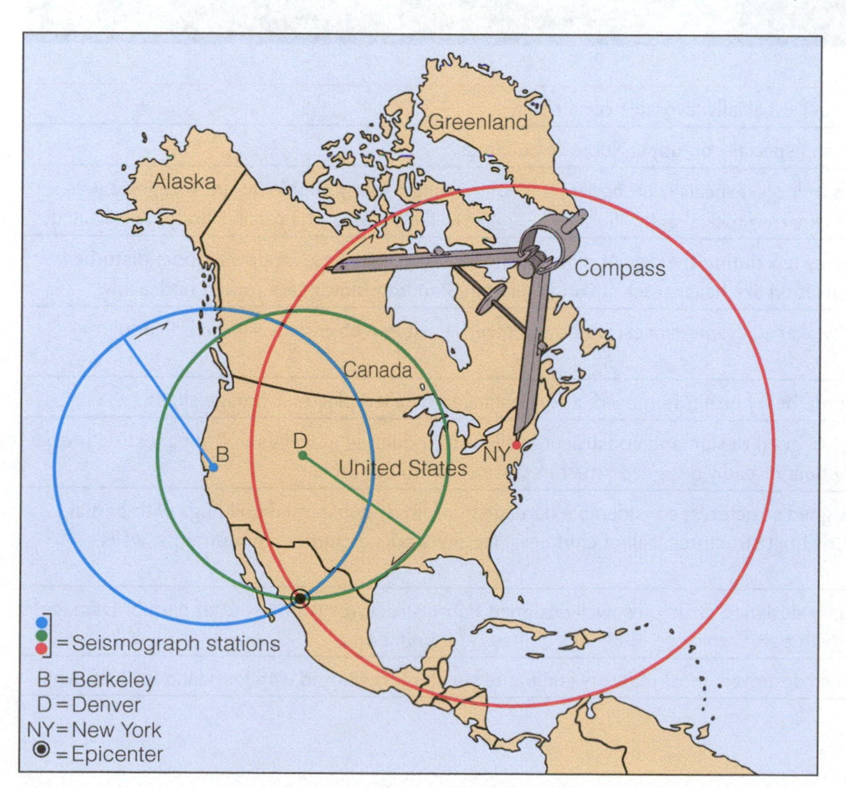

Figure 6.19 An earthquake epicenter can be closely approximated by trilateration from three seismograph stations.

6.2 Earthquake Measurement Scales (LO2)

Seismologists have developed several scales for measuring the sizes of individual earthquakes. These measurements have proven useful for researchers who study plate tectonic interactions, for emergency response planners, for building engineers, architects, contractors, home buyers, and many other people concerned about seismic safety.

6.2.1 Seismic Intensities

A sampling of the reactions of people who have been subjected to an earthquake can be put to good use in estimating the intensity of shaking. Numerical values can be assigned to the individuals' perceptions of earthquake shaking as well as local damage, which can then be contoured on a map.

One **intensity scale** developed for measuring these perceptions is the 1931 **modified Mercalli scale**. The scale's values range from *MM* = I (denoting not felt at all) to *MM* = XII (denoting widespread destruction), and they are keyed to specific U.S. architectural and building specifications. People's perceptions and responses are compiled from returned questionnaires; then lines of earthquake intensity, called **isoseismals**, are plotted on maps. Isoseismals enclose areas of equal earthquake damage and can indicate areas of weak rock or soil, as well as areas of substandard building construction (**Figure 6.20**). Such maps have proved useful to planners and building officials in revising building codes and creating safe construction standards (**Table 6.1**).

A method of plotting shaking intensities in the United States has evolved that is faster and compares favorably with the questionnaire method. This is to use Internet responses to create a community Internet intensity map (CIIM), an example of which is shown in **Figure 6.21**. The time needed to generate an intensity map is minutes, not months, and is particularly useful in areas with sparse seismograph coverage. Although perhaps less simple than the map generated by the traditional *MM* intensities obtained by using postal questionnaires, the CIIM values agree well and have actually proved more reliable in areas of low shaking. However, neither method allows comparison of earthquakes with widely spaced epicenters because of local differences in construction practices and local geology.

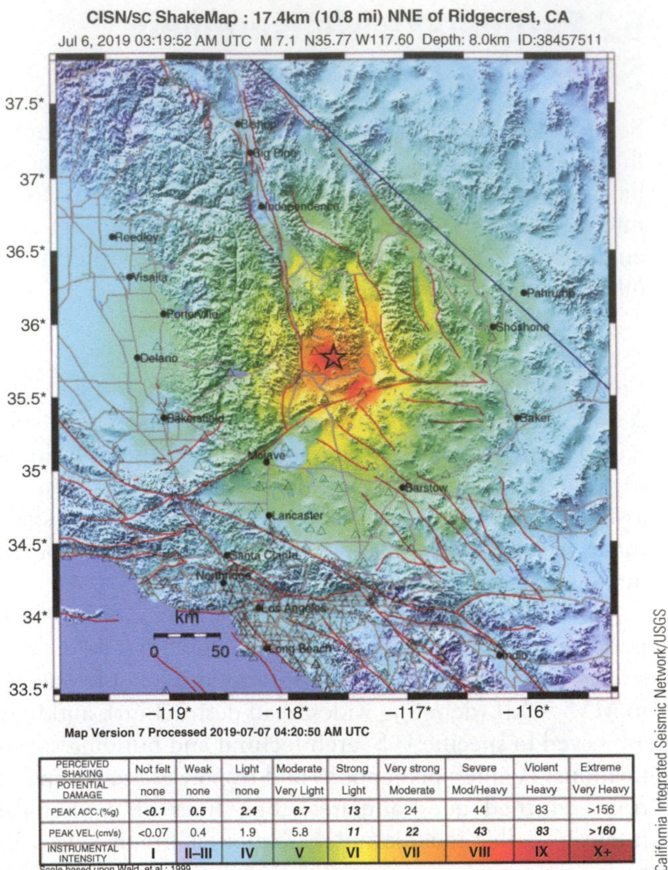

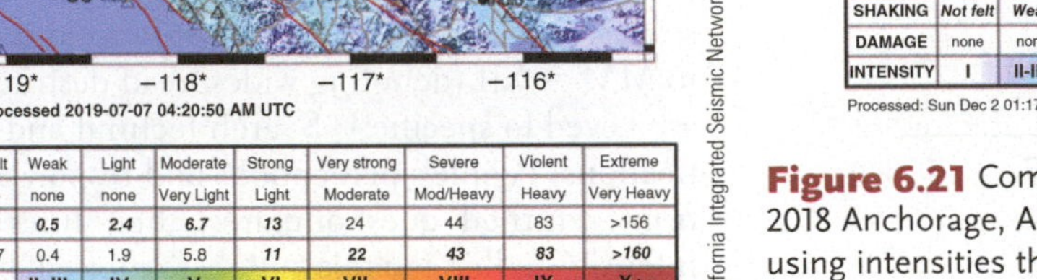

PERCEIVED SHAKING	Not felt	Weak	Light	Moderate	Strong	Very strong	Severe	Violent	Extreme
POTENTIAL DAMAGE	none	none	none	Very Light	Light	Moderate	Mod/Heavy	Heavy	Very Heavy
PEAK ACC.(%g)	<0.1	0.5	2.4	6.7	13	24	44	83	>156
PEAK VEL.(cm/s)	<0.07	0.4	1.9	5.8	11	22	43	83	>160
INSTRUMENTAL INTENSITY	I	II–III	IV	V	VI	VII	VIII	IX	X+

Scale based upon Wald, et al ; 1999

California Integrated Seismic Network/USGS

Figure 6.20 Mercalli seismic intensity map generated for the July 6, 2019, Ridgecrest, California, earthquake.

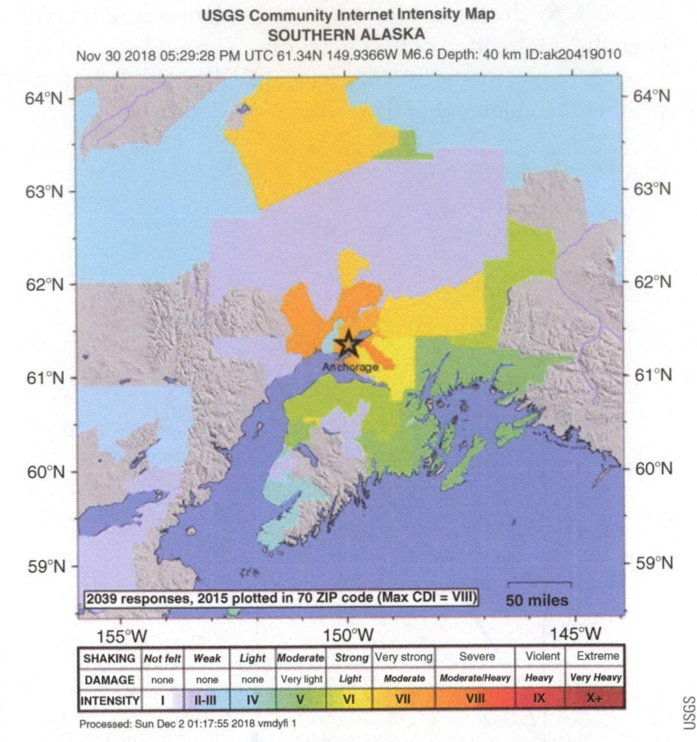

SHAKING	Not felt	Weak	Light	Moderate	Strong	Very strong	Severe	Violent	Extreme
DAMAGE	none	none	none	Very light	Light	Moderate	Moderate/Heavy	Heavy	Very Heavy
INTENSITY	I	II-III	IV	V	VI	VII	VIII	IX	X+

Processed: Sun Dec 2 01:17:55 2018 vmdyfi 1

USGS

Figure 6.21 Community internet intensity map (CIIM) for the 2018 Anchorage, Alaska, earthquake. This type of map is made using intensities that were reported by respondents within zip code areas.

Table 6.1 Mercalli Intensity Scale

Intensity	Shaking	Description/Damage
I	Not felt	Not felt except by a very few under especially favorable conditions.
II	Weak	Felt only by a few persons at rest, especially on upper floors of buildings.
III	Weak	Felt quite noticeably by persons indoors, especially on upper floors of buildings. Many people do not recognize it as an earthquake. Standing motor cars may rock slightly. Vibrations similar to the passing of a truck. Duration estimated.
IV	Light	Felt indoors by many, outdoors by few during the day. At night, some awakened. Dishes, windows, doors disturbed; walls make cracking sound. Sensation like heavy truck striking building. Standing motor cars rocked noticeably.
V	Moderate	Felt by nearly everyone; many awakened. Some dishes, windows broken. Unstable objects overturned. Pendulum clocks may stop.
VI	Strong	Felt by all, many frightened. Some heavy furniture moved; a few instances of fallen plaster. Damage slight.
VII	Very strong	Damage negligible in buildings of good design and construction; slight to moderate in well-built ordinary structures; considerable damage in poorly built or badly designed structures; some chimneys broken.
VIII	Severe	Damage slight in specially designed structures; considerable damage in ordinary substantial buildings with partial collapse. Damage great in poorly built structures. Fall of chimneys, factory stacks, columns, monuments, walls. Heavy furniture overturned.
IX	Violent	Damage considerable in specially designed structures; well-designed frame structures thrown out of plumb. Damage great in substantial buildings, with partial collapse. Buildings shifted off foundations.
X	Extreme	Some well-built wooden structures destroyed; most masonry and frame structures destroyed with foundations. Rails bent.

Sources: USGS

6.2.2 Richter Magnitude Scale: The Best-Known Scale

The best-known measure of earthquake strength is the **Richter magnitude scale**, which was introduced in 1935. It is a scale of the energy released by an earthquake; thus, in contrast with the intensity scale, it may be used to compare earthquakes in widely separated geographical areas. The Richter value is calculated by measuring the maximum amplitude of the ground motion as shown on the seismogram using a specified seismic wave, usually a surface wave.

Next, the seismologist "corrects" the measured amplitude (in microns) to what a "standard" seismograph would record at the station. After an additional correction for distance from the epicenter, the Richter **magnitude (M)** is the common logarithm of that ground motion in microns. For example, a M-4 earthquake is specified as having a corrected ground motion of 10,000 microns ($\log_{10}$ of 10,000 = 4), and thus can be compared with any other earthquake for which the same corrections have been made.

Because the scale is logarithmic, each whole number represents a ground shaking (at the seismometer site) 10 times greater than the next-lower number. Thus, a M 7 earthquake produces 10 times greater shaking than a M 6, 100 times that of a M 5, and 1,000 times that of a M 4. Total energy released, in contrast, varies logarithmically as some exponent of 30. Compared with the energy released by a M-5 earthquake, a $M = 6$ releases 30 times (30^1) more energy; an $M = 7$ releases 900 times (30^2) more; and an $M = 8$ releases 27,000 (30^3) times more energy (**Figure 6.22**).

The Richter scale is open-ended: theoretically, it has no upper limit. However, rocks in nature do have a limited ability to store strain energy without rupturing, and no earthquake has yet been measured directly with a Richter magnitude greater than 8.6 (the 1964 Alaska earthquake), although some geophysicists estimate that large meteorite or asteroid strikes on Earth's surface early in its Precambrian history, such as the Vredefort impact in South Africa, could have released energies scaling up to magnitude 14.

6.2.3 Moment Magnitude: The Most Widely Used Scale

Many seismologists today have abandoned Richter magnitudes in favor of **moment magnitudes** (Mw) for describing earthquakes. The reason is that Richter magnitudes do not accurately portray the energy released by large earthquakes (>M7) on faults with great rupture lengths. The seismic waves used to determine the Richter magnitude come from only a small part of the fault rupture, and hence cannot provide an accurate measure of the total seismic energy released by a very large event.

Moment magnitude is derived from a value termed the **seismic moment**, M_0 (in dyne-centimeters), which is proportional to the average displacement (slip) on the fault times the rupture area on the fault surface times the rigidity of the faulted rock. The amount of seismic energy (in ergs) released from the ruptured fault surface is linearly related to seismic moment by a simple factor, whereas Richter magnitude is logarithmically related to energy. Because of the linear relationship of seismic moment to energy released, the equivalent energy released by other natural and human-caused

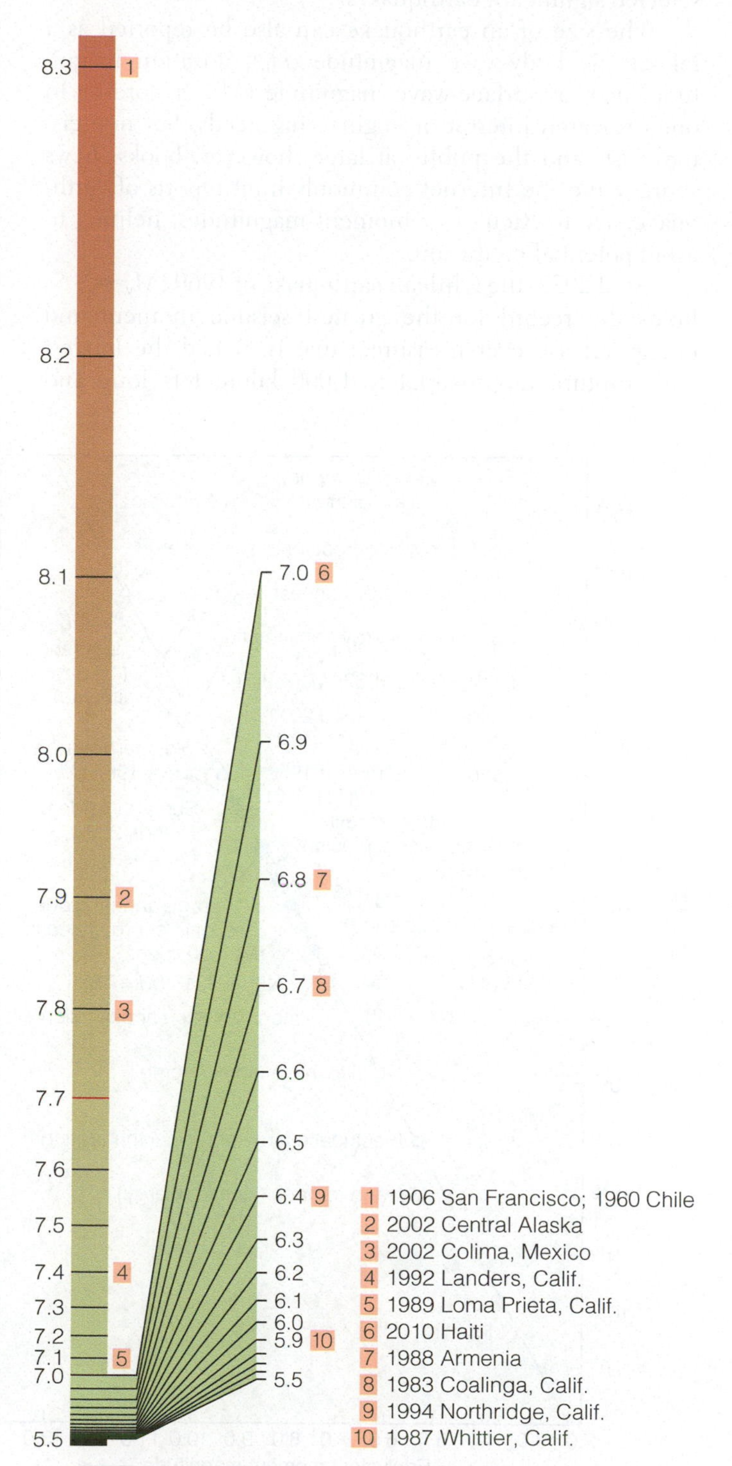

Figure 6.22 Earthquake magnitude and energy. The vertical bar to the left compares the relative energy releases of ten selected historical earthquakes of different Richter magnitudes. (Note how much more powerful an earthquake of M=8 is than one of M=7). The right bar expands the scale at the lower end of the range.

phenomena can be conveniently compared with an earthquakes' moment magnitudes on a graph (**Figure 6.23**).

Moment magnitudes (M_w) are derived from seismic moments by the formula: $M_w = (2/3 \log M_0 - 10.7)$. **Table 6.2** compares the two most commonly used scales for selected significant earthquakes.

The size of an earthquake can also be reported as a teleseismic body-wave magnitude (m_b), duration magnitude (m_d), or surface-wave magnitude (M_s), according to one's research interest or engineering needs. For nonseismologists and the public at large, however, books, news sources and the Internet commonly limit reports of earthquake size to Richter or moment magnitudes, helping to avoid potential confusion.

As of 2022, the Chilean earthquake of 1960 ($M_w = 9.5$) holds the record for the greatest seismic moment and energy release ever measured; that is, it had the longest fault rupture (approximately 1,000 kilometers long and 200 kilometers wide) and an upthrust displacement of as much as 20 to 40 meters (65–130 feet). The Alaskan earthquake of 1964 had the highest Richter magnitude of the twentieth century (M = 8.4), suggesting that this event may have had even more severe ground motion than the event in Chile.

Table 6.2 Magnitude of selected earthquakes

Earthquake, Year	Richter Magnitude (M)	Moment Magnitude (M_W)
Chile, 1960	8.3	9.5
Alaska, 1964	8.4	9.2
New Madrid, Missouri, 1812	8.7 (estimated)	8.1 (estimated)
Mexico City, 1985	8.1	8.1
San Francisco, California, 1906	8.3 (estimated)	7.7
Loma Prieta, California, 1989	7.1	7.0
San Fernando, California, 1971	6.4	6.7
Sumatra-Andaman, 2004	?	9.1
Northridge, California, 1994	6.4	6.7
Kobe, Japan, 1995	7.2 JMA*	6.9
Tohuko, Japan 2011	?	9.1
Haiti, 2021	?	7.2

*The Richter scale has not been used in Japan. The official earthquake scale in Japan is the scale developed by the Japanese Meteorological Agency (JMA).

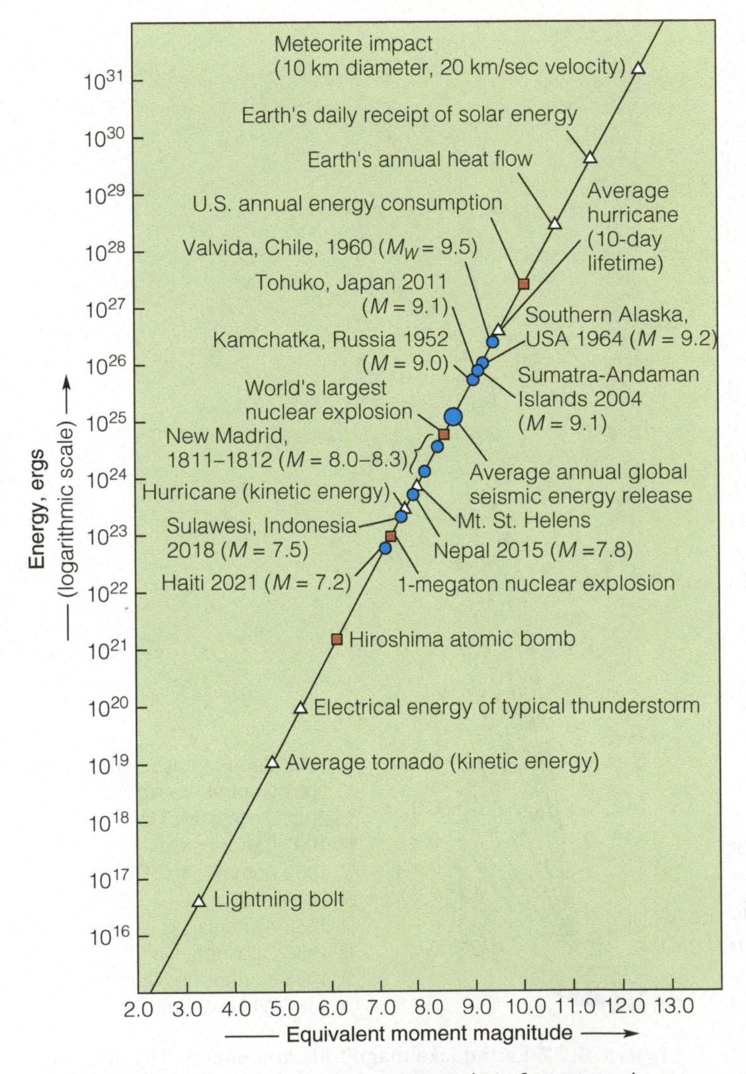

Figure 6.23 Equivalent moment magnitudes of energetic human-caused events (squares), non-seismic natural events (triangles), and large earthquakes (circles). The erg is a unit of energy, or work, in the metric (cgs) system. To lift a 1-pound weight 1 foot requires 1.4×10^7 ergs.

6.2.4 Foreshocks and Aftershocks, Doublets, and Triggered Earthquakes

The energy released by a large earthquake is only part of the seismic picture. Big earthquakes are sometimes preceded by foreshocks, and always followed by aftershocks, sometimes for years. The total crustal energy release associated with a major earthquake (the **main shock**) should take these numerous smaller, related shocks into account as well, though in practice this calculation is not often made, given the time required to acquire a complete set of data.

Foreshocks, the earthquakes that precede a much larger main shock, generally take place a few days to just minutes before the main shock. They have been identified retrospectively for about 70% of all main shocks exceeding magnitude 7.0, and for around 40% of all moderate-sized earthquakes (M_W 5–6 range). They can also be quite powerful in and of themselves. The M_W 9.5, May 22, 1960, Chilean earthquake was preceded by a damaging magnitude M_W 7.9 earthquake one day earlier in the same region. Foreshocks can be interpreted as cracking and slipping beginning along faults in the crust, which transfer stress to a nearby zone that in short order becomes the main shock rupture. Unfortunately for forecasters, not all main shocks are preceded by foreshocks, and it is not possible to identify foreshocks until a full sequence of "foreshock + main shock + aftershock" has taken place; too late to do emergency planners much good.

Aftershocks, the many smaller earthquakes that follow with diminishing intensity over time in the same area as a main shock, may be viewed as adjustments in the crust as it settles into a less stressful position. Think of the rebalancing you may do after taking a big jump (**Figure 6.24**). For seismologists, the foci of aftershocks are useful to map, because they generally outline the area of rupture taking place during the main shock. That in turn helps define the energy release, or moment magnitude of the main shock.

Unlike foreshocks, aftershocks tend to follow well-defined patterns. **Utsu-Omori's Law**, for instance, describes how the rate of aftershocks decreases quickly with time; if a certain number of aftershocks are recorded on the first day after a big earthquake, then one-third of that number are likely to occur the day following, and one-third of that number the day following (day 3). And so on. **Båth's Law** states that the moment magnitude of the largest aftershock will be on the order of 1.1–1.2 less than the moment magnitude of the main shock, regardless of the power of the main shock. The **Gutenberg-Richter Law** essentially states that the vast majority of aftershocks will be quite small in magnitude.

The time period of aftershocks following a major earthquake depends upon both the magnitude of the main shock and the structure of the crust in which it occurs. Generally speaking, the larger the main shock, the longer the aftershock sequence, ranging from mere weeks to several centuries. Main shocks taking place along divergent plate boundaries tend to be of lower magnitude than those related to convergent boundaries and plate interiors ("intraplate earthquakes," described further below).

Occasionally, main shocks occur in pairs called **doublet earthquakes**; two powerful shocks of nearly equal magnitude taking place in the same general area, or along two closely separated faults. Sometimes this happens almost at once, as during the August 14, 2021, Haiti earthquake described in the chapter opener. A single moment magnitude is published for this earthquake, although rupturing occurred in two parts along different, adjoining faults. In other instances, the time separating the shocks can be as long as several years. The Shumagin, Alaska quake of July 22, 2020 (M_w 7.8) was followed by a M_w 7.6 quake on October 19 as the zone of seismic slip expanded from the initial rupture. Likewise, the three powerful earthquakes (M_w 7–8 range) of the 1811–1812 New Madrid sequence, described below, could be regarded as a "multiplet" with individual main shocks occurring just a few months apart. The point is that people can't dismiss the possibility that an especially powerful earthquake will sometimes be followed by at least one other of nearly the same strength. Either way, the general pattern of foreshocks and aftershocks is the same, whether a main shock occurs an isolated event, or as part of a set.

Triggered earthquakes is a concept related to doublet earthquakes, though in this case, the magnitudes of the shocks are not closely similar, and the fault ruptures involved take place well separated from one another–even hundreds of kilometers. A triggered earthquake is typically much less powerful than the one that sets it off, though not so much as to classify it as a typical aftershock. The stresses relieved by the initial earthquake transfers through the crust to a distant fault, or part of the same large fault, which at a later point, independently snaps. A triggered earthquake is its own separate main shock. We consider several examples of triggered earthquakes (Sumatra, New Zealand) later in the chapter.

Of the three categories of earthquakes–foreshock, main shock, aftershock–the aftershocks are a special concern for safety reasons. Structures weakened during the main shock may be easily toppled in later shaking if they are not engineered to withstand earthquakes well. Aftershocks also tend to be psychologically very challenging for survivors because they are frequent and can take place any time without warning. They can generate and exacerbate depressive and anxiety disorders, keeping people–especially children–on edge for weeks to months following a main shock. Being aware that aftershocks occur and that their frequency decreases with time can empower victims of a major earthquake with knowledge, helping them control expectations of "what comes next."

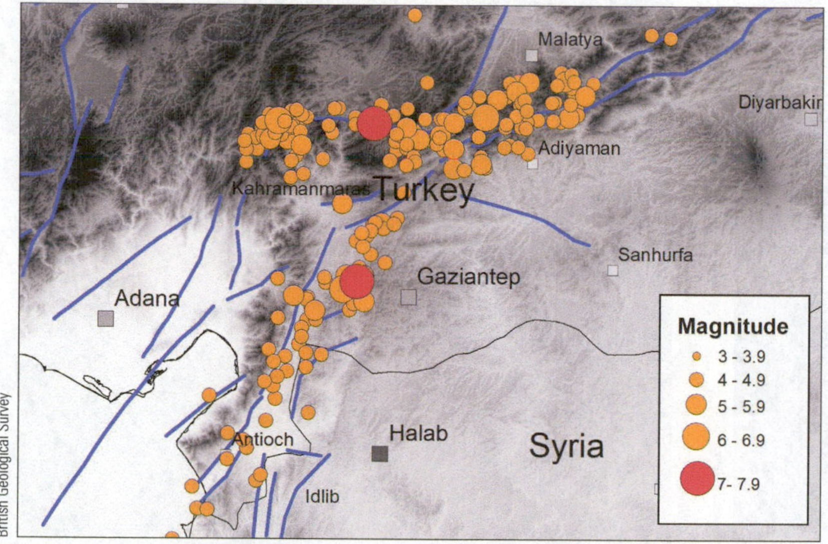

Figure 6.24 Map of aftershocks associated with the 6 February 2023 Turkish-Syrian earthquakes. The main shock to the south, shown in red, triggered another powerful earthquake on a fault to the north, also shown as a red circle. Orange circles are aftershocks, with the size of each circle scaled to show relative magnitudes.

6.2.5 Fault Creep, the "Non-Earthquake"

Not all fault movements involve measureable seismic shaking. This doesn't mean a fault is truly inactive, however. **Fault creep** is a slow, steady slipping that takes place along some faults, and is measured in terms of fault offset length per year.

The Hayward Fault, which is part of the San Andreas Fault system in California, is an example of a creeping fault. It is just east of the San Andreas Fault and runs through the cities of Hayward and Berkeley, California. The ruptured crust slides along the fault at a rate of millimeters per year and is one way that this transform plate boundary accommodates motion between the Pacific and North American plates. The San Andreas Fault is famously known as the "boundary" between these two plates but, in fact, the fault accounts for only about half the net relative plate motion. The rest is taken up by parallel or subparallel faults like the Hayward mostly to the east, some of them hundreds of kilometers away (**Figure 6.25**).

A good strategy in areas subject to fault creep is to map the surface trace of the fault so that it can be avoided in future construction (**Figure 6.26**). The University of California-Berkeley Memorial Stadium was built on the Hayward fault before the hazard of fault creep was recognized. The fault creep caused damage to the drainage system and masonry, both of which require periodic maintenance. It is ironic that creep damage occurs at an institution that installed the first seismometer in the United States.

The obvious benefit of a creeping fault–lack of earthquakes–comes with a price, as the residents of Hollister, California, can testify. The creeping Calaveras Fault runs through their town and, as in Hayward, it gradually breaks and offsets curbs, sidewalks, residences, and at least one winery (**Figure 6.27**). The U.S. Geological Survey installed a "creepmeter" in the winery, reportedly safely hidden behind wooden aging barrels. The data are automatically transmitted back to the survey. One of the geologists has been quoted after a field visit to the winery, "Instead of a dusty, nasty place, you get to come here."

Why do some faults continuously, silently creep, whereas others exhibit more normal stick-slip, earthquake-generating behavior? The answer seems to lie in the condition of the bedrock that they cross. The Hayward and

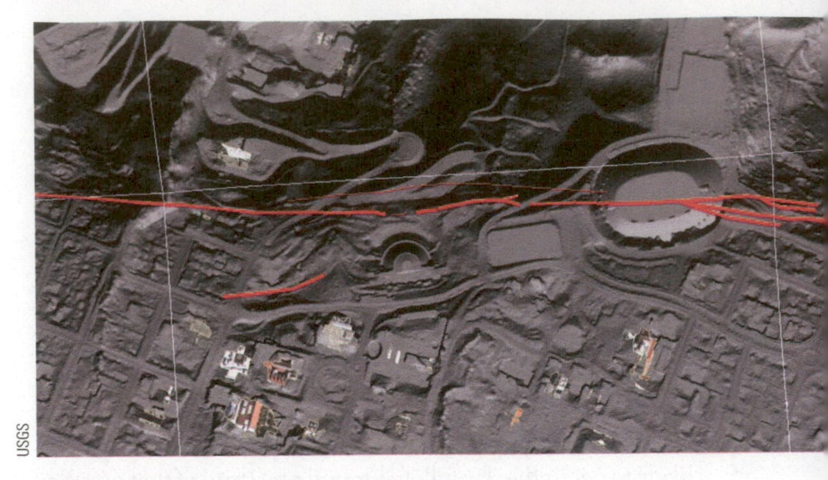

USGS

Figure 6.26 Lidar image of the Hayward Fault as is cuts right through the University of California-Berkeley Memorial Stadium.

Kelly Blake/US Navy

Figure 6.25 Measuring fault offset on Highway 178 in Ridgecrest, California, after the 2019 magnitude-6.8 earthquake. This is left-lateral motion.

DeRose Winery, Inc

Figure 6.27 Creeping offset of the Calaveras Fault going right through the DeRose winery. The offset is visible on a drainage channel out side the winery. This is right lateral offset.

Calaveras faults slice through large masses of dark gray-green serpentinite, a type of low-grade metamorphic rock that is notoriously weak and deformable, whereas the San Andreas Fault generally cuts through masses of much stronger granite, gneiss, and sedimentary rock prone to "catch" before breaking. Sometimes knowledge of the bedrock geology can provide important clues to expected long-term fault behavior.

6.3 Seismic Design Considerations (LO3)

The disasters in Haiti, and many other lesser resourced countries, could be averted by improving the strength and design of buildings, as has been done in other, better resourced parts of the world that are also prone to severe shaking. Several factors are important to do this well, quite apart from having the economic ability to build more quake resistant structures in the first place.

6.3.1 Ground Shaking

Ground movements, particularly rapid horizontal displacements, are most damaging during an earthquake. S- and short-period surface waves are the culprits, and the potential for them must be evaluated and then design specifications established. The design objective for earthquake-resistant buildings is relatively straightforward: structures should be built to withstand the maximum potential horizontal ground acceleration expected in the region. Engineers call this acceleration **base shear**, and it is usually expressed as a percentage of the acceleration of gravity (g).

On Earth, g is the acceleration of a falling object in a vacuum (9.8 meters per second^2, or 32 feet per second^2). In your car, it is equivalent to accelerating from a dead stop through 100 meters (320 feet) in 4.5 seconds. An acceleration of 1 g downward produces weightlessness, and even a fraction of 1 gram in the horizontal direction can cause buildings to separate from their foundations or to collapse completely. An analogy is to imagine rapidly pulling a carpet on which a person is standing; most assuredly, the person will topple.

The effect of high horizontal acceleration on poorly constructed buildings is twofold (**Figure 6.28**). Flexible-frame structures may be deformed from cube-shaped to rhombus-shaped, or they may be knocked off their foundations completely (**Figure 6.29**). More rigid, multistory buildings may suffer "story shift" if floors and walls are not adequately tied together. The result is a shifting of floor levels and the collapse of one floor on another like a stack of pancakes (**Figure 6.30**). Such structural failures are generally not survivable by inhabitants, and they clearly illustrate the adage that "Earthquakes don't kill people; buildings do."

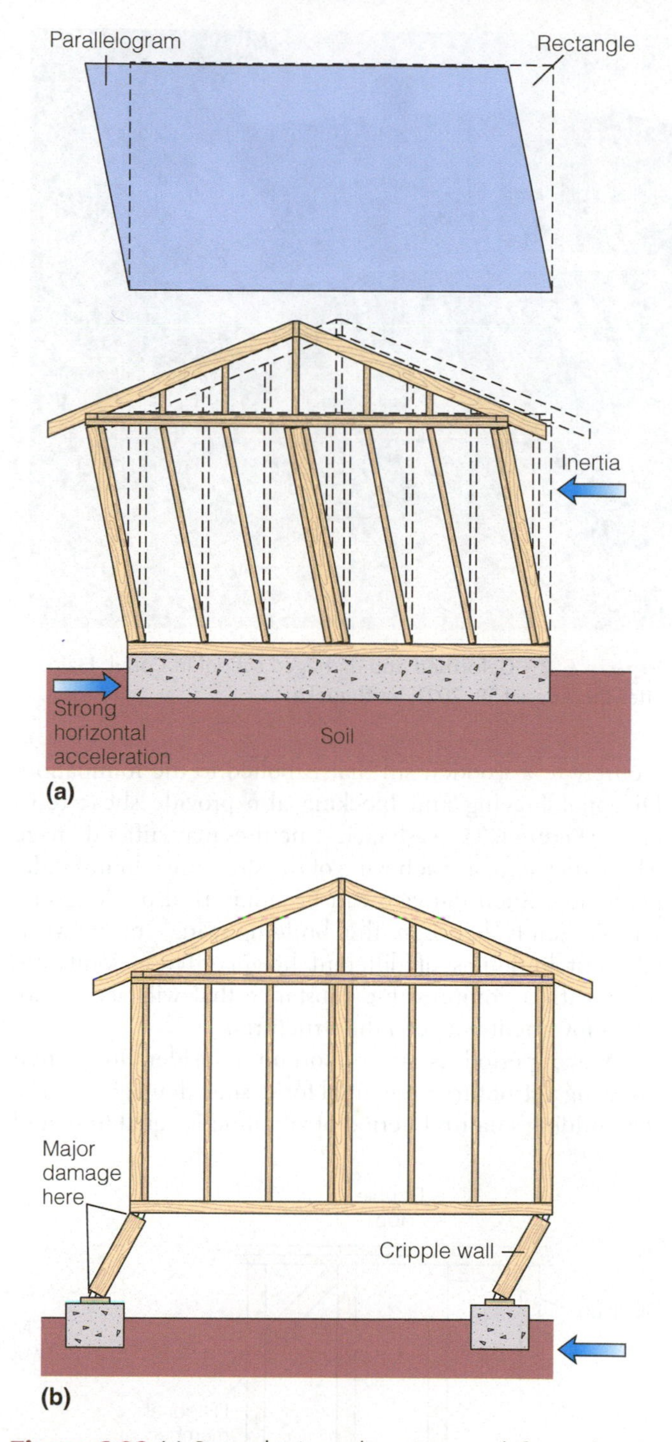

Figure 6.28 (a) Strong horizontal motion can deform a house from a cube to a rhomboid or knock it from its foundation completely. (b) A cripple-wall consists of short, vertical members that connect the floor of the house to the foundation. Cripple-walls are common in older construction.

Damage caused by shearing forces can be mitigated by bolting frame houses to their foundations and by **shear walls**. An example of a shear wall is plywood sheeting nailed in place over a wood frame, which makes the structure highly resistant to deformation by shearing. Wall framing, usually two-by-fours, should be nailed very

El Nuevo Herald/Tribune News Service/Getty Images

Figure 6.29 Deformed and damaged school in Corail, Haiti, after the August 19, 2021, earthquake.

Pacific Press/LightRocket/Getty Images

Figure 6.30 Pancaking collapse of building in Jaya, Aceh Province, Indonesia, from a Richter magnitude 6.5 quake.

securely to a wooden sill that is bolted to the foundation. Diagonal bracing and blocking also provide shear resistance (**Figure 6.31**). L-shaped structures may suffer damage where they join, as each wing of the structure vibrates independently. Such damage can be minimized by designing seismic joints between the building wings or between adjacent buildings of different heights. These joints are filled with a compressible substance that will accommodate movement between the structures.

Wave period is an important consideration when assessing a structure's potential for seismic damage because if a building's natural period of vibration is equal to that of seismic waves, a condition of resonance exists. **Resonance** occurs when a building sways in step with an oscillatory seismic wave. As a structure sways back and forth under resonant conditions, it gets a push in its direction of sway with the passage of each seismic wave. This causes the sway to increase, just as pushing a child's swing at the proper moments sends it higher with each push. Scale models are often used to test building designs for resonance with the largest-scale facilities in Japan, a country on a subduction zone and thus not a stranger to massive earthquakes (**Figure 6.32**).

In some instances, buildings of a certain size suffer serious damage, whereas smaller or larger structures in the surrounding area ride through an earthquake intact because of differences in resonance. Low-rise buildings have short natural wave periods (0.05 to 0.1 second), and high-rise buildings have long natural periods (1 to 2 seconds). Therefore, short-period (high-frequency) waves affect single-family dwellings and low-rise buildings, and long-period (low-frequency) waves affect tall structures. Close to an earthquake epicenter, high-frequency waves dominate; thus, more extensive home and low-rise damage can be expected. With distance from the epicenter, the short-period wave energy is absorbed or dissipated, resulting in the domination of longer-period waves, much to the chagrin of people who happen to be inside skyscrapers at the time.

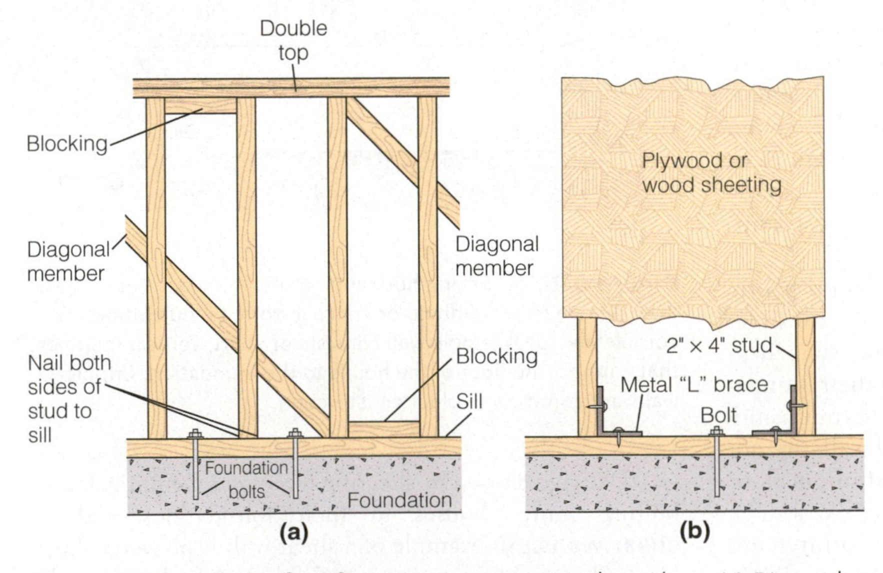

Figure 6.31 Methods of reinforcing structures against base shear. (a) Diagonal cross-members and blocks resist horizontal earthquake motion (shear). (b) Plywood sheeting forms a competent shear wall, and metal L braces and bolts tie the structure to the foundation.

6.3.2 Landslides

Earthquakes are a major cause of landslides, which we'll explore further in Chapter 9.

There were an estimated 17,000 landslides in mountainous and hilly terrain around the City of Los Angeles during the M_W6.7 1994 Northridge earthquake alone. These were not particularly newsworthy,

Reuters/Alamy Stock Photo

Figure 6.32 A building design being tested for seismic response on a giant shake table.

given the circumstances, but they had a significant, unintended consequence that became apparent in the weeks to come: The dislodged, windblown soil and dust raised by landsliding caused an outbreak of *coccidioidomycosis*, commonly known as Valley Fever. Endemic to the southwestern United States, the disease causes persistent flu-like symptoms and in extreme cases can be fatal. Exposed to dust released by the slides, many persons breathed in airborne *Coccidioides immitis* spores without knowing it. Between 24 January and 15 March, 166 people were diagnosed with Valley Fever symptoms in Ventura County, up from only 53 cases in all of 1993 (**Figure 6.33**). Most of the cases were reported from the Simi Valley, former home of President Ronald Reagan, an area where only 14% of the county's population lived.

In part, the disease outbreak was also weather and bedrock related: large clouds of dust hung over the Santa Susana Mountains for several days following the earthquake, promoted by a lack of winter rains. During this period, pressure-gradient winds known locally as Santa Ana winds (Chapter 2), of 5 to 7.5 meters per second (11 to 17 miles per hour) blew into the Simi Valley, carrying in spore-laden dust from the nearby mountain. Researchers believe that the more "coherent" metamorphic rocks of the neighboring San Gabriel Mountains northeast of the San Fernando Valley probably account for the lack of cases reported in the epicentral region.

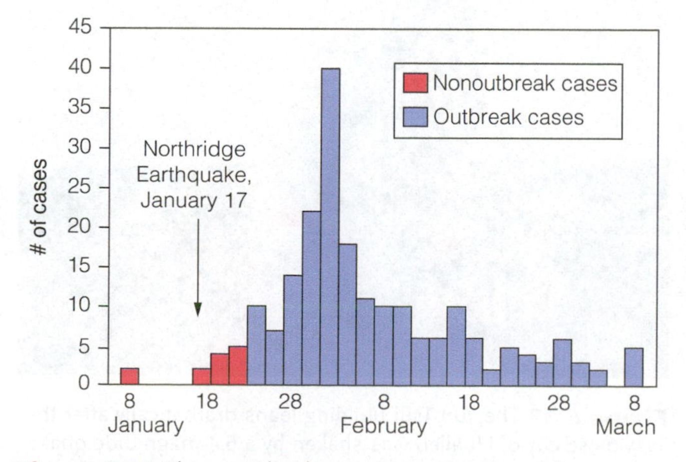

Figure 6.33 The coccidioidomycosis outbreak and the Northridge earthquake.

By far the most destructive quake-related landslide in recorded history took place on May 31, 1970, when a subduction-related M_w = 7.9–8.0 earthquake struck Peru. Strong shaking was felt over a vast area (the size of Belgium and the Netherlands combined). A great avalanche of rock, snow, and ice measuring 900 meters (3,000 feet) wide and 1.6 kilometers (1 mile) long, detached from the north face of 6,768-meter (22,200-foot)-tall Nevado Huascarán, an ice-covered Andean peak (**Figure 6.34**).

This broken mass roared down the steep flank of the mountain at speeds as high as 355 kilometers per hour (220 miles per hour), mobilized, in part, by a cushion of trapped air. A canyon funneled it toward the towns of Yungay and Ranrahirca (**Figure 6.35**), 11 kilometers (7 miles) away, which were buried just a few minutes later as it exited the base of the mountain and spread across the plain below (**Figure 6.36**). Of the 20,000 people in Yungay, only 400 survived, mostly children who happened to get to a higher part of the town and so weren't smothered and crushed (**Figure 6.37**). The Andes region has the greatest seismic-related landslide risk in the world because of high mountains, active tectonics, and heavily settled mountain valleys.

Nothing can be done to prevent seismically induced landslides or stop them after they've started. Geologists can

Dave Stamboulis/Alamy Stock Photo

Figure 6.34 Snow and glacier-clad Nevado Huascarán towers over the valley. This is a view of Campo Santo, a memorial and cemetery built on the deposit of the debris avalanche that wiped out the town of Yungay in 1970.

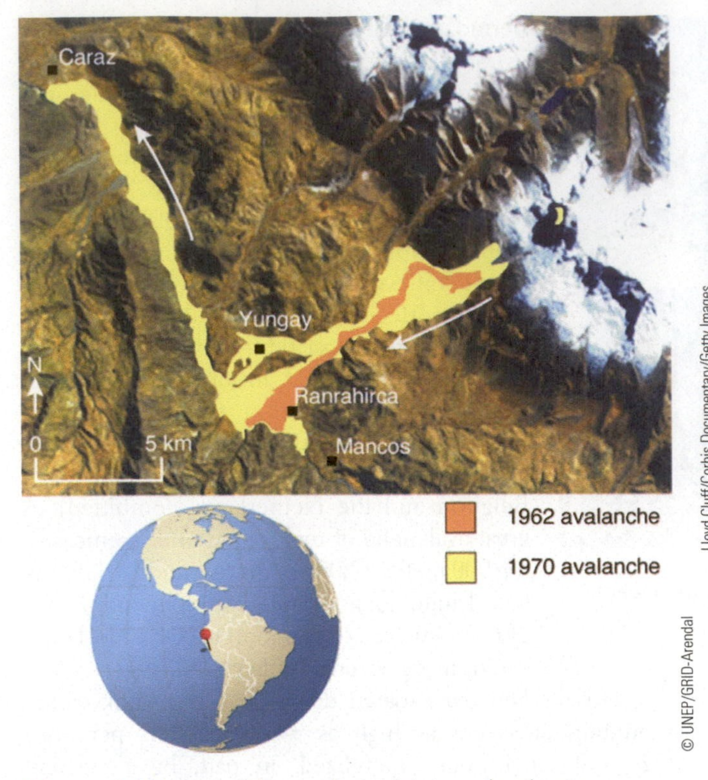

Figure 6.35 Map of the two large debris avalanches to descend Nevado Huascarán in recent times.

identify slide-prone areas and conditions in seismically active areas; then it is simply a matter of people not building structures that lie in harm's way, perhaps enforced by zoning ordinances. Out of necessity in a world with an ever-increasing population, however, such a precautionary relationship to the land is difficult to achieve.

Figure 6.36 Much of the town of Yungay was buried under the massive debris avalanche. The debris which covers the foreground originated from the peak in the background (Nevado Huascarán). If you look closely, you'll see the dark scar showing where the debris avalanche originated.

Figure 6.37 Referred to as Cemetery Hill, this area of Yungay damaged by the earthquake was above the level of the debris avalanche and so people there survived despite the damage all around them from the strong shaking.

6.3.3 Ground or Foundation Failure

Strong ground shaking during an earthquake can weaken soil, causing it to fail. Buildings on that weakened soil can suffer a catastrophic demise even if they are built to withstand the shaking (**Figure 6.38**). **Liquefaction** is the sudden loss of strength of water-saturated, sandy soils resulting from shaking during an earthquake. It can open large ground cracks, perhaps the origin of ancient myths that the Earth opens up to swallow people and animals during earthquakes.

During liquefaction shaking causes saturated sand and silt to consolidate, and thus to occupy a smaller volume (**Figure 6.39**). The process of consolidation squeezes water out from the spaces between the settling grains. The water can erupt at the surface, carrying with it a slurry of silt and sandy material. Buildings may sink into this weakened ground,

Figure 6.38 The Yun Tsui building leans dramatically after the Taiwanese city of Hualien was shaken by a 6.4-magnitude quake late on February 6, 2022. The building is intact; the ground beneath failed from strong shaking.

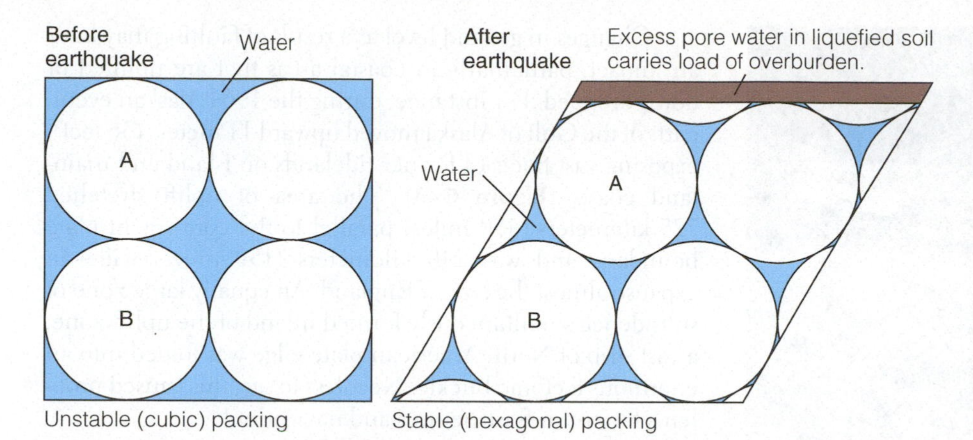

Figure 6.39 Liquefaction (lateral spreading) caused by repacking of spheres (idealized grains of sand) during an earthquake. The earthquake's shaking causes the solids to become packed more efficiently, and thus to occupy less volume. A part of the overburden load is supported by water, which has no resistance to lateral motion.

sometimes at crazy angles (**Figure 6.40**). Liquefaction at shallow depths may also result in extensive lateral movement or spreading of the ground, leaving large cracks and openings.

The ground areas most susceptible to liquefaction are those that are underlain at shallow depth–usually less than 10 meters (about 30 feet)–by layers of fine, water-saturated sediment. With subsurface geological data obtained from water wells and foundation borings, liquefaction-susceptibility maps have been prepared for many seismically active areas in the United States.

Similar failures occur in certain clays that lose their strength when they are shaken or remolded. Such clays, specifically called **quick clays**, are natural aggregations of fine-grained clays and water (Chapter 9). They have the peculiar property of turning from a solid (actually, a gel-like state) to a liquid when they are agitated by an earthquake, an explosion, or even vibrations from pile driving. It doesn't necessarily take much. Such clays occur in deposits of glacial-marine or glacial-lake origin, and are therefore found mostly in northern latitudes, particularly in Scandinavia, Canada, and the New England states. The failure of shallow quick clays underlying Anchorage, Alaska, produced extensive lateral spreading throughout the city in the M_W 9.2 1964 earthquake (**Figure 6.41**).

6.3.4 Ground Rupture and Changes in Ground Level

Structures or landforms that straddle an active fault may be destroyed by actual ground shifting and the formation of a **fault scarp** (**Figure 6.42**). By excavating trenches across fault zones, geologists can usually locate past rupture surfaces that may reactivate (**Figure 6.43**). In 1970, the California State legislature enacted the Alquist-Priolo Special Studies Zones Act, which was renamed the Earthquake Fault Zones Act in 1995. It mandates that all known active faults in the state be accurately mapped and the

The Asahi Shimbun/Getty Images

Figure 6.40 An apartment building tilted as a result of soil liquefaction beneath its inadequetely designed foundation, Niigata, Japan, just after the 1964 earthquake. Many residents of the building in the center exited by walking down the side of the structure. Look closely and you can see the people.

Everett Collection Historical/Alamy Stock Photo

Figure 6.41 Lateral spreading on quick clay destroyed Anchorage's Turnagain Heights neighborhood in the 1964 earthquake.

Stacy, J. R./USGS

Figure 6.42 Prominent fault scarp left by M_w 7.2 August 19, 1959, Hebgen Lake earthquake, near Yellowstone National Park.

land around them managed for seismic safety. The act provides funds for state and private geologists to locate the youngest fault ruptures and requires city and county governments to limit land use adjacent to identified faults within their jurisdictions, including construction of new schools, hospitals, fire and police departments, and other related public facilities.

USGS

Figure 6.43 Trench excavated across a fault scarp in the Wynoochee River Valley (Washington State, Olympic Peninsula, subduction zone setting). By observing the layers and how they are offset, geologists can map evidence of past earthquakes and, if the right materials are present, date when a quake ruptured the ground.

Changes in ground level as a result of faulting may have an impact, particularly in coastal areas that are uplifted or down-dropped. For instance, during the 1964 Alaskan event, parts of the Gulf of Alaska moved upward 11 meters (36 feet), exposing vast tracts of former tidelands on island and mainland coasts (**Figure 6.44**). The area of uplift stretched 725 kilometers (450 miles) parallel to the convergent plate boundary, and was 240 kilometers (150 miles) wide–an expanse almost the size of England. An equally large zone of subsidence simultaneously formed inland of the uplift zone; a vast strip of North American plate edge was folded into an enormous tectonic buckle. Needless to say, this caused problems for port infrastructures and navigation!

6.3.5 Tsunamis

Certain large undersea earthquakes generate **tsunamis** (pronounced "soo-nah-mee"), giant sea waves that pour across coastal areas, often causing a great amount of destruction and loss of life (see Chapter 10). Not all tsunamis are caused by undersea earthquakes, however, nor do all large undersea earthquakes cause tsunamis. Those that do are termed **tsunami earthquakes**. They most often take place at convergent plate boundaries, where thrust faults develop along the edge of the overriding plate. An earthquake caused by thrusting can raise the seabed sometimes by many meters, as shown by the 1964 Alaska example just described (see Figure 6.44). Such uplift displaces an enormous amount of overlying water, like a giant paddle being lifted out of the ocean.

Undersea movement is a prerequisite for tsunamis, but something more seems to be required for a really large tsunami to appear. Tsunami-generating earthquakes also are characterized by slower fault rupture speeds (1 kilometer per second, or 2,240 miles per hour) than measured for ordinary thrust-related earthquakes (2.5 to 3.5 kilometers per second, or 5,600 to 7,800 miles per hour). This possibly allows energy

USGS

Figure 6.44 The 1964 earthquake uplifted Cape Cleare on Montague Island in Prince William Sound, Alaska. The almost flat rocky surface with the white coating which lies between the cliffs and the water is about 400 meters (a quarter of a mile) wide. Before the quake, it was underwater!

Mainichi Shimbun/Reuters

Figure 6.45 Massive tsunami wave overtops seawall during the 2011 M_w 9.0 Tohoku-Oki earthquake in Japan. This was a slow rupture speed quake.

to become concentrated more intensely at one point, or to one side, of a rupture zone; the center of generation for a particularly large seismic sea wave (**Figure 6.45**).

6.3.6 Fires

Fires caused by ruptured gas mains or fallen electric power lines can add considerably to the damage caused by an earthquake. In fact, most damage attributed to the San Francisco earthquake of 1906, and much of that in Kobe, Japan, in 1995, was due to the uncontrolled fires that followed the earthquakes (**Figure 6.46**). The Kobe quake hit at breakfast time. In neighborhoods crowded with wooden structures, fires erupted when natural gas lines broke and falling debris tipped over kerosene stoves. Broken water mains made firefighting efforts futile.

In the first few days following the Tohoku-Oki, Japan earthquake in 2011, fires erupted in many places where stored fuel had escaped from cylinders and containers (ranging from ruptured gas tanks in automobiles to giant municipal tank farms). Sparks from metal colliding with metal during the ensuing tsunami set much of this spreading fuel on fire, igniting the wreckage of wooden houses and other combustible debris that firefighters could not reach. One of the principal "do's" for citizens immediately after a quake is to shut off the natural gas supply to homes and other buildings to prevent flammable gas leaks into the structure. This action alone can save many lives.

6.3.7 Dam Failures

In a number of locations (California, New Zealand, India), dams are located directly atop active, or potentially active faults. The low ground commonly found where faults cross landscapes, especially in mountainous areas, is an attractive site for reservoirs. Liquefaction is an especially important concern. During the $M_W = 6.6$ Sylmar earthquake on 9 February 1971 near Los Angeles, horizontal accelerations of 1.25 g (g meaning gravitational acceleration) took place–twice as strong as any observed up to that date. People would be thrown off their feet by the shaking. The 600-meter (2,100-foot)-long, 42-meter (140-foot)-tall Lower San Fernando Dam, filled by its reservoir to within 7.5 meters (25 feet) of its rim, experienced severe liquefaction beneath one of its abutments, and on the adjoining, clay-rich reservoir bottom. The abutment collapsed as 3 million cubic meters (4 million cubic yards) of dam material slumped into the reservoir, spreading across the submerged floor as far as 75 meters (250 feet) upstream. The dam came within 1.5 meters (5 feet) of being overtopped, giving way altogether–a close call! With strong aftershocks taking place, authorities took no chances and hastily evacuated 80,000 residents from 65 kilometers^2 (24 miles^2) of crowded city downslope (**Figure 6.47**).

To a certain degree, dams can be constructed to survive faults rupturing directly beneath them: They can be packed with ductile cores of rock and clay, which will retain some impermeability even if sheared from below. Drainage systems and fabric filters to control soil movements in the earthen core of a dam can prevent major leaks from appearing in concrete faces, and in one extreme example, a giant rubber plug sheathed in steel has been designed to wedge into a dam fracture, should one open. The concern persists in any case. Almost 30% of the world's 90,000-plus dams lie near, or actually on, known active faults, including several major reservoirs that continue to serve the metro areas of San Francisco and Los Angeles.

Sankei Archive/Sankei/Getty Images

Figure 6.46 Fires ravage Kobe, Japan, after the 1995 earthquake damaged the city.

Figure 6.47 Owing to basal liquefaction, the Lower San Fernando Reservoir dam came very close to failing. If the dam had failed, it would have released a disastrous flood across a crowded Los Angeles suburb in the minutes following the 1971 Sylmar earthquake.

6.4 Example Earthquakes That Make Important Points (LO4)

Most earthquakes are not destructive; they occur in remote locations or have great focal depths; much of their energy is dissipated before their seismic waves reach the surface. Like dodging the proverbial bullet, however, it's simply a matter of chance and time before a strong earthquake strikes an area with intensive human development. The recent Mw 7.8 Turkey Syria earthquake (February 2023) which killed nearly 60,000 persons, underscores how this can happen unexpectedly, both in terms of time and place. Let's investigate some examples of past major earthquakes with lessons that remain highly relevant today. We summarize what we learn from them at the end of this section. (See if you can anticipate what's important as you read along!).

As the world becomes more and more crowded with people and human infrastructure, statistics suggest that the destructiveness of earthquakes will increase–at least in monetary terms–despite greater understanding and sophisticated monitoring technologies. On the other hand, as investment in public safety and construction has improved, the total proportion of earthquake fatalities relative to total population impacted has declined over the past 100 years. Let's plan for this trend to continue if we can.

6.4.1 Northridge, California, 1994

The largest earthquake in Los Angeles's short history occurred at 4:30 a.m. Monday, January 17, 1994, on a unknown fault below the San Fernando Valley in the sprawling Northridge district. The M_w = 6.7 earthquake started at a depth of 18 kilometers (11 miles) and propagated upward in about 8 seconds to a depth of 5 to 8 kilometers (3 to 5 miles; **Figure 6.48**). There were thousands of aftershocks, and their clustered pattern indicated the fault to blame was a blind thrust (buried and hidden below the surface) with a shallow inclination of 3° to 5° to the south (**Figure 6.49**). A similar type of fault was responsible for the catastrophic 2010 Haiti earthquake, described in the chapter opener.

The earthquake did not last long; only 10 to 20 seconds. However, 13,000 buildings were severely damaged (**Figure 6.50**); 21,000 dwelling units had to be evacuated; 240 mobile homes were destroyed by fire; and 11 major

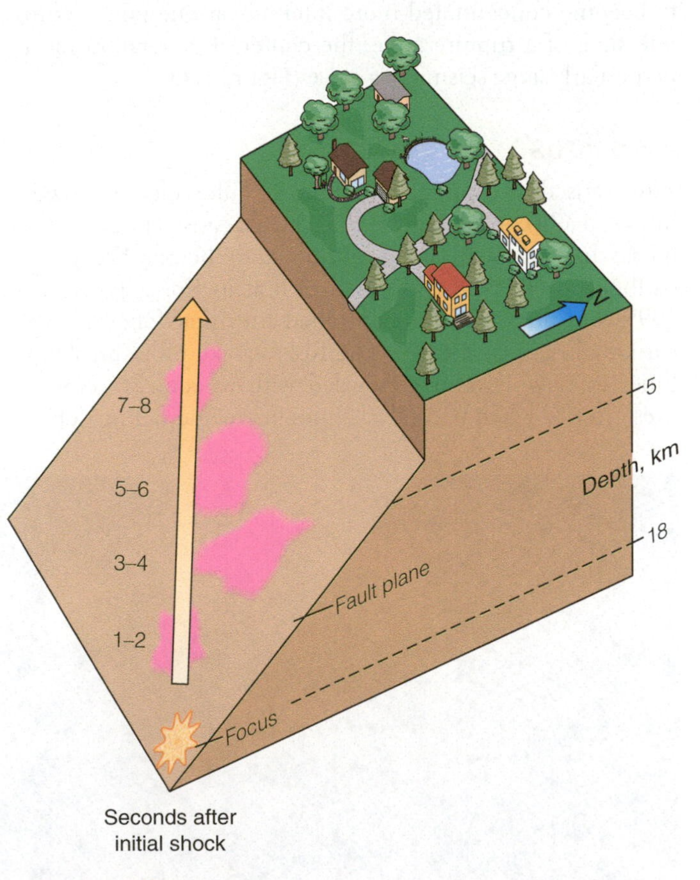

Figure 6.48 In the 1994 Northridge, California, earthquake, the fault rupture progressed up the fault plane from the focus at the lower right of the figure to the upper left in 8 seconds. Rather than rupturing smoothly, like a zipper opening, it moved in jerks along the fault plane, as shown by the pink patches. Total displacement was about 4 meters (12 feet).

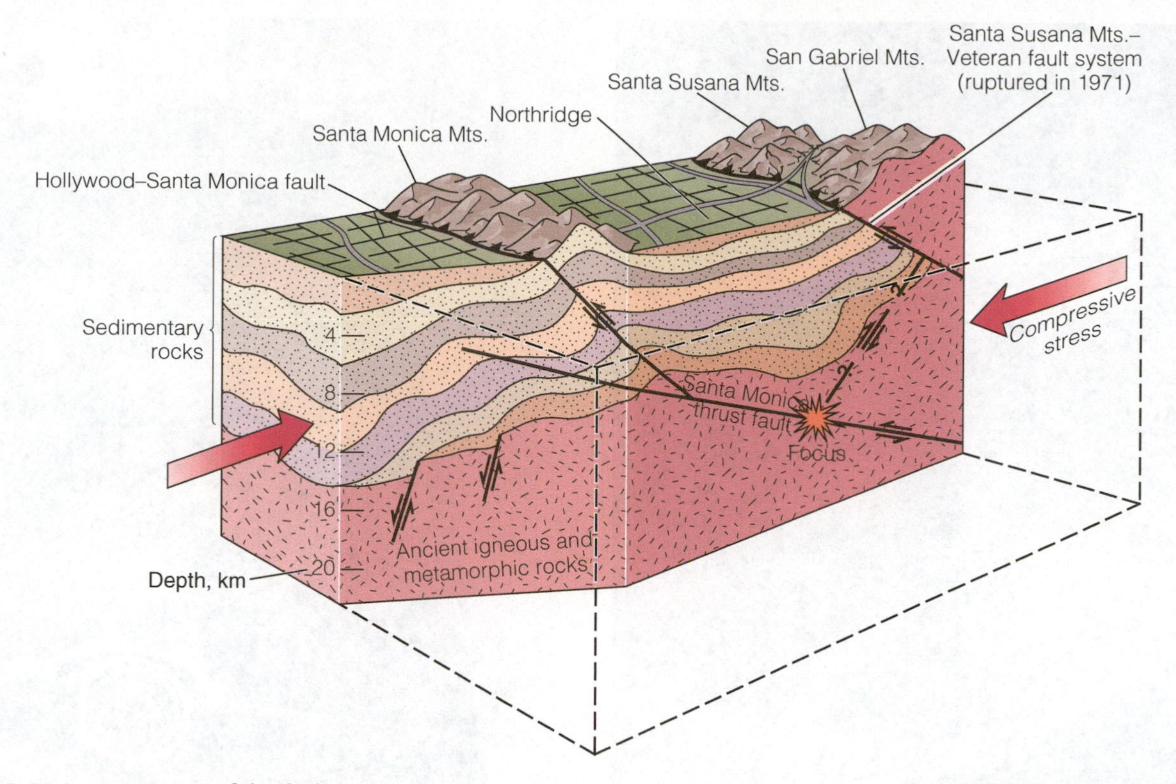

Figure 6.49 Interpretation of the fault system in the San Fernando Valley and surrounding area of Southern California. Compressive stresses built up over a long period, causing the subsurface blind thrust fault to rupture. The ensuing Northridge earthquake impacted the entire valley and extended from the Santa Susana Mountains on the north to the city of Santa Monica on the south. (Based on USGS data)

freeway overpasses were damaged (**Figure 6.51**). The reason for the extensive damage was the high horizontal and vertical accelerations generated by the earthquake. Accelerations of more than 0.30 g are considered dangerous, and a vertical acceleration of 1.8 g was measured by an instrument bolted to nearby bedrock.

The high ground acceleration explains why many people were thrown out of their beds, and objects as heavy as television sets were thrown several meters from their stands. Brick chimneys, both those reinforced with steel bars and unreinforced, came down; houses moved off their foundations; concrete-block walls crumbled; gaps broke open in plaster walls; and windows shattered. The horizontal ground motion was directional; that is, it was strongest in the north-south direction. With few exceptions, block walls oriented east-west tipped over or fell apart, whereas those oriented north-south remained standing. Many two- and three-story apartment buildings built over garages collapsed onto the residents' cars due to the lack of shear resistance in the open garage space, leading to failure of vertical supports (**Figure 6.52**).

California State University at Northridge literally sustained "textbook" earthquake damage. The university's library automated book-withdrawal program, Leviathan II, recorded a record withdrawal of about 500,000 books on the day of the earthquake–all of them onto the floor of the library. Bottled chemicals fell off their storage shelves (a major concern at any university) and caused a large fire in the chemistry building. In addition, one wall of a large, open parking structure collapsed inward to produce the stunning "art deco" architecture shown in **Figure 6.53**.

USGS

Figure 6.50 The January 17, 1994, Northridge earthquake severely damaged this reinforced concrete structure. Note that it remained standing, in contrast to the adjacent brick wall.

6.4.2 Sumatra-Andaman Islands, 2004

The third deadliest and one of the most powerful earthquakes in recorded history struck the northwest coast of Sumatra around breakfast time on December 26, 2004. The

Hum Images/Alamy Stock Photo

Figure 6.51 Freeway overpass collapse from strong shaking.

Reuters/Alamy Stock Photo

Figure 6.52 Open garages on the first floor turned out to be not such a good idea.

quake had an M_w of 9.2. Because each whole number increase in magnitude equals 30 times more energy released, the energy released in this earthquake was about 1,000 times greater than in the destructive Haiti 2010 event.

In a terrifying 6 to 8 minutes, the edge of the Indo-Pacific plate boundary flexed and lifted the coral-rimmed coasts of offshore islands closest to the 600-kilometer (350-mile)-long fault rupture as much as 90 centimeters (35 inches) (**Figure 6.54**). Meanwhile, the shoreline of the main island of Sumatra sank below sea level, inundating important farmlands, roads, and built-up areas. An enormous tsunami radiated outward from the focal area and battered coasts as far as eastern Africa, Sri Lanka, Myanmar, and Thailand. Because there was no early warning system in place to alert people of the approaching waves, nearly a quarter of a million people drowned (Chapter 10). In one instance, the sea waves derailed and washed away a crowded passenger train that happened to be traveling in the wrong place at the wrong time along the coast of Sri Lanka (**Figure 6.55**).

For geologists, the Great Sumatran earthquake was geophysically interesting because it set the whole Earth ringing like a giant bell struck with a hammer. Every location on the planet, including the remote poles and the flat plains of Kansas, rose and fell at least a centimeter in slow, gentle oscillations, with periods ranging from every 20.5 to 54 minutes, too slow for people to notice without instrumental assistance. This wiggling continued with diminishing strength for months. The quake was also strong enough to trigger aftershocks as far away as Alaska. Only a few months later, in March 2005, another great earthquake with a magnitude of 8.5 to 8.7, struck the same area of Sumatra, evidently triggered by the December 26 catastrophe. Although it did not generate a large tsunami, 1,300 people lost their lives.

6.4.3 Tohoku-Oki, Japan (2011)

The M_w Tohoku-Oki earthquake caused massive damage and revealed a lot about tsunami-generation for large subduction-zone earthquakes. A set of M_w 6.0 to 7.7 foreshocks preceded the magnitude 9.0 earthquake by a couple of days, with an ensuing pause in seismic activity. A calm before the storm. Then a modest magnitude-4.9 quake struck the epicentral region, followed by the gigantic rupture nobody anticipated based on the historical record (**Figure 6.56**). Only in retrospect do we appreciate that that record was too short, although other geological clues existed to suggest we hadn't learned enough about the seismicity of northeastern Japan. These included the natural debris left by tsunamis occurring centuries ago that could only have resulted from similarly powerful events.

iStock.com/Pandapix

Figure 6.53 Parking garage that collapsed in the Northridge quake.

© Earth Observatory of Singapore/Kerry Seih

Figure 6.54 *Pontes* coral heads at the global positioning system (GPS) survey station NGNG on one of the islands off the west coast of Sumatra. The coral heads that are normally submerged were uplifted about 90 centimeters (35 inches) by the December 26, 2004, Great Sumatran earthquake.

Reuters/Alamy Stock Photo

Figure 6.55 A Buddhist monk examines a train destroyed by the 2004 Indonesian tsunami wave 90 kilometers (56 miles) south of Colombo, Sri Lanka. Over 1,200 persons died in this train.

Seismograph data indicate that an enormous segment of lithosphere, measuring 400 kilometers (240 miles) horizontally and extending 20 to 200 kilometers (12 to 120 miles) into the Earth, ruptured along the marine trench marking the convergent plate boundary in the sea northeast of Honshu, Japan's main island. The shaking lasted a terrifying 7.5 minutes, far longer than most earthquakes. The fault displacement was by far the largest ever recorded, an estimated 40 to 80 meters (130 to 260 feet).

Enough energy was released in those 7 minutes to power the entire city of Los Angeles for a year. Because the rupture reached so close to Earth's surface, the overthrust plate experienced tremendous vertical uplift as high as 3 meters (10 feet). A giant tsunami resulted (see Chapter 10). As with the 2004 Sumatra-Andaman earthquake, most fatalities resulting from the earthquake were associated with these powerful sea waves. The quake also threw Earth off its rotational axis by 10 to 25 centimeters (4 to 10 inches),

Aclosund Historic/Alamy Stock Photo

Figure 6.56 Parts of Iwate Prefecture in Japan were utterly devastated by the 2011 Tohoku-Oki earthquake and tsunami. Street patterns are all that's left in much of this devastated area. The buildings are gone.

shortening the length of a day by 1 millionth of a second, a microsecond. (It took the planet several years to recover its ordinary polar rotation.)

Only about 5% of the destruction caused by the earthquake was due to the physical shock itself. The rest was due to the tsunami, for which inadequate coastal seawall protection existed (Chapter 1), even though some walls loomed as high as 10 meters (more than 32 feet) over the coast. Thanks to modern construction, outstanding engineering, and emergency response technology, structures held up to the strong shaking remarkably well. For instance, two dozen of Japan's famous bullet trains were operating on tracks in the region racing at speeds as high as 190 kilometers per hour (120 miles per hour) when the ground beneath began rumbling. Within seconds, automatic braking systems slowed the trains before they derailed and they stopped with no significant harm done.

The national earthquake warning system, operated by the Japan Meteorological Agency (JMA), also provided good warning that strong shaking would soon reach people living hundreds of kilometers away, perhaps saving hundreds, if not thousands, of lives. The JMA warning system includes not only a national network of more than 1,000 seismometers, but undersea sensors between the coast and the trench. In the crowded Tokyo area, people had a minute to prepare themselves, including legislators in the National Diet (parliament) then in session as alerts sounded. The alarms went off about 30 seconds after the earthquake started far to the north (**Figure 6.57**).

On the other hand, the size of the earthquake-in-progress was woefully underestimated by observers in charge of the early-warning system–and with it the size of the associated tsunami–because of incompletely processed information about the seismic energy being released. The fact that the earthquake lasted so long, mixing S- and P-wave signals together, further confused data processors.

Retrospectively, improved means of interpreting seismic and geodetic data in real-time were needed. Better ways of getting the word out to everyone also had to be considered. One can argue that a heavier loss of life occurred than would have been the case, had more accurate information been immediately available. Some citizens felt that they were safely protected by their local seawalls because tsunami heights were initially underestimated, or at least unexpected. Many survivors also blamed themselves, however, for not taking tsunami warnings more seriously. The critical window of opportunity–5 to 30 minutes in which to flee (depending upon location)–was widely squandered.

One positive outcome of this catastrophe has been the Japan-World Bank Program on Mainstreaming Disaster Risk Management in Developing Countries. A report from this program, "Learning from Megadisasters," is especially important. The report not only allowed Japan to take

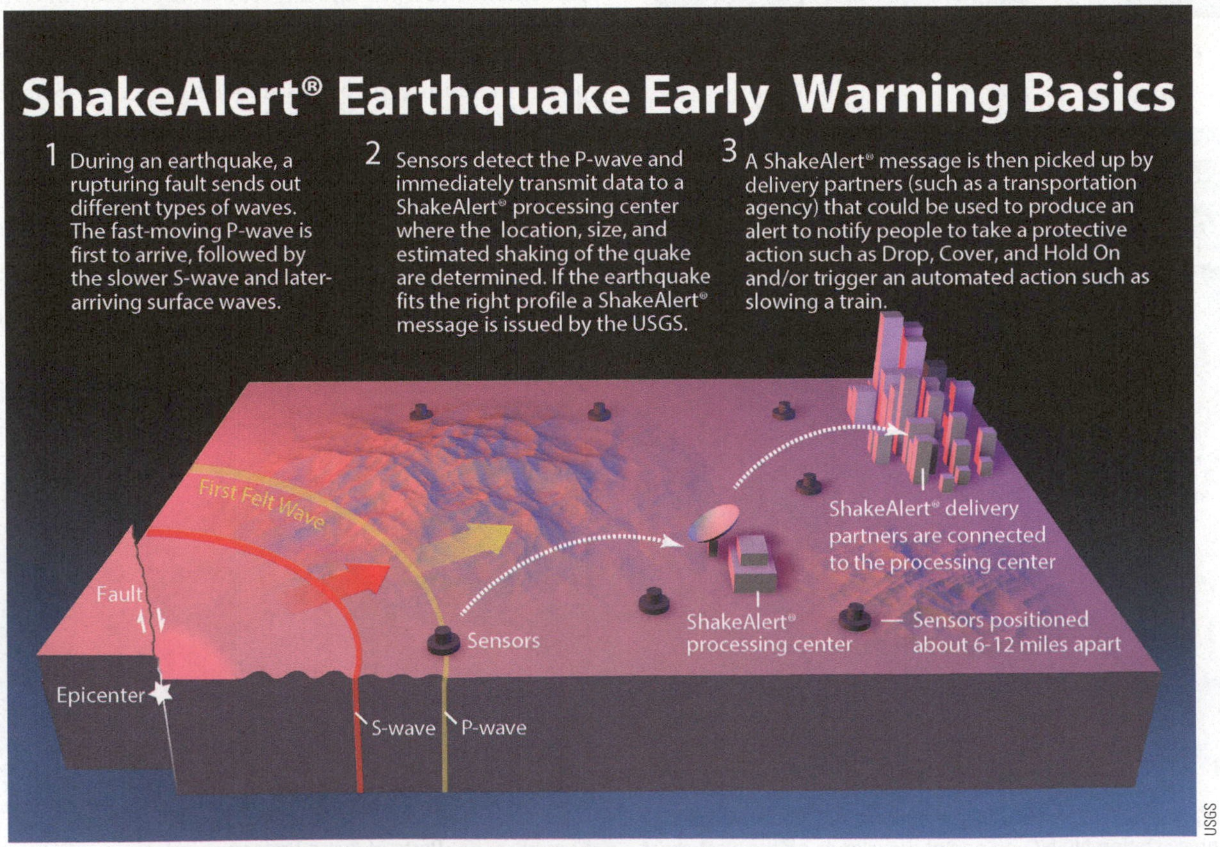

Figure 6.57 U.S. Geological Survey diagram showing how an earthquake early-warning system, in this case the U.S. ShakeAlert, can give people up to tens of seconds warning before the onset of severe shaking. A similar approach was used by the Japanese during the Tohoku-Oki earthquake.

stock of the lessons from the Tohoku-Oki quake, but to share what they learned internationally. Recommendations include improved, integrated earthquake planning between all levels of government and agencies that may be involved in relief efforts. Redundancy, the duplication of systems to keep services going in the case the ordinary providers fail, is an important goal. This includes the need to maintain safe, adequate water supplies and sanitation, transportation, and telecommunications, so that isolation, blackouts, and social breakdown don't take place in the critical hours following a massive earthquake. These are key factors needed for people to recover resiliently from disaster.

6.4.4 Christchurch, New Zealand, 2011

On the early morning of September 4, 2010, a M_W 7.1 earthquake rocked the central Canterbury Plain with an epicenter about 40 kilometers (25 miles) west of Christchurch, the second-largest city in the country. Casualties were minimal despite the strong shaking because of low population density, the modern construction of most buildings, and the fact that so many people were home asleep in timber-framed homes when the quake struck. Apart from some foundation issues and chimney damage, wood-framed dwellings generally respond well during earthquakes because of their flexibility, contrary to unreinforced stone and masonry buildings.

Many commentators compared the effects of this tremor with the one that occurred in Haiti shortly before, where loss of life was orders of magnitude greater despite lower magnitude shaking because of poorly designed construction and engineering. Long-lasting social and psychological impacts nonetheless were significant. Following the quake, the numbers of heart attacks greatly increased, overall property crimes in Christchurch greatly decreased, whereas incidence of domestic violence increased by more than 50%.

But the worst was yet to come. An unusually strong set of aftershocks developed. They spread eastward along blind faults into the Christchurch town area, showing a transfer of strain in that direction from the original rupture. Their magnitudes and frequency exceeded what was expected based on observations of similar seismic events elsewhere in the world.

Many buildings in the region had unnoticed or underestimated pre-weakening from the September 2010 earthquake. When over the lunch hour on February 22, 2011, a M_W 6.3 quake struck 10 kilometers (6 miles) southeast of central Christchurch, numerous structures were seriously damaged that might have withstood the shaking had it occurred without a precursor (**Figure 6.58**). This quake was so strong in comparison with ordinary aftershock statistics that some seismologists characterized it as an independent event–a triggered quake, one that resulted from stress transferred by earlier quakes to this section of the fault. None of this mattered to the victims, of course. The combination of pre-weakening and localization at a very shallow depth–only about 5 kilometers (3 miles) down–meant that the city suffered terribly. To compound the problem, the peak ground acceleration in downtown Christchurch reached 1.8 g, and at one local school 2.2 g.

Greg Ward NZ/Shutterstock.com

Figure 6.58 Emergency responder examines a building front that has shed masonry during heavy shaking in Christchurch, New Zealand (2011). Exiting a building once a quake has started risks serious injury.

What preconditioned the ground to whip around so violently? The shallowness of the earthquake focus was certainly partly responsible. But perhaps the presence of abundant shallow groundwater played a role as well. A water-saturated mass of sediment will shake like a giant bowl of gelatin during an earthquake, whereas solid bedrock can transmit seismic energy readily without amplified shaking. The presence of the large amounts of groundwater right beneath metro Christchurch could be easily seen even during the September 2010 earthquake. Sandy, silty liquid–expelled groundwater–erupted from the earth in many districts during the shaking, causing roads, foundations, and fields to crack and settle (**Figure 6.59**). The resulting liquefaction features are called **mud volcanoes** or **sand blows**, which in this case were especially spectacular.

Nearly 200 people died in the February earthquake, victims not only of pre-weakened structures and strong seismic accelerations, but of a time of day that engaged

Kyodo/Reuters

Figure 6.59 Dazed residents examine fissures and boils (cratered mounds formed by erupting groundwater) during liquefaction in a suburban field, Christchurch, New Zealand, 2011.

many people in activities that made them vulnerable, including using sidewalks along streets where they were exposed to falling masonry, and doing office work in non-wood frame structures that could not withstand the shaking. Important to note, however, is that most fatalities occurred at only a few building sites. The Canterbury TV building alone, a fairly modern structure, took 115 lives when it collapsed. This is typical of many earthquakes—human loss is concentrated in a relatively few key structures. Such underscores once again the importance of building well in earthquake country.

6.4.5 Earthquakes Far from Plate Boundaries

While the vast majority of earthquakes are associated with plate interactions at their boundaries, some very large ones have taken place historically far from such zones. Called **intraplate earthquakes**, they can be especially dangerous because they are so infrequent that people believe they live in areas that are not susceptible to seismic activity. In these areas, building codes and construction practices are generally not well established to withstand strong shaking, even in well-resourced economies. Intraplate earthquakes have several characteristics in common:

1. The faults that cause them are ancient–even hundreds of millions of years old–and deeply buried with little, if any, evidence of surface rupture before slipping.
2. Because the rocks of plate interiors are structurally more homogeneous than those at plate boundaries, which are laced with faults, they more effectively transmit seismic waves, causing ground motion over a huge area (**Figure 6.60**).
3. Their origins relate to strain that, in most cases, accumulates much more slowly than it does at plate boundaries.

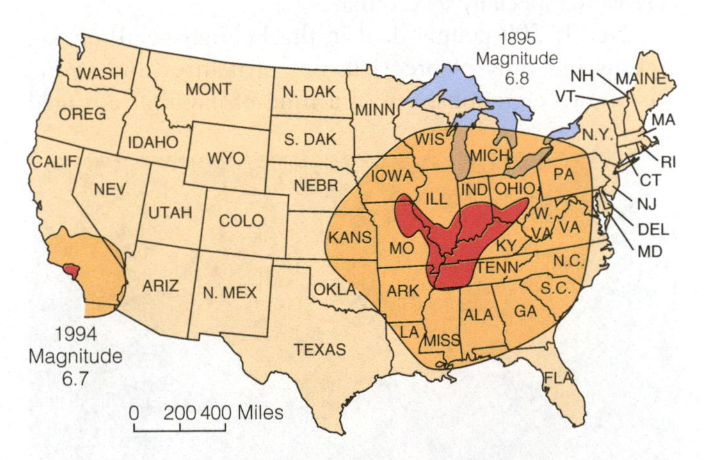

Figure 6.60 Comparison of the effects of two earthquakes of similar magnitude. The 1895 Charleston, Missouri, mid-continent earthquake impacted a large area in the central United States, whereas the 1994 Northridge earthquake's effects were limited to Southern California. Red indicates areas of minor-to-major damage to buildings and their contents. Yellow indicates areas where shaking was felt but there was little or no damage to objects, such as dishes.

Principal causes of strain in these settings include the twisting of deep plate interiors by interactions that occur on distant plate boundaries, the loading by sediment over millions of years atop ancient rift-related faults at former divergent margins, and vertical adjustments in the lithosphere caused by deglaciation or heat loss. Ancient faults are fleetingly reactivated by these seemingly innocuous, "everyday" processes.

One of the most famous intraplate earthquakes in history took place in the central United States: the New Madrid (pronounced "MAD-rid") earthquake swarm of 1811–1812. New Madrid is a small community in southeastern Missouri in the northern Mississippi River Valley. It lies within the Midwest Earthquake Zone (MEZ), a belt of mostly low-level seismic activity that roughly outlines a down-dropped block of crust largely buried by sedimentary material, called the Reelfoot Rift (**Figure 6.61**). The rift probably formed

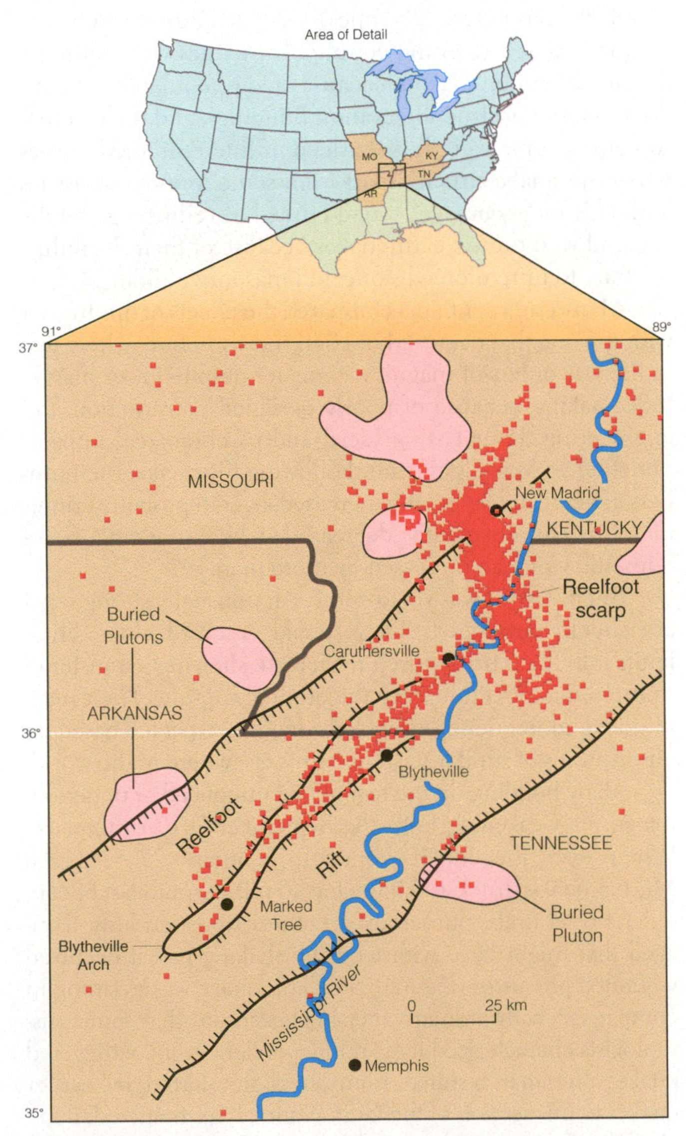

Figure 6.61 The New Madrid seismic zone covers parts of four states and consists of an ancient, fault-bounded depression (rift) buried below 1.5 kilometers (almost a mile) of sedimentary rocks. An area of major seismic hazard, it is studied intensively. (Based on USGS data)

around a half-billion years ago as the young continent of North America began to split apart. The new divergent plate boundary did not fully develop, however, and the continent rejoined before any through-going seaway could form. The zone of weakness represented by the ancient rift faults nonetheless remained. Perhaps it is no accident that the Mississippi at least partly follows the path of this failed rift system, too, inherited from a past topographic influence that no longer exists.

Shortly before Christmas in 1811, the MEZ suddenly generated a great earthquake (**Figure 6.62**). At that time, few people lived in this frontier region, but the shaking spread so widely across the plate interior that it swayed Washington, D.C., 1,600 kilometers (1,000 miles) away. Church bells rang and scaffolding around the Capitol, then under construction, collapsed. Perhaps no earthquake in recorded history has been more widely felt.

The initial shock triggered two additional great tremors in the following 3 weeks. Nobody knows how many people died. An amateur observer in Louisville, Kentucky, recorded almost 2,000 individual quakes with a homemade pendulum. Liquefaction was widespread along the Mississippi River, and the town of New Madrid, Missouri, was destroyed as it abruptly subsided 4 meters (13 feet). For a short while during the third main event, near the later Civil War battlefield of Island Number Ten, the Mississippi River actually ran backward–even spilling over upstream-facing waterfalls! Subsidence caused some swamps in its floodplain to drain, whereas others became new water bodies–the most famous being Reelfoot Lake in northwest Tennessee, which today is more than 15 meters (50 feet) deep. It seems that the old Reelfoot Rift had suddenly sprung back to life.

Although these events are long forgotten by most Americans, seismic trouble remains part of the popular imagination of people in the Mississippi River Valley (**Figure 6.63**). Iben Browning, a scientist with a Ph.D. in physiology but who is best known for his work on climate, predicted there would be a repeat of the 1811 New Madrid earthquake on December 3, 1990. His prediction was based on alignment of the planets and consequent gravitational pull, which he believed would be sufficient to trigger an earthquake on that date. Although other scientists discounted the prediction, it created considerable anxiety in Arkansas, Tennessee, Alabama, and Missouri. Months before "the day," earthquake insurance sales boomed, moving companies were booked up, and bottled-water sales rose dramatically, as did sales of quake-related souvenirs. One church sold "Eternity Preparedness Kits," and "Survival Revivals" were held. On the date of the scientifically discredited prediction, nothing earthshaking occurred.

USGS

Figure 6.62 Since the New Madrid quakes happened decades before photography became widespread, there are no images from the time, but art from 1851 shows cabins tilted by the tremors.

The Riverfront Times

Free | St. Louis' Best... Every Week | 76 Pages

December 5–11, 1990 | Number 635

EARTH QUACKS

The great quake of 90 scored a zero on the Richter scale, proving that Iben was as terrible as the hypesters who took him seriously. The Midwest had gotten stupid for nothing. And it wasn't New Madrid's fault

BY RICHARD BYRNE JR.

So did the earth move for you?

It did for the media and for local government, from the moment that climatological crackpot Iben Browning found a Midwestern audience for his bizarre seismology.

For them, the prediction of a Dec. 3 quake (give or take 48 hours) on the New Madrid fault and the panic that ensued were nothing short of titillating. There were scores of earthquake-preparedness videos, full-dress drills and pull-out sections. Guns and batteries and bottled water sold like there was, literally, no tomorrow. The boys in the fire departments parked the big red trucks outside to keep the firehouses from crumbling upon them. John Ashcroft trudged grimly along the fault line, while scores of satellite dishes bounced the message of doom from the still earth to

continued on page 10

USGS

Figure 6.63 Papers in the Mississippi Valley poked fun at the failed earthquake forecast.

Figure 6.64 Paleoseismic evidence clearly shows that the Reelfoot Rift has produced large earthquakes in the past. (a) Trench excavated to find evidence of prior earthquake preserved in sediment. (b) Sand blows found in trenches dug across the fault scarp. These show strong shaking prior to the most recent (1811–1812) earthquake sequence.

More recently, detailed analysis of the geology shown by trenching across the Reelfoot Fault scarp reveals evidence of three large prehistoric earthquakes within the past 2,000 years, giving an average recurrence time of one quake every 600 to 700 years (**Figure 6.64**). The recurring earthquakes were not necessarily of the same magnitude as the 1811 to 1812 events. Nonetheless, the evidence indicates that the citizens of St. Louis or Memphis should be as concerned about earthquake preparation as those living along plate boundaries. Citing field evidence and statistics, some geologists have argued, in fact, that there is a 90% probability of a M_W 6.7-scale earthquake in the MEZ before the year 2050.

On August 23, 2011–a very "quaky" year worldwide–a magnitude 5.8 earthquake struck the United States East Coast, centered in Virginia. This was not the strongest tremor to strike North America, but no earthquake on the continent has ever been felt by more people (**Figure 6.65**). Buildings were damaged as far as 80 kilometers (50 miles) from the epicenter. The shaking crumbled the spire of the National Cathedral in Washington D.C. and damaged the Washington Monument and Smithsonian Institution. Total damage ranged from $200–300 million. No one was killed, though many people were traumatized by the shock of something so unexpected. This quake was not alone, on August 9, 2020, a M_w 5.1 quake hit Sparta, North Carolina. It ruptured the ground, leaving a reverse fault scarp up to 25 centimeters high that extended about 2 kilometers–a very unusual occurrence along a tectonically passive continental margin.

In contrast, a much stronger intraplate earthquake (~M_W7.2) struck Charleston, South Carolina in 1886, killed 60 persons and seriously damaged over 2,000 structures (**Figure 6.66**). As in the case of New Madrid and Christchurch, New Zealand, widespread liquefaction took place, with the "eruption" of boils and flows of mud pouring out of cracks in the ground. These were triggered by strong shaking and relatively high lateral acceleration that also

Figure 6.65 U.S. Geological Survey commemorative post regarding the 2011 Virginia earthquake, which caused widespread unease on the East Coast.

USGS

Figure 6.66 Extensive damage to unreinforced masonry buildings in downtown Charleston, South Carolina from the August 31, 1886 earthquake.

severely damaged structures (**Figure 6.67**). The temblor lasted around 40 seconds and was felt as far north as Boston. People were so frightened by the unexpected quake, most initially refused to sleep in their homes (**Figure 6.68**).

In the 30 years following, over 300 aftershocks rattled the region, keeping the population on edge. One historical consequence of the Charleston quake is that it prompted the first-ever seismic retrofitting campaign on America's "placid" East Coast; the installation of reinforcement bolts in buildings. In purely monetary terms, it was less destructive than the much weaker 2011 Virginia tremor, but only because there was far less infrastructure and construction to damage there in the late-nineteenth century.

Historical/Corbis Historical/Getty Images

Figure 6.67 Locomotive derailed by strong shaking during the August 1886 Charleston earthquake. Ten Mile Hill, Berkeley County, South Carolina, CA. September 1886.

Everett Collection Historical/Alamy Stock Photo

Figure 6.68 After the earthquake of 1886, African Americans sheltered in a Charleston park to avoid their damaged homes.

Potential seismic hazards across the 48 contiguous United States are not uniform (**Figure 6.69**), Note that as much of the eastern half of the country, far from a plate boundary, can be impacted, as can locations in the West, where vigorous mountain building and volcanic activity are taking place. There is a noteworthy difference in the intensities, however: broader swaths of western North America are subject to severely damaging shaking than seismically vulnerable regions farther east. Large quakes in the West are much more common, with shorter recurrence intervals. Large eastern quakes are rare, so hazardous areas reflect only the quakes we know of. There are likely more seismically active places than we realize–a scary thought, for sure.

6.4.6 What We Can Learn

A number of important lessons derive from these examples of seismic calamity. For instance, in the Los Angeles urban area, the real earthquake threat is not from the giant San Andreas Fault 35 kilometers (20 miles) to the northeast, but rather from reverse faults that slant unseen directly beneath the city. Not only shallow, such faults also tend to generate unusually strong ground accelerations. The damage illustrated some weak links in modern architecture, including chimneys and apartment complexes.

Both the Sumatra-Andaman and Tohoku-Oki earthquakes set the whole earth ringing like a bell and wobbling on its rotational axis for months (a phenomenon called the Chandler wobble). Although impressive and maybe even unnerving, this displacement is not threatening, because planetary rotation is self-correcting. More importantly, the Sumatra-Andaman earthquake illustrated the potential for earthquake triggering–something reiterated in the Christchurch example. Both Sumatra-Andaman and Tohoku-Oki also make the point that although it's nice to live next to the beach, residents must be ready to leave home quickly after a sharp tremor. When they return, they could well find that seafloor and shoreline aren't quite positioned where they

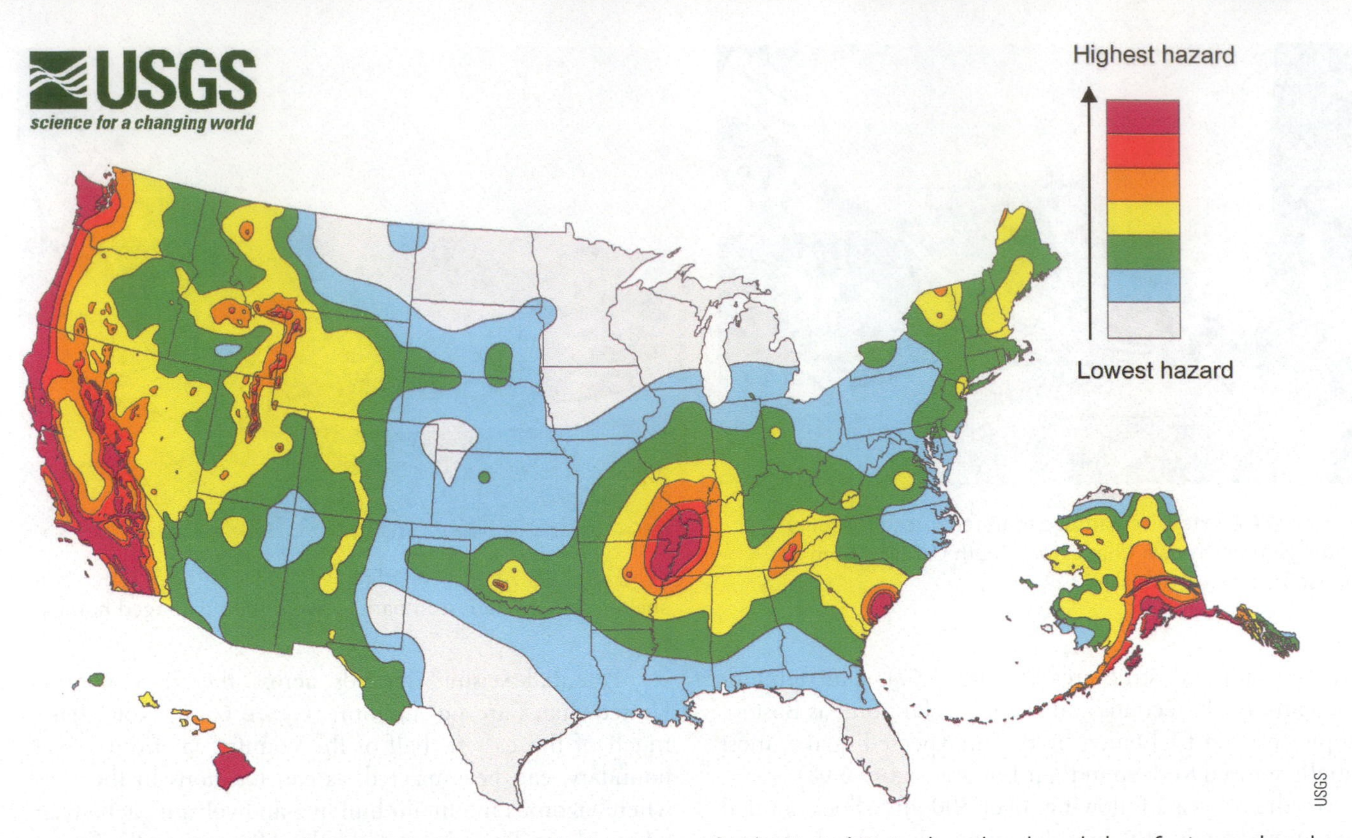

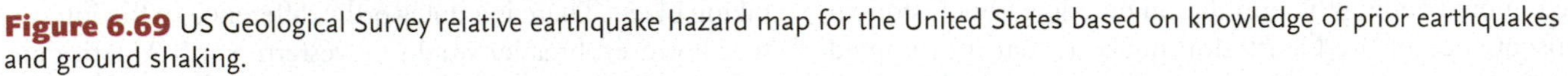

Figure 6.69 US Geological Survey relative earthquake hazard map for the United States based on knowledge of prior earthquakes and ground shaking.

Human-Induced Earthquakes

Most earthquakes are the result of tectonic forces at energy levels far beyond anything that humanity can match; and yet the ordinary activities of people can also trigger earthquakes! Every year there are 3 million earthquakes that naturally rattle the planet; that is 8,000 a day, most too small to be felt by people. To this number may be added, however, hundreds of additional **human-induced earthquakes (HIE)** that also take place annually, according to HIQuake–the Human-Induced Earthquake Database in the United Kingdom–and the U.S. Geological Survey. How is this possible?

Earth's crust is everywhere under stress, and a delicate balance exists in many places between the strength of rocks, the directions and magnitudes of stresses acting upon them, and the pressure of fluids that they contain. If this balance is upset, then the rock will fracture or rock on either side of an existing fracture will move; that is, an earthquake occurs. In many places, faults become inactive as the balance of forces inside Earth is restored. Inactive faults can remain so for thousands, or even millions of years. But people can reactivate them in several ways, in most cases simply by adding a proverbial straw to the camel's back (**Figure 6.70**).

About half of all HIE are a result of mining, oil/gas extraction, or reservoir impoundment. This changes the stress level on the surrounding rocks, in some places critically. For example, China reported in 2007 over 100 mines experiencing earthquakes with $M_W > 4$. Even a large, open pit mine can induce seismicity in the local crust as it is excavated. The causes vary. Groundwater percolating through the bottoms of newly filled reservoirs, or deliberately injected by fracking (Chapter 11), and petroleum or municipal wastewater disposal operations (Chapter 12) all are triggers. The resulting earthquakes are mostly in the 3 to 4 magnitude range, strong enough to be felt though not significantly damaging. Some are quite powerful indeed, though.

Prior to development of the Konya Hydroelectric Project in the State of Maharashtra, western India, in 1962, no seismic activity was reported regionally. However, the project filled four reservoirs and added a tremendous amount of new groundwater to the underlying crust (Chapter 8). Seismic activity developed over an area of 745 km^2 (288 mi^2), culminating in a M_W 6.3 temblor that killed 180 persons and injured another 1500 in December 1967. Earthquake activity has continued to the present, with most tremors taking place in June and July–wet months related to annual onset of the southwest monsoon. Over 20 have exceeded M_W 5, and over 200 have been stronger than M_W 4.

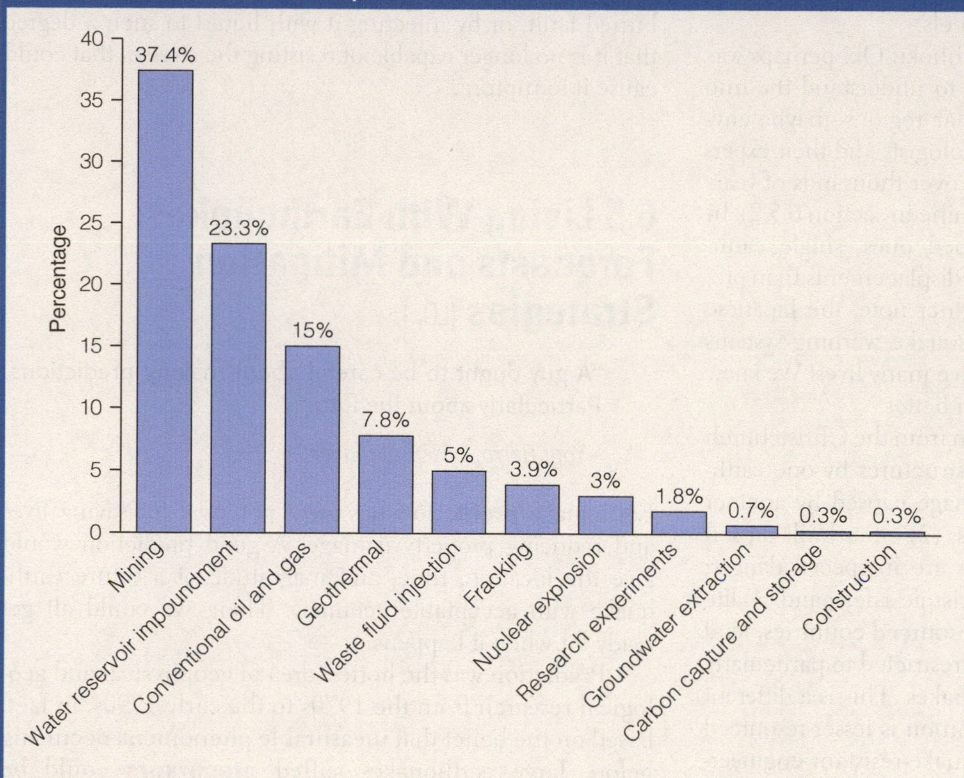

Figure 6.70 Drivers of human-induced earthquakes. (Data from The Human-Induced Earthquake Database)

To date, probable HIE have been reported at over 70 dam and reservoir sites worldwide. The most controversial case is filling of Zipangu Reservoir in Sichuan, China, in 2008. That earthquake measured M_W 7.9 and killed 80,000 persons (**Figure 6.71**)! A fault had ruptured 14 to19 kilometers (8.7 to 12 miles) deep, directly below the young reservoir. This focal area is so deep, however, that a correlation with changing groundwater conditions isn't clear.

Fracking is the process of injecting high-pressure solutions into existing cracks within the crust, causing them to open (fracture) in such a way that petroleum and natural gas seep into them and concentrate so that they can be easily extracted (Chapter 11). The solutions under pressure enable rocks making up the crack walls to slide past one another. Fracking can cause earthquakes, in other words, and these are not necessarily trivial. The strongest fracking-related earthquake to date took place near Pawnee, Oklahoma in 2016 (**Figure 6.72**). It measured M_W 5.7, and seriously damaged over a dozen homes. A similar earthquake in China killed two persons in 2019. In general, however, fracking-related HIE are less common and powerful than those related to wastewater injection disposal, which has gone on for much longer and is more widespread across the United States (Chapter 12). Furthermore, many areas experiencing fracking operations have not been seismically activated, at least in terms of felt shocks.

Overall, the U.S. Geological Survey estimates that some 7 million Americans are at risk of suffering potential damage from HIE. Those facing losses from natural tectonic quakes remains far greater, though, mostly concentrated in the plate boundary regions of the West Coast.

Figure 6.71 Crazily tilted buildings, some partly collapsed, in the aftermath of the 2008 Sichuan, China, earthquake. Some geologists have suggested that this shake related to the filling of a large, nearby reservoir.

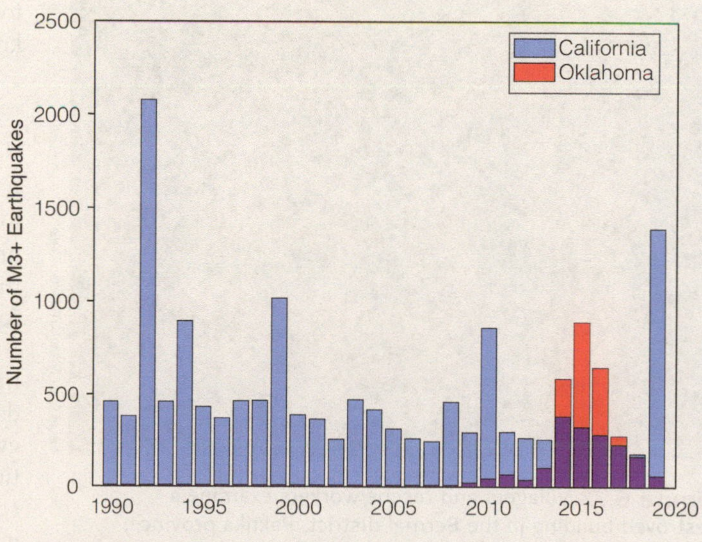

Figure 6.72 For several years, due to fracking, Oklahoma had more earthquakes than California! (Based on USGS data)

were before the earthquake occurred, requiring emplacement or construction of new coastal facilities, including roads, seawalls, piers, and boat channels.

The most important lesson of Tohoku-Oki perhaps was that our human history is too short to understand the true seismic risk and patterns of particular regions–maybe anywhere in the world. That's where geologists and their expertise in deciphering Earth's behavior over thousands of years become really important (as we examine in section 6.5.2). In some regions, even highly developed ones, single earthquakes may be capable of far greater displacements than previously thought possible. On a brighter note, the Japanese example showed that modern earthquake warning systems and technologies do work and can save many lives! We know now how to make these systems even better.

A number of key messages stream from the Christchurch example: (1) the pre-weakening of structures by one earthquake will greatly increase the damage caused by another large earthquake that shortly follows; (2) cities built on soft sediment with shallow groundwater are in special danger; (3) time of day matters in terms of seismic safety; and (4) life and property loss, at least in more resourced countries, tend to be concentrated in pockets–even restricted to particularly weak buildings–during large earthquakes. This is a different pattern of damage than what is common is lesser-resourced countries. Without access to earthquake-resistant engineering and building techniques, such countries and their people are threatened by even modest earthquakes. The 2021 Haitian earthquake and the 2022 Afghan quake are examples where weak structures, many of which were unreinforced masonry, collapsed from shaking (**Figure 6.73**).

Finally, as the Box above illustrates, human activity can cause destructive earthquakes. We must learn to anticipate this when undertaking certain new projects, especially when these involve changing the weight of the crust on top of a buried fault, or by injecting it with liquid to such a degree that it is no longer capable of resisting the stresses that could cause it to rupture.

Ahmad Sahel Arman/AFP/Getty Images

Figure 6.73 Villagers and rescue workers examine a destroyed building in the Bermal district, Paktika province, Afghanistan, on June 23, 2022. Numerous unreinforced masonry buildings, like this one made of mudbricks, suffered severe damage from this moderate sized quake (M = 5.9). More than 1,000 people died.

6.5 Living With Earthquakes: Forecasts and Mitigation Strategies (LO5)

> "A guy ought to be careful about making predictions. Particularly about the future."
>
> –*Yogi Berra, baseball player*

Earthquake **prediction** has great potential for saving lives and reducing property damage. A good prediction would give the location, time, and magnitude of a future earthquake with acceptable accuracy. Then, we could all get ready for when it happens.

Prediction was the hottest area of geophysical and geological research from the 1970s to the early 1990s, in fact, based on the belief that measurable phenomena occurring before large earthquakes called **precursors** could be identified. Researchers studied earthquakes and seismograms where presumed precursors were seen. Such things as changes in the ratio of P- and S-wave speeds, ground tilt, water-well levels, and emissions of noble gases in groundwater were measured. Unfortunately, the hope of finding precursors that would lead to reliable predictions seems to have evaporated. In fact, the U.S. Geological Survey is on record as saying that the prospect of earthquake prediction is slim to none. In short, earthquakes cannot be predicted because the mechanics of earthquake generation are too complicated to evaluate given our present state of knowledge.

6.5.1 Forecasts

A new approach that seems more promising, if less exact, is to evaluate the probability of a large earthquake occurring on an active fault or in a particular region during a given time period. This falls under the heading of **long-term forecasting**.

By compiling geological evidence pertaining to the timing of past earthquakes in a region, we acquire basic data for calculating the statistical probability for future events of given magnitudes (**Figure 6.74**). These calculations may be done on a worldwide scale or on a local scale, such as the example in **Figure 6.75**. Analysis of the graph indicates that, for the particular area in Southern California, the statistical **recurrence interval**–that is, the length of time that can be expected between events of a given magnitude–is 1,000 years for a M_W 8.0 earthquake

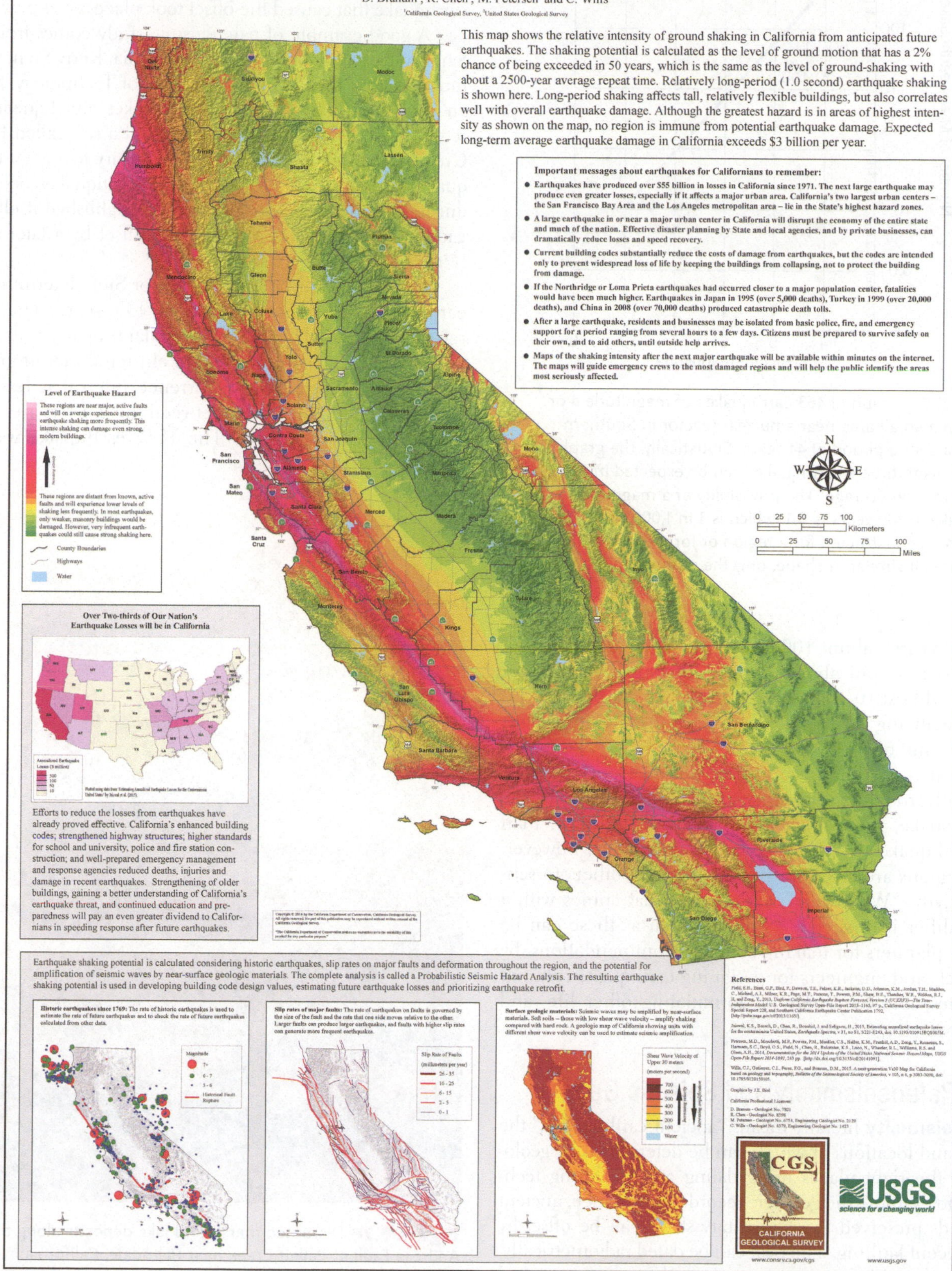

USGS

Figure 6.74 Large map represents seismic hazard level (color-coded) for the State of California based on mapped faults, their known frequency of movement, and the strength of surface materials when shaken (the smaller maps).

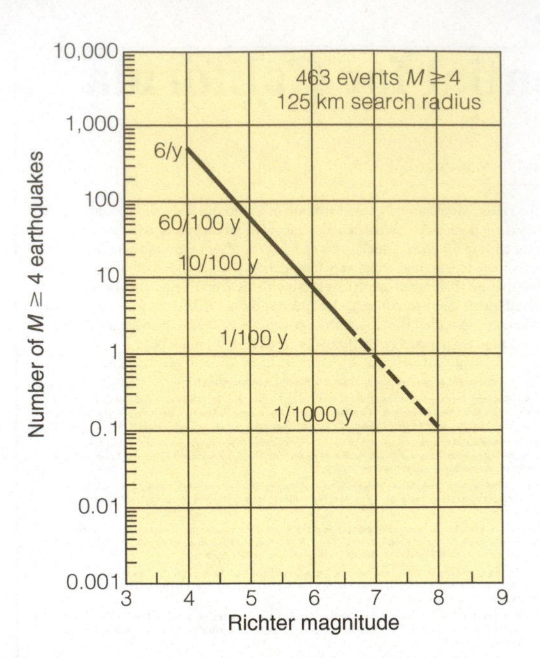

Figure 6.75 Graph of 463 earthquakes of magnitude 4 or greater in a small area near a nuclear reactor in Southern California over a period of 44 years. Statistically, the graph shows that 600 magnitude-4 earthquakes can be expected in 100 years, or 6 per year on average. The probability of a magnitude-8 earthquake in 100 years is 0.1, which is 1 in 1,000 years. Such plots can be constructed for a region or for the world. The graphs are all similar in shape; only the numbers differ.

(0.1/100 years), about 100 years for a M_W 7.0 earthquake (1/100 years), and about 10 years for an earthquake of M_W 6.0 (almost 10/100 years). The probability of a M_W 7.0 event occurring in any one year in Southern California is thus 1%, and of a M_W 6.0, 10%.

On an annual basis worldwide, we can expect, on average (with roughly a century of data), at least 1 M_W 8 earthquake, 17 M_W 7 earthquakes, and no fewer than 134 earthquakes of M_W 6.0. As you now know, however, some regions are far more susceptible than others to seismic activity. "Worldwide" is a concept that comes with a big qualifier in this case! Numbers such as these can be used by planners for making zoning recommendations, by architects and engineers for designing earthquake-resistant structures, and by others for formulating other life- and property-saving measures.

6.5.2 Paleoseismicity and Seismic Gaps

Paleoseismicity literally means "ancient earthquakes," the timing and locations of which can be determined by geologists in the right places using dating and trenching techniques to examine the geologic record. For instance, ancient peat beds preserved in sedimentary strata can be offset by more recent faulting. The peat can be dated radiometrically, using carbon-14 from ancient plant tissues or preserved wood. The peat may be exposed naturally in a stream cut, accidentally in a road cut, or deliberately by seismologists making a trench at a right angle across an active fault, as is often done. We know that any offset along such a fault must have taken place after the peat bed formed. The date obtained provides a maximum possible age for when the earthquake that caused the offset took place.

A good example of paleoseismic study comes from the San Andreas Fault in Southern California. Kerry Sieh, a seismologist from the California Institute of Technology, found an intriguing history of past earthquakes and liquefaction recorded in disrupted marsh deposits at a site called Pallett Creek, extending from the seventh century to a great earthquake in 1857 (~M 7.9). Ten large earthquakes occurred during this interval. A fresh marsh re-established itself after each quake, only to be disturbed or offset by a later shake (**Figure 6.76**).

Using carbon-14 dating, Professor Sieh determined an earthquake recurrence interval of 135 years. The dates reveal that they did not occur at regular time intervals, however; instead, the earthquakes are clustered in four groups. Within each cluster, the recurrence interval is less than 100 years, and the intervals between the clusters are two to three centuries in length. The 1857 earthquake was the

Figure 6.76 Disrupted marsh and lake deposits along the San Andreas Fault at Pallett Creek near Palmdale, California. Sediments range in age from about 200 C.E. at the lower left to 1910 C.E. at the ground surface. Several large earthquakes are represented by broken layers and buried fault scarps.

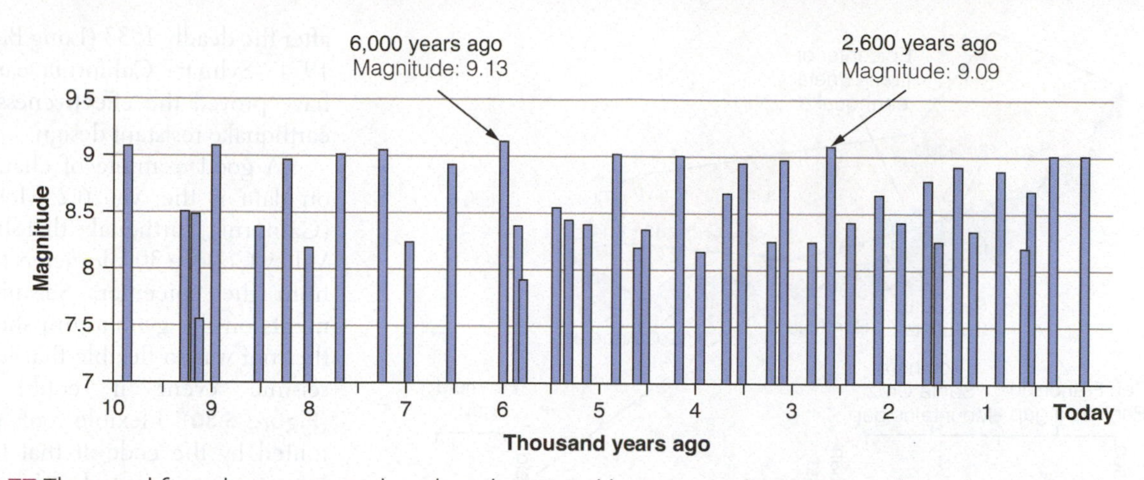

Figure 6.77 The record from deep ocean muds and sands retrieved by marine sediment coring shows that some 40 large quakes have shaken the Pacific Northwest in the last 10,000 years. The quakes leave tell-tale sandy deposits from erosion along the coast–a contrast to the mud that typifies deep sea sediment. Quakes occur every 246 years on average. (Data from C. Goldfinger, USGS)

final one in the latest cluster. Paleoseismic data suggest that this section of the San Andreas may remain dormant until late in the twenty-first century or beyond, which could be good news, at least in the short term, for the nearby City of Los Angeles, which draws much of its water supply from expensive aqueducts constructed across the fault.

Another example of valuable information coming from paleoseismic research pertains to the Pacific Northwest (British Columbia, Washington, Oregon, and northern California). In years past, earthquake hazards were considered minor in Oregon and Washington–something only people in California and Alaska, sharing the same coastline, had to worry about. During the 1980s, however, research changed this perception, revealing geological evidence that major or great subduction-zone earthquakes occurred repeatedly in the region prior to Euro-American settlement. Based on ocean cores and on buried salt marsh sediments, these quakes appear to recur every few hundred years. About 40 of these large subduction zone quakes have struck in the last 10,000 years (**Figure 6.77**). Seismologists warned on this basis that giant earthquakes will continue striking the region–the next perhaps sooner than we think, given after all that an average implies a range.

The last large Northwestern shake occurred in 1700, approximately 300 years ago. It generated a tsunami that traveled all the way to Japan, where destruction by the sea wave was recorded by baffled survivors - baffled because they felt no quake before the wave struck. The widespread distribution of tsunami deposits along the Pacific Northwest Coast are consistent with a very large quake; the same general range as the Tohoku-Oki temblor. The source doubtlessly was the Cascadia subduction zone, which stretches from northern California to Vancouver Island in British Columbia (**Figure 6.78**). Although moderate-sized quakes have rattled Seattle, Eureka, and several other communities in the brief recorded history of the region, nothing catastrophic has taken place yet. But the Tohoku-Oki earthquake reminds us that our recorded histories are, perhaps as a rule, imperfect measures of actual risk when it comes to very large events.

Much of the early research on the Cascadia quakes was done in Johnson Hall, the Geology Department building on the University of Washington campus. Ironically, while this work was being done, a list was circulated around campus. Johnson Hall was one of the three most dangerous buildings on campus in terms of earthquake risk. Years later, it was renovated with a full seismic retrofit to make it a safer place to be when the Big One strikes.

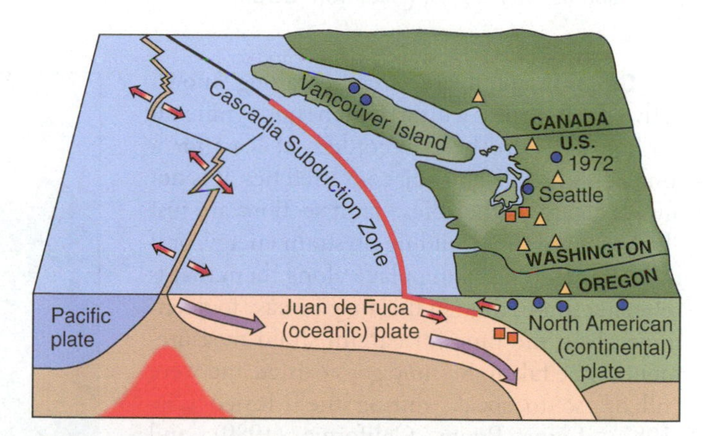

■ Deep earthquakes (>65 km or 40 mi deep)* occur when the oceanic plate descends beneath the continental plate. The largest deep earthquakes in recent times were in 1949 (M = 7.1) and in 1965 (M = 6.5).

● Shallow earthquakes (<17 km or 10 mi deep)* are caused by faults in the North American continent. Magnitude 7+ earthquakes have occurred along the Seattle fault in 1872, 1918, and 1946.

— Subduction earthquakes are huge quakes that occur when the subduction boundary ruptures. The most recent Cascadia Subduction Zone earthquake was in 1700 and it sent a tsunami as far as Japan.

Figure 6.78 Faults and historic earthquake epicenters in the Seattle, Washington, and Vancouver, British Columbia, area.

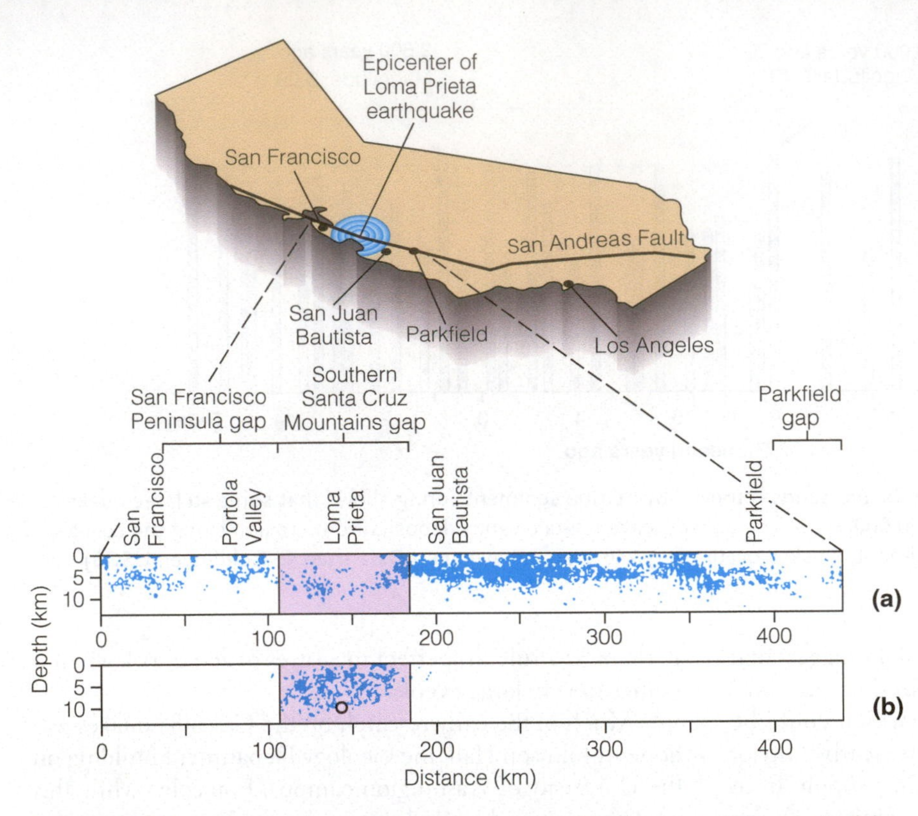

Figure 6.79 (a) Historical seismic activity of about 450 kilometers (275 miles) of the San Andreas Fault south of San Francisco before the 1989 Loma Prieta earthquake. The southern Santa Cruz Mountains seismic gap, where Loma Prieta is located, is highlighted. Other seismic gaps apparent in this historical record at the time were one between San Francisco and the Portola Valley, and one southeast of Parkfield. (b) Seismicity record of the Southern Santa Cruz Mountains portion of the fault after the 1989 earthquake. The former seismic gap had been "filled" by the earthquake (open circle) and its aftershocks. (Data from USGS)

Seismic gaps are stretches along known active fault zones within which no significant earthquakes have been recorded, at least for a long period relative to adjacent stretches. It is not always clear whether these fault sections are just "locked," and thus building up strain energy, or if motion (creep) is taking place along them that is relieving strain, though this may be easy to determine given the time for careful strain measurements to be taken. Seismic gaps existed and were "filled," so to speak, during the Mexico City (1985), Loma Prieta, California (1989), and Izmit, Turkey (1999) earthquakes (**Figure 6.79**). They serve as warnings of possible future events and can be used as forecasting tools. They provide no information about when such events will take place but can tell us where to expect them.

6.5.3 Mitigation

Reducing earthquake risks is an admirable goal of scientists and lawmakers. Building codes provide the first line of defense against earthquake damage and help ensure the public safety. Laws passed after the deadly 1933 (Long Beach) and 1971 (Sylmar) California earthquakes have proved the effectiveness of strict earthquake-resistant design.

A good example of change based on data is the M_W 6.2 Morgan Hill (California) earthquake that shook West Valley College 30 kilometers (20 miles) from the epicenter. Seismic instruments on the gymnasium showed that the roof was so flexible that in a strong seismic event it could collapse (**Figure 6.80**). Flexible roofs were permitted by the code at that time, and many gyms and industrial buildings were built that way. As a result of the experience at West Valley, the **Uniform Building Code** was revised–this is the code used by hundreds, if not thousands, of municipalities across the United States. The revision requires that the roofs be constructed to be less flexible and thus able to withstand nearby or distant strong earthquakes.

Most large cities in strong earthquake zones have their own building codes, patterned after the Uniform Building Code, that require construction to meet modern seismic standards. For instance, ground response due to differing soil types is now more appreciated than previously as a factor

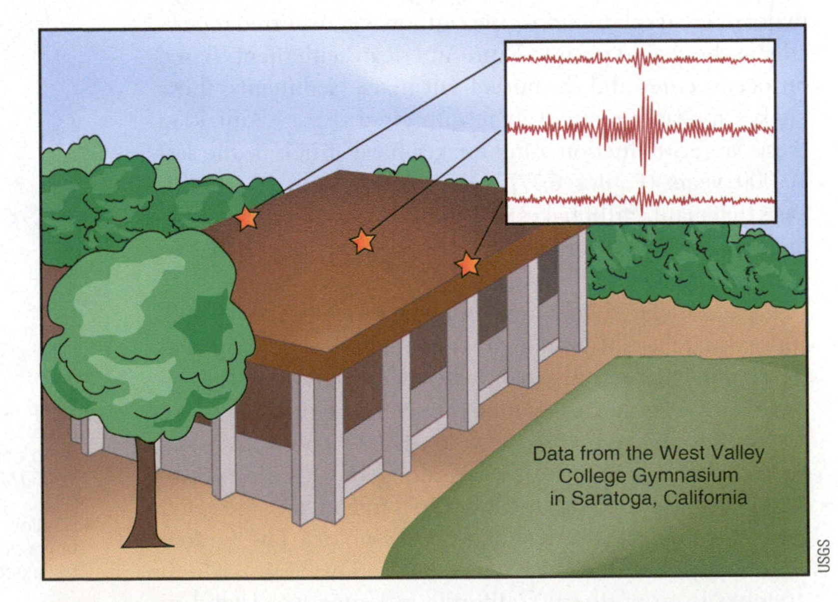

Figure 6.80 Seismic records (upper right) obtained during the 1984 Morgan Hill, California, earthquake led to an improvement in the Uniform Building Code (a set of standards used in many states). The center of the gym roof shook sideways three to four times as much as the edges. The code has since been revised to reduce the flexibility of such large-span roof systems and thereby improve their seismic resistance.

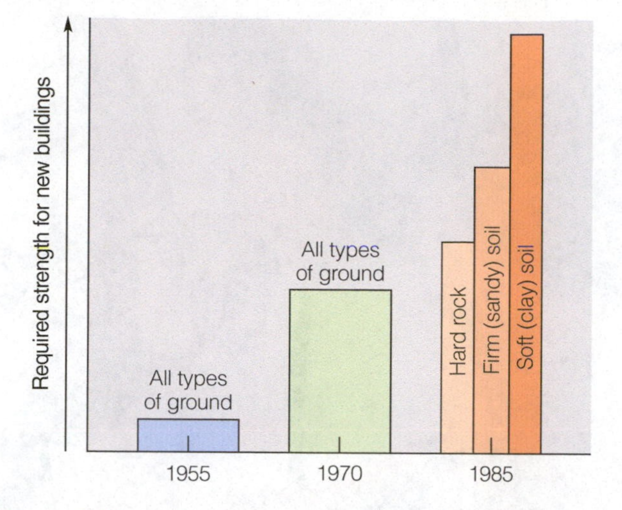

Figure 6.81 Earthquake requirements in building codes have increased over time as scientists and engineers have obtained new information. Note that recent codes specify separate criteria for different ground types.

Figure 6.82 The Space Needle in the foreground is the ninth highest building in Seattle. Mount Rainer, a composite volcano, the direct result of subduction, towers over the city from a distance.

contributing to quake damage. **Figure 6.81** shows how the code has evolved from 1955 as more information has been incorporated. For example, soft clays shake more violently with earthquake waves than does solid bedrock, and this difference in response is taken into consideration when structures are designed.

A primary objective in earthquake design is to incorporate resistance to horizontal ground acceleration; that is, base shear. As noted earlier, strong horizontal motion tends to topple poorly built structures and deform more flexible ones. In California, high-rise structures are built to withstand about 40% of the acceleration of gravity in the horizontal direction (0.4 g), and single-family dwellings are built to withstand about 15% (0.15 g). **Base isolation** is now a popular design option. The structure, low- or high-rise, is placed on Teflon plates, rubber blocks, seismic-energy dissipaters that are similar to auto shock absorbers, or even springs, which allows the ground to move but minimizes building vibration and sway.

Seattle's Space Needle, built for the 1962 World's Fair, is nearly 200 meters (605 feet) tall (**Figure 6.82**) It's also a classic example of seismic design. The needle was constructed with more mass below ground (6350 tons; 7000 short tons) than above ground (3630 tons, 4000 tons) (**Figure 6.83**). It's center of gravity is 1.5 meters (5 feet) above the ground surface and it's flexible–in other words,

Figure 6.83 Deep footings were excavated to anchor the Space Needle. The footings are steel-reinforced concrete.

it's meant to be bent without breaking, unlike the earlier mentioned gymnasium roofs. It's so flexible that during the 1965 Seattle earthquake (6.7 on the moment magnitude scale, 20 seconds of strong shaking), the rotating restaurant at the top moved side to side about 5 meters (16 feet). The restaurant manager reported that it was just like riding the top of a flagpole but in the end they lost only a few bottles of booze when they fell from the bar. Rumor has it that the wait staff working that day never came back to work.

There was a radio studio at the top of the Space Needle. It's where Frosty Fowler broadcast for Radio KING (**Figure 6.84**). He reported that structure continued to sway for at least a few minutes after the quake stopped. Structural engineer Dave Swanson when asked to comment on what will happen during a big Cascadia earthquake responded that, "The Space Needle is going to want to act like an inverted pendulum and sort of bend left and right, kind of like wheat in a wheat field during a windstorm." It will be wild ride for sure if you happen to be visiting when the Cascadia subduction zone next lets go.

Figure 6.84 The view from the Space Needle radio studio overlooking Seattle from 200 meters above the ground. What a place to be when a magnitude 6.7 earthquake hits!

6.5.4 Survival Tips

Knowing what to do before, during, and after a severe earthquake is of utmost importance to your survival should you experience one. Many governmental agencies (e.g., U.S. Federal Emergency Management Agency) provide guidelines that can be helpful in planning for "possible futures," especially for persons who live (or commute) within a few hundred kilometers of convergent and transform plate boundaries. During an earthquake:

- Remain calm and consider the consequences of your actions.
- If you are indoors, stay indoors and get under a desk, a bed, or a strong doorway.
- If you are outside, stay away from buildings, walls, power poles, and other objects that could fall. If driving, stop your car in an open area.
- Do not use elevators; if you are in a crowded area, do not rush for a door.

After the shaking stops:

- Turn off the gas at the meter if you have one available.
- Use portable radios for information; cell phone service could be unreliable.
- Check water supplies, remembering that there is water in water heaters, melted ice, and toilet tanks–all generally potable in many countries. Do not drink waterbed or pool water.
- Check your home for damage.
- Do not drive. Roads should be kept clear for emergency services, stoplights may be out, and streets may be blocked with rubble and downed, potentially live power lines.

Much less urgently, it could be helpful to take notes and photos; there's no telling if observations you make when you have the time to record them might turn out eventually to be valuable, both to others (e.g., for determining Mercalli intensities) and to yourself as a stress-reliever and personal log of a memorable experience.

Case Study

6.1 Earthquakes and Nuclear Reactors

South Africa, as it moves past apartheid, needs more electrical power as its majority Black population improves its social-economic standing. The country has large coal reserves: in 2021, it mined about half as much coal as the United States. But it has an interest in energy not sourced from fossil fuels, and so is exploring a new nuclear power facility in addition to the two reactors currently running north of Cape Town at Koeberg. The new facility would be situated at Thyspunt, on the southern coast. South Africa, with 7% of the world's uranium reserves, seeks to be energy self-sufficient by powering its own reactors.

South Africa sits on a tectonically passive continental margin, far away from subduction zone and mid-ocean ridges. One would think the chance of a strong, damaging quake is low, but nuclear power plant design mandates data collection and the creation of a **probabilistic seismic hazard analysis**–a document that lays out, based on geologic field data and laboratory measurements, the frequency of earthquakes and their magnitude in the past. This document is used to design the reactor, its building, plumbing, and pumps, to resist strong shaking.

Previous mapping by geologists showed a bedrock fault system that stretched 600 kilometers (400 miles) across the southern tip of Africa, about a hundred kilometers (60 miles) north of the proposed power plant. Intensive fieldwork by a dozen geologists scoured the length of the fault closest to the proposed plant. In some places they found beds of gravel, cemented to rock, draped over the landscape and covering the fault (**Figure 6.85**). Was the fault a dead relic of past tectonic activity or was it still potentially active? The answer was complex. Turns out, it was both.

Most of the fault zone was found to be stable, but part had reactivated–there was clear evidence of fault scarp development showing geologically recent movement along about 84 kilometers (52 miles) of the fault trace. Many other parts of the fault had gravel beds that were not offset by faulting, however, suggesting little or no recent earthquake activity in those areas. Dating the unfaulted gravel and the fault scarp were critical for evaluating seismic hazard for the ongoing power plant project.

Using isotopes of beryllium and aluminum, the gravel beds were dated. They had been sitting at or near the surface undisturbed by faulting, excepting one short stretch, for at least a million years (and maybe as much as 4 million years). That was good news; 1 to 4 million years is a long time for tranquility in human terms. Those sections of more recently active fault trace showed an offset of 1.5 meters (5 feet) about 25,000 years ago (**Figure 6.86**), with another section likely active between 10,000 and 15,000 years ago. These were nothing to worry about, however, given the distance from the planned reactor site–though earthquakes strong enough to feel (without causing damage) were certainly a possibility within the lifetime of the new facility.

Paul Bierman

Figure 6.85 The black cemented gravel in the foreground has been stable for between 1 and 4 million years. It makes up the flat surface in both the foreground and at a distance. The underlying tan quartzite bedrock in the distance is faulted, but that faulting does not offset the gravel here.

Paul Bierman

Figure 6.86 Sampling a 1.5-meter-high fault scarp for dating. The fault last moved in an event about 25,000 years ago.

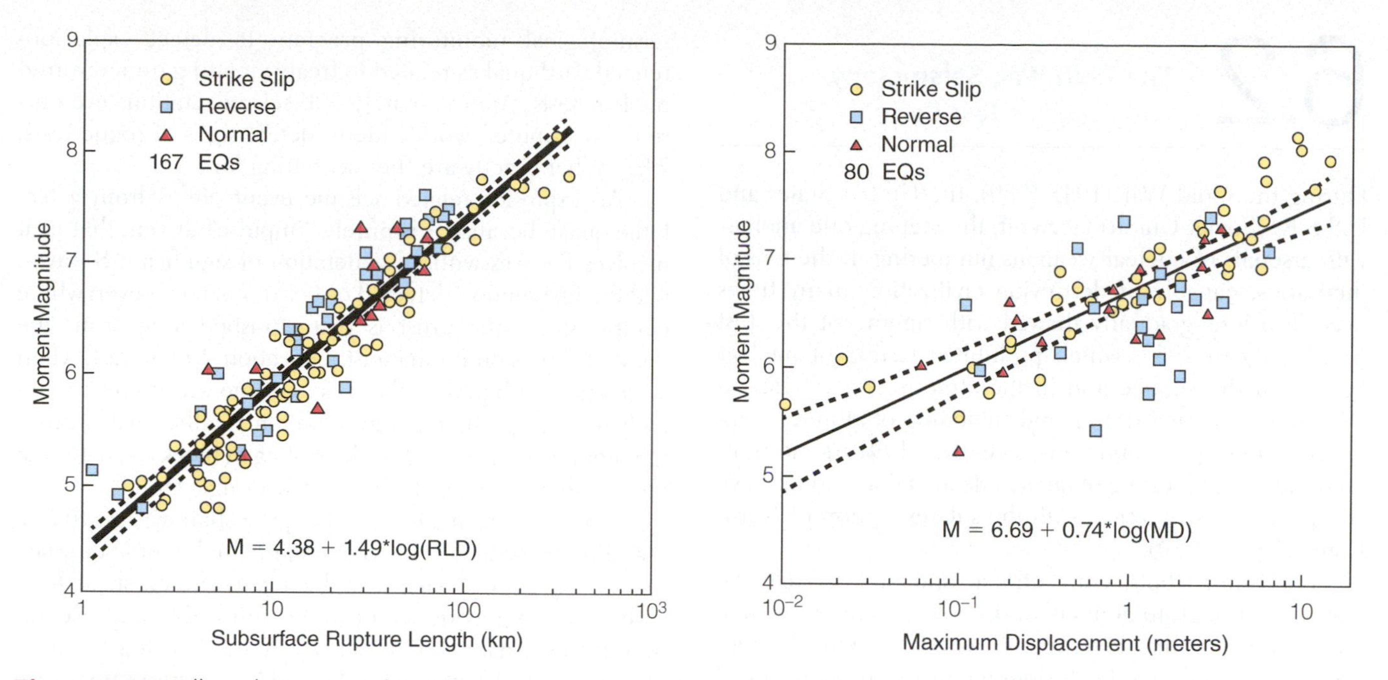

Figure 6.87 Wells and Coopersmith's 1994 graphs compile hundreds of historical earthquake and clearly show that longer fault rupture lengths and greater fault scarp displacements generate larger earthquakes. These empirical relationships are very useful for interpretting geological information in terms of paleo-earthquake strength. (Based on Donald L. Wells; Kevin J. Coppersmith Bulletin of the Seismological Society of America 1994 84 4: 974–1002.)

How does one go from geologic mapping of fault length and offsets to estimate earthquake shaking intensity and energy release–the information of greatest importance for building engineers? In 1994, two geologists, Wells and Coopersmith, analyzed hundreds of historical earthquakes for which the moment magnitude had been measured. They compared the magnitude of each earthquake to the length of its fault rupture and amount of offset (**Figure 6.87**). The graphs they created allow rough estimates of earthquake size from the geologic field information. For the dated 1.5 meter vertical offset, the resulting quake would have a moment magnitude of about 6.7. If all 84 kilometers of the potentially acitve South African fault had ruptured at once (unlikely because the dating suggests that different parts ruptured at different times), then the resulting quake would have had a moment magnitude of about 7.0. With this information in hand, engineers created an appropriate design that keeps the nuclear plant and the people around it safe from future earthquakes.

But there are other issues with the proposed plant and its location that engineering can't solve. The cost of building nuclear plants has skyrocketed as the cost for renewable energy, including wind and solar, has plummeted. The Thyspunt site holds a million-year record of human occupation and is home to massive dune fields that have preserved an exceedingly long record of climate and biologic change. These were long native African lands before colonial settlement. The proposed plant has been on the drawing board for almost two decades, but as of 2023, it's nowhere near being built. It faces steep political resistance and suggestions that it's time in some way for the country to move beyond nuclear power (**Figure 6.88**).

Figure 6.88 Local political opposition to the new nuclear powerplant is significant. Even the rubbish bins on the beach near the proposed site at Thyspunt are stickered by those who oppose the plant. Yet, South Africa faces electricity shortages and needs more power.

6.2 The Cold War, Seismology, and Nuclear Test Bans

During the Cold War (1945-1989), the United States and U.S.S.R. (Soviet Union) faced off, threatening one another with arsenals of nuclear weapons numbering in the tens of thousands, capable of destroying civilization many times over. Tensions gradually thawed with signing of the Test Ban Treaty in 1963, which prohibited testing of nuclear bombs on the surface and in the atmosphere; a treaty in 1967 that restricted testing and stationing of atomic weapons in Outer Space; and more recently, a 1991 moratorium on underground testing of nuclear devices that nations have accepted for many years, with the sole exception of North Korea (**Figure 6.89**).

Explosions (both conventional and nuclear) can be powerful enough to generate distinctive seismic wave patterns. Knowing this, governments worldwide have kept an eye on one another by becoming involved in a global seismological monitoring program to detect explosion-related earthquakes related to treaty-violating underground nuclear tests. Approximately 300 seismic stations are currently distributed worldwide to detect signs of rogue tests, 24/7. What exactly are they searching for?

Everett Collection/Shutterstock.com

Figure 6.89 The Grable atomic bomb detonation, southern Nevada, May 25, 1953. The bomb was fired as a projectile from the artillery piece in the right foreground, several kilometers away, known as the atomic cannon. Radiation from the explosion cloud spread over much of the American West in the following hours. Generally, nuclear testing of this sort was not done when the wind was blowing toward Las Vegas or Los Angeles, the largest regional metro areas.

An explosion-related seismic event differs from a tectonic quake because it is purely "impulse" driven; that is, it involves P-waves without generation of significant S-waves, and the first motion detected on seismometers is everywhere compressional–the crust is being pushed away from the point of explosion no matter the location. In contrast, when a tectonic earthquake takes place, some seismometers initially record a pulling, or tensional force, rather than a pushing motion. This represents the sliding of blocks of crust past one another (in opposite directions) along a fault.

Explosion events tend to be quite shallow; less than a few kilometers deep. This is due to the high cost of emplacing an explosive device any deeper in the crust, at least without leaving evidence of intensive drilling and excavation at the surface which can easily be detected by satellites. A sneaky country might consider escaping this issue in part by testing underground beneath the ocean; but even this is problematic. Large seabed explosions generate a distinctive type of seismic energy in the overlying water termed a T-wave.

Big underground nuclear tests also carve out large chambers that tend to collapse immediately following explosions (**Figure 6.90**). These collapses, termed chimneying, also generate seismic waves; a chaos of signals that seismologists find quite easy to identify. Sometimes multiple episodes of collapse follow a test, which illustrate its occurrence even more clearly. Strong underground blasts may exceed Richter magnitude 5, setting off a series of ordinary aftershocks along surrounding faults stressed by the explosion. In any case, the main shock itself is clearly human-induced.

International collaboration among seismologists has helped keep nuclear powers "honest," for the most part. That is not to say that these powers haven't from time to time considered ways of getting around treaty commitments or moratoriums. In the middle of the Cold War, proposals were made to test designs for nuclear bombs on the far side of the Moon or even on the far side of the Sun relative to Earth's orbital position. The presence of seismometers on the Moon's surface, and heliocentric satellites (orbiting the Sun), put an end to these ideas.

Explosions can also be done in large underground cavities filled with air that absorbs a lot of the explosion's energy. These decoupled nuclear tests still cause the surrounding crust to shake, allowing some energy to get through to the rocky walls enclosing the cavities. But the overall detectable

Science History Images/Alamy Stock Photo

Figure 6.90 Craters caused by some of the more than 1,000 underground nuclear tests at Yucca Flat on the Nevada test site, where the United States tested many nuclear weapons from 1951 to 1986. About 80% were underground tests.

strength of an underground nuclear explosion could be reduced by as much as an order or magnitude under ideal conditions, making such explosions difficult to detect, especially for smaller weapons (less than 5 to 10 kilotons of TNT; or one-third to two-thirds of the strength of the Hiroshima atomic bomb).

The United States conducted decoupled nuclear tests in southern Mississippi only about 160 kilometers (100 miles) from New Orleans, at a place now known as the Salmon Site, beginning in 1964. These were the only nuclear tests to have taken place in the eastern part of the United States. (America tested almost all of its bombs in the southern Nevada desert, and, previously, on remote atolls in the southwest Pacific, as well as one in Alaska). The Mississippi blasts took place in a salt dome, a large mass of upwelling halite similar to those found in the main nuclear test range of the Soviet Union at the time. The seismographs were ready (**Figure 6.91**).

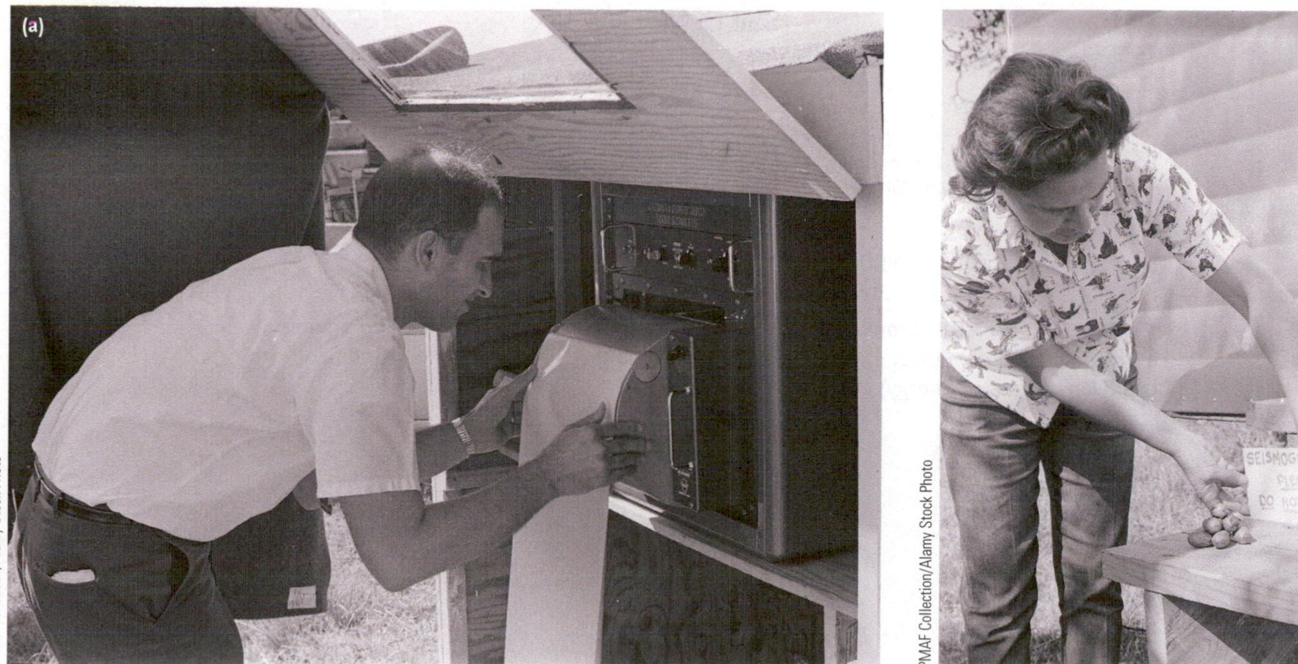

PMAF Collection/Alamy Stock Photo

PMAF Collection/Alamy Stock Photo

Figure 6.91 The seismographs near the blast site were checked and loaded with paper to make sure they accurately recorded the nuclear detonation. (a) The official instrument. (b) The homemade version.

Case Study

Technicians lowered a 5.3 kiloton nuclear explosive into a 820-meter (2,700-foot)-deep drill hole, plugged the upper 180 meters (600 feet) of the hole with concrete, and detonated it to carve out a cavity 34 meters (110 feet) wide in the weak, malleable salt (**Figure 6.92**). A local resident, Steve Thompson, standing 16 kilometers (10 miles) away, commented that the ground felt like "being in a boat" when the explosion took place. "There was [*sic*] two big waves and a bunch of ripples" in the solid earth. Reports of cracked plaster and masonry came from kilometers around, leading to some 400 legal claims for damages (**Figure 6.93**).

Not long after this initial nuclear preparation, project managers detonated a second, much less powerful explosive (equivalent to 380 tons, or 0.4 kilotons of TNT) within the

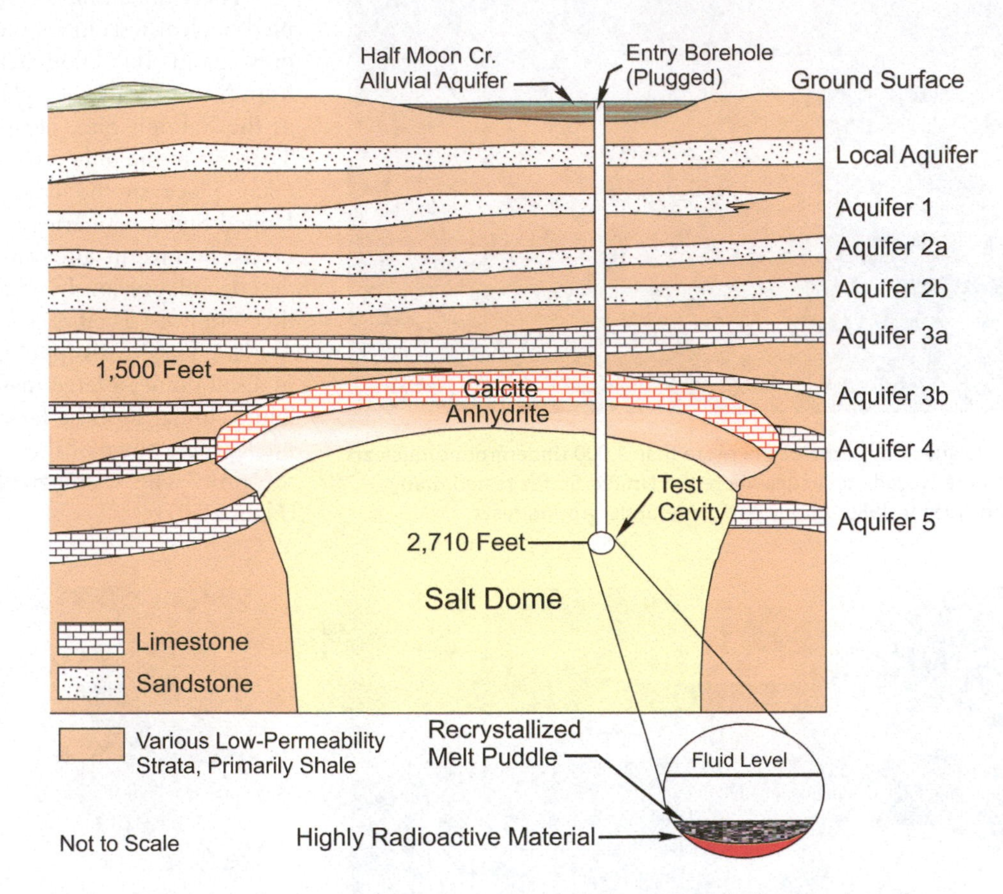

Figure 6.92 Cross-section of the nuclear test site in shows the salt dome and aquifers in the shale bedrock. Inset shows the cavity where the test was conducted. (Modified from U.S. Department of Energy)

cavity. Follow-up tests using methane-oxygen explosive mixtures took place in 1969 and 1970. The results in all instances were ambiguous; complete decoupling did not seem achievable as predicted. But these studies contributed to the Threshold Test Ban Treaty of 1974, signed by the United States and U.S.S.R. This restricted underground nuclear testing to explosives less powerful than 150 kilotons (TNT equivalence).

Present technologies make it possible to predict the yield (explosivity) of new weapons designs without actually testing them. Unannounced nuclear tests, as in North Korea or potentially Iran (where salt domes are common), remain a concern, in a world that has not yet discovered how to eliminate these immensely destructive weapons.

PMAF Collection/Alamy Stock Photo

Figure 6.93 A man inspects the damage to his kitchen from the strong shaking caused by the October 22, 1964, nuclear test in Mississippi.

Photo Gallery

Quakes Have Little Respect

Earthquakes have a way of taking apart just about everything people can build. They do that with both strong shaking and by displacing the ground on which we stand.

1 One-hundred-yard dash? The September 21, 1999, earthquake (M_w 7.8) in Taiwan was caused by movement on the Chelungpu fault. This running track at Kuang Fu High School was built across the fault, which here had a vertical offset of 2.5 meters (8.1 feet). Today, the offset track is preserved under cover as part of a park memorializing the earthquake that killed 2,415 persons.

Sam Yeh/AFP/Getty Images

Archive Holdings Inc./The Image Bank/Getty Images

2 Homes in San Francisco leaned vertiginously after the 1906 quake jostled their foundations.

Cars crushed by a building that collapsed in the 2011 Christchurch earthquake.

Reuters/Alamy Stock Photo

3 Cars and unreinforced masonry don't go well together in an earthquake. In rural northern Italy, 2012, a pair of closely spaced quakes did this damage and killed 26 people.

Zcdebala/E+/Getty Images

iStock.com/Arkanto

4 On Saturday, November 1, 1755, at about 9:50 in the morning, a massive earthquake struck Lisbon, Portugal, destroying the city, killing tens of thousands, and generating a massive tsunami. Igreja do Carmo, a medieval convent, was ruined as the roof collapsed.

Dinodia Photos/Alamy Stock Photo

5 Even if buildings remain standing, damage on the inside can be devastating. Here, an Indian family's home was badly damaged by the 2001 quake in Bhuj Gujarat (Mm = 7.7) that killed more than 10,000 people and destroyed more than 300,000 buildings.

Summary Earthquakes and People

1. The Nature of Earthquakes

LO1 Describe why and how earthquakes take place

a. Types of Seismic Energy Waves

Primary waves (P-waves) are longitudinal waves (push and pull) that produce ground motion in the same direction that the wave moves. Secondary waves (S-waves) are transverse waves that produce ground motion perpendicular to the wave direction. These waves move more slowly that P-waves, but can result in much more significant structural damage.

b. Seismic Wave Periods

A wave period is the time it takes for two successive wave crests (or troughs) to pass a particular point. Wave frequency is the number of wave crests or troughs that pass a point in an interval of time. Seismic waves are classified as short and long period; short-period waves result in more violent earthquakes.

c. Locating the Epicenter of an Earthquake

To locate an earthquake epicenter, seismographs are used to detect, measure, and record seismic activity. The average time interval between the arrival of P-waves and S-waves indicates how far the epicenter is from the seismograph. Thus, the epicenter could be anywhere on the circumference of the circle around the seismograph. To find the exact location, this same method is used on two other seismograph stations, where all three circles meet is the earthquake epicenter. This strategy is call trilateration.

Study Questions

i. What are the different types of waves and how do they each behave physically?

ii. How do you locate the epicenter of an earthquake using the trilateration method?

2. Earthquake Measurement Scales

LO2 Detail how the strength of an earthquake is measured

a. Seismic Intensities

One intensity scale, the modified Mercalli scale, gathers people's perceptions of an earthquake's intensity from "not felt at all" to "widespread destruction." These maps have been used to revise building codes and create safe construction standards in earthquake prone areas. Community internet intensity maps (CIIMs) are easy and quick to generate, using internet responses to gauge local earthquake intensity.

b. Richter Magnitude Scale: The Best-Known Scale

Richter magnitude is the common logarithm of the ground motion in microns. Because the scale is logarithmic, each whole number represents a ground-shaking 10 times greater than the next-lower number. This scale has no upper limit; however, no earthquake has been measured with a Richter magnitude greater than 8.4 in known history (the 1964 Alaska Earthquake).

c. Moment Magnitude: The Most Widely Used Scale

Moment magnitudes more accurately portray the energy released by large earthquakes on faults with great rupture lengths. This scale is derived from the seismic moment, a value proportional to the average displacement on the fault multiplied by the rupture area on the fault surface and the rigidity of the faulted rock. The Chilean earthquake of 1960 has the highest recorded moment magnitude at 9.5.

d. Foreshocks and Aftershocks, Doublets, and Triggered Earthquakes

Foreshocks, the earthquakes that precede the much larger main shock, typically occur anywhere from a few days to just moments before the main shock. Aftershocks are the many smaller earthquakes that follow the main shock with diminishing intensity. Doublets are two equally powerful main shocks taking place at nearly the same time and in the same general area. Triggered earthquakes are caused by other earthquakes taking place some distance away, on unrelated sections of a fault or on other faults altogether. They are stronger than expected aftershocks.

e. Fault Creep, The "Non-Earthquake"

Unlike foreshocks and aftershocks, fault creep is a slow and steady slipping that takes place along some faults. Fault creep can cause structural damage over time; however, it may also help prevent more violent and sudden slips, resulting in earthquakes. Areas which experience fault creep typically have weak and deformable bedrock (e.g., low-grade metamorphic rock), where areas with stronger bedrock (e.g., granite and gneiss) are more likely to catch and slip.

Study Questions

i. What are the differences between the Moment and Richter magnitude scales?

ii. What is fault creep and how is it different than a foreshock or aftershock?

3. Seismic Design Considerations

LO3 Identify some important ways that earthquakes can damage or destroy buildings and harm people

a. Ground Shaking

Rapid horizontal displacements, caused by shear and short-period surface waves, are the most structurally damaging aspects of an earthquake. Flexible frame structures can be deformed or knocked off their foundations. Rigid or multistory buildings may shift, resulting in the collapse of floor levels on top of one another. Resonance is another important consideration, as some tall, flexible structures may suffer serious damage from the increased sway they experience. All of this needs to be considered when building structures in earthquake-prone areas to reduce risk of damage.

b. Landslides

Earthquakes are a major cause of landslides and significant secondary effects from disease, destruction of buildings and communities, and death. While nothing can be done to prevent seismically induced landslides, the best mitigation strategy is to not build structures that lie in harm's way.

c. Ground or Foundation Failure

Liquefaction is the sudden loss of strength of water-saturated, sandy soils resulting from shaking during an earthquake. Shaking consolidates saturated soil and silt such that it occupies a smaller volume, resulting in openings at the ground surface, substantial settlement, extensive lateral movement of the ground, and sometimes large cracks and openings. Quick clays are aggregations of fine-grained clays and water, and seismic shaking turns this solid aggregate to liquid with agitation.

d. Ground Rupture and Changes in Ground Level

Structures that lie on an active fault are at risk of being torn apart by a developing fault scarp. Farther away, uplift or subsidence of the ground can cause additional damage, in some cases even submerging structures as a shoreline shifts inland.

e. Tsunamis

Undersea earthquakes can generate tsunamis that often flood coastal areas causing massive destruction.

f. Fires

Fires are a common earthquake hazard as a result of ruptured fuel tanks, downed power lines, and sparking of debris during shaking. A strong recommendation is to immediately shut off the natural gas supply to homes and other buildings following an earthquake.

g. Dam Failures

Many dams are located directly on top of active fault lines. For these dams, liquefaction and other ground movements can be a serious threat to dam stability. There are, nevertheless, ways of building dams to reduce risk of their rupture from direct fault motions.

Study Questions

i. List the three earthquake hazards and explain why each occurs.

ii. How are structures destroyed by earthquakes? Give three specific examples.

4. Example Earthquakes That Make Important Points

LO4 Explain the most important lessons that we've learned from past earthquake disasters

a. Northridge, California, 1994

This earthquake is the largest in Los Angeles's recorded history, resulting in thousands of aftershocks. Lasting only 10 to 20 seconds, 13,000 buildings were severely damaged and thousands more so damaged that they had to be evacuated. The extensive damage and strong feeling of the earthquake was the result of high horizontal and vertical accelerations generated by a hidden ("blind") thrust fault.

b. Sumatra-Andaman Islands, 2004

Lasting 6 to 8 minutes, with a magnitude of 9.2, the Sumatra earthquake sent waves worldwide and temporarily threw the Earth off its axis of rotation. Every seismograph, and presumably every location on the planet, rose and fell at least a centimeter in slow, gentle oscillations.

c. Tohoku-Oki, Japan

This 9.0 magnitude earthquake was preceded by a set of 5 to 8 magnitude foreshocks. The major shock was unexpected based on the incomplete historical record. Only a small percentage of the structural damage was caused by seismic waves. The majority of the structural damage and human fatalities were a result of the following tsunami. While this event was devasting, Japan's technological preparation and national earthquake warning system perhaps saved a substantial number of lives.

d. Christchurch, New Zealand, 2011

The first earthquake occurred in September 2010, with a magnitude of 7.1, resulting in minimal damage due to low population density and modern construction. In February 2011, a magnitude 6.3 earthquake resulted in serious widespread damage as it was closer to urban development in an area with soft, water saturated ground. Though a lower magnitude, the pre-weakening, shallow focus, and proximity to bustling city with a shallow water table resulted in a much greater amount of damage.

e. Earthquakes Far From Plate Boundaries

Intraplate earthquakes strike far from plate boundaries and can be especially dangerous due to occurring in underprepared regions with infrequent earthquakes. Caused by strain that accumulates slowly in ancient, deeply buried faults. The New Madrid, Missouri, earthquakes of 1811–1812 are examples of these types of earthquakes.

f. What We Can Learn

There are many lessons to be derived from previous earthquakes. The damage and life lost due to an earthquake can be related to types of construction, ground conditions, time of day, type of fault movement, proximity to the ocean, economic situation, even lack of general concern given an incomplete historical record or social memory of past quakes.

Study Questions

i. How do earthquakes occur far from plate boundaries?

ii. Describe an earthquake discussed in this chapter. What about it was informational for future earthquake mitigation?

5. Living With Earthquakes: Forecasts and Mitigation Strategies

LO5 Discuss what we can do to forecast and prepare for future earthquakes

a. Forecasts

Long-term forecasting can be done by compiling historical earthquake data to calculate statistical probability for future events of given magnitudes. These data allow geologists to estimate recurrence intervals, or the length of time that can be expected between earthquakes of a given magnitude.

b. Paleoseismicity and Seismic Gaps

Paleoseismicity refers to the study of the timing and locations of ancient earthquakes. Oftentimes, carbon-14 dating is used to examine ancient peat beds offset by faulting. Seismic gaps are stretches along known fault zones where no significant earthquakes have been recorded for a long period of time.

c. Mitigation

Buildings constructed according to earthquake-safe codes are a very important strategy to reduce structural damage (and loss of life) during an earthquake. Base isolation, a popular design option, places structures on Teflon plates, rubber blocks, and seismic-energy dissipaters to minimize building vibration and sway.

d. Survival Tips

If you're indoors, stay indoors and find shelter under a desk or strong doorway. If you're outside, stay away from buildings and other structures that could fall. Don't drive or use elevators. Turn off gas supply, locate and conserve potable water supplies, and check your home for damage.

Study Questions

i. How are long-term earthquake forecasts made?

ii. How are prehistoric earthquakes studied?

Key Terms

Aftershocks
Base isolation
Base shear
Båth's Law
Body waves
Directional bracing
Displacement
Doublet earthquakes
Elastic rebound theory
Epicenter
Fault creep
Fault plane
Fault scarp
Focal mechanism
Focus
Foreshocks
Fracking
Gutenberg-Richter Law
Human-induced earthquakes (HIE)
Hypocenter
Incompressibility
Intensity scale
Intraplate earthquakes
Isoseismals
Left-lateral
Liquefaction
Long period
Long-term forecasting
Longitudinal waves
Love waves
Main shock
Modified Mercalli scale (MM)
Moment magnitudes
Mud volcanoes
Normal faults
P-waves
Paleoseismicity
Precursors
Probabilistic seismic hazard analysis
Quick clays
Rayleigh waves
Recurrence interval
Resonance
Reverse fault
Richter magnitude scale
Right-lateral
Rupture zone
S-waves
Sand blows
Seismic gaps
Seismic moment
Seismicity
Seismograms
Seismograph
Seismology
Seismometer
Shear walls
Strain
Stress
Strike-slip fault
Subduction zone earthquakes
Surface waves
Teleseisms
Thrust faults
Transverse (shear) waves
Triggered earthquakes
Trilateration
Tsunami earthquakes
Tsunamis
Uniform Building Code
Utsu-Omori's Law
Wave frequency
Wave period

"A few inches between humanity and starvation..."

—Anonymous

7 Soil

Learning Objectives

LO1 Explain the processes that weather rock and sediment

LO2 Describe how soils form, including processes that establish soil profiles

LO3 Outline how soils are classified and explain how certain soil orders affect human uses of the land

LO4 Explain where and why the four most common soil problems occur

LO5 Discuss how humans degrade soils and the problems such degraded soils create

Early morning ploughing of a field on the island of Moen in Denmark.
iStock.com/ClarkandCompany

The Wind Blew and the Soil Flew

Figure 7.1 Clouds of dust rising over the fields of Texas Panhandle in 1936; this is the Dust Bowl.

The Great Depression of the 1930s was exacerbated by a protracted drought in the Great Plains of Oklahoma, Colorado, Texas, New Mexico, and Kansas. This part of the Great Plains was totally dependent on rainfall for crop production, and because the government had guaranteed wheat prices, farmers tilled as many acres as they could. A long-standing agricultural practice was to plow the fields after the fall harvest, crop stubble and all, and let the land lie fallow all winter. This was good practice as long as it didn't rain too much—or too little. If too much rain fell, the water running over the saturated surface eroded the soil. If there were droughts and the soil dried out, the result was the same: extreme soil erosion, only this time by wind.

During droughts in the mid-1930s, the wind blew huge quantities of topsoil away in the southern Great Plains (**Figure 7.1**). Fine particles were lifted 5 kilometers (16,400 feet) in the air and carried eastward as far as New York City and beyond. The five states that were the source of most of this airborne dust became known as the "Dust Bowl." Farming was impossible, and farm families, burdened by debt for equipment, seed, and supplies, left their farms to become migrant workers. Many traveled to California and became derisively known as "Okies." These unfortunate victims of severe soil erosion were the subject of John Steinbeck's touching novel *The Grapes of Wrath*. Even today, wind erosion exceeds water erosion in many parts of the Dust Bowl states. With climate changing rapidly and predictions for summer drought in the northeastern United States, could New England be the next dust bowl (**Figure 7.2**)?

question to ponder

1. What, if anything, do you think the farmers or the government could have done to prevent the Dust Bowl in light of the fact that climate conditions vary so much from year to year in the Midwest?

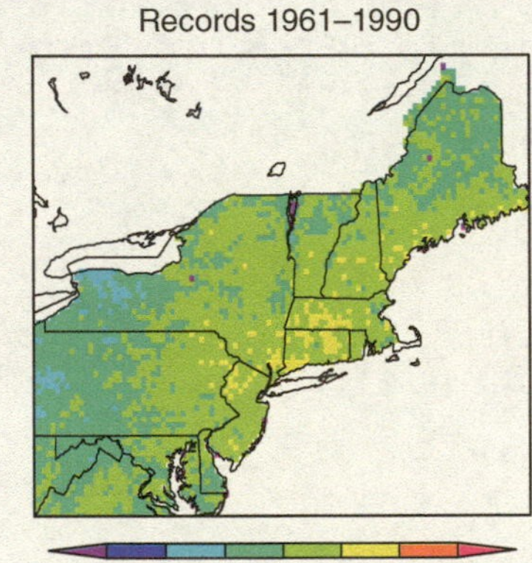

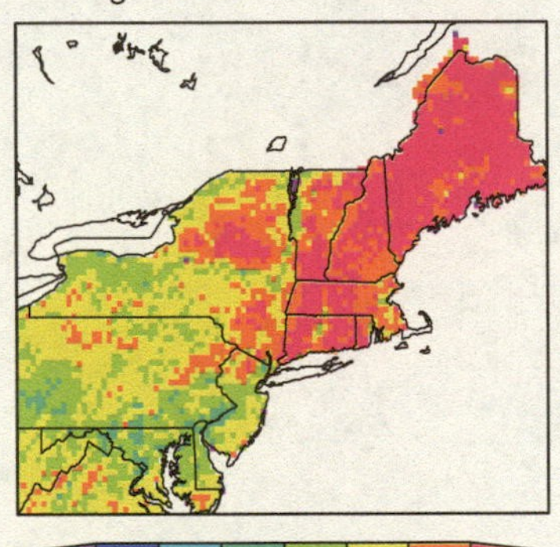

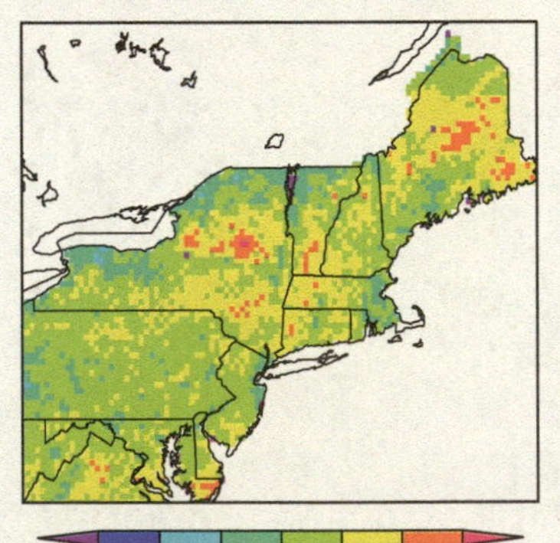

Figure 7.2 With continued global warming, the northeastern United States could be the next dust bowl. Models show that summers will get hotter and drier; in this image, red indicates an increase in likelihood of summer drought (the middle panel presumes that carbon dioxide emissions increase, and the lower panel suggests that emissions stabilize); with drought common, crops may fail. If the wind blows, the soil will erode unless farming practices change. Scale below the images shows the number of short (1- to 3-month) droughts (soil moisture deficit > 10%) over the 30-year window represented by each map.

7.1 Weathering (LO1)

Soil, the thin mixture of weathered rock and organic material below our feet, is Earth's most fundamental resource. We live on it, and through it we produce much of our food. Soil supports forest growth, which gives us essential products, including paper and wood. Soil material, organisms within the soil, and vegetation constitute a system critical to life on this planet (see Chapter 2). It is through soil that the four ingredients needed for plant growth are recycled: water, air, organic matter, and dissolved minerals (**Figure 7.3**).

Many definitions and classification schemes have been developed for soils. To the engineer, a soil is the loose material at Earth's surface–that is, material that can be moved about without first being dynamited and upon which structures can be built. The geologist and the soil scientist see soil as a mixture of weathered rock, mineral grains, and organic material that is capable of supporting plant life. A farmer, in contrast, is mostly interested in what crops a soil can grow and whether the soil is rich or depleted with respect to organic material and minerals.

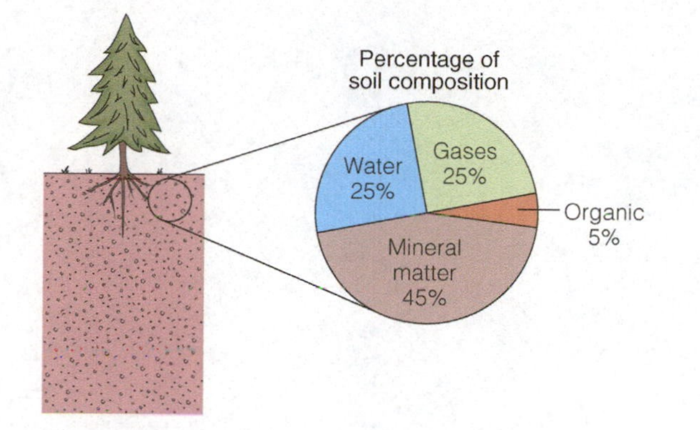

Figure 7.3 Composition of a typical soil. The organic component includes humus (decomposed plant and animal material), partially decomposed plant and animal matter, and bacteria. Mineral matter is what is left of the rock that weathered to form the parent material for the soil.

Because soil is made up in large part of weathered rock, we start by approaching the question, how does rock weather and erode to produce parent materials on which soils develop? **Weathering** is the process by which rocks and minerals are broken down by exposure to atmospheric agents. Specifically, weathering is the physical disintegration and chemical decomposition of earth materials at or near Earth's surface. **Erosion**, in contrast, is the removal and transportation of weathered or unweathered materials by wind, running water, waves, glaciers, groundwater, and gravity (**Figure 7.4**).

The carrying capacity of our planet–the number of people Earth can sustain–depends on the availability and productivity of soil. For this reason, understanding how soil is formed and how it can be best cared for is critical. Soil erosion removes this precious resource, and for all practical purposes, once productive topsoil is removed it is lost to human use forever. The goal of this chapter is to give you an appreciation of how important soil is to our environment by providing the information you need to understand how soil is formed, how it can be impacted by human actions, and how society can better care for this invaluable resource.

7.1.1 Physical Weathering

Physical weathering makes little rocks out of big rocks by many processes. Rocks are wedged apart along planes of weakness by thermal expansion and contraction as they heat and cool repeatedly, and by the activities of organisms. **Frost wedging** occurs when water freezes in joints (cracks) or other rock openings and expands. This expansion, which amounts to almost a 10% increase in volume, can exert tremendous pressures in irregular openings and joints, but only if the joints are somehow sealed. As one would suspect, this process operates only in temperate or cold regions that have seasonal or daily freeze-and-thaw cycles. Water is not the only substance to solidify in

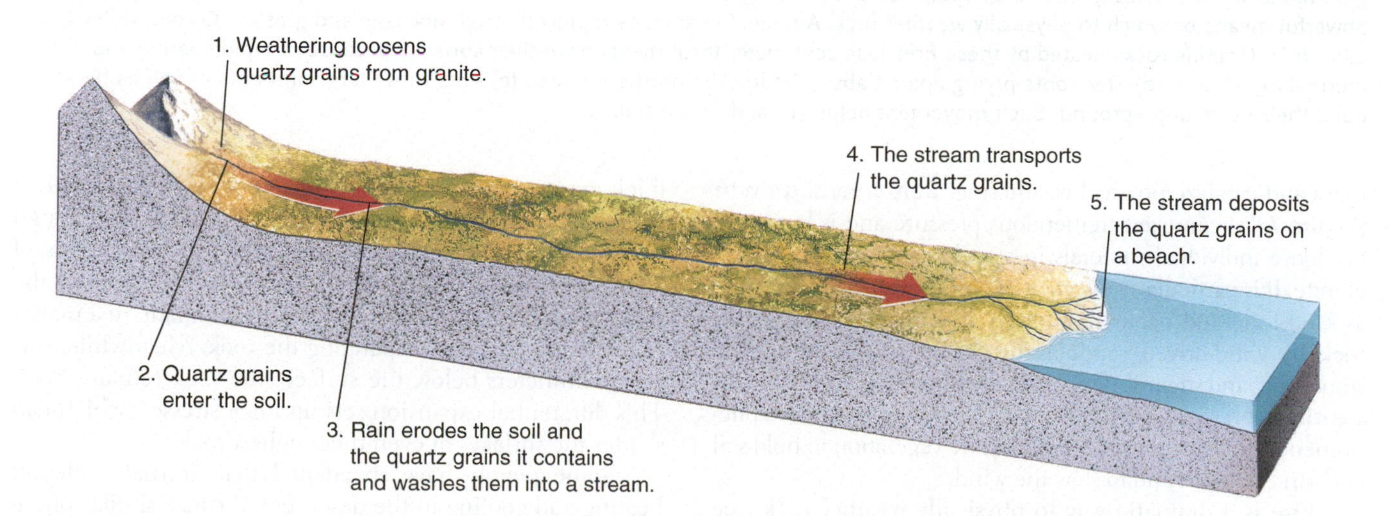

Figure 7.4 Weathering is catalyzed by water and atmospheric gases, which work together to loosen the quartz grains in the host rock, granite. The loosened grains are then incorporated into soil, eroded, and transported (in this example, by a stream) to a beach, which may be at a lake or at the ocean.

Figure 7.5 Physical weathering happens in many differnt ways. (a) Crystallizing salt can help pry rock apart. This granite in the hyperarid desert of Namibia in southern Africa is spalling in onion skin–like layers as salt crystallizes in small cracks and splits the rock apart. A pen (on the rock) is used for scale. (b) Wind-driven sand can effectively erode rock. In the Namibian desert the hard granite has been physically weathered (polished and shaped) by sandblasting, otherwise known as ventifaction. (c) Range fire is a powerful means by which to physically weather rock. A range fire spreads rapidly through dry sage and grass in Owens Valley, California. Granitic rocks heated by these fires lose centimeter-thick sheets from their surfaces because of rapid heating and thermal expansion. (d) Tree roots prying apart slabs of bedrock in northern Israel. (e) Ants move large amounts of soil as they build their nests underground. Such movement helps stir and aerate soil.

rocks and catalyze physical weathering. **Salt-crystal growth** (**Figure 7.5a**) can exert tremendous pressure and is known to pry loose individual minerals in a rock, even in solid, strong granite. It is particularly effective in porous, granular rock such as sandstone and occurs mainly in arid regions. Abrasion of rock by wind-driven sandblasting leaves telltale hints (the smoothing and streamlining of abraded surfaces) and is termed **ventifaction** (see **Figure 7.5b**). Ventifaction is a common phenomenon in deserts where there is little vegetation to hold soil and sand in place and to slow the wind.

Fire is a dramatic way to physically weather rock (see **Figure 7.5c**). It has long been known that intense forest fires can shatter the surface of rocks, but only recently have people observed similar physical weathering when dry shrubs and thick grasses burn in deserts and semiarid areas. How is it that range fires passing in mere minutes can damage tough rocks such as granite? Rocks are very poor conductors of heat; therefore, when the fire passes by, the surface of the rock may heat to hundreds of degrees centigrade in a matter of just a few minutes, expanding the rock. Meanwhile, several centimeters below the surface, the rock remains cool. This differential expansion sets up huge stresses, which can shatter the surfaces of even the toughest rock.

Some people have speculated that diurnal cycles of heating and cooling in the desert could cause similar physical weathering. Although short-term laboratory experiments to test this hypothesis have never yielded conclusive results, the orientation of cracks in rocks around the world suggests that, over millennia, solar heating and the resulting daily expansion and contraction do damage rocks.

Biota are important agents of physical weathering and help prepare rock for its conversion to soil and sediment. Roots of even the smallest plants can act as wedges in rock in any climate (see **Figure 7.5d**). The most dramatic example of root wedging is probably the damage to paving and structures caused by tree roots. Animals also contribute significantly to weathering and soil formation by aerating the ground and mixing loose material. Insects, worms, and burrowing mammals, such as ground squirrels and gophers, move and stir soil material and carry organic litter below the surface. Some tropical ants' nests (see **Figure 7.5e**) have tunnels hundreds of meters long, for example, and common earthworms can thoroughly mix soils to depths of 1.3 meters (4.25 feet). Earthworms are found in all soils where there is enough moisture and organic matter to sustain them, and it is estimated that they eat and pass their own weight in food and soil minerals each day, which amounts to about 22.5 tonnes per hectare (10 tons per acre) per year, on average. An Australian earthworm species is known to grow to a length of up to 3.3 meters (11 feet) and is a real earthmover! Charles Darwin was the first person to calculate just how much soil earthworms process.

7.1.2 Chemical Weathering

Chemical reactions alter rock and mineral debris at Earth's surface. These reactions are complex and involve many steps, but the fundamental processes are reactions between earth materials and atmospheric constituents such as water, oxygen, and carbon dioxide. The products of chemical weathering are new minerals and dissolved elements and compounds. The most important chemical weathering reactions are **solution**, the dissolving of minerals; **oxidation**, the "rusting" of minerals; **hydration**, the combining of minerals with water; and **hydrolysis**, the complex reaction that forms clay minerals and the economically important aluminum oxides. **Table 7.1** shows the formulas for the typical chemical-weathering reactions that are explained in the following discussion.

Solution—Carbon dioxide released from decaying organic matter and from the atmosphere combines with water to form carbonic acid. This natural, weak acid attacks solid limestone, dissolving it and yielding a watery solution of calcium and bicarbonate ions. Solution in limestone terrain creates caverns such as Mammoth Caves in Kentucky and Carlsbad Caverns in New Mexico. Acid rain, which forms when sulfur dioxide or nitrogen oxides combine with water droplets in the atmosphere to form sulfuric and nitric acids, takes its toll on rock monuments (**Figure 7.6**).

Oxidation and hydration—Chemically, oxidation occurs when an atom or molecule loses an electron and increases oxidation state; in other words, becomes more positive electrically (see Chapter 4). Rusting is a common oxidation reaction that occurs to metal objects that are left outdoors and exposed to precipitation and oxygen. Iron is a strong reducing agent, whereas oxygen is a strong oxidizing agent; thus, iron will readily give up an electron in the presence of oxygen. This reaction is furthered by water forming iron hydroxides.

Once dehydrated, oxidation produces iron-oxide minerals (hematite and limonite), which constitute rust. This process occurs in well-aerated soils, usually in the presence of water. Pyroxene, amphibole, magnetite, pyrite, and olivine are the minerals most susceptible to oxidation because they have high iron contents. The oxidation of these and other iron-rich minerals provides the bright red and yellow colors seen in many of the rocks of the Grand Canyon and the Colorado Plateau.

Hydrolysis—The most complex weathering reaction, hydrolysis, is responsible for the formation of clays, the most important minerals in soil. A typical hydrolytic reaction occurs when orthoclase feldspar, a common mineral in granites and some sedimentary rocks, reacts with slightly acidic carbonated water to form clay minerals, potassium ions, and silica in solution. The ions released from silicate minerals in the weathering process are salts of sodium, potassium, calcium, iron, and magnesium, which become important soil nutrients. Clay minerals are important in soils, because their

Table 7.1 Examples of Chemical-Weathering Reactions

Solution	
Step 1: A natural acid forms.	$CO_2 + H_2O \rightarrow H_2CO_3$ carbon dioxide, water, carbonic acid
Step 2: Acid dissolves limestone, placing calcium ions and bicarbonate ions in solution.	$CaCO_3 + H_2CO_3 \rightarrow Ca^{21} + 2(HCO_3)^2$ limestone, carbonic acid, calcium ions, bicarbonate ions
Oxidation and Hydration	
Oxygen and water combine with iron, producing hydrated iron oxide (rust).	$4Fe^{2+} + 3O_2 + 6H_2O \rightarrow 2(Fe_2O_3{\cdot}3H_2O)$ iron minerals, oxygen, water, limonite (rust)
Hydrolysis	
Orthoclase feldspar combines with acid and water, forming clay minerals and potassium ions.	$2KAlSi_3O_8 + 2H^1 + 9H_2O \rightarrow$ orthoclase feldspar, acid ions, water $Al_2Si_2O_5(OH)_4 + 2K^+ + 4H_4SiO_4$ clay minerals, potassium ions, soluble silica

Figure 7.6 Compare the weathering of granite obelisks, both believed to have been shaped 3,500 years ago (1500 B.C.E.). (a) Obelisk amid the ruins of Karnak near Luxor, Egypt, a hyperarid climate. The hieroglyphs are sharp, and the polished surface is still visible. (b) The red-granite obelisk known as "Cleopatra's needle," which stands in New York's Central Park, was a gift from the government of Egypt in the late nineteenth century. New York City's humidity and acidic pollution have caused chemical weathering of the incised hieroglyphs, which had endured several thousand years in Egypt's clean air and arid climate.

extremely small grain size–less than 2 microns (0.002 mm)–gives them a large surface area per unit weight. Soil clays can adsorb significant amounts of water on their surfaces, where it stays within reach of plant roots. Also, because the clay surfaces have a slight negative electrical charge, they attract positively charged ions, such as those of potassium and magnesium, retaining these nutrients and making them available to plants. Other soil components being equal, the amounts of clay and humus in a soil determine its suitability for sustained agriculture. Sandy soils, for example, have much less water-holding capacity, less humus, and fewer available nutrients than do clayey soils. But overly clay-rich soils drain poorly and are difficult to farm and till.

7.1.3 The Rate of Weathering

The rate of chemical weathering is controlled by the surface environment, mineral or sedimentary particle grain sizes (surface areas), and climate. The stability of the individual minerals in the parent material is determined by the pressure and temperature conditions under which they formed. Quartz is the last mineral to crystallize from a silicate magma, and the temperature at which it does so is lower than the temperatures at which olivine, pyroxene, and the feldspars crystallize. For this reason, quartz has greater chemical stability under the physical conditions at Earth's surface. The relative resistances to chemical weathering of minerals in igneous rocks thus reflect the opposite order of their crystallization sequence in magma. Most stable to least stable, the sequence is quartz (crystallizes at lowest temperature), orthoclase feldspar, amphibole, pyroxene, and olivine (crystallizes at highest temperature).

Micas and plagioclase feldspars form over a range of temperatures and thus vary in weathering stability. Note that ferromagnesian (iron and magnesium) minerals–olivine, pyroxene, and amphibole–are least stable and most subject to chemical weathering. Igneous rocks, such as gabbro, are made up of such weathering-susceptible, iron-rich minerals, and thus weather readily in humid climates where plenty of water is available.

A monolithic mass of rock is less susceptible to weathering than an equal mass of fractured rock. This is because the monolith has less surface area on which weathering processes can act. A hypothetical cubic meter (about 35 cubic feet) of solid rock would have a surface area of 6 square meters (65 square feet). If we were to cut each face in half, however, we would have eight cubes with a total surface area of 12 square meters (130 square feet) (**Figure 7.7**). Thus, smaller fragments have more area upon which chemical and physical weathering can act.

It is important to remember that chemical weathering dominates in humid areas, and physical weathering is most active in arid or dry alpine regions. A sixteenth-century monastery in Mexico that was built of volcanic tuff has been ravaged by physical and chemical weathering to the extent that some of the carved statues on its exterior are barely recognizable (**Figure 7.8a**). Likewise, a limestone rock wall protecting Chicago's Lake Shore Drive has suffered as water overtopping the wall slowly but surely dissolves clearly visible pits in a pattern known as "honeycomb" weathering (see **Figure 7.8b**). You may be able to see examples of chemical weathering in your own community by visiting a cemetery. Compare the legibility of the inscriptions on monuments of various ages and materials. Some rock types hold up better than others, and younger

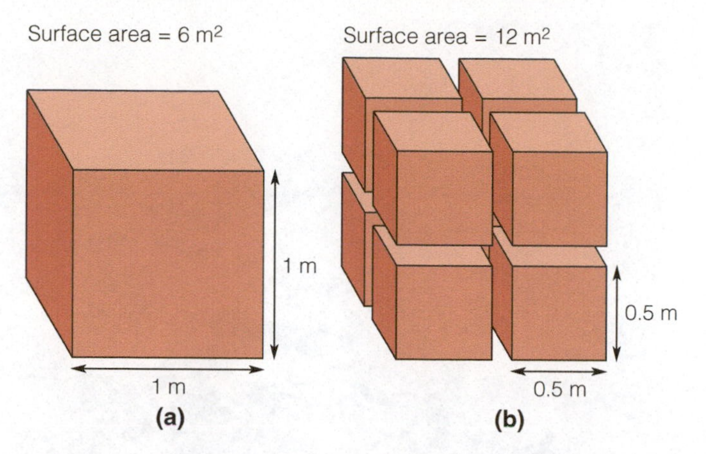

Figure 7.7 Surface area increases with decreasing particle size. A cube that is 1 meter on each side has a surface area of 6 square meters. If the cube is cut in half through each face, eight cubes result, with a total surface area of 12 square meters.

inscriptions are easier to read than those that have been exposed to the elements for a century or two.

In the past 40 years, geologists teaming up with high-energy particle physicists have come up with a new way to measure rates of erosion of bare rocky surfaces. How fast are these exposures of hard rock weathering and eroding down? It is possible to find out. The trick is the ability to count rare atoms of several elements, including beryllium (Be), chlorine (Cl), and helium (He) present within the rock. Earth is constantly being bombarded by cosmic radiation (mostly neutrons) that originates outside the solar system. Most of this radiation passes through the rocks that make up our planet, but occasionally there is a direct hit and something like the nucleus of an oxygen atom gets slammed so hard that it breaks apart. One of the pieces left is likely to be a Be atom, with 4 protons and 6 neutrons, giving it a mass of 10 atomic mass units–otherwise known as ^{10}Be. These rare creations are known as **cosmogenic isotopes**, because they are isotopes made (*genesis*) by cosmic radiation (*cosmo*).

With expensive and exotic tools, the number of ^{10}Be atoms can be counted (see **Figure 7.9a**). Because the neutrons that make ^{10}Be penetrate only a meter or two into rock, the number of ^{10}Be atoms reveals how quickly the rock is eroding; for example, a high concentration of ^{10}Be indicates slow erosion, and less ^{10}Be indicates fast erosion. Using this technique, researchers have shown that rocky exposures in the southern continents of Australia, Antarctica, South America, and Africa (see **Figure 7.9b**) are exceptionally stable, eroding only a few tens of centimeters (5 to 10 inches) over a million years. In contrast, parts of the active mountain ranges, such as the Himalayas, lose a kilometer or more (perhaps up to a mile) of rock over the same length of time.

7.1.4 Geological Features of Weathering

Spheroidal weathering is a combination of chemical and physical weathering in which concentric shells of decayed rock are separated from a block bounded by joints or other fractures (**Figure 7.10a**). Imagine a rocky block that initially

Figure 7.8 (a) A sixteenth-century monastery in Mexico shows the ravages of weathering, mostly from wind and wind-driven rain. The rock is volcanic tuff. (b) A Lake Michigan seawall built of Indiana limestone is chemically weathered by waves overtopping the structure. Solution of the limestone has caused "honeycomb" weathering.

has a rectangular shape embedded in the shallow crust. The spheroidal form results when the block is weathered from three sides at the corners and from two sides along its edges, while the block's plane faces are weathered uniformly. The block weathers most rapidly at the corners, less rapidly at the edges, and slowest on its planar faces. In time will acquire a spheroidal shape. This phenomenon is also called "onionskin" weathering because of superficial resemblance to the peeling layers of an onion.

Exfoliation domes (Latin *folium*, "leaf," and *ex-*, "off") are among the most spectacular examples of physical weathering. They result from the unloading of rocks that have been deeply buried. As erosion removes the overlying rock, it reduces pressure on the underlying rocks, which, in turn, begin to expand. When the rocks expand, they crack

© Lawrence Livermore National Laboratory

Paul Bierman

Figure 7.9 (a) At Lawrence Livermore Laboratory, best known for designing nuclear weapons, is a machine of exquisite sensitivity, an atom sorter and counter known as an accelerator mass spectrometer. (b) In the highlands of Namibia, two geologists collect samples of eroding rock to analyze on the accelerator seen in (a). The data showed that this is a very stable outcropping of rock; it is eroding only about 3 meters (10 feet) every million years.

Paul Bierman

Carol M. Highsmith/The Library of Congress

iStock.com/Julie Caron

Figure 7.10 (a) Spheroidal weathering in jointed granite; Alabama Hills, California. (b) Sheet jointing has produced these slabs of granite in Yosemite Valley, California. (c) Stone Mountain, one of the largest granite domes in the world.

and fracture along **sheet joints** parallel to the erosion surface. Slabs of rock then begin to slip, slide, or spall (break) off the host rock, revealing a large, rounded, domelike feature. Sheeting along joints occurs most commonly in plutonic (intrusive igneous) rocks, such as granite, and in massive sandstone. North Dome in Yosemite National Park in California and Stone Mountain in Georgia are well-known granite exfoliation domes (see **Figure 7.10b and c**).

7.2 Soil Formation (LO2)

Soil is the backbone of civilization because it is the medium in which plants grow. Thus, you can think of soil as providing food to people and animals alike. Soil also supports trees that provide lumber for the homes in which we live. Soils also store large amounts of carbon, keeping it out of the atmosphere. As climate change has intensified and means to address it have become clearer, soils have gained new respect and importance as storehouses of carbon. For practical purposes, soil scientists (pedologists) have put tremendous effort into understanding how soils form and devising ways of classifying their great diversity. Some soils are thin, while others are thick; some soils are young, others old, their ages not necessarily related to their thicknesses. Some soils are fertile, and some lack the nutrients plants need to grow.

7.2.1 Soil Development

Soils form by the weathering of **regolith**, the fragmental rock material at and just below Earth's surface. Regolith may consist of sediment that has been transported and deposited by rivers or wind, for example, or it may be rock that has decomposed in place. Regolith is known as the **parent material** of the soil. The composition of regolith depends strongly on the rock from which it was derived. Regolith derived from limestone will be very different than that formed from weathered basalt or granite. Limestone regolith will be rich in calcium, whereas regolith derived from basalt will be rich in iron. Soils developed on different parent materials will have different chemical and physical characteristics. The fundamental difference between a regolith and a soil is that soils involve biological activity, from the services provided by bacteria and thread-like mycelia to the decay of leaf litter and downed logs that add carbon to the topsoil. Life is not only present within soil, but as noted earlier, is essential to its full development. Other bodies in our solar system, such as Mars and the Moon, feature thick regoliths. But these are certainly not soils.

The transition from regolith to soil begins almost at once, as microorganisms "colonize" the regolith and infiltrating water further breaks it down to make clays and other minerals rich in nutrients capable of supporting life. The activity of organisms in turn further changes the overall chemical make-up and physical character of the soil. **Soil development** describes the physical and chemical changes that regolith undergoes to become soil. The longer a soil has been exposed near the surface of the Earth, the more chemical and physical changes will alter the parent material. Many soils contain organic matter which sequesters carbon that might otherwise be in the atmosphere and holds both nutrients and water that living organisms, including plants, animals, and bacteria need to survive.

Five environmental factors determine the rate and intensity of soil development: (1) climate, (2) organic activity, (3) relief (steepness) of the land, (4) parent material, and (5) length of time that soil-forming processes have been occurring. These factors make up the "CLORPT equation," a memory aid which is as follows: Soil $= f\{CL, 0, R, P, T\}$, where f is read "a function of" and CL, 0, R, P, and T represent climate, organic activity, relief, parent material, and time, respectively.

Soils change over time as iron-rich materials oxidize and grow redder (given a humid, warm climate), organic matter accumulates, various mineral crusts termed **hardpans** develop, and clays that may form by weathering elsewhere are transported into the soil by gravity and infiltrating water, a process called **illuviation**. Where rainfall is heavy, soils are deep, acidic, and dominated by chemical weathering. Soluble salts and minerals are leached (removed) from the uppermost soil, iron and aluminum compounds accumulate lower in the soil, and the soil supports abundant plant life. In arid regions, in contrast, soils are usually alkaline and thinner, coarse-textured, or rocky, and dominated by physical weathering. Because of infrequent precipitation in arid regions, salts are not leached from soils; when soil moisture evaporates, it leaves additional salts behind. These salts may form crusts and lenses on and in the soil, a serious problem for agriculture.

The most highly developed soils are found in warm, wet, flat-lying tropical areas with heavy vegetation where weathering proceeds rapidly and erosion rates are low (because slopes are not steep). In contrast, cold, dry areas or those with very steep slopes and little vegetation have thin, weakly developed soils. Climate is generally the most important factor determining the rate of soil development.

The influence of the parent material is most clearly apparent in young soils. In older soils, the parent material has been so extensively weathered that the original mineral may be unrecognizable. Hints may remain, however. Consider granite, which contains feldspar grains. After sufficient weathering, all of the feldspar will have been altered to clay minerals. As a result, quartz, one of the minerals most resistant to weathering, will dominate soil derived from granite.

Highly developed soils are vulnerable to changes in land use. In places like the Amazon rain forest, many of the nutrients that plants need are either in the uppermost layers of the soil or in the rainforest vegetation itself. When the rainforest is clear cut and the logs hauled away, the nutrients go with them. Farmers might get a crop or two out of the denuded soil before it's depleted and useless for farming; but that is it, and what remains often quickly erodes away. Even the original vegetation may have trouble regrowing in the barren, nutrient-poor soil that contains little more than clay, quartz, and iron.

7.2.2 Soil Profiles

A developed **soil profile** has several recognizable **soil horizons**, layers roughly parallel with the ground surface that are products of soil-forming processes (**Figure 7.11**). The uppermost soil horizon, the "**A horizon**," is the **zone of leaching**. Here, mineral matter is most strongly dissolved by downward-percolating water. It may be capped by a zone of organic matter (**humus**) of variable thickness called the "**O horizon**," which provides carbon dioxide and organic compounds that make the percolating water slightly acidic. As the dissolved chemicals move downward, some of them are redeposited as new minerals and other compounds in the "**B horizon**," also called the **zone of accumulation**. The B horizon is usually redder and harder than the A horizon. Below this is the "**C horizon**," the weathered transition zone that grades downward into fresh parent material, either regolith or rock.

The thickness and development of soil profiles is profoundly influenced by climate and time. In desert regions of the American Southwest, the A horizon may be thin or nonexistent, and the B horizon may be firmly cemented by a white, crusty calcium carbonate hardpan known as **caliche**. Caliche forms when downward-percolating water evaporates and deposits calcium carbonate that has been leached from above. Also, if the water table is close to the ground surface, upward-moving waters may evaporate into the layers, leaving deposits of caliche. Some of these deposits are so thick and hard that unusual measures are required to till or excavate them (**Figure 7.12**).

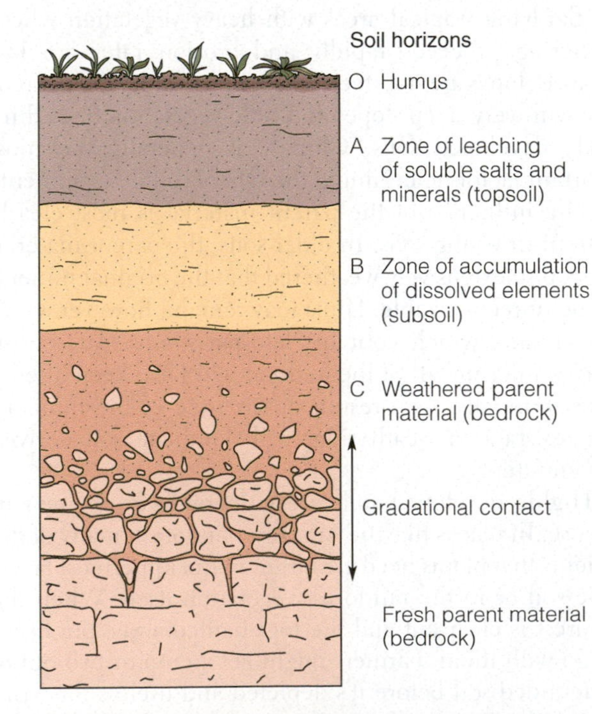

Figure 7.11 An idealized soil profile. Heavily irrigated soils and very young ones exhibit variations of this profile.

Figure 7.12 Lydia Staisch, a U.S. Geological Survey Research geologist, stands next to a very thick caliche deposit in arid eastern Washington state.

7.3 Soil Classification (LO3)

Soils are classified in a variety of different ways: color, texture, and origin, to name a few. These classifications are useful because they allow soils of similar types to be mapped. These maps are very useful for both agriculture and for determining where soils might be optimal for some uses and risky for others.

7.3.1 Soil Color and Texture

Color is a good indicator of a soil's humus (organic material) and iron content. As the proportion of humus increases, soil color darkens from brown to nearly black. Oxides of iron impart a reddish or yellowish color and are found mostly in semitropical and tropical areas. Bright red soils occur where there is aeration (oxygen present) and good drainage, whereas yellow hydrated-iron compounds form where there is poor drainage. Gray-green soil colors indicate the absence of oxygen and are most common in water-saturated soils. In arid climates, white B horizons and white-coated soil fragments indicate calcium carbonate or other salts, including gypsum (calcium sulfate) and halite (sodium chloride). Because both gypsum and halite are water soluble, their presence in soils indicates very low amounts of rainfall. Such soils are found in the driest deserts of the world, like the southern Negev in Israel.

A soil's texture–the sizes of the individual grains (particles) composing it–is a major determinant of its utility. A sample of any particular soil, excluding any gravel or boulders in it, can be divided into three grain sizes: sand, silt, and

GRAIN SIZE

mm	inch	classification	
2.0	0.08		
1.0	0.04	very coarse	sand
0.500	0.02	coarse	sand
0.250	0.01	medium	sand
0.100	0.004	fine	sand
0.050	0.002	very fine	sand
0.002	0.008	silt	
		clay	

(a)

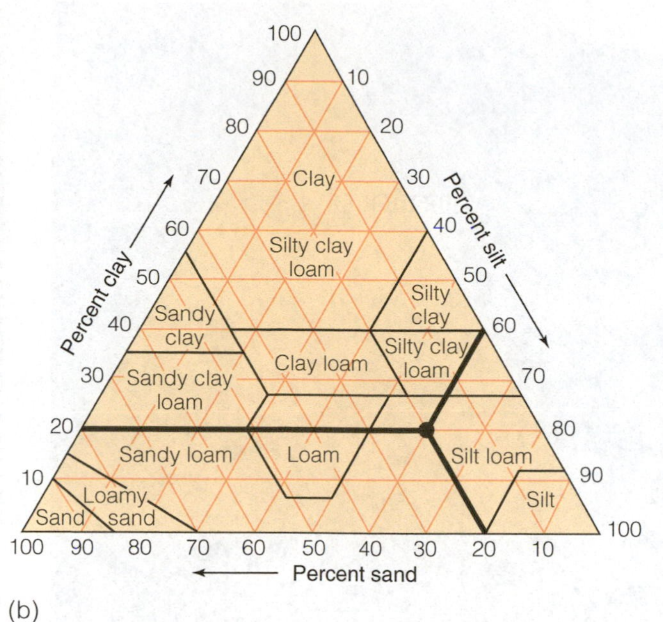

(b)

Figure 7.13 (a) The grain-size scale adopted by the U.S. Department of Agriculture and used by most soil scientists. (b) Soil texture classification triangle. Each side of the triangle is a scale for the percentage content of one of the three textural grades—sand, silt, or clay. A hypothetical soil sample (•) is 60% silt, 20% sand, and 20% clay. Using this triangle, thick black lines are followed inward from the respective percentages to learn the textural class of the soil, "silt loam." (Data from U.S. Department of Agriculture)

clay. The grain-size range used by the U.S. Department of Agriculture for each texture category is shown in **Figure 7.13a**. Based on the percentages of sand, silt, and clay in a soil sample, the soil can be classified as loam, sandy loam, and so forth (see **Figure 7.13b**).

The texture of a soil has practical significance because it determines the soil's properties; that is, its looseness, workability, and drainage. For example, coarse- textured soils are easily worked, readily penetrated by plant roots, and easily eroded. Cemeteries in formerly glaciated regions are almost always dug in sand and gravel, not heavy, wet clay. Fine-textured (clayey) soils tend to be heavy, harder to work, and sticky when wet.

Water retention and infiltration are also functions of soil texture. Loose, sandy soils tend to allow water to pass through them readily, but they have low water retention capacity, meaning that plants growing there need to be watered often or be drought tolerant. The opposite is true of fine-textured soils; they have large water-retention capacity but low infiltration rates. This condition leads to poor drainage and excessive water runoff. Loamy soils are generally considered most favorable for agriculture because they provide a balance between the coarse- and fine-textured soils, as shown in **Figure 7.13b**.

7.3.2 Residual Soils

Residual soils have developed in place on regolith derived from underlying bedrock. For example, in humid-temperate climates, rock decomposes into **saprolite**, rock that has lost some of its volume and much of its strength as chemical weathering leaches soluble elements. If the parent rock were granite, the saprolite would contain quartz, weathered feldspar, and clay minerals formed from the weathering of feldspar. At first glance, saprolite that has just begun to develop can easily be mistaken for its parent rock, except that one can often stick a shovel into it and the density of saprolite can be tens of percent less than the density of the rock from which it was derived (the mass is lost by solution of rock into moving groundwater and the transport away of the dissolved material). Residual soils take tens to hundreds of thousands of years to develop, and they are always subject to erosion and removal by one of many geological processes.

Engineers and geologists call slightly weathered granite **decomposed granite (d.g.)**. It is also known as grus, a word of German origin translating to "grit" or "pebble-sand." Decomposed granite is an excellent foundation material for structures and roads. In contrast, heavy clay soils develop from the weathering of shale bedrock. The expansive nature of some of these clays makes them troublesome for foundations as they alternately swell and contract.

7.3.3 Transported Soils

Many soils have formed on transported regolith. These **transported soils** are classified according to the geological agencies that are responsible for their transportation and deposition: *alluvial* soils by rivers, *eolian* soils by the wind, *glacial* soils by glaciers, *volcanic* soils by volcanic eruptions, and so on.

Volcanic soils are initially barren but as soil development proceeds, chemical weathering releases from minerals and accumulated organic matter the nutrients that plants need (Chapter 5). **Figure 7.14** illustrates a soil on the Island of Hawai'i that is slowly forming from beds of volcanic ash overlying an older basalt lava flow. Known locally as the *Pahala ash*, soil developed on this parent material supports such exotic and valuable crops as macadamia nuts, coffee, and papaya. In Italy, volcanic soils support citrus groves and vineyards. In northwestern North America, soils formed on weathered ash beds support forests filled with massive fir and cedar trees (**Figure 7.15**).

B. Pipkin

Figure 7.14 Volcanic ash (horizontal layers) overlying weathered basalt; Pahala, Hawai'i. This volcanic soil supports rich crops of coffee and macadamia nuts.

As Pleistocene glaciers expanded and moved over northern North America, they scraped off soils and regolith over millions of square kilometers. As the ice melted back, it left a wide blanket of fine clay, silt, sand and gravel, in places mixed with boulders in a band across much of Canada and the United States from Montana to New Jersey. Known as glacial drift, this material evolved into fertile soil for farmers--where it isn't too bouldery at least! (This is one reason that pastures for grazing dairy cattle, rather than cultivated grain crops, are today the agricultural mainstay of such states as Wisconsin and New Hampshire). After a rainstorm, roads on clay-rich glacial soils become a quagmire, trapping automobiles. Spring in New England, which has many such clay-rich soils, is known as mud season for this reason (**Figure 7.16**).

iStock.com/Siculodoc

Figure 7.15 An orange grove in Sicily thrives on volcanic soils courtesy of snow-covered Mount Etna, seen in the distance.

Vermont State Archives and Records Administration

Figure 7.16 A car from the early 1900s is swallowed by spring mud in Vermont.

Alluvial soils are found around the world adjacent to streams and rivers. The parent material for these soils was deposited by running water and is often rich in sand and gravel. Most alluvial soils are permeable and drain well. They are often useful for agriculture with the caveat that such farm fields are often flooded during storm events and even spring snowmelt season. Alluvial soils are often aggradational, which means that during each flood, sand and silt are deposited on top of the existing soil and soil-forming processes start anew. A soil pit dug into alluvial soil is likely to find multiple buried A and B horizons (**Figure 7.17**).

Loess (pronounced to rhyme with *cuss*) is the most widespread transported soil. Loess is windblown silt typically composed of very small grains of feldspar, quartz, calcite, and mica. Covering about 20% of the United States and about 10% of Earth's land area (**Figure 7.18**), loess is arguably Earth's most fertile parent material. The sources of loess are glacial deposits and deserts. Strong winds blowing off the ice-age glaciers that covered most of Canada and the northern United States swept up nearby fine-grain meltwater deposits and redeposited the dust far away as loess in valleys and on ridge tops.

In the United States, loess has produced especially rich agricultural lands in the Palouse region of eastern Oregon and Washington, and in the central Great Plains, sometimes

Figure 7.17 Buried soil horizons can be well preserved and indicate rapid aggradation.

called the "breadbasket" of the United States. The fertility of Ukraine is due to a favorable climate and its loess-rich soils. During World War II, Ukraine's soil was so highly regarded that Adolf Hitler had large quantities of it moved to Germany. Loess derives its fertility from its loamy texture, which allows roots to penetrate easily and retains water well. In addition, many of the minerals in loess are small and not yet weathered. When they do weather in the soil, they release nutrients that plants can use.

The largest loess deposits are in China, where they cover 800,000 square kilometers (310,000 square miles) and are as thick as several hundred meters (1000 feet). These deposits were derived from the Gobi and Takla Makan deserts to the north and west. On windy days, airborne silt from these deserts impacts air quality in Beijing, hundreds of kilometers away. Similarly, the loess in Africa's eastern Sudan was carried there by winds from the Sahara Desert. Even more impressive is the wind transport of dust from Asia to Hawai'i, where small grains of windblown quartz can be found dropped by wind on basaltic volcanic rocks and basaltic soils that should contain no quartz at all.

Loess has several noteworthy physical properties. It is tan in color and rather loosely packed (low density), which makes it easily excavated and eroded by running water. Loess has cohesive strength and thus is capable of standing in vertical slopes without falling. During the Civil War, these unusual properties served the Union and the Confederacy well in Mississippi. Both sides excavated tunnels, trenches, revetments, and other earthworks in the loess before the siege and battle at Vicksburg, which ultimately was won by

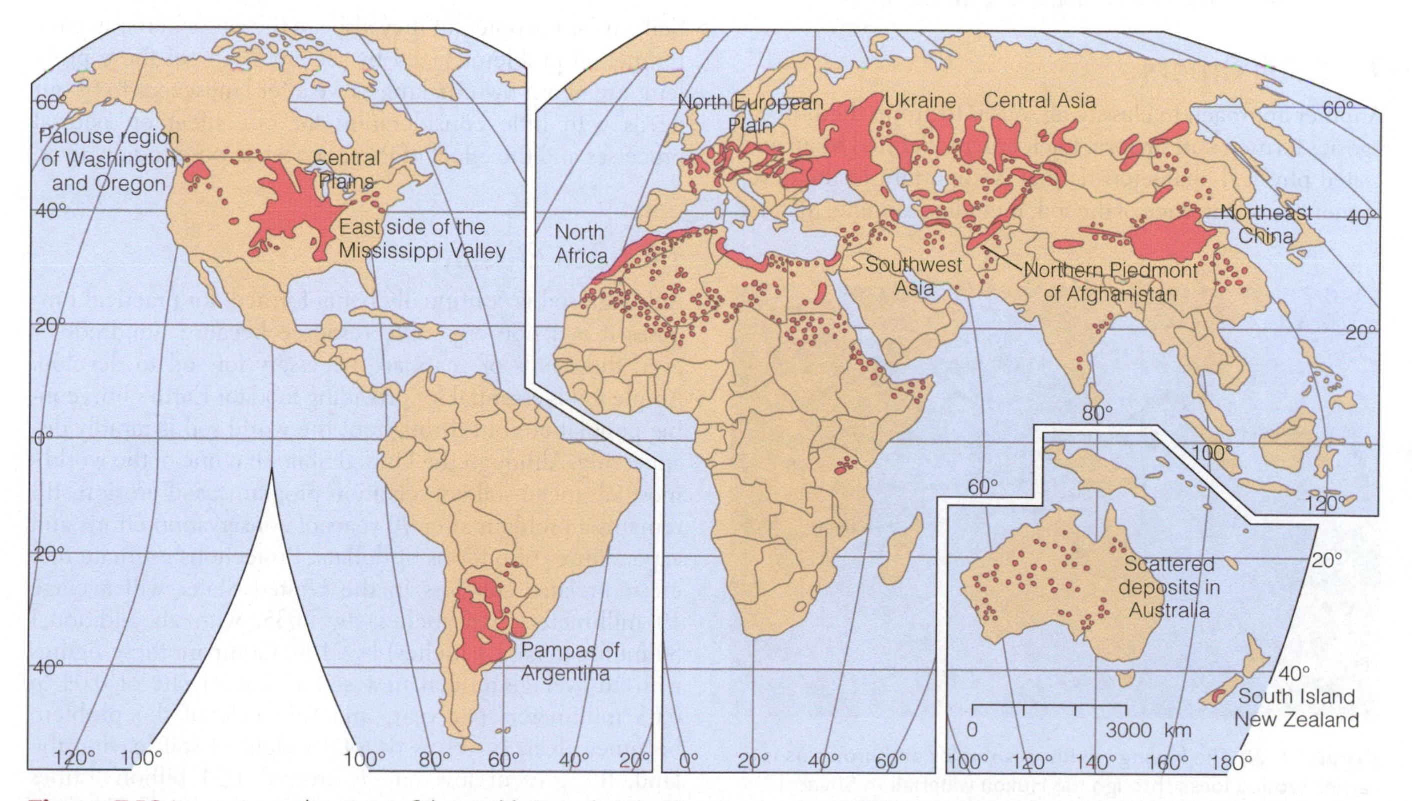

Figure 7.18 Loess-covered regions of the world. Note that the largest areas are in China, central and western Asia, and North America.

iStock.com/Pengyou91

Figure 7.19 Homes and businesses cut into loess in China.

Ulysses S. Grant's Union forces. That loess was deposited when wind blew silt from the Mississippi River 25,000 years ago as it was choked by sediment from the great Laurentide Ice Sheet upstream.

In northern China's Shansi Province in the 1500s, elaborate caves scraped out of loess provided refuge for more than a million people (**Figure 7.19**). Unfortunately for the cave residents, loess is subject to collapse under dynamic stresses, such as those of an earthquake, and an estimated 800,000 people were buried in their homes during the earthquake of 1557. The load of loess carried by the Huang He River as it flows through this area gave the river its western name, the Yellow River, which empties into the Yellow Sea. However, yellow is a mistranslation; Huang is better translated as brown, the color of loess (**Figure 7.20**).

7.3.4 Soil Orders

Another approach to classifying soils is to group them in categories termed **soil orders**. Each order is defined by an associated physical characteristic that is important to the development or occurrence of the soil, including climate, material content, permeability, even vegetation type. The orders are given Latin names, such as "oxisol" which means "highly oxidized soil." There are 13 soil orders (**Figure 7.21a** and **b**), which in turn are subdivided into suborders, families, and series. The number of soil entries increases astronomically at the lower levels (illustrated by the fact that greater than 14,000 soil series are recognized in the United States).

The soil order classification scheme is useful to geologists working in the field who need to understand environments and landscape history (**Figure 7.22**). Consider for instance a common geological project; locating a site for a septic system for human waste disposal. Such systems require well-aerated soils that are not saturated by water. After meeting with the landowner, the first thing a geologist will do is go to a library, government office (such as the Natural Resources Conservation Service in the U.S.) or a website to track down a soil survey map of the area. If the land is dominated by spodosols (**Figure 7.23**), the soils are probably well drained a necessity for treating human wastes. However, if there are histosols, they are wetland soils and less likely to work well for a septic system.

Digging soil pits will verify if there is a problem or not. If it's revealed that the soil on the wall of the pit is mottled (identified by alternating red and grey coloration), this is characteristic of poorly drained, unwanted histosol. The landowner will have to find another way of managing the sewage! (**Figure 7.24**).

iStock.com/Gionnixxx

Figure 7.20 The Huang He River is muddy and brown as is carries eroded loess through the Hukou waterfall in Shaanxi province, China.

7.4 Soil Problems (L04)

Soils sustain people but they also are the cause of many environmental problems faced by society. Many of these problems are of our own making as we alter landscapes to fit our needs with little consideration for our effect on natural processes and the effect of these processes on our daily lives.

7.4.1 Soil Erosion

Although soil is continually being formed, for practical purposes it is a nonrenewable resource because hundreds or even thousands of years are necessary for soil to develop. Ample soil is essential for providing food for Earth's burgeoning population, but throughout the world soil is rapidly disappearing. Although the United States has one of the world's most advanced soil-conservation programs, soil erosion still remains a problem after 90 years of conservation efforts and expenditures of billions of dollars. Projections estimate that erosion-related soil loss in the United States will average 10 millimeters (0.4 inches) by 2035, with an additional 80 millimeters (3.1 inches) by 2100. Compare these figures with an average annual new soil-formation rate of 0.04 to 0.08 millimeters per year, and the scale of the problem becomes clear. In terms of total weight of soil leaving the land, the present loss rate is around 1.54 billion tonnes (1.7 billion tons) per year across the United States. Some of

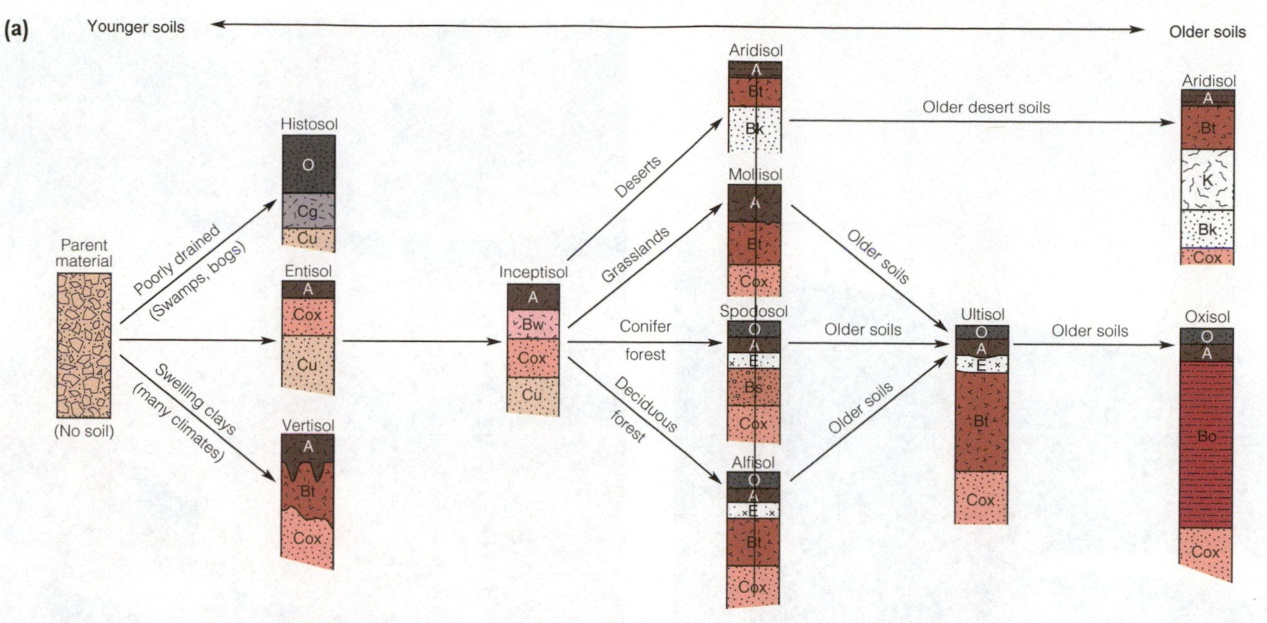

O horizon: Surface horizon dominated by organic material, dark
A horizon: Surface horizon or beneath O horizon, dominated by mineral material but with enough organic material to be dark
E horizon: Subsurface horizon, leached gray by acids from conifer needles
Bw horizon: Subsurface horizon, young and slightly reddened by iron oxide
Bs horizon: Subsurface horizon, reddened where organic matter, aluminum, and iron accumulate
Bk horizon: Subsurface, lightened by the accumulation of calcium carbonate
Bt horizon: Subsurface horizon that is reddened by iron and where clay has accumulated
Bo horizon: Subsurface horizon, deeply weathered and very red with iron, found in very old, tropical soils (laterite)
Cox horizon: Subsurface horizon of oxidized parent material, red from iron
K horizon: Subsurface horizon similar to Bk but so enriched in calcium carbonate that it is white
Cg horizon: Subsurface horizon that has patchy colors of grey and green because it is usually saturated with water
Cu horizon: Unweathered parent material

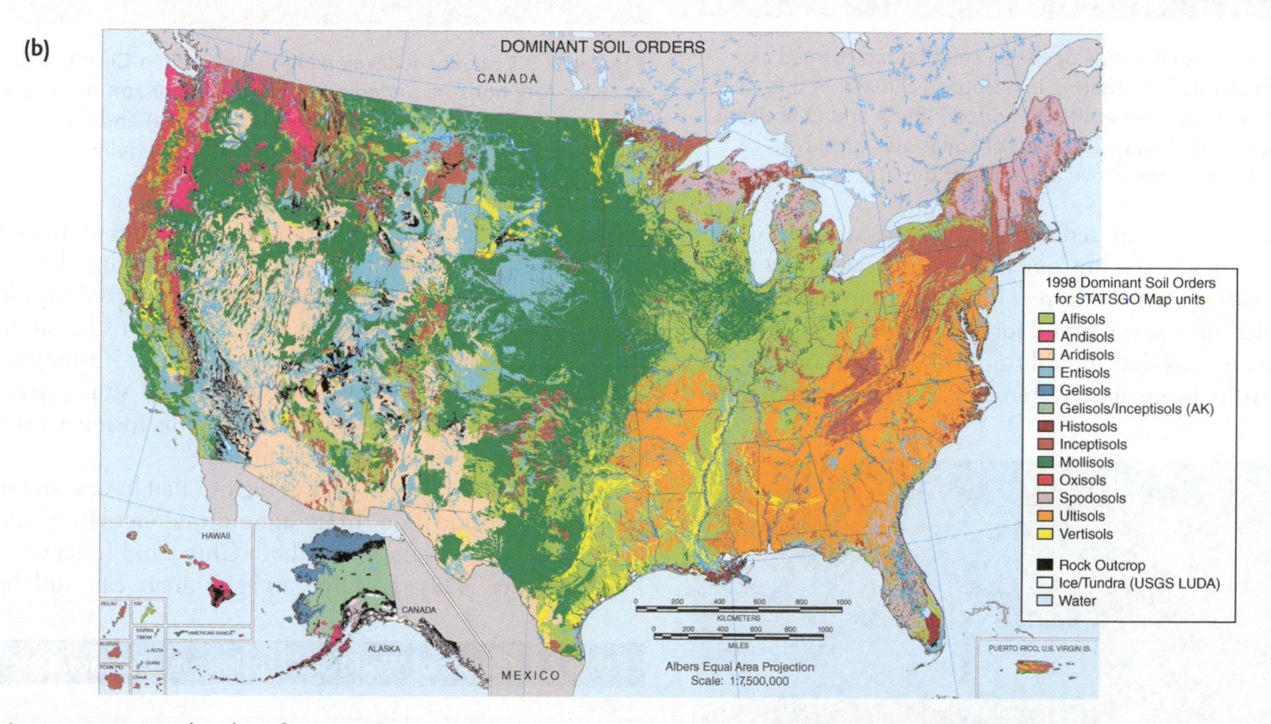

Figure 7.21 (a) The classification of 13 soil orders is an organized way to understand where different types of soils form and how soils change as they age. (Modified Birkeland, P. W. 1999. Soils and Geomorphology, 3rd ed. Figure 2.7, Oxford University Press, Inc. Courtesy of Dennis Netoff) (b) Wonder where to find your favorite soil order? Check the map! If you are in the United States, can you figure out what soil order likely underlies where you are sitting right now, reading this book? (Created using USDA and University of Idaho data)

Paul Bierman

Figure 7.22 An oxisol, a deep red soil that has formed as a result of extended weathering in a tropical climate, in this case, Brazil. You are looking at the Bo horizon; the A and O horizons have been eroded away. This type of soil is also known as a laterite. On the slope above are fields of tomatoes.

SoilPaparazzi/iStock/Getty Images

Figure 7.23 Spodosol developed in sandy soil in Germany has a classic A/O horizon, underlain by leached E horizon (a silt and sand-rich layer that lacks clays, iron, and alumina) and then a bright-red B horizon. This is a well-drained soil useful for a septic system.

it is on the move as sediment in streams and rivers and ultimately out to sea but much remains near where it was eroded, sitting at the bottom of hillslopes or in small streambeds, piled up and not useful for agriculture.

This means that in addition to reduction of precious crop fertility by erosion, waterways and drainages in many places experience pollution from soil-associated pesticides and nutrient overloads, clogging, and flooding. Losses from a farm of only 2 to 3 centimeters (1 inch) of topsoil represent about 180 tonnes per hectare (70 tons per acre) the disappearance of which generally goes unnoticed except by the farmers themselves. **Table 7.2** summarizes the recognized causes of accelerated soil erosion worldwide and assesses their relative impact.

On a global scale, the area of regions that have suffered irreparable degradation from agricultural activities and overgrazing exceeds the size of both China and India combined (**Figure 7.25**). Lightly damaged areas can still be

Minnesota Pollution Control Agency

Figure 7.24 Soil mottles characteristic of histosols in Minnesota suggest this is not a good location for installing a septic system. The soil does not drain well. The scale bar is in centimeters.

Table 7.2 Causes of World Soil Erosion and Deterioration

Cause	Estimated % of Total
Overgrazing	35
Deforestation	30
Bad agricultural practices	28
Other causes, natural and human	7

Source: World Resources Institute, United Nations Environment and Development Program. 1992. *World Resources: A Guide to the Global Environment*. New York: Oxford University Press.

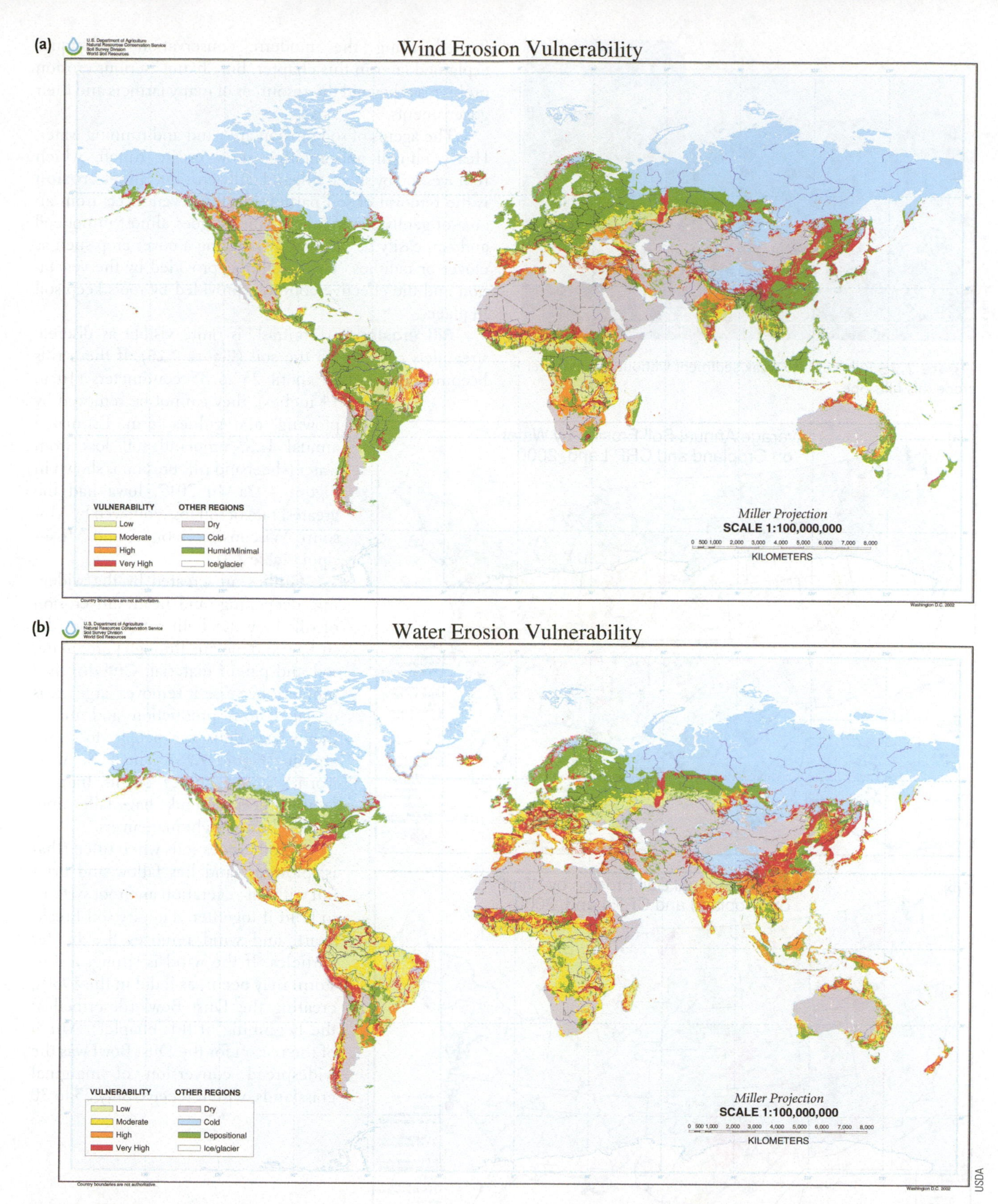

Figure 7.25 Soil is vulnerable to erosion by both wind and water. These maps show where the high-risk and low-risk areas are in the world for soil erosion. These are inverted. map a is wind, map b is water. Please change wording here to (a) Map of soils vulnerable to wind erosion. (b) Map of soils vulnerable to water erosion. Look hard at the maps. Are the high-risk areas for wind erosion the same as those for water erosion? What do you think is controlling the distribution of erosion by wind and water?

Figure 7.26 Rills cut into weak sediment without groundcover erode this hillslope.

farmed using the modern conservation techniques explained later in this chapter, but "fixing" serious erosion problems is beyond the resources of many farmers and their governments.

The agents of soil erosion are wind and running water. Heavy rainfall and melting snow create runoff, which removes soil by sheet, rill, and gully erosion. **Sheet erosion** is the removal of soil particles in thin even layers from an area of gently sloping, bare land. It goes almost unnoticed and can easily be stopped by planting a cover crop such as clover or radishes. The roughness provided by the vegetation and the effective cohesion provided by roots keep soil in place.

Rill erosion, in contrast, is quite visible as discrete streamlets carved into the soil (**Figure 7.26**). If these rills become deeper than about 25 to 35 centimeters (10 to 14 inches), they cannot be removed by plowing, and gullies form. Estimated annual U.S. cropland soil loss from water (sheet and rill) erosion is shown in **Figure 7.27a**. In 2017, Iowa had the greatest rate of soil loss, followed by Missouri, Wisconsin, Georgia, and Mississippi (**Table 7.3**).

Gullies are created by the widening, deepening, and headward erosion of rills by water both running over the surface and moving through pores in the soil and parent material. **Gullying** is a problem because it removes large areas of land from production and makes them hazardous for people to move around (**Figure 7.28**). The U.S. Army worries about gullies on its training grounds because tanks have fallen into gullies during night maneuvers.

Wind erodes soils when tilled (that is, plowed) land lies fallow and dries out without vegetation and root systems to hold it together. The dry soil breaks apart, and wind removes the lighter particles. If the wind is strong, a dust storm may occur, as it did in the 1930s, creating the Dust Bowl (described at the beginning of this chapter). Much of the reason for the Dust Bowl was the widespread conversion of marginal grasslands, which receive only 25 to 30

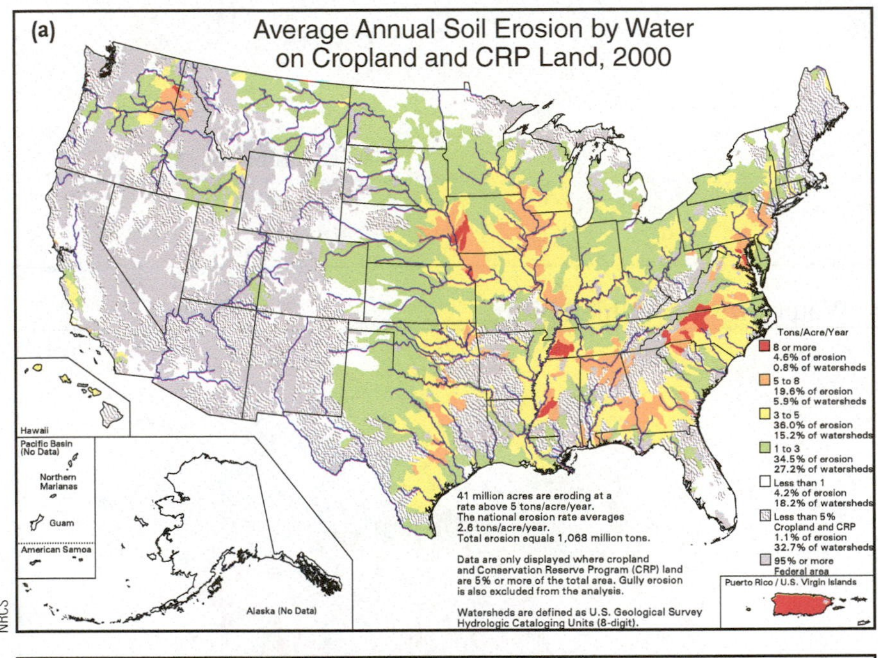

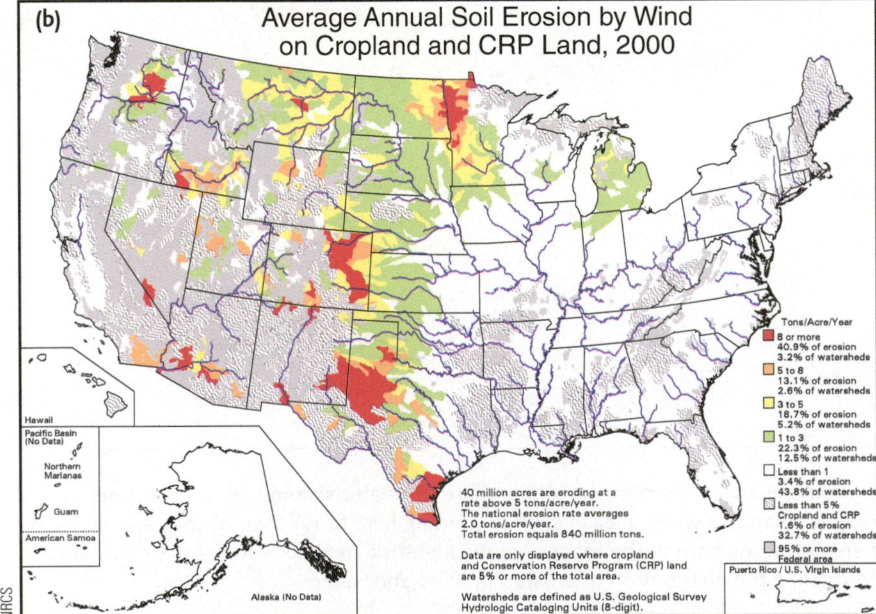

Figure 7.27 Estimated annual rates of cropland soil loss in the United States. (a) Sheet and rill erosion losses by water; (b) wind erosion losses.

Table 7.3 Significant U.S. Soil Erosion by State, 2017

Rank	State	Metric Tons per Acre per Year
Sheet and Rill Erosion		
1	Iowa	5.1 ± 0.16
2	Missouri	4.2 ± 0.22
3	Wisconsin	4.1 ± 0.21
4	Georgia	4.1 ± 0.3
5	Mississippi	3.9 ± 0.21
Wind Erosion		
1	New Mexico	20.0 ± 2.6
2	Arizona	10.9 ± 3.27
3	Texas	6.3 ± 0.67
4	Colorado	5.4 ± 0.49
5	North Dakota	4.6 ± 0.16
Total Soil Erosion, Water, and Wind*		
Rank	**State**	**Metric Tons per Year**
1	Iowa	135.4 million
2	Minnesota	125.6 million
3	Kansas	115.0 million
4	Illinois	92.0 million
5	Nebraska	74.9 million

*The U.S. average annual rate of total soil erosion on cropland is estimated at 4.2 metric tons/acre. (Data from NRI)

centimeters (10 to 12 inches) of rainfall per year, to wheat, owing to high market demand. Unfortunately, land conversion to agriculture is still occurring in the United States and the rest of the world. **Figure 7.27b** presents estimates of the amount of soil lost each year to wind erosion in the United States. Wind-caused soil losses can be reduced by planting windbreaks of trees and shrubs ("shelterbelts") near fields, and by planting row crops perpendicular to the prevailing wind direction. Wind losses are almost zero with cover crops in place.

In addition to the soil erosion caused by overgrazing, deforestation, and unsustainable agricultural practices, recreational and military pursuits cause soil erosion, particularly those that involve off-road vehicles (ORVs). A 1974 study indicated that recreational pressure on the semi-arid lands of southern California by ORVs was "almost completely uncontrolled," and that those desert areas had experienced greater degradation than had any other arid region in the United States. In the half century since that study, much has been accomplished in managing damaging off-road traffic in California and other states, but ORVs continue to exert severe impacts on areas established for their use (**Figure 7.29**).

Although significant soil erosion has gone on for millennia, starting with the beginning of intensive agriculture in the Middle East, it has not always been clear where the eroded soil goes and how far and fast it travels. More recent work–for example, that of geographer Stanley Trimble–suggests, as mentioned earlier, that most eroded soil does not go far at all. Most of it found today sits at the bottom of the slopes from which it came (**Figure 7.30**). Streams make a big difference. In Pennsylvania, Dorothy Merritts and Bob Walter found that small mill dams, built by colonial settlers to power the mills that were the backbone of their economy, have trapped much of the soil eroded from the then heavily logged slopes not far upstream. Today, these dams are an environmental disaster waiting to happen because of potential collapse during flooding. Many are failing or being removed, and all that sediment, eroded more than 200 years ago but in storage ever since, is headed now downstream to Chesapeake Bay, one of the most important estuaries in the country, and a place where millions of dollars are being spent to improve water quality and fisheries (**Figure 7.31**).

Gastateparks

Figure 7.28 Providence Canyon is a gully cut into the deeply weathered saprolite in Georgia as the result of poor farming practices and excessive run off in the 1800s.

iStock.com/Jtstewartphoto

Figure 7.29 Heavy offroad vehicle traffic erodes these public lands in Utah at the Factory Butte recreation area. Vehicle tracks are everywhere. The soils are thin and the rock underneath, the Mancos Shale, is very weak.

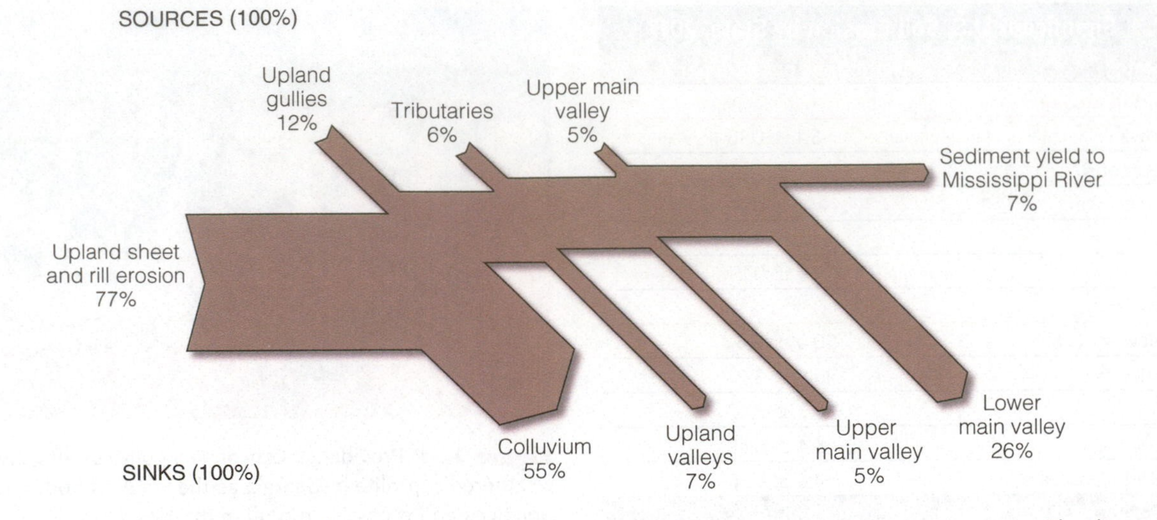

Figure 7.30 A "Trimblegram" for Coon Creek, a stream that is a tributary of the Mississippi River in Wisconsin. The diagram shows sources (top and left side) of eroded soil and sinks (bottom and right side) where the eroded soil has gone between 1938 and 1975. Most of the eroded soil ends up as colluvium, the poorly sorted mass of material at the base of hillslopes. Only 7% of the eroded soil makes it to the Mississippi River. (Adapted from Trimble, S. W. 1999. Decreased rates of alluvial sediment storage in the Coon Creek Basin, Wisconsin, 1975-93. Science 2855431: 1244–1246)

iStock.com/CynthiaAnnF

Figure 7.31 Pennsylvania Mill dam, along with the mill.

Added to the issue of soil loss due to erosion is the equally destructive transformation of cropland to nonagricultural uses. Increasingly, the U.S. population is moving into urban areas and, to accommodate them, cities are expanding outward onto agricultural land. In 2017, about 16% of the United States was classified as cultivated cropland (**Figure 7.32**), and each year more of that land is urbanized or converted to forestland, particularly in the northeastern and southeastern states. This trend is also seen in India and other nations with high growth rates. Between 2018 and 2019, the total land in farms in the United States declined by 2.1 million acres. World cropland decreases were not regarded as a major problem until the late twentieth century because increases in productivity had been great, especially following the introduction of synthetic fertilizers and pesticides. Increases in productivity leveled off in 1990, however, and in the face of a rapidly changing climate, it is questionable whether future increases will keep pace with the increasing population and decreasing cropland soils.

It is also important to consider that much of the increase in agricultural productivity has been leveraged from the use of fossil fuels to power large-scale industrial agriculture in developed nations and to maintain supplies of much-in-demand synthetic fertilizer. This cheap energy also allows for the shipment of food products long distances, so that in the dead of winter, people in Maine or Wisconsin can eat tropical fruits from the Philippines or lettuce from southern California. If energy prices continue to increase and fossil-fuel supplies tighten, one needs to ask: Is there still enough fertile and available land nearby to support the food needs of the local population? The answer in many intensively developed areas may well be, no.

Thomas Jefferson was among the first people in the United States to comment publicly on the seriousness of soil erosion. He called for such soil-conservation measures as

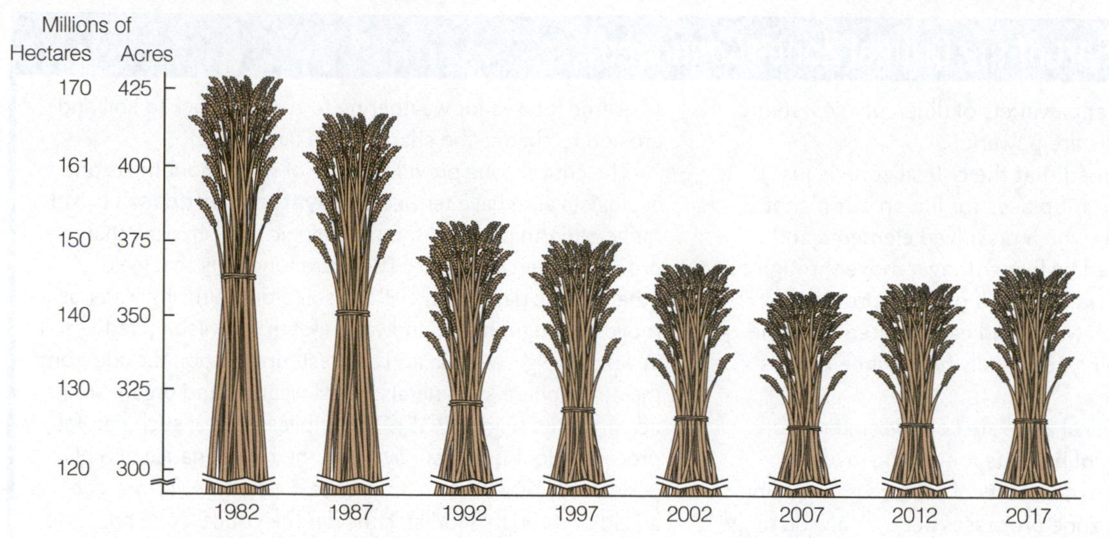

Figure 7.32 Total U.S. cropland, 1982–2017. Reductions were due to urbanization, conversion to other uses, and soil erosion. (Data from USDA and AGI)

crop rotation, contour plowing, and planting of grasses to provide soil cover during the fallow season. These practices are now increasingly common. Proper choices of where to plant, what to plant, and how to plant are primary components of soil conservation. The major erosion mitigation practices are:

- **Terracing**–creating flat areas, terraces, on sloping ground–is one of the oldest and most efficient means of conserving soil and water in steep, easily eroded terrain (**Figure 7.33**).
- **Strip-cropping**, where ground-covering plants are alternated with widely spaced ones, efficiently traps soil that is washed from bare areas and supplies some wind protection. An example of this is alternating strips of corn and alfalfa.
- **Crop rotation** is the yearly alternation of crops. Rotating corn and wheat with clover in Missouri, for example, has been found to reduce soil erosion from 44.5 tonnes/hectare

iStock.com/Marktucan

Figure 7.33 Terracing steep terrain is an ancient means of reducing soil erosion. At Machu Picchu in Peru, terraces that are many hundreds of years old hold soil for agriculture just as they did for the Inca who built them.

Soils: A Critical Part of the Critical Zone

Soils are where the atmosphere, terrestrial hydrosphere, lithosphere, biosphere, and people all interact in complex ways. As such, soils are an important component of what is now termed the **critical zone**, which extends from the top of the tallest trees into the bedrock below and incorporates life of all sizes and shape (**Figure 7.34**).

People and much life on Planet Earth depend on the critical zone for survival. It's where our food is grown and where land plants thrive that help to oxygenate our atmosphere. To better understand the critical zone, the U.S. National Science Foundation supported a decade-long investigation by scientists with a broad range of interests at Critical Zone Observatory sites scattered around the United

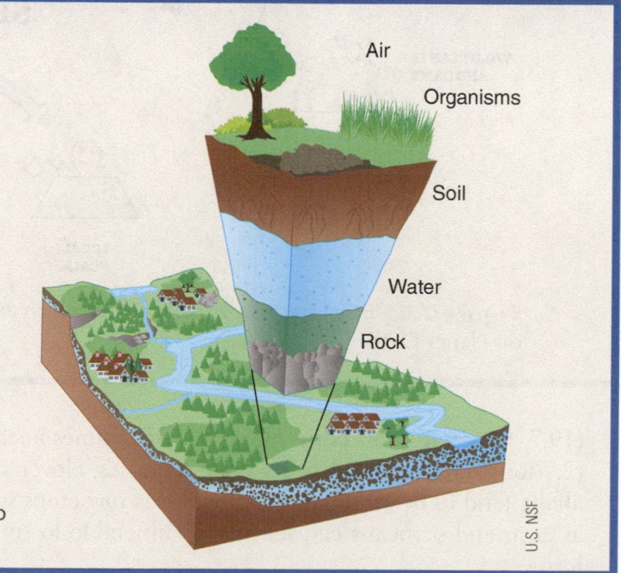

U.S. NSF

Figure 7.34 The critical zone stretches from the tops of the trees into the lithosphere, the rock that underlies our planet's landscapes.

(Continued)

States. The sites represent a variety of different ecosystems and climates. The results are powerful.

The research confirmed that the critical zone is just that: One of the most important places for life on our planet. Water is the medium by which dissolved elements and molecules, many created by living things, move through the critical zone by both physical and biological processes. Organisms interact with water and earth materials in the critical zone and by doing so literally help shape Earth's dynamic surface.

A hallmark of the critical zone are complex interactions that regulate the quality of habitats for organisms and determine the natural sustainability of resources, including food and water. Critical zone processes act over a wide range of time scales, from a small part of a second (the time it takes for chemical reactions to occur) to millions of years (the time it takes for weathering to turn bedrock to soil and erosion to change the shape of the landscape).

The critical zone provides some of what economists and ecologists alike have termed **ecosystem services**–a broad range of Earth processes, both biologic and physical, that provide benefit to people. These are functions that we'd otherwise be paying for, such as soil biota purifying water as it moves through the groundwater system, furnishing fertile soil in which we grow food, and sequestering carbon dioxide from the atmosphere as minerals in soil weather and organic matter accumulates (**Figure 7.35**). The implications of such natural processes for human survival and the maintenance of 8 billion people on this planet are immense. Every time you walk across a field or through a forest, consider the critical zone processes moving mass and energy from the tops of the trees to the bedrock below and vice versa, and know that soil is a key link.

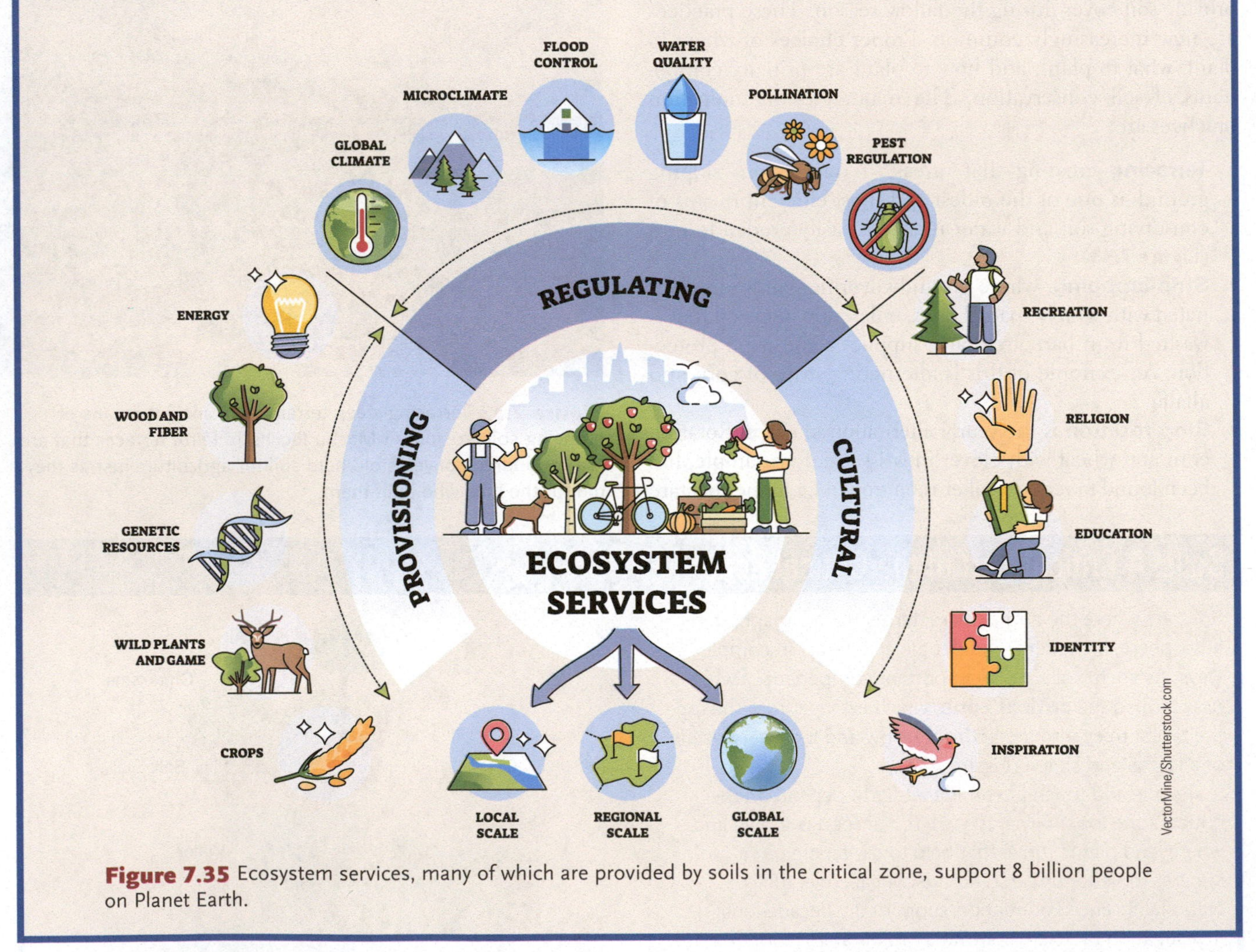

Figure 7.35 Ecosystem services, many of which are provided by soils in the critical zone, support 8 billion people on Planet Earth.

(19.7 tons/acre) for corn alone to 6.3 tonnes/hectare (2.7 tons/acre). Groundcovers such as grasses, clover, and alfalfa tend to be soil-conserving, whereas row crops such as corn and soybeans can leave soil vulnerable to runoff losses.

- **Conservation/reduced-tillage** practices minimize plowing in the fall and leave at least 30% of the soil surface covered with plant residue. As of 2017, U.S. farmers used conservation-tillage practices on over 40 million hectares (97.8 million acres) of farmland–about 39% of the land used for cropping. That's about the same as 2004 but much better than 1990, when only 26% of farmland was managed with conservation tillage. Retaining crop residues such as the stubble of corn or wheat on the soil surface after harvest has been found to reduce erosion and increase water retention by more than half.

- **No-till** and **minimum-till** practices are the most effective ways of planting to conserve soils. With no-till farming, seeds are planted into the soil through the previous crop's residue and weeds are controlled solely with chemicals. Specialized no-till equipment is required. No-till has been especially successful in soybean and corn farming. In 1990, U.S. farmers practiced no-till agriculture on only 6% of their land. By 2017, no-till practices were used on 260 million hectares (105 million acres) in the United States. The adaption of no-till and conservation-tilling practices has led to the decrease of intensive-tilling practices by 35% since 2012. No-till agriculture is most beneficial during times of drought, when plowed fields dry out and topsoil is more vulnerable to wind erosion. No-till practices saved much of the agricultural soil that would have otherwise been eroded by floodwaters of the Great Flood of 1993 and those that followed (2011, 2019) in the Mississippi River Valley.

7.4.2 Desertification

Grasslands and savannahs border most of the world's deserts. These landscapes are desirable for grazing animals, and if provided sufficient water can sustain crops. They are also very fragile environments. Especially sensitive to drying out from climate change, they can also be severely eroded by winds following plowing, and overgrazed by livestock. Such human-related degradation of these semiarid and subhumid lands, called **desertification**, is occurring at alarming rates, especially in countries that can least afford to lose arable land. According to the United Nations Convention to Combat Desertification, as many as 250 million people are threatened by this process, with 150 million likely to be displaced in the next few decades as their grassy steppe and savannah homelands transform into uninhabitable deserts. The United Nations regards on-going desertification as one of the severest environmental problems facing humanity. Accelerating climate change only worsens the situation.

Desertification is not a new phenomenon for people. Much of the lands bordering the southern and eastern Mediterranean Sea suffered from it terribly during Classical times. Even Plato referred to it in his writings. Many early conservationists and soil scientists cited the Mediterranean example as a cautionary tale. Until the last half of the twentieth century, however, desertification was often seen as a local problem, affecting only a few people, and it was ignored because new lands always seemed available to exploit and develop for food production. Currently, with an exploding world population, that's all changing. Globally, as much as 5% of all land area (above sea level) is undergoing or seriously threatened by desertification.

Figure 7.36 In southern Brazil, intensive agriculture and land clearance has allowed stable sand dunes to become mobile again, erode, and cover otherwise fertile land downwind.

In our modern world, human impact on fragile arid, semi-arid and subhumid environments has taken on new forms beyond overgrazing of sparse vegetation and poor farming practices (**Figure 7.36**); These include rapid growth of large desert cities such as Phoenix and Las Vegas in the southwestern United States and in other countries such as Egypt and Saudi Arabia; depletion of groundwater for irrigation and urban growth; replacement of well-adapted native vegetation with resource-intensive cultivated crops; and soil salinization caused by evaporation of irrigation water. Even inappropriate economic policies—such as those practiced in the United States with agricultural subsidies, which averaged $16.4 billion per year from 2010 to 2020, are cited as a cause. Such policies result in increased production for export, which encourages the placement of more land under cultivation, resulting in corresponding increases in erosion and soil loss.

The need for rethinking and improving land use in regions subject to to desertification grows significantly stronger as Earth's climate warms. World Wide Attribution (WWA), a consortium of international research scientists, conclude that the severe Northern Hemisphere droughts in 2022 should ordinarily occur once every 400 years. The American West was not the only region severely impacted in 2022. China hadn't seen such drying in 60 years, and its Yangtze River, one of the world's largest, dwindled to only half its average width. Modelling by WWA scientists indicates that droughts of this degree and intensity are now more likely to take place once every 20 than every 400 years, given present warming trends.

Africa—It was the desertification of Africa's sub-Saharan Sahel region in the late 1960s that first brought the world's attention to the problem of desertification and prompted thorough analyses by several climate specialists. The **Sahel** is a belt of semiarid lands along the southern edge of North Africa's Sahara Desert that stretches from the Atlantic Ocean on the west to the high mountains of Ethiopia on the east—300 to 1,500 kilometers (180 to 940 miles) wide

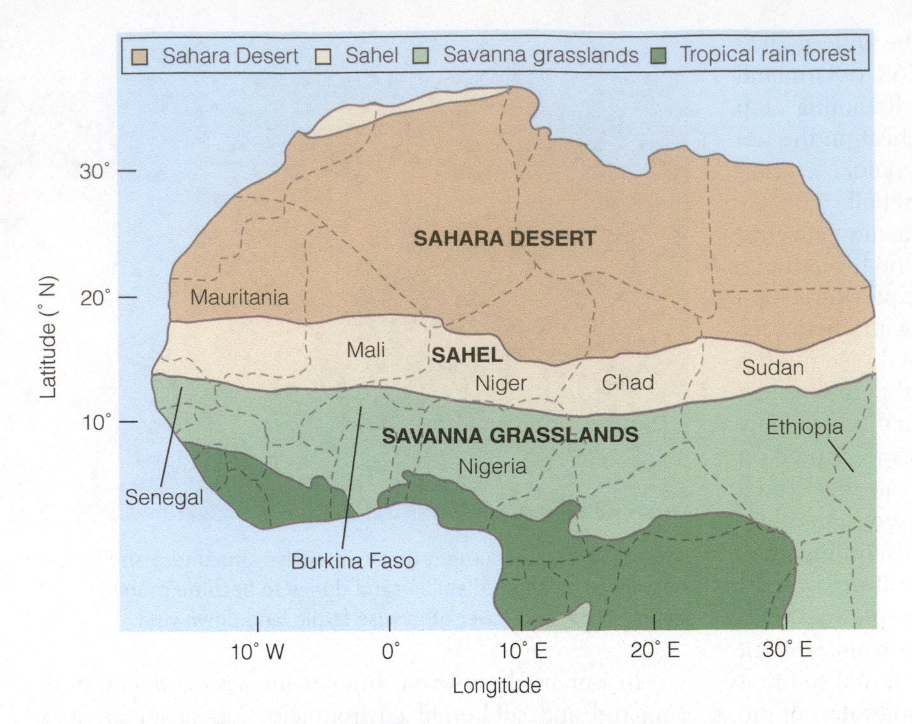

Figure 7.37 North Africa's semiarid Sahel is bordered by grasslands on the south and by the Sahara Desert on the north.

(**Figure 7.37**)—in which a large and rapidly expanding population resides. In 2022, the population was estimated to be 135 million, and projections suggest it could reach 330 million by 2050.

Typically, the Sahel region has dry winters and wet summers, when the intertropical convergence zone (a rather weak trough of low atmospheric pressure near the equator, where the trade winds converge) moves into the region, bringing rain. Rainfall amounts in the past have varied from year to year, averaging 10 to 20 centimeters (4 to 8 inches), but in some years it has been as much as 50 centimeters (20 inches) in the southern part, where the Sahel blends into a more humid biome (**Figure 7.38**). In 1968, the summer rains began decreasing, exacerbating the human-induced vegetation changes and initiating drought conditions, which captured the international community with grim images of skeletal mothers and starving children having bloated bellies. With dry soils and wind came dust storms (**Figure 7.39**).

Since the 1990s, the rains have returned somewhat, but not before 250,000 people and 12 million head of livestock died in one of the world's most devastating famines. The region is still experiencing a rainfall deficit, and the shortfall has extended to the Horn of Africa—Somalia, northern Kenya, and the lowland area of eastern Ethiopia (**Figure 7.40**). After the drought in 2016-2017, the Horn of Africa has had persistent below-average rainfall, leading to the loss of an additional 3 million livestock in Ethiopia and Kenya and 30% of household's herds in Somalia since mid-2021.

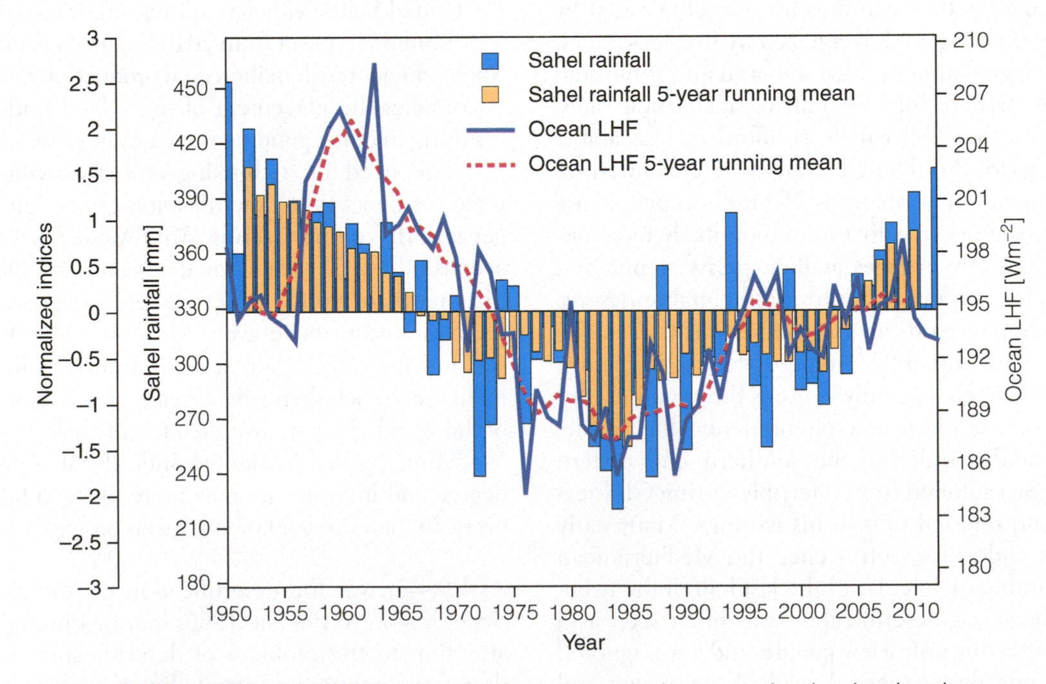

Figure 7.38 Rainfall in the Sahel varies greatly over time and appears closely related to the amount of evaporation in the global world oceans (here called LHF, latent heat flux). (Adapted from Diawara, A., Tachibana, Y., Oshima, K., Nishikawa, H., and Ando, Y.: Synchrony of trend shifts in Sahel boreal summer rainfall and global oceanic evaporation, 1950–2012, Hydrol. Earth Syst. Sci., 20, 3789–3798)

iStock.com/Guenterguni

Figure 7.39 A 2020 dust storm in Chad, deep in the Sahel.

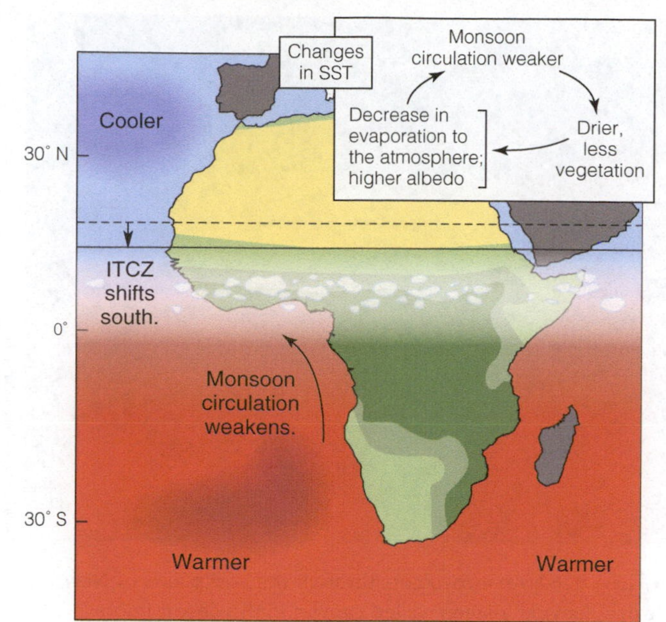

Figure 7.41 The Sahel drought since the 1960s was likely initiated by a change in global ocean temperatures, which reduced the strength of the African monsoon, and exacerbated by land-atmosphere feedbacks through loss of natural vegetation and land cover change. Land use by humans played a role. ITCZ = intertropical convergence zone; SST = sea surface temperature. (Adapted from Zeng, N. 2003. Drought in the Sahel. Science 302: 999–1000)

The drought that struck the Sahel region may be associated with a natural cycle of change in sea water surface temperatures off the west coast of Africa, known as the Atlantic Multidecadal Oscillation. A drop in sea temperatures and associated evaporation rates beginning in 1968 led to a decrease in life-sustaining monsoon rains onshore (Figure 7.38). The presence of aerosols from industries and other anthropogenic sources possibly enhanced this effect. In the Sahel, crops withered with the loss of rain, and vegetation elsewhere could not recover after people continued harvesting their firewood supplies. This triggered a positive feedback: evapotranspiration declined reducing rainfall further and drying accelerated, worsening the drought (Figure 7.40). It was a combination of natural factors, land use practices, and possibly even global human activities that made this situation so bad.

Another aspect of drought in arid and semiarid regions is dust, and the amount of dust originating from the Sahel during wet periods was markedly less than during the drought. Because dusts originating from the Sahel's windstorms are now known to account for about half of the world's total dust load, and dust reduces the amount of sunlight reaching Earth's surface, the drought in the Sahel may have an unexpected and widespread influence on Earth's climate. A dusty atmosphere probably cools the climate slightly (a negative feedback), but it will also mask the impact of accumulating greenhouse gases, meaning that the potential warming caused by these gases may actually be greater than what we measure.

iStock.com/Mustafa Olgun

Figure 7.40 In this drought-stricken region of Somalia, a woman carries water to her home.

The United States—In the United States, overgrazing has left an imprint on the arid lands of the West, along with the salinization and waterlogging that is common in many irrigated valleys. By the mid-1800s, overgrazing in the Southwest was already a problem, but when the railroads arrived in the 1880s, the explosion of cattle grazing that followed resulted in the carrying capacity of the land being exceeded, as shown by increased erosion and **arroyo** (gully) deepening (**Figure 7.42**). Consequently, the southwestern United States is a textbook example of the conversion of fertile, semiarid grassland to desert. The Rio Puerco Basin of central New Mexico is one of the most severely eroded regions.

Much of the American West has been in drought conditions since around the year 2000—conditions that have not been experienced for about 1200 years. The heavy rains and snowfalls in the winters of 2009-2010 and 2022-2023 in parts of the region offered some relief, but one wet season

Figure 7.42 An arroyo cuts through the arid plains of New Mexico, leaving a steep walled canyon in the desert floor.

amongst many does little to ease the net deficit of so many dry years. Furthermore, the temperature increase throughout the West since the 1950s has reduced the average snowpack and produced an earlier snowmelt (**Figure 7.43**). Since the global temperature has increased at an average rate of 0.08 degrees Celsius annually since 1881, and at twice that rate since 1980, it's likely that the connection between high temperatures and low rainfall is not a coincidence.

Southern Hemisphere—In the Southern Hemisphere, Australia has experienced unprecedented drought, especially in the continent's southwest. During the first 150 years of European settlement in southwestern Australia, winter rainfall was reliable. Now, however, the region experiences a rainfall deficit that is credited at least in part to global warming. Reduction of ozone has cooled the stratosphere over Antarctica, which has drawn Australia's zone of southern rainfall farther south over the ocean. The region of former winter wheat farming in southwest Australia is now turning to desert, with the condition exacerbated by increasing summer rainfall. Because of the historically erratic nature of the summer rains, farmers have never raised summer crops, so the rain now falls on vacant wheat fields and the water percolates down to the water table, where it encounters salt, which the prevailing westerly winds have been blowing in from the Indian Ocean for millennia.

Under natural conditions, the native plants utilized the summer rains and the salt remained untouched. Currently, however, as the rains fall on bare fields, evaporation draws salty water upward to the surface, poisoning all the plants that it touches (**Figure 7.44**). The former wheat-growing region is the world's worst case of dryland salinization, and it comes at a cost of billions of dollars. In addition to the lost

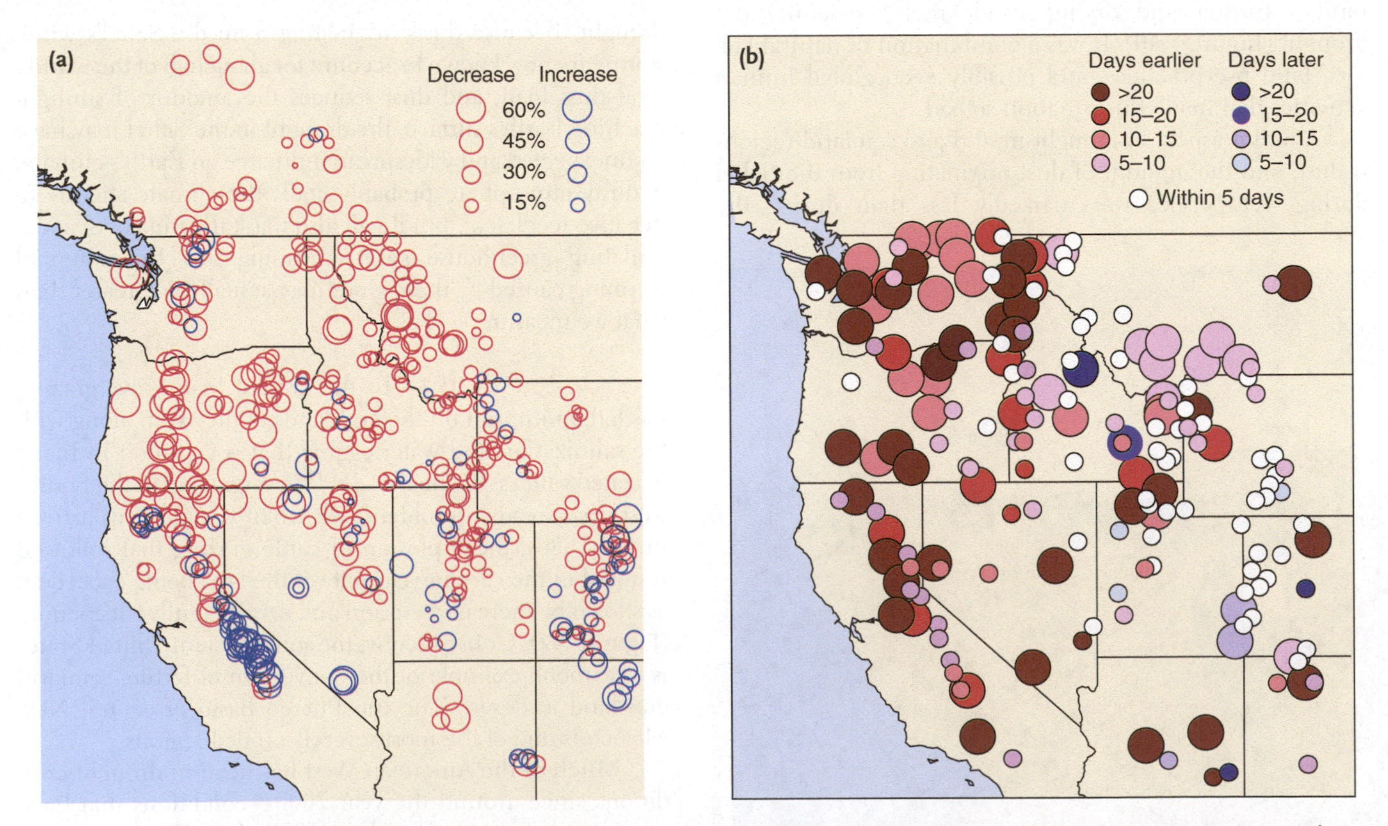

Figure 7.43 A modest temperature increase in the western United States from 1950 to 2000 (a) reduced the snow-water equivalent of the winter snowpacks and (b) shifted the peak snowmelt to earlier in the summer. (Adapted from Robert Service. 2004. Science 303: 1124–1127)

iStock.com/Basya555

Figure 7.44 Salt crust forming in an irrigated field.

agricultural production, houses, roads, and railroads are being damaged by the salt. There appears to be no solution to the problem.

What Do We Do?—As mentioned above, scientists have good evidence to suggest that our warming climate will intensify desertification over many parts of the globe, but where desertification is driven by land-use change, it can still be reversed or halted. An example of reversing desertification occurred in parts of the Great Plains of the United States impacted by the Dust Bowl in the 1930s (**Figure 7.45**).

US NARA

Figure 7.45 A black blizzard of dust hits Rolla, Kansas, in 1935. Night appeared to fall at 3 in the afternoon that April. This image, sent to President Roosevelt, was taken from the top of a 30-meter-tall water tower.

Political action was taken to address the problem not only because of the high cost to the thousands of farmers struggling to survive there, but also because dust blown from the Midwest seriously impacted distant cities downwind–with many concerned voters–including Washington, DC (**Figure 7.46**). Vastly improved methods of land and water management and farming methods have prevented the Dust Bowl disaster from recurring.

Unfortunately, disturbed desert and marginal grassland ecosystems are slow to recover. This is because those ecosystems are characterized by slow plant growth, little bacterial activity in the soil for recycling nutrients, low species diversity, and scarce water. Already sparse in vegetation, if damaged by overgrazing and off-road vehicle use, such lands may require decades, perhaps even centuries, to be restored to their natural state.

The best program for reversing desertification is to reduce the intensity of land use and to adopt measures such as planting drought-tolerant trees and shrubs (**Figure 7.47**), developing drip irrigation systems for effective water use, establishing protective screens (e.g., grass belts or strips of trees) as windbreaks that will help reestablish natural ecosystems, building sand fences, keeping grazing stock moving, and in extreme cases covering drifting sand dunes with boulders or wood, and paving areas to interrupt the force of wind near the face of dunes to prevent sand from shifting.

A good example of natural recovery is seen in Kuwait as an unexpected consequence of the 1991 Gulf War. Thousands of unexploded land mines and bombs are scattered across the desert in the western part of the country. These have discouraged off-road driving by sport hunters of the region's animals. The native vegetation is reestablishing itself, and the desert now resembles an undisturbed U.S. prairie. The number of birds in the area is much greater than it was before the war.

USDA

Figure 7.46 In March 1935, dust from the Midwest reached Washington, DC, obscuring the sun and surrounding the Lincoln Memorial with dust. Ships in the Atlantic reported a centimeter or more of dust on their decks.

Figure 7.47 Creosote bush, a common desert shrub, planted to restore eroded dry lands. The cages protect the small plants from animals.

Figure 7.48 This well-built concrete-block wall in Colorado was uplifted and tilted by stresses exerted by the underlying expansive soil.

7.4.3 Expanding Soils

As noted earlier, certain clay minerals have a layered structure that allows water molecules to be absorbed between the layers, which causes the soil to expand. This process differs from the adsorption of water to the surfaces of nonexpansive soil clays. Soils that are rich in these minerals are said to be **expansive soils** and fall in the soil order vertisol. Although this reaction is reversible, as the clays contract when they dry, soil expansion can exert extraordinary uplift pressures on foundations and concrete slabs, with resulting structural and cosmetic damage (**Figure 7.48**). Damage caused by expansive soils costs about $2.3 billion per year in the United States, exceeding the nation's estimated annual costs accrued by tornados, floods, earthquakes, and hurricanes combined. Expansive soils are mostly in the Rocky Mountain states, the Southwest, and Texas and the other Gulf Coast states (**Figure 7.49**). These states have sedimentary rocks that contain large amounts of clay.

The swelling potential of a soil can be identified by several tests, usually performed in a soil engineering laboratory. A standard test is to compact a soil in a cylindrical container, soak it with water, and measure how much it swells against a certain load. Clays that expand more than 6% are considered highly expansive; those that expand 10% are considered critical. Treatments for expansive soils include: (1) removing them, (2) mixing them with nonexpansive material or with chemicals that change the way the clay reacts with water, (3) keeping the soil moisture constant, and (4) using reinforced foundations that are designed to withstand soil volume changes. Identification and mitigation of damage from expansive soils are standard practice for soil engineers.

7.4.4 Settlement

Settlement occurs when a structure is placed upon a soil, a rock, or other material that lacks sufficient strength to support it. Settlement is an engineering problem, not a geological one. It is treated here, however, because settlement and specific soil types are inextricably associated. All structures settle, but if settlement is uniform–say, 2 or 3 centimeters (about an inch) over the entire foundation–there is usually

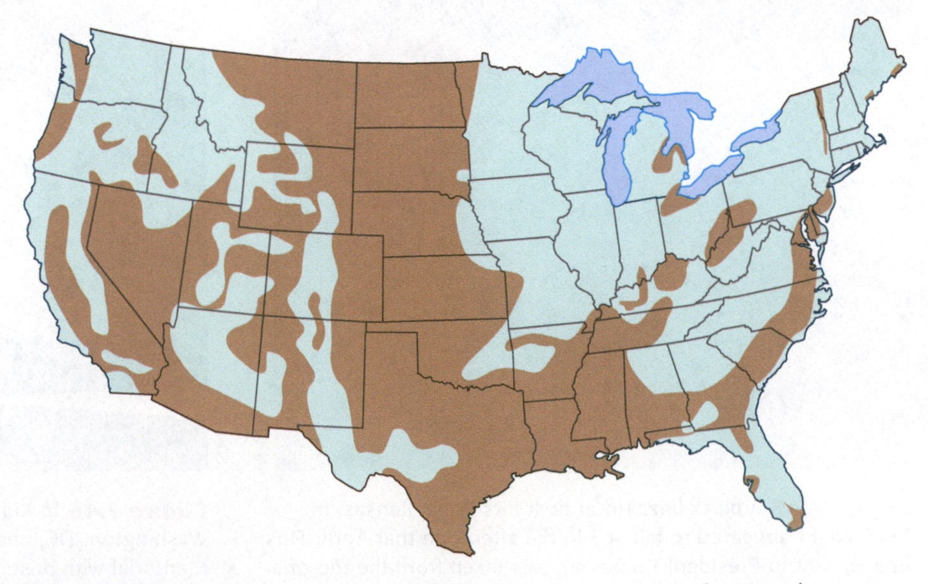

Figure 7.49 Distribution of expansive soils (shown in brown) in the United States.

no problem. When differential settlement occurs, however, cracks appear or the structure tilts.

The Leaning Tower of Pisa in Italy is probably the best known, most-loved example of poor foundation design. It may have scientific importance as well, as Galileo is said to have conducted gravity experiments at the 56-meter (183-foot) structure. In 1174, the tower started to tilt when the third of its eight stories had been completed, sinking into a 2-meter-thick layer of soft clay just below the ground surface. At first, the tower leaned north, but after the initial settlement and throughout the rest of its history, it has leaned south. To compensate for the lean, the engineer in charge had the fourth through eighth stories made taller on the leaning side. The added weight caused the structure to sink even farther. At its maximum lean, it tilted toward the south about 5.2 meters (17 feet) from the vertical, about 5.5 degrees, which gives one an ominous feeling when standing on the south side of the tower (**Figure 7.50**).

In 1989, a similarly constructed leaning bell tower at the cathedral in Pavia, Italy, collapsed, causing officials to close the tower at Pisa to visitors. In 1990, the Italian government established a commission of structural engineers, soil engineers, and restoration experts to determine new ways to save the monument at Pisa. The commission's goal was to reverse the tilt 10 to 20 centimeters (4 to 8 inches), which they believed would add 100 years to the lifetime of the 800-year-old structure. After considering many plans, it was decided to remove soft sediment from the north side of the tower, causing it to reverse its lean to the south.

This method was simple in concept but difficult to implement. It was done with drills 200 millimeters (about 8 inches) in diameter, which removed cores of soil about 15 to 20 liters (4.8 to 5.2 gallons) in volume. The weak sediment under the north side foundation quickly filled the voids left by the drilling and the tower started to straighten–that is, tilt back toward the north. The concept was less risky than installing stabilizing cables or constructing enlarged and stronger foundations. After being shut down for 11 years, the tower was reopened, still leaning, but 40 centimeters (16 inches) and 0.5 degree less. Since this correction was made, the tower has continued to straighten. As of 2018, the tilt has further reduced by 4 centimeters (1.5 inches). It is now estimated that the realignment will add 300 years to the life of the tower. The project cost about $30 million, but that is dwarfed by the enthusiasm of the over 5 million visitors to the tower each year.

iStock.com/F11photo

Figure 7.50 Pisa's famous leaning tower, the result of differential settlement on soft soils.

Tilting towers are not exclusive to Pisa, however. Few of the 170 campaniles (bell towers) still standing in Venice are vertical (see Chapter 1), and many other towers throughout Italy and Europe tilt. These graceful towers are certainly testimony to past architects' ability and talent, but also to their failure to understand that a structure can remain truly vertical only if local geology permits.

7.4.5 Thawing Soils

The term permafrost (Chapter 2), a contraction of "permanent" and "frost," was coined by Siemon Muller of the U.S. Geological Survey in 1943 to denote soil or other surficial deposits in which temperatures below freezing are maintained for several years (see Chapters 2 and 3). More than 20% of Earth's land surface is underlain by permafrost, and it is not surprising that most of our knowledge about frozen ground has been derived from construction problems encountered in Siberia, Alaska, and Canada. Permafrost becomes a problem for people when we change the surface environment by our activities in ways that thaw the near-surface soil ice. Such melting results in soil flows, landslides, subsidence, and other related phenomena.

Permafrost forms where the depth of freezing in the winter exceeds the depth of thawing in the summer. If cold ground temperatures continue for many years, the frozen layer thickens until the penetration of surface cold is balanced by the flow of heat from Earth's interior. This equilibrium between cold and heat determines the thickness of the permafrost layer. Thicknesses of as much as 1,500 and 600 meters (almost 5,000 and 2,000 feet) have been reported in Siberia and Alaska, respectively. Such thick permafrost is very old; it must have formed over hundreds of thousands, if not millions, of years as Earth cooled during repeated Pleistocene glaciations.

The top of the permanently frozen layer is called the **permafrost table** (**Figure 7.51**). Above this is the **active layer**, which is subject to seasonal freezing and thawing and is always unstable during summer. When winter freezing does not penetrate entirely to the permafrost table, water may be trapped between the frozen active layer and the permafrost table. These unfrozen layers and lenses, called **talik** (originating from the Russian word *tayat'*, meaning "melt"), are of environmental concern because the unfrozen groundwater in the talik may be

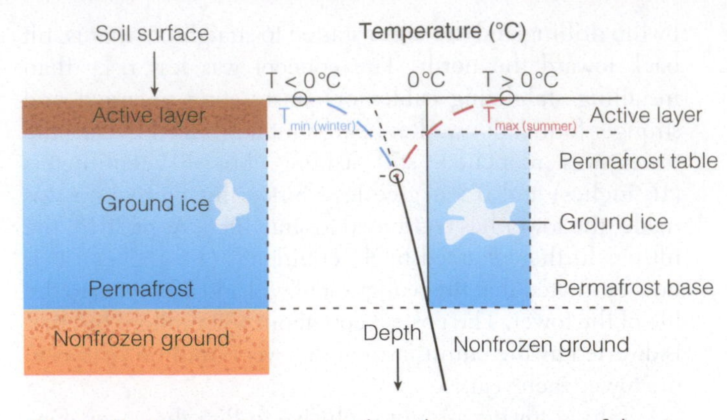

Figure 7.51 Permafrost responds to the temperature of the air above it and heat from the Earth below. In the summer, the surface warms and there is melting near the surface, creating the active layer. In the winter, the active layer freezes. Because of geothermal heat, permafrost gets warmer with depth, although some permafrost in the coldest parts of Earth is many hundreds of meters deep.

under pressure. Occasionally, pressurized talik water moves toward the surface and freezes, forming a dome-shaped mound 30 to 50 meters (100 to 160 feet) high, called a **pingo** (**Figure 7.52**).

The presence of ice-rich permafrost becomes evident when insulating tundra vegetation (the hearty, stunted shrubs and trees that cover much of the arctic) is stripped away for a road, runway, or building. Thawing of the underlying soil ice results in subsidence, soil flows, and other gravity-induced mass movements (**Figure 7.53**). Even the casual crossing on foot or in a Jeep can upset the thermal balance and cause thawing. Once melting starts, it is impossible to control, and a permanent scar is left on the landscape.

Structures that are built directly on or into permafrost settle as the frozen layers thaw. Both active and passive construction practices are used to control the problem. Although active methods involve the complete removal or thawing of the frozen ground before construction, passive methods build in such a manner that the existing thermal regime is not disturbed. For example:

- A house can be built with an open space beneath the floor, so that the ground can remain frozen (**Figure 7.54**).
- A building can be constructed on foundation piers whose temperature is held constant by heat dissipaters or coolant (**Figure 7.55**).
- Roads may be constructed on top of packed coarse-gravel fill to allow cold air to penetrate beneath the roadway surface and keep the ground frozen.

The most ambitious construction project on permafrost terrain to date is the 1200 kilometer (800 mile)-long oil pipeline from Prudhoe Bay on Alaska's North Slope to Valdez on the Gulf of Alaska. About half the route is across frozen ground, and because oil is hot when it is

iStock.com/Rlesyk

Figure 7.52 A pingo rises from the flat permafrost landscape of Canada's Northwest Territory.

iStock.com/Prospective56

Figure 7.53 Thawing permafrost near Fairbanks, Alaska, caused utility boxes near a highway to tilt as the ground below them deformed.

Danita Delimont/Alamy Stock Photo

Figure 7.54 Russian boys walk past houses in a Chukchi village that have been raised above the permafrost so that heat from the homes does not melt the ground below.

B. Pipkin

Figure 7.55 Heat-dissipating foundation piers on a modern structure; Kotzebue, Alaska.

pumped from depth and during transport, it was decided that the pipe should be aboveground to avoid the thawing of permafrost soil (**Figure 7.56**). The 1.2 meter (4-feet)–diameter pipeline also crosses several active faults, three major mountain ranges, and a large river. The pipeline's aboveground location allows for some fault displacement and for easier maintenance if it should be damaged by faulting. Placing the pipeline aboveground also accommodates the longstanding migratory paths of herding animals such as caribou and elk.

iStock.com/Sarkophoto

Figure 7.56 The Alaska pipeline is engineered to survive harsh winter cold, summer heat, and earthquakes. Pipe supports contain radiators to keep them cool and avoid melting the permafrost below.

Permafrost is likely to be one of the most troublesome victims of global climate change. All climate models and much recent observational data indicate that the warming of Earth as carbon dioxide levels continue to rise will not be a uniform affair. Climate change is predicted to have some of its most extreme impacts in the polar regions, with warming several times greater than at mid-latitudes. The effects will be dramatic, with the disappearance of areas of permafrost that may have existed for hundreds of thousands of years. Once permafrost is gone, traditional building techniques can be used, but the real problems will come during the decades-long transition when once-solid frozen ground melts, and structures, roads, and other engineering works that depend on that solid, frozen base are damaged and destroyed.

7.5 Humans and Soils (LO5)

In previous sections, we have discussed many ways in which humans negatively interact with soils, generally to their detriment (and our own). To recap, these include:

- direct exposure of soils through plowing and grazing to wind and water erosion
- continued harmful forms of land-use during droughts
- hazardous location of structures on soft or weak soils, or soils with expansive clays
- inappropriate construction of structures on soils that thaw (permafrost)
- offroad vehicle traffic

Positive interactions, including no till farming and useful identification of soil types to manage appropriate development (e.g.–septic tanks, foundations), also abound, but in general humanity has far to go in order to live wisely with soils, our most important natural resource.

In this final section we consider two selected issues in greater detail as examples of why this last statement is so important:

7.5.1 Salinization and Waterlogging

As mentioned earlier, **salinization**, a concentration of salts near the soil surface that is toxic to plants, has vexed farmers and ranchers throughout much of Australia and the American southwest, where it is an especially damaging aspect of desertification. In fact, salinization is the oldest soil problem known to humans, dating back at least 4,400 years. At that time, a highly civilized culture was dependent on irrigation agriculture in the southern Tigris-Euphrates Valley of Iraq, the "Cradle of Civilization." Within a few centuries of first being recognized as a problem, the soils had become so salty that wheat, a mainstay crop, could no longer be grown, and ultimately the area and their dependent cities had to be abandoned. Soils in parts of California, Pakistan, Ukraine, and Egypt are now suffering the same fate.

Salinization is human induced by intensive irrigation, which raises underground water levels very close to the soil surface. Capillary action then causes the underground water, containing dissolved salts, to rise in the soil, just as water is drawn into a paper towel or sugar cube. The soil moisture is then subject to surface evaporation, leaving behind salts that render the soil less productive and eventually barren (**Figure 7.57**).

For millennia, annual flooding of Egypt's Nile River added new layers of silt to the already rich soil of the agricultural floodplain and removed the salts by dissolving and flushing them downstream. After completion of the Aswan Dam in 1970, however, the yearly flooding and flushing action stopped, and salts have now accumulated in the soils of the floodplain, reducing productivity of the soil.

Salinization is particularly evident in the arid Imperial Valley of California. Soils of the Colorado Desert–or any desert, for that matter–can be made productive if enough water is applied to them. Because deserts have low rainfall, irrigation water must be imported, usually from a nearby river. The All-American Canal brings relatively high-salinity water from the Colorado River to the Imperial Valley, where it is used to irrigate all manner of crops, from cotton to lettuce. In some places, the water table has risen to such a degree that the cropland lies barren, its surface covered by white crusts of salt (**Figure 7.58**). Salinization matters, because the productivity of about 20% of all U.S. cropland is totally dependent on irrigation, and crops grown on irrigated cropland accounted for 54% of all U.S. crop sales.

The Imperial Valley's Salton Sea is one of the largest saltwater lakes in the United States, and it grows saltier and larger every year by the addition of saline irrigation water. This internally drained basin has no outlet, so all the salt that comes in stays in, unless, of course, it dries and blows away (**Figure 7.59**). The return of degraded irrigation water in canals to the Colorado River in southern Arizona has required that a desalinization facility be established near Yuma to make the water acceptable for agricultural use in Mexico.

Soil salinization can be reversed by lowering the water table through pumping and then applying heavy irrigation to flush the salts out of the soil (a measure unknown to the early Mesopotamian civilizations, and in any case too energy

USDA

Figure 7.58 Salt saturated soils in a Riverside, California, field.

iStock.com/Nuttaya99

Figure 7.57 Saline soil in Thailand kills rice seedlings.

Derrick Rivera/Shutterstock.com

Figure 7.59 The salt-encrusted shoreline of the Salton Sea.

intensive to have attempted at that time). Salinization can also be mitigated, or at least delayed, by installing perforated plastic drainpipes in the soil to intercept excess irrigation water and carry it off-site. Neither remedy is an easy or inexpensive solution to this age-old problem. Irrigation is an amazing thing, allowing the desert to bloom and many hungry mouths to be fed. But irrigation can adversely impact soils. When irrigated, some soils get so salty that few if any plants can grow.

7.5.2 Pedestrian and Vehicle Compaction

Have you ever hiked on a trail through the woods that became a stream after an afternoon thunderstorm passed? If so, you have seen the effect of soil compaction firsthand. In the adjacent woods, there is no water flowing over the soil surface; it has all been absorbed by the soil. People and vehicles moving across soils change soil structure by collapsing the pores that allow water and air to pass between grains. Some of these pores are microscopic, not visible to the naked eye. Others are large enough to put your hand in–perhaps the abandoned burrow of a gopher or the cavity left when an old root rotted away. When these pores are crushed, there is nowhere for the water to go, except to run off the surface.

The pounding of hiking boots is one way to compact soil. The tires of automobiles are an even more effective way (**Figure 7.60**). Park on a lawn a few times. The soil's ability to soak up rainwater might go down significantly. Park in the same place until the grass is dead and the soil packed tight, and what was once a lawn now behaves like asphalt and rainwater runs right off rather than soaking in. Repeat this process across a city, and storm sewers that used to be able to handle all the runoff in a major rainstorm no longer can, because so much more water is spilling into the streets every time it rains. However, with a pick axe, some compost, some fencing, and a handful of grass seed, the problem can be fixed. Soil, once aerated and protected from future compaction, will regain its ability to absorb water.

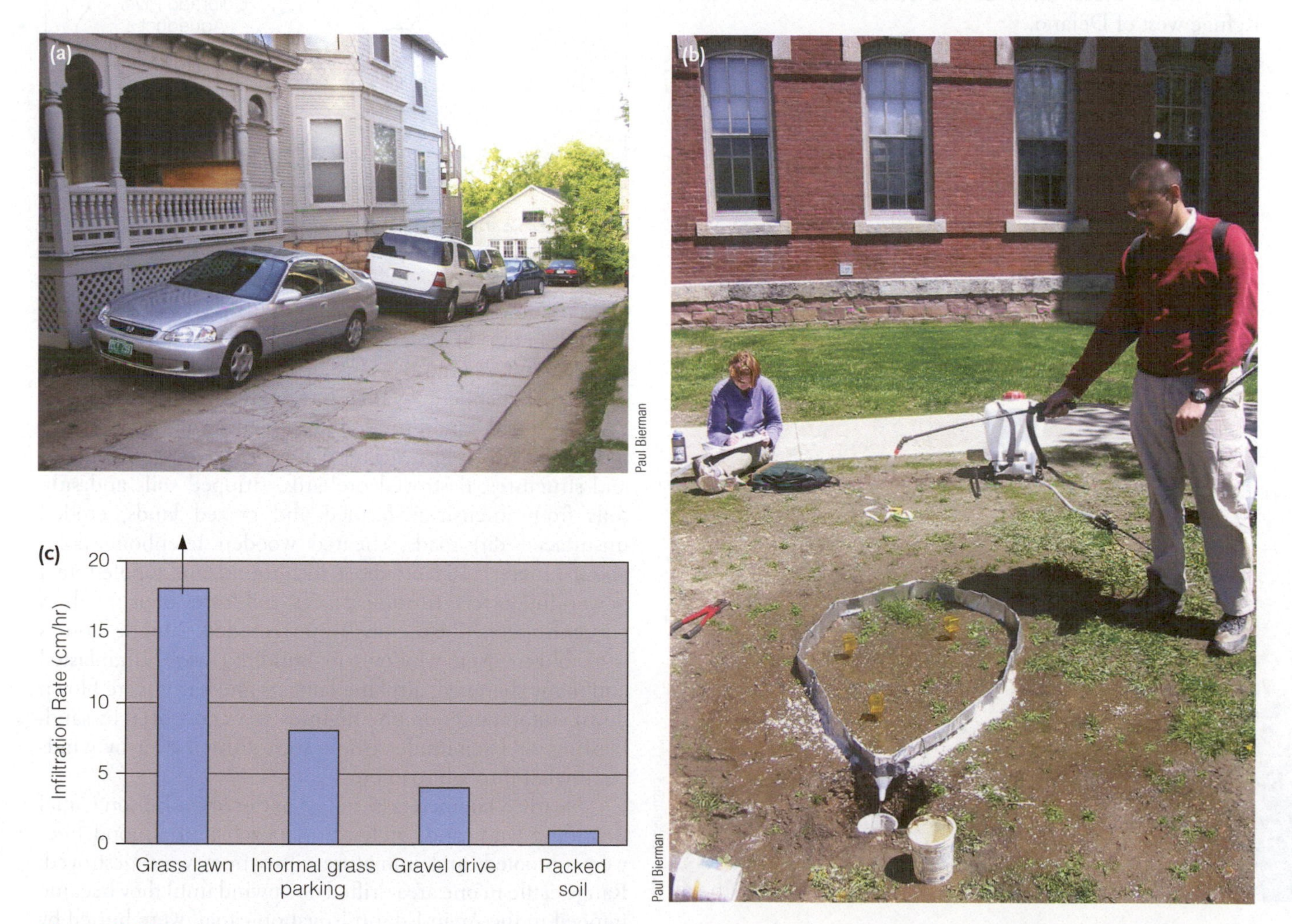

Figure 7.60 Soil compaction is a significant urban and suburban problem caused, in large part, by large objects that we drive. (a) Cars parked on what used to be lawn near college student apartments in Burlington, Vermont. (b) University of Vermont undergraduate research students measure how little water infiltrates on a part of campus where hundreds of people walk over the soil every day. Here, they are simulating a rainstorm and measuring runoff. (c) The results of many such infiltration tests show how well lawns soak up rainfall and how packed soil has almost no ability to infiltrate rainfall.

Case Study

7.1 A Dust Bowl in California

Historically, millions of acres of wetlands in California's San Joaquin Valley, including two lakes, sustained a rich ecosystem, home to countless migrating birds along the Pacific Flyway, large Tule elk, and some of the biggest grizzly bears in the world. Jedidiah Smith, who, with his company of beaver trappers, traversed the valley in the winter and spring of 1827, wrote about the fertility of this landscape. Today, only about 5% of the wetlands reported by Smith remain. They exist as publicly owned reserves, parks, and wildlife areas, such as the Kern National Wildlife Refuge west of Delano.

On the morning of December 20, 1977, an extraordinary windstorm generated an enormous dust plume (**Figure 7.61**), which swept out of the Tehachapi Mountains into the southern end of the San Joaquin Valley, today among the most productive agricultural regions of the United States (**Figure 7.62**). The dust cloud, rising as high as 1,500 meters (5,000 feet) above the valley floor, vividly illustrated how dry lands can be affected by human activities. The winds, with recorded speeds as much as 300 kilometers/hour (186 miles/hour), mobilized some 25 million tons of loosened soil from grazed lands, resulting in the extensive plume of dust, which extended as far as the northern Sacramento Valley hundreds of kilometers away. Wind-stripped agricultural lands added similar amounts of soil to the dust plume. The windstorm damaged vehicles and structures, destroyed orchards, stripped soils and subsoils from intensively farmed and grazed lands, eroded unsurfaced dirt roads, sheared wooden telephone poles about 3 meters (10 feet) above the ground, and toppled steel power-line towers. In some places, soil losses of as much as 60 centimeters (24 inches) were recorded. Mobile homes were blown over, windows in buildings were sandblasted and many shattered, airplane hangars and barns were blown down, automobiles on one highway were destroyed by sandblasting, and semitrailer trucks were turned over by winds channeled through road cuts.

Figure 7.61 Aerial photograph of the enormous dust plume that rose 1,500 meters (5,000 feet) above the southern San Joaquin Valley on 20 December 1977. Winds reached speeds up to 300 kilometers/hour (186 miles/hour). The Tehachapi Mountains are in the lower left.

Figure 7.62 Index map to localities in California.

Nearly 8 kilometers (5 miles) of the Arvin-Edison Canal was filled with sand, orchards of peach and almond trees were uprooted, and immature citrus trees were destroyed. Range cattle in one area drifted downwind until they became trapped in the Arvin-Edison irrigation canal, were buried by the drifting sand, and died of suffocation. More than 2,000 square kilometers (772 square miles) of once-productive lands were affected. Some areas of the valley have not yet recovered and may not do so for thousands of years.

A soil fungus, *Coccidioides immitis*, endemic to much of the southwestern United States, was carried aloft by the wind and spread valley fever, a respiratory ailment, throughout the region, even causing a widespread increase in reported cases in the San Francisco Bay area and the northern Sacramento Valley (see also Chapter 6). One known fatality from valley fever was a gorilla in the Sacramento Zoo, more than 480 kilometers (300 miles) to the north. Five people were killed in automobile accidents caused by poor visibility. The storm was very unusual because the sizes of sediments carried in traction (dragged along the ground surface), saltation (bouncing), and suspension by the wind greatly exceeded any previously reported for wind events.

The very high winds were generated by a large and strong high-pressure system positioned over the Great Basin states and a low-pressure system over the eastern Pacific Ocean. The high-elevation Sierra Nevada blocked the westward movement of the airflow, but the air mass was able to move across the lower-elevation Tehachapi Mountains and sweep into the San Joaquin Valley. The exceptional sediment transport caused by the storm was typical of lands subject to desertification: overgrazing, nearly 2 years of drought, recent plowing of the agricultural land in preparation for planting, recent stripping of natural vegetation in preparation for agricultural uses, and the general absence of windbreaks in agricultural areas. Less important contributing factors were the stripping of natural vegetation from the numerous oil fields of the region, extensive development near Bakersfield, and the local destabilization of land by recreational vehicles.

The San Joaquin Valley windstorm was not unique. In February 1977, a similar storm hit eastern Colorado and New Mexico. The setting was entirely different but involved similar human disturbances (agricultural development on relict sand dunes) and resulted in a dust plume, which was tracked eastward all the way across the Atlantic Ocean by satellite.

7.2 General Patton's Soils, Decades Later

Most people know General George S. Patton (**Figure 7.63**) as a legendary military commander who used his tank corps to defeat the Germans in North Africa during World War II; however, few know that he was a great compactor of soil.

In preparation for the North Africa campaign, Patton ordered the building of a dozen training camps in the desert U.S. Southwest, camps with names such as Iron Mountain. There, tens of thousands of recruits, most of whom had never seen a desert, lived and trained for months without rain or trees. Imagine the shock, going from the woods of Minnesota or the bayous of Louisiana to the Mojave Desert, where the biggest plant was a sagebrush bush and there wasn't anything green for miles. The camps were spartan places with rows of tents and roads bulldozed onto the gently sloping expanse of the desert. The soldiers, whom we can only assume were more than a bit homesick at times, did what they could to make the place home, building stone walls and intricate stone designs around their tents. Only the roads and low stone walls remain today. Toward the end of the war, the camps were abandoned and never reoccupied. For 80 years, they have been an ongoing experiment in how quickly desert soils recover from the pounding of soldiers' boots and tonnes of force that tank treads impose on the landscape.

Figure 7.63 General George S. Patton, tank commander and desert soil compactor.

The impacts of Patton's men are still very much affecting the area today. Soil compaction in deserts is a long-lasting insult. The soil beneath the abandoned roads is still compacted, and water flows along these roads during rainstorms, rather than through natural drainage in places carving deep ruts along their sides.

In the late 1990s, geologists from the University of Vermont and Duke University reoccupied one of Patton's camps to determine how the desert soils had recovered over the more than 50 years since Patton and his men had packed up and gone home. They used all sorts of tools to understand how the landscape and soils behaved, including high-precision surveys (**Figure 7.64**), low-tech soil pits, and measurements of ^{10}Be (see the explanation earlier

Figure 7.64 A survey rod equipped with a prism reflects laser light, allowing precise measurement of distance. In the background, a specially designed Global Positioning System (GPS) unit is used to measure locations to within a centimeter or two. Hundreds of locations will be measured in a day, and the data will be used to map the walkways and channels that carry water over this desert surface.

in this chapter). One of them set outlines of small, painted pebbles, 1,600 in all, to track how pebbles move (**Figure 7.65**).

Some of the pebbles moved in the several years that they were tracked. The pebbles in channels moved the farthest and fastest as they were swept away by moving water (it rained several times during the 3-year experiment). Even the pebbles away from channels moved a bit, probably by the action of gophers or other animals. The rock berms that the soldiers made to outline the walkways might seem insignificant, but they were a major influence on which way the water moved.

What can be done? At present, nothing, because this camp is a protected area of historical significance; but this experiment suggested some simple ways in which the impact of future camps on desert soils might be minimized. When a camp is abandoned, plow up the roads and disturb the soils to reduce compaction. This will accelerate erosion initially but allow the system to more quickly reestablish natural drainage patterns in the long run.

All of this should make us wonder about the impact people have had on the extensive deserts of the Middle East, where so much of our petroleum exploration and military activity has taken place.

Paul Bierman

Figure 7.65 University of Vermont geologists lay out lines of numbered and painted pebbles at General Patton's Iron Mountain camp in the Mojave Desert. Several years of surveying showed that most of these pebbles move only slowly or not at all—until massive rain from thunderstorms carries them downstream.

Photo Gallery

Soil Erosion Around the

Erosion is a natural process, but human actions almost always increase the rate of erosion and focus erosive activities in specific areas. Humans do this by removing vegetation for farming or clear-cutting for timber, destroying root networks that hold soils together. Tilled fields are vulnerable to erosion by wind and water, while grazing animals can compact soil and strip vegetation, making the soil more vulnerable to erosion.

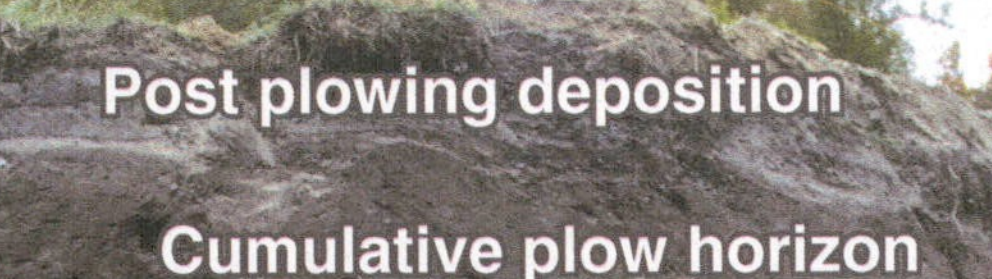

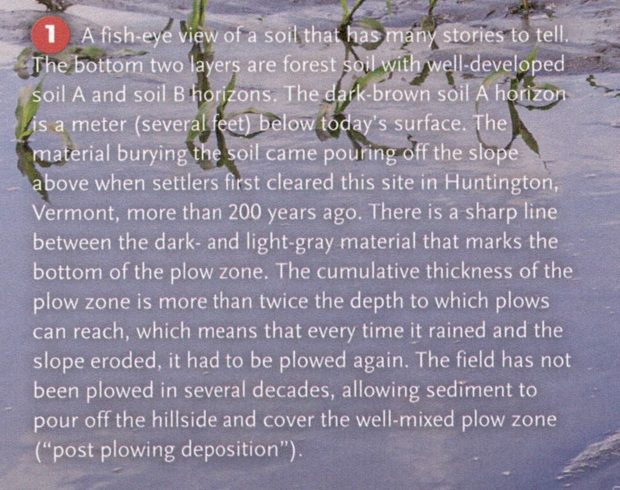

1 A fish-eye view of a soil that has many stories to tell. The bottom two layers are forest soil with well-developed soil A and soil B horizons. The dark-brown soil A horizon is a meter (several feet) below today's surface. The material burying the soil came pouring off the slope above when settlers first cleared this site in Huntington, Vermont, more than 200 years ago. There is a sharp line between the dark- and light-gray material that marks the bottom of the plow zone. The cumulative thickness of the plow zone is more than twice the depth to which plows can reach, which means that every time it rained and the slope eroded, it had to be plowed again. The field has not been plowed in several decades, allowing sediment to pour off the hillside and cover the well-mixed plow zone ("post plowing deposition").

2 Dirt road cuts through the grassland of Mongolia. Repeated passage by wheeled vehicles has eroded the road deep into the plain.

Erosion caused by heavy rain strips soil from an Illinois corn field in early summer.
iStock.com/JJ Gouin

iStock.com/Master2

3 Herding cattle across bare soil in the Serengeti of Tanzania, Africa. The lack of vegetation and the impact of the animals' hooves allow the soil to erode. Mount Kilimanjaro in the background is covered in snow.

iStock.com/totaila

4 In the outback of Western Australia (the Kimberley), termite mounds dot the landscape. The termites stir the soil, bringing materials from depths to the surface.

5 Remnants of a stone wall in an area that was once cleared for farming in Massachusetts. Today, soil erosion rates in these old, abandoned farm fields are very low. Leaves absorb the energy of rainfall, preventing erosion and allowing water to soak into the forest soil. The roots of the trees, which have regrown since the field was abandoned, hold the soil in place.

Summary Soil

1. Weathering

LO1 Explain the processes that weather rock and sediment

a. Physical Weathering

Physical processes which catalyze the breakdown of rocks contribute to physical weathering. In cold regions, frost wedging is common when water fills cracks, freezes, expands, and exerts pressure on rocks, resulting in fractures. The process of salt-crystal growth similarly creates pressure in rock fractures, resulting in weathering. Intense heating and cooling cycles, fire, and plant roots are other ways that rocks are physically weathered.

b. Chemical Weathering

The reactions which occur between earth materials and atmospheric constituents result in the formation of new minerals and dissolved compounds. Carbon dioxide and water react to form carbonic acid. Acidic solutions like this dissolve minerals. Materials interact with water and oxygen in oxidation reactions, which result in new forms of minerals, as when iron is oxidized to rust. Hydrolysis is a type of chemical weathering that forms clays, which provide some of the most important minerals in soils.

c. Rate of Weathering

Several factors impact the rate at which earth materials weather. Climate, surface environment, and grain size control the prevalence of physical or chemical weathering, as well as the area in which these types of weathering can impact materials. Parent material also determines the relative stability of the material itself, with minerals like quartz being very stable on Earth and olivine being the least stable.

d. Geological Features of Weathering

Where rocks are jointed (fractured), solutions can enter the joints and begin weathering the rock chemically while it is still underground. Corners made by three intersecting joints weather fastest, edges made by two intersecting joints somewhat less fast, and the planar faces of inidvidual joints most slowly. The result is called spheroidal weathering. At a larger scale, the expansion of large masses of rock as erosion takes place above them may cause the rock to split into sheets parallel to surface, which spall off once the rock mass is exposed, creating exfoliation domes.

Study Questions

i. What types of weathering are there, and how do they impact rocks and sediment?
ii. What environmental features impact the type and rate of weathering?

2. Soil Formation

LO2 Describe how soils form, including processes that establish soil profiles

a. Soil Development

Regolith, or freshly decomposed parent material, weathers away by both physical and chemical processes to form soil (once biological processes become involved). The intensity with which soil development occurs depends on several factors that impact the rate of weathering and decay, including climate, organic activity, relief of the land, parent material stability, and length of time that soil-forming processes have been occurring (think: soil = f(CL, O, R, P, T)).

b. Soil Profiles

A typical soil profile contains an O-horizon, the topmost layer of organic matter over the A-horizon, known as the zone of leaching because materials are highly weathered in this horizon by percolating water. The B-horizon, or the zone of accumulation, is deeper and is where dissolved minerals from the A-horizon accumulate. The final layer is the C-horizon, a transition zone in contact with the parent material. Soil profiles become more developed over time.

3. Soil Classification

LO3 Outline how soils are classified and explain how certain soil orders affect human uses of the land

a. Soil Color and Texture

Soil color depends mostly on the organic material of which it is composed and its iron content. Iron oxides make soils redder, whereas organic material makes soils darker. In reduced soils (oxygen depleted), gray and green colors are common. Soil texture varies from fine-grained loam to coarser-grained sand and gravel, and impacts drainage, workability, and looseness.

b. Residual Soils

Residual soils have developed on the parent material. Weathering of rocks decomposes into saprolite, which is decomposing rock that has lost some of its volume and strength but not necessarily its general overall shape and form as its most soluble minerals have weathered.

c. Transported Soils

Alluvial soils, transported by rivers and deposited along them, tend to be permeable and drain well. Glacial soils develop from glacial deposits and vary significantly depending on the sizes of the parent sedimentary particles they contain, ranging from fine clays to coarse boulders. Volcanic soils form on parent materials deposited by volcanoes and are initially barren. They make nutrients available to plants as they are chemically weathered.

d. Soil Orders

The 13 soil orders are based on major physical characteristics associated with each, including climate, degree of oxidation, permeability, vegetation, parent material, etc. This highest form of classification for soils is followed by suborders, families, and series. There are 14,000 soil series recognized in the United States. These classifications are extremely useful for many reasons, including building infrastructure safely and securely.

Study Questions

i. What criteria are used to categorize soils?
ii. Why are soil classifications important/how are they used?

4. Soil Problems

LO4 Explain where and why the four most common soil problems occur

a. Soil Erosion

Sheet, rill, and gully erosion are processes which erode soil by running water. Sheet erosion is the loss of soil particles in thin, even layers from a gently sloping area. Rill erosion begins with small channels carved into soil by runoff. Once rills become deep enough, they cannot be treated with plowing and form larger gullies. Wind erosion occurs mainly on dry, tilled soils in the absence of vegetation and root systems. Without these stabilizing forces, wind can pick up light soil particles. Erosion can be caused and worsened by overgrazing, deforestation, and agriculture, as well as by recreational and military pursuits.

b. Desertification

The degradation of land in arid regions caused by climactic and vegetation changes induced by human activities. Overgrazing of sparse vegetation, compaction of soils by livestock, land clearing, and urbanization all lead to the loss of productive soils and the formation of desert-like environments.

c. Expanding Soils

In the vertisol soil order, expansive soils made up of clay minerals with a layered structure allow water molecules to be absorbed between the layers and causes the soil to expand. The swelling potential of these soils results in significant infrastructure damage every year.

d. Settlement

Infrastructure settlement occurs when a structure is built on soil, rock, or other earth materials that lack sufficient strength to support it. The Leaning Tower of Pisa in Italy is a famous example of a structure's settlement into softer underlying soil beds.

e. Thawing Soils

Permafrost soils are frozen year-round for at least several years at a time and are common in arctic regions like Alaska, Canada, and Siberia. Issues arise when infrastructure is built on top of permafrost, which melts to create an active layer, leading to mass movements and subsidence.

Study Questions

i. What are the three examples of soil problems, where are they prevalent, and why?
ii. What led to the Dust Bowl in the 1930s throughout the Great Plains of the United States?

5. Humans and Soils

LO5 Discuss how humans degrade soils and the problems such degraded soils create

a. Salinization and Waterlogging

Salinization is caused by irrigation when capillary action on the surface draws groundwater toward the soil surface. The moisture at the top of the soil surface then evaporates, leaving salts in the water behind.

b. Pedestrian and Vehicle Compaction

Soil compaction, caused by trampling herds of grazing animals, vehicles, and people hiking, leads to the collapse of pores that would otherwise allow water and air to pass through soils. Compacted soils lead to greater surface runoff because water cannot infiltrate the soils.

Study Questions

i. In what ways can humans help reverse the effects of soil problems like erosion or settlement?

ii. How have humans contributed to soil problems like erosion, desertification, compaction, and settlement?

Key Terms

Active layer
A horizon
Arroyo
B horizon
Caliche
C horizon
Conservation/reduced-tillage
Cosmogenic isotopes
Critical zone
Crop rotation
Decomposed Granite (G.G.)
Desertification
Ecosystem services
Erosion
Exfoliation domes
Expansive soils
Frost wedging
Gullying
Hardpans
Humus
Hydration
Hydrolysis
Illuviation
Loess
Minimum-till
No-till
O horizon
Oxidation
Parent material
Permafrost table
Pingo
Regolith
Residual soils
Rill erosion
Sahel
Salinization
Salt-crystal growth
Saprolite
Sheet erosion
Sheet joints
Soil
Soil development
Soil horizons
Soil orders
Soil profile
Solution
Spheroidal weathering
Strip-cropping
Talik
Terracing
Transported soils
Ventifaction
Weathering
Zone of accumulation
Zone of leaching

“In an age when man has forgotten his origins and is blind even to his most essential needs for survival, water along with other resources has become the victim of his indifference.”

—Rachel Carson, Silent Spring

8

Water

Learning Objectives

LO1 Describe the primary controls on the distribution, types, frequency, and intensity of precipitation and what becomes of precipitation when it lands on the Earth

LO2 Summarize the distribution of fresh water on Earth and its sensitivity to environmental impacts

LO3 Explain why and how floods happen and their distribution over time and space

LO4 Predict where you will find groundwater and how it will move

LO5 Outline how planners manage water quality and water supply

LO6 Describe how water both supports and endangers human society

Rain pours down over the wetlands of Louisiana.

iStock.com/Konoplytska

The High Plains Aquifer and Water Sharing

Mark Twain, a writer who could twist a phrase, is credited with saying "Whiskey is for drinking, water is for fighting." There is much truth in the saying, as demonstrated in the conflict over water in the early settlement of the western United States. History may record that many twenty-first century conflicts, especially in the Middle East, were not over terrorists, territory, or even oil, but water. Most of the water in this region comes from three rivers: the Jordan, the Tigris-Euphrates, and the Nile. By 2050, the populations of water-scarce countries in the Middle East and North Africa are projected to double in size from their 2015 populations. Syria, Iraq, Israel, Jordan, and Egypt will be no exception to the population increase, and they will need more water for their people. You can bet that water sharing will be high on the agenda of any future peace talks between the leaders of these countries.

The same might be true, without the threat of military action, in the "water wars" over the High Plains **aquifer**, sedimentary formations from which abundant water can be removed for use. The High Plains aquifer contains as much water as Lake Huron and underlies 480,000 square kilometers (147,000 square miles) of parts of Kansas, Colorado, New Mexico, Wyoming, South Dakota, Nebraska, Oklahoma, and Texas (**Figure 8.1**). Water distributed by the irrigation system in **Figure 8.2** is from a well drilled into the aquifer's water-bearing strata. Indeed, 97% of the water drawn from this aquifer is used for irrigation. Congress, worried about the sustainability of such withdrawals, ordered the U.S. Geological Survey to monitor the aquifer and report on its condition every other year. The reports are not promising.

Water levels in the High Plains aquifer have been dropping dramatically, an average of more than 5 meters (15.8 feet) since pumping began for all the states, or a decline of 336 cubic kilometers (273.2 million acre-feet) in stored water since the 1930s. The latest data suggest that water levels are continuing to fall. When the amount of water withdrawn from an aquifer exceeds the amount replenished to it by rainfall, surface water, and groundwater,

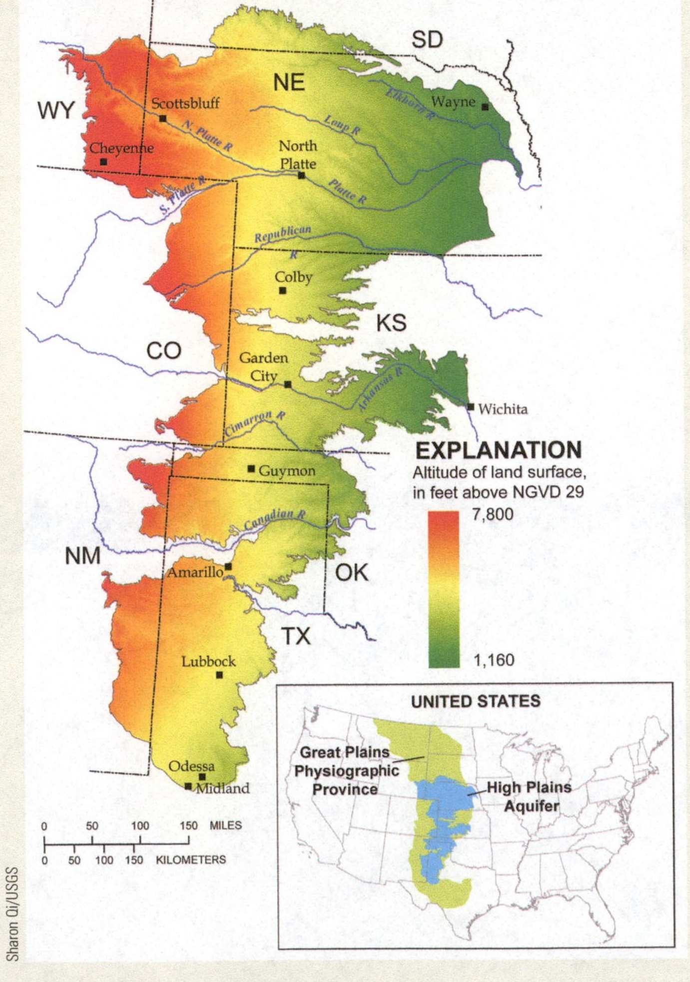

Figure 8.1 Location map showing the boundary of the High Plains aquifer, major cities and roads, and altitude of land surface.

Figure 8.2 For decades, groundwater from the High Plains aquifer has irrigated our nation's "breadbasket." This irrigation system draws water from a well that taps the aquifer. The well's potential yield is estimated at 4,000 liters per minute (about 1,000 gallons/minute) 24 hours a day. But the real question remains: Is that rate sustainable over the long term when rates of water withdrawal outstrip rates of groundwater recharge?

an overdraft condition exists, a shortfall; as with shortfalls of all kinds, it will have to be reckoned with, later if not sooner. This overdraft condition is called **groundwater mining** because the resource is being exploited without adequate replenishment. Models indicating the impact of human-induced climate change on rainfall suggest that the southern Great Plains (including Texas, where much of the water is used) will get much drier and the northern plains will get wetter. That means we'll see major shifts in where food is grown in North America. Political and economic upheaval are likely to follow, forced in large part by water–or better stated, the lack of water.

There are glimmers of hope. In one new approach, groundwater conservation easements are paid to farmers to abandon their rights to groundwater; in other words, to stop pumping. This leaves water in the ground for others to use while providing some income for the farmer. Other approaches include growing crops that require less water. Such drought-tolerant plants, common in arid regions around the world but not so common in the United States, include cassava (yuca), millet and black-eyed peas which are legumes and so fix their own nitrogen–a major advantage.

questions to ponder

1. With water levels in the High Plains aquifer still dropping precipitously, what should be done? Should pumping be strictly limited for everyone, or should water be allocated according to who is willing to pay the most for it?
2. How would you solve this tragedy, where the actions of each person end up destroying a resource used by many (known as the Tragedy of the Commons)?

8.1 Precipitation and Infiltration (LO1)

People who go into space and view Earth from afar cannot help but be struck by our planet's heavy cloud cover and its expanse of ocean, clear indications that Earth is truly "the water planet" (see Chapter 2). Abundant surface water distinguishes our planet from most other bodies in our solar system. Three-quarters of Earth's surface is covered by water. Water provides humans with pathways of transportation, and much human recreation is in water (swimming)–and *on* water, where that water exists in solid forms as ice and snow.

Without water, Earth would be a dead planet, its surface pock-marked with craters, as is the Moon's. Perhaps that's obvious, but less obvious are some really surprising things about the importance of water worth considering. For example, it takes 3000 liters (800 gallons) of water to grow a day's food for a family of four; it takes 75 liters (20 gallons) of water to make a half liter (pint) of beer; half the water used in the United States is required for electricity production; and ~200 million work hours are needed every day by people, mostly women and girls, to collect water for their families.

Most days, water is a good thing. Moving water supports whole ecosystems. Rivers are tapped for drinking water and to irrigate fields, allowing crops to grow in the desert. But sometimes, in some places, water can turn deadly. Floodwaters inundate riverside towns, storm surges rip homes from their foundations, and tsunami waves sweep away entire coastal communities. Water and its agents–rivers, streams, and oceans–perform most of the work of erosion and deposition, which shape and modify Earth's landscape.

Precipitation, the term for rain, snow, hail, and sleet delivers water to Earth's surface. Different types of weather systems produce precipitation (**Figure 8.3**). Fronts–boundaries between air masses–encourage precipitation. Recall from Chapter 2 that a cold front separates incoming cold air from warm air and tends to generate short-lived (minutes to hours) but

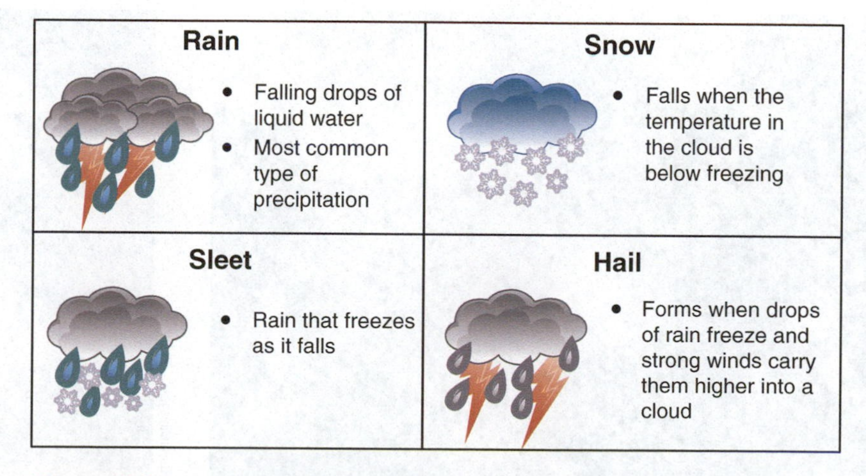

Figure 8.3 Types of precipitation (Adapted from Anbreen).

occasionally intense rainfall, including severe thunderstorms. A warm front, which separates incoming warm air from cold air, generates longer (by many hours) and gentler rainfall, which farmers greatly welcome.

Storms, which are organized disturbances in the atmosphere having low atmospheric pressure, come in several different types. **Tropical storms**, including hurricanes and typhoons, feed on warm tropical ocean waters. These typically compact storms (up to several hundred kilometers wide) can generate massive amounts of rain and high winds (**Figure 8.4**). Extra-tropical storms such as nor' easters are even bigger atmospheric disturbances up to 1000 kilometers wide which bring heavy precipitation and strong winds, although not of the speed generated by hurricanes (**Figure 8.5**). Convective rainstorms are localized events that result from the daytime heating of the land surface and the air above it (see Chapter 2). That air, being less dense, rises, cools, and results in precipitation, often accompanied by lightning and thunder (**Figure 8.6**).

8.1.1 How Precipitation Varies Over Space and Time

Precipitation is not equally spread over Earth's surface; rather, there are wet spots and dry spots predictably arranged around the world (**Figure 8.7**; Chapter 2). In the tropics, such as the islands of Hawai'i or Puerto Rico, moist air evaporated from warm ocean waters generates intense downpours. In contrast, downwind of cold ocean currents, such as in Peru (South America) and Namibia (southern Africa), hyperarid deserts are found where rain may fall only several times a decade.

Precipitation does not fall regularly over time, either. Storms, or **low-pressure systems**, are disturbances in Earth's atmosphere that come and go day by day and with the seasons. In contrast, extended winter rains commonly cause floods along the west coast of North America as Pacific storms come ashore. On the East Coast, hurricanes can drop 30 centimeters (1 foot) of rain in just a few hours during the summer and fall, whereas nor'easters batter the Atlantic Coast with rain, wind, and sometimes very heavy snow, mostly in the winter and spring. In the winter of 1888, much of New England got more than 1 meter (4 feet) of snow in March from a single nor'easter (**Figure 8.8**).

NOAA/NWS

Figure 8.5 Sprawling nor'easter covers the entire east coast of North America in 2014.

iStock.com/Elen11

Figure 8.4 Hurricane Willa swirls over Mexico in 2018.

iStock.com/BeyondImages

Figure 8.6 Convective thunderstorm in Australia.

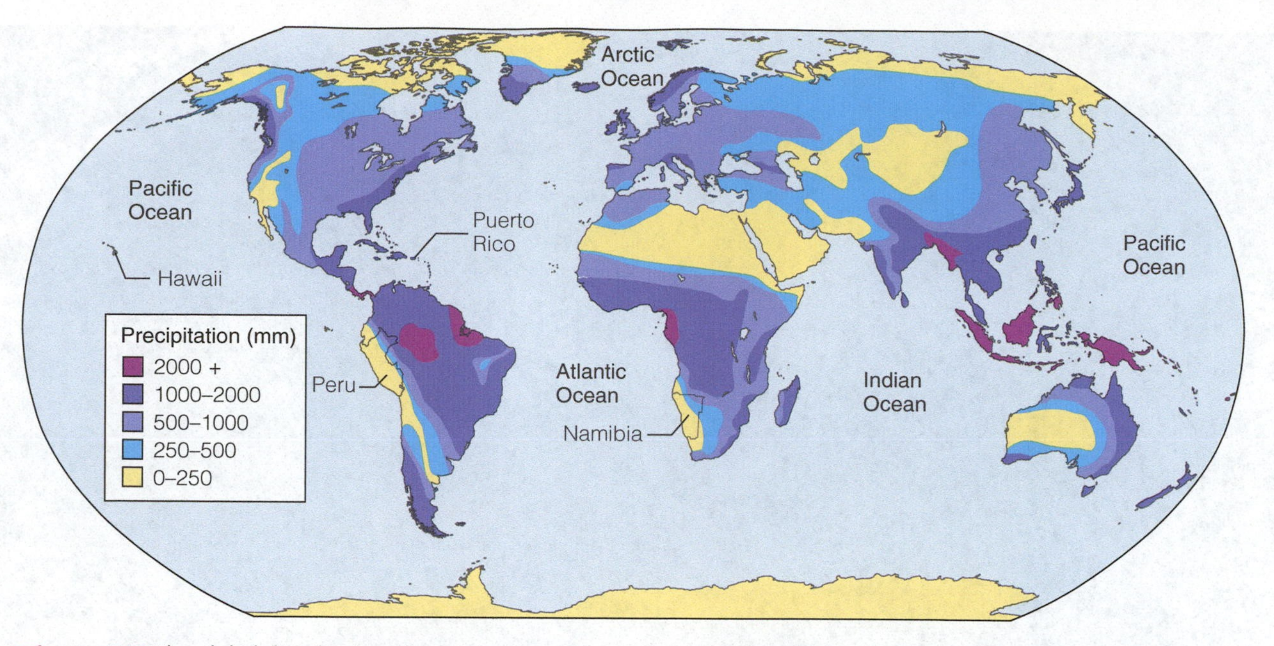

Figure 8.7 The global distribution of precipitation. Can you find areas near tropical oceans where it is very wet? Can you find areas downwind of major mountain ranges where it is very dry?

8.1.2 The Effects of Topography on Precipitation

Mountains are almost always wetter than adjacent valleys. Air masses forced up and over high terrain cool, and the moisture in their air condenses, falling as rain or snow (**Figure 8.9**). This "wringing out" of moisture by lifting air over mountaintops is termed the orographic effect, and it is why many floods get their start in highlands (introduced in Chapter 2).

Pictures Now/Alamy Stock Photo

Figure 8.8 The blizzard of 1888, a classic nor'easter, took the East Coast by surprise in mid-March. Temperatures plummeted from the 50s to a little above zero, and snow fell from Maryland to Maine. Here, in Brattleboro, Vermont, several feet of snow blanketed the city.

There are two sides to the orographic effect. Downwind of mountainous terrain, the rung-out air descends. Many lowland deserts of western North America, such as the Mojave are the result of mountains to their west having stolen the Pacific moisture before it could move inland. The Pacific Northwest in the United States is another great example. The Cascade Mountains get drenched with over 2 meters (7 feet) of precipitation yearly, while 100 kilometers (62 miles) farther east the Yakima Valley is lucky to get a few tens of centimeters (10 inches). The Cascades are lush and green. Yakima is dry and brown, except for the irrigated fields (**Figure 8.10**).

Some of the most dramatic rainfall occurs when storms and topography interact. Southeastern Texas is perhaps not what you would think of as a flooding hot spot, but it is. Near Austin, the Balcones Escarpment rises up to 160 meters (500 feet) above the coastal plain. Moisture streaming north from the warm Gulf of Mexico on southerly and southeasterly winds is forced to rise above the escarpment and onto the Edwards Plateau, a rural farming area. Incredible rains can result, swelling normally dry riverbeds into raging torrents in a matter of minutes.

During a 1954 flood, the Pecos River, often a dry stream, carried more than 28,000 cubic meters (more than a million cubic feet) of water per second–the average discharge of the Mississippi River–yet the area from which that

USGS

Figure 8.9 Moist trade winds blow southwestward over the summits of 5,480-foot Kohala and the 13,796-foot Mauna Kea. On the upwind (right) sides, the slopes are covered in green vegetation. Downwind (left) of the peaks, desert conditions prevail.

water came is only 0.3% of the area of the Mississippi watershed (**Figure 8.11**). Some of the highest rainfall rates in the continental United States, as well as the greatest 18-hour rainfall total, more than 90 centimeters (36 inches), have been measured on the Balcones Escarpment. There, in 1987,

iStock.com/KarenMassier

Figure 8.10 Green rows of irrigated wine grapes with the dry, brown desert hills of Yakima, Washington, in the distance.

Bob Daemmrich/Alamy Stock Photo

Figure 8.11 Very muddy floodwaters pour over the spillway at the Wirtz Dam in the Texas Hill Country during the record-breaking October 2018 floods. Between 20 to 30 centimeters (8 to 12 inches) of rain fell in 48 hours. The river feeding the reservoir carried over 150 times its average flow. All of this occurred after a record-breaking wet fall.

Figure 8.12 In July 1987, this bus was swept down the Guadeloupe River when rapidly rising floodwaters, caused by nearly 28 centimeters (11 inches) of rain bore down upon it. Forty three people ended up in floodwaters and ten teenagers drowned.

10 people died when their church bus was swept downstream by flooding (**Figure 8.12**). There are dramatic images of students being plucked from trees by helicopter rescue teams as floodwaters swirled around their bus.

8.1.3 Infiltration and Runoff

When a raindrop hits the ground, one of several possibilities may happen (**Figure 8.13**). If the ground is not already saturated with water and rain is not falling too heavily, the raindrop will be absorbed by the soil; this is called **infiltration**. Much of the water that infiltrates accumulates as groundwater, filling large and small pores between soil grains or along natural pipes made as roots rot away. Some water is taken up by plants and transpired back into the atmosphere. Other groundwater moves deeply through the underlying rock or sediment and may eventually emerge as springs or seep back out into the beds of streams and rivers. In forested areas all over the world, almost all rainwater follows these paths. This is all part of the hydrologic cycle introduced in Chapter 2.

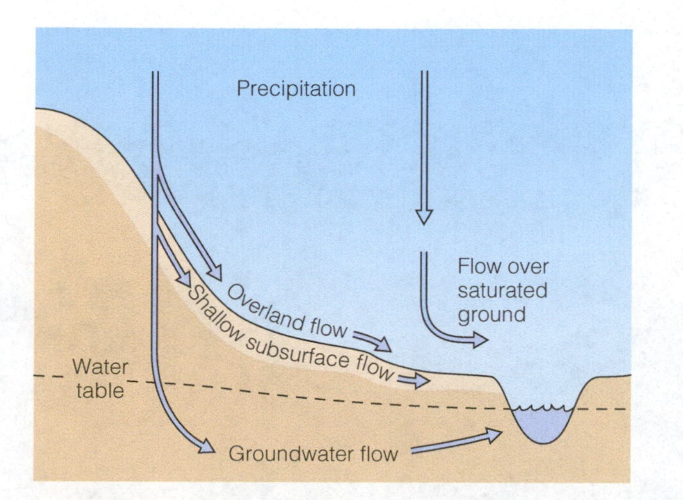

Figure 8.13 Water, falling from the sky as rain, can move various ways when it hits the ground. Development and other human impacts usually reduce the ground's ability to absorb water, forcing more rainfall downslope as overland flow.

However, if the conditions are right, water can run off the ground surface and directly into stream and river channels (**Figure 8.14**). Such runoff happens only if the water cannot infiltrate. Water won't infiltrate if the ground is already saturated by prior rain or melting snow, or if the rain is falling at a rate greater than water can be absorbed into the soil, or if the raindrop falls on something impermeable–perhaps clay-rich soil, a rocky outcrop, a hiking trail compacted by hundreds of footsteps, or a paved parking lot (see Chapter 7). Instead, the rainwater moves along the ground surface as **runoff**. It pours off the slopes and quickly enters streams and rivers.

Unless rainfall rates are exceptionally high or the ground has been compacted and there is little vegetation, most of the rain that falls after a dry period infiltrates the uppermost layers of soil and rock; little water runs off the land. As the surface materials become saturated or if rainfall rates remain extremely high, some rainfall cannot infiltrate and runoff begins. Runoff initially flows into small tributary streams, and finally consolidates into a well-defined channel of the main ("trunk") stream in the drainage basin. Whether a flood occurs is determined by several factors: the intensity of the rainfall, the amount of prior rainfall soaking the ground (known as **antecedent moisture**), the amount of snowmelt (if any), the topography, and the vegetation.

What do land-use change and development have to do with floods? As forests and grasslands are replaced by

Figure 8.14 Runoff after heavy spring rain, unable to infiltrate saturated ground in this midwestern North American farm field, is headed to a small stream.

impermeable asphalt and buildings, water that used to soak into the ground now pours into storm drains and rapidly enters rivers and streams. When forests covered the land, there were deep pits and high mounds where toppled trees had tipped up and roughened the forest floor. These pits, plus downed branches and leaves, captured rainwater, allowing it to infiltrate slowly. When trees are cleared for new development and 5-acre lots, bulldozers smooth the former forest floor and uniform grass lawns. Drainage pipes buried in the soil take the place of rotting logs and other forest detritus.

Even in the city, the conversion of grassy lawns to parking areas can make a big difference (**Figure 8.15**). Water levels in streams and rivers rise rapidly and attain higher peaks as rainwater moves quickly and efficiently off the paved landscape. Urban floods may not last long, but they can be devastating. The slug of floodwater, moving at high velocity, tends to erode and deepen channels, causing other unintended consequences such as overflowing culverts and flooded streets.

8.1.4 Frequency, Intensity, and Duration of Precipitation

The environmental effects of precipitation depend on how frequently it occurs, the intensity of the precipitation, and the duration of precipitation events. These three characteristics are controlled in large part by climate and thus geographic location.

In some parts of the world, like the equatorial, maritime tropics, precipitation is a daily event as the heating of the day causes warm, humid air to rise and condense (**Figure 8.16**).

Paul Bierman

Figure 8.15 When cars park on lawns, the mud starts quickly. In this photo, several vehicles in Burlington, Vermont, have repeatedly driven over the grass, killing it, compacting the soil beneath, and forming deep ruts. When it rains, that compacted soil can't absorb the rainfall. Its infiltration rate is so low that muddy puddles result. The good news is that, with fencing, rototilling to break up the compaction, and time, the soil will once again be able to absorb rainfall.

iStock.com/Itsskin

Figure 8.16 Tropical afternoon shower, Indian Ocean.

Afternoon showers are the rule and vegetation covers the land surface, making the soil permeable and runoff almost nonexistent. In contrast, the Atacama Desert of western Peru can go years without a drop of rain because so little moisture evaporates from the cold ocean offshore. There is no vegetation to protect the land surface. When it does rain, as it did in 2015, when severe thunderstorms struck the region, much of the precipitation runs off and flooding is common (**Figure 8.17**). The heavy 2015 rains triggered mudflows that buried towns and filled streets with overturned cars buried in thick, brown mud (**Figure 8.18**).

Rainfall intensity, the rate at which precipitation falls, in large part determines the impact of rainfall on the landscape and population. Low-intensity rainfall is absorbed

Alex Fuentes/AFP/Getty Images

Figure 8.17 In the hyperarid Atacama Desert of Chile in March 2015, the Copiapo River overflowed following heavy rainfall.

Figure 8.18 Mudflows, triggered by the heavy rains on barren slopes in the Atacama Desert, filled streets in the towns of western Chile in March 2015.

by the ground. High-intensity rainfall runs off because the rate at which water is falling from the sky exceeds the rate at which it can infiltrate the soil. Rainfall intensity and **rainfall duration** (how long a rainfall event of a certain intensity continues) are inversely related (**Figure 8.19**). The highest intensity rainfalls last only a short time (typically minutes), whereas low-intensity rainfalls can last for many hours, even days.

Summer thunderstorms and hurricanes are known for generating high-intensity, short-duration rainfall. In contrast, many of the storms that pummel western North America for days on end generate long-duration, low-intensity rainfall,

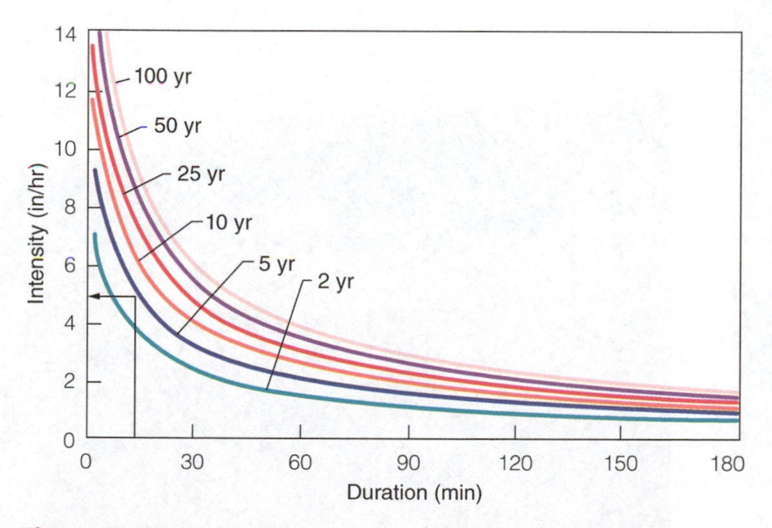

Figure 8.19 Precipitation intensity and duration are inversely related. High-intensity rain events don't last long! The lines on the graph represent the average time between events of a certain intensity and duration. The arrow marks an event that lasts 12 minutes with an intensity of 5 inches of rain per hour. That would occur, on average, once every 5 years.

easily infiltrated into the soil. These storms make for damp hiking but they don't result in flooding because they don't generate large amounts of runoff over short periods of time, and they are wonderful for replenishing surface and groundwater supplies and watering crops.

8.2 Surface Water (LO2)

Fresh surface water is used for drinking, recreation, waste disposal, and industrial purposes. Here, we consider the importance, behavior, and environmental sensitivity of lakes and rivers. It's important to remember that most of Earth's water resides in the oceans and is salty, making it much less useful for many human uses (Chapter 2; **Figure 8.20**). Most fresh water is locked up in the Greenland and Antarctic ice sheets (at least for now; Chapter 2).

8.2.1 Lakes

Lakes are bodies of water in landlocked basins that are formed by different geological processes. For example, the geological origin of the Great Lakes (except Lake Superior) is glacial, Crater Lake is volcanic, and the Great Salt Lake is a downfaulted basin. Lakes are not important quantitatively in the global water balance, holding less than 0.4% of all continental freshwater. However, they are of great local importance for agriculture and recreation and as a source of water. About 80% of all the water in lakes worldwide is held in fewer than 40 large lakes.

Lakes can be either fresh or saline (salty). Freshwater lakes are especially important as a source of water for drinking irrigation, and recreation. Saline lakes can be useful for mineral extraction. Lake basins can be filled with water, or dry for some or nearly all of the time. Whether a lake is fresh, salty, or dry depends on the balance between precipitation (P) and runoff (R) into the lake, and evaporation (E) and outflow (O) from the lake.

If a condition exists where $P + R = E + O$, the result is a freshwater lake. If $P + R = E$ and there is no outflow, the result is a salt lake. However, in the southwestern United States, the condition exists where $P + R < E + O$, producing in time a dry lakebed called a **playa**.

Dry lake beds, especially the salt flats of Utah's ancient Lake Bonneville have been used for automobile speed trials, because they are among the most level surfaces above sea level. Rogers Dry Lake in California is the home of Edwards Air Force Base and had the distinction of being used as the landing strip for space shuttles before they were retired. Playas are also known for a variety of salts and useful clays, products found in their sediments. Some lakes have their source of water cut off by humans, usually for agriculture, occasionally with catastrophic results–the Dead and Aral Seas are prime examples (**Figure 8.21**).

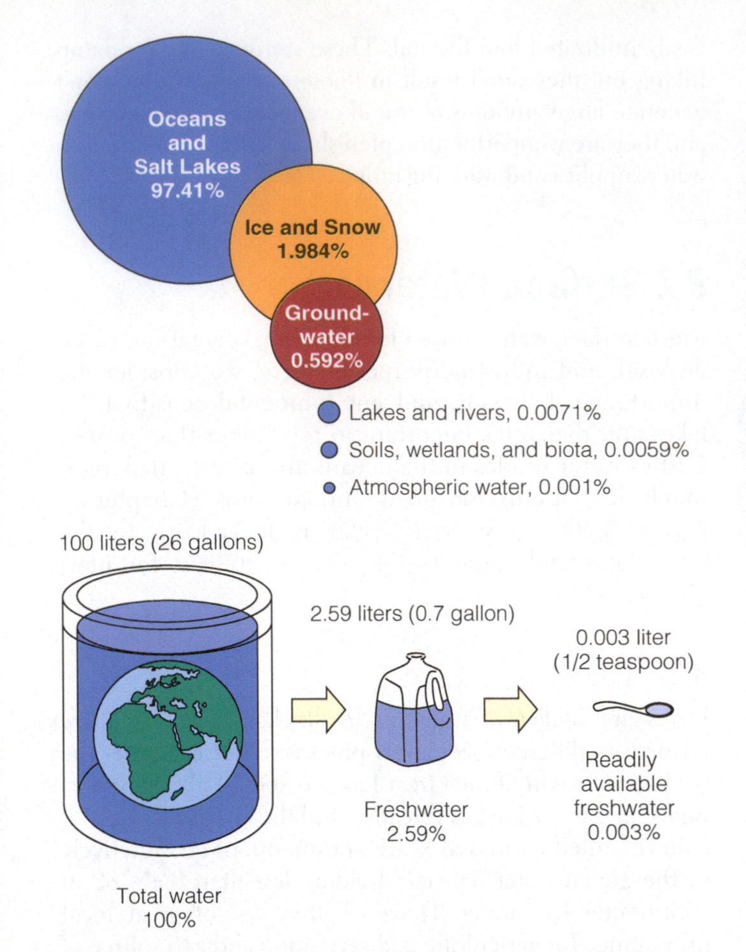

Figure 8.20 The planet's water budget. Only a tiny fraction by volume of the world's water supply is freshwater available for human use.

Changing climate can dramatically alter the water balance in lakes. Consider this: if you had visited the southwestern United States 15,000 years ago, you would have seen lakes in the Mojave Desert. Those lakes would have been connected by rivers with rapids and waterfalls, places to swim and take a drink of freshwater. The glacial climate was cooler and cloudier, so evaporation was lower, and precipitation may have been higher in some places. Even Death Valley, one of the hottest and driest places in the United States today, had a giant lake in it filling its floor; Lake Manly. How do we know? Because the shorelines and lake-bottom sediments of this and other now-vanished lakes can easily be found and their ages determined using carbon-14 and other age dating techniques. (**Figure 8.22**).

Many lakes are far more dynamic than they look thanks to an unusual property of water. The temperature of maximum density of freshwater is 4°C (about 40°F). Water at this temperature is heavier per unit volume than water that is warmer or colder. In the summer, the lake is stratified, and the warmer, less dense surface water is separated from the cooler, denser bottom waters by the thermocline (see Chapter 2), which acts as a floor below the surface water and a ceiling above the bottom waters. As it gets colder in the fall, the surface water temperature can decrease to 4°C and the thermocline weakens; that is, the temperature of the lake becomes about equal throughout, or **isothermal**. The surface water sinks and the bottom waters that are oxygen-deficient rise to replace them.

This is called **overturning** and it occurs in lakes in temperate and cool climates where surface waters are subject to near-freezing temperatures. Overturning keeps the bottom waters in lakes from becoming stagnant.

Figure 8.21 The Dead Sea in Israel. The "bathtub" rings of salt near the water's edge show the gradual lowering of the water level in as more water evaporates than is replaced by flow from (perdominately the Jordan River) feeding the Dead Sea. Damming and diversion of the river has reduced incoming water flow over the last century, and the Dead Sea has gotten smaller and saltier.

Figure 8.22 Horizontal shorelines, eroded into what is now called Shoreline Butte in Death Valley, testify to the presence of a now-vanished lake, nearly 160 km (100 mi) long, that filled the valley 15,000 years ago when the climate was cooler and wetter.

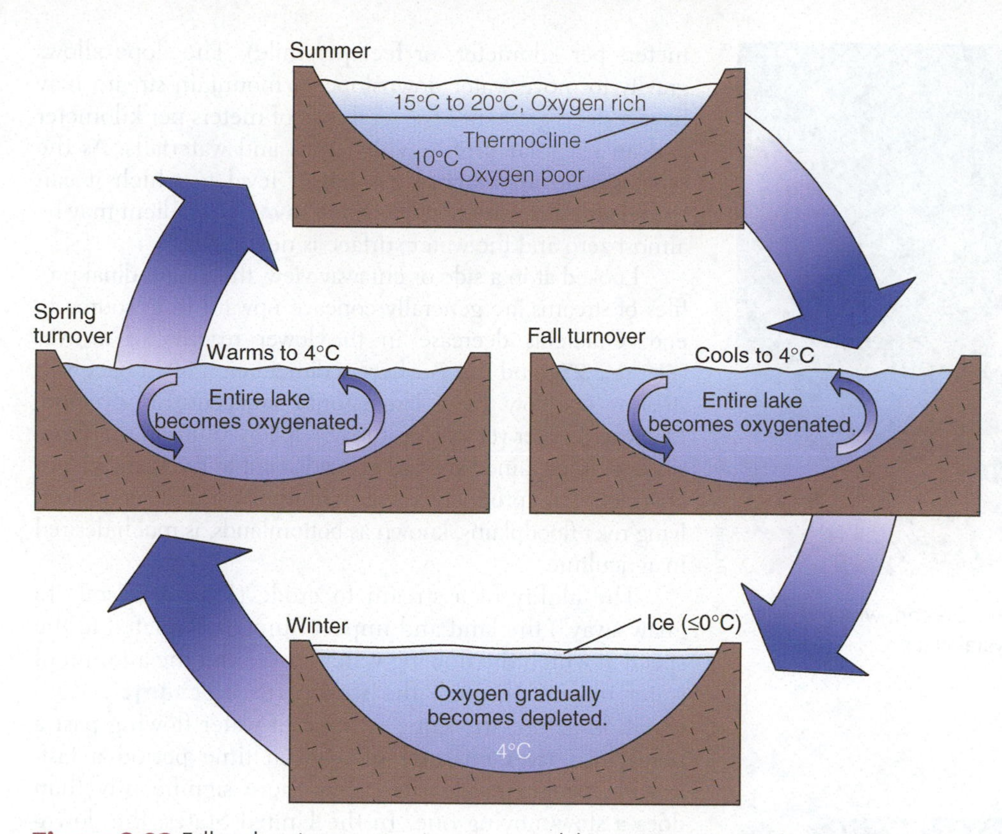

Figure 8.23 Fall and spring overturning occurs in lakes in temperate and cold regions. Overturning is a mechanism whereby deep lake waters are reoxygenated and do not become stagnant.

As **Figure 8.23** illustrates, the bottom waters stagnate during the winter but are refreshed again in the spring when the surface-water temperature increases to 4°C and surface water sinks, as the bottom waters simultaneously rise. The net result is the oxygenation of the lake waters and recycling of nutrients, which are transported from the bottom to the surface. This makes for abundant plankton and good fishing.

Another factor in the well-being of lakes and their suitability as a source of water is how well "nourished" they are. Nourishment consists of inorganic compounds (nitrate, phosphate, and silica) that are carried into the lake by streams and released by decaying organic matter on the lake bottom. **Oligotrophic** lakes are usually deep, with clear water, and low in nutrients, with a relatively low fish population (**Figure 8.24**). They are a source of good drinking water and are usually visually appealing; the water is crystal clear. Crater Lake on Oregon is an example of an oligotrophic lake (**Figure 8.25**). A **eutrophic** lake is "well fed" with nutrients. Such lakes are often shallow, have plants, contain a population of fish such as carp, have a low oxygen content, and usually support a rich growth of algae or other plants at the surface which give the water a light greenish color (**Figure 8.26**). Between the two extremes are **mesotrophic** lakes, which have a well-balanced nutrient input and output. The result is a healthy fish population, a variety of shore and bottom plants, and an abundance of plankton.

8.2.2 Rivers

You can think of rivers as natural conduits moving water from uplands to lowlands and in most cases to the world's oceans. Rivers flow in channels, and every river has a **drainage basin** (the land area that contributes water, which at some point in time fell as precipitation, to a particular stream, river or lake). Drainage basins are the basic unit into which hydrologists, scientists who study water, divide river systems. Individual drainage basins are separated by **drainage divides**. A drainage basin may have only one stream, or it may encompass many streams and all their tributaries. Drainage basins are also known as watersheds.

Every stream channel has a slope, known as its **gradient**, which can be expressed as the ratio of its vertical drop to the horizontal distance it travels (in

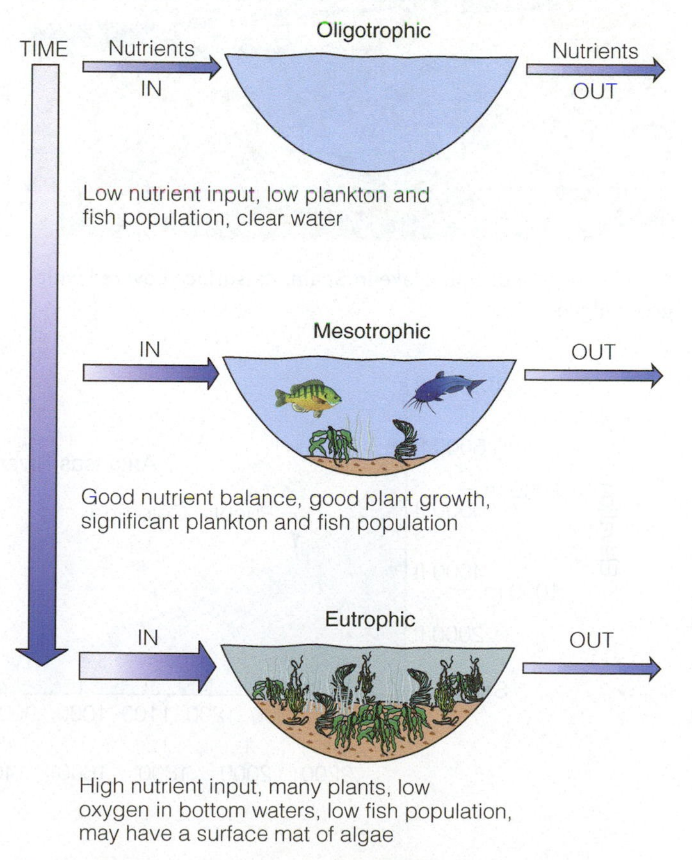

Figure 8.24 Over time, lakes go through a nutrient cycle, which can be accelerated by waste products from humans living around the lake or by agricultural chemicals.

D Currin/Shutterstock.com

Figure 8.25 The waters of Crater Lake, Oregon, which is oligotrophic, with few nutrients, are crystal clear.

IStock.com/Jon Benedictus

Figure 8.26 Eutrophic lake in Spain, its surface covered with green algae.

meters per kilometer, or feet per mile). The slope allows gravity to move water downslope. A mountain stream may have a steep gradient of several tens of meters per kilometer and an irregular profile with rapids and waterfalls. As the same stream approaches the lowest level to which it can erode, an elevation known as **base level**, its gradient may be almost zero and the water surface is nearly flat.

Looked at in a side or cutaway view, the longitudinal profiles of streams are generally concave upward from source to end. Gradients decrease in the lower reaches of a river (**Figure 8.27**), and this is where sedimentation often occurs as streams overflow their banks onto adjacent **floodplains**. Decreased water velocity on the floodplain results in the deposition of fine-grained silt and clay adjacent to the main stream channel. This productive soil, commonly found along low-lying river floodplains, known as bottomlands, is much desired in agriculture.

The ability of a stream to erode (Latin *erodere*, "to gnaw away") the land and impact humans is related to the speed at which the flowing water moves and the amount of water moving through the stream (its **discharge**). With equal discharge–the same volume of water flowing past a point in a river channel in a given time period–a fast-flowing stream erodes its banks more significantly than does a slow-moving one. In the United States, the downstream velocity (V; the maximum speed the water is moving directly down-channel) is usually expressed in feet per second (ft/sec), and discharge (Q), the amount of water, in cubic feet per second (cps). A cubic foot contains about 28 liters (7.4 gallons) of water. With increased velocity and discharge, there are increases in the stream's erosion potential and its sediment-carrying capacity (**Figure 8.28**). Stream erosion is a significant geologic hazard, eating away at banks and sometimes, destroying homes and bridges.

Planners recognize several categories of river-water use:

- *Instream use*–all uses that occur in the channel itself, such as hydroelectric power generation and navigation.

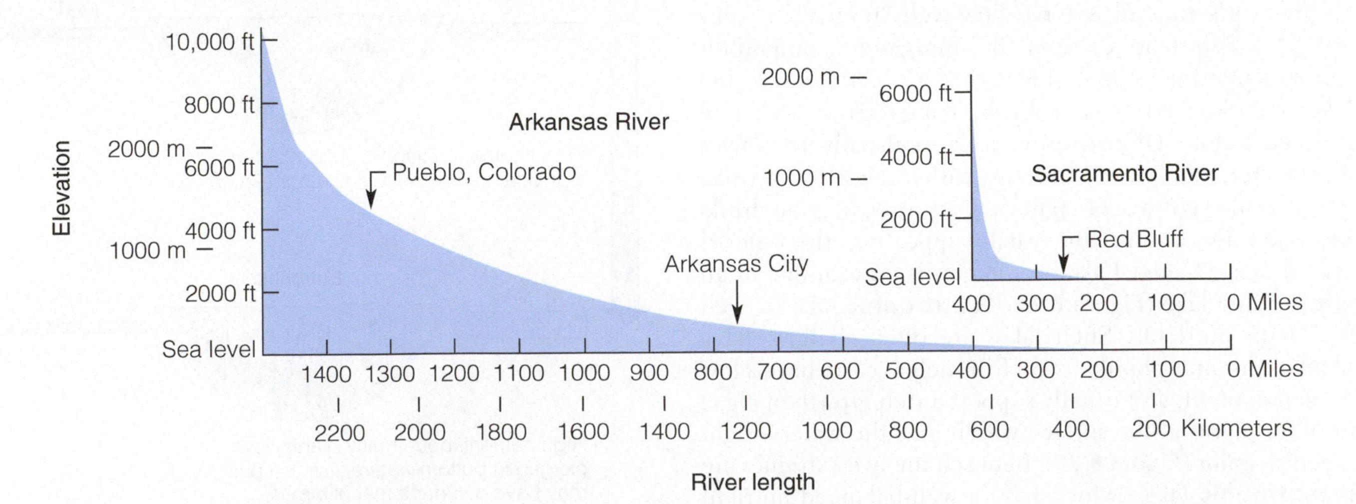

Figure 8.27 Cross sections, or longitudinal profiles, of the Arkansas River in Colorado and the Sacramento River in California. The vertical scale is exaggerated 275 times, meaning that these diagrams make the rivers look far steeper than they really are. Which river is steeper?

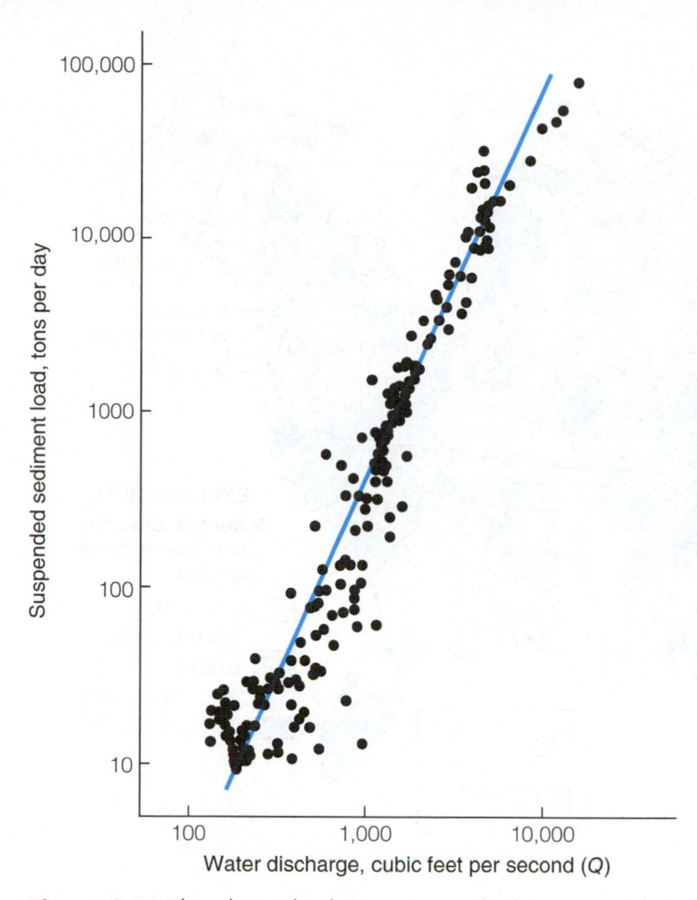

Figure 8.28 The relationship between water discharge *(Q)* and the suspended sediment load (the mass of sediment carried by the stream water) for a typical stream. Each point represents a separate measurement. A 10-fold increase in water discharge results in almost a 100-fold increase in sediment load. Look carefully at the axes of this graph. Can you see that the units increase in multiples of 10 because these are logarithmic scales? Consider how variable these measurements are. For any one discharge, suspended sediment load can vary by a factor of more than 50.

- *Off-stream use*–the diversion of water from a stream to a place of use outside it, such as to your home. The amount of water available for off-stream use along any stretch of a river is the total flow minus the amount required for in-stream purposes and for maintaining water quality.
- *Consumptive use*–water that evaporates, transpires, or infiltrates and cannot be used again immediately.
- *Nonconsumptive use*–water that is returned to streams with or without treatment so that it can be used again downstream. Domestic (household) water is used nonconsumptively: It is returned to the hydrological cycle through sewers and storm drains.

In many areas, particularly the western United States, consumption exceeds local water supply. This necessitates importing water from other areas or mining groundwater. Aqueducts, artificial river channels, are used to move water across the landscape. For example, about 75% of Southern California's water is imported from sources more than 330 kilometers (200 miles) away. Two aqueducts import water from west and east sides of the Sierra Nevada

Figure 8.29 The Los Angeles Aqueduct in Owens Valley California just after it was built (about 1913). In the background is Owens Lake, which is now dry because the water that once filled the lake flows to Los Angeles, hundreds of kilometers away.

(**Figure 8.29**), and a third aqueduct brings water from the Colorado River. The construction of the Los Angeles aqueduct, which siphoned off the streams of Owens Valley in the eastern Sierra, was a particularly contentious affair. Valley residents, furious that their water was being diverted to the big city, spent years protesting, sabotaging, and even bombing the aqueduct in a failed attempt to stop the export of the water that kept their farms alive (**Figure 8.30**).

Total world withdrawals of water for off-stream purposes amount to about 2,400 cubic kilometers per year (1.7 trillion gallons/day), of which 82% is for agricultural irrigation.

Figure 8.30 Dynamite for use in sabotaging the Los Angeles Aqueduct in Owens Valley. Local farmers spent years disrupting the flow of water in an attempt to save their livelihood.

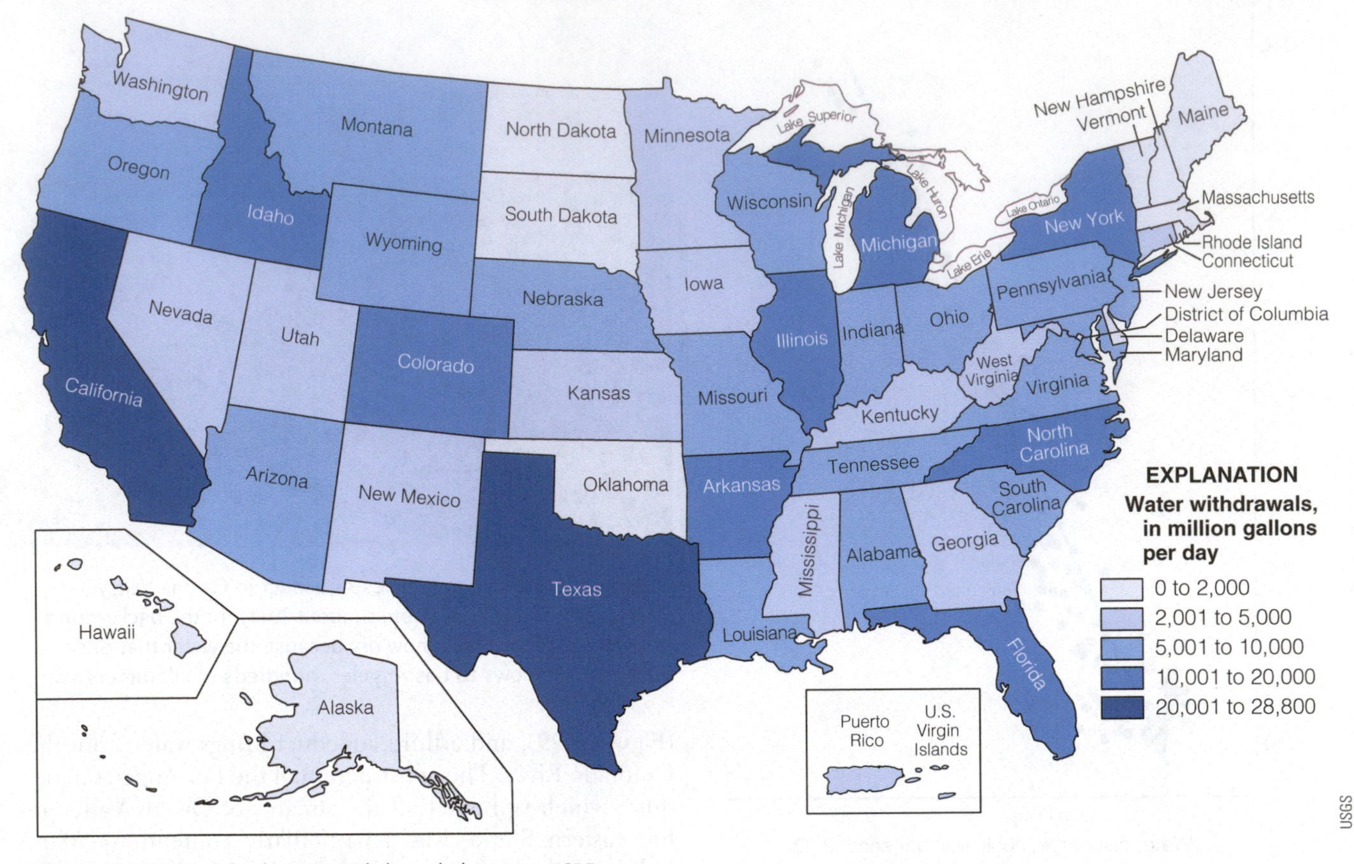

Figure 8.31 Total freshwater withdrawals by state, 2015.

This amount is expected to increase to 20,000 cubic kilometers (14.5 trillion gallons/day) by the year 2050, which is beyond the dependable flow of the world's rivers. Irrigation uses about 40% of the total water withdrawn in the United States, a startling 80% in California. It is not surprising, then, that California, with the largest population and a huge irrigation system, is the largest U.S. consumer of water, followed by Texas then Idaho (**Figure 8.31**).

The availability of water will limit population growth in the twenty-first century, either by statute in developed countries or by natural means, most likely drought and famine, in developing nations. Repetition of the terrible droughts experienced in Africa in the 1970s and 1980s, compounded by overtaxed water supplies, would be disastrous. Climate change is likely to make the situation even worse as droughts are predicted to increase throughout much of the world, particularly the semi-arid regions of the Global South (see Chapter 7).

8.2.3 Measuring Flow

A **hydrograph** is a graph of stream or river discharge or stage (height) over time; it is the basic tool of hydrologists (**Figure 8.32**). It represents graphically the flow of water over time. The "synthetic" hydrographs in **Figure 8.33** illustrate the basic difference between upland floods of short duration and high discharge, and lower discharge but longer lasting lowland floods. The timing of flood crests on the mainstream–whether they arrive from the tributaries simultaneously or in sequence–in large part determines the severity of lowland flooding.

Flood studies begin with hydrographs for a particular storm on a river within a given drainage basin. These graphs are generated from data obtained at a stream-gauging

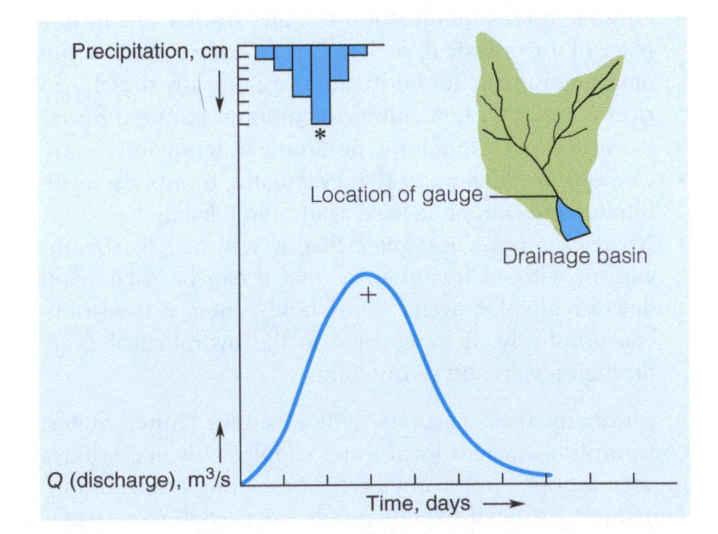

Figure 8.32 A synthetic hydrograph that relates precipitation and discharge to time. Note that peak discharge (+) occurs after the peak rainfall intensity (*). The drainage basin includes the main stream and all its tributaries.

Figure 8.33 Synthetic hydrographs of an upland flash flood and lowland riverine flood, both of which are carrying the same amount of water; therefore, the area under the curves is the same.

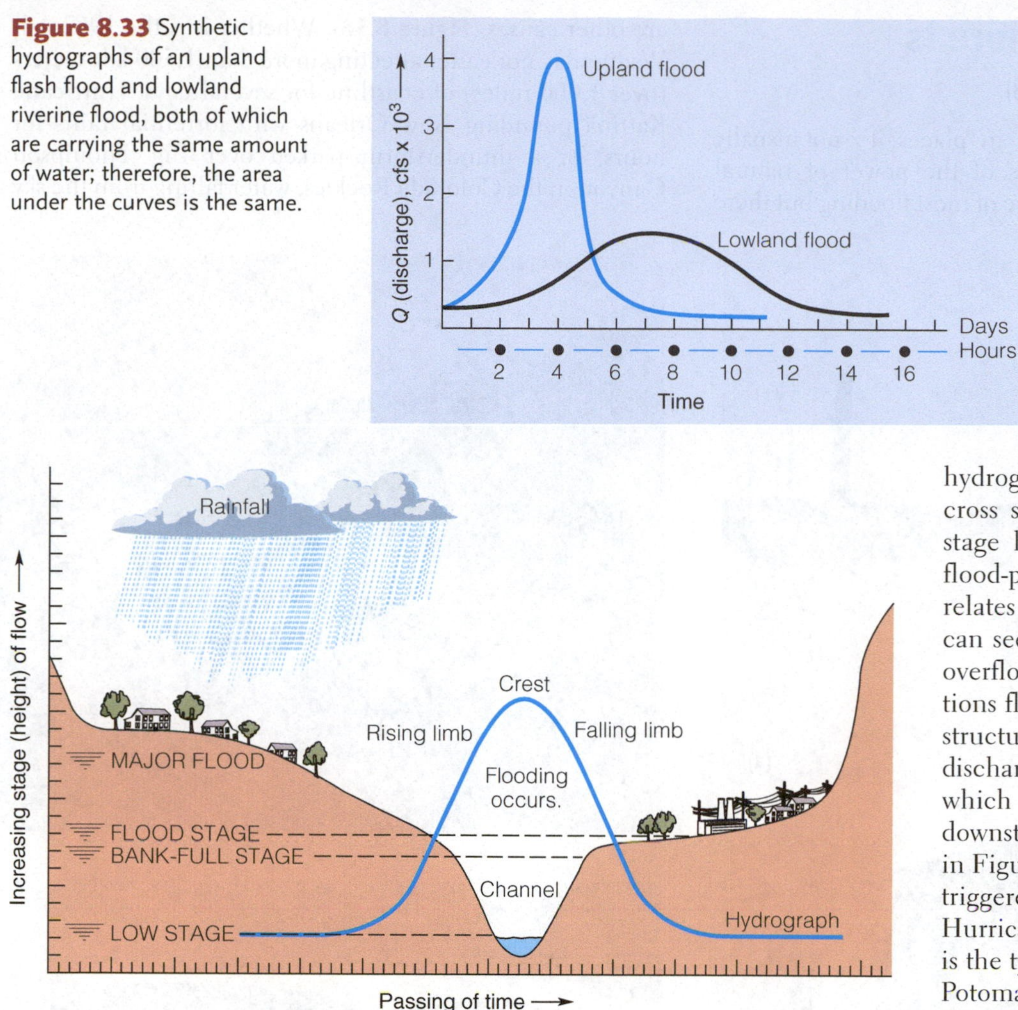

Figure 8.34 A stage hydrograph superimposed on a schematic stream valley illustrates the impact of increasing stage height on the flooded area.

station, where stage (the water elevation) is measured and related to discharge. Rapid runoff in a drainage basin is accompanied by high stream discharge and high water velocity. As discharge increases, the water rises to the **bank-full stage** and then spills onto the adjacent floodplain at **flood stage**. **Figure 8.34** shows such a condition. It is a hydrograph superimposed on the cross section of a stream valley; the stage hydrograph is most useful for flood-planning purposes because it relates water height over time. One can see at what point the stream will overflow its banks and at what elevations floodwaters will contact specific structures. Note that both stage and discharge are shown in **Figure 8.35**, which is for the Potomac River, just downstream of where the photographs in Figure 8.48 were taken. This flood, triggered by the torrential rains of Hurricane Isabel several days before, is the type of event that occurs on the Potomac River once or twice every decade.

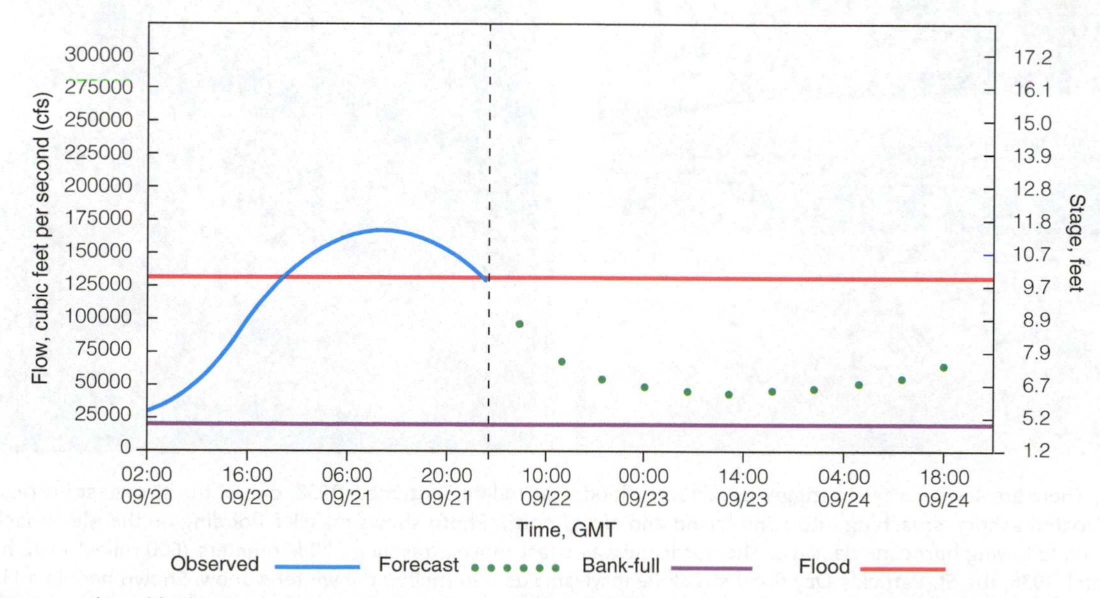

Figure 8.35 A combined hydrograph showing both stage and discharge for the Potomac River at Little Falls after the passage of Hurricane Isabel in fall of 2003. Note that the flood exceeds both the bank-full and flood stages. The blue line represents data collected at the gauging station. The green dots are predicted stage values calculated by NOAA's advanced hydrologic prediction service based on rainfall information and gauge data from upstream.

8.3 Floods: When There Is Too Much Water (LO3)

Floods, where water ends up in places it's not usually found, are dramatic reminders of the power of natural forces. Precipitation is the source of most flooding but there are other causes (**Figure 8.36**). Whether it is the 1962 Ash Wednesday nor'easter affecting more than 1,500 kilometers (over 1,000 miles) of coastline for several days, Hurricane Katrina pounding New Orleans with torrential rains for hours, or a thunderstorm parked over Big Thompson Canyon in the Colorado Rockies, water falling from the sky

Figure 8.36 There are several ways to trigger devastating floods. (a) In late September 1938, one of the strongest hurricanes to hit New England roared ashore, smashing into Long Island and racing north. Photo shows massive flooding on the Merrimack River, New Hampshire, following hurricane passage. The hurricane was a fast mover, traveling 950 kilometers (600 miles) in 12 hours. (b) In mid-March 1936, the St. Patrick's Day flood struck Pennsylvania as rain melted the winter's snow. Shown here is a Harrisburg, Pennsylvania, city bus nearly submerged. (c) The Johnstown, Pennsylvania, flood swept huge amounts of debris downstream when a private earth-fill dam failed. Trees and homes were piled high against a stone bridge. A crane had to be used to remove the debris. (d) The failure of the earth-filled Teton Dam in 1976 sent a massive wave of water downstream killing 14 people. Water seeping through the dam weakened it so much that one side failed.

faster than it can be absorbed by the ground or transported downstream propels rivers out of their banks.

Floods from all causes are the number-one natural disaster in the world in terms of loss of life. Some of the most disastrous U.S. floods are listed in **Table 8.1**. Although by no means a complete listing of severe floods, the table gives an idea of the distribution of river floods, coastal floods, and dam-break floods in the United States. Coastal flooding caused by hurricanes and typhoons is the greatest single killer. The record-holding river floods are due to dam failures. The 1889 Johnstown dam failure flood is legendary for killing about 3,000 people, and the 1928 St. Francis Dam failure in Southern California killed 450 people.

8.3.1 Types of Floods

Some floods occur in small drainage basin areas of moderate or high topographic relief. They may occur with little or no warning and are generally of short duration, hence the name **flash floods**. Flash floods are more common in areas where little rainwater infiltrates and most runs off the landscape. They result in extensive damage because of the high speed of the flowing water. During flash floods, the level and volume of water in streams rises rapidly during and shortly after rainfall (**Figure 8.37**). Homes and bridges are ripped from their foundations. Lowland or **riverine floods**, in contrast, occur when broad floodplains adjacent to the stream channel are inundated. The water in such lowland floods typically moves quite slowly; this results in damage by soaking and siltation–peoples' homes are left wet and full of mud but usually intact (**Figure 8.38**).

The difference in these types of floods is due to drainage basin morphology (shape). Mountain streams occupy the bottoms of V-shaped valleys and cover most of the narrow valley floor; there is no adjacent floodplain (the flat areas adjacent to the river channel) over which flood waters could spill. Heavy rainfall at higher elevations in the drainage basin can produce a wall of fast-moving water that descends through the canyon, destroying structures and endangering lives. Lowland streams, in contrast, have low gradients, adjacent wide floodplains, and meandering paths. Here, floods tend to move slowly and be less deep, but inundate large areas.

Table 8.1 Selected U.S. Floods Since May 1889

Type	Date	Location	Lives Lost	Estimated Damage $, Millions (at time of flood)
Regional flood	1927	Mississippi River Valley	Hundreds	1000
	March 1938	Southern California	79	25
	21 September 1938	New England	600	306
	September/October 1983	Tucson, Arizona	<10	50–100
	Summer 1993	Upper Mississippi River drainage basin	50	12,000
	January/February 1996	Oregon, Washington, Idaho, Montana	1	89
	October 1998	South-Central Texas, San Antonio to Houston, Guadalupe River	29	400
Hurricane	8 September 1900	Galveston, Texas	6,000	30
	17–18 August 1969	Mississippi, Louisiana, and Alabama (Hurricane Camille)	256	1,421
	September 1999	North Carolina (Hurricane Floyd and post-hurricane rains)	47	5,000+
	August 2005	Louisiana, Mississippi (Hurricane Katrina)	1,840	125,000+
	August 2017	Southeastern Texas (Hurricane Harvey)	100	125,000
Flash flood	July 1976	Big Thompson River, Colorado	139	30
Dam failure	May 1889	Johnstown, Pennsylvania	3,000	17
	12–13 March 1928	California, St. Francis Dam	450	14
	February 1972	Buffalo Creek, West Virginia	125	10
	June 1976	Southeast Idaho, Teton Dam	11	1,000

Source: Various sources, including the U.S. Geological Survey.

Figure 8.37 Flash flood in North Carolina overtops a bridge near Charlotte in 2018.

Figure 8.38 Village in Thailand inundated by a slow-moving, lowland flood of silty water.

Monsoon floods are yearly events that result from large-scale flow of air and water vapor in the atmosphere (**Figure 8.39**). The sun heats the land and the land warms more quickly than adjacent bodies of water. The warm land heats the air above which becomes less dense and rises drawing in more air from the ocean, air charged with ocean moisture. As the rising air cools high in the atmosphere, the moisture in it begins to condense and rain falls. That rain is critical for people and the environment but it can also trigger flooding. Monsoons occur predictably in the summer and the winter in different places as the amount of energy from the sun changes with the season (**Figure 8.40**). In North America, the southwestern monsoon is the most important, bringing welcome rain and less welcome summer humidity to Arizona, California, New Mexico and Mexico. The Asian monsoon is critical for the farmers of India and Indochina but can generate devastating floods such as those that routinely impact Pakistan (see Chapter 1).

Rain-on-snow events are big flood makers anywhere there is a substantial snowpack. Warm winds and early spring rains on deep snowpacks are particularly devastating as melting snow adds to the runoff from rain fall. Melting snow has often saturated the ground before the rain arrives, causing more water to run off and less to infiltrate. Rainfall on frozen soils has also become a problem, as more erratic freeze-thaw cycles have caused a loss in snowpack throughout the winter, exposing soils to the frigid air. When rain falls on these frozen soils, there is little capacity for infiltration, leading to a large amount of runoff in midwinter. These **rain-on-snow events** have caused devastating floods all over the United States, from southern Pennsylvania to the Cascades of Washington State (**Figure 8.41**). They are most common in the springtime, but climate change has increased their frequency in midwinter as well.

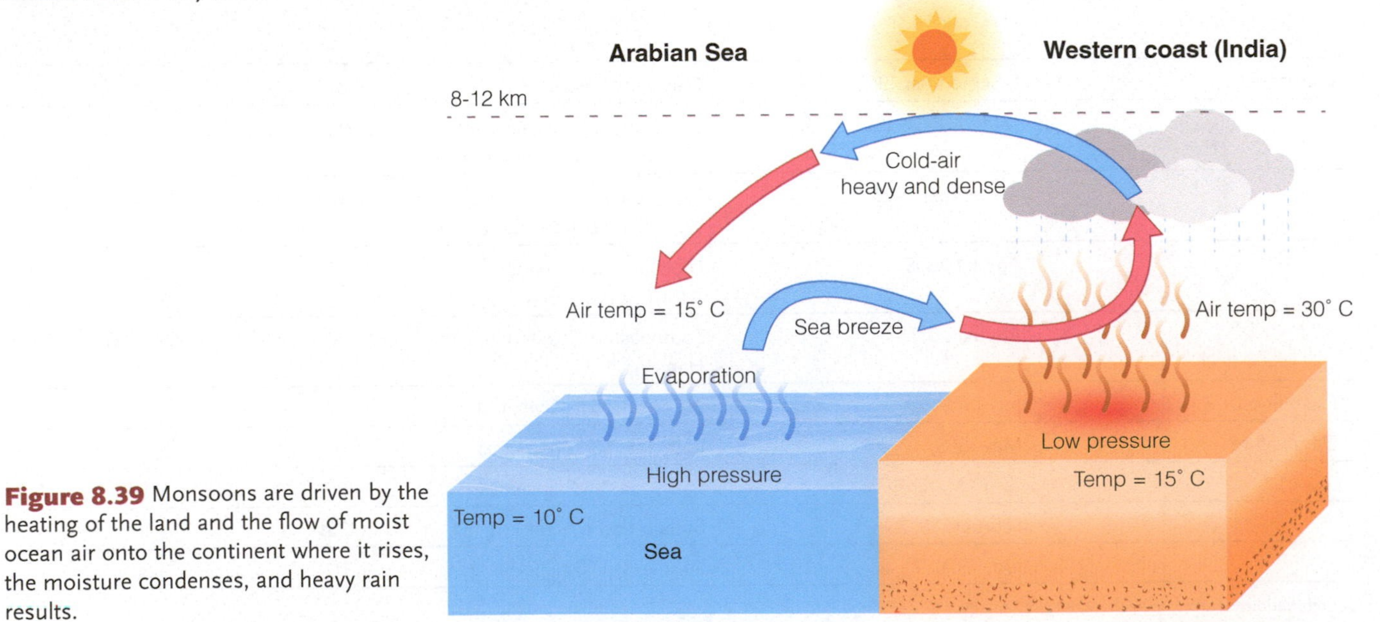

Figure 8.39 Monsoons are driven by the heating of the land and the flow of moist ocean air onto the continent where it rises, the moisture condenses, and heavy rain results.

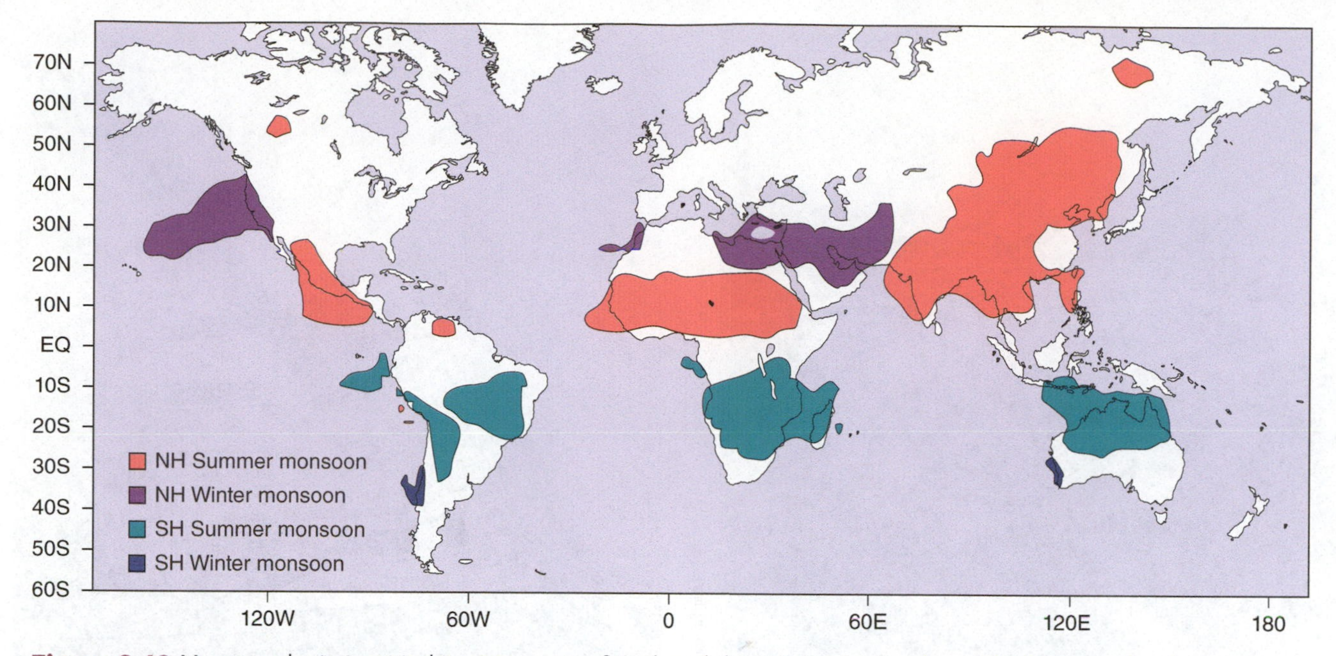

Figure 8.40 Monsoonal rain is critical to many parts of Earth and the people who live there. The location of such rains changes with the seasons.

Ice-jam floods occur when river ice breaks up, moves downstream, and wedges against obstacles in the channel. These short-lived ice dams quickly raise water levels upstream. Ice-jam floods are common anywhere rivers freeze over and often occur during rain-on-snow events. One moment, it's a pleasant, early spring day by the river; the first robins have appeared, and it's warm enough to pull off your wool hat for the first time since December. The snow around your feet is soggy; it's a good thing you are wearing high rubber boots. It rained hard last night, and the river should be rising, but you can't see it through the cover of gray ice.

Figure 8.41 Rain on snow event in 2009 in Minnesota caused major flooding. This homeowner sandbagged their home in an attempt to keep the water out.

Suddenly, there's a deafening crash, and the ice on the river breaks up before your eyes. Blocks of ice, half a meter (20 inches) thick and the length of small cars, move downstream in a crazy jumble, grinding past one another, ripping trees from the riverbanks, and jamming against the bridge piers just downstream. You watch in amazement as more and more ice stacks up against the bridge, ignorant of the water rising over your boot tops until it cascades down to your toes, bathing you and your wool socks in an extension of the icy river. The rising water and your soaking feet are telling you that an ice-jam flood has started. It's time to move to high ground, and fast!

Ice-jam floods can happen anywhere that streams freeze in the winter; that means high latitudes and high elevations. These floods occur when warm winds melt the snowpack, rain pummels icy slopes, and river ice weakens. Rising floodwaters can tear ice cover apart, rapidly adding to the hazard. When ice jamming occurs, water levels rise fast as dams of ice block downstream flow and thus quickly start to pond water behind them, upstream. Unlike most river floods during which the flood stage rises gradually and predictably, ice-jam floods are fast, devastating, and hard to predict.

Many times, ice jams and the floods they trigger happen in the same place where natural or human-made obstacles force ice blocks to pile up (**Figure 8.42**). Imagine a bridge abutment or a tight meander bend. Can you see blocks of ice, stacking one behind the next? What's the remedy? If you can be patient, adequate time to melt the ice is the answer. If people or their homes are in danger, though, dynamite and heavy equipment will do the trick, destroying the ice dam and letting the river flow freely again. Ice-jam floods may well become more common in cold climates as midwinter thaws accompany global warming; conversely, in mid-latitude locations, the climate will likely become warm enough that the river ice no longer forms.

UVM Howe Library

Figure 8.42 Ice jams have been damaging human constructs for as long as people have lived near frozen rivers. As shown in this 1863 stereoview, an ice jam diverted the White River right down Main Street in Hartford, Vermont.

City of Boston, MA/U.S. Census Bureau

Figure 8.44 Destruction in the wake of the Johnstown dam failure.

During a storm, water need not come from the sky to cause a flood; it might come from the ocean–a phenomenon called a **storm surge.** When hurricanes or nor'easters make landfall, they push before them a wall of water driven by wind and accentuated by low atmospheric pressure near the storm. These surges of water can rapidly raise sea level by meters, inundating low-lying coastal areas within an hour or two (**Figure 8.43**). The effect of storm surges is not limited to the coast. Storm surges can move inland, up coastal rivers. When they do, floodwaters moving downstream are blocked and flooding further intensifies.

Dams are designed in part to control floods. Most of the time they do, but devastating floods have been caused by the failure of dams, both natural and constructed. The 1889 failure of an earth-fill dam in Johnstown, Pennsylvania, swept its victims downstream in a wall of water 12 meters (40 feet) high (**Figure 8.44**). The stories are horrific. Riverside homes lifted from their foundations were carried intact and wedged up against a massive stone bridge in such a pile that no one could escape the fires that started when wood-fired cookstoves overturned. Victims were burned alive while their houses floated. Natural dams, formed when landslides or glacial moraines have blocked streams, can fail catastrophically (see Figure 6.3).

Earthquakes can do more than shake, rattle, and roll: They can cause floods in several ways. The Indian Ocean tsunami of 2004 generated rapid but short-lived flooding as massive waves washed ashore (see Chapter 6). But earthquakes can cause flooding in another way: by changing land level. In the Pacific Northwest, buried marsh soils and standing dead trees with their roots drowned by saltwater are compelling evidence that giant quakes have changed the land level over the past hundreds to thousands of years, intermittently flooding the coast when the land subsided (see Chapter 10). In Alaska, after the 1964 magnitude-9.1 earthquake, land level fell far enough that some settlements were partially submerged, and there was no choice but to abandon them (**Figure 8.45**).

Not all flooding is so obvious or rapid. Consider changing climate and its impact on sea level. At the peak of the last glaciation, 21,000 years ago, wooly mammoths roamed across North America and sea level was about 135 meters (440 feet) lower than today. As the climate warmed and the ice sheets melted, water flowed back into the ocean and sea level rose, rapidly at

iStock.com/Moorefam

Figure 8.43 Large waves and a storm surge batter the southern English coast near Devon in 2014.

Figure 8.45 The 1964 Alaska earthquake dropped the village of Portage about 2 meters as the ground subsided.

some times, more slowly at others. The camping sites of the first Americans along the Alaskan coast are now deep underwater, and Chesapeake Bay, once a broad river valley, is now drowned by the sea. Today, with the climate warming and most glaciers retreating, the scientific consensus is that sea level will continue to rise, likely quite rapidly as melting of the Greenland and Antarctic ice sheets accelerates (see Chapter 3). It takes only a few meters of sea-level rise to inundate parts of many major and important world cities such as New York, Miami, Venice, London, Hong Kong, and New Orleans.

8.3.2 Effects of Floods

Imagine a flood and its effects. If you are thinking about rising, muddy water covering fields, homes, and cars, you are imagining the most common impact of floods, **inundation**, flood waters covering area that is normally dry land. Because floods tend to carry large amounts of fine sediment, not only do things get wet but very muddy as well. The aftereffects of inundation can be more devastating than the flood itself. After the inundation of New Orleans in 2005, hundreds of thousands of automobiles were damaged, their engines and electrical systems filled with floodwater. Many homes needed to be gutted or demolished, not because they were structurally unsound but because their flood-soaked walls couldn't be dried and had mildewed so badly that safely inhabiting the buildings was impossible (**Figure 8.46**).

Floods do more than soak things. As floodwaters move down river channels, they pick up sediment, both loose material from the channel bottom and solid material from the banks. At bridge abutments, this tendency of moving water to entrain material from the river bottom can lead to disaster as support is removed and the bridge eventually collapses (**Figure 8.47**). This removal of material from beneath abutments, known as **bridge scour**, is a hazard worldwide. In the past, not everyone realized that bank erosion and channel migration, processes that are most active during floods, are critical and normal aspects of river function. Human attempts to stabilize rivers using engineering solutions, with concrete and straightened channels at their core, have typically failed over time, because these "solutions" ignore the dynamic nature of flowing water and moving sediment. Today, the emphasis is on river-corridor management and river-channel restoration to accommodate the natural behavior of flowing water.

In the long term, rapidly moving water is a dynamic force, shaping Earth's surface. Large, rare floods can cause river channels to shift dramatically, abandoning their old course and establishing a new channel, sometimes with disastrous results for people living along river banks. Some of the most spectacular features on our planet are deep canyons cut into rock. Research suggests that it takes massive and, therefore, rare flooding events to generate enough power to rip rock away and carve canyons with moving water. In the Himalayas, raging rivers fed by Asian monsoons can erode millimeters of rock per year. Surprisingly, the Potomac River, just outside Washington, DC, cut through hard rock at rates almost this fast during the last glaciation, driven presumably by major floods (**Figure 8.48**).

8.3.3 Flood Frequency

One of the most useful relationships that can be derived from long-term flood records is **flood frequency**, or **recurrence interval**–that is, how often, on average, a flood of a given magnitude can be expected to occur at a particular location. With this kind of information, we are better equipped to design dams, bridge clearances over rivers, the sizes of storm drains, and the like.

Figure 8.46 Inundation is a devastating effect of floods. In New Orleans, after the floodwaters of hurricane Katrina receded, mold was a devastating problem in the warm, moist Louisiana environment.

© Daily Gazette Sid Brown/Schenectady Gazette

USGS

Figure 8.47 Bridge scour can be devastating. (a) A bridge on the New York State Thruway collapsed on April 5, 1987, claiming 10 lives as raging floodwaters tore away the support under its footings. (b) The view after the collapse; it's hard to get from here to there anymore.

Paul Bierman

Paul Bierman

Figure 8.48 Only big floods can erode rock. At Great Falls on the Potomac River just 12 miles from the U.S. Capitol in Washington, DC, most flows spill gracefully over the rocks. However, when hurricanes or rain-on-snow events strike, the river becomes a raging torrent that can erode rock. (a) A calm November day with flow of several hundred cubic meters per second. (b) After Hurricane Isabel in 2003, the river channel is filled as almost 5,000 cubic meters (160,000 cubic feet) per second of water pour over the falls. The red arrows point to the same rock outcrop at low and at high water.

Floods are normal and, to a degree, predictable natural events. Generally, a stream will overflow its banks every year or two, gently inundating a portion of the adjacent floodplain. Over long periods of time, catastrophic high waters can be expected. The more infrequent the event, the more widespread and catastrophic the inundation (**Figure 8.49**). Depending on their severity, these events are referred to as 50-, 100-, or 500-year floods, denoting the average recurrence interval between events of similar magnitude.

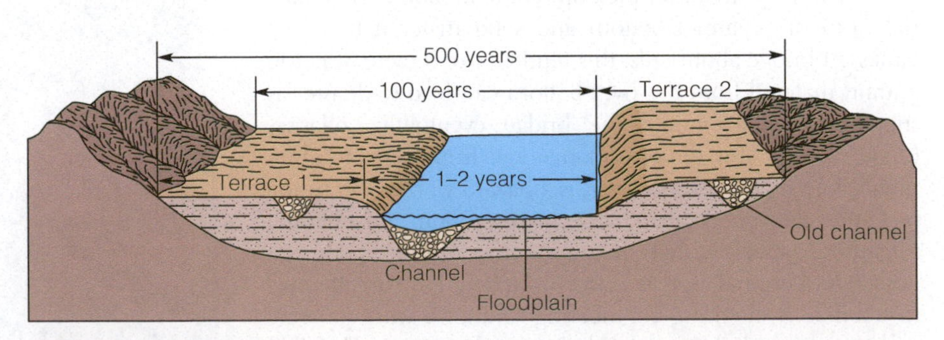

Figure 8.49 The probable extent of flooding over long time intervals on a floodplain. The 1- to 2-year flood just goes over the stream bank onto the adjacent floodplain. The 100-year flood inundates the low terrace (terrace 1), which formed when the river stood at a higher level. The 500-year event inundates an even higher terrace (terrace 2), and all lower terraces and the floodplain.

Figure 8.50 Stream gauge measuring discharge of a stream in the Sangre de Cristo Mountains, Colorado.

Figure 8.51 The stream gauge is in the small green building adjacent to the river.

The information needed for determining recurrence intervals is a long-term record of **annual peak discharge** (the largest flow for each year) for a particular location on a river. Such data are often gathered by the U.S. Geological Survey using stream gauges to monitor the flow on thousands of rivers and streams in the United States and its territories (**Figure 8.50**). The longer the period of record, the more reliable the statistics.

To calculate a recurrence interval, the annual peak discharges are ranked according to their relative magnitudes, with the highest annual peak discharge being ranked as 1, the second largest as 2, and so forth. To serve as an example, **Table 8.2** lists selected annual peak discharges on the Rio Grande near Lobatos, Colorado (**Figure 8.51**), with their ranks and recurrence intervals. A recurrence interval, T, is calculated as follows:

$T = (N + 1)/M$ where N is the number of years of record (122 years in our example) and M is the rank of the annual peak discharge.

The analysis predicts how often floods of a given size can be expected to occur in a drainage basin if the landscape and climate conditions have not changed. In this example, a flood the size of the 1941 flood ($M = 12$, discharge of 229 cubic meters per second) would be expected to occur, on average, about once every 10 years [(122 + 1)/12].

If 100 years of records are available, the largest flood to occur during that period ($M = 1$) has a recurrence interval of about 100 years, on average. This is the so-called 100-year flood. The 100-year flood can also be obtained by projecting stream-flow data beyond the period of record, although such extrapolation is extremely uncertain.

How often can the "20-year flood" occur? The term is misleading, because it implies that a flood of that size can happen only once in 20 years. The 20-year flood is really a statistical statement of the probability that a flood of a given *rank* will occur in any one year; that is, $1/T$. For the 20-year flood, there is a 5% chance (1/20) of its occurring in any one year; for the 50-year flood (1/50), a 2% chance. The U.S. Geological Survey suggests that a better term would be "the 1-in-20 *chance* flood."

There is, however, no meteorologic or statistical prohibition against more than one 100-year flood occurring in a century, or even in 1 year, for that matter. Recurrence intervals are not exact timetables; they merely indicate the statistical probability of a given-size flood. However, even rough estimates of the scale of flood discharge that can be expected in a river system every 25, 50, or 100 years enable us to plan better against the flooding that is certain to occur eventually. When this concept is viewed as a probability phenomenon, it is more understandable that we might get big floods in successive years or once every few years.

Keep in mind that, the longer the rainfall or stream-flow record, the better and more robust the statistical inference.

Table 8.2 Annual Peak Discharges on the Rio Grande Near Lobatos, Colorado, 1900–2021

Year	Discharge, Cubic Meters/Second	Rank (M)	Recurrence Interval (T)
1900	133	30	4.1
1901	103	43	2.9
1902	16	112	1.1
1903	362	2	61.5
1904	21	103	1.2
1905	374	1	123
1906	237	11	11.2
1907	249	8	15.4
1908	65	63	2.0
1909	216	15	8.2
1974	22	102	1.2
1975	70	58	2.1
1985	177	21	5.9
2007	44	82	1.5
2008	83	52	2.4
2021	27	97	1.3

Source: USGS

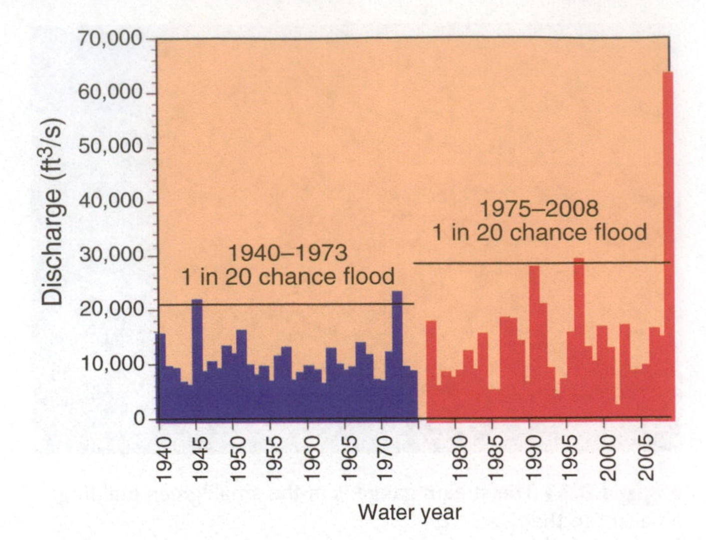

Figure 8.52 Annual maximum discharge data for the Chehalis River near Doty, Washington. Note the difference in the predicted value for the 20-year flood (the horizontal lines on the graphs) based on the first 34-year period and the second 34-year period. (USGS data.)

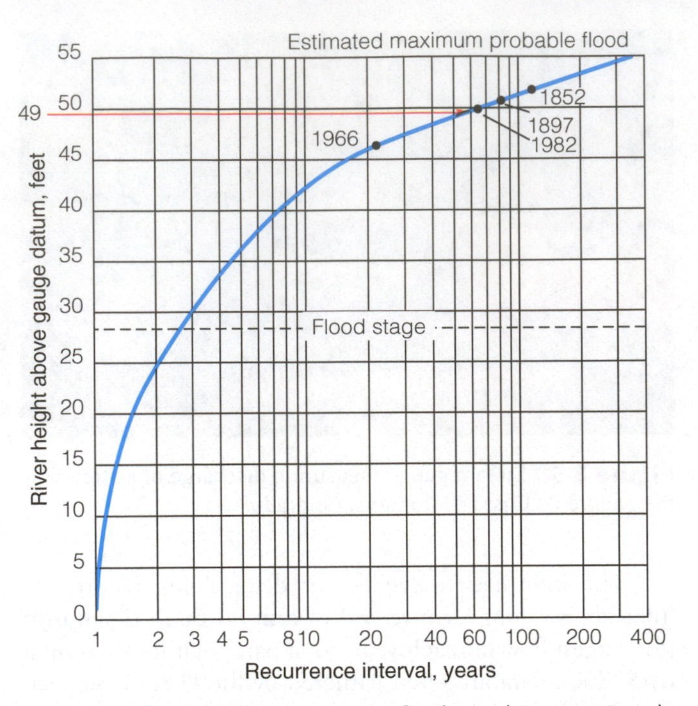

Figure 8.53 Flood-frequency curve for the Red River at Grand Forks, North Dakota, using water elevation as the standard.

Look at the stream-flow data collected between 1940 and 2008 on the Chehalis River in Washington State in **Figure 8.52**. Note that the 1-in-20 chance flood calculated for data from 1940 to 1973 is lower than the same 1-in-20 chance flood based on data for the period 1975 to 2008. A similar relationship occurs for the 1-in-10 and 1-in-5 chance floods (**Table 8.3**). This illustrates the importance of long-term data collection in assessing the probability of floods and illustrates a phenomenon called **nonstationarity**. That is the observation, noted on some rivers, that flow characteristics change over time. Sometimes the driving force is land-use change; other times, it is climate change. Whatever the cause, it is clear that over the last 35 years, the Chehalis River has produced more frequent, larger floods than it did previously. Most of these large floods occur in December, January, and February, and are presumably related to rain-on-snow events. Might this be a signal of a warming climate, more rain on snow, and/or a more active hydrologic cycle as predicted by global circulation models? Only time will tell.

Once we have determined recurrence intervals from the historical record, it is possible to construct flood-frequency graphs like those in **Figure 8.53**. Such graphs allow us to estimate how often a flood level or discharge of a particular magnitude is likely to occur. For example, if we wish to build a warehouse with a useful life of 50 years on the floodplain of the Red River at Grand Forks, North Dakota, we see in Figure 8.53 that the river reaches a reference elevation of about 49 feet (15 meters) every 50 years, on average. We would probably want to build the warehouse above that elevation, because there is a significant chance of a flood within the 50-year window of interest (and our chances for insurance would be better, too). However, urbanization, artificial levees, and other human-generated changes can alter the amount and timing of runoff and cause flood events to become more common. In urbanizing areas, a 10-year flood can become a 2- or 3-year flood as paved area expands and the ability of water to infiltrate the soil decreases.

Land-use changes, engineering modifications of the river channel, and long-term climate changes require that recurrence intervals, flood levels (stages), and floodplain management be updated periodically. Indeed, changing climate and climate cycles influence flood frequency, mandating that we continue to collect flow data from as many stream gauges as possible.

Table 8.3 Flow Statistics for the Chehalis River, Doty, Washington

1940–1973	Return Interval	Annual Probability	1975–2008
12,600	5 years	1:5	17,400
15,400	10 years	1:10	21,000
21,600	20 years	1:20	29,000
10,200	Mean annual maximum flow		13,400

8.4 Groundwater (L04)

Some of the water in the hydrologic cycle infiltrates underground and becomes **groundwater**, one of our most valuable natural resources (Chapter 2). In fact, most of Earth's liquid freshwater exists beneath the land surface; its geological occurrence is the subject of many misconceptions. For example, it is commonly believed that groundwater occurs in large lakes or pools beneath the land. The truth is that almost all groundwater is in small pore spaces and fractures in rocks.

8.4.1 Groundwater Supply: Location and Distribution

Groundwater supplies about 50% of the drinking water in the United States, with 140 million people relying on groundwater for their drinking water supply in the nation as of 2021. It is by far the cheapest and most efficient source of municipal water because obtaining it does not require the construction of expensive aqueducts and reservoirs. Furthermore, groundwater usually requires far less cleaning and purification than surface water. Many U.S. cities rely heavily on groundwater for their freshwater (and sometimes, saltwater) needs; San Francisco, New Orleans, Dallas, and Los Angeles are just a few. Groundwater provides over 40% of all irrigation water used in the United States, with about 70% of all freshwater groundwater withdrawals being used for irrigation. It is not surprising that some of the largest agricultural states, such as Texas, Arkansas, Nebraska, Idaho, and California, are the largest consumers of groundwater, accounting for almost half of all the groundwater produced and used. Groundwater is important even to communities that import water from distant areas, because, in almost every case, local groundwater provides a significant, low-cost percentage of their water supply.

Groundwater maintains stream flows during periods when there is no rain. This process, by which water from the surrounding landscape seeps into the bottom and sides of streambeds, creates what is known as **baseflow** (the flow in a stream when it's not raining). Baseflow is critical for maintaining the ecological integrity of streams because it keeps water flowing long after rains have ceased. In arid climates where there are permeable, water-saturated rocks beneath the surface, groundwater seeping into stream channels may be the only source of water for rivers. The fabled oases (wet, green places in the desert) occur where underground water is close to the surface or where it intersects the ground surface, forming a spring or watering hole.

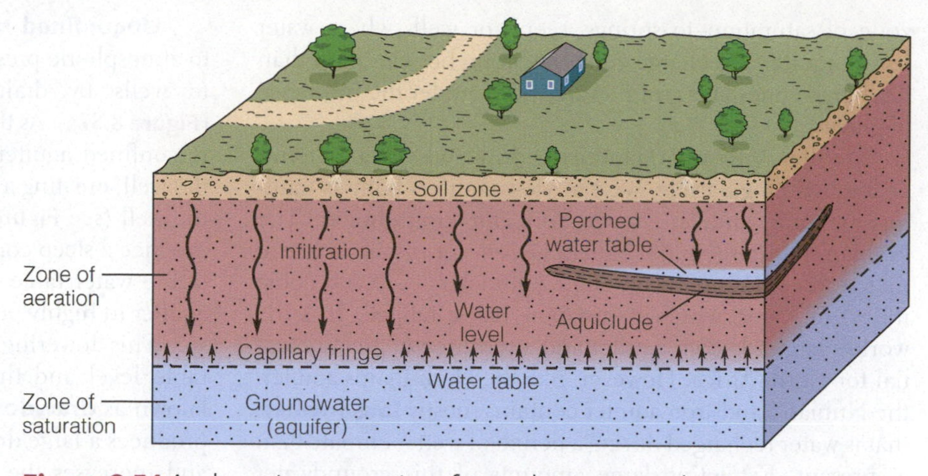

Figure 8.54 A schematic cross section of groundwater zones, showing a static water table with a capillary fringe. A **perched water table** exists where a body of clay or shale acts as a barrier (otherwise known as an aquiclude) to further infiltration.

8.4.2 Water Table and Flow Directions

Recall from Chapter 2 that the zone of aeration lies between the land surface and the depth at which we find all pore spaces saturated with groundwater. This is the zone where voids in soil and rock contain only air and water films. This is also known as the **unsaturated zone**. The contact between this zone and the zone of saturation below it is the **water table** (Figure 8.54). Above the water table is a narrow, moist zone, the **capillary fringe**, where pores are filled with water, but the water is held by surface tension and cannot freely drain from the soil.

In the zone of saturation, free (unattached) water fills the openings between grains of clastic sedimentary rock, the fractures or cracks in hard rocks, and the solution channels, including caves, in limestones and dolomites (Figure 8.55).

A rock's **porosity** (root word *pore*) is a function of the volume of openings, or void spaces, in it and is expressed as a percentage of the total volume of the rock. Without such voids, a rock cannot contain water or any other fluid. **Permeability** is the ease with which fluids can flow through the ground and is thus a measure of the connections between pore spaces.

Aquifers are important because they transmit water from **recharge areas**–areas where water is added to the

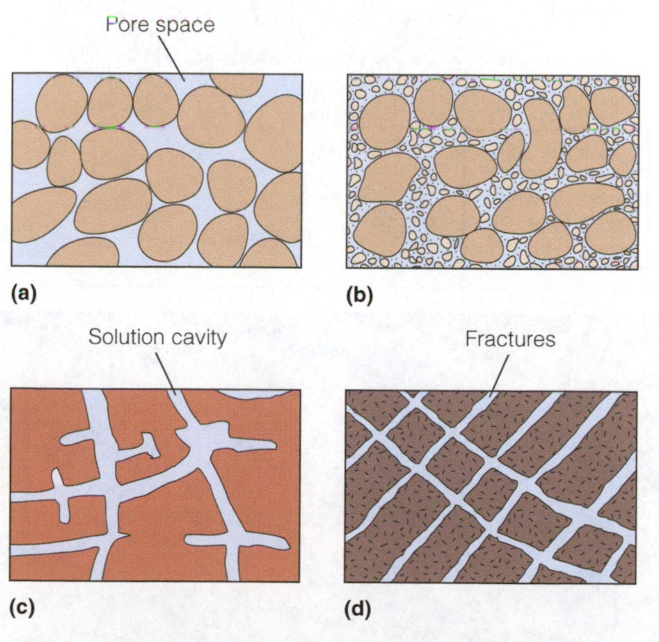

Figure 8.55 A rock's porosity is dependent on the size, shape, and arrangement of the material composing the rock. Whereas (a) a well-sorted sedimentary rock has high porosity, (b) a poorly sorted one has low porosity. (c) In soluble rocks, such as limestones, porosity can be increased by solution, whereas (d) crystalline metamorphic and igneous rocks are porous only if they are fractured.

zone of saturation–to springs, seeps, or wells where water leaves aquifers. Recharge zones are at higher elevation than discharge zones and groundwater flows under the influence of gravity.

Aquifers store vast quantities of drinkable water. In terms of the hydrologic cycle, we can view aquifers both as transmission pipes and as storage tanks for groundwater. The Nubian sandstone in the Sahara Desert, for instance, is estimated to contain 600,000 cubic kilometers (130,000 cubic miles) of water, which is just now being tapped. It is the world's largest known aquifer and a resource of great potential for North Africa. However, like the High Plains aquifer, the Nubian Sandstron aquifer contains mostly **fossil water**, that is water recharged during a period of wetter climate than at present. Extracting large amounts of this groundwater amounts to unsustainable mining. Started in 1984, the Great Man-Made River Project, a brainchild of former Libyan leader, Muammar Al Qadhafi, now carries 6.5 million cubic meters of water each day from desert wells to the cities and agricultural fields of this arid nation (**Figure 8.56**). This extraction is unsustainable as aquifer is not being replenished by rainfall.

Unconfined aquifers are formations that are exposed to atmospheric pressure changes and that can provide water to wells by draining adjacent saturated rock or soil (**Figure 8.57a**). As the water is pumped and removed from an unconfined aquifer, a **cone of depression** forms around the well, creating a gradient that causes water to flow toward the well (see **Figure 8.57b**). A low-permeability aquifer will produce a steep cone of depression and substantial lowering of the water table in the well. The opposite is true for an aquifer in highly permeable rock or soil.

This lowering, or the difference between the water-table level and the water level in the pumping well, is known as **drawdown**. Stated differently, low permeability produces a large drawdown for a given amount of pumping and increases the possibility of a well "running dry." As pointed out, a pumping well in permeable materials creates a small drawdown and a gentle gradient in the cone of depression. Some closely spaced wells, however, have overlapping "cones," and excessive pumping in one well lowers the water level in an adjacent one (**Figure 8.58**). Many lawsuits have arisen for just this reason. One neighbor's well dries out another's.

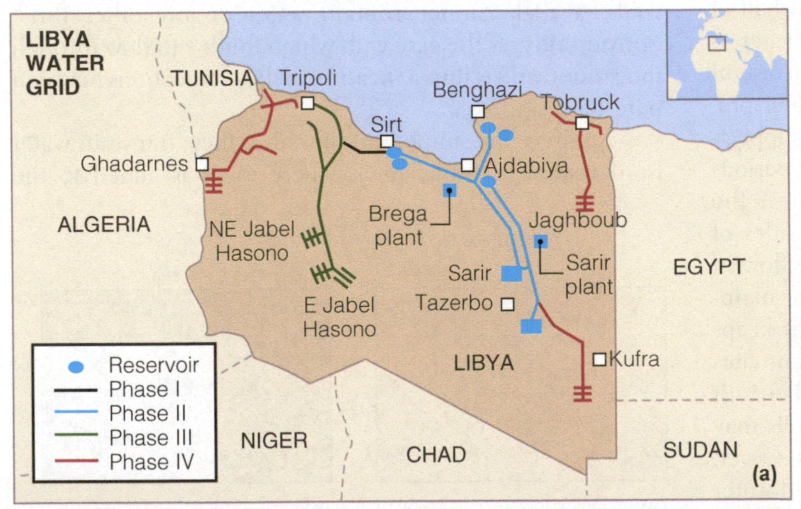

Yves Gellie/Gamma-Rapho/Getty Images

(b)

Gary Cook/Alamy Stock Photo

Figure 8.56 Before he was deposed in 2011, former Libyan leader, Muammar Al Qadhafi, had overseen an ambitious plan to water his nation with fossil water from the Nubian sandstone aquifer. (a) A map of Libya and the system that takes ancient water (recharged during cooler, moister glacial times) to the cities of today. (b) A billboard showing Qadhafi and the water system. (c) One of the 4-meter-wide (14 foot) pipes that carry the water.

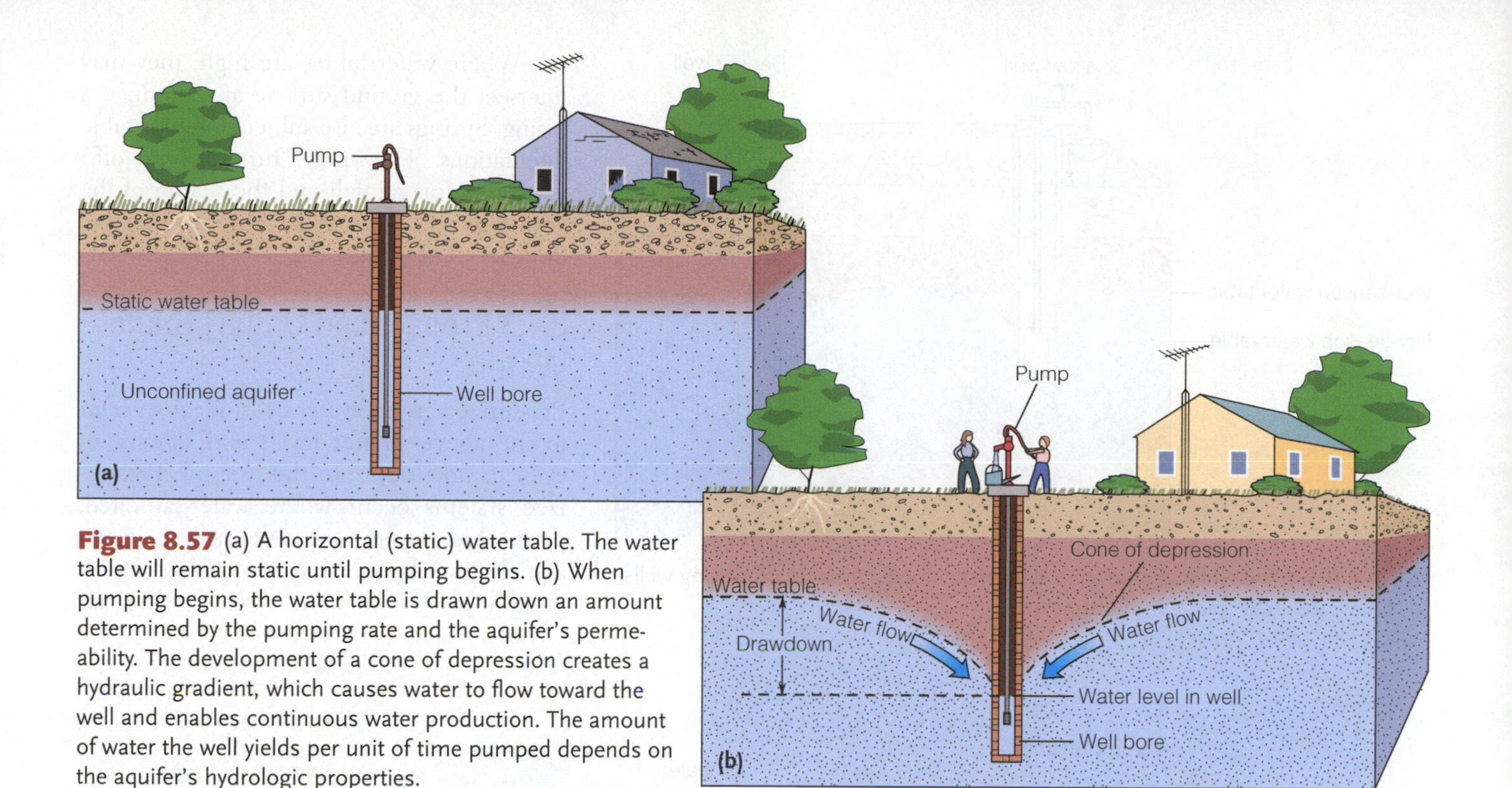

Figure 8.57 (a) A horizontal (static) water table. The water table will remain static until pumping begins. (b) When pumping begins, the water table is drawn down an amount determined by the pumping rate and the aquifer's permeability. The development of a cone of depression creates a hydraulic gradient, which causes water to flow toward the well and enables continuous water production. The amount of water the well yields per unit of time pumped depends on the aquifer's hydrologic properties.

A good domestic (single-family) well should yield at least 11 liters (2.6 gallons) per minute, which is equivalent to about 16,000 liters (3,700 gallons) per day, although some families use as little as 4 liters per minute (1,350 gallons/day). Many people get by with domestic wells that yield considerably less water than this. The reason is simple. If the well casing is long, large amounts of water can be stored in the well itself. For a 6-inch well, which is typical for a residence, each foot of casing stores about 5.8 liters (1.5 gallons) of water, so a deep well of 60 meters (200 feet) gives people a supply buffer of nearly 1200 liters, good enough for very long showers but not enough to fill the family swimming pool.

The water table rises and falls with consumption and the season of the year. Generally, the water table is lower in late summer and higher during the winter or the wet season. Trees use a great deal of groundwater, extracting it with their roots mostly from the unsaturated zone. When deciduous trees lose their leaves in the fall and cease **evapotranspiring** (biologically pumping ground water), water tables rise. Thus, water wells must be drilled deep enough to accommodate seasonal and longer-term climatic fluctuations of the water table (**Figure 8.59**). If a well "goes dry" during the annual dry season, it needs to be deepened–an expensive proposition.

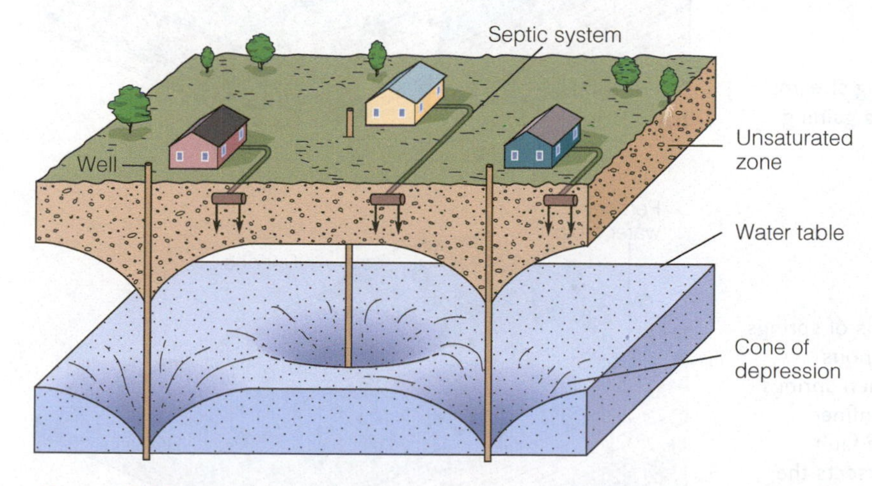

Figure 8.58 Cones of depression overlap in an area of closely spaced wells. Note the contributions to groundwater from septic tanks.

Streams located above the local water table are called **losing streams** because they contribute to the underground water supply (**Figure 8.60a**). Beneath such streams, there may be a mound of water above the local groundwater table, known as a **recharge mound**. Losing streams are commonly found in the desert, where the water table is usually far below the ground surface and stream water frequently originates from runoff in high mountains towering above the desert floor. A stream that intersects the water table and is fed by both surface water and groundwater is known as a **gaining stream** (see **Figure 8.60b**). These streams are most common in humid regions where precipitation is regular and abundant, keeping groundwater tables near the surface. A stream may be both a gaining and a losing stream, depending on the time of year and variations in the elevation of the water table. This gain of water by streams is what makes up baseflow.

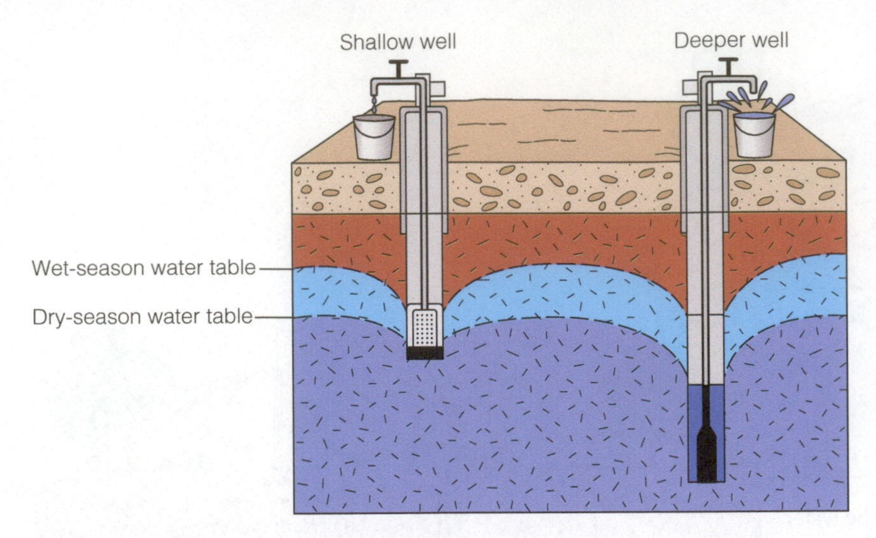

Figure 8.59 The effect of seasonal water-table fluctuations on producing wells. A shallow well may go dry in the summer.

Where water tables are high, they may intersect the ground surface and produce a spring. Springs are also subject to water-table fluctuations. They may "turn on and off," depending on rainfall and the season of the year. **Figure 8.61** illustrates the hydrology of three kinds of springs.

8.4.3 Pressurized Underground Water

Pressurized groundwater systems cause water to rise above aquifer levels and sometimes even to flow out onto the ground. Pressurized systems occur where water-saturated, permeable layers are enclosed between aquicludes (layers that do not transmit groundwater) and for this reason they are

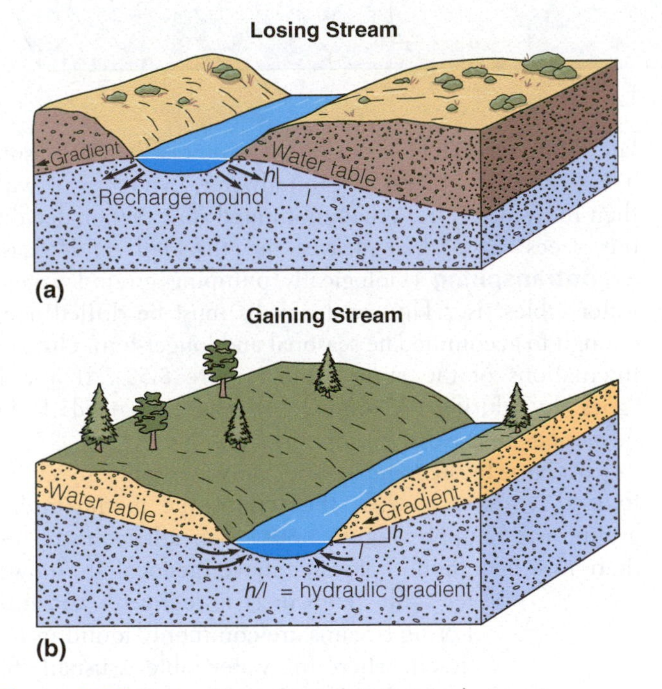

Figure 8.60 Hydraulic gradient (*h*/*l*) in (a) a losing stream, which contributes water to the water table, and (b) a gaining stream, which gains water from the water table.

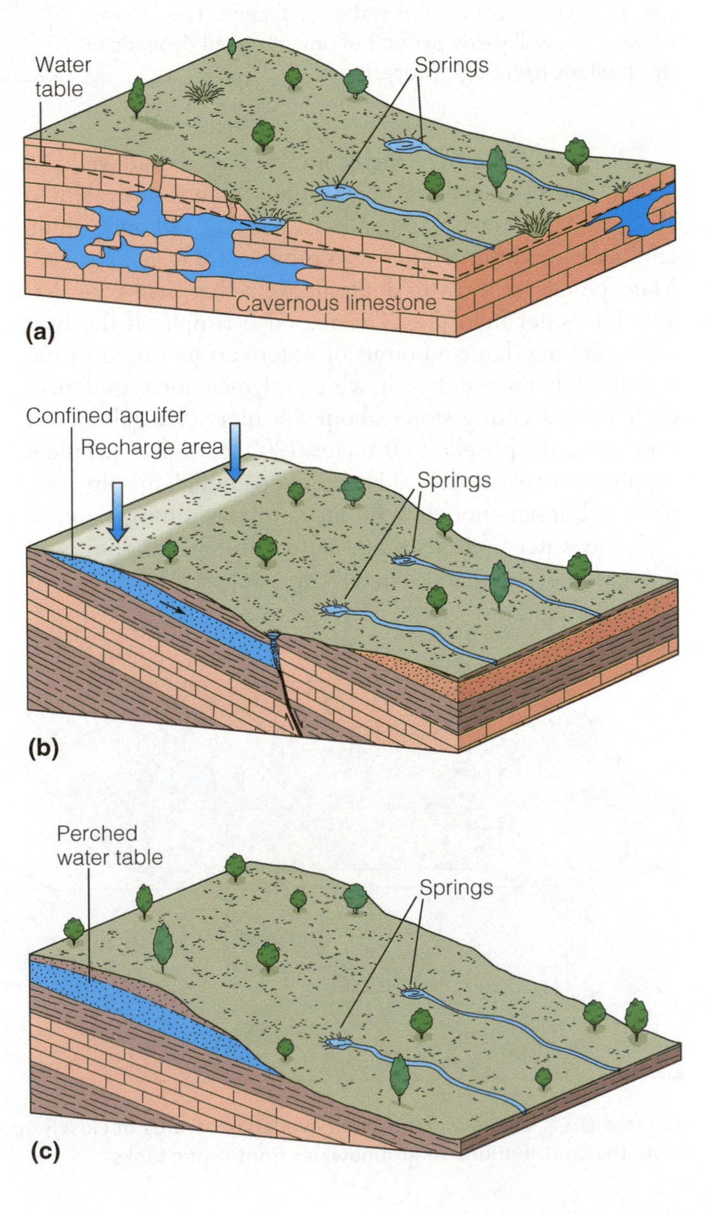

Figure 8.61 Geology and hydrology of three kinds of springs. (a) Springs form where a high water table in cavernous limestone intersects the ground surface. (b) Artesian springs. A fault barrier can cause pressurized water in a confined aquifer to rise as springs along irregularities in the fault. (c) Springs form where a perched water table intersects the ground surface. Such springs are intermittent because the small volume of perched water is quickly depleted.

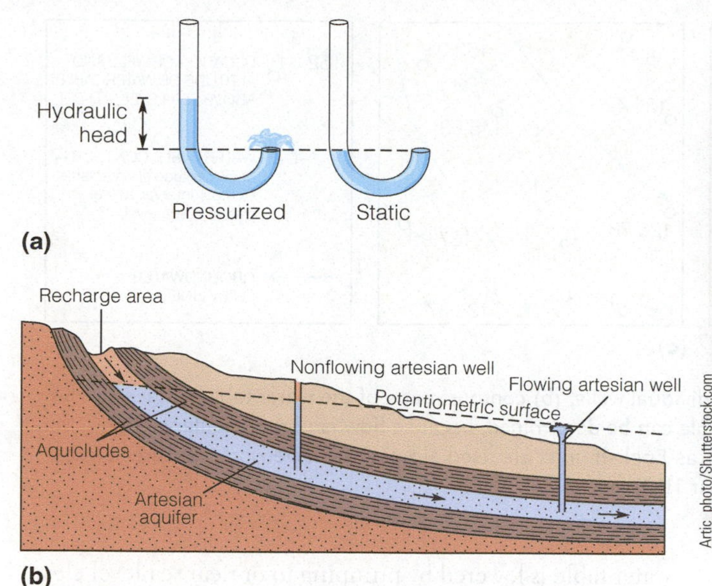

Figure 8.62 Water under pressure. (a) The hydraulic head in a hose lowers as water flows out the low end. Eventually, a static condition prevails. (b) A confined aquifer, commonly called an artesian aquifer. The potentiometric surface is the level to which water will rise in wells at specific points above the aquifer. Why is one well flowing, whereas the other is not?

Figure 8.63 This artesian well on the valley floor flows continuously, its water recharged in the hills that rise above the valley.

called **confined aquifers** (**Figure 8.62**). A confined aquifer acts as a water conduit, very similar to a garden hose that is filled with water and situated so that one end is lower than the other. In both cases, a water-pressure difference known as a **hydraulic head** is created. Water gushes from the low end until there is no longer a pressure difference. At that point the water level stabilizes, becoming static.

Wells that penetrate confined aquifers are called **artesian wells**, after the province of Artois, France, where they were first described. Water in some artesian wells is under enough pressure that it flows out onto the ground (**Figure 8.63**). In other cases, water simply rises through the well casing above the top of the aquifer. Artesian wells differ from water-table wells in that the latter depend on pumping or topographic differences to create the hydraulic gradient that causes the water to flow. Because the energy required to raise water to the surface is the greatest expense in tapping underground water, artesian systems are much desired and exploited. Not all artesian wells flow to the ground surface; water may rise naturally only partway up the well bore. Nevertheless, a natural rise of water any distance reduces the cost of its extraction.

Many bottled-water and beverage producers attempt to convince us that deep-well or artesian water is purer than other kinds of water. Artesian refers only to an aquifer's hydrogeology and has no water-quality connotation except that recharge zones for artesian wells are often in high-elevation, mountainous zones, which are less densely settled than valleys and less likely to be impacted by industrial waste dumping and polluting agricultural practices. Because artesian aquifers are under pressure where they are tapped as water supplies, water leaks out of them rather than in, protecting water quality. Do not be misled by advertisements attributing greater purity to artesian or deep-well water.

8.4.4 Groundwater Flow: Darcy's Law

Groundwater flows from an area of higher head (potential energy) to lower head. For confined aquifers, head can be considered pressure. For unconfined aquifers, head can be thought of as water table elevation. Both are called the **potentiometric surface**. Thinking about this imaginary surface and a contour map of head values can be very useful because it indicates the direction of groundwater flow. Water will always flow down gradient (**Figure 8.64**). If an aquifer's hydraulic conductivity can be estimated or measured, the rate and volume of groundwater flow can be determined using Darcy's law, which we discuss below.

Permeability is a measure of how fast fluids can move through a porous medium and is thus expressed in a unit of velocity, such as meters or feet per day. Knowing the permeability of an aquifer allows us to determine in advance how much water a well will produce and how far apart wells should be spaced. In the case of water, permeability is referred to as the **hydraulic conductivity**. Hydraulic conductivity, K, is defined as the quantity of water that will flow in a unit of time under a hydraulic gradient (head change over distance, the slope of the water table or potentiometric surface) of 1 through a unit of area measured perpendicular to the flow direction. K is expressed in distance of flow with time, usually meters (feet) per day (**Figure 8.65**).

To estimate flow volume, we use **Darcy's law**, which states that the rate of flow (Q in cubic meters/day) is the product of hydraulic conductivity (K), the hydraulic gradient (S in meters per meter), and the cross-sectional area

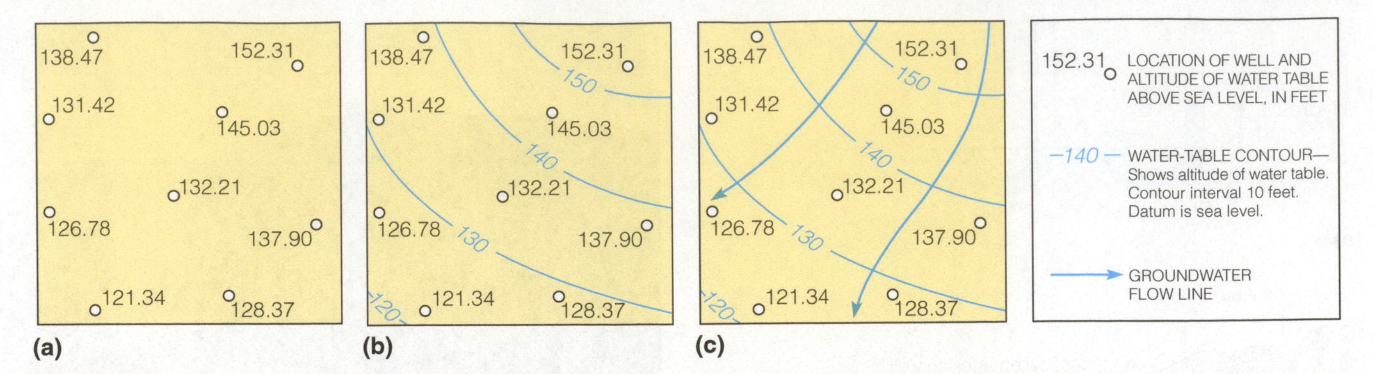

Figure 8.64 Using (a) known altitudes of the water table at individual wells, (b) contour maps of the water-table surface can be drawn and (c) directions of groundwater flow along the water table can be determined, because flow is perpendicular to the contours. Elevations are in feet above a datum, usually sea level, as English units are used almost exclusively on pressure-surface contour maps in the United States. (Modified from USGS circular 1139)

through which the groundwater is moving (A in square meters). Thus,

$$Q = K * A * S$$

If we use a unit hydraulic gradient, then Darcy's law reduces to

$$Q = K * A$$

Unit hydraulic gradient is the slope of 1-unit vertical to 1-unit horizontal ($S = 1/1 = 45°$). This it allows us to compare the water-transmitting ability of aquifers composed of materials with different hydraulic conductivities. Flow rates may be less than a millimeter per day for clays, and many meters (feet) per day for well-sorted sands and clean gravels. In general, groundwater moves at a snail's pace.

8.4.5 Groundwater-Saltwater Interaction

Near the coast, the aquifers people rely on for drinking water are fragile and can easily be contaminated by **saltwater intrusion**, when saltwater from the ocean mixes with freshwater from the land. Such intrusion happens when the

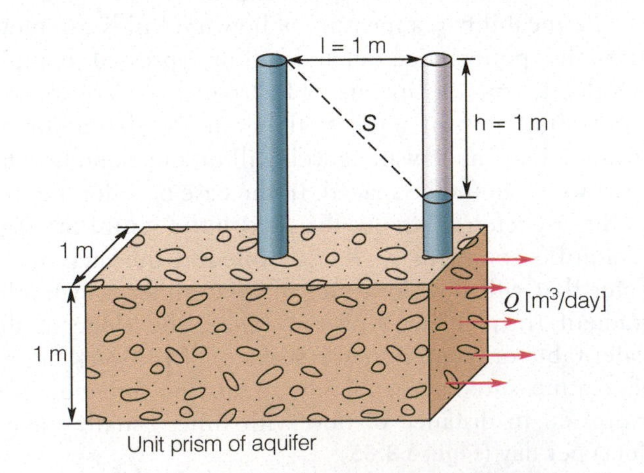

Figure 8.65 The unit hydraulic gradient (*I*) and unit prism (1 square meter) of a sandstone aquifer. When the hydraulic conductivity (*K*) is known, then the discharge (*Q*) in cubic meters per day per square meter of aquifer can be calculated.

water table is lowered by pumping to or near sea-level elevation, and saltwater invades the freshwater body, causing the water to become saline (**brackish**) and thus undrinkable.

The direction of flow between the two water masses is determined by their density differences. Seawater is 2.5% (1/40th) denser than freshwater. This means that a 31.2 meter (102.5 foot)-high column of freshwater will exactly balance a 30.5 meter (100-foot)-high column of seawater (**Figure 8.66a**). Freshwater floats on seawater under the Outer Banks of North Carolina, Long Island, and many other coastal areas.

This relationship is known as the *Ghyben-Herzberg lens*, named for the two scientists who independently discovered it and for the lens-like shape of the freshwater body (see **Figure 8.66b**) that floats on the denser saltwater. In theory, the mass of freshwater extends to a depth below sea level that is 40 times the water-table elevation above sea level. If the water table is 75 centimeters (2.5 feet) above sea level, the freshwater lens extends to a depth of 30.5 meters (100 feet) below sea level.

Any reduction of water-table elevation causes the saltwater to migrate upward into the freshwater lens until a new balance is established. If the water table drops below the level necessary for maintaining a balance with denser seawater, wells will begin to produce brackish (slightly salty) water and eventually saltwater (**Figure 8.67**). Thus, it is important to maintain high water-table levels in coastal zones, so that the resource is not damaged by saltwater intrusion. High water tables are maintained by good management combined, when necessary, with artificial recharge using local or imported water (water spreading), or with the injection of imported water into the aquifer through existing wells.

Long Island, which lies just northeast of New York City, depends heavily on its aquifers for drinking water and has struggled with saltwater intrusion. For thousands of years, rainwater infiltrated the sandy soil and flowed to the sea. Then, groundwater was pumped intensively as the island was urbanized, causing the water table on the western side of the island to drop below sea level. By 1936, saltwater had invaded the freshwater aquifer. Something had to be done. The response was a mix of lowering demand by

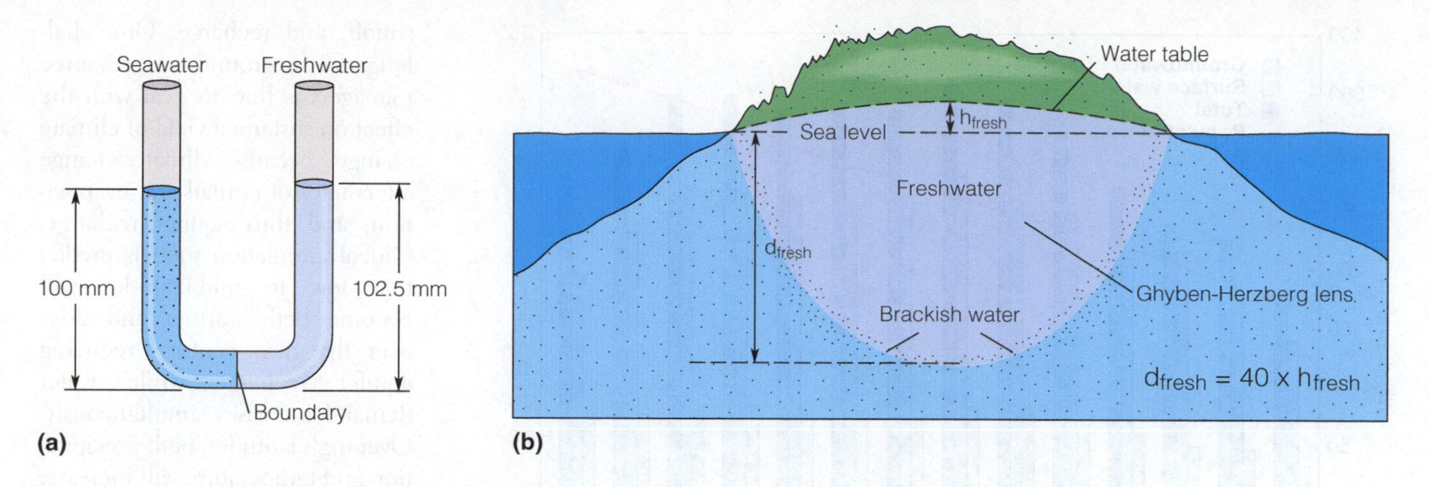

Figure 8.66 (a) The density contrast between freshwater and seawater. Because seawater is 2.5% denser than freshwater, a 102.5-millimeter column of freshwater is required to balance a 100-millimeter column of saltwater. (b) A hypothetical island and Ghyben-Herzberg lens of freshwater floating on seawater. The freshwater lens extends to a depth 40 times the height of the water table above sea level (not to scale).

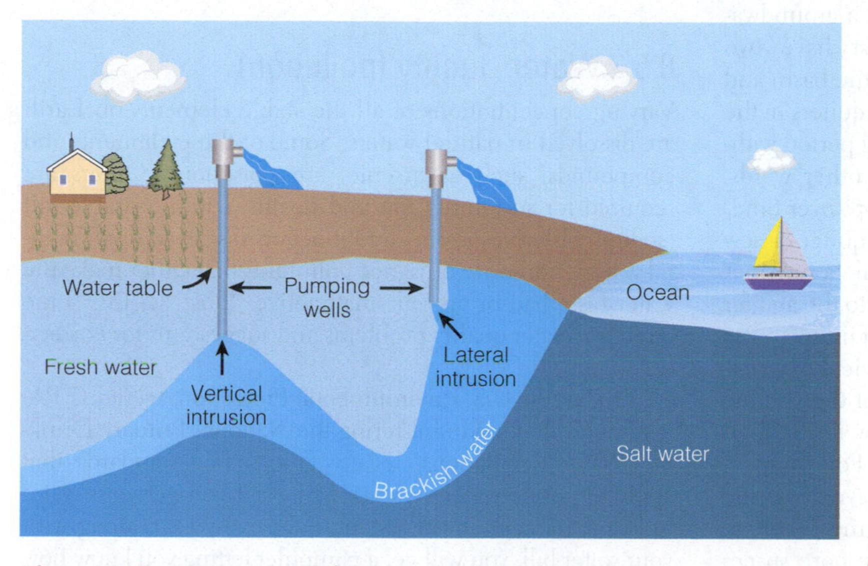

Figure 8.67 Pumping groundwater can induce saltwater intrusion.

importing water and artificial recharge to the aquifer using wells and leaky ponds that allowed sewage, treated to drinking water standards, to refill the aquifer. This approach built a freshwater barrier against saltwater intrusion.

8.5 Water Quality and Quantity (LO5)

Surface water and groundwater are not separate entities. Nearly all surface-water features–lakes, streams, rivers, artificial lakes, and wetlands–interact with water underground. Some surface-water bodies gain water and acquire a different chemistry from groundwater, and other surface waters contribute to underground water and may even pollute it. Thus, in water planning for quality and quantity, it is important to have a clear understanding of the linkages between groundwater and surface water, and of the geology that controls both.

8.5.1 Water Quantity

We use freshwater for many different things. The greatest volume of freshwater is used for irrigation, and the greatest total volume of water, both fresh and saline, is used for power generation. However, there is a difference in how this water is used: 98% of the water used for power generation is returned to streams or other reservoirs; only 2% is consumed. Irrigation water is almost entirely consumed.

How water is used affects the reuse potential of return flows. For example, irrigation water may be too salty or contaminated by pesticides and insecticides to have any reuse potential unless it is purified naturally or artificially. In contrast, water used in power generation has great reuse potential, because the principal change is usually an increase in its temperature and location (or head) as it is released downstream.

Since 1950, U.S. water-use statistics have been compiled every 5 years. There was a steady increase in per-person water usage from 1950 to 1980, the all-time high. By 2015, the average water use at home had fallen to 310 liters (82 gallons) a day per person. Between 2005 and 2015, there was a steady decrease in water withdrawals in the United States, most notably due to decreases in withdrawals for cooling power generation stations. However, the U.S. population grew by 50 million people between 2000 and 2020 (**Figure 8.68**), so withdrawals for use in public supply have increased steadily since 1950 to keep up with population demand. Some of the reasons cited for this decrease were the development and use of more efficient irrigation methods, conservation programs, increased water recycling, and a downturn in farm economies.

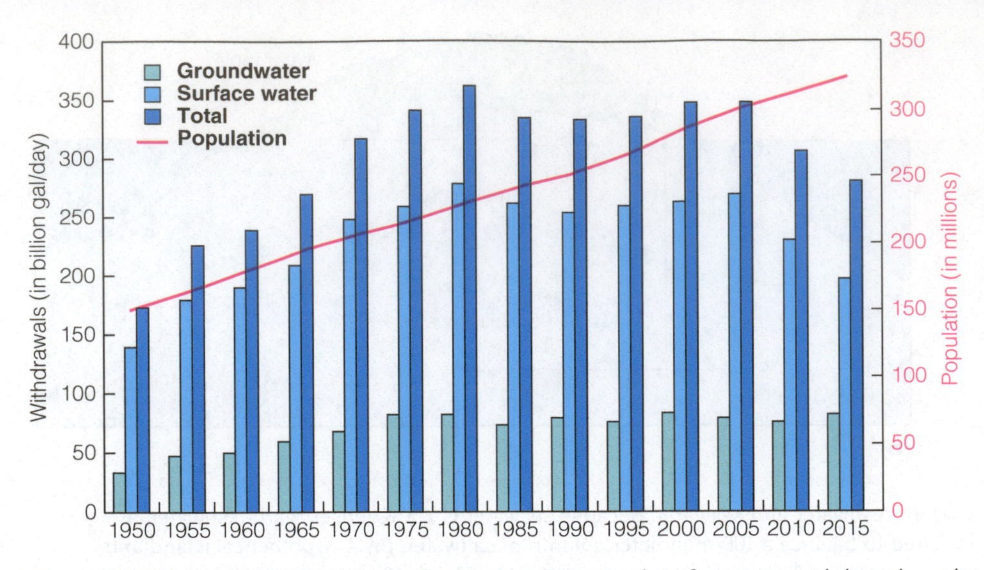

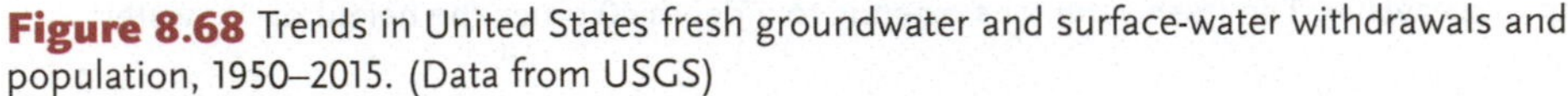

Figure 8.68 Trends in United States fresh groundwater and surface-water withdrawals and population, 1950–2015. (Data from USGS)

To manage underground water supplies in a groundwater basin, the hydrologist or planner must have two measurements: the quantity of water stored in the basin and the **sustained yield**, the amount of water the aquifers in the basin can yield on a day-to-day basis over a long period without overdrafting the ground water supply–in other words, mining groundwater so that the water table drops over time.

A groundwater basin consists of a single aquifer or several aquifers that have well-defined geological boundaries and recharge areas, places where water seeps into the aquifer (see Figure 8.61b). To determine the amount of usable water in the basin, we rely on the concept of specific yield of water-bearing materials. **Specific yield** is the ratio of the volume of water an aquifer will give up by gravity flow to the total volume of material, expressed as a percentage. For example, a saturated clay with a porosity of 40% may have a specific yield of 3%; that is, only 3% of the clay's volume is able to drain into a well. The remaining water in the pore spaces (37%) is held by the clay-mineral surfaces and between clay layers against the pull of gravity. This "held" water is its **specific retention**. Similarly, sand with a porosity of 35% might have a specific yield of 23% and a specific retention of 12%. Note that the sum of specific yield and specific retention equals the porosity of the aquifer material.

Thus, the total available water in an aquifer can be calculated by multiplying the specific yield by the volume of the aquifer determined from water-well data. The process is a bit more complicated for artesian aquifers, but it follows the same procedure after accounting for the compaction of the aquifer and the expansion of the water when the pressure is reduced. The amount of water available in a groundwater basin can be determined accurately if the basic data are available from numerous wells.

As noted earlier, sustained yield is the amount of water that can be withdrawn on a long-term basis without depleting the resource. Sustained yield is more difficult to assess than aquifer storage capacity because it is affected by precipitation, runoff, and recharge. One challenge facing groundwater resource managers is how to deal with the effect on sustained yield of climate change, because climate change alters rates of rainfall and evaporation, and thus aquifer recharge. Global circulation models predict that low to mid-latitudes will become both warmer and drier over the next century, reducing aquifer recharge while water demand increases simultaneously. Over high latitudes, both precipitation and temperature will increase. In the face of changing climate, it's more important than ever to estimate sustained yield, so that groundwater resources can be managed intelligently.

8.5.2 Water Quality (pollution)

Varying concentrations of all the stable elements on Earth are dissolved in natural waters. Some of these elements and compounds, such as arsenic, are poisonous. Others are required for sustaining life and health; common table salt (sodium chloride) is one, and it occurs dissolved in all natural waters. Some are nuisance pollutants that either make the water taste bad or appear unattractive. **Table 8.4** lists some common water quality problems and suggestions for correcting them.

Today, the U.S. Environmental Protection Agency (EPA) is responsible for administering the National Primary Drinking Water Regulations, legally enforceable standards that apply to public water systems. In fact, if you get your water from such a centralized system, once every year, along with your water bill, you will get a pamphlet letting you know how well the water you drink meets these standards and what your water company is doing to help make the water even cleaner. These primary standards protect your health by limiting the levels of potentially harmful chemicals in your drinking water. These concentrations are expressed in weight per volume (milligrams/liter). The standards can also be expressed in a weight-per-weight system. For example, in the English system, 1 pound of dissolved solid in 999,999 pounds of water is one part per million (ppm).

The quality of groundwater can be impaired either by a high total amount of various dissolved salts or by a small amount of a specific toxic element. High salt content may be caused by solution of minerals in local geological formations, seawater intrusion into coastal aquifers, contamination by road salt, or the introduction of industrial wastes into groundwater. Water with a high salt content, particularly bicarbonate and sulfate contents, can have a laxative effect on humans. Dissolved iron or manganese can give coffee or tea an odd color, impart a metallic taste, and stain laundry.

Table 8.4 Common Water Quality Problems

Problem	Probable Cause	Correction
Mineral scale buildup	Water hardness	Water softener or hardness inhibitor (such as Calgon)
Rusty or black stains	Iron and/or manganese	Aeration/filtration; chlorination; remove source of Fe/Mn
Red-brown slime	Iron bacteria	Chlorination/filtration unit; remove source of slime
Rotten-egg smell or taste	Hydrogen sulfide and/or sulfate-reducing bacteria; low oxygen content	Aeration; chlorination/filtration unit; greensand filters
Gastrointestinal diseases, typhoid fever, dysentery, and diarrhea	Coliform bacteria and other pathogens from septic tanks or livestock yards	Remove source of pollution; disinfect well; boil water; chlorinate; abandon well and relocate new well away from pollution source

Source: Daly, D. 1994. Groundwater quality and pollution. Geological Survey of Ireland Circular 85–1, 1985. In Moore, J. E., et al. *Ground water: A primer*. American Geological Institute.

Concentrations greater than 3 ppm of copper can give the skin a green tinge, a condition that is reversible. Organic contaminants, mainly animal wastes, solvents, and pesticides, also degrade water quality.

Several types of water quality standards have been set by the federal government to protect public health. The most important enforcement standard is the **maximum contaminant level (MCL)**, which is the highest level of a contaminant that is allowed in drinking water. **Maximum contaminant level goals (MCLGs)** are the level of a pollutant at and below which there is no known or expected risk to health. MCLs are set as close to MCLGs as is politically and scientifically feasible, while considering the available treatment technology and taking cost into consideration. **Table 8.5** lists EPA limits for some of the inorganic substances that occur in groundwater and impair its safety.

A **persistent pollutant** is one that builds up (or bioaccumulates) in the human system because the body does not metabolize or excrete it. The Romans, for example, used lead plumbing pipes, and skeletal materials exhumed from old Roman cemeteries retain high lead concentrations. Lead concentrations in the water supply greater than about 0.1 ppm can build up over a period of years to debilitating concentrations in the human body. Arsenic is also persistent, and the standards for arsenic and lead are similar, less than 0.015 ppm, preferably absent. Nitrate in amounts greater than 10 ppm is known to inhibit the blood's ability to carry oxygen and to cause "blue baby syndrome" (*infantile methemoglobinemia*) during pregnancy. The element with the greatest impact on plants is boron, which is virtually fatal to citrus and other plants in concentrations as low as 1.0 ppm.

Table 8.5 Standards (Maximum Contaminant Level or Action Level, 2022) and Health Effects of Significant Concentrations of Common Inorganic Environmental Pollutants

Inorganic Pollutant	Standard, Milligrams/Liter	Health Effects
Arsenic	0.01	Dermal and nervous system toxicity effects, paralysis
Barium	2.0	Gastrointestinal effects, laxative
Boron	1.0	No effect on humans; damaging to plants and trees, particularly citrus trees
Cadmium	0.005	Kidney effects
Copper	1.3	Gastrointestinal irritant, liver damage; toxic to many aquatic organisms
Chromium	0.1	Liver/kidney effects
Fluoride	4.0	Mottled tooth enamel
Lead	0.015	Nervous system/kidney damage; highly toxic to infants and pregnant women
Mercury	0.002	Central nervous system disorders, kidney dysfunction
Nitrate	10.0	Methemoglobinemia ("blue baby syndrome")
Selenium	0.05	Gastrointestinal effects
Silver	0.1	Skin discoloration (argyria)
Sodium	20–170	Hypertension and cardiac difficulties
Sodium	70.0	Renders irrigation water unusable

Source: U.S. Environmental Protection Agency. 2022, January. *National primary drinking water regulations*

A new persistent pollutant has come to the public's attention in the past few years: PFAS, which is short for perfluoroalkyl substances. They go by the alias **forever chemicals (PFAS)**. These are a class of organic compounds made by people. They are mostly carbon and hydrogen but also incorporate the element fluorine. They are remarkably stable and don't degrade in nature. They have some nifty properties, like being really slick (think nonstick kitchen pans or water beading off your raincoat). The problem is that they are toxic and bioaccumulate. Some of the hundreds of different PFAS compounds are water soluble and have ended up in people's drinking water. The extent of PFAS contamination is unknown but appears to be widespread. As recent rainfall data show, these chemicals are literally everywhere on Earth now (see also Chapter 12).

Dissolved calcium (Ca) and magnesium (Mg) ions cause water to be "hard." In hard water, soap does not lather easily, and dirt and soap combine to form scum. Hard water is most prevalent in areas underlain by limestone ($CaCO_3$) and dolomite ($CaMgCO_3$). A water softener exchanges sodium ions (Na) for Ca and Mg ions in the water using a zeolite mineral (a silicate ion exchanger) or mineral sieve. Water with more than 120 ppm of dissolved Ca and Mg is considered hard (**Table 8.6**). Because sodium is known to be bad for persons with heart problems (hypertension) and water softeners increase the sodium content of water, people on low-sodium diets should not drink softened water.

The good news is that fluoride (F^-) ions dissolved in water reduce the incidence of tooth cavities when fluoride is present in amounts of 1.0 to 1.5 ppm. Tooth enamel and bone are composed of the mineral apatite, $Ca_5(PO4)_3$(F, Cl, OH), whose structure allows the free substitution of fluoride (F^-), chloride (Cl^-), or hydroxyl $(OH)^-$ ions in the mineral structure (see Chapter 4). In the presence of fluoride, the crystals of apatite in tooth enamel become larger and more perfect, which makes them resist decay. Too much fluoride in drinking water, however–concentrations greater than 4 ppm–can cause children's teeth to become mottled with dark spots. Indeed, it was this cosmetic blemish, among children of Colorado Springs, Colorado, in particular, that led to the discovery of the beneficial effect of fluoride ions and the establishment of concentration standards for fluoride.

Table 8.6 Water Hardness Scale

Concentration of Calcium, Milligrams/Liter	Classification
0–60	Soft
61–120	Moderately hard
121–180	Hard
>180	Very hard

Note: Dissolved calcium and magnesium in water combine with soap to form an insoluble precipitate that hampers water's cleansing action.

Hardness is expressed here as milligrams of dissolved Ca per liter.

In 1908, Dr. McKay, a Colorado Springs dentist, described a malady he called **Colorado brown stain**, a discoloration of the tooth enamel that we now know results from ingesting high levels of fluoride (**Figure 8.69**). Not only did nearly 90% of the schoolchildren in Colorado Springs have stained teeth, but they also had far fewer cavities than their peers in nearby Boulder, Colorado. It took decades and many studies to prove that fluoride is a very effective way to dramatically reduce tooth decay. Now much of the nation drinks fluoridated water and gets far fewer fillings. Dissolved fluoride is also believed to slow osteoporosis, a bone-degeneration process that accompanies aging.

Residence time, the average length of time a substance remains (resides) in a system, is an important concept in water-pollution studies. We may view it as the time that passes before the water in an aquifer, a lake, a river, a glacier, or an ocean is totally replaced and continues its way through the hydrologic cycle. Residence time can be calculated simply by dividing the volume of the water body (cubic meters) by the flow rate into and out of it (cubic meters/second), assuming everything is in balance.

The average residence time of water in rivers is hours to weeks; in large lakes, several decades; in shallow gravel aquifers, days to years; and in deeper, low-permeability aquifers, perhaps thousands or even hundreds of thousands of years. This knowledge allows us to estimate the approximate time a particular system will take to clean itself naturally and indicates those cases in which a pollutant must be removed from a water body by human means because it would otherwise cause harm over a long residence time.

Most pollution is due to the careless disposal of waste at the land surface by humans. Groundwater pollution occurs in urban environments because of the improper disposal of industrial waste and leakage from sewer systems, old fuel-storage tanks, wastewater settling ponds, and chemical-waste dumps (**Figure 8.70a**). In suburban and agricultural areas, septic-tank leakage, fecal matter in runoff and seepage from

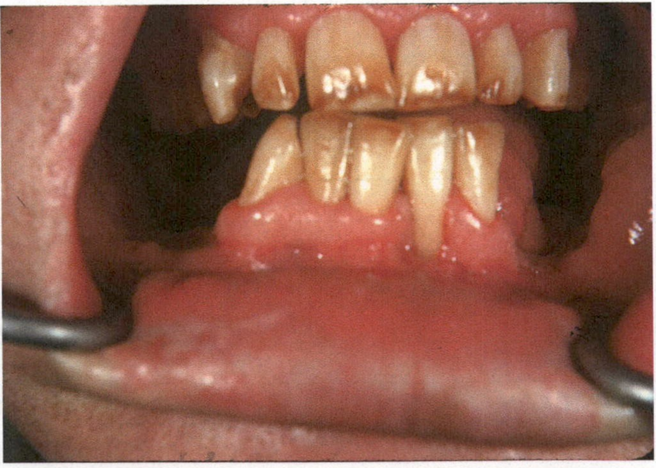

Flickr/National Museum of Health and Medicine

Figure 8.69 These teeth, from a person living in an area where the water has high levels of fluoride, are stained brown but have few cavities.

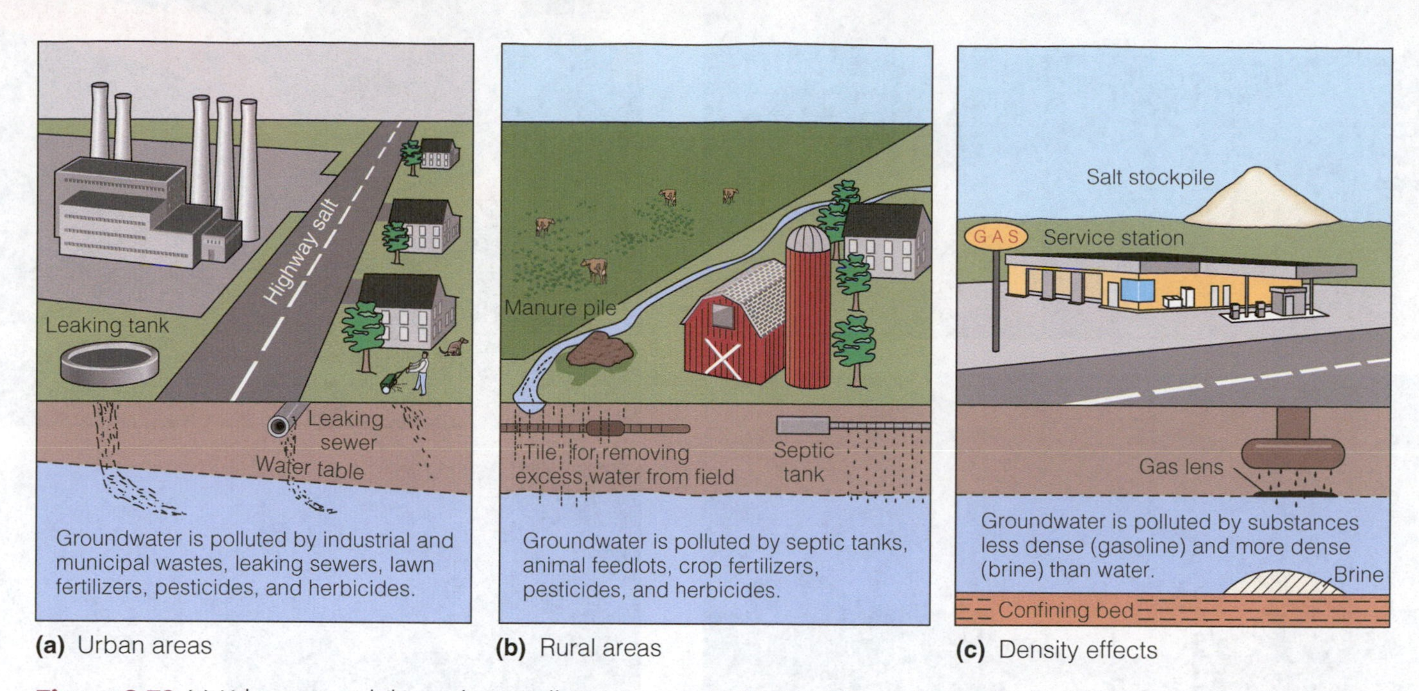

Figure 8.70 (a) Urban-area and (b) rural-area pollution sources. (c) Some pollutants float and others sink because of differences in density between the groundwater and the pollutants. Brine, being denser than water, sinks. Gasoline, being less dense than water, floats.

animal feedlots, the use of inorganic fertilizers including nitrates and phosphates, and the application of weed- and pest-control chemicals can all degrade water quality (see **Figure 8.70b** and Chapter 12).

In the past few decades, leaking underground fuel-storage tanks have been recognized as a major cause of soil and groundwater pollution (see **Figure 8.70c**). By federal law, all underground tanks now must be inspected for leakage, and all single-wall tanks are being replaced with double-wall tanks so fuel leaking from the inner tank is contained rather than being released to the environment; if fuel is found in the adjacent soil, the tanks must be removed and the soil cleaned.

Density differences between groundwater and particular pollutants often govern the remedial measures that can be used; for example, gasoline and oil float on groundwater, whereas salt brines from industry or agriculture and some organic chemicals sink to lower levels (see Figure 8.70c). In some cases, it is possible to skim a layer of gasoline from the top of the water table. Gasoline or fuel oil that is floating on the capillary fringe may yield vapors that can rise and be trapped beneath or in the walls of buildings. Gasoline fumes contain BTX (*b*enzene, *t*oluene, and *x*ylene), and are thus carcinogenic. "Air-stripping" is currently the preferred method of removing volatile organic pollutants such as BTX compounds and trichloroethylene (TCE) and perchloroethylene (PCE) cleaning solvents from groundwater. (The origin of these substances is usually connected to refinery, industrial, or military activities. See also Chapter 1.) The polluted water is brought to the ground surface, where the contained volatile pollutants are gasified (air-stripped) and either vented to the atmosphere or captured in charcoal filters and then destroyed.

In the past decade, increasing attention has been paid to the role of sediment as a pollutant. Increased loads of fine sediment can smother in-stream biota, including plants and fish. Silty, turbid water reduces the amount of light available for photosynthesis and clogs the gills of fish. Many pollutants, such as phosphorus and lead, have affinities for sediment and thus can be carried into streams attached to solid particles. Sediment pollution is often the result of poorly constructed roads, agriculture, excavation, and mining (**Figure 8.71**), as well as rapid land development without proper engineering controls.

There are simple things one can do to improve water quality. Increasing scarcity of clean water and a realization that actions taken in one part of the hydrologic system affect other parts of the system have spawned an increasingly active citizen-driven movement to protect the quality of surface and groundwater. For example, many storm drains lead directly to water bodies; used motor oil someone pours down the drain might go directly to the local trout stream (**Figure 8.72**). If you have unused pesticides, paints, or fertilizers, don't dump them down the drain but instead take them to a community hazardous waste collection facility. Most towns and cities have hazardous waste drop-off days and some even have mobile collection vans that might set up in a parking lot near you to collect waste and keep it out of the environment.

8.5.3 Cleanup, Conservation, and Alternative Water Sources

Many regions of the United States are or have repeatedly experienced drought conditions. Excessive groundwater withdrawals have led to water mining in many areas, particularly

iStock.com/WeiseMaxHelloween

iStock.com/Ysbrandcosijn

iStock.com/Avalon_Studio

iStock.com/Ilbusca

Figure 8.71 Sediment pollution can result from (a) muddy roads like this one in the Carpathian Mountains of Romania; (b) this freshly tilled field; (c) excavating a construction site in Poland; (d) hydraulic mining of gold-rich stream gravel in the western United States with jets of pressurized water.

in the desert southwest. Proposed remedies include desalinizing seawater, recycling wastewater, tugging Antarctic icebergs up the coast of South America to drier areas in the Northern Hemisphere, and floating 35-ton water bags down from the northwestern United States to Southern California. However, in the long run, as population increases, there is no alternative to conserving water. Voluntary water conservation has been quite successful, reducing consumption in American households by as much as 15%, and even greater reductions have been achieved where rationing is mandatory. The advent of Energy Star appliances such as water-efficient dishwashers and front-loading clothes washers lets many of us contribute while often earning rebates from local utilities.

Despite methods for water conservation, global water use is increasing at a rate of 1% per year since the 1980s. Global water demand for the world's 8 billion people is currently approximately 4,600 kilometers^3 per year. By 2050, global water demand is expected to rise by 20% to 30%, increasing use to 5,500 to 6,000 kilometers^3 per year. If total consumption increases at the current rate, more than 6 billion people will suffer from clean water scarcity by 2050.

iStock.com/Helt2

Figure 8.72 Many storm drains lead directly to streams, rivers, and lakes, so citizens often stencil them in the hope of dissuading dumping of polluting liquids.

Ironically, the combination of pumping, floodwalls, and levees has caused the city to sink even farther below sea level as dried, organic-rich soils decay and compact, and the river's floods, which used to deposit sediment and raise the land, are funneled south, beyond the city's reach. Indeed, by walling in a city that now exists below sea level, engineers and politicians made it possible for people to live in a flood-hazard zone and created the potential for disaster. The levees and floodwalls were designed to withstand flooding from a rapidly moving category 3 hurricane, a moderately severe storm. But larger storms lurked.

Several times in the last few decades, the levee system was threatened, but each time the hurricanes changed course at the last minute, sparing New Orleans their full fury. In general, the levees held and flooding of the city was not catastrophic, until Katrina hit in August 2005. The strong category 3 storm hammered the city with wind and rain, but the real flooding came a day later, when the storm surge breached the flood-control system in numerous places, allowing water from Lake Pontchartrain to pour into the city. Nearly 80% of New Orleans was flooded, more than 1,800 persons died, and nearly a million were homeless. This was flooding and destruction on a scale the United States had never seen.

The levee and floodwall system failed because some sections were underdesigned and others poorly installed. The levees failed in several ways (**Figure 8.79**). Those that were not built high enough were overtopped by water. As this water poured over the levees, it eroded and undermined them, leading to collapse. Other flood-control features, built on soft soils, were simply pushed aside by the pressure of the water they were meant to contain as the soils beneath them gave way. Sometimes water moved under the levees in what are termed "piping failures," eroding material under the levees until they lost support and caved in. These engineering failures can be remedied by improved design and construction techniques; however, what really needed to be fixed was the administrative

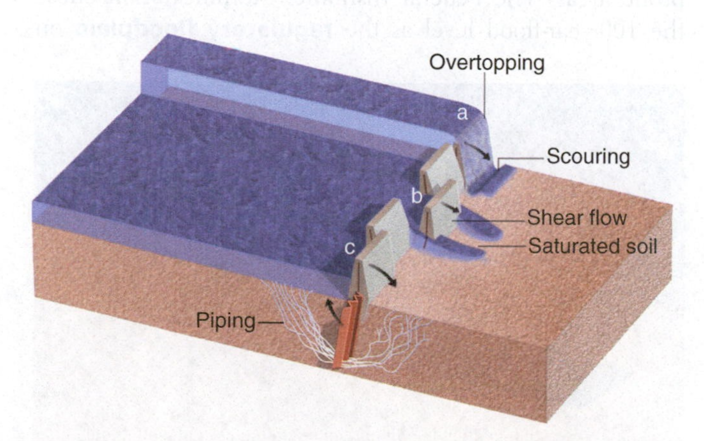

Figure 8.79 Levees and floodwalls can fail in several ways. (a) Water can pour over the top, scouring away the base. (b) Weak soils beneath can fail, allowing the levee to be pushed outward. (c) Water can move under the levee, called "piping", a process by which the levee is demolished by erosion from below.

system that allowed the design and construction of substandard flood control when millions of people's lives were at stake.

Ironically, flood-control structures may increase flood risk because floodplain development occurs in response to the assumptions that floods have been controlled. Population and property values adjacent to and below flood-control structures increase, placing people and their property at risk when the levees fail. For example, many artificial levees were built along the Mississippi River to protect farmland. The floodplain was then urbanized, and the existing levees, although adequate to protect the farms (which can tolerate occasional flooding), were not adequate for protecting homesites. Is "flood control" the solution or the problem?

8.6.2 Dams

In the United States, there are more than 91,000 dams (2022), and nearly every major river is controlled by at least one dam. Dams provide such societal benefits as cheap electricity, flood control, recreation, and drought mitigation. However, dams destroy and fragment habitats, disturb natural cycles of flooding and sediment deposition, change aquatic and riverine biology, and destroy important spawning grounds for migratory fish such as salmon. The days of big dam construction in the United States appear to be over. Although more large dams are being built in developing countries, only a few sites in the United States, Canada, and Europe are seriously being considered.

The water emerging from a dam has different physical properties than the water upstream. Sediment-free outflow water scours the riverbed below the dam; because the temperature is different, native fish are stressed and some die. Even with fish ladders, adult fish have trouble migrating upstream to spawn, and many fry cannot survive the trip back downstream. Different **riparian** aquatic communities arise in the downstream area as the frequency and magnitude of disturbance change (most floods are now controlled). In addition, some dams fail, and dam-failure floods have caused loss of life.

For all these reasons, some dams are being dismantled and others are being altered to make them more ecologically friendly. The Kennebec River in Maine now flows freely because the 162-year-old Edwards Dam in Augusta, Maine, was breached in 1999. Other dams that have been removed (to make way for spawning Pacific salmon) are on the Elwha River in Washington State, which backs up water into Olympic National Park, and 10 hydroelectric dams on the Deerfield River in Vermont and Massachusetts (to allow passage by Atlantic salmon and other migratory species). This dismantling is occurring because a 1996 federal law requires that environmental concerns be addressed before dams can be relicensed. Dam removal is by no means straightforward. Large sediment loads after removal must be considered, as decades of trapped sediment are released, sometimes in just a few days. Some of these sediments can be rich in nutrients, toxic metals, and organic compounds.

Figure 8.80 Location map of the Aral Sea. (Redrawn from Philip Micklin, Western Michigan University)

Dams have downstream effects as they divert water from rivers that is then used consumptively for agriculture. Consider the Aral Sea, once the fourth-largest lake in the world, second only to the Caspian Sea in the former Soviet Union (**Figure 8.80**). The level of the sea in 1960 was about 53 meters (170 feet) above sea level, and it contained a fish population that supported more than 60,000 people in fishing and fish processing. In 2003, the level of the Aral Sea was only 30 meters (100 feet) above sea level, and it ranked as the world's sixth-largest lake, having lost more than 74% of its surface area since 1960. The result of this shrinkage is that the fishing industry is no more (**Figure 8.81**). In addition, the salinity of the Aral Sea has increased from 10 grams/liter (about 1%) to more than 100 grams/liter (10%) in some areas, making it inhospitable to most marine life. This story leaves a question that begs an answer: What caused this water body to disappear?

Agricultural demands of the former Soviet Union deprived this great salt lake of water inflow sufficient to overcome evaporation losses. Two major rivers feed the lake: the Amu Darya (*darya*, "river") flowing into the lake from the south, and the Syr Darya, which reaches the sea at its north end. The Soviet leaders wanted exportable crops, and cotton was high on their list. Irrigated agriculture was expanded, and the Kara Kum Canal was opened in 1956, which diverted water from the Amu Darya into the deserts of Turkmenistan. Millions of hectares of arid lands came under irrigation.

By 1960, the level of the sea had begun to drop, and where in the past it received 59 cubic kilometers (about 13 cubic miles) of water annually, this amount had fallen to zero by the early 1980s. As the lake area receded, the salinity increased, and by 1977, the fish catch had dropped to one-fourth its level before water diversion. By 1990, the area of the Aral Sea was cut in half, and its water volume was but a mere one-third its historic amount. The lowering of the sea exposed 30,000 square kilometers (about 11,000 square miles) of salt- and chemical-filled lakebeds (**Figure 8.82**).

Blowing dust, particularly south of the lake, has resulted in air pollution, producing a high incidence of throat cancer, eye and respiratory diseases, anemia, and infant mortality. In addition, the shoreline habitat was lost and delta wetlands desiccated. Animal populations shifted because conditions favored animals better adapted to drought and high salinity. By any measure, the desiccation of the Aral Sea is an ecological disaster. The central Asian republics that surround the sea–Kazakhstan, Turkmenistan, Uzbekistan, and the little-known Karakalpak Republic (within Uzbekistan)–are the poorest of the former Soviet bloc and have no money to invest in rebuilding irrigation systems. In addition, Afghanistan and Iran, with small areas within the drainage basin, have little interest in the problem.

As a sad Uzbek poem goes, "You cannot fill the Aral Sea with tears."

8.6.3 Insurance, Flood-Proofing, and Floodplain Management

The National Flood Insurance Act of 1968 provides federally subsidized insurance protection to property owners in flood-prone areas. The Federal Insurance Administration chose the 100-year-flood level as the **regulatory floodplain** on

Figure 8.81 Abandoned Aral Sea fishing boats, left stranded when the water level fell.

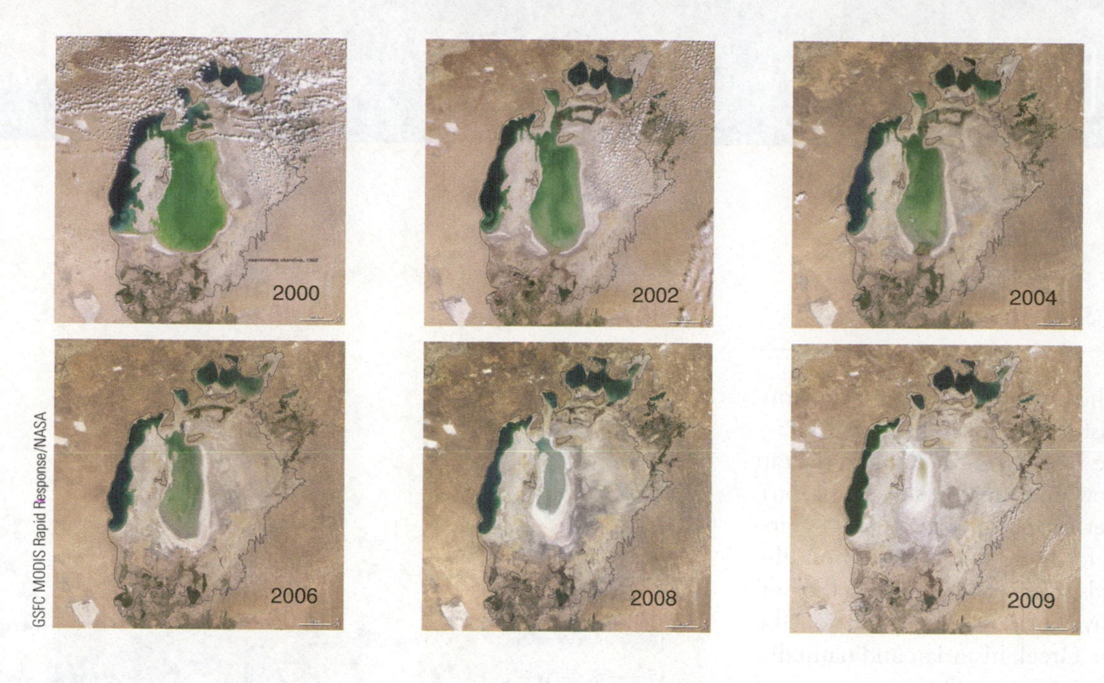

Figure 8.82 The shrinkage in area of the Aral Sea from 2000 to 2009 is dramatic. By 2003, compared with 1960, the sea had lost 74% of its area and 90% of its volume. In 2005, Kazakhstan built a dam and cut off the southern Aral Sea, keeping all of the water to the north. After the dam was closed, the water levels in the south plummeted. Can you see the changes?

which to establish the limits of government management and set insurance rates. Maps delineate regulatory floodplains in thousands of communities in the United States. Good floodplain management limits the uses of floodplains in chronically flood-prone areas, the **floodway**. The subsidized flood insurance is not available on new homes within the regulatory floodway (**Figure 8.83**). Pre-1968 houses in the floodway were "grandfathered" (allowed to stay and to be insured under the government program), but they cannot be rebuilt at that site with flood-insurance money if they are more than 50% damaged by flooding.

Desert regions provide a particular challenge to flood-hazard management. There, the problem is less inundation and more the migration of channels during large flood events. In some cases, channels have moved laterally hundreds of feet without overflowing as banks were undercut by high flows. The National Academy of Sciences recently came up with a different flood-hazard evaluation strategy more appropriate for arid regions, which explicitly considers the likelihood of channel movement during floods.

Development has literally poured onto flood-prone areas in the United States. Annual flood damage (in constant, inflation-adjusted dollars) appears to have increased over the last 100 years to an average of several billion dollars per year, but damage amounts are extremely variable on a year-to-year and decade-to-decade basis. For example, a National Oceanographic and Atmospheric Administration (NOAA) compilation of flood damage since 1903 suggests that, as of 2020, the average cost per year in flood damage in the United Sates was $32.1 billion, adjusted for inflation. By comparison, the periods 1905-1914 and 1950-1959 had $3.2 billion and $4.4 billion worth of flood damage (adjusted for inflation). Floods are natural and inevitable, as demonstrated in this chapter. A congressional report perhaps summed it up best in noting that "floods are an act of God. Flood damages result from acts of people."

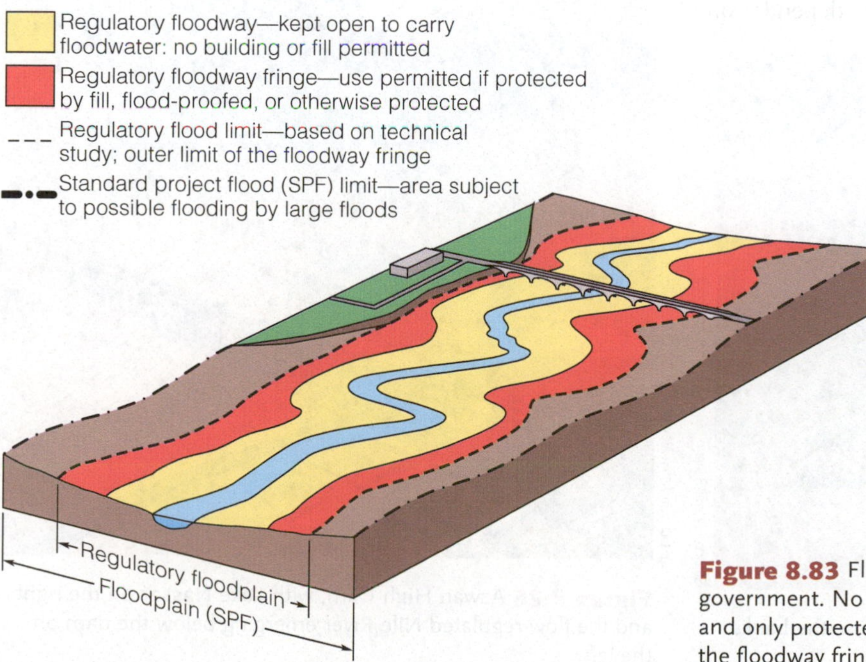

Figure 8.83 Fluvial flood- hazard areas as defined by the U.S. government. No building is allowed in the regulatory floodway, and only protected or raised structures are allowed to be built in the floodway fringe area.

8.1 The Nile River–Three Cubits between Security and Disaster

Flooding is a natural part of the hydrologic cycle and, as farmers, hydrologists, and geologists know, it can be beneficial.

The annual flood on the Nile River, the longest river in the world (depending on how one measures the Amazon), enabled civilization in Lower Egypt for at least 5,000 years. Eighty percent of the water for these floods came from the faraway Ethiopian Highlands, and it replenished the soil moisture and deposited a layer of nutrient-rich silt on the floodplain (**Figure 8.84**). The Greek historian and naturalist Herodotus, in 457 B.C.E., recognized the importance of annual flooding when he spoke of Egypt as "the gift of the Nile." The priests of ancient Egypt used principles of astronomy to calculate the time of the river's annual rise and fall. Records of the height of the flood crests date back to 1750 B.C.E. (some say to 3600 B.C.E.), as measured with stepped wells connected to the river and known as "nilometers" (**Figure 8.85**). Water height was measured in cubits, 1 cubit being the distance from one's elbow to the tip of the middle finger, approximately 0.5 meter or 20 inches.

The annual flooding cycle has changed since completion of the Aswan High Dam on the Upper Nile in southern Egypt in 1970 (**Figure 8.86**). The dam's purpose was to increase the extent of arable land, generate electricity, improve navigation, and control flooding on the Lower Nile River. It is accomplishing all of those goals. Now that the flooding is controlled, crop maintenance depends on

Figure 8.85 The Nile outside of Cairo as it flows past the pyramids. Inset is a schematic view of a Nilometer, a stepped well that allowed the measurement of river stage or height. Water height was measured in cubits, a cubit being the distance between one's elbow and the tip of the middle finger (usually about 18 inches or half a meter). The environmental interpretations shown here were used during the time of Pliny the Elder (23–79 C.E.) of Mount Vesuvius fame.

Figure 8.84 Rich, irrigated, agricultural fields line the banks of the Nile River in Egypt.

Figure 8.86 Aswan High Dam, with Lake Nassar on the right and the flow-regulated Nile River emerging below the dam on the left.

irrigation. Severe environmental consequences have resulted, including depletion of the soil nutrients that are beneficial to farming (necessitating the use of artificial fertilizers) and the proliferation of parasite-bearing snails in irrigation ditches and canals. *Schistosomiasis* is a human disease found in communities where people bathe or work in irrigation canals that contain snails serving as temporary hosts to parasitic flatworms known as schistosomes (**Figure 8.87**).

These creatures, commonly called "blood flukes," invade humans' skin when people step on the snails while wading in the canals. The flukes live in their human hosts' blood, where the females release large numbers of eggs daily and cause extensive cell damage. Next to malaria, schistosomiasis is the most serious parasitic infection in humans. Its outbreak in Egypt is a result of controls on the river that prevent the generous floods that formerly flushed the snails from canals and ditches each year.

Another negative consequence is the loss of a huge shrimp fishery off the Nile delta; the Aswan Dam traps the nutrients that encouraged the growth of plankton, upon which the shrimp larvae fed. Every dam has a finite useful lifetime, which is determined by how fast the lake it impounds fills with sediment. Lake Nasser, behind the Aswan Dam, is rapidly filling with the (precious) sediment that formerly enriched the floodplain of the Lower Nile and the waters off the delta.

iStock.com/Sinhyu

Figure 8.87 Schistosomes, known as blood flukes, viewed in a petri dish using a microscope. Adult worms can exceed a centimeter in length.

8.2 A Very Wet Fall

New England is not the kind of place one imagines floods that take out a thousand bridges, destroy towns of 6,000 people, and remove miles of roadway. Instead, people think of Vermont and New Hampshire as pastoral places where snow blankets the landscape in winter and gentle rains nurture the maple trees that provide sap for syrup in the spring and flaming orange hillslopes every fall. Not so in the fall of 1927.

October 1927 was dreary and wet in New England. It rained and rained. Many towns got twice their average rainfall. By early November, the ground was sodden and the trees, having lost their leaves, weren't pumping water from the ground. Groundwater levels were on the rise. Then, on November 4, 1927, the heavy rain began, driven by moist, tropical air from the south trying to override cold air from the north. As the warm air rose over the Green Mountains of Vermont, the moisture was squeezed out of the saturated atmosphere. Over the next 48 hours, mountainous parts of Vermont saw more than 20 centimeters (8 inches) of rain fall from the sky; even the lowlands recorded 10 to 15 centimeters (4 to 6 inches) of rain. Because the ground was already soaked from the precipitation in October, much of this new rainfall ran off into creeks and streams, driving river levels up. Riverside homes that had stood for 150 years were swept away, railroads were twisted into pretzels, and bridges were ripped from their piers and swept downstream (**Figures 8.88** and **8.89**).

That flood is now the legend to which every other flood in this region is compared. On the Winooski, the

Fairbanks Museum and Planetarium

Figure 8.88 The 1927 flood picked up houses and dropped them in some unusual places. This robust post and beam farmhouse survived the half-mile trip from its foundation to the railroad tracks intact. How do you think today's balloon-framed homes might do in a similar situation?

© Fletcher Free Library Burlington VT

© Vermont State Archives and Records Administration

Figure 8.89 Rampaging flood waters carried away more than 1,000 bridges in the tiny state of Vermont, devastating the transportation network. (a) This photo was taken just before the bridge connecting the cities of Burlington and Winooski washed away. (b) After the flood, a temporary pontoon bridge linked the mills of Winooski (shown in the background) with the city of Burlington until a permanent bridge could be rebuilt the next year.

main river draining northern Vermont, the flood of 1927 was the flood of record, the largest to affect the basin since western settlement. At the flood's peak, 3,200 cubic meters (113,000 cubic feet) per second of water was flowing downstream, 2.5 times the second largest flood on this river, and almost 70 times the average flow. Simple calculations suggest that 1927 was the 100-year flood, but this is misleading because it does not plot in line with other floods on the flood-frequency curve. This was an odd event, a perfect combination of widespread, heavy rain on saturated ground. Miraculously, only 84 Vermonters and 111 people in New England died in the flood, but the region was forever changed. It took years to rebuild the lost bridges, and in many places rivers found new paths, leaving fields as islands separated from their farms.

How rare was the 1927 flood? Lake mud answers that question: It was not that rare. Geologists from the University of Vermont ventured onto frozen lakes for several winters, bringing back cores of lake sediment that had been secreted on lake bottoms for the last 14,000 years since the glaciers left (**Figure 8.90**). When they cut open those cores, they found lots of black, sticky, stinky mud. However, every once in a while, they cut open a core to find a layer of bright gray

© Anders Noren

Figure 8.90 Collecting 10,000 years' worth of lake sediment is serious business easily accomplished off the winter ice. Here, one author (Bierman, right) and University of Vermont graduate student Josh Galster (left), working on a warm spring day, struggle to loosen a piece of coring equipment before the ice melts.

Figure 8.91 Once the cores are cut open, storm layers clearly stand out from the black, organic-rich, and foul-smelling pond sediment. Photo shows a 10-centimeter-thick, light-colored layer of sediment likely washed into Ritterbush Pond by a big storm about 6,800 years ago. At the top of the image are two very thin storm layers.

sand carried into the lake when big storms struck the drainage basin (**Figure 8.91**). By dating these sand layers from lakes all over New England, the geologists, led by Anders Noren, determined that very big storms hit New England, on average, every few hundred years. What's even more interesting is that the frequency of these storms has changed over time, and if the past is a clue to the future, the forecast for the next 600 years is an increase in big storms.

The New England flood of 1927 followed an even more destructive multiday flood in the Mississippi River Valley in April, perhaps the worst flood in U.S. history. The country as a whole suffered a very bad wet-weather year in 1927. Before the Mississippi flood, the Federal government did little to help natural disaster victims. Afterward, Federal relief programs began.

Everyone Needs Water

Water is a basic human need. Without freshwater, you can live only a few days. Water slakes our thirst and cleans our skin. It cools us in the heat of summer and keeps the plants that feed us from wilting. The people of the world get the water they need in very different ways. Some gather untreated water from hand-dug wells, whereas others lower buckets down more formal, masonry wells.

iStock.com/Hadynyah

1 Indian woman drawing water from a well in the Thar Desert, Rajasthan, India.

iStock.com/Kavram

2 The Pont du Gard Roman aqueduct as built 2000 years ago to carry water over 50 kilometers (31 miles) to the city of Nîmes in Provence, France, for drinking, public baths, and fountains.

Flooding from a 2016 typhon in Japan carries large amounts of eroded soil downstream.
iStock.com/Rssfhs

USGS

4 Dam removal is becoming more common. The Glines Canyon Dam on the Elwha River in Washington State was one of the first large dams to be removed in the United States. An excavator breaches the dam as sediment erodes from the reservoir.

Paul Bierman

3 Groundwater under pressure soaks two drillers as they try to cap an artesian well in western Massachusetts. The water is rising over 100 feet (30 meters) from a deep, glacial, gravel aquifer fed from the highlands and capped by nearly impermeable silt and clay.

iStock.com/Matt Hunt

5 Sometimes life keeps going right through the flood. In Bangkok, Thailand, a delivery person rides a motorbike through streets flooded in the rainy season.

Summary Water

1. Precipitation and Infiltration

LO1 Describe the primary controls on the distribution, types, frequency, and intensity of precipitation and what becomes of precipitation when it lands on the Earth

a. Precipitation Variation

Precipitation (rain or snow) is the result of the atmosphere being saturated with water, which condenses then falls to Earth. In tropical regions, water evaporates from warm oceans and generates intense downpours. Desert regions, downwind from cold ocean currents, may experience precipitation only once every few years. Storm systems control the intensity of precipitation.

b. Effect of Topography on Precipitation

The orographic effect describes why the upwind side of mountains experience more precipitation than downwind areas. Air masses rise upwards over high terrain, cool, condense their moisture and precipitate. This effectively dries the air, creating arid lowland regions downwind of mountains.

c. Infiltration and Runoff

Some precipitation infiltrates the soil and fills pores between soil grains, accumulating as groundwater. Soil conditions that prevent infiltration lead to runoff, during which precipitation flows across the ground surface into nearby receiving waters. Conditions that lead to runoff include saturated soils, frozen soils, and precipitation landing on impermeable surfaces like concreate, compacted hiking trails, or clay-rich soils.

d. Frequency, Intensity, and Duration of Precipitation

Precipitation frequency varies greatly depending on the region, with some tropical regions experiencing daily rainfall, versus desert regions that go long stretches of time without any rain. When rainfall does occur in these desert regions, the impact may be severe (flash floods, debris flows) due to intense precipitation events and lack of vegetation to help slowly infiltrate water to the underlying ground. Rainfall intensity and duration also impact how precipitation affects the land. For example, flooding is more common when there is high-intensity rainfall.

Study Questions

i. Give the three reasons why some regions experience more rain than others.

ii. List the three conditions that result in reduced soil infiltration.

2. Surface Water

LO2 Summarize the distribution of fresh water on Earth and its sensitivity to environmental impacts

a. Lakes

Several types of lakes exist, including salty and freshwater. Some persist only during wetter climates. The type of lake depends on the balance between precipitation and runoff into the lake and evaporation and outflow from the lake. Lakes are very dynamic, sometimes overturning because of changing water temperatures and thus density differences between deep and shallow water layers. The health of lakes is described based on nutrient concentrations, water clarity, and ability to sustain life.

b. Rivers

Rivers are pathways (channels) in which water moves from highlands to lowlands. Ranging from streams with only small trickles of water to rivers raging with high discharge, rivers change the shape of Earth, and the way humans live. The Grand Canyon was carved by the Colorado River over millions of years, while hydroelectric dams harness the power of moving water to create electricity for humans.

c. Measuring Flow

Flow is measured by a stream's discharge or stage, as a function of time. Hydrographs plot flow or stage over time, allowing hydrologists to study and predict flooding within drainage basins. In the United States, the USGS monitors a large network of stream gauges that measure discharge, velocity, and stage at sites across the country.

Study Questions

i. Consider water balance and describe how this helps explain the different types of lakes that exist.

ii. In what ways do rivers contribute to human society?

3. Floods: When There Is Too Much Water

LO3 Explain why and how floods happen and their distribution over time and space

a. Types of Floods

The type of flood (flash, riverine, ice-jam) depends largely on the environmental conditions in which the flood is occurring. Flash floods occur most commonly in small drainage basins with moderate or high topographic relief (large changes in elevation, steep slopes). Ice-jam floods, on the other hand, are most likely to occur during late winter and early spring in regions where temperatures below freezing are sustained long enough for river ice to form. The type of flood determines how hazardous the event may be.

b. Effects of Floods

Inundation is the most common impact of floods. However, floods not only carry water but sediment and large woody debris as well. The effects of this kind of disturbance can be disastrous. Bridge scour can lead to bridges collapsing, and flood-soaked drywall can make homes uninhabitable. Lateral channel erosion can undermine channel banks leading to collapse and erosion.

c. Flood Frequency

Flood frequency or recurrence interval is how often a flood of a certain magnitude can be expected to occur, on average, at a particular location. This information is derived from long-term flood records and is valuable for engineering and for infrastructure design that can withstand natural disasters expected in a particular location over a given amount of time. In the face of rapidly changing climate, such calculations are increasingly uncertain.

Study Questions

i. Give the three examples of types of floods, define each, and explain where and when they are most likely to occur.
ii. How are reoccurrence intervals calculated?
iii. What is the meaning of a 20-year flood?

4. Groundwater

LO4 Predict where you will find groundwater and how it will move

a. Groundwater Supply: Location and Distribution

Groundwater is by far the cheapest and most efficient source for municipal water supply because it is easy to obtain, available even during periods of drought, and less likely to be contaminated than surface water.

b. Water Table and Flow Directions

In the subsurface, lies the zone of aeration, or the unsaturated zone, where voids in soil and rock contain only air and water films. During rainfall, water infiltrates through this zone into the zone of saturation. The boundary of these two zones is the water table. In the zone of saturation, if the porosity, volume, and permeability of the material are sufficient, there is an aquifer. Aquifers recharge surface springs and wells.

c. Pressurized Underground Water

Confined aquifers are under pressure, known as the hydraulic head. Wells that penetrate confined aquifers, called artesian wells, can flow without being pumped.

d. Darcy's Law

Groundwater flows through aquifers along the potentiometric surface, or from high pressure/elevation to low pressure/elevation. Darcy's Law estimates the rate of flow in volume per time as the product of hydraulic conductivity, hydraulic gradient, and the cross-sectional area of an aquifer.

e. Groundwater-Saltwater Interaction

Saltwater intrusion occurs due to the density difference between saltwater and freshwater. In coastal regions, when the elevation of the groundwater table drops, saltwater migrates upward, pushing the freshwater/saltwater interface upward. Saltwater is more dense than freshwater, so it sits below the freshwater "lens."

Study Questions

i. Explain Darcy's Law and the variables within it.
ii. What defines an artesian well?

5. Water Quality and Quantity

LO5 Outline how planners manage water quality and water supply

a. Water Quantity

The way water is used affects the reuse potential of return flows to streams, rivers and lakes. For example, power generation and irrigation are two of the largest users of freshwater globally; however, 98% of water used for power generation is returned to streams and other reservoirs, while almost all water used for irrigation is consumed. Increases in population also affect freshwater use, making it important to manage groundwater resources responsibly (i.e., maintaining a withdrawal below or at sustained yield) to preserve the resource.

b. Water Quality

Water quality considers the concentration of many elements, molecules, and organic compounds that can impact the utility of water as a resource for humans and other life. The U.S. EPA is responsible for administering the National Primary Drinking Water Regulations and Maximum Contaminant Levels, drinking water quality standards for all public water supply systems.

c. Cleanup, Conservation, and Alternative Water Sources

Global water use increased at a rate of 1% per year since the 1980s due to growing population and consumption. Six billion people globally will likely suffer from water scarcity by 2050. To avoid this disastrous outcome, alternative water sources, like the desalinization of saltwater, are being explored. Policies to protect and conserve the freshwater we have on Earth are crucial.

Study Questions

i. What makes water-use consumptive versus nonconsumptive?

ii. Name the three common water pollutants and the danger they present to human health.

6. Human Interactions with Water

LO6 Describe how water both supports and endangers human society

a. Levees

Levees, both natural and artificial, are raised channel margins that help prevent flooding when water stage increases. While some artificial levees are useful to prevent flooding in adjacent regions, levees funnel this floodwater downstream to regions with no artificial levees, harming those areas. As the frequency of intense storms (such as stronger hurricanes) increases, the risk of levees failing becomes greater.

b. Dams

Dams have many useful functions to society, including cheap electricity, flood control, and drought mitigation. However, dams destroy habitats, disturb natural cycles of flooding and sediment deposition, and disrupt spawning pathways for migratory fish. Additionally, poorly maintained dams increase the risk of dam failure, which could lead to major flooding, loss of reservoirs, and disruption of drinking water supplies.

c. Insurance, Flood-Proofing, and Floodplain Management

The average cost annually of flood damage in the United States exceeds $32.1 billion; nonetheless, development in flood-prone areas continues. While flood insurance is subsidized for those who live in homes built before 1968 in floodways, floodplain management is the most sustainable way to avoid future flood damage.

Study Questions

i. How are levees both helpful and harmful to mitigating flood damage in communities?

ii. Besides inundation, what are three examples of the impacts of floods and the damage they cause?

Key Terms

Antecedent moisture
Artesian wells
Artificial levee
Base level
Baseflow
Bridge scour
Capillary fringe
Colorado brown stain
Cone of depression
Confined aquifers
Darcy's law
Discharge
Drainage basin
Drainage divides
Drawdown
Eutrophic
Evapotranspiring
Flash floods
Flood frequency
Floodplains
Floods
Floodway
Fossil water
Forever chemicals (PFAS)
Gaining stream
Gradient
Groundwater mining
Hydraulic conductivity
Hydraulic gradient
Hydraulic head
Hydrograph
Ice-jam floods
Infiltrates
Infiltration
Inundation
Maximum contaminant level (MCL)
Maximum contaminant level goals (MCLGs)
Mesotrophic
Natural levees
Oligotrophic
Overturning
Perched water table
Permeability
Playa
Porosity
Potentiometric surface
Precipitation
Rain-on-snow events
Rainfall duration
Rainfall intensity
Recharge areas
Recharge mound
Recurrence interval
Regulatory floodplain
Residence time
Riparian
Riverine floods
Runoff
Saltwater intrusion
Specific retention
Specific yield
Storm surge
Sustained yield
Tropical storm
Unconfined aquifers
Unsaturated zone
Water table

"I grew up in a family of peasants, and it was there that I saw the way that, for example, our wheat fields suffered as a result of dust storms, water erosion and wind erosion; I saw the effect of that on life–on human life."

—Mikhail Gorbachev

9

Erosion

Learning Objectives

LO1 Explain how the balance of driving and resisting forces controls the occurrence of mass movements

LO2 Compare the processes leading to rockfalls, landslides, and lateral spreads and the effects of these mass movements on people

LO3 Explain why, when, and where debris flows will occur

LO4 Explain what conditions make snow avalanches likely and thus how to mitigate their impact

LO5 Reflect on why and where subsidence happens and how it can be managed and/or avoided

LO6 Explain how the risks of mass movements to people and society can be reduced using your understanding of how mass movements work

Debris flow after heavy rains in 2015 buries steep village road in Sicily, Italy.

iStock.com/LuckyTD

A Bad Slice on the Golf Course

Golf's explosion in popularity in the late 1990s led to many new golf courses. Some are in resort communities, but others are in urban and suburban areas, like the beautiful Ocean Trails Golf Course south of Los Angeles on the Palos Verdes Peninsula, since renamed the Trump National Golf Club-Los Angeles. Geologists know the peninsula well for its many landslides, due to several factors, including: it is uplifted well above sea level, the sea constantly batters and erodes the cliffs' base with waves, and weak rock underlies the area.

It is easy to see in 30- to 50-meter (100- to 165-foot) cliffs, where the layers are exposed, that the stratified sedimentary rock underlying the course inclines gently toward the sea. Just as a slice of bread slides down a breadboard when tilted, some weak layers have slid toward the sea in the past, carrying houses and roads with them. The area is dotted with landslides up and down the coast.

Ocean Trails Golf Course was almost ready for its early summer 1999 opening when, without warning, a fissure opened parallel to the cliff, and 300 meters (985 feet) of the 18th fairway slid into the ocean along with walking and biking trails, as well as part of a Los Angeles County sewer line (**Figure 9.1**). This slide should have been no surprise. The 18th fairway was built on an ancient landslide, well known to local geologists. The area was underlain at depth by a very thin layer (only a few inches) of weak clay known as bentonite, which when wet loses much of its strength. Some believed that the sewer line had been leaking prior to the slide.

Local resident Tony Baker and his dog were temporarily stranded on a precarious, 215-meter (700-foot)-long island between fissures. "I heard crumbling earth and started seeing dust rising," Baker said later. "The trail started cracking up. I was doing a little running around, jumping over big cracks. I just found a place and hunkered down. I wasn't sure if I was going to make it back."

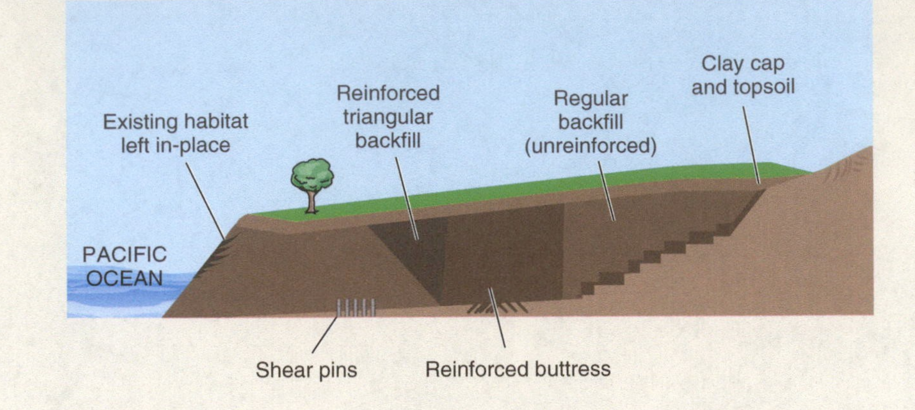

Figure 9.2 Cross-section of the slice repair. Clay cap keeps out rainwater, and shear pins anchor the repair to stronger material below. Geofabric-reinforced buttresses resist driving forces.

Landslide remediation ensued. The plan was ambitious. Leave the landslide block by the coast intact and strengthen the area between it and stable ground (**Figure 9.2**). Geologists and geotechnical engineers first installed 116 steel shear pins. Each of these was 6 meters (20 feet) long and 1 meter (3 feet) wide, and filled with concrete. They anchored the slide at its base to stable material below. Then, engineers dug out the slide area and backfilled the hole with fill reinforced by exceptionally strong cloth materials known as geofabrics (**Figure 9.3**). This patch cost \$61 million at the time, about half the original cost of building the courses (\$126 million).

But after 3 years of wrangling, the developers and lenders reached an impasse, and the landslide stabilization work stopped. The beautiful golf course, now only 15 holes long, was blemished by the scar of the landslide and stabilization work. Enter Donald Trump, the New York entrepreneur, real-estate magnate, enthusiastic golfer, and now former president. He bought the troubled project (at a bargain price, \$27 million) in 2002 and vowed to finish the project by the summer of 2003 (not quite a promise fulfilled). On January 20, 2006, the Trump National Golf Club officially opened with all 18 holes. The "buttress fill" designed to stabilize the "slice"

Figure 9.1 The slice took not only the golf course but a large excavator with it.

of the safest coastal places to be standing as a result of the many layers of high-strength geosynthetics sitting below their feet. When asked about this reinforced hole by a reporter, course owner Donald Trump remarked, 'If I'm ever in California for an earthquake, this is where I want to be standing.' ("Golfing Atop a Landslide: A Signature Hole Is Born at Trump National Golf Course," Rich)

MediaNews Group, Inc

Figure 9.3 Remediation of the slide (on the right side of this aerial photograph) was a massive undertaking.

created by the landslide is buried under the last two holes. The total cost for 18 holes was a mere $300 million. The most expensive golf course in the world, at least at the time.

As golfers take in the breathtaking view from the 18th hole, few will ever know that they are playing on the most expensive hole ever built. Even fewer golfers will realize that this hole is one

questions to ponder

1. If they had been careful, what geological and landscape clues might the designers of the golf course have noticed that would have alerted them to potential slope stability issues before they built the course?
2. If rainfall greatly increases or decreases due to climate change, how do you imagine this will impact potential for future landslides along this part of the California coast?

9.1 The Basics of Mass Movements (L01)

Although it appears stable, Earth's surface is constantly in motion. Most motion is subtle, slow, and unless you have a lot of patience, difficult to see and measure. But sometimes the Earth beneath us gives way quickly and the result is disastrous. Earth movements come in all shapes and sizes, from single grains moving downslope when a gopher moves sand from a burrow to the failure of entire mountainsides during volcanic eruptions. All of these are known as **mass movements**, downslope movements of soil and rock under the direct influence of gravity.

Mass movement processes include landslides, rapidly moving debris flows, slow-moving soil creep, and rockfalls of all kinds. Annual damage from landslides alone in the United States is estimated at between $2 and $4 billion, up from an estimated annual cost of $1 billion in 1980. This damage is widespread, affecting much of the country (**Figure 9.4**). It is estimated that landslides affected 4.8 million people and caused 18,000 deaths between 1998 and 2017 worldwide. If we include other ground failures, such as subsidence,

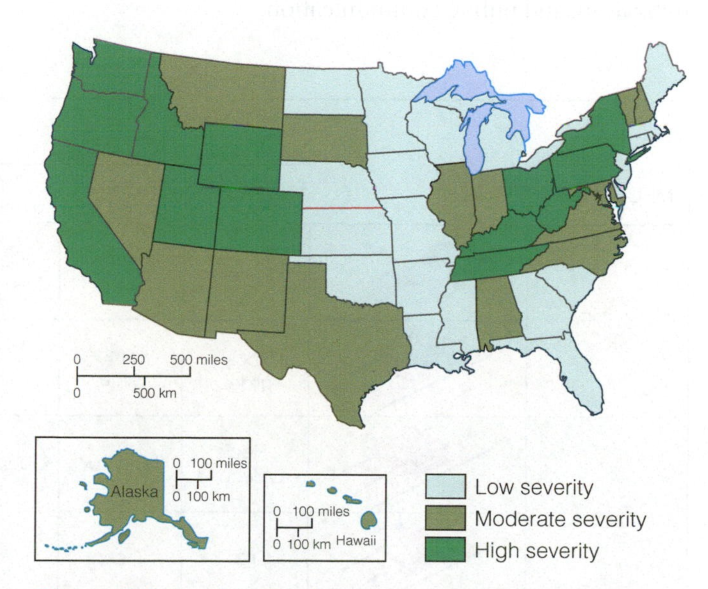

Figure 9.4 Severity of landsliding in the United States. Note that the most serious problems are found in the Appalachian Mountains, the Rocky Mountains, and the coastal mountain ranges of the Pacific Rim. (Based on data from J. P. Krohn and J. E. Slosson, 1976. Landslide potential in the U.S. California Geology 29(10))

expansive soils, and construction-induced slides and flows, total losses are, on average, many times greater than the annual combined losses from earthquakes, volcanic eruptions, floods, hurricanes, and tornadoes–unless, of course, it happens to be the year that Hurricane Katrina strikes New Orleans or an earthquake devastates Japan.

Landslides, debris flows, and avalanches are terrifying environmental phenomena that test our certainty whether the Earth beneath us will always be stable. As climate changes so will the distribution and intensity of rainfall. With an increased frequency and duration of hurricanes and other strong storms, some parts of the world will see more landslides and debris flow, the result of a landscape responding to a new and different hydrological regimen.

Mass movement and subsidence are global environmental problems. Earthquake-triggered landslides in Gansu Province in China killed an estimated 200,000 people in 1920, and debris flows left 600 dead and destroyed 100,000 homes near Kobe, Japan, in 1938. These events pale in comparison to the death toll from a volcanic mudflow (a lahar) in Columbia at the Nevado del Ruiz Volcano. That one event killed over 23,000 people when their town was buried in mud (see Chapter 1).

The goal of this chapter is to reveal the variety of ways in which mass movements occur and place people and their property in harm's way. We consider the relevant physical processes and means by which different types of Earth-movement hazards can be effectively mitigated or avoided.

Mass movements are driven by gravity and its interaction with earth materials including soil, rock, and snow. Understanding what drives these movements is key to reducing hazards and remediation when the time comes. Classification is an important tool for hazard evaluation, mitigation, and public communication.

The classification of mass movements used in this textbook is similar to the one widely used by engineering geologists and soil engineers (**Figure 9.5**). The bases of the classification are:

- The type of material involved, such as rock, soil, or snow
- How the material moves, such as by sliding, flowing, or falling
- How fast the material moves; its speed.

For instance, soil creep occurs at imperceptibly slow rates; debris flows are water saturated and move swiftly; and slides are coherent masses of rock or soil that move along one or more discrete failure surfaces. Rock falls endanger those below steep cliffs in part because they happen so quickly. Where you are located matters when evaluating the hazards posed by mass movements.

9.1.1 The Mechanics of Mass Movements

Let's examine the physics of mass movements, because by understanding how mass movements work you will be better able to understand how they can be predicted and how the hazards they present are best managed.

Mass movements occur when the forces always present within sloping ground become unbalanced (**Figure 9.6**). There are only two forces to consider. One is **driving force**, the gravity pulling rock and soil downslope. The other is **resisting force**, the strength of the material making up the slope.

Failure occurs when the force that is pulling the potentially unstable material downslope (gravity) exceeds the strength of the earth materials that compose the slope. Because the force of gravity is constant, either the earth materials' resistance to sliding must decrease or the steepness of the slope must increase for sliding to occur. The presence of water is key because it has the potential to diminish the resisting force that makes slopes stable. The strong shaking of a major earthquake can also momentarily change the balance of forces on a sensitive (barely stable) slope and send much of it rushing downhill.

Now, let's delve deeper (no pun intended) into this **force balance**. First, consider the driving force. Gravity is directed downward, toward the center of Earth. Indeed, it's what keeps your feet planted solidly on the ground (see Figure 9.6a). Now imagine a block of material on a 15° slope (see Figure 9.6b). We can partition that gravity force into two separate component forces: one is oriented downslope, trying to get the block to slide; the other is pushing onto the slope. The latter, called the **normal force**, helps anchor the

MECHANISM		MATERIAL: Rock	Fine-grained Soil	Coarse-grained Soil	VELOCITY
SLIDE		Slump	Earth slump	Debris slump	SLOW
		Block glide	Earth slide	Debris slide	RAPID
FLOW	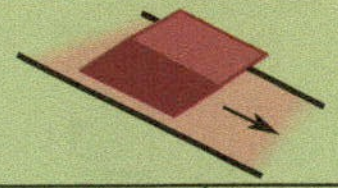	Rock avalanche	Mudflow	Debris flow	VERY RAPID
		Creep	Creep	Creep	EXTREMELY SLOW
FALL		Rockfall	Earthfall	Debris fall	EXTREMELY RAPID

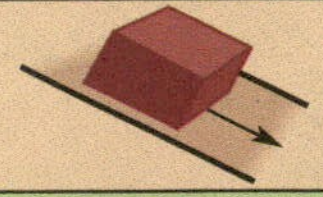

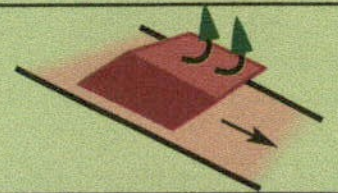

Figure 9.5 Classification of mass movements by mechanism, material, and velocity.

The resisting force also has two components, both familiar through life experiences. One component is cohesion, or the material's tendency to stick together. Consider clay. It has **cohesion** and can hold a vertical face; you can mold and shape clay in many ways and it will still stick together.

Now consider sand, a common **granular material** with no cohesion. It has what is termed **frictional strength**, the result of frictional interaction–largely just the catching of tiny bumps and corners on individual sand grains where they come in contact with one another. The harder you push the grains together, the stronger sand becomes, at least up to moment you begin crushing them. As a sandy slope steepens and the normal force holding grains together decreases, you soon reach a critical point. For sand, that critical point, where the friction between grains is no longer able to resist the downslope pull, is about 35°. This is termed the **angle of repose**. Try as you might, you'll never be able to stack dry sand much steeper than 35°, nor can nature (**Figure 9.7**).

Many materials that experience landslides have both cohesion and frictional strength. Even granular materials such as sand and gravel can have small amounts of **effective cohesion** where networks of tree roots bind the grains together. The strongest earth materials have large amounts of both frictional and cohesive strength. Rocks are stronger than soils because they have very high cohesive strengths. For sedimentary rocks, cohesion is provided by the natural cement holding grains together. For igneous and metamorphic rocks, a range of possible factors make them strongly cohesive, including rapid cooling from molten conditions to form homogeneous groundmass or glass, or the complete interlocking of crystal grains in more slowly cooling rocks. For snow, it's ice holding flakes together. For sediment and soil, the clay they often contain provides some cohesion.

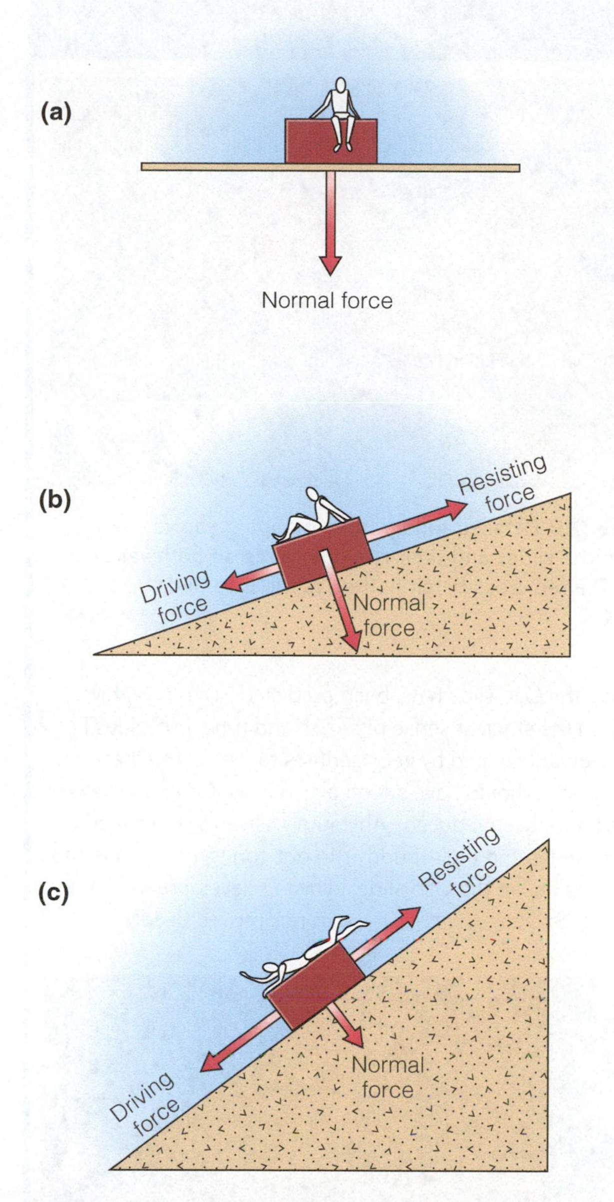

Figure 9.6 Landslide physics is all about balance. In the diagrams, the red block represents the landsliding material. (a) In the middle of the Great Plains, all of the gravity force is directed straight down and there are no landslides. Long arrows mean big forces. Small arrows mean small forces. (b) As slopes steepen, some of the gravity force is directed downslope and makes up the driving force. Some of the gravity force is directed into the slope as the normal force. (c) Here's a slide about to happen on a steep slope. Look carefully at the length of the arrows, compared with (b). What happens to the driving, normal, and resisting forces? How do you know this slope is about to fail?

slide block. Now, let's crank up the slope to 45° (see Figure 9.6c). The partitioning of forces has now changed. More of the gravity force is oriented downslope and less into the slope. Imagine what the partitioning must look like on a 70° slope. The steeper the slope, the more the gravity force is directed downslope, and the more likely the landslide will fail.

Figure 9.7 Pile of sand at a construction site sits at the angle of repose, about 35°.

When a Force Balance Matters

Mark Reid/USGS

Figure 9.8 The Oso landslide from the air. The slide began in glacial sediment underlying the terrace on the right and moved downhill and across the river, damming the river and burying dozens of homes on the far bank in mud and debris.

Ralph A. Haugerud/USGS

Figure 9.9 Detailed topographic mapping shows many older slides and their run-out zones on both valley walls. The red cross-hatch pattern is the Oso Slide of 2014.

Of the 43 people who died when the Oso landslide swept across the North Fork Stillaguamish River in Western Washington at 10:37 a.m. on March 22, 2014, it's a good bet that none of them were thinking about the force balance on the slope of glacial sediment above them (**Figure 9.8**). The development in which they lived sat on the valley bottom alongside the river. Its construction had been permitted by the local government. But all was not as safe as it seemed. The area had a history of landslides and geologists had repeatedly warned of landslide hazards in the area (**Figure 9.9**). The Oso landslide was not an isolated event but one in a series of many in the Stillaguamish River Valley, all underlain by the same geology.

Sand and clay, deposited at the end of the last glaciation by ice that once filled the valley, failed in a massive landslide when exceptional rainfall caused the factor of safety to dip below 1. February and March were record-breaking wet months, with 150% to 200% of mean annual precipitation. Groundwater levels were high, reducing the resisting force by increasing the pore pressure.

Once the hillside above Oso began collapsing, the excess water had another deadly effect. It increased the mobility of the slide mass, which on its trip downslope liquefied to a very mobile debris flow, smothering homes and their residents in thick, viscous mud (**Figure 9.10**).

Could the Oso slide have been predicted? That is highly unlikely in the strictest sense of a date and time, but slides like it had been anticipated by geoscientists for years and the record rainfall should have set off alarm bells. Could the hazard it posed have been reduced? Absolutely. Given the number and extent of slides and their muddy run out zones, mapped in the valley years before the Oso slide, allowing development there was a risk. Sadly, in March 2014, that risk proved deadly.

US Air Force Photo / Alamy Stock Photo

Figure 9.10 Rescue workers move through the dense, wet mud trying to locate victims buried in the liquified slide debris at Oso.

Geologists and engineers have devised many ways to quantify this balance of forces, most of which involve physical measurements and mathematical models. All of this experimentation and computation can be reduced to one number, referred to as the **factor of safety**, which is a ratio between the resisting and driving forces. When the two forces are equal, the factor of safety equals 1.0 and the slide is teetering on the brink of failure. When resisting forces are greater than driving forces, the factor of safety is greater than 1.0 and all is well. The moment the resisting forces are less than the driving forces, the factor of safety is less than 1.0 and a landslide starts. Most modern building codes require that manufactured and natural slopes have a factor of safety of 1.5 or greater to provide a margin of error in the measurements and calculations.

9.1.2 Triggering Mass Movements

To trigger a mass movement, the factor of safety must decline to less than 1.0 and driving forces must exceed resisting forces. A short-term change in this balance is most often driven by water, as in the case of the Oso landslide (see box on prior page). In wet years, mass movements are more common. Why is this?

Let's go back to force balance and your personal experience in swimming pools and at the beach. When rainfall saturates earth materials with water, some of the weight of the grains is buoyed up by that water, just as when you are floating in water. This buoyancy, caused by the pressure of groundwater in the soil and rock pores and referred to as **pore pressure**, partially offsets the gravity force holding the material on the slope.

Because the water pressure supports a portion of the block's weight, it reduces the normal force, thus reducing the frictional strength of the material and the all-important resisting force. Sometimes this reduction in strength can be just enough to tip the balance, drop the factor of safety below 1.0, and trigger the slide. The greatest effect of heavy rains, raising groundwater tables and increasing pore pressure, is on mass movements that are made up of low-cohesion material such as sand, gravel, weathered rock, and snow. Water usually saturates slopes after heavy rain, but occasionally human errors (such as leaky swimming pools and sprinkler systems) are the culprit.

Other human triggers of mass movements include cutting away the toe or base of the slide block (which is equivalent to steepening the slope) and adding weight at the top of a slope, such as by building a house, which increases the driving force. Snow avalanches are often triggered just by adding mass to a slope. That mass could be windblown snow, another snowstorm, or the weight of a skier on a slope just hanging in balance. Over the longer term, on geological time scales of millennia or more, minerals in rock and soil will weather, weakening the material, especially if the weathering products are clay. This is the underlying factor in many landslides in basaltic terrains such as those that plague parts of the Oregon Coast Range.

Earthquakes and the shaking they cause can trigger mass movements in several different ways. The shaking alternately increases and decreases the effective force of gravity. Shaking can also cause sediments to weaken as the internal structure collapses and frictional strength is overcome. The 2018 Alaska earthquake, M_w 7, caused numerous mass movements, including a failure that took out a highway north of Anchorage (**Figure 9.11**).

iStock.com/John Pennell

Figure 9.11 Mass movement triggered by the 2018 Alaska earthquake took out the highway north of Anchorage.

9.1.3 Subtle but Important Mass Movement

Most people think of mass movements as catastrophic events–landslides taking out entire hillslopes and rock falls burying roads in massive boulders–but most mass movement is far more subtle. It moves along at a pace you can't see with your eyes, but if you know what to look for and how to measure these slow changes on Earth's surface, you'll realize how important they are! And despite moving slowly, such ubiquitous mass movements have real consequences.

Creep, the gradual downslope transport of mass, is the most important, yet for most people the least appreciated, form of mass movement. Creep typically proceeds only a few millimeters per year, perhaps a tenth of an inch at most; almost imperceptibly. Snow can creep; soil can creep; weak rocks can creep. Creep involves any number of specific and different processes that are all typically lumped together in this one term. Creep is most prevalent and easiest to detect on steeper slopes.

Creep involves deforming or changing the shape of the slope materials. Some deformation of weak materials, such as soft clay, may be driven completely by gravity. For other materials, freezing and thawing or alternate wetting and drying of a hill slope, which causes upward expansion of the ground surface perpendicular to the face of the slope, causes creep (**Figure 9.12**). As the slope dries out or thaws, the soil surface drops vertically, resulting in a net downslope

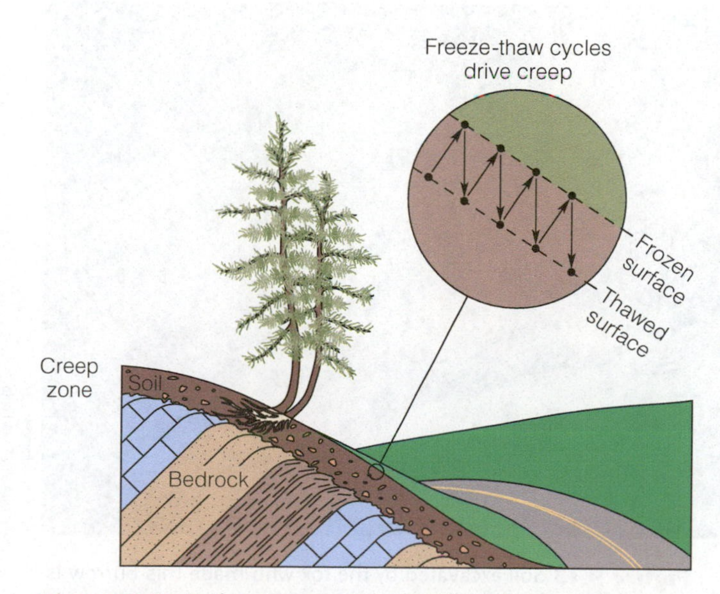

Figure 9.12 Soil creep moves material (including trees) slowly downslope

movement of the soil. Biota of all sorts contribute to soil creep, and burrowing animals can move soil onto steep slopes. More soil moved by animals ends up moving downslope than upslope; the net result is downslope transport (**Figure 9.13**). Trees are also an important catalyst for soil creep. Every time one blows over and its root wad tips up (**Figure 9.14**), soil moves, usually downslope.

How do we know soil creep is active? Bent trees, leaning fence posts, tombstones, and telephone poles, as well as the bending of rock layers downslope, are all evidence of soil creep (**Figure 9.15**). Homes built with conventional foundations 45 to 60 centimeters (18 to 24 inches) deep may develop cracks caused by soil creep. This is seldom catastrophic; typically, it's an expensive cosmetic and maintenance problem. The influence of soil creep can be overcome by placing foundations through the creeping soil into bedrock or more solid soil at depth. Soil creep matters to the Earth System. It delivers large amounts of sediment to streams and rivers. Think about creep as a conveyor belt of soil moving down just about every slope in the world.

Paul Bierman

Figure 9.14 Soil creep on an Oregon Coast Range scale. Here, in the Drift Creek Wilderness Area, geomorphologist Tom Dunne (1.8 meters [5 foot 10 inches]) shows just how large tip ups and root wads can be. This is an old-growth forest; when trees fall on the 25° slopes, soil moves.

9.2 Landslides, Rock Falls, and Lateral Spreads (LO2)

Landslides, rock falls, and lateral spreads are the mass movements people think about most often. These mass movements occur in different places, have different triggers, and cause different hazards and impacts on people and society. Below we consider each in turn.

iStock.com/Jmrocek

Figure 9.13 Soil excavated by the fox who made this burrow is headed downslope (to the right), an excellent example of soil creep that the pups don't seem to notice.

J M Barres/Alamy Stock Photo

Figure 9.15 Soil creep, the slow downhill movement of soil, is shown by bent trees in the Sierra de Guadarrama, Madrid, Spain.

9.2.1 Rockfalls and Rock Avalanches

Rockfalls, which occur when a mass of rock detaches from a cliff and falls to the bottom, are the simplest type of mass movement. They usually occur very rapidly and are among the most common mass-wasting processes. Road signs alerting motorists to watch for falling rocks are a frequent sight. Chunks of rock of all sizes fall from vertical or very steep cliffs, having separated from the main mass along joints, faults, foliation planes, and other weaknesses in the rock mass. Rocks are loosened by root growth, frost wedging as water freezes in cracks and expands, and heavy precipitation. A pile of rock debris may build up at the base of a cliff, forming a slope called **talus** (**Figure 9.16**). Two huge rockfalls in Yosemite Valley in 1996 killed 1 visitor and injured 14 others. The force of the shockwave from one of these rockfalls toppled nearby trees and stripped bark from more distant ones.

Some rockfalls can mobilize, crumble, and fan out downslope. Termed **rock avalanches**, they can leave lobate-shaped deposits on a valley floor (**Figure 9.17**). Some geologists speculate that massive rockfalls, especially those which begin with a steep vertical drop, can move exceptional distances because they are supported on a cushion of air trapped beneath the rock, rather like an air hockey puck sliding across the table when the blower is on. The Blackhawk rock avalanche in southern California failed many thousands, perhaps even tens of thousands of years ago. It got its start near the top of the mountains and dropped over 1300 meters (4000 feet) to the valley floor, travelling more than 8 km (5 miles) before coming to a stop. You can clearly see it today on the ground, but it's much more impressive from the air (**Figure 9.18**).

iStock.com/Евгений Харитонов

Figure 9.17 Rock avalanche in the Caucasus Mountains of Russia has run out onto the flat valley floor.

iStock.com/Raclro

Figure 9.16 Talus extends toward the photographer in this image from Rocky Mountain National Park in Colorado.

© Doc Searls

Figure 9.18 The Blackhawk rock avalanche (which is often called a rockslide or landslide) where it ran out across the valley bottom. It started as a rockfall in the mountains visible to the upper right. The run out is 8 kilometers (5 miles) long!

Figure 9.19 With climate warming, melting glaciers have left crumbling rock faces unsupported. Soon after the glacial ice melts, the rock gives way. Here, hikers point to the massive fractures on the face of Eiger Mountain, where hundreds of tons of rock fell soon after; above Grindelwald, Switzerland, July 2006.

Rockfalls may increase as a side effect of global warming. As the Earth warms, alpine glaciers are retreating ever more rapidly (most alpine glaciers have been in retreat since the Little Ice Age began waning several hundred years ago). As these glaciers melt away, they leave steep bedrock walls in the canyons they used to occupy. Without the buttressing of ice, these walls are often unstable. In summer 2006, the famous Eiger face in the Swiss Alps, a result of glacial retreat, collapsed, dumping 20 million cubic feet of rock on the remaining glacier below, equal to about half the volume of the Empire State Building (**Figure 9.19**).

9.2.2 Landslides

The term **landslide** encompasses all moderately rapid mass movements that have well-defined boundaries and move downward and outward from a natural or artificial slope. In other words, mass movements that move as a unit. Landslides can range widely in size from a small translational failure no bigger than a car to one of the largest landslides on record, the sector collapse of Mount St. Helens, Washington, in 1980, during which the entire north side of the mountain slid away (see Chapter 5).

Landslides occur in all 50 U.S. states and are an economically significant factor in more than 25 of them (see Figure 9.4). Areas with the highest slopes are at greatest risk because gravity is the force that drives landslides. For example, more than 2 million landslide scars dot the steep slopes of the Appalachian Mountains between New England and Alabama.

The costliest landslide in U.S. history occurred in Thistle, Utah, in 1983, completely wiping out the town below and creating an estimated $200-400 million worth of damage. Colorado's steep terrain makes it prone to frequent landslides, often in remote areas of the state which are difficult to monitor; the Colorado Geological Survey estimates that landslides create millions of dollars in damage in Colorado alone each year. Slope stability problems like this have such an impact that a national landslide loss reduction program was implemented by the U.S. Geological Survey in the mid-1980s. A high priority of the program is to identify and map landslide-prone lands.

Geologists recognize two kinds of landslides, distinguished by the shape of the slide surface. They are **slumps**, or **rotational slides**, which move on curved, concave-upward slide surfaces and are self-stabilizing, and **block glides**, or **translational slides**, which move on inclined slide planes (see Figure 9.2). A block glide moves as a unit until it meets an obstacle, the slope of the slide plane changes, or the block disintegrates into a debris avalanche or debris flow.

Slumps, the most common kind of landslide, range in size from small features a few meters or feet wide to huge failures that can damage structures and transportation systems. They are spoon-shaped as viewed looking down from above, and their concave-upward basal slide surfaces cause them to rotate backward as they move (**Figure 9.20**). The rotation aspect is key to the definition, and is observed in the compressed lower part of the landslide, near its end.

If you cleaned away the landslide and looked just at the **failure plane** (the surface on which the landslide moves), it

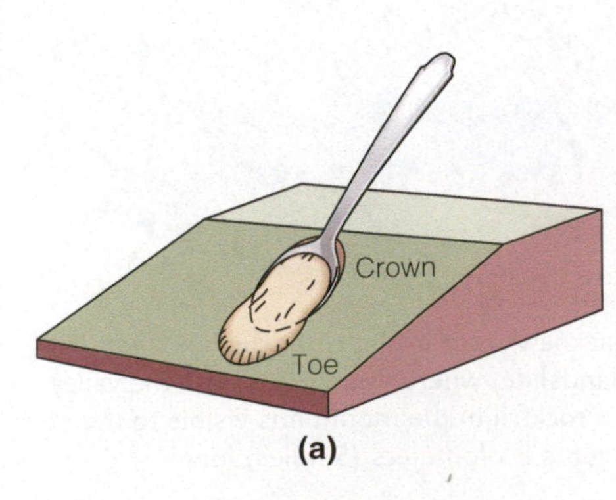

Figure 9.20 A rotational landslide. (a) An oblique stylized view showing the ideal spoon shape and crown and toe of the landslide. (b) An oblique view of a rotational landslide (slump), showing toe with earth flow and potential headward growth of the landslide.

would be curving and cupped. Slumps occur in geological material that is homogeneous, such as soil or intensely weathered or fractured bedrock or clay, and the slide surface cuts across geological boundaries. Slumps move about a center of rotation; that is, as the toe of the slide rotates upward, its mass eventually counterbalances the downward force, causing the slide to stop (see Figure 9.20b).

Slumping produces repeated uniform depressions and flat areas on otherwise sloping ground surfaces. A slump typically propagates headward as it develops and critically weakens the remaining slope uphill. This is because, as one slump mass forms, it removes support from the slope above it. Slump-type landslides are the scourge of highway builders and repairing or removing them costs millions of dollars every year (**Figure 9.21**).

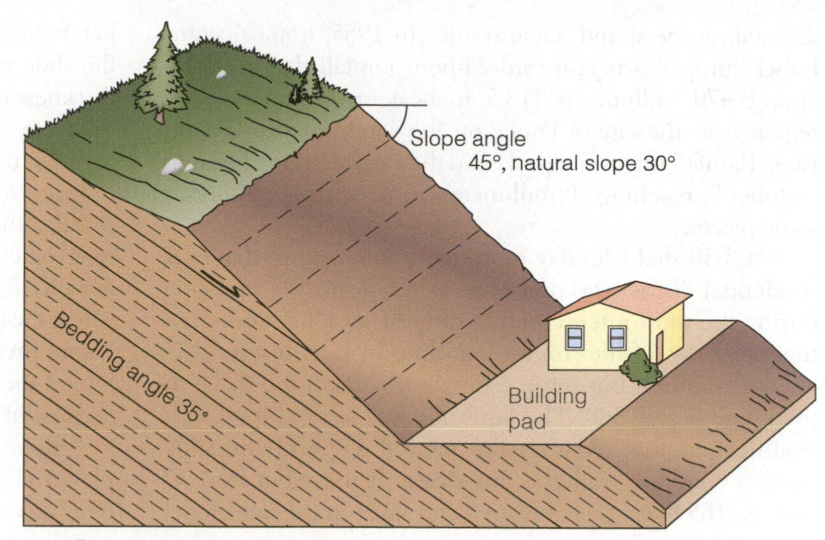

Figure 9.22 Cutting a natural slope for homesites can lead to landslides on bedding or foliation planes. After cutting a natural slope for a building pad, the resulting slope exposes potential slide planes.

Block glides are coherent masses of rock or soil that move along relatively planar sliding surfaces (failure planes), which may be sedimentary rock bedding planes, metamorphic rock foliation planes, faults, or fracture surfaces. Block glides are translational landslides: imagine the landslide block moving down a plane, just like sliding your cell phone down the sloping of your notebook. For a block glide to occur, it is necessary that the failure plane be inclined less steeply than the inclination of the natural or manufactured hill slope. Slopes may be stable with respect to block glides until they are steepened or their bases are cut vertically by excavation for building subdivisions or roads, thus leading to landsliding (**Figure 9.22**).

A classic block glide of about 10 acres is seen along a sea cliff undercut by wave action at Point Fermin in San Pedro, California (**Figure 9.23**). Movement was first detected there in January 1929, and by 1930, the landslide had moved 2 meters (7 feet) seaward. It has been intermittently active ever since. The rock of the slide mass is coarse sandstone, but the slide plane is a thin layer of **bentonite** clay sloping 15° seaward. Bentonite is volcanic ash that has chemically weathered to clay minerals; it becomes very weak and slippery when wet. Bentonite is commonly involved in slope failures; addition of water is all that is needed to initiate a landslide where the slope of the bentonite layer is as little as 5°. Remember that just a few inches of bentonite helped the 18th hole of the Trump golf course end up in the Pacific Ocean at the start of this chapter.

Some translational slides (a synonym by which fast-moving block glides are also known), are large and move with

iStock.com/Bespalyi

Figure 9.21 Stair step landslide surface caused by slump blocks. The road is no more.

Trekkerimages/Alamy Stock Photo

Figure 9.23 Known today as the Sunken City slide, pavement mangled by the ongoing Point Fermin landslide (right side of image) has been painted bright colors.

devastating speed and tragic results. In 1985, tropical storm Isabel dumped a near-record 24-hour rainfall that averaged almost 470 millimeters (18.5 inches) on a mountainous region near the city of Ponce on the south coast of Puerto Rico. Rainfall intensities peaked in the early morning hours of October 7, reaching 70 millimeters (2.8 inches) per hour in some places.

At 3:30 that Monday morning, much of the Mameyes residential district was destroyed by a fast-moving rockslide during the most intense period of rainfall. This resulted in the worst loss of life from a landslide in U.S. history, 129 known fatalities, but some estimate the death toll to be in the range of 200-300. The landslide was in sandstone with stratification that parallels the natural slope of the slide mass, a condition called a **dip slope**. It moved at least 50 meters (165 feet), probably on a clay layer in the sandstone, before breaking up into large blocks. The scarp at the top of the slide was 10 meters (32 feet) high, and the maximum thickness observed at the toe of the landslide was 15 meters (49 feet).

A similar landslide occurred in the Alps of northeastern Italy in 1963 at Vajont Dam, the highest thin-arch dam in the world at the time (275 meters [900 feet]). The geology at the reservoir consists of a sedimentary-rock structure that is bowed downward into a concave-upward **syncline,** a U-shaped fold where the rock layers are deformed, with its axis parallel to the Vajont River canyon. Limestones containing clay layers slope toward the river and reservoir from both sides of the canyon because of this structure (**Figure 9.24**).

Slope movements and slippage along clayey bedding planes above the reservoir had been observed before the dam was constructed. This condition gave engineers and

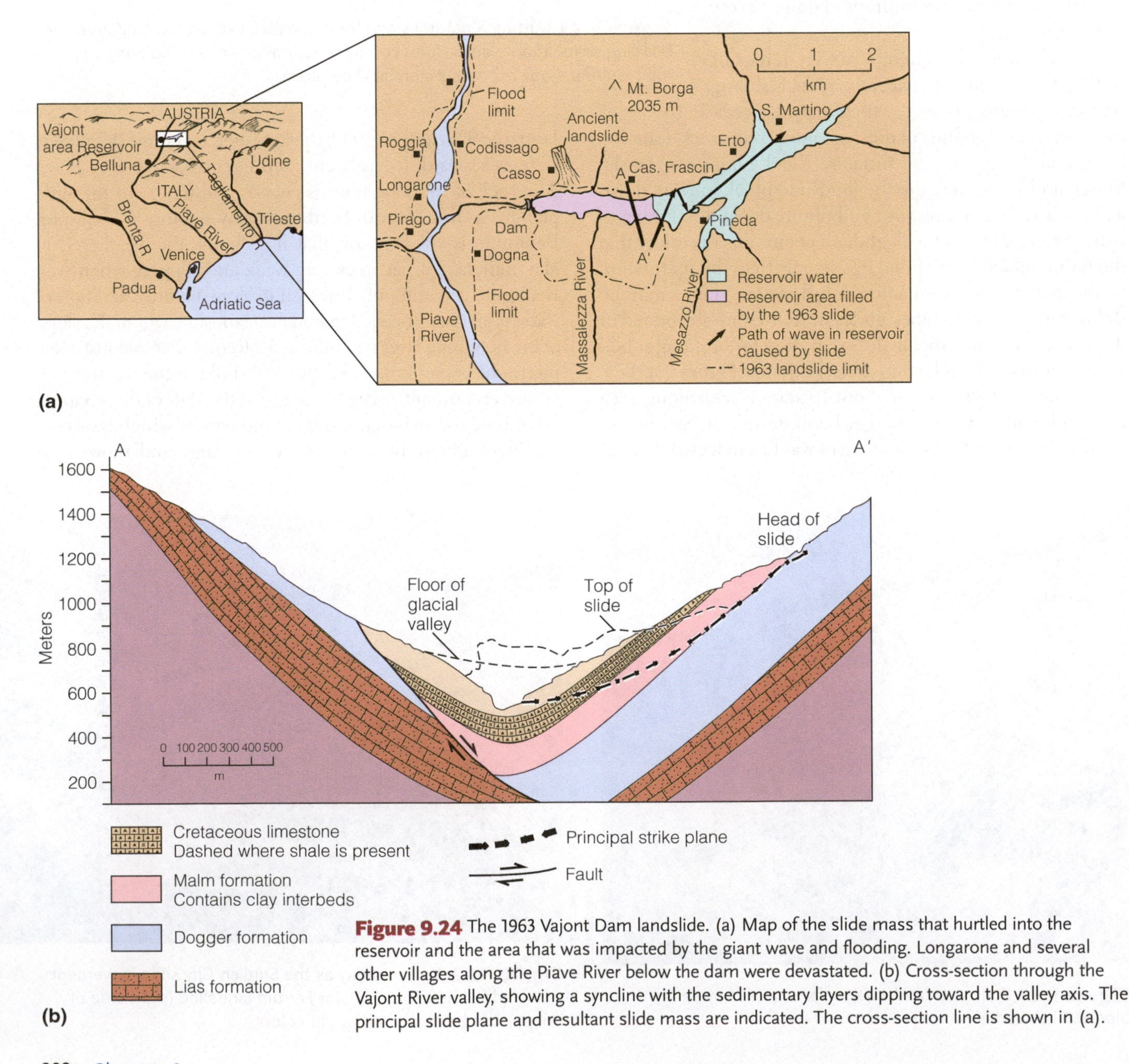

Figure 9.24 The 1963 Vajont Dam landslide. (a) Map of the slide mass that hurtled into the reservoir and the area that was impacted by the giant wave and flooding. Longarone and several other villages along the Piave River below the dam were devastated. (b) Cross-section through the Vajont River valley, showing a syncline with the sedimentary layers dipping toward the valley axis. The principal slide plane and resultant slide mass are indicated. The cross-section line is shown in (a).

The History Collection / Alamy Stock Photo

Figure 9.25 The mass of the landslide occupies most of the reservoir in this 1963 photograph. The top of the Vajont Dam remains visible.

iStock.com/Iurii Buriak

Figure 9.26 Today, the Vajont Dam stands empty, its reservoir drained. This is a similar view to the 1963 photograph but more than 50 years later.

geologists sufficient concern that they placed survey monuments on the slope above the dam for monitoring such movement. Heavy rains fell for 2 weeks before the disaster, and slope movements as large as 80 centimeters (31 inches) per day were recorded as slow-moving block glides began.

Then, on the night of October 9, without warning, the speed of the glides dramatically accelerated and a huge mass of limestone slid into the reservoir so quickly it generated a wave 100 meters (330 feet) high. The wave burst over the top of the dam and flowed into the Piave River valley, destroying villages in its path and leaving 2,500 people dead. The speed of the landslide into the reservoir was so great and the slide block so large that the slide mass almost emptied the lake (**Figure 9.25**). The dam did not fail and is still standing today–a monument to excellent engineering but poor site selection (**Figure 9.26**). The reservoir has never been refilled.

There are some exceptionally large mass movements on Earth. Composite volcanoes suffer sector collapse (see Chapter 5). These are dramatic, oversized rock avalanches where entire parts of the volcanic edifice fail. A sector collapse was first witnessed at Mt. St. Helens by two photographers who just happened to be lucky enough to document the eruption (they were close enough to see well and far enough away to survive). The entire north flank of the volcano peeled off (**Figure 9.27**). At first there was a gaping hole in the mountain (**Figure 9.28**), but since 1980 that hole has been largely filled by a dome of new volcanic rock. The size of the slide was extraordinary. The U.S. Geological Survey reported that "The total avalanche volume is about 2.5 km3 (3.3 billion cubic yards), equivalent to 1 million Olympic swimming pools."

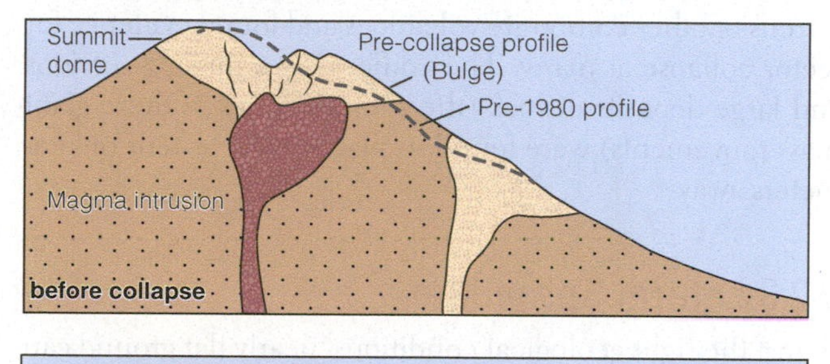

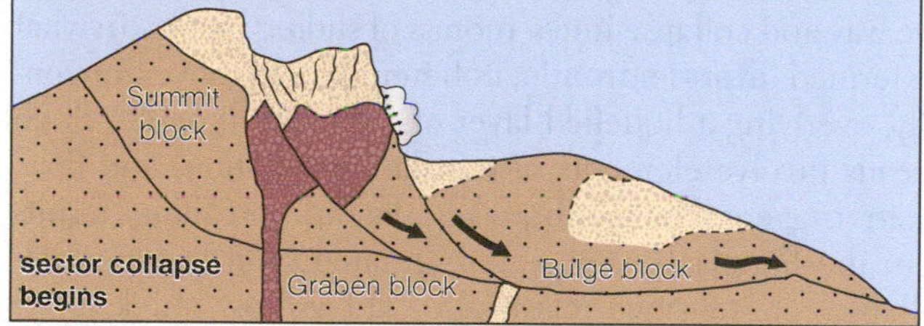

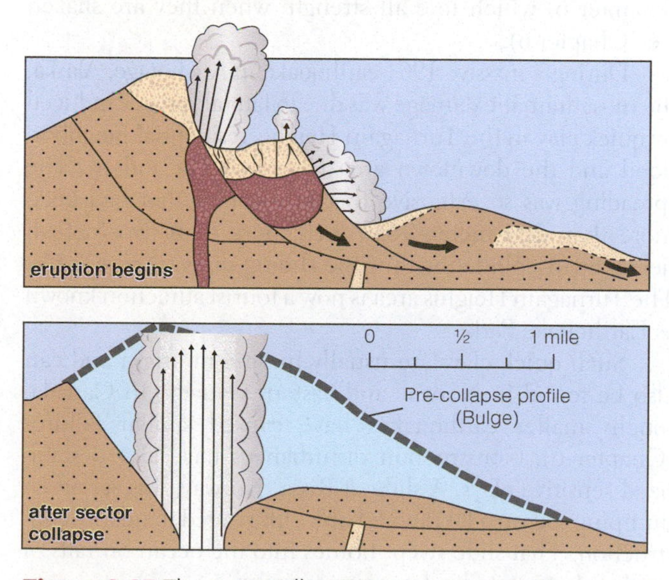

Figure 9.27 The sector collapse over a minute on the morning of May 18, 1980–a giant rock avalanche that removed the north face of Mount Saint Helens. (Adapted from USGS)

Adison pangchai/Shutterstock.com

Figure 9.28 Before the 1980 eruption, Mount Saint Helens had a classic, symmetrical cone. After the eruption and sector collapse, the entire north flank of the volcano (to the right side) was gone.

Once geologists knew what to look for, they reevaluated dozens of other composite volcanoes and found evidence for sector collapse at many. The edifices were missing sections and large deposits of rock (the run-out zones of these giant mass movements) were found in places as far as tens of kilometers away.

9.2.3 Lateral Spreading

In just the right geological conditions, nearly flat ground can give way and collapse into a morass of sliding blocks. In what are termed **lateral spreads**, coherent blocks move horizontally, overlying a liquefied layer of sediment at depth. Such spreads involve elements of translation, rotation, and flow. Often triggered by earthquakes, lateral spreading results from the liquefaction of water-saturated sand layers or the collapse of sensitive clays–also known as quick clays (Chapter 6) which lose all strength when they are shaken (see Chapter 6).

During a massive 1964 earthquake in Anchorage, Alaska, the most dramatic damage was due to lateral spreads induced by quick clay in the Turnagain Heights residential neighborhood and the downtown area (**Figure 9.29a** and **b**). The spreading was so extensive that two houses that had been more than 200 meters (650 feet, more than two football fields) apart collided within the sliding mass (**Figure 9.30**). The Turnagain Heights area is now a tourist attraction known as Earthquake Park.

Such quick clays are usually marine in origin and can also be found in Norway and eastern Canada. In Canada, much smaller earthquakes have caused similar failures (Chapter 6). Construction disturbances have also destabilized sensitive clays. A slide at Rissa, Norway, was set off by dumping soil on a lakeshore bank-cliff related to nearby construction. That slide swept homes into the ocean on rafts of mud and generated a damaging tsunami.

Quick clay beds result from deposition of fine sediment in salt water; the charged ions of salt helped develop the clay

USGS

USGS

Figure 9.29 Turnagain Heights. (a) Overview of the Anchorage area showing the extent of lateral spreading in Turnagain Heights. (b) The Bootlegger Clay (a marine sediment) underlying the Turnagain Heights housing development failed during strong shaking from the 1964 Alaska earthquake.

Shutterstock.com

Figure 9.30 The catastrophic demise of Anchorage neighborhoods underlain by the Bootlegger Clay formation in the 1964 Alaska earthquake.

layer structure, which weakens significantly after postglacial isostatic uplift permits fresh-water infiltration that flushes out the salt "glue." In this case, climate-related environmental changes precondition areas underlain by quick clays, like Rissa, Norway, to failure.

Lateral spreading is not restricted to natural materials but can happen in fill placed by people. After the 1906 San Francisco earthquake on the San Andreas Fault, much of the rubble of the city was pushed into marshland that became the Marina District. Homes and stores were built on this new land. When the 1989 Loma Prieta earthquake shook the area strongly, the filled ground failed. Much of the damage was due to lateral spreading on the fill–ironically, the rubble of one earthquake haunted the residents of the same city nearly a century later (**Figure 9.31**). This is a sad example of how nature controls Earth processes and people create the resulting hazards.

9.3 Debris Flows (LO3)

Debris flows are most often the result of landslides. As some groundwater-saturated landslides move downslope, the material in them loses strength and coherence, transitioning to a muddy mix of water, soil, and rock that flows rather than slides down valleys, overwhelming and often incorporating everything in its way–cars, homes, and boulders. Debris flows can be mostly muddy and fluid or thick and chock full of boulders and debris. Imagine rivers of wet concrete thundering down the street in front of your home and you will know what it's like to be next to an active debris flow. The ground itself can rumble and shake as they pass by.

The recipe for debris flows is short. Take a steep slope, provide plenty of loose material (including some clay and silt), add just enough water to get everything wet but not too wet, mix well, and let the mixture loose down a stream channel or steep canyon. Once initiated, debris flows usually follow pre-existing drainages often bulking up as the flows move by incorporating loose rocks and surface water. Small flows may end at the bottom of their track on a debris fan (**Figure 9.32**). Large flows can keep going for kilometers until the slope is too low or they get too thin to flow.

9.3.1 Behavior of Debris Flows

Debris flows are dense, one-phase fluid mixtures of rock, sand, mud, and water. The one-phase part is important. Add too much water and the sediment settles out. Don't put enough water in the recipe and the debris flow won't flow. Steep slopes help the driving force of gravity exceed the strength of the debris flow material, so the flow keeps moving.

Figure 9.31 Apartment building in the Marina District damaged beyond repair by the 1989 Loma Prieta earthquake.

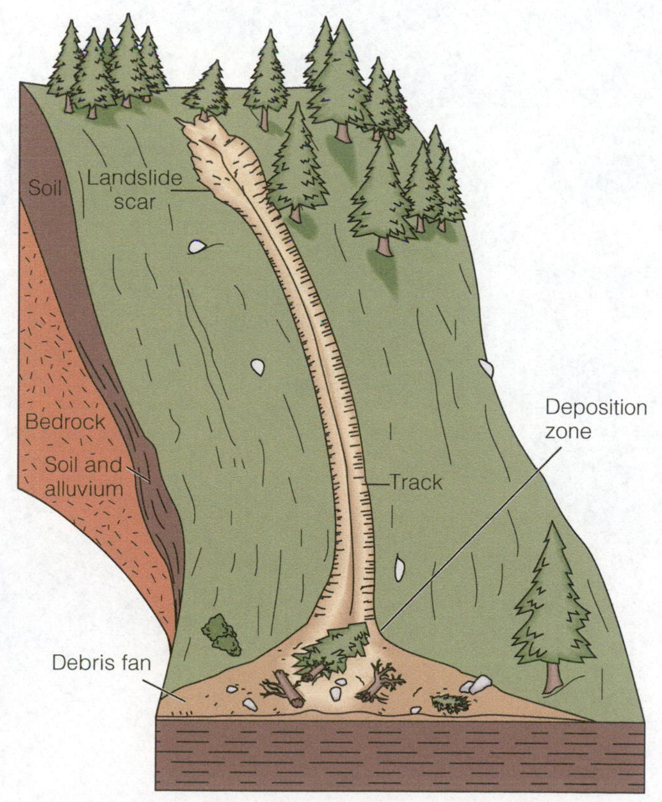

Figure 9.32 Debris flow track and zone of deposition. A debris flow may travel thousands of feet beyond the base of the slope.

When debris flows reach flat valley bottoms, the driving force lowers and the flow comes to a stop as the strength of the debris flow material exceeds the driving force.

Debris flows can be very destructive for two reasons. First, they can travel at high speeds (many meters (tens of feet) per second) and second because unlike clear water, which flows around objects, the "bulked up" debris flows full of rock and mud simply mow down and flow over objects. Because of their high viscosity and high density, commonly 1.5 to 2.0 times the density of water, debris flows are capable of transporting large boulders, automobiles, and even houses in their mass (**Figure 9.33**). Depending on the slope on which they are flowing, and the amount of water and type of sediment, debris flows can just creep along, or they can thunder down canyons at speeds in excess of 50 kilometers per hour (30 miles per hour). Several flows may merge and travel many kilometers from their source area.

9.3.2 Debris Flow Triggers

Areas most subject to debris flows are characterized by sparse vegetation and intense seasonal rainfall. Debris flows are relatively common in steep terrain. They occur in tropical, temperate, and desert environments worldwide. They can

Figure 9.33 Debris flows can make quite a mess of the landscape. (a) Two debris flows in 1.5 hours hit and moved a farmhouse and destroyed several other buildings. The farm structures were on an alluvial fan without a well-defined main channel. The flow occurred in Madison County, Virginia, which shows that debris flows are not limited to the southwestern United States.
(b) These are old boulders on a debris flow fan in Madison County. The boulder in the background is the size of a small house; a debris flow dropped these two where they sit today. The soils that partially bury them suggest that repeated debris flows have covered this fan over the past tens of thousands of years.
(c) In 2011, very heavy rains triggered widespread landslides that spawned debris flows in the mountains west of Rio de Janeiro, Brazil. Here, Nelson Fernandes, a Brazilian hydrologist, examines a swimming pool filled with boulders in front of vacation homes destroyed by one of the flows.

USGS

Figure 9.34 The Tuttle River mudflows were many meters deep. Can you find the person for scale? Compare their height to the mudlines on the trees.

happen in the arctic when permafrost melts or in the desert when monsoonal rains pour from the sky. Debris flows often cascade down barren slopes when rain falls after wildfires have stripped away the vegetation.

Debris flows associated with volcanic eruptions (lahars) are potential hazards in all volcanic zones of the world. They pose a particular danger in populated areas near stratovolcanoes and are the most significant threat to life of all the mass-movement hazards (see Chapter 5 opener). Some lahars form when hot eruptive products (like pyroclastic flows) land on glaciers and snowfields common on volcanoes (even in the tropics) because volcanic edifices are so high and thus cold. This was the scenario at Mt. St. Helens in 1980. The debris flows there were delayed until many hours after the main eruption on May 18, 1980 because it took that long for the snow and ice buried by the hot eruption materials to melt sufficiently. When enough melting occurred, mudflows thundered down the Tuttle River on the north side of the mountain submerging everything in their way (**Figure 9.34**).

Shutterstock.com

Figure 9.35 In 1995, 4 years after the main eruption, typhoon rains unleashed massive lahars from Mount Pinatubo. The mud buried the town of Bacolor and the first floor of the Catholic church. People walked across the mud in front of the church carrying what they could salvage from their muddied homes.

Other lahars are the result of heavy rains (in some cases from tropical storms) falling on the weak, noncohesive layers of volcanic ash that cover the flanks of many volcanoes; this was the case in at Mount Pinatubo in the Philippines, where massive lahars followed the 1991 eruption for months to years (**Figure 9.35**). The mountain gets massive amounts of rainfall during the monsoon season. Two months after the 1991 eruption, the monsoon struck. Up to 160 inches (4,000 millimeters) of rain fell on the volcano's peak and the flows began. According to the U.S. Geological Survey, "0.7 cubic miles (3 cubic kilometers) equivalent to 300 million dump-truck loads" of ash and debris moved off the volcano in debris flows since 1991, mostly during the first few years after the eruption. It took many years before the volume of lahars decreased as the easiest to erode material had been moved and vegetation was regrowing on the slopes.

Composite volcanoes generate repeated massive lahars even when they have not been recently active. Mt. Rainier in Washington State is a great example. The three largest lahars have been well mapped (**Figure 9.36**) and provide the basis for hazard maps going forward. Material from Mt. Rainer has

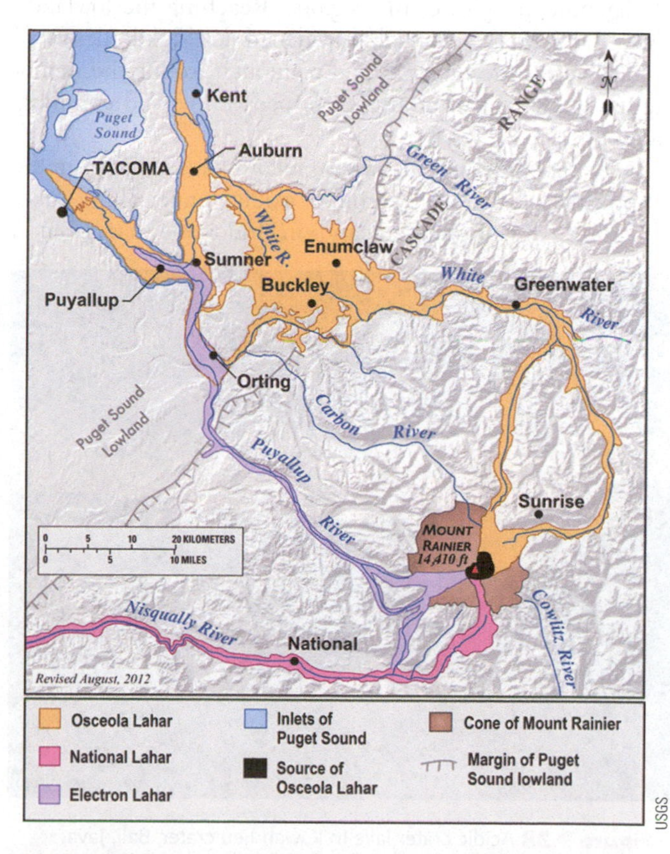

USGS

Figure 9.36 Map showing the extent of major Mount Rainer mudflows.

John, Dave/USGS

Figure 9.37 The Osceola lahar looks very typical in this section exposed by river erosion. It's a mix of volcanic boulders, clay, and sand with no discernable layering. Wood collected from outcrops likes this allows the flow to be dated using radiocarbon.

traveled many tens of kilometers reaching to the suburbs of Tacoma and Seattle and overwhelming valleys now filled with condominiums, car dealers, and shopping malls with meters of mud and boulders (**Figure 9.37**). The Osceola lahar flowed about 5,600 years ago. The Electron lahar is much younger, only about 500 years, and is described in oral stories told by Indigenous peoples of the region. Reaching the lowlands, Mt. Rainier's Osceola lahar widened to 12 kilometers (7 miles) across the Puget Sound plain after exiting the mountains. Many tens of thousands of people live and commute atop this deposit today.

Some of the worst lahars on record have come from the rupturing of the walls around acidic, crater-filling lakes (**Figure 9.38**). Kelud volcano in Indonesia has been a particularly nasty source of catastrophic crater-lake discharges. To deal with these, the Dutch colonial government in 1919 began constructing tunnels in the volcano's crater wall to keep the lake level inside at a safer, lower level, following a lahar that killed 5,000 people. Authorities undertook a second tunnel project in 1966 to further maintain drainage.

iStock.com/Bicho_raro

Figure 9.38 Acidic crater lake in Kawah Ijen crater, Bali, Java, Indonesia. The yellow rock on the left side of the image is a sulfur mine.

NPS Photo / David Swanson

Figure 9.39 Thaw slump erodes into Alaskan landscape as the permafrost thaws. Debris flows carry much of the thawed sediment to the river.

Climate change is triggering debris flows. In the arctic, these flows result as permafrost melts. **Thaw slumps** are high-latitude landslides triggered by the loss of ground ice that gave the frozen soil strength (**Figure 9.39**). When the soil thaws enough that the driving force exceeds the resisting force, the thaw slump starts to fail and because there is so much water around, immediately becomes a debris flow. At Tahoma Creek, in Mount Rainier National Park, the road and campground have been taken out by repeated debris flows generated by the rapidly melting glacier upstream. The Park Service is often issuing warnings to hikers (**Figure 9.40**).

Taylor Kenyon/National Park Service

Figure 9.40 Tahoma Creek is filled with mud and boulders from repeated debris flows originating on Mount Rainier. The trees along the floodplain have been killed, the campground closed, and the road damaged.

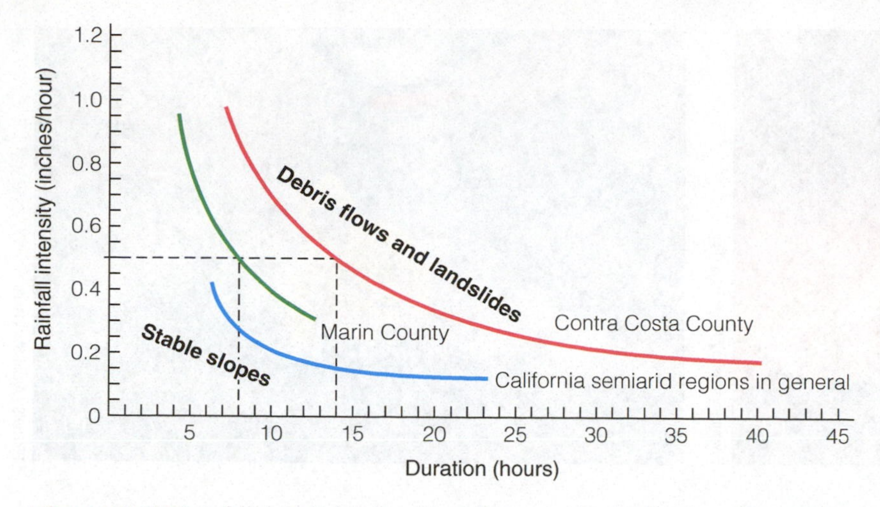

Figure 9.41 Rainfall thresholds for debris flows and landslides for two central California counties and for California semiarid regions in general. The dashed lines show that rainfall intensity of 0.5 inch per hour for 8 hours is sufficient to initiate flows in Marin County. The same rainfall intensity for 14 hours is required to trigger debris flows in Contra Costa County. (Data from California Division of Mines and Geology)

9.3.3 Debris Flow Prediction

Although predicting the timing and location of specific debris flows is not possible, the conditions that lead to debris-flow initiation are well known. Researchers have collected the field data required to create what are called **intensity/duration envelope diagrams** (**Figure 9.41**). Simply put, these are graphs that show the combination of how long it needs to rain and at what rainfall intensity before debris flows will begin. In reading the graphs, it becomes clear that very heavy rainfall need occur for only short periods of time to trigger flows, whereas moderate rain needs to fall for much longer before slopes will give way. **Antecedence**, or the amount of rain that has fallen over the past week or two, is also important. As you might suspect, saturated or nearly saturated soils are more likely to fail than those that have plenty of dry pore space available to soak up the rainfall.

Intensity/duration data have been compiled for flows and slides in the San Francisco Bay area of northern California (see Figure 9.41). These curves show that, for a rainfall intensity of 0.5 inch per hour, the threshold time for the onset of debris flows is 8 hours in Marin County and 14 hours in Contra Costa County. The differing rainfall thresholds in these relatively close areas are due to the variability of geological materials and topography. The lower curve of Figure 9.41 shows that less-intense rainfall will produce debris flows in semiarid areas of California than in more humid areas. This is because these dry areas have little vegetation and abundant loose surface debris; thus, there are few root systems to retain the weak soils. Although these curves are regional, they are useful because they allow authorities to alert residents in critical areas once the threshold conditions for debris flows have been attained.

Not all debris-flow seasons are created equal. Heavy rains (**Figure 9.42**) associated with El Niño events (the change in Pacific Ocean water temperature distribution) have resulted in exceptional landslide and debris flow activity, particularly along the west coasts of North and South America. Studies show that weather and environmental conditions from El Niño more strongly impact mass movement hazards (slumps, slides, and flow) than typical seasonal rainfall variability in Southeast Asia and Latin America. The effects of El Niño have been noted in the United States as well. The 1997–1998 El Niño, possibly the largest of the twentieth century, heavily impacted the San Francisco Bay area, where some of the 85,000 landslides were reactivated. This is an excellent example of interaction among the atmosphere, hydrosphere, and solid-earth systems. Climate change could exacerbate landslide and debris-flow hazards in the Bay area because, with warming climate, some models suggest that El Niño-like conditions will become more common.

Mountain slopes burned by range and forest fires during the dry season are very susceptible to debris flows during the wet season. John McPhee, in his book *The Control of Nature,* does a marvelous job of describing the fire/storm/flow continuum in the hills overlooking the Los Angeles basin, as well as the societal response (or lack thereof). High in those very steep hills, wildfires (**Figure 9.43**) scorch the scrubby chaparral vegetation (see **Figure 9.43b**), releasing waxy chemicals that essentially waterproof the ground. When the winter rains come, the soils are impermeable and, with

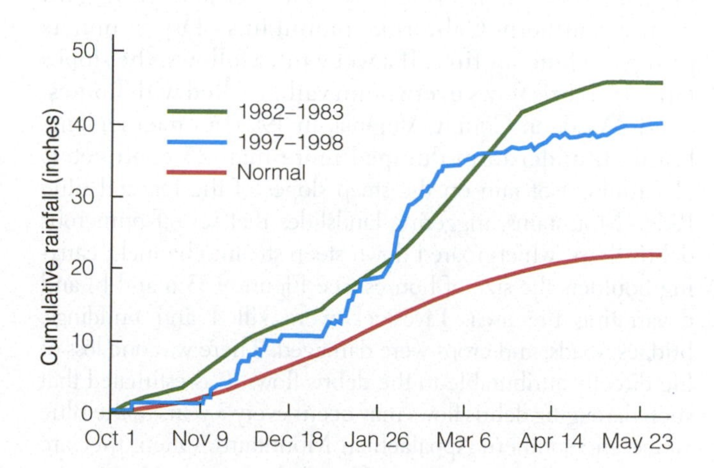

Figure 9.42 Rainfall was plainly in excess of normal in the San Francisco Bay area during El Niño years 1982–1983 and 1997–1998. These heavy rainfalls translated directly into a greater frequency of landslides and debris flows as saturated and nearly saturated soils failed and came roaring down slopes.

Figure 9.43 Fires are a key element of debris flow hazards in Southern California. (a) Fire strikes dry and flammable chapparal. (b) The fire-denuded landscape above a suburban southern California neighborhood. It's like looking up the barrel of a loaded gun. (c) When the rains come, the slopes fail in shallow landslides generating debris flows that run down the streets. (d) Homes are overwhelmed with mud and destroyed.

the vegetation burned off, slopes quickly lose strength, letting loose debris flows that sweep through canyons (see **Figure 9.43c**) with people and their homes and vehicles in the way (see **Figure 9.43d**). This cycle plays out over and over again in the southern California mountains. Dry summers prime the land for fires. If a wet winter follows, the slopes fail and debris flows overwhelm valleys filled with homes.

In Madison County, Virginia, in 1995, an intense, long-lasting thunderstorm dumped more than 25 centimeters (10 inches) of rain on the steep slopes of the forested Blue Ridge Mountains, triggering landslides that set off numerous debris flows, which roared down steep stream channels, carrying boulders the size of houses (see Figure 9.33 a and b) and devastating the area. Livestock were killed and buildings, bridges, roads, and crops were damaged. There was one loss of life directly attributable to the debris flows. It is estimated that such damaging debris flows may occur every 3 years in the Blue Ridge and southern Appalachian Mountains. Often, they are triggered by the intense rains of tropical storms and hurricanes as warm, moist air is lifted over the steep mountains.

Other geologic events can cause debris flows. On May 30, 1983, a debris flow was generated on Slide Mountain near Reno, Nevada. The flow killed one person, injured many more, and destroyed a number of homes and vehicles, all in less than 15 minutes. How did this happen without any rainfall? About 720,000 cubic meters (about a million cubic yards) of weathered granite gave way from the mountain's steep flank and slid into Upper Price Lake. This mass displaced the water in the lake (about 26,000 cubic meters or 7 million gallons), causing it to overflow into a lower lake, which, in turn, overflowed into Ophir Creek gorge as a water flood. Picking up sediment as it went, it became a debris flow that emerged from the gorge, spread out, destroyed homes, and covered a major highway (**Figure 9.44**).

Modern remote-sensing techniques clearly show this was not the first debris flow at Slide Mountain nor is it likely to be the last. Lidar (light detection and ranging), a technique that relies on using laser light to measure millions of points on the landscape (often from an airplane or helicopter), makes incredibly detailed maps, allowing geologists to see the landscape in whole new ways (**Figure 9.45**). The result is clear–there are numerous lobes of sediment left by debris flows at the mouth of the canyon leading away from Slide Mountain. The 1983 flow was a repeat performance. With lidar, many former slides and flows, for which there is no written record, are now being identified.

Figure 9.44 The Slide Mountain debris flow in 1983 made quite a mess. A horse trapped in the mud was saved by an airlift (a) but the house, truck, and school bus were not so lucky (b).

Figure 9.45 Slide Mountain Canyon viewed from above.

9.4 Snow Avalanches (LO4)

Snow avalanches have been with us ever since humans ventured into areas of the world with snow. There's more than 2,300 years of recorded avalanche history. Hannibal crossed the Alps in 218 B.C.E. on his way to Rome accompanied by his elephants. Histories written decades after his military campaign suggest he and his men were caught in a massive, early-season avalanche and that many of the troops perished as they were buried in snow. Modern generals may have learned from Hannibal. During World War I, 6,000 troops were buried in one day by avalanches in the Dolomite Mountains of northern Italy as opposing armies intentionally triggered avalanches with the hope of burying their opponents.

The mechanics of snow avalanches are similar to those of other mass movements like landslides, the differences being in the material and the speed. Although snow avalanches have been around as long as there have been mountains and snow, the relatively recent emergence of the recreational skiing industry as big business and the growing interest in backcountry over-snow travel has made avalanche awareness and control all the more critical (**Figure 9.46**). Avalanches can occur anywhere there is snow and at any season there is snow on the ground.

Avalanches in high mountain regions have wiped out villages, disturbed railroad track alignments, blocked roads, and otherwise made life difficult for millennia (**Figure 9.47**). In western North America, several dozen people die yearly from avalanches. Even eastern North America is affected. In many years in New Hampshire, on Mount Washington, above the tree line, a skier or climber dies in Tuckerman Ravine when an avalanche buries that person alive.

9.4.1 Types of Snow Avalanches

There are several kinds of snow avalanches. Each has a different character, tends to happen at different times, and has

iStock.com/Filip Majercik

Figure 9.46 Snowboarder triggering small sluff avalanche in Slovakia.

Kirill Skorobogatko/Shutterstock

Figure 9.48 Avalanches begin in the starting zone, move though the track, and come to rest in the run-out zone.

a different risk profile. All snow avalanches have a **starting zone** where the failure begins, often high up where the slope is the steepest and the driving force is greater. After leaving the starting zone, many avalanches follow well-established **avalanche tracks**, valleys where the snow moves confined by valley walls. Avalanches come to rest at the base of slopes in a **run-out zone** which is less steep; here the driving force is less. This is where avalanche victims are typically found (**Figure 9.48**)

A **slab avalanche** is a coherent mass of snow and ice that hurtles down a slope–much like a magazine sliding off a tilted coffee table (**Figure 9.49**). Slabs are made of compact, dense, dry snow with high cohesion. This type of avalanche is most frequent on slopes between 35° and 40°, just about black-diamond ski run territory for those who are expert skiers (**Figure 9.50**). Slab avalanches account for most avalanche deaths because of their size and their tendency to solidify when they stop. They often release on slopes where windblown snow has accumulated, typically near ridgelines and under **cornices**, accumulations of windblown snow on the downwind side of ridges (**Figure 9.51**). Slabs leave a clear scarp behind just above where they originate.

Climax avalanches, often the biggest and thickest of the season, usually occur in spring when the snowpack melts and weakens. It is then that the snowpack warms to the freezing point and is saturated with water from melting snow. Because of this melting, there is no longer ice holding snow grains together into a strong solid capable of resisting the force of gravity. With climate change increasing, extreme midwinter freezing and thawing cycles are more often resulting in spring-like conditions and the massive, wet avalanches that result. Climax avalanches often run over the ground, snapping off trees and burying anything in their run-out

iStock.com/Pattonmania

Figure 9.47 Best to avoid having your truck buried in an avalanche because it just won't be the same in the spring.

iStock.com/Nicolamargaret

Figure 9.49 Small slab avalanche released from a cornice where there remains a clear scarp. Note the blocks of snow that survived the trip downslope intact.

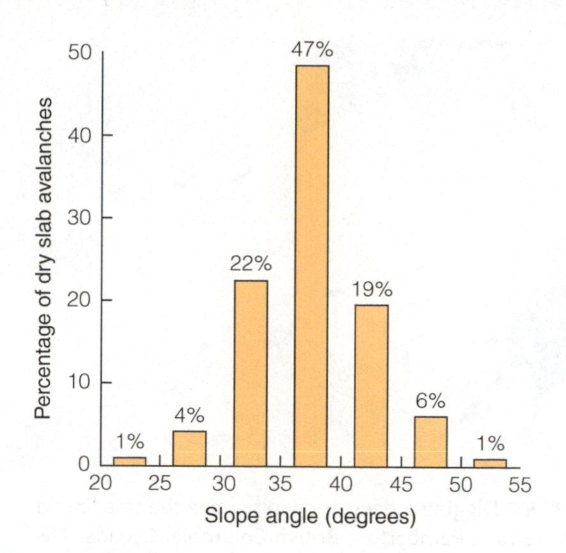

Figure 9.50 Percentage of slab avalanches versus slope angle in degrees. (Data from Knox Williams, October 2000, "Avalanche formation and release," Rock Talk 3(4)).

Figure 9.51 Skier looking cautiously over a large cornice deposited as wind blew the snow toward the right side of the image. He'd be smart not to go any farther out because cornices like this are often unstable and prone to collapse.

zones with thick, wet, heavy snow. Climax avalanches can be devastating, wiping out stands of old growth trees, taking down buildings, and covering roads in thick, slushy snow difficult to remove (**Figure 9.52**).

Powder-snow avalanches are most common during and after cold storms (**Figure 9.53**). These "dry" avalanches are dangerous because they entrain a great amount of air which because of cushioning makes them highly fluid, traveling far and fast. Furthermore, the strong shockwave that precedes them can uproot trees and flatten structures. Victims of dry avalanches tend to be buried in light snow close to the surface, which makes their chances of being rescued alive better than those of wet-avalanche victims.

Sluff avalanches occur most frequently during storms, when slopes are loaded with newly fallen snow either falling from the sky or by wind. They are typically small and frequent and lower the subsequent hazard. Sluff avalanches are most common on steep slopes.

Avalanches are most often triggered by snow loading during and just after heavy storms. Rockfalls and cornice collapse are other natural avalanche triggers, as is the weakening of once stable dense, icy snow as spring comes and the snowpack warms to the melting point. People and their machines can be very effective avalanche triggers. Backcountry skiers, or skiers going out of bounds at ski areas routinely set of slopes that are just on the threshold of stability as do snowmobilers, especially those **high-marking**, a game of going up very steep slopes and turning down just below over hanging cornices–a really risky endeavor (**Figure 9.54**).

9.4.2 Snow Stability Analysis and Avalanche Predication

Snow is a fickle substance. Because snow is water close to its freezing point, its strength characteristics can change rapidly. In minutes to hours a once stable slope can turn deadly.

Figure 9.52 This massive avalanche blocked the Richardson Highway near Valdez, Alaska, in January 2014. It was the result of almost 12 inches of rain and abnormally high temperatures that week–very spring-like conditions in midwinter.

Figure 9.53 Powder snow avalanche roars downslope in the Caucasus Mountains of Russia.

iStock.com/CynthiaShepherd

Figure 9.54 Snowmobile high-marking at Mount Baker volcano in Washington State.

iStock.com/StockstudioX

Figure 9.55 Digging a deep snow pit above the tree line in avalanche terrain, Pemberton, British Columbia, Canada. The skiing looks great, but is it safe?

Snow stability is determined by the same factors that determine the stability of landslides, the driving and resisting forces. In the case of snow, strength changes can happen in just a few hours as it warms and weakens or wet snow freezes into a strong, solid mass.

Snow pits, weather forecasts, and experience are key to predicting avalanche hazard levels. Snow pits are dug near avalanche-prone slopes by experts trained in reading the snowpack. (**Figure 9.55**). They look for weak layers in the snowpack because most avalanches occur on such weak layers. For example, an ice crust to which the snow above is not well bonded is a dangerous set up: the snow above can easily slide on the ice. Ice layers and ice crusts can form from freezing rainstorms or when bright sun melts the snow, which then refreezes. Other weak layers are **surface hoar**, the delicate crystals of ice that form on top of the snow during cold, moist nights, and **depth hoar**, which forms when the ground beneath the snowpack is moist and warm, and the air above the snowpack is very cold. Water vapor moving upward forms a layer of very large, weak crystals, which can fail catastrophically. Depth hoar is common at high altitude, in thin snowpacks, and on north-facing slopes in places such as the Rocky Mountains and the Wasatch Range. It's also common in the arctic (**Figure 9.56**). In these situations, the gradient in temperature and moisture is extreme between the ground and the air.

Most avalanches are caused by an addition of new snow (an increase in driving force). Indeed, the most hazardous time to be on the mountains is when it's snowing and the wind is blowing strongly. During such high avalanche-hazard conditions, snow is rapidly loaded onto slopes and avalanches may occur constantly. Snow avalanches have one resisting force that landslides do not–**terrain anchors**. Trees are probably the most effective terrain anchor, but large rocks can help, too. These roughness elements on the landscape hold snow and minimize avalanche hazard by providing an additional resisting force.

9.4.3 Mitigating Snow Avalanches

The best personal defense against snow avalanche hazard is to stay away from steep, open slopes, especially when

Cambridge University Press

Figure 9.56 Snow pit reveals a slab of wind-blown snow over depth hoar. The insets are photographs of the snow grains magnified. The small windblown grains bond to each other. The large depth hoar crystals do not and thus are weak.

Figure 9.57 There many warning signs for avalanches. (a) Ski areas post places you might want to avoid based on their knowledge of the terrain. (b) Nature does the same; here in the summer is a view down an avalanche track. The dead trees map out the flow in the run-out zone of the avalanche that killed them. The trail at the bottom is safe during the summer but not the best place to be in the winter.

conditions are hazardous. Avalanche forecasts and hotlines exist for most mountain areas near population centers. If you spend time in the backcountry or ski out of bounds, take an avalanche safety course to learn how to read the snow. It could very well save your life. Near ski areas and in places where avalanche danger is known to be high, warning signs are often posted (**Figure 9.57a**). Ignore them at your own risk! If you spend summer hiking in the same area you ski in the winter, you can find valuable clues as to where avalanches have run in the past. Trees will be knocked down in a telltale pattern–nature's warning sign (see **Figure 9.57b**)

Where buildings already exist, structures can be built to divert avalanches. In some areas in the Alps, the uphill sides of chalets are shaped similarly to a ship's prow, so that they will divert avalanching snow (**Figure 9.58**). Railroads and highways can be protected by building snow sheds along stretches adjacent to steep slopes (**Figure 9.59**). On some slopes high in the Alps, miles of large concrete, metal, and wood barriers have been built to help anchor snow to the slopes. They work even if they are thought by some to be an eyesore in the summer (**Figure 9.60**). It's a tradeoff for winter safety.

Avalanche control is usually achieved by triggering an avalanche with explosives tossed by ski patrollers, dropped from helicopters, or fired from cannons when dangerous snow conditions exist in areas frequented by people, such as ski areas and roads (**Figure 9.61**). The explosions collapse the depth hoar, break loose slabs, or detach powder snow, causing a "controlled" avalanche. These controlled releases are done when the run-out zones of the avalanches they set off are empty, minimizing hazard to people. Although roads close and people may need to be evacuated or ski days cut short, avalanche control has saved countless lives. In recent years, drones have been used both to scout avalanche starting zones and deliver explosives to trigger controlled avalanches.

Figure 9.58 What a great place to have your home, until the avalanches from the bare slope above come roaring down. Look carefully and you will see the rock prow that defends the upslope side of this home in the Swiss Alps above Davos.

Figure 9.59 In Davos, intensive avalanche-control measures are used to keep the road open. This is an avalanche shed designed to let the traffic go under and the snow go over.

Figure 9.61 Ski patroller at Crested Butte Ski Resort in Colorado has just tossed a charge and set off a controlled avalanche.

9.5 Subsidence (LO5)

Subsidence, the sinking of the land surface, happens for different reasons in different places. Sinkholes dot the limestone landscape of Florida. Walls of soil and rock (known as dikes or embankments) prevent the sea from flooding the sinking land in some places in the Netherlands, and in Mexico City, the famous Palacio de Bellas Artes (Palace of Fine Arts) rests in a large depression in the middle of the city. These areas suffer from natural and human-induced **subsidence**, a sinking or downward settling of Earth's surface.

Figure 9.60 Snow anchors high in the mountains above the Austrian ski town of Zillertal provide artificial terrain anchors for the snow, reducing the chance of avalanches.

Subsidence is not usually catastrophic and loss of life due to it is rare, but land subsidence is currently observed in 45 U.S. states, affecting an estimated 44,000 square kilometers (17,000 square miles). This is greater than the area of Vermont and New Hampshire combined. Subsidence was estimated to cost $125 million annually in 1991 (the equivalent of nearly $270 million in 2022) because of property damage and increased flood frequency.

9.5.1 Human-Induced Subsidence

Human-induced subsidence occurs when humans extract underground water, when they engage in mining or oil and gas production, and when they cause loose sediments at the ground surface to consolidate (see Chapter 1). The effects can be local or regional in scale. In the United States and Mexico alone, an area the size of Vermont has slowly subsided 30 centimeters (1 foot) because of withdrawal of underground water (**Figure 9.62**). Sinking of the land changes drainage paths, and it is particularly damaging to coastal areas and lands adjacent to rivers, because it increases flood potential. Subsidence can also result from the compaction of sediments rich in organic matter. Parts of the city of New Orleans, built on cypress swamps of the delta of the Mississippi River and now isolated from the sediment-laden floods of that river, are below river level and sea level (Chapter 7).

In Texas, areas with concerning rates of human-induced subsidence are monitored as "subsidence districts." These management regions have the goal of reducing groundwater consumption to slow the rate and negative effects of subsidence. The Harris-Galveston Subsidence District along the Texas Gulf Coast, for example, was created in 1975 as concerns about subsidence grew amongst the community and local governments. This region represents one of the largest areas of land subsidence in the United States, the result of groundwater withdrawals primarily for the municipal drinking water supply. Monitoring of the subsidence and groundwater levels by the USGS in this region found that, as of 2017, some parts of the region were experiencing dramatic changes in groundwater levels. The water table is some areas rose more than 28 meters (61 feet), while other parts of the region groundwater levels declined as much as 30 meters (65 feet). Since 1974, a little more than a meter (3.7 feet) of subsidence has been measured in some areas of the district.

Figure 9.62 Perhaps the most famous subsidence photograph. Extensive pumping of groundwater for irrigation in the San Joaquin Valley of California caused the land surface to drop more than 9 meters (30 feet) between 1925 and 1977. The dates on the pole mark the former land surface elevation.

One of the most unusual subsidence stories comes from the United Kingdom. The area around Cheshire is famous for salt production (the rocks underlying the area include beds of pure salt). For many years, shallow tunnels mined into the salt would collapse, damaging buildings. When those in control of the mines tried a new way of removing salt by pumping out brine (groundwater full of salt), the problem got much worse. Large areas subsided, leaving roads impassable and buildings tipped at crazy angles (**Figure 9.63**).

9.5.2 Natural Subsidence

Natural subsidence is caused by earthquakes, volcanic activity, and the solution of limestone, dolomite, and gypsum. Severe ground shaking that leads to liquefaction–when soil and sediment lose their strength and flow like liquids–can also lower the ground surface (see Chapter 6). Volcanic activity that empties magma chambers can cause collapse and subsidence over much larger areas when it forms calderas (see Chapter 5). Non-seismic subsidence caused by tectonic (mountain-building) processes occurs so slowly that it is not considered an environmental hazard (see Chapter 2). Sinkhole development is the most prevalent form of sudden natural subsidence.

The first sign of impending disaster in Winter Park, Florida, occurred one evening when Rosa Mae Owens heard a "swish" in her backyard and saw, to her amazement, that a large sycamore tree had simply disappeared. The hole into which it had fallen gradually enlarged, and by noon the next day her home had also fallen into the breach (**Figure 9.64**). So did five Porsches, part of the city swimming pool, and two residential streets. Deciding to work with nature, the City of Winter Park stabilized and sealed the sinkhole and converted it to a beautiful urban lake. Few people who read of

Figure 9.63 Subsidence fractures from salt mining make a road impassable in Northwich, Cheshire, England, about 1910.

this notorious sinkhole in May 1981 realized the extent of the sinkhole problem in the southeastern United States.

Sinkholes are most commonly found in carbonate terrains (limestone, dolomite, and marble), but they are also known to occur in rock salt and gypsum. They form when carbonic acid (H_2CO_3) dissolves carbonate rock, forming a near-surface cavern. Continued solution leads to collapse of the cavern's roof and the formation of a sinkhole. About 20% of the United States and 40% of the country east of the Mississippi River are underlain by limestone, dolomite, or carbonate rocks. The states most impacted by surface collapse are Alabama, Florida, Georgia, Tennessee, Missouri, and Pennsylvania.

The dissolution (a chemical reaction discussed in Chapter 7 and listed in Table 7.1) of limestone and gypsum by underground water leads to the formation of caves, caverns, disappearing streams, and high-yielding, extensive groundwater sources such as the Edwards aquifer in Texas that supplies drinking water to the city of Austin. Land area exhibiting these features is called **karst terrain** (German *karst*, for the Kars limestone plateau of northwest Yugoslavia) (**Figure 9.65**). Urbanization on karst terrain typically results in problems such as the contamination of underground water and collapse of cavern roofs beneath structures.

Tens of thousands of sinkholes exist in the United States, and the formation of new ones can be accelerated by human activities. For example, excessive water withdrawal that lowers groundwater levels reduces or removes the buoyant support of shallow caverns' roofs. In fact, the correlation between water-table lowering and sinkhole formation is so strong that "sinkhole seasons" are designated in the southeastern United States. These seasons are declared whenever groundwater levels drop naturally because of decreased rainfall and in the summer when the demand for water is heavy. Vibrations from construction activity or explosive blasting can also trigger sinkhole collapse. Other mechanisms that have been proposed for sinkhole formation are fluctuating water tables (alternate wetting and drying of shallow rock reduces its strength) and water tables that are so high that they erode cavern roofs.

Although the development of natural sinkholes is not predictable, it is possible to assess a site's potential for collapse resulting from human activities. This requires extensive geological and hydrogeological investigation before site development. Under the best of geological conditions, good geophysical and subsurface (drilling) data can tell where collapse is most likely to occur, but they cannot tell exactly when a collapse may occur. Armed with this information, planners and civil engineers can implement zoning and building restrictions to minimize the hazards to the public.

9.5.3 Classification of Subsidence

One method of classifying land subsidence is based on the depth at which the subsidence is initiated. **Deep subsidence** is initiated at hundreds of meters below the surface generally when oil, or gas are removed. **Shallow subsidence**, in

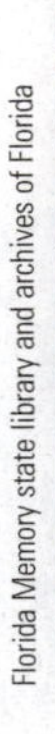

Anthony S. Navoy/USGS

Figure 9.64 The Winter Park sinkhole took out the city's public swimming pool (a) and five Porsches at a repair shop specializing in German automobiles. (b) Look closely at the image and see if you can find the sixth Porsche (it is blue) that was rescued by a crane before it landed in the drink.

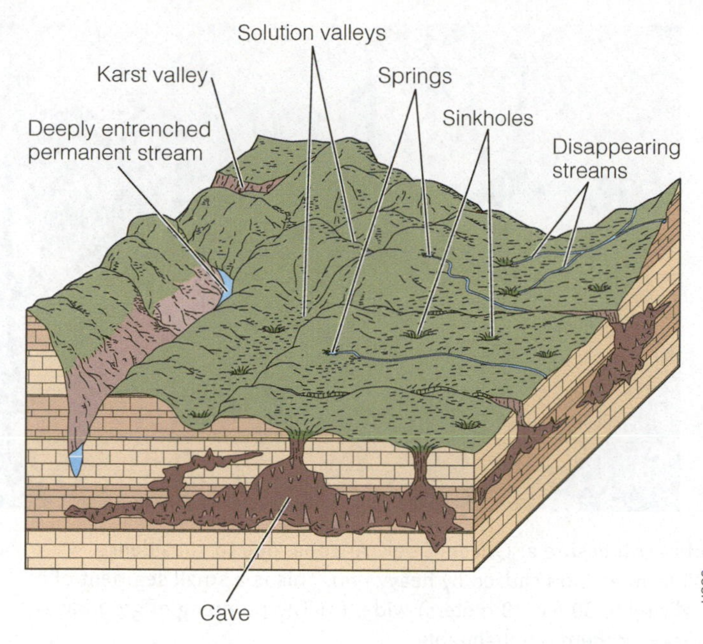

Figure 9.65 Surface and subsurface features, as well as discontinuous surface drainage, are indicative of karst topography.

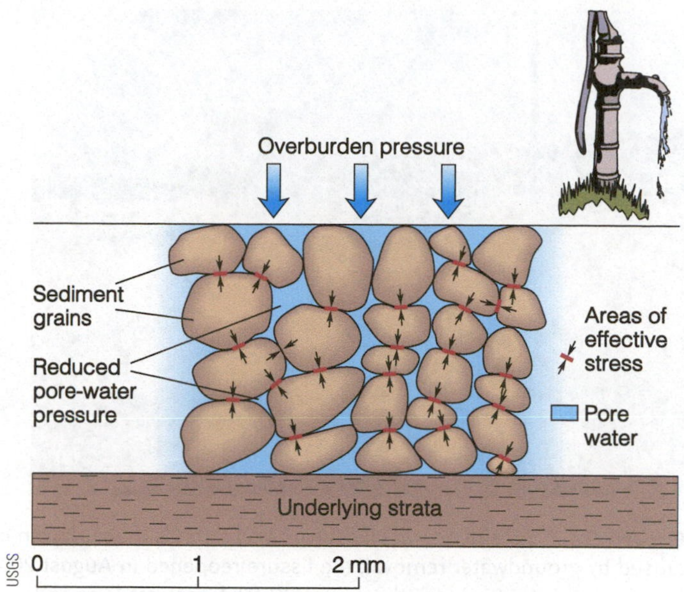

Figure 9.66 A reduction of pore pressure, as groundwater is pumped away, causes increased effective stress and the rearrangement and compaction of sediment particles.

contrast, takes place nearer the ground surface when shallow groundwater or solid material are removed by natural processes or by humans. In addition, many poorly consolidated, shallow deposits such as peat are subject to compaction by the pressure of overlying material and/or groundwater withdrawal.

Settlement differs from subsidence. It occurs when an applied load, such as that of a structure, is greater than the bearing capacity of the soil onto which it is placed. Settlement is totally human-induced, and it is a soils- and foundation-engineering problem. Italy's Leaning Tower of Pisa is a classic example of foundation settlement (see Chapter 7).

The removal of oil or gas and associated groundwater confined in the pore spaces in rock or sediment causes deep subsidence. Pressure from these substances helps support the overlying material. As pressure is reduced by resource extraction, the weight of the overlying sediment gradually transfers entirely to mineral-and-rock-grain boundaries. If the sediment was originally deposited with an open structure, the grains reorient into a closer-packed arrangement, thus occupying less space, and subsidence ensues (**Figure 9.66**). Because clays are more compressible than sands, most compaction takes place in clay layers.

At least 22 oil fields in California have subsidence problems, as do many fields in Texas, Louisiana, and other oil-producing states. A near world record for subsidence is held by the Wilmington Oil Field in Long Beach, California, where the ground has dropped 9 meters (about 30 feet). This field is the largest producer in the state, with more than 2,000 wells tapping oil in an upward-arched geological structure called an **anticline**. The arch of the Wilmington field's anticline gradually sagged as oil and water were removed, causing the land surface to sink below sea level in one place (Chapter 1).

In Arizona, groundwater withdrawal in some places has caused subsidence of 5 meters (more than 15 feet) and the development of cracks and earth fissures, particularly along the edges of some basins (**Figure 9.67**). The subsidence has changed natural drainage patterns and caused highways to settle and crack. The magnitude of the subsidence caused by groundwater withdrawal is similar to oil-field subsidence.

The Houston suburb of Baytown was another victim of groundwater extraction. It subsided almost 3 meters in the 1900s, and 80 square kilometers (31 square miles) of this coastal region is now permanently underwater (see **Figure 9.67**). Flooding became so frequent some people built concrete levees and seawalls; the rest moved out. Hurricane Alicia finished off most of what was left in 1983. The Brownwood subdivision in Baytown is no more. The buildings are gone and the area is now a public park.

Mexico City sits in a down-dropped fault valley that has accumulated nearly 2,000 meters (6,600 feet) of lake sediments, mostly of pyroclastic (volcanic) origin. Subsidence occurred there because of the withdrawal of groundwater from these sediments. Total subsidence varies throughout the city, but almost 6 meters (20 feet) of subsidence had occurred by the 1970s in the northeast part of the city, mostly in the water-bearing top 50 meters (160 feet) of sediment. In 1951, an aqueduct was completed that brought water to the city; as a result, many wells were closed.

The effect on subsidence rates was dramatic, and by 1970, subsidence had decreased to less than 5 centimeters (2 inches) per year. Flash forward to 2022, and subsidence rates have skyrocketed in Mexico City, with studies then estimating

© Ray Harris/Arizona Geological Survey USGS

Figure 9.67 Subsidence wreaks havoc on Earth's surface. (a) An old earth fissure at Queen Creek, Arizona, due to subsidence caused by groundwater removal; the fissure reopened in August 2005 from erosion caused by heavy rain. This is a small segment of a fissure system that is more than a mile (1.5 kilometers) long and locally up to 30 feet (9 meters) wide. (b) The pumping of groundwater in Baytown, Texas, has caused the earth to subside under homes leaving them uninhabitable.

annual subsidence to be 50 centimeters (20 inches) per year. These found that subsidence in this region is less linked with groundwater pumping and more a result of compaction of the ancient Texcoco lake beds on which the heavy city has grown. At this rate, "total compaction" of the subsiding ground will be reached in approximately 150 years in Mexico City, resulting in an additional 30 meters of subsidence.

Subsidence can occur for a variety of other reasons as well, related to tectonics and sediment supply. Great subduction earthquakes, such as those that affected Alaska in 1964 and the Pacific Northwest in the 1700s, changed land levels by meters. Much of the subsidence was along the coast, and the subsided area is now regularly flooded at high tide. In fact, the land actually tilted and a large area of the Gulf of Alaska was uplifted several meters. In Alaska, whole villages were submerged by earthquake-induced subsidence (**Figure 9.68**).

There are natural and human markers of subsidence. In the Pacific Northwest, the last great subduction quake occurred in 1700. In this case, prior land level is clearly shown by the submergence of an entire forest into saltwater, killing the trees (**Figure 9.69**). In the Mississippi River Delta a Civil War era fort is a graphic indicator of subsidence as it sinks beneath the waves (**Figure 9.70**). The cause: a greatly diminished sediment supply resulting from dams upstream and river levees funneling muddy water to the sea, bypassing the delta.

Shallow subsidence can also cause trouble. Subsidence that results from the heavy application of irrigation water on certain loose, dry soils is called **hydrocompaction**. On initial saturation with water, such as for crop irrigation or the construction of a water canal, the open fabric between the grains collapses and the soil compacts. The in-place densities of such "collapsible" soils are low, usually less than 1.3 grams per cubic centimeter, about half the density of solid rock and just 30% more dense than fresh water.

Subsidence of monumental proportions has occurred because of hydrocompaction. In the United States, it is most common in arid or semiarid regions of the West and Midwest, where soils are dry and moisture seldom penetrates below the root zone. Known areas of hydrocompaction are the Heart Mountain and Riverton areas of Wyoming; Denver, Colorado; the Columbia Basin in Washington; southwest and central Utah; and the San Joaquin Valley of California.

Collapsing soils were a major problem for the designers of the California Aqueduct in the San Joaquin Valley. It was necessary to identify the areas of loose soil that

USGS

Figure 9.68 The town of Portage, Alaska was very close to sea level before the 1964 earthquake. After the 1964 quake, the town was underwater at high tide because the land level dropped.

USGS

Figure 9.69 A ghost forest on the coast of Washington State was the victim of submergence when the 1700 Cascadia subduction zone earthquake dropped land levels several meters along the coast on January 26, submerging the tree roots in saltwater and killing the trees (see Chapter 6).

Marli Bryant Miller, University of Oregon

Figure 9.70 Fort Proctor, built in the 1850s on the Mississippi Delta, is now surrounded by water as the delta subsides, largely because the sediment supply from the river has been cut off by channelization.

could not be avoided, to maintain a constant ground slope for the flow of water. Sediments were field-tested by saturating large plots and measuring subsidence at various levels beneath the surface. Subsidence was found to average 4.2 meters (close to 14 feet) over a period of 1.3 years, during which thousands of gallons of water had been applied to the test plots. Based on the field tests, 180 kilometers (112 miles) of the aqueduct alignment in the valley were "presubsided" by soaking the ground 3 to 6 months before construction. This assured that the slope of the canal would be maintained after construction and water would continue to flow where it was intended.

9.5.4 Mitigating Subsidence

The best strategy for minimizing losses is to restrict human activities in areas that are most susceptible to subsidence and collapse. Unfortunately, most instances of land subsidence caused by fluid withdrawals were not anticipated. But knowledge of the causes, mechanics, and treatment of deep subsidence has increased with each damaging occurrence, so that we can now better implement measures for reducing the incidence and impact of these losses. Some of the measures taken to minimize subsidence include controlling regional water-table levels by monitoring and limiting water extraction, replacing extracted fluids like oil by water injection, and importing surface water for domestic use to avoid drawing on underground supplies.

9.6 Reducing Losses from Mass Movements (LO6)

Mass movements are a part of the Earth System, transferring mass from hillslopes to lowlands. But they won't cause as much grief to people if we plan and take steps to mitigate these hazards. New technologies that let us see past mass movements are a powerful first step to reducing damage and loss of life–but only if people pay attention and act on the data.

The basic principles behind all methods of preventing or correcting mass movements relate to strengthening the earth materials involved (the resisting forces) and reducing the stresses within the system (decreasing the driving forces). For a given slide mass, this requires halting or reversing the factors that promote instability, which will probably be accomplished by one or more of the following: draining water from the slide area, excavating and redistributing the slide mass, or installing retaining devices. Debris flows require large structures that either divert the path of the flow away from buildings or capture the debris in basins.

9.6.1 Landslide Hazard Zonation

The first step in reducing losses caused by mass movement over the long term is to make a careful inventory or map of known landslide areas. Ancient landslides have a distinct topography characterized by hummocky (bumpy) terrain with knobs and closed depressions formed by backward-rotated slump blocks. On topographic maps, slide terrain has curved, closely spaced, amphitheater-shaped contour lines indicating scarps at the heads of landslides. In the field, one may see

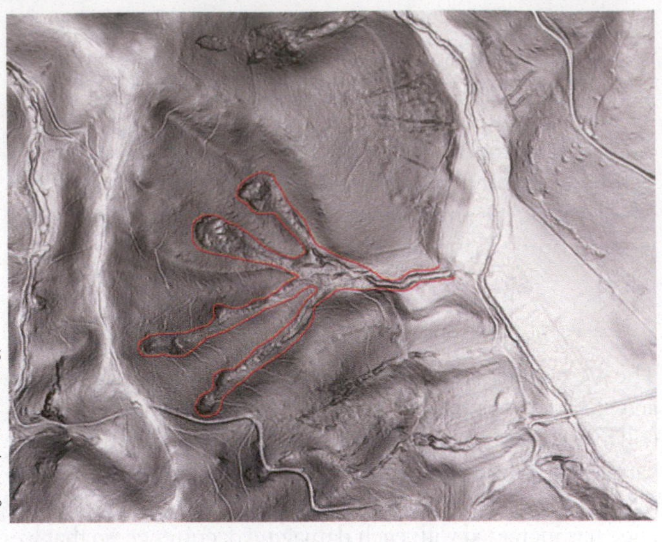
Virginia Department of Energy

Figure 9.71 Lidar image clearly reveals landslide scars in heavily forested Appalachian Mountains in Virginia. Slides are outlined in red. Fortune's Cove in Nelson County, Virginia.

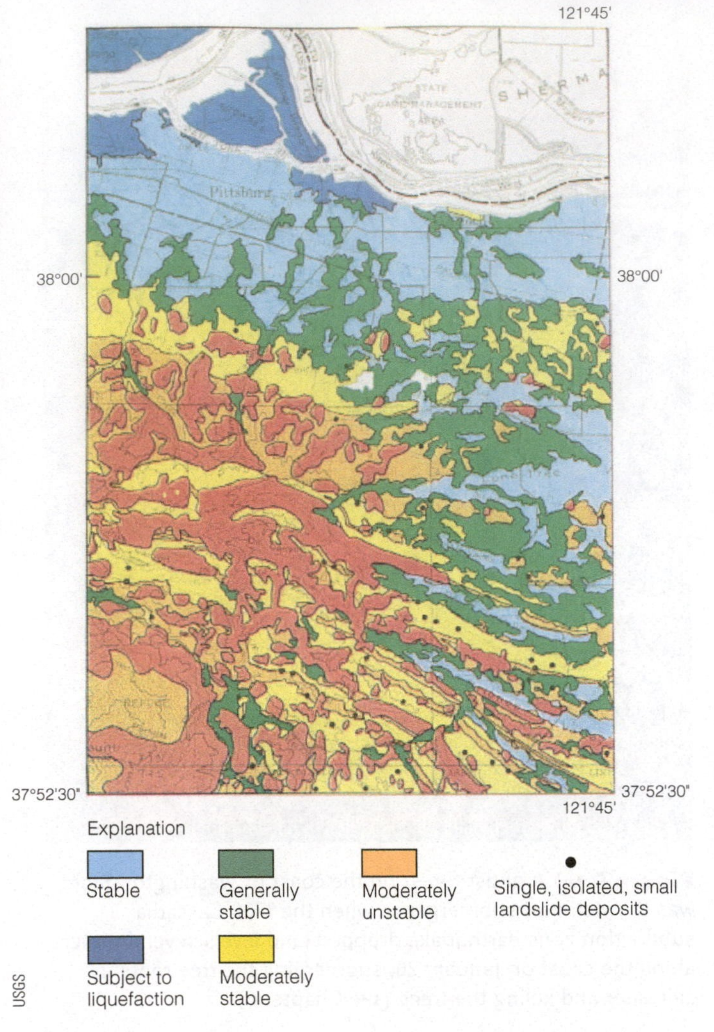

Figure 9.72 Map of the relative slope stability of parts of Contra Costa County and adjacent counties near San Francisco, California, derived by combining a slope map, a landslide-deposit map, and a map of susceptible geological units. The area shown here is 15 kilometers (9 miles) wide.

scarps, trees, or structures that are tilted uphill; unexpected anomalous flat areas; and water-loving vegetation, such as cattails, where water emerges at the toe of a landslide. Laser-based surveys from drones and planes now give us unprecedented data and even allow geologists and planners to "see through" heavy tree cover to find old slide deposits. Such lidar imagery has truly revolutionized the mapping of landslides and thus areas of potential hazard (**Figure 9.71**).

With an adequate database provided by maps and field observations, it is possible to define landslide hazard zones. Maps can be generated that show the landslide risk in an area; then these areas can be avoided, or work can be done to stabilize the landscape (**Figure 9.72**). Mapping programs do not prevent slides, but they offer a means for minimizing the impact on humans. The human tragedy of landslides in urban areas is that they cause the loss of land, as well as of homes and other infrastructure. **Table 9.1** summarizes

Table 9. 1 Summary of Mass-Wasting Processes and Their Mitigation

Landslides		
Cause	**Effect**	**Mitigation**
Excess water	Decreased strength	Horizontal drains, surface sealing
Added weight at top	Increased driving force	Buttress fill, retaining walls, decrease slope angle
Undercut toe of slope	Decreased resistance	Retaining walls, buttress fill
"Daylighted" bedding	Exposure of unsupported bedding planes	Buttress fill, retaining walls, decrease slope angle, bolt

Other Mass-Wasting Processes	
Process	**Mitigation**
Rockfall	Rock bolts and wire mesh on the slope, concrete or wooden cribbing at bottom of the slope, cover with "shotcrete"
Debris flow	Diversion walls or fences and catch basins
Soil creep	Deep foundations
Lateral spreading	Dewater, buttress, abandon

mitigation measures that can slow or stop landsliding and other mass-wasting processes.

9.6.2 Building Codes and Regulations

Building codes dictate which site investigations geologists and engineers must perform and the way structures must be built. Chapter 70 of the *Uniform Building Code,* which deals with slopes and alteration of the landscape, has been widely adopted by cities and counties in the United States. It specifies compaction and surface-drainage requirements, and the relationships between such planar elements as bedding, foliation, faults, and slope orientation. Code requirements for manufactured slopes in Los Angeles resemble the specifications found in codes elsewhere (**Figure 9.73**). The maximum slope angle allowed by code in landslide-prone terrain is 2:1; that is, 2 feet horizontally for each foot vertically (27°). The rationale for the 2:1 slope standard is that granular earth materials such as sand and gravel have a natural angle of repose of about 34°; 2:1 slopes allow for a factor of safety.

Losses have been cut dramatically by the enforcement of grading codes, and they will continue to decrease as codes become more restrictive in hillside areas. For instance, before grading codes were established in Los Angeles in 1952, 1,040 building sites were damaged or lost by slope failures out of each 10,000 constructed, a loss rate of 10.4%. With a new code in effect between 1952 and 1962 that required minimal geological and soil-engineering investigation, losses were reduced to 1.3% for new construction. In 1963, the city enacted a revised code requiring extensive geological and soils investigations. Losses were reduced to 0.15% of new construction in the following 6 years. These codes were updated and tightened again in 2011, as populations continue to increase in the greater Los Angeles region.

9.6.3 Water Drainage and Control

Water is the major culprit in land instability. If surface water can be prevented from infiltrating the potential slide mass, stability is the likely result. Water within the slide mass can be removed by drilling horizontal drains, called **hydroauger** (water bore) holes, and lining them with perforated plastic drainpipe that weeps water. This is commonly done on hillside terrain where groundwater presents a problem. In some areas, the land surface of a building site above a suspect slope is sealed with compacted fill to help keep surface water from percolating underground. Also, plastic sheeting is widely used to cover cracks and prevent water infiltration and erosion of the slide mass (**Figure 9.74a**). The efficacy of drainage control and planting is shown in Figure 9.67b.

When mass movements dam rivers, they create a whole new hazard: catastrophic dam failure. At Mt. Saint Helens, the sector collapse dammed Spirit Lake, resulting in the lake becoming several hundred feet deeper–a major concern for the Army Corps of Engineers. Four years after the eruption, they spent millions of dollars drilling a tunnel to partly drain the lake and reduce the risk of water rising overtop the dam and leading to catastrophic failure (**Figure 9.75**).

Landslides in tectonically active, mountainous regions often dam rivers flowing in deep valleys eroded into the rapidly uplifting landscape such as the Himalayan Mountains (**Figure 9.76**). Lakes dammed by glacial moraines are similarly fragile (**Figure 9.77**). When moraine and landslide dams fail, they cause outburst floods that can be deadly to people living along the river channel downstream.

Snow avalanches can also create temporary dams, which are much weaker than landslide dams and have the bad habit of melting and then failing. Such dams are often removed before the ephemeral lakes behind them can fill and over top the snow and ice blockage. Spring climax avalanches are particularly hazardous in this regard because they can block valleys during flood season (**Figure 9.78**). The Alaskan name for these beasts is suitably descriptive: "damalanche."

9.6.4 Excavation and Redistribution

Recontouring is one method of stabilizing a slide mass. Material is removed from the top of the landslide and placed at the toe. Compacted-earth structures called **buttress fills** are often designed to retain large, active landslides and known inactive ones believed to be vulnerable. They are constructed by removing the toe of a slide and replacing it, layer by layer, with compacted soil. Such fills can buttress huge slide masses, thus reclaiming otherwise unusable land. The finished product is required to have concrete terraces (for intercepting rainwater and preventing erosion) and downdrains for conveying the intercepted water off the site (**Figure 9.79**). Buttressing unstable land has aesthetic and economic advantages, in that the resulting slope can be landscaped and built upon. The cost of the additional work needed to stabilize

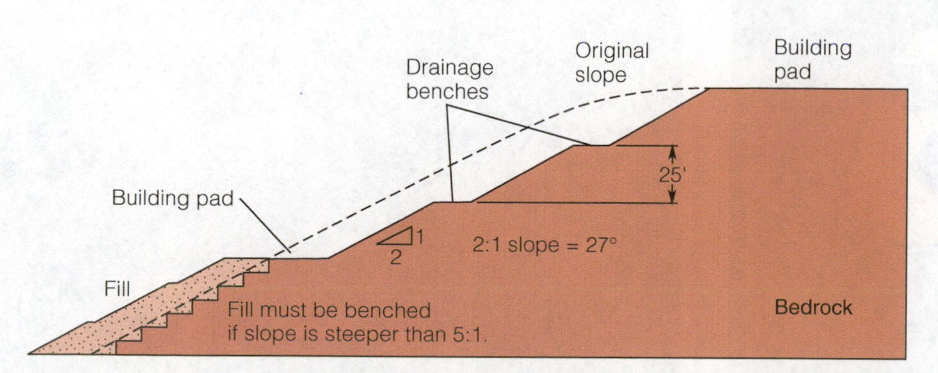

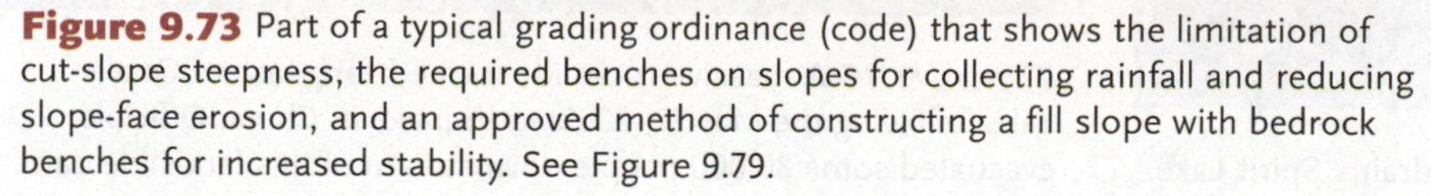
Figure 9.73 Part of a typical grading ordinance (code) that shows the limitation of cut-slope steepness, the required benches on slopes for collecting rainfall and reducing slope-face erosion, and an approved method of constructing a fill slope with bedrock benches for increased stability. See Figure 9.79.

Pam Irvine CDMG

Siang Tan CDMG

Figure 9.74 Excess water is the culprit in initiating many landslides. (a) Plastic sheeting placed over the head scarp of a landslide. This is, perhaps, closing the barn door after the horse has escaped but is common practice in landslide-prone areas. (b) Surface-water-erosion-prevention measures on a graded slope, San Clemente, California. Although slope planting and terracing have kept the middle slope in good condition, the slope on the left has gullied badly and will need maintenance in the future.

unstable land is usually offset by a relatively low purchase price, which can make it economically buildable. Of course, the ultimate mitigating measure is total removal of the slide mass and reshaping of the land to buildable contours. One has to wonder about the wisdom of such dramatic reshaping of our planet or whether it makes better sense to build someplace more stable.

9.6.5 Retaining Devices

Many slopes are over-steepened by cutting back at the toe, usually to obtain more flat building area. The vertical cut can be supported by constructing steel-reinforced concrete-block retaining walls with drain (weep) holes for alleviating water-pressure buildup behind the structure (**Figure 9.80**). Retaining devices are constructed of a variety of materials, including rock, timber, metal, and wire-mesh fencing. An interesting, if not particularly attractive, means of stabilizing weak slopes is the application of a cement cover. This material, known as **shotcrete**, is sprayed onto rock slopes, where it hardens into a skin (**Figure 9.81**). Steel nets can be used to catch falling rocks and protect people and roads below steep cliffs (**Figure 9.82**).

Steep or vertical rock slopes that are jointed or fractured are often strengthened by inserting long rock bolts into holes drilled perpendicular to planes of weakness. This binds the planes of weakness together and prevents

USFS

Figure 9.75 Inspecting the tunnel that now drains Spirit Lake.

China Photos/Getty Images News/Getty Images

Figure 9.76 A massive rockslide formed Tangjiashan Quake Lake following the May 12, 2008 earthquake in China. Officials evacuated some 80,000 people downstream of this landslide dam.

Figure 9.77 Lake Laguna Paron in the Cordillera Blanca, Peru, is dammed by a moraine from the rock glacier descending from the tributary valley to the left.

Figure 9.78 A lake formed outside of Valdez, Alaska, after a January 2014 "damalanche" blocked the Lowe river and backed water up over the only highway to town.

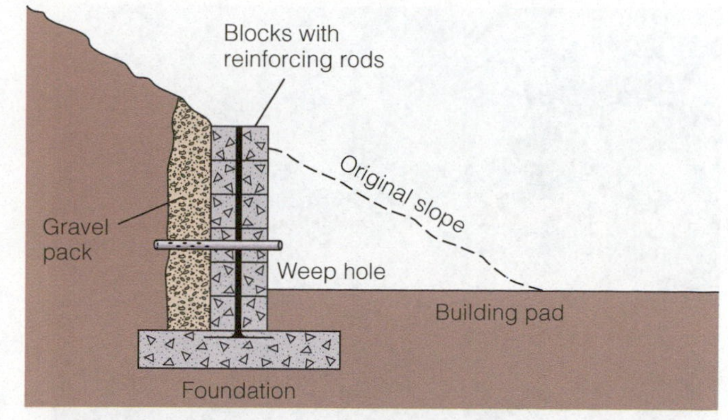

Figure 9.80 Typical retaining wall with drain (weep) holes to prevent water retention and the buildup of hydrostatic pressure behind the wall. The wall is constructed of concrete blocks with steel reinforcing rods and concrete in their hollow centers.

slippage. Rock bolts are used extensively to support tunnel and mine openings. They add considerably to the safety of these operations by preventing sudden rock "popouts." They are also installed on steep roadcuts to prevent rockfalls onto highways (**Figure 9.83**).

9.6.6 Mitigating Debris Flows

Mitigating debris flows presents a different challenge than landslides, because flows can originate some distance away from the site of interest. How can one avoid debris-flow hazards? The most obvious answer is to live away from steep slopes, and if you choose to live in a mountainous area, stay away from canyon bottoms and the mouths of canyons, which act as conduits for debris flows. Home sites by streams seem attractive until that stream is filled with a raging mixture of rock and mud. This is particularly true in areas of high mean annual rainfall, areas that are subject to sudden cloudbursts, and areas that have been

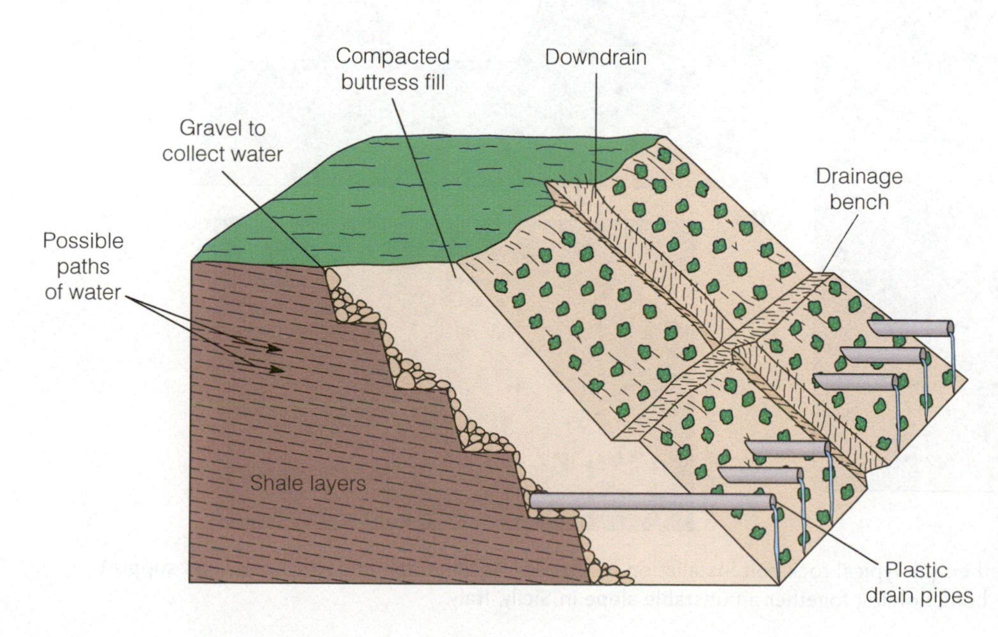

Figure 9.79 Buttress fill with horizontal drains for removing water and surface drains for preventing erosion. The fill acts as a retaining wall to hold up unstable slopes, such as those that result from excavation.

Figure 9.81 Shotcrete being sprayed onto a fragile rock wall to prevent lose rock from falling to the road below.

Figure 9.82 Along the steep, rocky sea coast just outside of Cape Town, South Africa, is an amazing road cut into steep, rocky sea cliffs. Rockfalls here are common. In addition to rock bolting, rock fall hazard is mitigated by the use of strategically placed steel nets that catch falling rocks before they land on the road or cars below.

burned, such as in the tragic October 1993 firestorms in Southern California. As long as people continue to build houses in the "barrels" of canyons, as is done in the foothills of Southern California, debris flows will continue to take their toll.

Debris-flow insurance is expensive, but the cost can be lowered if **deflecting walls** are designed into house plans (**Figure 9.84**). These walls have proved to be effective in protecting dwellings, but they can have adverse consequences if they divert debris onto neighbors' structures, in which case neither an architect nor a geologist would be as helpful as a lawyer. Los Angeles has taken a large-scale approach to debris-flow mitigation. It has built a series of large catch

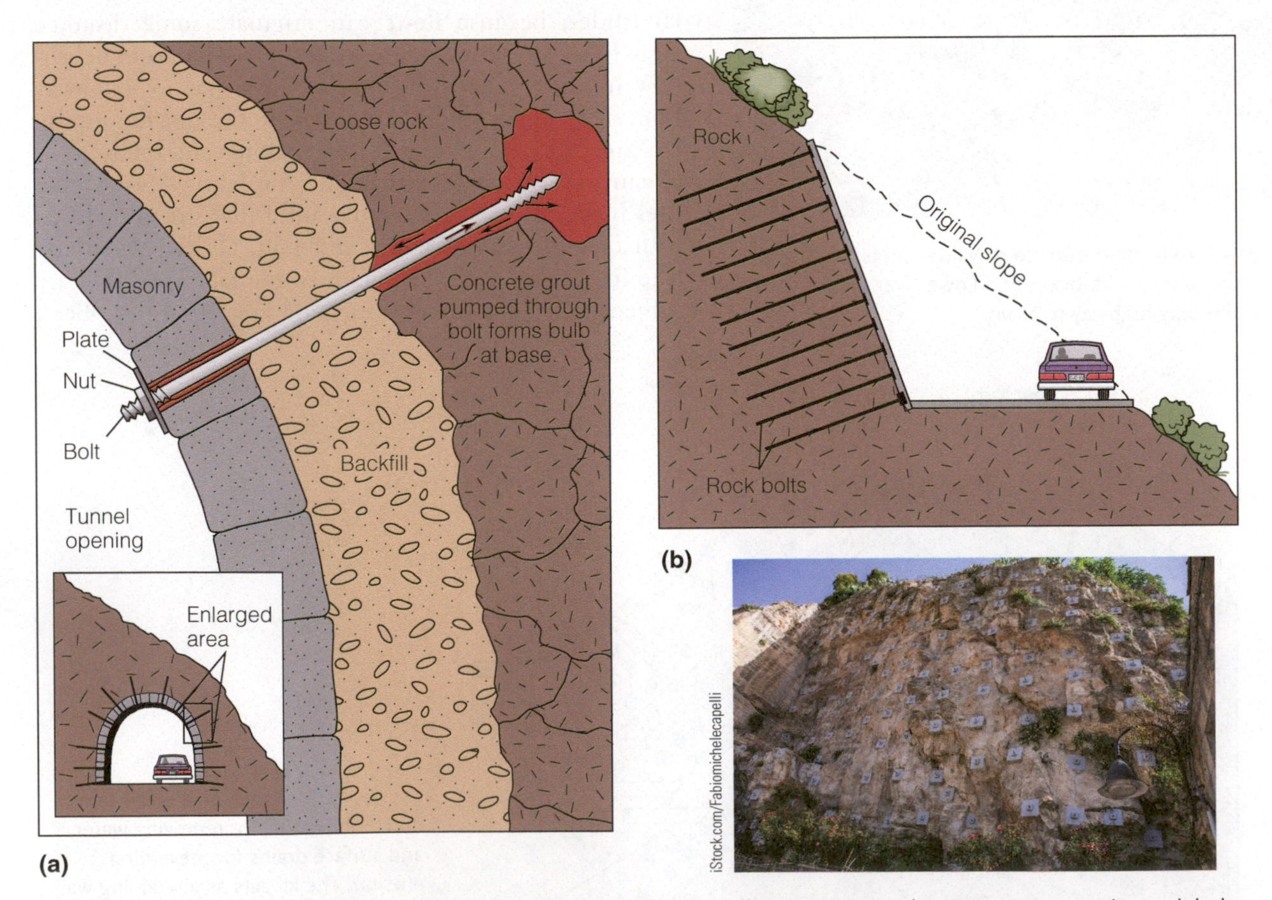

Figure 9.83 Keeping rock slopes together. (a) Typical rock bolt installation in a tunnel or mine opening. (b) Rock bolts support jointed rocks above a highway. (c) Rock bolts holding together an unstable slope in Sicily, Italy.

Figure 9.84 Mitigating debris flow hazards. (a) Debris flow diversion structures are designed to deflect flows moving downslope, and thus protect property from being damaged or destroyed by debris flows. (b) One way to deal with debris flows is to try to capture them before they can harm people and property. Los Angeles attempts this with a series of debris basins. Sometimes this approach works well, but other times the flows are so large that they overtop the basins and cover the houses below. Can you find the result of just such an overtopping in this photograph?

basins designed to trap debris flows before they can harm property (see Figure 9.84). These basins are situated in the upper reaches of deep canyons that drain the mountains and, until they fill, are effective means of capturing flows. Some have argued that the debris basins, similar to flood-control levees, provide a false sense of security for those living beneath them and encourage development in what are actually very hazardous places to live. John McPhee's book *The Control of Nature* has wonderful descriptions of these basins, including stories of how they do and do not work.

Of course, the basins need to be cleaned out with great frequency and the material dumped somewhere else. That does not always happen. Montecito, in southern California is a prime example of deferred maintenance. Reporting in the *Los Angeles Times* (December 20, 2018) suggests that failing to maintain these basins likely costs lives and millions of dollars of damage. In late 2017 and on into early 2018, the Thomas Fire charred more than 200,000 acres in Ventura and Santa Barbara counties near and above prime Southern California real estate (see Photo Galley at chapter end). When the rains came soon after, mud and boulders poured off the slopes (**Figure 9.85**), but the catch basins were full or nearly so. They hadn't been cleaned out in years. The result was 23 dead, 130 homes destroyed, and hundreds of millions of dollars in damage.

Alison Hardey, who owned a café in Montecito, summed it up well. "You feel how enormous this thing was … We're like ants against these mountains. We probably should have never been here to begin with."

Figure 9.85 Boulders surround and partially bury a home in Montecito, California, January 2018.

Case Study

9.1 When the Trees Come Down, So Do the Hills

Trees do so much for us. Their wood frames our homes, and their trunks and leaves store carbon, which would otherwise warm our climate. They give us oxygen to breathe and shade under which to relax on a warm day. But trees do something you may not have given much thought to before. They hold up many of our hillslopes. Trees have a dense network of roots, which binds the soil together. In terms of resisting forces, these roots provide cohesion until the tree dies and the roots rot, or maybe the stumps are pulled so that sheep can graze on open slopes. On dry days and for soils with high natural cohesion, the loss of roots and their assistance in holding the soil together matters little. But let it rain for a long time on sandy soils and the hillside will fail in landslides, perhaps sending debris flows hurtling down steep slopes to the valley below.

If the trees are grown for wood products, then the slope will soon be replanted for the next rotation of logging. The window is short between when the old roots rot and the new roots are numerous and large enough to hold up the slope—perhaps a decade or less (**Figure 9.86**). If no big storms hit in this window, everything's okay. But if the slope is kept clear and not replanted, it's only a matter of time before the mass movements will start. We know this because the experiment has been run over and over again, not by scientists but by the people who settled North America. First they cleared the Mid-Atlantic states and the slopes failed, then New England, then the Southeast, and finally the Pacific Northwest. Everywhere the loggers went, the slopes responded. The erosion is well documented in writings of the time (see the work of George Perkins Marsh, writing *Man and Nature* in the 1850s, who lamented the clearing of hillslopes and the erosion that followed). More recently, scientists such as Lee Benda, David Montgomery, and Beverley Wemple, geomorphologists who specialize in debris flows and erosion, have directly observed the effects of land clearance.

There is another data source, often overlooked, but available just about everywhere: historic images. Chances are, no matter where you live, with a little detective work, you can find 100-year-old images of where you live today and, with those images, do your own research about landscape change and response. Let's look at what historic images of Vermont can tell us about the results of that clear-cutting experiment our ancestors ran more than a century ago.

Study **Figure 9.87**, an attractive image of a stream in one of the most picturesque towns in Vermont, a town best known for having the longest-running agricultural fair in the state.

Paul Bierman

Figure 9.86 A logging road in southwestern Washington goes through a recently clear-cut stand of timber. In the distance are forests of different ages, each replanted after it was cut.

SERC

Figure 9.87 The deforested slopes of Tunbridge, Vermont, in the late 1800s. The sandy river terraces are being ripped apart by gullies and small landslides. All the sand is landing on the alluvial fans below. A horse stands at the apex of one of these fans.

First, find the river terraces, the three flat surfaces. Then, see if you can find the alluvial fans. Here's a hint: find the horse standing at the apex of one of the fans. Then, look above the horse and you'll see the eroding hillslope and the gully that is feeding the sandy sediment to the alluvial fan. Can you find a second fan? Can you find more erosion dumping sediment on the terrace below the alluvial fan? Geologists, graduate students, and faculty have dug up more than a dozen alluvial fans like the one in the photograph. Every one of them was covered with at least a meter (3 feet) of sediment deposited after land clearance; today, all the fans and the slopes above them are stable, covered in trees. But in the photograph is the smoking gun– evidence that the cleared slopes eroded and fed sediment to the alluvial fans below.

This is not an historical problem. It's relevant today. Forest clearance is a issue in developing nations where trees are removed for timber and so that agriculture can feed burgeoning populations and in some places, the American need for low price beef. (**Figure 9.88**).

iStock.com/Marcio Isensee e Sa

Figure 9.88 Deforestation in Brazil leads to erosion of thin jungle soils and the deeply weathered rock below.

9.2 Water, Water Everywhere

New Orleans has the flattest, lowest, and youngest geology of all major U.S. cities. The city's "Alps," with a maximum elevation of about 5 meters (16 feet) above sea level, are natural levees built by the Mississippi River, and the city's average elevation is just 0.4 meter (1.3 feet). No surface deposits in the city are older than about 3,000 years. The city was established in 1717 on the natural levees along the river. The levees' sand and silt provided a dry, firm foundation for the original city, which is now called *Vieux Carre* (French, "old square"), or the French Quarter. Farther from the river, the land remained undeveloped because it was mainly water-saturated cypress swamp and marsh formed between distributaries of the Mississippi River's ancient delta. These areas are underlain by as much as 5 meters (16 feet) of peat and organic muck.

When high-volume water pumps became available in the early 1900s (**Figure 9.89**), drainage canals were excavated into the wetlands to the north. Swamp water was pumped upward into Lake Pontchartrain, a cutoff bay of the Gulf of Mexico. About half the present city is drained wetlands lying well below sea and river level. The lowest parts of the city, about 2 meters (6.5 feet) below sea level, are also the poorest parts of the city and the parts most devastated by the flooding of Hurricane Katrina (**Figure 9.90**).

Subsidence is a natural process on the Mississippi River delta because of the great volume and sheer weight of the sediment laid down by the river. The sediment compacts, causing the land surface to subside. The region's natural subsidence rate is about 12 centimeters (4 inches) per century for the past 4,400 years. This estimate is based on carbon-14 dating of buried peat deposits and does not take into account any rise in sea level during the period. Urban development in the 1950s added to the subsidence.

© Alexey Sergeev

Figure 9.89 Hundred-year-old pumping station does its best to keep New Orleans dry by removing groundwater and rainwater from the city.

The compaction of peaty soils in reclaimed cypress swamps coincided with the construction of drainage canals and the planting of trees, both of which lowered the water table. Peat shrinks when it is dewatered, and the oxidation of organic matter and compaction contribute to subsidence. Most homes constructed on reclaimed swamp and marsh soils in

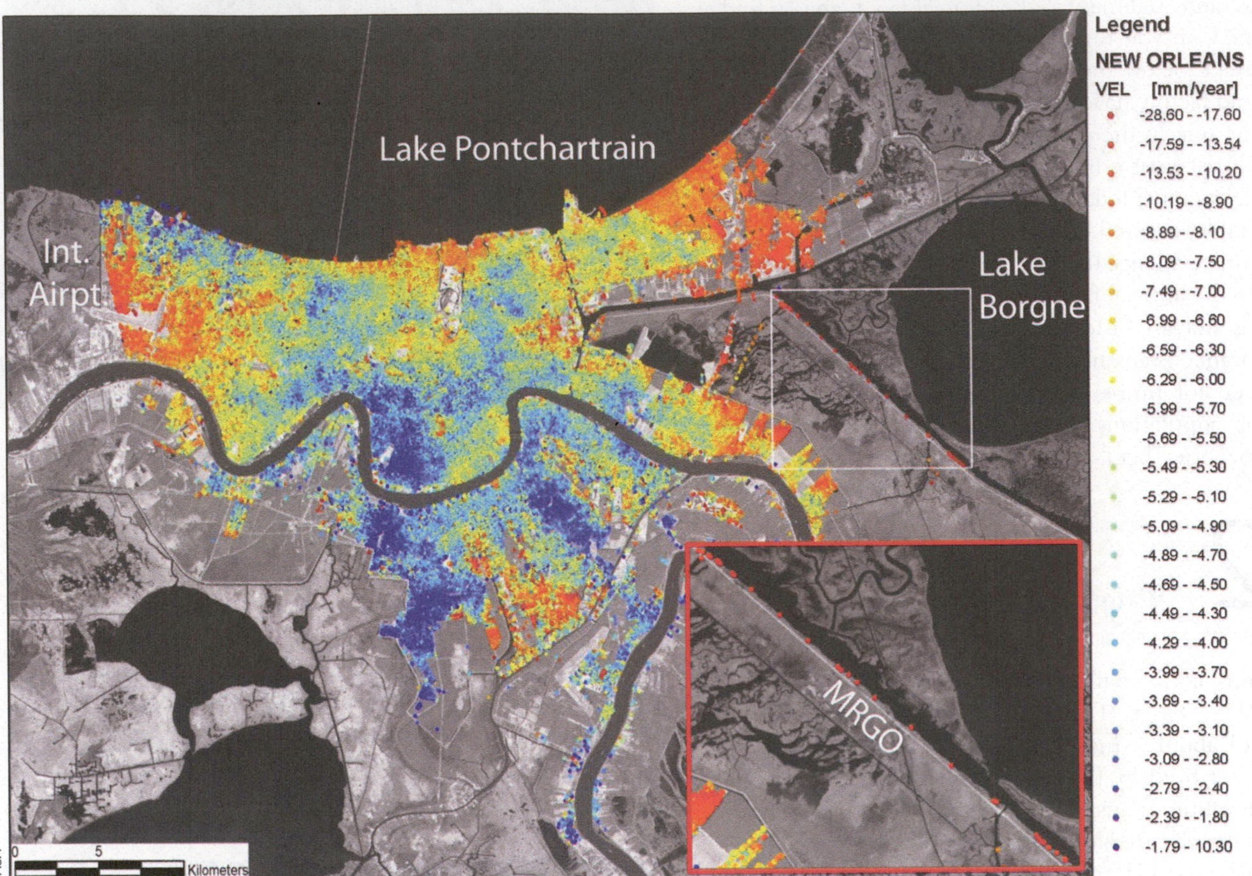

Figure 9.90 A map of flooding in New Orleans shows just how bad the situation was 5 days after Hurricane Katrina swept ashore. The map is based on combination aerial photography and laser elevation data. The color coding conveys water depth in the flooded areas. Most of the city, which is below sea level, is at least partially submerged.

the 1950s were built on raised-floor foundations. These homes are still standing, but they require periodic leveling. Unfortunately, homes built on concrete-slab foundations, a technique that had just been introduced, sank into the muck and became unusable.

Other home foundations were constructed on cypress-log piles sunk to a depth of 10 meters or more (at least 30 feet). These houses have remained at their original level, but the ground around them has subsided. This **differential subsidence** requires that fill dirt be imported to make the sunken ground meet the house level. It is very common to see carports and garages that have been converted to extra rooms when subsidence has cut off driveway access to them (**Figure 9.91**). One also sees many houses with an inordinate number of porch steps; the owners have added steps as the ground has sunk. In 1979, Jefferson and Orleans parishes (counties are called "parishes" in Louisiana) passed ordinances requiring 10- to 15-meter (33- to 50-foot)-deep wooden-pile foundations for all houses built on former marshland and swampland. This has been beneficial, but differential subsidence continues to damage New Orleans' sewer, water, and natural gas lines, as well as its streets and sidewalks. This subsidence will require ever-higher levees to protect New Orleans against storm surges and Mississippi River floods.

NOAA

Figure 9.91 This new Orleans home was built on pilings, which have kept it from settling; but not so the lawn and driveway.

Slip, Slide, and Fall

Mass-wasting processes are tragic, devastating, and ongoing. Both the geological and photographic records are clear: this hazard is not going away. It's not clear we have learned much as a society despite repeated tragedies and the loss of life and property. Houses are still built on the edges of unstable cliffs and on the deposits of old debris flows. The past is a clue to the future.

© University of Vermont Special Collections Howe Library

1 A man stares out at clear-cut slopes sometime in the late 1800s. Can you find the shallow, translational landslide at which he's looking? The slope failed because the trees were cut and the effective cohesion their roots provided was lost.

iStock.com/Atlantic-kid

2 Rockfall is a hazard wherever steep bedrock slopes tower over roads and buildings. In Spain, a rockfall blocked the road and sent a car, parked in the wrong place, to the junkyard.

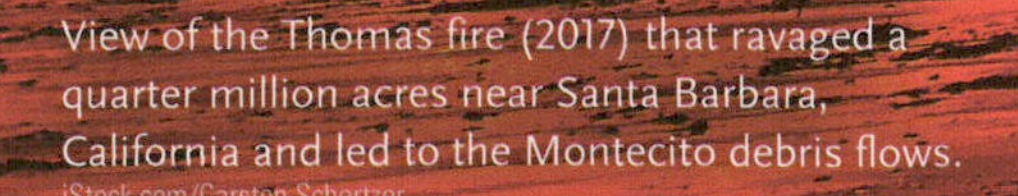

View of the Thomas fire (2017) that ravaged a quarter million acres near Santa Barbara, California and led to the Montecito debris flows.
iStock.com/Carsten Schertzer

Windoast/Shutterstock.com

3 In 1903, the side of Turtle Mountain in Alberta collapsed in the massive Frank rock slide, killing almost 100 coal miners and causing Canada's worst mining disaster. The slide path and the debris apron of broken rock are clearly visible today, more than 100 years later.

iStock.com/Matthew J Thomas

4 Landslides along steep seacoasts are common. On the East Yorkshire Coast of Britain waves have undercut the sea cliff and the road is now a difficult drive, unless you are on a bicycle.

Paul Bierman

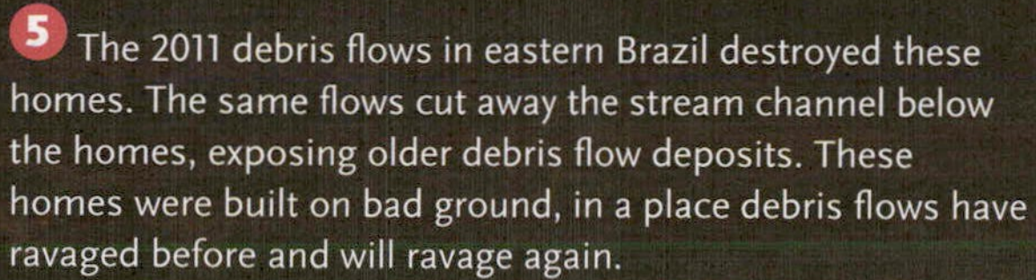

5 The 2011 debris flows in eastern Brazil destroyed these homes. The same flows cut away the stream channel below the homes, exposing older debris flow deposits. These homes were built on bad ground, in a place debris flows have ravaged before and will ravage again.

Summary Erosion

1. The Basics of Mass Movements

LO1 Explain how the balance of driving and resisting forces controls the occurrence of mass movements

a. The Mechanics of Mass Movements

A mass movement occurs when the driving force (a the component of gravity acting down a potential slip plane is greater than the resisting force (the force which strengthens a slope including friction). The ratio of these forces is called the factor of safety. When the factor of safety is 1.0, the forces are equal and the slope is close to failure but not moving. When the factor safety is less than 1.0, the driving force is greater than the resisting force and mass movement occurs.

b. Triggering Mass Movements

A mass movement is triggered when the factor of safety falls below 1.0 and the driving forces exceed the resisting forces. There are many examples of mass movement triggers; heavy rainfall increases pore pressure, earthquakes change the force balance and human activities like mining and building may increase the driving force or reduce the resisting force.

c. Subtle But Important Mass Movement

Not all mass movements are sudden or rapid. Creep is the gradual downslope transport of mass and can be as slow as a few millimeters of movement per year. Over time, creep deforms and changes the shape of slope materials and trees.

Study Questions

i. What forces determine if a mass movement will occur?

ii. Give three specific examples of how mass movements are triggered.

2. Landslides, Rock Falls, and Lateral Spreads

LO2 Compare the processes leading to rockfalls, landslides, and lateral spreads and the effects of these mass movements on people

a. Rockfalls

Rockfalls are a simple, rapid mass movement in which rock detaches from a cliff and falls. Rocks can be loosened by root growth, frost wedging, and heavy precipitation.

b. Landslides

A landslide refers to any mass movement that involves a sloping mass with well defined boundaries. Slumps are the most common kind of landslide, occurring in homogenous material, moving about a center of rotation as they head downslope, and leaving a concave-upward, spoon-shaped surface on the slopes from which they slide. Block glides typically occur on planar sedimentary and metamorphic rock bedding/foliation planes. They take place where planes of failure are less steeply inclined than the adjoining slopes.

c. Lateral Spreading

On flat ground, lateral spreading occurs when coherent blocks of earth move horizontally over a liquefied layer of sediment at depth typically made of water-saturated sensitive clays. This phenomenon is often triggered by earthquakes.

Study Questions

i. What are the distinguishing characteristics of the two types of landslides?

ii. What can trigger a lateral spread?

3. Debris Flows

LO3 Explain why, when, and where debris flows will occur

a. Behavior of Debris Flows

Debris flows are generated quickly during heavy rainfall, snowmelt, and volcanic eruptions. The water combines with loose rocks, sand, and mud to form a one-phase fluid. Depending on the slope and its composition, debris flows may travel rapidly and merge, creep along, or stop.

b. Debris Flow Triggers

A debris flow is a phenomenon that can occur anywhere, typically in regions with sparse vegetation and intense seasonal rainfall. Volcanic eruptions cause debris flows, called lahars.

b. Debris Flow Prediction and Occurrence

Field data collected over time have been used to create intensity/duration envelope diagrams, which show the combination of how long rainfall needs to last and at what intensity before a debris flow will begin. Other factors which effect the likelihood of a debris flow are antecedent moisture conditions, slope steepness, and amount of vegetation.

Study Questions

i. What does it mean that a debris flow is a one-phase fluid?

ii. How can the impact of debris flows be mitigated?

4. Snow Avalanches

LO4 Explain what conditions make snow avalanches likely and thus how to mitigate their impact

a. Types of Snow Avalanches

Slab avalanches slip as a coherent block, oftentimes released on slopes were windblown snow has accumulated to form cornices. Climax avalanches occur when snowpack begins to melt and weakens. These can be some of the most devasting avalanches due to the amount of the wet snow accumulating at the base of the slope. Powder-snow avalanches are dry, fast-moving avalanches which typically occur after cold storms. Sluff avalanches are the result of large volumes snow-fall, are typically smaller-scale and short lived.

b. Snow Stability Analysis and Avalanche Prediction

Analyzing layers of snow and ice in snow pits, observing terrain anchors, and being aware of heavy snowstorms are all useful ways of forecasting avalanches.

c. Mitigating Avalanches

The best strategy to mitigate the hazards from avalanches is to avoid avalanche-prone areas during hazardous conditions. It's best to avoid steep slopes after heavy snow, or in areas where there are few terrain anchors. Avalanches can be triggered in a controlled environment when conditions suggest they are likely, oftentimes in fully evacuated ski areas. Measures are taken to maintain safe ski areas, so stay on groomed runs to avoid dangerous avalanches.

Study Questions

i. What are two examples of safety equipment developed to help save humans in an avalanche?

ii. List three types of avalanches and their defining characteristics.

5. Subsidence

LO5 Reflect on why and where subsidence happens and how it can be managed and/or avoided

a. Human-Induced Subsidence

Humans can cause subsidence by extracting large volumes of groundwater or by engaging in mining or oil and gas production, or when loose sediments are compacted at the ground surface.

b. Natural Subsidence

Subsidence happens naturally, often caused by earthquakes, volcanic activity, and the dissolution of limestone, dolomite, and gypsum. Large-scale subsidence occurs when earthquakes occur on large fault lines, lowering land surface, or due to liquefaction.

c. Classification of Subsidence

Subsidence is often defined by the depth at which the subsidence is initiated. Deep subsidence is typically a consequence of oil or gas extraction many hundreds of meters below the ground surface. Shallow subsidence results from groundwater withdrawals and also compaction from construction and other human activities.

d. Sinkholes

A sinkhole is formed when carbonic acid dissolves carbonate rock, forming a near-surface cavern. As the dissolution continues, the ground-surface (cavern's roof) can collapse, resulting in a sinkhole. Sinkholes are a natural phenomenon, but their formation can be accelerated by human activities like groundwater pumping and removal.

e. Mitigating Subsidence

Monitoring and controlling regional water table levels in regions known to be heavily impacted by subsidence is the most powerful way humans can mitigate subsidence.

Study Questions

i. What are two natural and two human causes of subsidence?

ii. List three hazards created by subsidence.

6. Reducing Losses from Mass Movement

LO6 Explain how the risks of mass movements to people and society can be reduced using your understanding of how mass movements work

a. Landslide Hazard Zonation

The best way to mitigate landslide hazard is to predict when, where, and under what conditions mass movements are most likely to occur. Research and construction of databases and landslide hazard maps help humans make conscious decisions regarding safety and stability when building new infrastructure.

b. Building Codes and Regulations

Building codes and regulations provide standards by which infrastructure must be built, where, and on which materials to improve the safety of buildings. Building codes greatly reduce losses from mass movements.

c. Water Drainage and Control

Water is a main culprit in land instability. To prevent surface water from infiltrating into a potential slide mass, hydrauger holes help drain water away from unstable areas. Mass movements (avalanches, landslides, rockfalls.) can also dam rivers, which increases the risk of catastrophic dam failure or outburst floods.

d. Excavation and Redistribution

Recontouring and buttress fills are two examples of excavation and redistribution of materials within a landslide to provide stability. There are advantages to this type of mitigation for safety, aesthetics, and land use; however, it also requires significant human intervention to reshape the land.

e. Retaining Devices

Retaining devices are infrastructure composed of rock, timber, metal, and wire mesh fencing as a means of stabilizing weak slopes and fractured steep or vertical slopes that endanger human safety.

f. Mitigating Debris Flows

The safest way to mitigate the hazards of debris flows is to build infrastructure away from the bases of steep slopes, within or below canyons, or other places at high risk for debris flows. Deflecting walls can protect homes, but can also deflect flows to other, unprotected buildings. Los Angeles has built large catch basins to capture debris flows on a large scale. While effective, these catch basins introduce their own problems, including the need for frequent maintenance.

Study Questions

i. What are two examples of policies in place to mitigate mass movement hazards?

ii. What are two examples of infrastructure used to mitigate mass movement hazards?

Key Terms

Angle of repose
Anticline
Avalanche tracks
Bentonite
Block glides
Climax avalanche
Cohesion
Creep
Debris flows
Deep subsidence
Depth hoar
Differential subsidence
Dip slope
Driving force
Effective cohesion
Factor of safety
Failure plane
Force balance
Frictional strength
Granular material
High-marking
Hydrocompaction
Ice crust
Intensity/duration envelope diagrams
Karst terrain
Landslide
Lateral spreads
Mass movements
Normal force
Pore pressure
Porewater pressure
Powder-snow avalanche
Resisting force
Rockfalls
Rock avalanche
Rotational slides

Run-out zone
Settlement
Shallow subsidence
Shotcrete
Slab avalanche
Sluff avalanche
Slumps
Starting zone
Subsidence
Surface hoar
Talus
Terrain anchors
Thaw slumps
Translational slides

"But where, after all, would be the poetry of the sea were there no wild waves?"

—Joshua Slocum, the first person to sail single-handedly around the world, taking three years between 1895 and 1898

10

Coasts and Oceans

Learning Objectives

LO1 Explain how waves change as they move toward shore and the water shallows

LO2 Describe how shorelines respond to changing seasons and the passage of large storms

LO3 List the three phases of a tsunami and identify which one causes the greatest hazard

LO4 Explain how hurricanes work and why they occur most often during the summer and fall

LO5 Contrast nor'easters with hurricanes, explaining how the storms are similar and different

LO6 Explain how human actions can change the flow of sediment to and along the world's coastlines

Waves and storm surge of Storm Imogen, the most intense storm of the winter 2015/16 season, batter the south coast of England at Porthleven in Cornwall.
iStock.com/Moorefam

The 2004 Indonesian Tsunami: A Quarter Million Perish

Indonesia is no stranger to earthquakes, volcanic eruptions, and other spasms of our restless planet. However, within hours after the largest earthquake to strike Earth in 40 years (since the Alaskan and Chilean quakes of 1964) shook the region on December 26, 2004, the shoreline was unrecognizable and 230,000 people lay dead, mostly drowned–among the greatest losses of human life from a single event in world history. What happened that December day? Why was the death toll so staggering?

The moment magnitude-9.2 quake originated 10 kilometers (6 miles) below the surface, where the Indian and the Burma plates meet in a locked subduction zone–locked, that is, until the quake struck. When more than 1,000 kilometers (600 miles) of the fault slipped, the seafloor rose as much as 15 meters (50 feet) in places. Areas close to the quake had little warning as waves, moving in deep water at the speed of an airplane, followed the earthquake within minutes.

One coastal city, Banda Aceh on the island of Sumatra, was hit especially hard and fast. Nearly a third of its residents were dead or missing as a result of the repeated tsunami waves that surged through the city. The waves, and the debris they carried, smashed most everything in their path. Field measurements show that the tsunami waves, when they came ashore, were astonishingly high; in some places, run-ups exceeded 30 meters (100 feet). Homes, businesses, and people were simply swept away (**Figure 10.1**).

The damage was not limited to the area near the quake. As the waves spread across the Indian Ocean, so did the tragedy (**Figure 10.2**). In just a few hours, beach resorts in Thailand were demolished as water surged over hotels and homes. Next, the coast of Sri Lanka was assaulted by the waves and finally, hours later, the east coast of Africa, thousands of kilometers across the Indian Ocean. Tsunami waves were even detected in the Atlantic Ocean, where tide gauges recorded waves exceeding 0.3 meters (1 foot) in height along U.S. and Canadian coasts. Farther from the quake, fewer people died–in part because the waves decreased in height and in part because tsunami warnings allowed people to respond.

What can be done to prevent this and similar tragedies from reoccurring (**Figure 10.3**)? Simple measures such as public education about what to do when the Earth shakes can save lives. A sudden withdrawal of water from a shoreline that widely exposes the seabed is a dangerous development requiring people to flee inland as fast as possible. This must be publicly recognized from an early age. A coordinated, region-wide warning system that relays information (much like that already in place to warn of cyclone hazards) would likely save even more lives. Establishing a direct tsunami monitoring system in the Indian Ocean was a top priority after 2004. Currently, the Indian Ocean Tsunami Warning System (IOTWS) uses pressure recorders placed on the seafloor to measure how the weight of the water changes. The data are recorded in buoys at the water's surface and picked up via satellite signal to receiving stations. Thus, when a tsunami is detected, a communication network can alert potentially impacted communities before impact. Some in India have suggested replanting mangroves along developed coastlines to diffuse tsunami energy. The coastal trees, and the swamps they create, moderate tsunami waves in the same way coastal wetlands tame large waves and storm surges. Some people are taking matters of tsunami safety literally into their own hands. If you live in a region that could suffer tsunamis and own a cell phone, you can download an app that uses real-time earthquake data to warn of potential tsunami hazards.

Illinoisphotos.com/Philin A. McDaniel/U.S. Navy

Figure 10.1 A week after the December 26, 2004 tsunami, flooding and damage to a village near the coast of Sumatra, Indonesia, were on a scale so massive that it is hard to comprehend. In this image, can you find the limit of inundation that indicates how far the tsunami ran inland?

questions to ponder

1. Do you live in an area that could be affected by a tsunami? If so, do you know how to reduce your risk?
2. Study the map in Figure 10.2. Why were the west coast of India, and the shoreline of the Arabian Peninsula (across the narrow sea north of Somalia) little impacted by the tsunami waves, whereas coasts not far away from them (e.g.—the east coast of India) so seriously affected by the waves?

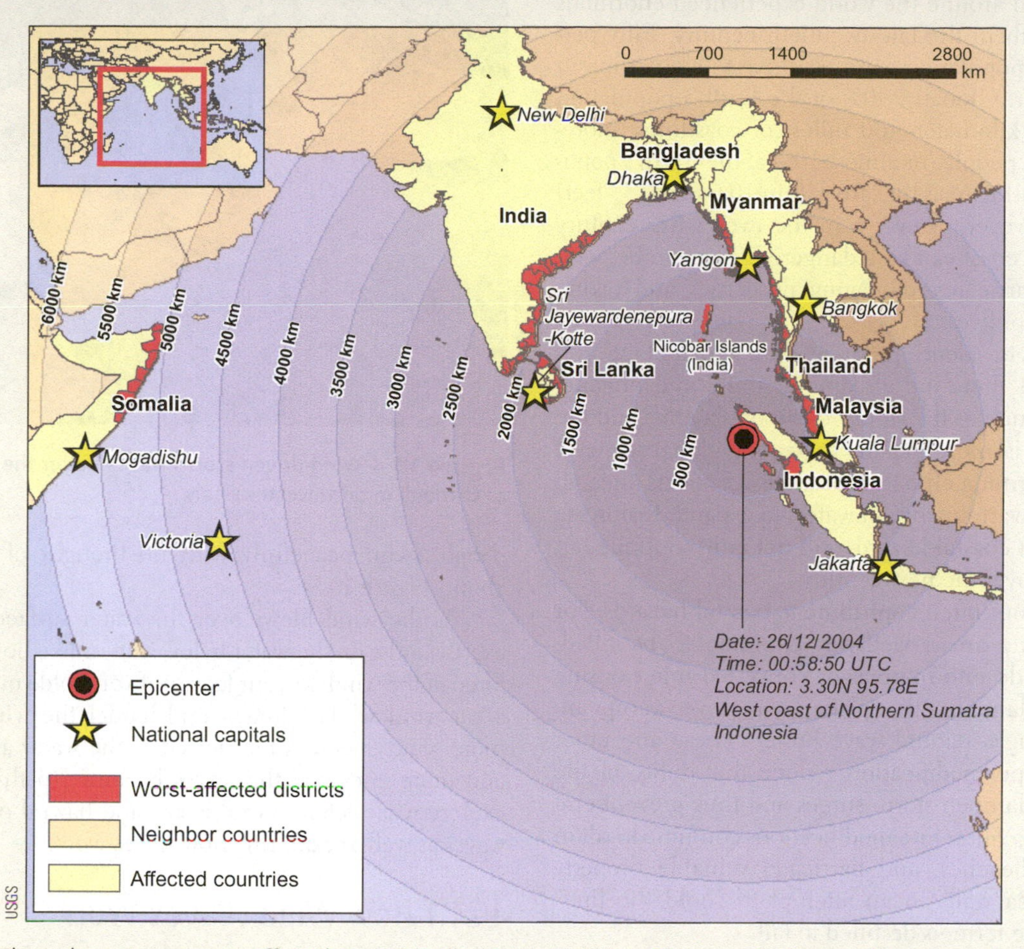

Figure 10.2 The Indonesian tsunami affected countries all around the Indian Ocean basin. Some of the waves traveled thousands of kilometers before crashing ashore.

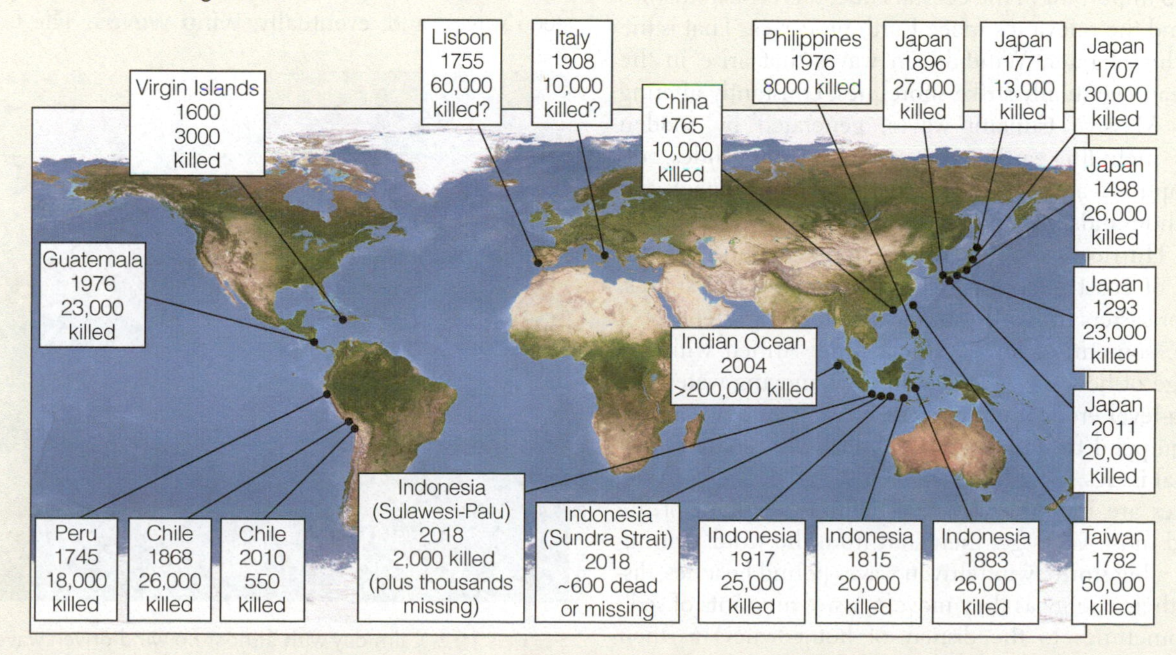

Figure 10.3 The Indonesian tsunami may be the most deadly we know of, but it is only one of many that have struck in recent history. (Adapted from Bryant, Edward. 2001)

10.1 Wind Waves (LO1)

Many people want to live near a beach. Coastal areas in the United States and around the world experienced enormous population growth in the late twentieth century. Fifty percent of the U.S. population can now drive to a coastline in less than an hour. Globally, 40% of the Earth's population lives within 100 kilometers (60 miles) of coastlines; moreover, 600 million people (or almost 10% of the Earth's population) live in coastal areas that are within 10 meters (33 feet) of sea level. However, as events in the twenty-first century clearly show, the coast can be a dangerous place to be, with tsunamis and hurricanes damaging properties and taking many thousands of lives.

These discrete events produce dramatic footage and photographs, but there are also more subtle and ongoing coastal hazards such as the slow erosion of sea cliffs and the retreat of barrier islands. Steadily increasing global sea level is driven by a warming climate that melts ice sheets and glaciers and causes warming ocean waters to expand. Rising sea level ensures that coastal hazards will not only continue but also grow worse over the next century.

Human actions often contribute to coastal hazards. For example, in some areas, wetlands continue to be filled, drained, and made into marinas or converted into housing and agricultural land. In other places, mangrove swamps are cleared, removing a natural wave buffer. These and other types of landscape modification reduce the ability of the coastal zone to dampen storm surges and thus prevent erosion and flooding. Shoreline modifications continue to silt in harbors, erode beaches, and barricade valuable property with concrete sea walls in an attempt to "hold the line" which in the long term, is destined to fail.

To understand and approach the environmental issues that are so important in the coastal zone, it is critical that we understand the relevant surface Earth processes. That is the goal of this chapter. Wind-driven waves that arise in the open ocean eventually strike shorelines, constantly altering beaches. Massive tsunami waves, generated by sudden impulses, including earthquakes, submarine landslides, volcanic eruptions, and bolide (asteroid or comet) impacts can change thousands of kilometers of coastline in a matter of minutes. Hurricanes, borne in the tropics, pound smaller stretches of coast for hours, whereas extra-tropical storms, called nor'easters in North America, batter long stretches of coastline, sometimes for days at a time. Armed with the knowledge of how waves work, and how moving water and rising sea-level erode and flood the coastal zone, we are all better equipped to understand and deal with coastal environmental issues.

Waves are in some ways the defining element of the coast and most waves get their start from the wind. During and after a big storm, wind-driven waves pound beaches, dissipating their energy as they move massive amounts of sediment, sometimes to the dismay of homeowners as their homes fall victim to coastal erosion (**Figure 10.4**). On a calm day, waves lap gently onto the shore, children play, and people swim peacefully with no thought of being tossed about (**Figure 10.5**).

iStock.com/Antonioscarpi

Figure 10.4 Wind-driven storm waves batter the coastal town of Camogli in northwestern Italy.

As the wind blows over the water surface, moving air exerts a force on the water below (not unlike how you feel the force of the wind on your face or if you hold your hand out of a car window. The longer and harder the wind blows, the more wind energy is transferred to the water and the bigger and more energetic the waves become. In this section, we explore what determines the size and hazard of wind-driven waves as well as exploring how waves work.

10.1.1 How Wind Waves Work

When wind blows over smooth water, surface **ripples** are created that enlarge with time to form local **chop** (short, steep waves) and, eventually, **wind waves**. The following

iStock.com/Sungwoo Kim

Figure 10.5 Calm day with almost no wind-driven waves on a South Korean beach. The dark patch in the center is likely a seaweed farm.

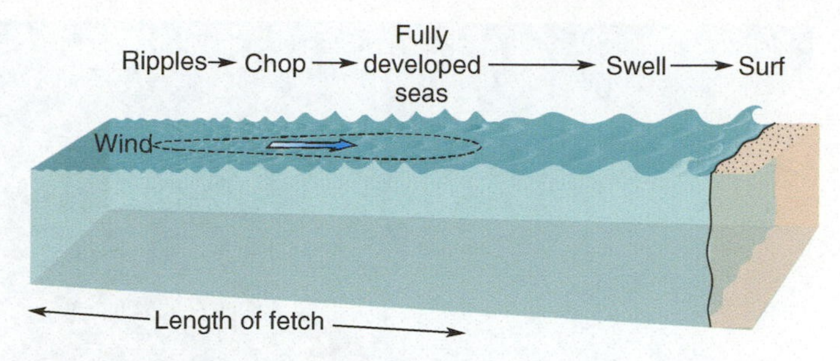

Figure 10.6 The evolution of wind-generated sea waves from a rippled water surface to fully developed seas (storm waves). Long waves with very regular spacing and height that outrun the storm area are called *swell.* Surf only occurs when the waves closely approach and strike land.

factors determine the size of waves at sea and ultimately the size of the surf that strikes the shoreline:

- Wind speed
- The length of time the wind blows over the water
- The **fetch**, the distance along open water over which the wind blows (**Figure 10.6**)

For a given wind speed, wind duration, and fetch, wave dimensions will grow until they reach the maximum size that they can. This condition is known as a **fully developed sea**, and such a sea growing during a severe storm can take as long as 10 hours to form. The sea becomes "fully developed" when the energy added to it by wind energy becomes equal to the energy lost as the waves break into foam. This knowledge is empirical, the result of years of measurement. For example, wave heights of 32 meters (100 feet) are possible for a 50-knot (93-kilometers/hour or 58 mph) wind blowing for 3 days over a fetch of 2,700 kilometers (1600 miles). Wind waves will move out of a stormy area as regularly spaced, long waves known as **swell** and eventually become huge breakers on a distant shoreline. In other words, wind waves can not only outlast the wind that spawned them but move far away from where they originated! Some of the largest, finest surfing waves in the world are related to incoming swells.

Important dimensions of wind waves are their height (also known as **amplitude**), **wavelength** (the distance between wave crests), and period (the length of time between the passage of equivalent points on the wave; see also Chapter 6). These dimensions are measures of the amount of energy the waves contain–that is, their ability to "do work" such as lifting boats and transporting sediment (**Figure 10.7**). Although waves with long wavelengths and long periods are the most powerful, both long and short period and wavelength waves are capable of eroding beaches.

You may imagine a wave moving through deep water as analogous to the rippling of a rope that you snap at one end. (Perhaps you have a friend holding the other end of the rope to keep it still). Note that the individual particles of rope do not rush down to the far end as they race along. It is only the form of the ripples themselves–the "wave energy"–that advances along the rope. Perhaps if you snap the rope hard enough a wave will return to you, but only as a diminished reflection of the wave energy off your friend's hand holding on at the far end.

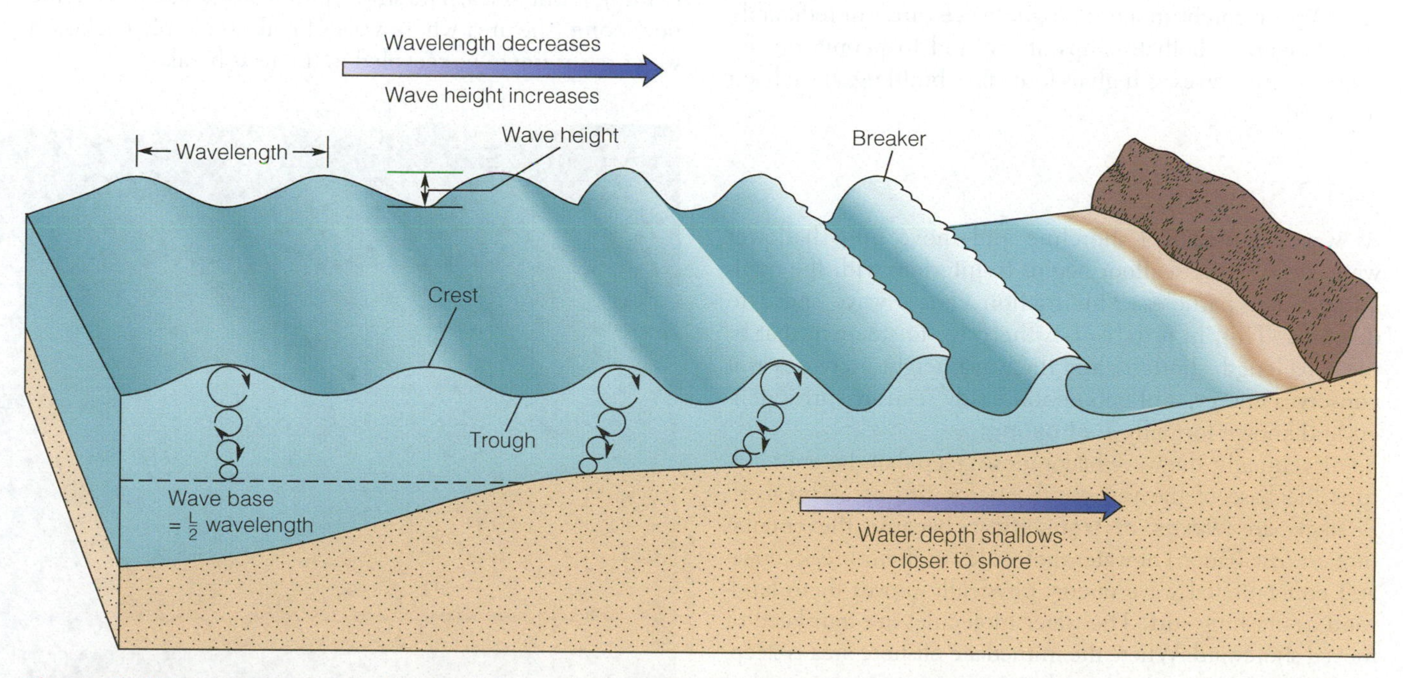

Figure 10.7 Wave motion is complex and changes as the wave approaches shore and begins to "feel" the bottom, or shoal. Small circles with arrows show the orbital motion of water in a wave. The terminology of wave geometry is shown, including wavelength and wave height. Note that the orbital water motion decreases with depth until it is essentially nil at a depth of half the wave length ($L/2$), a depth called *wave base.*

Similarly, the actual motion traced by individual water molecules in a body of water through which a wave passes describes a circular path, the diameter of which at the sea surface is equal to the wave height. You can see this clearly if you toss a cork or other float into the water and watch what happens to it as the wave passes by. The cork, a proxy for all the surrounding water molecules, moves up-and- down and back-and-forth, but the wave keeps on going, as though ignoring it, in one direction. The circular motion diminishes with depth and becomes essentially zero at a depth equal to half the wavelength ($L/2$), the depth referred to as **wave base** (see Figure 10.7). Many marine creatures such as fish and whales (and people in submarines) escape the fury of storms; by hovering just a short distance beneath the wave base.

Another way of looking at this is that the wave base is the maximum depth to which waves can disturb or erode the seafloor. The form of the wave has particular components we can more easily sea at the surface: **Wave crests** are the ridge like tops of waves, and **wave troughs** are the valley or gulley-like depressions at their low points. Mariners in a rough sea often try to aim toward and keep their boats at a right angle to oncoming waves both to see what's coming and to avoid having their craft rolled over and capsized as crests and troughs pass by. This is especially important when the wavelengths of waves exceed the length of the boat. This is, as you can imagine, a pretty rough ride, but the margin of safety is generally worth it.

Sometimes a **rogue wave**, a solitary wave much larger than others nearby, appears. The formal definition is a wave twice as tall as the mean of the top 30% of waves for the state of the sea that day. Such exceptional waves appear to form when two waves come together; they combine their energies. As you might imagine, rogue waves are unpredictable and dangerous both to ships at sea and to people on the coast. Rogue waves as high as four-story buildings have been recorded.

10.1.2 Shoaling

As waves approach a shoreline and move into shallower water depths, the seafloor begins to interfere with the oscillating water particles. This friction causes wave speed to decrease. This, in turn, causes the wave to steepen; that is, the wavelength shortens, and the wave height increases as it conserves energy, a phenomenon easily seen by surf watchers and known formally as **shoaling**.

Eventually, the wave becomes so steep that the crest outruns the base of the wave and jets forward as a "breaker." Breakers are **waves of translation**, because the circular motion (oscillation) of water molecules in them breaks down and concentrated wave energy physically moves them landward as crashing **surf**. The water molecules are translated or moved shoreward. Where the immediate offshore area is steep, we find the classic tube-shaped **plunging breakers** typical of Sunset Beach and Waimea Bay on the north shore of O'ahu and many other Pacific Islands (**Figure 10.8**). This is because the wave peaks rapidly on the steeply sloping seafloor.

iStock.com/Helivideo

Figure 10.8 Plunging (tubular) breakers form where there are steep offshore slopes. The front of the wave steepens until the top of the wave "jets" forward and down. This one is in French Polynesia. Such waves only break during the biggest winter storms, which occur a few times each winter.

On a flat nearshore slope, in contrast, waves lose energy by friction, so they build up slowly, and the crest simply spills down the face of the wave. This type is called a **spilling breaker** and is typical of wide, flat, nearshore bottoms, such as those around much of the United Kingdom (**Figure 10.9**).

Regardless of the type of wave, **swash**–water that rushes forward onto the foreshore slope of the beach when waves come ashore–returns seaward as **backwash** to become part of the next breaker. A common misconception is the existence of a bottom current, called **undertow**, that pulls swimmers out to sea. The strong pull seaward one feels in the **surf zone** (the area where waves break) is simply backwash water returning to be recycled in the next breaker.

iStock.com/Alphotographic

Figure 10.9 Paddle boarders and surfers enjoy spilling breakers on a shallow, gently sloping beach in northeastern England, at Tynemouth.

Figure 10.10 Waves refract (bend) around a headland in Australia.

10.1.3 Wave Refraction

Where waves approach a shoreline at an angle, they are subject to **wave refraction**, or bending, which ends up making the wave crest approach the shore in a nearly parallel fashion. The bending of waves occurs because waves slow first at their shallow-water ends, whereas their deeper-water ends move shoreward at a higher speed. Thus, a fixed point on a wave crest follows a curved path to the shoreline (**Figure 10.10**).

Another consequence of wave refraction is the straightening of irregular shorelines. This occurs because the concentration of wave energy at headlands causes shoreline erosion, and the dissipation of wave energy in embayments allows sand deposition (**Figure 10.11**).

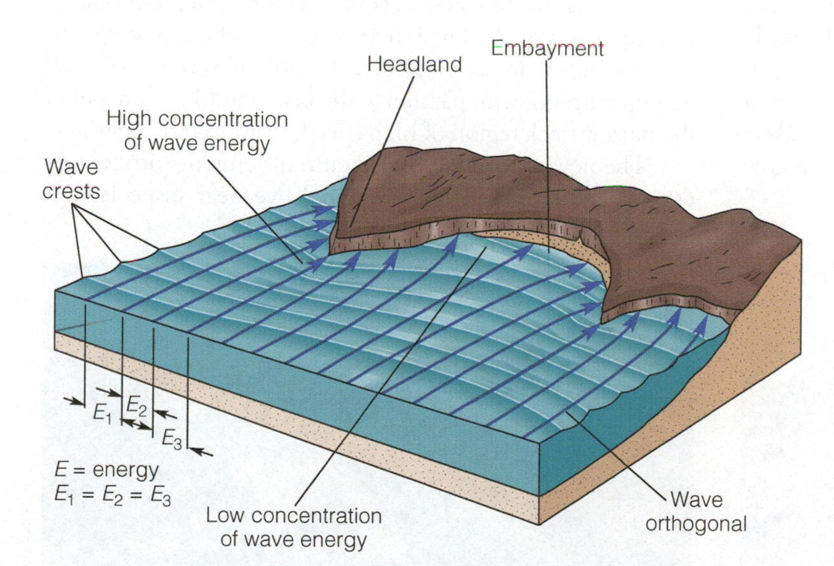

Figure 10.11 Wave refraction leads to a concentration of energy on headlands and dissipation in embayments. The arrowed lines, called orthogonals, are constructed perpendicular to the wave crest and show the wave path. E_1, E_2, and so on represent equal parcels of wave energy and, therefore, equal ability to erode. Note that energy is focused toward the headlands and spreads out in the bays. This is not abstract knowledge! Experience shows that erosion hazards are highest on the headlands and lowest in the embayments.

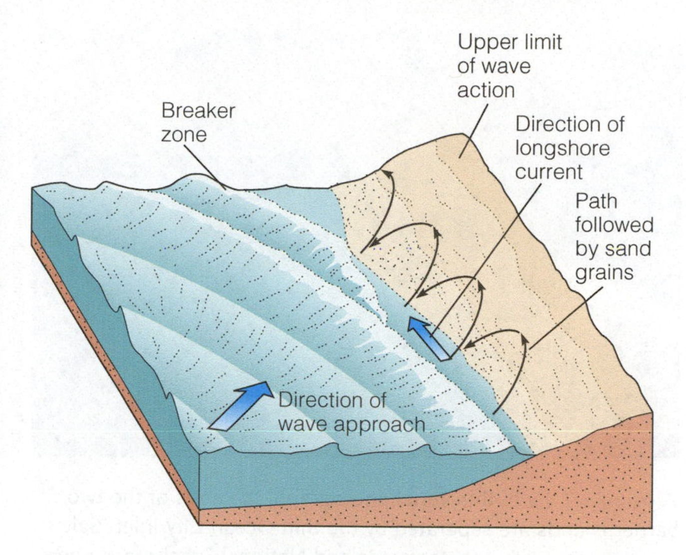

Figure 10.12 Waves approaching the shoreline at an angle generate a longshore current parallel to the beach. The current transports sand, resulting in longshore drift.

Because waves seldom approach parallel to the shore, some water moves along the beach as a weak current, called a **longshore current**, within the surf zone. The greater the angle at which the wave approaches the shoreline, the greater the volume of water that is discharged into the longshore-current stream. This current is capable of moving swimmers, as well as sand grains, down the beach. Sand becomes suspended in the turbulent surf and swash zone and moves in a zigzag pattern down the beach's foreshore slope (**Figure 10.12**). Beach-sand transport parallel to the shore in the surf zone is known as **longshore drift**, and its rate is usually expressed in thousands of cubic meters or cubic yards of sand per year. Beaches are dynamic. The sand grains present on the foreshore today will not be the same ones there tomorrow.

Longshore currents are likened to rivers on land, with the beach and the surf zone constituting the two banks. Just as damming a river causes sedimentation to occur upstream and erosion to occur downstream, so does building structures perpendicular to the beach. Such a structure slows or deflects longshore currents, causing sand accretion upcurrent from the structure and erosion downcurrent (**Figure 10.13**).

Groins (French *groyne*, "snout") are one of those structures placed on the beach. They are constructed of sandbags, wood, stone, or concrete, ostensibly to trap sand as it moves down-beach, carried by longshore drift. The idea is to create a wider beach. For groins of equal length and design, the magnitude of the erosion-deposition pattern is a function of the strength of the waves and the angle at which they strike the shoreline. Most groins are short and are intended to preserve or widen a beach in front of a private home or resort. Unfortunately,

Figure 10.13 In this aerial photograph, the ends of the two barrier islands are separated by the thin Ocean City Inlet. Salt marshes separate Assateague Island National Seashore (upper right) and the densely populated beach resort of Ocean City, Maryland (lower left), from the mainland. Sand erosion on Assateague, caused by jetties that keep the Ocean City Inlet open to navigation, poses a severe threat to the island.

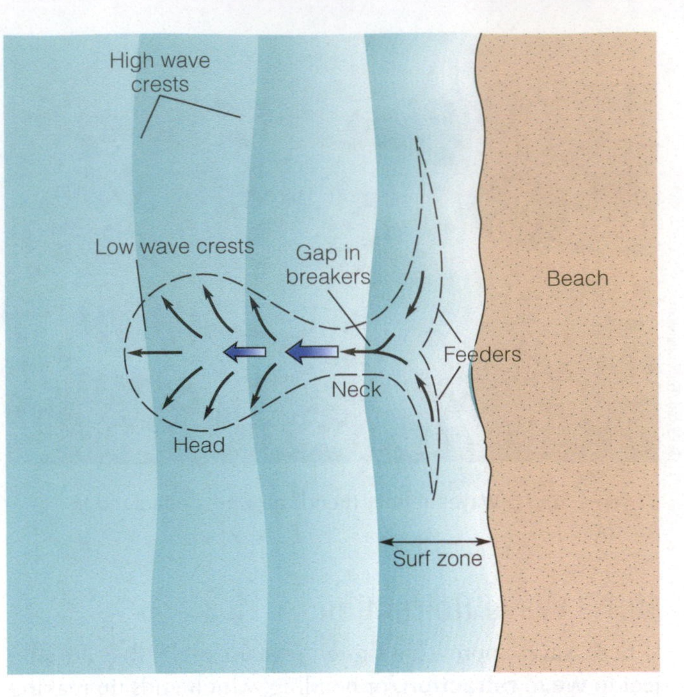

Figure 10.15 A rip current is characterized by relatively weak feeder currents, a narrow neck of strong currents, and a head outside the surf zone, where the rip dissipates.

because this causes neighbors' beaches downcurrent from the groin to narrow, many of those neighbors build groins to preserve and widen their beaches (too bad, neighbors!) and so on. Like artificial levees alongside a river, groins tend to generate more groins, and we generally see fields of them along a given stretch of beach instead of just one (**Figure 10.14**).

Just as rivers overflow their banks, longshore currents may flow out through the surf zone, particularly during periods of big surf. Outflow to the open ocean occurs because the water level within the surf zone is higher than the water level outside the breakers, making for a hydraulic imbalance. The higher water inshore flows seaward to lower water offshore as a **rip current** through a narrow gap, or neck, in the breakers (**Figure 10.15**). Because the location of rip currents, erroneously called rip tides, is controlled by bottom topography, they usually occur at the same places along a beach.

Rip currents can become a problem to swimmers under big-surf conditions, when speeds in the neck may reach 8 kilometers/hour (5 miles/hour)–faster than an Olympic swimmer. Awareness can save your life, so be alert to a gap in the breakers, white water beyond the surf zone, and objects floating seaward, all of which are evidence of a strong rip current flowing out to sea (**Figure 10.16**). Should you find yourself in a rip current, swim parallel to the beach until you are out of the narrow neck region of high-speed, seaward-moving water.

The newest research on rip currents employs drones and dye. Scientists drop brightly colored dye near suspected rip

Figure 10.14 Groin field at Eastbourne on the southern coast of England. Can you determine which way the longshore drift is moving sand (hint: observe the groins)?

Figure 10.16 A large rip current is headed away from shore in the center of the photograph. Dunedin, Otago Peninsula, New Zealand.

© Rob Brander

Figure 10.17 In Sydney, Australia, rip currents from the southern ocean are a risk. Here, purple dye traces the current flowing from the beach to offshore.

currents and watch from above as the dye mixes with the ocean water and starts moving (**Figure 10.17**). If it heads out to sea, the rip current is mapped out in fluorescent colors. However, the only way to protect yourself from the dangers of rip currents is to heed warnings (**Figure 10.18**). They could save your life.

iStock.com/Cosmonaut

Figure 10.18 If rip currents have happened once on a beach, they will happen again.

iStock.com/Gyro

Figure 10.19 The Pacific Ocean erodes the shoreline near Hokkaido, Japan. In front of the rocky sea cliff is a discontinuous bedrock platform eroded by waves during storm surges and high tides.

10.2 Shorelines (LO2)

Where the sea and land meet, we find **shorelines**, some of which are shaped primarily by **erosional processes** and others by **depositional processes**. Shorelines that are exposed to open-ocean, high-energy wave action are likely to be erosional and display such erosional features as sea cliffs or broad, wave-cut platforms (**Figure 10.19**). Shorelines that are protected from strong wave action, by offshore islands or by the way they trend relative to the direction of dominant wave attack, are characterized by depositional features such as sandy beaches. Coastal areas supplied with large amounts of sediment, such as near river mouths and deltas, are often depositional settings (**Figure 10.20**).

iStock.com/ianmcdonnell

Figure 10.20 Sand from San Juan Creek, where it flows into the Pacific Ocean in Orange County, southern California, helps build the beach.

iStock.com/MNStudio

Figure 10.21 Black sand beach at Punalu'u on Hawai'i's Big Island. Can you find the sea turtle?

Of course, not all sand beaches are stable, as we'll discover in the next few pages.

The **shoreline** is the distinct boundary between land and sea that changes with the tides, whereas the **coast** is the area that extends from the shoreline to the landward limit of features related to marine processes. Thus, a coast may extend inland some distance, encompassing estuaries and bays.

10.2.1 Beach Types

In addition to their usefulness as recreational areas, beaches also protect the land from the erosive power of the sea. They are where the energy of waves is dissipated. Beaches are composed of whatever sediment rivers and waves deliver to them. For example, in some places in Hawai'i, black sand forms from weathered basalt and from hot lava that disintegrates when it flows into the sea (**Figure 10.21**). In contrast, reef-fringed oceanic islands are characterized by beautiful white coral sand and shell beaches (**Figure 10.22**).

iStock.com/BryttaBrytta

Figure 10.22 White coral sand beach, Zanzibar island, Tanzania.

iStock.com/Vadiar

Figure 10.23 A pebble beach in Monterey, California.

Continental beach materials are largely quartz and feldspar sand grains derived from the breakdown of granitic and sedimentary rocks. **Pebble beaches** are composed of gravel; they typically occur where the beach is exposed to high-energy surf (**Figure 10.23**). Amusing or not, there's also a Glass Beach at Fort Bragg in northern California–it used to be a dump and now waves have smoothed the discarded glass into small treasures (**Figure 10.24**).

Beaches change shape with the season. As a rule, winter beaches are narrow, because high-energy winter-storm waves erode sand from the upper part of the beach (the **beach berm**) and deposit it in offshore sandbars parallel to the beach (**Figure 10.25**). You may have found such an offshore sandbar while wading in waist-deep water that abruptly shallowed as you walked onto the bar. Winter

Wollertz/Shutterstock.com

Figure 10.24 Instead of pebbles, rounded bits of broken glass, once bottles, make up this beach at Fort Bragg in northern California. Decades of wave action have made this unique pebble beach.

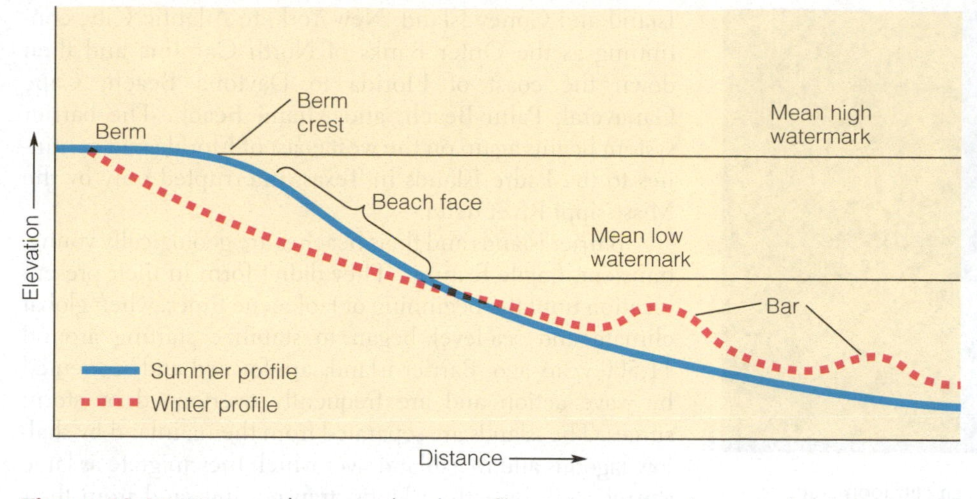

Figure 10.25 Winter and summer beach profiles and some beach terminology. (Modified from Wicander, R., and J. Munroe. 1999. Essentials of Geology Brooks/Cole)

beaches are also more likely to be covered in gravel, rather than sand, as the sand is more easily moved offshore. With the onset of long, low-energy summer waves, the sand in the offshore bar gradually moves back onto the beach, and the beach berm widens. Thus, the natural annual beach cycle is from narrower and coarser in the winter to wider and sandier in the summer. On the U.S. East Coast, beach narrowing is related to waves generated by long-duration storm events that may occur at any time of the year but are more likely in the winter.

Many beaches are backed by sea cliffs (**Figure 10.26**). Such cliffs are dynamic places, eroded both by wave action at the toe and by mass-wasting processes on the face and the top. An **active sea cliff** is one where erosion is dominated by wave action. The result is a steep cliff face with large, angular blocks of rock (talus) accumulating in the booming surf at the base of the cliff. Much of the south shore of the Island of Hawaiʻi, where lava flows continue to pour into the ocean, has become an active sea cliff as much as 35 meters (115 feet) tall as swells from stormy southern seas ceaselessly pound against the newly formed coast (**Figure 10.27**).

In contrast, the erosion of **inactive sea cliffs** is dominated by running water and mass-wasting processes. They can be much taller than active sea cliffs, but in many places have somewhat gentler, irregular slopes, depending on their age. One of the tallest inactive sea cliffs in the world lines the northeastern coast of Molokai, another Hawaiian Island. It formed when a gigantic landslide, originating undersea just offshore, tore away half of the East Molokai shield volcano around 1.5 million years ago, no doubt creating an equally giant tsunami with impacts all around the Pacific. Since then, streams, weathering, and smaller mass-wasting events have modified the massive slide scarp, creating a rugged, dramatic coast. Wave action at the base of the sea cliff contributes little to its diminishment. It towers as much as 900 meters (3,000 feet) above sea level (**Figure 10.28**).

Active sea cliffs can also become inactive if they are isolated from the ocean by the formation of wide beaches at their base plus other depositional and tectonic processes. Their original steep profiles become smoother, gentler, and soil covered as they age, a process that typically takes thousands of years.

Most people enjoy ocean views, particularly from atop a sea cliff. Of course, that view disappears when the cliff erodes and their lawn, deck, or even their home falls onto the beach below.

iStock.com/Bloodua

Figure 10.26 Pocket beach in a cove backed by active bedrock sea cliffs in Nusa Penida Island, Bali, Indonesia.

J. Wei/NPS

Figure 10.27 Active sea cliff, southeastern coast of the Island of Hawaiʻi. The sea arch in the middle distance is 27 meters (90 feet) tall. Sea arches are hallmarks of actively eroding, potentially unstable coastlines, in this case due to strong swells that frequently beat against the shore.

Figure 10.28 The giant East Molokai sea cliff looms above rough North Pacific waters. The sea cliff began forming 1.5 million years ago and continues to erode today, but wave action plays only a secondary role in creating this landscape.

10.2.2 Barrier Islands

Barrier islands, ribbons of sand parallel to the coast and extending just above sea level, are common around the world where there is little tectonic activity ("passive margins") and the continental shelf is wide, there's ample sediment supply, and the water offshore does not deepen quickly. These islands stretch along about 10% of the world's coastlines and are found fringing every continent except Antarctica (**Figure 10.29**).

Barrier islands form the most extensive beach system in the United States. They extend southward from Long Island and Coney Island, New York, to Atlantic City, continuing as the Outer Banks of North Carolina and then down the coast of Florida to Daytona Beach, Cape Canaveral, Palm Beach, and Miami Beach. The barrier system begins again on the west coast of Florida and continues to the Padre Islands in Texas, interrupted only by the Mississippi River delta.

Figure 10.29 A 10-kilometer-long barrier island in the Canadian Arctic. This satellite image was taken in June before the sea ice melted. This island is only active a few months each year when the ice is gone and waves strike its beaches.

Barrier islands and their beaches are geologically young, transient, fragile features. They didn't form in their present location until the beginning of Holocene times, when global climate and sea-level began to stabilize starting around 11,500 years ago. Barrier islands are formed and reoriented by wave action and are frequently overtopped by storm surges. The islands are separated from the mainland by shallow lagoons and marshland over which they migrate as large storms wash over the islands, transporting sand from their seaward sides to their landward sides (**Figure 10.30**). What's key to remember is that barrier islands are close to sea level and "by design" wash over during big storms; yet, people choose to build homes on them and live by the sea anyway, willing and able to take their chances. This sets up an inevitable clash between how the Earth and people behave.

Sometimes, barrier islands and the communities on them simply disappear when a big enough storm comes along–a storm with waves and water high enough to sweep completely over the island. Hog Island, Virginia, was a barrier island that was a popular hunting and fishing spot of the rich and famous in the late 1800s. Its town of Broadwater had 50 houses, a school, a cemetery, and a lighthouse. After a 1933 hurricane inundated the island, killing its pine forest, Broadwater was abandoned. Today, Broadwater is underwater a half-kilometer (1600 feet) offshore. How's that? Barrier island erosion and sea level rise conspired to change the landscape regardless of what people did and thought.

Typical shoreward migration rates for an Atlantic coast barrier island are 2 meters (6.5 feet) per year, but extraordinarily rapid displacements have been reported. The rate of barrier island overwash, flooding, and migration will increase as sea level rises in response to global warming and the resultant melting of glaciers and thermal expansion of seawater. Although heavily developed in many parts of the United States, barrier islands are hazardous places to be in the face of any major coastal storm. They are routinely ravaged by hurricanes and nor'easters and will only become more hazardous places to live over the next century (**Figure 10.31**).

10.2.3 Beach Accumulation and Erosion: The River of Sand

Beaches are dynamic, changing by the minute, hour, and day as water and wind move sand and rock. A beach is stable when the sand supplied to it by longshore current, rivers, and erosion of sea cliffs replaces the amount of sand removed by waves. Where the amount of the supplied sand is greater than the available wave energy can

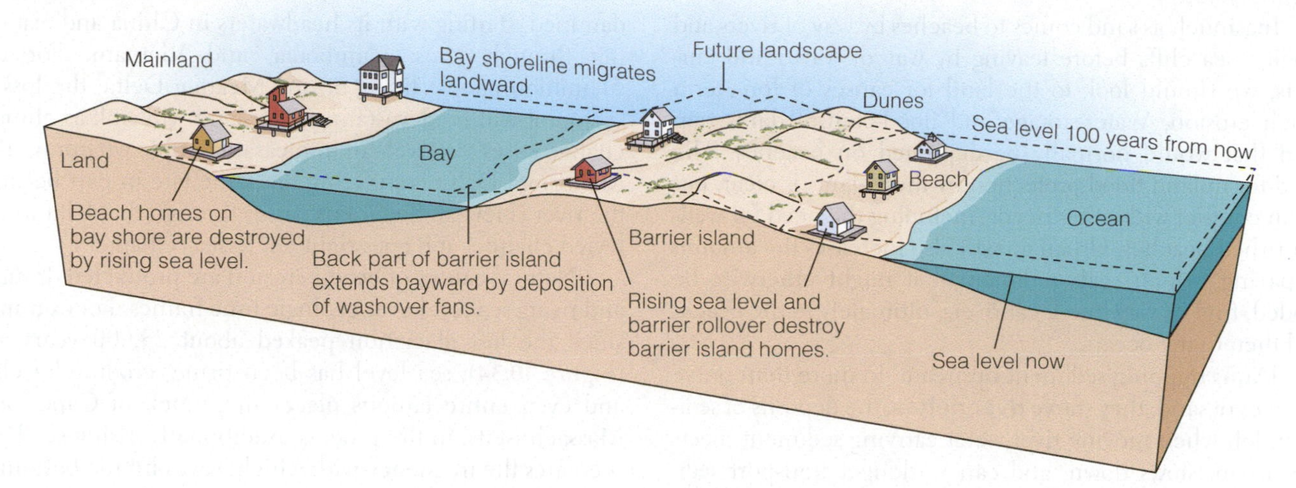

Figure 10.30 This diagram shows the landward movement of barrier islands and the likely effects of sea level rise on homes.

remove–typically near the mouth of a river or rapidly eroding cliffs–the beach widens until wind action forms sand dunes.

Significant areas of coastal dunes occur in Oregon, on the south shore of Lake Michigan in Indiana, near the Nile River in Egypt, on the tip of Cape Cod near Provincetown, and on the Bay of Biscay on France's west coast, to name just a few (**Figure 10.32**). When wave action and longshore currents move more sand than is supplied, we find eroding beaches or no beaches at all. All 30 of the U.S. coastal states have beach-erosion problems of varying magnitudes

Figure 10.32 Coastal dunes are common and the movement of their sand can become a geologic problem. (a) Oregon has some very large coastal dune fields including those shown here near the Pistol River with sea stacks in the distance. (b) Someone's beach home was partly buried when sand was blown to the lee side of 25-meter (80-foot)-high coastal dunes along the southeastern coast of South Africa.

Figure 10.31 Dauphin Island, Alabama, a barrier island, was devastated by Hurricane Katrina in 2005. The high water and waves from the storm washed over the island, carrying away many homes and moving sand landward.

resulting from both human activities and natural causes (**Figure 10.33**).

Inasmuch as sand comes to beaches by way of rivers and eroding sea cliffs before leaving by way of waves and currents, we should look to the land for causes of long-term beach erosion. Water-storage and flood-control dams trap sand that would normally be deposited on beaches. The need for upland flood protection by these dams is clear, but it can conflict with other needs, including the need for well-nourished beaches. Urban growth also increases the amount of paving, which seals sediment that might otherwise be eroded, find its way into local rivers, ultimately to the beach, and then to the ocean.

Dams trapping sediment upstream do more than starve beaches of sand: they starve **river deltas**, the deposits of sediment left when moving river water carrying sediment meets the ocean, slows down, and can no longer transport sediment. These deltas are home to tens of millions of people around the world (e.g., Egypt, China, Italy, and the United States). For them, losing this sediment supply is causing a slow but steady disaster. The sediment in large river deltas is so weighty that it causes the Earth beneath the delta to slowly sink (subside). The sediment also compacts under its own weight (see Chapter 1).

When the sediment supply is cut off, the delta keeps sinking, taking people's homes and fields with it. The result of this is flooding and sometimes massive loss of life. The Mekong River is one of Earth's longest and it's rapidly being dammed, starting with its headwaters in China and extending through Laos, Cambodia, and Vietnam. For the 21 million people living on the Mekong Delta, the loss of sediment will be devastating. The land will sink as climate change raises sea level–an unpleasant double whammy. The Mississippi Delta is suffering the same fate in part because the river is levied. Sediment passes through the delta in the levied channel and goes right out to sea (Chapter 8).

Natural causes of beach erosion are protracted drought and rising sea level over geologic time frames. For example, since the last glaciation peaked about 25,000 years ago (**Figure 10.34**), sea level has been rising, eroding beaches and even entire famous places like much of Cape Cod, Massachusetts, in the process. Additionally, rising sea level increases the frequency with which waves hit the bottom of sea cliffs, eroding them more significantly. In drought conditions, little runoff occurs; therefore, less sand is delivered to beaches when the rains fail to fall, as they have in parts of Africa or California for extended periods of time which could last months, years, or even decades.

Rising sea level erodes beaches and has catastrophic results. On a human scale, it has been estimated that, as Earth warms through the next century, sea level could rise as much as 1 to 3 meters (3 to 10 feet) because of thermal

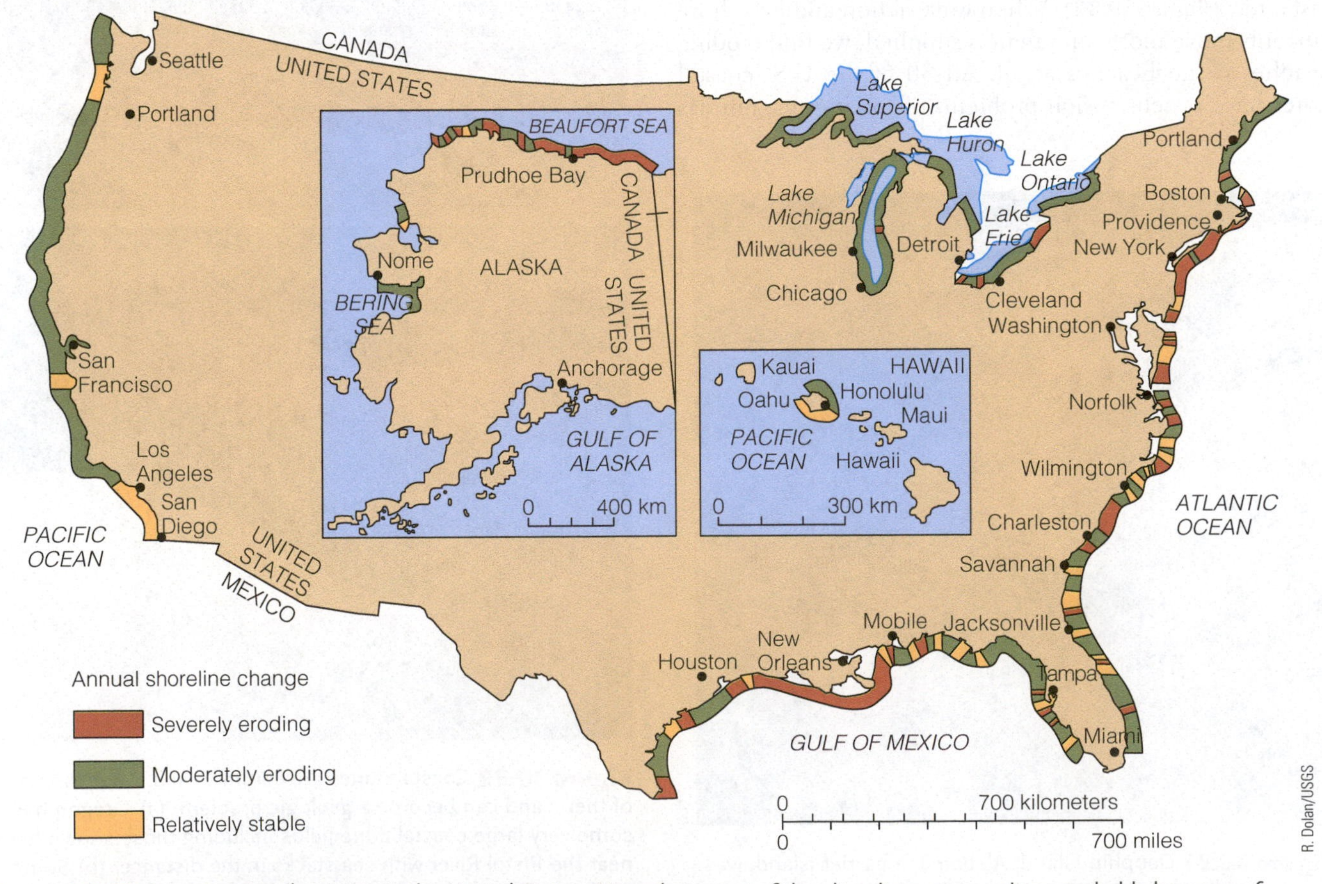

Figure 10.33 Present coastal erosion in the United States. Note that most of the shorelines are eroding, probably because of sea-level rise both in the long term (in response to deglaciation) and in the short term (in response to global warming). Uncolored coastal areas on the map represent shorelines for which data are not available.

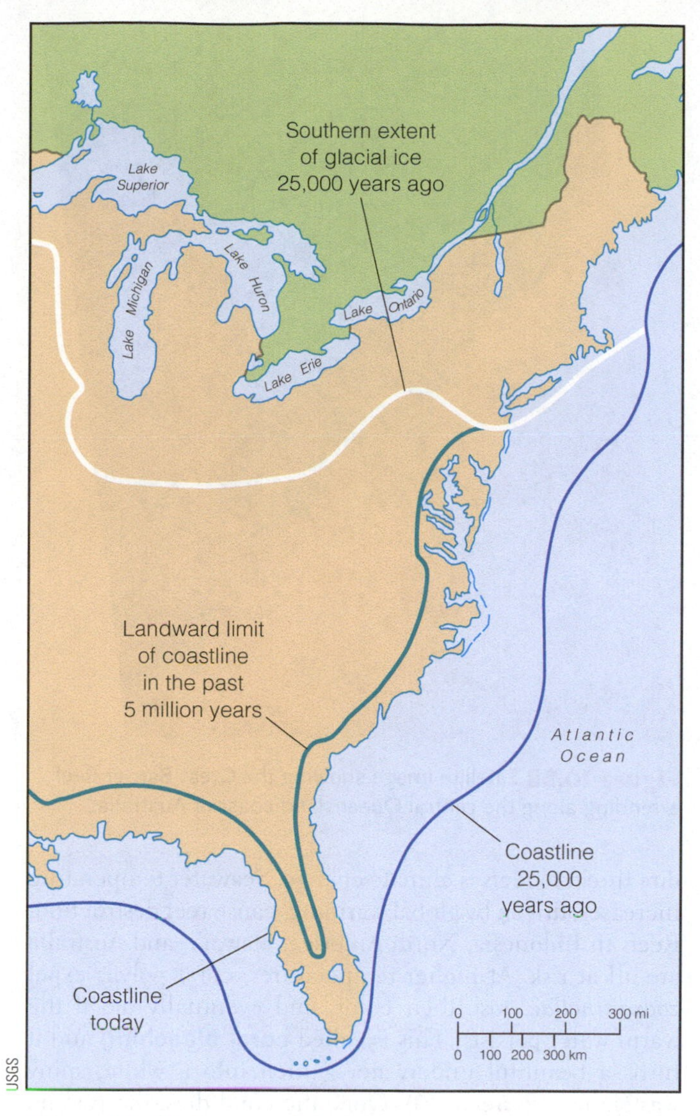

Figure 10.34 Position of the coastline in the eastern United States in late Pleistocene time (25,000 years ago), Miocene time (5 million years ago), and today.

expansion of warming seawater and melting of glacial ice, both adding volume to the ocean. On flat coastal plains, where slopes are as gradual as 20 to 40 centimeters per kilometer (13 to 25 inches per mile), a sea-level rise of 2.5 centimeters (1 inch) would cause the shoreline to retreat between 62 and 125 meters (200 and 400 feet). If the coast is also subsiding from compaction or groundwater withdrawal, this shoreline migration landward would be magnified. Careful tidal-gauge measurements at New York City, where the shoreline is tectonically stable, indicated a sea-level rise of 29 centimeters (almost 1 foot) in the last 100 years, with a rate of rise of 2.89 millimeters (0.1 inch) per year (**Figure 10.35**).

10.2.4 Coral Reefs

Coral reefs are unique coastal landscapes. Corals are animals, related to jellyfish, that secrete a hard skeleton of calcium carbonate (limestone). They grow only in warm water (above 20°C [approximately 68°F]) within the zone of light. Corals require light because they live in symbiosis with the tiny marine algae, *zooxanthellae*. These plants live within the coral polyp and give the calcareous skeleton its color (**Figure 10.36**). Hard corals build their own landforms on which they live; we call them reefs. Reefs are part of the shoreline system in most tropical areas of our planet (**Figure 10.37**).

Reefs come in all different sizes. The Great Barrier Reef stretches along the east coast of Australia for over 2,300 kilometers (1430 miles) (**Figure 10.38**). There are fringing reefs that surround coastal islands (**Figure 10.39**), and at a much smaller scale the Pacific Ocean has many atolls (isolated coral islands), some built on extinct sea mounts. All of these reefs are exceptionally vulnerable to climate change. Imagine being here for a typhoon and consider what this landscape might look like when sea level rises a meter or two (as much as 6 feet) by the year 2100, the

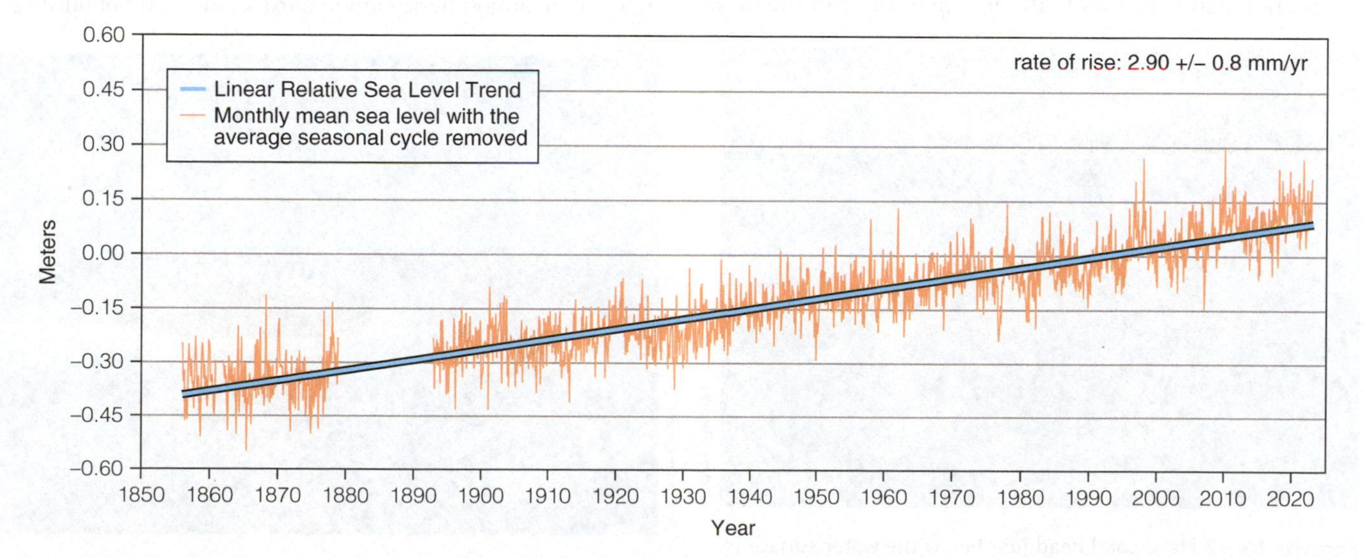

Figure 10.35 Tide gauge record from New York City shows the steady rise of sea level over the past century. This is one of the longest tide gauge records in the United States. (Modified from NOAA Tides & Currents)

Figure 10.36 A colorful Red Sea coral reef, with a diver for scale.

Figure 10.38 Satellite image showing the Great Barrier Reef extending along the central Queensland coast of Australia.

prediction of most climate scientists. You'll note that future sea-level rise is predicted to be much faster than current rates of rise. Why? Because ice sheet melt and ocean heating are both accelerating due to climate change (see Chapter 3).

You might think sea-level rise is uniform around the world, but it's not. In places where the land is sinking (river deltas are one) sea-level rise is faster because not only is the water level going up but the land is going down. On the other hand, sea level could actually be falling near large ice sheets because the land, freed of the burden of ice, is rising more quickly than the ocean.

Coral reefs help people in diverse ways. They form barriers to beach erosion, attract tourism, support fisheries, and build atoll islands upon which people live. Unfortunately, reefs are being destroyed worldwide at a phenomenal rate. Mining of reefs for limestone, dumping of mine tailings on reefs, construction of airplane runways, fishing by use of explosives, land reclamation, and polluted and sediment-laden river runoff all take their toll. But the most dire threat to reefs is climate change. Seawater temperature increases, driven by global warming, cause reef destruction. Reefs in Indonesia, North America, Hawai'i, and Australia are all at risk. At higher temperatures, coral polyps expel *zooxanthellae*, lose their color, and eventually die if the warm water persists. This is called **coral bleaching** and it turns a beautiful underwater garden into a white, spiny wasteland (**Figure 10.40**). Once the coral dies, the reef no longer grows and waves begin to tear it apart.

The newest threat to coral is the acidification of near-surface ocean waters (**Figure 10.41**), caused by human-induced increases in atmospheric carbon dioxide, the result of burning

Figure 10.37 Hard coral head just below the water surface is a reef builder. In the distance is a coral atoll, built of dead coral and covered with coral sand (French Polynesia).

Figure 10.39 Fringing reef surrounds small coral island in the Maldives.

Figure 10.40 Scuba-diving marine scientist uses giant calipers to measure the diameter of a bleached coral in the Seychelles Islands, Indian Ocean.

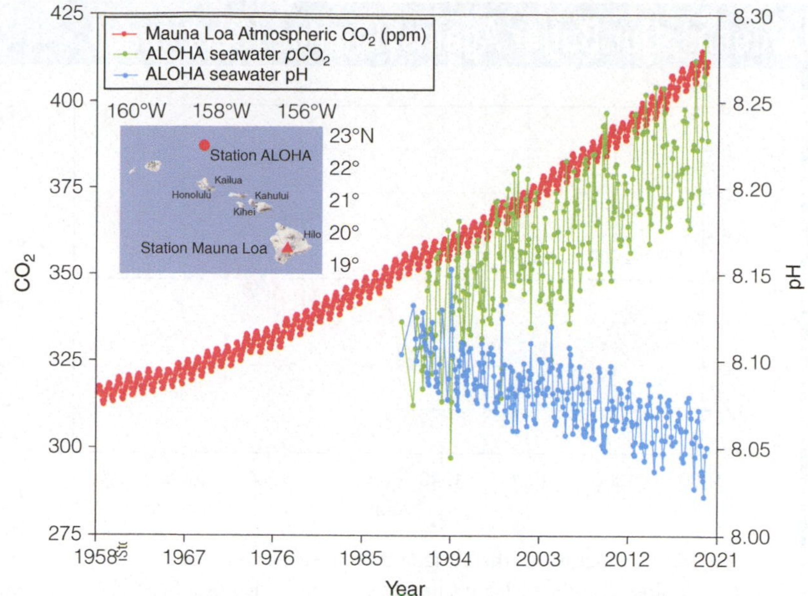

Figure 10.41 Much of the carbon added to the atmosphere by human burning of fossil fuels ends up dissolved in ocean waters. When CO_2 dissolves in water, it forms carbonic acid (HCO_3), which lowers the pH of the water and makes it more acidic. Since 1958, atmospheric CO_2 has increased by over 100 ppm at Mauna Loa, Hawaii. As a result, the surface ocean is growing more acidic, being a decrease of pH from ~8.15 in 1990 to ~8.05 in 2021. This acidification makes it more difficult for organisms to incorporate calcium carbonate in their shells. (Modified from NOAA PMEL Carbon Program)

fossil fuels. Carbon dioxide from the atmosphere dissolves in the ocean water, making carbonic acid which in turn makes the ocean more acidic. The growing acidity of ocean waters makes it increasingly difficult for marine creatures to fix calcium carbonate because calcium carbonate dissolves in acidic waters.

Oceans, Our Buffer Against Change

Oceans, which cover more than 70% of Earth's surface, have massive inertia. The great mass and volume of ocean water buffers changes to the Earth System which would otherwise happen far more quickly. NASA has determined that 90% of the heat from human-induced global warming is now held in the oceans, as is about a third of the carbon dioxide emitted by the combustion of fossil fuels.

Water has a high heat capacity, which means it takes a lot of energy to warm it. This physical property, along with the massive volume of water in the ocean and the slow rate at which water masses move from the surface to the deep oceans, means that heat from our warming planet is increasingly stored in the ocean.

Since 1970, the heat content of the ocean has steadily risen (**Figure 10.42**). On average, the additional heat stored in the ocean each year represents about ten times more energy than that used annually by all the people on earth. As you might be thinking, the temperature of the ocean is also rising

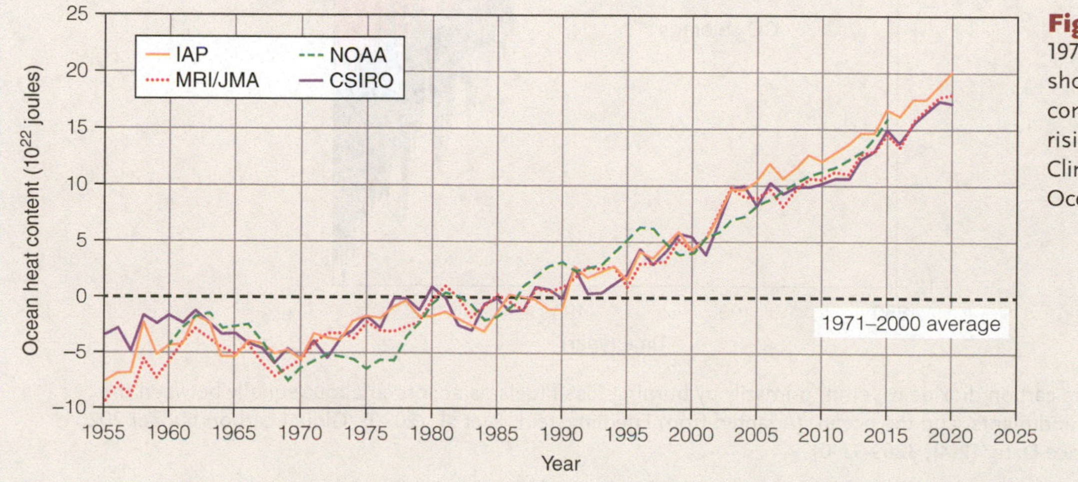

Figure 10.42 Since 1970, numerous data sets show that ocean heat content has been steadily rising. (Modified from Climate Change Indicators: Ocean Heat, U.S. EPA)

(*Continued*)

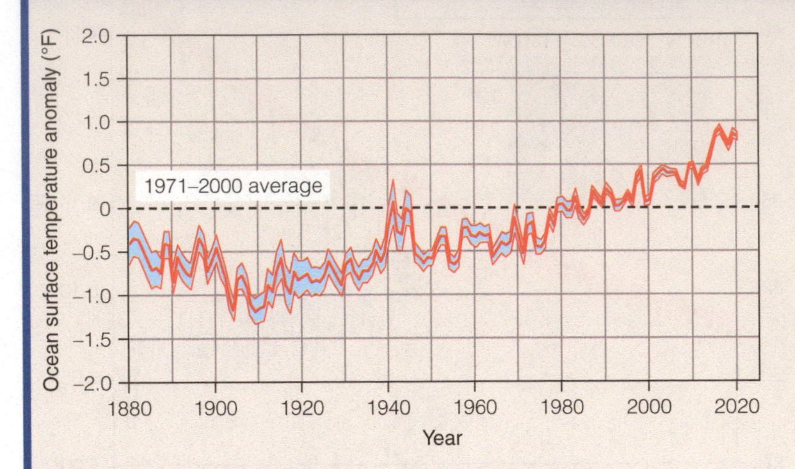

Figure 10.43 Since 1970, the surface ocean temperature has also been rising. (Modified from Climate Change Indicators: Sea Surface Temperature, U.S. EPA)

(**Figure 10.43**). All of this is buffering the rise in temperature we experience at Earth's surface.

Carbon dioxide levels in the ocean are also rising quickly. Because the oceans and the atmosphere exchange gas, when carbon dioxide concentration rises in the atmosphere, it also rises in the ocean. But the oceans have another trick up their sleeves. The currents that move water across the ocean flow in three dimensions and as water moves from the surface ocean to depth, it carries dissolved carbon dioxide into the depths, sealing that carbon off from the atmosphere (see Chapter 2). This provides a sink for carbon, removing it from the atmosphere for hundreds or even thousands of years. On average, individual carbon atoms are stored in the ocean for approximately 500 years. Contrast that to the residence time of a carbon atom in the atmosphere, less than 5 years! Once again, the ocean is keeping carbon out of the atmosphere, thus buffering the warming effects of fossil fuel emissions (**Figure 10.44**).

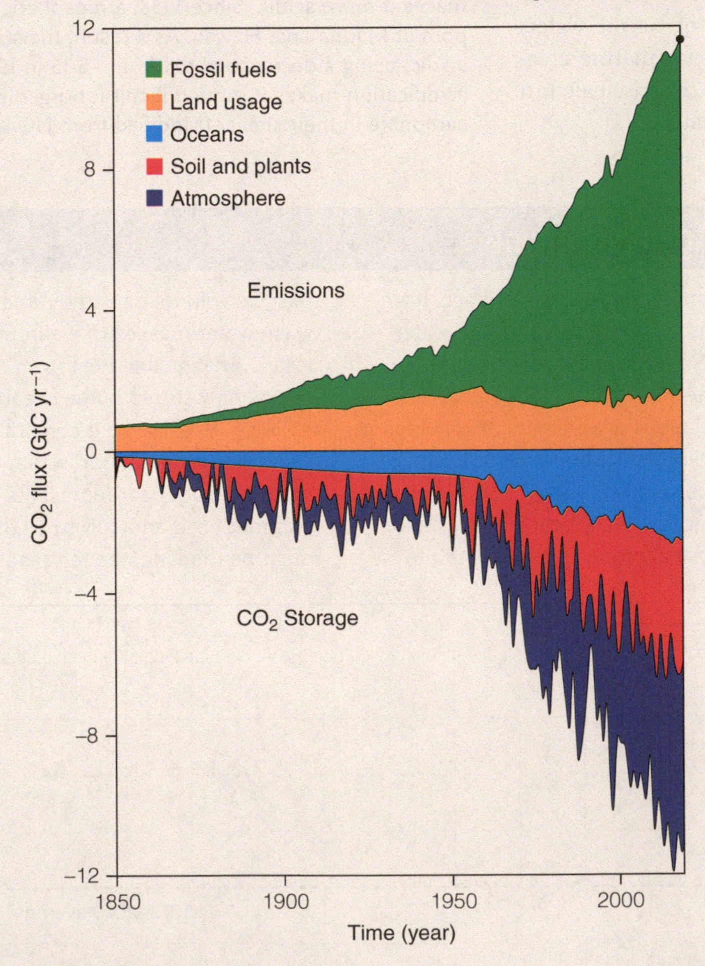

Figure 10.44 The carbon dioxide we emit, primarily by burning fossil fuels, is absorbed about equally between the atmosphere, soils and plants, and the ocean. (Adapted from Friedlingstein, P. et al. (2020). Global carbon budget 2020. Earth System Science Data, 12(4), 3269-3340)

Table 10.1 Causes of Tsunami in the Pacific Ocean Region over the Last 2,000 Years

Cause	Number of Events	Percentage of Events	Number of Deaths	Percentage of Deaths
Landslides	65	4.6	14,661	2
Earthquakes	1,171	82.3	>625,000	89
Volcanoes	65	4.6	51,643	8
Unknown (including possibly bolides)	121	8.5	5,364	1
Total	1,422	100	>700,000	100

10.3 Tsunami (L03)

As we illustrated at the beginning of this chapter, tsunami can be one of the most devastating environmental hazards, causing water levels to rise dramatically in just minutes, sweeping away people, buildings, and entire communities. Tsunami (pronounced "soo-nah-mee") is a Japanese word meaning "harbor" *(tsun)* and "wave" *(ami)* because harbors are where these waves are most violent and observable. The plural is the same as the singular word, by the way, based on Japanese language practice.

On 11 March, 2011, another massive quake and tsunami shocked the world (see Chapter 6). No matter that the disaster struck Japan, probably the best prepared country in the world in terms of natural hazard mitigation; the Tohoku-Oki earthquake and tsunami were soon both well-known, not only because of the size of the waves and the scale of the damage, but also because the tsunami overtopped undersized seawalls and devastated a cluster of nuclear reactors along the shoreline (see Chapter 1). Few natural phenomena can do so much harm so quickly as a tsunami.

There are three phases to tsunamis: **initiation** (a massive impulse of energy that moves ocean water and initiates the wave(s)), **propagation** (the wave's movement across the ocean), and **inundation** (when the wave gets to land). The following sections discuss each phase in turn.

10.3.1 Initiation

Something needs to move ocean water at a massive scale in order to set off a tsunami. Four main geologic culprits can lead to tsunami: underwater fault movement, volcanic eruptions, submarine landslides, and meteorite impacts (**Table 10.1**). These impulses give rise to waves that move out from their source in all directions, analogous to those produced when you drop a rock into a pond.

Fortunately, most underwater earthquakes do not generate tsunami. Scientists estimate that approximately ten major tsunami occur each century, even though tens to hundreds of thousands of earthquakes occur each year. The west coast of South America, a seismically active area, experienced 1,100 earthquakes in the same period and only 20 tsunami. One reason is that many tsunami have such small amplitudes that they go unnoticed when they are travelling far out at sea. In addition, earthquake-induced tsunami require shallow-focus quakes with surface-wave magnitudes greater than 6.5 (see Chapter 6).

Earthquake depth is critical. For example, even though the 2010 Chilean subduction earthquake was huge (magnitude > 8.5), it was deep, and thus the tsunami it generated, although devastating to local communities, was smaller than tsunami generated by some lower magnitude quakes that were shallower. The Atlantic Ocean experiences only 2% of all tsunamis, whereas the Pacific Ocean experiences 80% of them. Why do you think the Pacific Ocean is so tsunami-rich? (Hint: Consider what you learned about plate tectonics from Chapter 2.)

Most major tsunami are generated by large-scale reverse faulting in subduction zones that fringe the Pacific and Indian Ocean basins. Japanese studies have shown that strike-slip faulting seldom produces tsunami. In the late twentieth century, earthquake-generated tsunami around the Pacific Rim damaged Chile (1960, 2010), Alaska and the Pacific Northwest (1964), Nicaragua (1992), New Guinea (1998), Indonesia (2004, 2018), and Japan (2011). The Hawaiian Islands are very tsunami-prone due to multiple tsunami source areas all around the Pacific. On average, the Islands experience a damaging tsunami once every 12 years.

Submarine landslides are a significant source of tsunami, the probable East Molokai tsunami mentioned earlier being an extreme example that involved masses of dry land falling into the sea as well. Keeping aware of this may save lives in population centers around both the Pacific and Atlantic margins that are currently under the false impression that tsunami is not a hazard. San Francisco, Santa Barbara, Los Angeles, San Diego, and New York are some of the cities that could be affected by tsunami-generating landslides on nearby continental slopes. Known landslide-generated tsunami resulted from seafloor instability after the Alaskan earthquakes in 1946 and 1964, the Papua New Guinea earthquake in 1998, and the Grand Banks earthquake in 1929.

While relatively quiescent, the Atlantic Ocean is nevertheless subject to the occasional giant tsunami. Mapping of prehistoric submarine landslide complexes on both sides of the Atlantic hints of what could happen. About 8,000 years ago on the eastern edge of Norway, in a place called Storegga, a giant underwater landslide let loose (**Figure 10.45**). The tsunami it generated reached heights of greater than 25 meters (80 feet) in parts of the Shetland Islands off the coast of Scotland.

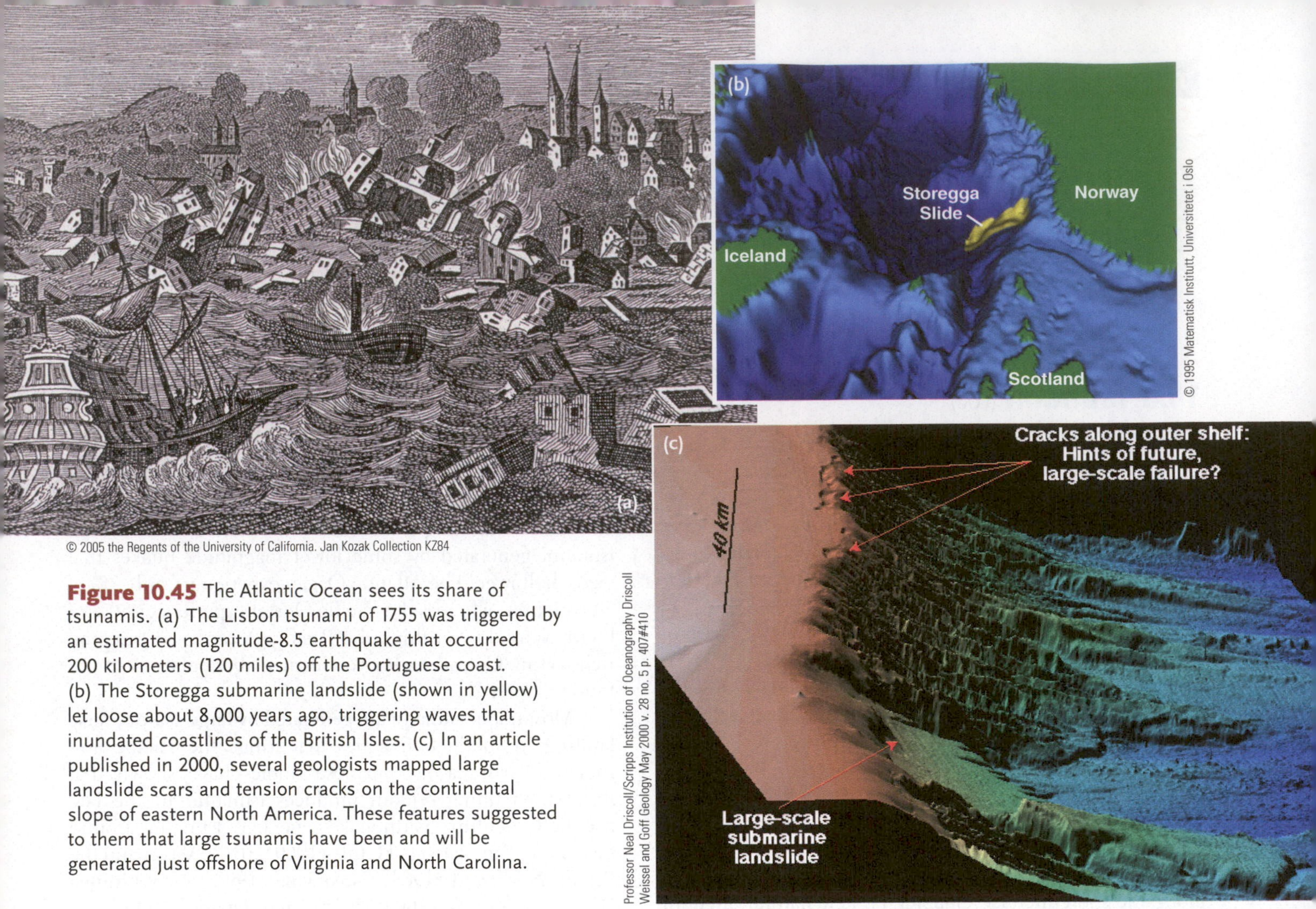

Figure 10.45 The Atlantic Ocean sees its share of tsunamis. (a) The Lisbon tsunami of 1755 was triggered by an estimated magnitude-8.5 earthquake that occurred 200 kilometers (120 miles) off the Portuguese coast. (b) The Storegga submarine landslide (shown in yellow) let loose about 8,000 years ago, triggering waves that inundated coastlines of the British Isles. (c) In an article published in 2000, several geologists mapped large landslide scars and tension cracks on the continental slope of eastern North America. These features suggested to them that large tsunamis have been and will be generated just offshore of Virginia and North Carolina.

Perhaps the most worrying landslide source for triggering an Atlantic Ocean tsunami is the sector collapse of volcanoes such as those making up the Canary Islands off the east coast of Africa. Side-scan sonar maps of the ocean floor clearly shows that these big ocean volcanoes have shed large parts of their towering cones in the past–hummocky topography is all over the ocean floor around them. Field mapping on the islands shows that waves transported boulders 240 meters (800 feet) above sea level about 73,000 years ago. That's quite the wave!

On the other side of the Atlantic, off the coast of Virginia and North Carolina, there is evidence of a gigantic submarine landslide that occurred some 18,000 years ago. This slide, which contained more than 33 cubic miles (150 cubic kilometers) of material, came off the edge of the continental shelf and must have generated a substantial tsunami. North of this ancient slide are tension cracks in the crust along the edge of the shelf, suggesting another slide may be "imminent" (geologically speaking that is; see Figure 10.45).

Another cause of tsunami is powerful volcanic eruptions at or slightly below sea level. The explosive eruption of Krakatoa (Indonesia) on August 27, 1883, was one of the largest volcanic eruptions and subsequent tsunamis ever recorded. The volcano collapsed into its own magma chamber and because it was an island in the Sunda Strait between Java and Sumatra, the ocean flowed in where the volcano had once been. The eruption was disastrous, but the death toll of more than 37,000 from the tsunami was the worst effect.

The Krakatoa tsunami had a maximum run up height of 43 meters (140 feet) and traveled 5 kilometers (3 miles) inland over low-lying areas (**Figure 10.46**). The largest wave struck the town of Merak, because of the funnel shape of its harbor. There, the 15-meter (45-foot) wave increased to >40 meters (135 feet) because of the concentration of energy, and coral blocks weighing more than 600 tons were transported onshore. About 5,000 boats were destroyed, and for months afterward bloated bodies washed ashore, along with blankets of pumice.

Meteors and comets striking Earth are a much less frequent but still possible source of tsunami. Most asteroids come from the Asteroid Belt between Jupiter and Mars. Smaller bits of space debris incinerate during their trip through the atmosphere before reaching the surface; only the big ones make it through. Comets have a greater variety of orbits, some of which are extremely elongate, taking them away from Earth for decades or even hundreds to thousands of years. When their impacts do occur, which

Figure 10.46 The Krakatoa 1883 tsunami wave carried the Dutch gunship *Berouw* 3 kilometers miles inland, almost 20 meters (66 feet) above sea level. A report from the time reads "She lies almost completely intact, only the front of the ship is twisted a little to port, the back of the ship a little to starboard. The engine room is full of mud and ash. The engines themselves were not damaged very much, but the flywheels were bent by the repeated shocks. It might be possible to float her once again." (From *Krakatoa: The Day the World Exploded*, by Simon Winchester)

thankfully is not often, they can be severe. Hollywood has picked up on this particular tsunami-generating process and produced such popular movies as *Deep Impact* and *Armageddon*, both of which feature devastating and massive tsunami created when bolides landed in the ocean (**Figure 10.47**).

Indeed, there is good evidence in the geologic record of a major bolide impact and a truly exceptional associated tsunami 65 million years ago at the end of the Cretaceous period (see Chapter 2). Boulder beds are an indication of the great size of the end-Cretaceous waves, though the extinction of the dinosaurs and about 70% of all other creatures that lived at that time was due to much more than simply giant walls of water. Fire, air pressure changes, acid in the atmosphere, and much more combined to kill off much of the world's life. The 180-kilometer (110 mile)-wide crater created by this impact has been found in the northern Yucatan Peninsula, and is named for a town there called Chicxulub. Near there, models indicate that the wave was over 1500 meters high (5000 feet)! It swept inland more than 100 kilometers (60 miles) in places and propagated around the world.

10.3.2 Propagation

As soon as they are initiated, tsunami waves are on the move. In deep water, wave velocities can exceed 800 kilometers per hour (480 miles/hour) and could easily keep pace with a jet airliner. With wavelengths more than 160 kilometers (100 miles) and wave heights of only a meter or two (3 to 6 feet) in the open ocean, however, tsunami waves

Solarseven/Shutterstock.com

Figure 10.47 An artist's view of a large bolide striking Earth. Because about 71% of Earth's surface is covered by ocean water, the chance of splashdown and the resulting tsunami is high.

have such gentle slopes that they go unnoticed at sea (**Figure 10.48**).

Tsunami waves travel readily across ocean basins. As mentioned earlier, the Hawaiian Islands, perched in the center of the Pacific Ocean, have been assaulted by tsunami

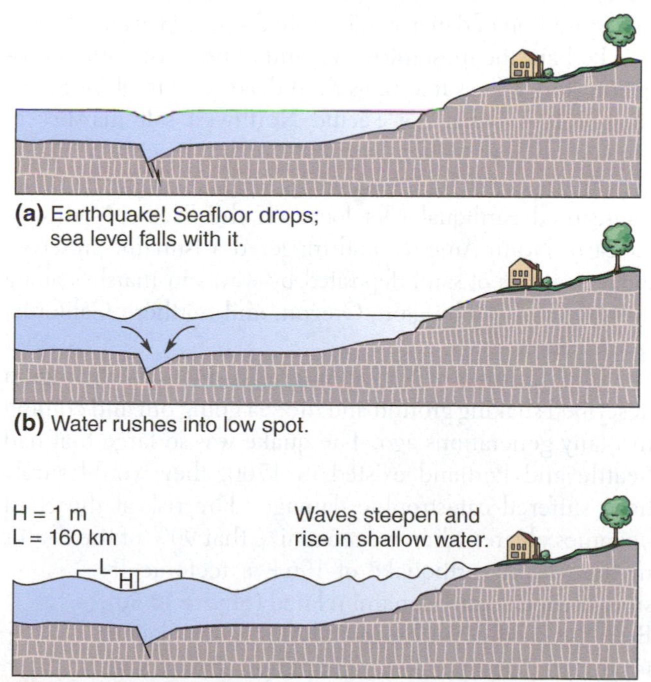

Figure 10.48 Formation of a tsunami when a portion of the seafloor is down-faulted and the water surface over the fault is lowered. Water rushes in to fill the low spot and creates an impulsively generated wave that moves out in all directions with a very high speed (hundreds of kilometers per hour). As the wave approaches shore, it shoals, steepens, breaks, and moves onshore, flooding the land.

Brian Atwater/USGS

Figure 10.49 Ghost forest on the Copalis River in Washington State bears witness to the massive subduction earthquake that inundated the tree's roots with salt water as the land sank.

NOAA

Figure 10.50 Seward, Alaska, was devastated by waves after the 1964 Alaska subduction quake. The waterfront was littered with destroyed vehicles and railcars tossed about by the waves. Strong shaking from the quake lasted several minutes, triggering an underwater landslide that generated local tsunami waves as much as 10 meters (33 feet) high. These were followed by additional waves from farther away.

waves from all sides. In 1946, an earthquake in the Aleutian Islands of Alaska spawned a wave that killed more than 170 Hawaiians. The wave from the massive 1960 Chilean quake took 15 hours to reach Hawai'i, where it caused 61 deaths. In 2011, the Tohoku-Oki quake spawned a tsunami that did more than 8 million of damage as it washed up on Hawaii's shores.

Perhaps the most intriguing and unusual tsunami propagation story relates tree rings from dead trees in ghost forests (**Figure 10.49**) that dot Pacific Northwest salt marshes to impeccably kept written Japanese tsunami records, a catalog of giant waves that hit the island. In 1700 C.E., a great subduction-caused earthquake let loose on the Pacific Northwest Coast of North America and triggered a tsunami preserved today as layers of sand deposited by waves in marshes along the shores of Washington, Oregon, and northern California (see Chapter 6).

Oral histories of Native Americans living in the region described shaking ground and the sea going out and coming in many generations ago. The quake was so large that had Seattle and Portland existed in 1700, they would surely have suffered catastrophic damage. The risk at the coast becomes clearer when we recognize that 90% of the deaths in the Alaska earthquake of 1964, a tectonically similar event, were tsunami-related (**Figure 10.50**). Both the Pacific Northwest and Alaska lie adjacent to linear, coast-parallel, subduction zones with active seismic histories.

The 1700 quake and the tsunami it spawned were dated by counting rings in trees killed as saltwater flooded their roots, the result of coastal subsidence accompanying the shaking (**Figure 10.51**). Pinning down the year using tree biology opened the door for an amazing discovery. In the Japanese records was an "orphan" tsunami; that is, a "wave without a preceding felt quake" that left the leaders of flooded villages perplexed (**Figure 10.52**). According to the records, an orphan wave began flooding coastal Japan as midnight approached on January 26, 1700. The Japanese did not realize, at least at first, that the source of the wave was from an earthquake too far away for them to feel. Computer models show that a wave generated in a the Pacific Northwest quake would most likely take about ten hours to cross the Pacific Ocean from North America to Japan. So an international team of scientists reached the conclusion that a very large earthquake, struck the Pacific Northwest at about 9 pm on January 24, 1700. It is not often that geologists can be so exact about the timing of events that occurred hundreds of years ago.

10.3.3 Inundation

As the leading edge of the tsunami wave moves into shallower water near the coast or an island, it slows from the speed of a jet plane to more like the speed of a car on the freeway, and the water behind that leading edge begins to pile up–the process of shoaling that we've already described. The wave increases in height and steepness and then runs

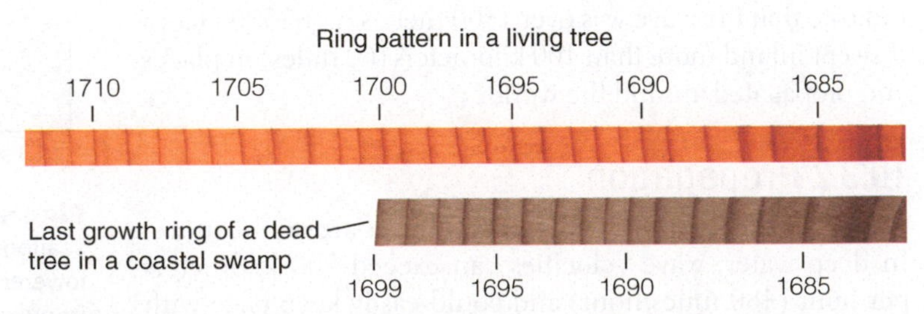

Figure 10.51 Tree rings ending at 1699 on ghost tree sampled in a marsh along the Washington coast. (Modified from Steven Earle, Physical Geology (2ed), 2019)

USGS

Figure 10.52 This painting shows the village of Miho as it was flooded by the orphan tsunami in January 1700. The village leader urged the thirty people who lived there to flee to a shrine on the hilltop, thus saving their lives. Mount Fuji towers in the background.

ashore as a wall of white, foamy water; a surging, sweeping flood. The "sweeping flood" part of this description should be emphasized. It is not that a wave comes in and moments later goes back out, like an ordinary wave crashing on a beach. Rather, given the immense wavelength and energy of tsunami, the incoming wave continues to come ashore sometimes for many minutes before receding. It is relentless. This is one reason that a tsunami can be so damaging.

To make matters worse, a tsunami consists of multiple surges; not just a single, sweeping flood but a series of waves that may continue in some cases over a period as long as 24 hours. The arrival times between crests of incoming tsunami waves may range from 5 minutes to as long as 2 hours, and the first crest may not be the largest. The trough of the initial wave (the low point) may arrive first, and many lives have been lost by curious people who wandered into tidal regions exposed as water withdrew, some in search of an easy seafood meal. For example, in Lisbon, Portugal, on November 1m 1755, when many people were in church commemorating All Saints' Day, an immense earthquake (magnitude ~8.5) struck in three distinct shocks (see Figure 10.45a). The worshipers ran outside to escape falling debris and fire, many of them joining a group seeking safety on the waterfront. The length of time of strong shaking has been estimated from historical accounts of how many "Ave Maria's" and "Paternosters" the churchgoers recited during their fearful experience.

After the quake, there was a quiet withdrawal of water followed by a huge wave minutes later. Sixty thousand people died, about two-thirds of them in the quake and the rest in the aftershocks and fires that followed. The quake was felt throughout Europe and northern Africa, and the tsunami is reported to have crossed the Atlantic, raising the level of the ocean 6 meters (20 feet) in Antigua, Martinique, and Barbados.

Water withdrawal before the first high wave also contributed to fatalities in the 1946 Hawaiian tsunami; some people walked offshore to collect exposed mollusks just before the first wave crest arrived. Most people in Hawai'i died as they were sucked back to sea by the first receding surge. Survivors found flotsam to cling to out at sea while awaiting rescue, hours later. Perhaps more would have lived if they had been warned of the impending danger, but the Scotch Cap, Alaska (on Unimak Island in the Aleutian Island Chain), was swept away completely before it could transmit a single message, there had been many false alarms before, and it probably didn't help that the wave struck on All Fool's Day, April 1.

The tsunami wave that destroyed the Scotch Cap facility peaked at 45 meters (148 feet) above sea level (**Figure 10.53**). The wave got its start when a massive subduction zone

(a)

lighthousefriends.com

(b)

NOAA/NGDC

Figure 10.53 The Scotch Cap lighthouse, on the coast of Alaska, was destroyed by the 1946 tsunami. All five men there were killed by the wave. (a) The lighthouse station before the wave. (b) What little was left of the lighthouse station after the wave.

iStock.com/RyuSeungil

Figure 10.54 Not quite where you expect to find a large ship–high and dry, intruding on a road. The *Asia Symphony* was washed into the town of Kamaishi, Japan, by the March 11, 2011 tsunami.

earthquake occurred below the Aleutian Islands, a volcanic island arc. That strong shaking is thought to have triggered a submarine landslide, which displaced more ocean water and thus lifted the local wave even higher.

The inundation phase of a tsunami does damage in several ways. The incoming wave floats and moves anything that's not firmly attached–that includes houses, cars, people, and boats (**Figure 10.54**). The flood of water moving inland quickly fills with debris, creating floating battering rams. That moving debris makes the wave even more dangerous as it slams into buildings, causing far more damage than water alone. What is most striking about tsunami damage is how spotty it can be. Homes above the high-water mark are untouched (except of course if the earthquake was centered nearby); while ten meters away, a neighbor's home is flattened by the debris-filled wave (**Figure 10.55**).

Stocktrek Images, Inc. / Alamy Stock Photo

Figure 10.55 In rural northeastern Japan, most of a seaside village was destroyed by the 2011 tsunami but homes and trees outside of the waves' reach remain standing.

AFP/Getty Images

Figure 10.56 Cleaning up after the 2011 tsunami generated massive piles of waste along much of Japan's western coastline.

The impact of a major tsunami continues long after the waves have passed. In areas damaged by the waves, there are immediate issues of people without housing, disrupted public utilities (water, sewer, and gas), and trash–amazing volumes of debris from destroyed and damaged buildings (**Figure 10.56**). Debris carried out to sea may linger afloat and wash onto distant shores even years later. NOAA suggests that 4.5 million tonnes (5 million tons) of debris entered the ocean because of the tsunami following the 2011 Tohoku-Oki earthquake. That's more than 3000 times the usual amount of trash Japan discharges to the ocean (not by design!) annually. Tsunami debris can carry many different species with it and introduce them to new habitats: Some Japanese marine creatures have found their way to the northwestern coast of North America, surviving on debris for several years at sea (**Figure 10.57**).

Buddy Mays / Alamy Stock Photo

Figure 10.57 A Japanese floating dock, set loose by the 2011 tsunami, landed on an Oregon beach a year later and became an instant tourist attraction.

10.3.4 Mitigating Tsunami Risk

How do we avoid terrible losses of life and property from a tsunami, knowing that they impact practically any coastline in the world? Of course, public recognition of warning signs as a tsunami approaches matters greatly; stay away from a shoreline with receding water. Government action is important too. Thousands who tried to flee the tsunami during the Sendai earthquake, well aware of its approach, were caught in traffic jams on roads with stoplights out as the water reached them. The highway grid was unable to accommodate such a speedy evacuation. In another instance, school children drowned as they lined up for a lengthy evacuation drill, sorting themselves by height to be counted. Only the class "renegades" survived by ignoring their teachers and bolting at once to nearby hillsides.

Uncertainty about where to go can also be a problem. Realistic spur-of-the-moment evacuation plans and designated points of assembly must be made clear to everyone. The Japanese knew this for centuries prior to the advent of modern industrial life; ancient Buddhist temples on high ground near frequently threatened shores served as "tsunami shelters." Respect for tsunami became part of the national religion. The best defenses against tsunami risk are an educated public that knows how to respond, using maps showing high-hazard areas to make wise land-use decisions (**Figure 10.58**), and an early warning system.

Technology can help, too. International tsunami warning systems are far better integrated and widespread than they were in 1946, when Scotch Cap was destroyed. The goal now is to place on the seafloor more real-time,

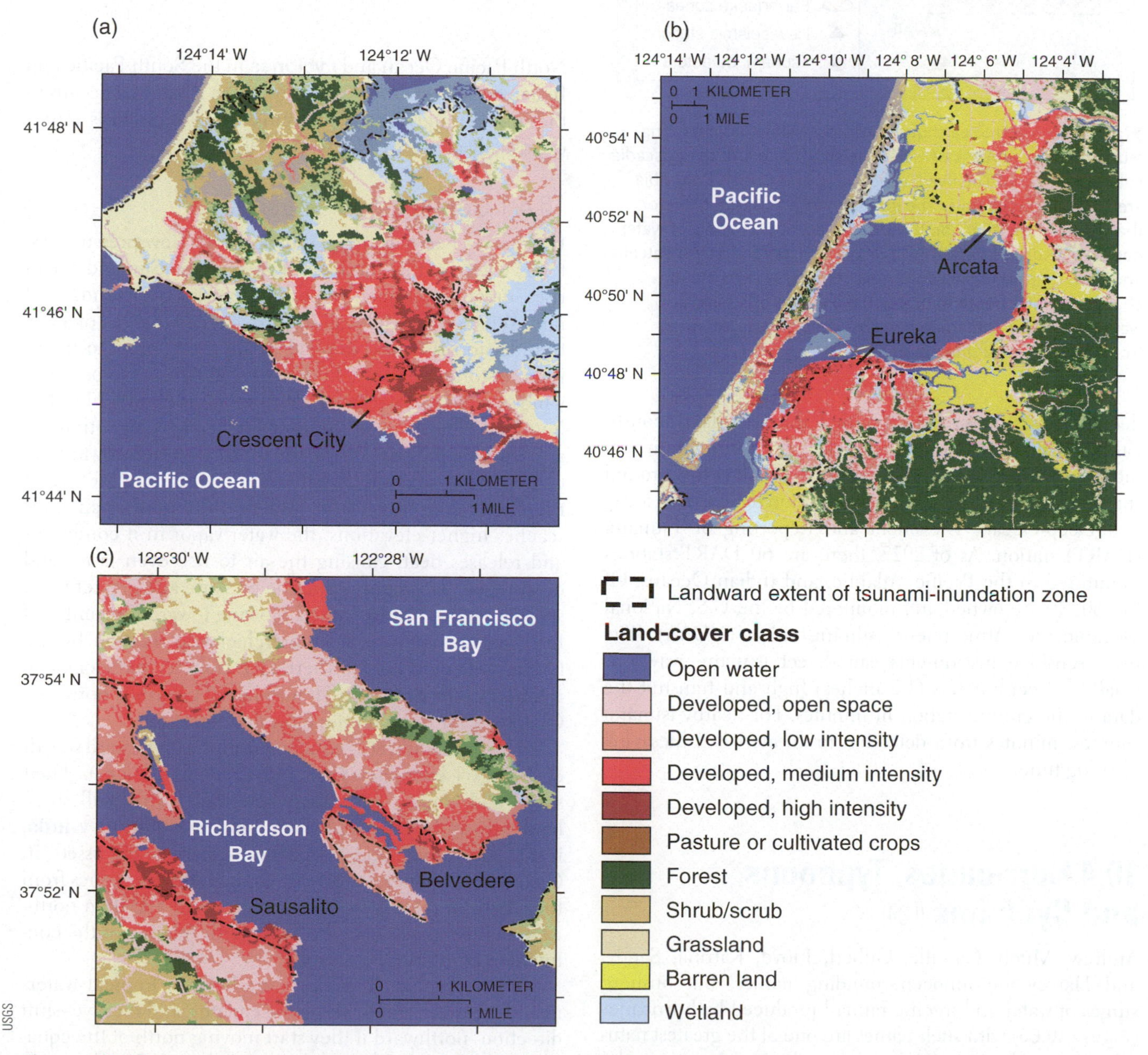

Figure 10.58 Tsunami hazard maps for three cities in California showing land use and the zone of maximum predicted inundation.

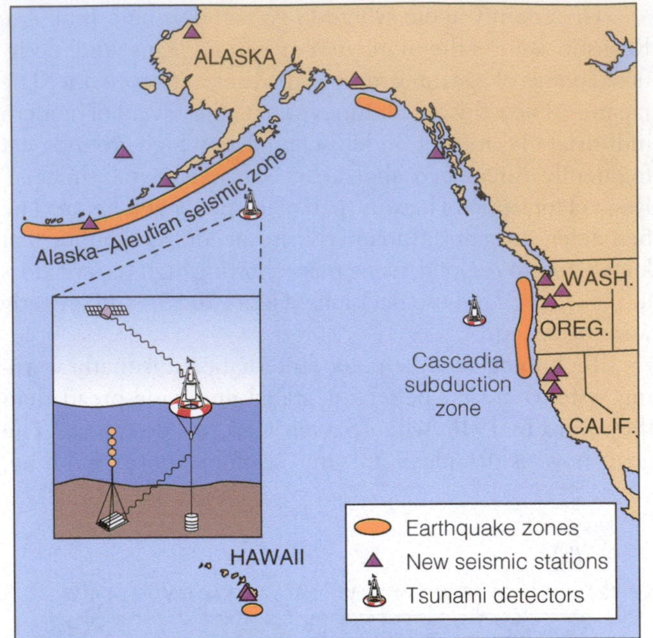

Figure 10.59 U.S. earthquake zones capable of generating tsunamis are the Alaska-Aleutian seismic zone and the Cascadia subduction zone. The five Pacific states and British Columbia are subject to tsunamis. The inset shows a tsunami detector that rests on the seafloor. It transmits acoustic signals of water depth to a surface buoy, which relays signals to a NOAA satellite and then to ground-based warning centers. A slow, steady change in water depth indicates the passage of a long-wavelength, low-amplitude tsunami wave at sea. (Adapted from *EOS 79*, no. 2, June 2, 1999)

deep-ocean tsunami detectors that transmit data acoustically to a floating buoy, which relays the information to a satellite; in turn, the satellite transmits the data to a ground station (**Figure 10.59**). The most common device used today is a Deep-Ocean Assessment and Reporting of Tsunami (DART) station. As of 2022, there are 60 DART stations monitored in the Pacific, Atlantic, and Indian Oceans. Of the 60, 39 are owned and monitored by the U.S. National Oceanic and Atmospheric Administration (NOAA). The most sensitive instruments can detect tsunami waves as small as 3 centimeters (1.2 inches) high and transmit the data to the ground station in minutes. For nearby tsunami sources, minutes from detection to an alert is a lifesaving warning time.

10.4 Hurricanes, Typhoons, and Cyclones (LO4)

Andrew, Mitch, Camille, Gilbert, Floyd, Katrina, Sandy, and Harvey are innocent-sounding names, but onshore surges of water and intense rainfall produced by hurricanes (**Figure 10.60**) with such names are one of the greatest natural threats to humans. Hurricanes (called **typhoons** in the North Pacific Ocean and **cyclones** in the South Pacific and Indian oceans) get their energy from the equatorial oceans in the summer and early fall months, when seawater is warmest, usually above 25°C (about 80°F).

Figure 10.60 Wind, waves, and rain of Hurricane Sandy come over the Winthrop, Massachusetts, seawall in October 2012.

10.4.1 Genesis and Seasonality

Hurricanes begin as tropical depressions (low pressure systems) when air that has been heated by the Sun and ocean rises, creating reduced tropospheric pressure, clouds, and rain. What starts as a localized tropical disturbance or a few thunderstorms in the tropical Atlantic can grow to be an intense storm, up to 800 kilometers (500 miles) across, with wind speeds far in excess of 120 kilometers / hour (74 miles/hour), the lower speed limit for hurricane designation. As atmospheric pressure drops, the warm air moves toward the center of low pressure, what later may become the eye of the impending hurricane. When the moist, rising air reaches higher elevations, the water vapor in it condenses and releases heat, causing the air to rise even faster and creating even lower atmospheric pressure and greater wind speeds (**Figure 10.61a**). Wind blowing toward the center of low pressure is given a counterclockwise rotation by the Coriolis effect in the Northern Hemisphere, and clockwise rotation in the Southern Hemisphere. A storm is born (see Chapter 2 for more on the Coriolis effect).

Hurricanes travel as coherent storms with forward speeds of 20 to 65 kilometers/hour (12 to 40 miles/hour). Their speed and course, and thus where and if they will strike land, are set by large-scale, atmospheric **steering winds**; first trade winds and then the westerlies discussed in Chapter 2. In general, steering winds guide hurricanes from their birth in the eastern Atlantic westward and then northward to their graves over the cool North Atlantic or the continent of North America.

In the Pacific, hurricanes are born in tropical waters near the eastern rim or along the equator, and recurve–shift direction–northward if they start moving north of the equator, or southward if they are born to the south. Northeastern

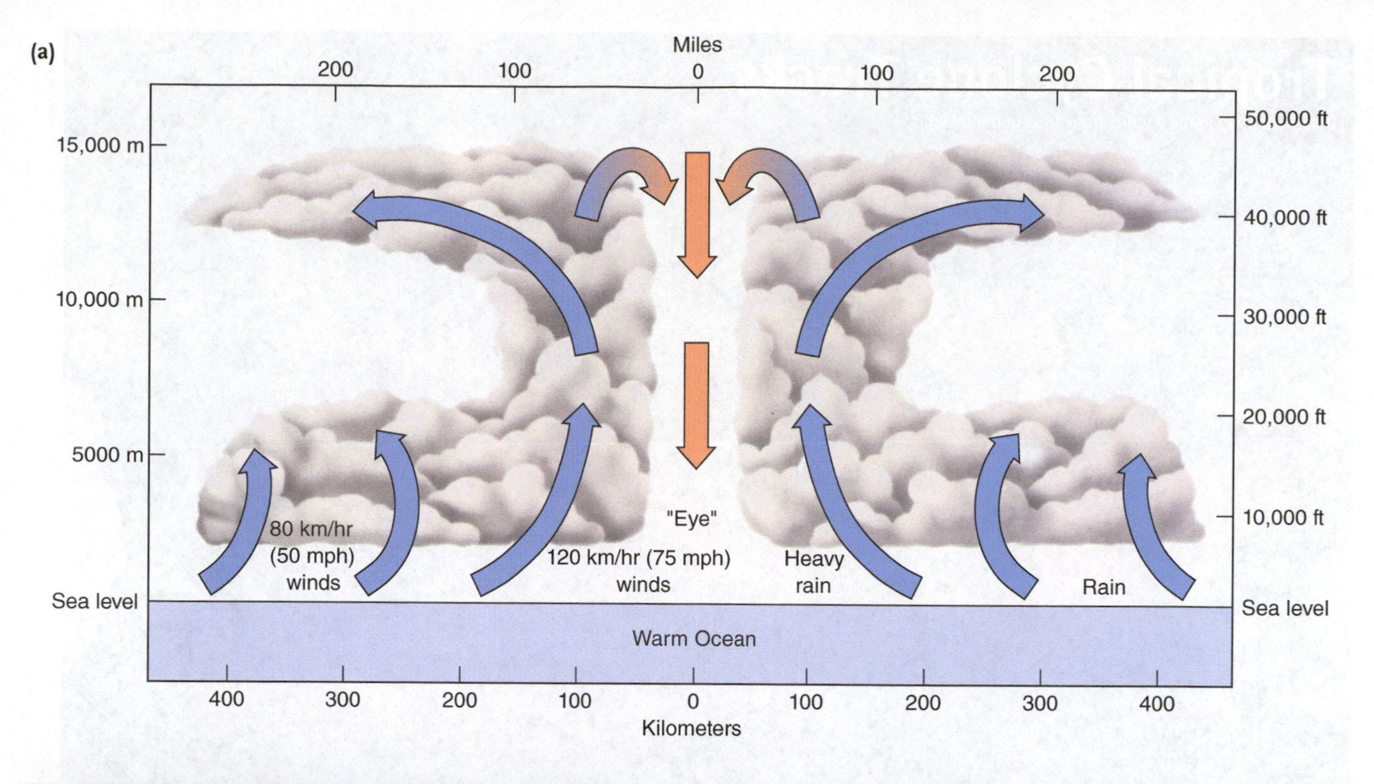

Figure 10.61 (a) Circulation patterns within a hurricane, showing inflow of air in the spiraling arms of the cyclonic system, rising air in the towering circular wall cloud, and outflow in the upper atmosphere. Subsidence of air in the storm's center produces the distinctive calm, cloudless "eye" of the hurricane. (b) A satellite photograph of Hurricane Katrina coming ashore near New Orleans on August 29, 2005. Katrina's rotation, bands of rain-producing clouds, and central eye are clearly visible.

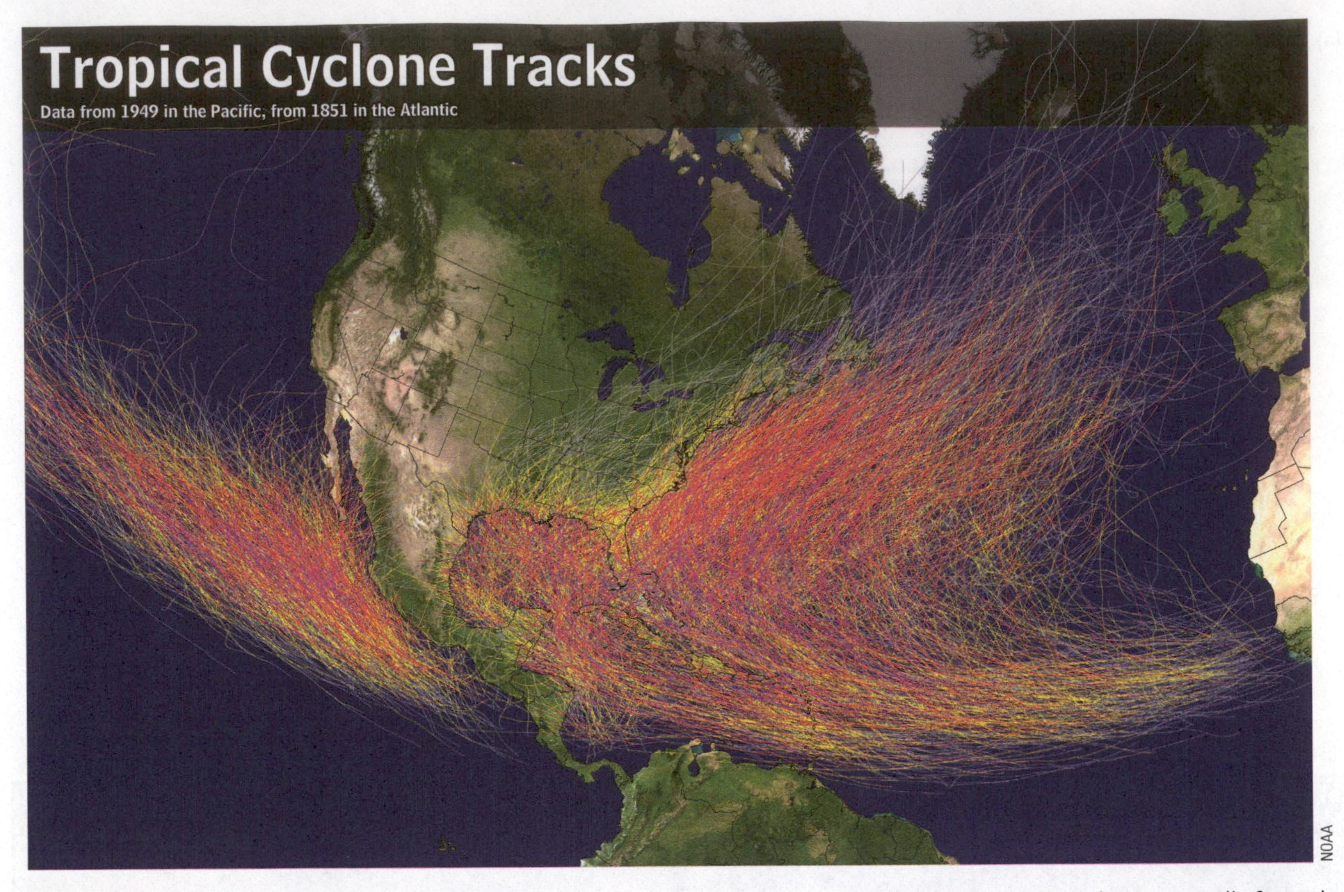

Figure 10.62 Hurricanes follow many different paths from their birthplace in the tropical Atlantic to their demise, usually far to the west and north. Shown are the tracks of all hurricanes from 1851 to 2017.

Australia, the Philippines, China, and Japan are battered by some of the world's strongest storms. Similar east-to-west cyclone tracking impacts the coasts of the Indian Ocean. However, hurricane tracks in detail vary as widely as the irregularities of the steering winds that guide them (**Figure 10.62**). Hurricanes occur in the late summer and early fall because they need the energy contained in warm ocean water to grow strong (**Figure 10.63**).

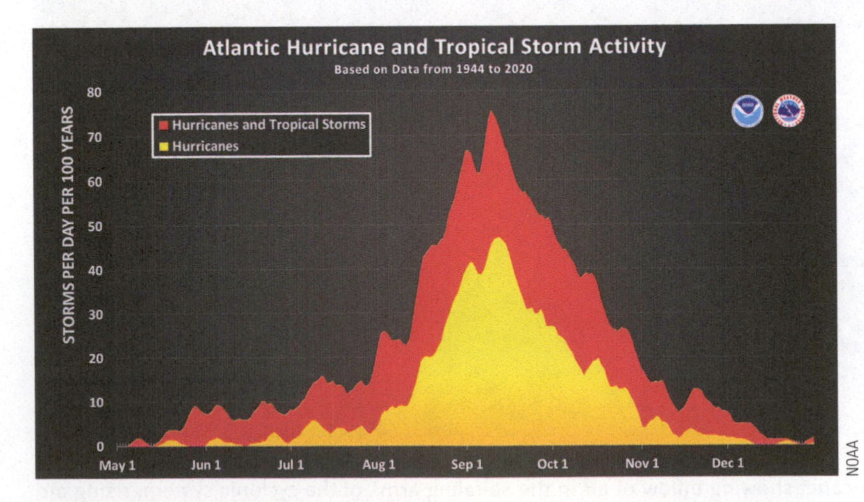

Figure 10.63 Hurricanes and tropical storms are most common in the late summer and early fall, after the ocean has warmed in the strong summer sun.

10.4.2 Circulation

To understand the distribution of wind-induced damage from hurricanes, you must know something about where and how quickly hurricane winds blow. Let's start with the hurricane itself. In the Northern Hemisphere, the winds are wrapping counterclockwise around the eye (**Figure 10.64**). But this isn't the entire story because the hurricane isn't standing still; it's moving along, as noted above, being guided by the steering winds.

Imagine yourself as a television reporter stationed to the east of New Orleans as Hurricane Katrina comes ashore in 2005. The hurricane has winds of 225 kilometers/hour (140 miles/hour), and its rotating eye is moving toward you from the south at 20 kilometers/hour (12 miles/hour). What wind speed do you feel (**Figure 10.65**)? Because both the hurricane and steering winds are blowing in the same direction (from the

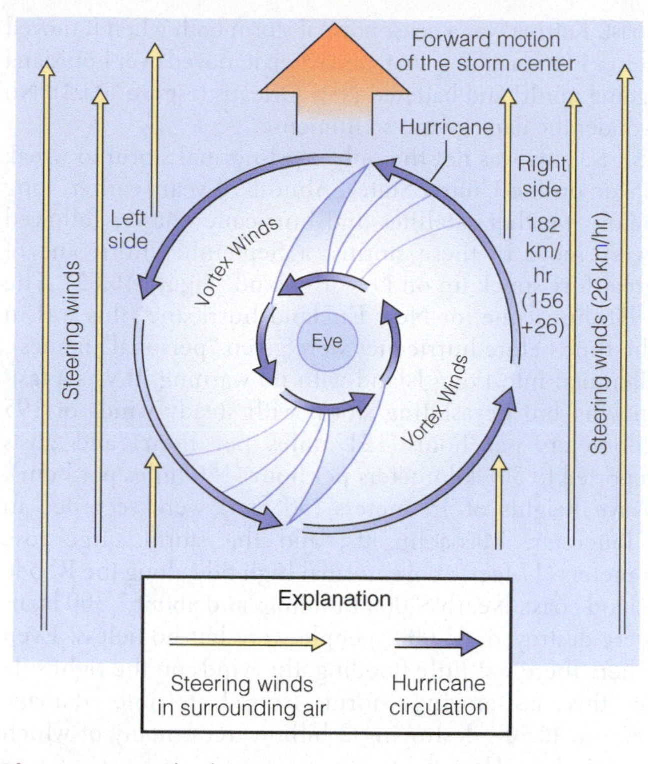

Figure 10.64 This hurricane is being steered by winds of 26 kilometers/hour. The storm winds are 156 kilometers/hour. On the right side of the storm, people on the ground feel 182 kilometers/hour winds (156 + 26). What wind speed do people on the left side of the storm experience? (Modified from Coch, N. K. 1994, August. Geological effects of hurricanes. *Geomorphology* 10(1–4): 37–63)

south), you add them and feel winds of 245 kilometers per hour (152 miles per hour) on your face, if of course you can stand up! Now your colleague, working for the other network, happens to be stationed west of New Orleans. There, the steering winds are blowing from the south, but the

Figure 10.65 Wind speed in hurricanes can be so high that it is difficult to stand—unless, of course, you lean into the wind, as this fellow is doing on August 25, 2005, in south Florida, where hurricane Katrina first made landfall.

Figure 10.66 Hurricane Katrina's wind-driven storm surge was massive and severely damaged low-lying areas along the Gulf Coast. Here the surge is coming over Highway 90 near Gulfport, Mississippi, on August 29, 2005.

hurricane winds are blowing from the north. As a result, your colleague feels only 205 kilometers per hour winds (127 kilometers per hour)–still fierce, but 40 kilometers per hour (25 miles per hour) less than what you are feeling and showing your viewers. When a roof is about to fail, the 30- or 40- kilometer-per-hour wind-speed difference from one side of the hurricane really matters.

Hurricane winds also drive **storm surges**, the mound of water pushed ahead of the hurricane that can cause sea level to rise many meters above the normal high-tide level, as at Galveston in 1900 (**Figure 10.66**). The storm surge is highest where the winds are moving most quickly. Katrina's storm surge, which overtopped levees and flooded New Orleans, came from east of the city, where the right side of the hurricane and its strongest winds passed.

The reduction in atmospheric pressure associated with a hurricane also causes sea level to rise, like fluid rising by suction in a giant soda straw. Average sea-level atmospheric pressure is 1,013 millibars, and the rise in sea level is about 1 centimeter (0.4 inches) for every millibar of pressure drop. Hurricane Camille in 1969, the strongest storm to hit the U.S. mainland, had a near-record low barometric pressure of 905 millibars (26.61 inches of mercury). Camille's storm surge and winds devastated the entire Gulf Coast from Florida to Texas. Just the drop in pressure alone raised local sea level by over 1 meter (>3 feet).

10.4.3 Track and Intensity

Hurricane tracks determine where and how severe the damage will be. Consider a hurricane approaching the U.S. East Coast (**Figure 10.67**). If the hurricane moves parallel to the coast, the strongest winds and the storm surge stay offshore. These are referred to as **coast-parallel**, or CP, hurricane tracks. A hurricane that hits the coast at a right angle has a **coast-normal**, or CN, track. The full force of the storm surge is pushed ashore on the right side and the highest winds (remember where they are and why) slam into the

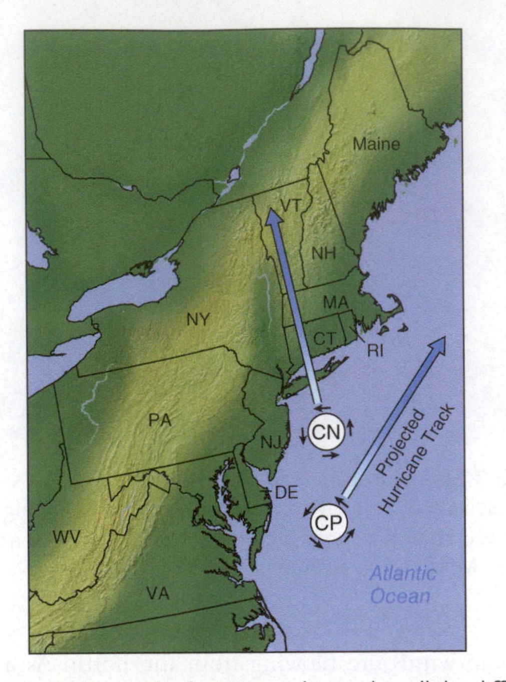

Figure 10.67 The track a hurricane takes makes all the difference. Coast-parallel (CP) hurricanes do little damage to the east coast of North America, because their strong winds and storm surge remain offshore. However, when a hurricane hits the coast straight on—a coast-normal (CN) storm—the damage can be immense. Witness the storm surge damage from both Hurricane Katrina and the 1938 New England hurricane. (Adapted from Coch, N. K. 1994, August. Geological effects of hurricanes. Geomorphology 10(1–4): 37–63, Elsevier)

coast. Katrina was a coast-normal storm both when it moved across Florida (going west) and when it moved over Louisiana (going north) and battered New Orleans (**Figure 10.68**). No wonder the damage was so immense.

Katrina was not the only coast-normal storm to wreak havoc on the United States. Almost 70 years earlier, long before weather satellites and hurricane chasers followed every move of these storms, a September hurricane of great fury snuck up on New England (**Figure 10.69**). The 1938 hurricane, or New England hurricane (this was in the time before hurricanes were given "personal" names), slammed into Long Island with no warning. It was a fast-moving but devastating storm with steady winds of 195 kilometers per hour (121 miles per hour) and gusts reported to 300 kilometers per hour (186 miles per hour). Wave heights of 15 meters (50 feet) were recorded in Gloucester, Massachusetts, and the storm surge rose 5 meters (17 feet) above normal high tide along the Rhode Island coast. Nearly 9,000 buildings and about 3,300 boats were destroyed; 63,000 people were left homeless. Even where there was little flooding, the winds on the right side of this fast-moving storm caused terrible damage (**Figure 10.70**), destroying 2 billion trees, many of which were in New Hampshire.

Hurricane intensity is ranked using the Saffir-Simpson scale (**Figure 10.71**). Herb Saffir was an engineer specializing in wind and Bob Simpson was a meteorologist. The scale considers maximum sustained wind and ranks

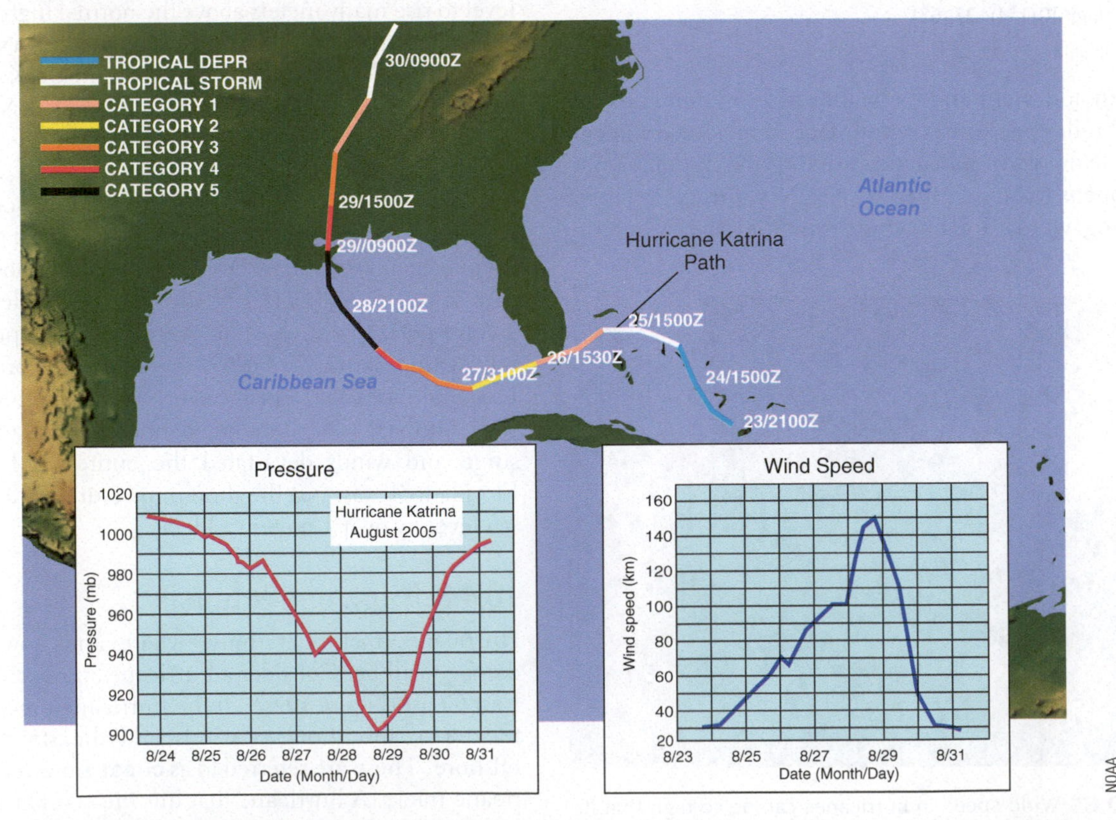

Figure 10.68 When Hurricane Katrina hit land (in Florida and Louisiana), it was a coast normal (CN) storm pushing its highest winds and storm surge (to the right of the eyewall) ashore. Can you see from the graphs how wind speed and barometric pressure are related? What happened to both of these indications of storm strength after Katrina went ashore?

hurricane strength from one to five where the weakest hurricanes are given a category 1 ranking and the strongest hurricanes are assigned a rank of 5. This scale only considers wind speed. It does not explicitly consider storm surge or flood and tornado risk, which often increase along with wind speed. Hurricanes with a ranking of category 3 and above are considered "major."

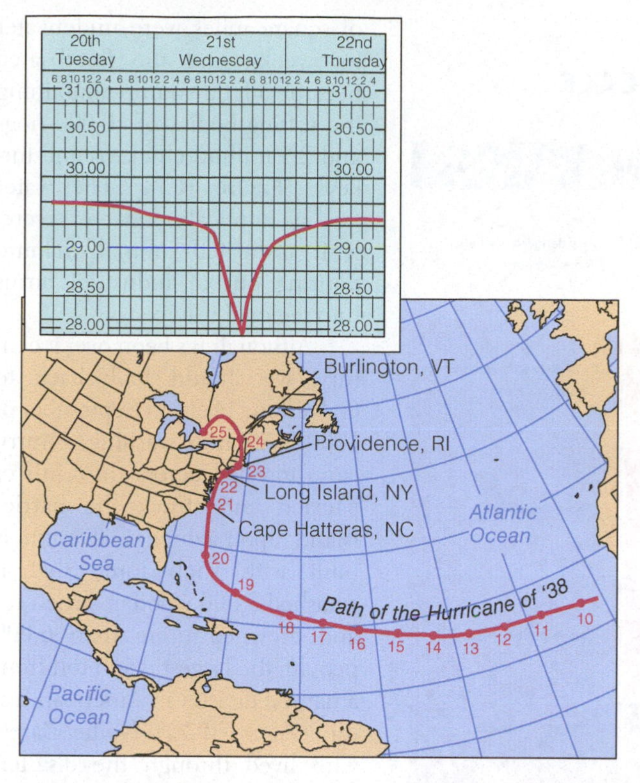

Figure 10.69 The 1938 hurricane roared into New England, slamming into Long Island and taking just about everyone by surprise. The red numbers on the map indicate the hurricane's location on sequential dates in September. It was a fast-moving, coast-normal hurricane when it hit Long Island, which meant high wind speeds and a massive storm surge pounded communities to the right, or east, of the eye. Can you tell from the barometer trace (from a weather station on Long Island) how long the storm lingered and how fast it was moving? (Adapted from The Path of the Hurricane of '38 and Hendrickson, Richard G. 1996. Winds of the Fish's Tail. Mattituck, NY: Amereon Ltd., pp. 33–62)

10.4.4 Hurricane Impacts on Society

What makes hurricanes so dangerous? Certainly, the heavy rains they spawn cause massive flooding and can trigger landslides and debris flows, but the worst actor for many storms is wind. Wind alone rips roofs off buildings and throws debris through the air, but wind also drives waves and storm surges, which cause extensive and costly flooding damage.

In 1970, tropical Cyclone Bhola slammed into East Pakistan (now Bangladesh) and resulted in perhaps the deadliest coastal flooding of all time (**Figure 10.72**). The death toll may have been as great as 500,000 people. Flooding inundated much of the area and people had nowhere to go. So great was the dissatisfaction with the central government of Pakistan's response to the tragedy that the storm led to popular rebellion and the formation of Bangladesh, a new country separated from Pakistan.

Another similar cyclone tragedy in the same area in 1991 killed 60,000 people and prompted formation of a tropical-cyclone forecasting center. The real issue for Bangladesh is not cyclones but elevation–or lack of it. Most of the country is within a few meters of sea level; it's the delta of the Ganges River. Experts at the center recommended the construction of raised concrete bunkers above storm-surge level for refuge as there are no hills to climb and escape floodwaters (**Figure 10.73**).

Some hurricanes veer eastward and follow a track up the U.S. East Coast, causing flooding and beach erosion from the Carolinas to Maine. In 1989, Hurricane Hugo struck the coast

Courtesy of the Technical Association Inc/New York New Haven and Hartford Railroad

Hartford Historical Society

Figure 10.70 Damage from the 1938 New England hurricane differed, depending on location. (a) Near the coast, huge waves and the storm surge ripped apart infrastructure and tossed ships around like toys, as shown here in eastern Connecticut. In New London, Connecticut, the lightship tender TULIP was thrown out of New London harbor and deposited on the New Haven's main rail line. Restoration of passenger service on the Shore Line between New Haven, Connecticut, and Boston, Massachusetts, took nearly 2 weeks. (b) Farther inland, wind was the primary agent of destruction. In Hartford, Vermont, homes were damaged as massive trees were felled by the hurricane's wind.

THE SAFFIR-SIMPSON HURRICANE SCALE

CATEGORY	WINDS (mph)	WINDS (km/h)	DAMAGES
1	74 - 95	119 - 153	Very dangerous winds Some damage
2	96 - 110	154 - 177	Extremely dangerous winds Extensive damage
3	111 - 130	178 - 209	Devastating damage
4	131 - 155	210 - 249	Catastrophic damage
5	> 156	> 250	Catastrophic damage

Artellia / Alamy Stock Photo

Figure 10.71 The Saffir-Simpson hurricane category ranking scale.

just north of Charleston, South Carolina, with sustained winds of more than 130 miles per hour and a storm surge of 6 meters (20 feet). Hurricane Floyd was the 1999 storm in North Carolina that wouldn't go away: it dropped 55 centimeters (22 inches) of rain. Although not a strong hurricane, it created flooding of massive proportions in the eastern part of the state, aided by Hurricane Dennis, which had dropped 25 centimeters (10 inches) of rain 2 weeks earlier. Thousands of square miles were underwater for weeks, and the flood area became a fetid swamp containing the rotting bodies of 10,000 hogs and 2.5 million chickens and turkeys. (**Figure 10.74**) The water quality implications were severe, with untreated animal manure pouring out of industrial farms into regional waterways.

Although it's been over a century, we should look back to Galveston, Texas, in September of 1900. With little warning, a hurricane traveled across the Gulf of Mexico and struck the barrier island upon which Galveston is built with such force that it smashed 3,600 wooden structures and drowned more than 6,000 people, the largest death toll from a natural disaster in American history (**Figure 10.75**). Nellie Carey, who lived through the disaster, wrote in a letter: "Thousands of dead in the streets–the gulf and bay strewn with dead bodies. Not a drop of water–food scarce. The dead are not identified at all–they throw them on drays and take them to barges, where they are loaded like cord wood, and taken out to sea to be cast into the waves." The letter was reprinted in its entirety in a book published shortly after the disaster. Today, Galveston is among the cities most likely to be evacuated in case of a hurricane.

In September 2008, Hurricane Ike, a 1,000-kilometer (600-mile)-wide category 2 hurricane, pummeled Texas, coming ashore over Galveston, where 20,000 people ignored evacuation orders (but more than 2 million people did leave

Harry Koundakjian/Shutterstock.com

Figure 10.72 In 1970, Cyclone Bhola slammed into Bangladesh (then Pakistan). Much of the low-lying country was underwater.

iStock.com/Saikat Bhadra

Figure 10.73 In 2021, Cyclone Yaas roared ashore. Its track, to the west of Bangladesh, put the nation on the right side of the storm where wind speed on the ground and storm surge were the highest.

© Dove Imaging

Figure 10.74 The corpses of hundreds of hogs that drowned in the floodwaters of Hurricane Floyd in 1999.

NOAA

Figure 10.76 The aftermath of Hurricane Ike (2008) on the Bolivar Peninsula just east of Galveston. The homeowners knew the water was coming: their homes are on stilts, but that wasn't enough for many homes which are gone. Only the stilts remain.

the Texas and Louisiana coasts). The storm surge was between 3 and 5 meters (10 to 15 feet), just overtopping the seawall in some places and deeply flooding low-lying neighborhoods (**Figure 10.76**). Quoted in the *New York Times*, one resident who had tried to ride out the storm at his Galveston home with his family said, "I know my house was dry at 11 o'clock, and at 12:30 a.m., we were floating on the couch putting lifejackets on." Once the water reached the television, a meter (3 feet) off the floor, he retrieved his boat from the garage and loaded his family into it. "I didn't keep my boat there to plan on evacuating because I didn't plan on the water getting that high, but I sure am glad it was there."

You might ask how often is someone living along the coast likely to experience a hurricane. Well, that question has been answered at least for the past using a century of hurricane data for 45 cities along the eastern and southern coast of the United States. If you live in Florida, your chances of being impacted by a hurricane are many times higher than if you live in New England. Hurricanes are so frequent in Miami, it would be almost impossible to grow up there without experiencing one (**Figure 10.77**).

10.4.5 Hurricane prediction

We know that hurricanes are part of the normal tropical weather pattern and that they will occur seasonally every year. Meteorologist William Gray at Colorado State University (1929–2016) developed an empirical prediction system that can estimate the severity of the upcoming hurricane season. His work is based on understanding the two most important ingredients needed to make strong hurricanes: warm sea-surface temperatures and an atmosphere with little **wind shear** (winds blowing in different directions at different elevations), so that hurricanes, once they start to form vertical circulation, are not ripped apart by winds aloft. Gray's model, now being implemented by others at Colorado State University where he taught, uses the following data:

- Tropospheric winds up to an altitude of 12,000 meters (40,000 feet). Strong winds in the troposphere work against hurricanes by shearing off the tops of their circulation.
- West African climate. Dry years in Africa promote high-altitude winds from the west above the tropical Atlantic, increasing the shear between the upper and lower levels of the atmosphere and diminishing the strength of hurricanes. During wet years, upper atmosphere winds blow from the east, greatly reducing wind shear and thus promoting hurricane-strength storms.

Library of Congress

Figure 10.75 Men attempt to remove a body from the wreckage of Galveston after the 1900 hurricane. This is a stereo view–a popular medium at the time that allowed people to see the world three dimensionally using a small viewer.

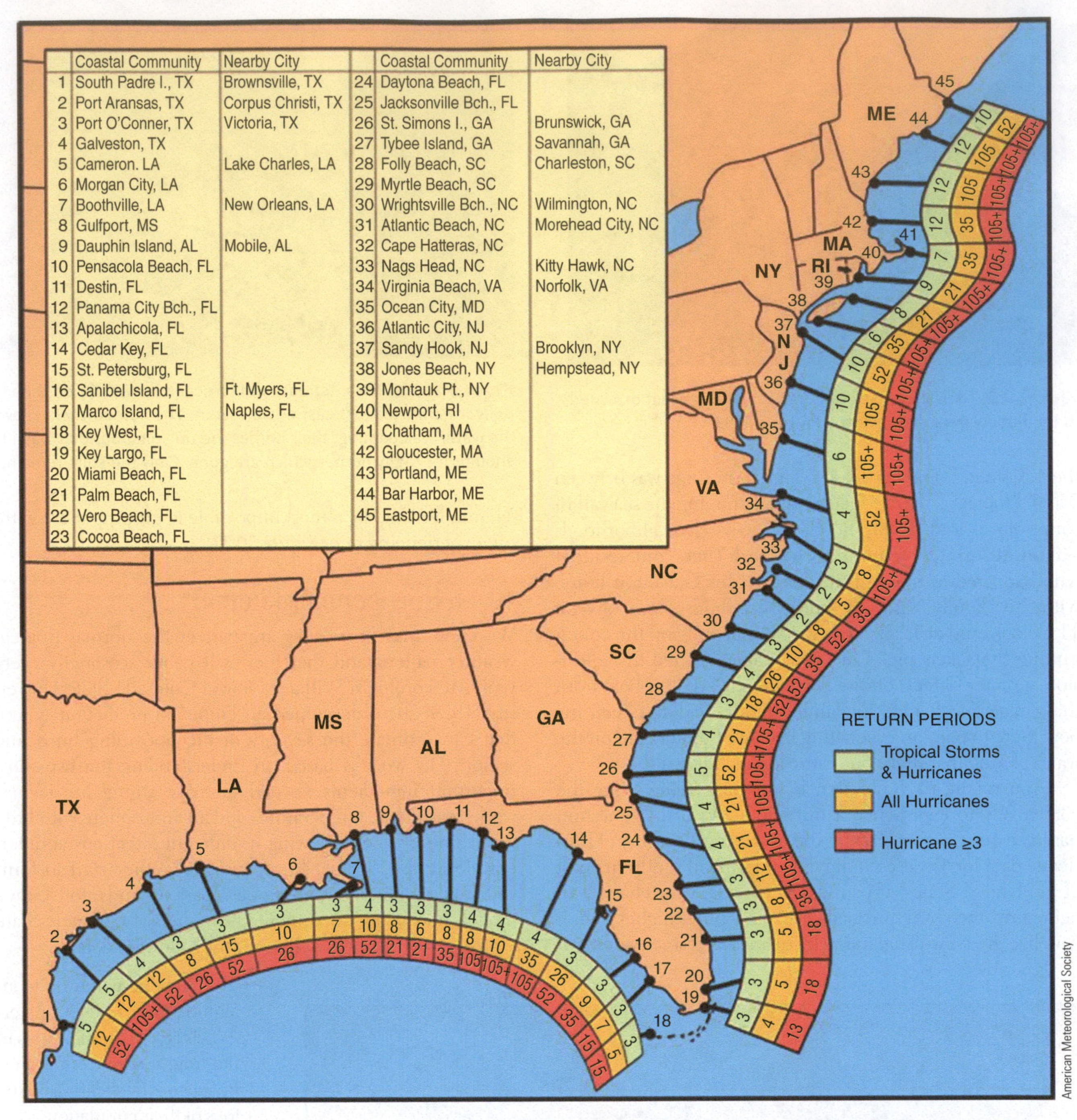

	Coastal Community	Nearby City		Coastal Community	Nearby City
1	South Padre I., TX	Brownsville, TX	24	Daytona Beach, FL	
2	Port Aransas, TX	Corpus Christi, TX	25	Jacksonville Bch., FL	
3	Port O'Conner, TX	Victoria, TX	26	St. Simons I., GA	Brunswick, GA
4	Galveston, TX		27	Tybee Island, GA	Savannah, GA
5	Cameron. LA	Lake Charles, LA	28	Folly Beach, SC	Charleston, SC
6	Morgan City, LA		29	Myrtle Beach, SC	
7	Boothville, LA	New Orleans, LA	30	Wrightsville Bch., NC	Wilmington, NC
8	Gulfport, MS		31	Atlantic Beach, NC	Morehead City, NC
9	Dauphin Island, AL	Mobile, AL	32	Cape Hatteras, NC	
10	Pensacola Beach, FL		33	Nags Head, NC	Kitty Hawk, NC
11	Destin, FL		34	Virginia Beach, VA	Norfolk, VA
12	Panama City Bch., FL		35	Ocean City, MD	
13	Apalachicola, FL		36	Atlantic City, NJ	
14	Cedar Key, FL		37	Sandy Hook, NJ	Brooklyn, NY
15	St. Petersburg, FL		38	Jones Beach, NY	Hempstead, NY
16	Sanibel Island, FL	Ft. Myers, FL	39	Montauk Pt., NY	
17	Marco Island, FL	Naples, FL	40	Newport, RI	
18	Key West, FL		41	Chatham, MA	
19	Key Largo, FL		42	Gloucester, MA	
20	Miami Beach, FL		43	Portland, ME	
21	Palm Beach, FL		44	Bar Harbor, ME	
22	Vero Beach, FL		45	Eastport, ME	
23	Cocoa Beach, FL				

Figure 10.77 Location matters! The return period (or average time between hurricane strikes in years) for the east and south coasts of the United States, based on data from 1901 until 2005. Different storm strengths are considered separately. If you live in Bar Harbor, Maine, you are unlikely ever to see a strong hurricane and perhaps you'll see a tropical storm once a decade. Contrast that to Miami, where you'll see on average a tropical storm every few years and a major hurricane every decade.

- El Niño. The warming of the equatorial Pacific Ocean promotes westerly winds that blow into the Caribbean and Atlantic basins, where trade winds blow at lower levels from the east. This promotes wind shear that knocks down hurricane buildups. For example, between 1991 and 1994, repeated El Niños resulted in few hurricanes. In 1995 and 1999, El Niños gave way to a cooling of the equatorial Pacific Ocean (La Niña) and hurricane activity greatly increased.
- Atmospheric pressure. Low atmospheric pressure in the Atlantic Ocean and Caribbean Sea indicates areas of warm water, which fuel hurricanes, promote the convergence of air masses, and stimulate vertical buildup of moisture-laden clouds.

Although Gray's model has a good record of predicting storm frequency for any one hurricane season, it is up to seasoned forecasters and increasingly sophisticated computer models to predict both the track and the strength of

403RD Wing/UAF

Figure 10.78 1st Lt. Ryan Smithies, 53rd Weather Reconnaissance Squadron pilot, flies in the eye of Hurricane Dorian off the coast of Savannah, Georgia, in 2019. Dorian was a category 2 hurricane.

any individual hurricane after it is born. In the United States, the National Weather service is charged with this duty, and it has made great advances in the accuracy of its predictions over the past two decades.

Forecasters are aided in their work by several sources of data. First, they use sophisticated atmospheric models to predict the behavior of steering winds that help determine both the track of the hurricane and the degree of shearing that could tear the storm apart. Second, they use data collected in real time by hurricane-hunter planes that fly right through the storms, collecting wind and pressure data (**Figure 10.78**). Third, they rely on more than 100 years of hurricane history to understand how such storms behave along different parts of the Gulf and Atlantic coasts.

The work of many forecasters at the National Hurricane Center in Miami is condensed into track and intensity prediction maps (**Figure 10.79**). The improvement in

NOAA

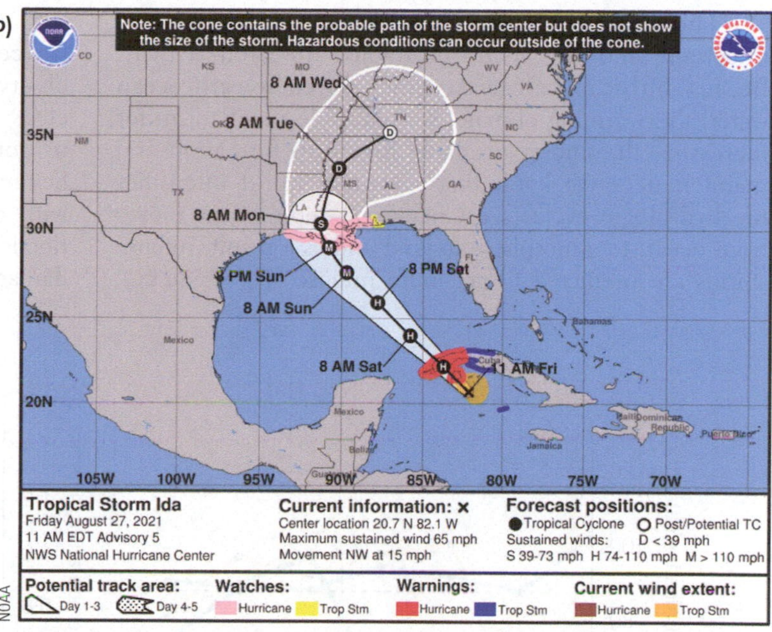

NOAA

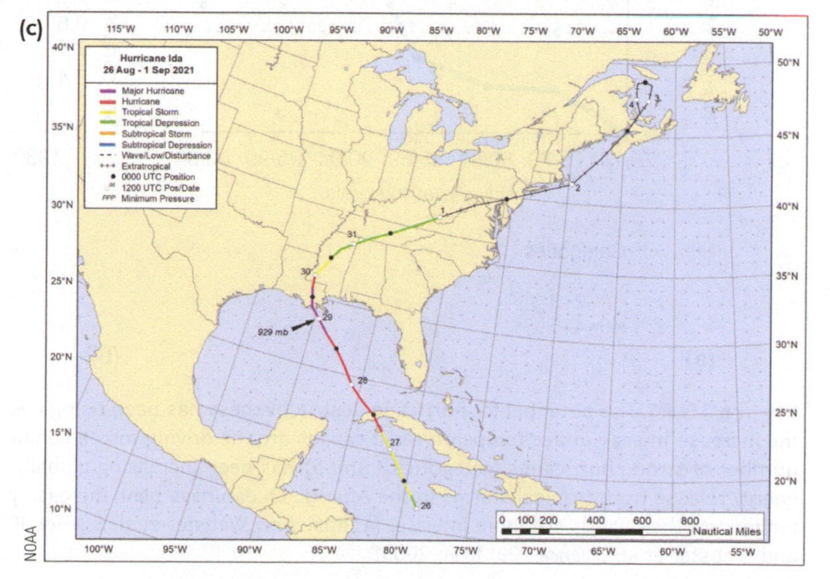

NOAA

Figure 10.79 In August 2021, Hurricane Ida struck the U.S. mainland near New Orleans after crossing over the western tip of Cuba. (a) Satellite image of Ida at 4:10 a.m., just before landfall. The lights of a sleeping nation clearly highlight cities and towns. (b) The 5-day forecast of Ida's track made on August 27. (c) Ida's actual track. How well did the forecasters do?

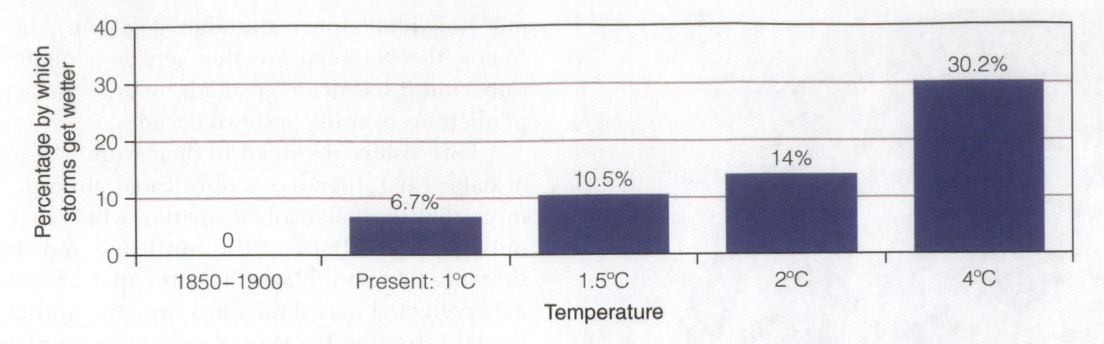

Figure 10.80 Both current data and model predictions show that the biggest storms will get wetter as the climate warms. The percentage by which storms get wetter is shown above each bar. (Adapted from Mathew Barlow (July 29, 2022). Climate change is intensifying the water cycle, bringing more powerful storms and flooding – here's what the science shows, The Conversation)

hurricane forecasting accuracy is impressive. Today, the accuracy of 5-day hurricane forecasts matches that of the 3-day forecasts only a decade ago.

The effect of climate change on hurricanes remains uncertain and the focus of intense research activity. Because hurricanes are heat engines, driven by warm ocean water, the warming climate is a plausible mechanism for increasing the energy available for storm formation and maintenance. We know that a warmer world intensifies the hydrologic cycle (see Chapter 2)–more water moves from ocean to atmosphere to land and as a result, intense storms are predicted to get more intense (**Figure 10.80**). Consistent with these findings, two studies suggest that the total energy released by hurricanes, as well as the frequency of strong hurricanes, is increasing (**Figure 10.81**). Yet, when overall hurricane frequency data over the last century are analyzed and corrections made for fewer observations in the time before weather satellites, it's not clear that hurricane frequency is actually increasing–it could be observation bias. We've simply gotten better and finding all the hurricanes. What we do know with certainty is that with sea level rising, storm surge flooding driven by hurricanes will indeed become more damaging.

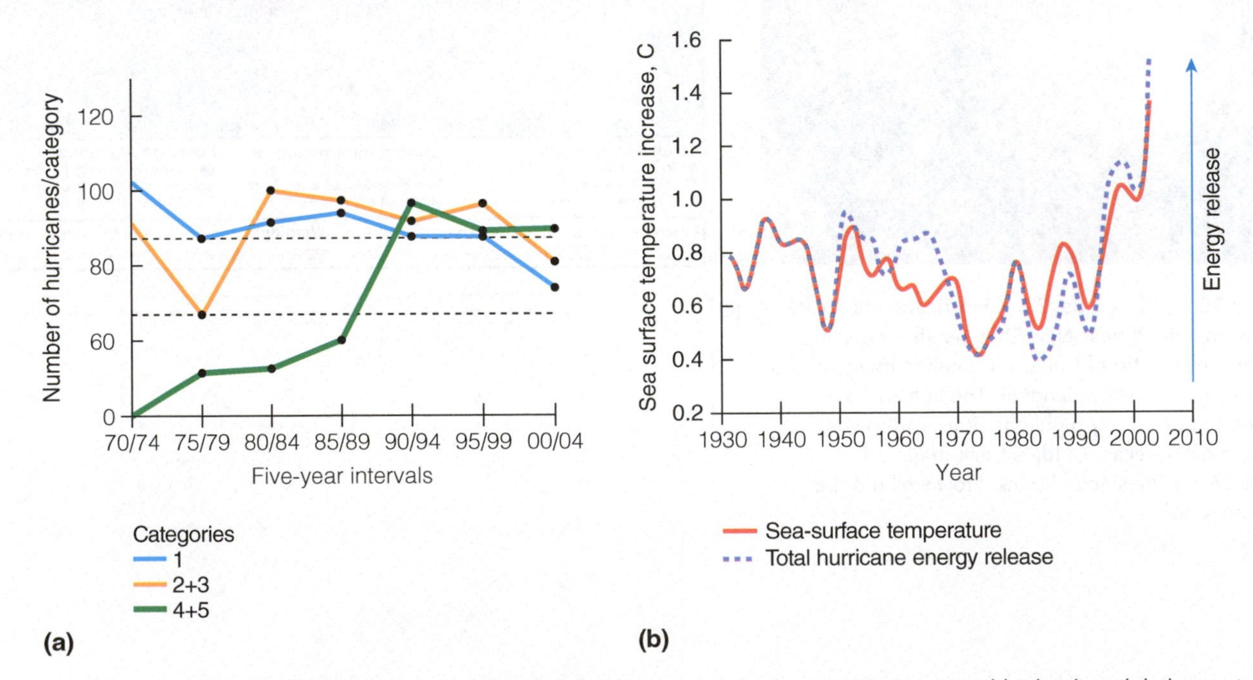

Figure 10.81 The potential for hurricane destructiveness has been rising over the past 30 years. Could it be that global warming and the increase in sea-surface temperatures it causes are the driving forces behind changing hurricane dynamics? Time will tell. (a) The number of strong hurricanes (categories 4 and 5) has been increasing globally since 1970. (b) Both sea-surface temperature and total energy release from tropical storms in the Atlantic Ocean have been increasing in lockstep over the past 50 years. Storms are lasting longer and their winds are more intense. (a: Based on Webster et al. *Science* 309: 1844; b: Emanuel, Kerry. 2005, August. *Nature* 436(4), and Webster et al. *Science* 309: 1844. 2005)

10.5 Nor'easters (L05)

The east coast of North America is affected by major coastal storms pretty much year-round. Although the media focuses heavily on the hurricanes that threaten eastern North America in the summer and fall, during the winter and spring, other powerful but different storms, termed **nor'easters**, threaten the Atlantic Coast. The name "nor' easter," comes from the direction that their winds most frequently blow onshore as the storms track north just off the U.S. East Coast.

Nor'easters are predominately winter and spring storms that differ not only in season but also in many other ways from hurricanes.

- While hurricanes are born and thrive over warm, tropical waters, nor'easters thrive in the winter cold.
- Hurricanes are torn apart by the jet stream, a rapidly moving current of air high in the atmosphere (at the altitude jet planes fly, the boundary between the stratosphere and troposphere; Chapter 2). Nor'easters thrive on the jet stream and the contrast it sets up between colder, drier air to the northwest of the storm, and warmer, moister air to the southeast of the storm.
- Nor'easters are massive storms. The largest ones cover the entire east coast of North America. In contrast, hurricanes are compact storms, perhaps a few 100 kilometers across.
- Most hurricanes move quickly, rarely lingering more than half a day in one location. Nor'easters can batter the coastline for days on end.

10.5.1 Growing a Monster

Nor'easters are cold storms born over North America, where they begin as nothing more than average low-pressure systems, moving from the west to east. Once the parent storm hits the western Atlantic, it rapidly intensifies, exploding into a **bomb cyclone**, as some weather forecasters term such storms. In other words, the atmospheric pressure in the storm drops rapidly, the storm expands dramatically, and the winds increase. A small, weak storm becomes a monster in a matter of hours, or literally overnight.

Nor'easters develop explosively along the eastern coast of North America, often getting their start off Cape Hatteras, North Carolina. From there, the storms move toward the north and northeast, most often tracking along and near the coast. On their right side are moist ocean waters, warm only in comparison to the frigid arctic air to their west. This contrast of air masses makes nor'easter's a menace and allows them at the same time to produce torrential rain on the eastern side of the storm and massive snow fall on its western side (**Figure 10.82**).

Figure 10.82 Massive, swirling nor'easter batters the East Coast of the United States in March 2018.

10.5.2 Impacts on Society

The impacts from nor'easters are the combined result of how large these storms are, how slow-moving they can be, and how strong they can become. Their strength and persistence create sometimes extreme coastal hazards, as wave heights can build over days and those waves relentlessly pound beaches and sea walls. The long duration of the storm also ensures that the storm surge will correspond with at least one, or sometimes several, high tides, increasing the potential for flooding and damage to buildings close to sea level (**Figure 10.83**). Many large nor'easters dramatically alter coastlines, breaching barrier islands and creating new inlets. When homes are in the way, this does not end well (**Figure 10.84**).

Inland, nor'easters create different hazards. They routinely bring deep winter snowfall, blanketing cities with so

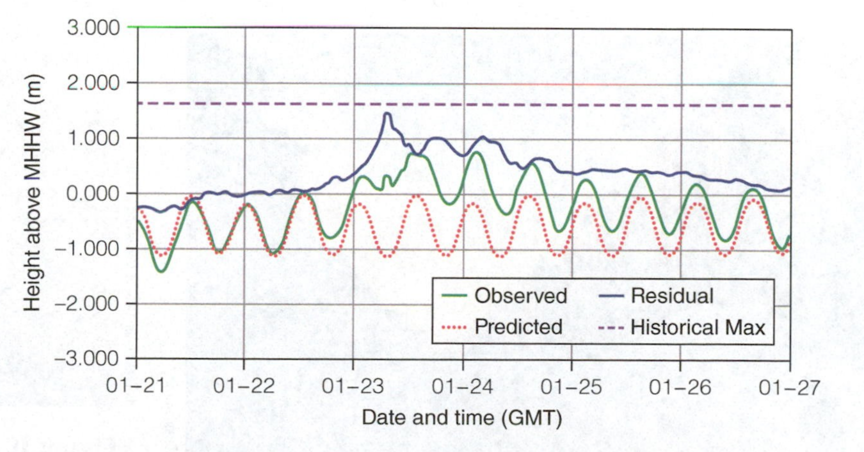

Figure 10.83 Ocean water levels at Money Point, Virginia during a 2016 nor' easter. The red line shows the prediction of tides rising and falling. The green line is the measured sea level. The difference is the storm surge (the blue curve). The effect of the surge lasted from January 22 to January 27, 5 days. (Modified from NOAA, Center for Operational Oceanographic Products and Services)

Figure 10.84 Before and after images of a barrier island breach caused by Superstorm Sandy at Mantoloking, New Jersey.

much snow it can't be plowed and rather must be trucked away to open the roads (**Figure 10.85**). The snow is often accompanied by very high winds that lead to blowing, drifting, and whiteout conditions, shuttering airports and closing highways, sometimes for days. If the air is not cold enough to produce snow, then inches of rain can trigger landslides and debris flows.

Perhaps the best-known nor'easter is the Halloween storm of 1991, which pummeled New England with massive waves and high seas and is featured in the book and movie *The Perfect Storm*. The Halloween storm was massive and devastating, affecting much of the northeastern United States and Atlantic Canada. It took the lives of fishermen at sea (yes, George Clooney went down with the ship) and demolished many shorefront homes that had stood for a century or longer. The Bush family compound in Kennebunkport, Maine, suffered serious damage from the pounding surf.

Strong nor'easters are common, occurring several times each winter, but severe nor'easters are much less common, occurring perhaps once every few years. The biggest nor'easters are events people remember by name and there are only a few each century.

Figure 10.85 Winter after a strong nor' easter in the South Boston (Southie) neighborhood of Boston.

10.6 Human/Ocean Interactions (LO6)

The coast, because it is a dynamic, sometimes rapidly changing system, can be a risky place to build structures and to live one's life. Waves erode sea cliffs, dropping homes with a view onto the beach below (**Figure 10.86**). Storm surges flood low-lying areas near the coast and routinely wash over barrier islands, carrying buildings and roads into the adjacent lagoon. While these hazards are typically thought of as ocean-related, they can also affect large lakes, like the Great Lakes.

The Earth preserves a long record of humans interacting with the ocean. Shell middens, where humans left the shells of clams and oysters they harvested from the coastal zone, date back thousands of years. Some middens are high above the waves or far from today's coast due to sea-level change since they were active (**Figure 10.87**). Tectonics can play unexpected games with coastal infrastructure. In 2012, the Kaikoura quake shook the South Island of New Zealand. Up went the land along the coast, including the intertidal zone where an intertidal ecosystem all of a sudden found itself high and dry (**Figure 10.88**).

Today, we measure sea level using a variety of tools. Satellites track sea level in real time. **Tide gauges** record the

Figure 10.86 Rapid sea-cliff retreat near Big Lagoon, Humboldt County, California, has left this home in a precarious position. These cliffs near the Oregon border are composed of uplifted marine sediments (sand and gravel). As shown by the photograph, the cliffs are subject to direct wave attack. Drastic undercutting by wave action and the near-vertical cliff indicate that this area is actively eroding.

Figure 10.87 A shell midden, left by people gathering clams and oysters for food, covers the ground well above the shoreline near Playa de Cofete Jandia, Fuerteventura, Canary Islands. Note the volcano in the background.

ups and downs of the ocean surface as the water rises and falls, day in and day out. After decades and in some places more than a century, these tidal records clearly show sea-level rise when the average over time is considered–the slope of the sea-level line. Consider for example San Francisco, California. Its tide gauge has been taking data since 1854! There's much year-to-year variability but the trend is up at about 2.1 millimeters per year (**Figure 10.89**).

Figure 10.88 The intertidal zone, where these kelp plants normally live, is now high and dry thanks to uplift from the massive Kaikoura quake in New Zealand. Without daily immersion in the seawater, the kelp is dying.

Many shoreline-impact studies use scale models to evaluate various remediation schemes before they are built. Physical modeling makes sense, because beaches are complex systems with several feedbacks that might not be obvious at first glance or are difficult to handle in a mathematical model. Such physical models are an exact replica of the project but require careful scaling of the materials and flows to compensate for differences in size. The U.S. Army Corps of Engineers at its facility in Vicksburg, Mississippi, has been at the forefront of such modeling efforts (**Figure 10.90**).

10.6.1 Sea- and Land-Level Changes

Mavra Litharia is an ancient harbor, built by the Romans, in the northern Peloponnese of Greece. It had a large jetty built of stone. Today, much of that jetty is well above today's sea level. Archeologists excavating the area worked with geoscientists and determined that two different earthquakes elevated the jetty above sea-level. The first occurred around 264 C.E. and lifted the harbor by 4.35 meters (14.3 feet). One thousand years later, the harbor rose another meter. Today, the jetty is 40 meters (130 feet) inland from the shoreline. Try mooring your boat there now.

Land can go down as well as up. Another spectacular example of a Roman coastal work impacted by sea-level change (relatively speaking) is the port of Pozzuoli (Puteoli), just west of Naples. This was one of the most important harbors in late Republic/early Empire times; it is now deeply submerged, sinking in the third and fifth centuries CE as the floor of the surrounding volcanic caldera began to deform. It is preserved today as the Baia Underwater Archaeological Park. You'll need scuba or snorkel gear to visit, unless you can find a glass bottom boat (**Figure 10.91**).

Then, there is Doggerland, the now-flooded land bridge that connected Brittan to Europe when ice sheets lowered global sea level more than 120 meters (400 feet) (**Figure 10.92**). As the Earth warmed and the ice melted, Doggerland began to shrink, going entirely below the waves about 7000 years ago. But its history lives on. In 1931, a fisherman working 25 miles offshore in water 120 feet deep pulled up a haul that included a spectacular, carved harpoon tip encased in freshwater peat (**Figure 10.93**). The tip, carved in deer antler, was dated to about 11,500 years ago by measuring radiocarbon in the peat. Whoever dropped that weapon was walking on dry land! Since then, spectacular artifacts have been found by fishermen and people walking on the area's beaches. For thousands of years, people thrived on what is now the bottom of North Sea. Slowly, the lands on which they lived and hunted fell victim to a rising sea. Doggerland is a preview of our future if we continue to warm the climate, melting the planet's ice sheets and raising sea level.

10.6.2 Protecting Harbors

Harbors are valuable along the coast and that value leads people to take some less than wise actions. Ocean City, Maryland, is at the south end of Fenwick Island, one island

(a)

(b)

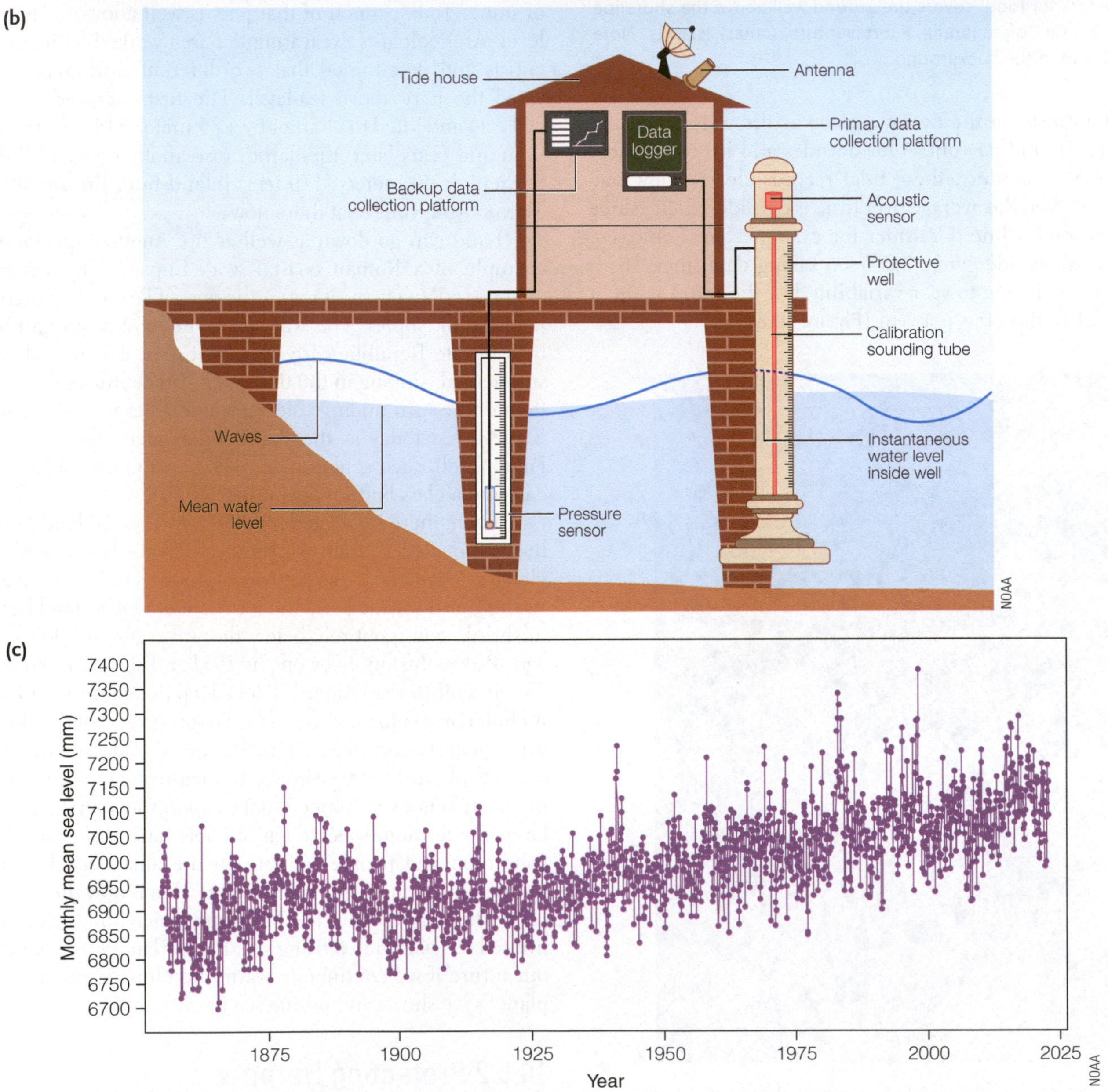

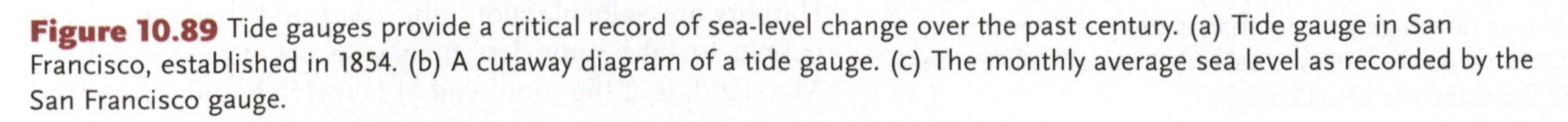

Figure 10.89 Tide gauges provide a critical record of sea-level change over the past century. (a) Tide gauge in San Francisco, established in 1854. (b) A cutaway diagram of a tide gauge. (c) The monthly average sea level as recorded by the San Francisco gauge.

Engineer Research and Development Center

Figure 10.90 A scale model of Chicago Harbor built by the U.S. Army Corps of Engineers. The physical model, built at a scale of 1:120, depicts more than 5,500 meters (18,000 feet) of the city's shoreline. Along with a wave simulator, the model is useful for assessing potential damage from waves of varying height and source direction. Data from the model are used to make future design improvements.

in the chain of barrier islands extending down the U.S. East Coast from New York to Florida. It is a major and highly developed destination beach resort for the millions of people who live in the Baltimore/Washington/Philadelphia corridor. Long jetties were constructed near Ocean City in the 1930s to maintain the inlet between Fenwick and Assateague Island to the south (see Figure 10.13). As a result, the beach was stabilized at Ocean City–the north jetty blocked the southward longshore transport of sand–and the Assateague Island beach has been displaced 500 meters (1,650 feet) shoreward because of sand starvation. Now, as sea level rises, the beach at Ocean City is washing away. The solution, beach nourishment. Sand from offshore is pumped on the the beach (**Figure 10.94**).

In the 1920s, Santa Barbara's civic leaders decided to mold their community into the "French Riviera" of California.

BIOSPHOTO / Alamy Stock Photo

Figure 10.91 Drowned ancient mosaics at Campi Flegrei, west of the Gulf of Naples. The Underwater Park of Baia, Italy.

Claus Lunau/Science Source

Figure 10.92 Doggerland is the region between the British Isles and mainland Europe. It was dry land during the last glacial period but it began to flood as the ice sheets melted and sea level rose. The green regions show the last time land in that area was above sea level.

Norwich Castle Museum and Art Gallery

Figure 10.93 Deer-bone harpoon brought up by a fisherman over now-flooded Doggerland.

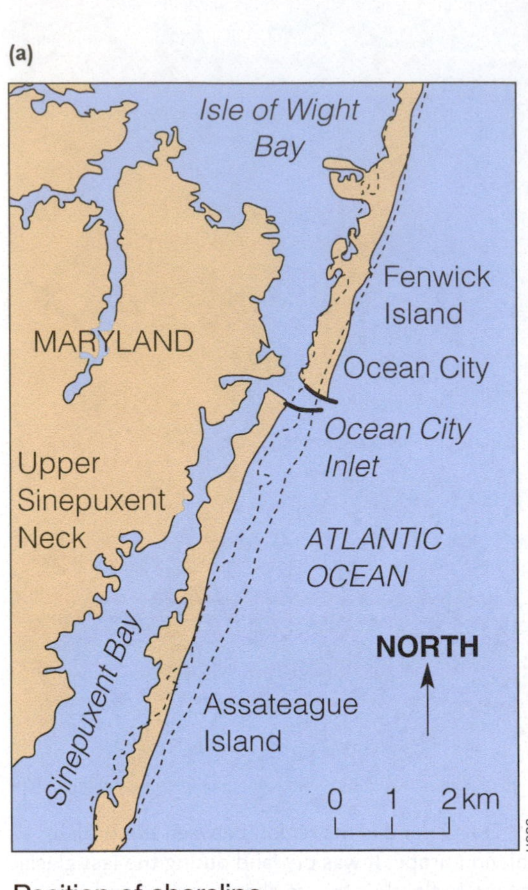

Figure 10.94 Ocean City Maryland has a beach erosion problem. (a) Jetties built at Ocean City Inlet (thick black lines on map) in the 1930s cut off the supply of sand from the north (see Figure 10.13). In response, the north end of Assateague Island has now shifted 500 meters (1650 feet) shoreward of the south end of Fenwick Island. (b) Using large pipes to pump sand from offshore onto the beach at Ocean City, Maryland. (c) Beach replenishment project at Ocean City, which is doing what it is designed to do: protect the buildings and give beachgoers more room to play.

Figure 10.95 The Santa Barbara, California, breakwater in 1988. Sand accumulation in the harbor is clearly visible.

Despite unfavorable reports by the Corps of Engineers, city officials authorized the construction of a long, L-shaped breakwater for a yacht anchorage in 1929 (**Figure 10.95**). The predominant drift direction here is to the southeast, and as the beach west of the breakwater grew wider, near-record erosion occurred on the beaches east of the structure. Summerland Beach, about 16 kilometers (10 miles) downdrift (east), retreated 75 meters (250 feet) in the decade following construction of the breakwater–not the preferred outcome.

Eventually, the southeast-moving littoral drift traveled around the breakwater, forming a sandbar in the harbor entrance. The bar built up, diminishing the size of the harbor and creating a navigational hazard which remains to this day. The harbor inlet must be frequently dredged at great expense to keep it from clogging with sediment. When the Santa Barbara city fathers asked Will Rogers, the noted

American humorist, what he thought of the harbor in the 1930s, he is alleged to have replied, "It would grow great corn if you could irrigate it."

10.6.3 Protecting Beachfront Property: Hard Stabilization

For decades, the most common approach to protecting beachfront property was the installation of coastal engineering works built to protect the land from the sea. These are collectively known as **hard stabilization**, the construction of metal, stone and concrete sea wall, jetties, and groins designed to preserve the shoreline in its current configuration–no flexibility for dynamic coastal processes is allowed (**Figure 10.96**). Geoscientists know from experience the long-term folly in building walls to protect us from the rise of sea level or withstand enormous storm surges, but that does not keep people from trying. In the short term, hard stabilization is buying time and is a way to mitigate flooding during moderate storm events. Below, we provide some examples of what has been done and what has not gone well.

Rock revetments and concrete seawalls are built landward of the shoreline and parallel to the beach, not to protect the beach but rather to protect the land and property behind them. Commonly, these "hard" structures reflect wave energy downward, removing sand and undermining the structure (**Figure 10.97**). There is a question among coastal engineers and geologists that goes something like "Do you want a revetment, or do you want a beach?" Because waves reflecting off hard shoreline structures remove the beach material in front of them, the best-designed rock revetments have sufficient permeability to absorb some of the wave energy. Another problem is that wave action erodes around the ends of revetments. When that happens, they must be extended to prevent damage to structures located there.

Figure 10.96 Large boulders, each weighing several tons, are moved into position to armor the shoreline against wave attack on the Palm Jumeirah Island in Dubai, United Arab Emirates.

Communities along the shores of the Great Lakes also have erosion problems. Early in the twentieth century, protective structures were built on the Lake Michigan shoreline. Lake levels began dropping around 1960, and there was little concern at the time about shoreline erosion as this happened. Since 1964 levels have been rising at an irregular pace, threatening the older, decaying lakeshore protection works. Between 2013 and 2019, the lake water level rose 2 meters (6 feet). In 2020, the high-water levels in Lake Michigan surpassed monthly high-water level records. With high water levels comes an increased likelihood of shoreline erosion and lakeside flooding, increasing public pressure to put more protective structures on the shoreline (**Figure 10.98**).

10.6.4 Soft Stabilization

The drawbacks of hard stabilization are well known and some states, such as North Carolina, have largely prohibited its use in the coastal zone. In its place, soft stabilization has been implemented, a less invasive way of managing a changing coastline.

Sand pumping (**beach nourishment**) is a soft stabilization favorite (see Figure 10.94). It's expensive, temporary, and disturbs marine ecosystems while using large amounts of energy, almost all of which is generated by fossil fuels. The premise is simple: pump sand from offshore onto the beach. Unfortunately, artificially nourished beaches are generally short-lived. One reason is that the imported sand is often finer than the native beach sand and is thus more easily removed by the prevailing waves. The other reason is that artificial beaches usually have a steeper shore face and are therefore more subject to direct wave attack than are gently sloping, natural shorelines, which better dissipate wave energy offshore.

At present, the designs of most new, long structures built perpendicular to the shoreline allow for the pumping of sand from the upcurrent side of the structure to nourish the beaches downcurrent. In addition, downcurrent beaches can be artificially nourished by pumping in sand from offshore areas or by transporting it to them from land sources (see Figure 10.94).

As a last resort, some homeowners reason that, if you can't save the beach, then how about just building above tide level? All around the country, on barrier islands from New York to Texas, homes have been jacked up and placed on stilts to keep them above rising sea level (**Figure 10.99**). This approach works until a large enough storm comes along that the pilings are swept away and the house collapses, or until the barrier island underneath migrates landward, leaving the house still on stilts but over water instead of land. Raising houses on stilts is at best a short-term approach to coast hazard mitigation. The best solution is to minimize coastal construction altogether.

10.6.5 Managed Retreat

As sea level rises due to climate change and rates of coastal erosion increase, talk has begun of **managed retreat**, the idea that it makes more sense to move than to fight rising sea

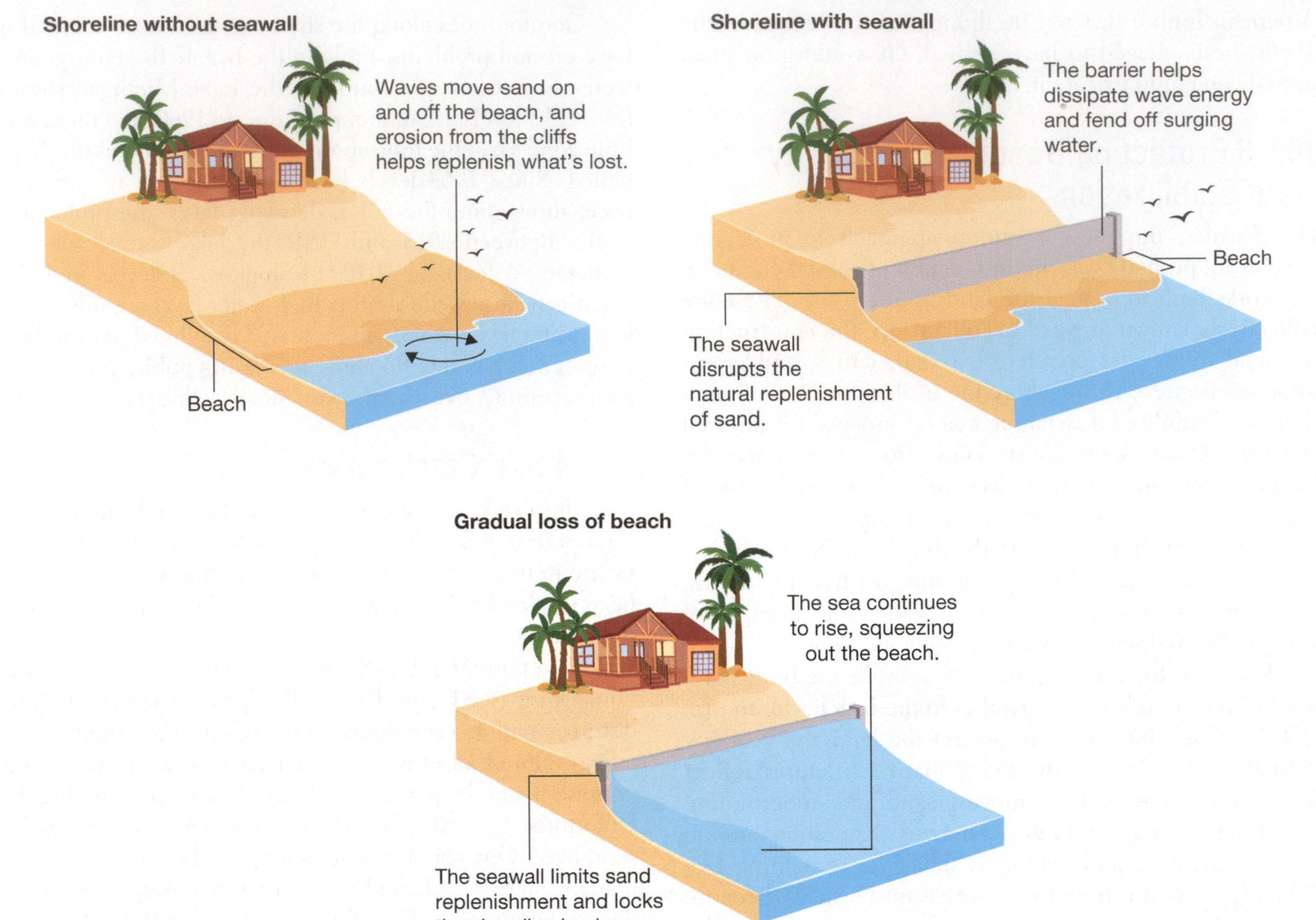

Figure 10.97 Seawalls, a form of hard stabilization, may save houses but at the expense of beaches, which disappear along with the sediment supply. (Adapted from Stefanie Sekich (2021). Seawalls Are Stealing Our Sandy Beaches, Surfrider Foundation)

Figure 10.98 Concrete seawall protect a Chicago park's bike path from erosion by waves on Lake Michigan, Montrose Point, Illinois, 2019.

Figure 10.99 Shoreline erosion at Nags Head on the Outer Banks of North Carolina is a problem. It's not clear that stilts are the long-term solution for this home.

level and retreating coastlines. Rather than invest in hard stabilization that is expensive and doomed to failure in the end, buildings could be abandoned, or moved back from the coast if they are sufficiently valuable.

The Cape Hatteras Lighthouse, with its black-and-white, candy-cane paint job, is the tallest brick lighthouse in the world and an excellent example of managed retreat. Built in 1870 with state-of-the-art lenses, the 63-meter (208-foot) structure is so distinctive that it is on the National Register of Historic Places. Besides saving lives, the lighthouse has hosted millions of visitors. It is one of the most enduring landmarks in North Carolina and the nation. The lighthouse was located on the Outer Banks, a string of barrier islands along the coast on North Carolina (**Figure 10.100**).

Fixed structures built on the shifting sands of barrier islands are inevitably at the mercy of beach erosion and, ultimately, direct wave action. This is because barrier islands "migrate" shoreward. When it was built in 1870, the lighthouse was 460 meters (1,500 feet) from the ocean. By 1919, the sea had come to within 100 meters (310 feet) of its base and, by 1935, to within 30 meters (100 feet). When it comes to a battle of humans against the sea, the sea eventually wins.

Illustrating the futility of shore-protection efforts during times of rising sea level, the National Park Service and the Corps of Engineers spent millions of dollars attempting to save the beloved lighthouse–constructing groins to trap sand, replenishing beaches, artificially depositing a barrier-sand-dune system in front of the structure, installing offshore wave-energy dissipaters, and, finally, constructing hard structures (seawalls) to reflect wave energy. They all failed. By the 1980s, the sea was lapping at the base of the lighthouse and more extraordinary measures were required to save the structure (**Figure 10.101**). The National Park Service decided to move the lighthouse in 1990.

Moving structures of great weight has become easier with the development of modern hydraulic systems. The massive Cape Hatteras Lighthouse was moved by using horizontally mounted "pusher" jacks, which transported it along a track system in 5-foot increments. The move took just 22 days, 19 of which were "pushing" days–a remarkable engineering feat, indeed (**Figure 10.102**).

The powerful light that had been out for 9 months was relit on Saturday, November 12, 1999–a prime example of managed retreat from a retreating coastline.

Figure 10.100 North Carolina's Outer Banks are a chain of islands composed of sand that form a barrier between lagoons and the ocean. Cape Hatteras, near the center of the photo, juts the farthest out into the Atlantic.

Figure 10.101 The Cape Hatteras lighthouse stands alone, waves getting ever closer to its base, just before the plan was implemented to move it landward to safety.

Figure 10.102 Overview of the lighthouse's new home and the move path it took. The Cape Hatteras lighthouse sits by the beach, ready for its managed retreat to a new, safer home.

Figure 10.103 A New Orleans home is jacked up in the hope that the next time floodwaters come, it will remain high and dry.

10.6.6 Coastal Restoration

What can be done in the face of rising sea level and increasing storm frequency and energy release? Proposals for New Orleans include raising and strengthening the levee system, raising the land with fill before rebuilding, elevating homes (**Figure 10.103**), abandoning low-lying neighborhoods, and restoring the marshlands that help dissipate storm energy.

A 2009 study, published in *EOS*, a widely read weekly newspaper for geologists, suggests that diverting muddy Mississippi River water into the delta downstream of New Orleans could make a significant difference by rebuilding riverside marshlands, which effectively temper storm surges and dissipate wave energy (**Figure 10.104**). Today, extensive levee systems efficiently transport Mississippi River sediment and water directly to the Gulf of Mexico, bypassing the delta. This bypassing is in part responsible for the Mississippi River delta's losing an average of 44 square kilometers (17 square miles) of land to the sea every year since 1940.

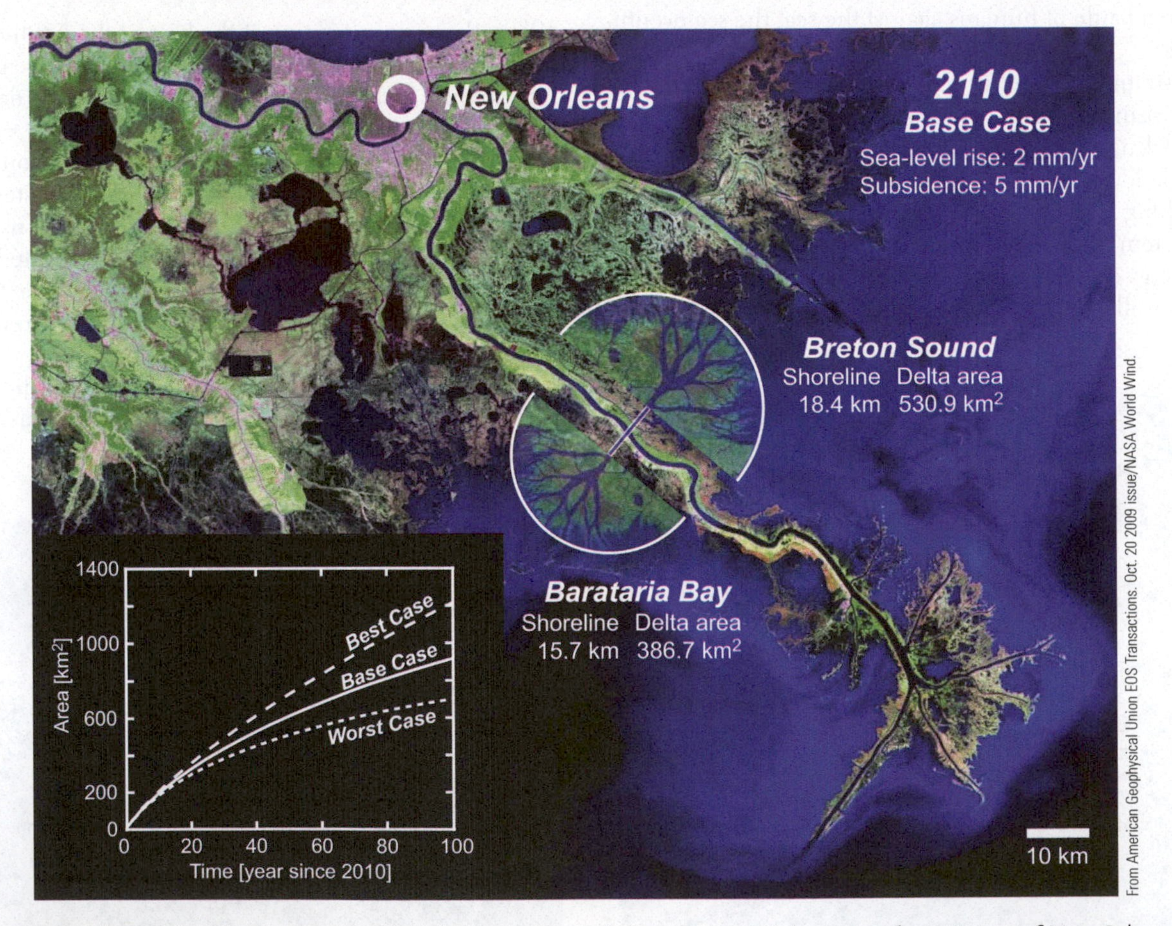

Figure 10.104 The results of a computer model, showing what is likely to happen if the levees downstream of New Orleans were partially breached, allowing sediment to again spill out onto the Mississippi River delta. Inset shows area of new marshland created under different model scenarios.

Using a verified computer model, the authors of the *EOS* article found that, by partially breaching the levees, they could reduce by half the amount of marshland loss predicted for over the next century, even considering delta subsidence and global sea-level rise. Breaching the levees to allow the delta to grow again seems like reasonable, low cost protection against rising sea level. Only time will tell if there is sufficient political will to act responsibly on the scientific data that already exist.

Plants can help ameliorate the effects of sea level rise. Mangrove restoration is a hot topic in the tropics. Mangroves thrive in the coastal zone, where they grow together to form a nearly impenetrable thicket of roots and stems. As mentioned at the beginning of this chapter, while this pile of biomass may be difficult to get through, it provides a very effective means of dispersing wave energy and thus helps protect coasts from both storm surge and wave erosion. For decades, mangroves have been aggressively cleared from the coastal zone. Today, many communities are restoring them. As an added benefit, mangroves, and the ecosystems they support, are effective sinks for carbon in the nearshore realm–think black, gooey and somewhat stinky marsh muck (**Figure 10.105**)

iStock.com/Akarawut Lohacharoenvanich

Figure 10.105 Newly planted mangrove trees stand ready to protect the coastline of Thailand.

Case Study

10.1 Superstorm Sandy

If there were a Hall of Fame for storms, it would feature Superstorm Sandy and its destruction of New York and the New Jersey coast. Sandy would join Hurricane Katrina and New Orleans in 2005, Hurricane Camille and the southeastern United States in 1962, the Ash Wednesday nor'easter that pummeled the east coast of North America in March 1962, and the 1938 hurricane that flattened much of Long Island and then raced north through New England. Hurricane Sandy produced the highest storm surge ever recorded in New York harbor, the first wholesale flooding of New York City's subway tunnels, and an unprecedented early-season snowfall in the Appalachian Mountains (**Figure 10.106**).

Superstorm Sandy began as a tropical depression in the southwestern Caribbean Sea on October 22, 2012; 10 days later, it fell apart in western Pennsylvania. Within 2 days, the tropical depression had become Hurricane Sandy, threatening Jamaica, then the Bahamas and Cuba on October 25 before moving over Puerto Rico, Haiti, and the Dominican Republic. Sandy now had winds over 170 kilometers (100 miles) per hour near its center and tropical storm force winds extending out more than 330 kilometers (200 miles).

One day later, on October 26, things started to change. Sandy was weakening as it moved north, and a strong cold front was approaching from the west. Forecasters were predicting that Hurricane Sandy would transition from a tropical, warm core system to an extratropical, cold core system, a nor'easter. Jokingly, NOAA scientists termed this hybrid storm "Bride of Frankenstorm," in honor of its approach near Halloween. Turns out they were not far off the mark.

As Sandy moved north and then west and approached the coast, its maximum winds slowed to only 130 kilometers (80 miles) per hour but the storm exploded, rapidly growing larger. Its cloud deck and circulation extended 3200 kilometers (2000 miles) into the Atlantic (**Figure 10.107**). Cold air was infiltrating the storm from the west, changing its character. Two days later, Sandy was turning to the west, a very unusual pattern for coastal storms, which usually move toward the northeast, away from the mid-Atlantic coast. The westward movement was caused by high pressure over southern Greenland, which blocked Sandy from moving to the east. Sandy was becoming extratropical; that is, a massive storm characteristic of higher latitudes. On October 28, storm surge warnings were issued: 1- to 2-meter (3- to 6-foot) surges in North Carolina, 1.3- to 2.5-meter (4- to 8-foot) surges between Maryland and Connecticut, and 2- to 3.5-meter (6- to 11-foot) surges in New York harbor.

When the storm moved ashore on October 29, it covered most of the eastern United States and Canada as it became extratropical, losing its warm core but keeping all of the moisture that could be provided by a tropical hurricane. High wind warnings were posted and flood warnings covered much of the eastern United States. Inland, winter storm warnings suggested that feet of snow would fall. Sandy was a huge storm; when it made landfall, it affected 4.7 million square kilometers (1.8 million square miles) north to Canada, south to the Mid-Atlantic states, and west to the Ohio Valley. Parts of New Jersey received over 27

615 collection / Alamy Stock Photo

Figure 10.106 Superstorm Sandy left up to 1 meter (3 feet) of heavy, wet October snow in the West Virginia mountains, toppling trees and closing roads. National Guard soldiers helped manage the disaster, here clearing a road.

Trading Ltd/Shutterstock.com

Figure 10.107 Superstorm Sandy covers much of the east coast of North America as it comes ashore in New York in October 2012.

Michael Bocchieri/Getty Images News/Getty Images

Figure 10.108 Dozens of cabs were submerged as the storm surge flooded much of coastal New York.

centimeters (11 inches) of rainfall while the central Appalachian Mountains saw up to 100 centimeters (3 feet) of heavy, wet snow.

Coastal flooding during Sandy caused massive damage. Over 380,000 housing units were destroyed or left uninhabitable. Almost a quarter-million cars were soaked in saltwater, leaving them as total losses providing an unanticipated stimulus to the automobile industry and a major headache for insurance providers as these vehicles had to be replaced (**Figure 10.108**)

Most flooding was due to the storm surge, at its highest and most devastating in northern New Jersey and the New York City area. As the storm surge pushed up rivers and estuaries surrounding the city and its harbor, water poured over coastal roads and walkways and into city streets. Low-lying areas were flooded for days. Homes near the coast became islands, subway and train tunnels flooded (**Figure 10.109**), and underpasses became fetid lakes, impassable for cars. The transportation network was crippled. Boats floated into front yards and were stranded for months before they were removed. The coast was ravaged by wind, waves, and water.

U.S. Coast Guard/Getty Images News/Getty Images

Figure 10.109 Sandy's storm surge floods the coastline on October 30, 2012, at Tuckerton, New Jersey.

10.2 Distant Tsunami: The Silent Threat

Ample warning had been given the residents of Hawai'i as the Chilean tsunami raced across the Pacific Ocean at jetliner speeds. A warning had been issued about 6:45 p.m. on May 22, 1960, that large waves were expected to reach Hilo on the island of Hawai'i at about midnight. Coastal warning sirens had wailed at 8:30 p.m. and continued for half an hour.

At midnight, a wave arrived, but it was only a few feet high. Hundreds of people had stayed home, and those who had evacuated assumed the danger was over and returned home. Then, another wave appeared, and another. The highest wave struck at 1:04 a.m. (**Figure 10.110**). Sixty-one people died, and another 282 suffered severe injuries. A

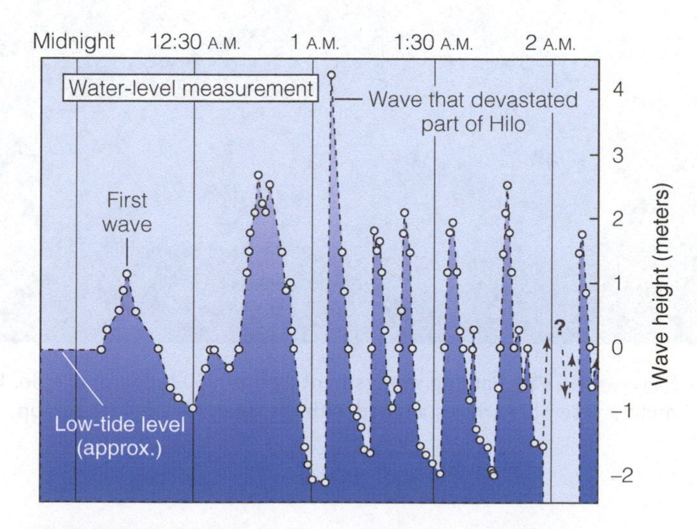

Figure 10.110 Water-level measurements beneath the Wailuku River bridge during the first hours of the tsunami of May 23, 1960.

restaurant memorialized the event with a waterline on its window (**Figure 10.111**).

The city of Hilo, with its bay facing northeast toward the stormy Gulf of Alaska lies along a section of coast where high waves are expected. These higher than average wave heights are due to the orientation of the coast and the funnel shape of the bay, which focus tsunami energy. An engineering study in the aftermath of the 1960 tsunami resulted in some interesting recommendations for mitigating tsunami damage in that city. For example, parking meters in the run-up area were bent flat in the direction of debris-laden wave travel (**Figure 10.112**), with the direction varying throughout the affected area. It was recommended that new structures be situated with their narrowest dimension aligned in the direction indicated by nearby bent parking meters. Several waterfront structures had open fronts that allowed water to pass through them; these structures sustained less damage during the tsunami. Waterfront buildings are now designed with open space on ground level, so that damage from the impact of water and debris can be minimized. Most of all, stretches of packed housing and businesses near

B. Pipkin

Figure 10.111 A line painted on the window of a rebuilt restaurant commemorates the height to which the tsunami waters rose. Can you explain why the tsunami waves are not truly tidal waves, as the notation on the window suggests?

USGS

Figure 10.112 Parking meters bent by the 1960 tsunami at Hilo, Hawaii. The meters resemble arrows aligned with the direction of wave run-up.

the shore have been replaced by lovely public parks, golf courses, soccer fields, and picnic areas–open spaces that won't be seriously impacted by the next big tsunami. People have permanently evacuated most parts of the historical tsunami impact zone. Hawai'i is simply in a tough place when it comes to tsunami. There are major subduction zones surrounding the islands (**Figure 10.113**).

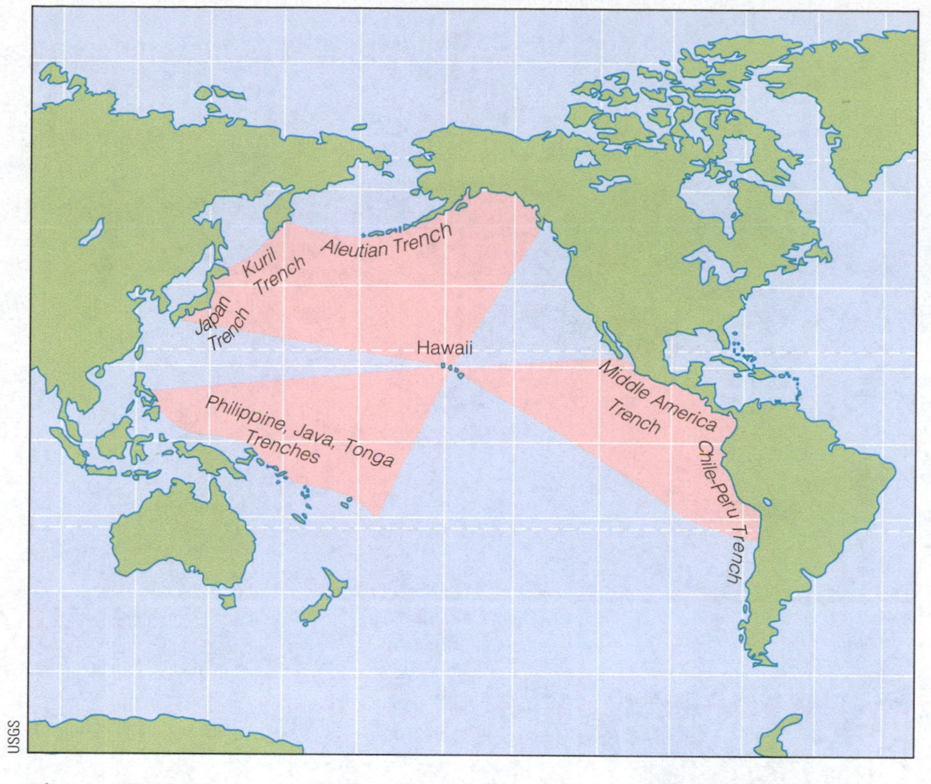

Figure 10.113 Tsunami sources and travel paths to the Hawaiian Islands.

Photo Gallery In Harm's Way

Looking at these photographs, one might wonder, "What were they thinking?" Each of these devastating situations could have been avoided. But it's not that simple. Although scientists know why and how coastal environments can be risky places to live, they have not communicated that well to the broader community and there are competing interests.

iStock.com/TSSnowImages

1 A low-lying area near New Orleans, Louisiana, flooded by Hurricane Katrina. The area is below sea level; no need for a map to determine its elevation.

2 In Laguna Beach, California, hundreds of homes are perched just above an unstable sea cliff. Landslides threaten.

iStock.com/LaurenSimmons

High tide floods low-lying road in the United Kingdom. Climate change and the sea level rise it's causing are increasing nuisance flooding like this near coastlines.
iStock.com/Astrid860

iStock.com/Summerseretrievers

3 Vacation homes along the Outer Banks of North Carolina survive on stilts. What's next for these barrier island buildings?

4 People who survived the 2018 tsunami walk among debris in the coastal town of Ujung Jaya village, Sumur district Banten province, Indonesia.

iStock.com/Heyfajrul

Coasts and Oceans

1. Wind Waves

LO1 Explain how waves change as they move toward shore and the water shallows

a. How Wind Waves Work

Wind waves are formed when wind blows over smooth water, creating surface ripples that enlarge with time, growing to local chop and eventually to wind waves. Wind speed, length of time that wind blows over water, and fetch influence the size of wind waves formed.

b. Shoaling

Shoaling is the phenomenon in which the seafloor interferes with the oscillating water particles as the waves approach the shoreline, resulting in decreased wave speed and wavelength, steepened waves, and increased wave height. Shoaling leads to the formation of many types of waves you see at the beach, spilling breakers, plunging breakers, and backwash.

c. Wave Refraction

Wave refraction is the bending of waves that results in waves approaching the shoreline at an angle. This phenomenon occurs because water speed slows at the shallow end of a wave while water speed remains high at the deep end. There are several effects of wave refraction, including the straightening of irregular shorelines due to shore erosion, the formation of a longshore currents, and formation of rip tides.

Study Questions

i. What are the differences between plunging and spilling breakers?

ii. How and why does wave refraction impact shorelines?

2. Shorelines

LO2 Describe how shorelines respond to changing seasons and the passage of large storms

a. Beach Types

Beaches dissipate the energy of waves and protect land from erosion. They are composed of whatever is delivered to them by waves or rivers entering the sea, varying from sand composed of quartz and feldspar to pebble-size rocks. The energy of the sea drastically impacts the beach, so beaches change with seasons, usually most narrow and pebbly during the winter and widest in the summer.

b. Barrier Islands

Barrier islands are ribbons of sand parallel to the coast that sit just above sea level. They are considered transient features because they are formed and reoriented by wave action and are separated from the mainland by shallow lagoons. Though barrier islands are home to many large communities of people, like the Outer Banks, North Carolina, these landforms are often overtopped by storm surges, and are at risk of disappearing completely as sea level rises.

c. Beach Accumulation and Erosion: The River of Sand

A beach is stable when the sand supplied to it by longshore currents, rivers, and erosion of sea cliffs replaces the sand removed by waves. When the volume of sand supplied is greater than that carried away by waves, the beach will widen and sometimes form sand dunes. When more sand is removed than supplied, beaches erode or do not form. Other causes of beach erosion are reduced sediment supply from river damming and protracted drought, as well as rising sea level.

d. Coral Reefs

Corals, animals that secrete a hard skeleton of calcium carbonate, grow near the surface within warm regions of the ocean. Hard corals build their own landforms where they live, called reefs. Coral reefs provide a barrier to beach erosion, attract tourism, support fisheries, and build atoll islands upon. Acidification of ocean waters caused by increased concentration of atmospheric carbon dioxide threatens coral reefs, as acidic water dissolves calcium carbonate.

Study Questions

i. What natural and unnatural circumstances lead to eroding beaches?

ii. How does atmospheric carbon dioxide lead to the acidification of oceans, and how does this impact coral reefs?

3. Tsunamis

LO3 List the three phases of a tsunami and identify which one causes the greatest hazard

a. Initiation

Tsunami occur when something moves ocean water on a massive scale, in all directions around the source. Most large tsunami are triggered by large-scale earthquakes in subduction zones.

Volcanic eruptions, submarine landslides, and bolide impacts can also cause tsunamis.

b. Propagation

Tsunamis travel hundreds of kilometers per hour across deep water, but usually pass unnoticed until they reach shallow water and shoal.

c. Inundation

Like other types of waves, tsunami experience shoaling as they approach the shoreline. The energy contained in the wave is condensed into a smaller volume of water, and the wave increases in height and steepness. When it hits the shore, the wave breaks into a tidelike flood. This period of a tsunami is the most dangerous, picking up homes, cars, people, and boats as it flows inland. Beyond flooding and the force of moving water, floating debris acts as a battering ram, increasing damages.

d. Mitigating Tsunami Risk

All coastlines can experience tsunami, so educating the general public about tsunami hazards and safety is important Moving inland and getting to higher ground are the best ways people can keep safe; tsunami warning systems help give communities time to react.

Study Questions

i. What are the geologic phenomena that trigger tsunami?

ii. What are the most effective ways to reduce tsunami hazards?

4. Hurricanes, Typhoons, and Cyclones

LO4 Explain how hurricanes work and why they occur most often during the summer and fall

a. Genesis and Seasonality

A hurricane is formed from a tropical storm when atmospheric pressure drops and warm air moves toward the center of low pressure. This moist, rising air condenses at higher elevations and releases heat, causing the air to rise even faster and creating an even lower atmospheric pressure at a well-defined center. As this occurs, wind speeds increase and a storm is formed.

b. Circulation

The damage caused by hurricanes varies depending on the direction of the steering winds and the side of the hurricane. If the steering winds and hurricane winds are aligned, wind speed and damage will be greater. Hurricane winds also drive storm surges raising sea level far above normal high tide.

c. Track and Intensity

Hurricane tracks, either parallel or normal to the coast, directly impact the severity of damage. Coast normal (CN) storms are more hazardous than coast parallel (CP). The intensity of a hurricane is ranked between category 1 and 5 using the Saffir-Simpson scale, where category 5 is the strongest and most severe.

d. Hurricane Impacts on Society

Hurricanes can be extremely dangerous events with many effects to society, including landslides, debris flows, wind damage, storm surges, and flooding. Humans are at risk for injury and death during these powerful storms. The power and energy of hurricanes can also contribute to existing, long-term problems like beach erosion.

e. Hurricane Prediction

Gray's method is commonly used to predict the severity of an upcoming hurricane season using data collected about tropospheric winds, West African climate, El Niño, and atmospheric pressure. Predicting the track and strength of hurricanes occurs on an individual storm basis, using advanced computer models and the experience of seasoned forecasters.

Study Questions

i. Why might climate change impact hurricane frequency and intensity?

ii. How does a hurricane's track impact its severity?

5. Nor'easters

LO5 Contrast nor'easters with hurricanes, explaining how the storms are similar and different

a. Growing a Monster

Like a hurricane, nor'easters begin as low-pressure systems that intensify as central atmospheric pressure drops rapidly. As these storms move to the north or northeast, moist ocean water to the east and frigid winter air masses to the west are juxtaposed fueling the storm.

b. Impacts on Society

Nor'easters are slow-moving storms, meaning that they can grow quite large and intense with time. They bring high wind speeds, storm surges, and floods to coastlines. Inland, nor'easters bring deep and sudden winter snow fall and high winds.

Study Questions

i. What are the differences between hurricanes and nor'easters?
ii. Why was Superstorm Sandy so unusual.

6. Human/Ocean Interactions

LO6 Explain how human actions can change the flow of sediment to and along the world's coastlines

a. Sea- and Land-Level Changes

Beach erosion and sea-level rise are pressing concerns for coast communities around the world. Local attempts to reduce impacts of sea-level rise and erosion are often failures and can worsen the problem.

b. Protecting Harbors

The construction of barriers to protect harbors can have unintended negative consequences. Barriers can limit sand transport, causing some beaches to be sand-rich and others to be sand-starved.

c. Protecting Beachfront Property–Hard Stabilization

Hard stabilization of beachfront property has for a long time been the most common approach to protecting some coastal properties. The construction of sea walls, jetties, and groins are examples of hard stabilization meant to protect property from storm surges, sea-level rise, and other major coastal storms.

d. Protecting Beachfront Property–Soft Stabilization

Soft stabilization is an alternative approach to addressing coastal hazards. Sand pumping or beach nourishment replenishes eroded beaches by moving sand from offshore onto the beach. Another common approach is to elevate buildings on stilts so they stand above the rising sea. The problem with these and other soft stabilization techniques is that they are expensive, energy intensive, and typically short-term solutions.

e. Managed Retreat

Managed retreat suggests that the best way to reduce coastal hazards over the long term is to move communities and infrastructure away from the shifting shoreline.

f. Coastal Restoration

Other potential mitigation efforts to address sea-level rise include mangrove restoration, breaching coastal levee systems, and wetland restoration.

Study Questions

i. What are the pros and cons of hard and soft stabilization?
ii. Describe one coastal restoration/mitigation strategy and explain why or why it may not be effective to address problems caused by sea-level rise.

Key Terms

Active sea cliff
Amplitude
Backwash
Barrier islands
Beach nourishment
Bomb cyclone
Chop
Coast-normal
Coast-parallel
Coral bleaching
Cyclones
Depositional processes
Erosional processes
Fetch
Fully developed sea
Groins
Hard stabilization
Inactive sea cliff
Initiation
Inundation
Longshore current
Longshore drift
Managed retreat
nor'easters
Pebble beaches
Plunging breakers
Propagation

Ripples
River deltas
Rogue wave
Shoaling
Shorelines
Spilling breaker
Surf zone
Steering winds
Swash
Swell
Typhoons
Undertow
Wave base
Wave crests
Wave refraction
Wavelength
Waves of translation
Wind shear
Wind waves

"The future is green energy, sustainability, renewable energy."

—Arnold Schwarzenegger

11

Energy and the Environment

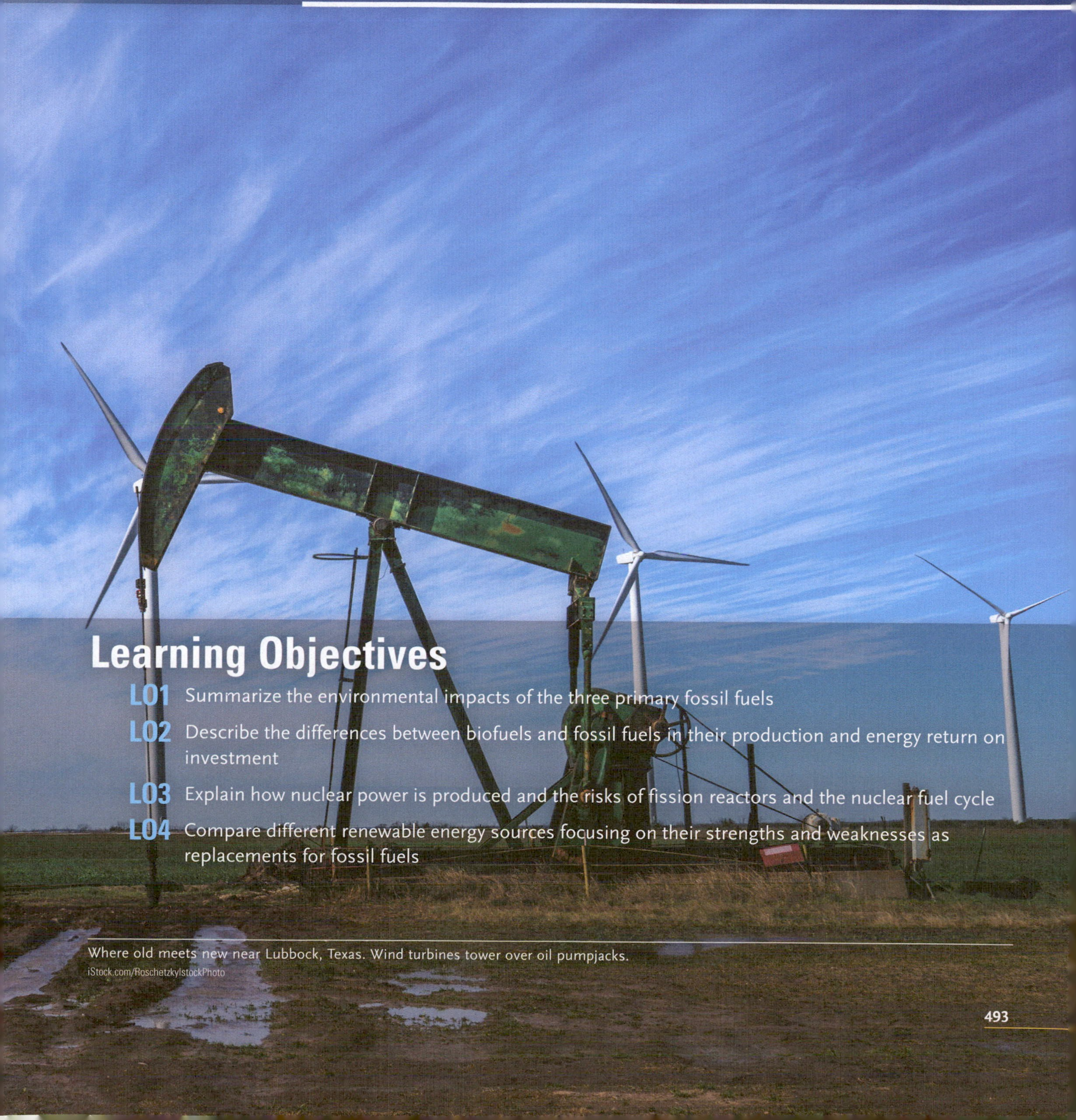

Learning Objectives

LO1 Summarize the environmental impacts of the three primary fossil fuels

LO2 Describe the differences between biofuels and fossil fuels in their production and energy return on investment

LO3 Explain how nuclear power is produced and the risks of fission reactors and the nuclear fuel cycle

LO4 Compare different renewable energy sources focusing on their strengths and weaknesses as replacements for fossil fuels

Where old meets new near Lubbock, Texas. Wind turbines tower over oil pumpjacks.
iStock.com/RoschetzkyIstockPhoto

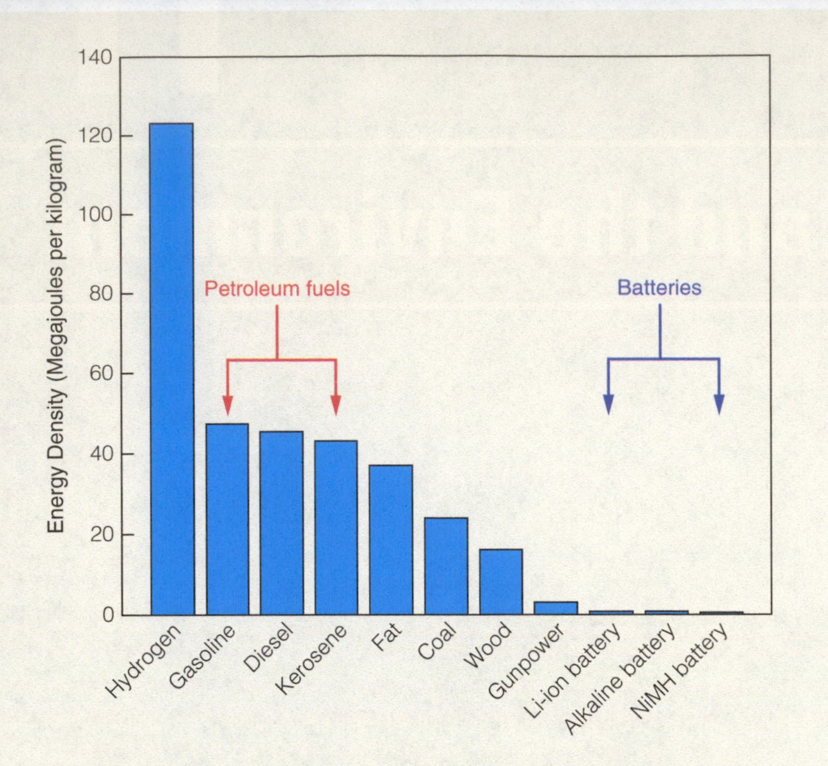

Figure 11.1 Energy density varies widely between different fuels and energy storage devices.

From the simplest algal scum to the most complex ecosystem, energy is essential to all life. Derived from the Greek word *energia*, meaning "in work," **energy** is defined as the capacity to do work. The challenge with energy is twofold–concentrating energy so it's useful to us as humans, and storing it in a way that makes it safely available when we need it.

It's no surprise that **fossil fuels** such as coal, oil, and gas have become attractive energy sources. They are **energy dense** (a small mass provides a lot of useful energy) and easy to store (**Figure 11.1**). Fossil fuels have been storing ancient solar energy (as the potential energy of combustion in chemical bonds) deep in the lithosphere for millions of years. Add a little oxygen and that energy is released as the carbon-hydrogen bonds or carbon-carbon bonds break and are replaced with oxygen, a process called **oxidation** that, when it happens quickly, we know as **combustion**. For example, a barrel of oil, which is 42 American gallons or 159 liters, is the equivalent of 1,700 kilowatt hours of electricity. To put that in perspective, in 2021 the average American home used 10,600 kilowatt hours of electricity, the equivalent of just more than six barrels of oil.

There are of course plenty of other ways to store energy. Dams hold water behind them that has potential energy which is converted to kinetic energy as it flows downstream. If that water encounters a turbine in a hydroelectric dam, then that potential energy is converted to electricity (**Figure 11.2**). Storing electricity is trickier. Until recently batteries have been heavy, didn't last long, and weren't particularly energy dense. That's been changing rapidly as new technologies evolve and enter the market. Good batteries are a big deal, because they buffer the ups and downs of renewable energy sources which by their nature are intermittent; that is, they don't generate power consistently. The wind blows and then it doesn't. The sun doesn't shine at night.

The shift in use from one form of energy to another began long before today's transition from petroleum to renewable source, such as wind turbines, solar panels, and geothermal generation. For thousands of years, raw animal (and human) muscle, along with the burning of wood, were the major sources of energy needed by people. By 200 B.C.E., simple windmills were pumping water in China, and vertical-axis windmills were milling grain in the Middle East (**Figure 11.3**). Later, the Crusaders carried the knowledge of windmills back to Europe, where the new technology was adapted for draining swamps and lakes in Holland.

In the late 1600s, steam engines were developed and, by the early 1800s, significant improvements by James Watt and George Stephenson gave rise to the wood- and coal-fired steam engine. Steam engines were more convenient than water or wind for generating power (even if they did have a bad habit of exploding every now and then, **Figure 11.4**). Soon, steam engines

iStock.com/Edsongrandisoli

Figure 11.2 The Ilha Solteira hydroelectric power plant on the Paraná River in Brazil.

iStock.com/Abdolhamid Ebrahimi

Figure 11.3 These vertical windmills are a thousand years old. Built in the Iranian town of Nashtifan, close to Afghanistan border, they are used to mill grain.

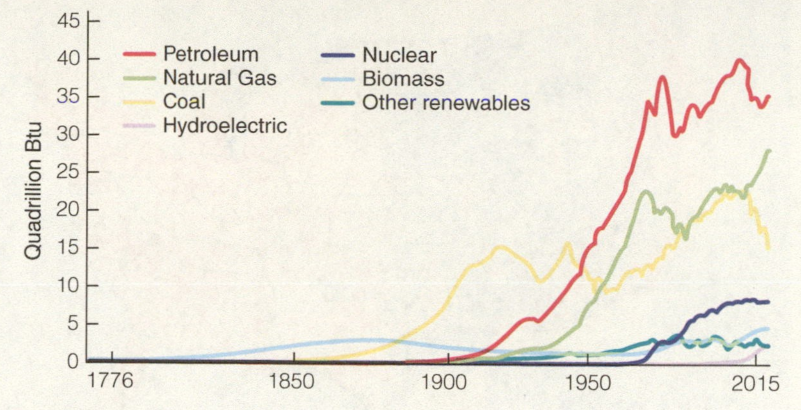

Figure 11.5 Energy sources in the United States since the country's founding. Not only has energy use risen over time but the source of energy has changed dramatically. (Modified from U.S. Energy Information Administration, Monthly Energy Review)

became less expensive than maintaining stables of draft animals. By the middle of the nineteenth century, steam engines were powering factories, farm machinery, and railroad locomotives and had replaced water-pumping windmills and sent many working animals to pasture. The **Industrial Revolution** was well under way, with coal, a fossil fuel, soon displacing wood as the primary energy source. In 1899, a coal-powered steam engine was attached to an electric generator and the modern world of electricity began.

This would not be the last energy transition. During the twentieth century, the demand for energy derived fossil fuels grew quickly, doubling every 10 years. Automobiles and airplanes became commonplace. Along with this dramatic uptick in fossil fuel combustion came problems –air pollution, smog, acid rain, and climate change. Once again, our dependence on specific energy sources began to transition in response. Since about 2010, coal use has diminished in the United States, and natural gas use increased (**Figure 11.5**). A map of carbon emissions is instructive. It clearly shows that the most carbon is emitted from heavily urbanized areas including much of eastern North America and the western cities (**Figure 11.6**)

Our current transition in energy sources will eventually reduce our reliance on fossil fuel-based energy systems. But for now, we find ourselves in a world of contrasts (**Figure 11.7**). While the shift toward renewable energy, including that derived from the wind, the sun, and moving water, is slowly gaining momentum, fossil fuels remain the dominant source of energy for humans. Change is clearly coming, driven by falling costs for renewable energy and now government subsidies to encourage the transition away from fossil fuels (**Figure 11.8**).

Energy, how we get it and the pollution it causes, is one of the most pressing environmental issues of our times. Human-induced climate change is largely driven by the combustion of fossil fuels that powers the twenty-first century world and the resulting release of carbon dioxide. Extracting fuels from

Interfoto/Alamy Stock Photo

Figure 11.4 Satirical drawing as horse-power started being replaced by steam engines around 1820 near London. People were clearly nervous of this new technology.

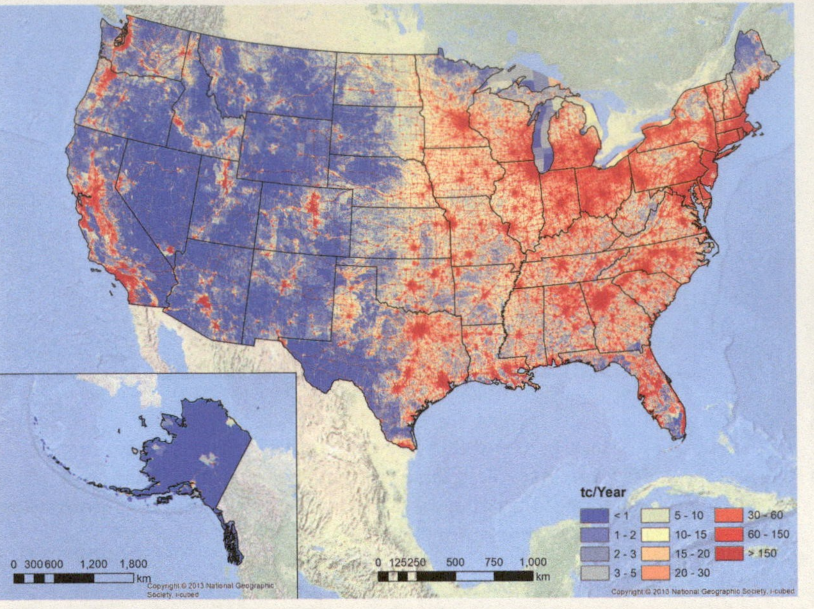

NASA

Figure 11.6 Carbon emission map for the United States from 2010 to 2015. Red areas have high emissions. Blue areas have low emissions.

iStock.com/Jonathanfliskov-photography

Figure 11.7 Wind turbines in the ocean stand in contrast to the ship powered by fossil fuel in this photograph taken from the beach in Copenhagen, Denmark, a country quickly moving away from fossil fuels.

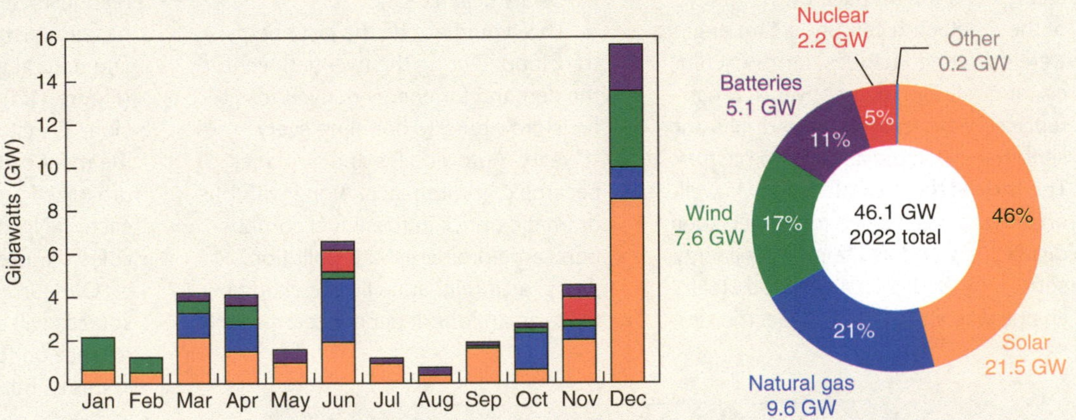

Figure 11.8 In 2022, almost half the new generating capacity installed in the United States was solar. (Modified from U.S. Energy Information Administration, Preliminary Monthly Electric Generator Inventory, October 2021)

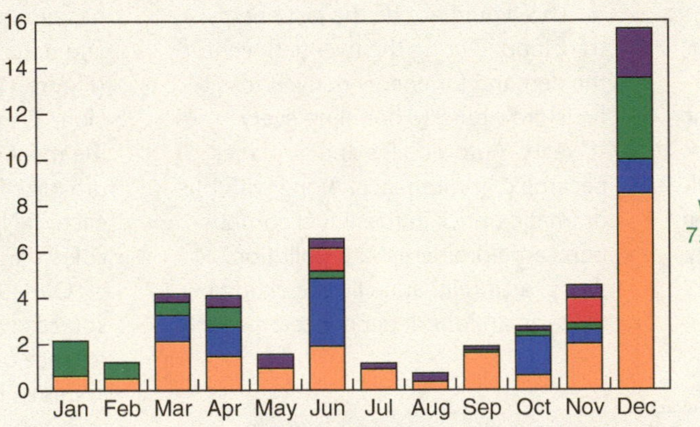

the Earth, be they fossil fuels like coal and gas or nuclear fuels like uranium, have significant impacts on air, water, and landscape quality and in many cases, human and ecosystem health.

It's commonplace to characterize energy sources as renewable or nonrenewable resources. **Renewable resources** are replenished at a rate equal to or greater than the rate at which they are used. Examples include solar, water, wood, geothermal, wind, ocean and lake thermal gradients, and wave/tidal energy. The energy in all these resources—except for tidal and some geothermal (gravitational) energy—was originally derived from the Sun. **Nonrenewable resources** are not replenished as fast as they are used and, once consumed, they are in human terms, gone. Crude oil, oil shales, tar sands, natural gas, coal, and fissionable elements are nonrenewable energy resources. The quantities are finite.

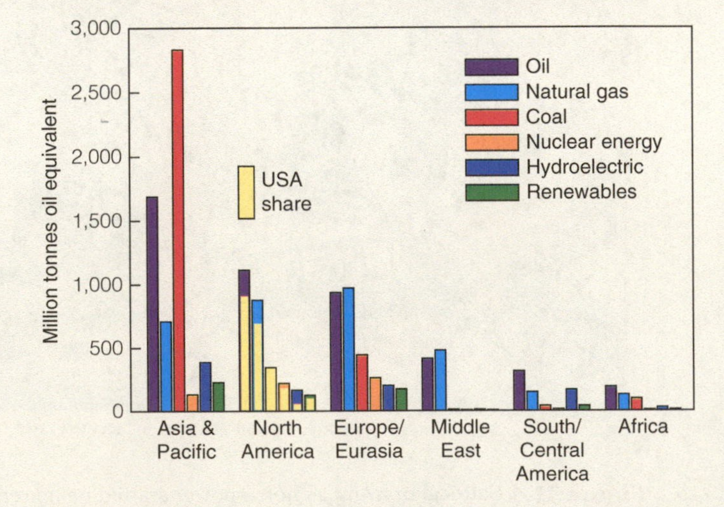

Figure 11.9 World energy use by continent and by fuel in 2018 shows great disparities.

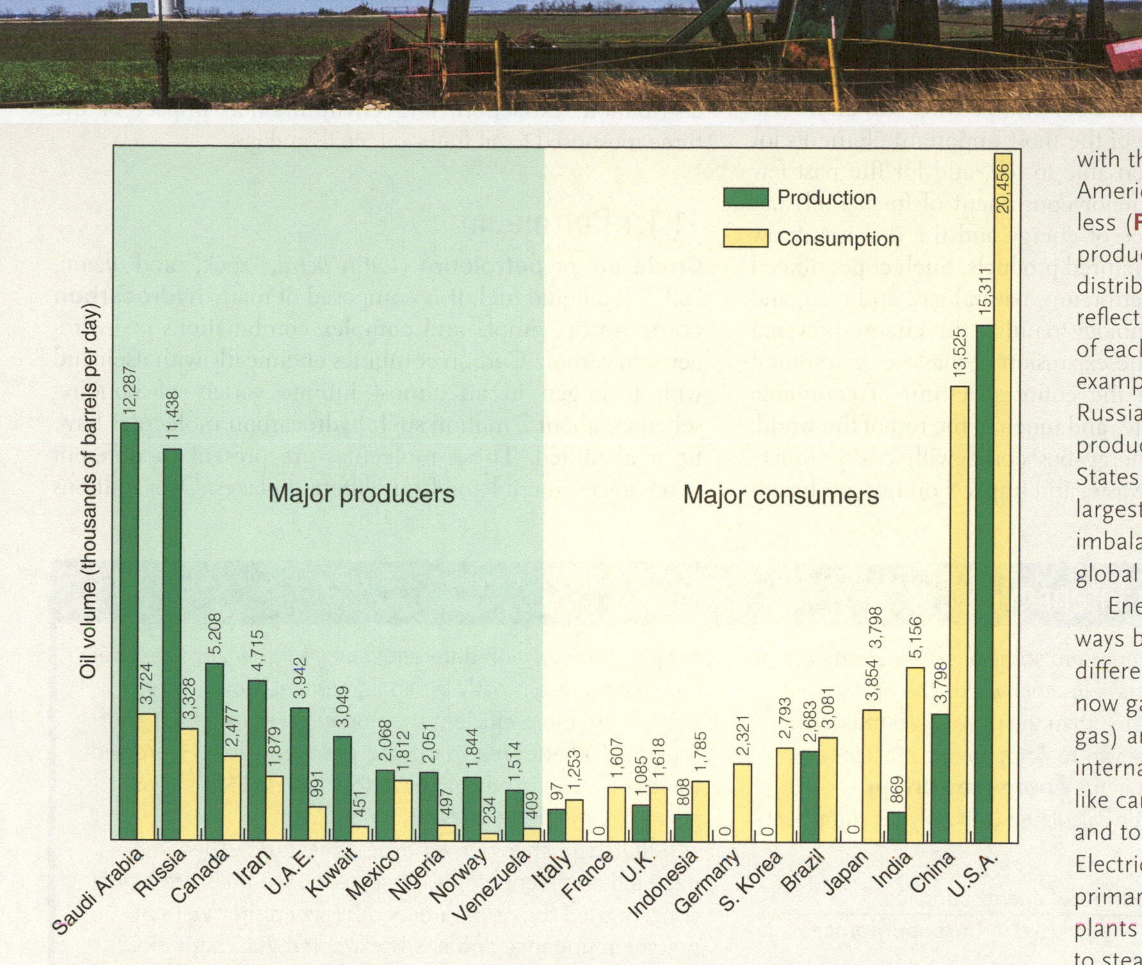

Figure 11.10 The world divided into net energy producers and net energy users. (Modified from The Global Education Project. Data from BP Statistical Review of World Energy, June 2019)

Geothermal has also been historically, non-renewable – but not because heat from the Earth is greatly limited. Rather, geothermal fields have a finite lifetime, generally constrained by groundwater overdrafts.

Energy resources and energy use are not evenly distributed around the world. Asia is now the greatest consumer of energy, much of which is coal. Europe and North America (mostly the United States) use similar amounts of energy with the Middle East, South America, and Africa using far less (**Figure 11.9**). Energy production is also not evenly distributed (**Figure 11.10**) and reflects the underlying geology of each country. Take for example oil. Saudi Arabia and Russia are the largest global producers, but the United States and China are the largest consumers. This imbalance drives much of global politics.

Energy is used in different ways by different people in different places. Liquid and now gaseous fuels (natural gas) are used to power internal combustion vehicles like cars, busses, and trucks, and to heat buildings. Electricity is generated primarily in thermal power plants where water is heated to steam and used to drive turbines, that in turn spin generators. The water in these plants can be heated by burning fossil fuel (coal, oil, gas), biomass (wood, trash), solar energy, or nuclear reactors. Although the energy source is different, the remainder of the power plant is very similar (**Figure 11.11**).

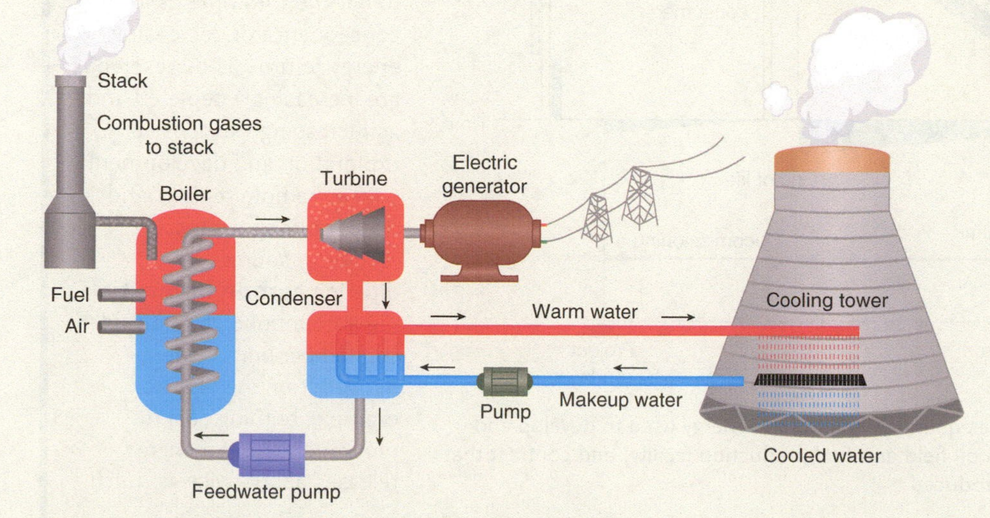

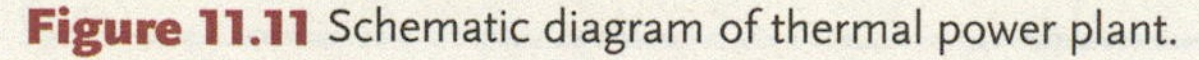
Figure 11.11 Schematic diagram of thermal power plant.

questions to ponder

1. What do you think will be the dominant source of energy in the United States in the year 2100?
2. What could be done to make low energy density fuels competitive with higher energy density fuels?

11.1 Fossil Fuels (L01)

Although the carbon content of Earth's crust is less than 0.1% by weight, carbon is one of the most important elements for humankind. It is indispensable to life, and for the past few centuries carbon (as a major constituent of fossil fuel) has been the principal source of energy and the dominant raw material of many manufactured products. Such concentrated energy, available from petroleum, natural gas, and coal, and the development of technology to utilize this energy, powered the growth of industry, the expansion of massive agricultural production, and indeed the entire spectrum of economic growth of the United States and much of the rest of the world. But, using this fossil energy has come with costs–climate change, acid rain, toxic waste, and impacts on human health from air pollution. It has also become increasingly expensive to extract (see box). In this section, we examine the formation, distribution, extraction, and environmental impact of the three most-used fossil fuels–oil, coal, and gas.

11.1.1 Petroleum

Crude oil, or **petroleum** (Latin *petra*, "rock," and *oleum*, "oil"), is a liquid fuel. It is composed of many **hydrocarbon compounds**, simple and complex combinations of hydrogen and carbon. Carbon combines chemically with itself and with hydrogen in an almost infinite variety of bonding schemes; about 2 million such hydrocarbon molecules have been identified. These molecules are present in different percentages in crude oil from different places. Over millions

EROI and Carbon Intensity

It takes energy to get energy and some energy systems are far more efficient than others – in other words the payback is greater, often much greater, than the initial investment. An imperfect but interesting way to compare the efficiency of different energy sources is the **energy return on investment (EROI)**. It's calculated with only two numbers– energy in and energy out.

$$\text{EROI} = \frac{\text{quantity of energy supplied}}{\text{quantity of energy used in the supply process}}$$

While simple to calculate in theory, calculating actual EROIs is tricky (**Figure 11.12**). The final EROI value is sensitive to the assumptions used in calculating the energy inputs required to extract the energy – and thus different people come up with different values for the same technology. Yet, there is mostly broad agreement. Some energy sources are more efficient than others (**Figure 11.13**) and thus yield a better return on energy investment. Hydroelectric, wind power, and coal have the highest EROIs. Corn ethanol is by far the worst.

EROIs can change over time as resources become depleted and technologies change. Fossil fuels are highly concentrated forms of energy and, when they were in greater abundance and easily extracted, had high EROIs. For example, in 1930, oil wells were relatively shallow and easily drilled, and the energy in one barrel of oil could be used to produce the energy in 100 barrels of oil; thus, oil at that time had an EROI of 100:1. By 1970, the EROI of U.S. petroleum had dropped to 30:1, and since then the EROI of U.S. petroleum has fluctuated between 11:1 and 18:1 (**Figure 11.14**). The EROI for petroleum is likely to continue to decline with time as a consequence of decreasing energy returns as oil reservoirs are increasingly depleted and as increasingly expensive exploration and development shift to remote regions and deeper offshore sites.

Energy sources have differing **carbon intensity**, the amount of carbon dioxide created per unit of energy derived from the fuel. For example, burning coal to recover one unit of energy releases about twice as much

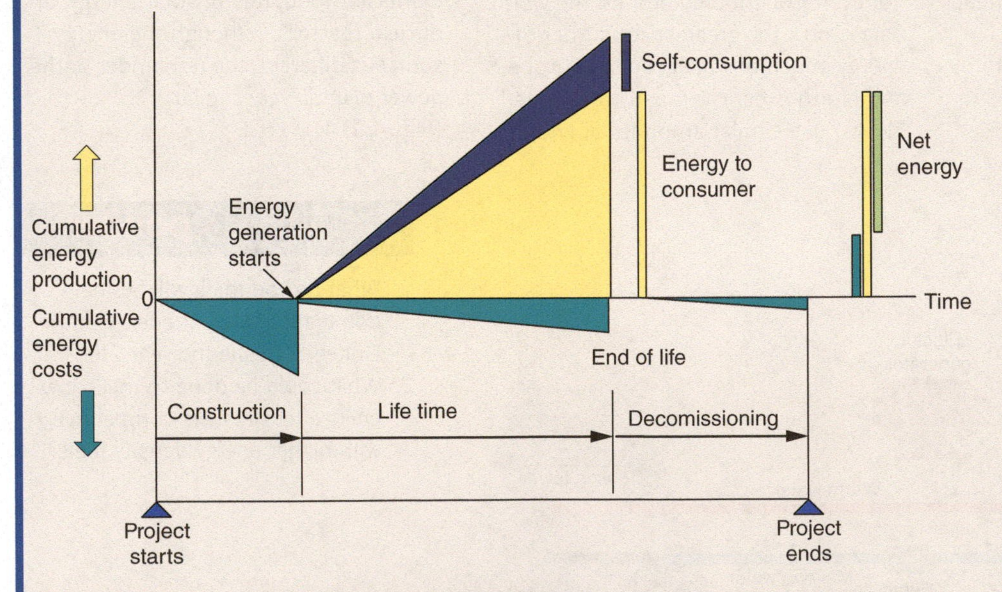

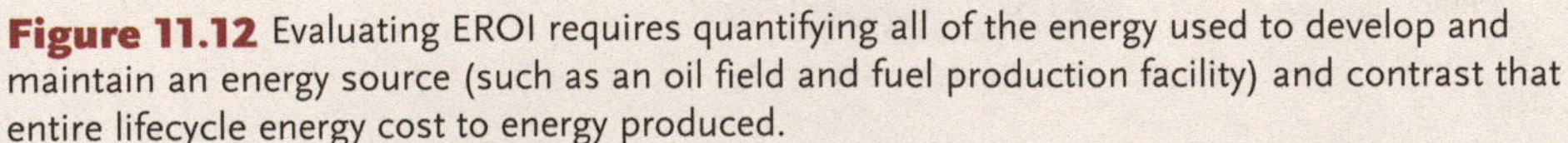
Figure 11.12 Evaluating EROI requires quantifying all of the energy used to develop and maintain an energy source (such as an oil field and fuel production facility) and contrast that entire lifecycle energy cost to energy produced.

EROI and Carbon Intensity (*Continued*)

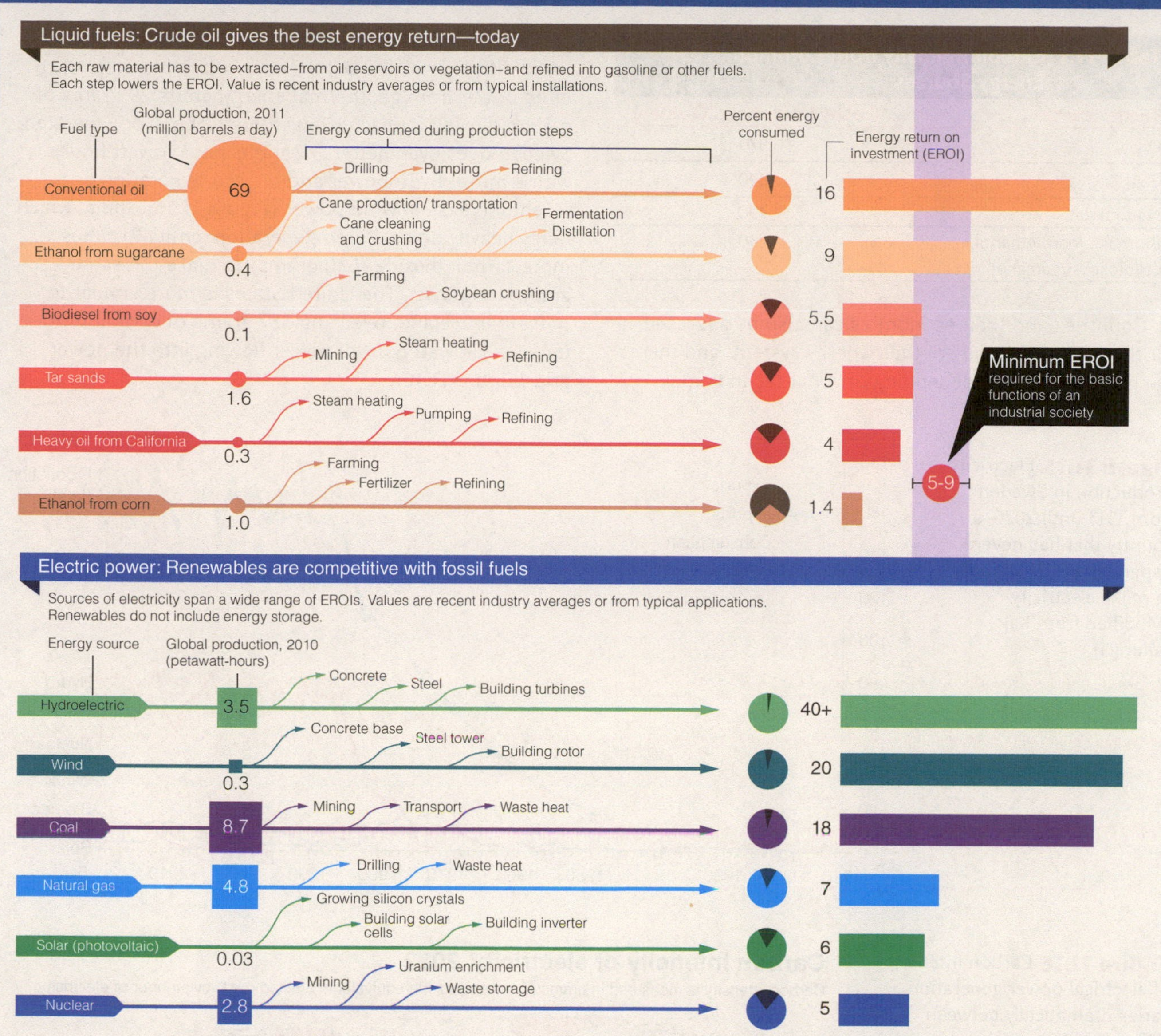

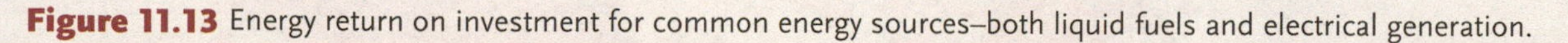

Figure 11.13 Energy return on investment for common energy sources–both liquid fuels and electrical generation.

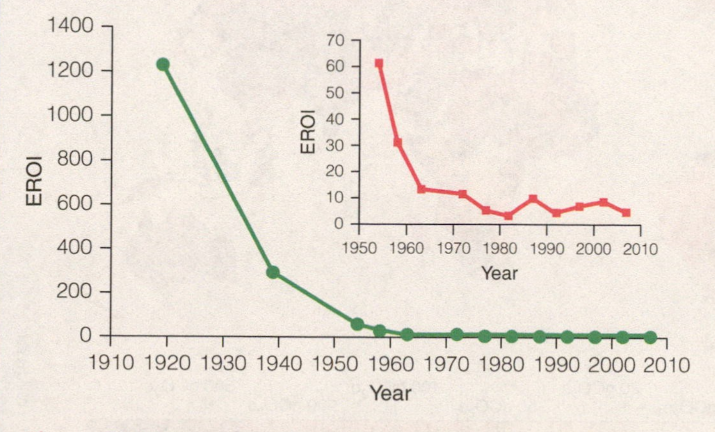

Figure 11.14 EROI for U.S. oil and gas extraction has changed dramatically over time. Inset diagram is same data since 1950 at different scale.

carbon dioxide as burning natural gas (**Table 11.1**). This is because coal is composed entirely of carbon atoms; thus, carbon-carbon bonds are broken to release energy. In contrast, methane (one carbon and four hydrogen atoms) is the primary ingredient of natural gas. Methane has four carbon-hydrogen bonds to break (and thus release energy during combustion). The result, the same amount of energy generated from methane releases less carbon dioxide than if it were generated from coal. The recent (since 2010) coal to natural gas transition is one reason U.S. carbon emissions, when considered on a per capita basis, have declined substantially since 2000. Even with the decline, the United States remains one the world's top emitters of carbon dioxide per capita.

(*Continued*)

Table 11.1 Carbon Intensity from Common Energy Sources

Energy Source	Carbon Intensity (gCO_2/KWh)
Oil	970
Coal	820
Natural gas	490
Biomass (corn ethanol, cellulose, sugarcane)	230

Countries and their economies can also be assessed for the carbon intensity of their energy system—and there are huge differences depending on what technologies and energy sources are used. Electricity generation shows these differences clearly. Iceland, which generates much of its power from geothermal energy, emits 29 grams of carbon dioxide for each kilowatt hour of power generated. Sweden does even better (12 grams per kilowatt hour) with a mix of hydropower and nuclear generation and recent increase in wind power (**Figure 11.15**). India, which relies heavily on coal-fired generation, emits 21 times more carbon dioxide (630 grams) to make the same amount of power. The United States is more similar to India than Iceland, releasing 357 grams of carbon dioxide for each kilowatt hour of power flowing into the power grid (**Figure 11.16**).

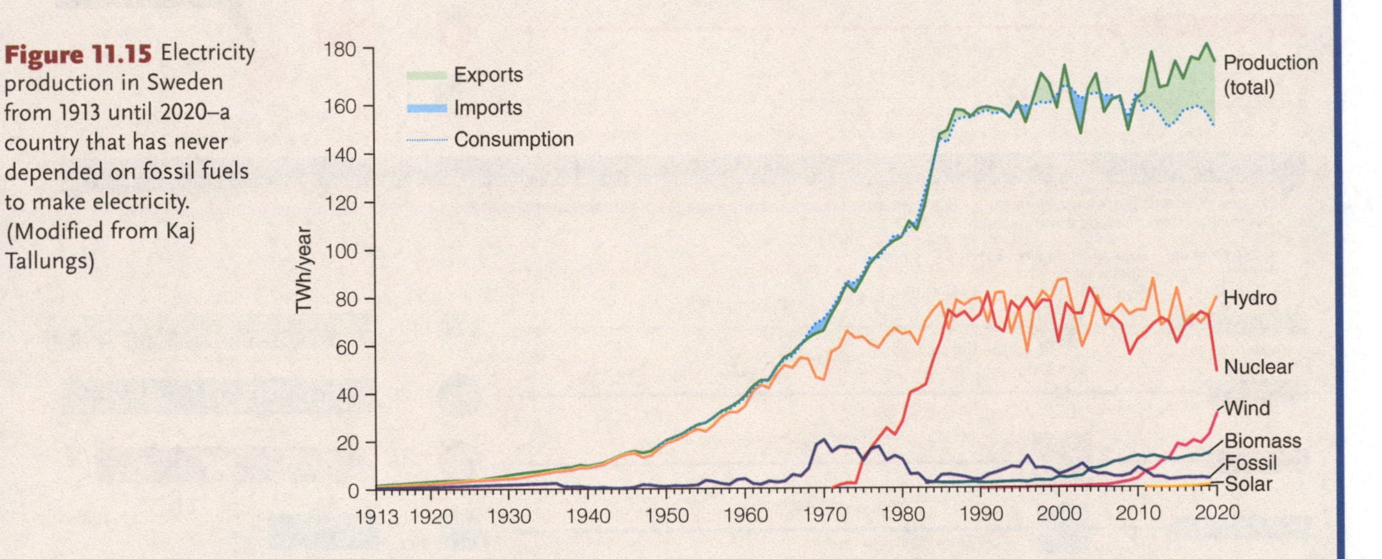

Figure 11.15 Electricity production in Sweden from 1913 until 2020—a country that has never depended on fossil fuels to make electricity. (Modified from Kaj Tallungs)

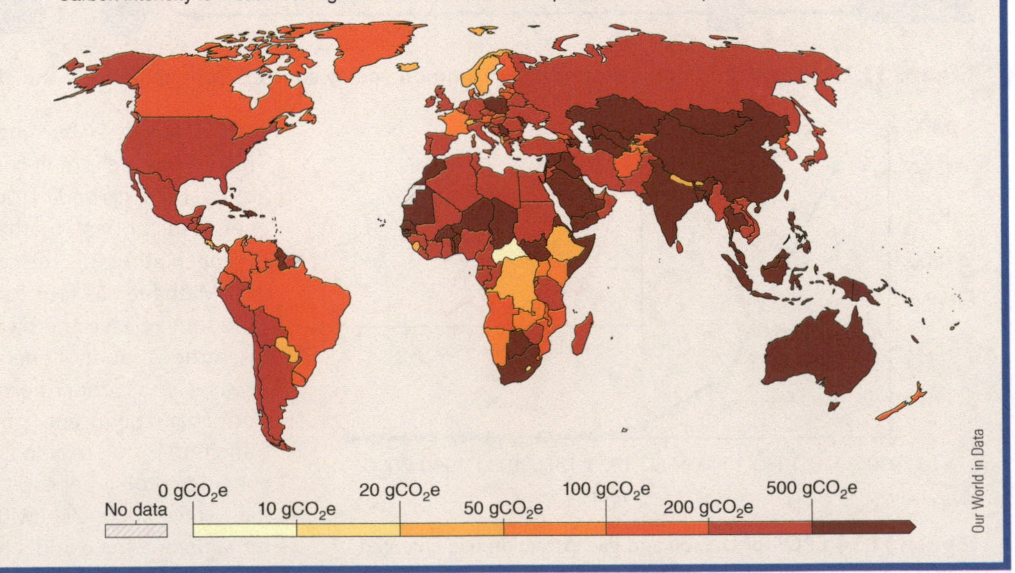

Figure 11.16 Carbon intensity of electrical power generation varies dramatically between different nations.

Figure 11.17 Crude oil reserves in billion barrels (Gbbl). Not your average map! The size of the country corresponds to the amount of crude oil believed to underlie its lands in 2019. (Modified from The Countries With the Most Oil Reserves, map by the Visual Capitalist)

of years, as the rocks in which they reside are warmed and deformed, some compounds are lost and others are created. The result is a unique blend of hydrocarbons that give every oil field in the world a characteristic mix of hydrocarbons.

Oil is not uniformly distributed around the world–a few countries hold much of the world's **reserves** (the amount of oil thought to remain in the ground) and thus control much of the world's supply (**Figure 11.17**). Many of the oil-rich countries are in the Middle East. Nigeria and Libya hold most of Africa's oil. Venezuela dominates South American oil and Canada dominates North American oil reserves. Russia has most of the oil in Asia.

Petroleum is found in **reservoir rocks,** rocks that have sufficient porosity and permeability that the oil can flow between the grains and can be extracted. The discovery, drilling, and extraction process can take years to complete. Field geologists start by mapping rocks and determining how they are folded and faulted. Those data are compiled at a local and regional level before areas are targeted for test drilling. Finally, wells are completed and oil pumped to the surface. Once out of the well, crude oil is moved to refineries where it is separated into its different hydrocarbon components.

The manufacturing process of separating crude oil into its various components is known as **refining** or **cracking** and involve separating hydrocarbon molecules of different sizes and weights. Petroleum occurs beneath Earth's surface in liquid and solid forms, and at or near the surface as oil seeps and tar sands. In addition to their use as a fuels (including gasoline, diesel fuel, and heavy oil for ships) hydrocarbon compounds derived from petroleum are used in producing paints, plastics, fertilizers, insecticides, soaps, synthetic fibers (nylon and acrylics, for example), pharmaceuticals and synthetic rubber.

Petroleum extraction and transport can have severe environmental consequences. Drilling for oil can go badly wrong, most often because crude oil is mixed with natural gas under pressure in many geologic formations. When a well is drilled into such a formation, the pressure can blow out the well causing an unintended and uncontrolled release of crude oil to the environment. Crude oil is transported across the ocean primarily by tankers which have a bad record of both hitting things (like rocks) and then leaking and even sinking. On land, oil pipelines can rupture and leak.

April 20, 2010, was not a good day for the oil drilling industry. At 9:56 pm, the *Deepwater Horizon*, a British Petroleum (BP) drilling rig operating in the Gulf of Mexico about 64 kilometers (40 miles) off-shore of Louisiana, suffered a massive blowout on the sea floor followed by an explosion and fire on the rig. Flames from the fire roared up to almost 100 meters (300 feet) and could be seen from 90 kilometers (56 miles) away (**Figure 11.18**). Eleven workers were killed, the drilling rig sank, and more than 750 million liters (200 million gallons) of crude oil flowed into the Gulf of Mexico–the largest off-shore oil spill in history.

At the time of the explosion, the rig had just completed drilling a 3,960-meter (13,000- foot)-deep well in 1,520 meters (5,000 feet) of water. An investigation by BP and government authorities revealed that the immediate cause of the explosion was gas shooting up the drill column, bursting through the blowout preventer as well as various safety barriers and seals. The emergency response was rapid in an attempt to minimize environmental damage. Concentrated slicks of oil on the ocean surface

USCG

Figure 11.18 The Deepwater Horizon oil rig burns after the blow out despite the efforts of multiple fireboats. Soon after, the rig sank.

NOAA

Figure 11.19 Floating oil from the Deepwater Horizon is burned in an attempt to keep it from damaging Gulf of Mexico ecosystems.

NOAA

Figure 11.20 A skimmer vessel with a boom attempts to corral crude oil from the Deepwater Horizon floating in the Gulf of Mexico.

were burned in place to remove them–the ocean was literally on fire (**Figure 11.19**). In other places, oil was skimmed off the surface of the water (**Figure 11.20**).

Nonetheless, the rapid response was not enough. Wildlife was coated with oil as were the marshes and beaches of the Gulf Coast (**Figure 11.21**). The loss of the Gulf's bird and marine life was immense, with oceanographers recording evidence of long-range damage from the oil spill to marine life. Dead and dying deep-sea corals coated in "brown gunk" (raw crude oil) were found as far away as 11 kilometers (7 miles) from the well at a water depth of 1,311 meters (4,300 feet). A study of bottlenose dolphins in Barataria Bay, Louisiana, revealed that many of the animals were underweight and suffering from liver and lung disease attributed to seawater contamination by oil. The use of **chemical dispersants** (chemicals that allow oil slicks to break up into small globules) in the clean-up effort, may have had adverse biological impacts.

The ultimate cost of the blowout and resulting spill was estimated at $146 billion dollars in 2018. The value of the drilling platform exceeded a billion dollars, and the operation of such a deep-water operation, with support boats, numerous services, and helicopter support, amounted to about $1 million per day. Although the exact toll will never be known, when BP settled with the government for violations of the Clean Water Act, $8.8 billion was earmarked for ecological restoration. The funds have been used to reduce pressure on fisheries, to rebuild populations of coastal animals, and to restore wetlands.

NOAA

NOAA

Figure 11.21 Crude oil coated animals and their habitats. (a) A NOAA veterinarian prepares to clean an oil-covered marine turtle. (b) Oil and tar foul the beaches and marshes of southern Louisiana after the spill.

The second largest oil spill in United States history happened in a very different environment–the cold waters of southeast Alaska. On March 24, 1989, the *Exxon Valdez*, a massive tanker, was headed to the refineries at Long Beach, California, carrying 53 million gallons of crude oil. That oil had been drilled on the north slope of Alaska, near Prudhoe Bay and moved across the state through the trans-Alaska pipeline before being loaded on the tanker. At 12:04 a.m., the tanker struck shallow rock reef, the hull ruptured, and 37,000 tonnes (10.8 million U.S. gallons) of crude oil spilled into the frigid waters (**Figure 11.22**).

The cleanup was difficult and hampered by logistical issues as well as a lack of people and supplies available immediately after the spill. Because the area was so remote, crews and equipment had to be delivered by helicopter and by ship. Skimming the oil failed when kelp, prevalent in the sound, clogged the skimmers. Dispersants failed to work effectively in cold water while raising public concern about impacts of these chemicals on local fisheries and wildlife. Rough seas and wet weather meant that burning the floating slick was not an option. Soon, the oil coated the rocky beaches ringing Prince William Sound. The primary tool used to clean oil-coated rocks was high pressure washing and then skimming of the liberated oil (**Figure 11.23**).

The cleanup was incomplete and the impacts of the spill on the local ecosystem were dramatic. The U.S. Geological Survey estimates that a quarter of a million sea birds were killed by the spill (**Figure 11.24**). The cold water limited biologic degradation of the oil and the rocky shoreline made recovering the oil that washed ashore challenging. In the end, more than 2,000 kilometers (1,300 miles) of shoreline were fouled by oil. Although difficult to measure, it appears that only about 10% of the spilled oil was recovered and removed from the sound. Exxon and government employees, as well as 11,000 Alaska residents, assisted in the clean-up, which cost Exxon over 2 billion dollars. In addition, Exxon eventually spent almost $2 billion more to restore damaged habitat and to compensate people (many in the fisheries industry) for damages related to the spill.

Alaska and the oil industry learned a lot from the Exxon Valdez spill (**Figure 11.25**). Tankers leaving the port of Valdez are now guided through the riskiest section of the harbor and channel by two tugboats. More emergency response equipment is in place. Most importantly, the United States Congress passed the Oil Pollution Act of 1990, which mandated that all new oil tankers be double hulled and mandated that all single hull tankers would be retired by 2015. Having two layers of steel offers greater protection when a vessel hits something it should not.

US Coast Guard Photo/Alamy Stock Photo

Figure 11.23 Six months after the spill, shorelines in Prince William Sound were still covered by crude oil. Here, cleanup workers were testing oil dispersant to see if it would loosen oil from the rocks.

Bettmann/Getty Images

Figure 11.22 Oil is pumped off the grounded Exxon Valdez into the Exxon Baton Rouge in the middle of Prince William Sound, Alaska. Over 150 million liters (40 million gallons) of oil were removed from the damaged tanker.

Alan Levenson/Alamy Stock Photo

Figure 11.24 A dead, oil-soaked seabird with the snowy mountains of Prince William Sound in the background.

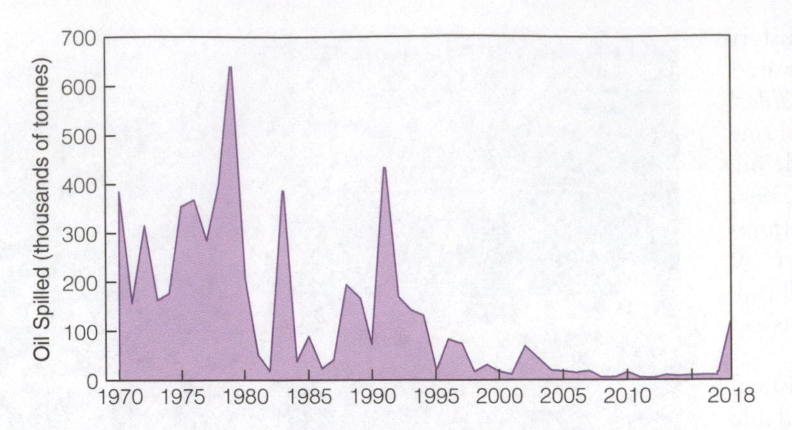

Figure 11.25 After double hulls on most large oceangoing tankers were mandated in the 1990s, the mass of oil lost to spills decreased dramatically. (Modified from Earth, The Global Education Project)

The Norwegian oil tanker, the SKS Satilla, provides the evidence (**Figure 11.26**). On March 8, 2009, the nearly 300-meter (900-foot)-long tanker struck a submerged (and unmarked) oil rig in the Gulf of Mexico. The rig had been blown free of its mooring during Hurricane Ike in September 2008 and had gone missing. The tanker found it. That impact ripped open the outer hull, but the inner hull remained intact. The double-hulled tanker was carrying 140 million tonnes (41 million gallons) of crude oil. Not a gallon of the oil leaked. The double hull saved the day.

11.1.2 Coal

Coal is the carbon-rich residue of ancient plant matter that has been preserved and altered by heat and pressure over millions of years (**Figure 11.27**). It is a solid fossil fuel easily stored and transported. Following oil and oil shales, coal is Earth's most abundant reserve of fossil energy. Coal deposits are known from every geological period since the appearance of widespread terrestrial plant life some 390 million years ago. The large coal fields of North America, England, and Europe were deposited during the Carboniferous period (360 to 299 million years ago), so named because of the extensive coal deposits in rocks of that age. Permian coal (299 to 252 million years ago) is found in Antarctica, Australia, and India–pre-continental drift Gondwanaland (see Chapter 2). Younger coals (66 to 3 million years ago) are found in such diverse locations as Spitsbergen Island in the Arctic Ocean, the western United States, Japan, India, Germany, and Russia.

Forming coal requires the accumulation of large amounts of plant debris under conditions that encourage preservation. This requires high plant productivity in a low-oxygen depositional environment; conditions usually found in brackish water swamps. The accumulated plant matter must then be buried to a depth sufficient for the heat and pressure to expel water and volatile matter, at least several kilometers below the surface of the Earth. The degree of metamorphism (conversion) of plant material to coal is denoted by its **rank**. From lowest to highest rank, the metamorphism of coal follows the sequence **peat**, **lignite**, **subbituminous**, **bituminous**, **anthracite** (**Figure 11.28**). From peat to anthracite, the sequence of coal rank is accompanied by increasing amounts of fixed carbon and heat content and a decreasing amount of quickly burned volatile material and water.

Different types of coal are found in different places. In the Appalachian coal basin, high-volatile bituminous coals occur in the western part of the basin and increase in rank to low-volatile bituminous coals to the east and then to anthracite. When the Pennsylvania Anthracite was formed, about 275 million years ago, the eastern United States was a convergent plate margin. Anthracite, the highest rank of coal, is formed when coal-bearing rocks are subjected to intense heat and pressure–a situation that occurs in areas of plate collision which prevailed when the Appalachian Mountains and their counterparts in the British Isles formed. The interior coal

NOAA

Figure 11.26 Despite a massive gash from colliding with a sunken oil rig, the double-hulled *SKS Satilla* didn't leak any of its 41 million gallon cargo of oil.

iStock.com/Philip_hens

Figure 11.27 A shovel full of coal. The fossil fuel that stoked the Industrial Revolution remains critical for electricity generation worldwide.

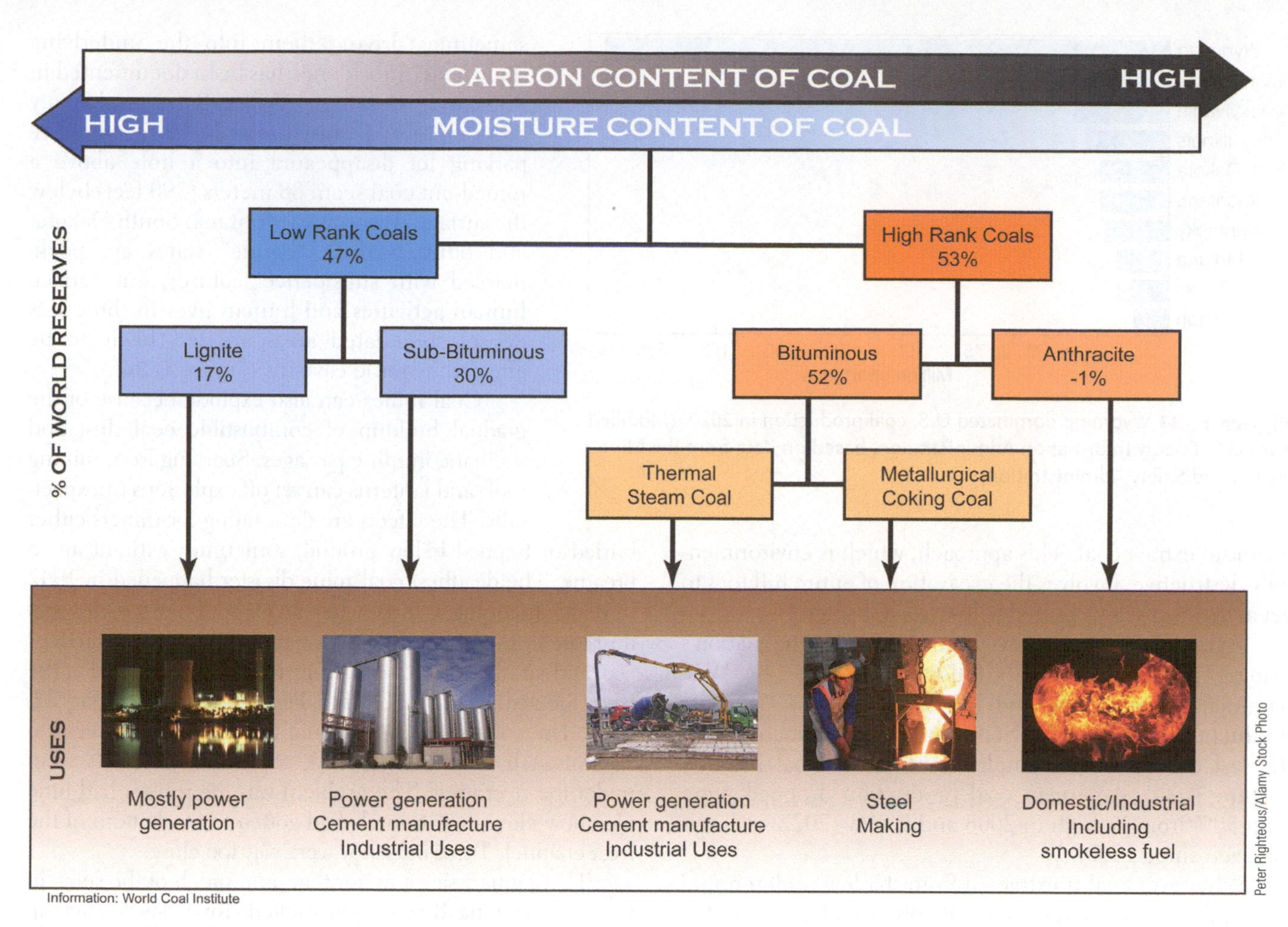

Figure 11.28 Coal rank and the uses of different types of coal.

basins of the Midwest are characterized by high-sulfur bituminous coal, whereas the younger western deposits in Wyoming are low-sulfur lignite and subbituminous coal.

There are many different ways to mine coal (see Chapter 4). Peat, lignite, and bituminous, or soft, coal usually occur in flat-lying beds at shallow depths that are amenable to surface-mining techniques (**Figure 11.29**). Higher rank coal is often more deeply buried and thus mined using more traditional tunneling techniques (**Figure 11.30**). More recently, mountaintop removal has become a popular

Figure 11.29 Near surface lignite mine in Germany. Most of Germany's coal resources are lignite. With the start of war in Ukraine and the loss of Russian oil and gas supplies, Germany has stepped up its coal mining, much to the dismay of climate activists.

Figure 11.30 Abandoned #9 Coal Mine tunnel in Pennsylvania's Carbon County. This tunnel is about 2,000 feet (625 meters) underground.

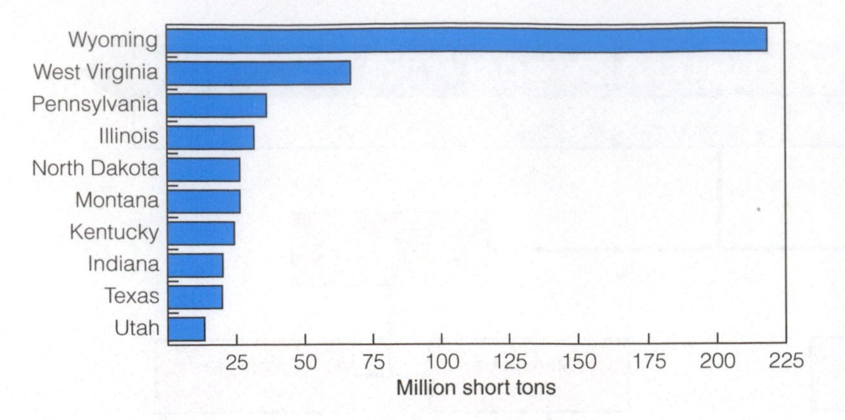

Figure 11.31 Wyoming dominated U.S. coal production in 2020. (Modified from U.S. Energy Information Administration, based on data from the Mine Health and Safety Administration)

means to extract coal. This approach, which is environmentally destructive, involves the excavation of entire hill tops to get at the coal seams buried below (see Chapter 4).

Wyoming, with its low-sulfur coal, was the nation's leading producer in 2020 followed in order by West Virginia, Pennsylvania, and Illinois (**Figure 11.31**). Coal production in the United States has fallen dramatically in the last decade as has employment in the coal industry (**Figure 11.32**). American coal production dropped more than 50% from its high in 2008 and is now (2022) at levels not seen since 1965.

Whenever coal is extracted from shallow underground seams, there is a risk for mine collapse–in other words, the roof falls in. Lignite and bituminous (soft) coals are most susceptible to collapse, because they are exploited from near-horizontal beds at relatively shallow depths. The "room and pillar" method recovers about 50% of the coal, leaving the remainder as pillars for supporting the roof of the mine (**Figure 11.33**). As a mine is being abandoned, the pillars are sometimes removed or reduced in size, increasing the yield of coal but also increasing the possibility of collapse sometime in the future.

Even when the pillars are left, failure is possible. The weight of the overlying rock can cause the pillars to rupture, sometimes driving them into the underlying shale beds. Subsidence has been documented in Pittsburgh, Scranton, Wilkes-Barre, and many other areas in Pennsylvania. In 1982, an entire parking lot disappeared into a hole above a mined-out coal seam 88 meters (290 feet) below the surface. Wyoming, Montana, South Dakota, and other western "lignite" states are pockmarked with subsidence features, but surface human activities and human lives in those less densely populated areas are less likely to be affected by mine cave-ins (**Figure 11.34**).

Coal mines can also explode because of the gradual buildup of combustible coal dust and methane in mine passages. Sparking from mining tools and lanterns can set off explosions unexpectedly. The effects are devastating for miners either buried or trapped below ground, sometimes without air to breathe. The deadliest coal mine disaster happened in 1942 in Benxi, Liaoning, China, when an explosion of gas and coal dust tore through the mine. More than 1,500 miners died.

Perhaps the most unusual mine disaster happened at the River Slope mine in Port Griffith, Pennsylvania, on January 22, 1959. The coal was anthracite and the mine was adjacent to the Susquehanna River–indeed, some mine tunnels went under the river itself! The problem was the miners had little idea how close their tunnels had gotten to the bottom of the river channel. Turns out, they were way too close.

That January saw a major thaw and much of the snow in the Susquehanna River Basin melted. River levels shot up quickly, almost 7 meters (22 feet) in just a few days. Despite the flooding, mining continued. Just before lunch, several workers heard a loud sound and hurried to escape. A whirlpool was forming at the surface of the river as 10 million liters (2.7 million gallons) of water per minute poured into the mine below. To put that number in context, it's about three cubes, 50 meters (160 feet) on each side, going down the hole, every minute. Eighty-one men were working in the mine that day. Twelve lost their lives. Their bodies were never found.

Plugging the hole in the riverbed was a challenge. Workers dumped in mine cars and railroad cars, but the river kept flooding the mine. Finally, they built a small diversion dam and plugged the hole in the drained river bed. Investigations after the accident showed that the mine had taken out so much coal that only 50 centimeters (19 inches) of rock remained between the bottom of the river and the top of the mine. Why did the mine get so close to the river bottom? Because the mine was not well mapped; the miners were "mining blind." A safe thickness of rock separating the river and the mine should

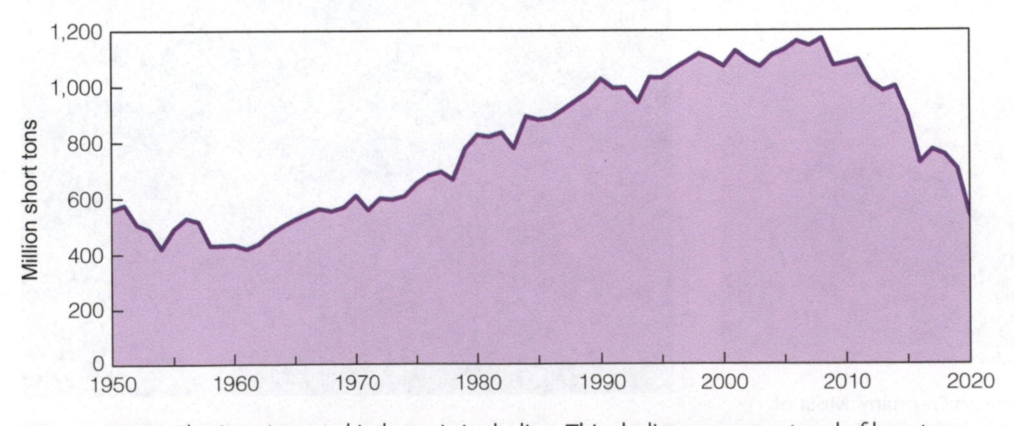

Figure 11.32 The American coal industry is in decline. This decline reverses a trend of long-term growth in coal production from 1960 until 2008. (Modified from U.S. Energy Information Administration)

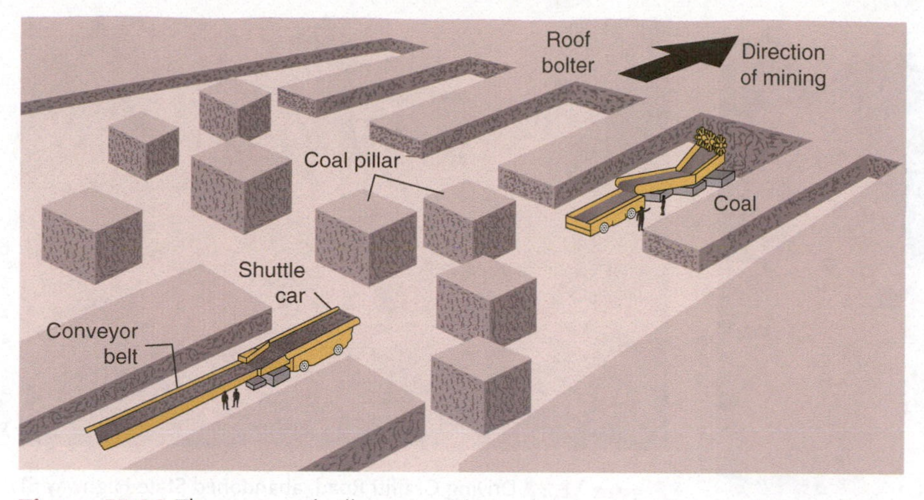

Figure 11.33 The room and pillar approach to mining coal underground. (Modified from Ach Coal Company)

have been at least 10 meters (33 feet). The disaster led to a tightening of mining laws in Pennsylvania including a requirement to map accurately all mining activities as they progress.

Coal is combustible not only in a coal stove or a power plant but also when it is still in the ground. Long-lasting coal seam fires are burning at most places coal is and has been mined, threatening communities unfortunate enough to live above the mines. In 1980, Centralia was a town of 1,000 people in the Appalachian Mountains of eastern Pennsylvania. The rock below Centralia is laced with tunnels of the now-abandoned coal mine and the coal in those tunnels is on fire (**Figure 11.35**). Rumor has it that the fire started in 1962, when someone was burning trash near an old mine tunnel entrance. The impacts to the community have been devastating. As the fire consumed the coal, the abandoned mine tunnels began to collapse into sinkholes of a very different sort than those that form in limestone from dissolution (see Chapter 9). Soon, people were overcome by carbon monoxide and sulfur-rich gases, the result of incomplete underground combustion.

In 1981, a young boy almost lost his life when the ground gave way below him. That was the beginning of the end of Centralia. Soon after, the state of Pennsylvania condemned buildings in the town, moved people out, and properties were demolished. Today, Centralia has four residents, three cemeteries, three homes left standing, and one fire department (**Figure 11.36**). Otherwise, it's a ghost town. What's left of the town does have an unusual tourist attraction, "Graffiti Road" or what used to be State Highway 61. The highway was abandoned because the land below it was deforming as the burning coal collapsed. Now the crumbling road is a destination not only for those who enjoy driving all-terrain vehicles but for artists who have covered the pavement with graffiti (**Figure 11.37**). The fires beneath Centralia may take hundreds of years to burn out naturally.

Coal seam fires are not just a Pennsylvania problem. Such fires haunt coal beds throughout the American West, including major fires presently ongoing in Colorado and Wyoming. Coal fields in China and India, both of which are heavily dependent on coal for electricity generation, have been beset with coal seam fires for more than a century. In Jharia, India, alone, there are nearly 100 coal seam fires burning and a million people are at risk (**Figure 11.38**). For India, this is a billion-dollar problem covering an area about three times larger than the District of Columbia.

All of that uncontrolled burning coal is contributing carbon dioxide to the atmosphere at a time when we least need more emissions. Some estimates suggest that more 33 million tonnes (37 million tons) of coal, worth billions of dollars, have burned in the Jharia coal fields. If that's the case, then the fires have released over 80 million tonnes (88 million tons) of carbon dioxide, equal to the annual emissions of countries like Colombia, Chile, and Kuwait, or the American states of Washington, Colorado or West

Science History Images/Alamy Stock Photo

Figure 11.34 Subsidence above now-abandoned coal mines near Sheridan, Wyoming, resulting from roof collapse.

Michael Brennan/Hulton Archive/Getty Images

Figure 11.35 Steam and smoke rise from collapsing ground in Centralia, Pennsylvania, engulfing Lamar Mervine, the town's last mayor.

Reuters/Alamy Stock Photo

Google Earth

James Schaedig/Alamy Stock Photo

Figure 11.36 Centralia, Pennsylvania, is a modern-day ghost town. (a) Main Street sure has changed from 1983 when the buyout of homes started until 2000, when it was largely complete. (b) The view of Centralia from above is haunting. It's nearly empty. (c) If you visit Centralia, you'll see signs like this.

Virginia. It could be worse in China, where coal mining is far more distributed with families and communities mining small pits and shafts which are frequently abandoned and thus prone to ignition. Some estimate that up to 1% of global carbon emissions may come from coal seam fires.

Michael Ventura/Alamy Stock Photo

Figure 11.37 Driving Graffiti Road, abandoned State Highway 61 in Centralia, Pennsylvania.

Coal can have significant health impacts on miners, on people who burn coal, and on ecosystems downwind from coal-burning power plants. Black lung (*pneumoconiosis*) is a disease caused by inhaling coal dust which irritates and scars the lungs. Coal dust, an inevitable byproduct of mechanized mining has led to the slow, miserable deaths of tens of thousands of miners (**Figure 11.39**).

Downwind of coal-fired power plants there are additional environmental impacts. Coal, because it naturally contains sulfur and mercury, releases both when it's burned. The combusted (or oxidized) sulfur creates sulfuric acid when mixed with rainfall (Chapter 4). Coal combustion is the most important source of acid rain, which in the 1960s and 1970s, decimated the forests of eastern North America, especially on mountain tops and where the soil did not have much ability to neutralize the acid falling from the sky (**Figure 11.40**). The worst acid rain had an acidity between that of lemon juice and vinegar (pH between 2 and 3). But it's not all bad news. Greatly improved pollution control devices on coal-fired power plants have dramatically reduced both the acidity of rain and toxic mercury emissions between 1990 and 2020 (**Figure 11.41**).

Jonas Gratzer/LightRocket/Getty Images

Figure 11.38 In Jharia, India, coal seams burn 24/7.

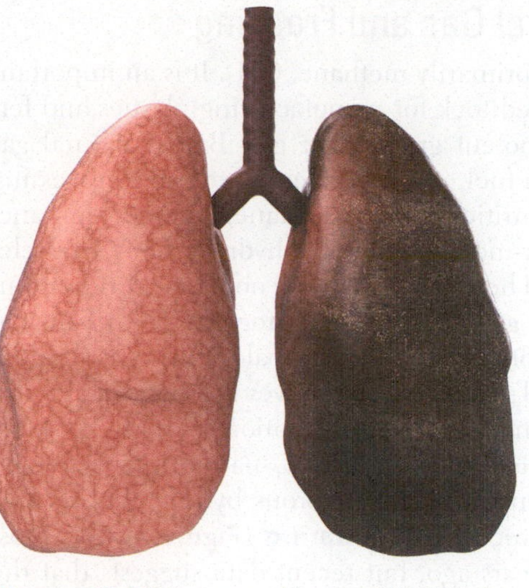

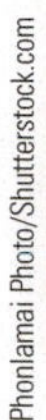

Figure 11.39 Lungs archived at NIOSH, the National Institute for Occupational Science and Health; it's easy to tell which one belonged to a coal miner.

Some coal contains arsenic, a toxic metal, and fluorine, a very reactive element that binds with calcium in living organisms and above certain concentrations is toxic. In many rural areas of Asia, coal is burned in unvented stoves for cooking, space heating, and water heating (**Figure 11.42**). When the weather is too cool and damp to

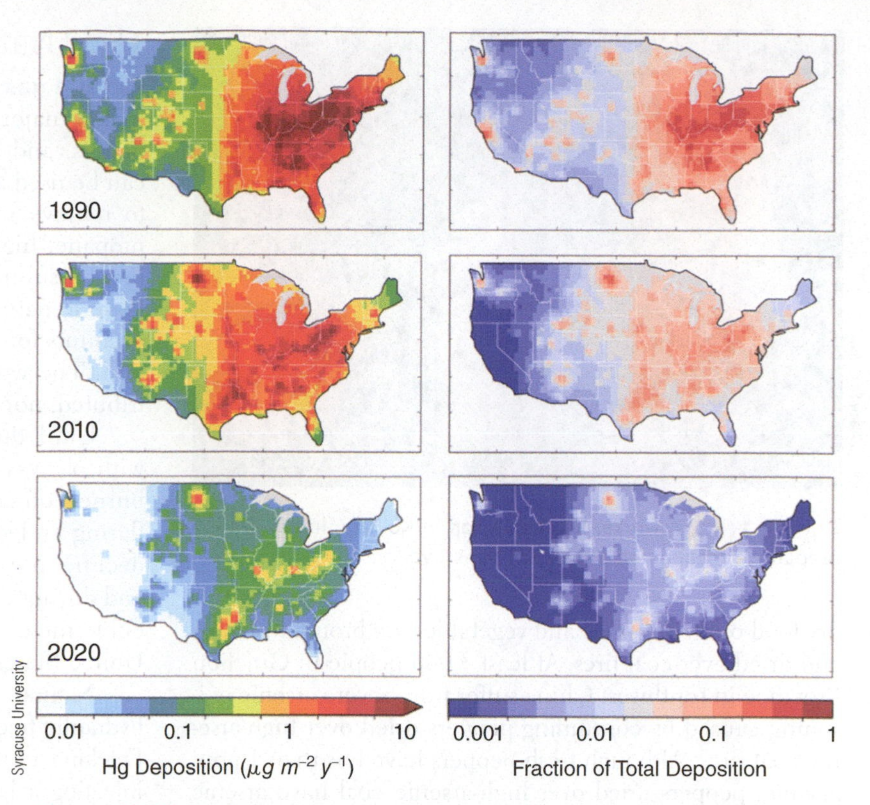

Figure 11.41 Model results showing dramatic decrease in mercury emissions from power plants (left side) and fraction of total mercury emissions attributable to power plants on the right. Most of these power plants burn coal as their primary fuel. Emissions reductions are due both to improved pollution control and the retirement of coal-fired power plants.

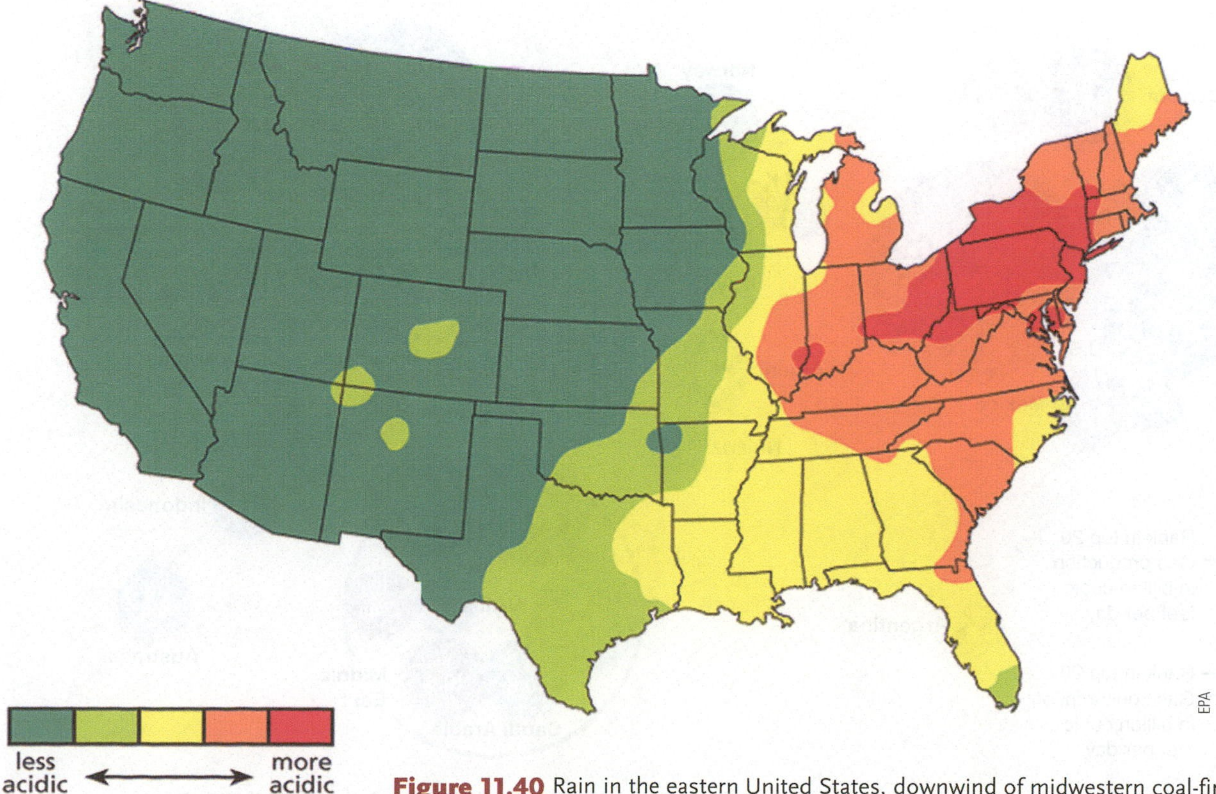

Figure 11.40 Rain in the eastern United States, downwind of midwestern coal-fire power plants, is far more acid than rain elsewhere in the country.

Figure 11.42 A worker in a coal shop in Kolkata, India, prepares a meal on an unvented coal stove.

dry food outdoors, meat and vegetables are brought indoors and dried over coal fires. At least 3,000 people in Guizhou Province in southwest China suffer from severe arsenic poisoning caused by consuming peppers dried over high-arsenic coal fires. Although fresh peppers have less than 1 ppm arsenic, peppers dried over high-arsenic coal have arsenic levels up to 500 ppm. More than 10 million people in the same province suffer dental and skeletal fluorosis caused by eating corn that has been cooked over coal containing large amounts of fluorine. As many as 3.5 billion people worldwide are exposed by indoor coal burning to toxic fumes and fine particulate matter.

11.1.3 Natural Gas and Fracking

Natural gas is primarily methane, CH_4. It is an important fuel, a major feedstock for manufacturing plastics and fertilizers, and a potent greenhouse gas. Before natural gas can be used as a fuel, it must undergo extensive processing to remove impurities, such as ethane, butane, pentane, propane, higher-molecular-weight hydrocarbons, and elemental sulfur. The United States is now the world's number-one natural gas producer, and, together with Canada, accounts for more than 25% of global natural gas production. The world's natural gas reserves are not equally distributed, nor is natural gas consumption (**Figure 11.43**).

Until the last several decades, natural gas extracted with oil was considered a dangerous by-product and was burned off–a process called **flaring** (**Figure 11.44**). Most flaring ended years ago, but recent data suggests that the decline in gas flaring has plateaued in years shortly before and during the COVID-19 pandemic era, with 144 billion cubic meters of gas flared in 2021–about 20% of what the United States consumes annually. That's waste!

Natural gas production has been revolutionized by hydraulic fracturing, also known as fracking (see Chapter 6). Fracking improves the production of oil and gas wells by injecting a high-pressure fluid, infused with chemicals and sand, deep into "tight" formations like shale, to fracture the formation. These fractures increase the rock's porosity and permeability, promoting the release of oil and/or natural gas into well bores drilled to recover them (**Figure 11.45**). Fracking, in concert with improvements in horizontal drilling technology, was one of the most significant energy production innovations

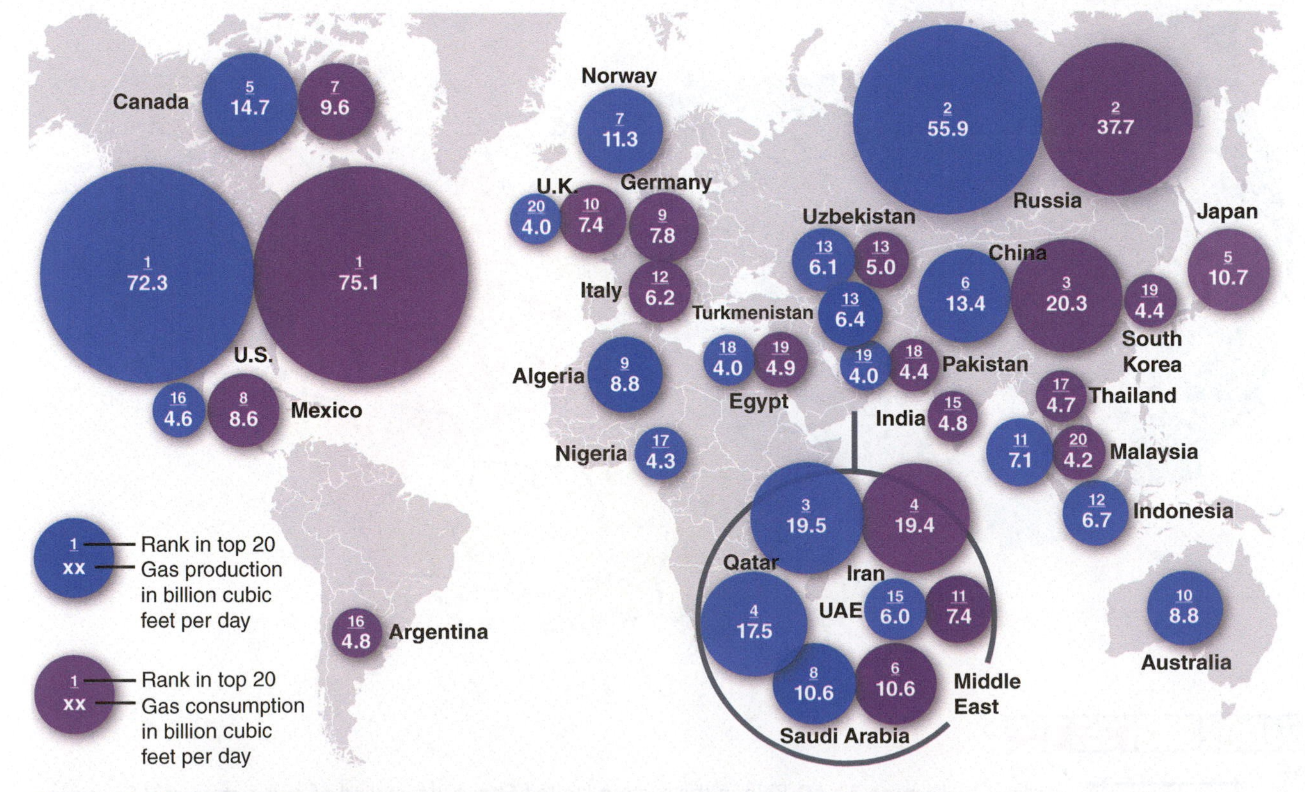

Figure 11.43 Map of natural gas production (blue) and consumption (purple) in units of billions of cubic feet per day. (Modified from map by BP, 2016 data)

Figure 11.44 Natural gas being flared at the Mossmorran processing plant in Fife, Scotland, the United Kingdom. The wind turbines in the background stand in stark contrast.

of the twentieth century. For example, the results of fracking have been dramatic in Pennsylvania. The state has experienced a phenomenal increase in drilling since 2000. By 2021, the state had more than 71,000 fracking-related active gas wells. But there are downsides. Fracking can have large environmental and public health costs. The fracking process produces toxic wastewater that has the potential to contaminate surface water and drinking water wells.

Fracking treatments may require up to 23 million liters (6 million gallons) of fluid and 100,000 pounds of **proppants** (sand that holds open the fractures) for a single large well. The fluid includes less than 2% of chemical additives, rust inhibitors, microorganism-killing agents, and friction reducers. Although only 2% may seem a small amount, over the life of a typical producing well, this may amount to 380,000 liters (100,000 gallons) of additives. A list of fracking chemicals includes emulsifiers, surfactants, biocides (such as used as disinfectants in swimming pools), and viscosity modifiers that produce a fluid the drillers call slickwater. The additives vary widely in toxicity. Many of the compounds are found in ordinary household products such as soap, paint, floor wax, cosmetics, with some even used in food products. Others, however, are toxic: carcinogens; neurotoxins (e.g., benzene, toluene, xylene, ethylene glycol [antifreeze]); butoxyethanol and methanol. The interaction of fracking water with rock at depth can also liberate natural radioactivity present in the fracked shales (derived mostly from radon; see Chapter 12) and bring that radioactivity to the surface as wastewater.

Disposal of wastewater is perhaps the most significant environmental impact of fracking. Somewhere between 10% and 40% of the fracking fluid pumped down a well returns to the surface as **flowback**, a murky, highly salty water enriched in drilling fluids and other contaminants dissolved from the rock into which the fracking fluid was injected and circulated. The flowback is recovered and stored in aboveground ponds until removed by tanker trucks

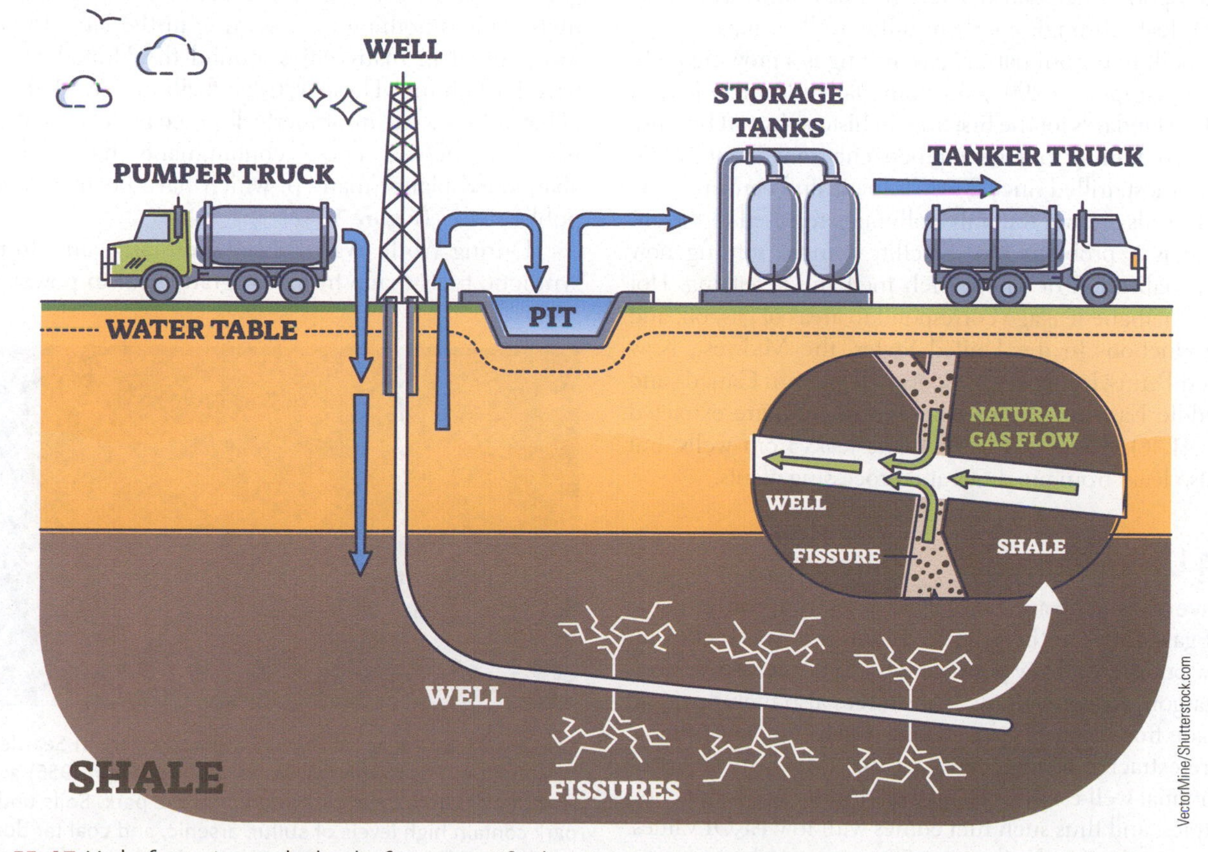

Figure 11.45 Hydrofracturing, or hydraulic fracturing ("fracking"), requires injecting over a million gallons of water, chemicals, and sand per well at high pressure to crack the rock and release oil and/or gas into the well.

Figure 11.46 Methane data sensed by European Copernicus Sentinel-5P satellite in 2019. Circle size equals size of methane plume. Darker circles show areas with higher methane concentration.

for disposal. The remaining fracking fluid stays underground. Contamination of groundwater by flowback leaking through a cracked pond liner is also a very real possibility as is flowback that leaks through poorly installed well casings.

Air pollution from natural gas drilling is a growing problem. For example, in 2009, Wyoming failed to meet federal air quality standards for the first time in history, in part because of fumes containing toluene and benzene from about 27,000 gas wells, most drilled since 2004. Toluene and benzene leaking from wells are not the only pollutants released to the air. Methane is a problem too. Satellite remote sensing now allows visualization of where such methane is leaking. Hot spots for methane leakage correspond to areas of gas, oil, and coal production. In the United States, the Midwest, New York, and Pennsylvania are hotspots as is western Canada and the Middle East–all regions where fossil fuels are extracted (**Figure 11.46**). Some of this methane leaks from wells, but some also leaks from pipelines and processing plants.

11.1.4 Unconventional Fossil Fuels

Unconventional oil and natural gas differ from conventional oil and gas. They can be produced from reservoir rocks with very low porosity and low permeability–"tight" reservoirs in oil patch jargon. Unconventional oil sources also include liquid fuel made from natural gas and coal. Often, unconventional fuels are extracted using technology that is less efficient than conventional well-established oil well drilling and production techniques, and thus such fuel comes with low EROI values. Many unconventional oil extraction processes have greater environmental impacts than conventional oil production.

Some unconventional fuels have been proposed as sources of motor vehicle fuel for transportation; synthetic gasoline and corn-based ethanol are in limited use as vehicle fuels. This is nothing new, starting in the later 1800s, coal was gasified in many cities around the United States and used for lighting. These **syngas** facilities have left a legacy of hazardous waste improperly disposed at the time the facilities were active. These contaminants haunt the now-abandoned plants, many of which have been reclaimed as public spaces (**Figure 11.47**).

During World War II, Nazi Germany came to rely on synthetic fuels made from low-grade coal to power its war

iStock.com/Seastock

Figure 11.47 An aerial view of Gasworks Park in Seattle, Washington. The former gasworks (shut down in 1956) are fenced off from access in the center of the park. Soils under the park contain high levels of sulfur, arsenic, and coal tar dumped on the site when it was active. The site has been capped to keep the waste away from the people picnicking on the grass.

College Park, MD/NARA

Figure 11.48 A B-17 flying fortress bombs the I.G. Farben syngas facility at Monowitz, Poland, in 1944.

machine as the country had large coal reserves but little oil. During the first several years of the war, synfuels accounted for 60% of Germany's demand for petroleum products, including almost all aviation gasoline (over 90%) and about half of the vehicle gasoline supply. German synthetic fuel plants were a frequent target of Allied bombing during the war. At some plants, such as Buna in Poland, slave labor to make the fuel was provided by inmates held at the nearby Auschwitz and Monowitz concentration camps (**Figure 11.48**).

Tight oil flows readily at room temperature if it can be released from the rock in which it is hosted. Extraction requires the extensive well drilling and fracking technology–some wells even go sideways to follow the oil-bearing rocks (**Figure 11.49**). Among the important tight oil deposits in the United States are the Bakken Shale in North Dakota and adjacent parts of Montana and Canada, the Barnett Shale, and Eagle Ford Shale of Texas. Tight oil also occurs elsewhere in the world, for example, in the Persian Gulf, West Siberia, Oman, and Mexico. In general, tight oil reservoirs, especially found in shale, tend to be depleted much more quickly than big conventional oil reservoirs. Prospecting for fresh supplies is a constant need.

iStock.com/MajaPhoto

Figure 11.50 Fracking in action, Dawson Creek, Canada to release more natural gas.

Coal-bed methane (CBM) is natural gas trapped in coal beds. It is a hazard to coal mining, as mentioned above, and CBM extraction was originally undertaken simply to make coal mining safer. Commercial development of CBM in the United States has taken off and in 2021 accounted for about 3% of total natural gas production. Australia, Canada, and China also produce CBM. CBM is easy to extract by a drilled well as opposed to extracting coal, which involves mining. But gas production, as with all the nonconventional fuels, comes at a price, because it requires disruptive wells, a network of roads, pipelines, and methane leakage, all of which have environmental impacts. The completion of CBM wells requires enhancement by hydraulic fracturing (or "fracking") as described above (see Figure 11.45). With that comes the need for large amounts of water and a means to dispose of water used and contaminated by fracking (**Figure 11.50**).

Methane hydrate is natural gas trapped in ice. Some believe that methane hydrates may eventually supply some of the world's future fuel needs. Methane hydrates are widely distributed in deep-sea sediments in all oceans (**Figure 11.51**) and are present in permafrost in the Arctic and

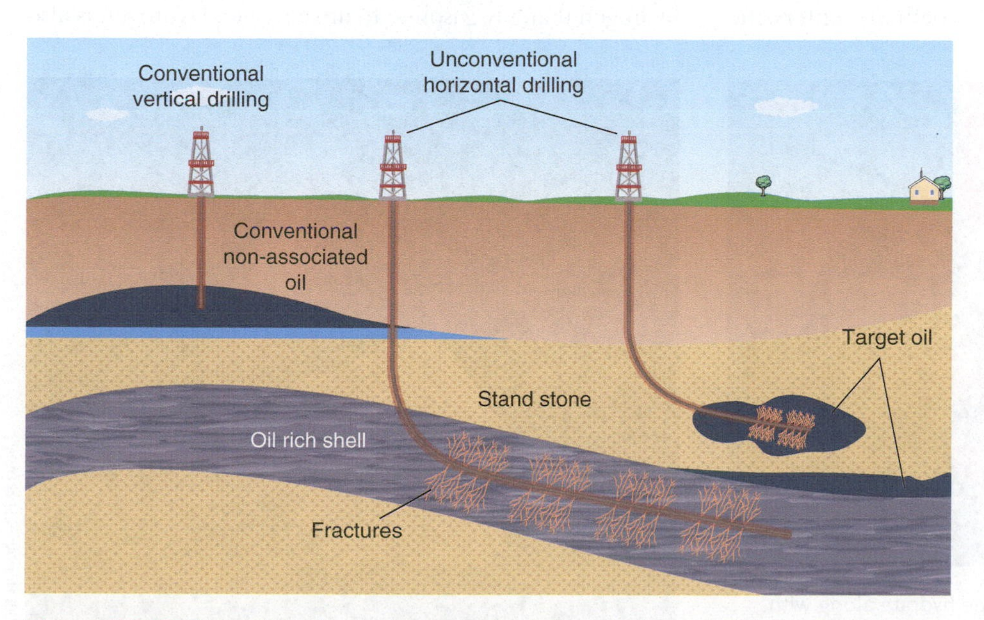

Figure 11.49 Tight oil requires technologically challenging horizontal drilling techniques remove the crude oil.

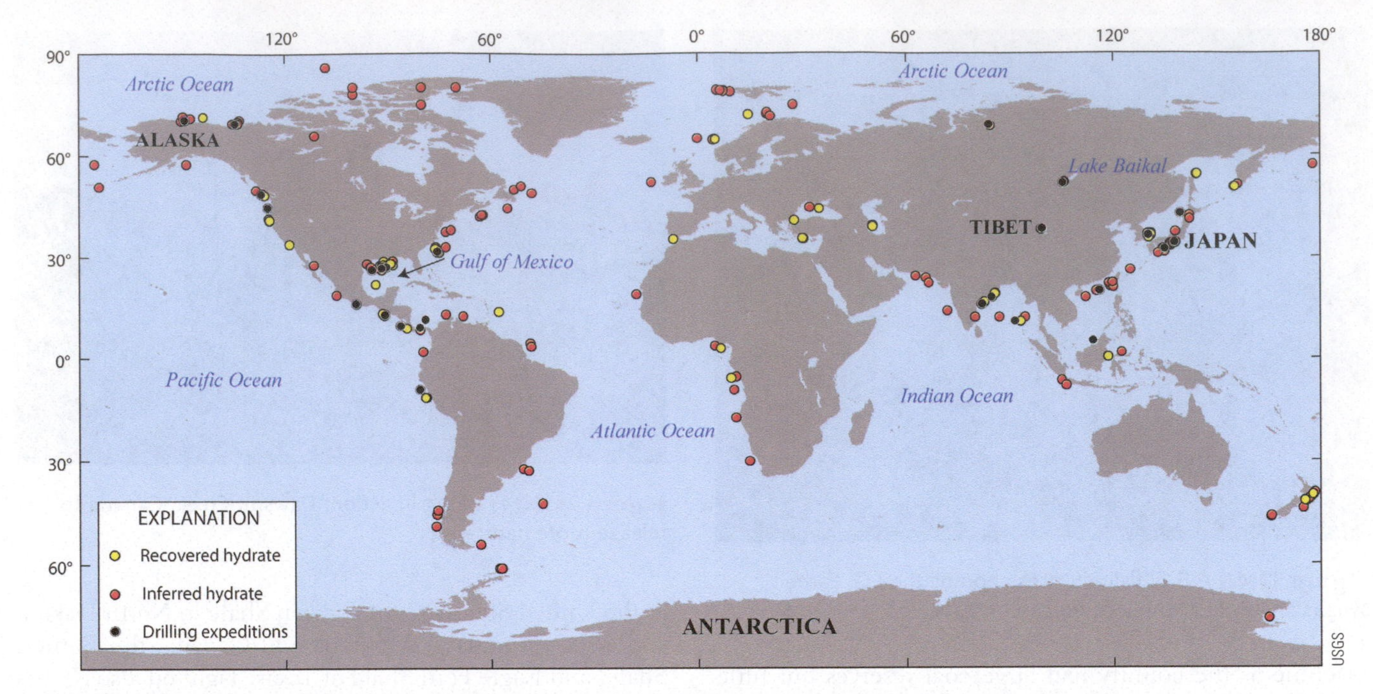

Figure 11.51 Map of methane hydrate deposits either identified by drilling or inferred from seismic data.

subarctic. The exact origin of the gas is unknown, but it is strongly suspected that sea-bottom bacteria consume organic-rich detritus and generate methane as a waste product. Under appropriate temperature and pressure, the gas becomes trapped in ice within the sediment instead of being released to the overlying water column (**Figure 11.52**). The bad news is that these deep-sea hydrated molecular structures can release their caged methane into the atmosphere if there is even a slight decrease in pressure (if sea level lowers, for example) or a slight warming of the surrounding seawater. Methane hydrates buried in permafrost can be released by arctic warming sufficient to melt the permafrost. Once released, the methane enters the atmosphere, a dangerous positive feedback for global warming (**Figure 11.53**).

Hydrogen has been touted as a fuel of the future. It gives the promise of combustion that releases only water (the oxidation of hydrogen gas, H_2, generates H_2O, water). Unfortunately, there are no natural reservoirs of hydrogen to mine. It's better to think of hydrogen as an energy carrier. It can be manufactured from fossil fuels, but this just shifts the burden of carbon dioxide pollution. More promising is the use of solar power to split molecules of water to produce hydrogen–a way to store solar energy for later use. Hydrogen has other liabilities. You need a much larger volume of hydrogen than say gasoline to drive a mile. Hydrogen is also

Figure 11.52 White chunks of methane hydrate along with gray sediment were retrieved using a sediment core from a water depth in the ocean of about 2,400 meters (8,000 feet).

Figure 11.53 Methane hydrate, the ice that burns!

GL Archive/Alamy Stock Photo

Figure 11.54 On May 6, 1937, the hydrogen-filled *Hindenburg* explodes while attempting to land at Lakehurst, New Jersey, Naval Air Station.

explosive. It's important to remember the German airship *Hindenburg* and its fiery demise in 1937, when the hydrogen which gave it lift caught fire explosively and destroyed the airship, killing 35 people (**Figure 11.54**).

Tar sands, or heavy-oil sands, consist of layers of sandstone, claystone, and siltstone containing water-coated sediment grains surrounded by films of thick, tar-like **bitumen** (**Figure 11.55**). Tar sands contain oil that is too thick to flow at normal temperatures. These deposits, found in Alberta (Canada), Venezuela, Madagascar, the United States, and elsewhere, are surface-mined like coal, using enormous diesel-powered shovels and haul trucks, or drilled using steam to mobilize and recover the oil (**Figure 11.56**). To get one barrel of synthetic oil from tar sands requires large amounts of natural gas and hot water and 4 tons of ore.

The size of Canadian and Venezuelan deposits tar sands deposits is mind-boggling. Canada's Athabasca sands alone constitute the largest oil field in the world, covering

iStock.com/Francisblack

Figure 11.55 A handful of tar sands from Alberta, Canada.

iStock.com/Dan_prat

iStock.com/Dan_prat

Figure 11.56 Tar sands are removed from (a) open pits over large areas and (b) transported for oil extraction by massive trucks.

an area about the size of North Carolina. Canada claims that 1.73 trillion barrels of recoverable oil could potentially come from the deposits, with oil reserves (unconventional and conventional) eight times that of the United States. Thanks to the prodigious Athabasca tar sands, Canada is second only to Saudi Arabia in oil reserves.

Tar sands that are too deep for surface exploitation, such as those that underlie Alberta's Cold Lake area, are extracted differently. Boreholes produce this oil by a process called **in situ extraction**. Steam at 300°C (575°F) is injected through boreholes at high pressure and allowed to "soak in" for a week or more to make the bitumen more fluid. The oil released from the heated sand is then brought to the land surface by "rocking horse" pump jacks as seen in conventional oil fields (**Figure 11.57**). In 2021, in Canada alone, 4.04 million barrels of crude oil were produced daily from the tar sands in Alberta and Saskatchewan.

Exploitation of the tar sands comes with a huge environmental footprint. As of 2022, the tar sand mines had converted 760 square kilometers (294 square miles, an area about the size of New York City), or about 0.01%, of Canadian boreal forest into dirt, dust, and toxic tailings ponds. Eventually, 4,790 square kilometers (1,850 square miles) ultimately could be impacted. The Canadian Association of Petroleum Producers reports that reclamation

iStock.com/Dszc

Figure 11.57 Pump jack extracting crude oil adjacent to an oil refinery in Seminole, West Texas.

(restoration of soils and vegetation) of mined land is an ongoing process since mining began in the 1960s. As of 2019, the Alberta provincial government reports that 11% of total disturbed land has been reclaimed. In cooperation with the Fort McKay First Nation, Syncrude, the tar sand mining company–a joint venture of several major oil companies–has successfully developed wood bison habitats on reclaimed lands. More than 300 bison now graze on what were once the sites of tar sand mining and tailings operations (**Figure 11.58**), but it will take decades to completely restore mined landscapes.

Another environmental concern is the large quantity of water used in processing tar sand. Two to four barrels of water (in addition to energy to heat the water and make steam) are needed to make one barrel of synthetic crude oil. Although 80% to 95% of process water is recycled, some ends up in highly contaminated tailings ponds. In April 2008, several hundred migrating waterfowl died after landing on one of the Alberta ponds. Since 2014, 11 million liters (around 3 million gallons) of toxic leakage from tar sands tailings seeps into the Athabasca River daily, causing First Nation (Indigenous Canadian) communities downstream to worry about the purity of their drinking water and the fish they eat.

Gunter Marx/Alamy Stock Photo

Figure 11.58 Bison grazing in the foreground on reclaimed tar sands tailings. The Syncrude extraction works, where oil is recovered from tar sands, is in the distance. Fort McMurray, Alberta, Canada.

Greenhouse gas emission is another issue. The Syncrude refinery at Fort McMurray is one of Canada's major emitters of greenhouse gases, as it requires about 5,650,100 cubic meters (196 million cubic feet) of natural gas per day, around 40% of Alberta's total usage, to produce the steam required to extract this heavy oil. Carbon emissions related to extracting oil from the tar sands are 20% higher than average emissions from using an equivalent amount of conventional oil. The EROI value for synthetic oil extracted from tar sands is low, somewhere between 2:1 to 6:1.

Oil shale is dark, fine-grained sedimentary rock that like tar sands yields petroleum when heated. Some oil shale is not even shale; it is a clay-rich rock containing **kerogen**, a solid, waxy hydrocarbon (**Figure 11.59**). A ton of oil shale may yield 38 to 570 liters (10 to 150 gallons) of high-quality oil. Found on all the continents, oil shales were originally organic-rich sediment deposited in lakes, marshes, or the ocean. The most extensive U.S. deposit, the Green River Formation, was formed during Eocene time (56 to 34 million years ago) in huge freshwater lakes in the present-day states of Colorado, Utah, and Wyoming.

Shallow oil shales can be mined in open pits, but deeper deposits can be mined by underground methods. Converting solid kerogen into gasoline requires heating it to about 370°C (698°F). Adding hydrogen and other chemical treatment is needed to produce a finished product. There are environmental issues. Processing oil shale is

Department of Energy

Figure 11.59 Kerogen-rich oil shale from the Green River Formation in Wyoming and a beaker of the heavy oil that can be extracted from it.

energy intensive and requires three to four barrels of water to produce one barrel of oil, conditions that do not bode well for oil shale development in arid lands, such as the western United States. At its very best, fuel extracted from oil shale has a low estimated EROI of perhaps 1.5:1 to 2:1.

11.2 Biofuels (LO2)

Efforts are underway to develop a suitable alternative for liquid fuels currently derived from fossil fuels. One promising option is biofuels which are any fuels derived directly from **biomass** (straw; agricultural waste; sugarcane; food leftovers; waste from the wood, paper, and forestry industries corn; switchgrass; and any organic waste including manure). Contenders include **ethanol**, **methanol**, and **biodiesel** fuel. As with petroleum and coal, biomass is a form of stored solar energy that has been captured by growing plants using the process of photosynthesis–the difference is one of timing. Biofuels have been manufactured in modern times and so their energy-dense carbon-bearing molecules have only recently formed. Fossil fuels contain fossil carbon, stored millions of years ago.

For a long time, ethanol (ethyl alcohol) derived from corn led the biofuel pack (**Figure 11.60**). Ethanol made from sugarcane has been used in Brazil for many years as an auto fuel, and at one time almost 90% of the cars there burned it. Ethanol derived from corn has received the most attention in the U.S. where sugarcane is hard to grow, but it's EROI is hard to pin down and probably quite low. Corn-based ethanol may even have a negative EROI, although some experts suggest a low but positive EROI of about 1.3:1. Sugarcane ethanol's EROI ranks higher, about 5:1 or 6:1 because waste from the cane is used to fuel the distilling process rather than fossil fuel.

Corn-based ethanol is also not an efficient biofuel, although it is popular among farmers for the high price the corn fetches and the idea of energy self-sufficiency. Initially, farmers growing corn benefitted from federal subsidies of about $6 billion a year, supporting ethanol as a gasoline additive. Those subsidies ended in 2012, when the companies producing ethanol lost a tax credit of 46 cents per gallon. Ethanol fuel use does have one positive impact on the environment; it cuts ozone-forming gases and is a safe substitute for the fuel-oxygenating, gasoline additive, MTBE, which has poisoned groundwater around the world.

In 2022, the United States had 200 ethanol plants in operation. The feedstock for almost all of these plants is corn, although 10 ethanol plants use beer waste, wood waste, or food-processing waste as feedstock. The plants use different energy sources to drive the distillation. Ethanol distilled in natural gas-powered plants increases greenhouse gas emissions by 5%, compared with petroleum-based gasoline. For ethanol distilled in coal-powered plants, the gain in greenhouse gas emission (GHG) is a whopping 34%.

Biodiesel is another alternative fuel produced from animal fats, plant oils, or algae. It's a great way to recycle used cooking oils (**Figure 11.61**)–although the exhaust from cars using biodiesel from French Fries cookers have a certain smell you might recognize (a bit like fast food). Biodiesel contains no petroleum, but it can be blended at various levels with petroleum-derived diesel. Biodiesel blends can be used as a fuel in most compression-ignition (diesel) engines with no modification. Biodiesel has a low EROI, in the range of 1.5:1, but that is offset by its high marks in cutting GHG emissions: carbon dioxide is cut by about 40%, and nitrogen oxides are reduced by up to 85%.

Pond scum (algae) may have a role to play in producing biofuels, especially biodiesel (**Figure 11.62**). As of 2022, at least 50 small startup companies are working on the idea of algae cultivation of ethanol and biodiesel using CO_2 from coal-fired power plant emissions. Algae cultivation looks attractive because it produces four times as much ethanol as the same weight of corn. Emissions from the coal-fired power plants are routed to an algae bioreactor, where carbon dioxide is absorbed and utilized by the algae to grow at a rapid rate.

iStock.com/Jmichl

Figure 11.60 Ethanol plant in the midwestern United States. Corn that is soon to be distilled is piled in front of the plant.

Martin Bond/Alamy Stock Photo

Figure 11.61 An English bus running on the biodiesel that might have cooked last week's fish and chips.

Ashley Cooper/Alamy Stock Photo

Figure 11.62 A commercial algae growing system with clear pipes so algae can photosynthesize. The algae is used to make ethanol and/or biodiesel.

Once harvested, the algae can be processed to produce a variety of biofuels: gases, such as methane; liquid fuels, such as biodiesel and ethanol; and a variety of useful solids, such as bioplastics and proteins. Although most ongoing research focuses on controlled industrial production of algae as a crop to harvest for biofuel production, some companies are experimenting with collecting coastal algae blooms. These are the oxygen-sucking, fish-killing masses of algae in the ocean, whose growth is stimulated by fertilizer runoff from farm fields (see Chapter 8).

Most early studies of biofuel as a substitute for gasoline show a reduction of GHG, because biofuels sequester carbon through the growth of feedstock. But these studies fail to account for the carbon emissions that occur as farmers worldwide respond to higher prices by converting grassland and particularly forest to new croplands, needed to replace the grain diverted to biofuels. In addition, there are ethical issues to converting land that can produce food into land producing energy when almost 10% of the world population is food insecure and 50 million people live in areas where famine is happening now (2022) or likely to happen soon in the future.

11.3 Nuclear Energy (LO3)

Nuclear energy is fundamentally different than other energy sources used by humans (except perhaps geothermal, the heat in which is derived in part from radioactivity deep in the Earth). Nuclear energy is not extracted from chemical reactions (like the burning of fossil fuels), nor did it originate in the Sun. Nuclear energy is derived from a fundamental property of matter; that being the splitting of an atom or the combining of two atoms into one, which releases massive amounts of energy–if the reaction can be controlled. And there lies the rub!

During the 1950s and 1960s, nuclear energy was considered the solution to the world's future energy needs, generating power that some thought would be "too cheap to meter." It never did work out that way. So pervasive was the drive to put nuclear power to use, however, that the U.S. military installed a nuclear reactor and power plant in the Greenland Ice Sheet and another in Antarctica (**Figure 11.63**). Reactors were placed on submarines in battle fleets all over the world; the power they produced could permit the submarines to remain underwater for very long periods of time, but accidents at sea and the release of radioactivity were always a possibility (see Chapter 12). On land, large amounts of nuclear power were generated with little visible footprint on the landscape other than an occasional cooling tower and containment dome–the icons of the nuclear industry (**Figure 11.64**). There were accidents, a few of them fatal

Bettmann/Getty Images

Figure 11.63 Installation of parts of the nuclear reactor that powered a 200-person military camp (Camp Century) inside the Greenland Ice Sheet 138 miles from the coast.

iStock.com/RelaxFoto.de

Figure 11.64 Grohnde nuclear power station in Germany, for 6 years, held the record for the most power produced by any power plant. As Germany moved toward more renewable energy and away from nuclear power, the plant's decommissioning began on December 31, 2021, largely in response to the 2011 Fukushima nuclear accident.

PA Images/Alamy Stock Photo

Figure 11.65 Protests outside a British nuclear power plant on the first anniversary of the 2012 Fukushima disaster.

and a few that released some radioactivity. But, for the first several decades that nuclear technology was used (the late 1950s, the 1960s, and the 1970s), nuclear power seemed to hum along, with new plants opening regularly around the world.

Then three accidents (1980 at Three Mile Island, the United States; 1986 at Chernobyl, former USSR; and 2011 at Fukushima, Japan) changed the public perception of nuclear power and the associated risks (**Figure 11.65**). In 2011, about 25% of German electricity was generated by nuclear power. After the Fukushima disaster that year, German Chancellor Angela Merkel indicated that Germany would phase out nuclear power by 2022. It almost happened, but the 2021 war in Ukraine changed the German approach: although many German reactors were shut down, at least two reactors have been given a new lease on life and will continue operating past their scheduled December 2022 shutdown date.

More than 400 nuclear power plants have been built worldwide, with 54 commercially operating nuclear power plants and 32 decommissioned plants in the United States as of 2022. Nuclear energy accounts for 14% of the planet's electrical power (**Figure 11.66**). As of 2022, the United States generates 91.5 gigawatts of nuclear energy, contributing nearly 20% to the country's total electrical energy generation. Some countries have come to rely even more on nuclear power. France, which has a nuclear capacity of 61.3 gigawatts, obtains more than two-thirds of its power from nuclear sources; 17% of this energy is generated from recycled nuclear fuel.

The fuel for nuclear reactors is far more energy dense that any other fuel. The energy stored in a tiny pellet of uranium is equal to many barrels of oil or a truckload of coal (**Figure 11.67**). This means that large amounts of power can be generated from small amounts of fuel–an advantage. But, the waste resulting from today's nuclear power plants is extremely radioactive and

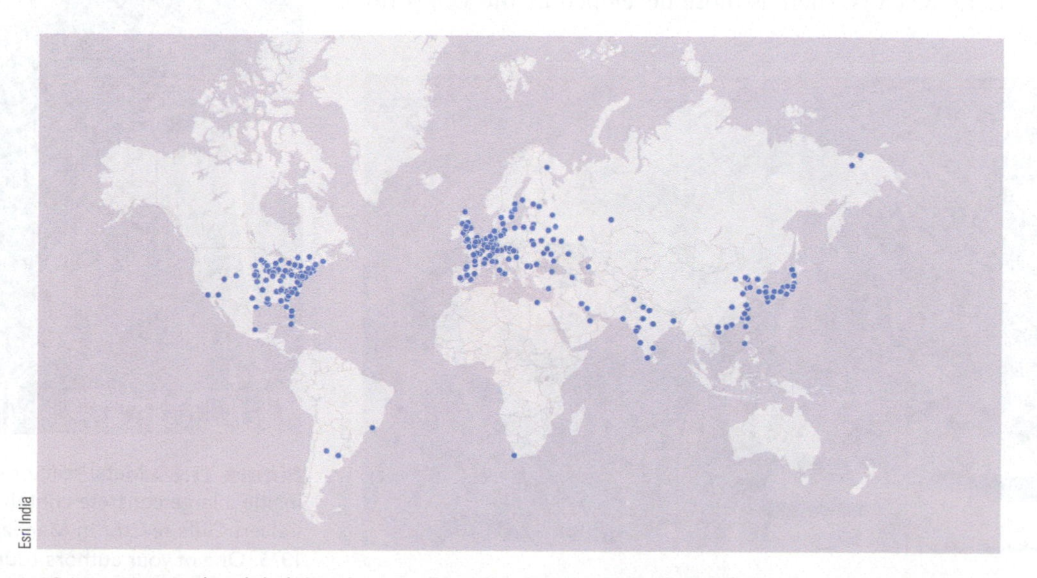

Esri India

Figure 11.66 The global distribution of nuclear reactors used for power generation in 2019. The Australian continent has no nuclear reactors used for power generation.

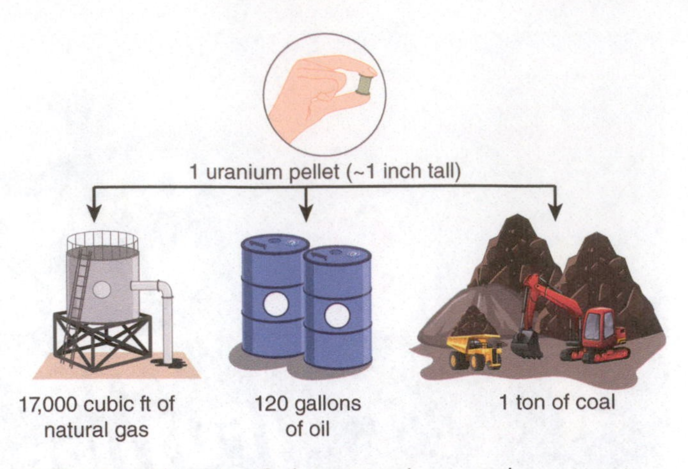

Figure 11.67 Nuclear fuel is extremely energy dense. Uranium used to power nuclear reactors is far more energy dense than fossil fuels. (Adapted from U.S. Department of Energy)

thus presents a significant hazard. While some believe that nuclear power is an important, low-carbon energy source for our future, others point to a growing and potentially toxic waste stream for which we have not seriously considered safe long-term storage and disposal. Below we examine how nuclear power works and the risks and benefits it presents. Nuclear waste disposal and storage is considered in greater depth in Chapter 12.

11.3.1 Fission Energy

Nuclear reactors and fossil-fuel plants generate electricity by similar processes. Both heat a fluid, which then directly or indirectly makes steam (Figure 11.16). The steam spins turbine blades to drive an electrical generator (**Figure 11.68**). Nuclear power plants are large, sprawling faculties but the reactor itself is a very small part of the plant held inside a concrete and steel containment structure (**Figure 11.69**). Early reactors, such as those developed by the U.S. Army for the arctic and the U.S. Navy for submarines, had cores of the nuclear fuel so small that you could nearly wrap your arms around them (**Figure 11.70**).

Uranium is the fuel of atomic reactors because its nuclei are so packed with protons and neutrons that the element is capable of sustained nuclear reactions. All uranium nuclei have 92 protons and between 142 and 146 neutrons. The common isotope uranium-238 (^{238}U) has 146 neutrons and is barely stable (remember, isotopes of an element have differing numbers of neutrons but the same number of protons). The next most common isotope, ^{235}U, is so unstable that a stray neutron penetrating its nucleus can cause it to split apart completely, a process known as **fission**–a term particle physicists borrowed from cellular biologists. We take advantage of ^{235}U's instability under controlled conditions to extract the huge energy potential of atom splitting, 90% of which is heat.

Nuclei that are split easily, such as those of ^{235}U, ^{233}U, and plutonium-239 (^{239}Pu), are called **fissile isotopes**. What results from the reaction are called **fission products** (**Figure 11.71**). The lighter fission products are elements that recoil from the split nucleus at high speeds. This energy of motion and subsequent collisions with other atoms and molecules creates heat, which raises the temperature of the fluid used to cool the reactor and eventually to drive the turbines and generator.

industryviews/Shutterstock

Figure 11.68 Steam in thermal power plants spins turbine rotors like this one.

Everett Collection Historical/Alamy Stock Photo

Figure 11.69 Metal reactor vessel being raised into position inside a large concrete containment building. This is at the Calvert Cliffs reactor in Maryland in 1971. The plant opened in 1975. One of your authors toured this plant when he was in elementary school as it was being built, although he didn't get to see the reactor vessel!

Figure 11.70 Two soldiers standing a model of the core of one of the U.S. Army's first nuclear reactor–the PM-1, deployed to a radar installation in Sundance, Wyoming in 1962. It ran until 1968. The core holds enough fuel for the reactor to run for 2 years providing 1.25 megawatts of power as well as hot water for heating.

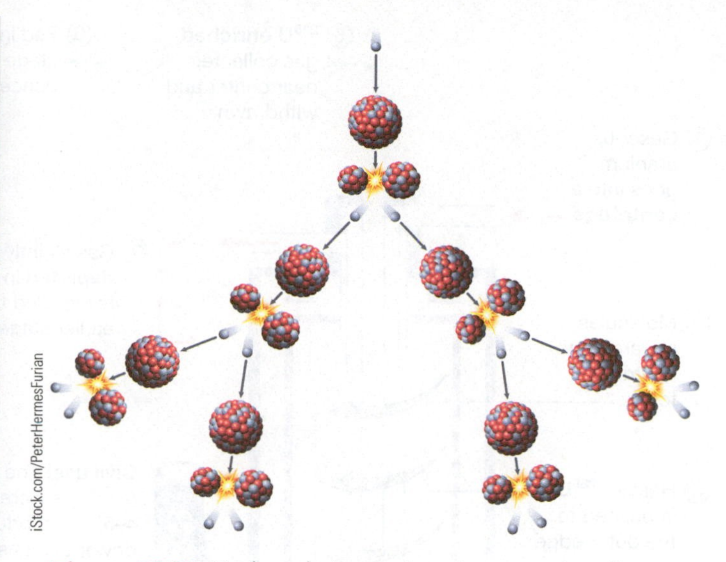

Figure 11.72 Nuclear chain reaction. A neutron (small grey sphere) hits the first 235-U nucleus, splitting and releasing two additional neutrons. They go on to split two more 235-U nuclei and the reaction grows exponentially.

Meanwhile, the stray neutrons collide with other nuclei, repeating the process in what is called a **chain reaction**. A controlled chain reaction occurs when one free neutron (on average) from each fission event goes on to split another nucleus. If more than one nucleus is split by each emitted neutron (on average), the rate of reaction increases rapidly, and the reaction quickly goes out of control (**Figure 11.72**). Keeping the reaction controlled is critical for nuclear power generation and involves moderating (slowing and absorbing) the neutrons using compounds like water and elements like boron, both good neutron absorbers. If the reaction goes out of control, life-threatening energy and radiation releases occur. A nuclear weapon works because of an uncontrolled nuclear reaction.

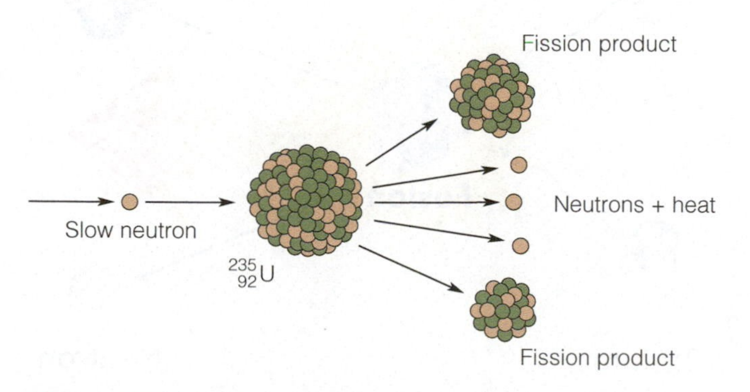

Figure 11.71 Nuclear fission. A slow neutron penetrates the nucleus of a U-235 atom, creating fission products (also called **daughter isotopes**), two or three fast neutrons, and heat.

Because ^{235}U makes up only 0.7% of all naturally occurring uranium, the rest being ^{238}U, uranium ore as mined can't sustain fission though it is radioactive. But once it's enriched, by removing some of the ^{238}U, the resulting fuel is ready for use in commercial nuclear reactors. The enrichment process is technically demanding (it most commonly involves hundreds, even thousands of gas centrifuges) and relies on the slight mass difference between ^{238}U and ^{235}U (**Figure 11.73**). Enrichment also produces large amounts of waste including the rejected ^{238}U. Some of this **depleted uranium** (so named because it's depleted in ^{235}U) sees reuse in both armor for fighting vehicles (because it's so dense and thus resistant to piercing by bullets and shells) and in munitions designed to pierce that armor.

There are major environmental concerns associated with nuclear energy, including radioactive contamination during fuel processing and transportation, as well as the disposal of radioactive waste from reactor operation. Other problems include the potential for nuclear weapons proliferation, the decommissioning of antiquated reactors, radiation leaks from nuclear power plants, and nuclear fuel supply limitations.

The Nuclear Regulatory Commission, the nuclear power industry, the media, and even some environmental groups promote nuclear energy as a clean and safe alternative to fossil fuels. Climate change has spurred many people (including some environmentalists) to argue that nuclear

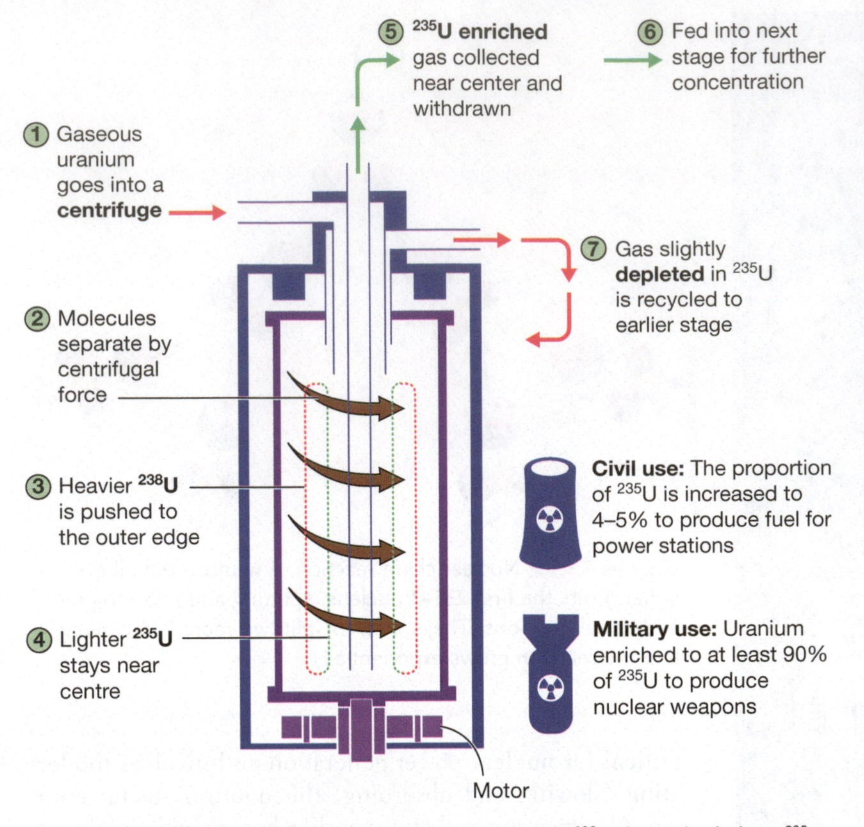

Figure 11.73 A gas centrifuge separates the heavier ^{238}U from the lighter ^{235}U. Nuclear energy is produced from ^{235}U, which makes up just 0.7% of naturally-occurring uranium, the rest being ^{238}U. The enrichment process increases the proportion of ^{235}U by separating it from ^{238}U. (Adapted from AFP, data from USNRC)

power is critical to reducing greenhouse gas emissions and that at least currently running plants should remain open. Nuclear power proponents see it as an established energy technology that can help slow global warming. The EROI of nuclear power ranges from 5:1 to about 18:1, a ratio high enough that some argue that we should build modern nuclear power stations.

11.3.2 Fusion Energy

Nuclear fusion brings two atoms together to form a third–a process that releases massive amounts of energy. Fusion reactions power the Sun (and we all know how well that works) but people have never been able to harness the power of a fusion reaction to generate power. We have designed, built, and exploded fusion bombs, but those involve uncontrolled, runaway reactions. Power generation requires just the opposite–a stable, ongoing, controlled reaction. There's a saying in the energy community that power generation from nuclear fusion is just 20 years away and has been that way for decades.

Designs for fusion reactors are centered on bringing together two isotopes of hydrogen. **Deuterium** is hydrogen with one neutron and one proton. **Tritium** is hydrogen with two neutrons and one proton. Getting these two isotopes to fuse into helium (two protons, two neutrons) requires massive amounts of energy because the closer the two atoms get, the stronger the repulsion caused by the interaction of the two positively charged protons (**Figure 11.74**). But, if that energy barrier can be overcome and the atoms fuse, immense amounts of energy are released and the byproducts, helium gas and neutrons, are relatively benign.

Many designs have been proposed for fusion reactors. One now seems to have the best chance of generating more power than is put in (and EROI>1): a **tokamak** (this is a Russian acronym for a fusion reactor design) uses powerful magnets to contain an energetic plasma (a highly energized and different state of matter than we encounter day to day) in a donut-shaped containment container (**Figure 11.75**). Here, deuterium and tritium are accelerated as they race around inside the tokamak. When they collide, the energy is sufficient for some of them to fuse. So far, a tokamak has achieved an EROI of 0.66–a power sink, not a power source!

A new machine may change this. The result of an international collaboration that began with leaders of the United States and Soviet Union over several decades: Nixon, Brezhnev, Reagan, and Gorbachev. Together with European scientists, Soviet (now Russian) and American researchers worked together to design and now build the **International Thermonuclear Experimental Reactor (ITER)**. The ITER is currently under construction in southern France (**Figure 11.76**). It's a tokamak on a scale of no other: 23,000 tonnes (25,300 tons) and 30 meters (100 feet) tall. The ITER is scheduled for its first test run to generate a plasma in 2025 with the goal of making power (with an EROI>10) by 2035. Time will tell if this instrument lives up to its design potential.

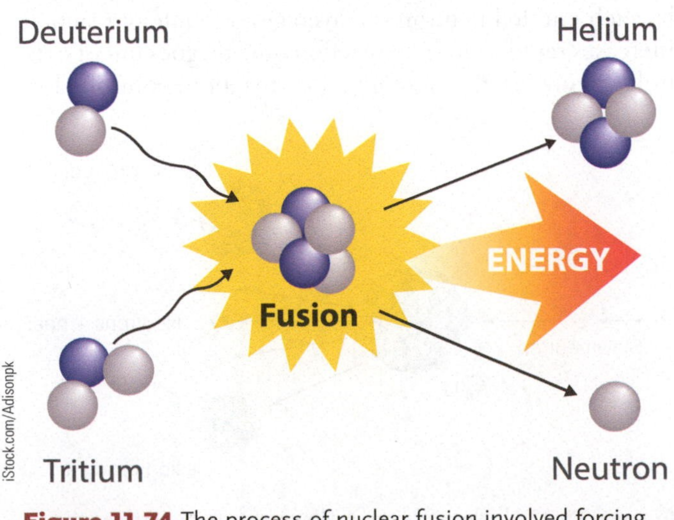

Figure 11.74 The process of nuclear fusion involved forcing two types of hydrogen together to create helium. This reaction releases large amounts of energy and a neutron.

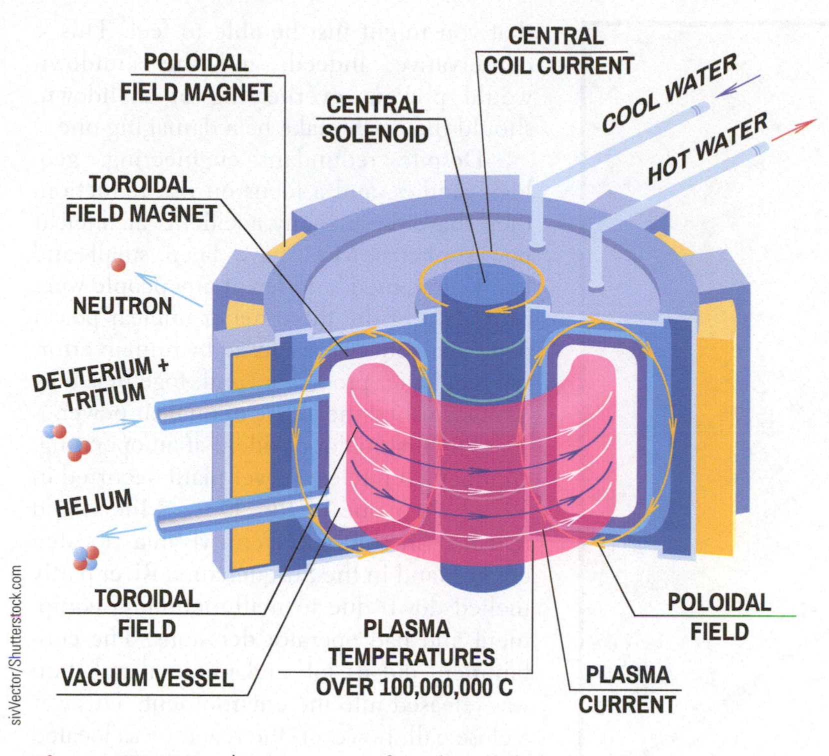

Figure 11.75 A schematic view of a Tokamak fusion reactor designed for heating water that could be used for power generation.

11.3.3 Nuclear Safety, Accidents, and Contamination

The standards for building, licensing, and maintaining nuclear power stations are strict. In the United States, such standards are set by the U.S. Nuclear Regulatory Commission, the federal government body that licenses nuclear reactors for public utilities. Its standards are often adopted by other nations.

Geologists and geology are key to safely siting a nuclear power plant. One of the many criteria for siting nuclear reactors is geological stability. Ideally, the selected site is free from landslides, tsunamis, earthquake activity, active volcanoes, and flooding. This is not always possible or practical. The Nuclear Regulatory Commission requires that all active and potentially active faults within 320 kilometers (200 miles) of a nuclear power plant be located and mapped. Reactors must be built at a distance from an active fault, determined by the fault's earthquake-generating potential. The NRC defines an active fault as one that has moved once within the last 30,000 years or twice in 500,000 years. This mapping and dating of fault movement keeps geologists busy around the world.

Sites that meet these strict seismic hazard criteria are difficult to find along active continental margins where faults are common, such as southern California, where the San Andreas Fault extends from the Mexican border almost to Oregon. For example, California had several nuclear power plants located along fault zones, San Onofre and Diablo Canyon (**Figure 11.77**) being the best

Figure 11.76 The ITER under construction in October 2022.

Figure 11.77 Diablo Canyon, a nuclear power plant in California, is located less than 1,000 meters from an active fault. It sits on a marine terrace about 26 meters (85 feet) above the Pacific Ocean.

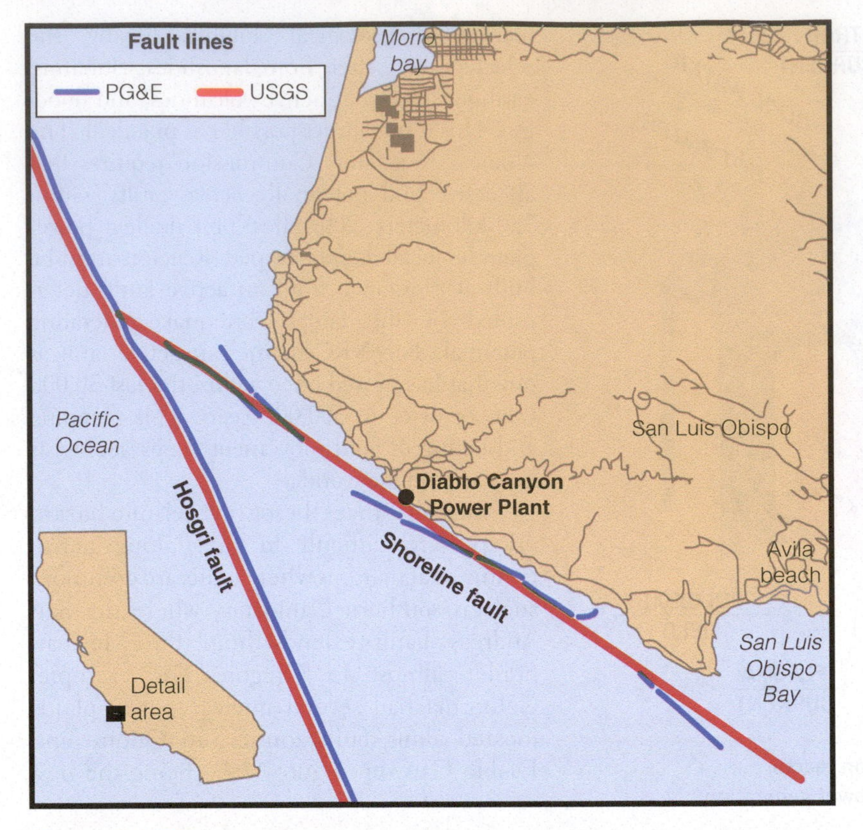

Figure 11.78 The Diablo Canyon nuclear plant sits close to two faults. Geologists differ over the potential impact of two offshore faults – one of which lies directly offshore and could potentially generate a magnitude 7.2 earthquake.

known. San Onofre was shut down in 2013, but Diablo Canyon, which was completed in 1986, remains on the electrical grid adjacent to the San Andreas Fault System. The plant sits within 1,000 meters of the Shoreline Fault and about 3,000 meters from the Hosgri Fault (**Figure 11.78**). The Hosgri fault was discovered in 1971, after the plant had been designed but before it was built. Redesign to a higher seismic standard did not come cheaply.

In 2016, Pacific Gas and Electric announced that Diablo Canyon would close by April 2025. This was a business decision. The company was facing increasing public pressure to close the plant, and the cost of upgrades needed to relicense the plant were high–but the real decision-maker might just have been the falling cost of renewable power in California and a state mandate for 60% of all power in the state to be produced by renewable sources by 2030. When Diablo Canyon closes, California will have no nuclear power plants.

Faults can be much more difficult to find along tectonically quiescent continental margins, where there are many nuclear power plants. This is because faults along such older passive margins are rarely active. For example, one of this book's authors worked in South Africa mapping and dating faults several hundred kilometers from a proposed nuclear power plant (see Case Study, Chapter 6). To provide additional safety and allow for geological uncertainty, reactors are programmed to shut down immediately at a seismic acceleration of 0.05 *g*–a small quake that you might just be able to feel. This is conservative, indeed, and the shutdown would prevent overheating or meltdown, should the earthquake be a damaging one.

Despite redundant engineering, geologic studies, and a focus on risk reduction, there have been many accidents at nuclear power plants. Most have been small and easily contained, and few if any people were hurt. Two of the three major nuclear power plant accidents were caused by human error, not geologic processes, and together they have tarnished the allure of nuclear power.

The first major accident at an operating, commercial nuclear power plant occurred in 1979, when one of the Three Mile Island reactors (in southern Pennsylvania) located on an island in the Susquehanna River partly melted down due to malfunctioning equipment and bad operator decisions. The containment did its job and minimal radiation was released into the environment. This was a close call; however, the reactor was located on an island in the Susquehanna River upstream from Chesapeake Bay and close to Baltimore, Philadelphia, and Washington, DC (**Figure 11.79**). President Jimmy Carter, a nuclear engineer by training, visited the damaged plant 4 days after the accident to reassure a nervous nation (**Figure 11.80**).

The world was not so lucky in April, 1986. In northern Ukraine (then in the USSR), close to the Belarusian border, an error by plant operators had disastrous consequences in part because of flawed reactor design. The reactor exploded, killing two operators immediately and

Figure 11.79 Aerial view of the Three Mile Island nuclear power plant on an island in the middle of the Susquehanna River, southeastern Pennsylvania.

Everett Collection Inc/Alamy Stock Photo

Figure 11.80 President Carter visiting the crippled Three Mile Island nuclear plant only 4 days after the accident that caused the core of one reactor to melt down.

spewing about 5% of the radioactive core material into the environment (**Figure 11.81**). After the explosion, graphite in the reactor core ignited and flames carried radioactivity aloft. Many radioactive particles settled around the plant, but some kept going, carried by the wind, contaminating hundreds of square kilometers of land. Some radioactivity lofted into the atmosphere, circulating around the world. Much of northern Europe had detectable levels of radiation falling out of the sky, contaminating fields, and as a result, the milk supply when cows ate the grass (**Figure 11.82**).

Over the next few weeks, hundreds of workers trying to put out the fire at the reactor and contain the damage were sickened by radiation. Nearly 30 died of acute radiation poisoning from overexposure. Upward of a quarter million people were evacuated. Most were never allowed to return to their homes. Fields and forests were severely contaminated with radioactive materials from the damaged reactor. A 4,300-square-kilometer (1,660-square-mile) exclusion zone was set up and monitored. Today, nearly 40 years later, the nearby City of Prip'yat' remains abandoned (**Figure 11.83**). A few people, mostly older, never left, and today live quiet lives in the exclusion zone. In the absence of people and disturbance, wildlife abounds, but a sampling shows that much of it carries a body burden of radioactivity.

The Soviet Union brought in over 200,000 **liquidators**, people from across the nation, whose job it was to assist in recovery and cleanup (**Figure 11.84**). The reactor was entombed in a concrete sarcophagus in 1986 to keep the remaining melted and radioactive nuclear fuel isolated from

AP Images/Igor Kostin

Figure 11.81 View of the damaged Chernobyl reactor from a helicopter soon after the explosion that ripped it apart.

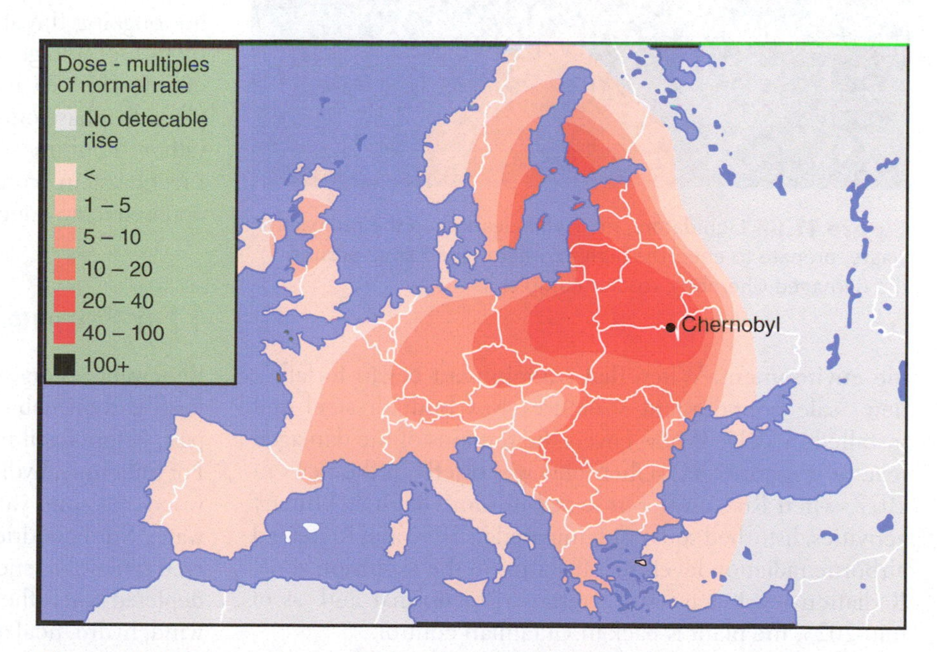

Figure 11.82 The wind carried fallout from the damaged reactor across Europe. People closest to the accident were the most heavily exposed to radiation. (Adapted from the BBC)

NST/Alamy Stock Photo

Figure 11.83 The dodge-em cars at Hotel Polisya Park, in Pryp'yat', Kyivs'ka oblast, Ukraine, Chernobyl, were abandoned in 1986. The city is now a ghost town.

Francois Lochon/Gamma-Rapho/Getty Images

Figure 11.84 Liquidators, men who cleaned up the radioactive waste, prepare to enter the highly contaminated zone around the damaged Chernobyl reactor complex.

the environment. When that containment began to fail, a new "safe containment structure" was built of steel and installed in 2017. It now covers the remains of the damaged reactor (**Figure 11.85**). Chernobyl was briefly in the news in 2022, when Russian troops seized it from Ukraine. Military activities disturbed soils contaminated in 1986 and increased airborne radiation levels set off alarms in the exclusion zone. Radiation levels have since returned to normal and, as of mid-2023, the plant is back in Ukrainian control.

The Fukushima Daiichi, Japan, nuclear power plant disaster, the largest nuclear disaster since Chernobyl, exacerbated concerns and further raised the specter of danger from nuclear electricity generation. The disaster was the result of the giant seismic tsunami of March 11, 2011 (see Chapter 6), though fundamental design flaws in the placement of generators and water pumps, described in Chapter 1 played a role, showing that even here some human error was involved.

iStock.com/Diegograndi

Figure 11.85 The new safe confinement structure (as of 2021) covering the damaged reactor at the Chernobyl nuclear power plant.

It's worth emphasizing that in each of these three historical nuclear accidents, the circumstances were both unexpected and unique. They took place despite extensive safety measures taken in reactor design and operation. But these measures proved insufficient. Humanity is still on a "nuclear learning curve." With military action taking place around nuclear power plants quite recently (e.g.–Zaporozhye in Ukraine), reactor safety becomes an even greater concern.

In addition to these nuclear disasters there remains the challenge of radioactive waste disposal (see Chapter 12) and the ongoing threat of terrorist attacks on power plants and spent fuel storage. These concerns have led to cancellation of orders for new nuclear plants in the United States, and the plans to phase-out nuclear power in Germany and Japan. Other countries–Australia, Austria, Denmark, Ireland, Liechtenstein, Portugal, Israel, Norway, and others–remain opposed to any nuclear power.

11.4 Renewable Energy (LO4)

Renewable energy is the future if the world wants to kick the fossil carbon habit. Ironically, it was also the energy of the past, before fossil fuels came to dominate our energy supply. For millennia, hydropower drove mills along rivers as they cut wood and spun yarn. Windmills ground grain and pumped water. Sunlight dried food and warmed homes. To be considered renewable energy, the source of power is not measurably depleted when the energy is harvested. That holds for solar, wind, hydro, tidal and some geothermal energy sources.

11.4.1 Hydroelectric

Flowing water, our largest renewable resource next to wood, has been used as an energy source for thousands of years. First used to generate electricity in 1882 on the Fox River near Appleton, Wisconsin, falling water today provides a fourth of the world's electricity. Hydroelectric facilities are particularly useful for providing power during times of peak demand in areas where coal, oil, or nuclear generation provides the **base load** (the minimum amount of power that must be supplied to the electrical grid at a steady rate) Hydroelectricity can be turned on or off at will to provide peak power. Some smaller hydroelectric power stations can be run remotely, and even larger stations require far less staffing than a fossil fuel or nuclear power plant.

The principle is simple: impound water with a dam and then cause the water to move under the force of gravity through a system of turbines and generators to produce electricity (**Figure 11.86**). Because of this simplicity, and the fact that hydroelectric dams and the turbines within them can last for a century or more, electricity coming from existing facilities is one of the cheapest sources of power. Hydropower can generate a lot of energy where there is a lot of water and steep terrain. In Norway, 1,690 hydropower plants and 1,000 storage reservoirs account for 88% of the country's electricity consumption.

China's Three Gorges Dam on the Yangtze River is the world's largest producing electrical power station (**Figure 11.87**). At peak capacity it generates 22,500 megawatts of electricity (1 megawatt is enough electricity to power 400 to 900 homes for a year). The world's second-largest source of hydropower, the Itaipu Dam on the Paraná River between Paraguay and Brazil, completed in 1984, has the capacity to produce 14,000 megawatts of electricity. However, though the Itaipu Dam has a smaller generating capacity, seasonal variations in water availability in China means that both dams actually generate approximately the same amount of electricity per year!

Although hydroelectric energy is clean and has a high ROI, there are other costs, primarily to people and river ecosystems. Fish are perhaps the species most affected.

Imaginechina Limited/Alamy Stock Photo

Figure 11.87 The Three Gorges Dam was completed in 2004 after 12 years of construction. It is 185 meters (600 feet) tall and impounds water hundreds of kilometers upstream, drowning over a thousand known historical and archaeological sites of importance.

The Columbia and Snake Rivers provide an example of the impacts of large dams and the costs of trying to mitigate those impacts. The Columbia River watershed is home to more than 60 dams that together generate 36,000 megawatts of electricity every year. There are 14 dams on the Columbia River and 20 on the Snake River. Together, they have changed the rivers from a series of rapids where millions of salmon migrated upstream every year to a series of lakes. The impact on fisheries has been devastating, with the number of fish returning to spawn as well as the number of juvenile fish making it downstream far lower than before the dams were built.

The solution: fish ladders and fish trucking; yes, they truck fish around dams (**Figure 11.88**). They solve (in part) the problem of fish trying to get upstream to spawn but they do nothing to help the young fish get downstream. What used to be a journey along flowing waters is now interrupted by long, still lakes that force small fish to swim exhaustingly long distances. Add to that, changes in the temperature of river water (the water coming out of dams is from the bottom of the lake and therefore cold) and the sediment on the bed of the river (much of it is trapped in reservoirs) and the fish, which have evolved over millions of years to succeed in a specific river environment, now need to survive in waters that are very different.

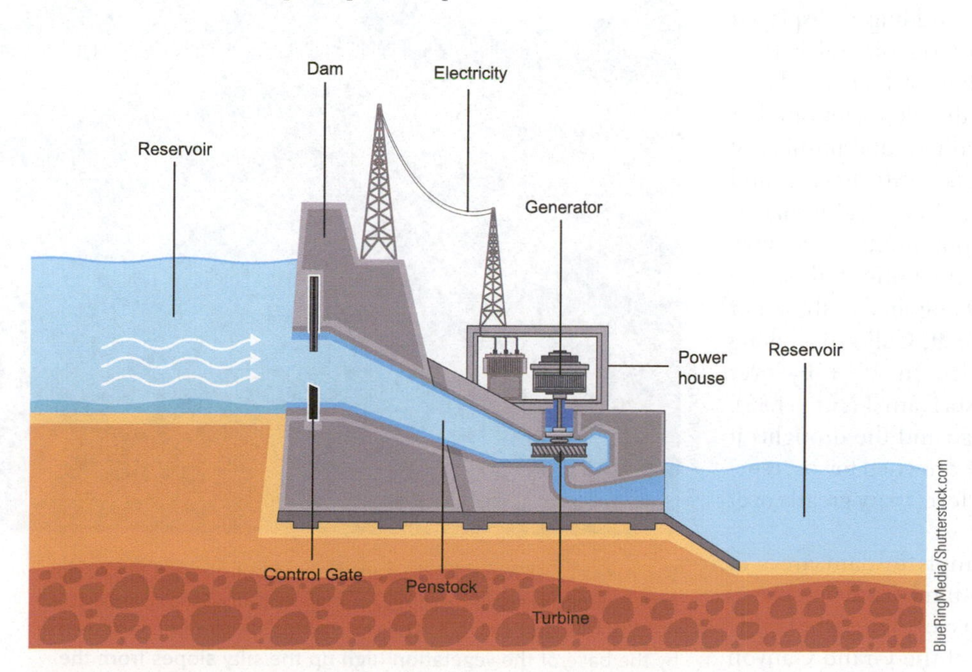

BlueRingMedia/Shutterstock.com

Figure 11.86 Schematic diagram of hydroelectric power plant.

iStock.com/JerryPDX

Figure 11.88 A fish ladder allows salmon to continue up the Columbia River, bypassing the Bonneville Dam.

US Bureau of Reclamation

Figure 11.89 Only the upper floor of Kennett's last buildings peek out over the rising waters of Shasta Lake in 1944. The town is now 400 feet below the reservoir's surface (when it's full).

One social consequence of dam building is displaced persons and lost towns and cities. As reservoirs fill behind newly completed dams, towns, emptied of their residents, flood. China's Three Gorges Dam set the record for number of people displaced (more than 1.2 million) and number of communities flooded (1,350 villages, 140 towns, and 13 cities). Across the United States, from California to Massachusetts, dozens of former communities are now underwater. Sometimes buildings were bulldozed before they were flooded; others were simply engulfed as the water rose. The former was the fate of Kennett, California, along the Sacramento River. It was flooded in 1944 by over 400 feet of water by impounded by Shasta Dam (**Figure 11.89**). In an odd twist of fate, climate change and the droughts it causes are now exposing some of these drowned towns, especially the arid regions where reservoir levels vary greatly over years and decades (**Figure 11.90**).

There have been repeated attempts to dam areas of great aesthetic value, such as the Grand Canyon. David Brower, who led the Sierra Club for years, fought valiantly (and successfully) to keep dams out of the Grand Canyon driven in large part by his failure to protect Glen Canyon, a less well-known site, from damming. Brower used the power of the press and advertising to sway public opinion. He took out a full page advertisement in the *New York Times*. The title read "Should We Also Flood the Sistine Chapel So Tourists Can Get Nearer the Ceiling?", a not-so-veiled dig at the Army Corps of Engineers claim that if the canyon were in part flooded by lakes (150 river kilometers, 92 river miles), tourists could get a better view of the walls. With climate change-induced drought leaving

Carmelo Alen/AFP/Getty Images

Figure 11.90 Aceredo, a Spanish town submerged below the Alto Lindoso reservoir in 1992, is high and dry (abandoned to all but the tourists) in 2022. The normal high-water line is shown by the base of the vegetation high up the silty slopes from the present reservoir shore.

Ian Dagnall/Alamy Stock Photo

Figure 11.91 Lake Mead's bathtub ring, or stain of mud plaster on steep slopes, left as the reservoir level has dropped. Parts of the Grand Canyon would have looked this way if it too had been dammed. It's not easy to see the rocks through the thick, light-colored mud.

T.O.K./Alamy Stock Photo

Figure 11.92 The Glines Canyon Dam blocked the Elwha River in Washington State from 1927 until 2011 when it was removed. The photograph taken in 2020 shows what's left of the dam. Upstream the sediment impounded by the dam has mostly revegetated, and the channel has cut though the sand and silt back to its previous depth.

Colorado River reservoirs almost dry, Brower's view to the future looks more and more reasonable. If those dams had been built, the canyon would have had some serious bathtub rings left by sediment-laden water that once filled the reservoir (**Figure 11.91**).

Because of land costs and environmental considerations, no large dams have been built in the United States for decades, and some dams are being removed as they come up for periodic relicensing. Dam removal brings its own set of challenges. Decades of sediment stored behind dams is released quickly; much of the easily eroded materials moves downstream in weeks to months. As the sediment moves downstream, it changes stream channels and can impact the ecosystems living in them. So far, however, monitoring of dam removal shows that ecosystems and the landscape recover well and quickly (**Figure 11.92**).

From an efficiency point of view, hydropower is very attractive, with EROI estimates ranging from 20:1 to 40:1. This places hydropower in a favorable position relative to conventional (fossil fuel) power-generation technologies, but there is limited opportunity for expanding hydropower, as the world's most favorable sites have already been developed. Those remaining, like much of the Mekong River basin in southeast Asia, could produce additional power but at the expense of ecosystem health and societies that have developed along the riverbanks.

11.4.2 Geothermal Energy

Earth's interior is an enormous reservoir of heat produced by the decay of natural radioactive elements that occur in all rocks. When the heat rises to shallower depths, often as magma, this **geothermal energy** can be accessed for human use (**Figure 11.93**). Geothermal energy can be used for electricity generation, space heating, and heat pumps. Unlike other renewable energies, such as solar or wind, geothermal can function 24 hours a day, 7 days a week. Geothermal is ideal for providing nearly zero-emission renewable base-load electricity.

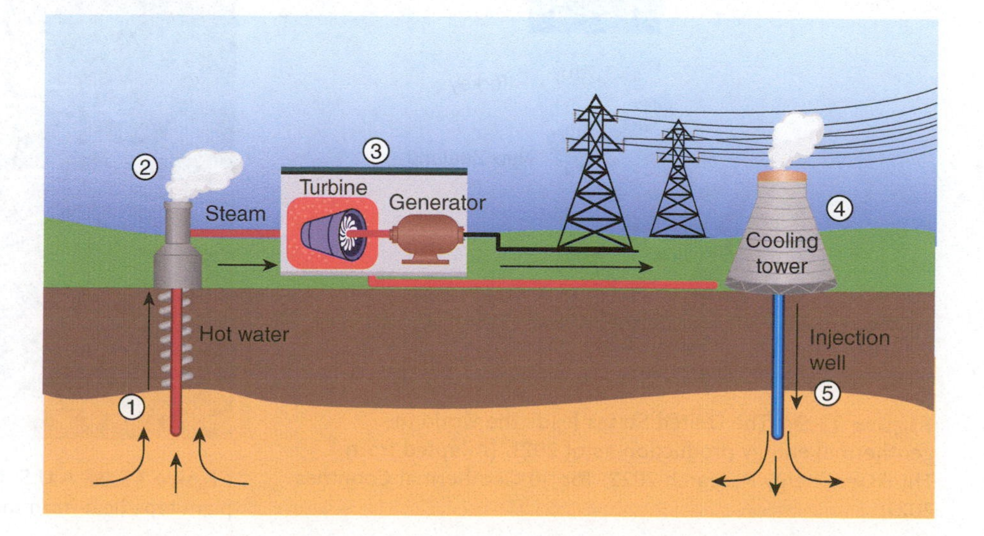

Figure 11.93 Schematic view of a geothermal electricity generation plant. (Adapted from U.S. EPA, Global Climate Change)

Electricity was first produced from geothermal energy in 1904 in Larderello, Italy, where the initial output illuminated four lightbulbs. Over 100 years later, geothermal energy production grew to nearly 95,000 gigawatt hours around the world. As of 2021, the United States represents 25% of the global online geothermal electricity generation capacity of 3.7 gigawatts, or a power generation over 17,000 gigawatt hours. This amounts to 1% of the U.S. energy capacity, with the potential to exceed 8% by 2050. Together, California and Nevada generate almost 95% of the United States' geothermal electricity. Geothermal operations contribute about 5% of California's electricity and contribute just under 10% of Nevada's electrical grid. Globally, geothermal energy is used commercially in over 80 countries, with Iceland, China, and Turkey among the leading geothermal generators (**Figure 11.94**).

Geothermal is cleaner than many other energy sources, and there are fewer negative environmental consequences of using it. Not surprisingly, the prospects for using geothermal energy are best at or near plate boundaries where active volcanoes and high heat flow are found. The Pacific Rim (the Ring of Fire), Iceland on the Mid-Atlantic Ridge, and the Mediterranean belt offer the most promise. Today, the United States, Japan, New Zealand, Mexico, and countries of the former USSR are utilizing energy from Earth's heat. Iceland uses geothermal energy directly for space heating (one geothermal well there actually began spewing lava), and many other countries have geothermal potential. The versatility of Earth heat ranges from using it to grow vegetables in the winter to driving steam-turbine generators (**Figure 11.95**).

Geothermal energy fields may be found and used for energy production where magma is present at shallow depths and where sufficient underground **hydrothermal** (hot-water) systems exist to form steam. These systems are classified as follows:

- *Hot-water hydrothermal systems*, with temperatures of about 200°C (390°F) or more, produce steam at sufficient pressure to drive turbine generators. A classic example is the four power plants operating at the Coso geothermal field in south-central California (**Figure 11.96**).
- *Vapor-dominated hydrothermal systems* are those in which the pore spaces in rocks of the high-temperature system are saturated with steam rather than liquid water. In this instance, only steam is produced and is routed directly to turbine generators. Vapor-dominated systems–the most

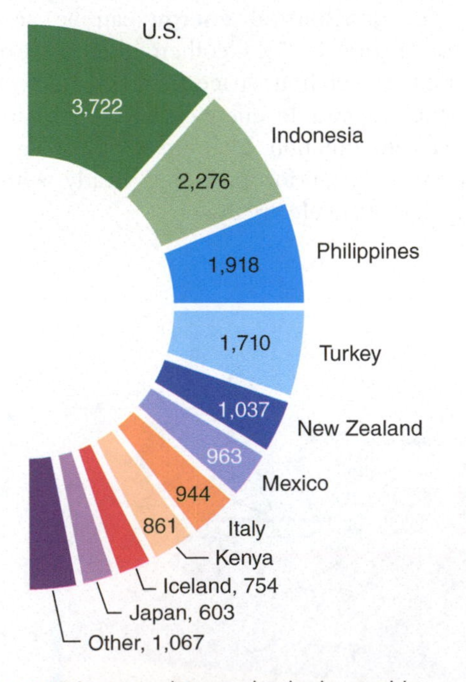

Figure 11.94 The United States leads the world in geothermal energy production as of 2021. (Adapted from ThinkGeoEnergy Research 2022, Top 10 Geothermal Countries 2021)

Figure 11.95 Geothermal heat warms greenhouses near Husavík, Iceland, where lettuce grows year round without the use of fossil fuels.

Figure 11.96 A U.S. Navy built and run geothermal power plant taps heat from the Coso Volcanic Field in California.

efficient, the simplest, and the most desired for electrical power production–are rare. The Geysers geothermal area, near Santa Rosa, California, with 21 operating power plants, supplies about 20% of California's renewable electrical energy and is the largest vapor-dominated system in the world (**Figure 11.97**). There's not sufficient water in the system, so water is brought in from outside and injected into the hot rock.

- *Moderate-temperature hydrothermal systems* are not capable of producing steam at high enough pressure to drive a turbine generator directly. They utilize geothermal heat to vaporize a second, "working" fluid (e.g., isobutene) whose boiling temperature is lower than that of water. The vapor from the second fluid drives the turbine generator. Such a system is called a **binary-cycle system** or, simply, a binary system. A successful binary-cycle system operates near Mammoth Lakes, California.

Geothermal energy is low-carbon energy. The amount of CO_2 emitted is about one-tenth that emitted by coal-fueled plants per megawatt of electricity produced, and about one-sixth that of relatively clean-burning natural gas. Similar contrasts exist for emissions of sulfur and sulfur oxides. The wastewater from most geothermal energy systems is injected back into the subsurface to help extend the useful life of the hydrothermal system or in Iceland, and some is used for a massive outdoor swimming pool and spa complex (**Figure 11.98**).

The problems associated with geothermal energy production are related to groundwater withdrawal, water quality, and the longevity of geothermal fields and fluid-handling equipment. Geothermal waters contain toxic elements such as arsenic, boron, and selenium, and heavy metals, such as gold, silver, and copper. In some geothermal areas, the removal of underground water can cause surface ground lowering. At the Geysers geothermal area in California, subsidence of 13 centimeters (5 inches) has been measured, without noticeable impact. Reinjecting the cooled wastewater into the geothermal reservoir decreases the subsidence risk and helps ensure a continued supply of steam to the electrical generators. Geothermal plants in California are monitored for possible seismic activity related to steam production because changes in pore pressure in the rock, as water and vapor are removed and reinjected, can cause small, old faults to slip. Small quakes have been reported near the Geysers geothermal area, but none of significant magnitude–the largest according to the United States Geological Survey is approximately a moment magnitude of 4.5.

Kim Steele/The Image Bank/Getty Images

Figure 11.97 Turbine run by geothermal energy at the Geysers is used to generate electricity.

iStock.com/Narvikk

Figure 11.98 The power plant from which waste water is piped to the Blue Lagoon tourist site in Iceland.

The future of geothermal energy looks promising. A 2019 U.S. Department of Energy report claimed that the generation of electricity by geothermal methods could increase 26-fold by 2050, producing 8.5% of the U.S. electricity. Accessing geothermal energy will be expensive, but could pay off substantially, especially with an estimated EROI of perhaps 7:1, it is competitive with photovoltaic and sugarcane ethanol. As of 2021, there are 23 geothermal fields in the United States that generate electricity, and 58 developing projects from currently active geothermal companies.

Geothermal energy does not need to come from hot magma. It can come, at shallow depths, from the Sun, which warms the air and the ground. A meter or two into the ground, the temperature of rock or soil is similar to the average annual air temperature and varies little over the year. That energy can be extracted for heating during the winter. In the summer, the ground functions as a heat sink. In this way, geothermal energy can help cool or heat buildings (**Figure 11.99**). To do so requires the burial of piping a few meters deep, where there is little seasonal temperature change through the year. By using a ground-sourced heat pump to circulate water or some other fluid through the

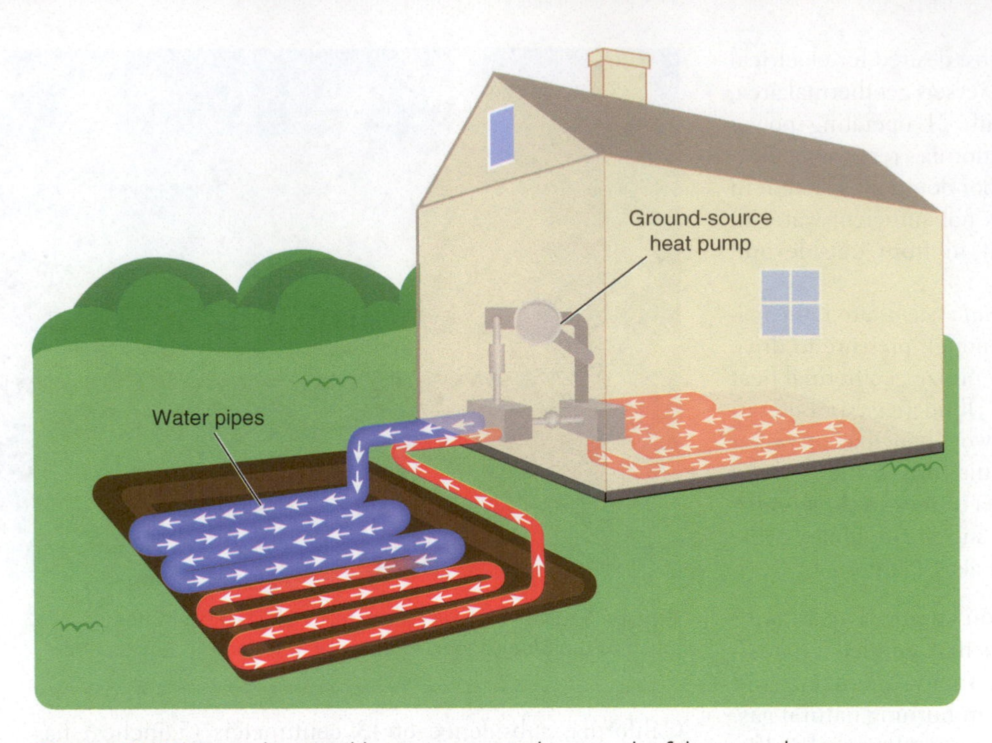

Figure 11.99 Ground-sourced heat pumps use the warmth of the ground to extract energy for heating.

piping, thermal energy can be transferred between the ground and the structure. A heat pump and Earth's thermal energy can effectively and economically provide a viable cooling and heating system. As of 2021, more than 50,000 geothermal heat pumps are installed in homes and buildings each year in the United States.

11.4.3 Wind Energy

Wind energy is a big success story in renewable power. Wind generates electricity when the wind turns rotors (windmills) attached to a turbine. Globally, wind power has grown dramatically since 2000 (**Figure 11.100**). China is the global leader in wind-energy development, with and installed capacity of 236 GW followed by the United States with 105 GW.

Wind energy investments in the United States have resulted in immense growth in wind energy generation. In 2000, total wind capacity from installed turbines was only 2.53 gigawatts across four states. By 2022, installed wind generation capacity increased by a factor of more than 55, with an installed wind capacity of 14 gigawatts. Globally, wind generation is the fastest-growing energy segment. The cost of wind power dropped significantly, and present production pales in comparison with what electrical wind generation will be in 20 years. Current estimates suggest that 35% of U.S. electricity will be generated from wind by 2050. The U.S. DOE estimates that current wind energy potential in the United States exceeds 10,640 gigawatts - enough to power over a billion people's homes. The wind potential of the lower 48 states, shown in **Figure 11.101**. is huge and mostly untapped.

The energy that can be tapped from the wind is proportional to the cube of the wind speed: a slight increase in wind speed yields a large increase in electricity produced. Longer blades requiring a higher tower yield greater production. Increasing the blade diameter from 10 meters (32.8 feet) (1980s technology) to 50 meters (164 feet) (2000 technology) gives a 55-fold increase in yearly electrical production. State-of-the-art "smart" wind turbines (2022 technology) may stand more than

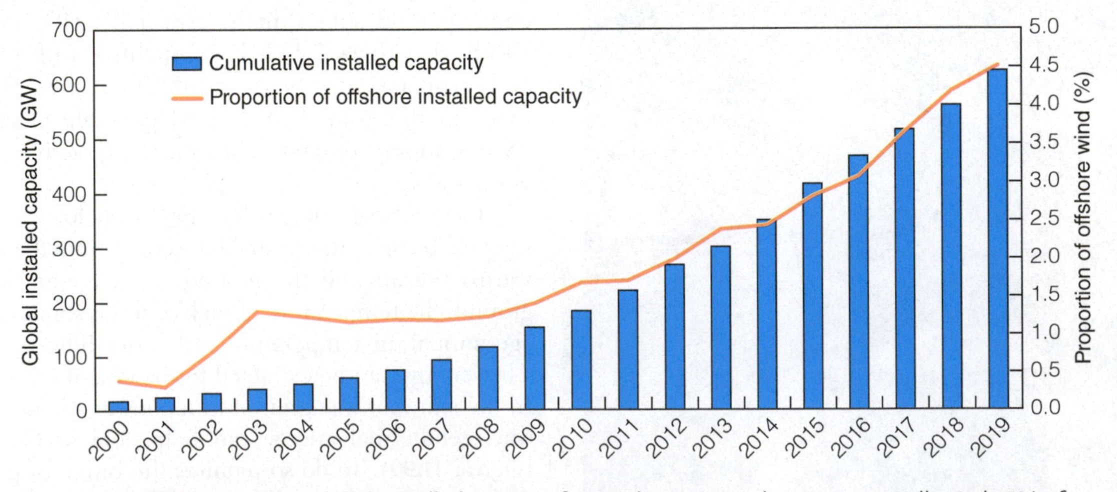

Figure 11.100 Since 2000, global installed capacity for wind generation has grown rapidly. Only 5% of wind generation is offshore. (Data from IRENA, 2020)

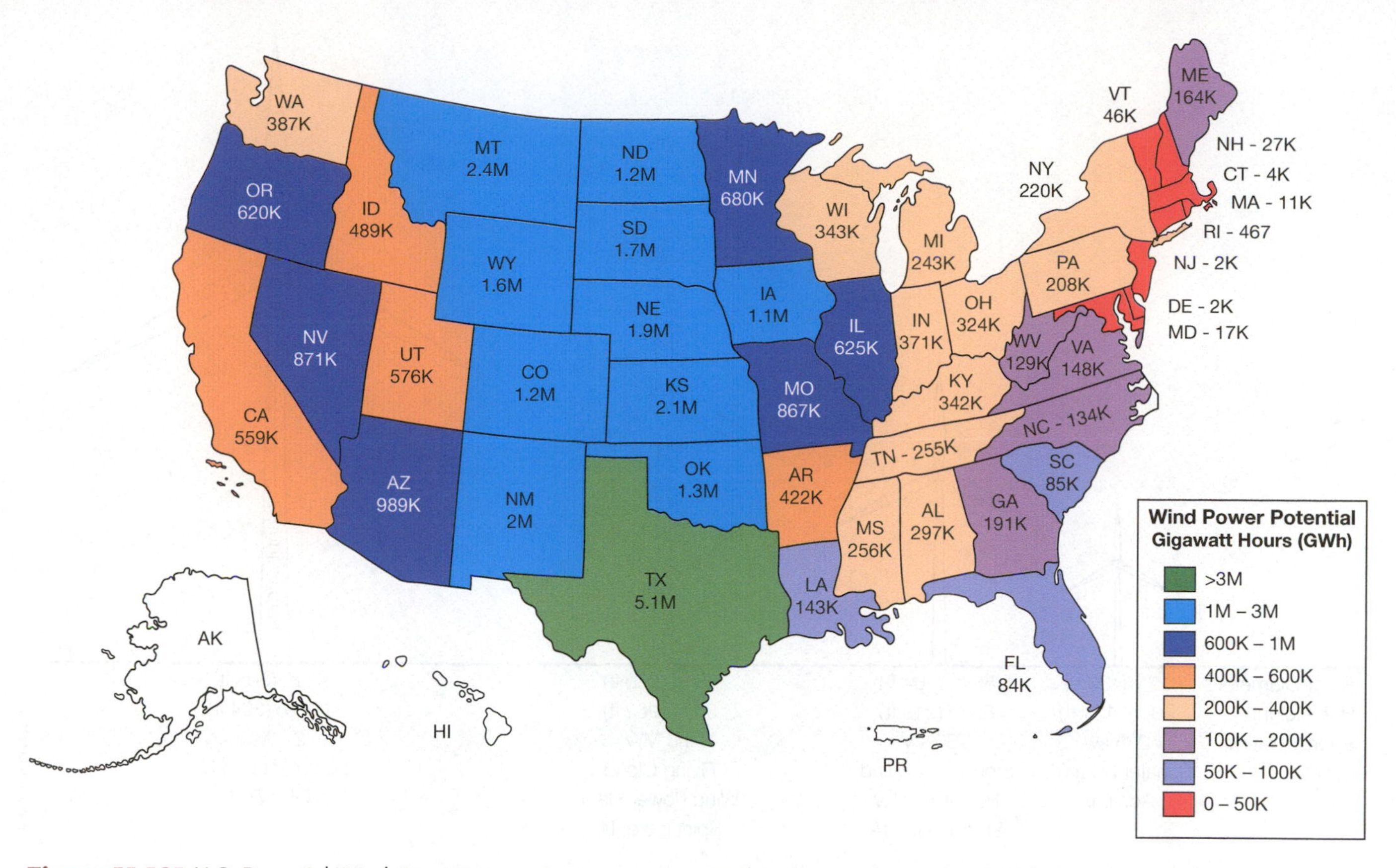

Figure 11.101 U.S. Potential Wind Generation in Gigawatt Hours (GWh). There is the potential for far more wind generation (at 80 meters above the ground, the height of a large turbine) than is currently installed in the United States. Texas, Montana, and Kansas lead the nation in potential power production. (Modified from US Office of Energy Efficiency and Renewable Energy)

120 meters (394 feet) tall, are rated at 2.3 megawatts each, and have rotor diameters that exceed the length of a football field (**Figure 11.102**).

The latest model turbines are programmed to adjust automatically for varying wind conditions, and they produce electricity at prices competitive with other sources (**Table 11.2**). Depending on location and other factors, the estimated EROI for wind power ranges from 10:1 to 30:1, and EROIs for windy regions, such as Denmark's islands, may be even higher. The direct costs of wind generation are a function of how much and how strong the wind blows, turbine design, and the diameter of the rotor. There are also indirect costs related to siting surveys and permitting which in many communities can be quite difficult and time consuming.

Until 2005, California led the United States in wind power, and the world's first large wind farms were constructed in the state at three locations: Tehachapi Pass, southeast of Bakersfield; San Gorgonio Pass, near Palm Springs; and Altamont Pass, east of San Jose (**Figure 11.103**). California is no longer the leading state for wind power. Texas (37,422 MW), Iowa (12,427 MW), Oklahoma (11,991.9 MW), Kansas (8,244.7 MW), and Illinois (7,037.2 MW) now surpass California (6,117.3 MW). These states combined, excluding California, generated 56% of the U.S. total wind electricity in 2021. The amount of power generated by wind keeps increasing in the United States (**Figure 11.104**) and wind power is generated in many states (**Figure 11.105**).

In 2021, California was second in the nation for total electricity generation from renewable energy sources, falling only behind Texas. California's negligible wind-power growth in the last few decades has been replaced by a greater prevalence of other renewable energy sources including solar, geothermal, and hydroelectric. Nevertheless, wind provided about 8% of California's energy needs in 2021. Wind accounts for greater than 20% of all energy used in Texas, and the state even exports wind power to other states. In 2021, Texas produced electricity from more than 40 wind farms and Texas ranchers are delighted with wind turbines. They can graze their cattle while receiving land-use royalties.

There are environmental issues that accompany wind generation, such as noise, land acquisition, TV interference, and most important, visual impact. Modern generators have largely solved the problem of noise and TV reception, but a couple of hundred windmills are hard to ignore. Some people feel that the presence of wind towers on ridgelines lowers scenic values; others find the towers attractive. Other environmental issues are that wind farms

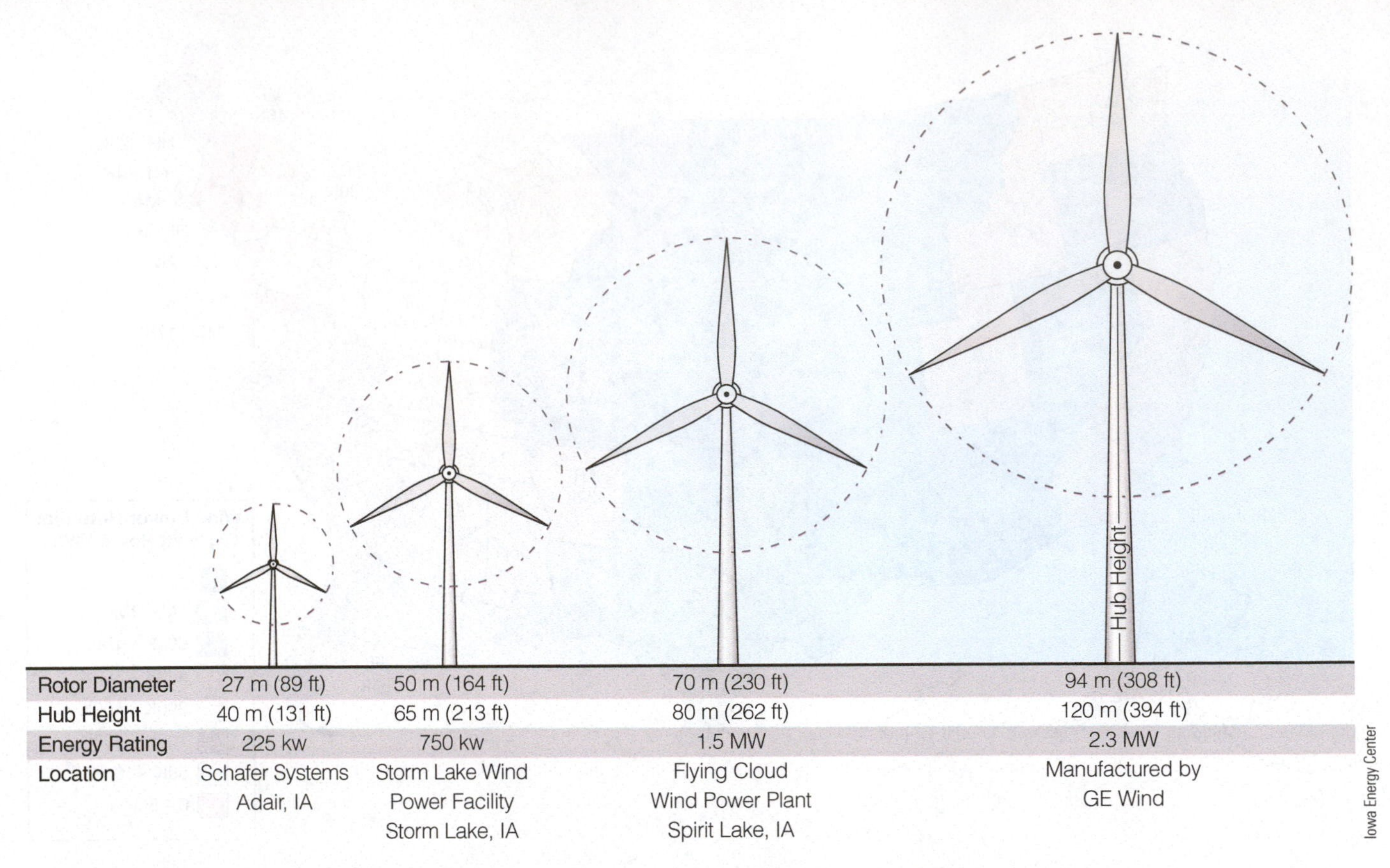

Figure 11.102 How big are wind turbines? Some, including rotor radii, are as tall as a 55-story skyscraper.

require road-building often in steep terrain or wildlands. If not done well, such roads can catalyze erosion. Constructing turbine pads disturbs large acreages and sometimes requires blasting.

Wildlife, too, is affected. Fences may block game migration routes, and bird deaths occur at wind farms, but at rates far less than urban areas where cats and tall buildings are deadly to birds. A study by the American Birding Association shows that an estimated 680,000 birds are killed by collision with wind turbines annually as of 2021. That sounds like a lot until you consider some estimates of bird deaths from collisions with buildings–which could kill as many as a billion birds annually. Bats, too, succumb to wind turbines, especially those species that migrate seasonally. Because bats are nocturnal, the easy solution is to feather the rotor blades and shut down the turbines at night during migration periods. Energy demands are lower at night, so the economic loss is not great.

Table 11.2 Cost of Electricity by Energy Source

Energy Source	Approximate Cost ($USD/MWh)
Offshore wind	104
Nuclear	61
Coal	52
Hydroelectric	47
Biomass	41
Onshore wind	30
Solar photovoltaic cells	27
Geothermal	22

Sources: Data from Levelized Costs of New Generation Resources in the Annual Energy Outlook 2022, U.S. Energy Information Administration.

Figure 11.103 Wind turbines at sunset on Altamont Pass in California (90 minutes east of San Francisco).

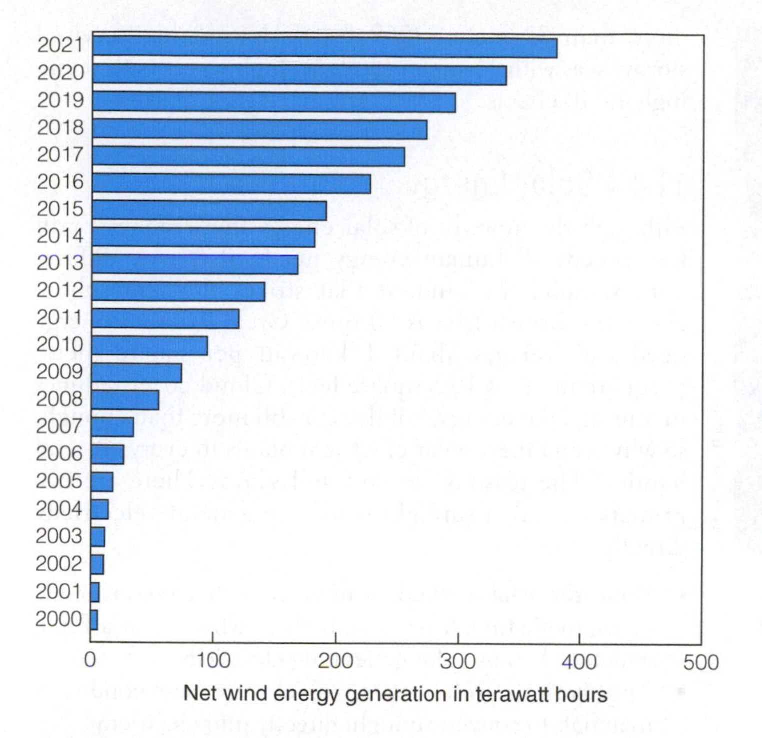

Figure 11.104 Wind energy generation in the United States grew rapidly between 2000 and 2021. (Data from Monthly Energy Review - March 2023, U.S. Energy Information Administration)

Offshore wind farms are not yet common but are a promising new source of wind energy. In April 2009, Interior Secretary Ken Salazar suggested that shallow-water offshore wind farms were the greatest opportunity for offshore wind energy in the United States and that the Atlantic Coast holds the potential for 1,000 gigawatts (GW) of electricity generation–one-quarter of national demand. He stated, "the wind potential off the coasts of the lower 48 states actually exceeds our entire U.S. energy demand."

Yet, there are few such offshore wind farms more than a decade later. Why? **NIMBY**–Not In My Back Yard–sentiment has blocked many coastal wind farms (**Figure 11.106**). People who own homes along the beach often suggest that wind turbines offshore will interfere with their views and lower property values. There have been numerous complaints from people who fish in coastal waters about danger to navigation and fishing gear from the turbine platforms. As a result, while Denmark has had coastal wind turbines since 1991, the first United States offshore wind farm (in Block Island Sound, off the coast of Connecticut) did not become operational until 2016. Many more projects are in different stages of development up and down the east coast, but only one is fully operational.

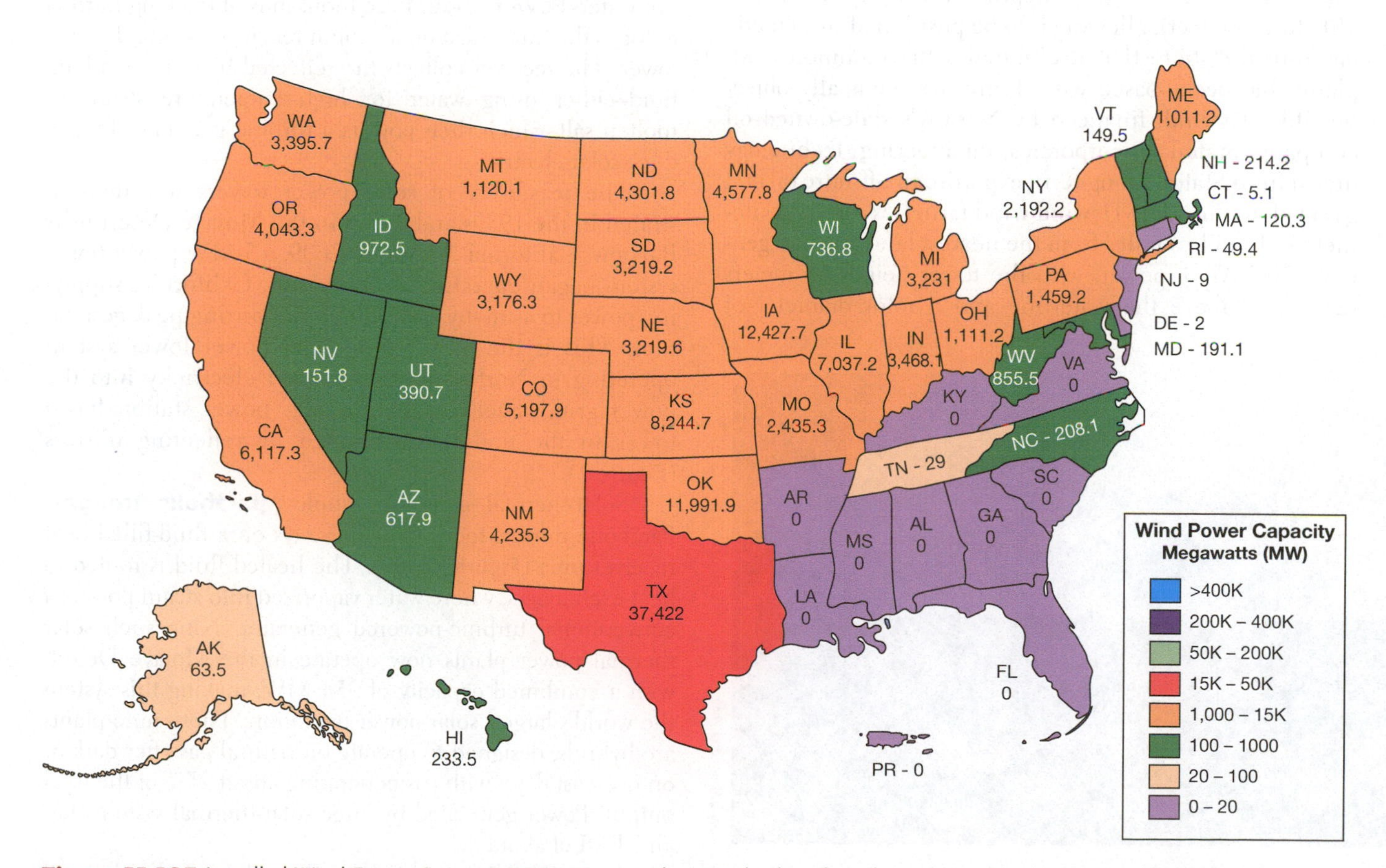

Figure 11.105 Installed Wind Power Capacity (MW). Texas is the state leader of wind energy in the U.S. Only nine states had no installed wind generation capacity in 2022. (Modified from US Office of Energy Efficiency and Renewable Energy)

Boston Globe/Getty Images

Figure 11.106 Cape Wind was a proposed wind farm off the southern coast of Cape Cod. It was never built, in part because of popular protests and complaints by influential landowners including the Kennedy family.

In 2017, the first deep-water, full-scale floating wind turbine farm, Hywind, went into operation off the coast of Scotland (**Figure 11.107**). A key factor of Hywind is that it can be anchored in water depths of 122 to 701 meters (400 to 2,300 feet), allowing it to be positioned sufficiently far from the shore that it eliminates the common complaint that ocean-based wind farms are scenically objectionable. Hywind, financed by Norway's state-owned oil company Statoil, incorporates engineering technology drawn from Statoil's long-term expertise in offshore oil and gas exploitation. The Hywind wind farm lies some 25 kilometers (15 miles) miles from the nearest land and can generate 30 MW of power, with the tower height 65 meters (213 feet) above the waterline and a rotor diameter of more than 82 meters (269 feet). Hywind has survived stormy seas with 11-meter (36-foot)-high waves with seemingly no ill effects.

Xinhua/Alamy Stock Photo

Figure 11.107 The world's first floating wind farm, in Scotland. It is operated by the Norwegian energy company Statoil and generates enough power for 20,000 homes.

11.4.4 Solar Energy

Although the amount of solar energy that reaches Earth far exceeds all human energy needs, it is very diffuse. For example, the amount that strikes the atmosphere above the British Isles is 80 times Great Britain's energy needs; it averages about 1 kilowatt per square meter (a square meter is 10.8 square feet). Cloud cover reduces incoming solar energy, but there is still more than enough, so why aren't there solar-electrical plants in every city and hamlet? The reasons are cost and space. There are two primary ways that sunlight is used to generate electricity directly.

- **Solar-thermal** methods, which use either power towers or parabolic-trough systems, both of which heat a fluid, which is then used for generating electricity
- **Photovoltaic (PV) cells**, which use semiconductor materials to convert sunlight directly into electricity

Solar thermal generation is primarily an industrial, large-scale process of power generation although some homes use solar thermal approaches for heating domestic hot water. **Power towers** use thousands of tracking mirrors to focus the Sun's heat on a central receiver mounted atop a tower. The receiver collects the reflected heat in a working fluid–either using water for high-temperature steam or molten salt–which then powers a turbine generator to produce solar electricity.

The feasibility of solar power towers was demonstrated in the 1980s and 1990s in the Mojave Desert near Barstow, California. In August 2009, a 5-MW power tower system began operating in Lancaster, California, supplying power to as many as 4,000 homes during peak generation. This is the only commercial power tower system operating in North America to feed electricity into the power grid. Israel's Aschalim solar power station has a tower in the middle surrounded by reflecting mirrors (**Figure 11.108**).

Solar-thermal can also employ **parabolic troughs**. Parabolic mirrors focus the Sun's rays on a fluid-filled heat receiver pipe (**Figure 11.109**). The heated fluid is routed to heat exchangers, where water vaporized into steam powers a conventional turbine-powered generator. Nine such solar thermal power plants now operate in the Mojave Desert, with a combined capacity of 354 MW, making this system the world's largest solar-power operation. These nine plants are hybrids, designed to operate on natural gas after dark or on overcast days, with gas generating about 25% of the total output. Power generated by large solar-thermal systems has an EROI of about 4:1.

As potentially attractive as these large solar-thermal systems are, they come with environmental baggage. Industrial-scale solar thermal installations take up large

iStock.com/Barbara Gabay

Figure 11.108 In the Negev Desert, near Beer-Sheva, Israel, the 260-meter (850-foot) tower of the Aschalim solar power station rises above the town.

iStock.com/Ollo

Figure 11.110 Large industrial building with a flat roof–now covered by solar panels that generate power without taking up otherwise useful land.

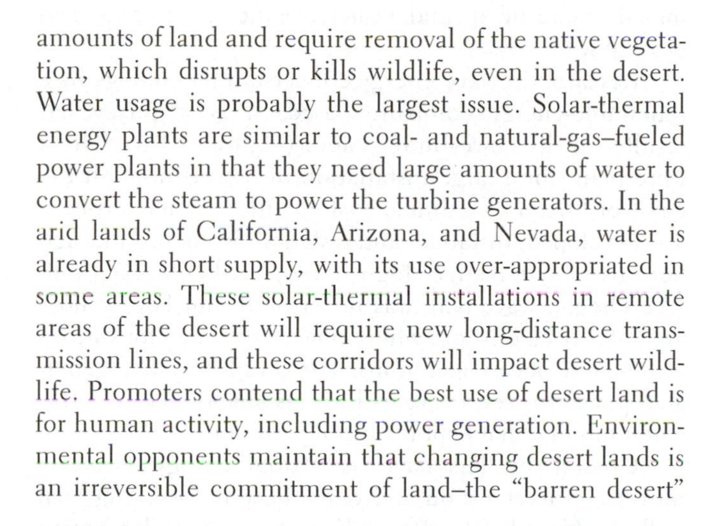

amounts of land and require removal of the native vegetation, which disrupts or kills wildlife, even in the desert. Water usage is probably the largest issue. Solar-thermal energy plants are similar to coal- and natural-gas–fueled power plants in that they need large amounts of water to convert the steam to power the turbine generators. In the arid lands of California, Arizona, and Nevada, water is already in short supply, with its use over-appropriated in some areas. These solar-thermal installations in remote areas of the desert will require new long-distance transmission lines, and these corridors will impact desert wildlife. Promoters contend that the best use of desert land is for human activity, including power generation. Environmental opponents maintain that changing desert lands is an irreversible commitment of land–the "barren desert" may seem lifeless, but it supports a complex and fragile ecosystem.

Solar PV is very different than solar thermal power production. It has no moving parts, requires no water, and is easily scalable from a home installation to large, industrial-scale generation (**Figure 11.110**). Light falling on semiconductor materials generates an electric current. Solar PV is used widely as the power source for satellites, calculators, watches, and remote communications and instrumentation systems. Millions of homes around the world have solar panels on their roofs (**Figure 11.111**). The cost of solar PV has plummeted, and with that cost drop, businesses and institutions such as universities and colleges have begun generating their own power, covering flat roofs and fields with solar panels. One of the best uses of solar panels is as shade structures in parking lots (**Figure 11.112**). The cars stay cool and the power flows.

William Warren/Alamy Stock Photo

Figure 11.109 Parabolic troughs at sunrise in Kramer Junction, California, deep in the Mojave Desert.

iStock.com/RoschetzkyIstockPhoto

Figure 11.111 New development in Austin, Texas, where many of the homes are equipped with solar panels.

Figure 11.112 Solar parking canopies are not just for the desert. This one in Stourton, Leeds, the United Kingdom, keeps cars dry and provides power to the grid even on cloudy, showery days like this one.

In 2020, the electricity produced from PV cells cost an average of about 16 cents per kWh for residential scale installations (the cost per kWh is less for commercial and utility scale systems)–still more costly than electricity produced by fossil fuels. The costs of installed PV systems and solar panels have been decreasing at about 15% per year, with the price of solar panels dropping 90% since 2010 because of increasing efficiency of solar cells and major improvements in manufacturing technology. PV electricity costs will continue to decline with technological improvements and PV electricity is already cheaper than electricity generated by fossil fuels and nuclear reactors in some countries (see Table 11.2).

PV cells are not environmentally benign. They use toxic elements such as gallium and cadmium in their manufacture. Although cells last decades, eventual disposal of used cells could become a significant environmental problem unless the panels are properly recycled. The early PV cells utilized silicon as the energy converter, but cadmium telluride (CdTe), copper indium gallium selenide (CIGS), and other, newer types of PV cells are more efficient and have an EROI of about 7:1.

11.4.5 Energy from the Sea

As the need to decarbonize our energy supply becomes more urgent, scientists, engineers, and entrepreneurs are searching for new technologies to achieve clean, renewable energy. One possible source of renewable energy is one of Earth's largest: its oceans. **Marine renewable energy** can include **wave energy** and **tidal energy** as well as ocean thermal energy conversion. Because these new technologies are in the research and early developmental stages, EROIs cannot be estimated yet.

The U.S. Electric Power Research Institute estimates that marine renewable energy in the United States offers an energy-generation potential equal to that currently generated in the United States by hydropower. Capturing marine energy and converting it to electricity is not a new idea, but the technology faces challenges. Various technologies are currently being tested and developed but few have come to market yet. The marine environment is aggressive–marine organisms grow on submerged parts of power plants and cause fouling, intense storms can damage generating facilities, and seawater is corrosive. Marine energy projects experience environmental, policy, and permitting issues.

There have been many attempts to harvest **wave energy,** but none have yet succeeded at commercial scale. The idea is simple in theory but complicated in practice. The rising and falling of waves causes a floating buoy to move freely up and down. The resultant mechanical motion is converted to electricity by an electrical generator, and the power is transmitted ashore by an underwater electric cable or used in place at sea.

Research in wave energy conversion technology by a renewable-energy company based in New Jersey has resulted in its PowerBuoy system (**Figure 11.113**). There were plans for a large installation of PowerBuoys off the coast of Oregon. A similar system (the Sotenäs Project) was developed, installed, and tested in Sweden. Although both these systems had high hopes, neither worked out. The Oregon wave park was never built, but PowerBuoys are today used offshore where power is needed for various stand-alone ocean operations. The Swedish system generated power for several years but was abandoned in 2017, when funding that kept it running ran out.

Coastal dwellers have used **tidal energy** to power mills for at least a thousand years. Restored working tidal mills in New England and Europe are popular tourist

Figure 11.113 PowerBuoy ready for offshore installation.

attractions (**Figure 11.114**). Both flood (landward-moving) and ebb (seaward-moving) tidal currents are used for power. It is possible that as much as 10% of U.S. electricity could be supplied eventually by tidal power, a potential equaled in the United Kingdom and surpassed at some coastal sites, such as Canada's Bay of Fundy. Compared with solar and wind energy, tidal-energy generation has the advantage of predictability. Tides are reliable, being driven by the gravitational pull of the Sun and Moon, as opposed to the weather, and their timing and strength can be determined in advance with a high level of certainty. Cosmetically, the underwater machinery is less obtrusive than wind turbines.

There are two basic types of tidal-energy-generating technology. The first type uses a **barrage** (a damlike barrier) on a bay or in an estuary with a large tidal range–the height difference between high and low tides–with a large area behind the barrier for storing water at the elevation of the high tide. Power is usually generated during the ebb (outgoing) tide, as the barrage creates a substantial head of water, much as a hydroelectric dam does. The second type uses currents and requires no damming or interruption of water movement.

There are environmental impacts of harvesting tidal energy, particularly when a barrage is needed. Damming the estuary changes circulation patterns and often traps silt and sand coming from upstream, eventually filling the impounded area of the estuary. Marine ecosystems are affected by longer periods of standing water upstream of the barrage.

The first and best-known large-scale tidal power plant is on the Rance River near St. Malo on the English Channel in northwestern France (**Figure 11.115**). The tidal range there varies from 9 to 14 meters (30 to 40 feet), and the plant generates power at both the flood and ebb tides. Twenty-four 10-MW generators with a total capacity of 240 MW are used. The plant has been in operation since 1966.

Hemis/Alamy Stock Photo

Figure 11.115 The tidal power station on the Rance River, Ille et Vilaine, Saint Malo, France.

Annapolis Royal in Nova Scotia, Canada, is the only modern tidal-generating plant in North America (**Figure 11.116**). It generates electricity to support 4,500 homes. An island in the mouth of the Annapolis River was selected as the site for the powerhouse, which opened in 1984 after 4 years of construction. It is an ideal location, because the Bay of Fundy has the highest tides in the world, and there was an existing causeway on the river with sluice

iStock.com/RolfSt

Figure 11.114 Tidal mill on Ile de Brehat, Saint-Brieuc, Brittany, France. As the tide rises, it flows right to left filling the estuary behind the dam with water. When the tide falls, water pours through the dam, powering the mill.

Ron Garnett/All Canada Photos/Alamy Stock Photo

Figure 11.116 As the tide falls, water pours back into the Atlantic Ocean, generating power at Annapolis Royal, tidal power generating station in Nova Scotia, Canada.

gates that dammed the river. On the incoming tide, the sluice gates are opened and the incoming seawater pours upriver and fills the pond. The sluice gates are then closed, which traps the seawater upstream of the generating turbine. As the tide recedes, the level of the water below the pond drops and a hydraulic head develops. When the difference is 1.6 meters (5 feet) or more, the gates open and water flows through the turbine to generate electricity.

Fast-moving marine currents caused by tidal action, termed **hydrocurrents**, can also be used for tidal energy generation. Shallow or narrow constrictions that restrict the tidal flow produce the fastest and most powerful tidal currents, from which energy is extracted using submerged turbine generators or other designs. The submerged turbines are very much like windmills, but because water is more than 800 times denser than air, the blades can be quite small and more compact, so that a relatively small device can create a relatively large amount of energy. Because tides vary in time and intensity from day to day, but do so in a predictable manner, tidal power can be integrated more easily into the local power grid.

A recently concluded advanced demonstration of hydrocurrent power generation was Verdant Power's Free Flow Kinetic Hydropower System, in New York City's East River, which is actually a tidal channel. In 2006, five turbines were moored to the riverbed in 10 meters (30 feet) of water and supplied power to the grid. Each turbine had a 5-meter (16.4-foot) diameter rotor turning at 32 rpm. In January 2012, Verdant Power was given the first commercial license from the Federal Energy Regulatory Commission to install up to 30 MW of permanent of tidal power in the East River. In 2020 and 2021, a three-turbine 630 kW system was installed and operated, providing 312 MWh to the grid in 9 months (**Figure 11.117**)

Figure 11.117 Three Verdant turbines await installation in the East River in October 2020.

Ocean thermal energy conversion (OTEC) utilizes the temperature difference between warm surface water and colder deep waters. OTEC relies on heat pump technology that uses a working fluid (sometimes ammonia), which passes between gas and liquid phases to extract energy with a turbine that drives a generator (**Figure 11.118**). The ideal OTEC location is a coastal area where water depth increases rapidly enough to achieve the appropriate temperature differential and close enough to shore for power transmission.

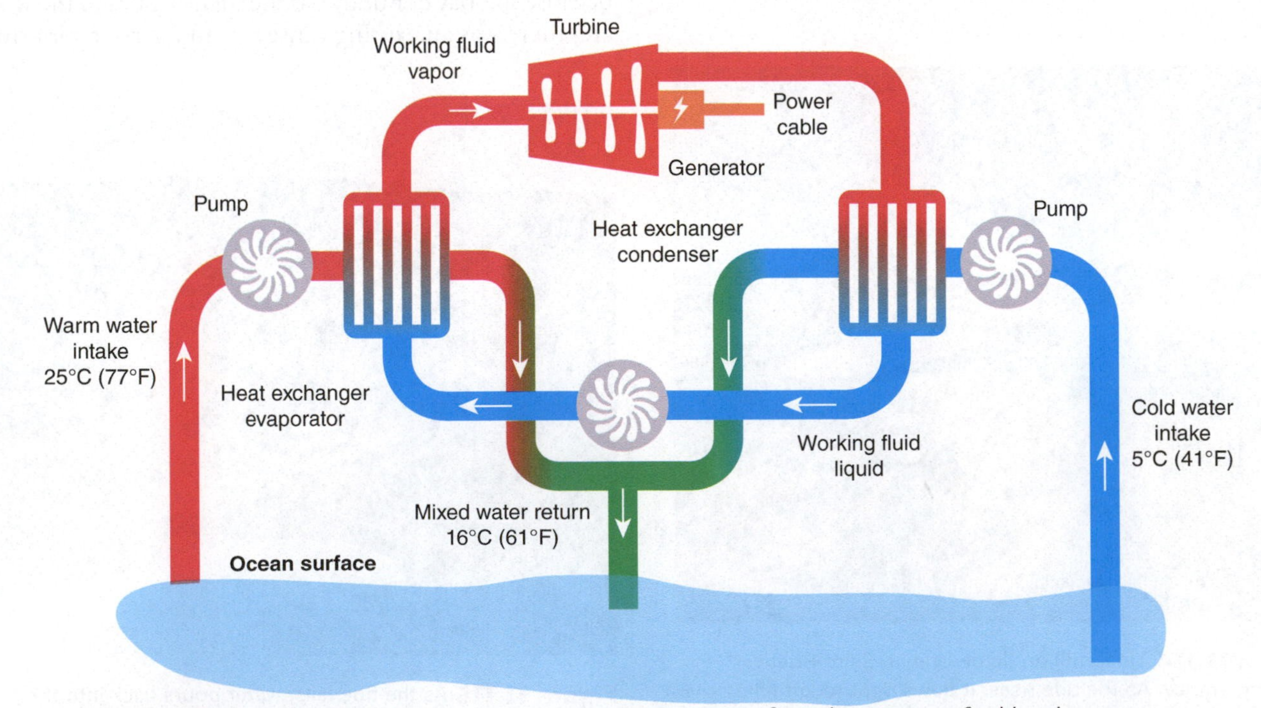

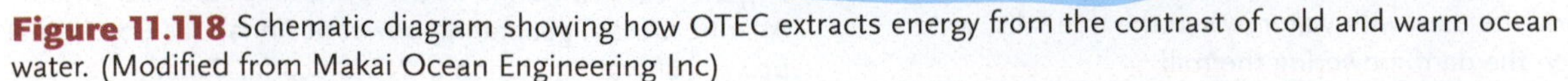

Figure 11.118 Schematic diagram showing how OTEC extracts energy from the contrast of cold and warm ocean water. (Modified from Makai Ocean Engineering Inc)

Design Pics Inc/Alamy Stock Photo

Figure 11.119 Otec-Ocean Thermal Energy Conversion open-cycle plant at Kona, Hawaii.

The Hawaiian Islands have the population, need, and oceanographic conditions for this form of energy conversion. The temperature differential between the surface water and the deep ocean is 20°C (36°F), ideal conditions for OTEC. In 1979, scientists at the National Energy Laboratory of Hawaiʻi Authority (NELHA), located at Keahole Point near Kailua on the Big Island (Hawaiʻi), successfully produced 10 to 15 kW of net electrical power by OTEC.

Since then, the NELHA has experimented with a so-called open-cycle OTEC system, which not only can produce 100 kW electricity but also produces 600,000 gallons of pure freshwater per day as a byproduct of the process (**Figure 11.119**). The freshwater is condensate from the chilling part of the cycle and is a valuable commodity on the arid western side of Hawaiʻi Island.

In January 2022, Mark Betancourt, writing in *EOS*, a weekly newsletter for geologists, was optimistic about OTEC:

> "The theoretical potential of OTEC is vast. It could produce at least 2,000 gigawatts globally, rivaling the combined capacity of all the world's coal power plants on their best day. And unlike many renewables, OTEC is a baseload source, which means it can run 24/7 with no fluctuation in output. But the conditions necessary to make the process viable–at least a 20°C difference between surface and deep water–occur only near the equator, far from most of the world's power demand and much of its wealth."

Large parts of the global south, central America, and ocean island nations have the deep/shallow water temperature difference sufficient that they will benefit greatly from this energy technology as it matures (**Figure 11.120**).

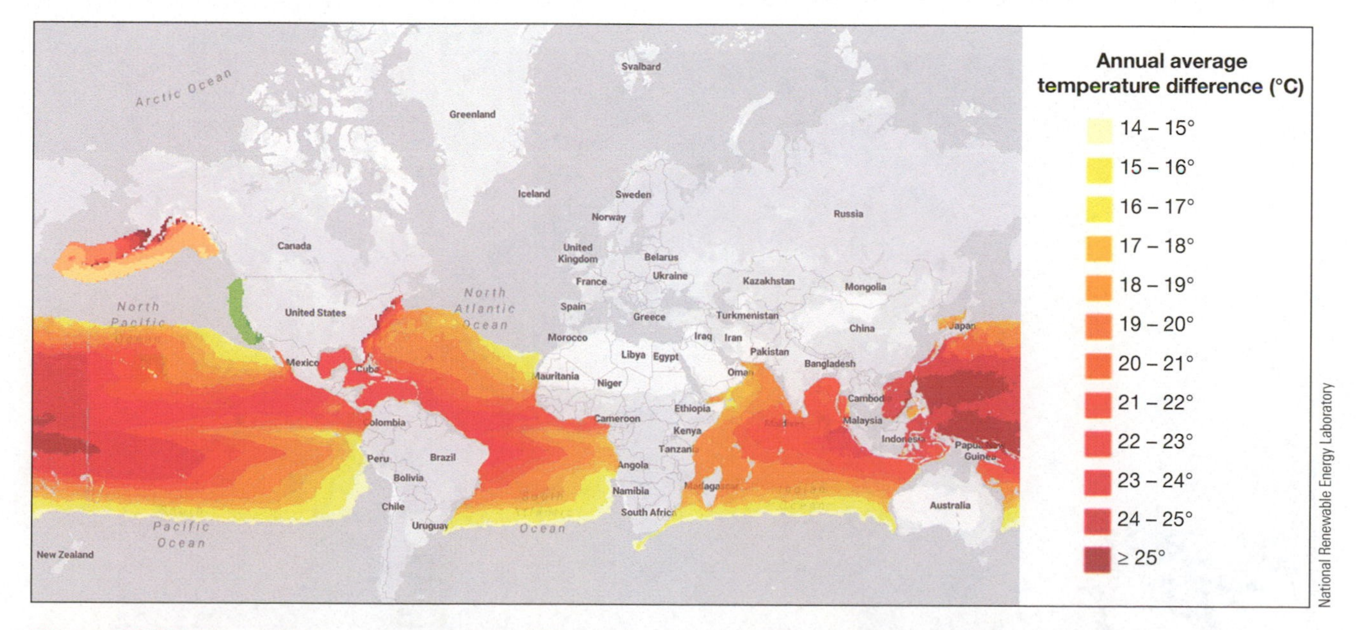

National Renewable Energy Laboratory

Figure 11.120 The redder areas have a greater difference between the surface and deep sea water temperature. A 20°C difference is optimal for OTEC power generation. The tropics of the western Pacific and the east coast of South America are well positioned for this form of energy development.

11.4.6 Intermittency

Most renewable energy has the advantage of not requiring the combustion of fossil fuels, but it has the disadvantage of not being a consistent generation source. In other words, sometimes it produces power and sometimes it does not. This phenomenon, termed **intermittency**, is a challenge that utilities have tried in various ways to overcome. The bottom line is that the power grid must supply exactly the amount of power people demand at any one time–thus, generation needs to be flexible.

To address intermittency, some utilities have added large sets of batteries that store power when demand is low and release power when demand is high. As battery technology improves and cost drops, battery storage is increasing across much of the United States (**Figure 11.121**). Such energy storage is often paired with solar and wind installations to buffer their inherently variable outputs of energy (**Figure 11.122**). As of 2019, there were 163 large-scale battery systems in the United States. Together, they could store 1,688 megawatt hours of electricity.

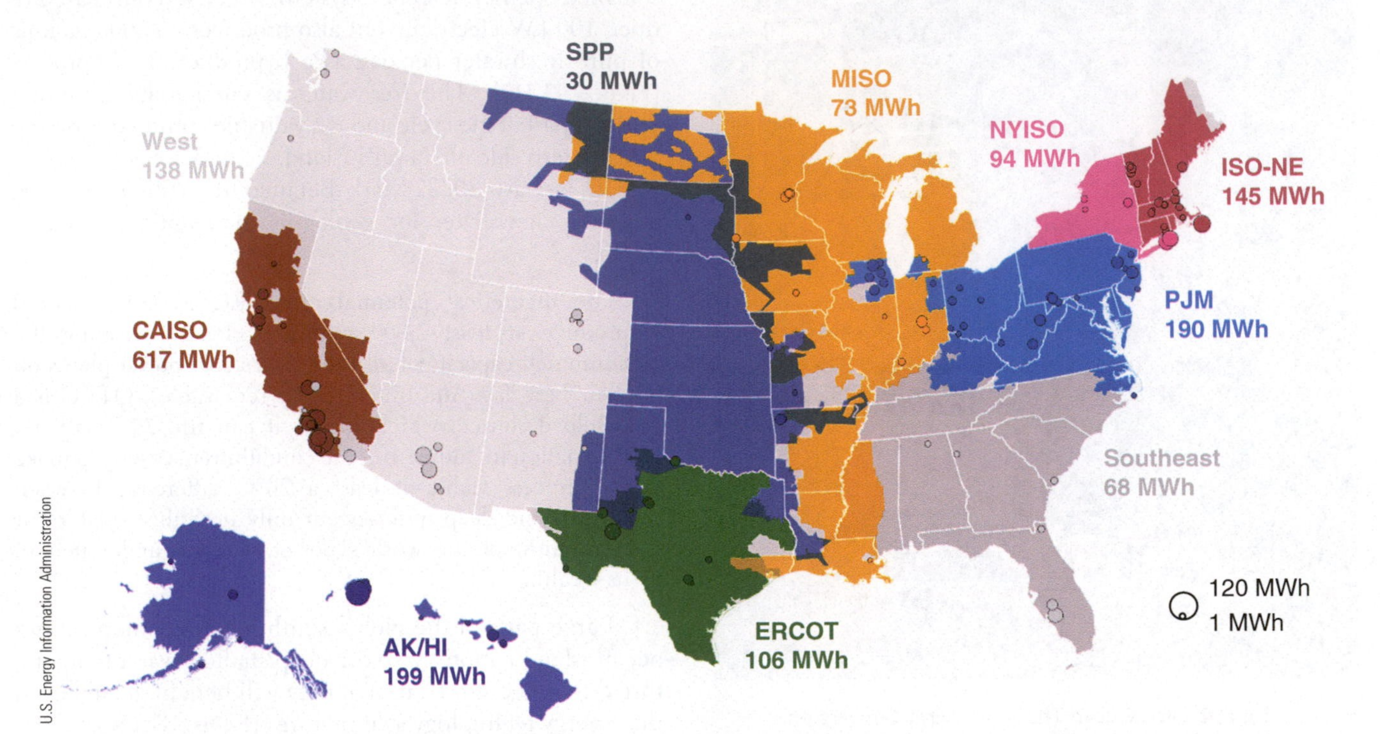

Figure 11.121 Battery storage capacity for the energy grid as of 2019. Storage capacity shown in Megawatthours (MWh).

Figure 11.122 Battery storage paired with intermittent wind and solar generation makes sense.

Another approach to grid-scale energy storage involves moving water–the creation of what are termed **pumped-water-storage facilities** or **water batteries**. During off-peak hours, when plenty of power is available, water is pumped from an aqueduct or another source to a reservoir at a higher elevation (**Figure 11.123**). During peak demand, the water is allowed to fall to its original level, fulfilling the temporary need for added electricity by generating hydroelectric power. This is not a new concept. The first water battery was built in New Milford, Connecticut, in 1930. As of 2021, 18 states have water batteries in operation. California, Virginia, and South Carolina have the largest capacity, and three other states will come online soon.

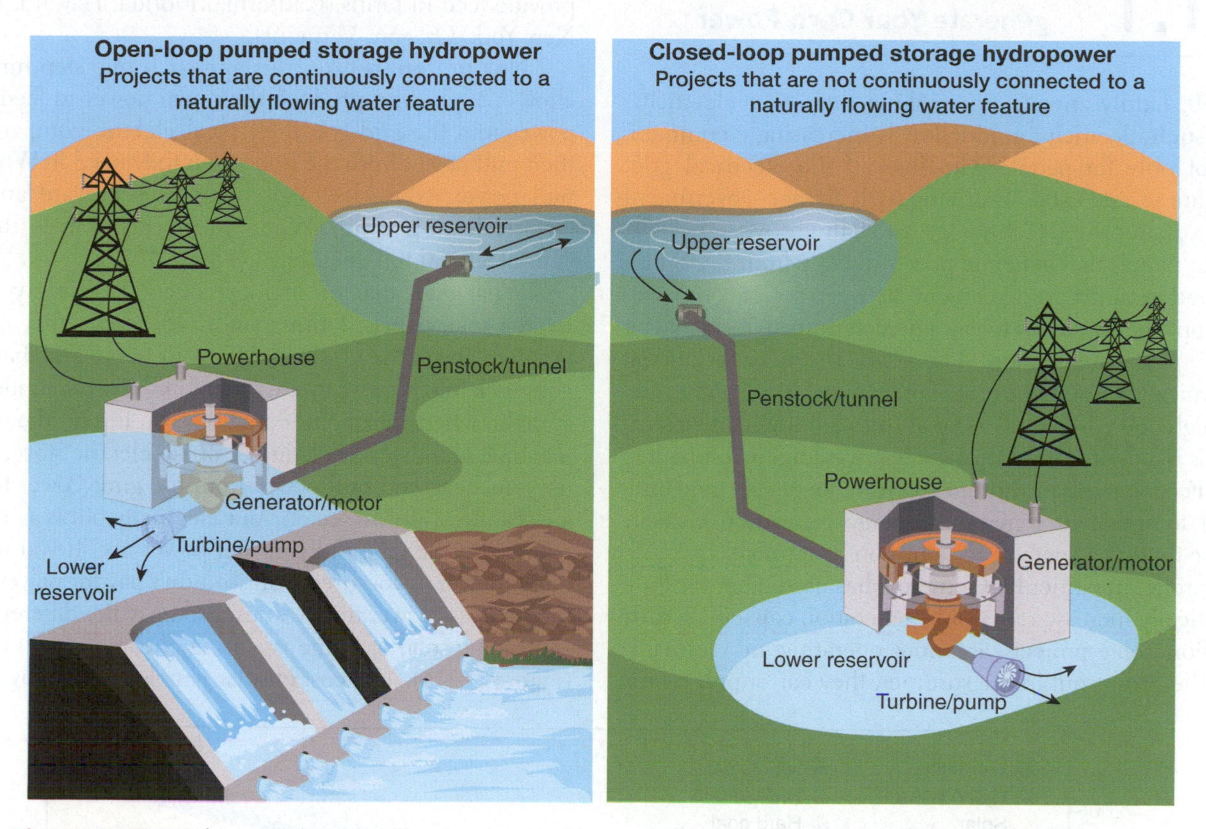

Figure 11.123 Schematic diagram of pumped storage between two reservoirs (closed loop) and between a river and reservoir (open loop). (Modified from U.S. Department of Energy, Pumped Storage Hydropower, Water Power Technologies)

Case Study

11.1 Making It Possible to generate Your Own Power

In 1950, hardly anyone was making their own electricity; today, such electricity production is increasingly common, and not only for people who live off the electrical grid. There are several catalysts driving such change–government incentives, improved technology, and falling costs–especially for solar panels that generate power directly from the Sun.

Feed in tariffs are agreements between power companies and people or businesses to pay a predetermined, fixed rate for "green" power production. The payment for each kilowatt hour typically exceeds the cost of other generation sources and is set high enough to send a signal (through the marketplace) that this is a desired (and financially rewarding) investment to make. Feed in tariffs provide an incentive for people (and businesses) to invest in renewable energy generation capacity because there is a guaranteed return on investment.

Europe, in particular Germany, has used these tariffs to dramatically increase renewable generation capacity. Nearly 2 million solar projects are now operating in Germany; when the Sun is out in the summer, they can supply nearly 40% of the country's electrical needs (**Figure 11.124**). The approach is less used in the United States; only seven states provide feed in tariffs (California, Florida, Hawaiʻi, Maine, New York, Oregon, Vermont).

Net metering takes this approach one step further. It allows people who produce their own power to feed excess power into the grid (via their local electric company) and then pull power from the grid when they need it. When you pull power, your electrical meter runs forward and your electrical bill goes up. When you feed power to the grid, your electrical bill goes down (**Figure 11.125**).

What does that mean in practice? Let's say you have rooftop solar panels. During sunny afternoons they produce far more power than you need for your home and that power goes back into the grid to run other folks' air conditioners. But at night, when your lights come on, you flip on the washing machine and dryer, cook dinner on an electric stove, or turn up your heat, you pull power from the gird. Power transfers are seamless–all you see on your electricity bill is a charge to connect to the grid and your net power use. If you made as much power as you used, you wouldn't pay for power. If you produce less power than you use, you pay the difference.

Going solar at home is getting cheaper–much cheaper (**Figure 11.126**). The cost of solar panels dropped by a factor

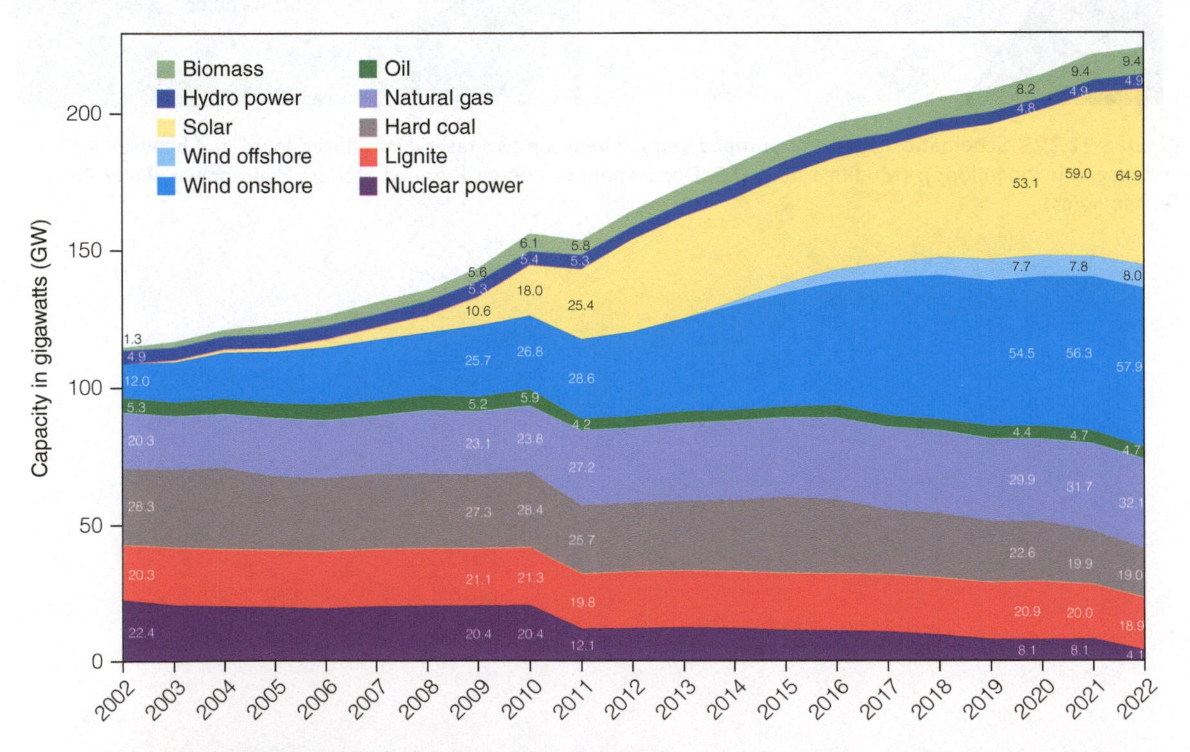

Figure 11.124 Changes in public policy encouraged the development of renewable energy sources in Germany starting in the early 2000s. The results are very clear–public policy can change energy generation strategies. Solar and wind power increased dramatically. Fossil fuel derived power declined. (Adapted from Clean Energy Wire (2023), Installed net power generation capacity in Germany 2002-2022)

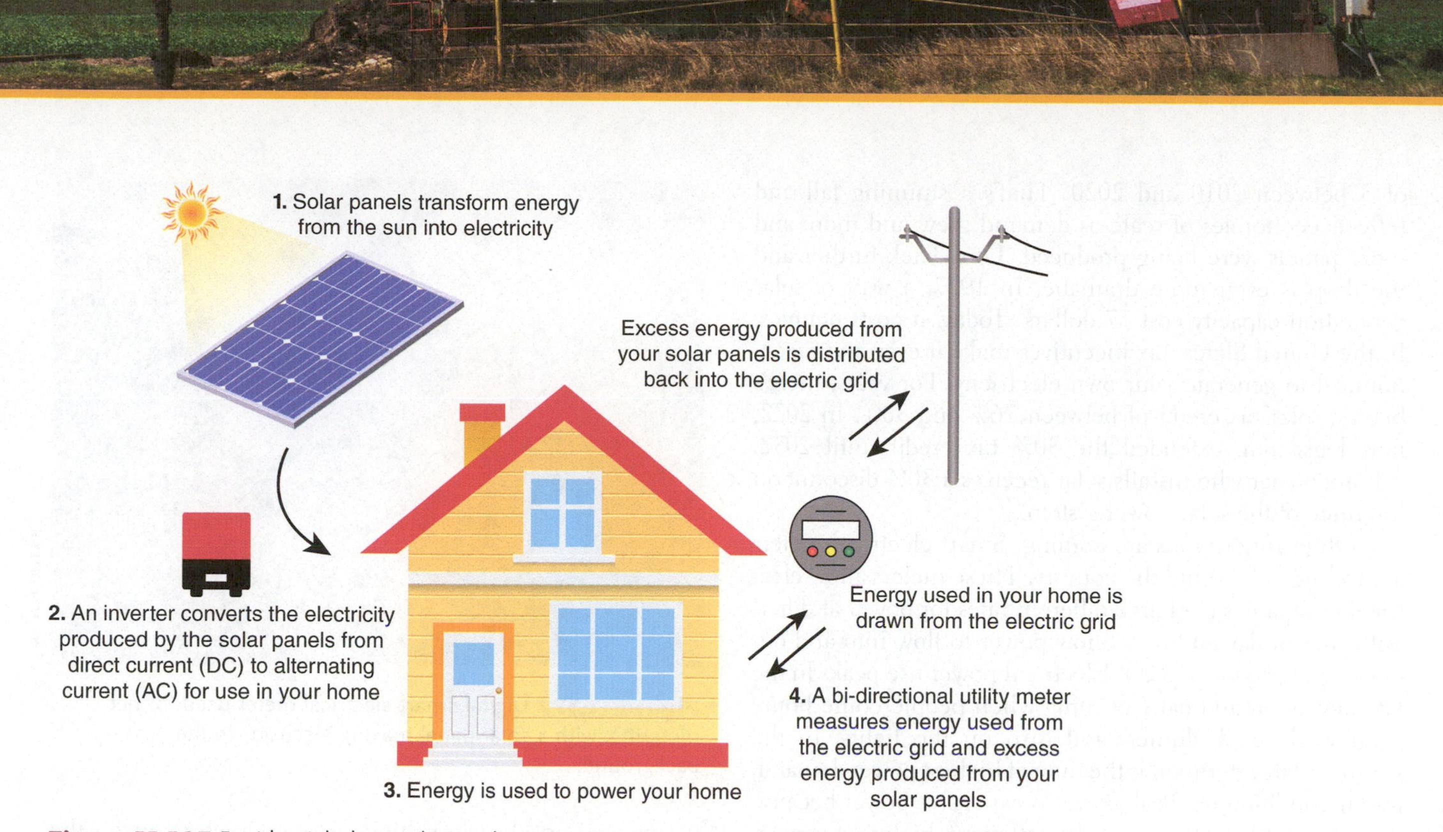

Figure 11.125 Residential photovoltaic solar generation is an example of how net metering works. (Modified from State of Vermont Public Utility Commission, Net-Metering)

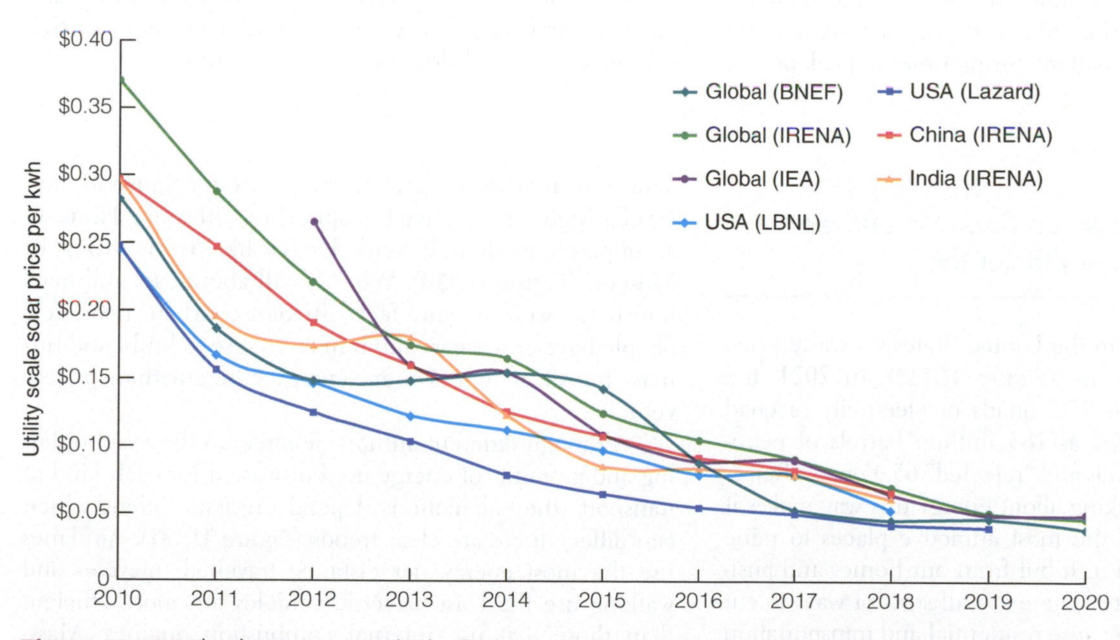

Figure 11.126 The drop in cost per kWh of solar power is stunning. These data are for large-scale generation from seven different estimates. Home solar has shown a similar drop in cost. Most important perhaps is that, starting about 2015, solar power was as cheap as power from burning fossil fuels!

of 5 between 2010 and 2020. That's a stunning fall and reflects economies of scale as demand grew and more and more panels were being produced. Look back further and the drop is even more dramatic. In 1977, a watt of solar generation capacity cost 77 dollars. Today, it costs pennies. In the United States, tax incentives make it even more economical to generate your own electricity. For years, there's been a solar tax credit of between 26% and 30%. In 2022, new legislation extended the 30% tax credit until 2032. A homeowner who installs solar receives a 30% discount on the price of the solar power system.

Other innovations are coming. Smart electrical meters are rolling out around the country. These meters allow electrical companies to charge different rates for power at different times of day and they allow power to flow into and out of the grid (**Figure 11.127**). Electrical power use peaks in the late afternoon and early evening when people come home from work, cook dinner, and turn on the lights. In the summer, late afternoon is the time of highest power demand, for air conditioning. Peak power is expensive power because it requires utilities to activate less-efficient, high-cost generation systems many of which emit large amounts of carbon dioxide and other pollutants.

There are various ways of cutting the peak load. Smart meters can help. Some utilities subsidize home batteries with an agreement that the utility can use your smart meter to pull power from your battery during times of peak power.

iStock.com/onurdongel

Figure 11.127 Digital smart electrical meter used for net metering, with a solar panel making electrons in the background.

If you own an electric vehicle, a smart meter can do the same. This is a win-win-win. You get financial incentives, the power utility gets peak power, and dirty, peaking power sources are replaced with cleaner power.

Decarbonizing our electrical system is not only possible. It's happening; not yet at the rate it needs to so that climate change will slow, but we are on our way.

11.2 How You Can Use Energy More Efficiently

It's hard to believe, but in the United States we waste twice as much energy as we use (**Figure 11.128**). In 2021, the United States generated 97.3 quads of electricity (a quad contains as much energy as 183 million barrels of petroleum). We used 31.8 quads and "rejected" 65.4 quads. Rejection means waste. Thinking about energy this way makes it easy to see that one of the most attractive places to mine energy is not from the Earth but from our homes and businesses by reducing waste. There are all sorts of ways to cut energy consumption. Because residential and transportation make up nearly half of U.S. energy use, they are a great place to start considering how to lower your energy footprint and save some money (**Figure 11.129**).

Where you live matters. As early as 1989, urban planners (Peter Newman and Jeffrey Kenworthy) presented data showing that per-person energy use for transportation was far higher if one lived in sprawling cities like Houston as opposed to densely settled cites like Hong Kong or Moscow (**Figure 11.130**). Why? It's all about cars and mass transit (as well as your feet). In dense urban areas, few people have or use cars and it makes sense to build and run mass transit, which is more energy efficient than private vehicles.

Although different authors disagree on the exact ordering and amounts of energy used or wasted for each kind of transport (the calculations depend on assumptions which can differ), there are clear trends (**Figure 11.131**). Airplanes use the most energy per distance travelled; bicycles and walking use the least. Electric vehicles are more efficient than those that use internal combustion engines. Mass-transit efficiency depends on the system and is higher if electricity is used to power the system.

Your home offers many ways to reduce energy consumption. Simple, easy changes like replacing incandescent light bulbs with LEDs are no-brainers, especially if

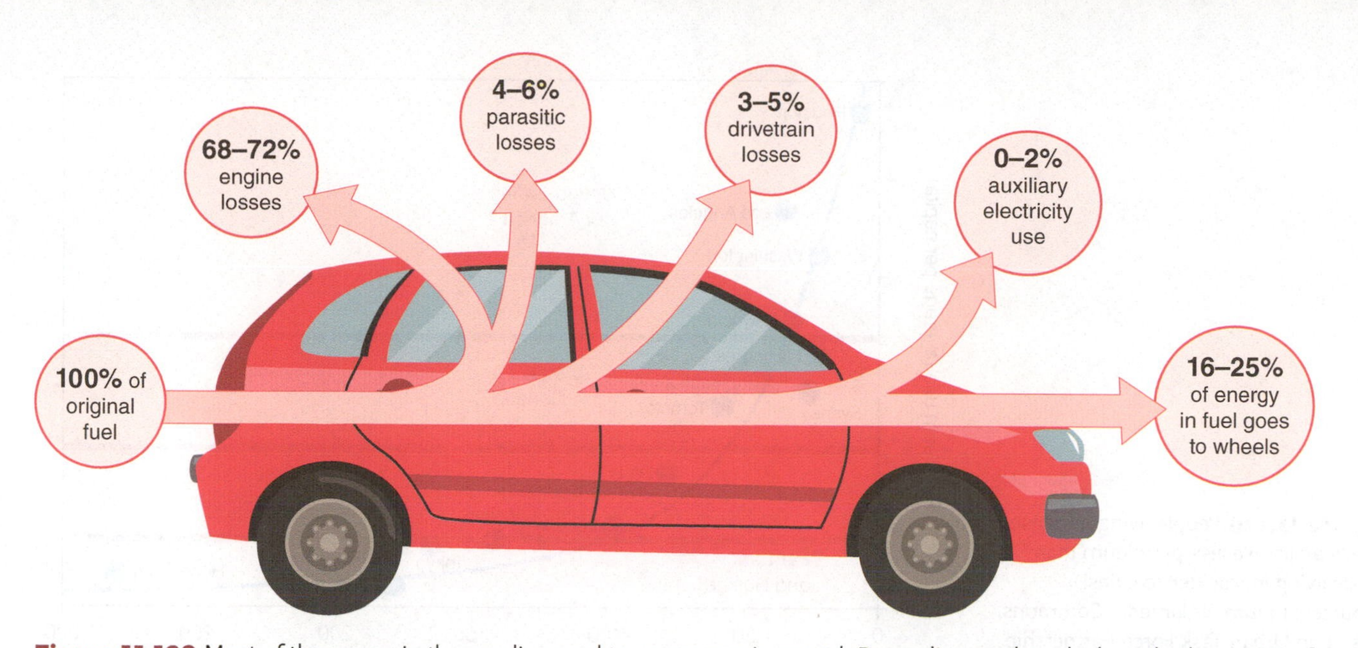

Figure 11.128 Most of the energy in the gasoline used to power cars is wasted. Depending on the vehicle, only about 20% of the energy goes to the wheels, the rest is lost as heat in the engine, friction in the drivetrain, and electronics in the car. In contrast, in an electric vehicle, about 90% of the energy goes to the wheels! (Modified from Yale Climate Connection)

your utility (like many) supplies bulbs free or at cost. Next time you replace an appliance, check the Energy Star label and find one that uses less electricity. The savings in energy use (and cost on your bill!) over a decade or more that your refrigerator, dish washer, or clothes drier lasts will be substantial.

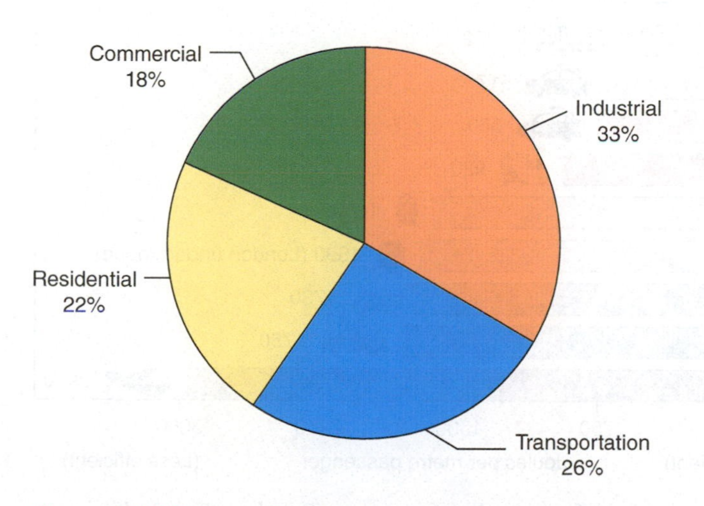

Figure 11.129 Share of total U.S. energy consumption by end-use sectors, 2020. Cuts to energy consumption can come from a variety of sources. For the average citizen, transportation and residential uses are the places where it's easiest to make change. Sum of individual percentages does not equal 100 because of rounding. (Modified from Energy Information Administration)

Home weatherization can have a quick and substantial payback, whether you live in Phoenix, Arizona, or Duluth, Minnesota. Insulation keeps your home warm in the winter and cool in the summer and lowers both your heating and cooling loads–meaning you use less energy, are more comfortable, and spend less (**Figure 11.132**). Air sealing, which keeps warm air in during heating season and out during cooling season, is an economical way not only to reduce energy bills but increase comfort: no more drafts. Many energy utilities subsidize weatherization and will provide energy audits so you know the most cost-effective ways to tighten up your house.

Innovative ways of heating and cooling your home have the potential to greatly reduce residential energy use whether you live in cold or hot climates. Heat pumps, which extract energy from the air even when it's cold, can be used for both heating and cooling (**Figure 11.133**). They are several times more efficient than electric resistance heating and double as efficient air conditioners in the summer. Federal legislation passed in the summer of 2022 will make heat pumps even more economical, with support to expand both manufacturing and rebates at installation.

Case Study

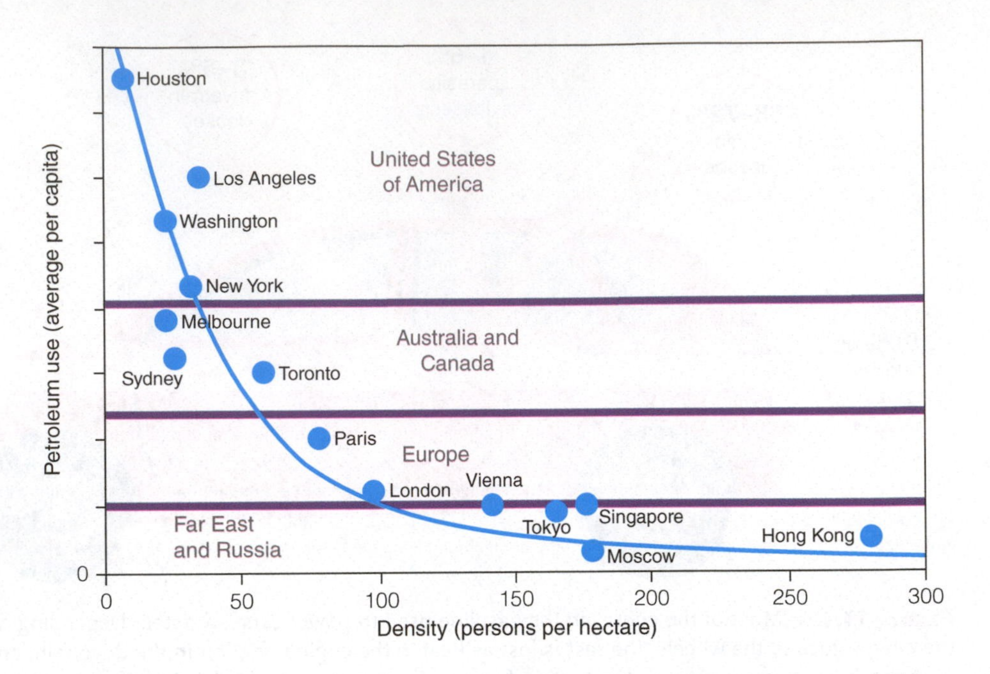

Figure 11.130 People living in dense urban areas use less petroleum than those living in less dense cities. (Source: Hohum, Wikimedia Commons. Based on Urban Task Force Partnership, DETR, 1999)

New construction can be done in ways to dramatically reduce energy use. In many climates, **passive solar design** along with weather sealing and insulation can be used to create buildings that need little if any heating or cooling using fossil-fuels or fossil-fuel derived energy. Such buildings are designed to absorb sunlight in the morning and not at mid-day. Passive solar buildings have large thermal mass, meaning they heat up and cool down slowly. In some climates, buildings may be built partially underground to buffer temperature changes (**Figure 11.134**).

Other cutting-edge buildings employ **biomimicry**, using ideas found in nature, particularly living things, to keep spaces warm or cool, dry or moist. This could include shading, natural ventilation, and, in a design influencing one of Qatar's 2022

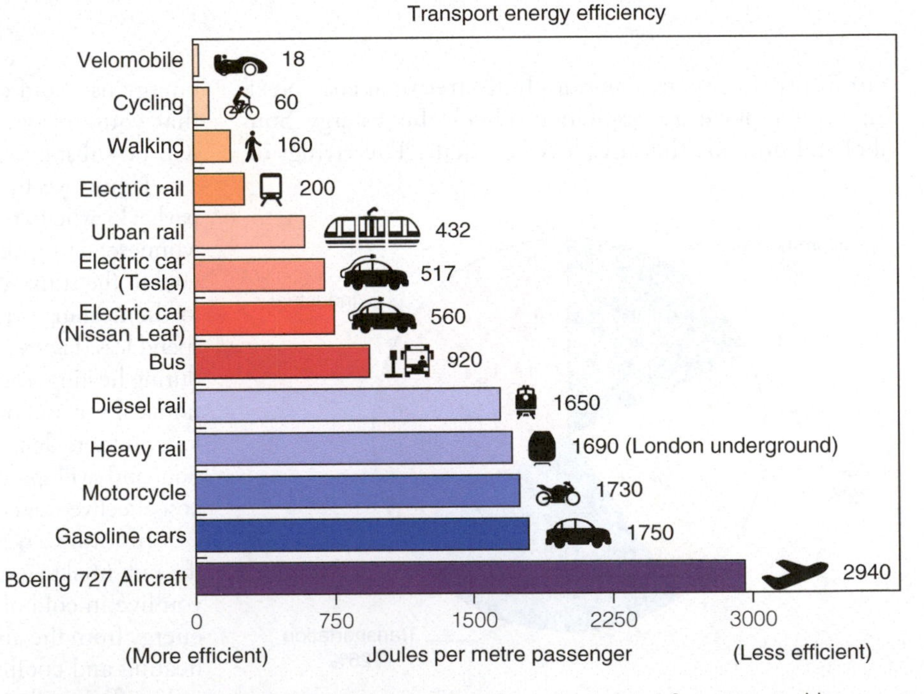

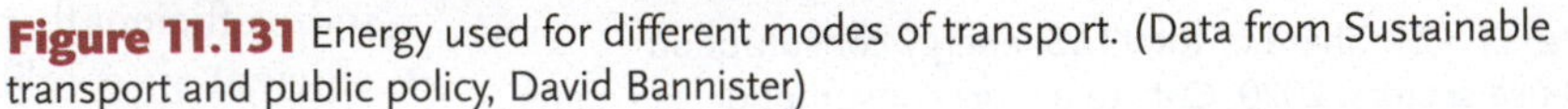

Figure 11.131 Energy used for different modes of transport. (Data from Sustainable transport and public policy, David Bannister)

Figure 11.132 Spray foam insulation reduces both heat loss from conduction and air leakage and is standard on most new construction in the United States. It expands to fill voids but emits hazardous vapors when it's applied (not after it's cured), so workers need protection.

Paul Chauncey/Alamy Stock Photo

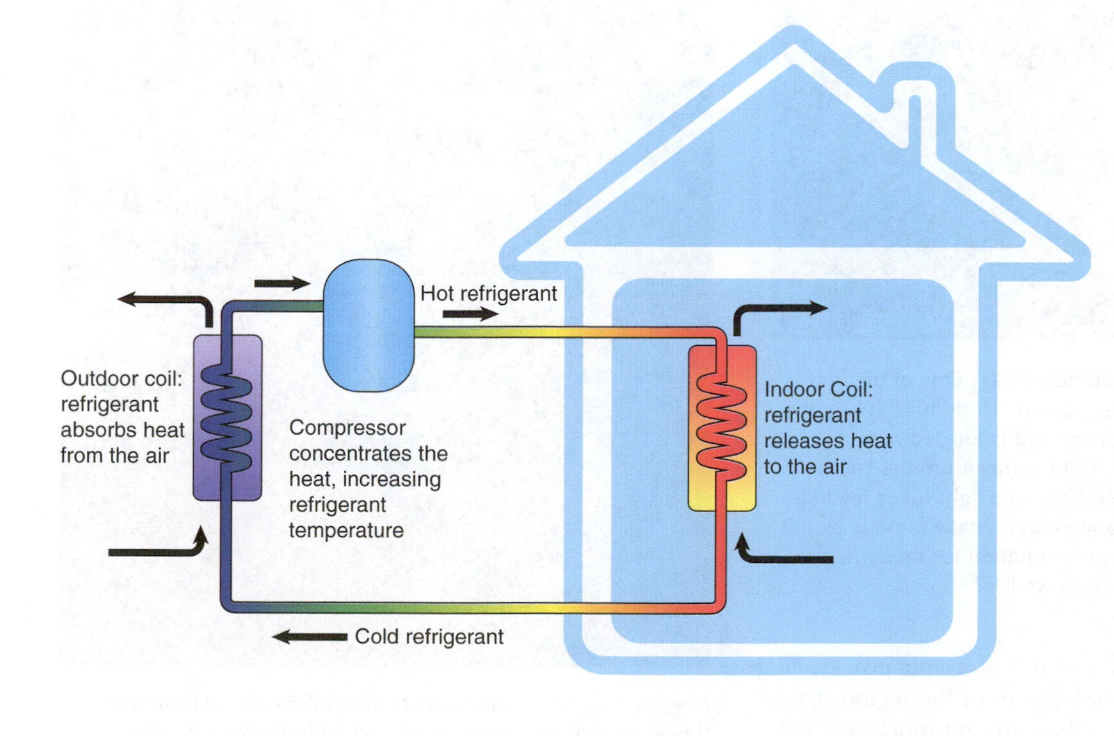

Figure 11.133 A heat pump harvests heat from outside and releases it into homes. Cold climate heat pumps now work well below zero degrees Fahrenheit. (Modified from David Pill, Pill-Maraham Architects)

Martin Bond/Alamy Stock Photo

Howard Harrison Photography/Alamy Stock Photo

Figure 11.134 Earth-sheltered homes. (a) One of the first modern, earth-sheltered houses. Called "Underhill," it was the home of architect Arthur Quarmby, and is located near Holmfirth, Yorkshire, England. Light comes from the south-facing windows, the massive skylight, and light tubes feeding each room. (b) Could the Hobbit House (made famous by J. R. R. Tolken) in New Zealand be an inspiration for modern, energy-efficient adaptations of how we live?

World Cup soccer stadium, the use of evaporative cooling that relies on traditional knowledge from the region. Under the stadium will be **qanats**, subterranean tunnels that bring in groundwater that, when it evaporates from open pools under the stadium, will cool the air (**Figure 11.135**). People in

Hemis/Alamy Stock Photo

J Marshall-Tribaleye Images/Alamy Stock Photo

Figure 11.135 Ancient qanats allowed desert civilizations to thrive. (a) Surface entrances to a qanat in Kerman Province, Iran, listed as a World Heritage Site by UNESCO. (b) The qanat channel cut into rock deep below the surface.

the Middle East have used this technology form over 2,000 years to bring water from wet highlands to distant farms in arid lowlands. The water, travelling underground, remained cool, fresh, and did not evaporate as much as it would have flowing at the surface. The qanats had a great side benefit: cooling. People in the Middle East so mastered natural cooling that they could keep ice through the summer in specially designed, naturally air-conditioned structures (**Figure 11.136**)

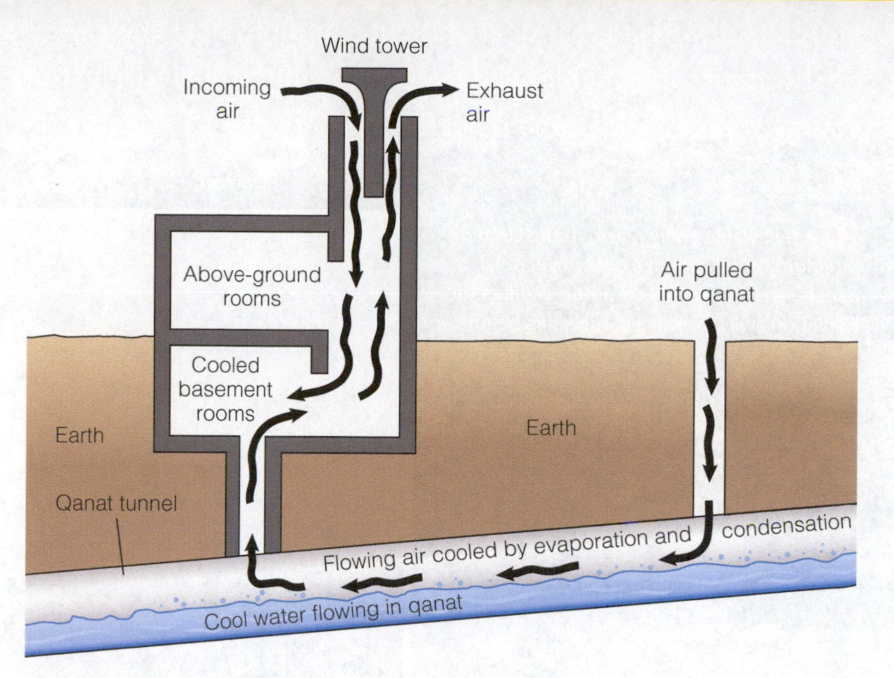

Figure 11.136 The design of a qanat passive air-conditioning system.

Photo Gallery

Energy Is Where You Find It

Humans devote a great deal of mental and physical energy to harnessing Earth's energy. Energy, like freshwater, is invaluable. The five photos in this gallery illustrate some of Earth's nonconventional energy sources and some relics of fossil energy.

1 A renewable energy farm in the United Kingdom. Wind turbines tower over solar panels, an increasingly common sight as the transition away from fossil fuels has begun.

iStock.com/A-Shropshire-Lad

2 Drilling for oil in Pennsylvania in the 1890s. There were no pipelines then. Look closely for the barrels of oil between the oil rigs.

iStock.com/RockingStock

The fun side of geothermal energy. Bathers enjoy the Blue Lagoon, a hot-water pool 40 kilometers (25 mi) from Reykjavik, Iceland. It uses geothermal power plant outlet water. The lagoon attracts about 100,000 visitors annually.

Barry Lewis/Alamy Stock Photo

Joern Sackermann/Alamy Stock Photo

3 New, vertical-axis wind turbines (VAWT) are starting to augment more traditional designs. VAWTs are 43% to 45% efficient, in contrast to conventional propeller-type turbines, which have 25% to 40% efficiencies. The VAWTs can function at wind speeds up to 110 kilometers per hour (68 miles per hour), which would shut down conventional turbines. This one is in Germany—a leader in renewable energy.

Jonas Gratzer/LightRocket/Getty Images

4 Mining coal by hand and using excavators at the Jharia coal mine in India. The area, rich in bituminous coal, has been mined since 1894; it's the largest coal deposit in the country. The area around the mine has been affected by underground coal seam fires for more than a century. There are frequent sinkhole collapses as the coal burns, and the air can become toxic.

iStock.com/T-lorien

5 Sixteen offshore wind turbines, between Copenhagen, Denmark, and Sweden. Their bases are covered in fog, and if you look closely, you'll see the ocean beneath.

Summary Energy and the Environment

1. Fossil Fuels

LO1 Summarize the environmental impacts of the three primary fossil fuels

a. Petroleum

Petroleum, a fuel composed of hydrocarbon compounds, is unequally distributed around the world. Found in reservoir rocks, petroleum occurs in both liquid and solid forms. In addition to their use as fuel, hydrocarbon compounds derived from petroleum are used in paints, plastics, soaps, and synthetic fiber. Oil spills are a significant threat from extracting this type of fuel, threatening local ecosystem health and difficult to clean up quickly and effectively.

b. Coal

Coal is formed when carbon-rich residue of ancient plant biomass is preserved and altered by heat and pressure over millions of years. Coal can be found and mined around the world, using different mining techniques depending on where the coal is located. Mountaintop removal excavates entire hilltops to access coal seams buried below. Coal shallowly buried can be mined traditionally but this introduces many risks, from mine collapse to mine fires.

c. Natural Gas and Fracking

Natural gas is primarily composed of methane, a potent greenhouse gas. The production of natural gas has increased with the development of hydraulic fracturing, a process which injects high-pressure fluids and sand into tight formations to fracture the rock. This increases rock's porosity and permeability, thus promoting the release of trapped oil and natural gas. Fracking involves many environmental risks, including water pollution, air pollution, and wastewater disposal.

d. Unconventional Fossil Fuels

Unconventional oil is typically produced from reservoir rocks with very low porosity and low permeability. Often, unconventional oil is more difficult to extract, has lower EROI values, and poses greater risk for environmental impacts than conventional oil. Other unconventional fossil fuels include tight oil, coal-bed methane, methane hydrate, hydrogen, oil shale, and tar sands.

Study Questions

i. What is EROI, and how does it differ between fossil fuels?
ii. What are the environmental risks of energy generation (from mining to processing) for two different fossil fuels in this section?
iii. What is the general process used to generate electricity using fossil fuels?

2. Biofuels

LO2 Describe the differences between biofuels and fossil fuels in their production and energy return on investment

a. Biofuels

Biofuels are any fuels derived directly from biomass–they are recently produced fuel sources. Some common examples are ethanol, methanol, and biodiesel. Ethanol derived from corn for a long time received quite a bit of attention; however, this type of ethanol is not efficient, with a low EROI. Sugarcane ethanol often has a higher EROI than corn. Biodiesel is produced from animal fats, plant oils, or algae, but like ethanol it has a low EROI.

Study Questions

i. How do biofuels differ from fossil fuels?
ii. What is a positive effect of adding ethanol to gasoline?

3. Nuclear Energy

LO3 Explain how nuclear power is produced and the risks of fission reactors and the nuclear fuel cycle

a. Fission Energy

Uranium is used as fuel in nuclear reactors because its nuclei are packed with protons and neutrons, making it potentially unstable and thus capable of sustained nuclear reactions. The uranium isotope ^{235}U is the most important fissile isotope, meaning it can undergo fission reactions which split the nucleus. Stray neutrons produced in this reaction collide with other nuclei, creating a chain reaction. Fission releases heat, which is transferred to a working fluid, usually water, that spins turbines to produce electricity (like fossil-fuel generated electricity).

b. Fusion Energy

Nuclear fusion brings two hydrogen atoms together to form helium producing massive amounts of energy in the process. Current designs for fusion reactors are unable to generate power.

c. Nuclear Safety, Accidents, and Contamination

The standards for building, licensing, and maintaining nuclear power stations are strict. One important criterion is geological stability, such that the site is not a risk for landslides, tsunamis, seismic activity, and flooding. Only one major nuclear disaster has been caused by a geologic phenomenon (2011, in Fukushima, Japan), while others have in large part been caused by operator mistakes (1986, in Chernobyl, Ukraine).

Study Questions

i. How do fission and fusion reactions release energy and how do they differ?

ii. What are the environmental risks of nuclear energy generation, and what is one example of a real-world nuclear disaster?

4. Renewable Energy

LO4 Compare different renewable energy sources, focusing on their strengths and weaknesses as replacements for fossil fuels

a. Hydroelectric

The principle of hydroelectric energy is impounding water with a dam and then allowing water to move under the force of gravity through a system of turbines and generators to produce electricity. Hydroelectric power generation is low carbon and has a high EROI, but it impacts ecosystem function. Fish populations have been reduced, with dams obstructing migration patterns critical to fish spawning. Hydroelectric power can impact communities due to inundation of towns and villages, as the building of a dam and the formation of a reservoir are required to generate hydroelectric power.

b. Geothermal Energy

Geothermal energy is not only harnessed for electricity generation but can also be used directly for space heating and heat pumps. Geothermal energy is low carbon and has few negative environmental impacts, but this type of energy is usually harnessed along plate boundaries where there is high heat flow from the Earth (not available everywhere). Additionally, this type of energy can create environmental problems due to groundwater withdrawal and water contamination. Geothermal fields have production capacity limited by the availability of water supply needed to keep them going.

c. Wind Energy

Wind turns rotors attached to a turbine and generates electricity. Globally, wind is one of fastest-growing energy segments, with a great deal of potential energy currently unharnessed worldwide. Depending on the location, wind EROI estimates vary between 10:1 and 30:1. There are environmental drawbacks of wind energy generation, including noise, land acquisition and the death of birds and bats that hit the turbines as they are spinning.

d. Solar

There are two main ways that energy can be generated directly from solar energy; solar-thermal methods heat fluid to spin turbines and generate electricity; photovoltaic cells convert sunlight directly to electricity using semiconductor materials. Solar towers have proven to be a reliable source of electricity. On large scales, this type of energy generation requires large amounts of land, removal of native vegetation, and high-water use. Photovoltaic cells require no water and moving parts but can use large spaces if not placed on rooftops, and use toxic chemicals in their manufacture, which makes them difficult to dispose of safely.

e. Energy from the Sea

Marine renewable energy includes wave and tidal energy, as well as ocean thermal energy conversion. Though this type of energy generation is currently in the research and early developmental phase, some argue that the marine renewable energy generation potential could be equal to that of currently installed hydropower in the United States.

Study Questions

i. What are two examples of renewable energy sources and how is energy generated by each?

ii. What are the environmental drawbacks of two different types of renewable energy generation?

Key Terms

Anthracite
Barrage
Base load
Binary-cycle system
Biodiesel
Biofuels
Biomass
Biomimicry
Bitumen
Bituminous
Carbon intensity
Chain reaction
Chemical dispersants
Coal-bed methane (CBM)
Combustion
Cracking
Crude oil
Daughter isotopes
Depleted uranium
Deuterium
Energy
Energy dense
Energy return on investment (EROI)
Ethanol
Feed in tariffs
Fissile isotopes
Fission
Fission products
Flaring
Flowback
Fossil fuels
Geothermal energy
Hydrocarbon compounds
Hydrocurrents
Hydrogen
Hydrothermal
In situ extraction
Industrial Revolution
Intermittency
International Thermonuclear Experimental Reactor (ITER)
Kerogen
Lignite
Marine renewable energy
Methane hydrate
Methanol
Natural gas
Net metering
NIMBY
Nonrenewable resources
Ocean thermal energy conversion (OTEC)
Oil shale
Oxidation
Parabolic troughs
Passive solar design
Peat
Petroleum
Photovoltaic (PV) cells
Power towers
Qanats
Rank
Refining
Renewable resources
Reserves
Reservoir
Reservoir rocks
Solar-thermal
Subbituminous
Syngas
Tar sands
Tidal energy
Tight oil
Tritium
Unconventional oil
Wave energy
Wind energy

"Use it up, wear it out, make it do, or do without."

—New England proverb

12

Waste Management

Learning Objectives

LO1 Characterize types of waste

LO2 List the three primary ways that we take care of wastes

LO3 Explain how a landfill is properly designed to handle waste, including liquid and gaseous by-products

LO4 Summarize the importance to waste management of recycling and composting

LO5 Describe the ways that we deal with hazardous wastes, and the shortcomings of the approaches people have taken so far

Aerial view of metal waste ready for recycling.

iStock.com/Bfk92

Our Plastic Planet

The world is in love with plastic. Strong, lightweight, and moldable into a variety of forms for many purposes, this product (mostly derived from petroleum) finds its way into items as diverse as lawn chairs, water bottles, test tubes, spacecrafts, shopping bags, and even stockings.

British metallurgist Alexander Parkes introduced plastic to the world in 1862, calling it "Paper of Parkesine," although it did not really capture the imagination of consumers until the DuPont Corporation invented nylon in 1939. There are many types in use today, though seven principal categories dominate the market (**Table 12.1**).

Many plastics contain potentially harmful additives, including bisphenol A (BPA) and phthalates, that interfere with the normal estrogen-producing functions of the human body–and those of wild animals too. Problems with brain development, reproduction, and other organ functions have been attributed to chronic exposure to these substances. Such toxicity would not be such a serious environmental concern, though, if plastics did not enter the natural world so easily.

Plastics are the principal form of long-lasting litter in the world today. Being lightweight and easily disposable, much of this waste simply comes from casual discards or articles falling off of vehicles along roads and highways (**Figure 12.1**). Wind blows discarded plastic packaging for kilometers across landscapes. The United States alone uses more than 100 million packages a year (as of 2022), little of which is saved or reused for long. Runoff from rain and snowmelt sweeps up plastic litter and carries it to the ocean–the world's ultimate sump. Some estimates warn that the mass of plastics floating at sea could exceed the mass of all the world's fish by 2050. The principle source of plastic appears to be urbanized areas because of their concentrated numbers of consumers and centralized surface water drainage systems, including urban streams (**Figure 12.2**).

Fishing vessels and larger container ships also contribute to the plastic problem. Severe storms and high waves toss 2,000 to 10,000 cargo containers from the decks of ships each year. The unrecoverable debris floating at sea (flotsam) can range from plastic toothbrushes to entire prefab homes (**Figure 12.3**). In 1990, the massive cargo vessel the *Hansa Carrier* was halfway across the Pacific when a storm hit. Numerous shipping containers were ripped off the ship and 61,280 Nike shoes (using plastic to make them more comfortable) were cast into the Pacific Ocean. About a year later, shoes started to appear along beaches in

Table 12.1 The seven types of plastics

Plastics are categorized by numbers corresponding to general chemical compositions and physical properties. The numbers are labeled in triangular recycling logos on many products; in some instances, on their bases. Some types can be recycled and some cannot.

Number of Plastic	Composition	Notes	Recyclable?
1	Polyethylene terephthalate (PET or PETE)	Clear in color. Makes up most soda and water bottles. Safe, but should not be reused because bacteria and flavoring will accumulate after initial use.	Most curbside recycling programs will pick up PET/PETE.
2	High-density polyethylene (HDPE)	Most milk jugs, detergent and juice bottles, containers for butter, and toilet products are HDPE. Often opaque.	The most often recycled plastic
3	Polyvinyl chloride (PVC)	Food wrapping, plumbing pipes, and cooking oil bottles commonly use PVC. Contains potentially harmful phthalates (don't cook with food wrap, especially in microwave ovens!).	Generally not accepted by recycling programs. Must be separated when thrown out.
4	Low-density polyethylene (LDPE)	LDPE goes into grocery bags (where still legal and in use), some food wrapping (notably breads and baked goods), and squeezable food bottles.	Not harmful to human health, but still not accepted by most curbside recycling programs; some markets accept in specialized recycling collectors.
5	Polypropylene	Goes into medicine bottles, ketchup and syrup bottles, straws, yogurt cups, and many water bottles (with cloudy finishes)	Not harmful to human health but not universally accepted by curbside recycling programs
6	Polystyrene or Styrofoam	Disposable cups, plates, picnic coolers. Will release potentially harmful chemicals, especially when heated	Not accepted by most curbside recycling programs
7	All other types of plastics	Includes some potentially harmful substances, including bisphenol A in polycarbonate products. Includes DVDs and CDs, baby and many water bottles, and medical and dental devices	Not accepted by most curbside recycling programs

the Pacific Northwest in Oregon and Washington and along Vancouver Island.

This is one case where science benefited from plastic pollution. Curtis Ebbesmeyer, an oceanographer who studied the spill and meticulously tracked the time and place at which the shoes washed ashore, used the data to map ocean currents (see Chapter 2). The shoes moved with the water; he knew exactly where and when they went into the ocean. The results validated an important model of how the oceans work. He recalls:

> "The shoe story... had real implications for the study of the world's oceans. Building on the sneaker research, I would eventually use flotsam to calculate, for the first time, the orbits of the 11 oceanic gyres–the giant, circular currents that are, after the oceans and the continents, the Earth's largest geographical features."

Figure 12.1 Litter in a street, Toulouse, France. This is mostly plastic and paper waste; easily discarded into the environment after use, whether intentionally or not.

Plastic is structurally complex. At the microscopic level it consists of foams, films, fibers, and beads. As randomly discarded plastic disintegrates due to sunlight, atmospheric oxygen, organisms, acidic water, hot days, and other factors in the wild, it gradually releases swarms of particles to the environment–unnoticed without careful study and virtually impossible to recover. These are termed **microplastics** (**Figure 12.4**). The U.S. Geological Survey estimates that in 2022, 12% of freshwater fish swimming in American waters contain substantial concentrations of fine plastic particles. You consume 50 tiny plastic particles per serving of oysters, and embedded in every square foot of river sediment there are on average 1,285 microscopic bits of plastic, many waiting to enter the food chain (**Figure 12.5**).

Microscopic plastic bits are deliberately manufactured in some cases. Known commercially as **microbeads**, these tiny particles have been intentionally included in some toothpastes and lotions, and abundantly so in cosmetics, helping to activate important ingredients and extend the shelf lives of products (**Figure 12.6**). Recognizing the problem, the U.S. Congress passed the Waterbead-Free Waters Act, which went into effect in 2018, prohibiting the use of microplastics in the making of cosmetics. There is other good news, too: of all plastics consumed annually in the United States, 85% presently ends up going into landfills (2022), one of our

Figure 12.2 Most of the waste floating this urban stream in the Philippines is plastic that will eventually enter the ocean.

Figure 12.3 One of thousands of Nike sneakers that washed ashore on U.K. beaches in 2019, after a massive shipping container of shoes washed off a cargo vessel in the western Atlantic. These shoes, which are mostly plastic, had a long trip across the ocean.

iStock.com/pcess609

Figure 12.4 Microplasticcs on a beachgoer's hand.

Imagegallery2/Alamy Stock Photo

Figure 12.6 Those exfoliating microbeads (look closely at the label and the lotion, they are green) might just end up in the ocean and in marine life.

most important sites of waste management, as you'll learn later in the chapter. That is 25 million tonnes (27 million short tons). Less than 10% ends up recycled, but these data indicate that we can push back against this all-too-human problem. As of 2021, Eight U.S. states have prohibited the single use of plastic bags in sales, and 80 countries–about 40% of the world total–have already partly or wholly restricted plastic packaging (2021).

There are new plastics not made of petroleum that have the potential to solve the waste problem in interesting ways. Many are compostable and some are even edible. Plastic cups and thin films made from seaweed extract are designed to be eaten as one drinks their contents (**Figure 12.7**). And they come in a variety of delicious flavors, from lemon to rosemary-beet.

questions to ponder

1. How much plastic waste do you generate in a typical week, and which two kinds of plastic do you most often use (if any)?
2. What could be an easy way for people to reduce their generation of plastic wastes, without causing great personal inconvenience?

Dotted Yeti/Shutterstock.com

Figure 12.5 Microplastic particles floating in an ocean current. Individual pieces look tiny but summed up this plastic debris is a major environmental problem.

Milenova Elena/Shutterstock.com

Figure 12.7 Some coffee and tea cups are made of edible plastic. Here's with roasted coffee beans for scale. These have been marketed as a way of reducing waste.

12.1 Types of Waste (LO1)

Our civilization generates a multitude of different waste types. It is convenient to categorize them into four overall categories: municipal solid waste, industrial waste, agricultural waste, and wastewater (**Figure 12.8**). There is overlap between these categories. Plastics, for instance, are present in municipal, industrial, and agricultural waste depending upon circumstances. A framework for classifying waste has proved useful in terms of management and regulation, so we'll follow such an approach in this chapter.

12.1.1 Municipal Solid Waste

Municipal solid waste (MSW) is primarily the stuff that we toss away on a day-to-day basis at home, school, or work; in short, "garbage." Municipal solid waste includes uneaten food, plastics, glass, wood and paper products, yard cuttings, construction waste, and discarded appliances such as refrigerators (**Figure 12.9**). MSW also includes consumer electronics, some household medical waste, and construction as well as demolition waste. Some of this material is recyclable and some can be composted, but in much of the world, most municipal solid waste is simply discarded.

On average, in 2018, each American threw away about 2.2 kilograms (4.9 pounds) of MSW per day. About 0.7 kilograms (1.5 pounds) ended up in recycling bins or compost piles (that's 32% of the total MSW produced). At this rate, this means that during a year's time the typical American generates approximately 750 kilograms (1,650 pounds) of MSW, or 265 million tonnes (292 million tons) summed up for the entire country (**Table 12.2**). That is about 12% of the world total, even though the United States makes up only 4% to 5% of the total global population (**Figure 12.10**).

In comparison, individual Europeans typically produce around 450 kilograms (1,000 pounds) of garbage annually, reflecting differences between U.S. citizens and Europeans that have to do with culture and lifestyle. The difference might also reflect legislation. The European Union (EU) regulations mandate that manufacturers make products from recyclable materials and offer ways to recycle products when they are no longer usable.

Figure 12.9 Municipal solid waste at a transfer station (before it heads to a centralized landfill) includes furniture, plastic, and household garbage.

Plastics, despite the concerns raised in the chapter opener, are not the top-tier MSW waste product when ranked by weight (Table 12.2). The environmental problem lies not in the mass of plastic products but in the mobility of plastic waste. Wood, though it dominates the waste stream, is generated by fewer numbers of people, in contrast to plastic. Because wood is biodegradable, it alone is not harmful if left to degrade in the environment, unlike plastics. Indeed, termites, beetles, and fungi love digesting wood in one way or another. The environmental impact of wood is typically the coatings that cover or impregnate it, including paint containing lead and wood preservatives that include toxic organic chemicals and arsenic (**Figure 12.11**).

Figure 12.8 The four categories of waste.

Table 12.2 EPA List of Selected Solid Wastes in Select Categories by Weight per Person in the United States per Annum (2018 Datum)	
Waste Type (ranked by weight discarded)	**Kilograms/Pounds of Waste Thrown Away Averaged per Person per year**
Wood (old furniture, home improvements)	387/854
Food waste	100/221
Corrugated boxes	85/188
Nonbeverage glass storage bottles and jars	42/93
Yard trimmings	41/91
Newspapers, magazines, books	31/68
Junk mail and other paper not used for packaging	29/65
Plastic bags, wraps, sacks, and packaging	24/54
Glass, beer, wine, liquor, and soft drink bottles	21/47
Rubber products, including old tires	19/41
Office paper, including computer printer paper	13/30
Metal cans	10/23
Tissue paper and paper towels	10/22
Paper plates and cups	4/8
Paper bags and sacks	3/6
Old clothing and footwear	2/5

Figure 12.11 Waste resulting from the demolition of an older wooden building in New York. Wood from buildings built before the 1960s is almost certain to be coated by lead paint which was not banned in the United States until 1978. Demolition waste is often mixed with municipal solid waste and landfilled.

12.1.2 Industrial Waste

Industrial wastes are produced not only by industry, particularly manufacturing, but also by lumber mills, mining, and powerplants. Examples include scrap metal, oils, varnishes, chemical solvents, greases, dust, crushed stone, ash, and radioactive waste largely from nuclear power plants but also from industrial processes and medical facilities (**Table 12.3**). These wastes may be disposed of on site (e.g.,

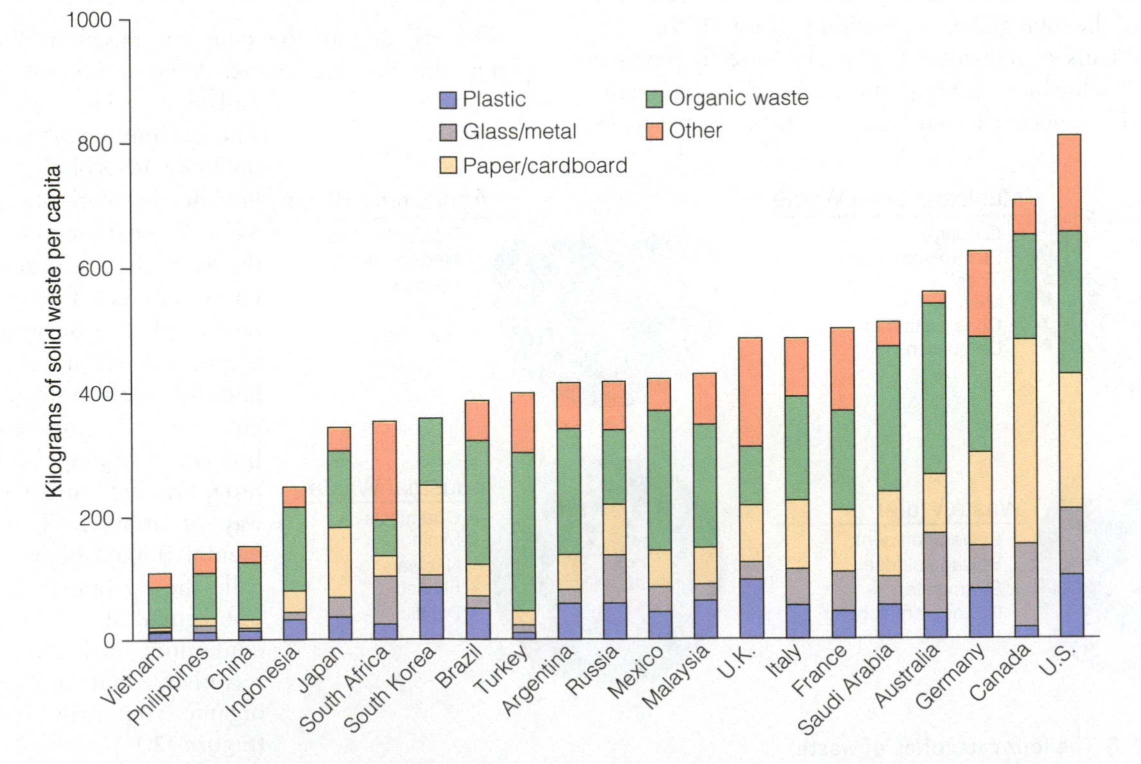

Figure 12.10 The amount of solid waste of different types generated per person varies greatly between countries.

mining waste; see Chapter 4) or directed to special facilities for treatment and disposal if the wastes are toxic or flammable. In the past, industrial waste was often disposed of alongside MSW–a practice that led to widespread contamination, human exposure to toxins, and costly cleanups.

Not all industrial wastes are solids, unlike MSW. Liquid industrial wastes can include toxic and reactive substances. Before such wastes were well regulated, they often were disposed of using municipal sewer systems and thus entered streams, rivers, and the ocean degrading water quality and endangering the public (**Figure 12.12**). The Cuyahoga River, which flows through Cleveland on its way to Lake Erie, became a national symbol for poorly managed industrial waste. The river caught fire many times (**Figure 12.13**); well, not actually the river but slicks of oil released by industry along its banks. A small fire in 1969 hit the press (*TIME* magazine and *National Geographic*) and is credited by some with catalyzing creation of the U.S. Environmental Protection Agency (EPA) soon after, as well as the 1972 expansion of the Clean Water Act, which regulates discharges to waters of the United States.

12.1.3 Agricultural Waste

The production of food generates **agricultural waste** of various kinds. Industrialized agriculture relies on toxic pesticides, herbicides, and fungicides to grow monocultures (single crops) with high yields. Large amounts of plastic are used both to protect crops (e.g., as mats to suppress weeds) and to store farm products like hay. The most ubiquitous agricultural waste is **manure**, fecal matter that is a mix of living organisms and undigested material. All animals produce manure, which is rich in bacteria and viruses as well as plant nutrients such as nitrogen and phosphorous.

Manure from farms of all sizes is a useful fertilizer but it can have negative environmental impacts. Free-ranging herds of animals spread their manure across the landscape, where it mixes with soil, returning nutrients to the ecosystem

Table 12.3 Various Sectors of Industry with Wastes of Concern to the U.S. EPA (2022)

Industry Category	Common Solid Wastes	Common Toxic Liquid Wastes and Outgassing*
Construction, demolition, and renovation	Ignitable or toxic wreckage and debris, lead pipe, asphalt wastes (partly recycled)	Petroleum distillates, used oil, acetone, adhesives, coatings, pigments, solvents, striping compounds, kerosene
Dry cleaning (e.g., for clothes, sheets)	Spent filter cartridges and residues,	Spent solvents and various hazardous substances including perchloroethylene, outgassing of chlorofluorocarbons harmful to ozone layer (see Chapter 2) and trichloroethane.
Educational and vocational shops (auto repair, graphics, woodworking)	Paint wastes	Solvents, various hazardous acids, bases
Equipment repair	Paint wastes	Solvents, various hazardous acids and bases
Furniture manufacturing and refinishing	Wood, plastic, and metal scraps	Acetone, alcohol, various hazardous liquid wastes including petroleum distillates, mineral spirits; outgassing of volatile organic compounds
Laboratories	Rags, gloves, wraps; various contaminated solids, including plastics and biological matter.	Spent solvents and contaminated liquids
Leather manufacturing	Residual suspended solids	Alkaline waste water, hazardous sulfides, chromium, acid and alkaline salts and acids, dye overspray
Motor freight and rail transportation	Residues from shipping products, various sludges and paint pigments, cadmium, nickel, scrap metal, batteries, used oil filters	Alkaline and acid cleaners, oils, greases, acetone, radiator flushing solutions
Pesticide end-users and application services	Ignitable wastes, contaminated soil, empty containers	Used and unused pesticides
Photo processing	Silver, plus various corrosive, ignitable chemicals	Bleaches, system cleaners
Printing	Chromium, lead	Waste ink; cleaning solvents including various hazardous chemicals used for etching plates, etc.
Textile manufacturing (e.g., clothes, fabrics)		Hydrogen peroxide and various other bleaches; alkali and sodium hydroxide alcohol; various hazardous chemicals used in equipment maintenance

*Outgassing is the slow release of gases, often toxic, from some solids and liquids. Think of that "new-car smell" as one example.

Science History Images/Alamy Stock Photo

Figure 12.12 A slurry of mine tailings (in this case, taconite, an iron ore) pour into Silver Bay on Lake Superior, Minnesota, in 1968. This was unregulated and unprocessed industrial waste.

so they can be used again by plants (**Figure 12.14**). When animals become concentrated in barns and feeding areas, manure management becomes a necessity. Small farms have for centuries spread the manure back on land. When and where this spreading happens matters. Manure spread on

iStock.com/Amed-in-bali

Figure 12.14 Manure left on a pasture by a herd of grazing cows in Indonesia–the natural way to spread animal waste.

frozen or saturated ground is more likely to be carried by runoff into streams (**Figure 12.15**). For that reason, some states restrict where and when manure spreading can happen.

When agriculture is scaled up, so is the waste stream. **Concentrated animal feed operations (CAFOs)** are giant feedlots for livestock (**Figure 12.16**). They produce meat, dairy, and/or egg products. A single CAFO may contain as many as 125,000 broiler chickens, 80,000 hens, or 100,000 cows, or 10,000 pigs all crammed together. In 2019, Americans consumed 66-million CAFO-related turkeys for Thanksgiving and Christmas. Because CAFOs can include thousands of animals gathered in small areas, the animals are prophylactically dosed with antibiotics to keep them healthy, intensively fed, and grown with industrial efficiency, but not without significant environmental impact.

The biggest CAFOs cover a hundred hectares or more (a few hundred acres), but most livestock spend a large part of their lives confined within single structures. As many as 2,500 pigs or 50,000 chickens can be housed under one expansive roof (**Figure 12.17**). For cows, the entire existence of an individual may be restricted to a space measuring no more than 2 to 2.3 square meters (22 to 25 square feet). As many

Bettmann/Getty Images

Figure 12.13 In November 1952, an oil slick on the Cuyahoga River caught fire and destroyed three tugboats, three buildings, and ship repair facilities.

iStock.com/SimplyCreativePhotography

Figure 12.15 Spreading liquid manure on fields in Ontario, Canada. The brown manure covers the greening grass showing clearly where it has been applied.

Jim West/Alamy Stock Photo

Figure 12.16 Nearly 100,000 cattle are fed and fattened at a CAFO in Kersey, Colorado, awaiting slaughter.

as 10 billion land animals are raised and slaughtered in the United States each year to nourish us, and 80 billion animals are eaten annually worldwide.

Why do we need CAFOs? Two reasons: (1) Earth's population of 8 billion people, and (2) our choices of diet. On a strictly vegetable diet, 0.02 hectares (0.05 acres) of land can provide a person with 730,000 kilocalories a year, enough to live on. To produce this many calories of meat takes about 20 times more land area. Animals need room to forage without overgrazing, while plants only need soil, sunlight, and water in one specific location. Earth has about 1.5 billion hectares (3.7 billion acres) of arable land for food production. To feed 8 billion of us via a meat-based diet would require 3.2 billion hectares (or almost 8 billion acres). Our planet simply isn't big enough to support such a diet by devouring free-range livestock. Our livestock must be concentrated if we keep doing what we are doing now.

Animals in confinement in the United States produce about a half-billion tonnes annually of manure, or three times the total of all human excrement nationwide. The dust from feedlot operations and gaseous manure emissions includes fungal spores, animal skin cells, antimicrobial agents (such as the antibiotic tetracycline, used in animal feed), hormones, bacteria, respiratory pathogens, and 168 different kinds of gases, including hydrogen sulfide (the "rotten-egg"-smelling H_2S) and ammonia (NH_3), which can cause respiratory problems for people, including throat irritation and distinctive nasal burning. As much as 80% to 90% of the ammonia pollution in Earth's atmosphere today is from CAFOs and the fertilization of crops, some of which are needed to feed animals.

The prodigious amount of manure derived from CAFOs is stored, at least temporarily, in **feedlot lagoons**, sealed reservoirs with no inlets or outlets (**Figure 12.18**). Bacteria turns the liquid in these lagoons a purple or pink color. The waste may later be spread to fertilize fields elsewhere in a district or held indefinitely. Some lagoons fail, however, owing to faulty dam construction and overflows during periods of heavy rainfall–especially hurricanes. The effluent enters rivers and the waste, being high in the plant nutrients nitrogen and phosphorus, can stimulate choking eutrophication (see Chapter 8).

In one 2012 incident, manure waste issuing from a hog farm along Beaver Creek, Illinois, killed an estimated 150,000 fish and 20,000 mussels. One observer described the polluted aquatic habitat as looking like ink. Ecological recovery was slow; even 2 years later, 9 species of fish still seemed to be missing. The EPA estimates that 56,000 kilometers (35,000 miles) of United States rivers suffer pollution related to CAFO wastes that have escaped control. Twenty-two U.S. states report serious surface water pollution from CAFOs, and 17 states have CAFO-related groundwater contamination problems.

Jim West/Alamy Stock Photo

Figure 12.17 The Vande Bunte Egg Farm, in Martin, Michigan, houses keeps 2.7 million laying hens and 750,000 young chickens. That's a lot of eggs and a lot of chicken poop.

© Waterkeeper Alliance

Figure 12.18 Wastewater lagoon at a pig-raising CAFO. Drains enter the lagoon from feed houses at bottom of the image. Hoses spray water to agitate and oxygenate it near the center of lagoon. Bacteria give it the dark-pink color.

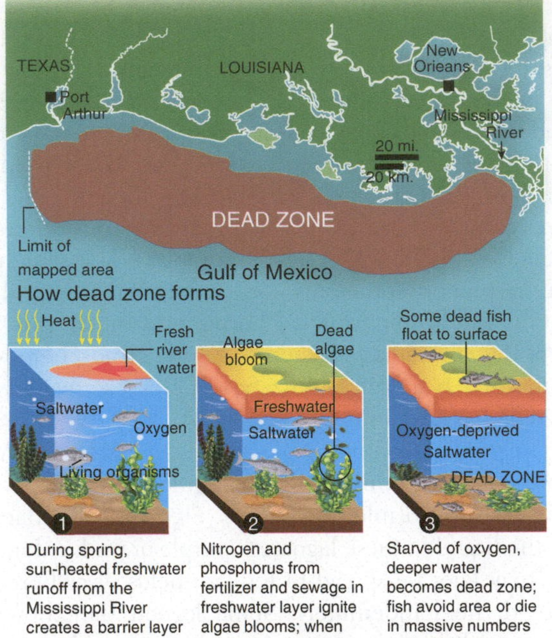

Figure 12.19 How dead zones form from excess nutrients in surface water, including phosphorus and nitrogen. (Based on MCT, Louisiana Universities Marine Consortium, NOAA, St. Louis Post-Dispatch)

Nitrogen pollution (in the form of nitrate and nitrite, -NO_3 and -NO_2) is not restricted to CAFO operations; it is a universal problem related to the heavy use of artificial nitrogen fertilizers on crops practically everywhere (recall the Haber-Bosch process of making nitrogen fertilizer from Chapter 2). Runoff from fields brings nitrogen waste to streams and river sides, lakes, and ultimately to the sea, where vast marine **dead zones** are the result of algal blooms that stretch beyond river mouths and deltas (**Figure 12.19**). When these algae die, decomposition uses much of the oxygen in the water. The resulting **hypoxic** (oxygen-deprived) dead zone at the mouth of the Mississippi River, which drains America's Midwestern agricultural heartland, has seasonally grown as large as 22,800 square kilometers (8,800 square miles)–larger than the state of New Jersey (**Figure 12.20**). The dead zone has severely harmed Gulf of Mexico fisheries.

Nitrate contaminated water comes back to haunt us directly. Drinking it can lead to a condition called methemoglobinemia ("blue-baby syndrome")–so named because oxygen transport in the blood is reduced and the skin turns blue (see Chapter 8). Groundwater, now enriched in nitrate by farming, underlies many agricultural areas and is tapped by drinking water wells.

Agriculture often increases erosion and the transport of soil and sediment from fields to rivers and streams, which eventually enter lakes and oceans (see Chapter 8). Many pesticides and some nutrients, like phosphorus, are bound to that sediment and move with it into bodies of water where the chemicals can be released into the water column. Some of the sediment eroded by agriculture is stored on the floodplains of rivers, only to be moved again by floods decades and even centuries later. This **legacy sediment** complicates management of both toxins and nutrients because it provides an ongoing source of both, even if contemporary nutrient loads have been reduced and pesticides banned.

12.1.4 Wastewater

Cities have always struggled with disposal of human waste and with the resulting outbreaks of disease. For centuries, people didn't associate illness with the mixing of human waste and water supplies; rather, they blamed all sorts of other environmental factors, like **miasmas**, vapors that arose from swamps, garbage, graves, and other foul-smelling sites where organic material decayed. In the 1800s, London physician Dr. John Snow spent years tracking the cholera outbreaks across the city. Cholera is a bacteria that kills by dehydrating its victims with massive diarrhea. He was not a believer in miasmas.

Mapping the cases of cholera in the 1854 London outbreak, Snow noticed most people getting sick got their drinking water from the same shallow public well that was serviced by a pump on Broad Street (**Figure 12.21**). Nearby was a shallow **cesspool**, an unlined, covered tank into which people deposited their feces and urine. Biological

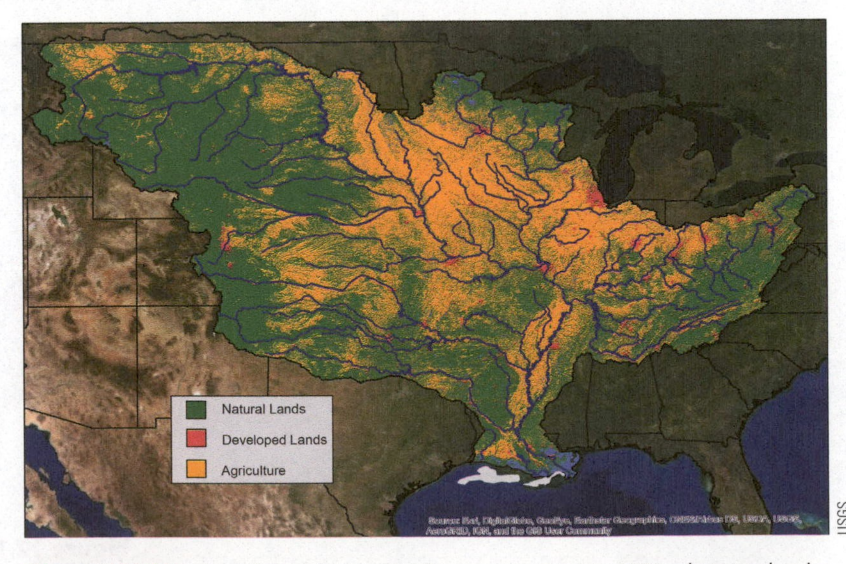

Figure 12.20 Map of the Mississippi River watershed in 2022, showing land use patterns, with a prominent marine dead zone (white patches) at the mouth of river, on the Gulf of Mexico.

Figure 12.21 Replica of the Broad Street well that John Snow correctly identified as the source of cholera-bearing water that sickened Londoners in 1854.

pathogens in the liquid (including cholera germs from sick people) infiltrated the surrounding ground, seeping into the aquifer and then into the well. Londoners were drinking a cocktail of sewage mixed with their water. Soon after the pump was disabled, the outbreak ceased.

Despite being based in firm geographic evidence, Snow's hypothesis was rejected for years and ridiculed in the press (**Figure 12.22**) His work has stood the test of time: we know now that both surface water and groundwater are vectors for disease-causing organisms. Just as importantly, we know how, through good geologic and environmental science, to reduce the risk of drinking contaminated groundwater.

In more recent times, an added layer of protection has developed to protect the environment from the impacts of accumulating human waste: the **septic tank**. A well-sealed septic tank separates solids from liquid wastes; the solids stay within the tank, where anaerobic bacteria break them down. Meanwhile, the liquids travel from the tank via shallow, buried **leach lines** that disseminate it into the surrounding soil or rock (**Figure 12.23**). Periodically, the tank must be pumped clean and the solid waste taken elsewhere for disposal, usually a landfill or a sewage treatment plant.

Figure 12.22 Satirical cartoon from 1866 lampooning John Snow's theory of groundwater contamination from human waste as the cause of London's cholera epidemics.

The liquid from septic tanks flows into an underground **leach field**. The leach field consists of rows of tile pipe surrounded by gravel, through which the effluent percolates before seeping into the ground. As the effluent infiltrates the soil and rock surrounding the leach field, it is further filtered and oxidized, and aerobic bacteria get to work decomposing it. Studies have shown that the movement of effluent through a few feet of unsaturated, fine-grained soil is sufficient to render it harmless bacteriologically, but nutrients like nitrogen keep moving through the soil, as can chemicals dumped down the drain, including solvents, paints, and drain cleaners. Septic systems work well if they are properly sited and maintained and most importantly - only if receive biological or other non-toxic wastes.

Geologists are often involved in the siting of septic systems; they know that if they choose the wrong site and

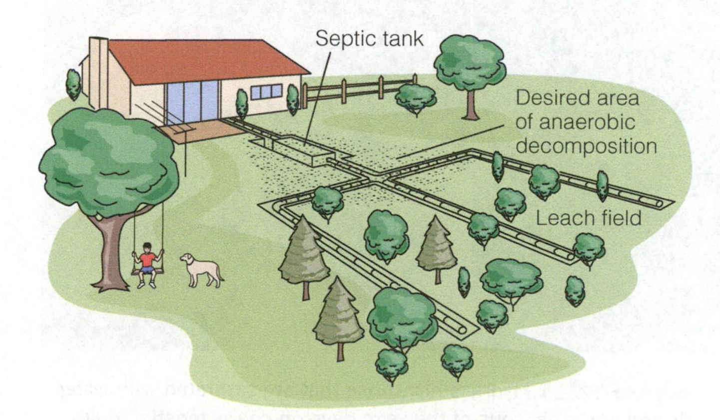

Figure 12.23 A septic tank near a house drains into a leach field of drain tiles that are laid in trenches and covered with a filter gravel and soil.

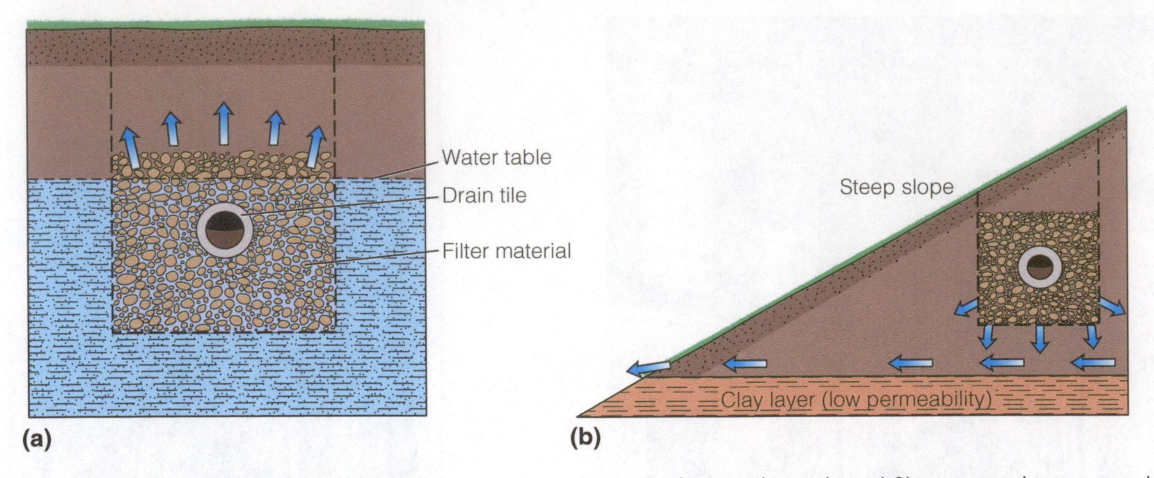

Figure 12.24 Conditions under which raw effluent can seep into the ground. (a) A drain tile and filter material intersect a high-water table. When this occurs, sewage rises, and the growth of anaerobic bacteria is promoted in the leach field. This leads to biological plugging of the filter material. (b) A leach line is underlain by low-permeability layers adjacent to a steep slope.

groundwater comes to the surface, the result is raw sewage effluent seeping out onto the ground (**Figure 12.24**; Chapter 1). This is usually avoided by doing two things. First, they examine shallow soil excavations for evidence that the soil is frequently and seasonally saturated by ground water; such saturation, which often happens in the spring, leaves a characteristic mottling pattern (**Figure 12.25**; Chapter 7). Then, they conduct **percolation tests** (perc tests for short) to make sure the soil is permeable enough that the water from the leach field will flow through it (**Figure 12.26**)

Figure 12.25 Hydric soils, those that are saturated with water during some seasons of the year, develop characteristic, multi-colored, red and grey mottles. In the background is a mottled hydric soil profile. The inset shows a specimen of a hydric soil. Hydric soils are poor choices for siting septic systems.

A **sewer**, or central waste disposal system, collects the wastes from many sources and deliberately flushes them via pipes, drains and tunnels to rivers, the ocean, or injection wells, and optimally via treatment plants that reduce the health hazard before release to the environment.

The ancient Mesopotamians introduced the concept of the sewer, using clay drainage pipes over 6,000 years ago to help keep areas clean around important temples. Much later, the ancient Romans constructed the 1.6-kilometer (1-mile)-long Cloaca Maxima, one of the world's first metropolitan sewers (**Figure 12.27**). It discharged raw human waste directly into the Tiber River for centuries and is still in use 2,500 years later to channel storm water away from Rome. It wasn't until 1843 that the world's first modern, urban sewer system came into existence, in Hamburg, Germany.

Figure 12.26 Navy engineers measure how quickly water soaks into soil during a perc test. They are building a septic system for an indigenous community in Colombia.

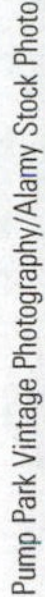

Figure 12.27 A 1880's view of the Cloaca Maxima (the semi-circular tunnel at water level) as it enters the River Tiber in Rome.

Figure 12.29 Giant circular clarifying tanks for removing fats and greases from wastewater at a sewage (wastewater) treatment facility.

The development of sewage treatment technologies has greatly reduced the impact of human waste not only on the environment–but only where treatment technologies are used. A sewage treatment plant achieves its purpose in a multistep process (**Figure 12.28**). In an early stage, termed **primary treatment**, the liquid and solid components of the waste stream are separated, and the liquid part clarified by removal of fats and greases using clarifiers (clarifying tanks) that skim floating debris while the solids sink and are removed from the bottom (**Figure 12.29**). The solids are disposed of in landfills and/or composted.

Then, in many plants, what follows is **secondary treatment**. Secondary treatment uses a mix of living bacteria to reduce the level of nitrogen and phosphorus in the water. The bacteria feast on the sewage and when they die, sink to the bottom of treatment tanks and are removed from the water, taking with them the nutrients. The resulting organic-rich material (largely the microscopic bodies of dead bacteria) is called **sewage sludge**. In **tertiary treatment**,

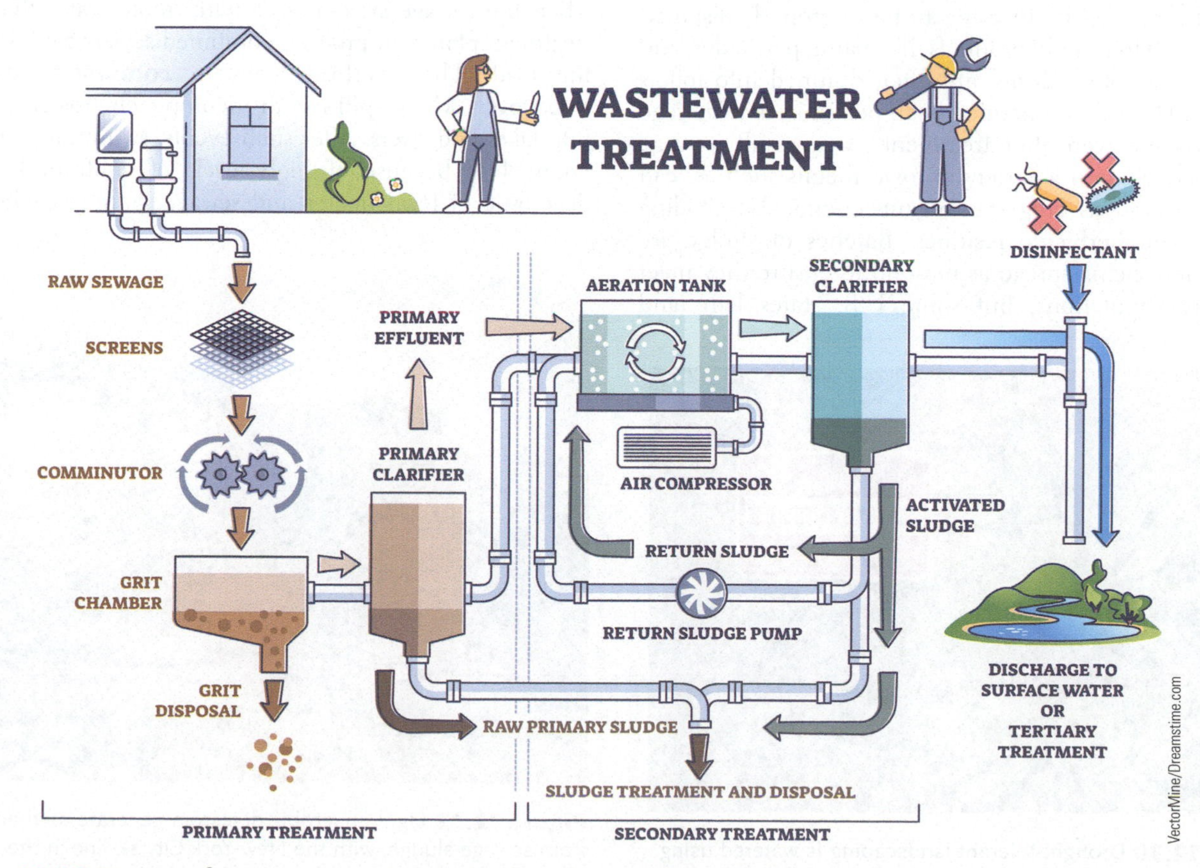

Figure 12.28 Diagram of sewage treatment plant showing primary and secondary treatment.

nutrient levels in the clarified liquid are reduced even further with chemical and biological approaches.

Secondary and tertiary treatment of sewage both reduce the potential for eutrophication or overfertilization, resulting algal blooms and oxygen depletion of water bodies into which the sewage is released at the end of the treatment process (Chapter 8). Before treated sewage is released, it is typically chlorinated to kill bacteria and viruses that may have survived the treatment plant. Treated wastewater can then be discharged into lakes and rivers or used as **greywater** to water lawns and trees (**Figure 12.30**). Some treated wastewater is injected into the ground or allowed to infiltrate into aquifers to recharge the groundwater system.

Sewage sludge–the solid or semisolid part of sewage–can be managed in different ways. About half of all sewage sludge ends up going to landfills or incinerators; the rest is converted into **biosolids** for mixing into soils or use as crop fertilizers (**Figure 12.31**). Composted sludge is applied as fertilizers to fields and forests. In some places, it is anaerobically digested (without oxygen) and used to generate methane (**Figure 12.32**). That methane can be used to run electrical generators. At Seattle, Washington's main sewage treatment plant at West Point, sewage sludge supplies 18,000 megawatt (MW) hours to the power grid worth $1.4 million annually. This waste-to-power project helps Seattle City Light meet its goal to generate at least 15% renewable power.

Sewage sludge carries risks. If not properly composted, it can still host unwanted bacteria and viruses. As mentioned above, industrial waste can be improperly disposed of in sewers. Household products, like paint, pesticides, and leftover prescription drugs, are often dumped into toilets and sinks. Hazardous chemicals can persist in both sewage and greywater even after treatment. As a result, sewage sludge may contain a variety of toxic metals such as lead and silver as well as dangerous organic chemicals including pesticide and herbicide residues. Batches of sludge are tested before being spread as biosolid to ensure they meet regulatory stipulations; but some U.S. states ban land application of sludge for crops that people will eat. (**Figure 12.33**). The presence of contaminants in sludge has triggered protests and pushback in many communities.

Justin Kase zsixz/Alamy Stock Photo

Figure 12.31 Composted biosolids derived from sewage sludge in the United Kingdom have been pelletized and are designed for bulk application to agricultural fields as a soil amendment and fertilizer.

Sewage and storm water are often carried by the same sewers, leading to further problems. During heavy storms, when the sewage system is carrying more water than the treatment plant can process, combined sewers can't handle the load. This overload leads to **combined sewage overflows**, which spill raw or incompletely treated sewage into lakes and rivers. After such events, swimming beaches often close because of high levels of coliform bacteria diagnostic of fecal matter and sewage. Some beaches are

iStock.com/Simone Hogan

Figure 12.30 Drought-tolerant landscaping is watered using reclaimed (grey) water in California.

Robert K. Chin - Storefronts/Alamy Stock Photo

Figure 12.32 Eight anaerobic digestors generate methane from sewage sludge, with the New York City skyline in the background. Their shape lead to their nickname, digester eggs.

Figure 12.33 Warning sign: biosolids have been spread here on a cattle ranch in Florida. The goal is to grow better grass for the cows.

Figure 12.35 Composting toilet in the New Zealand wilderness. (Who could ask for a more comfortable setting?)

affected so frequently by combined sewage overflows that permanent warning signs are posted (**Figure 12.34**).

At present about 72% of all sewerage in well-resourced countries is treated by centralized systems, as described above, while only around 4% of sewerage in less resourced countries is collected and treated. Inexpensive, small-scale, domestic sewage treatment is especially appealing to consider in these largely rural regions (**Figure 12.35**). Alternatives to centralized sewage treatment include the use of composting toilets using wood chips and earthworms to speed up decomposition, and aerobic treatment systems (small scale bioreactors) so that domestic sewage, including most kitchen wastes, can be taken care of onsite, at home.

Hundreds of millions of people lack basic, clean drinking water in large part because human and animal waste enters aquifers and surface water bodies. One cost-free and easy way to implement local approach for treating contaminated drinking water is solar disinfection. Untreated water is placed in transparent containers, such as inexpensive plastic water bottles, and exposed to sunlight for a prolonged period (**Figure 12.36**). The combination of high temperature and ultraviolet (UV) radiation in the sunlight disinfects the bottled water–as simple as that! One disadvantage, however, is the time required; as much as 6 hours of exposure on a sunny day, 2 days on a cloudy one. The system also works best only at lower latitudes (<35°), where sunlight is strongest. Addition of catalysts such as titanium oxide can speed up disinfection and adding citric acids or riboflavin will make it more effective. In any case, low-cost, easily implemented innovations such as this could provide safe drinking water in many regions where waterborne diseases are endemic.

12.1.5 Hazardous Waste

When someone mentions hazardous waste, you might visualize 55-gallon drums leaking a foul-smelling liquid–which is not wrong, but **hazardous waste** is more than that (**Figure 12.37**). According to the U.S. EPA:

Figure 12.34 The Potomac River as it flows through Washington, DC, is subject to repeated combined sewage overflows so frequently that there are permanent warning signs.

Figure 12.36 Solar disinfection in Mukuru-Kayaba, Nairobi, Kenya, a settlement where the piped water is often contaminated by wastewater.

Bettmann/Getty Images

Figure 12.37 In 1983, the U.S. EPA brought a mobile filtration unit to the Enviro-Chem Plant north of Indianapolis to capture hazardous waste after it leaked during heavy rain.

> … a hazardous waste is a waste with properties that make it dangerous or capable of having a harmful effect on human health or the environment. Hazardous waste is generated from many sources, ranging from industrial manufacturing process wastes to batteries and may come in many forms, including liquids, solids gases, and sludges.

While it's tempting to think of hazardous waste as strictly having an industrial origin, some hazardous waste is related to residential use (think of the cleaning agents and insecticides found under a sink or in the cupboard of a typical kitchen), and some is generated by agriculture (pesticides, fungicides, and herbicides). Hazardous waste can also be flammable, chemically reactive, and corrosive. Improper disposal means that hazardous waste finds its way into municipal solid waste and wastewater, too (**Figure 12.38**).

Even though hazardous waste is regulated around the world, some escapes into the environment. Every second, industries worldwide release around 300 kilograms (660 pounds) of known toxic chemicals. Some waste escapes despite pollution-control regulations and management policies–other releases are intentional as discharges permitted at levels thought to minimize harm to people and the environment.

Independent Picture Service/Alamy Stock Photo

Figure 12.38 Typical household hazardous waste that finds its way into the municipal solid waste stream.

Much hazardous waste contains synthetic chemicals–those made by people, which otherwise do not exist in nature. As of 2020, the Chemicals Abstracts Service had catalogued nearly 160,000 individual commercial chemicals in daily use around the world, and the number could be much higher–as much as 350,000 according to Chemical Engineering News. The U.S. National Toxicology Program notes that this number increases by about 2,000 every year in the United States alone (as of 2022). Many of these substances are not thought to be harmful, but because there is so little testing, we simply don't know whether they are or not–at least, not yet.

Some of the most hazardous chemicals humans have made and released into the environment are pesticides, herbicides, and flame retardants. They belong to a group of compounds referred to as **persistent organic pollutants (POPs)** because they persist in the environment long after their initial use. Characteristically, they are nearly insoluble, often clinging to soils and organic material. Most are large molecules with high molecular masses, and are resistant to chemical breakdown in the environment. The so-called **Dirty Dozen** are of greatest international concern (**Table 12.4**).

Some of the Dirty Dozen chemicals are pervasive in the environment; others travel long distances after they are released through a process called **global distillation**. Global distillation involves chemicals that evaporate where it is warm, then are transported in the air to cooler regions, generally at higher elevations and latitudes. There, they condense and return to the surface in rain and snow (**Figure 12.39**). These chemicals end up in soil and water bodies, then in plants and animals and from there, reach people. Global distillation explains why arctic people and the animals that they consume have higher levels of some POPs (PCBs in particular) than those living at mid latitudes.

The world still lacks an adequate monitoring system for evaluating the entry and spread of chemical wastes through the environment. Highly sensitive analytical techniques can now detect the presence of hundreds of these pollutants in the blood of ordinary persons living today. Measured concentrations are usually very low but the pace of synthetic chemical creation has outpaced the ability of scientists to quantify the effects of these POPs. Lower income minority populations in the United States appear especially vulnerable relative to the general American population because these people are more likely to live near sites where POPs are used, processed, and disposed (**Figure 12.40**) (see Case Study 1.1 in Chapter 1).

Steps are being taken to address impacts of hazardous wastes on people and the environment. The United Nations

Table 12.4 The Dirty Dozen (POPs)

Aldrin–An insecticide and a potent POP; banned in most countries beginning in 1970s; still found lingering in the environment and in the blood of most people in developed and developing countries like India in trace amounts today.
Chlordane–A pesticide no longer in use in United States; implicated in testicular and prostate cancers, migraines, diabetes onsets, anxiety, and depression.
DDT–An insecticide; 40,000 tons per year applied worldwide 1950-1980, before bans put a stop to much production; very hard on birds, causing lethal thin-eggshell syndrome. May impact human endocrine systems and cause genetic harm. Brought to the attention of the public dramatically by Rachel Carson's important book, *Silent Spring*.
Dieldrin–Introduced as an alternative to DDT following the political fallout from *Silent Spring*, but is a POP and readily enters the food chain as a bioaccumulator–a substance that, once inside the body, stays there. Linked to Parkinson's disease and breast cancer, and causes damage to immune, reproductive, and nervous systems. Leads to deformities in male babies.
Endrin–Used as a pesticide attacking insects, particularly in fields of cotton. Other crops include rice, sugar cane, grain, and sugar beets. It is also applied to kill rats in orchards. This is an organochloride and a POP that bioaccumulates in fatty tissues. It readily attacks the central nervous system. Now banned in the U.S., but used still in some lesser developed countries.
Heptachlor–One of the pesticides along with DDT questioned by Rachel Carson in *Silent Spring*. A POP organochloride that bioaccumulates and persists in soils for decades. Still legal in the U.S. for control of fire ants in the South and Midwest. Gets into breast milk readily and is probably carcinogenic. It can be very hard on the central nervous system of exposed humans.
Hexachlorobenzene–Once used as a fungicide in wheat fields to control a plant disease called "bunt." A POP, now banned worldwide. Carcinogenic for lab animals. People ingesting it in Turkey in the 1950s suffered liver damage and skin lesions
Mirex–An insecticide and fire retardant in plastics, rubber, paint paper, and electrical goods. No human incidents of negative exposure to date, but the study of laboratory animals indicates that it could be quite carcinogenic–hence banned.
Polychlorinated biphenyls (PCBs)–Used in transformers, capacitors, fluorescent bulbs, oil in motors, and cable insulation (See Case Study 1.1, Chapter 1). Highly carcinogenic, leading to diseases ranging from brain, breast, and liver cancer to melanoma. One of the most highly toxic of hazardous wastes if not contained.
Polychlorinated dibenzofurans–Related closely to PCBs, with similar potential impacts. Used as fire retardants, dielectric fluids in cables, transformers, and added to some plastics. Burning of MSW and oil can produce dibenzofurans. A common product of pulp and paper mills.
Polychlorinated dibenzo-p-dioxins–A group of 75 related chemicals ("dioxins"), among the most poisonous known compounds. Used to make chlorinated pesticides. Mostly human-made, but in small amounts also produced by volcanoes and forest fires. Very carcinogenic: it attacks the central nervous, reproductive, and immune systems. Identification of a dioxin concentration forced evacuation of the town of Times Beach, Missouri, in 1983. Times Beach has been abandoned ever since then.
Toxaphene–An insecticide widely used in the American South on cotton fields in the 1960s and 1970s. Remains in soil without breaking down for 1 to 14 years. Spreads naturally and rapidly through the natural environment. No human harm incidents known, but tests on lab animals show that it attacks central nervous systems, the thyroid gland, liver, and kidneys.

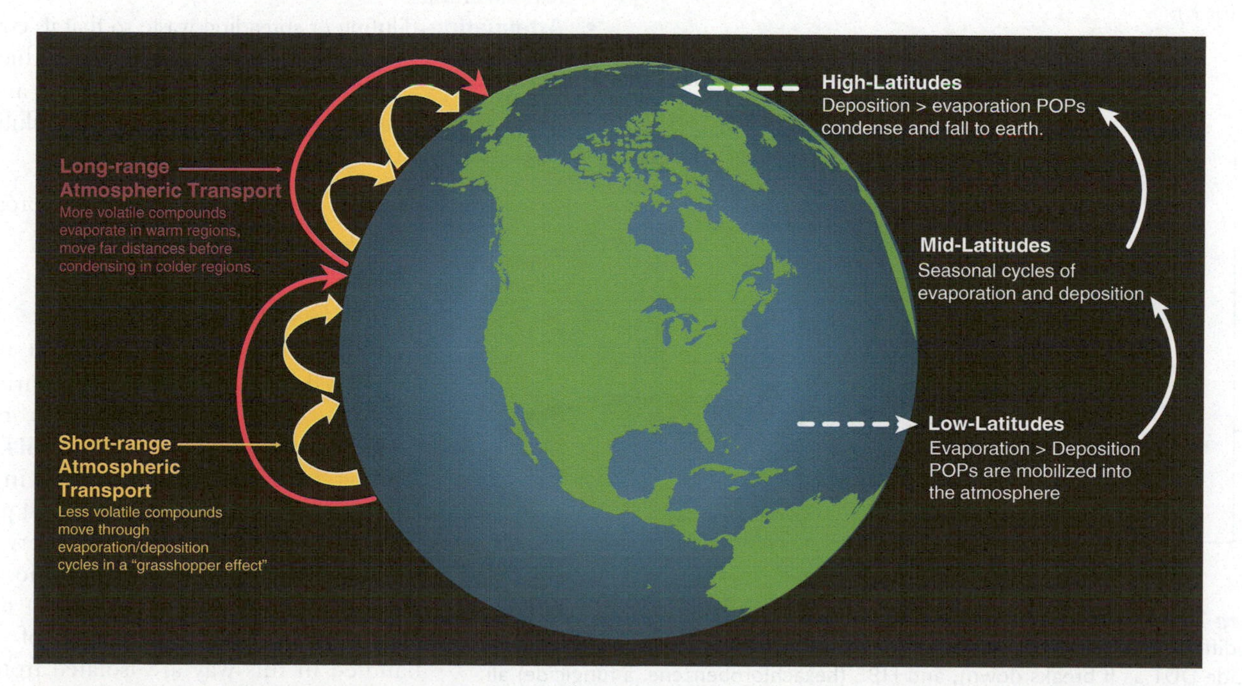

Figure 12.39 Diagram of global distillation that brings POPs to the arctic, thousands of kilometers from where these chemicals are used.

Andrew Lichtenstein/Corbis News/Getty Images

Figure 12.40 Exxon oil refinery and chemical production plant adjunct to a school in an African American community along the Mississippi River, south of Baton Rouge, Louisiana–an area informally known as "Cancer Alley."

Stockholm Convention (signed in 2001 and effective 2004) banned the generation of certain wastes including 12 POPs. The treaty has proven to be effective and levels of most POPs on the banned list have started to drop in soil and water–but not always in people. Because these chemicals persist so long in the environment, they continue to be available and ingested by living organisms (**Figure 12.41**). In 2009, 16 more POPs were added to the banned list by the Stockholm Convention, those included perfluorinated alkylated substances (PFAS)–so-called forever chemicals (because they resist degradation) that were used in firefighting and water- and stain-resistant clothing (see Chapter 8).

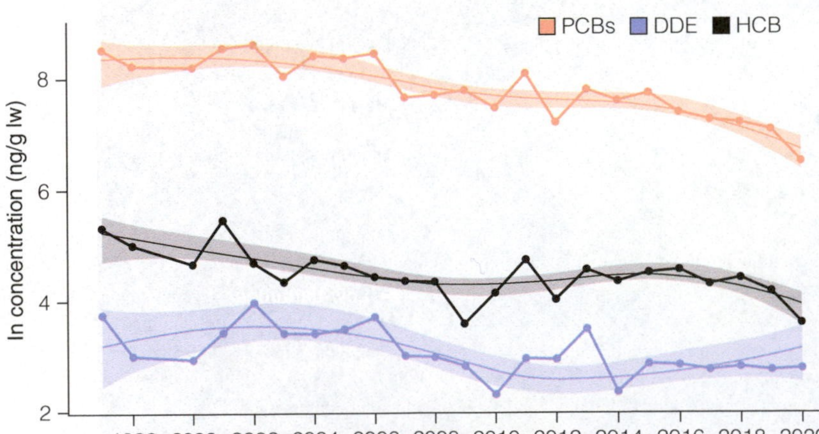

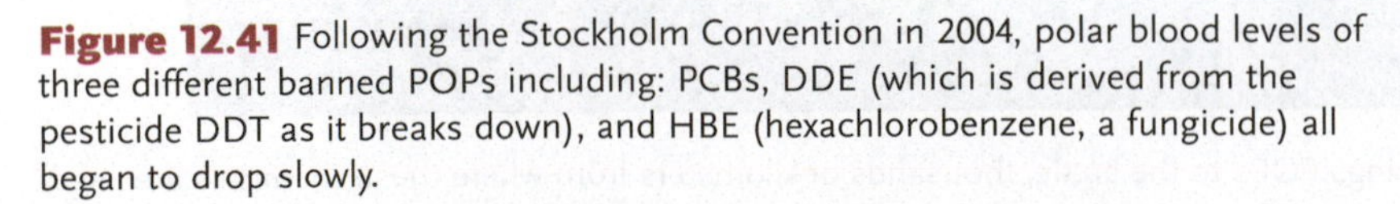

Figure 12.41 Following the Stockholm Convention in 2004, polar blood levels of three different banned POPs including: PCBs, DDE (which is derived from the pesticide DDT as it breaks down), and HBE (hexachlorobenzene, a fungicide) all began to drop slowly.

12.2 Waste Management Strategies (LO2)

Successful strategies for dealing with waste require considering the pathways by which people and other organisms could be exposed to wastes. The U.S. EPA explicitly considers exposure pathways in its analysis of waste management strategies, particularly at sites that contain materials to which exposure could prove dangerous to people. Waste becomes a problem only when there are exposure pathways to people and the environment.

There are various media through which people and other organisms may be exposed to waste or chemical constituents of that waste: air, water, soil and dust, food, aquatic biota, or manufactured consumer products. Conceptual diagrams can be useful for understanding the movement of such contaminants through both the Earth System and the biosphere (see Chapter 2). For example, hazardous waste can be released from a landfill or a spill and move into groundwater that people might drink or air that people might breathe. Farm animals could ingest the material and then transfer it to people once the animals are eaten. If the waste is volatile, fumes might seep into people's homes (**Figure 12.42**). Waste management strategies are designed to minimize such exposures.

Approaches to dealing with solid and liquid wastes, regardless of how and where these wastes are generated, fall into three main categories:

- **Isolation**–encapsulating, burying, or in some other way isolating waste from people and the surface environment.
- **Incineration**–the burning of waste, leaving behind a much smaller volume residue that then can be attenuated or isolated and, in the case of some materials, rendered less hazardous.
- **Attenuation**–diluting or spreading waste so that its concentration is reduced to low levels at which the waste has little impact on people and the environment, in other words, "Dilution is the solution to pollution."

Let's consider each of these approaches in turn:

12.2.1 Isolation

In well-resourced countries, most wastes are sorted, concentrated, and buried at carefully designed disposal sites called sanitary landfills, or simply **landfills**. The method involves spreading trash in thin layers, compacting it to the smallest practical volume with heavy machinery, and then covering the day's accumulation with at least 15 centimeters (6 inches) of soil (**Figure 12.43**). Compartments of waste handled in this way are isolated from the environment in rhomb-shaped units, or

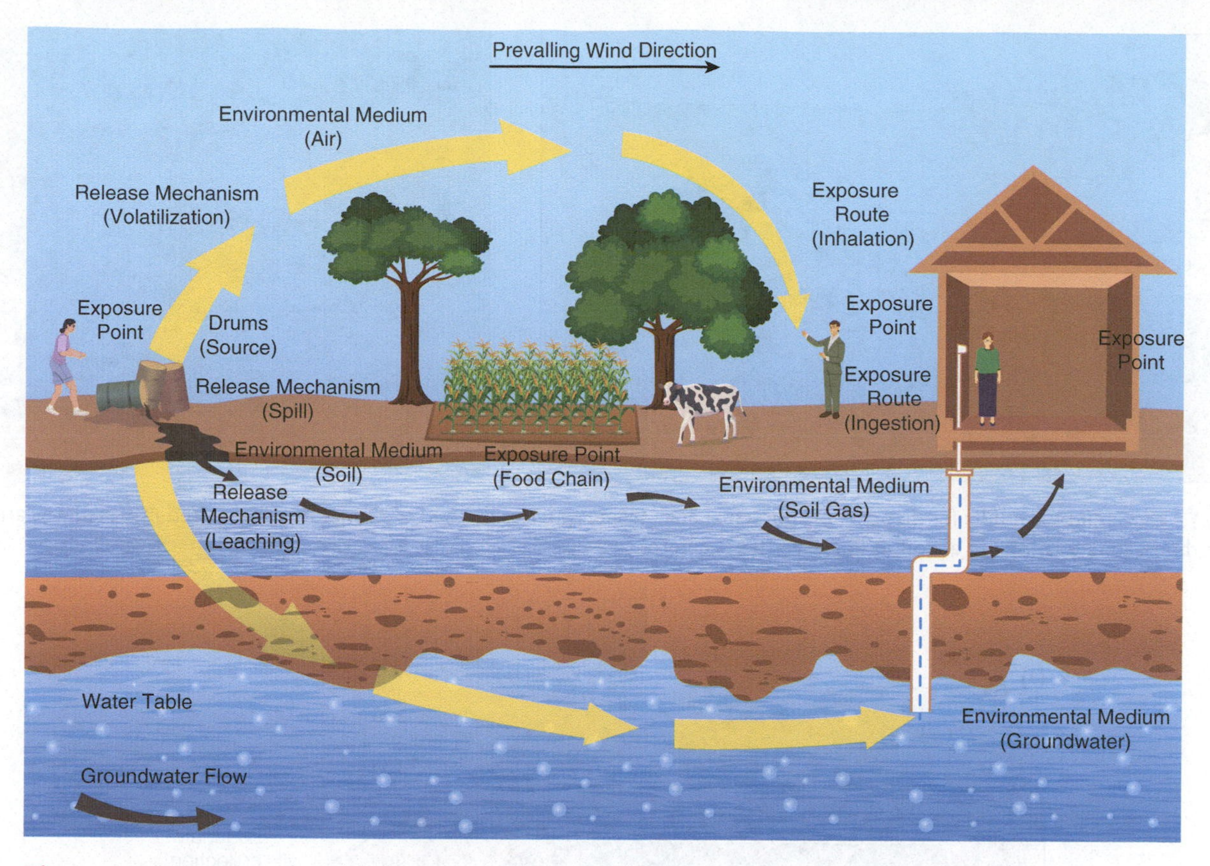

Figure 12.42 Conceptual model of exposure pathways to a hazardous waste release. (Modified from Agency for Toxic Substances and Disease Registry)

landfill cells (**Figure 12.44**). When finished, the landfill is sealed by 50 additional centimeters (20 inches) of compacted soil and is graded so that water will drain off the finished surface. This reduces water infiltration and the production of potentially toxic fluids within the fill (**Figure 12.45**). A well-operated sanitary landfill is relatively free of odors, blowing dust, and debris.

Unfortunately, most landfills in lesser-resourced countries have not achieved this level of waste management: over 90% of disposal sites are open dumps, according to the World Bank (2022). Changing this is a major priority for international development agencies, because the amount of waste generated per capita in the lesser-developed world is projected to increase by 40% between now and 2050, compared to just 19% in wealthier countries. Cost is a major concern. Operating expenses for sanitary landfill waste management typically run around \$100 per tonne (\$110 per ton) of garbage. For open dumps, the cost is as low as \$35 per tonne (\$40 per ton). It is not a matter of will, but a question of funding (**Figure 12.46**).

Figure 12.43 Bulldozer covering municipal solid waste with a layer of soil.

Leachate is the water, primarily from precipitation, which filters down through a landfill, acquiring (leaching out) dissolved chemical compounds and fine-grained solid and microbial contaminants as it goes (**Figure 12.47**). The content of the waste will determine the content of the leachate. Leachate quality is assessed using groundwater monitoring wells, and if it is found to be contaminated, leachate is often collected and treated onsite or piped to a sewage treatment plant. Protecting the local environment from groundwater leachate is a major consideration in designing and operating waste-disposal sites.

Landfills or individual sections within them are classified in three ways based on their underlying geology and the potential impact of leachate on local groundwater or surface water (**Table 12.5**). Each of the three classes is certified to accept specific wastes grouped according to differing degrees of chemical reactivity. Some landfill sites encompass all three classifications within their boundaries (**Figure 12.48**).

Figure 12.44 Sanitary landfills for disposal of municipal solid waste. (a) The method of dump and cover used in a sanitary landfill. (b) An urban sanitary landfill in action in Palos Verdes, California.

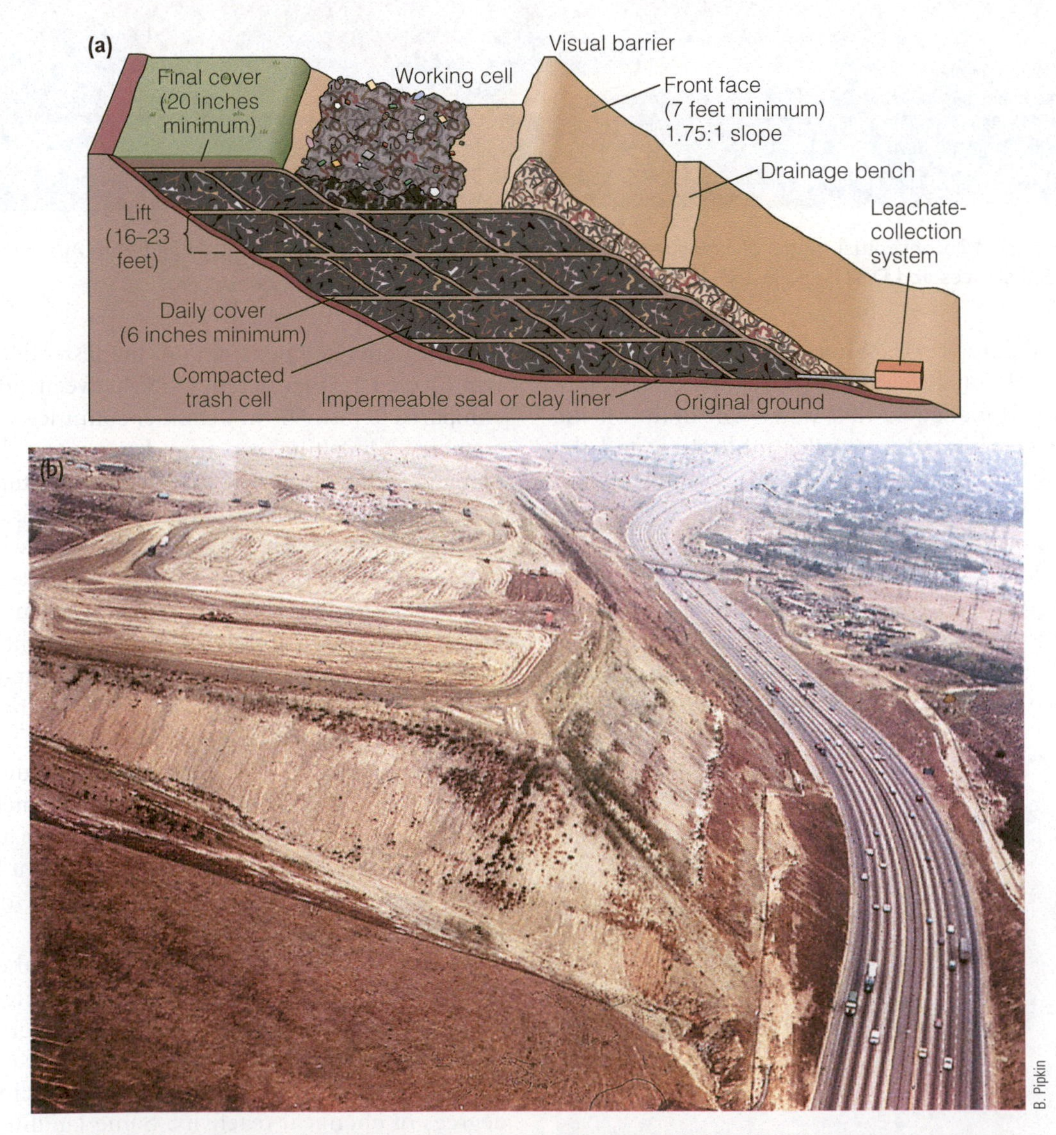

Figure 12.45 Landfills can be massive, engineered structures. (a) Local government landfill specifications for cells, final cover, and leachate-collection system. Note the barrier for reducing the visual effect of the landfill. (b) A full, graded landfill before planting in Puente Hills, Los Angeles County, California. Final grades and slope benches for intercepting runoff are visible.

Picture alliance/Getty Images

Figure 12.46 Open dump with a mixture of municipal solid waste, metal goods, and electronics in Accra, Ghana.

The world's first solid-waste sanitary landfill was established in Great Britain in 1912. Known as "controlled tipping," this method of isolating wastes expanded rapidly and had been adopted by thousands of municipalities by the 1940s. The city of Fresno, California, claims to be the first city in the United States to use a sanitary landfill for waste disposal. Perhaps counterintuitively, the number of U.S. landfills has declined in the last few decades, from more than 7,900 in 1988 to just around 2,000 in 2022. Despite the drop, total national landfill capacity has

International Development Issues/Alamy Stock Photo

Figure 12.47 Leachate seeping uncontrolled from a landfill.

Table 12.5 Groundwater Quality Measured Near One Landfill

	Measurement Site		
Water Characteristics	**Local Groundwater**	**Leachate**	**Monitoring Well**
Total dissolved solids, parts per million (ppm)	636	6,712	1,506
BOD,* milligram/liter	20	1,863	71
Hardness, ppm	570	4,960	820
Sodium content, ppm	30	806	316
Chloride content, ppm	18	1,710	248

Source: Brunner, D. R., and D. J. Keller. 1972. *Sanitary landfill design and operations.* Washington, DC: Environmental Protection Agency.

*Biological oxygen demand (BOD): the amount of oxygen per unit volume of water required for total aerobic decomposition of organic matter by microorganisms.

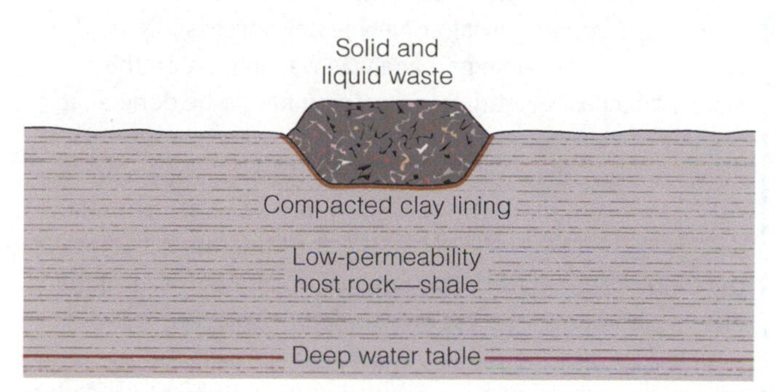

(a) Class I landfill

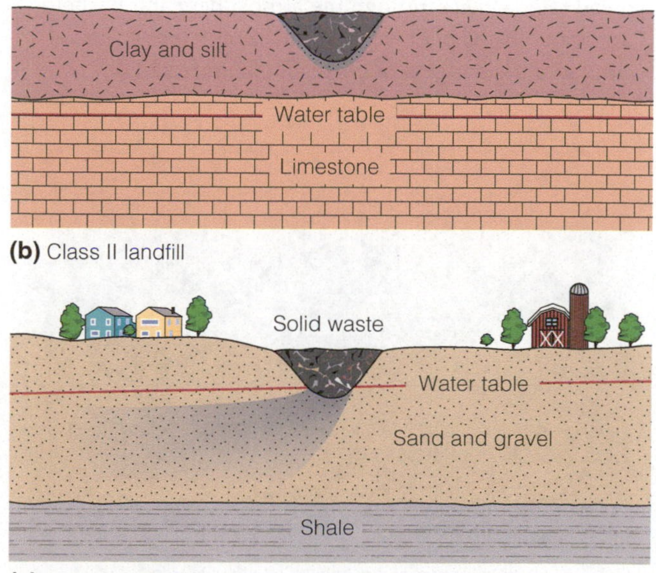

(b) Class II landfill

(c) Class III landfill

Figure 12.48 The geology of the three classes of landfills (a–c). Note that a Class I fill offers maximum protection to underground water and a Class III fill offers no protection.

Environmental Legislation Helping the United States Manage its Wastes

Until recent decades, the prevailing approach toward dealing with wastes in the United States, indeed most of the world, was to dispose of it any way one saw fit on one's personal property or in the local environment. Wood wastes were burned, and kitchen wastes often went out the window for pigs and other livestock to eat. Pigs served as a major solution to accumulating garbage in most rural areas. Some communities designated open-air dumps to control the spread of wastes, usually just gullies or shallow excavations at the edge of a town. Odors, vermin, and air pollution increased as human populations grew.

The situation was far worse in big cities. Residents grew sick from the waste accumulating in streets and in water supplies, contributing to cholera, diarrhea, typhoid fever, and other diseases. Beginning in 1881, New York City, began collecting its waste using hired labor. Between 1900 and 1931, the city simply dumped its garbage into the adjacent ocean (**Figure 12.49**). This created a slimy mire along shorelines that thickened ominously over time. Free-for-all ocean dumping stopped only after New Jersey successfully sued New York City because garbage was washing up on the state's beaches. Eventually something had to be done at an organized government level.

While the urban garbage crisis took place, the environmental movement in the United States was already underway, as shown by the writings of John Muir, the activism of the Sierra Club, the photography of Ansel Adams, and the efforts of other conservation-minded people, primarily directed toward the preservation of wilderness. Cities drowning in their own waste were simply no great cause for them. More inclusive political activism blossomed after World War II, however, when surface nuclear weapons tests spread radioactivity through the atmosphere and Rachel Carson published her seminal book, *Silent Spring* (1962) warning of the dangers of pesticides (**Figure 12.50**).

Everett Collection Historical/Alamy Stock Photo

Figure 12.49 Loading a garbage barge in New York. People are scavenging the garbage at the foot of Beach Street.

Everett Collection Historica/Alamy Stock Photo

Figure 12.50 Rachel Carson, author of *Silent Spring*.

Suddenly, we needed to be concerned about more than simply protecting "wild Nature." A key "Ah-hah!" moment was the first photo of Earth taken by astronaut William Anders as Apollo 8 orbited the Moon on December 24, 1968 (**Figure 12.51**). The public impact of the image showing a small, blue-and-white ball floating in the black emptiness of space galvanized public imagination as nothing had before. It had both scientific and spiritual meaning: it jolted many citizens out of their immediate neighborhood mentalities, showing how everything and everyone are interconnected. To preserve raw wilderness, we had to be mindful of what we did with our wastes. It is no coincidence that the first Earth Day soon followed, and the political establishment took note (**Figure 12.52**).

The Nixon Administration, working with Congress, passed the **National Environmental Policy Act (NEPA)** of 1969, which requires an in-depth field study and the issuance of an **Environmental Impact Statement (EIS)** on the consequences of all projects on federal land, or even on privately owned or municipal lands, whenever any federal money is used to help support them. If deemed environmentally impactful, proposed projects are stopped before breaking ground, or taken back to the drawing board. The EPA came into existence to enforce this act.

Environmental Legislation Helping the United States Manage its Wastes (*Continued*)

Figure 12.51 Earthrise over the moon.

Figure 12.52 Protests at the first Earth Day in Welland, Ontario, Canada, on April 22, 1970.

Landfill and other waste disposal projects fall under the purview of NEPA. Questions to be answered include: "Will a proposed landfill be large enough to collect and safely store all of the waste that it receives?" "Will it be designed sufficiently to keep wastes from contaminating air and groundwater supplies?" NEPA compels the careful siting and planning, operation, and closures of all major waste disposal facilities.

In 1976, the U.S. government passed the **Resource Conservation and Recovery Act (RCRA)**. The RCRA regulates the generation, transportation, treatment, storage, and disposal of hazardous wastes, including steps to be taken in the case of accidental releases to the environment (**Figure 12.53**). It also provides a framework used for the management of nonhazardous wastes. The Act encourages solid-waste planning by the individual states.

In 1980, a third key piece of waste-related legislation passed into law, the **Comprehensive Environmental Response, Compensation and Liability Act (CERCLA)**, meant to provide funding (known as a **Superfund**) to clean up major waste sites, particularly old mining and industrial activities no longer in operation (e.g., Love Canal; see Chapter 1). Unfortunately, expenses required to complete cleanup have generally exceeded the ability of CERCLA to operate fully in a timely way, but the act has nonetheless done much to remediate environmental damage.

These laws sent a message to future generations that the then-prevailing "free for all" approach to disposing of wastes of all kinds, hazardous and benign, was unacceptable. The laws mandated cradle-to-grave systems for impounding wastes and monitoring them to ensure their low potential for migrating into freshwater supplies and otherwise contaminating the natural environment. State governments and many other national governments have since come up with parallel approaches to managing humanity's ongoing environmental impacts. The world is a far cleaner and healthier place because of such legislation.

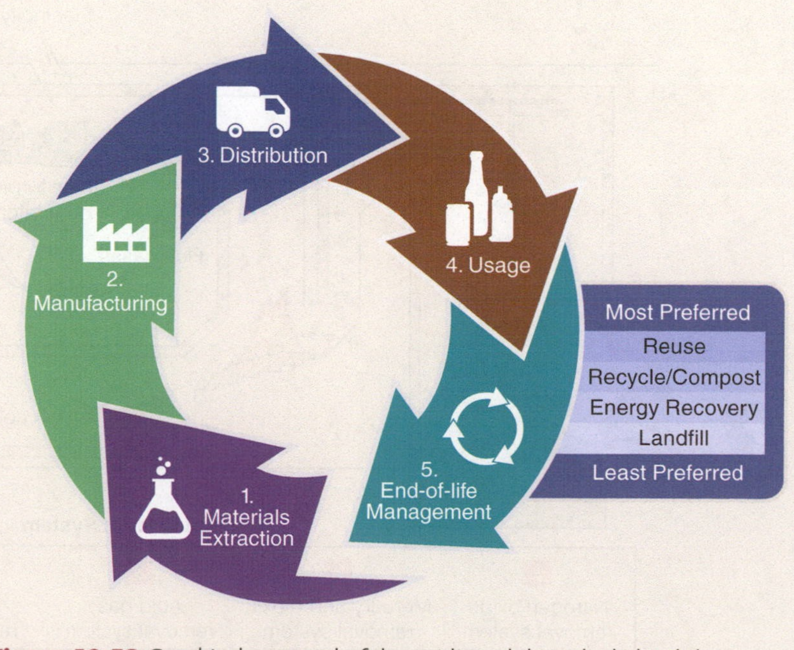

Figure 12.53 Graphical portrayal of the guiding philosophy behind the Resource Conservation and Recovery Act (RCRA). (Modified from Sustainable Material Management's Life-cycle Perspective, U.S. EPA)

increased as individual landfills have grown larger. Most landfill closures have occurred because they had reached their capacity, but many others could not meet the stringent EPA standards for protecting groundwater (see box feature "Environmental Legislation Helping the United States Manage its Wastes" on page 580).

12.2.2 Incineration

Incineration, or burning, significantly reduces the volume of waste, sometimes by as much as 90%. In addition to reducing volume, combustors, when properly equipped, can convert water into steam to generate electricity or fuel heating systems (**Figure 12.54**). **Scrubbers** (devices that use lime slurries to neutralize acid gases) and **filters** (devices that remove tiny ash particles and if equipped with activated carbon, toxic chemicals) significantly reduce gases and other pollutants emitted to the atmosphere during combustion so that emissions must meet the applicable government standards. The use of incineration is greatest in areas with high population density, little open land, or shallow water tables. In the United States, Massachusetts, New Jersey, Connecticut, Maine, Delaware, and Maryland, for instance, burn more than 20% of their waste (**Figure 12.55**). New York City is a leader in trash incineration and energy recapture (known as **waste to energy** plants), burning as much as a million tons, or 30% of all its waste, at a single centralized facility; enough to power 46,000 homes. That facility is located across the river in New Jersey. New York banned municipal incinerators within the city limits in 1990.

Visions of America, LLC/Alamy Stock Photo

Figure 12.55 The Wheelabrator waste to energy incineration plant in Baltimore, Maryland. The heat here is recaptured and used to generate enough electricity to power about 40,000 homes from 2,250 tons of Baltimore's municipal solid waste (post-recycling). It also provides steam for heating 255 downtown businesses. Recycling removes most eligible materials before combustion.

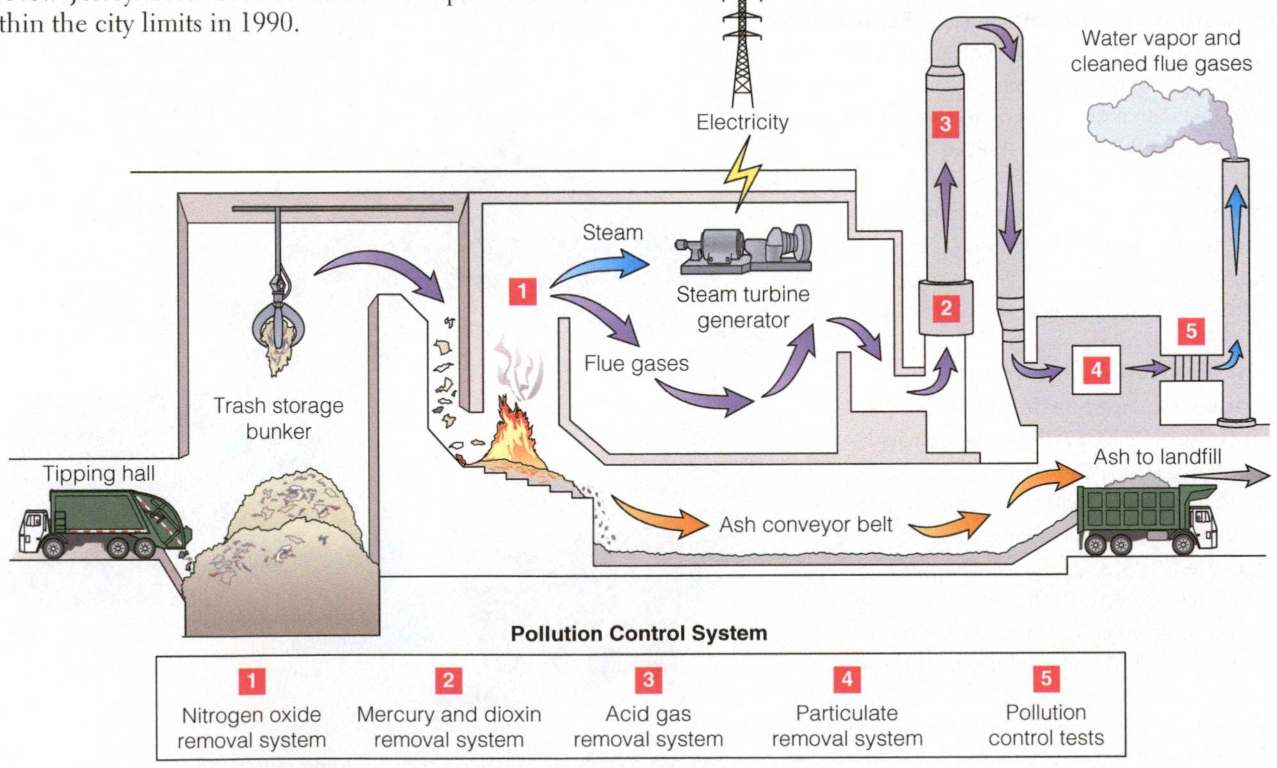

Figure 12.54 Diagram showing optimal industrial trash incinerator operation. Advantages of trash incineration include significant reduction of waste volume that otherwise would enter landfills, and generation of electricity. Air pollutants can be reduced using emission control systems. (Adapted from Ecomain Industry Solutions, Siemens Corporation)

To reduce dangerous emissions, waste is combusted at very high temperatures (1100°C; 2000°F), and incinerator exhausts are fitted with sophisticated scrubbers. The gases that ultimately emerge are largely benign; nitrogen, carbon dioxide, water vapor. The emission of methane, a highly potent greenhouse gas, is significantly less with incineration of trash than what is released from slow decay in landfills.

Incinerator ash presents another problem, however, because it can be hazardous waste, containing high concentrations of toxic elements like lead, silver, and cadmium. The ash is shipped largely to landfills after it is produced, while scrap metal is drawn out using giant magnets and recycled for use typically before the waste is burned. Three or 4 days of New York City garbage can yield enough metal to make 21,000 auto bodies!

More recently, incineration has fallen into disfavor, with only 72 U.S. incinerators remaining in 2022 from the 186 operating in 1990. The reasons for the reduction include more stringent federal and state air pollution emission requirements, public opposition, and expense. Direct disposal of MSW in landfills for instance, is less expensive than disposal of the residual ash from burning.

12.2.3 Attenuation

The old adage that "dilution is the solution to pollution" is based on the idea that if something harmful is dispersed widely in very low concentrations (attenuated), it won't hurt anyone. This has been the operating philosophy behind emissions of waste pollutants into the atmosphere for centuries; a major reason for tall smokestacks on factories (e.g., Sudbury; see Chapter 4). Today, innovations such as those already described for incinerators assure that most toxic emissions are captured before leaving smokestacks, although some still escapes.

Another attenuation approach is to mix large amounts of clean water from streams, rivers, or the ocean, with liquid wastes–including treated wastewater–to make contaminants from the waste less harmful to the environment. This approach presumes that nature has a certain ability, termed **assimilative capacity**, to break down well-diluted waste pollutants through bacterial activity, the effect of sunlight, and/or other environmental factors. In practice, however, we've learned that ecosystems are far more complex and dynamic than is often presumed. We can't always count on some presumed "assimilative capacity" to save the day. Potentially harmful contaminants, even in trace amounts, may persist. Moreover, attenuation does not prevent bioaccumulation of contaminants in organisms to harmful levels. A level of precaution is needed before undertaking any attenuation scheme; what we don't know can hurt us!

Time of year also matters. During the late summer months, when river and stream levels are low, dilution decreases and water quality may become a problem as concentrations of contaminants build up. In the winter, when rivers and streams may freeze over, oxygen levels in natural waters drop, and the organic matter in admixed waste can easily use up what oxygen remains as it decomposes, causing fish and other organisms to die. Where direct dumping of liquid wastes is permitted in small or modest amounts, government regulations strictly limit legal discharge amounts based upon concerns such as these.

12.2.4 Ocean Dumping

The ocean has been used for waste disposal for millennia. Until recent decades, ocean dumping was legal everywhere and all sorts of waste was hauled out to sea and dumped. Some waste, as in the case of New York City sewage and sewage sludge, was dumped with attenuation in mind. Other waste was sealed in drums or otherwise isolated before dumping. If any leakage took place after settling to the seabed, it was presumed that attenuation and the ocean's assimilative capacity would take care of it.

Ocean dumping of radioactive waste began during the Cold War. Between 1946 and 1970, the U.S. government sank nearly 50,000 210-liter (55-gallon) drums of low-level radioactive waste, mostly from laboratories, across 1,400 square kilometers (530 square miles) of ocean floor. This dumping site is now the Farallon National Wildlife Refuge, a sanctuary 50 kilometers (30 miles) west of San Francisco, famous for its abundance of marine mammals and great white sharks. Laboratory wastes weren't the only radiation sources deposited there.

In 1951, the Navy scuttled a 10,000-ton World War II aircraft carrier, the *U.S.S. Independence* (**Figure 12.56**) at the dumping site. The ship had been irradiated during two South Pacific atomic bomb tests in 1946 (**Figure 12.57**) and was then

Figure 12.56 The *U.S.S. Independence* being towed out from San Francisco to its watery grave.

© University of Washington Libraries

Figure 12.57 The *U.S.S. Independence* was one of 22 ships moored near the "Baker" test of a 21-kiloton nuclear weapon (about 40% larger than the bomb that destroyed Hiroshima) at Bikini Atoll in the South Pacific, 1946. The ships and the animals on them (pigs and mice) were supposed to be monitored for after effects. But almost all of the animals died from acute radiation poisoning. Because the bomb was detonated underwater, radioactive spray coated the ships and exposed the animals to large doses of very radioactive (but short-lived) elements.

towed back to San Francisco and used for testing radiological decontamination procedures before it was scuttled in 1951. A 1996 biological study revealed an order of magnitude greater concentration of plutonium isotopes 239 and 240 (nuclear weapons leftovers) in the Farallon Refuge fish relative to average values measured elsewhere in California waters–a sure indicator that contamination had spread from the waste.

The United States was not alone in dumping radioactive wastes at sea during the Cold War. The Soviet Union sank a large number of decommissioned nuclear submarines and reactor parts in arctic waters. In the Barents Sea alone, 8 scuttled submarine hulls and 16 nuclear reactors (6 still containing fuel), plus 8,200 tonnes (9,000 tons) of discarded fuel assemblies lie on the shallow seabed which hosts a large commercial fishery. Large amounts of low- to intermediate-level radioactive waste were routinely dumped at more than 50 sites in the North Pacific and Atlantic oceans before an international moratorium in 1982.

Today, ocean pollution in the United States is governed by the Marine Protection, Research and Sanctuaries Act of 1972, better known as the **Ocean Dumping Act**. Amended in 1988, this legislation requires anyone dumping waste into the ocean to have an EPA permit and to provide proof that the dumped material will not degrade the marine environment or endanger human health. In U.S. waters, the only legal dumping now consists of relatively harmless sediment dredged from coastal shipping channels and ports. The only potential risk this poses is the possible reintroduction of contaminants into the water column stirred up by dredging and settling.

12.3 The Challenges of Landfilling (LO3)

Landfills remain the most common destination for solid wastes, and even some liquid wastes stored in containers. But landfills only work well through careful design and keen understanding of what takes place within them over time. They are projects requiring the inputs of geologists, hydrologists, chemists, microbiologists, and civil engineers. Functioning at their best, landfills are an example of humans working with nature to control wastes and render them harmless to the surrounding environment.

12.3.1 Landfill Maturation Phases

Landfills may be regarded as giant biological reactors (**bioreactors**); they host decomposition of organic wastes, their most common type of garbage, through multiple phases defined by the ability of oxygen and water to access the buried, compacted garbage. In Phase 1, a period of initial adjustment, the packed refuse contains abundant oxygen, trapped in void spaces. Microbes that use oxygen to begin breaking down the refuse thrive, a process called **aerobic biodegradation**.

In Phase 2, a transition takes place. Oxygen decreases as it is used up by the aerobic microbes, and a new ecosystem of microbes develops, utilizing nitrates and sulfates (derived from decaying food wastes). This is the onset of oxygen-poor **anaerobic biodegradation**. Carbon dioxide is a by-product, and it begins to leak out of the landfill–a concern for scientists who monitor the sources of greenhouse gases, but by no means, as you'll soon see, their biggest concern.

In Phase 3, moisture and infiltrating water stimulates microbial decomposition of waste to produce acidic leachates with pH as low as 5.6 (around the acidity of normal rainwater), and hydrogen gas. The acids leach metals from the waste, and if toxic metals such as lead, cadmium, and silver are present (discarded batteries and electronics are a source of these metals), the leachate can carry these hazardous materials away from the landfill.

In Phase 4, microorganisms convert some of the acids produced in Phase 3 (acetic, propionic, and butyric) to CO_2 and methane (CH_4), which is an environmental worry given its greenhouse potency (see Chapter 3) and an explosion hazard (because it's flammable). Inevitably, the methane escapes the landfill–it is virtually impossible to keep confined. Altogether, around 11% of human-produced methane worldwide is landfill-derived, and landfills are the third-largest source of methane generated in the United States. Livestock and the natural gas industry are two other major contributors. The greenhouse warming equivalent of annual methane releases from U.S. landfills is equivalent to driving 20.3 million passenger vehicle miles, while the annual CO_2 release has the same climate impact as the

© Sally Brown

Figure 12.58 This magazine had been buried in a Seattle landfill for 51 years and is still readable (even if it might not smell very nice).

Table 12.6 Natural Decomposition Times for Various Types of Waste

Rates are based on observed decompositions in landfills, at sea, and in laboratories (e.g., plutonium half-life). Data: EPA; Mote Marine Laboratory, Florida; U.S. National Park Service.

Type of Waste (ranked in order of disintegration time	Time It Typically Takes to Fully Decompose Under Natural Conditions
Pesticides	Hours to several decades
Waxed paper cartons	3 months
Paper	2 weeks to 1 year
Chewing gum	5 years
Cigarette butts	2–25 years
Nylon	30–40 years
Aluminum cans	50–250 years
Disposable diapers	550 years
Monofilament fishing line	600 years
Plastics	250–1,000 years, depending on type and product
Batteries	500–1,000 years
Glass	Thousands of years
Plutonium-239	Hundreds of thousands of years because of 29,000 year half-life
Styrofoam	Unknown; years to hundreds of thousands of years

energy consumption of 11.9 million homes (2022 EPA data). Phase 4 is the longest and potentially most consequential stage of landfill maturation.

In Phase 5, microbial activity slows because waste decomposition had reached a stage where there are few nutrients, especially phosphorus, left to supply the landfill ecosystem. Remaining waste will degrade into material resembling the humus that enriches many soils (see Chapter 7).

As landfills become starved for oxygen, materials degrade far more slowly. There are many stories of landfills being excavated and newspapers decades old being easily readable because they are preserved in the low-oxygen environment (**Figure 12.58**). Different wastes take different lengths of time to degrade (**Table 12.6**).

12.3.2 What to Do About Landfill Gases?

A well-managed waste operation makes an effort to capture at least some of its landfill gas produced onsite (**Figure 12.59**). About 50% of this gas is methane, almost all the rest CO_2 with traces of hydrogen sulfide and other volatiles. Perforated pipes may be laid as the landfill builds up, cell-by-cell, or wells drilled to extract the gas with the aid of blowers (**Figure 12.60**). The excess methane can be combusted (flared) to produce water vapor and CO_2, which is less impactful on climate than methane. If not flared off, methane can be separated from the other gases to generate electricity by spinning turbines. Some methane may be burned to make steam and hot water in addition to electrical generation. Heat from combusting methane can be used to evaporate leachates collected from the base of a landfill. The harmful metals in the leachate will precipitate out as salts that can be easily collected, protecting the environment.

iStock.com/ValterCunha

Figure 12.59 Captured methane from the landfill to the left is piped to the generation station in the distance and burned.

Michael Vi/Shutterstock.com

Figure 12.60 Former landfill in Sunnyvale, California, capped and converted into a park-like public greenspace, with wells for capturing **landfill gas**.

Inga Spence/Alamy Stock Photo

Figure 12.61 High Density Polyethylene Geomembrane being installed in new California landfill as a liner to keep waste and leachate separated from the underlying groundwater.

12.3.3 What to Do About Leachates?

The composition of leachate in and leaving landfills is highly variable; routine sampling, mandated by governmental regulation in many nations, determines whether it is hazardous to people and the environment and how it is treated. Left unconfined and unaddressed, leachate from landfills holding toxic materials can pollute groundwater aquifers used for drinking. In one of the author's home state (Vermont), there are three former sanitary landfills that have been declared EPA Superfund sites, all because groundwater around them has been contaminated by leachate tainted with industrial chemicals. All three landfills accepted industrial waste when they were operating in the 1950s, 1960s, and 1970s. Some was permitted for disposal, much was not.

Leachates typically show high oxygen demand, meaning that they have the potential to consume large amounts of oxygen either by chemical reactions or the decomposition of organic material dissolved in the leachate. Some leachates contain high concentration of dissolved metals, some of which are toxic including lead, mercury, cadmium, and chromium. Other leachates contain tiny fragments of plastics, generally in the 100 to 1,000 micron (0.1 to 1 millimeter) range. They often carry organic pollutants including phthalates (plasticizers), resulting from the degradation of plastics including polyethylene and polypropylene–common packaging materials (Table 12.1). Leachates can off-gas ammonia, produced as nitrogen-rich waste degrades.

Many different geological circumstances determine if and how quickly leachates will mix with underlying groundwater, including the structure and permeability of the rock and sediment under the landfill, the amount of rainfall, and the depth of the water table. The most effective way to isolate leachate from the underlying geologic materials and the groundwater table is to line the base of the landfill with impermeable material knows as **geomembranes**. These rubber or plastic sheets keep leachates confined within landfills where they can be pumped out, stored and evaporated in collecting tanks (**Figure 12.61**). The tightly woven, higher-density geomembranes underlying landfills, up to several millimeters thick, have service lives lasting as long as 400 years, meaning that it takes that long for 50% the membrane to degrade; long enough for at least organic waste to completely break down. Beds of impermeable clay can also be spread beneath landfills during initial construction to strengthen potential for confinement (**Figure 12.62**).

If leachates do escape, their progress is tracked by drilling sampling wells downslope from a landfill (**Figure 12.63**). Pumping draws up groundwater for analysis to detect signs of contamination. In this way, the outline of a contaminant plume can be defined (see Chapter 1) and the groundwater pumped and treated if it is contaminated enough to present a hazard.

Leachates may be treated physically (e.g., evaporated, as mentioned above) and chemically. One promising new avenue of research in leachate treatment is **bioremediation**, the use of fungi, and bacteria, and other to convert "garbage juices" into less harmful waste. No two leachate situations are identical, however, and the approaches taken to deal with this problem vary accordingly. In any event, extraction, confinement by plastic liners, and bioremediation with follow up physical and chemical treatments, are all successful ways to manage this potentially hazardous liquid.

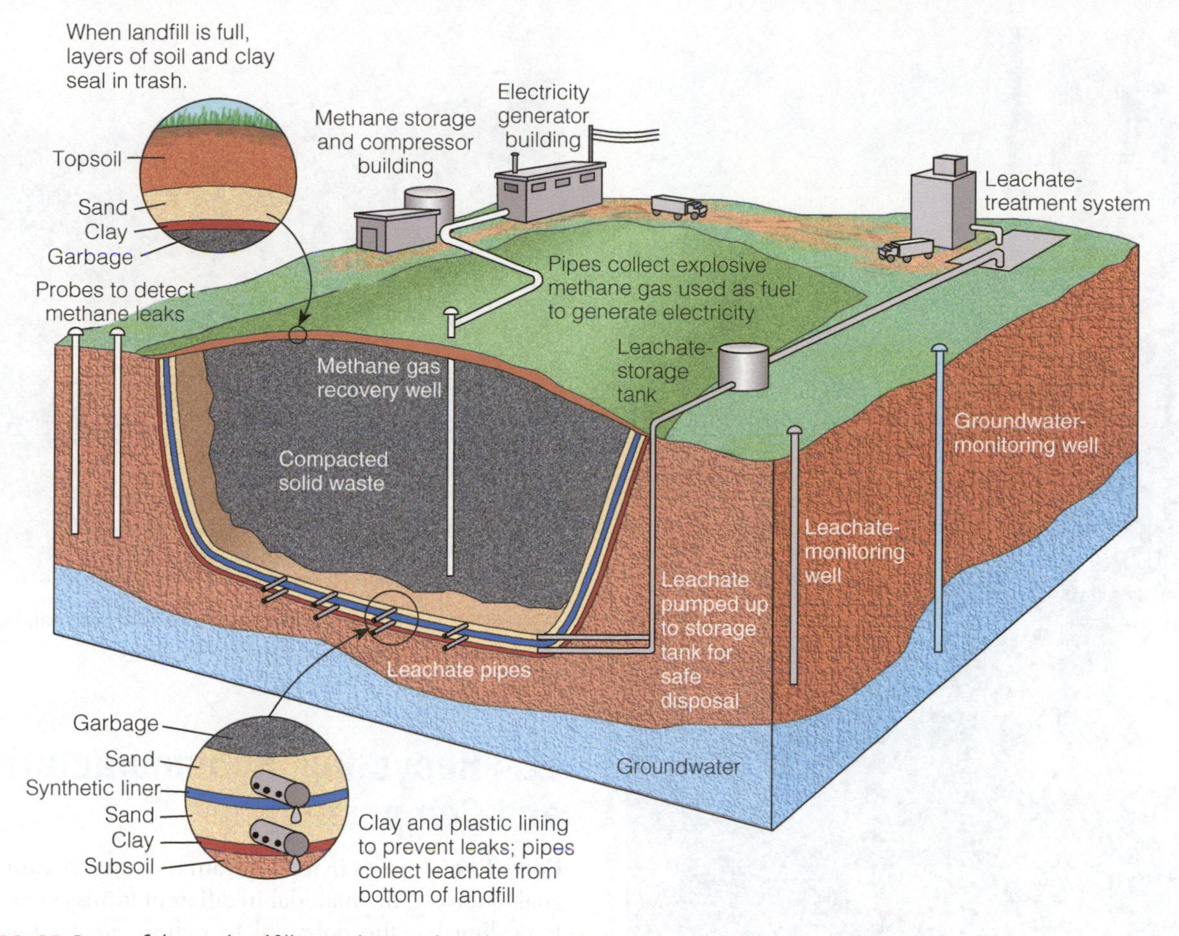

Figure 12.62 State-of-the-art landfills are designed to eliminate the problems of toxic leachate and methane escaping into the environment. Note the double lining (clay and plastic layers) to prevent leakage.

12.3.4 Landfill Subsidence

As landfills age, they settle, compact, and lose mass; their volumes decrease and their surfaces subside. Typically, surface sagging in a freshly filled landfill is rapid at first and diminishes to a much slower rate after the first 10 years (**Figure 12.64**). The rate of subsidence depends on many factors, including the climate, the kind of waste undergoing decomposition, and the degree of compaction and covering with fresh material that took place during filling. Organic matter decomposes more slowly in cold or dry climates than in warm and wet ones, accounting for some of this variability. Expulsion of gases and leachates from void spaces, and the increasing compressibility of particular waste items as they degrade, also account for settlement-rate variability. Some landfill thicknesses have decreased as much as 40% due to internal decomposition, meaning that a landfill that begins with a thickness of 15 meters (50 feet) may thin to as little as 9 meters (30 feet), possibly over just a few decades. In one case, a 6-meter (20-foot)-thick landfill in Seattle settled 23% in the first year of operation, while a 23-meter (76-foot)-thick fill in Los Angeles settled only 1% during the 3 years following completion—a two order of magnitude difference in settlement rates.

No matter the rate, any settlement on a landfill can result in cracks and fissures, allowing water to pond and percolate downward, creating additional leachate. Gases can escape more easily too, while open fissures also allow animals access to decaying garbage, posing another kind of public health hazard. To avoid these problems, properly operated landfills are compacted daily, and then recompacted, on a daily basis, using heavy tractors weighing up to 27 tonnes (30 tons).

At the completion of a fill, a surface geomembrane will be laid across the landscape to keep water out altogether—optimally for a period of decades. Surface geomembranes tend to be thinner than those liners at the bottom, and do not last as long as they are more exposed to the elements. On the other hand, their condition is easier to monitor. Compost, wood chips, greenwaste, and additional soil added atop the geomembrane promote the growth of grasses and other vegetation as a final cap atop closed landfills. The vegetation not only helps stabilize sites, but intercepts and transpires much precipitation, reducing the amount that can infiltrate to reach the geomembrane.

© Andrea Melendez/USA Today Network/Imagn Content Services

Figure 12.63 Drillers install monitoring well at a landfill.

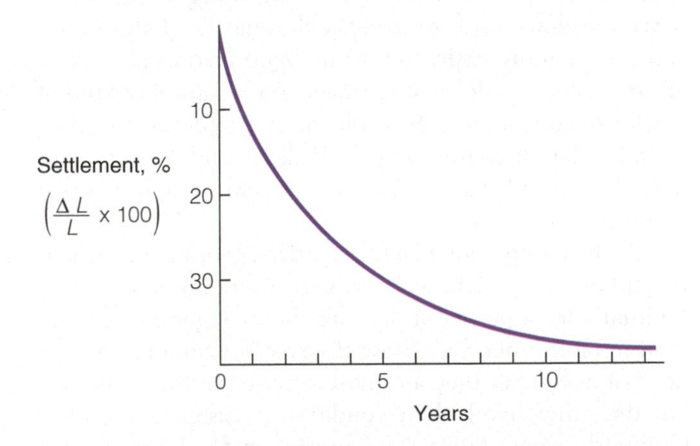

Figure 12.64 Landfill settlement over time. The percentage of settlement is calculated by dividing the amount of settlement (ΔL) by the original thickness of the landfill (L) and multiplying by 100. The settlement values here are for illustration only and are not real values.

Zoonar GmbH/Alamy Stock Photo

Figure 12.65 Paper and cardboard carton material, bundled and stacked in preparation for recycling.

12.4 Recycling, Remanufacturing, and Composting (LO4)

Recycling removes materials from the waste stream with the goal of reusing the material in different forms (Figure 12.65). Recycling has the potential to reduce greatly the volume of waste but recycling rates vary greatly around the world (Figure 12.66). Not everything that goes into garbage can be recycled owing to technological limitations and expense. Over 18 billion disposable diapers end up in landfills every year, for instance–the third-largest consumer item entering garbage disposal sites, worldwide. About 4% of the volume of all municipal solid waste in the U.S. consists of diapers. Diapers cannot be recycled for many reasons, both because they contain human waste and because they are composites made up of many different materials.

The recycling process varies widely according to the product being recycled. For example, once paper is collected and taken to a recycling center, it must be shredded and water added with other chemicals (hydrogen peroxide, sodium peroxide, and sodium silicates typically) to break down and separate fibers. A pulp results from which inks and dyes are then removed. This is typically done in a flotation tank through which various chemicals and air bubbles percolate in order to whiten the pulp. Finally, the pulp is passed through massive rollers to squeeze out the water. Virgin wood fiber and potato starch may be added to give the resulting press–recycled paper–smoothness and an ability to take fresh ink without blotting. The whole process takes about a week. In this way, paper may be recycled as much as seven times before its fibers weaken. In other words, recycled material itself can be recycled over and over.

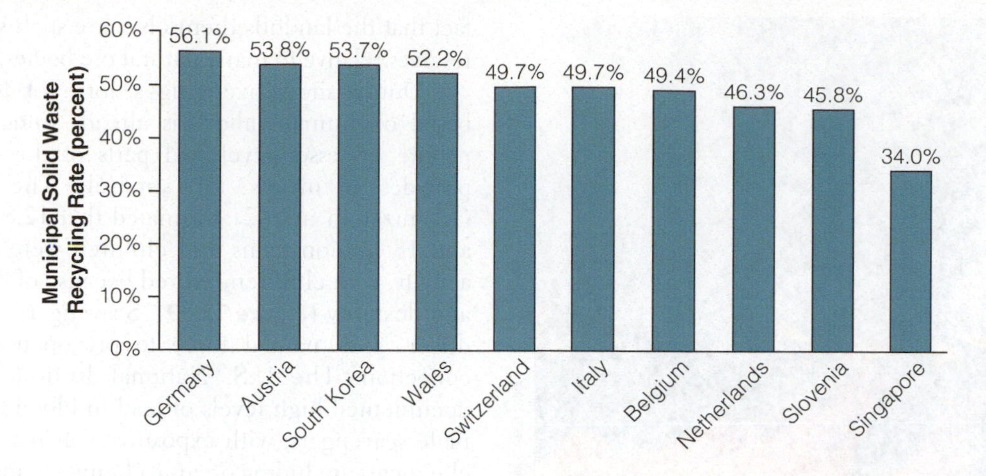

Figure 12.66 Recycling rates vary dramatically around the world. Several countries recycle more than 50% of their waste. (Adapted from mapofworld)

Recycling not only saves materials and reduces the strain on waste disposal facilities, but also saves energy and reduces resource consumption. Each 0.9 tonnes (1 ton) of recycled paper can save 17 trees, 1,400 liters (380 gallons) of oil, 4,000 kwh of electricity, 26,500 liters (7,000 gallons) of water, with 64% energy savings, and 58% water savings. The recycling also prevents 27 kilograms (60 pounds) of pollution from entering the atmosphere from the making of an equivalent amount of fresh paper. Likewise, simply remanufacturing old, worn-out products by cleaning and replacing their parts and reassembling them–a "proactive" type of recycling–can provide important energy savings. Annually, remanufacturing worldwide is roughly equal to the electricity generated by five nuclear power plants and saves enough raw materials to fill 155,000 railroad cars, forming a train 1,170 kilometers (1,100 miles) long.

In 2021, 63 million tonnes (69 million tons) of MSW were recycled in the United States. Of this, paper made up the vast majority–42 million tonnes (46 million tons)–while 7.9 million tonnes (8.7 million tons) of metal were recycled. With limited ability for centers to handle plastics, only 1.8 million tonnes (2 million tons) of this waste were recycled. But doing so prevented the equivalent of almost 30 million barrels of oil being freshly pumped out of the ground. Because of a growing awareness that mineral and other resources are finite, recycling increased from 10% to 35% of municipal waste between 1985 and 2018. Apart from certain batteries, there is presently no national law in the U.S. that mandates recycling. However, as of 2020, about half of all U.S. states and some city governments have at least one mandatory recycling requirement.

Curbside recycling programs have grown enormously since 1980. In 2022, approximately 60% of homes in the United States had access to curbside recycling. Many communities require separation of glass, metal, paper products, and plastics from other trash materials. The City of San Francisco reported an 80% decrease in MSW in 2021 through a program combining recycling with waste reduction and reuse.

In the past several decades, the volume of electronic waste, or "**e-waste**" entering the waste stream has grown dramatically (**Figure 12.67**). Computers, printers, fax machines, and cell phones have short lives; only about 2.5 years for a phone, with portable computers being replaced every few years. As of 2021, 36 million tonnes (40 million tons) of electronic waste were being discarded annually worldwide; or put another way, the

Bloomberg/Getty Images

Figure 12.67 Recognize that phone or computer? Most of us are e-waste generators and much of that waste ends up in Asia or Africa. Here, workers in New Delhi, India sort waste.

Picture alliance/Getty Images

Figure 12.68 The 2005 Indonesian tsunami left massive amounts of waste. Here, in 2005, a year later, a scrap metal recycler burns the rubber off electrical wire so that he can sell the copper and aluminum in Banda Aceh.

equivalent of 800 laptop computers thrown away every second!

Only about 17% of global electronic waste is recycled. Most of it ends up in landfills and incinerators, which is especially problematic. Cell phones typically contain around 40 chemicals including toxic metals, and some computer monitors may contain more than 100 chemicals, including cadmium, mercury, and lead. Much of the world's e-waste is shipped to lesser developed nations where people disassemble it in ways, including burning in open pits, that release harmful substances (**Figure 12.68**). As waste, these items are a significant environmental burden: the EPA estimates that they contain about 70% of all toxic material that we discard. But they also represent a valuable potential resource: In the United States alone, the gold and silver found in landfill-bound cellphones each year amounts to around $60 million. According to the Nickel Institute (2022), the total value of raw materials contained in the world's annual stream of e-waste is around $55 billion!

We don't think of landfills ordinarily as potential future mines, but there they sit, seemingly just waiting for their time to come. Attempts have already been made to tap this resource. In 1953, enterprising Israelis scooped up fertilizer for local orchards by collecting decomposing landfill wastes mixed with soil. A pilot project documented by the University of Southern Maine in 2011 recovered $9.72 million (2022 value) in metals from a landfill holding incinerator ash. Of special interest are hard-to get materials such as platinum-group metals and rare earths, vital for reducing pollution from car engines, and for emerging technologies such as electric vehicles. The downside is the high expense of separating and cleaning desired metals (see Chapter 4), and the fact that the landfills themselves are such small, short-lived reserves relative to many natural ore bodies.

Dump site scavenging–a form of low-tech mining based on human labor– is already undertaken by some people in lesser-developed parts of the world, where it provides a means for survival. The World Health Organization in 2021 estimated that 12.8 million women and 18 million teens and children were involved in this activity, with children favored because of their small hands and dexterity (**Figure 12.69**). Scavengers simply use screw drivers and manual force to pry open components for collection. The U.S. National Institute of Health has documented high levels of lead in blood taken from adult male scavengers, with exposure to as many as 1,000 toxic chemicals, including organic chemicals and lead. Exposure levels well exceed the safe legal standards set by the countries that produce most of this waste.

Electronics manufacturers including Dell, HP, Intel, and Apple, offer "e-cycling" programs to their customers, including batteries as well as hardware. To date, this is a more efficient and economical approach than landfill mining. Programs may require their customers to buy a new computer to recycle the old one for free. Most large cities also have commercial recycling firms that accept all e-waste, regardless of origin.

Automobile tires are easily recycled. At least 2 billion used tires are stockpiled in the United States, a number that is growing at a rate of 280 million per year (**Figure 12.70**). When placed in landfills, they tend to add bulk very rapidly, a fact that has caused 38 states to ban landfilling of tires. Most states also levy a tax on each new tire sold to contribute to tire-recycling programs. Tires aren't remanufactured

Nicolas Aguilera/EyeEm/Getty Images

Figure 12.69 People scavenging from a municipal solid waste dump.

Gabe Palmer/Alamy Stock Photo

Figure 12.70 A Connecticut tire dump.

Rob Anscombe/Alamy Stock Photo

Figure 12.72 Simple, small-scale domestic composting operation with an early stage (left) and advanced stage of decomposition (right).

for automotive purposes; the rubber degrades too much. But incidental uses of old tires are many, including the construction of small dams, planting stacks for potatoes, erosion control, houses (e.g., the Greater World Earthship Community in Taos, New Mexico), fencing, rifle ranges, playgrounds, and grain-storage structures (**Figure 12.71**). Used tires are being used experimentally as bedding for livestock corrals, as a replacement for stone in septic systems, as reinforcement in flood-control structures, even (ironically) as a rubber modifier to make asphalt for new roads. Such roads may contain as much as 20% by weight tire scraps and provide a generally quieter ride. Unfortunately, tires contain a

Susan E. Degginger/Alamy Stock Photo

Figure 12.71 Old glass bottles and tire waste put to use artistically in a wall. New Mexico's Earthship residential community.

number of potentially toxic substances that can be released to the environment through burning or ordinary weathering, including synthetic organic chemicals selenium, lead, and cadmium.

Composting is the microbial breakdown of organic wastes that can be done on a small-scale in backyards, on farms, golf courses and other outdoor operations (**Figure 12.72**). Residential compost piles consist of vegetable matter such as food scraps, lawn trimmings, and some paper waste, which is stacked in layers. Kitchen wastes should not include meat products because they nourish maggots and attract other vermin. The wastes are generally interlayered with cut grass or other yard cuttings. Successful decomposition is measured by taking the temperature of a pile. Aerobic microbial activity is exothermic (releases heat). Too much heat, though, and the pile will steam and could even burst into flame, though this rarely happens. The compost may easily be spread as a mulch and fertilizer after it breaks down. A garden hose that is passed through maturing compost can also provide hot water, free of charge, for outdoor baths and showers.

On a larger scale, commercial composting operations gather compostable wastes from numerous sources, including grocery stores, restaurants, greenwaste bins and college campuses, for centralized management (**Figure 12.73**). The wastes are piled in sets of parallel rows, large silos, or giant, carefully maintained aerated mounds to dry. Internal temperatures are raised and held above 50 to 60°C (122 to 140°F) for as long as a week to break down pathogenic microorganisms. The resulting compost then may be sold to farms, nurseries, or individuals for spreading. Challenges abound. Plastics (even so-called biodegradable plastics) and pet manures must be removed, and other hazardous materials separated before composting; a time-consuming process. Even then, toxic chemicals, such as

Jim West/Alamy Stock Photo

Figure 12.73 Commercial composting operation in Michigan, United States.

herbicides and motor oil coating plant leaves collected from driveways, occasionally make it through to be included in the finished product.

12.5 Disposal of Hazardous Wastes–The Nastiest Stuff (LO5)

The EPA estimates that by 2021 U.S. industries were generating 360 million tones (400 million tons) of hazardous waste per year; more than 80% from the chemical, petroleum, and coal industries. The RCRA (see Box Environmental Legislation Helping the United States Manage its Wastes) in the United States strictly regulates the hazardous wastes from these major sources, but does not concern itself with smaller businesses and manufacturers each producing less than 1 tonne (1.1 tons) of waste per month; nor from households.

Regulated hazardous wastes, primarily solids, are shipped to dedicated, well-secured landfills in 55-gallon **hazmat drums**, designed to be corrosion-resistant, inflammable, and impermeable (**Figure 12.74**). Nuclear waste is classified as "hazardous" too, of course, and is the greatest long-term disposal headache of them all (see Chapter 1). As such, it is handled differently than most industrial hazardous wastes, as we'll discuss further below.

Noomcpk/Shutterstock.com

Figure 12.74 Inspecting drums of hazardous waste.

12.5.1 Deep-Well Injection

Some hazardous waste can be recycled, some incinerated or undergo **pyrolysis**–that is, broken down in the absence of oxygen at high temperature. This has proven effective for destroying some very toxic compounds, including PCBs. About 80% of the hazardous waste we generate, though, is in liquid form; and while some of this toxic soup can be managed in landfills, a different approach toward dealing with it has had to be considered, given concerns about limited space and storage. Although it might seem like a desperate measure, pumping the waste deep underground via specialized injection wells is a common practice.

Deep-well injection of hazardous liquids takes place in permeable rocks far below, or otherwise isolated from freshwater aquifers (**Figure 12.75**). Essentially, we use the crust as a gigantic garbage can. Before such a well is approved, thorough studies of subsurface geology must be made to determine the location of earthquake faults, the state of rock stresses at depth, and the impact of fluid pressures on the receiving formations. The injection boreholes are lined with steel casing to prevent the hazardous fluids from leaking into freshwater zones above the injection depth.

Multiple classes of injection wells are legally permitted in the United States by the EPA's Underground Well Injection Program; each class is based on the types and depths of waste disposed (**Table 12.7**). The mandate for legal

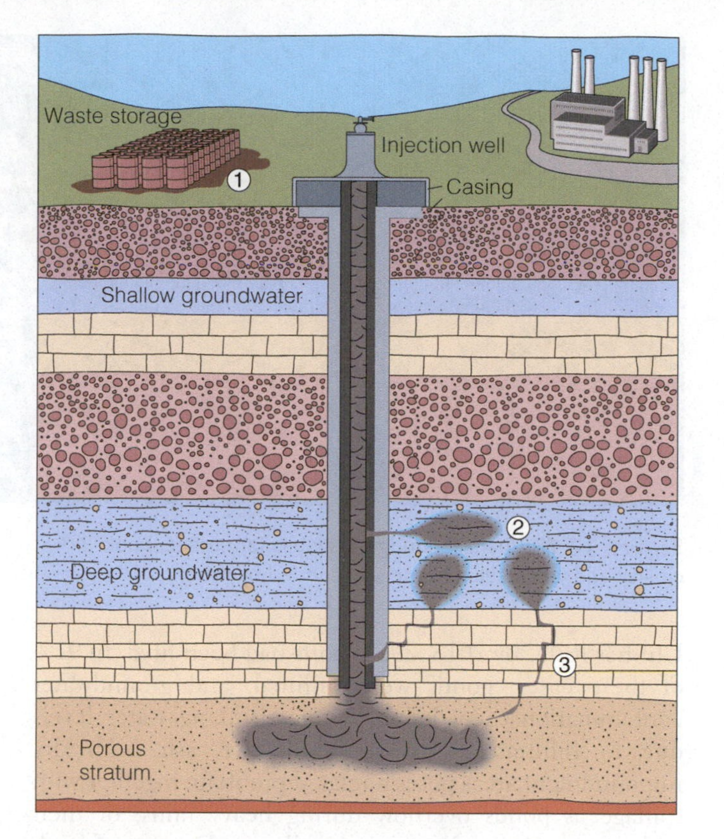

Figure 12.75 Deep-well injection of hazardous waste. This method presumes that wastes injected into strata of porous rock deep within the ground are isolated from the environment forever. What could go wrong? (1) Toxic spills may occur at the ground surface. (2) Corrosion of the casing may allow injected waste to leak into an aquifer. (3) Waste may migrate upward through rock fractures into the groundwater. (Modified from American Institute of Professional Geologists)

well operation is that underground sources of drinking water (USDW) will not be impacted–or *if* potentially impacted, not for at least 10,000 years. About 850,000 injection wells operate annually across the United States. Some wells inject nonhazardous municipal waste and sewage. On the order of 34 billion liters (9 billion gallons) of liquid hazardous wastes–almost 90% of it–disappears into the nation's lithosphere through other wells. To treat and render harmless these substances otherwise would be extremely expensive; too much so given existing funding.

12.5.2 The Problem with Coal

Coal combustion in powerplants produces around 37% of the world's energy (see Chapter 11). A major by-product of this, besides air pollution, is **solid combustion residue**–hydrocarbon ash, rich in arsenic, cadmium, mercury, and even some radioactive materials. About 80% of the ash remains in the plant combustion chamber that provides the heat for producing power. Some of it is wet and gloopy, **coal-ash sludge**, produced as scrubbers (sprayers) and filters in stacks partly remove the ash billowing up the inside of smokestacks (**Figure 12.76**).

In an effort to reduce the accumulation of this waste, as much as a third of coal ash may be mixed into cement and marketed as a construction material that is easy to pour and weather-resistant. Coal-ash cements are used in making dams all around the world in large part because of these additive properties. The toxicity of the ash in this application is inconsequential because of attenuation and mixing and because there are few exposure routes for people and the environment. Likewise, some coal ash ends up going with wood pulp, plaster and other material into making wallboard.

Table 12.7 Classes of Waste Injection Wells in the United States

These classes are divided into two groups here; shallow injection (above the water table or USDW supplies), and deep injection (in rock layers below or remote from USDW supplies—typically 1,700 to 10,000 feet deep).

Category	Number of Wells in United States	Notes
Shallow-Injection Wells (above the water table or USDW supplies)		
Class V	500,000–650,000	Injection of nonhazardous fluids only. Includes cesspools and septic tanks. Fluids tend to lose their harmfulness by natural breakdown during percolation.
Class IV	40	Hazardous and radioactive wastes injected. These wells are no longer legal but are still used to return treated contaminated water back into original aquifers as part of cleanup efforts.
Deep-Injection Wells (in rock layers below or remote from USDW supplies—typically 1,700–10,000 feet deep)		
Class III	17,000	Used by mining industry to inject water into ground to dissolve valuable minerals that are extracted when water with solutes is returned to surface. Fifty percent of all salt and 80% of all uranium is produced this way in the United States. After minerals are extracted, the remaining fluid is then reinjected.
Class II	147,000	Used by oil and gas industry to extract fossil fuels. Brine (highly saline water) brought up with the oil and gas is separated and re-injected into oil and gas reservoirs through these wells.
Class I Nonhazardous	366	Used for injection of municipal liquid wastes (sewage, etc.); mostly in Florida and Texas.
Hazardous	163	Used for deep injection of toxic liquids, which may be radioactive.

USDW = underground sources of drinking water.

DuxX/Shutterstock.com

Figure 12.76 Coal ash sludge spread for kilometers across the landscape surrounding a coal plant.

Getting rid of the rest of the ash, about 60% of it, is a problem. It cannot simply accumulate in combustion chambers, or no room for burning fresh supplies of coal would be available. Of the 7.8 billion tonnes (8.6 billion tons) of coal burned worldwide each year, according to Duke University, about 12% ends up as coal ash–that's a billion tonnes (1.1 billion tons).

As a "temporary" storage fix, most of this waste, both scrubber-related liquid and solid, ends up going directly into landfills, or to **coal impoundment ponds** not far from the plants where it is produced (**Figure 12.77**). In 2022, 500 such impoundment sites of environmental concern existed in the United States, according to the EPA. American coal companies are required to report groundwater quality beneath these ponds, following the 2015 Federal Coal Combustion Residuals from Electric

Jim West/Alamy Stock Photo

Figure 12.77 Bulldozers spread coal ash on side of dam at a coal ash impoundment pond.

Dot Griffith

Figure 12.78 Coal ash sludge disaster, Tennessee. Look closley and you'll see a home surrounded by mud.

Utilities Rule. By 2018, 292 reports had been filed. In 91% of these cases, ponds were unlined, and groundwater contamination exceeded EPA safe drinking water levels due to infiltrating toxins. Windblown ash from dry impoundment areas poses a related problem, together with spillage as ponds overflow during heavy rains, or their embankments deteriorate and collapse. Dozens of such spill sites have been reported by the EPA and public-interest groups throughout the United States, mostly in the humid, wet, and coal-rich eastern United States. Together with groundwater contamination, these appear to contribute to higher cancer rates in local populations: a person living within 1.6 kilometers (1 mile) of an unlined ash pond has a 1 in 50 chance of contracting cancer during a lifetime–more than 2,000 times greater than EPA goals.

A major case that helped trigger EPA oversight of coal wastes took place on December 22, 2008, when a 21-hectare (84-acre) ash sludge impoundment reservoir near Tennessee's Kingston Fossil coal-fired powerplant burst open and sent a slurry downslope into the nearby Emory River, destroying houses and covering more than 120 hectares (300 acres) of land (**Figure 12.78**). The sticky flood poured into two watersheds: that of the Emory and the Clinch rivers, both major tributaries of the Tennessee River–a total of 4.2 billion liters (1.1 billion gallons) of dark, contaminated mud.

A few months after this disaster, a scientific team from Duke University traveled to these rivers and took 200 water samples, collecting them from streams and rivers as well as from sediment pores at depths of 10 to 50 centimeters (4 to 20 inches) beneath each riverbed. They searched for high arsenic concentrations (see Chapter 11) that might remain from the slurry. On a positive note, they were pleased to discover that just a few months of precipitation and snowmelt were sufficient to cleanse the rivers where they were actually flowing–no problem for the fish swimming in them. But levels of arsenic measured as high as 2,000 parts per billion (ppb) in the associated soil and sediment pore waters.

A.P.S. (UK)/Alamy Stock Photo

Figure 12.79 This British nuclear fuel reprocessing plant at Sellafield closed in July 2022 after 58 years of operation.

Lack of flow in the spaces between the silt, sand, and gravel particles meant that the arsenic simply stayed put at immediate post-slurry levels. The EPA's mandated safe level for drinking standard is only 10 ppb as of 2021. The level of arsenic regarded as "safe" for aquatic life is 150 ppb. River bottoms are where the food chain begins in waterways. Worms and other invertebrates burrow into the bottom sediments scavenging for algae, microbes, and one another. Plants send their roots into this substrate, too, and they incorporate arsenic and other metals, which, in turn, bioaccumulate within plant-eating invertebrates and fish–which may well end up on a dinner plate somewhere. Fortunately, in the 5 years following the Emory River disaster, various researchers found no substantial build-up of arsenic and other contaminants in potentially impacted predator fish species (bluegill, redear sunfish, and large-mouth bass).

12.5.3 Existing Approaches to Nuclear Waste Disposal

High-level nuclear wastes, by-products of nuclear-power generation and military weapons production, are the most intensely radioactive and dangerous wastes. About once a year, a third of a nuclear reactor's fuel rods are removed and replaced with fresh rods. Replacement is necessary because as the fissionable uranium in a reactor is consumed, the fission products begin to capture more neutrons than the remaining uranium. This causes the chain reaction to slow, and if rods were not replaced, the chain reaction would eventually stop. Rising fuel rod heat and increasing radioactivity are also concerns.

The used, radioactive rods, called **spent fuel**, are the major form of high-level nuclear waste. Currently, they are stored at the individual reactor sites, supposedly a temporary measure until a final repository is prepared for them. By 2022, more than 80,000 tonnes (88,000 tons) of heavy metal spent fuel, mostly uranium, were stored across the United States in carefully engineered pools of water and, at many of the reactors, pool storage capacity is nearly filled. Each year, we add 2,000 tonnes (2200 tons) of additional spent fuel to this amount. The number does not include other materials such as the tubes that contain the fuel and structural materials.

Concerns about pool storage of fuel rods include loss of water which would cause the rods to overheat (e.g., due to fracturing of pool walls from an earthquake, or even terrorist attack) and the corrosion of fuel rod jackets that could release highly radioactive pellets—sufficient to kill people located nearby in a matter of seconds. Although the fuel removed from a reactor every year on average weighs about 31,000 kilograms (65,000 pounds), because of its high density, it is only about the size of an automobile.

Alternatively, spent fuel rods can be **reprocessed** to recover unused uranium-235 and plutonium-239. The French and English do this routinely, but the United States has never reprocessed spent commercial reactor fuel (**Figure 12.79**) because of the worry that the extracted plutonium could be used for making nuclear bombs. Reprocessing reduces the volume of high-level waste and recycles uranium. However, the liquid waste remaining after reprocessing contains more than 50 radioactive isotopes, such as strontium (^{90}Sr) and iodine (^{131}I), which pose significant health hazards. Of course, if we don't reprocess, we still have the nuclear waste.

By U.S. Nuclear Regulatory Commission (NCR) definition, **low-level nuclear wastes** are those not designated as transuranic—that is, made up of the heaviest, most unstable radioactive isotopes (see Chapter 1). Low-level wastes are created in research laboratories, hospitals, and industry, as well as in nuclear reactors. Ocean dumping of these wastes was a common practice between 1946 and 1970, as noted earlier. Since then, shallow land burial has been the preferred disposal method. As of 2022, four such centralized sites existed in the United States; one in South Carolina, one in Utah, one at the Hanford nuclear reservation in Washington State, and one in Texas (**Figure 12.80**).

Michael Stravato/Redux

Figure 12.80 Low-level nuclear waste buried in a Texas landfill designated to receive this material.

Uranium-mill tailings, the finely ground residue that remains after uranium ore is processed, are typically found in mountainous piles outside mill facilities–around 210 million tonnes (230 million tons) of it. The tailings contain low concentrations of around a dozen radioactive materials, including thorium (^{230}Th) and radium (^{226}Ra), both of which produce radioactive radon gas (^{222}Rn), the so-called "hidden killer (see Chapter 1)." The Uranium Mill Tailings Radiation Control Act (1978) makes the U.S. Department of Energy responsible for 24 inactive sites, in 10 states, left from uranium milling operations (**Figure 12.81**). The act specifies that the federal government will pay 90% of the cleanup cost, with each state to pay the remainder.

Grand Junction, Colorado supplied uranium used in the first atomic bombs (1943–1946). Authorities were less knowledgeable and concerned about radioactivity in those early days than they are now. An Atomic Energy Commission policy permitted the use of uranium-mill tailings in the manufacture of building materials, such as concrete blocks and cement. Sidewalks and many houses in Grand Junction included concrete mixed with radioactive sand. The winner of Miss Atomic Energy pageant (a Grand Junction tradition) was awarded a truckload of uranium ore as her prize (**Figure 12.82**). The pageant is no more, and sidewalks and homes have been demolished because of their high radioactivity. The debris is presently isolated in a special landfill.

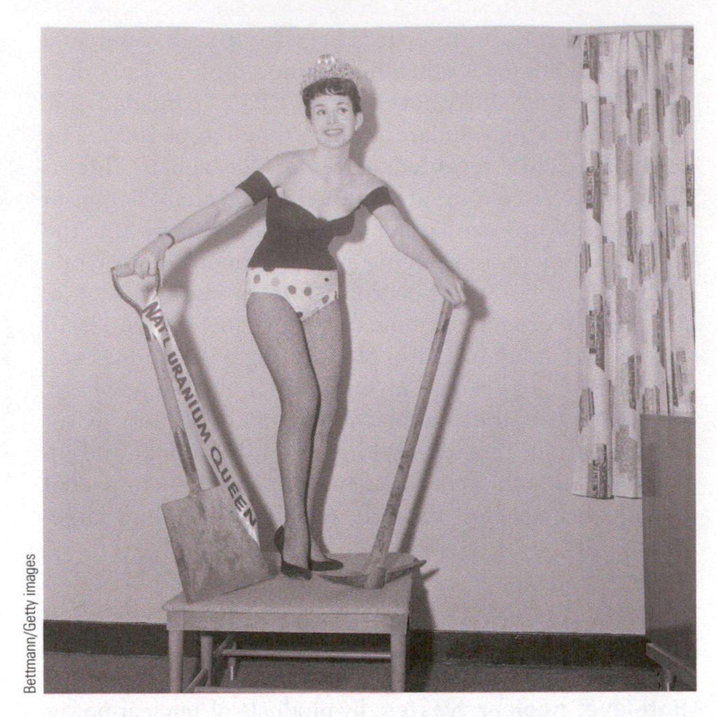

Bettmann/Getty images

Figure 12.82 Brooke Robin was the National Uranium Queen in 1956. After winning, she shows off her pick, shovel and miner's lamp crown. She reigned over National Uranium Week in festivities in Grand Junction, Colorado. Miss Robin said, "she would take the pick and shovel west with her and that she might try a little digging for the precious metal."

Much of the uranium procured at Grand Junction came from the Navajo Nation, Native American lands in northeastern Arizona and adjoining regions of Utah and New Mexico. From 1944 to 1986, approximately 500 mines on the Navajo land yielded around 27 million tonnes (30 million tons) of uranium ore. Lung cancers soon began to afflict former Navajo uranium miners, raising concerns about radiation exposure to the wider Native American population. The U.S. EPA has been working with the Navajo Nation since 2008. More than a thousand homes have been surveyed for radioactivity; 47 have been removed and disposed of because they were radioactive enough to harm those living inside. Uranium tailings mixed with soil have also been removed from 60 residential yards and along many roadsides. Substantial cleanup needs remain (**Figure 12.83**).

The Washington Post/Getty Images

Figure 12.81 Abandoned uranium mine with tailings on the Navajo Nation, Red Water Pond. This area is now undergoing EPA cleanup.

US EPA

Figure 12.83 EPA contractors removing uranium-contaminated soil along a road in the Navajo Nation. The sprayed water reduces airborne dust.

12.5.4 Repositories–A Long-Term Nuclear Waste Solution

The existing nuclear waste disposal approaches discussed above are temporary fixes to a problem that simply will not go away. A long-term approach is needed, scientists agree. Several methods of permanently storing wastes have been proposed, but any method must meet these conditions:

- Safe isolation of radioactive materials for at least 250,000 years and possibly as long as 500,000 years (designated durations of legal responsibility for warning the future, such as "10,000 years" for the WIPP site discussed in Chapter 1, are meaningless when it comes to how long radioactive material actually remains dangerous)
- Security from terrorists or accidental entry
- Complete protection from natural disasters (hurricanes, landslides, floods, volcanic eruptions, and even meteor impacts have been considered)
- Compete isolation to prevent contaminating nearby natural resources, especially air and water supplies.
- Failsafe handling and transport mechanisms

Ideas discussed for doing this have historically included:

1. *Dispose of waste in a convergent plate boundary (subduction zone).* The feasibility of this proposal has not been demonstrated because the rate of plate movement is so very slow. Presumably, however, sediment would envelop waste containers as they sink into the trench too remote to threaten humans on land far away. The tremendous displacement caused by some earthquakes at convergent plate boundaries (see Chapter 6) casts doubt on the wisdom of this approach as does the movement of warm water through rock, water that could transport dissolved radionuclides.
2. *Place containerized waste on the ice sheets of Antarctica or Greenland.* Radioactive heat would cause the containers to melt downward to the ice-bedrock contact surface, which could place the radioactive waste as much as 3 kilometers (2 miles) beneath glacier surfaces in either Greenland or Antarctica. This approach, however, given both measured rates of ice melting and the general trend of ongoing climate change is exceptionally risky and has been abandoned.
3. *Shoot the waste into the Sun or deep space.* This would be an excellent solution, except for the possibility of rockets blowing up on launch or as they traverse the atmosphere.
4. *Place waste in a geological repository (secured underground vault): salt mines, or deep chambers in granite, volcanic rocks, or some other very competent, relatively dry rock formation.* This approach requires detailed site analysis to ensure that all geological criteria are met, but all things considered, repositories seem to be the best way to go. Several have been established already and are receiving wastes, including the WIPP site in southeastern New Mexico (**Figure 12.84**; see Chapter 1). The United States has proposed a second repository at Yucca

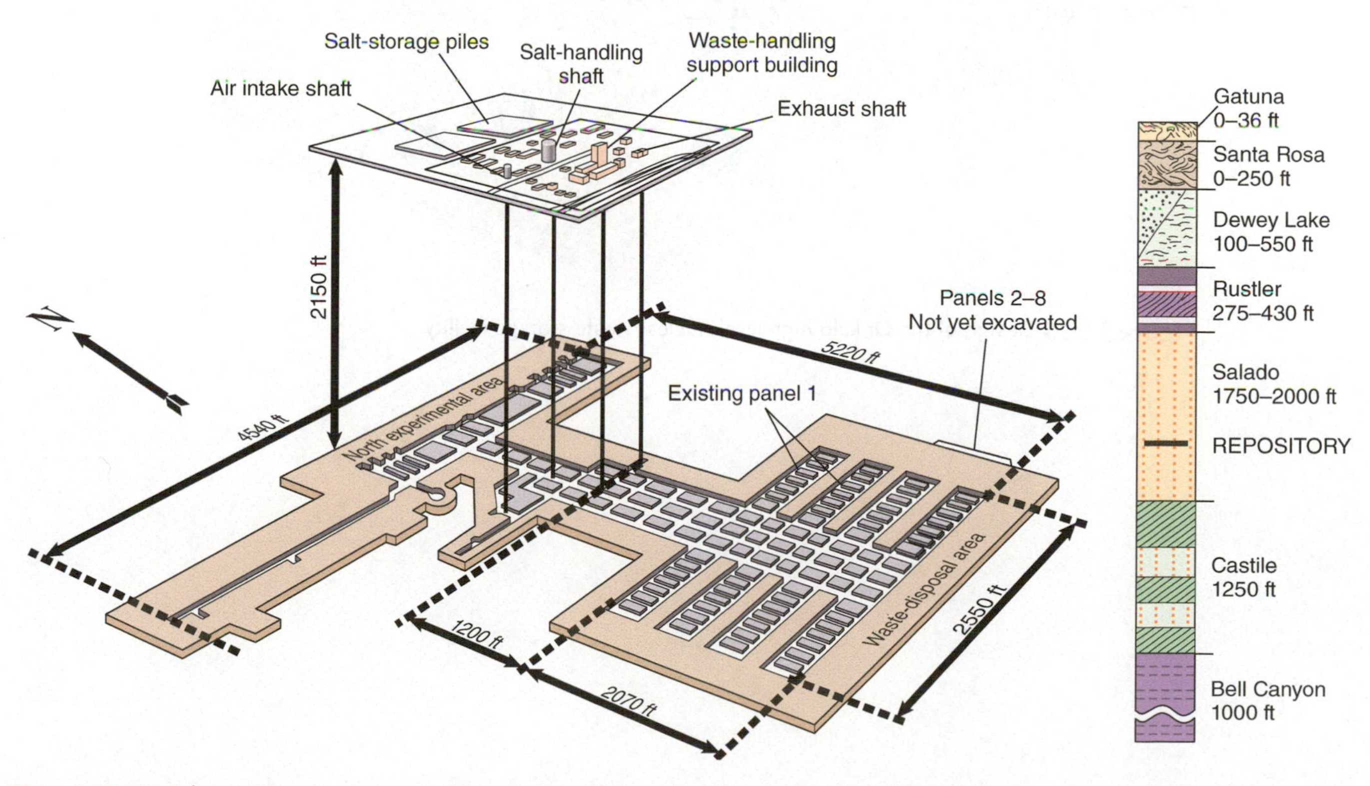

Figure 12.84 Schematic representation of the Waste Isolation Pilot Plant underground facility for storage of transuranic waste and the sedimentary rock sequence near Carlsbad, New Mexico. Only two of the storage panels have been mined. (Department of Energy)

Mountain, southern Nevada, but this project has been abandoned, at least for now. (see Case Study 12.2).

5. *Isolate waste in sealed containers on a deep abyssal plain.* Among the most stable places on Earth are the abyssal plains of the deep ocean floor. Currents are sluggish and the temperature near uniform; clay-rich sediment is capable of trapping and holding radioactive ions; and seismic and volcanic activity is nil, provided the seafloor is distant from any subducting trench, undersea fracture zone, or spreading center. The chances of human interference with storage are very low.

The Finnish government has constructed a repository at Onkalo, in a cavern undergoing excavation to a depth of 510 meters (1,710 feet). The tunneling was extremely difficult, taking place in hard, stable, granitic bedrock beneath a remote forest (**Figure 12.85**). Groundwater activity is minimal, and the landscape is geologically quite stable. Spent fuel rods from the nearby Olkiluoto nuclear power stations may be delivered beginning as early as 2023 or 2024. The repository has a capacity to accept wastes for about a century before it is closed and sealed (**Figure 12.86**)

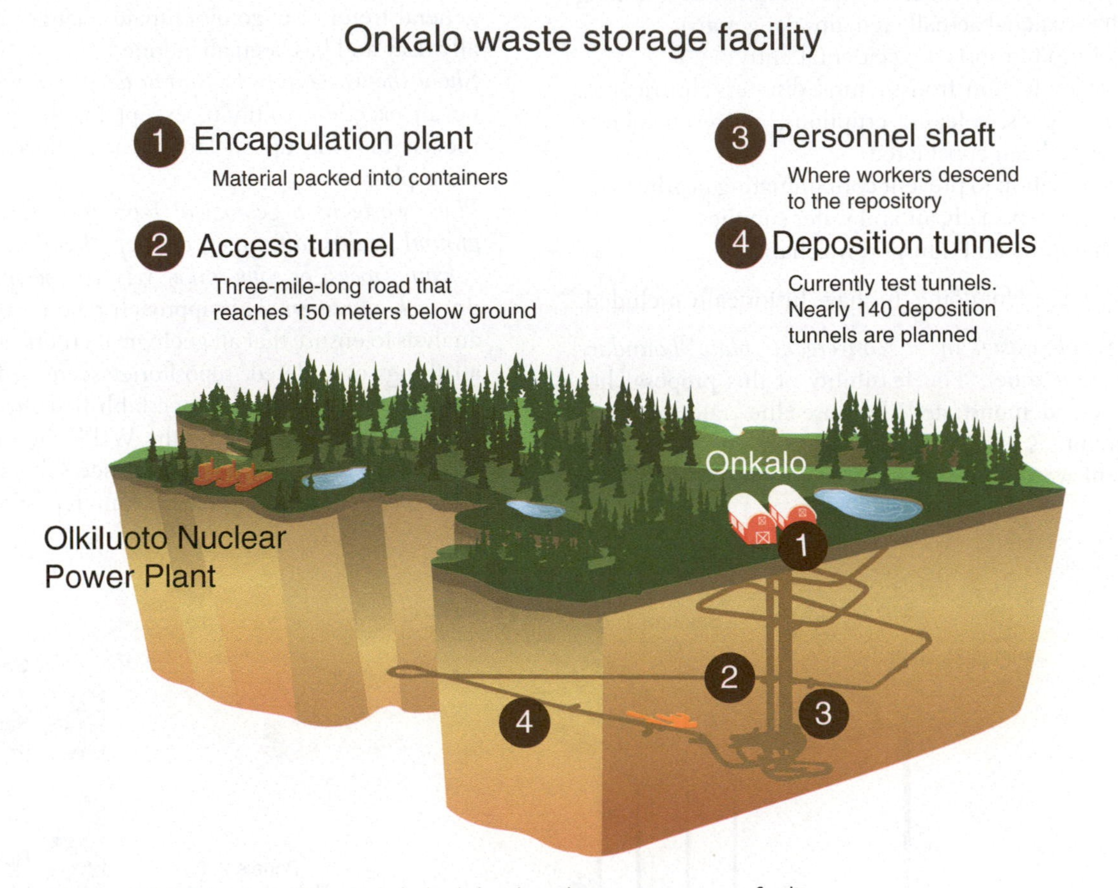

Figure 12.85 Design of the Onkalo high level nuclear waste storage facility.

NurPhoto/Getty Images

Figure 12.86 Tunnel inside the Onkalo nuclear waste repository, under construction in Finland and 420 meters below the surface.

Unlike the earlier described WIPP repository (see Chapter 1), it seems likely that no warning monuments or archives will be established at Onkalo to mark the site and keep people away. Instead, the entrance and shallow passages will be filled and made to look perfectly natural; never touched by human hands. The wager is that no one in the distant future is ever going to become curious or otherwise interested and disturb it. What could possibly attract them to the area? It has no urban, agricultural, or other current economic value apart from logging, which is a surficial activity that poses no threat. Why would anyone else wish to dig deeply into solid granite in such a practically featureless landscape. All the wars, political swings, climate changes, and other affairs that roil human life here at the surface have no impact on granite in ancient, stable continental crust that is hundreds of meters deep (**Figure 12.87**). What do you think?

Google Earth

Figure 12.87 The above-ground portion of the Onkalo facility is in the foreground. Olkiluoto's nuclear power plants are in the background. The landscape is flat and mostly tree covered.

12.1 Freshkills–From Landfill to Beloved City Park

Freshkills landfill on Staten Island, New York, can claim fame for two reasons. It stands 165 meters (500 feet) above sea level, the highest elevation on the eastern shore of the United States, and before closure it was the world's largest landfill, with 25 times the volume of the Great Pyramid of Giza, Egypt (**Figure 12.88**). Freshkills (from the Dutch *kil*, "stream") served the five boroughs of New York City, its suburbs, and parts of New Jersey. The landfill was established in 1948 on a salt marsh, with no provisions for constraining leachate. For more than five decades, more than 160,000 tonnes (175,000 tons) of municipal waste per week ended up at the site (some by barge; **Figure 12.89**), and up to 4.2 million liters (1 million gallons) of leachate leaked into the marsh and nearby waters each day.

Included in the waste stream to the fill was ash residue from the trash incinerators then operated by New York City. Because the ash contained toxic metals, specially designed "ash-only" fill areas were constructed, each with a double liner and a double leachate-collection system to keep toxic substances within the fill. A clay slurry barrier was built around the fill to further confine leachate if there were leaks. Landfill managers extracted methane and used it for heating and cooking in nearby homes, while construction debris was recycled.

Hum Images/Alamy Stock Photo

Figure 12.89 Garbage barges from Manhattan ready to unload at Freshkills Landfill in 1973.

The landfill was scheduled for closure in March 2001, because of public pressure with EPA support (**Figure 12.90**). However, following the attacks on the World Trade Center (WTC) on September 11, 2001, Freshkills was reopened to accommodate the disposal of debris from the destruction (**Figure 12.91**). Most of this debris ended up being sold for scrap and recycled, and the site was finally closed in 2003.

Richard Levine/Alamy Stock Photo

Figure 12.88 Freshkills Landfill when it was still active.

Matt Campbell/AFP/Getty Images

Figure 12.90 The closing of Freshkills Landfill after 53 years of waste.

Figure 12.91 The attack on the World Trade Center on 9/11 destroyed more than 1,000 vehicles. They, and other debris, were brough to the Freshkills landfill. There, police examined the debris as evidence and to separate the remains of victims.

At once, the New York City Department of Parks and Recreation proposed a 30-year plan for developing Freshkills as a 9-square-kilometer (3.4-square-mile) public park, and the Municipal Art Society collaborated with the City of New York to sponsor an international design competition, run by the Department of City Planning. Turning a massive pile of waste into a public park took planning and a cadre of engineers and geologists (**Figure 12.92**). The landfill had to be capped completely, the methane captured, and the leachate intercepted and piped to a treatment plant.

The approved plan included open grasslands for school wildlife studies, waterways for recreational sports, nature paths, public art displays, and birdwatching in a rolling terrain marked by four prominent, artificial hills bisected by a central valley with a tidal creek. Grasses thrived, red foxes moved in, rare grassland birds returned to Freshkills, and in 2017 the first section of the rapidly transforming landscape opened to the public (**Figure 12.93**). Freshkills is the largest new park added to New York City since the nineteenth century and is three times larger than Manhattan's famous Central Park.

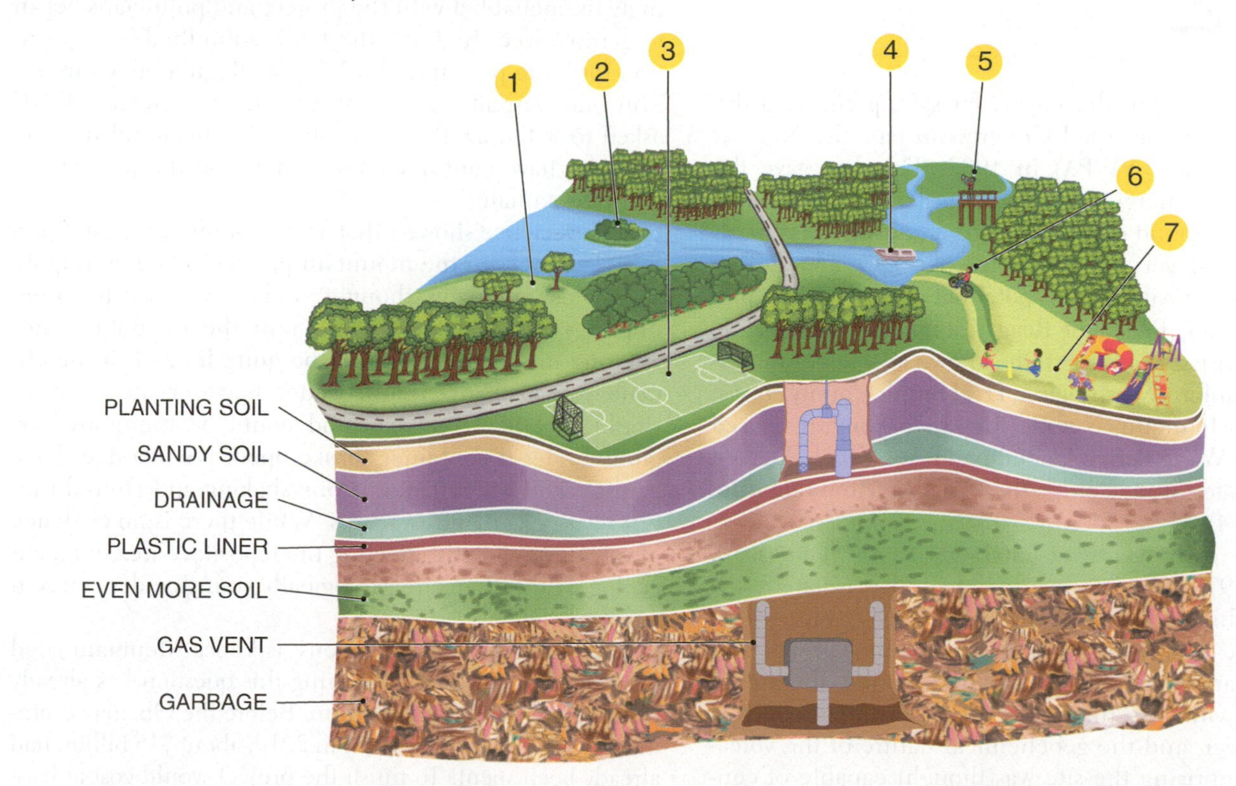

Figure 12.92 An artist's view of Freshkills' transformation from a dump to a park. (Based on illustration from Johan Thörngvist)

Case Study

Figure 12.93 The New York City skyline from Freshkills Park.

In 2018, New York University, New York City Parks, and the Freshkills Park Alliance sponsored the Reclaimed Lands Conference, with sessions on engaging residents in reclaimed lands stewardship and monitoring changes in biodiversity, at places like Freshkills. A studio and art gallery focusing on the restored wildland environment also opened. Meanwhile, a system of trenches, wells, and underground pipes continues to collect landfill by-products at the now closed Freshkills landfill, allowing environmental scientists to keep tabs on decomposition of the buried wastes.

12.2 Yucca Mountain, America's Troubled Plan for a Nuclear Waste Repository

The need to develop the means for safe, permanent disposal of nuclear waste led Congress to pass the Nuclear Waste Policy Act (NWPA) in 1982. This law gave the Department of Energy (DOE) responsibility for locating a failsafe underground disposal facility, a **geological repository**, for high-level nuclear waste. The DOE designated nine locations in six states as potential sites. Based on preliminary studies, President Reagan approved three of the sites for intensive scientific study (known as **site characterization**): Hanford, Washington; Deaf Smith County, Texas; and Yucca Mountain, Nevada. In 1987, Congress passed the Nuclear Waste Policy Amendments Act, which directed the DOE to study only the Yucca Mountain site (**Figure 12.94**).

In the early years of characterization (site work started in 1978), the site seemed ideal. It is isolated, it is located within the area of the Nevada Test Site (where atmospheric and underground testing of nuclear weapons had been carried out for more than 40 years), the area is dry desert with less than 15 centimeters (6 inches) of rainfall a year, and the geochemical nature of the volcanic tuff comprising the site was thought capable of containing any potential leakage of radioactive materials. Drilling began in 1994 and continued with geologic investigations for a decade, while Nevada became increasingly disenchanted with the project and politicians began to get involved. In 2008, the DOE submitted its application for a site permit to the Nuclear Regulatory Commission but soon after President Obama was elected, DOE asked to withdraw the application. Political and funding struggles have continued since then and the project has ground to a halt.

The science showed that Yucca Mountain wasn't perfect. The rocks in the mountain proved to be not as quite as dry as originally thought and they were fractured in places. Today, with very little rain, the amount of water entering the repository would be quite limited. If the climate were to change for the wetter, however, the waste too would very likely get wet and could be transported by groundwater if containers broke open. Nevada does have earthquakes, which cause strong shaking and ground ruptures with clearly offset scarps. While there is no evidence of volcanism directly affecting the repository area in recent geologic time, there are geologically young cinder cones in a nearby valley.

Despite these considerations, is Yucca Mountain good enough to do the job? Answering this question has already cost taxpayers an enormous sum. Before the Obama administration shuttered the project in 2010, about $15 billion had already been spent. To finish the project would cost at least $100 billion.

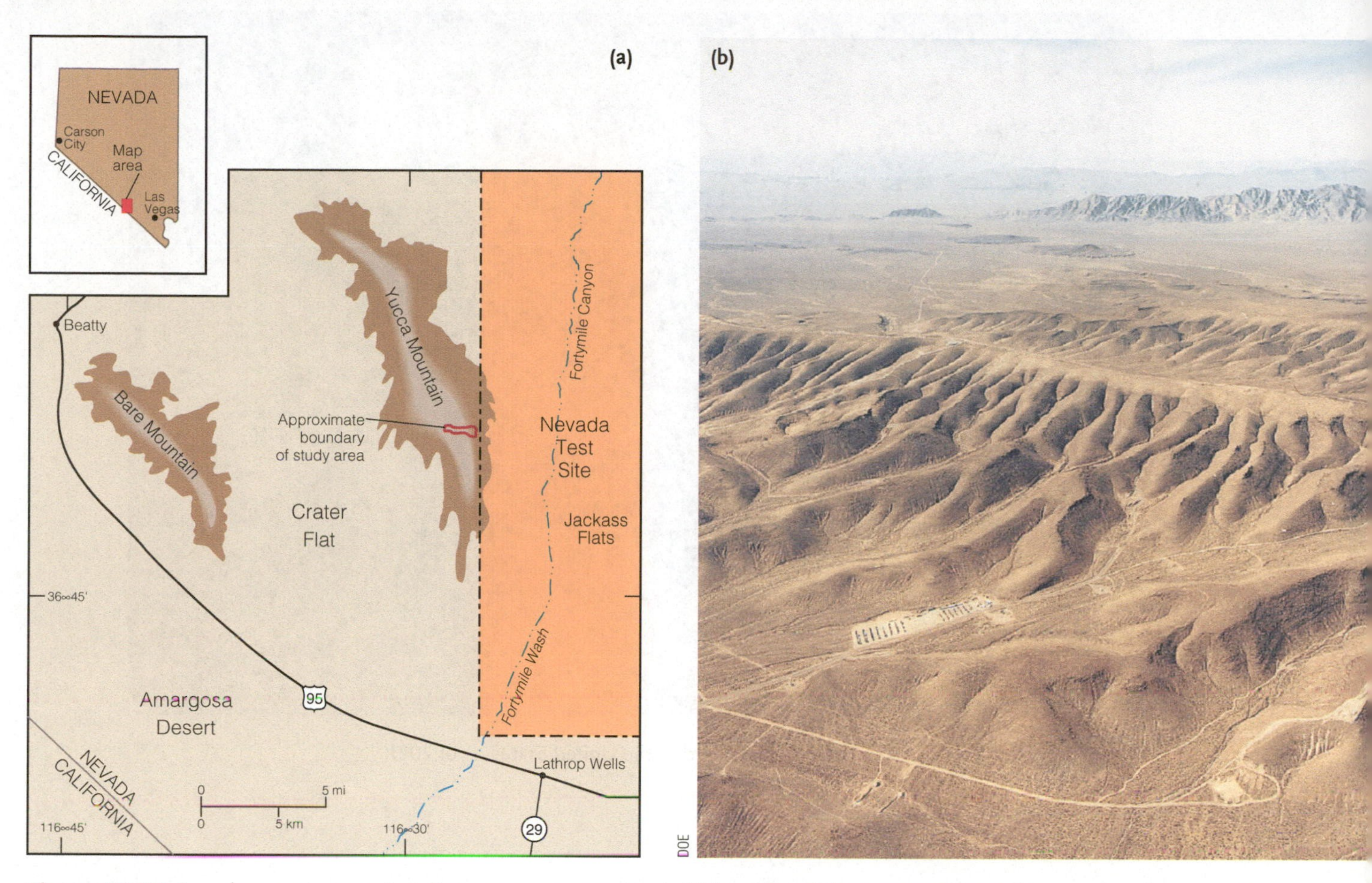

Figure 12.94 A nuclear waste repository that was never completed. (a) Location of the proposed Yucca Mountain nuclear repository. (b) Aerial view of the Yucca Mountain study area. The structures are at the site of the proposed nuclear-waste repository. The site was shelved in 2009. (Data from the Department of Energy)

So what now? While the United States wrestles with this important problem, nuclear waste remains in interim storage throughout the country (**Figure 12.95**). This includes not only the water-filled pools at nuclear power plants, but also high-level nuclear waste warehoused temporarily in **dry casks** consisting of steel encased in concrete. Each cylindrical cask stands roughly 6 meters (20 feet tall) and is 3 meters (10 feet) wide (**Figure 12.96**). The casks began to be introduced as a stopgap measure in the 1970s, when the water-filled pools at nuclear powerplants designed to store spent reactor fuel rods drew close to their storage capacities (**Figure 12.97**). Each cask is meant to retain the overflow waste for only a period of 20 years, with possible renewal authorized by the NRC for up to 40 years. But now it looks like dry cask storage will grow and continue indefinitely in the United States until the country decides what to do on a permanent basis.

Over 2,500 dry casks with spent fuel have been shipped across parts of the United States in the past 55 years (**Figure 12.98**). As of 2021, 65 sites called Independent Spent Fuel Storage Installations (ISFSIs) were in operation to receive them. Each resembles a large, well-guarded parking lot (**Figure 12.99**). ISFSIs may continue to grow in number over the next few centuries even if a new geological repository comes into operation, given the tremendous amount of nuclear waste in need of storage.

This perhaps is a good note on which to end this text. In stark terms, the topic of nuclear waste management brings us face to face with questions of what humanity owes the future, and with how our behaviors today can impact unborn generations for long spans of geologic time, not just the human time we deal with in our ordinary lives. Understanding environmental geology is a vital way of establishing "reality checks" in our frenetic, self-insulating world–sometimes when they are most urgently needed.

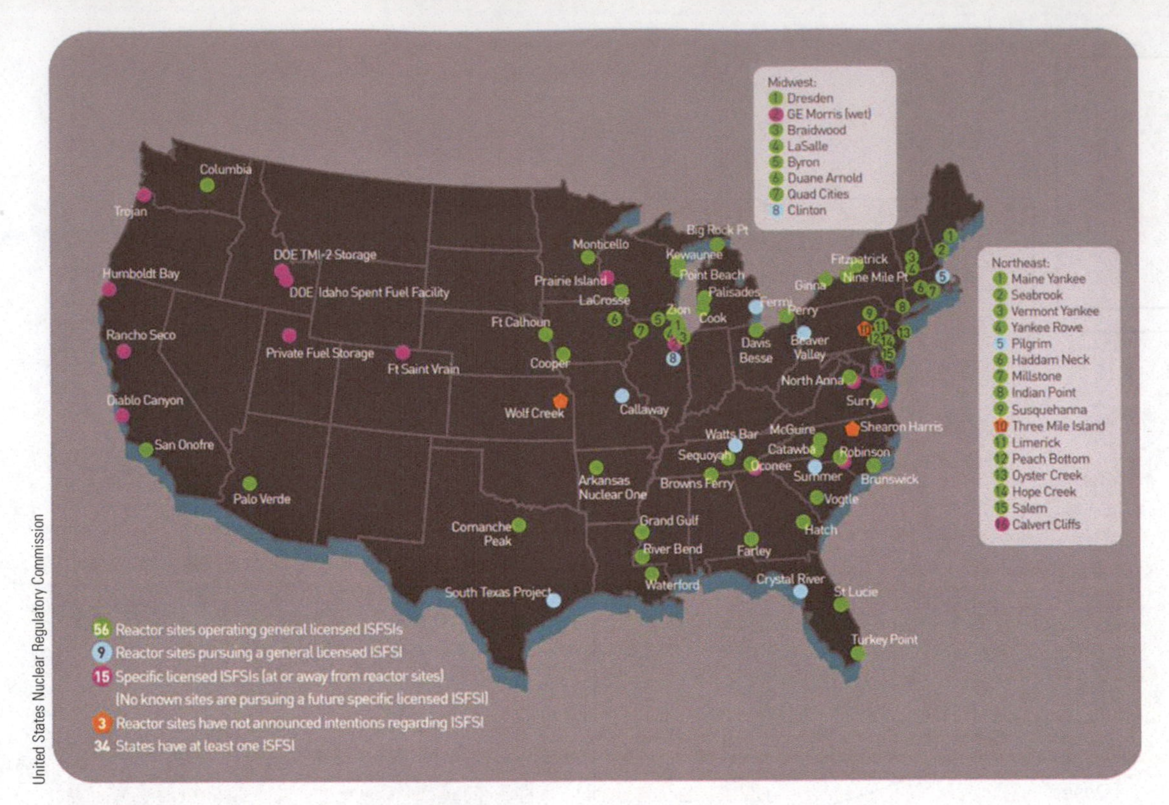

United States Nuclear Regulatory Commission

Figure 12.95 Where spent nuclear fuel is stored in the United States as of 2021.

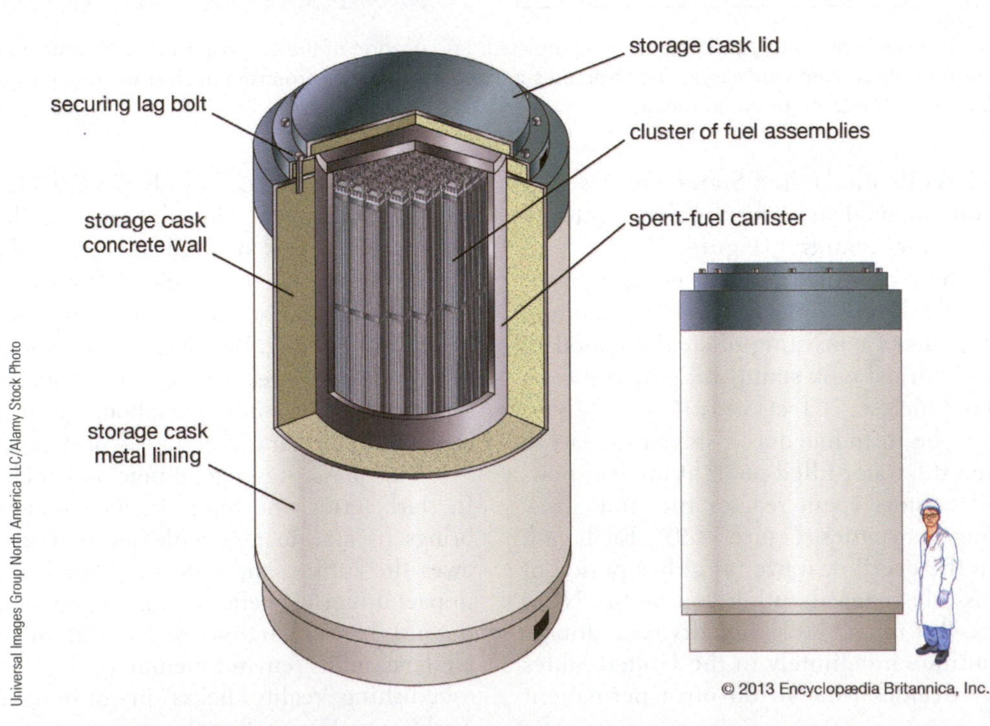

Universal Images Group North America LLC/Alamy Stock Photo

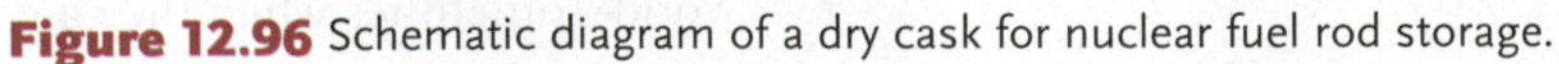

Figure 12.96 Schematic diagram of a dry cask for nuclear fuel rod storage.

Reuters/Alamy Stock Photo

Figure 12.97 Nuclear fuel assemblies in a storage pool as they await reprocessing at the Roan reprocessing plant in La Hague, France. The United States does not reprocess fuel, so storage facilities like this are located at every nuclear power-plant. The water absorbs much of the radiation emitted by the spent fuel and keeps the fuel cool.

Boston Globe/Getty Images

Figure 12.99 Dry cask storage is used to store spent fuel at the Vermont Yankee nuclear power plant in Vernon, Vermont. The power plant is now being decommissioned after 42 years of operation.

USDOE

Figure 12.98 Dry casks used by the Department of Energy for transport by rail of spent nuclear fuel from New York to the Idaho National Laboratory. These casks are about 7.1 meters (23.3 feet) long, 3.3 meters (10.8 feet) wide, and weigh 125 tonnes (275,000 pounds) when loaded.

Photo Gallery

Impacts of Waste

Waste in the air, water, and soil impacts the environment in which we live. Some wastes are unsightly, some have long-term consequences, and others are acutely toxic or harmful to life.

1 Smog, the brown haze enveloping Los Angeles, is the result of unburned gasoline and other hydrocarbons interacting with sunlight.

Daniel Stein/iStock/Getty Images

Allison Joyce/Getty Images News/Getty Images

2 A tributary to the Buriganga River in Dhaka, Bangladesh, is overwhelmed with plastic trash.

ZONE INTERDITE

3 A hazardous waste disposal team from France uses water to liquefy polluted soil before it is removed from landfills near Abidjan, Ivory Coast, where over 500 tonnes (550 tons) of toxic waste was dumped, killing seven people and injuring thousands.

Issouf Sanogo/AFP/Getty Images

Radiation contaminated soil from the reactor accident at Fukushima, Japan is stored awaiting disposal.

The Asahi Shimbun/Getty Images

4 A man bathes in a polluted section of the Yamuna River, near New Delhi, India in 2011.

Daniel Berehulak/Getty Images News/Getty Images

5 Recycling reduces waste disposal costs and impacts. Here workers sort mixed recycling into its various components.

iStock.com/Mindful Media

Summary Waste Management

1. Types of Waste

LO1 Characterize types of waste

a. Municipal Solid Waste

Municipal solid waste is the garbage that the average person generates on a daily basis, including food scraps, wood and paper products, plastic packaging, and yard waste. The average person in the United States generates approximately 750 kilograms (1650 pounds) of waste per year, adding up such that the United States accounts for 12% of global annual MSW, while only accounting for 4% to 5% of global population.

b. Industrial Waste

Industrial wastes account for the many byproducts of manufacturing, lumber mills, mining, and power plants. This includes scrap metal, oils, varnishes, chemical solvents, crushed stone, and radioactive waste. Improper disposal of industrial wastes has led to many significant and long-term environmental crises, and more visible ones like the Cuyahoga River fire in Cleveland, Ohio.

c. Agricultural Wastes

Agricultural wastes are all the disposed by-products of food production, from herbicides and pesticides to plastic storage packaging and manure. Manure is a major concern due its composition including bacteria and viruses, and nutrients like nitrogen and phosphorus. Confined animal feeding operations produce more livestock with less land use; however, this type of system creates other major problems. First, there is the huge generation of concentrated waste and pollutants in the form of fungal spores, antimicrobial agents, hormones, respiratory pathogens, and manure. Nutrient inputs into waterways from manure and fertilizer from the CAFOs and agriculture in general are also the primary cause of eutrophication (e.g., algal blooms). Additionally, CAFOs create incredibly poor living conditions for animals.

d. Wastewater

Proper disposal of human and storm wastewater is critically important to stop the spread of bacterial illnesses carried in water. Septic tanks and leach fields are one way to safely dispose of wastewater, in which liquid and solid wastes are separated and filtered or broken down by anaerobic bacteria. Sewage systems collect waste from many sources and transport it to treatment plants that process the waste and return water to the environment.

e. Hazardous Wastes

Hazardous waste is any waste with properties that could make it dangerous to human health or the environment, including household cleaning agents, insecticides, agricultural pesticides and herbicides, or any flammable, chemically reactive, or corrosive waste. Many hazardous wastes are human-made and do not naturally occur in the environment. These wastes typically persist in the environment for long periods of time and accumulate in soils, water, organisms, plants, animals, and humans.

Study Questions

i. What are the four main types of waste? Describe their characteristics.
ii. What threats do these types of waste pose to the environment (including human health)?

2. Waste Management Strategies

LO2 List the three primary ways that we take care of wastes

a. Isolation

Landfills are one widely used system to isolate waste. In well-resourced countries waste is sorted, concentrated, and buried at carefully designed disposal sites. These sites seal waste disposal cells to prevent water infiltration into the waste, and capture leachate from the waste before it gets into the environment. Lesser-resourced countries typically don't have advanced landfill systems like those described above, but rather have open dumps.

b. Incineration

Burning of waste can significantly reduce the volume of waste and in turn be used to generate electricity by heating water and using steam to spin turbines. Scrubbers and filters are used to reduce gases and other pollutants emitted in the incineration process from escaping into the environment. However, the operational cost, as well as the threat of air pollution from incineration is high, reducing the number of incinerators that remain today.

c. Attenuation

Attenuation is waste dispersal rather than disposal; it hinges on the concept that waste in low concentrations isn't harmful. Smokestacks are an example of the attenuation principal. Waste from factories is released into the atmosphere, where it travels widely such that it is detectable in a large area, but in small concentrations in any given place. This dispersal technique assumes that the environment has a certain assimilative capacity to break down well-diluted waste through bacterial activity, sunlight, or other environmental factors.

d. Ocean Dumping

Ocean dumping was a widely used waste disposal strategy until recently. Often, waste was believed to be attenuated in the ocean when dumped, while at other times waste was sealed into drums then dumped. During the Cold War, the United States dumped approximately 50,000 drums of radioactive waste across the ocean floor near the coast of San Francisco, California. Now, ocean dumping is prohibited without a permit from the EPA indicating that the waste material will not degrade the marine environment or endanger human health.

Study Questions

i. What waste disposal technique seems like the safest and most environmentally friendly option? Why?

ii. What are three examples of specific waste management strategies you've encountered in your life? Categorize each management technique into one of the four strategies outlined in this section.

3. The Challenges of Landfilling

LO3 Explain how a landfill is properly designed to handle waste, including liquid and gaseous by-products

a. Landfill Maturation Phases

Landfills operate much like bioreactors in the process of breaking down organic wastes in phases based on oxygen availability and water access. Phase one breaks down waste to carbon dioxide via aerobic biodegradation due to abundant oxygen concentration. Phase two is the onset of anaerobic degradation as oxygen is consumed and additional carbon dioxide is produced (respiration). Phase three is when infiltrating water produces acidic leachate, potentially carrying hazardous metals away from the landfill. Phase four is when microorganisms convert organic acids in leachate to carbon dioxide and methane, contributing to greenhouse gas emissions. Phase five occurs when microbial activity slows and nutrient levels in leachate are low. What is left at this stage will degrade into material resembling humus.

b. What to Do About Landfill Gases?

About 50% of the gas produced by landfills is methane, while the other 50% is mostly carbon dioxide with traces of other volatiles. Perforated pipes may be laid or wells drilled to extract gas that builds up in landfills. Excess methane can be combusted (flared) to produce carbon dioxide and water.

c. What to Do About the Leachates?

The composition of leachates is highly variable. Some carry toxic metals like lead and chromium, others contain microplastics or organic pollutants. Underlying geological features determine if leachates will mix with groundwater, dependent on the permeability of the underlying material. The most effective way to isolate leachates is by installing an impermeable membrane called a geomembrane. Leachates are regulated with routine sampling and monitoring mandated by local governments. If leachates do appear in groundwater (or anywhere outside of a landfill), they can be treated with bioremediation techniques.

d. Landfill Subsidence

Subsidence occurs as landfills age, settle, and lose mass and volume. Rate of subsidence depends on several environmental factors including climate. Regardless of the rate, settlement can result in cracks and fissures in the landfill, allowing water to pool and percolate downward through the landfill, creating leachate. Well-managed sites prevent subsidence by daily compaction and stabilizing the site with vegetation.

Study Questions

i. What are two examples of challenges faced by landfills, and how are these challenges addressed?

ii. What are the five phases of landfill maturation, and what risks are present at each phase?

4. Recycling, Remanufacturing, and Composting

LO4 Summarize the importance to waste management of recycling and composting

Recycling removes materials from the typical waste stream such that they can be re-used. The process varies widely depending on the material being recycled. The most common recycled products are paper and metal, and some plastics, though the ability for recycling centers to handle plastics is limited. An alternative to recycling in some instances is remanufacturing or taking worn-out products and replacing their parts. Electronic waste, or e-waste, is a dramatically growing portion of waste generation worldwide. This type of waste contains significant amounts of toxic materials as well as potentially valuable resources like gold and silver. Composting is the microbial breakdown of organic wastes that can be done at the small residential scale and at a large, utility-scale. Compost can result in nutrient rich soil after the organic materials break down.

Study Questions

i. What is the difference between recycling and remanufacturing?

ii. What are some of the shortcomings of recycling?

5. Disposal of Hazardous Wastes—The Nastiest Stuff

LO5 Describe the ways that we deal with hazardous wastes, and the shortcomings of the approaches people have taken so far

a. Deep-Well Injection

Deep-well injection disposes of hazardous liquids far below the surface of the Earth, far from freshwater aquifers. This disposal technique is very simple. Hazardous fluids are injected deep underground, and left there.

b. The Problem With Coal

A by-product of coal combustion is solid combustion residue, which is rich in arsenic, cadmium, and mercury. To avoid accumulation of this waste, some is mixed into cement and marketed as construction material, as well as wood pulp and plaster to make wallboard. A lot of waste is "temporarily" stored in coal impoundment ponds or landfills. However, these solutions create a problem for human and environmental health, as toxic and carcinogenic compounds contaminate groundwater when they leach from unlined ponds.

c. Existing Approaches to Nuclear Waste Disposal

Spent fuel, a high-level radioactive waste created in nuclear power generation, is currently stored at individual reactor sites in "pools." This type of disposal is considered temporary–permanent solutions aren't widely known yet. Reprocessing can recover unused uranium-235 and plutonium-239, reducing the volume of high-level radioactive waste. However, this creates liquid waste that contains many radioactive isotopes. Low-level radioactive wastes can be produced in research laboratories and industry, with shallow land burial being the most used disposal method.

d. Repositories: A Long-Term Nuclear Waste Solution

Longer-term approaches to nuclear waste disposal have been proposed, none have yet been adopted. Any method must create a safe isolation of radioactive materials for at least 250,000 years, secure from accidental entry, protected from natural disasters, isolated from natural resources, and including failsafe handling and transport mechanisms.

Study Questions

i. What are three examples of proposed long-term nuclear waste storage? What are the pros and cons of each approach?

ii. Besides greenhouse gas emissions, what are other types of waste generated by coal combustion, and how is this waste disposed of?

Key Terms

- Aerobic biodegradation
- Agricultural waste
- Anaerobic biodegradation
- Assimilative capacity
- Attenuation
- Bioreactors
- Bioremediation
- Biosolids
- Cesspool
- Coal Impoundment ponds
- Coal-ash sludge
- Combined sewage overflows
- Composting
- Comprehensive Environmental Response, Compensation, and Liability Act (CERCLA)
- Concentrated animal feed operations (CAFOs)
- Dead zones
- Deep-well injection
- Dry casks
- E-waste
- Environmental Impact Statement (EIS)
- Feedlot lagoons
- Geological repository
- Geomembranes
- Global distillation
- Greywater
- Hazardous waste
- Hazmat drums
- High-level nuclear wastes
- Hypoxic
- Incineration
- Industrial wastes
- Isolation
- Landfill cells
- Landfill gas (LFG)
- Landfills
- Leach field
- Leachate
- Legacy sediment
- Low-level nuclear wastes
- Manure
- Miasmas
- Microbeads
- Microplastics
- Municipal solid waste (MSW)
- National Environmental Policy Act (NEPA)
- Ocean Dumping Act
- Percolation tests
- Persistent organic pollutants (POP)
- Primary treatment
- Pyrolysis
- Reprocessed
- Resource Conservation and Recovery Act (RCRA)
- Scrubbers
- Secondary treatment
- Septic tank
- Sewage sludge
- Sewer
- Site characterization
- Solid combustion residue
- Spent fuel
- Superfund
- Tertiary treatment
- Uranium-mill tailings

Glossary

'a'ā A basaltic lava that has a very rough, rubbly surface.

A horizon The upper inorganic soil horizon capped by a layer of organic matter. It is the zone of leaching of soluble soil materials.

acid mining drainage (AMD) An environmental problem that occurs when sulfide-rich tailings from high-sulfur coal or metallic-sulfide ore mining react with drainage water to produce highly acidic sulfuric acid, which in turn releases toxic metallic ions, including iron, lead, and copper into waterways.

active layer The shallow underground zone above permafrost that is subject to seasonal freezing and thawing. Active layers contribute to forming unique arctic surface features.

active sea cliff A cliff whose formation is dominated by wave erosion.

active volcano A volcano with a historical record of eruptions, or one in a current state of eruption.

adaptation In the context of this course, steps taken to lessen the impact of climate change on people and societies.

adiabatic heating The heating that occurs in air that is compressed, as individual gas molecules are driven closer together and increase their collisions.

adits Horizontal or gently sloping tunnels that are drilled sideways into hill and mountainsides for sub-surface mining.

aerobic biodegradation Breakdown of organic matter by microbes in the presence of oxygen.

aftershocks The many smaller earthquakes that follow with diminishing intensity over time in the same area as a main shock.

aggregates Raw materials extracted from quarries including crushed stone, sand, and gravel (for building roads, foundations, and other constructions).

agricultural waste The wastes of industrial agriculture including toxic pesticides, herbicides, and fungicides, as well as large amounts of plastic that are used both to protect crops (e.g., weed mat) and to store farm products like hay. Large quantities of manure and fecal matter are also common agricultural wastes.

air masses Large approximately homogeneous bodies of air with distinct temperature, moisture, and density characteristics due to differences in heating of the Earth's troposphere.

albedo The reflectivity of surfaces, in other words the percentage of incoming shortwave solar radiation that is reflected back into space. High albedo surfaces are bright, low albedo surfaces are dark.

algal blooms Large colonies ("blooms") of algae that grow out of control due to the addition of nitrogen based fertilizers that run off into streams, lakes, and the ocean from farms and residential properties. The overproduction of algae creates toxic conditions for aquatic life by taking up all available oxygen in the water.

alpine glaciers Glacier that is confined by surrounding mountain terrain.

amalgam An alloy of mercury and gold.

amber-rat The urine of pack rats that crystalizes and makes a tough, resin-like substance that preserves vegetation such that it can be studied thousands of years later.

amplitude The height of a sea (or any other energy) wave.

anaerobic biodegradation The degradation of organic matter by microbes in the absence of oxygen.

andesite A gray to black volcanic rock that most often erupts from composite volcanoes at convergent plate boundaries. It contains abundant plagioclase phenocrysts, with 52% to 63% silica content by weight.

andosol Nutrient-rich, volcanically derived soil.

angle of repose The maximum stable slope angle that granular, cohesionless material can assume. For dry sand, this is about 34°.

anions Negatively charged ions are called anions.

annual peak discharge The largest water flow rate recorded in a stream or river each year.

antecedence The amount of recent rainfall prior to a given time

antecedent moisture The amount of moisture present in soil or the atmosphere at the beginning of a storm event.

anthracite The most highly metamorphosed form of coal. It contains nearly pure carbon and is a prized type of coal for energy production.

Anthropocene The proposed new epoch in the geological time scale defined by a world in which human activity creates great physical, biological, and chemical change; change sufficient to leave a long-standing trace in the geological record.

anticline A generally convex-upward fold that has older rocks in its core. (Contrast with syncline.)

apatite A group of phosphate minerals with the chemical formula [$Ca_5[PO_4]_3$ (OH, F, Cl)]. An important source of phosphorus in the natural world.

aphanitic A rock texture used to describe igneous rocks that are composed of small grained crystals of roughly uniform size.

aphyric Volcanic rocks that lack crystals because the magma that form them was too hot to crystallize before erupting.

aquiclude A layer of low-permeability rock or sediment that hampers or stops groundwater movement. Aquicludes provide the confining beds enclosing confined aquifers.

aquifer A groundwater bearing body of rock or sediment that will yield water to a well or spring in usable quantities.

aquitards An underground layer that slows, but does not altogether prevent the movement of groundwater.

Arctic amplification The enhancement of climate change due to global warming at high northern latitudes. It is largely a phenomenon that is driven by the ice-albedo feedback.

area mining The mining of coal in flat terrain (strip mining).

arroyo A steep-sided gully formed by the action of flowing water in an arid region.

artesian wells Wells that penetrate confined aquifers. Water under pressure rises up from the aquifers into these wells, and may upwell all the way to the surface.

artificial levee An embankment constructed to contain a stream at flood stage. See also levee.

asbestiform Minerals that can crystallize in a way that results in long, skinny fibers, such as chrysotile, amosite, crocidolite, and anthophyllite.

asbestos A term used to describe highly fibrous minerals that are resistant to heat and corrosion. Also, a well known carcinogen. See also mesothelioma.

asbestosis A debilitating, noncancerous, lung disease caused by scarring from the inhalation of asbestos fibers.

assimilative capacity The ability for pollutant to be absorbed by an environment without detrimental effects to the environment.

asthenosphere The plastic layer of the Earth, which underlays the lithosphere and consists of a part of the upper mantle.

Atlantic Meridional Overturning Circulation (AMOC) The sinking of cold and dense saltwater at the northern end of the Atlantic Ocean off the coast of Greenland. It is integral to the worldwide climate system. (See also global conveyor.)

atmosphere The layers of gases surrounding a planet.

atmospheric river Long and narrow plumes of moist air in the atmosphere that originate over the tropics and that bring plentiful rain to regions at temperate latitudes.

atoms The smallest part of a substance that cannot be broken down chemically, composed of a nucleus, protons, neutrons, and electrons.

atomic mass The sum of the number of protons and neutrons in the nucleus of an atom.

atomic number The number of protons (positively charged particles) in an atomic nucleus. Atoms with the same atomic number belong to the same chemical element.

atomic radius The size of an atom.

attenuation (one approach toward waste disposal). To dilute or spread out waste material, optimally so that it is not concentrated enough to be harmful.

attribution science Determining whether and how much climate change is to blame for weather disasters, particularly extreme weather events.

aurora Commonly known as the northern and southern lights, these are glowing displays of color in the sky caused by Earth's magnetism deflecting highly charged particles approaching from the sun.

avalanche tracks Paths taken by avalanches, focused by terrain (such as stream drainages on the steep sides of valleys).

B horizon A subsurface horizon present below the topsoil, or A horizon, that is relatively more compact and contains less humus, soluble minerals, and organic matter than the soil above.

base load The minimum demand for power that must be supplied by an electrical grid to consumers.

backwash The retreat of water that has been washed up onto a beach, helping build up the next wave approaching the shore.

Banded Iron Formation (BIF) A type of banded rock that was deposited on the floor of lakes and oceans before the Earth's atmosphere became enriched with oxygen during the Precambrian period. It consists of layers of red, iron-rich minerals that were deposited during summers when microbial oxygen production was high, and dark layers of chert deposited in the winters when aquatic systems were more anoxic.

bank-full stage The level of water in a river that has reached the top of the river banks. Any increase in water level causes overflow onto the surrounding floodplain.

barrage A dam-like barrier used to capture energy from masses of water moving in and out of a bay or river due to tidal forces.

barrier island A narrow sandy island rising slightly above high tide level that is oriented generally parallel to the coast. A barrier island is typically separated from the coast by a lagoon or a tidal marsh.

base isolation A building design option using a movable base in its foundation. This allows the building to move only a little or not at all, as the ground underneath shakes during an earthquake.

base level The level below which a stream cannot erode its bed. The "ultimate base level" worldwide is sea level.

base metals The more abundant "nonprecious metals" that are of high industrial importance, such as copper and tungsten.

base shear Caused by a rapid horizontal ground acceleration during an earthquake. It can cause a building to part from its foundation, thus the term.

baseflow Stream discharge that is sustained between precipitation events.

Båth's Law This Law states that the moment magnitude (Mw) of the largest aftershock will typically be on the order of 1.1 to 1.2 less than the moment magnitude of the main shock.

batholith A large body of phaneritic igneous rock that forms when magma rises through Earth's crust but does not erupt onto the surface.

beach berm A ridge of coarse materials that forms on a beach parallel to the shore by the wave uprush.

beach nourishment The addition of sand to a beach by trucking or pumping in order to build the beach back up after it has eroded.

bentonite A soft, plastic, and light-colored clay that swells to many times its volume when it gets wet. Named for outcrops near Fort Benton, Wyoming, bentonite is formed by chemical alteration of volcanic ash.

binary-cycle system A geothermal system that uses geothermal heat to vaporize a second fluid (e.g., isobutane) with a lower boiling temperature than water, in order to power a turbine generator.

biodiesel A diesel-fuel equivalent derived from plant or animal sources.

biofuel Any fuel derived from biomass, for example, sugarcane, food waste, straw, corn, or switchgrass.

biogenic sedimentary rock Rocks resulting directly from the activities of living organisms, such as a coral reef and shell-derived limestone or coal.

biogeochemical cycles The processes that allow matter and energy to flow throughout the atmosphere, hydrosphere, lithosphere, and cryosphere that are critical to life on Earth.

biomass Organic matter or matter from recently living organisms used for bioenergy production.

biomimicry Using natural processes carried out by living things to design spaces that keep desirable levels of temperature and humidity within them.

bioreactors An apparatus where biological reactions or processes are carried out in a confined space with optimal conditions.

bioremediation Use of bacteria, fungi, and other microorganisms to break down or consume environmental pollutants.

biosignatures The distinctive spectroscopic signs of life and Earth-like conditions on exoplanets.

biosolids Nutrient-rich organic materials that result from domestic sewage treatment.

biosphere the regions of the surface, atmosphere, hydrosphere, and crust of Earth that are occupied by living organisms.

bitumen A thick, semisolid, vicious mix of hydrocarbons obtained naturally and from petroleum distillation.

bituminous Referring to the second highest rank of coal (after anthracite). A product of metamorphism. Used widely in electricity generation and steel making.

block (volcanology) Large, angular fragments of lava rock, typically with smooth faces.

block lava A lava flow with a rough surface consisting of lava blocks. The blocks are not as jagged and spiny as they are in the similar-appearing a'a flows.

block glide (or translational slide) A landslide in which movement occurs along a well-defined planar surface or surfaces, such as bedding contacts, metamorphic foliations or faults.

body wave Earthquake wave that travels through the body (deep interior) of the Earth.

bomb cyclone A fast developing storm that results from sudden drop in atmospheric pressure at the storm's center.

bombs Large, rounded, or streamlined fragments of volcanic debris.

brackish Somewhat saline water produced when salt water mixes with freshwater (saltwater intrusion).

bread-crust structure A characteristic surface of volcanic blocks and bombs characterized by many superficial cracks caused by expansion of hot gases trapped inside them during eruption.

bridge scour The removal of material from beneath bridge abutments caused by moving water. This phenomenon is a leading cause of bridge collapse.

brine exclusion effect The generation of saltier, denser water where sea-ice, which is salt-free, forms.

buttress fills Compacted soil that is used to replace the loose, broken-up toe of a landslide to increase slope stability.

C horizon The bottom layer of a mature bed of soil, this horizon is a transition zone between the soil and unweathered parent material below.

caldera A large crater caused by collapse of an underlying magma chamber, sometimes accompanied by a highly explosive volcanic eruption. Calderas may be enclosed by cliff walls and are by definition more than 1.5 kilometers (1 mile) in diameter.

caliche In arid regions, sediment or soil that is cemented by calcium carbonate ($CaCO_3$). See also hardpans.

capillary fringe The moist zone in an aquifer above the water table. The water is held in this area by capillary action (adhesion to the surfaces of sedimentary grains).

carbon capture and storage Technology intended to trap CO_2 emissions and keep them out of the atmosphere.

carbon cycle The natural processes by which carbon cycles through the atmosphere, biosphere, geosphere, cryosphere, and hydrosphere.

carbon dioxide (CO_2) A colorless and odorless gas produced by respiration as well as the burning of carbon and organic compounds. Carbon dioxide is absorbed by plants during photosynthesis to produce plant tissues.

carbon intensity The amount of carbon dioxide (CO_2) created per unit of energy derived from using a particular fuel.

carbonate Any member of a family of minerals that contain the carbonate ion, $(CO_3)^{-2}$, including calcite, dolomite, and limestone.

carbonatites Products of rare, carbonate-based magmas erupted in continental rift zones. They are the source of most rare earth element deposits.

cations Positively charged ions.

cesspool An underground tank into which people deposit their feces and urine.

chain reaction A self-sustaining nuclear reaction that occurs when atomic nuclei undergo fission (split into two). Free neutrons are released that collide with and split other nuclei, causing more fissions with the release of more neutrons, and so on.

chemical dispersants Chemical agents used to break oil slicks into small globules in the water column.

chemical sedimentary rock Rock resulting from precipitation of chemical compounds from a water solution. See also evaporite.

Chinooks Warm, dry winds that occur on the downward slope of a mountain range when warm air has lost its moisture. (Named for this phenomenon observed in Colorado.)

chop Many small waves in a localized area causing the water surface to appear rough.

cinder cone Tall, smoothly sloping volcanic cones consisting mostly of pyroclastic material, or cinders, that pile up around explosive volcanic vents.

cinders Glassy and vesicular (containing gas holes) volcanic fragments, each only a few centimeters across, that are thrown from a volcano.

circulation cells Divisions of the global air circulation model in which each cell covers a region where the air circulates from high to low pressure.

circumpolar marine current An ocean current that flows from west to east around Antarctica which formed when the continent broke free from South America and became isolated from the climate-moderating influence of warmer ocean waters.

clastic sedimentary rock Rock composed of fragments of older preexisting rocks and minerals.

cleavage The characteristic way minerals break smoothly along certain crystallographic planes of weakness that reflects their internal atomic structure.

climate The general types and patterns of weather that characterize a place.

climate mitigation The effort to rein in the degree of climate change already well underway, primarily through technological upgrades ranging from small-scale (e.g., solar panels on rooftops) to the very large (e.g., capture of carbon dioxide, a greenhouse gas, then pumping it back deep underground where it can dissolve and be stored in saline groundwater).

climax avalanche A thick and heavy avalanche that tends to occur in late spring when the snowpack has warmed, partially melted, and weakened.

clinker A stony residue that is produced when limestone is heated in a kiln. It is to create "ready-mix concrete" when mixed with gypsum and additional limestone. ("Clinker" is also the term used for the spiny pieces of rubble atop a'a lava flows.)

cloud reflectivity enhancement A proposal for creating cloud condensation nuclei that would increase albedo and reflect more sunlight away from Earth.

cluster A series of small earthquakes that occur over a short period of time.

coal Carbon-rich residue from ancient plant matter that has been preserved over millions of years and metamorphosed by heat and pressure. A fossil fuel.

coal impoundment ponds The storage sites of coal combustion waste that is not sent to landfills.

coal-ash sludge The buildup of ash removed by scrubbers and filters in smokestacks at coal-fired power stations.

coal-bed methane (CBM) A form of natural gas, methane (NH_3), extracted from coal beds and seams. In the early twenty-first century, it has become an important source of energy in the United States, Canada, and other countries.

coast The area that extends from the shoreline to the landward limit of marine features such as beaches and related sand dunes, marshes, estuaries, and rocky shorelines.

coast-normal A hurricane that hits the coast at a right angle, such that the storm surge and strong winds slam into the coast with strongest intensity.

coast-parallel A hurricane that moves parallel to the coast, such that the strongest winds and the storm surge remain offshore.

cohesion The tendency of a material to stick to itself; the strength a material has when no normal force is applied.

cold front Where a cold air mass encroaches upon a mass of warm air, with the cold air wedging in beneath and easily lifting the warm air in its way.

color The result of sensations on the eye based on the way an object reflects light, a property that can be used to help identify minerals.

Colorado brown stain A discoloration of the tooth enamel that results from ingesting high levels of fluoride.

columnar jointing Polygonal columns in a solid lava flow or shallow intrusion formed by contraction on cooling. The joints (fractures) are oriented perpendicular to the cooling (e.g., air, older underlying bedrock). They are common in basaltic lavas.

combined sewage overflows Spills of raw or incompletely treated wastewater into lakes and river during heavy storms, when sewage systems are overwhelmed by more water (runoff) than they can process.

combustion A high-temperature, exothermic chemical reaction between some fuel and an oxygen.

composite volcano A type of volcano, also known as a stratovolcano, which typically forms along convergent plate margins. They are built up by successive pyroclastic and lava flows resulting in large, steep cones. See also stratovolcano.

composting Bacterial breakdown of organic material, optimally to produce useful material, such as garden fertilizer or soil amendments for agriculture and forestry.

Comprehensive Environmental Response, Compensation and Liability Act (CERCLA) An act established to coordinate cleanup at waste sites around the United States. See also Superfund and Superfund sites.

compressive A force that occurs at convergent plate boundaries when a physical force presses tectonic plates together.

concentrated animal feed operations (CAFOs) Giant feedlots for livestock that exist to produce meat, dairy, and egg products.

concentration factor The enrichment of a deposit of an element, expressed as the ratio of the element's abundance in the deposit to its average crustal abundance.

conduit A pathway of chemically altered, fractured rock that connects magma chambers to volcanic vents at Earth's surface.

cone of depression The downward-pointing cone shape formed by the water table undergoing extraction around a pumping well.

confined aquifer A water-bearing formation bounded above and below by impermeable beds (aquicludes) or beds of distinctly lower permeability (aquitards).

conservation/reduced-tillage A method used to reduce soil erosion on farmland by minimizing plowing in the fall and leaving never less than 30% of the soil surface covered with harvested crop residue.

contacts The upper and lower boundaries of each bed in a stratified sedimentary rock, marking when the environmental conditions in which sedimentation took place changed.

continental crust Earth's crust that underlies terrestrial surface features and consists largely of lighter-colored, silica-rich, igneous and metamorphic rocks (e.g., granite and gneiss) and related sedimentary cover. It typically ranges from 20 to 90 kilometers (12 to 56 miles) thick.

continental ice sheet The largest type of land-based glacier. Ice sheets can cover virtually an entire continent (e.g., Antarctica).

contour mining The mining of coal by cutting into and following the coal seam around the perimeter of the hill.

convection cell A complete cycle ("rollover") of overturning air.

convergent boundaries Tectonic plate boundary interactions where plates collide with tremendous compressive force.

core The dense, metallic center of the Earth composed of iron (Fe) and nickel (Ni) rich molten rock surrounding a solid metallic center.

coral bleaching The process of coral polyps expelling zooxanthellae at increased water temperatures which causes them to lose their color, and eventually die if the warm water persists.

Coriolis effect The deflection of wind and ocean currents by rotation of the Earth. Deflection is toward the right in the Northern Hemisphere and toward the left in the Southern Hemisphere.

cornices Accumulations of windblown snow on the downwind side of ridges.

cosmogenic isotopes Rare isotopes or types of an element, such as ^{10}Be or ^{36}Cl, formed as cosmic rays, such as neutrons, interact with minerals at and near Earth's surface.

cracking See refining.

creep (soil) The imperceptibly slow downslope movement of rock and soil particles caused by gravity.

cross-bedding Arrangement of thin stratigraphic layers within a sedimentary or volcanic ash bed, inclined at an angle to the bed's contacts.

crop rotation A method to limit soil erosion in agricultural land use by annually alternating crops.

crude oil A liquid fuel composed of many hydrocarbon compounds found in natural underground reservoirs. Unrefined petroleum.

crust The thin, solid, uppermost layer of the solid Earth composed of relatively lightweight aluminum and alkali-rich silicate rocks that enclose Earth's mantle.

crustal root Pendants of low-density crustal rocks extending deep into the mantle beneath many mountain ranges.

cryosphere That part of Earth's surface covered with permanent or long-lasting bodies of ice (e.g., ice sheets) and associated frozen groundwater (permafrost).

crystalline Describes the way atoms are arranged in minerals, such that they form geometrically regular and orderly patterns.

cyanide heap leaching A process of using cyanide to dissolve and recover gold and silver from ore.

cyclones A hurricane-strength tropical storm in the South Pacific and Indian Oceans.

Darcy's Law A derived formula for the flow of fluids, particularly groundwater, in a porous medium.

dead zones Hypoxic (oxygen-depleted) zones in water bodies that are result of the decomposition and respiration of algal blooms, consuming much of the oxygen in the water column.

debris flow A moving mass of a water, soil, and rock intimately mixed. More than half of the soil and rock particles are coarser than sand, and the mass has the consistency of wet concrete.

decomposed granite (d.g.) Slightly weathered granite that is well-suited for foundation material for structures and roads.

decompression melting A process that occurs when a portion of the Earth's mantle pushes or wells upward into the crust, where pressure drops significantly and the mantle partially melts as a result.

deep carbon sub-cycle The pathways that define how carbon dioxide (CO_2) has accumulated in Earth's atmosphere and recycled back into the planet via hydrosphere and lithosphere through billions of years.

deep ocean Includes many layers beneath the epipelagic ocean that are defined by the different lifeforms that can tolerate these great depths. Seawater temperature drops the deeper one goes in the deep ocean.

deep subsidence Subsidence is initiated at hundreds of meters below the surface when water, oil, or gas are extracted.

deep-well injection A liquid waste-disposal process that pumps liquids hundreds of feet underground into aquifers that have no potential to contaminate potential or existing potable water aquifers.

deflecting walls Reinforced concrete structures built to divert threatening flows away from structures and prevent damage.

denitrification A microbially driven process of reducing nitrate (NO_3^-) and nitrite (NO_2^-) to gaseous forms of nitrogen, primarily nitrous oxide (N_2O) and nitrogen (N_2).

depleted uranium (DU) The material left after most of the highly radioactive form of uranium, ^{235}U, is removed from the natural uranium ore.

depositional processes The ways that sediment can be moved and deposited by wind, flowing water, sea, or ice.

depth hoar Ice crystals formed in the snowpack by migration of water vapor. Forms a weak layer upon which avalanches can slide.

desertification The process by which semiarid grasslands are converted to desert.

deuterium A hydrogen with one neutron and one proton.

devitrification Also known as deglassing, a process in which crystallization occurs slowly in natural glass at the microscopic level and a mineral glass (e.g., obsidian) loses shiny luster.

differential subsidence Subsidence that occurs in areas adjacent to each other at different rates.

differentiation The separation of matter based on density (e.g., such as crystals settling in magma).

dip slope A sloping land surface that is inclined in the same direction and angle as the dip angle of the underlying rocks.

directed lateral blast A lateral (sideways) explosive volcanic eruption.

Dirty Dozen A group of 12 high persistent and toxic chemicals including DDT, aldrin, and polychlorinated biphenyls.

discharge (Q) The flow rate of water, usually expressed in cubic feet per second (cfs) or cubic meters per second (cms), passing a given point in a stream or river.

displacement The amount of slip taking place along a plane of rupture such as a fault.

divergent boundaries Tectonic plate boundary interactions where plates under tension move apart ("diverge") from one another.

dormant volcano A volcano that has no recent historical record of volcanic activity (if any record at all) but which may appear to have been active at some time in the recent geological past, including fresh appearing volcanic deposits, areas lacking vegetative cover, etc. A dormant volcano could erupt again, though implicitly—and perhaps deceptively—no time soon.

doublet earthquakes Two powerful shocks of nearly equal magnitude taking place in the same general area, or along two closely separated faults within a brief period of time.

drainage basin The tract of land that contributes water to a particular stream, lake, reservoir, or other body of surface water. Also known as a watershed.

drainage divide The boundary between two drainage basins.

drawdown The lowering of the water table immediately adjacent to a pumping well.

drifts Mine tunnels cut specifically to follow a valuable vein or mineral seam.

driving force The component of gravity that pulls rock and soil downslope.

dry casks A storage unit made of steel that is encased in concrete for spent nuclear fuel rods.

Earth System The various parts of our planet, including the oceans, ice cover, rocky interior, atmosphere, and living realm that are linked together and support one another naturally to sustain a habitable world.

earthquake swarm A series of earthquakes that occur in the same general area over a relatively short period of time.

eccentricity The degree to which the orbit of a body revolving around another body (e.g., Earth orbiting the Sun) deviates from a perfect circle. In other words, the degree to which it is elliptical.

ecosystem services Biological and physical Earth processes that provide benefits to people at no monetary cost.

eddies Looping side-currents that branch off oceanic gyres.

effective cohesion The ability of a material (e.g., slope) to resist shear (e.g., sliding), including the strength imparted by tree roots and capillary forces holding the material together.

effusive Volcanic eruptions that primarily produce flowing lava.

elastic rebound theory The theory that movement along a fault and the resulting seismicity are the result of an abrupt release of stored elastic strain energy between two rock masses on either side of the fault.

electrical resistivity A measure of how easily an electrical current can flow through a material, measured in ohm-meters ($\Omega \cdot m$).

electrostatic charges A deficiency or excess of electrons which occurs on ungrounded or insulating surfaces.

elements Substances that cannot be separated into different substances by usual chemical means. Each element consists of atoms having a distinctive number of protons in its nuclei.

end-members A mineral that is at the extreme end of a mineral series in terms of purity of its chemical composition.

energy The capacity of matter or radiation to do work.

energy dense A resource from which a small volume provides a lot of useful energy.

energy return on investment (EROI) The ratio of energy returned relative to the energy invested to discover and obtain that energy.

enhanced oil recovery Processes implemented to increase the ability of oil to flow to a well by injecting water, chemicals, or gases into the reservoir or by changing the physical properties of the oil.

environmental geologists Professionals who apply understanding of geological processes on or near Earth's surface in order to reduce the threat of natural hazards, pollution, and environmentally abusive land-use practices.

environmental geology An applied science involving the study and identification of geological hazards (e.g., landslides, volcanoes) and environmental degradation (e.g., erosion), including land abuse, the spread and containment of pollutants, resource depletion, and other processes relevant to human well-being.

Environmental Impact Statement (EIS) A comprehensive study of the potential consequences of a proposed project on the environment required by National Environmental Policy Act of 1969.

environmental justice The fair treatment and meaningful involvement of all persons and sections of society impacted by an environmental problem, irrespective of race, economic category, and gender.

environmental refugees People escaping drastic, long-lasting environmental change.

epicenter The point at the surface of Earth where the seismic waves released by an earthquake first emerge.

epipelagic ocean The shallow layer at the surface of the ocean that has a nearly uniform temperature, wind-driven circulation, and is home to approximately 90% of all ocean life.

erosion The scouring out and transportation of loose materials and exposed soils at Earth's surface.

erosional processes The ways that erosion takes place to sculpt and change a landscape, including the activity of runoff and streams, landslides, and impact of strong winds on loose materials.

eruption column A tall pillar of pyroclastic material and gases that rises above a volcanic vent.

ethanol A colorless liquid chemical compound, commonly called grain alcohol, with the formula CH_3CH_2OH.

eutrophic A type of lake that is typically shallow, has high nutrient input, low oxygen concentration in the bottom waters, small fish populations, and often has a surface mat of algae.

evapotranspiration The process by which water is transferred from the biosphere to the atmosphere by evaporation from plant leaves (transpiration).

evapotranspiring The verb form of evapotranspiration.

e-waste Electronic wastes including computers, printers, and cell phones that have been disposed of.

exfoliation dome A large, rounded dome resulting from exfoliation (breaking off shells of granitic rock by mechanical weathering); for example, Half Dome in Yosemite National Park.

expansive soils Clayey soils that expand when they absorb water and shrink when they dry out.

extinct volcano A volcano that has not erupted in such a long time that, typically, it has become significantly eroded and is well vegetated. Extinct volcanoes will not ever erupt again—though eruptions could recur in the same area or region.

fabric The structures and overall appearance of a rock.

factor of safety The ratio between resisting and driving forces in a landslide. Failure occurs when the factor of safety is less than 1—that is, when resisting forces are no longer greater than driving forces.

failure plane The buried fracture surface on which a landslide slips.

fault creep (tectonic) The gradual slip or motion along a fault without an earthquake.

fault plane The buried fracture surface of a fault.

fault scarp A steep slope that formed directly as a result of fault movement. It marks the location of where a fault intersects the surface.

feed in tariffs Agreements between power companies and people or businesses to pay a predetermined, fixed rate for "green" power production.

feedback A spontaneous, natural response to some process occurring in nature. See also positive feedback and negative feedback.

feedlot lagoons A temporary storage solution for manure consisting of sealed reservoirs with no inlets or outlets.

feldspars A silicate mineral with the chemical formula, $KAlSi_3O_8$.

felsic Descriptive of magma or rock with abundant light-colored minerals and a high silica content. The mnemonic term is derived by combining the words feldspar and silica.

ferromagnesian Minerals that contain high proportions of iron and magnesium.

fetch The unobstructed stretch of sea over which the wind blows to create wind waves.

filters Devices that remove tiny ash particles and toxic chemicals from gaseous pollutants emitted into the atmosphere during combustion.

firn A transitional material between snow and glacial ice, being older and denser than snow, but not yet transformed into glacial ice. Snow becomes firn after surviving one summer melt season; firn becomes ice when its permeability to liquid water drops to zero.

fissile isotopes Isotopes with nuclei that are capable of being split into other elements. See also fission.

fission The separation of one large atom into two smaller atoms when a neutron hits and the nucleus of the large atom splits.

fission products What results from a fission reaction.

fjord A narrow, steep-walled inlet of the sea between cliffs or steep slopes, excavated or at least largely shaped by the passage of a glacier. A drowned glacial valley.

flaring The process in which natural gas that is extracted as a by-product of oil pumping is burned off.

flash floods Floods which occur with little or no warning in small drainage basins and typically are of short duration. Extensive damage is possible due to the high velocity speed movement of water including homes and bridges ripped from their foundations.

flood basalt A lava flow resulting from the rapid outpouring of an exceptionally large volume of typically basaltic lava. These flows stack up to develop plains, plateaus, and submarine platforms. They originate from long fissure systems and represent a type of effusive eruption.

flood frequency The average time interval between floods that are equal to or greater than a specified discharge.

floodplain The portion of a river valley adjacent to the channel that is built of sediments deposited during times when the river overflows its banks at flood stage.

floods An overflow of water onto normally dry land.

flood stage The water level of a body of water at which water will overflow onto the floodplain (and beyond).

floodway Floodplain area under federal regulation for flood insurance purposes.

flowback Fracking fluid pumped down a well that returns to the surface as a murky, highly salty water enriched in drilling fluids and other contaminants.

fluidized When a non-fluid mixture of high-pressure gas and fine particles acquire the characteristics of a fluid and moves like a giant liquid flood.

fluids In the context of Earth's deep crust, fluids are substances that behave somewhat like liquids, somewhat like gases. They are an "intermediate" physical phase, with no equivalent under surface conditions.

fluorosis A condition caused by ingestion of fluorine with symptoms including bone and joint pain and deformation. Fluorosis has led to the deaths of many livestock following volcanic eruptions, which can produce abundant fluorine.

focal mechanism The type of slip that occurs during an earthquake.

focus (or hypocenter) Point within Earth where an earthquake originates.

foliation The planar or wavy structure that results from the flattened growth under high stress conditions of minerals in a metamorphic rock.

foraminifera Commonly known as forams, these are single-celled organisms with shells that can be easily fossilized after settling to the sea floor.

force balance The balance between the resisting force and the driving force that determines whether a landslide will slip or not.

foreshocks The earthquakes that precede a much larger main shock. They generally take place a few days to just minutes before the main shock.

forever chemicals (PFAS) Per-or polyfluoroalkyl substances that are stable, toxic chemicals which do not degrade in nature and thus bioaccumulate. These substances are widely used in consumer goods like nonstick pots and pans, water-resistant fabrics, and stain resistant upholstery due to their use in making fluoropolymer coatings that make products "nonstick."

fossil fuels Coal, oil, natural gas, and all other solid or liquid hydrocarbon fuels.

fossil water Water that has been contained and undisturbed over thousands of years, typically in aquifers.

Fournier d'Albe equation An equation that measures the economic aspects of risk. For example, volcanic risk = hazard × value × vulnerability, where hazard may be measured as the potential area threatened by a particular eruptive phenomenon in a specified period of time; value is the monetary value of property and infrastructure threatened; and vulnerability is the capacity of a particular area to withstand the impact of the hazard.

fracking Oil company jargon for improving oil or gas production by hydraulic fracturing.

fracture Break in a rock caused by tensional, compressional, or shearing forces. In some minerals that do not cleave, their distinctive fracture patterns are useful to help identify them. See also joint.

frictional strength The strength of a material that is proportional to the normal force applied.

fronts In terms of weather systems, fronts are the boundaries between air masses. See also cold fronts and warm fronts.

frost wedging The opening of joints (fractures) by the freezing and thawing of water.

fully developed sea The largest possible waves that form in a wind of a certain strength blowing over a set length of water.

fumaroles Vents in a volcanic area that issue volcanic gases.

fusain Fossilized charcoal.

gaining stream A stream that receives water from the zone of saturation.

garbage patches Large fields of floating trash that accumulate in the calm centers of gyre currents; primarily small bits of plastic, but also including much larger debris.

gas chromatography A sophisticated method for separating compounds in a mixture using a solvent, a mobile gas and an absorbent or inert liquid.

geoengineering Technological options that involve large-scale engineering of our environment in order to combat or counteract the effects of changes in atmospheric chemistry.

geologic timescale A timescale beginning with the consolidation of Earth ~4.6 billion years ago and ranging to the present day primarily using the classification of various fossil-bearing rocks into age groups based on type-localities/regions, life-forms, and ecological characteristics.

geological repository (nuclear waste) An underground vault in an area free from geological hazards. Repositories in salt mines, granite, and welded tuff have been constructed or proposed.

geomembranes Impermeable rubber or plastic sheets that line landfills and isolate leachate from the underlying geologic materials and the groundwater.

geosphere The rocks and minerals on Earth's surface and the solid and molten rock within the Earth's interior.

geothermal activity In areas of active or inactive volcanism, subsurface magma heats groundwater creating steam and hot water. Hot, less dense groundwater rises through cracks and fissures in the ground forming features such as geysers and hot springs. See also geothermal energy.

geothermal energy Energy derived from accessible thermal energy from Earth's interior.

geothermal gradient The change in temperature with depth inside Earth.

glacial periods Time intervals during ice ages, when Earth's climate is much colder and icier than today, and marked by glacial advances.

glacier A large mass of ice formed on land by the compaction and recrystallization of snow. It moves slowly downslope by creep or spreads outward in all directions because of its weight, and it survives from year to year.

global conveyor The integrated system of worldwide marine circulation initiated by the descent of large masses of cold water into the deep sea from the edges of Antarctica and Greenland.

global distillation A geochemical process by which persistent organic pollutants are transported from warmer to colder regions of the earth.

Global Positioning System (GPS) A navigational system of coded satellite signals that can be processed in a GPS electronic receiver, enabling the receiver to compute the exact positions and elevations of specific locations.

gold rushes Historical instances of urgent dashes to find treasure and fortune in the form of gold.

Goldilocks environment The environment on Earth due to the planet's position in the solar system that makes its habitable; just the right distance from the Sun for abundant liquid water to exist at the surface, a robust magnetic field that intercepts damaging high-energy cosmic rays, and an atmosphere that life not only can tolerate but also exploit to renew itself and evolve.

Gondwana The southern part of Pangaea which has since split into the continental "fragments" we now know as Antarctica, Africa, South America, India, and Australia.

GRACE experiment The Gravity Recovery and Climate Experiment, in which a pair of satellites traveling together measure minute by minute changes in the Earth's gravitational attraction as ice melts and glaciers retreat.

gradient The slope of a streambed usually expressed as the amount of drop per horizontal distance in meters per kilometer (m/km) or in feet per mile (ft/mi).

granular material A material with no cohesion between grains.

gravimeters An instrument that measures changes in the force of gravity in different locations.

greenhouse effect The added atmospheric temperature imparted by the trapping and reradiation of heat energy by greenhouse gases.

greenhouse gases (GHG) Gases responsible for the greenhouse effect including water vapor (H_2O), carbon dioxide (CO_2), and methane (CH_4). They absorb infrared radiation entering the atmosphere from space, from the surface below, and from neighboring gas molecules and reradiate it.

greenhouse warming Rise in average temperature of a body of air caused by an increase in the concentration of heat trapping gas— notably, carbon dioxide, methane, and water vapor in Earth's atmosphere.

greywater Wastewater generated in households or workplaces that does not have fecal contamination

groin A structure of rock, wood, or concrete built roughly perpendicular to a beach to trap sand.

ground shaking Ground movements that take place during an earthquake seismicity.

groundwater mining When the amount of water withdrawn from an aquifer exceeds the amount replenished to it by rainfall, surface water, and groundwater. The resource is being exploited without adequate replenishment.

groundwater Water that collects in the subsurface pore spaces and fractures within bedrock, sediment, and soils.

guano The excreta of birds and bats that is rich in phosphorus.

gullying The cutting of channels into the landscape by running water. When extreme, it renders farmland useless.

Gutenberg-Richter Law The vast majority of aftershocks will be relatively small in magnitude.

Haber-Bosch Process An industrialized process that fixes inert nitrogen gas (N_2) with hydrogen to produce ammonia based plant fertilizers, invented by the German chemists Fritz Haber and Carl Bosch.

half-life The period of time during which half of a given number of atoms of a radioactive element or isotope will disintegrate.

halocarbons Industrial gases that are carbon bearing that contain halogens such as chlorine (Cl), bromine (Br), and fluorine (F).

hardpans An impermeable solid layer of soil near the ground surface, typically composed of clay or silty-clay soils.

hard stabilization The construction of metal, stone and concrete sea wall, jetties, and groins designed to preserve the shoreline in its current configuration

hazardous waste Any wastes with properties that make it dangerous to human health or the environment.

hazards Phenomena or processes that could potentially harm people.

hazmat drums Impermeable, corrosion-resistant, and inflammable containers made to transport hazardous wastes.

heat capacity How much energy it takes to warm or cool a mass of material by a specific number of degrees.

high-level nuclear wastes Radioactive and otherwise dangerous by-products of nuclear-power generation and military weapons production.

high-marking When snowmobilers go quickly up very steep slopes and turn back down just below overhanging cornices.

highwall A cliff-like, excavated face of exposed overburden and coal that remains after mining is completed.

Holocene The name of the geologic epoch of the last 11,700 years characterized by climate stability and warmer, post-ice age conditions.

homeostasis A self-regulatory property of many both biological and physical systems in which processes occur to minimize long-term drastic change, or maintain equilibrium.

horse latitudes Latitudes between 30° and 35° north and south of the equator, at sea, in which winds tend to be mild or nonexistent and the air dry and warm for long periods of time.

hot spot A point (or area) on the lithosphere over a plume of magma rising from the upper mantle and crust, characterized by intensive volcanic activity.

human-induced earthquakes (HIE) Earthquakes caused by humans, most commonly due to the filling of reservoirs, mining activities and the extraction of groundwater, oil, and natural gas.

humidity Water content in the atmosphere.

humus Dark colored, organic material that forms in soil when plant and animal matter decays.

hydraulic conductivity A groundwater measurement expressed as the volume of water that moves in a specific period of time through a unit of area measured perpendicular to the flow direction. Commonly called permeability, it is expressed as cubic meters/(day/square meters) (or meters/day) or cubic feet/(day/square feet) (or feet/day).

hydraulic fracturing A method for improving oil and gas production by injecting high- pressure fluid infused with chemicals and sand into "tight" formations like shale to increase their porosity and permeability.

hydraulic gradient The slope of the water table underground, or measured another way, the slope between the water levels in two different wells, made by water rising into the wells from a confined aquifer below.

hydraulic head Pressure in a water body that influences flow; a body of water will flow from areas of high hydraulic head to low. The greater the pressure difference over a given distance, the faster the rate of flow.

hydraulic mining or hydraulicking A mining technique by which high-pressure jets of water are used to dislodge unconsolidated rock or sediment so that it can be processed.

hydroauger Drilled holes that behave as horizontal drains for water within a landslide.

hydrocarbon compounds Simple and complex organic chemical compounds composed only of hydrogen and carbon.

hydrocompaction The compaction of initially dry, low-density soils due to the heavy application of water.

hydrocurrents Fast-moving marine currents caused by tidal action.

hydrofracturing Also known as hydraulic fracturing, a method of extracting natural gas using high-pressure pumps that inject water into bedrock. See also fracking, hydraulic fracturing.

hydrogen A colorless, odorless, highly flammable gaseous element, the lightest of all gases and the universe's most abundant element. It is used in the production of synthetic methanol, in petroleum refining, and in many other uses.

hydrograph A graph of the stage (water level) or discharge of a stream, river, or related body of water over time.

hydrological cycle The circulation of water through the Earth System.

hydrolysis The chemical reaction between hydrogen ions in water with a mineral, commonly a silicate, resulting in the formation of clays, hydrated oxides and related alteration products.

hydroseeding Spraying seeds mixed with water, mulch, fertilizer, and lime onto regraded soil, used on steep reclaimed slopes to aid in establishing vegetation to help prevent erosion.

hydrosphere Water storage on Earth's surface, subsurface, and in the air including water vapor, oceans, lakes, rivers, and aquifers.

hydrothermal fluid In the context of the ocean floor, this is seawater that infiltrates fractures and faults overlying magma bodies, heats up, and reemerges.

hydrothermal Of or pertaining to naturally heated water, the action of heated water, or a product of the action of heated water—such as a mineral deposit that precipitated from a hot aqueous solution.

hypocenter See focus.

hypoxic A condition of having too little oxygen.

ice cap A perennial mass of ice that covers a large area, but covers an area smaller than a continental ice sheet. Ice caps can form where alpine glaciers merge together to smoother the crests of mountain ranges.

ice cores Cylinders of ice drilled from ice sheets and glaciers that are used to help reconstruct past climates.

ice crusts Thin ice layer on top or within a snowpack.

ice shelves Floating platforms of ice that are hundreds of meters thick and extend up to hundreds of kilometers from shore, formed where continental ice sheets meet the ocean.

ice-albedo effect A positive feedback in which snow and ice reflect sunlight directly back into space, keeping Earth cooler–and therefore more favorable to the development of snow and ice–than it would be if land and sea were exposed instead.

ice-jam floods Floods that occur when river ice breaks up, moves downstream, and wedges against obstacles in the channel to form an ice dam, which quickly raises water levels upstream.

igneous rocks A type of rock forms from the cooling of magma ("molten rock") which either hardens and crystallizes underground (plutonic igneous rocks) or erupts at the surface (volcanic rocks).

illuviation Movement of clay through a soil bed through the percolation of water.

in situ extraction Mining done by drilling wells to extract bitumen or ore with minimal disturbance to the surface environment or surrounding bedrock.

inactive sea cliff A cliff whose erosion is dominated by subaerial processes rather than wave action.

incineration A waste-treatment process that involves the combustion of waste at high temperatures to produce heat that can be used to generate electricity to make steam that is passed through turbines.

incompressibility A material's resistance to a change in volume.

Industrial Revolution A period of scientific and technological development in the eighteenth century that transformed many rural areas into industrialized cities, largely in North America and Europe.

industrial wastes The waste produced not only by industrial processes, particularly manufacturing, but also by lumber mills, mining, and powerplants.

infiltrates Particles or dissolved substances that enter the ground via infiltrating water from the surface.

infiltration The movement of water, including rain and snowmelt, into the soil surface.

infrared What energy from sunlight that is not reflected is absorbed, even by you standing under a bright Sun; then it is reradiated later, but mostly in a different and longer wavelength, the infrared, which we feel as heat.

infrared radiation Ultraviolet radiation from the sun that is absorbed by Earth's atmosphere and surface is transformed into this longer wavelength form of radiation which we feel as heat.

infrasound measurements Networks of instruments that detect unobserved eruptions of volcanoes using sound below the normal range of human hearing.

initiation A massive impulse of energy that moves ocean water and sets a tsunami wave in motion.

intensity scale An earthquake rating scale (I–XII) based on subjective reports of human reactions to ground shaking and on the damage caused by an earthquake.

intensity/duration envelope diagrams Graphs showing how rainfall durations and intensities may result in debris flows.

interferograms Interferometric Synthetic Aperture Radar (InSAR) measures the difference between two radar images acquired by orbiting satellites, producing an interferogram. Interferograms are used to show deformation in volcanoes and along active faults.

interglacial Intervals between cold glacial periods when Earth's climate is warm and stable.

intermediate Igneous rocks that are neither silica-rich nor silica-poor. Another way of looking at it is that their compositions lie between mafic and felsic.

intermittency Describes power that is derived from inconsistent generation, often true of renewable energy sources.

International Thermonuclear Experimental Reactor (ITER) An international nuclear fusion research project aimed at created energy through fusion reactions. Participating nations (as of 2023) include the United States, China, India, Japan, Russia, Korea, and 27 countries in the European Union.

intracontinental mountain chains Mountain ranges that are created where two plates consisting of continental lithosphere collide. Good examples are the Himalaya Mountains and Alps.

intraplate earthquakes Earthquakes that occur far from plate boundaries due to slipping of deep, ancient faults.

inundation Flooding of land by water.

ion An atom with a positive or negative electrical charge because it has lost or gained one or more electrons from the outer electron orbital shell.

island arcs Chains of islands created by explosive volcanoes which overlie subduction zones where two plates of oceanic lithosphere collide.

isolation (waste disposal) Buried or otherwise sequestered waste.

isoseismals Lines on a map that enclose areas of equal earthquake shaking based on a seismic (e.g., Mercalli scale) intensity scale.

isostasy The gravitational equilibrium between parts of the Earth's crust, the underlying mantle, and surface topography which helps explain why some parts of Earth's surface stand high above sea level and others are widely submerged. Just as ice floats in water, so too does the lithosphere "float" on Earth's mantle, according to this principle.

isostatic equilibrium The equilibrium between the Earth's crust and upper mantle.

isothermal In thermodynamics, a process in which the temperature of a system remains constant. In limnology, "isothermal" indicates a lake with an approximately equal temperature throughout.

isotope Forms of an element that have the same atomic number but varying numbers of neutrons, resulting in differing atomic masses.

jet streams Stratospheric winds in narrow, high altitude currents that blow mostly west-to-east due to Earth's rotation. Their location is determined by where atmospheric cells come together lower in the troposphere.

juvenile water Water that comes from the Earth's deep interior.

Karman Line The formal designation of where space "begins," at approximately 100 kilometers (60 miles) altitude, based less on atmospheric physics than our ability to rocket objects into Earth orbit above that altitude.

karst terrain Limestone land areas that are characterized by numerous springs, caves, and sinkholes. Karst topography is characteristically rugged.

kerogen An insoluble carbonaceous residue in certain fine-grained, clastic sedimentary rocks that has survived bacterial metabolism. It consists of large molecules from which hydrocarbons and other compounds are released on heating that yields petroleum when it is distilled.

kīpuka Hawaiian word referring to a patch of older land surrounded by younger lava flows.

Köppen-Geiger Classification An international classification scheme for world climates: tropical, dry, temperate, continental, and polar.

labradorescence Bright blue reflections caused by internal structures in translucent minerals such as labradorite.

lahar A debris flow or mudflow consisting of volcanic material.

land conversion The transformation of land from natural conditions to human use, primarily for agricultural and industrial development.

landfills A dedicated surface disposal site where wastes are sorted, concentrated, and to the extent possible isolated from the surrounding environment until they can naturally decay.

landfill cells The separate, sealed, rhomb-shaped concentrations of waste in a landfill.

landslide The downslope movement of rock and/or soil as a semicoherent mass on a discrete slide surface or plane. See also mass wasting.

lapilli Fragments only a few centimeters in size that are ejected from a volcano. Cinder and pumice are common types of lapilli.

llateral spreads See lateral spreading (above)

lava Molten rock (magma) that breaks through Earth's surface and flows as a thick liquid. See also magma.

lava caves Internal channels in lava flows that form during volcanic eruptions, develop crusts, drain, and leave cavernous tunnels inside the lava for people to explore.

lava lakes Accumulation of lava in pools above active vents or in craters. Lava lakes can take decades to cool depending on their thickness.

lava tubes See lava caves, above.

leach field An underground system of perforated popes adjacent to a septic tank, where liquid waste flows and percolates through the soil where aerobic bacteria finish breaking it down.

leach lines The underground pipes in a leach field that disperse septic effluent.

leachate The polluted water that percolates through landfills. If the leachate seeps to the surface or reaches the underlying groundwater, it becomes an environmental problem for the surrounding area.

left-lateral A classification of strike-slip faults based on the observation that features viewed on the opposite side of the fault are displaced to the left relative to features on the side where the viewer is standing. See also right-lateral.

legacy sediment Sediments deposited following human disturbances in a drainage basin.

lenticular clouds Smooth cloud caps that indicate fast-moving air above a mountain range.

light detection and ranging (lidar) Use of lasers from drones or airplanes to map topography.

lignite A low grade of coal that is brown in color and may include fossils of plants from which the coal formed. It is intermediate in composition between peat and bituminous coal, and is very polluting when burned.

lime A chemical compound, CaO, which results from the breakdown of limestone, $CaCO_3$, when exposed to high heat. Important for the production of cement.

limnic eruption Sudden eruption of gas, typically carbon dioxide, from a lake. Associated with aging volcanic craters filled with water.

liquidators A job in which people in this role are appointed to wind up the affairs of a company, firm, or project.

liquefaction The sudden loss of strength in water-saturated, sandy soils and sediment as a result of shaking during an earthquake.

lithification The conversion of sediment into solid rock through processes of compaction, cementation, and crystallization.

lithified Transformed into stone, as in the transition of loose sand to sandstone.

lithosphere A rocky layer of Earth's interior that is composed of the crust and upper mantle. It is Earth's outermost layer, which includes the crust.

lode A zone of valuable ore veins generally found in igneous and metamorphic rocks.

loess A blanket deposit of buff-colored silt that shows little or no stratification. It covers wide areas in Europe, eastern China, and the Mississippi Valley. It is primarily wind-blown glacial dust of Pleistocene age.

long period Energy waves with a long wavelength. They are lower in energy, and thus are less violent than short period waves

longitudinal wave A type of seismic wave involving particle motion that alternates in expansion and compression in the direction of wave propagation. Also known as P-waves or compressional waves, they resemble sound waves in their motion.

longshore current The current adjacent and parallel to a shoreline that is generated by waves striking the shoreline at an angle.

longshore drift The movement of sand along a beach caused by longshore currents.

long-term forecasting The use of compiled geological data regarding the timing of past earthquakes in a region to estimate the statistical probability for future events of a given magnitude.

losing streams A stream that lies above the water table and infiltrates water through its bed as it flows downstream gradually reducing stream discharge.

Love waves A surface wave that exhibits horizontal motion normal to the direction of travel.

low-level nuclear wastes Nuclear wastes generated in research laboratories, hospitals, and industry that are not designated as transuranic. Shallow burial in containers is sufficient to sequester this waste.

low-pressure systems A storm system that has lower pressure at its center than the areas around it. Wind blows towards the low-pressure area, air rises and water vapor condenses, resulting in the formation of clouds and precipitation.

LP-earthquakes Long-period earthquakes, caused by cracks resonating as magma and gases move toward the surface. These earthquakes often occur before volcanic eruptions but can also be a part of background seismic activity. See also VT-earthquakes.

luster The manner in which a mineral reflects light, described as metallic, resinous, silky, or glassy, to name a few.

mafic Descriptive of magma or rock rich in iron and magnesium. The mnemonic term is derived from magnesium and ferric.

magma Molten material within the earth that is capable of intrusion or extrusion (eruption) and from which igneous rocks form.

magma chamber A large, underground pool of accumulated magma that forms as the molten material rises above deep, dense host rock to a depth where the surrounding material is of equal density.

magnetic fields The region of magnetic forces surrounding a planet that shields the planetary surface from deadly, charged cosmic-ray particles like protons that make up cosmic radiation.

magnetic reversals Swapping of the positions of a planet's magnetic poles, which generally occurs when overall magnetic field strength greatly weakens.

magnetometers An instrument that measures the strength of a magnetic field.

main shock A major earthquake, preceded by foreshocks and followed by aftershocks.

managed retreat The idea that it makes more sense to move buildings away from the coast than to fight rising sea level and retreating coastlines.

mantle A thick, mostly solid shell of magnesium-silicate rich rock that surrounds and is less dense than Earth's core.

manure Animal fecal matter that is a mix of living organisms and undigested material.

mapping Graphical representations that depict features of particular interest ranging from roadmaps to detailed site studies of bedrock.

maria Flood basalt plains that formed 3–4 billion years ago and appear as dark spots on the face of the moon.

marine renewable energy Includes energy systems sourced from wave energy, tidal energy, and ocean thermal energy conversion.

marine sediment cores Cylinders of oceanic sediment drilled from the deep ocean floor that can be studied to help reconstruct past climates.

marine upwellings These occur as deep ocean waters rise to replace shallow waters that are pushed farther offshore by wind, bringing important nutrients for marine life to the surface.

mass movements Downslope movements of soil and rock under the direct influence of gravity.

massive sulfide deposit Beds of rich ore that develop at mid-ocean ridges on the deep ocean floor as a result of submarine volcanic activity.

maximum contaminant level (MCL) The highest level of a contaminant that is allowed in drinking water.

maximum contaminant level goals (MCLGs) The level of a contaminant present in drinking water. This level may be set somewhat below the maximum contaminant level (see the term in the left column) to provide a public margin of safety.

megacrysts Very large or more coarsely grained crystals that are scattered amongst smaller grained crystals in a phaneritic igneous rock.

melting point The temperature at which a given solid will melt.

mesosphere The layer of the atmosphere above the stratosphere that serves as a shield, vaporizing approximately 95% of space debris before it hits Earth's surface. The mesosphere is enriched in atoms of iron and other metals owing to this vaporization.

mesothelioma A serious cancer of the tissues lining the lungs, heart, abdomen, and chest caused by exposure to asbestos.

mesotrophic A type of lake with characteristics that are between oligotrophic and eutrophic, including a balanced nutrient input level, some plant growth, and significant plankton and fish populations.

metals Products of minerals that are typically hard, shiny, and easily formed substances used to make coins, electrical wires, and skyscrapers.

metamorphic grades Describes the extent and intensity of metamorphism.

metamorphic rocks Preexisting rocks that have been altered by heat, pressure, and often chemically active fluids.

metamorphism The process of a rock changing form due to heat, pressure, and fluid alteration.

metastable substance A material that appears stable for long periods of time but is not.

meteoric water The shallower water stored in aquifers that originates by percolation from the surface.

methane (CH_4) A naturally occurring gas as well as industrial by-product. Methane is a powerful greenhouse gas.

methane hydrate A solid compound in which a large amount of methane gas is trapped within a cage of ice, Methane clathrates and hydrates are widely distributed in deep-sea sediments and in permafrost in the Arctic and subarctic.

methanol A colorless liquid chemical compound, commonly called wood alcohol, that is volatile, flammable, and poisonous, with the chemical formula CH_3OH.

methylmercury An organic form of mercury which is readily incorporated into biological tissues and is highly toxic to humans.

miasmas Unpleasant or toxic vapors or smells such as those from swamps and garbage.

microbeads Deliberately manufactured microscopic plastic bits added to some cosmetic products to extend the shelf life.

microplastics Disintegrated plastic particles ranging in size from 1 nanometer to 5 millimeters found as mostly unnoticed waste products in water bodies.

mid-ocean ridges Underwater volcanic ranges that bisect the seafloor at divergent plate boundaries where lava erupts from fissures opened by the separation of plates.

mineral Naturally occurring, inorganic substances that have defined chemical compositions, orderly internal atomic structures (they are "crystalline"), and characteristic physical properties.

mineral reserves Known mineral deposits that are recoverable under present conditions but are yet to be developed.

mineraloid An amorphous (noncrystalline) substance that superficially show the attributes of a true mineral, such as opal.

mitigation Reducing the severity of climate change by lowering global emissions of greenhouse gases.

Modified Mercalli Scale (MM) Earthquake intensity scale from I (not felt) to XII (total destruction) based on damage to built structures and reports of human reactions.

Mohs' hardness scale A relative scale of mineral hardness that ranks all minerals on a scale of 1 to 10, with 1 being very soft (talc) and 10 being very hard (diamond).

moment magnitude A scale of seismic energy (M_w or M) released by an earthquake derived as the product of the rock rigidity along the fault, the area of rupture on the fault plane, and the amount of slip.

monitor The nozzle of the high-pressure jet of water used in hydraulic mining.

monocultures Farming that produces a single crop. Pesticides, herbicides, and fungicides are often intensively applied to maximize crop yields in monocultures, and biological diversity is very low.

monogenetic Referring to volcanoes whose full development takes place during a single eruption.

moraine A mound, arcuate ridge, hill, or other distinct accumulation of unsorted, unstratified glacial sediment, formed by the shoving and melting of ice at the edges or ends of glaciers.

mother lode A complex of igneous veins the erosion of which provides ore minerals to placer deposits found along streams or in steam deposits downslope.

mountain glaciers Glaciers that form on mountains as a result of snowfalls that accumulate year after year without completely melting.

mountaintop-removal mining The mining of coal that underlies the tops of mountains, involving clearing forests, stripping and removal of overburden, and the use of explosives to shatter rocks and expose coal beds.

mud volcanoes (or sand blows) Groundwater dense with sand and silt that erupts from the Earth during shaking as a result of liquefaction.

municipal solid waste (MSW) The materials that we toss away on a day-to-day basis at home, school, or work; garbage.

National Environmental Policy Act (NEPA) National Environmental Policy Act of 1969. Requires in-depth field study and the issuance of Environmental Impact Statements (EIS) for all proposed projects that are supported by U.S. federal funds.

native elements Minerals that are composed of a single element, for example, gold.

natural gas Gas, primarily methane (CH_4), that occurs naturally with liquid petroleum, in coal beds, and as methane clathrates. It is an important fuel, a feedstock for manufacturing fertilizers and plastics, and a potent greenhouse gas.

natural levee A ridge of sediment deposited naturally atop a river bank by overflowing water. See also artificial levee.

natural systems Systems that occur naturally and are products of physics and chemistry, operating according to principles such as the laws of thermodynamics, the structures of atoms, and the influence of gravity.

negative feedback A natural response to a process occurring in nature that lessens the effects of that process over time.

net metering A process that allows people who produce their own power to feed excess power into the grid (via their local electric company) and then pull power from the grid when they need it.

neutral buoyancy zone The level in the crust at which the density of magma matches that of the surrounding hardened crust, causing the magma to stall out as it ascends, and accumulate to form a magma chamber or reservoir. Neutral buoyancy can also refer to the atmosphere, where a growing cloud spreads out in the thinner air above.

NIMBY Not In My Backyard, a sentiment held by many people who reject new developments or projects near their homes or neighborhoods because of concern that they would negative impact their property values.

nitrification The microbially driven oxidation of reduced nitrogen compounds, primarily ammonia (NH_3), to form nitrate (NO_3^-) and nitrite (NO_2^-).

nitrogen cycle The nitrogen cycle is a complex exchange of the element nitrogen (N) in various forms between the atmosphere, hydrosphere, biosphere, and soils.

nitrogen fixation The oxidation of inert N_2 to biologically available forms of nitrogen, primarily nitrate (NO_3^-), nitrite (NO_2^-), and ammonia (NH_3) initiated by microbes and bacteria in soils.

non-precious (or base) metals More abundant metal ores that serve great industrial importance such as copper and tungsten.

nonrenewable resources Resources that are not replenished as fast as they are used.

nonstationarity The observation that natural processes impacting an area change over long periods of time.

nor'easters Low-pressure storm systems that develop over the northeastern coast of the United States year-round, but most frequently and violently during the winter. They are characterized by winds blowing from the northeast off the Atlantic Ocean.

normal faults A type of fault where the hanging wall (upper block) moves down relative to the footwall (lower block). This sense of motion is opposite that of a reverse fault.

normal force The force applied perpendicular to the failure plane of a landslide.

no-till/minimum-till A method to reduce soil erosion on farmland by not tilling farm fields and instead planting seeds into soil through the previous crops residue. Weeds are controlled using chemicals.

nucleus The positively charged central core of an atom, consisting of protons and neutrons. The nucleus contains nearly all the mass of an atom.

nuées ardenté French for "glowing cloud," it is a highly heated, almost incandescent cloud of volcanic gases and pyroclastic material that travels with great speed down the slopes of a volcano. It is typically produced by the explosive disintegration of viscous lava in a vent. Also known as pyroclastic flow.

numerical modeling Use of mathematical techniques to study physical processes in detail.

O horizon Zone of organic matter at the top of a soil profile.

obliquity The tilt of Earth's rotational axis.

obsidian A silica-rich, black, and glassy mineral formed by magma that is commonly recognized by its characteristic conchoidal fracturing.

Ocean Dumping Act Officially the Marine Protection, Research and Sanctuaries Act of 1972, American legislation that requires anyone dumping waste into the ocean to have an EPA permit and to provide proof that the material will not degrade the marine environment or endanger human health.

ocean fertilization Spreading finely ground iron particles into the ocean to stimulate phytoplankton growth and subsequently capture CO_2 via photosynthesis.

ocean thermal energy conversion (OTEC) A technology that utilizes the temperature difference between warm surface ocean water and colder deep water to generate electricity.

oceanic crust Crust underlying oceans composed largely of dark basaltic rocks. It is typically 5 to 10 kilometers (3 to 6 miles) thick.

oceanic gyres Great looping currents in shallow seawater. Each hemisphere has its own set of gyres, the sizes and shapes of which are determined by the adjoining continental coastlines. The directions of circulation reflect the Coriolis effect.

oil shale A group of fine-grained sedimentary rocks rich enough in organic material (kerogen) to yield petroleum on distillation.

oligotrophic A type of deep, clear water lake that is characterized by low nutrient inputs, as well as small plankton and fish populations.

open pit mines See quarries.

ores Metallic mineral reserves.

orographic This term refers to "mountains," especially with regard to their geographical locations and forms.

orographic effect The influence of a mountain range on a moving air mass: As the mass ascends the mountain range, it cools and condenses to form rain and snow. Descending the far side, the air is now dry, warms up adiabatically, and evaporates moisture in the landscape, potentially causing desert conditions.

outer core A mass of molten metal that makes up 15% of Earth's total volume but never erupts at Earth's surface due to its depth and enclosure by the mantle above.

overburden Any material that lies above an zone of mineral or fossil fuel resources in the crust.

overturning The process of sudden lake water mixing to homogenize water temperature and oxygen content. Lake water overturning takes place, for instance, in cool, temperate regions that are subject to near-freezing winter temperatures; the dense, colder surface water sinks and the bottom waters that are oxygen-deficient rise to replace them.

oxidation The loss of an electron (or a gain of oxygen) during a reaction by a molecule, atom, or ion.

hydration The combining of minerals with water molecules. One form of chemical weathering.

oxides Minerals that are formed by the combination of one or more cations with anions of oxygen.

ozone A heavy compound of three oxygen atoms, O_3, that is formed when atmospheric oxygen reacts with incoming sunlight (ultraviolet light, UV).

ozone layer The layer of ozone produced at the base of the stratosphere when oxygen reacts with incoming ultraviolet sunlight, forming ozone and filtering out UV-C radiation.

pack rats Pack rats are small, common desert creatures that scamper about the landscape collecting twigs and seeds that they use to make nests in rocky crevices. See also amber-rat.

pāhoehoe Basaltic lava typified by smooth, billowy, or ropy surfaces.

paleoclimatologists Scientists who study past climates.

paleoseismicity The rock record of past major earthquakes, preserved in the form of displaced beds and liquefaction features exposed in natural outcrops or the walls of excavated trenches.

Pangaea The supercontinent that resulted from plate convergence approximately 250 million years ago.

parabolic trough Parabolic mirrors that focus sunlight on a fluid-filled reservoir pipe to generate electricity.

parent material The geologic material from which soil horizons form.

passive solar design A construction technique that allows buildings to minimize the need for external heating and cooling, as for instance designing houses that capture warm morning sun but greatly reduce interior exposure to sunlight in the hot middle of the day.

peat A soil-like material formed by the partial decomposition of organic matter in wet, acidic conditions of bogs and fens. Metamorphism of peat produces coal (beginning with lignite).

pebble beaches Beaches composed of rounded pebbles and cobbles that occur where a shoreline is exposed to high energy surf.

pedologist (Greek: *pedo*, "soil," and *logos*, "knowledge") A person who studies soils.

pegmatite An exceptionally coarse-grained igneous rock with interlocking crystals, usually found as irregular dikes, lenses, or veins, especially at the margins of batholiths.

Pele's hair Fine golden threads of volcanic glass spun by high lava fountaining.

Peléean eruptions A type of eruption that is very similar to Vulcanian eruptions, but which involve pyroclastic flows and surges released from the summit of a composite cone capped by a plug of hardened lava.

perched water table The upper surface of a localized body of groundwater held up by a discontinuous impermeable layer. A perched water table lies at a shallower depth than the much deeper main water table below.

percolation tests Tests performed to make sure soil in a proposed leach field is permeable enough that waste water will flow through it.

peridotite A rock from Earth's mantle made up almost entirely of olivine and augite crystals.

period The length of time between the passage of equivalent points on a series of passing waves (for example, wave crests or wave troughs).

permafrost table The shallowest level underground of permafrost.

permafrost Permanently frozen ground, occurring in arctic, subarctic, and alpine regions.

permeability The degree of ease with which fluids flow through a porous medium. See also hydraulic conductivity.

persistent pollutant A substance that bioaccumulates in the human system because the body cannot metabolize or excrete it.

persistent organic pollutants (POP) Pollutants such as pesticides or flame retardants that persist in the environment long after their initial use.

petroleum A liquid mixture of hydrocarbons found in natural underground reservoirs, which can be extracted and refined to produce fuels such as gasoline and diesel oil See also crude oil.

petrologists Geologists who study the composition and origin of rocks.

phaneritic texture Fully crystallized plutonic igneous rocks that are coarse-grained, resulting from slow cooling of magma.

phenocrystic Volcanic rocks with widely scattered discernable crystals embedded in a noncrystalline groundmass.

phenocrysts Individual crystals in phenocrystic volcanic rocks.

photovoltaics (PV) cells A technology for generating electricity using various semiconductor materials, such as silicon or germanium arsenide. PV cells convert sunlight into direct current (DC) electricity.

phreatic Steam-driven explosions that occur when ground or surface-water is heated by erupting magma, flowing lava, or pyroclastic materials.

pingo Ice-cored round hills in arctic regions formed as groundwater migrates toward the surface and freezes.

playa Dried lake bed found in a desert region.

Plinian eruptions The most violent type of volcanic eruption which characteristically injects large amounts of volcanic gas and pyroclastic material into Earth's lower stratosphere.

plunging breakers A type of breaker loved by surfers that forms when waves approach a steep-sided sea bed near the shore and the crest of the wave falls forward as a well-defined curl with considerable energy.

plutonic Pertaining to igneous rocks that cool and fully crystallize underground, usually with a phaneritic texture.

plutons Any large igneous body that has congealed from magma underground.

point sources Any single identifiable sources of pollution from which wastes are discharged, such as a smoke stack or sewer outlet.

polar air The masses of frigid, dry air residing over Earth's poles.

polar front The zones of collision between warm, moist air heading poleward and polar air occurring at high latitudes encircling both of Earth's poles.

polarity Having a North and South magnetic pole, much like the ends of a bar magnet.

pollen Pollen is a fine powder given off by certain plants during reproduction in order to fertilize other plants. It is resistant to decay, thus appears in fossils and is useful in paleoclimate research.

polychlorinated biphenyl compounds (PCBs) Large, complex, synthetic molecules, made of carbon, hydrogen, and chlorine with a wide array of industrial applications. They are associated with multiple illnesses for humans who are exposed to them.

polygenetic Referring to volcanoes whose full development takes place from multiple eruptions, generally occurring over a period of many centuries or millennia.

polymerase chain reaction (PCR) A quick method of making multiple copies of a DNA sequence.

pore pressure The stress exerted by the liquids that fill the voids between particles of rock or soil.

porewater pressure See pore pressure, above.

porosity A material's ability to contain fluid. It is measured as the ratio of the volume of pore spaces in a rock or sediment to its total volume, usually expressed as a percentage.

porphyries Any igneous rock characterized by the presence of course grained crystals such as feldspar and quartz, dispersed in an otherwise fine-grained texture.

porphyritic A description of texture that occurs in volcanic and igneous rocks that is characterized by the presence of large crystals surrounded by considerably finer grained crystals or noncrystalline groundmass. (The term "phenocrystic" describes a volcanic rock with a porphyritic texture; there is some overlap in meaning in this case).

positive feedback A natural response to a process occurring in nature that enhances or increases the intensity of the effects of that process.

potentiometric surface The level to which water will rise in a well or aquifer. In confined aquifers this is based on pressure, and in unconfined aquifers is considered to be the water table.

powder-snow avalanche Avalanche of dry, recently fallen snow that has little cohesion.

power towers A system with a large number of sun-tracking, movable mirrors called heliostates, that focus sunlight onto a receiver at the top of a tall tower. The focused sunlight heats fluid to produce steam, which turns turbines and produces electricity.

precious metals Metals that are valuable because they are scarce, and have represented wealth throughout history including gold, silver, platinum, and palladium.

precipitation A term for all forms of water that form in the atmosphere including rain, snow, and sleet, and fall back down to the surface of the Earth.

precursors Observable non-seismic phenomena that occur before an earthquake that may be used (hypothetically) to predict it.

prediction A statement of what will happen in the future.

primary treatment The initial steps in wastewater treatment where the liquid and solid components are separated, the liquid part is clarified by removal of fats and greases, and the solids sink and are removed.

probabilistic seismic hazard analysis A document mandated for nuclear power plant design that analyzes the frequency of earthquakes and their magnitude in the past, and is used to design the reactor, its building, plumbing, and pumps, to resist the strongest probable shaking in the future.

proglacial A type of lake formed by the damming action of an ice sheet or moraine during the retreat of a melting glacier, and is filled with glacial melt water.

propagation A tsunami wave's movement across the ocean.

protolith The original rock that existed before pressure and/or temperature in the surrounding crust changed.

pumice Frothy-appearing rock composed of natural glass ejected from high-silica, gas-charged magmas. It is the only rock that floats.

pumped-water-storage facilities (or water batteries) Facilities that take water pumped from reservoirs during off-peak hours when plenty of power is available. The water is stored in these facilities (e.g., storage tanks), and is then allowed to fall back to the reservoir during peak hours to generate additional needed electricity.

pyroclastic ("fire broken") Descriptive of the fragmental material, ash, cinders, blocks, and bombs ejected from a volcano. See also tephra.

pyroclastic flows A dense cloud of hot volcanic gases, ash, and other pyroclastic materials, especially pumice and blocks, that pours across a landscape.

pyroclastic surge A dilute cloud of hot volcanic gases, ash, and other pyroclastic material that blows across a landscape at hurricane speeds.

pyroclasts Fragmental volcanic material exploded from a vent.

pyrolysis A process in which material is broken down in the absence of oxygen at high temperatures.

P-waves P(Primary)-waves are examples of longitudinal seismic waves that alternately compress and stretch the crust as they propagate through it. See also longitudinal waves, body waves, and S-waves.

qanats Subsurface tunnels excavated in alluvial fans and desert floors in arid climates (e.g., Algeria and Iran) to extract water. An ancient technology, qanats effectively transport water by gravity without the need for pumping.

quarries Pits or excavations in hillsides, open to the sky, from which rock is typically removed by explosions and the use of heavy hauling equipment.

quick clay A clay possessing a "house-of-cards" sedimentary structure that collapses when it is disturbed by an earthquake or other shock. Also known as sensitive clay.

radicals Complex ionized molecules.

radio detection and ranging (radar) Use of radio signals for determining elevations and mapping topography.

radioactive decay The process of an unstable, radioactive atom spontaneously decaying and releasing radiation to form a more stable form (or forms).

radiogenic Something that was produced by radioactivity.

radon Chemical symbol Rn, radon is a radioactive gas emitted as uranium (U) breaks down to stable, non-radioactive lead (Pb).

rainfall duration How long a rainfall event of a certain intensity continues.

rainfall intensity The rate at which precipitation falls.

rain-on-snow events Weather events that occur when rain falls on existing snowpack resulting in the melting of snow.

rank A coal's carbon content depending on its degree of metamorphism.

Rayleigh waves Surface waves that exhibit elliptical motion opposite to the direction of travel in a plane that is perpendicular to the ground surface.

recharge The process of groundwater being replenished when rain, stream, or meltwater percolates into soil, sediment, or rock cracks.

recharge areas Areas where water is added to the zone of saturation via infiltration.

recharge mound A mound of water that forms above the groundwater table, typically beneath losing streams or human-made stormwater drainage systems.

recurrence interval The return period (or frequency) of an event, such as a flood or an earthquake, of a given magnitude.

recycling The process of converting waste into reusable material.

reefs In terms of ore deposits, these are layers of valuable metalllic minerals found within the Bushveld intrusion (South Africa).

refining (or cracking) The manufacturing process of separating crude oil into its various useful (diesel fuel, etc.) components.

regolith Unconsolidated rock and mineral fragments at the surface that has not yet developed into soil.

regulatory floodplain The part of a floodplain that is subject to U.S. federal government regulation for insurance purposes.

renewable resources Resources that are replenished at a rate equal to or greater than the rate at which they are used.

reprocessed A process which separates unused plutonium and uranium from other nuclear wastes for further future applications.

reserves Those portions of an identified resource that can be recovered economically.

reservoir rocks A permeable, porous geological formation that will yield oil or natural gas.

residence time The average length of time a substance (atom, ion, molecule) remains in one part of a system.

residual soil Soil formed in place by decomposition of the rocks on which it lies.

resisting force Strength along a slide plane that resists failure of the rock and/or soil. It must be overcome for sliding to begin.

resonance The tendency of a system (a structure) to vibrate with maximum amplitude when the frequency of the applied force (seismic waves) is the same as the vibrating body's natural frequency.

Resource Conservation Recovery Act (RCRA) Resource Conservation and Recovery Act of 1976 that regulates the generation, treatment, transportation, storage, and disposal of hazardous wastes.

reticulite Pieces of foamy lava that resemble spun glass. They are formed by tall fountains of highly fluid basaltic lava.

retrograde metamorphism Changes that occur to mineral assemblage and composition during uplift and cooling of a metamorphic rock.

reverse fault A type of fault when the hanging wall (upper block) moves up along a fault plane relative to the footwall (lower block).

rhyolite A light-colored volcanic rock or lava that has the most siliceous composition (approximately 75% to 80% by weight) and tends to erupt explosively because of its very high viscosity and resistance to flow.

Richter magnitude scale The logarithm of the maximum trace amplitude of a particular seismic wave on a seismogram, corrected for distance to epicenter and type of seismometer. A measure of the energy released by an earthquake.

right-lateral A classification of strike-slip faults if the features viewed on the opposite side of the fault are displaced to the right relative to features on the side where the viewer is standing. See also left-lateral.

rill erosion The carving of numerous, roughly parallel small channels, up to 25 centimeters (10 inches) deep, in soil by running water.

riparian Pertaining to or situated on the bank of a river.

rip current A narrow, strong, and localized current of water that moves away from the shore and cuts between lines of breaking waves.

ripples Small waves that are created when wind blows over the surface of water.

risks The chances of being harmed by a hazard, usually measured as probabilities over certain periods of time. See also hazards.

river deltas Deposits of sediment left when quickly moving river water carrying sediment meets the ocean, slows down, and can no longer transport sediment.

riverine floods Lowland floods that occur when broad floodplains adjacent to a stream channel are inundated. Typically slow moving, they cause damage to basements and first floors of buildings by soaking into floors, walls, and foundations.

rock avalanche A type of mass movement of bedrock broken from a destabilized slope.

rock cycle The cycle or sequence of events involving the formation, alteration, destruction, and re-formation of rocks (igneous, sedimentary, and metamorphic) as a result of processes such as erosion, transportation, lithification, melting, and metamorphism.

rock texture A characterization of a rock based on the size and shape of its mineral grains.

rockfalls A type of mass movement when a mass of rock detaches from a cliff and falls to the bottom.

rocks Aggregates of minerals, rock fragments, natural glass, or solidified organic matter that make up the solid Earth.

rogue wave A solitary wave that is much larger than others around it.

rotational slides (or slumps) A type of landslide that moves on a curved, concave-upward slide surface.

runaway feedback A feedback that is not equally balanced and overwhelms those processes that oppose it.

runoff That part of precipitation falling on land that runs into surface streams.

run-out zone Typically at the base of slopes where topography is less steep and an avalanche slows to a stop.

rupture zone The area of a fault surface that slips during an earthquake. The rupture zone may be entirely underground, or reach to the surface to create a fault scarp.

Sahel Belt of populated semiarid lands along the southern edge of North Africa's Sahara Desert.

salinization A human-induced soil problem caused by intensive irrigation that causes increased concentration of salts near the soil surface toxic to plants.

salt-crystal growth The growth of salt crystals evaporated from solution, such as those of sodium chloride, which can generate pressures sufficient to break rock.

saltwater intrusion When salty groundwater from the ocean mixes with or displaces fresh groundwater on land.

sand blows Outpourings of wet sand, silt, and mud on the surface, resulting from liquefaction during a large earthquake.

Santa Ana winds Strong winds that originate over high pressure, dry, desert regions of the southwestern United States. They flow westward toward low pressure regions of the U.S. west coast.

saprolite Rock that has lost some of its volume and strength as chemical weathering leaches soluble elements.

satellites Instrumental platforms that orbit the Earth in space. They can examine the surface of our entire planet with great precision, collecting data ranging from air temperatures and ice cover to the uplift of landscapes following major earthquakes.

saturated zone The region of subsurface below the water table where the pores and voids in soil and bedrock are filled completely (saturated) with water.

schiller A variable optical phenomenon in minerals caused by light reflecting off tiny mineral platelets under the mineral surfaces.

scoria A pyroclastic, vesicular, dark colored volcanic igneous rock which forms from explosive lava eruptions. Also called cinder.

scrubbers Devices that use lime slurries to neutralize acid gases.

sea ice Thin covering of floating ice on an ocean.

seafloor spreading The development of new sea floor through eruption of lava as a result of rifting in the oceanic crust at divergent boundaries.

secondary treatment A phase of wastewater treatment that uses a mix of living bacteria to reduce the level of nitrogen and phosphorus in the water.

sector collapse Structural failure and subsequent collapse of a wedge-shaped side of a volcano that may accompany massive lateral blasts (as at Mt. St. Helens in 1980).

sediment Particulate matter derived from the physical or chemical weathering of the materials of Earth's crust and by certain organic processes.

sedimentary rock A type of rock that is formed from many pieces of preexisting rocks (sediments) or matter precipitated from water, either directly or via biological activity. The loose clasts and precipitates subsequently harden (lithify) to form coherent masses of sedimentary rock. Commonly identified by distinctive layering seen in the rock.

seismic gaps Stretches along known active fault zones within which no significant earthquakes have been recorded.

seismic moment A value that is proportional to the average displacement on the fault multiplied by the rupture area on the fault surface and the rigidity of the faulted rock. Used to derive an earthquake's moment magnitude.

seismicity The frequency of earthquakes, or other seismic events, in a region.

seismogram The recording from a seismograph.

seismograph A device for recording and measuring seismic energy waves.

seismologists Scientists who study the pattern of energy waves generated by earthquakes to determine how they took place and learn about the structure of Earth's interior.

seismology The study of earthquakes.

seismometer A device for detecting earthquake shaking. The data are sent to a seismograph for recording and measuring.

semidiurnal An event which occurs in a cycle once every 12-hours.

sensitive clay See quick clay.

septic tank An underground sewage waste chamber that digests solid waste with anaerobic bacteria and separates liquid waste which is discharged from the tank into a leach field.

sequestration The capture and storing of carbon dioxide (CO_2).

settlement This occurs when the applied load of a structure's foundation is greater than the bearing strength of the foundation material (soil or rock).

sewage sludge A mud-like residue left from the treatment of waste-water. It is often applied as a fertilizer, though it may be contaminated with heavy metals not removed during treatment.

sewer A system that collects wastes from many sources and deliberately flushes them via pipes, drains and tunnels to rivers, the ocean, or in many instances treatment plants designed to reduce wastewater contamination.

shafts Vertical tunnels from which adits branch at depth. See also adits.

shallow carbon sub-cycle The processes that define how human activity and atmospheric CO_2 concentration are interconnected and why it is such a major concern today.

shallow subsidence Subsidence that takes place near the ground surface when underground water or solid material is extracted.

shear walls A wall that is designed and constructed to resist damage from forces such as shearing.

sheet erosion The removal of thin layers of surface rock or soil from an area of gently sloping land by broad, continuous sheets of running water (sheet flow), rather than by channelized streams.

sheet joints Cracks more or less parallel to the ground surface that result from expansion because of deep erosion and the unloading of overburden pressure. See also joint.

sheet-silicate Also called phyllosilicates, a group of minerals that are characterized by planar structures made of SiO_4 tetrahedra, including mica, chlorite, and serpentine.

shells In classical models of the atom, the orbits taken by electrons as they revolve around the nucleus.

shield volcanoes Massive, wide volcanoes constructed mostly of basaltic lava flows. Some shield volcanoes are the largest mountains on Earth, including Hawaii's Mauna Loa.

shoaling The steepening of waves as they approach the shore and "feel" the bottom, eventually causing the wave to break.

shorelines Where water bodies meet land.

short period Energy wave with a short length and stronger overall energy than a long period wave.

shotcrete Concrete that is sprayed onto a rock slope to increase its stability.

silica tetrahedron An arrangement of ions that resembles a three-sided pyramid (four-sided–the "tetra"–if you include the bottom). Silicon and oxygen atoms combine in tetrahedral arrangements, which are the fundamental building blocks of silicate minerals, Earth's most common mineral group.

silicates The mineral group that is characterized by substances that contain both silicon (Si) and oxygen (O) atoms.

site characterization Understanding the geologic, hydrologic, and engineering properties of an area to determine if it would be an appropriate site for a potential project or use.

slab avalanche A coherent block of snow that fails as a translational slide and then moves downslope in chunks.

slag An artificial, lava-like, waste generated during smelting.

sluff avalanche A type of avalanche that occurs most frequently during storms, when slopes are loaded with newly fallen snow either falling from the sky or deposited by wind. These are typically small compared to other types of avalanches.

sluice boxes Long, open-ended boxes with transverse slats on the bottom over which a slurry of muddy water flows. Gold particles in the slurry may collect in the slats, enabling easy recovery.

slump See rotational slides.

smelting The process of using a furnace to extract pure metal from metallic minerals and rock.

soil development The physical and chemical changes that regolith undergoes to become soil.

soil efflux Emission of gases released from inside Earth via soils to the atmosphere.

soil horizon A layer of soil that is distinguishable from adjacent layers by properties such as color, texture, structure, and chemical composition.

soil order A means by which major soil types can be classified.

soil profile A vertical section of soil that exposes all of its horizons.

soil (a) Loose material at the surface of Earth that supports the growth of plants (pedological definition). (b) All loose surficial earth material resting on coherent bedrock (engineering definition).

solar-thermal A technology for converting solar energy into thermal energy (heat) that heats a fluid, which is then used to run a turbine and generate electricity.

solid combustion residue An ashy by-product of coal combustion, rich in arsenic, cadmium, mercury, and some radioactive materials.

solid earth tide The slow, planet-wide rise and fall of the Earth's crust due to the orbit of the Moon around the Earth.

solid solutions A homogenous mixture of two different kinds of atoms in solid state sharing a single crystal structure.

solution The dissolving of rocks and minerals by natural acids or water. One process of chemical weathering.

sound navigation and ranging (sonar) Use of acoustical signals to measure distances underwater. Sound waves are beamed beneath a ship and distance can be calculated based on the time it takes for the sound to bounce off the desired measurement location and back to the ship.

specific retention The ratio of the volume of water that a given body of rock or soil will hold despite the pull of gravity to the overall volume of the body.

specific yield The ratio of the volume of water drained from an aquifer by gravity to the total volume of the aquifer.

spectroscope A device that identifies the wavelengths of light characteristic of certain gases.

spent fuel Nuclear waste in the form of used radioactive rods.

spheroidal weathering Weathering of rock surfaces that creates a rounded or spherical shape as the corners of the rock mass are weathered faster than its flat faces.

spilling breaker A type of breaker wave that occurs on flat nearshore slopes, where waves lose energy by friction, and the crest spills down the face of the wave.

spoil Refuse rock material that results from mining, excavation, or dredging.

starting zone Where the failure that causes a snow avalanche begins.

steering winds Low to mid-level winds in the atmosphere that determine the travel direction of storms and rain cells.

stock An exposed area of phaneritic igneous rock that is less than 100 square kilometers (39 square miles).

storm surge The sudden rise of sea level on an open coast because of a storm. Storm surge is caused by strong onshore winds and results in water being piled up against the shore. It may be enhanced by high tides and low atmospheric pressure.

strain Deformation resulting from an applied stress (force per unit area). It may be elastic (recoverable) or ductile (nonreversible).

stratification The arrangement of sedimentary rocks in strata or layers.

stratosphere The layer of the atmosphere overlaying the troposphere, with relatively more stable, layered air that increases in temperature with increasing altitude.

stratovolcanoes A type of volcano, also known as a composite volcano, which typically forms along convergent plate margins. They are built up by successive pyroclastic, debris, and lava flows resulting in large, steep volcanic cones. See also composite.

stress The force applied on an object per unit area, expressed in pounds per square foot (lb/ft^2) or kilograms per square meter (kg/m^2).

strike-slip fault A type of fault where shearing causes horizontal displacement of one rocky mass against another,

strip-cropping A method to limit soil erosion by planting parallel strips of crop plants to reduce runoff and wind abrasion.

Strombolian eruption Characterized by pulsating bursts or explosions from a volcanic vent that throw out large quantities of cinders and volcanic bombs (blobs of lava).

structure-from-motion A process to produce 3-D images by merging hundreds of still photos.

subbituminous A mid-ranked metamorphism level coal between lignite and bituminous, with approximately 40% carbon content.

subduction zone earthquakes An earthquake that occurs along a subduction zone, where two tectonic plates collide and one plate is thrust beneath another.

subduction The sinking of one lithospheric plate beneath another at a convergent plate margin.

sublimates Water-soluble minerals, that may include toxic elements such as chlorine or fluorine, deposited by aerosols of volcanic gas on ash particles. These solutes can contaminate local water sources after volcanic eruptions.

subsidence Downward settling of Earth's surface caused by solution, compaction, withdrawal of underground fluids, cooling of the hot lithosphere, or loading with sediment or ice.

sub-surface mines Mines that are fully underground, involving sinking shafts and tunneling, to reach more concentrated mineral deposits otherwise too deep to quarry.

Suess Effect The build-up in Earth's atmosphere of isotope ^{12}C, from the burning of fossil fuels through transportation emissions and industrial activity.

sulfate aerosols A fine mist of aerosols consisting of water dissolved sulfates. Sulfate aerosols are produced naturally by large volcanic eruptions. The aerosols may be suspended in the upper atmosphere for months or even years.

sulfide minerals Minerals that are composed of one or more metals combined with sulfur.

sulfur scrubbers Devices that remove sulfur gases from emission plumes.

sumps Shafts in mines designed to collect drainage water.

Superfund site An area in the United States identified to be in need of priority environmental cleanup according to the Comprehensive Environmental Response, Compensation, and Liability Act (CERCLA). See also Comprehensive Environmental Response, Compensation, and Liability Act (CERCLA).

Superfund A spinoff of CERCLA insuring financial accountability for environmental polluters.

surf zone The near-shore area where waves are breaking.

surface hoar Deposit of ice directly from water vapor onto the top of the snowpack. Forms a very weak layer on which avalanches run.

surface mining A special type of quarrying that involves removing overburden that overlies the rock or mineral resource of interest.

surface wave Seismic waves that are trapped and spread along Earth's surface. Two types exist: Love and Rayleigh waves. They are the most damaging of seismic waves.

sustained yield The amount of water an aquifer can yield on a daily basis over a long period of time.

swash Water that rushes forward onto the foreshore slope of the beach after a wave breaks.

S-wave S (Secondary)-waves are an example of transverse seismic waves that produce ground motion perpendicular to their direction of travel. With P-waves, these energy waves are generated at the focus of an earthquake and travel through the interior of the Earth. See also transverse wave, body wave, and P-waves.

swell Water waves that travel out of the area in which they were generated. They tend to have long periods.

syncline A concave-upward, U-shaped fold where rock layers are deformed.

syngas "Synthetic gas," a mixture of hydrogen and carbon monoxide manufactured by the gasification of coal or other organic materials.

system A set of components that work together as parts of mechanism or an interconnected network or body.

tailings The rocky debris left over when the valuable materials have been extracted from an ore.

talik Layers of water trapped between the frozen, near-surface active layer and deeper permafrost that are produced when winter freezing does not penetrate all the way to the permafrost table.

talus Coarse, angular rock fragments lying at the base of the cliff or steep slope from which they were derived.

tar sand A sand body that is large enough to hold a commercial reserve of asphalt or other thick oil. It is usually surface excavated to remove the thick hydrocarbons.

tectonic exhumation The erosional unroofing of rocks once deeply buried by tectonically driven processes.

tectonic plates Large moving pieces of Earth that make up the lithosphere.

teleseisms The records on seismograms of the arrival of very long period seismic waves from large, distant earthquakes.

tension Stretching forces. In Earth's lithosphere, tension tears plates apart at divergent tectonic margins.

tephra A collective term for all pyroclastic material ejected from a volcano that travels aloft in the air for awhile before falling to Earth's surface.

terminus End of a glacier at any given time, where the melting takes place.

terracing A method of conserving soil and water in steep and easily eroded terrain by creating flat terraces.

terrain anchors Surface roughness, such as trees or rocks, which prevent snow avalanches from starting by holding back the snow.

tertiary treatment A final phase of wastewater treatment where nutrient levels in the clarified liquid are reduced even further with chemical and biological approaches.

thaw slumps A type of high-latitude landslide triggered by the loss of ground ice that gave the frozen soil strength.

thermocline The border zone that separates the epipelagic layer from deep ocean and is defined by changing water temperature.

thermohaline Ocean circulation that is driven by density-related differences in saltiness and temperature.

thermosphere The layer of the atmosphere that overlays the mesosphere, with widely spaced air molecules and extremely low air pressure.

thrust faults A low-angle (15 degrees or less) reverse fault.

tidal energy A technology for harnessing the natural energy of oceanic flood and ebb tides to generate electricity. Tidal energy has been used in the past to power grain mill water wheels.

tidal range The height difference between high and low tides.

tight oil Crude oil contained in petroleum-bearing formations of low permeability and porosity, such as shale.

tiltmeter A device designed to measure change in the angle of a slope.

tipping points Conditions in which small perturbations in a system will cause it to change significantly.

tokamak A machine that confines a plasma by using magnetic fields in a donut shaped container. The Tokamak is designed to enable controlled nuclear fusion for power generation, but so far has not worked.

trade winds Steady winds blowing from areas of high pressure at 30° north and south latitudes toward the area of lower pressure at the equator. This pressure differential produces the northeast and southeast trade winds in the Northern and Southern hemispheres, respectively.

transform boundaries Tectonic plate boundary interactions with mostly horizontal (lateral) movement, such that plates slide past one another.

transform fault Where tectonic plates slide past one another with minimal effect on the earth crust, but high potential to cause earthquakes. Transform faults are a type of strike-slip fault.

translational slide A type of landslide where a downslope movement of material occurs along a distinct, weak planar surface like a fault or bedding plane. See also block glide.

transported soils Soils developed on regolith that has been transported and deposited by a geological agent, hence the terms glacial soils, alluvial soils, eolian soils, volcanic soils, and so forth.

transuranic elements Artificially produced, radioactive elements that are members of the actinide series and have an atomic number between 93 (neptunium, Np) and 103 (lawrencium, Lr).

transverse (shear) waves A seismic wave propagated by a shearing motion (known as S-wave) that involves oscillation perpendicular to the direction of travel. It travels only in solids.

tree rings The result of the growth of new wood in a tree, new rings form approximately each year. They are thicker during years of good growing conditions and thinner during years of poor conditions.

triggered earthquakes A triggered earthquake is one set off when an earlier main shock transfers stress through the crust to a another distant fault, or potential rupture zone along the same fault, causing it to slip. Triggered earthquakes may be almost as strong as the shocks that cause them.

trilateration A triangulation method to locate an earthquake's epicenter by drawing circles around the location of three seismometers where the radius of each circle is equal to the distance between each respective seismometer and the epicenter. The location where all circles intersect on a map is the approximate location of the earthquake's epicenter.

tritium A hydrogen atom with two neutrons and one proton.

tropical storm Low-pressure storm and wind systems that form over warm tropical oceans, including hurricanes and typhoons.

tropopause The top of the troposphere, above the atmosphere is more stable.

troposphere The densest layer of the atmosphere that starts at the Earth's surface and extends 8 to 18 kilometers (5 to 11 miles) in altitude. This layer is characterized by turbulent upwelling and circulating winds due to decreases in temperature and density of air with increasing altitude. This layer contains 99% of all atmospheric water vapor and is the origin of evaporated moisture that forms clouds, storm systems, and rainfall.

tsunami A very large long-wavelength, sea wave produced by a submarine earthquake, a volcanic eruption, or a landslide.

tsunami earthquakes Earthquakes that trigger tsunamis. They typically occur along undersea convergent plate boundaries, where thrust faults develop along the edge of an overriding plate.

typhoon A tropical storm in the western Pacific Ocean.

unconfined aquifer Underground body of water that has a free (static) water table—that is, water that is not confined under pressure beneath an aquiclude.

unconventional oil and gas Petroleum and natural gas that is obtained using different techniques than production of conventional oil by oil wells. Unconventional oil and gas extraction is less efficient and has a greater environmental impact than producing conventional oil.

Underground Well Injection Program Multiple classes of injection wells are legally permitted in the United States by the EPA's Underground Well Injection Program; each class is based on the types and depths of waste disposed.

undertow A near-shore current of water just below the surface that moves in a different direction from any surface current.

Uniform Building Code The building code used by many municipalities across the United States, which has earthquake-resistant deisgn codes written into it.

unsaturated zone The subsurface region above the water table where voids in soil and rock contain air and water.

upwells Cold, deep, and nutrient rich water that rises to replace shallow shore waters that are pushed away from coastlines by winds on the western side of continents.

uranium-mill tailings Finely ground residue that remains after uranium ore is processed.

urban heat island effect Higher temperatures in urban areas packed with buildings, pavement, and little greenspace, compared to leafy suburbs and rural areas. The effect is caused by the low albedo of dark pavement, thermal mass, and lesser evaporation from greenspaces.

Utsu-Omori's Law This observation states that the rate of aftershocks decreases quickly with time; if a certain number of aftershocks are recorded on the first day after a big earthquake, then one-third of that number are likely to occur the day following, and one-third of that number the day following.

valence Ionic charge of an atom equal to the number of excess or missing electrons orbiting its nucleus.

ventifaction The process of making ventifacts, or wind eroded stones, by abrasion from windblown particles such as sand and silt.

vermiculite A phyllosilicate mineral with chemical formula, $(Mg,Ca,K,Fe^{+2})_3(Si,Al,Fe^{+3})_4O_{10}(OH)_2 \bullet 4H_2O$, used widely in the past century in construction and consumer products because of its absorption and heat resistant properties. More than half of this material was mined in Libby, MT, where it was found to be contaminated with asbestos.

viscosity A measure of a fluid's resistance to flow or "stiffness."

volatiles Substances that readily liquefy or become gaseous upon heating, including water vapor and the sulfurous fumes of fumaroles and virtually all volcanic vents.

volcanic arc A belt of volcanoes that form above a subducting oceanic tectonic plate.

volcanic ash Dust-like particles of exploded rock, crystals, and volcanic glass created during a volcanic eruption.

volcanic bombs Large globs of lava pitched out from a volcanic vent during an eruption that can land up to several hundred meters away.

volcanic rocks Rocks formed from magma which erupts and cools on the Earth's surface.

volcanic tremor A continuous seismic signal measured near active volcanoes.

VT-earthquakes Earthquakes that occur as a result of stresses building up within and around a volcano as it gradually fills with magma. "VT" stands for "volcano-tectonic." See also LP-earthquakes.

Vulcanian ash clouds Clouds of volcanic ash and pyroclastic materials that can rise kilometers above a vent and develop strong electrostatic charges because of the fine size of their particles resulting in lighting within the clouds.

Vulcanian eruptions This highly violent type of explosive eruption can last up to 12 hours at a time, clearing the throat of a vent while throwing out abundant older volcanic rocks as well as fresh ash and gases.

warm front A slow moving, warm air mass that rises gradually above any cooler air mass that it encounters.

waste to energy Power generation plants that incinerate garbage to generate energy.

water table The contact between the zone of aeration (unsaturated zone) and the zone of saturation.

wave base The depth at which water waves no longer move sediment across the bottom; equal to approximately half their wavelength.

wave crest The top or highest point of a wave.

wave energy Energy produced by ocean surface waves, which may be tapped to do work, such as generating electricity. The technology for capturing of wave energy is accomplished by a wave energy converter (WEC).

wave frequency The frequency with which individual seismic waves arrive at any given location.

wave period The time (T, time in seconds) between passage of equivalent points (e.g., crests or troughs) on two consecutive waveforms.

wave refraction The bending of wave crests as they move into shallow water.

wave trough The valleys, depressions, or lowest points of a wave.

wavelength The distance between two equivalent points on two consecutive waves; for example, crest to crest or trough to trough.

waves of translation Waves that are accompanied by a net movement of fluid in the direction of wave motion. Breakers approaching a shore are one example of waves of translation.

weather Weather is the day-to-day conditions of the atmosphere, including humidity, precipitation, cloudiness, and temperature.

weathering The physical and chemical breakdown of rocks by interaction with the atmosphere and biosphere.

welded tuff A glassy pyroclastic rock that hardens due to welding together of its particles at high temperature.

westerlies Steady winds blowing from areas of high pressure at 30° north and south latitudes toward the area of lower pressure at the poles. This pressure differential produces the northwest and southwest winds in the Northern and Southern hemispheres, respectively.

wind The movement of air from areas of high pressure to low.

wind energy Electrical power that is generated by wind spinning turbines.

wind shear Winds that blow past one another in different directions and typically at different elevations.

wind waves Surface waves that are created by wind blowing over the surface of water.

wobbles A change in the orientation of Earth's rotational axis with respect to surrounding space over a period of thousands of years.

Younger Dryas The time between 12,900 and 11,700 years ago when Earth's climate slipped at least part way back to intensive ice age conditions.

zone of accumulation The layer of soil where soluble nutrients leached out of above soil layers accumulate (B horizon).

zone of aeration The zone of soil or rock through which water infiltrates by gravity to the water table. Pore spaces between rock and mineral grains in this zone are only partially filled with water. Also known as the vadose zone.

zone of leaching The zone in a mature soil, generally near the top (the A-horizon), from which nutrients are leached by infiltrating water and flushed down to deeper levels in the soil.

zone of saturation The groundwater zone where voids in rock or sediment are filled entirely with water.

Index